International Conference on Mechanical Engineering, Automation and Control Systems 2016

IOP Conference Series: Materials Science and Engineering
Volume 177

Tomsk, Russia
27 – 29 October 2016

Editors:

Anna Zykova
Nikita Martyushev

ISBN: 978-1-5108-3715-7
ISSN: 1757-8981

Printed from e-media with permission by:

Curran Associates, Inc.
57 Morehouse Lane
Red Hook, NY 12571

proceedings.com

Some format issues inherent in the e-media version may also appear in this print version.

Printed by Curran Associates, Inc. (2017)

For permission requests, please contact the Institute of Physics
at the address below.

Institute of Physics
Dirac House, Temple Back
Bristol BS1 6BE UK

Phone: 44 1 17 929 7481
Fax: 44 1 17 920 0979

techtracking@iop.org

Additional copies of this publication are available from:

Curran Associates, Inc.
57 Morehouse Lane
Red Hook, NY 12571 USA
Phone: 845-758-0400
Fax: 845-758-2633
Email: curran@proceedings.com
Web: www.proceedings.com

Table of contents

Volume 177

International Conference on Mechanical Engineering, Automation and Control Systems 2016

27–29 October 2016, Tomsk, Russian Federation

Accepted papers received: 7 February 2017
Published online: 1 March 2017

Preface

011001
OPEN ACCESS
International Conference on Mechanical Engineering, Automation and Control Systems 2016

011002
OPEN ACCESS
Peer review statement

Integrated Computer Control Systems in Mechanical Engineering

012001
OPEN ACCESS
A comparative analysis of efficiency of nonlinear dynamics control methods for a buck converter

A I Andriyanov.....1

012002
OPEN ACCESS
Modeling startup and shutdown transient of the microlinear piezo drive via ANSYS

A V Azin, E P Bogdanov, S V Rikkonen, S V Ponomarev and A M Khramtsov.....10

012003
OPEN ACCESS
Microlinear piezo drive experiments

A V Azin, E P Bogdanov, S V Rikkonen, S V Ponomarev and A M Khramtsov.....16

012004
OPEN ACCESS
The method providing fault-tolerance for information and control systems of the industrial mechatronic objects

E V Melnik, A B Klimenko and V V Korobkin.....22

012005
OPEN ACCESS
Improving machine operation management efficiency via improving the vehicle park structure and using the production operation information database

V Yu Koptev.....28

012006
OPEN ACCESS
The algorithm of measuring parameters of separate oil streams components

A V Kopteva and I N Voytyuk.....34

012007
OPEN ACCESS
AC motor diagnostics system based on complex parametric analysis

N A Korolev and S V Solovev.....40

012008
OPEN ACCESS
Development and program implementation of elements for identification of the electromagnet condition for movable element position control

R I Leukhin, D V Shaykhutdinov, K M Shirokov, N D Narakidze and A S Vlasov.....47

012009
OPEN ACCESS
Combined finish machining of nonrigid parts by the brush electrode

V P Smolentsev, O N Kirillov, O N Fedonin and A Yu Ryazantsev.....52

012010
OPEN ACCESS
Intelligent fuzzy-neural pattern generation and control of a quadrupedal bionic inspection robot

D Sayfeddine and A G Bulgakov.....57

012011
OPEN ACCESS
The digital algorithm for fast detecting and identifying the asymmetry of voltages in three-phase electric grids of mechanical engineering facilities

O B Shonin, S B Kryltcov and N G Novozhilov.....63

012012
OPEN ACCESS
On potentiality of brake rigging unification for freight cars with separate braking

V V Sinitsyn and V V Kobishchanov.....69

012013
OPEN ACCESS
Operational control and diagnostics of the equipment by the parameters of the electric drive power supply by the example of deep well pump units

D I Shishlyannikov and M A Vasilyeva.....75

012014
OPEN ACCESS
Automated system for definition of life-cycle resources of electromechanical equipment

Y Zhukovskiy and N Koteleva.....81

Information technologies in Mechanical Engineering

012015
OPEN ACCESS
The application of mathematical modeling for the development of devices as an example of viscous fluid purification from magnetic impurity

V A Zhukov, E P Masyutkin and B A Avdeyev.....86

012016
OPEN ACCESS
Analysis of noisy signal restoration quality with Lanczos filter

A S Belyaev, I A Tutov and S A Maykov.....93

012017
OPEN ACCESS
Simulation model of load balancing in distributed computing systems

I A Botygin, V N Popov and S G Frolov.....99

012018
OPEN ACCESS
Feasibility of the Davidenkov method for investigation of hoop residual stresses in cold expanded cylinders

V F Skvortsov, A Yu Arlyapov, A O Boznak, I I Ogol and A B Kim.....105

012019
OPEN ACCESS
The algorithm of crack and crack tip coordinates detection in optical images during fatigue test

S V Panin, V O Chemezov, P S Lyubutin and V V Titkov.....111

012020
OPEN ACCESS
Optimization of the coherence function estimation for multi-core central processing unit

A G Cheremnov, V A Faerman and V S Avramchuk.....117

012021
OPEN ACCESS
Electronic load as part of the test complex of the power processing unit of electric and plasma propulsion

S V Chubov and A I Soldatov.....123

012022
OPEN ACCESS
Computer-assisted design of flux-cored wires

Yu N Dubtsov, I V Zorin, G N Sokolov, A A Antonov, A A Artem'ev and V I Lysak.....129

012023
OPEN ACCESS
The comparative analysis of different computations methods of strength of materials by the example of calculations of the axle beam

S I Evtushenko, I A Petrov, M N Shutova and A S Alekseeva.....137

012024
OPEN ACCESS
Evaluation of the vehicle state with vibration-based diagnostics methods

V E Gai, I V Polyakov, M S Krasheninnikov, A A Koshurina and R A Dorofeev.....143

012025
OPEN ACCESS
Designing a combined casting mold for manufacture of a gasoline centrifugal pump body using CAD/CAM-systems

N E Galin, I I Ogol, Yu B Chervach, V Kh Dammer and Jia Hong Ru.....149

012026
OPEN ACCESS
The analysis of the influence of the material antifrictional layer frictional properties on the parameters of the spherical bearing contact zone

A A Kamenskikh and N A Trufanov.....154

012027
OPEN ACCESS
A fuzzy set approach for reliability calculation of valve controlling electric actuators

D P Karmachev, A A Yefremov and E E Luneva.....160

012028
OPEN ACCESS
A simple algorithm for distance estimation without radar and stereo vision based on the bionic principle of bee eyes

A A Khamukhin.....166

012029
OPEN ACCESS
Automation of data processing and calculation of retention parameters and thermodynamic data for gas chromatography

A I Makarycheva and V A Faerman.....172

012030
OPEN ACCESS
Scheduling based on a dynamic resource connection

A E Nagiyev, I A Botygin, A I Shersntneva and P A Konyaev.....178

012031
OPEN ACCESS
Models of neural networks with fuzzy activation functions

A T Nguyen and A M Korikov.....184

012032
OPEN ACCESS
Research of thermal deformation of a kinematic wave reducerwith a modified tooth profile during the work in low temperature conditions

D L Stepanova and P A Nikulin.....189

012033
OPEN ACCESS
The method of optimization of neuro-based concurrent operations in neurocomputers

V A Romanchuk.....193

012034
OPEN ACCESS
The concept and the principle of the diagnostic observability of the object in problems of monitoring and non-destructive testing

A V Sedov.....199

012035
OPEN ACCESS
Designing of routing algorithms in autonomous distributed data transmission system for mobile computing devices with 'WiFi-Direct' technology

I A Nikitin, V S Sherstnev, A I Sherstneva and I A Botygin.....205

012036
OPEN ACCESS
A modern approach to storing of 3D geometry of objects in machine engineering industry

E A Sokolova, G A Aslanov and A A Sokolov.....210

012037
OPEN ACCESS
Development of an algorithm for automated enhancement of digital prototypes in machine engineering

E A Sokolova and G A Dzhioev.....216

012038
OPEN ACCESS
Energy efficiency improvement of electric drive of cold pilgering mill

D A Sychev, N V Savosteenko and A A Gryzlov.....221

012039
OPEN ACCESS
Application of Lucas–Kanade algorithm with weight coefficient bilateral filtration for the digital image correlation method

V V Titkov, S V Panin, P S Lyubutin, V O Chemezov and A V Eremin.....226

012040
OPEN ACCESS
Design and optimization of the micro-engine turbine rotor manufacturing using the rapid prototyping technology

R A Vdovin and V G Smelov.....231

012041
OPEN ACCESS
Anchoring of development workings in a zone of influence of mining in case of the level anchoring system

V F Demin, O B Fofanov, T V Demina and V V Yavorskiy.....237

012042
OPEN ACCESS
Deflected mode of marginal rock massif around mine working boundaries depending on anchoring parameters

V F Demin, O B Fofanov, T V Demina and V V Yavorskiy.....243

012043
OPEN ACCESS
A quality evaluation of stabilization of rotation frequency of gas-diesel engines when using an adaptive automatic control system

A A Zhilenkov and A A Efremov.....251

012044
OPEN ACCESS
Implementation of 14 bits floating point numbers of calculating units for neural network hardware development

I V Zoev, A P Beresnev, E A Mytsko and A N Malchukov.....256

012045
OPEN ACCESS
Information and diagnostic tools of objective control as means to improve performance of mining machines

I E Zvonarev and D I Shishlyannikov.....261

012046
OPEN ACCESS
Forming system of strategic innovation management at high-tech engineering enterprises

O T Ergunova, V G Lizunkov, E Yu Malushko, V I Marchuk and A Yu Ignatenko.....267

012047
OPEN ACCESS
New Methods of Three-Dimensional Images Recognition Based on Stochastic Geometry and Functional Analysis

N G Fedotov, A V Moiseev, A A Syemov, V G Lizunkov and A Y Kindaev.....274

012048
OPEN ACCESS
Hardware-software complex of informing passengers of forecasted route transport arrival at stop

V Yu Pogrebnoy, M I Pushkarev and A S Fadeev.....279

012049
OPEN ACCESS
Features of control systems analysis with discrete control devices using mathematical packages

E M Yakovleva and V A Faerman.....286

Numerical simulation of applied problems

012050
OPEN ACCESS
Metal cutting of tooth gears by a method of a nonlinear generating process of a profile

S K Ambrosimov and A V Khandozhko.....293

012051
OPEN ACCESS
Analysis of vibrational load influence upon passengers in trains with a compulsory body tilt

D Ya Antipin, V V Kobishchanov, V F Lapshin, A S Mitrakov and S G Shorokhov.....300

012052
OPEN ACCESS
The generalized mathematical model of the failure of the cutting tool

N I Pasko, A V Antsev, N V Antseva and V P Fyodorov.....307

012053
OPEN ACCESS
A statistical model of operational impacts on the framework of the bridge crane

V Yu Antsev, A S Tolokonnikov, A D Gorynin and A A Reutov.....313

012054
OPEN ACCESS
Test of the method for calculation of derating of workshop transformers on engineering plants

A Y Shklyarskiy and A I Bardanov.....317

012055
OPEN ACCESS
Numerical simulation of the non-Newtonian fluid flow using the indirect boundary element method

M P Bessonova and V A Yakutenok.....322

012056
OPEN ACCESS
Numerical simulation of dynamic channel-angular pressing of copper specimens

A S Bodrov and S A Zelepugin.....328

012057
OPEN ACCESS
Mathematical simulation of surface heating during plasma spraying

V I Bogdanovich and M G Giorbelidze.....333

012058
OPEN ACCESS
Results of the modal analysis of the underground mine hydraulic leg in various modes of its operation

G D Buyalich, K G Buyalich and S V Uvakin.....340

012059
OPEN ACCESS
Inductive sensor to detect metal impurities in non-metallic medium

N A Bykovsky, L N Puchkova and N N Fanakova.....346

012060
OPEN ACCESS
Modelling and control algorithms of the cross conveyors line with multiengine variable speed drives

M S Cheremushkina and S V Baburin.....351

012061
OPEN ACCESS
Interaction of the conical impactor with barriers containing an explosive

A N Ishchenko, M V Habibullin, S A Afanas'eva, A V Chupashev and A I Zykova.....357

012062
OPEN ACCESS
The investigation of the influence of thermomechanical treatment of the material of rotary cutter bit toolholders on its hardness

S A Chupin and V I Bolobov.....363

012063
OPEN ACCESS
Determination of size properties of the organomineral insulation nanofiller based on the wood matrix

V E Danilov, A M Ayzenshtadt, T A Makhova and M A Frolova.....369

012064
OPEN ACCESS
Mathematical modeling of the steel ingot teeming and the solidification process

S B Gamanyuk, N A Zyuban, D V Rutskii and S V Palatkin.....374

012065
OPEN ACCESS
Spherical gearing with intermediate ball elements: parameter ranges with a high contact ratio

M V Gorbenko and T I Gorbenko.....380

012066
OPEN ACCESS
Modeling a VR-type piston engine as the power plant

A A Gorshkalev, S S Kayukov, S S Korneev and V V Urlapkin.....384

012067
OPEN ACCESS
Modeling of the egress of a drilling liquid from the nozzle of a drill bit with Ansys Fluent

D V Smorkalov, A V Tyutyaev, A M Shterenber and A A Gorshkalev.....390

012068
OPEN ACCESS
Capacity calculation of the electrotechnical scheme of discharge gap replacement of the ozonizer in the COMSOL environment

Yu N Isaev, V A Kolchanova, A P Maltsev and V P Aksenov.....395

012069
OPEN ACCESS
The single-zone numerical model of homogeneous charge compression ignition engine performance

E A Fedyanov, E M Itkis, V N Kuzmin and S N Shumskiy.....400

012070
OPEN ACCESS
Simulation of switching overvoltages in the mine electric power supply system

D I Ivanchenko and N G Novozhilov.....406

012071
OPEN ACCESS
Fracture of cylindrical ampoules during solid phase synthesis under explosive loading

S A Zelepugin, O V Ivanova, A S Yunoshev and A S Zelepugin.....413

012072
OPEN ACCESS
The use of magnetic marks in steel wire ropes

A S Khoroshev, A V Pavlenko, A A Korotky, D A Tchoutchkin, V S Puzin and E V Khorosheva.....419

012073
OPEN ACCESS
Mathematical model for the power supply system of an autonomous object with an AC power transmission over a cable rope

V M Rulevskiy, V G Bukreev, E B Shandarova, E O Kuleshova, S M Shandarov and Yu Z Vasilyeva.....426

012074
OPEN ACCESS
Method of technological control of proportional electromagnets for control systems of their production

A M Lankin, M V Lankin and M Y Lankina.....433

012075
OPEN ACCESS
Approximation of dynamic characteristics of the magnetization for electromagnets

A M Lankin, M V Lankin and M Y Lankina.....439

012076
OPEN ACCESS
Modelling of flame propagation in the gasoline fuelled Wankel rotary engine with hydrogen additives

E A Fedyanov, E A Zakharov, K V Prikhodkov and Y V Levin.....443

012077
OPEN ACCESS
Topology optimization of a gas-turbine engine part

R N Faskhutdinov, A S Dubrovskaya, K A Dongauzer, P V Maksimov and N A Trufanov.....449

012078
OPEN ACCESS
Calculation by iterative method of linear viscoelastic plate under biaxial tension

A A Svetashkov, J Miciński, K K Manabaev and A A Vakurov.....454

012079
OPEN ACCESS
Simulation modelling of tribotechnologies system and its parametric reliability assessment on tribotechnical parameters of the joints of sliding friction

M N Nagorkin, V P Fyodorov and V V Nagorkina.....460

012080
OPEN ACCESS
Simulation of the plain milling process

A Yu Nikolaev.....466

012081
OPEN ACCESS
The frequency-domain method of calculation for the pulsed electromagnetic field in a conductive ferromagnetic plate

G V Nosov, E O Kuleshova, S Lefebvre, A A Plyusnin and D M Tokmashev.....471

012082
OPEN ACCESS
The simulation of the workpiece and saw teeth heating during the high-speed cutting process

L M Gurevich, A I Bannikov, D V Pronichev, R E Novikov and A G Serov.....478

012083
OPEN ACCESS
Implementation of inverse calculation method in diagnostics of digital electric substations

R G Oganyan, N D Narakidze, D V Shaykhutdinov, N I Gorbatenko and S O Maksuta.....484

012084
OPEN ACCESS
Diagnostic features of relief formations on the nanostructured titanium VT1-0 surface after laser shock-wave treatment

I V Lytvynenko, S A Lupenko, P O Maruschak, S V Panin and Yu I Hats.....489

012085
OPEN ACCESS
Numerical simulation of electron beam welding with beam oscillations

D N Trushnikov and G L Permyakov.....495

012086
OPEN ACCESS
Estimation of the vibration loading vehicle with pneumohydraulic suspensions

A S Gorobtsov, S K Kartsov and Yu A Polyakov.....500

012087
OPEN ACCESS
Using a specialized Markov chain in the reliability model of disk arrays RAID-10 with data mirroring and striping

P A Rahman.....506

012088
OPEN ACCESS
Analysis of the mean time to data loss of nested disk arrays RAID-01 on basis of a specialized mathematical model

P A Rahman.....512

012089
OPEN ACCESS
Modelling the behaviour of ingot axial defects under open die forging

D V Rutskii, A N Stjagin, N A Zyuban and S B Gamanyuk.....518

012090
OPEN ACCESS
Simulation and optimization of wind and diesel power supply systems

B V Lukutin, E B Shandarova, D L Matukhin, A A Igisenov and S M Shandarov.....524

012091
OPEN ACCESS
The choice of the principle of functioning of the system of magnetic levitation for the device of high-performance testing of powder permanent magnets

D V Shaykhutdinov, N I Gorbatenko, N D Narakidze, A S Vlasov and I A Stetsenko.....530

012092
OPEN ACCESS
Research of dynamic loading in a drivetrain by means of mathematical modeling

N S Sokolov-Dobrev, M V Ljashenko, V V Shekhovtsov and P V Potapov.....536

012093
OPEN ACCESS
GTE blade injection moulding modeling and verification of models during process approbation

I S Stepanenko and A I Khaimovich.....541

012094
OPEN ACCESS
Algorithm of iterative transformation for effective modules of multicomponent isotropic composite

A A Svetashkov, J Miciński, M S Pavlov and A A Vakurov.....548

012095
OPEN ACCESS
Algorithm of constructing hybrid effective modules for elastic isotropic composites

A A Svetashkov, J Miciński, N A Kupriyanov, V N Barashkov and A V Lushnikov.....554

012096
OPEN ACCESS
Methods of determining the margin of cyclic crack resistance of metal structures for hoisting machinery

V Yu Antsev, A S Tolokonnikov, A V Vorobev and V I Sakalo.....559

012097
OPEN ACCESS
Parameters assessment of the inductively-coupled circuit for wireless power transfer

Yu N Isaev, O V Vasileva, A A Budko and S Lefebvre.....564

012098
OPEN ACCESS
The power supply system model of the process submersible device with AC power transmission over the cable-rope

V M Rulevskiy, V G Bukreev, E O Kuleshova, E B Shandarova, S M Shandarov and Yu Z Vasilyeva.....570

012099
OPEN ACCESS
Studies in electromagnetic compatibility of optical and digital current and voltage transformers

V D Lebedev and A A Yablokov.....576

012100
OPEN ACCESS
Modelling of dredging process with vibrating digging bucket under subzero temperatures

S A Zenkov, D Y Kobzov and N A Balahonov.....585

Material Science in Mechanical Engineering

012101
OPEN ACCESS
Piercing mandrel strengthening by surfacing with nickel aluminide-based alloy

I V Zorin, Yu N Dubtsov, G N Sokolov, A A Artem'ev, V I Lysak and S N Elsukov.....591

012102
OPEN ACCESS
Preparation of the highly dispersed powder of titanium carbonitride by SHS azide technology with previous partial nitriding

A P Amosov, Yu M Markov, R A Dobrovolskaya and E N Nikolaeva.....595

012103
OPEN ACCESS
The study of mechanical effect on the characteristics of the PZT-system piezoelectric elements

V I Berg, A E Brand, V A Petryakov and A G Zakirzakov.....601

012104
OPEN ACCESS
Metal of cavitation erosion of a hydrodynamic reactor

A G Zakirzakov, A E Brand, V A Petryakov and E F Gordievskaya.....606

012105
OPEN ACCESS
The connection analysis between the dilution of the deposited Fe-Cr-V-Mo-C layer by the basic metal and the parameters of its microstructure

A S Degterev and S F Gnusov.....611

012106
OPEN ACCESS
Study of the Ti-5Al-5Mo-5V-1Cr-1Fe titanium alloy grain structure uniformity after bending and annealing

A V Balaykin, E A Nosova and N V Galkina.....618

012107
OPEN ACCESS
The effect of cooling speed on the structure and properties of the heat affected zone in welded compounds of ferrite-austenitic steels

I L Gonik, D N Gurulev and O P Bondareva.....623

012108
OPEN ACCESS
Tribological properties of hydraulic fluids modified by peat-based additives

V V Ionov, S A Larionov, Ju S Sarkisov, N O Kopanica, A V Gorchkova, N P Gorlenko, N A Tzevtkov and K V Ikonnikova.....628

012109
OPEN ACCESS
The structure and phase composition of hard alloys of the Cr_3C_2-Ti system produced by explosive compacting of powders

V O Kharlamov, A V Krokhalev, M A Tupitsin, S V Kuz'min and V I Lysak.....633

012110
OPEN ACCESS
The structure and the phase composition of copper aluminide-based coatings formed on the aluminum surface under contact melting in the Al-Cu system joints

V G Shmorgun, O V Slautin and V P Kulevich.....638

012111
OPEN ACCESS
Non-destructive testing of nanomaterials by using subminiature eddy current transducers

S F Dmitriev, A V Ishkov, A O Katasonov, V N Malikov, A M Sagalakov and L I Shevtsova.....644

012112
OPEN ACCESS
Development and implementation of the NMR-spectrometer on the basis of the National Instruments technologies

N D Narakidze, D V Shaykhutdinov, K M Shirokov, N I Gorbatenko and S G Yanvarev.....649

012113
OPEN ACCESS
Effect of heat treatment conditions on structure and properties of high-speed steel

E P Nikolaeva and D B Vlasov.....655

012114
OPEN ACCESS
Reasons for crack nucleation in welded joints of main gas-pipelines after a long-term operation

P O Maruschak, R T Bishchak, L S Shlapak and S V Panin.....661

012115
OPEN ACCESS
Development of SHS azide technology of silicon carbide nanopowder

Y V Titova, A Yu Illarionov, A P Amosov, D A Maidan and K S Smetanin.....665

012116
OPEN ACCESS
Contact stresses modeling at the Panda-type fiber single-layer winding and evaluation of their impact on the fiber optic properties

Yu I Lesnikova, O Yu Smetannikov, A N Trufanov and N A Trufanov.....671

012117
OPEN ACCESS
Influence of stress concentrator shape and testing temperature on impact fracture regularities of pipeline steel

I V Vlasov, S V Panin, P O Maruschak, D D Moiseenko and F Berto.....676

012118
OPEN ACCESS
Influence of soaking time of modifier in melt on microstructure of Al-12%Si alloys

N V Martyushev, V S Bashev and A P Zykova.....682

Mechanical Engineering Processes and Metals Treatment

012119
OPEN ACCESS
Research of the possibility of using an electrical discharge machining metal powder in selective laser melting

A A Golubeva, A V Sotov, A V Agapovichev, V G Smelov and V N Dmitriev.....687

012120
OPEN ACCESS
Influence of the direction of selective laser sintering on machinability of parts from 316L steel

V P Alexeev, A V Balyakin and A I Khaimovich.....693

012121
OPEN ACCESS
The investigation of the deformation wave hardening effect on the strength of the medium and low alloy steels

A V Kirichek, S V Barinov, A V Yashin and S E Kolontsov.....700

012122
OPEN ACCESS
Research of the process of axisymmetric forming of thin-walled flat blanks into the conical parts with minimal thickness variation

E G Demyanenko, I P Popov and V S Menshikov.....706

012123
OPEN ACCESS
Creation of wear-resistant frictional surfaces by implanting materials based on tungsten carbide

S V Davidov and A O Gorlenko.....714

012124
OPEN ACCESS
The influence of surface finishing methods on touch-sensitive reactions

M S Kukhta, A P Sokolov, P Y Krauinsh, A D Kozlova and C Bouchard.....721

012125
OPEN ACCESS
Application of the geometric immersion method based on the Castigliano variational principle for the axisymmetric problems of elasticity theory

Y S Kuznetsova, N A Vorobyev and N A Trufanov.....726

012126
OPEN ACCESS
The use of the rotating electromagnetic field for hardening treatment of details

V A Lebedev, A A Kochubey and A V Kiricheck.....732

012127
OPEN ACCESS
Thermodynamic aspects of the coating formation through mechanochemical synthesis in vibration technology systems

S U Shtyn, V A Lebedev and A O Gorlenko.....738

012128
OPEN ACCESS
Investigation of cutting process and process of surface quality formation during machining of fragile components

O N Fedonin and D V Levy.....744

012129
OPEN ACCESS
Dependence of the diffusion wear of the hard alloy surface on its fractal dimension

V P Nesterenko, A A Kondratyuk, E B Belousova and I A Shulepov.....750

012130
OPEN ACCESS
Effect of technological heredity on the fatigue strength in the manufacture of gas turbine engine blades

G V Smirnov, N D Pronichev and M V Nekhoroshev.....756

012131
OPEN ACCESS
Experimental and theoretical study of the hydriding behaviour in the pulse ecm of titanium alloys

G V Smirnov, N D Pronichev, M V Nekhoroshev and V I Bogdanovich.....763

012132
OPEN ACCESS
Surface profiling in mating parts by combined nonabrasive finishing

EV Smolentsev, ON Fedonin and VP Smolentsev.....771

012133
OPEN ACCESS
The application of finite element analysis during development of the Integral Strain Gauges calibration method for the study of the welded construction

K V Syzrantseva, V N Syzrantsev and M V Dvoynikov.....781

012134
OPEN ACCESS
The study of the chemical and phase composition of the diffusion interaction zone in layered composite Cr20Ni80-AD1

V G Shmorgun, L D Iskhakova, A I Bogdanov, A O Taube and R P Ermakov.....787

012135
OPEN ACCESS
Analysis of cavitation on the surface of steel under the ultrasonic cleaning

I K Tsybry and I L Vyalikov.....792

012136
OPEN ACCESS
The study of oscillations excitation patterns in the process of milling with portable equipment

A M Kozlov, E V Kiryuschenko and A V Khandozhko.....797

Electric Drive in Mechanical Engineering

012137
OPEN ACCESS
Friction self-oscillation decrease in nonlinear system of locomotive traction drive

D Ya Antipin, V I Vorobiyov, O V Izmerov, S G Shorokhov and D A Bondarenko.....804

012138
OPEN ACCESS
Multifactorial determination of the electric drive for the force compensating manipulator

G Ya Pyatibratov and A A Danshina.....811

012139
OPEN ACCESS
Methods for rotational speed reduction for alternating current electric motors

A V Kagan and D Y Glukhanich.....817

012140
OPEN ACCESS
Forecasting of the electrical actuators condition using stator's current signals

T N Kruglova, I V Yaroshenko, N N Rabotalov and M A Melnikov.....822

012141
OPEN ACCESS
Temperature control of power semiconductor devices in traction applications

A A Pugachev and N N Strekalov.....827

012142
OPEN ACCESS
Assessment of duration of the drive operation in the mode of kinetic energy recovery under power supply voltage sags in electrical grids of mechanical engineering enterprises

O B Shonin and N G Novozhilov.....833

012143
OPEN ACCESS
The experimental studies of operating modes of a diesel-generator set at variable speed

S G Obukhov, I A Plotnikov, M A Surkov and L P Sumarokova.....839

012144
OPEN ACCESS
Structural heredity influence upon principles of strain wave hardening

A V Kiricheck, S V Barinov and A V Yashin.....845

Preface

The **X International Conference on Mechanical Engineering, Automation and Control Systems (MEACS 2016)** was held in Tomsk, Russia, on October 27–29, 2016.

MEACS 2016 represents a continuation of the conferences on mechanical engineering, held in Russia since 2003, as well as seminars and conferences on the technological applications of modern scientific development, traditionally organized in Tomsk.

The four-day Conference brought together the specialists from different countries and organizations and provided an excellent opportunity to exchange knowledge, make oral contributions and poster presentations, and initiate discussions on the topics that are of interest to the Conference participants.

The selected papers of the Conference cover a wide range of technical areas and modern aspects of the integrated computer control systems in mechanical engineering, numerical simulation of applied problems, materials science in mechanical engineering, mechanical engineering processes, and metals treatment.

The Conference was hosted by Tomsk Polytechnic University.

Organizing Committee of MEACS 2016

Edited by:

Anna Zykova, PhD,
Tomsk Polytechnic University

Nikita Martyushev, PhD,
Tomsk Polytechnic University

Peer review statement

All papers published in this volume of *IOP Conference Series: Materials Science and Engineering* have been peer reviewed through processes administered by the proceedings Editors. Reviews were conducted by expert referees to the professional and scientific standards expected of a proceedings journal published by IOP Publishing.

A comparative analysis of efficiency of nonlinear dynamics control methods for a buck converter

A I Andriyanov

Department of Electronics, Radioelectronic and Electrotechnical Systems, Bryansk State Technical University, 7, 50-letiya Oktyabrya bul., Bryansk, 241035, Russia

E-mail: mail@ahaos.ru

Abstract. The paper presents a comparative analysis of efficiency of methods of nonlinear dynamics control with feedback: a method of linearization of Poincaré map, a method of time-delay feedback, a method of target-oriented control. Efficiency of each method is evaluated on basis of the achieved domain of the desired dynamic mode. Dynamics modes maps and diagrams allowing evaluation of a particular method efficiency are given. It is shown that a target-oriented control has the greatest efficiency, and when it is applied in the selected range of system parameters, undesired dynamic modes do not exist. The method of Poincaré map linearization and the method of time-delay feedback control have approximately equal efficiency. A comparative analysis is performed for the first time, and the obtained results are of great importance for practice.

1. Introduction

At present, pulse-width converters find a widespread application in systems that require conversion of DC voltage to another value of DC voltage [1].

Considered devices are closed-loop automatic control systems with voltage feedback that are prone to complex dynamics [2, 3]. In these systems, bifurcations occur with changing of one or several parameters, which may result in origination of undesired dynamic modes accompanied by large amplitude oscillations leading to an increase of electromagnetic interference during the load, current overload of power switches and appearance of acoustic effects during the device operation. To prevent the appearance of these modes, specific control algorithms for nonlinear dynamics are required.

In this case, it is necessary to provide a so-called desired dynamic mode. Under the desired dynamic mode, we mean a mode in which the period of switching processes coincides with the period of pulse width modulation (PWM). This mode is called a 1-cycle [2]. In addition, the undesired m-cycles can appear, when the period of the system switching processes is m times greater than the PWM period. Further, index m will be called the cycle ratio.

There is a large number of works on control of nonlinear dynamics of discrete systems [4-9]. Among control approaches, two main directions can be marked:

– methods with feedback: Poincaré map linearization method (PML) [4], time-delay feedback control methods (TDFC) [5, 6], a method of target-oriented control (TOC) [7, 8].

– methods without feedback, based on resonant perturbation of parameters [9].

Application of these methods in power electronics was widely reported in scientific literature, and the first group of methods has found the widest application. However, one of the main disadvantages of the existing works is the absence of systematic studies on efficiency of these methods. For example, authors can restrict themselves by demonstration of one timing diagram, or a bifurcation diagram, whereas the drift of two parameters can occur in the system, and at this point, 2-D dynamic modes maps are much more informative. In addition, there is no comparative analysis of efficiency of different control methods.

This paper contains a comparative analysis of efficiency of methods of nonlinear dynamics control with feedback by the example of an automatic control system with a buck converter. The evaluation will be conducted on the relative area of the desired dynamic modes in the space of two system parameters with the use of one or another method. The buck converter relates to switching converters of the first kind, that later on will allow the application of the obtained results in the design of a wide class of systems.

2. Nonlinear dynamics control systems of the buck converter

Three methods of nonlinear dynamics control will be briefly reviewed in this section. One of them (PML) is based on the introduction of perturbation of one parameter, calculated on the basis of Poincaré map linearization [4], and two others (TDFC, TOC) are based on the introduction of corrective actions, calculated on the basis of differential correction functions [5, 7].

The generalized functional diagram of the automatic control system (ACS) with switching converters is shown in Figure 1. The following designations are accepted here: SC – switching converter, U_L – load voltage, $x_1, x_2, \ldots x_n$ – state variables, U_{in} – input voltage, U_{ref} – reference signal, U_p – power switches control pulses, U_{con} – control signal, U_{rmp} – ramp voltage with amplitude $U_{rmp,m}$, U_{ck} – corrective action; Δp_k – perturbation parameter for PML, *NDCS* – nonlinear dynamics control system, *MC* – master oscillator, *RG* – ramp generator, *SH* – sample-and-hold device, *SB* – subtractor, *SM* – summator, U_{err} – error of ACS, U_{ld} – load voltage, I_{ld} – load current.

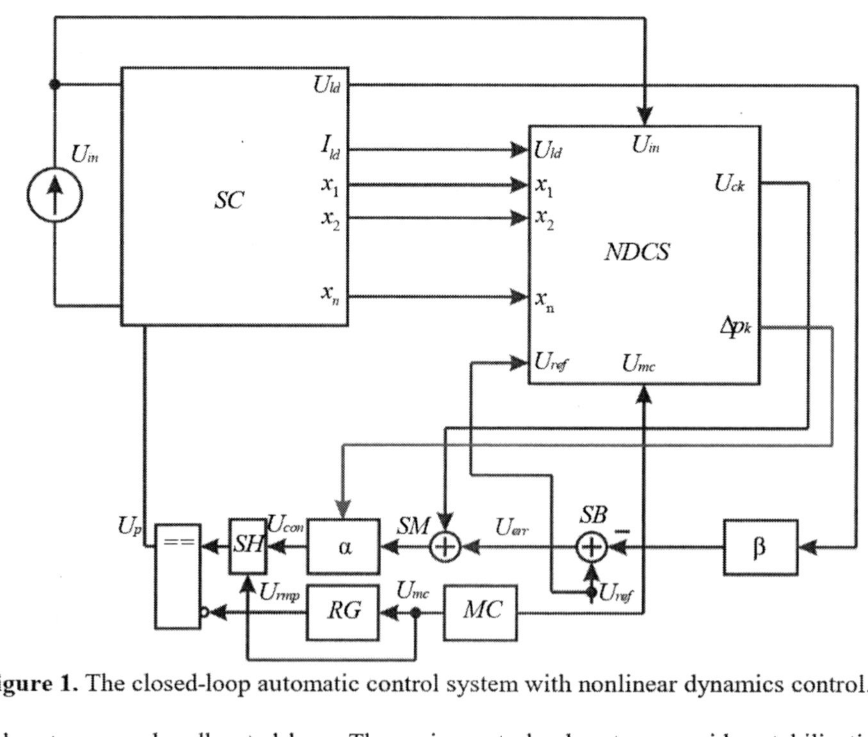

Figure 1. The closed-loop automatic control system with nonlinear dynamics control.

Two subsystems can be allocated here. The main control subsystem provides stabilization of the average value of the output voltage without considering system nonlinear dynamic properties. The nonlinear dynamics control system (*NDCS*) provides dynamic stabilization of the desired mode (1-cycle).

TOC and TDFC belong to methods with differential corrective functions. The essence of the methods with differential correction functions is in the following. Let there be given some abstract discrete system described by a stroboscopic map which is presented as

$$x_k = f(x_{k-1}), \tag{1}$$

where x_k – the value of the variable on the k-th iteration of the stroboscopic map.

Stabilization of an unstable fixed point of stroboscopic map (1) x^* can be accomplished by introduction of a stabilizing action, during which the stroboscopic mapping function of the nonlinear dynamic system takes the following form:

$$x_k = f(x_{k-1}) + f_{cf}(x_{k-1}),$$

where function f_{cf} is the correction function (CF).

In case of TDFC, the correction function has the following form [5]:

$$f_{cf}(x_{k-1}) = c_p(x_{k-1} - f(x_{k-1})), \tag{2}$$

In case of TOC, the correction function has form [7]

$$f_{cf}(x_{k-1}) = c_p(x^* - f(x_{k-1})),$$

where c_p is the adjustable coefficient.

Here, the main property of the correction function should be noted. As the representative point approaches the stabilized fixed point of stroboscopic map $f(x_{k-1})$, the value of the correction function tends to zero, and after reaching the desired mode, the correction is actually absent, which corresponds to $f_{cf}(x_{k-1})=0$. This approach guarantees that the use of the correction function has no effect on the position of the stabilized fixed point of original stroboscopic map $f(x_{k-1})$, which is a necessary condition for correct nonlinear dynamics control.

As can be seen from (2), in case of TDFC, the value of the variable in previous iteration x_{k-1} of the stroboscopic map is used to calculate the corrective action. In case of TOC, fixed point x* is used instead of x_{k-1} and is stabilized. This is a disadvantage of TOC, since x* should be calculated in advance, which requires higher computational resources.

Let us consider the mathematical description of ACS of the buck converter in the form of a stroboscopic map. The conventional automatic control system with output voltage average value feedback for DC-to-DC converters, operating in a discontinuous current mode, is described by a stroboscopic map as follows [10]:

$$\begin{aligned}\mathbf{X}_k &= \Psi(\mathbf{X}_{k-1}) = e^{\mathbf{A}_3(1-z_{k2})a} e^{\mathbf{A}_2(z_{k2}-z_{k1})a} e^{\mathbf{A}_1 z_{k1} a}\mathbf{X}_{k-1} + \\ &+ e^{\mathbf{A}_3\cdot(1-z_{k2})a} e^{\mathbf{A}_2(z_{k2}-z_{k1})a}(e^{\mathbf{A}_3 z_{k1} a} - \mathbf{E})\mathbf{V}_{AB1} + \\ &+ e^{\mathbf{A}_3\cdot(1-z_{k2})a}\left[e^{\mathbf{A}_2(z_{k2}-z_{k1})a} - \mathbf{E}\right]\mathbf{V}_{AB2} + (e^{\mathbf{A}_3(1-z_{k2})a} - \mathbf{E})\mathbf{V}_{AB3},\end{aligned} \tag{3}$$

where $\mathbf{X}=[x_1, x_2]^T=[i_L, u_c]^T$; i_L – the inductor current; u_c – the capacitor voltage; z_{k1}, z_{k2} – switching moments in the k-th clock period in relative time: $z_{k1} = \dfrac{t_{k1}-(k-1)a}{a}$, $z_{k2} = \dfrac{t_{k2}-(k-1)a}{a}$, where a – the duration of the clock period; $\mathbf{X}_{k-1}$ – the vector of system state variables at the beginning of the k-th clock period.

The circuit of a buck converter and the diagram of processes occurring in clock period [1, 10] are presented in Figure 2, where *VT* – power switch, *VD* – power diode, *L* – inductor, *R* – inductor resistance, *C* – capacitor, R_l – load resitance.

Analysis of Figure 2 shows that the clock period for the buck converter is divided into three sections of system structure constancy.

IOP Conf. Series: Materials Science and Engineering **177** (2017) 012001 doi:10.1088/1757-899X/177/1/012001

1. Section 1: $(k-1)a < z < z_{k1}$. Here the power switch is at the OFF-state, and inductor current i_L increases linearly.

2. Section 2: $z_{k1} < z < z_{k2}$. Here the power switch is in the ON-state, and the inductor current starts to decrease from moment z_{k1}; the energy is transferred from the input to the load.

3. Section 3: $z_{k2} < z < ka$. At this section, the inductor current is equal to zero, and the output is being discharged in this case, transferring energy to the load.

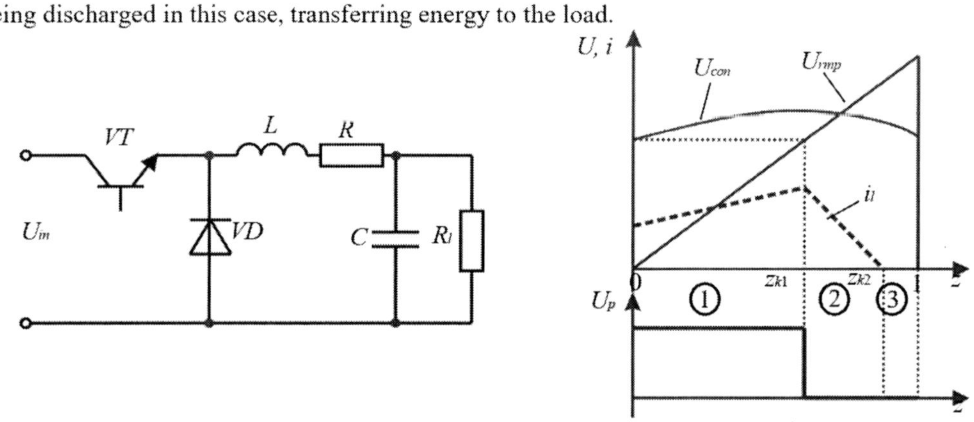

Figure 2. Illustration of a buck converter operating principle: a) the buck converter circuit, b) waveforms of processes in a clock period.

Switching moments z_{k1} and z_{k2} can be calculated on basis of switching functions [10]:

$$\xi_{k1}(\mathbf{X}_{tk1}, z_{k1}) = \alpha\left(U_з - \beta c_1{}^T \mathbf{X}_{tk1}(z_{k1})\right) - U_{rmp,m} z_{k1};$$

$$\xi_{k2}(\mathbf{X}_{tk2}, z_{k2}) = c_2{}^T \mathbf{X}_{tk2}(z_{k2}),$$

where $\mathbf{X}_{tki}$ – values of the vector of state variables at the i-th switching point in k-th clock period (i=1, 2); c_1=[0; 1] and c_2=[1; 0] – constant vectors that define the component of the vector of state variables that is used in expressions. At that, it is necessary to solve the so-called switching manifolds equations:

$$\xi_{k1}(\mathbf{X}_{tk1}, z_{k1}) = 0; \ \xi_{k2}(\mathbf{X}_{tk2}, z_{k2}) = 0.$$

When using NDCS with correction functions, additional control actions u_{ik} are introduced, which, for the k-th clock period, are determined by correction functions according to the following expressions:

– in case of TDFC,

$$u_{ik} = K_i(x_{ik-1} - x_{ik}); \tag{4}$$

– in case of TOC,

$$u_{ik} = K_i(x_{iref} - x_{ik}) \tag{5}$$

Therefore, the stroboscopic map for the ACS with the correction function takes the following form:

$$\mathbf{X}_k = \hat{\Psi}(\mathbf{X}_{k-1}) = e^{\mathbf{A}_3(1-z_{k2})a} e^{\mathbf{A}_2(z_{k2}-z_{k1}-\Delta z_k)a} e^{\mathbf{A}_1(z_{k1}+\Delta z_k)a} \mathbf{X}_{k-1} +$$
$$+ e^{\mathbf{A}_3 \cdot (1-z_{k2})a} e^{\mathbf{A}_2(z_{k2}-z_{k1}-\Delta z_k)a} (e^{\mathbf{A}_3(z_{k1}+\Delta z_k)a} - \mathbf{E})\mathbf{V}_{AB1} +$$
$$+ e^{\mathbf{A}_3 \cdot (1-z_{k2})a} \left[e^{\mathbf{A}_2(z_{k2}-z_{k1}-\Delta z_k)a} - \mathbf{E}\right] \mathbf{V}_{AB2} + (e^{\mathbf{A}_3(1-z_{k2})a} - \mathbf{E})\mathbf{V}_{AB3},$$

where Δz_k – the increment of a duty cycle in the k-th clock period.

The specified increment can be found on the basis of expression

$$\Delta z_k = \frac{\alpha U_{ck}}{U_{rmp,m}} = \frac{\alpha\left(\sum_{i=1}^{n} u_{ik}\right)}{U_{rmp,m}},$$

where U_{ck} is the corrective action (Figure 1), calculated as the sum of u_{ik}.

Switching function $\xi k1(\mathbf{X}, z_k)$, defining the switching moment of the power switch [11] for the ACS with correction functions, described by a system of differential equations of the n-th order, takes the following form:

$$\xi_{k1}(\mathbf{X}, z_{k1}) = \alpha\left(U_з - \beta c_1{}^T \mathbf{X}_{k-1} + \sum_{i=1}^{n} \Delta u_{ik}\right) - U_{ref,m} z_{k1},$$

where n – the order of the system of differential equations describing the ACS (in the case considered, when n=2).

The generalized functional diagram of NDCS (Figure 1) is presented in Figure 3.

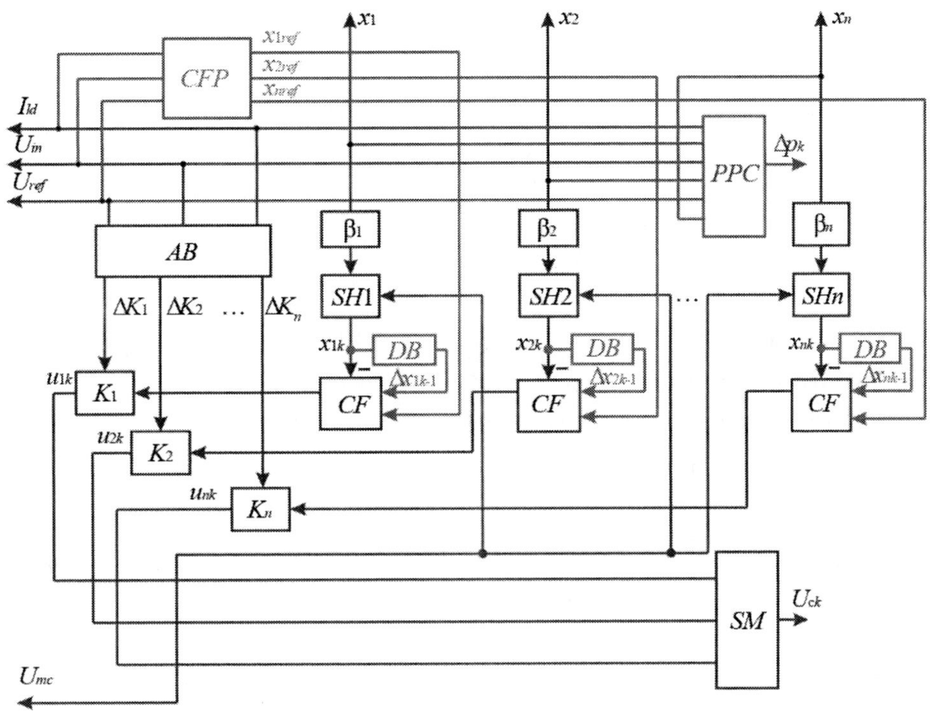

Figure 3. The generalized nonlinear dynamics control system.

PML blocks are shown with red lines, TOC and TDFC blocks – with blue and green ones, accordingly. Blocks that are common both for TOC and for TDFC are shown with black lines. It is evident that with application of a particular method only the corresponding blocks operate. The following designations are accepted in the Figure 3: β_i – a scale factor of the feedback circuit for the i-th state variable, SHi – a sample-and-hold device for the i-th state variable, DB – a delay unit, AB – an adaptation block, K_i – proportional coefficients for the i-th state variable, x_{ik} – the i-th state variable in the k-th stroboscopic moment, x_{iref} – the i-th state variable of the vector of reference signals for the fixed point, u_{ik} – a deviation component for the i-th state variable, SM – a summator, *PPC* – a parameter perturbation calculator, *CF* – a correction function block (function (4) or (5)).

The main task of the design of the ACS based on differential correction functions is selection of such coefficients K_i under which the desired dynamic mode is stable. This problem can be solved using methods of optimization with constraints. The author has used the method of Nelder-Mead with constraints $K_{min,i}$ and $K_{max,i}$ set empirically. In fact, this means that calculated optimal value K_i lies within $K_{min,i}< K_i<K_{max,i}$. In this case, the optimality criterion is the target value of the largest multiplier of the desired dynamic mode [5] that should be less than unity.

In case of the use of the method of Poincaré map linearization (PML) for stabilization of the unstable mode, the perturbation of one or another system parameter is performed. Therefore, in Figure 3, the relationship between NDCS and the proportional controller is marked by the dashed line. In this case, the NDCS forms increment $\Delta\alpha$ that adjusts the gain ratio of the proportional controller. In general, any parameter available for correction may be a subject of perturbation.

For calculation of the necessary increment, stroboscopic map (3) is replaced by a similar linearized map at point $(\mathbf{X}^*,p^*)$:

$$\mathbf{Y}_k = \mathbf{M}^*\mathbf{Y}_{k-1} + \mathbf{C}^* u_{k-1}, \tag{6}$$

where $\mathbf{M}^* = \dfrac{\partial\Psi(\mathbf{X}^*,p^*)}{\partial\mathbf{X}_{k-1}}$ – a monodromy matrix of the stabilized 1-cycle; $\mathbf{C}^* = \dfrac{\partial\Psi(\mathbf{X}^*,p^*)}{\partial p}$ – a stroboscopic map derivative with respect to the perturbed parameter; $\mathbf{Y}_{k-1} = \mathbf{X}_{k-1} - \mathbf{X}^*$, $u_{k-1} = p_{k-1} - p^*$ – the required parameter perturbation, where p_{k-1} – the value of the perturbed parameter in the k-th clock period.

As we have mentioned previously, gain α of the proportional controller is selected as parameter p.

For linear system (6), stabilizing control u_{k-1} is chosen in the form of linear state feedback

$$u_{k-1} = -\mathbf{K}^*\mathbf{Y}_{k-1}. \tag{7}$$

Taking into account (7), from (6), we obtain the following expression:

$$\mathbf{Y}_k = (\mathbf{M}^* - \mathbf{C}^*\mathbf{K}^*)\mathbf{Y}_{k-1}.$$

Thus, unstable point $\mathbf{X}^*$ of stroboscopic map (3) will be stabilized if matrix $\mathbf{K}^*$ is defined so that matrix $\mathbf{M}^* - \mathbf{C}^*\mathbf{K}^*$ has all the eigenvalues (multipliers) that are less than one.

To calculate desired matrix $\mathbf{M}^* - \mathbf{C}^*\mathbf{K}^*$, technique [12] is used:

$$\mathbf{M}^* - \mathbf{C}^*\mathbf{K}^* = \frac{\mathbf{M}^*}{\max(|\rho_i|)}\left(1 - \frac{\delta}{100}\right), \tag{8}$$

where i=1, 2, ρ_i – i-th matrix multiplier $\mathbf{M}^*$, δ=[%] – stability margin.

When using expression (8), the eigenvalues will always be less than one, at that the system properties do not change significantly because the multipliers of the desired matrix remain complex or real numbers, as they do in original matrix $\mathbf{M}^*$. Calculation of feedback matrix $\mathbf{K}^*$ based on known matrix $\mathbf{M}^* - \mathbf{C}^*\mathbf{K}^*$, as well as calculation of matrices $\mathbf{M}^*$ and $\mathbf{C}^*$, can be implemented using Ackermann's formula [13], widely known in automatic control theory. Calculation of required parameter perturbation u_{k-1} is carried out according to expression (7).

3. Study of nonlinear dynamics of the switching converter with PWM

The dynamic modes map of the ACS without nonlinear dynamics control based on a buck converter with PWM-I is shown in Figure 4, a.

The simulation was performed under the following system parameters: L=0.1 Hn; C=1 μF; R=10 Ohm; R_L=100 Ohm; α=56; β=0.01; U_{ref}=5 V; $U_{rmp,m}$=10 V; a=0.0001 s.

In Figure 4, symbols $D_{i,j}$ mark domains of existence of different dynamic modes (i stands for the m-cycle that characterizes a particular domain, j is the number of the domain on the dynamic modes map). Domains $D_{ch,j}$ correspond to nondeterministic modes of the converter operation, in which large amplitude quasiperiodic oscillations are observed in the system.

As can be seen from the Figure 4, in the whole range of variation of the input voltage, there are undesired modes that limit the possibility of reference voltage regulation in a wide range of values. As the analysis of Figure 4, a shows, there are multistability domains on the dynamic modes map. In particular, a certain part of domains $D_{4,1}$ and $D_{ch,1}$ coexists with the desired 1-cycle. The relative area of the domain of the desired 1-cycle in Figure 4, a is 52.74%.

Next, we will consider the results of the control system operation based on TDFC. The two-parameter diagrams for the system with TDFC-control with adaptation of parameters of the method are presented in Figure 4, b. TDFC parameters have the following values: K_{1min}=-3; K_{1max}=3, K_{2min}=-3; K_{2max}=3, β_1=0.08; β_2=0.01, the target value of the multiplier is ρ_{tg}=0.3. As follows from the figure, the domain of undesired modes at high control voltages decreases, which, in its turn, leads to an increase of the domain of desired mode $D_{1,1}$ (96.56% of the map total area).

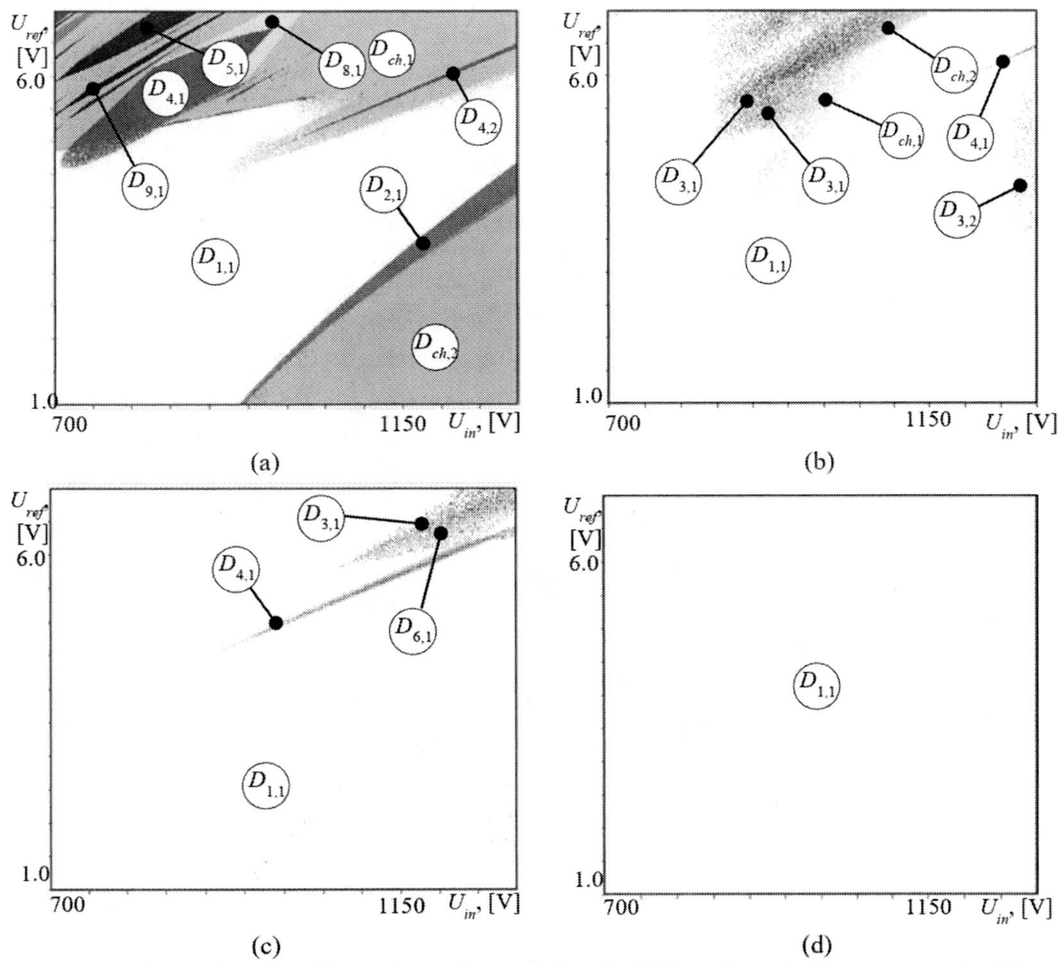

Figure 4. The results of nonlinear dynamics analysis of ACS based on a buck converter: a) without nonlinear dynamics control; b) using TDFC; c) using PML; d) using TOC.

With the introduction of adaptation domains, $D_{ch,2}$ and $D_{2,1}$, disappear completely, and the area of undesired modes at high reference voltages is replaced by domains, where, along with the stable 1-cycle, there are chaotic modes. Special attention should be paid to this case, because it shows the disadvantages of the adaptation algorithm based on linearization of the stroboscopic map. In the case considered, the author managed to ensure the stability of the desired mode within a given margin, but at the same time, the system properties changed in such a way that the undesired modes occurred in coexistence with the desired one. For this reason, it is necessary to carry out additional research at the design stage. Building a more complex objective function that is able to exclude such a situation is difficult for nonlinear dynamic systems.

The map of ACS dynamic modes based on PML, with selected stability margin δ=20% (see (8)) is presented in Figure 4, c. As is seen from the figure, the introduction of PML has led to a significant expansion of the domain of the desired 1-cycle, but the domains of undesired dynamic modes remained. At small values of the reference voltage, there is domain $D_{2,1}$ of a little area, and in case of large values of the reference voltage, there are domains $D_{4,2}$, $D_{4,1}$, $D_{3,1}$, $D_{6,1}$. A characteristic feature of the obtained map is the impossibility of regulation of the reference voltage throughout the whole range of values without appearance of undesired modes, but at this, the area of the domain of the desired mode on the map (96.63% of the map total area) is somewhat larger than on the dynamic modes map with the TDFC control method (Figure 5, b).

In case of TOC with parameters adaptation, the simulation has been performed for the system with algorithm parameters: K_{min}=-0.7; K_{max}=-0.7; β_1=0.08; β_2=0.01. Dynamic modes maps with the use of TOC are shown in Figure 4, d. As is seen from the figure, the use of TOC solves the problem of nonlinear dynamics control completely. The desired 1-cycle occupies 100% of the total area of the map of dynamic modes.

Bar charts for evaluation of efficiency of nonlinear dynamics control methods are shown in Figure 5. As shown in the figure, all control methods have led to the increase of the area of the domain of the desired mode, but the highest efficiency has been demonstrated by TOC.

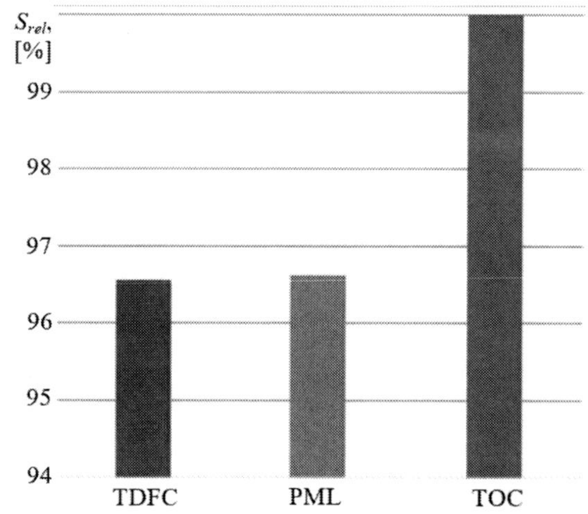

Figure 5. Bar charts of the relative area of the domain of the desired dynamic mode in percentage terms for a buck converter.

4. Conclusion

The conducted research allows drawing the following conclusions.

1. TOC demonstrates the highest efficiency, for, as a result, the area of the domain of the desired mode is maximal, but meanwhile, TOC requires calculation of the fixed stroboscopic map point for the stabilized dynamic mode that demands the use of high-performance MCUs.

2. TDFC has acceptable (but slightly worse than TOC) control efficiency, but at the same time, its application may cause appearance of multistability domains. This method has the minimum requirements for computing resources of MCU.

3. PML shows the worst performance, but with that, as follows from [12], its application is appropriate in multistability domains, where it provides an effective retention of the system in the desired mode and, therefore, ensures a minimal influence of system parameters deviation on the fixed stroboscopic map point. In the unstable area of the desired mode, the control efficiency may be low.

The obtained results can be used at the stage of schematic design of a wide class of switching DC-to-DC converters. On the basis of the results presented, the method of nonlinear dynamics control that is most suitable in each case can be chosen.

References

[1] Rashid M H 2011 *Power electronic handbook: devices, circuits, and applications handbook* (Burlington, MA, USA: Butterworth-Heinemann)
[2] Baushev V S, Zhusubaliev Zh T 1992 *Elektrichestvo* **8** 47–53
[3] Zhusubaliyev Zh T, Mosekilde E 2003 *Bifurcations and chaos in piecewise-smooth dynamical systems*. (Singapore: World Scientific Publishing Co. Pte. Ltd)
[4] Ott E, Grebogi C., Yourke G. 1990 *Phys. Rev. Lett* **64** 1796–1199
[5] Pyragas K 1992 *Phys. Rev. Lett.* A **170** 421–428
[6] Batlle C, Fossas E, Olivar G 1999 *Int. J. of Circuit Theory and Appl.* **27** 617–631
[7] Dattani J, Blake J C H, Hilker F M 2011 *Phys. Lett.* A **375** 3986–3992.
[8] Andriyanov A I, Krasnov N I 2013 *Proc. of the Higher Educ. Inst.. Instrumentation* **56** 33–38
[9] Zhou Y, Tse C K, Qiu S, Francis C M 2003 *Int. J. of Bifurcation and Chaos* **13** 3459–3471
[10] Andriyanov A I, Bulokhov N M 2013 *J. Eng.* **10** 30–39
[11] Andriyanov A I 2016 *Int. Siberian Conf. Control and Communications* (*SIBCON 2016*) **1** 1–6
[12] Andriyanov A I, Bulokhov N M 2013 J. *Elektrisity* **8** 41–49
[13] Kapitaniak T 1996 *Controlling Chaos – Theoretical and Practical Methods in Non-Linear Dynamics* (London: Academic Press)

Modeling startup and shutdown transient of the microlinear piezo drive via ANSYS

A V Azin[1], E P Bogdanov[2], S V Rikkonen[1], S V Ponomarev[1] and A M Khramtsov[1]

[1] Tomsk State University, 36, Lenina Ave., Tomsk, 634050, Russia
[2] Tomsk Polytechnic University, 30, Lenina Ave., Tomsk, 634050, Russia

E-mail: epbogdanov@mail.ru

Abstract. The article describes the construction-design of the micro linear piezo drive intended for a peripheral cord tensioner in the reflecting surface shape regulator system for large-sized transformable spacecraft antenna reflectors. The research target –the development method of modeling startup and shutdown transient of the micro linear piezo drive. This method is based on application software package ANSYS. The method embraces a detailed description of the calculation stages to determine the operating characteristics of the designed piezo drive. Based on the numerical solutions, the time characteristics of the designed piezo drive are determined.

1. Introduction

The problem of reducing the spacecraft (SC) system mass-dimension parameters is especially acute in the space field. One current solution consists in replacing electro-mechanical drives in different SC units with piezo drives which, in its turn, would decrease the mass-dimension parameters n-fold times [1, 2].

The peripheral cord tensioner unit (PCTU) in the reflecting surface shape regulator system for large-sized transformable spacecraft antenna reflectors provides the preliminary tension of the spacecraft reflector surface itself. The peripheral cord tension is F_{tens}– 300 N, the unit mass – 250 g, the total reflector spoke mass – 35 kg, cord displacement X_{tens}– 0.5 m. Cord displacement at 0.5m could be only in case of the animated regime of operating micro linear piezo drives (MLPD) [3, 4].

Studying the types of MLPD transient involves:

- establishing the animated regime of operating MLPD within PCTU to determine the operation transient behavior during MLPD startup and shutdown;
- preventing startup transient as a consequence of the impluse regime with temporary vibration displacement which is significantly exceeds the certified MLPD values and could result in the emergency mode of the system itself.

Modeling of startup and shutdown transient of the micro linear piezodrive method is described step-by-step via application software package ANSYS.

2. Problem statement

Generally, the MLPD system status is described by elastodynamics equations. Let us consider domain Ω with boundary S. Let us find displacement field $\vec{u}(x_1,x_2,x_3,t)$ as a function of spatial coordinates

$x = (x_1, x_2, x_3)$, changing in domain Ω of 3D Euclidean space and time t [5-9]. Differential continuum equations are obtained from static equilibrium equations, including mass inertia forces:

$$\frac{\partial \sigma_{i,j}}{\partial x_j} + \rho F_i = \rho \frac{\partial^2 u}{\partial t^2}. \quad (1)$$

The deformation and displacement connection is

$$\varepsilon_{ij} = \frac{1}{2}\left(\frac{\partial u_i}{\partial x_j} + \frac{\partial u_j}{\partial x_i}\right). \quad (2)$$

To stimulate the mechanical behavior in tension, Hooke's law is used:

$$\sigma_{ij} = \lambda\theta\delta_{ij} + 2\mu \cdot \varepsilon_{ij}, \quad (3)$$

$$\theta = \varepsilon_{11} + \varepsilon_{22} + \varepsilon_{33} = \vec{u},$$

where, δ_{ij} – Kronecker symbol; u_i, σ_{ij}, ε_{ij} – components of the displacement vector, second stress tensor of Piola-Kirchoff; deformation stress tensor; μ, λ – Lame parameters;ρ –material density; F_i– mass force.

Boundary conditions: if the body surface includes two parts:

$$S = S_1 + S_2,$$

$$u_i(\bar{x}) = 0, \bar{x} \in S_1, \quad (4)$$

$$n_k \sigma_{kj}\left(\delta_{ij} + \frac{\partial u_i}{\partial x_j}\right) = 0, \bar{x} \in S_2. \quad (5)$$

Initial conditions: at $t = 0$, the displacement field and velocity are assumed to be

$$u(x,0) = u_0(x), \quad (6)$$

$$\frac{\partial u}{\partial t}(x,0) = \vec{v}_0(x).$$

For piezo-electrical bodies, equation (1) is written as straight-line and reverse piezoeffect equation (7, 8) and constrained electrostatics equation (9):

$$\sigma_{ij} = C^E_{ijkl}\varepsilon_{kl} + \xi C^E_{ijkl}\varepsilon_{kl} - e_{lij}U_i, \quad (7)$$

$$D_i = e_{ijk}\varepsilon_{jk} + \varepsilon^S_{ij}U_j, \quad (8)$$

$$U = -\mathrm{grad}\varphi, D = 0, \quad (9)$$

where, D – the electrostatic field induction vector; U – the electrostatic field intensity vector; ϕ – electrostatic potential; C^E – the elastic piezo ceramic material modulus; e – the piezoelectrical modulus; ε^S – dielectric constants; ξ – the introduced coefficient to describe damping in the Rayleigh solid medium (viscoelastic losses).

Temperature components are excluded from the equation systems as operating conditions of the micro linear drive are considered to be isothermal.

3. The mathematical calculation model for MLPD

The basic steps in engineering analysis through the finite element method (FEM) are the following:

1. creating a geometrical model (CAD): an actual physical model of the investigated object is developed and parametrized;
2. creating an idealized model, i.e. moving from an actual physical model to a changed (simplified) mathematical model;
3. creating a discrete model (FEM), corresponding to constrained degree-of-freedom, i.e. idealized model discretization: physical material properties are given and splitting model grid parameters are set;
4. solving equation systems, corresponding to the selected analysis type (modal analysis, harmonic analysis, transient analysis);

5. analyzing obtained results.

Let us consider the following case study by applying the above-mentioned algorithm.

3.1 CAD model development

MLPD includes a framework, an adjustment screw, PZT Stack, a cheek with an oscillator element, a pushrod and a resilient spacer (Figure1a). The CAD model of the MLPD construction-design is illustrated in figure 1b. The framework impact on other MLPD elements was simulated by constraint conditions: attaching the lower PZT Stack boundary and the upper resilient spacer boundary. The load on MLPD is simulated by the point mass attached to the upper pushrod surface. As the described problem is axisymmetric, only ¼ of the MLPD construction is considered.

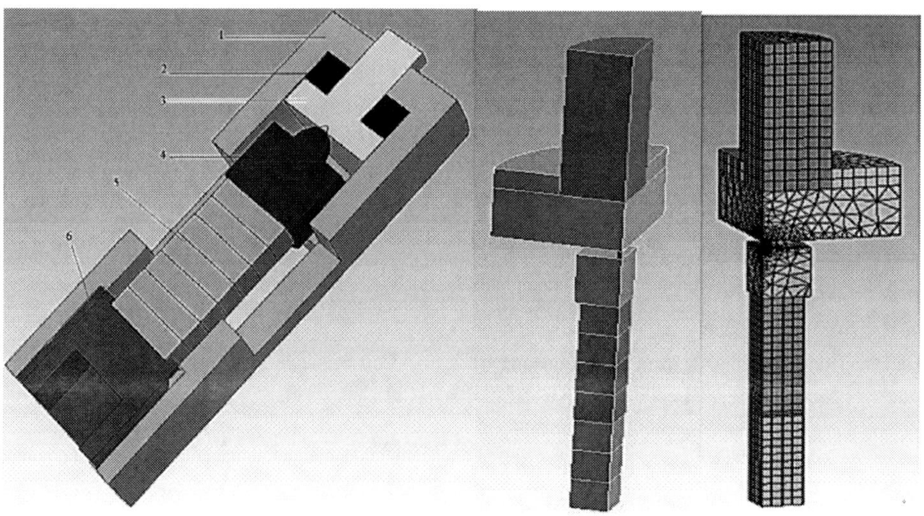

a b c

Figure 1. MLPD Modeling: a – overview; b – CAD model; c – FE model. 1– framework, 2 – resilient spacer, 3 – pushrod, 4 – cheek with oscillator element, 5 – PZT Stack, 6 – adjustment screw.

3.2 The finite element model

The finite element model (FEM) is designed on the basis of the geometrical model. To visualize the stress-strain state (SSS) of construction elements when subjected to loads, it is necessary to reduce the size of the elements in positions of likely maximum stress. In MLPD, it is the cheek-oscillator and the oscillator-pushrod contact zone. Such tools as line partitioning and boundary splitting offer this possibility. The finite element model of MLPD is illustrated in figure 1c showing the reduction of elements in the zone of likely maximum stress. The number of elements in presented FEM of MLPD is 5582.

The authors have developed a mathematical calculation model of the MLPD mechanical system in order to determine what materials are suitable for the mechanical cheek-oscillator-pushrod contact. Based on this model, the MLPD reinforcement element materials were determined to improve MLPD operation reliability and quality. In this case, the following materials were considered for the described MLPD construction model: cheek-steel, oscillator-steel, pushrod-plexiglas, spacer-rubber and PZT Stack-piezoelectric ceramics.

The material properties in calculation, presented in table 1, are reference data. The PZT Stack APM-2-7 was used in the MLPD construction. This PZT Stack embraces 7 PZT blocks. These PZT blocks are oriented towards the direction opposite to each other.

Table 1. Material properties in calculation

Material	Density, kg/m^3	Modulus of elasticity, Pa	Poisson ratio
Steel	7800	$2.1\cdot10^{11}$	0.29
Plexiglas	1190	$3\cdot10^{9}$	0.33
Rubber	1200	$5\cdot10^{6}$	0.5

Calculation data of piezo-ceramics (used in APM-2-7) have been received from the producer. Piezo-ceramics properties are depicted in table 2.

Table 2. Piezo-ceramics properties

Material properties	Value
Density, kg/m^3	7600
Planar Q-factor	70
Dielectric constant, F/m	ε^S_{xx}= 2100; ε^S_{zz}= 1300
Modulus of elasticity (anisotropic material), N/m^2	$C^E_{11}=1.1\cdot10^{11}$; $C^E_{21}= 6.1\cdot10^{10}$; $C^E_{22}= 1.1\cdot10^{11}$; $C^E_{31}=5.8\cdot10^{10}$; $C^E_{33}=1.01\cdot10^{11}$; $C^E_{11}=1.1\cdot10^{11}$; $C^E_{44}=1.8\cdot10^{10}$; $C^E_{55}=1.1\cdot10^{11}$; $C^E_{11}=2.45\cdot10^{10}$
Piezoelectric modulus	$d_{31}=-190\cdot10^{-12}$, $d_{33}=450\cdot10^{-12}$, $d_{15}=400\cdot10^{-12}$

The next stage in calculation model plotting is the application of loads and introducing constraint conditions.

In actual conditions, the pushrod presses the spacer which is attached to (glued on) the MLPD frame work surface. Accordingly, fastening the nodes on the spacer surface in all three axes simulates this situation. The lower PZT Stack boundary is also rigidly fixed so as the force arising from input voltage by law $U(t) = U_0{\cdot}\sin(f{\cdot}t)$ (where,U_0 – the amplitude applied to PZT Stack, equals 100 V; f – frequency; t – time), is transferred to the remaining MLPD elements.

The following factors are also considered: the load of displacement of reflector spokes and the spoke mass of 35 kg. This load is the point mass attached to the upper pushrod edge.

Equal values $\beta = 4\cdot10^{-5}$ were introduced into the damping MLPD system value. This value was determined according to the method described in ANSYS documents [10], applying the dominant oscillation frequency which, in its turn, involves the maximum ratio of effective mass to total mass.

3.3 Calculation results

The MLPD construction-design was considered. The following materials were used for the system elements: steel, piezoelectric ceramics, plexiglas and rubber. MLPD was examined under normal operation conditions, i.e. electrical connection and a sine wave signal of 100 V amplitude.

The modal analysis was conducted to determine the resonant frequencies at which MLPD executes the maximum displacement along a Z-direction. The following modes were obtained: 1 – 145 Hz and 6 – 13520 Hz.

The next calculation stage involved harmonic analysis. Based on the calculation results the following factor was determined: frequency at which the displacement of the pushrod is maximum: f_0=140 Hz, X = $28.1\cdot10^{-6}$ m. The calculation results are depicted in figure 2 – the dependence of the pushrod displacement amplitude value on the set signal frequency from PZT Stack.

The last mathematical calculation stage involves the transient analysis. The analysis results are displacement diagrams reflecting the information on the node displacements on the upper pushrod edge where voltage is applied to MLPD PZT Stack (Figure 3) and where voltage is applied to MLPD PZT Stack lining (Figure 4).To visualize MLPD start-up and shut-off transient more exactly, the voltage value applied to the PZT Stack is changed in three-time intervals:

1. First time interval (t = [0; 0.01] s) –stress on lining equals 0.

2. Second time interval (t = [0.01; 0.37] s) – the voltage being applied to PZT Stack lining $U = U_0 \cdot \sin(2 \cdot \pi \cdot f_0 \cdot t)$, where U_0 – the amplitude of applied voltage (U_0 = 100 V), f_0 – resonant frequency (f_0= 140 Hz). The MLPD start-up moment is simulated.
3. Third time interval (t = [0.37; 1.5] s) – stress on lining equals 0. The MLPD shut-off moment is simulated.

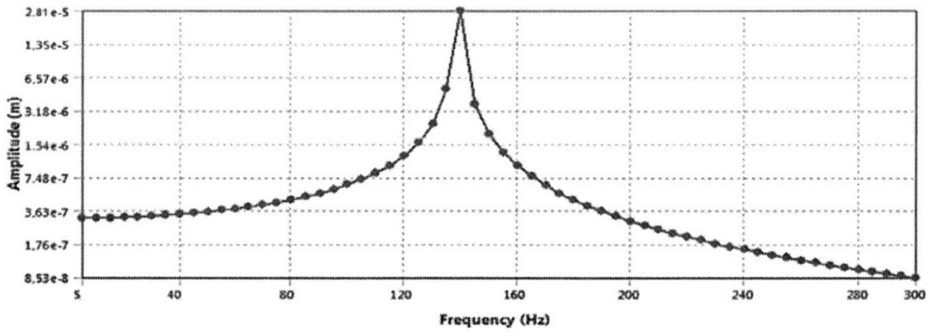

Figure 2. Dependence of the pushrod displacement amplitude value on the set signal frequency from PZT Stack.

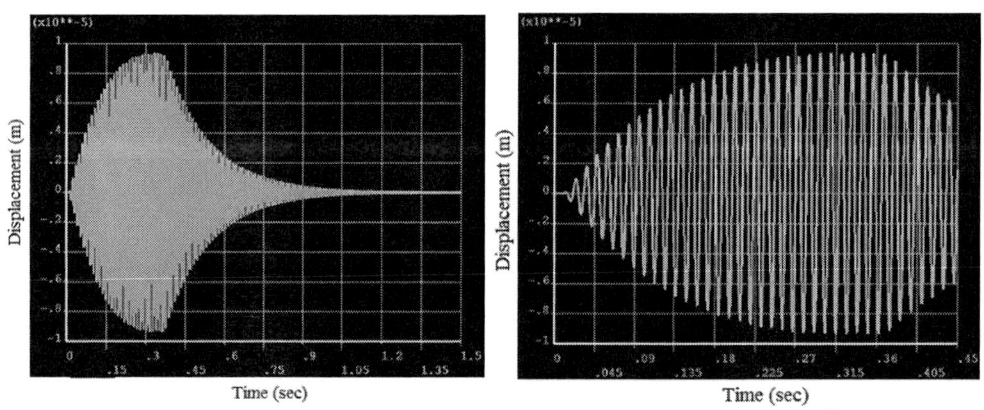

a b

Figure 3. Start-up and shut-off transient of MLPD. Displacement along the upper MLPD pushrod boundary: a– complete process; b – process start-up.

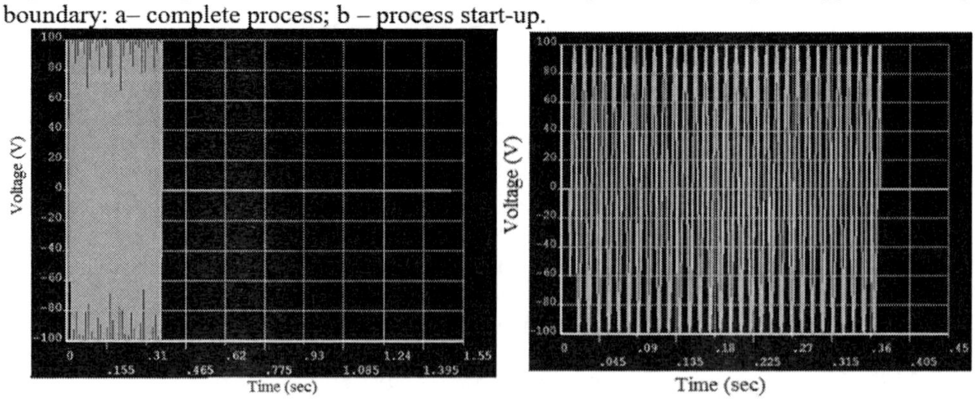

a b

Figure 4. Start-up and shut-off transient of MLPD. The voltage in MLPD PZT Stack lining: a – complete process; b – process start-up.

The calculation results of MLPD start-up showed that the above-mentioned MLPD at a load of 35 kg requires 0.23 s to operate in a steady-state mode (Figure 4). The calculation results of MLPD shut-off showed that 0.98 s is required in damping the oscillations (Figure 4).

4. Conclusion

The modeling procedure of the startup and shutoff micro linear piezo drive has been developed.

This procedure includes a detailed description of the calculation stages to determine the operating characteristics of the designed piezodrive.

Based on numerical equations, the time characteristics for the designed micro linear piezodrive were determined. This, in its turn, made it possible to select the optimal operating regime for the micro linear piezodrive under the animated operating regime.

Acknowledgments

This work was financially supported by the Ministry of Education and Science of Russia; unique identifier RFMEFI57814X0060.

References

[1] Park S 2011 *Single vibration mode standing wave tubular piezoelectric ultrasonic motor* (Toronto: Ryerson University)
[2] Wang Z, Li T and Cao Y 2013 *Aerospace Science and Technology* **26** 160
[3] Ponomarev S V, Rikkonen S V and Azin A V 2014 *News of Higher Educational Institutions. Physics* **2** 196
[4] Ponomarev S V, Rikkonen S V and Azin A V 2015 *Tomsk State Univ. J. of Mathem. and Mechanics* **2(34)** 86
[5] Abramenko T, Gorish A, Bogush M and Mitko V 2003 *Proc. of the Tenth Int. Congress on Sound and Vibration* (Stockholm) vol 3 (Stockholm: International Institute of Acoustics and Vibration) p 951
[6] Bogush M 2014 *Design of piezoelectric sensors based on spatial electrical and thermoelastic models* (Moscow: Technosphera publishing house)
[7] Davoudi S 2012 *Effect of Temperature and Thermal Cycles on PZT Ceramic Performance in Fuel Injector Applications* (Toronto: Department of Mechanical and Industrial Engineering University)
[8] Bogush M and Pikalev E 2008 *Izvestiya SFedU. Engineering sciences* **2(79)** 74
[9] Bogush M 2008 *Izvestiya SFedU. Engineering sciences* **6** 18
[10] 2013 Matrix resistance *Ansys Help Viewer* 15.0

IOP Conf. Series: Materials Science and Engineering **177** (2017) 012003 doi:10.1088/1757-899X/177/1/012003

Microlinear piezo drive experiments

A V Azin[1], E P Bogdanov[2], S V Rikkonen[1], S V Ponomarev[1] and A M Khramtsov[1]

[1] Tomsk State University, 36, Lenina Ave., Tomsk, 634050, Russia
[2] Tomsk Polytechnic University, 30, Lenina Ave., Tomsk, 634050, Russia

E-mail: epbogdanov@mail.ru

Abstract. The article embraces the experimental description of the micro linear piezo drive intended for the peripheral cord tensioner in the reflecting surface shape regulator system for large-sized transformable spacecraft antenna reflectors. The research target is the experimental investigation of the micro linear piezo drive to determine the stable oscillatory system operating modes which would include improved energy conversion parameters. The following points are briefly presented: test stand construction-design of the peripheral cord tensioner; the determined frequency characteristics and the identified resonant and actual frequencies of an oscillatory system under inertia load. A series of experiments has been conducted for both different preliminary voltages and inertia mass values.

1. Introduction

The problem involving the reduction of spacecraft (SC) system mass-dimension parameters is especially acute in the space field. One current solution is to replace electro-mechanical drives in different SC units for piezo drives which, in its turn, would decrease the mass-dimension parameters n-fold times [1–4]. The peripheral cord tensioner unit (PCTU), a system regulating the reflecting surface shape for large-sized transformable spacecraft antenna reflectors, provides the preliminary tension of the spacecraft reflector surface itself. The peripheral cord tension is F_{tens} – 300 N, the unit mass – 250 g, the total reflector spoke mass – 40 kg and the pushrod displacement with an increment of X_{tens}– 12 μm.

2. Problem statement

The problem targets involve the following: conducting micro linear piezo drive experiments based on PZT Stack APM-2-7 in inertia load modes under different inertia load values; identifying the characteristic oscillatory system frequencies; and determining oscillatory system (OS) resonant frequencies and vibration displacement amplitudes within resonant frequencies.

3. Test stand construction-design description of the peripheral cord tensioner

The oscillatory system of the test stand was calculated on the basis of 3D and one-dimension mathematical models [5, 6]. This test stand is designed to investigate the micro linear piezo drive (MLPD) operation modes (figure 1). Micro linear piezo drive (MLPD) operation modes depend on the following OS parameters: the type and the power of PZT Stack; the weight (loads) mass; preliminary stack voltage forces; voltage on PZT Stack; PZT Stack current; the frequency effect [7–10]. Operation mode parameters are recorded as electrical signals in the experiment testing of MLPD:

voltage on PZT Stack; PZT Stack current; signal strength from the PZT Stack sensing unit; vibration acceleration weight (loads). Another important parameter to be considered is preliminary tension force.

The following data were obtained during the MLPD experiment:

- weight (loads) frequency responses;
- weight vibration displacement frequency responses;
- frequency responses of forces acting on weight.

Based on the results of these frequency responses, the operating OS resonant frequencies and vibration displacement amplitudes were determined.

Varying the OS parameters and frequency exposure on the test stand, numerous system operation modes were investigated. Based on the test results, the optimal frequency exposure to given load was determined, at which the maximum forces acting on loads, maximum vibration displacement load and maximum power acting on loads were observed.

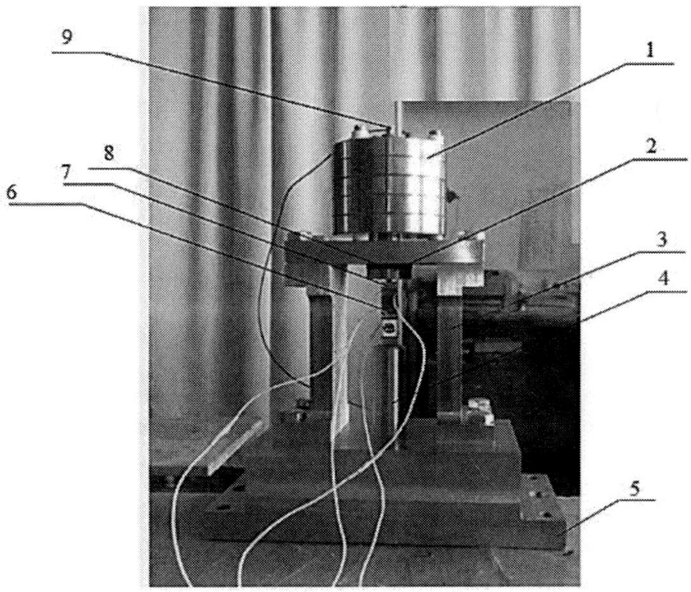

Figure 1. The test stand, operating MLPD under the inertia load: 1 – inertia load mass (weight); 2 – pushrod; 3 – support structures; 4 – adjustment screw; 5 – frame; 6 – PZT Stack APM- 2-7; 7 – force sensor; 8 – rigidity; 9 – accelerometer AR 1019.

Figure 1 illustrates the test stand. PZT Stack (6) is installed in frame (5) on adjustment screw (4), through which force sensor (7) rests upon pushrod (2), then pushrod (2) through rigidity (8) acts on inertia load (1) (weight). Accelerating inertia load (1) is measured by accelerometer (9). Through adjustment screw (4), the preliminary PZT Stack tension force is established. Force sensor PZT Stack calibration, accelerometer calibration and the preliminary tension setting procedure require a more detailed description which is impossible to include in the present paper.

The detailed diagram of the test stand is illustrated in figure 2, where minor details are included – energy cheeks and the MLPD oscillator, as well as the power source (alternating current 12 and direct current 13), data acquisition and storage system 11.

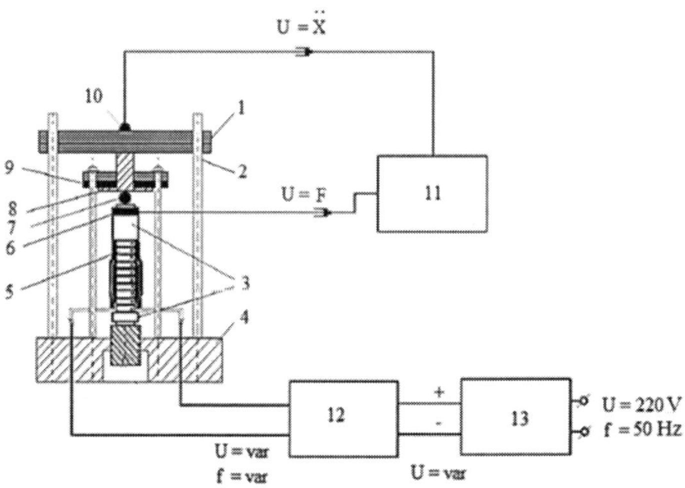

Figure 2. The layout diagram of the experimental test stand for prototype MLPD: 1 – inertia mass load (weight); 2 – guiding weight; 3 – energy cheeks; 4 – frame; 5 – PZT Stack; 6 – force sensor; 7 – oscillator; 8 – pushrod; 9 – preliminary tension rigidity; 10 – accelerator; 11 – data acquisition and storage system; 12 – AC power source; 13 – DC power source.

4. Experiment results

Varying the AC frequencies input to PZT Stack, the PZT Stack generates the disturbance force, which, in its turn, through the oscillator-pushrod acts on the inertia load imposing vibration at a given frequency. Oscillograms are presented in figure 3 a, b. Measuring the inertia load acceleration, the vibration displacement acceleration is determined. Based on the experiment results, frequency-response characteristics are plotted. PZT Stack current was maintained at 0.5 A, whereas the power supply decreased proportionally to PZT Stack capacitance. Parameters for the 1st operation mode (figure 3a): f =10.7 kHz,I = 0.5 A; preliminary tension force F_0 = 240 H; force sensor signal F_f = 2.82 H, $\ddot{X}$ = 57 m/s^2; weight force G_{wf} = 0.4 kg; X = 0.012 μm. Parameters for the 2nd operation mode (figure 3b): f = 1400 H, I = 0.5 A; preliminary tension force F_0 = 240 H, F_f = 2.82 H, $\ddot{X}$ = 57.4 m/s^2; weight force G_{wf} = 3 kg, X = 0.742 μm.

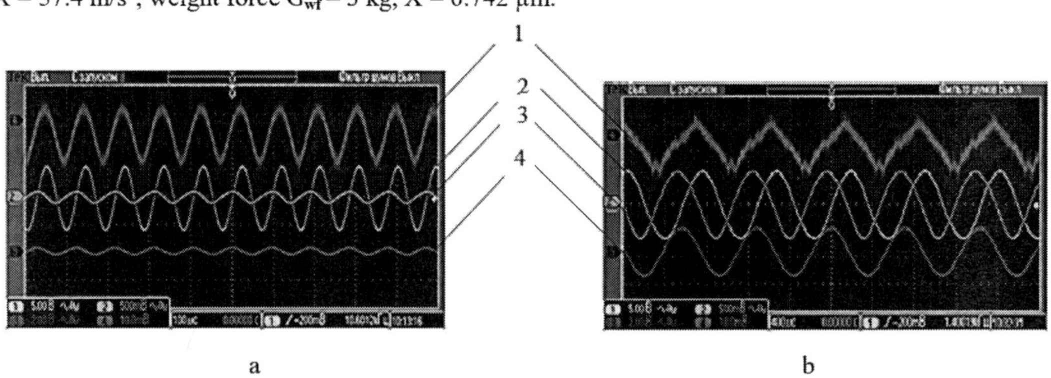

Figure 3. Signal distribution from the oscillatory system transducer to the electron oscillograph screen:a –1st operation mode; b –2nd operation mode.1 –acceleration; 2 – PZT Stack current;3 – PZTStack force; 4 – PZT Stack voltage.

MLPD frequency-response characteristics without preliminary tension and mass load of 0.5 kg are illustrated in figure 4. These characteristics have four distinct resonances. This effect could be based on the fact that separate PZT Stack elements in SC resonate differently and the system does not have a unified operation mode. The most energetic SC mode is at the frequency of 750 Hz (at current I = 0.5 A and load mass G_{wf}= 0.5 kg). It is within the range of this frequency that the vibration displacement and forces acting on the load are more significant (figure 4). In the specific frequency points, the vibration displacement has the following values: 1 – 29 μm, 2 – 9.32 μm, 3 – 0.735 μm, 4 – 0.002 μm, 5 – 0.0013 μm, 6 – 0.0013 μm.

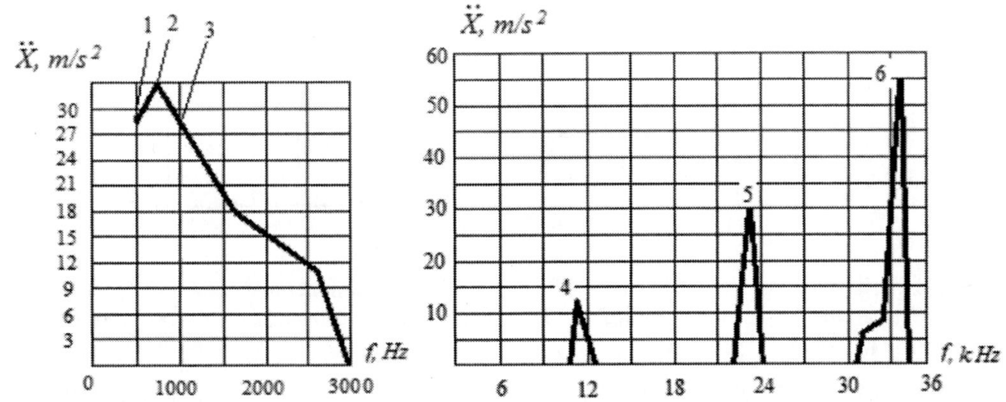

Figure 4. Frequency characteristics of MLPD vibration acceleration, operating under inertia load without preliminary tension

Increasing inertia load furthers decreasing resonant frequency and increasing vibration displacement (figure5), at current I = 0.5 A and preliminary tension F_0 = 240 N.

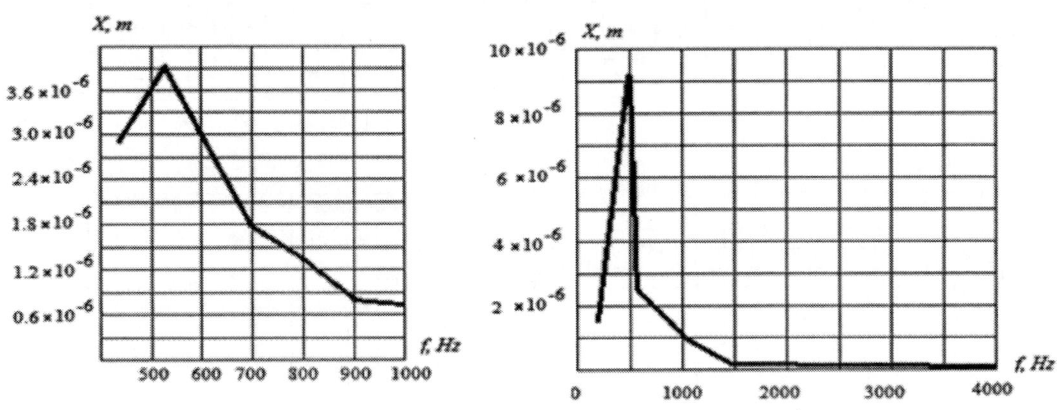

Figure 5. Frequency-response characteristics of operating MLPD at different inertia loads and preliminary tension: a – mass load G_{wf} = 0.4 kg; b – mass weight G_{wf} = 0.5 kg.

If inertia mass is 3 kg, the resonant frequency is 300 Hz, load force — F_0= 550 N (figure 6, curve 2) and vibration displacement — 12 μm (figure 7, curve 1). Experimental resonant frequency characteristics and vibration displacement amplitudes depending on inertia mass weight values are illustrated in figure 7 a, b.

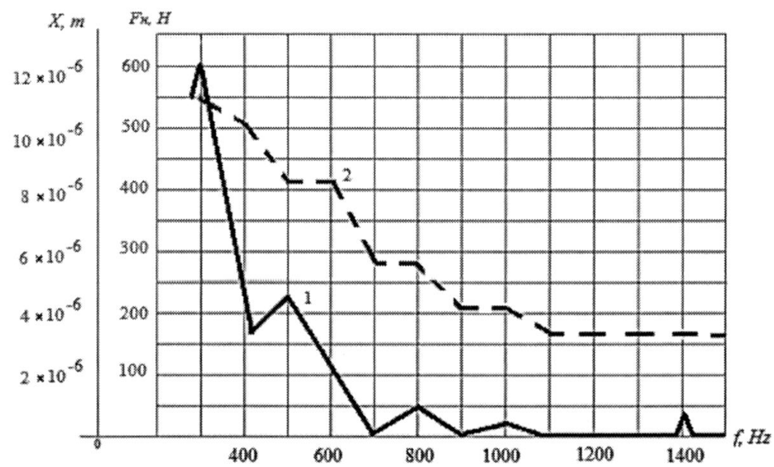

Figure 6. Experimental frequency characteristics of vibration displacement load and forces acting on load of operating MLPD: I = 0.5 A and G_{wf}= 3 kg. Preliminary voltage F_0 = 240 N. Frequency effect interval – 100 Hz

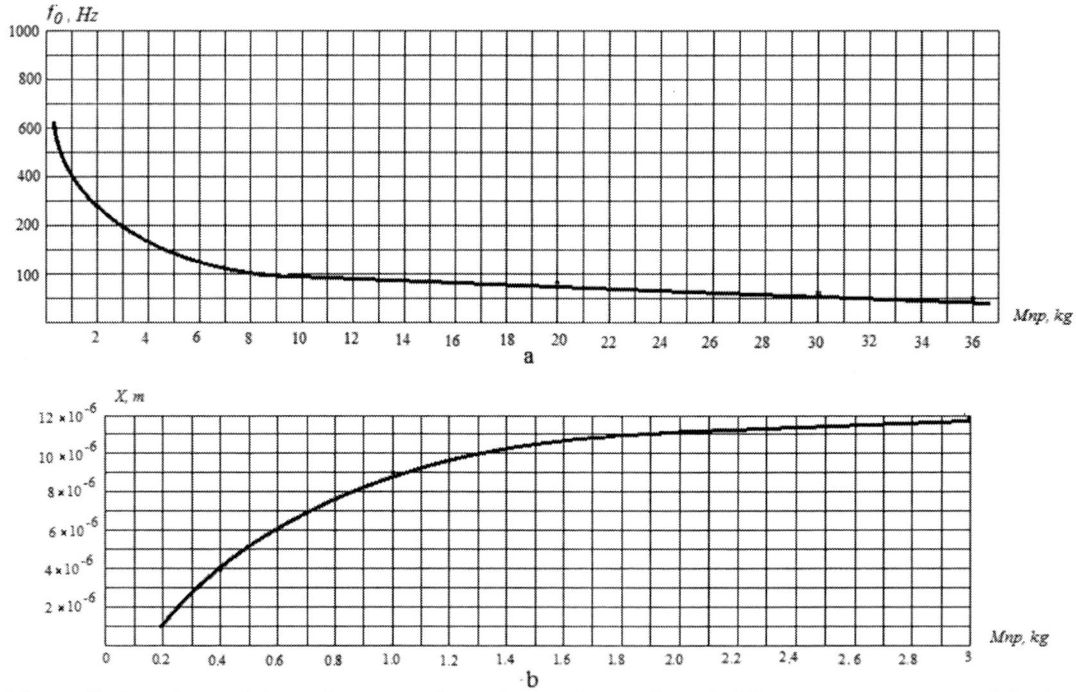

Figure 7. Experimental dependence: a – dependence of operating MLPD resonant on mass weight; b – dependence of weight vibration displacement of operating MLPD resonant on mass weight. Mass weight – up to 3 kg.

The basic MLPD operating frequency interval will be 50 – 100 Hz for a SC with the corresponding peripheral cord tensioner unit (PCTU) load parameters, as this is the most energetic operating interval of this frequency system. In this case, the excitation force on the load itself would be more than 550 Hz, the vibration displacement interval – more than 12 μm.

5. Conclusion

1. The experiment showed that the test stand is designed as a spacecraft (SC) involving only inertia load, however, the operating SC modes are quite similar to the operating modes with combined loads. SC resonant frequencies decrease with the increase of the inertia mass (typical for inertia load), while vibration displacement amplitudes increase with the decrease of the inertia load, which indicates the fact that the elastic component exists in the SC, i.e. preliminary tension elasticity, elasticity of stand supporting legs, PZT Stack elasticity which, in its turn, is characteristic of combined load.

2. The basic MLPD operating frequency interval will be 50 – 100 Hz for a SC with the corresponding peripheral cord tensioner unit (PCTU) load parameters, the vibration displacement interval of more than 12 μm.

3. Experimental data are in good agreement with the numerical experiment results on 3D and one-dimension mathematical models. In case SC has inertia load and is a resonant system, according to calculation results, it is possible to match the MLPD design to desired operating frequency resonant intervals.

4. The analysis of experimental data and numerical experiments showed that designed MLPD specifications could meet the requirements applicable to the peripheral cord tensioner unit in the reflecting surface shape regulator system for large-sized transformable spacecraft antenna reflectors.

Acknowledgments
This work was financially supported by the Ministry of Education and Science of Russia; unique identifier RFMEFI57814X0060.

References

[1] Park S 2011 *Single vibration mode standing wave tubular piezoelectric ultrasonic motor* (Toronto: Ryerson University)
[2] Wang Z, Li T and Cao Y 2013 *Aerospace Science and Technology* **26** 160
[3] Sanguinetti B and Varcoe B 2006 *Cryogenics* **9** 694
[4] Hsin-Jnag S, Faa-Jeng L, Li-Tao T and Po-Kai H 2006 *IEEE Trans. Ultrason. Ferroelec. and Freq. Contr. (Boston)* vol 53 (New York: IEEE Publisher) p 1649
[5] Ponomarev SV, Rikkonen SV and Azin AV 2014 *News of Higher Educational Institutions. Physics* **2** 196
[6] Ponomarev SV, Rikkonen S, Azin A, Karavatskiy A, Maritskiy N and Ponomarev SA 2015 *Advanced materials in construction and engineering* (Tomsk) vol 71 (Bristol: IOP Conference Series: Materials Science and Engineering) p 1-7
[7] Davoudi S 2012 *Effect of Temperature and Thermal Cycles on PZT Ceramic Performance in Fuel Injector Applications* (Toronto: Department of Mechanical and Industrial Engineering University) 99
[8] Henderson DA 2006 *10-th International conference on new actuators (Bremen)* vol 1 (Bremen: HVG Hanseatische Veranstaltungs-GmbH Division Messe Bremen) p 252
[9] Henderson DA 2007 *NSTI Nanotech (Santa Clara)* vol 1 (Washington: Nano Science and Technology Institute) p 17
[10] Henderson DA and Sheryl L 2006 *Electronic products (New York)* vol 1 (New York: New Scale Technologies) p 2

IOP Conf. Series: Materials Science and Engineering **177** (2017) 012004 doi:10.1088/1757-899X/177/1/012004

The method providing fault-tolerance for information and control systems of the industrial mechatronic objects

E V Melnik[1], **A B Klimenko**[2] **and V V Korobkin**[2]

[1] Southern Scientific Center of the Russian Academy of Sciences, 41, Chekhov Str., Rostov-on-Don, 344006, Russia
[2] Scientific Research Institute of multiprocessor computing systems, Southern Federal University, 2, gsp-284, Chekhov Str., Taganrog, 347900, Russia

E-mail: anna_klimenko@mail.ru

Abstract. The paper deals with the provision of information and control system fault-tolerance. Nowadays, a huge quantity of industrial mechatronic objects operate within hazardous environments, where the human is not supposed to be. So the question of fault-tolerant information and control system design and development becomes the cornerstone of a large amount of industrial mechatronic objects. Within this paper, a new complex method of providing the reconfigurable systems fault-tolerance is represented. It bases on performance redundancy and decentralized dispatching principles. The key term within the method presented is a 'configuration', so the model of the configuration forming problem is represented too, and simulation results are given and discussed briefly.

1. Introduction

Nowadays, industrial mechatronic objects (IMOs) are widespread within such areas as nuclear energy plants, autonomous production, hazardous industry, etc. IMOs are used frequently in industries with hazardous environments where IMOs have to operate without maintenance for a long period of time [1]. So, IMO reliability becomes the cornerstone of an important practical and scientific direction.

IMOs in general are considered as systems of sensors, actuators, communication environment, power and a control system. According to this IMO representation, the typologies of IMO faults are made: as an example, figure 1 represents the percentile values of the IMO subsystem impact onto the IMO mission success, while figure 2 contains the typology of IMO failure sources [2].

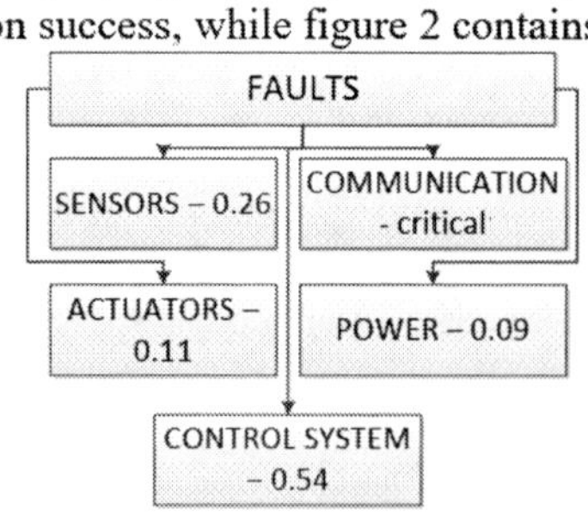

Figure 1. The IMO subsystem impact onto the mission success.

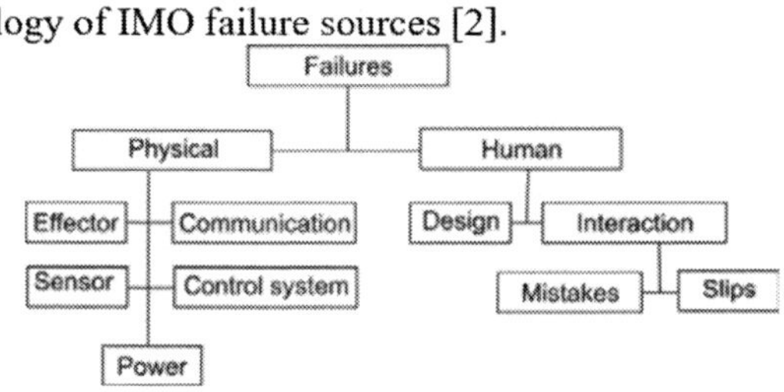

Figure 2. The IMO failures sources typology.

As it is shown in Figures 1 and 2, the faults of physical IMO subsystems have a great impact on the system operability. In the scope of this paper, the IMO control systems (more precisely - information and control systems (ICSs)) are under consideration: it is obvious, that the reliability of ICS is one of the most important aspects of the reliable IMO design and development.

Contemporary ICSs are the complexes of computational units, communicational environment and software. The ICS hardware components are developed, verified and tested; there are some prospective methodologies and techniques for very-large-scale integration (VLSI) devices development and verification: for instance, VHDL, Verilog, SystemC. Using appropriate coverage metrics, VLSI is verified, but it is useful to remember that, for example, for most coverage metrics, constrained random verification achieves 80% of coverage rapidly and then asymptotically approaches 100% with test redundancy and time overheads [3]. The slow convergence of random conditions to complete coverage is well known and referred to as the 'coupon collector problem' [4]. As the reliability of IMOs is incredibly important, and it is almost impossible to provide full verification coverage, it is necessary to provide additional methods and techniques for the ICS fault-tolerance providing.

The basis of the fault-tolerant reconfigurable IMO systems is the redundancy. Until recently, the structural redundancy was the key method for the fault-tolerance via reconfiguration provision, but contemporary hardware and software allow using another redundancy type – the performance one [5, 6]. Besides performance redundancy, decentralized dispatching methods for the net-centric ICSs were synthesized and used [5, 6], so the main bottleneck of the centralized dispatching was taken away.

Within this paper, the method of IMO ICS fault-tolerance provision is represented. The subsequent sections contain a brief review of the performance redundancy concept, the formalization of the configuration forming problem, an algorithmic description of the reconfiguration procedure and some experimental results and discussion.

2. Performance redundancy and decentralized dispatching principles

Structural redundancy is well-known and used widely in different types of ICSs. For instance, almost all modern aircrafts such as Boeing 777 and Airbus A320/330/340 have used triplex- or quadriplex-redundant activation systems, flight control computer and databus systems [7, 8]. Performance redundancy is an alternative way to organize a redundancy in the system: instead of reserve computational units (CUs), there is a performance redundancy within each CU. So, in case of CU fault, monitoring and control tasks can be re-launched on the operational nodes. To minimize the impact of faulted CU onto the system, the fault-tolerance providing method is enriched by the decentralized dispatching principles: every CU is controlled by its own software agent, which is a special type of the monitoring and control task. The problem to be solved by agents is to provide the reconfiguration via the cooperative action.

The basis of fault-tolerance provision presented within this paper is a combination of the performance redundancy and decentralized dispatching.

However, to reconfigure IMO ICS, there must be a set of pre-defined and verified system states described according to the fault-tree analysis – a set of configurations. Within this paper, a term 'configuration' means a control and monitoring tasks(CMTs) distribution among the CUs. The configuration forming problem is multiobjective and multiconstraint; the configuration set is designed and verified at the system design stage.

A generalized model of the configuration forming problem is represented in the next section.

3. The model of the configuration forming problem

Let the input data be the following:

- A set of CMTs $G=\{x_i\}$, $i=1...N$, where x_i – the size of task i, N – the number of tasks.

- Let $G = G_c \cup G_{nc}$, $G_c \cap G_{nc} = \varnothing$,where G_c – a subset of critical MCTs, G_{nc} – a subset of non-critical ones. Non-critical CMTs can be stopped or eliminated from the system during reconfiguration. The number of critical CMTs is N_c, the number of non-critical CMTs N_{nc};
- Let G_f be the set of CMTs from the faulted CU. $G_f \subseteq G$, G_p are the performing tasks, $G_p \subseteq G$, $G_p \cap G_f = \varnothing$.
- A planned completion time for set G is T_{plan}.
- The number of CUs is M with equal performance p.

Let us take into consideration the fact that we have to allocate the CMTs from set G_f within the system of operational CUs, on which the tasks from set G_p are allocated with the constraint of completion time T_{plan}. Let the resource allocated by CU j for subtask i be λ_{ij}. The tasks allocation before the failure is described by matrix R:

$$R = \begin{vmatrix} r_{11} & r_{12} & r_{1M} \\ \dots & \dots & \dots \\ r_{N1} & \dots & r_{NM} \end{vmatrix} \quad (1)$$

where $r_{ij} = f(\frac{x_i}{\lambda_{ij} p})$, $f(\frac{x_i}{\lambda_{ij} p}) = \begin{cases} \frac{x_i}{\lambda_{ij} p}, & if \quad x_i \text{ is running on CU } j, \\ 0, & otherwise. \end{cases}$

Let the failure occurred on the CU with number d. Column d of matrix R is deleted, so there are M-1 columns and N-$|G_f|$ lines in new matrix R_f. Renumber the elements of R_f in the following way, saving the indexes from matrix R in the upper positions:

$$R_f = \begin{vmatrix} r_{11}^{ij} & r_{12}^{ij} & r_{1(M-1)}^{iM} \\ \dots & \dots & \dots \\ r_{N-G_f}^{Nj} & \dots & r_{(N-G_f)(M-1)}^{NM} \end{vmatrix}. \quad (2)$$

R_f describes the system state before the reconfiguration and contains the allocation of the operational tasks among the operational CUs. R_r will be the allocation of task set G on the M–1 CUs. Formally, subset G_f will be added to G_p with the number of $CU_s = M - 1$:

$$R_r = \begin{vmatrix} r_{11}^{ij} & r_{12}^{ij} & r_{1(M-1)}^{iM} \\ \dots & \dots & \dots \\ r_{N1}^{Nj} & \dots & r_{(N)(M-1)}^{NM} \end{vmatrix}; \quad (3)$$

$$r_{ij}^{kl} = f(\frac{x_i}{\lambda_{ij} p}) g(x_i);$$

$$g(x_i) = \begin{cases} 0, if \ x_i \in G_{nc} \text{ and removed from the system}, \\ 1, otherwise. \end{cases} \quad (4)$$

Let us consider matrix Φ.

$$\Phi = \left[\varphi(x_{1j}^{kl}) \quad \varphi(x_{2j}^{kl}) \quad \dots \quad \varphi(x_{Nj}^{kl}) \right] \quad (5)$$

where $\varphi(x_{1j}^{kl}) = \begin{cases} 0, & l = j, \\ \xi, & otherwise \end{cases}$.

k, l – the saved indexes of matrix R, j – the number of CU in matrix R_r, ξ – the integer number. Matrix Φ describes whether the CMT x_i was relocated from CU l to CU j.

The first objective function can be written in the following manner:

$$F_1 = \sum_{i=1}^{N} \varphi(x_{ij}^{kl}) \to MIN \cdot \tag{6}$$

The maximum number of non-critical tasks running equals the maximum sum of all $g(x_i)$ in matrix R_r. The next objective function component can be presented as following:

$$F_2 = -\sum_{i}^{N} g(x_i) \to MIN \cdot \tag{7}$$

Herewith, the time constraint must be satisfied:

$$\forall j : \sum_{i=1}^{N} r'_{ij} \le T_{plan},\ j \in [1...M],\ r'_{ij} \in R_r\,. \tag{8}$$

Then, another criterion must be added to our model – a load-balancing objective function. As it was shown in [5-7], load balancing has a great impact onto the reliability function of CUs and, hence, onto the system reliability. It can be put in the following way:

$$F_3 = MAX(\sum_{i=1}^{N} \lambda_{ik} - \sum_{i=1}^{N} \lambda_{il}) \to MIN, \forall k, l, \lambda_{ik} \subset R_r \cdot \tag{9}$$

With expression (9), the multicriteria configuration forming problem can be written as a one-criteria optimization problem:

$$F = \sum_{i=1}^{N} \varphi(x_{ij}^{kl}) \to MIN\,; \tag{10}$$

$$MAX(\sum_{i=1}^{N} \lambda_{ik} - \sum_{i=1}^{N} \lambda_{il}) \le \gamma\,;\quad -\sum_{i}^{N} g(x_i) \le \mu\,;\quad \forall j : \sum_{i=1}^{N} r'_{ij} \le T_{plan},\ j \in [1...M]\,;$$

$x_i > 0;\ 0 < \lambda_{ij} < 1$, where $0 < \gamma < 1$ is the assumed level of load dispersing, μ is the constraint for the eliminated non-critical tasks.

4. The reconfiguration method and the configuration forming problem

The general reconfiguration method on the basis of performance redundancy and decentralized dispatching is described below. Every CU controlling agent gets the configuration tables through the system initialization. Also, we assume that all agents are fully interconnected and sent to the communicational network with a 'presence' sign. When the failure occurs, each agent is informed about the fault, and the new appropriate configuration is loaded from the agent's own configuration storage. If the new configuration is adequate to the system state (all agents are active), control and monitoring tasks from the faulted node are run on the other nodes according to the configuration due to the performance redundancy.

A set of configurations is a key of a reconfiguring procedure, and has to be designed according to the fault-tree analysis. As it was shown in the previous section, the configuration forming problem is multiobjective, has constraints and, in general, np-hard. So, the forming of the configuration set can take inacceptable periods of time.

In order to reduce this time-consuming procedure, the parallel stochastic search was used, more precisely – asynchronous parallel simulated annealing via parallel multistart. Simulated annealing in general has slow convergency, but some modified methods allow obtaining the solution within the satisfactory time period [9-11]. To solve the configuration forming problem described above, we used a parallel multistart search with a quenching temperature scheme. This kind of modification of SA allows us to obtain the local minimas in the acceptable time period, and the best solution among the local ones is chosen.

5. Simulation results

Within this paper, two types of simulation were made: the first one demonstrates the convergency speed of the SA with the quenching temperature scheme, and the second one contains aggregated results of the parallel SA search procedure.

During the first simulation, a set of 50 monitoring and control tasks with randomized laboriousness was taken. The objective function was minimizing the eliminated non-critical tasks (other objective functions were transformed into constraints). In figure 3, the tracing results of SA are represented: it is seen that the quenching temperature scheme allows us to obtain solutions with a satisfactory quality within 10-20 iterations. Such results make SA the prospective technique in the area of configuration forming.

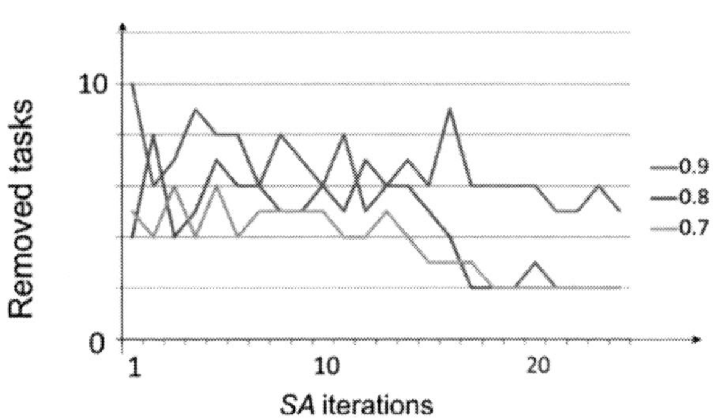

Figure 3. Simulation annealing convergency with quenching ratios 0.9, 0.8, 0.7.

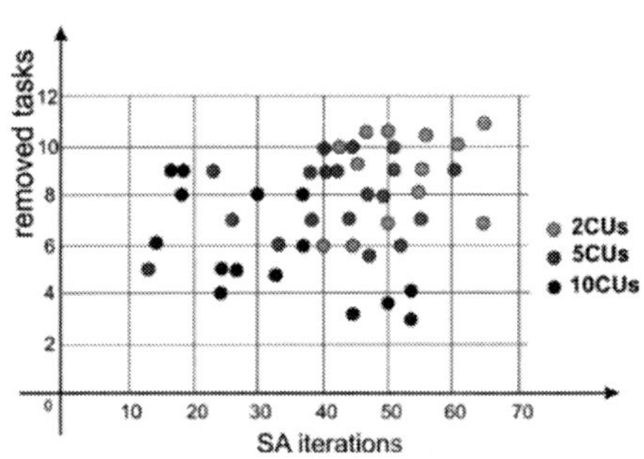

Figure 4. Solution quality for the 2, 5, 10 parallel search procedures.

The next simulation is for the comparison of different numbers of search processes of solution quality: 2,5 and 10 parallel searches. As shown in figure 4, the best results are obtained by 10 parallel searches on 10 CUs. These results are rather expectable: with the increasing of the initial search points number, the search results become better. With 50 control and monitoring tasks (30 critical, 20 non-critical), the task distribution with 3 eliminated tasks was obtained. It can be just an example of the local optimum, and, possibly, the better solution exists, but this type of solution was accepted by the authors of this paper as satisfactory.

6. Conclusions

A problem of fault-tolerance of IMO information and control systems is one of the contemporary fundamental science problems. Fault-tolerant reconfigurable systems, in general, use the structural redundancy, which can be inconvenient in many aspects: too expensive in development and maintenance, with the deficiency of proper dependability characteristics. Within this paper, the new fault-tolerance providing method based on performance redundancy and decentralized dispatching is presented.

The reconfiguration procedure of IMO ICS needs the set of configurations, which must be designed and verified at the design system stage. The configuration forming problem is non-trivial and time-consuming, so it is expedient to develop fast procedures to obtain the solutions. Parallel simulated annealing with the quenching temperature scheme allows one to obtain solutions of a satisfactory quality within the acceptable time periods, so these search techniques are prospective for the configuration forming.

As a conclusion, we can state that the area of IMO ICS research is developing intensively, so we look forward to the new principles, methods and techniques of IMO ICS fault-tolerance provision.

Acknowledgments

The reported study was funded by SSC RAS projects 0256-2014-0082, 0256-2015-0008 and by RFBR project 16-58-00191 bel-a.

References

[1] Zhang Y and Jiang J 2008 *Annu. Rev. Control* **32(2)** 229-52
[2] Carlson J, Murphy R 2005 *IEEE Trans. Robot.* **21(3)** 423-37
[3] Jerinic V, Langer J, Heinkel U and Miller D 2006 *Des. Aut. Test. Europe* **1** 1-6
[4] Blom G, Holst L and Sandell D 2004 *Probl. and Snapshots from the World of Probab.* **191** 85
[5] Melnik E, Klimenko A and Korobkin V 2015 *Proc. 10th Int. Conf. Application of Information and Communication Technologies* (Baku) (New York: IEEE) p 274
[6] Melnik E, Klimenko A and Korovin I 2016 *Lecture Notes in Comput. Sci.* **9812** 9-17
[7] Bartley G 2001 Boeing B-777: fly-by-wire flight controls *The Avionics Handbook* ed Spitzer C R (Boca Raton: CRC Press)
[8] Biere D, Favre C and Traverse P 2001 Electrical flight controls, from Airbus A320/330/340 to future military transport aircraft: a family of fault-tolerant systems *The Avionic Handbook* (Boca Raton:CRC Press)
[9] Kirkpatrick S, Gelatt C and Vecchi M 1983 *Science* **220 (4598)** 671-80
[10] Ingber L 1993 *Mathl. Comput. Modelling* **18** 29-57
[11] Azencott R 1992 *Simulated Annealing: Parallelization Techniques* (New York: John Wiley and Sons)

Improving machine operation management efficiency via improving the vehicle park structure and using the production operation information database

V Yu Koptev

Saint-Petersburg Mining University, 21-st line, Saint-Petersburg, 199106, Russia

E-mail: kvuy@mail.ru

Abstract. The work represents the results of studying basic interconnected criteria of separate equipment units of the transport network machines fleet, depending on production and mining factors to improve the transport systems management. Justifying the selection of a control system necessitates employing new methodologies and models, augmented with stability and transport flow criteria, accounting for mining work development dynamics on mining sites. A necessary condition is the accounting of technical and operating parameters related to vehicle operation. Modern open pit mining dispatching systems must include such kinds of the information database. An algorithm forming a machine fleet is presented based on multi-variation task solution in connection with defining reasonable operating features of a machine working as a part of a complex. Proposals cited in the work may apply to mining machines (drilling equipment, excavators) and construction equipment (bulldozers, cranes, pile-drivers), city transport and other types of production activities using machine fleet.

1. Introduction

Nowadays, mining, construction and other works use more than half a million of various models of machines. The most efficient way of using them is via machine complexes (comprehensive mechanization demands) that are established in industries: the mining industry deals with excavator vehicles, cargo handling complexes on transport; the construction industry deals with various construction machines, depending on the particular work to carry out.

A complex includes various machines for different purposes to be grouped, depending on the amount of work, into machine marks with a varying amount of similar equipment (i.e. excavators, bulldozers, dump trucks, cranes etc.). The parks include machines with different characteristics (i.e. tonnage, capacity, productivity, parts implementation). Since machines are expensive, while production and technical operation is costly either, the park structure must meet the demand for efficient operation of the machines with required productivity, safety guaranteed in combination with reduced time waste and expenses.

A basic element in any control system is a controlled object. Regarding mining facilities transport systems, a controlled object may be equipment for a technological process and the technological process itself. Depending on the intended purpose, the controlled object may possess input and output parameters, i.e. material, power and information flows incoming and outgoing, as well as it may influence the object by deflecting output values compared to the specified one. Creating complex parameters allows the most efficient selection of a particular system automatically, with dispatching open-cut transport, justifying specific equipment unit selection and the number of units in the fleet.

2. Materials and methods

Let us consider open-pit mining using open-pit dump trucks as an example of opportunities to manage the dump trucks park structure (operating certain amount of dump trucks from different producers, different models from the same producer, with different ages) to guarantee an efficient use of machines, material and labor resources.

The amount of resources mined in Russia constantly grows, with most resources mined via open-pit mining to be transported by vehicles (up to 70%). Developing an open pit further results in deepening the location and increases the transportation distance.

Automation and dispatching systems nowadays used in open pit resource transportation allows monitoring the mining works and managing it via transport vehicle redistribution. For example, should some dump trucks or excavators be rendered inoperable, the traffic situation deteriorate, the resource mined volume to load change, the system software (and the dispatcher, if necessary) allow selecting the solution to maintain transportation efficiency. It is back in 2000s, when the systems from the companies Micromine Pty Ltd. (Western Australia), Wenco International Mining Systems Ltd. (Canada), Modular Mining Systems Inc. (USA), LLC VIST Group and CJSC NPO Sojuztechnocom (Russia) etc. were implemented in large open pit mining sites to allow accumulating much information on dump trucks operation.

The mining transport complex dispatching system is a complex of vehicle-borne, navigational and radio equipment allowing automated accounting of technical and operational parameters of vehicle operation, it has a dispatching center and vehicle-borne systems mounted on dump trucks. Those systems continuously collect information: about a vehicle location in the coordinate system, the distance covered, the dump truck speed, loaded weight, residual fuel in fuel tanks, the engine and other units operation time. All the information collected is transferred over the radio (GSM antenna) to the dispatching center for further processing and storing in the database on the working shift basis or daily, or monthly.

A dump truck is routed to minimize time waste and to maximize excavator productivity.

In turn, the volume of information (the amount of parameters and the information update frequency) is defined by the formulated tasks of managing the excavator and vehicle complex. Major factors for determining the operative management tasks are the rock mass volume which should be transported, total demand for loading/unloading equipment etc.

It is proposed to use the information to form the best dump truck park management structure via selecting the model and determining the best number of machines.

When the amount of work changes, it is virtually impossible to quickly change the park of machines and replace the models the machine park comprises. Increasing the number of machines to operate is costly and time consuming. Reducing the number of machines results in losing the productivity and time waste on machines unused, and that means financial losses. Practically, when scheduling mining works, determining the number of working machines is followed by determining the reservation factor to increase the number of dump trucks in a park 1.3 … 2 times.

Today, there are models [1] and methodologies [2, 7] developed to form dump truck parks by profitability criteria that use operation criteria related to various types of dump trucks throughout their entire service life. A park is formed with minimum operation, ownership and service costs according to statistics available for various types of dump trucks used in Russia and abroad.

An improvement also counts the park of machines to be used for 5-10 years and operated on constantly changing routs. It is necessary to finish the park formation with the following criteria taken into account: stability [6] and transport flow to compare transport vehicles [3-5] used.

It is also necessary to include the machines from at least two producers into the park to provide a better service and easier repairs and to reduce the cost of accessories and spare parts, and also repair workers' and drivers' skills are thus improved. A dump truck park formation algorithm (figure 1) for a mining enterprise takes the following technological and production parameters: productivity, dimensions, cost, reliability, operating costs etc.

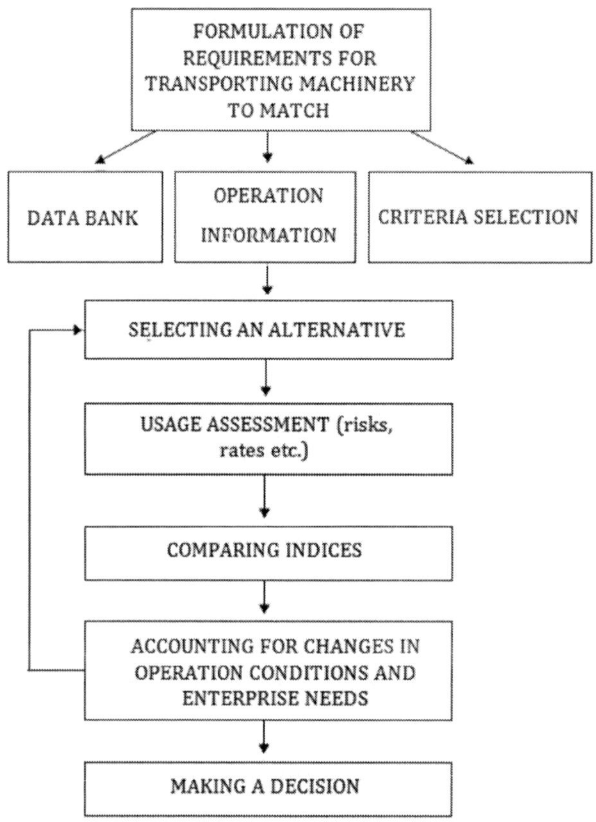

Figure 1. The dump truck park formation algorithm.

To improve production enterprise management efficiency at the level of the machine park structure, one needs, based on outsourcing, to establish centers of technics with a function of conducting technical operation and leasing service provision. The conditions that are favourable for launching these centers are present in mining clusters of Ural region, Yakut republic, Karelia, Krasnoyarsk region, Northwest region (Leningrad and Murmansk areas).

Cargo transportation from the production place (mining) to the delivery point (consumption or processing) is to be considered as a material flow motion within the space-time coordinate system. It has three characterizing parameters: volume Q[t], time T[h] and distance L[km]. And it is in one of 4 states: loading-transportation-unloading-storage. The value of transportation services represents the integral parameter, with the physical meaning of reflecting the value of the power spent by the transportation machine within the transportation time [8-10].

The total efficiency of the transportation service, Em, accounting for power spent, would be defined as the ratio between the transportation work carried out and the power spent:

$$Em = (Q \cdot L) / (Q \cdot L^2\, T^{-3}) = T^3 / L. \quad (1)$$

And the ratio between the transportation service provided and the work, Ep, carried out is:

$$Ep = (Q \cdot L)(L \cdot T^{-1}) / (Q \cdot L^2 \cdot T^{-2}). \quad (2)$$

Bigger values of Em and Ep mean better usage of transportation machines. Calculation results for comparable conditions depending on inclination are represented in figure 2.

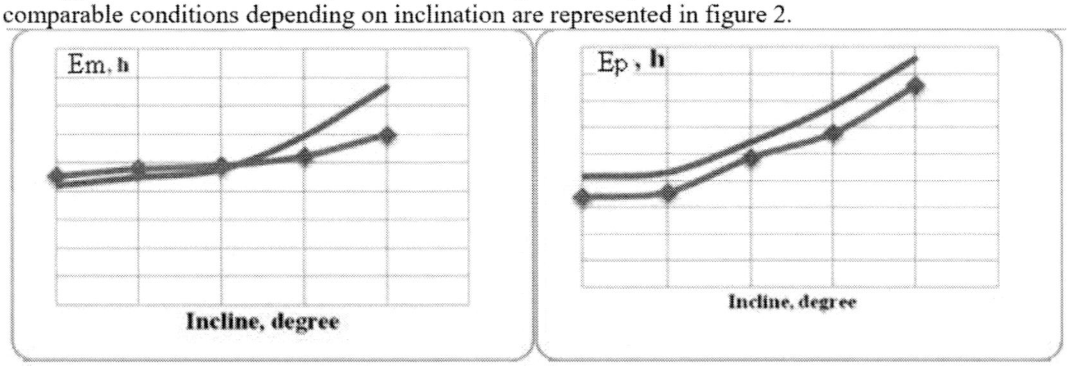

Figure 2. Changes in Em and Ep indicators depending on the angle of inclination: the red line for the dump trucks with load capacity of 55 120 tons; the blue line - railroad trains with dump cars, having load capacity of 60 and 120 tons.

3. Batch-continuous transport system parameters calculation

Designing transportation systems for transportation machine complexes with different principles of operation – batch and continuous ones – suggests that its logistics requires consideration of the transportation process in general. The complex operation efficiency is provided by the continuous transportation process implementing hopper storage devices [11].

For example, single-stall methods of mining potash utilize the complex of mining transportation equipment that includes a tunneling machine, a self-driving wagon, a reclaim hopper and a movable hopper (figure 1) [12]. The latter is not always used for unloading the wagon onto a drag-type conveyor. The complex consisting of a tunneling machine and a self-driving wagon only, is rarely used, mainly for short distances: during cutting, driving, short workings etc.

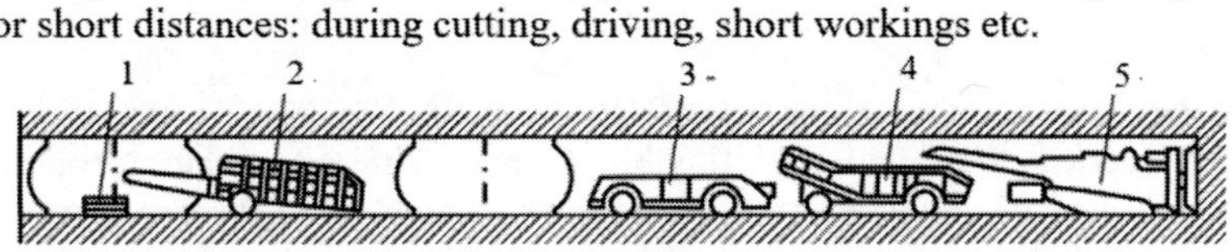

Figure 1. A tunneling complex for ore mining: 1 - a drag-type conveyor; 2 – a hopper; 3 – a self-driving wagon; 4 – a reclaim hopper; 5- a tunneling machine.

The main disadvantages of wagons is the impossibility to handle the ore in a chamber in case the mainstream transport fails, the performance depending on the delivery distance, and the presence of a cable that limits the transportation length [13, 14].

A distinctive feature of the discussed complex operation is the existence of the critical delivery distance, exceeding which results in the tunnel machine idle time waiting for a wagon, thus significantly compromising the performance of the complex. The critical delivery distance may be derived from comparing complex machine operation times during a single transportation cycle. The operation time of the reclaim hopper per cycle is:

$$A = t_{z.p} + t_{r.p}, \quad (1)$$

where $t_{z.p} = G_{b.p}/Q_k$ is the time of loading a reclaim hopper by the tunneling machine, in minutes; $G_{b.p}$ is the loading capacity of the reclaim hopper (the wagon), ton; $Q_к$ is the technical performance of the tunneling machine (in machine hours), t/min; $t_{r.p}$ is the unloading time of the reclaim hopper into the self-driving wagon per cycle, in minutes.

The hopper operating time is:

$$B = t_{z.v} + t_d + t_{r.v}, \quad (2)$$

where $t_{z.v}$ is the time of loading the wagon, in minutes; $t_d = 2L/\upsilon$ is the total time of motion for the loaded and empty wagon, in minutes; L is the delivery distance, in meters; υ is the equivalent speed of the wagon, m/min.; $t_{r.v}$ is the wagon unloading time, in minutes.

The equivalent speed of the wagon shall be:

$$\upsilon = \frac{2\upsilon_g \cdot \upsilon_p}{\upsilon_g + \upsilon_p}, \quad (3)$$

where υ_g , υ_p are the average speeds for the loaded and the empty wagons, correspondingly, m/min.

We take the reclaim hopper's and the wagon's operation time as equal during unloading, as the amount of the unloaded ore and the unloading method (by a bottom conveyor) is the same. Therefore, with accuracy good enough for calculations, we assume $t_{r.p} = t_{r.v} = t_r$. Then, believing that $t_{z.v} = t_{r.p}$, we have:

$$A = q/Q_к + t_p; \quad B = 2L/\upsilon + 2t_p. \quad (4)$$

If delivery distance L is such that $B < A$, then the wagon is idle waiting for the tunneling machine to fully load the reclaim hopper. If $B > A$, then the tunneling machine is idle, having finished with loading before the wagon arrives. Critical delivery distance L_{cr} is reached when $A = B$:

$$L_{cr} = \frac{\upsilon}{2}\left(\frac{q}{Q_к} - t_r\right). \quad (5)$$

The discussed methodology is applicable for other mining transportation complexes used in open and underground mining works to determine the required performance of batch and continuous operation machines establishing the complex, and to determine the capacity and the layout of accumulation tankages in combined transportation schemes.

4. Conclusion

Production transport network automation is complicated due to a vast range of technologies employed in production. It necessitates creating a flexible system with adjustable technologies, with the performance dependent on the abovementioned complex parameters of the transport unit types as well as operating features thereof, such as the total efficiency of the transportation service, Em; the operating time, B; the equivalent speed, v; the critical delivery distance, L_{cr}.

The proposals presented herein may be applied to mining machines (drilling machines, excavators) and road construction machines (bulldozers, cranes, pile driving machines), municipal public transport in cities, as well as other types of production activities that use vehicles and advanced dispatching systems.

References

[1] Anistratov K Yu, Stremilov V Yu and Teterin M V 2007 *Open pit technological transport vehicle enterprise functioning economic and mathematical model* vol 2, ed E Gorgelyak (Moscow: Gemos Limited Mining Industry) pp 16-18

[2] Anistratov K Yu 2010 *The dump trucks park structure justification for vehicle load classes 40-100 tons* vol 5, ed S Ivanov (Moscow: Gemos Limited Mining Industry) pp 26-29

[3] Koptev V Yu 2012 *Justification of Mining Enterprises Transport Machines Selection Mining equipment* (St Petersburg: Slavutich) pp 58-61

[4] Koptev V Yu 2013 *I Interuniv. Academical and Research Conf. Innovative Systems of Planning and Management in Transport and Heavy Machinery* (St. Petersburg: SPGGU) pp 158-165

[5] Radkevich J M and Skhirtladze A G 2003 *Metrology, standardization and certification* (Moscow: Mining University) p 58

[6] Kuleshov A A, Mariev P L, Egorov A N and Zyrjanov I V 2004 *Open pit vehicles: nowdays*

and in future (Saint-Petersburg: Science) p 398

[7] Schadov M I and Fedorov A V 2009 *Open pit equipment park structure formation method for operating enterprise* vol 5, ed A Ustraitchovskyi (Moscow: Gemos Limited Mining Industry) pp 10-13

[8] Kuleshov A A 2003 Economic evaluation of the impact of environmental factors when choosing the best type of loading and transportation vehicle complex *Week of the miner* vol 12, ed G Goryachev (Moscow: MSMU) pp 152-154

[9] Radkevich J M and Belyankina O V 2006 The mathematical expression for the basic target functions of mining vehicles and machines *Mining informational and analytical bulletin* vol 2, ed D Voloshko (Moscow: Scientific and Technical J.) pp 262-268

[10] Keerthika R and Jagadeeswari M 2015 Coal Conveyor Belt Fault Detection and Control in Thermal power plant using PLC and SCADA *Int.l J. of Advanced Research in Computer Engineering & Technology* vol 4, ed M V Ragahavendra (India: IJARCET) pp 1649-52

[11] Alexandrov V I and Koptev V Yu 2013 Modern transport schemes of mining *Aerology and Security* vol 3, ed L Gitis (Moscow: Mining Book) pp 202-211

[12] Tarasov Yu D and Koptev V Yu 2012 *Mining and transportation machinery* (St. Petersburg: SPGGU) p 380

[13] Balabaev O T 2015 *Special Courses Conveying* (Karaganda:CarGTU) p 178

[14] Puhov Yu S 1991 *Mine trucks* (Moscow: Nedra) p 240

The algorithm of measuring parameters of separate oil streams components

A V Kopteva and I N Voytyuk

Saint-Petersburg Mining University, 21-st line, Saint-Petersburg, 199106, Russian Federation

E-mail: Alexandrakopteva@gmail.com

Abstract. This paper describes a development in the area of non-contact measurement of moving flows, including mass flow, the number of components and their mass ratios in a multi-component flow, as well as measurement of flows based on algorithms and functional developed for various industries and production processes. The paper demonstrates that at the core of the proposed systems, there is the physical information field created in the cross section of the moving flow by hard electromagnetic radiation. The substantiation and measurement of the information parameters are performed by the hardware and the software of the automatic measuring system. A new way of statistical pulsation measurements by the radioisotope technique is described, being alternative to the existing stream control methods and allowing improving accuracy of measurements. The basic formula fundamental for the method of calibration characteristics correction is shown.

1. Introduction

The subject on which we work refers to the general theory of streams. Streams can be divided into two types: homogeneous (uniform or Newtonian) streams and multiphase multiunit streams. They all are characterized by the fact that parameters and features of streams are changed by random stationary or non-stationary laws which are characterized in time by functions of 'hite noise' type. These high-frequency changes of random values trouble quantitative measurement of any stream parameter.

Talking about homogeneous streams, most often it is required to measure the discharge and the quantity of substances and materials. In liquid streams (oil, mineral pulps, etc.), this problem is solved easily – by integral discharge measurement, for example, measuring volumetric discharge by a turbine flowmeter that has high-quality accuracy and benefits industry.

Multiphase and multiunit streams are controlled worse or even poorly when it is essential to measure quantitatively, for example, one of the components of the stream (the amount of non-associated gas in oil, the amount of dead rock on a conveyor) [1]. Until now this problem used to be transcendental [2].

2. The need for instrument development

However, demands of our time, new technologies, new economic relations put a question point-blank – for example, what is the volume of non-associated gas in oil transported by a pipeline? [3] When the crude-oil production in Russia was about 600 million tons, even insignificant volume inclusions of gas created big ambiguity and contradiction between producers and consumers [4]. In the eighties of the last century, this type of ambiguity between two big states resulted in the two million tons imbalance that forced scientists to search for explanation of that imbalance. In the first decade of the twenty first century, it was discovered and proved that the reason was in the presence of non-associated gas in the stream (about 2% of the volume), but oil producers did not admit that fact until the last minute [5, 6]. The scientists from the Central Research and Development Institute of Robotics and Engineering Cybernetics, and the scientists from Research and Production Company 'Complex Resource' proved that fact with the help of experimental facilities at the laboratories and under production conditions. Whereupon the fact was admitted by the oil producers with great difficulties, and the adequate government standards were adopted and the process of instrument development began. The instruments of Research and Production Company 'Complex Resource' have passed the metrological examination and are placed in the oil industry register. We should mention that postgraduates Moiseev, Bragin, Levashov, etc. of Saint-Petersburg Mining University took part in the development, examination and implementation.

3. Description of the measurement algorithm

The essence of the project (Figure 1) is the following: the statistical pulsation method of measurement of the separate components of the stream was invented on the basis of a highly reliable, simple and endurable radioisotope densimeter. Figure 1 presents: 1 – he pipe section with a multicomponent flow moving with speed V; 2 – bubbles of free gas, 3 – the transducer, 4 – the light source (chemical element ^{137}Cs); 5 – direct radiation of detection block N1; 6 – an intelligent system (single-chip computer), 7,8 – scattered radiation of detection blocks N2 and N3, respectively, 9,10 – direct and scattered γ-rays, respectively. Oil flow and its amount in a pipeline are controlled by the turbine flowmeter or by the densimeter as follows: 'speed — section' (certainly, there should be marking in the stream to calculate the speed, for example, ionization marking). The individual components of oil stream are registered by transducers (to measure the flux density), previously calibrated for the gas content, due to the 'knockout' of its hard electromagnetic radiation (in this case, γ-radiation of isotopes of chemical elements ^{137}Cs). Processing a large amount of statistic material, one determines that the analytical form of the measured signal is random, provoked, for example, by accumulation of bubbles of non-associated gas. That is usually characterized by the correlation function or by the probability density function [7].

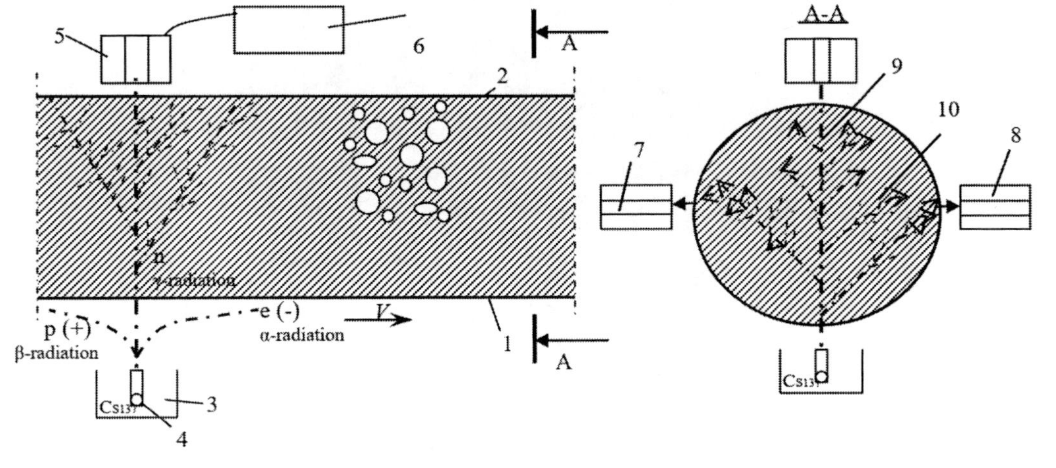

Figure 1. The structure of measurement parameters of separate components of streams.

The instrument that was previously calibrated on the metrological stand in the static gas content mode ought to be switched at the moment of bubbles passing (it is usually the length of 5-10 pipeline diameters) over calibrated 'gas' characteristics [8, 9]. This switching on a signal of the corresponding pulsing is produced by the single-chip computer of a secondary instrument. Calibration factors are corrected in a definite period of production activity.

4. Automatic updating of calibrated characteristics

We have found the way of enchancing the measuring accuracy of calibrating instrument, i.e. self-correcting transfer characteristics of the measuring system. The main point of the algorithm is as follows: the system on a main stream (oil, coal, gas, etc.) is calibrated in a static mode; according to sampling theorem, Δt_i – the time interval between measurements, that is important for exact reproduction of characteristics (of a random signal) of the measured parameter. For conveyor scales, Δt_i=0.02s; for an oil-trunk pipeline, Δt_i=0.2 s. According to the method of moving average during each measurement, the controller of a secondary instrument calculates the function value of the measured parameter according to the formula [10]:

$$y(t) = \frac{1}{2l}\int_{t-l}^{t+l} x(t)dt, \tag{1}$$

where y(t) – the value of the measured parameter; t – the time, s; l – the time period between measurements, s.

The value of measured parameter $y(t_i)$ is saved in the memory of instrument (basic calibrated characteristics) and is calculated according to the given formula (1). The value is improved by averaging and is recorded by the instrument memory unit as a new (improved) value of calibrated characteristics. During the whole period of the instrument work, new characteristics are averaged with previously iteratively averaged calibrated characteristics by the Monte-Carlo method. Figure 2 explains the process of averaging the calibrated characteristics, where 1 – calibrated characteristics which are recorded by the processor of the secondary instrument; 2 – calibrated characteristics which are calculated by the processor according to the method of moving average; 3 - improved calibrated characteristics obtained according to the Monte-Carlo method.

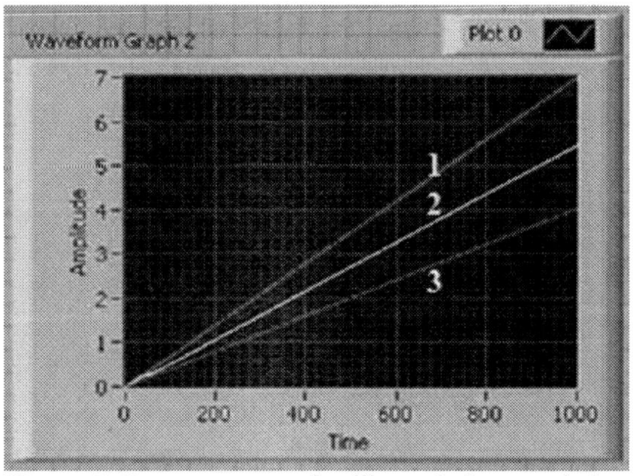

Figure 2. Improvement of calibrated characteristics.

5. Measurement methodology

With RAIS, we may obtain the information about:

– Pipeline integrity even for quite a long pipeline;
– The technical condition of the pipeline;
– The content of gas, water and other components in a large amount of oil;
– The type of the oil transported;

– The content of paraffin residing on the pipeline's inner wall;
– The flow rate.

For example, lying in the basement of the liquid and gas flow rate measurement, there is the area-speed method, developed in the Mining University. This method accounts for the rate of each component in a multiphase and multicomponent flow, determining it by multiplying the average flow rate per component by the area of a part of the cross-section of the flow occupied by this component to be measured with the proposed measuring system. Therewith, the averaged mass flow of oil shall be defined by the following expression:

$$Q_{mdl.qty} = Q_{qty}\,(1 - W_{mdl}), \tag{1}$$

where $Q_{mdl.qty}$ – the mass flow of oil averaged by time T_{mdl}; W_{mdl} – the mass fraction of water in the crude oil averaged by time T_{mdl}.

Oil weight m_o shall be defined by the following expression:

$$m_o = T_{mdl} \cdot Q_{mdl.qty} \tag{2}$$

The flow rate of the crude oil flow is measured via the mark method. Impurities, for example, free gas bubbles in liquid, are used as natural flow marks. The speed of gas in this case is assessed by the average flow speed of all bubbles (bubbles with various sizes). The speed of liquid is thus assessed by the average flow speed of the bubbles within, which is less than some threshold value, and which, therefore, behave as if they had been "frozen" into the liquid, i.e. with no sliding relatively the liquid. Figure 3 presents the portion of a pipeline with two radioactive isotope measuring converters (RMC).

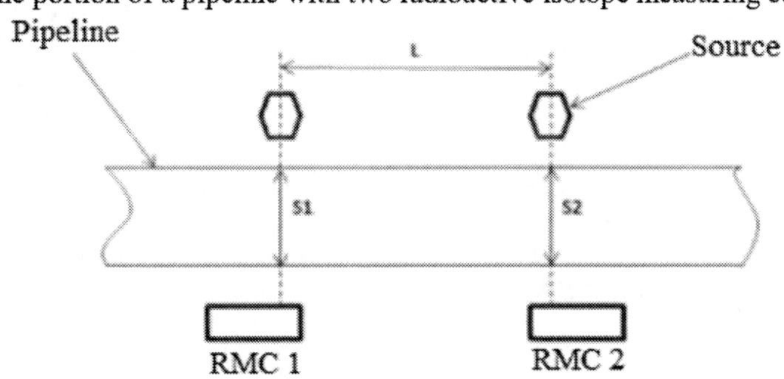

Figure 3. The portion of a pipeline with two radioactive isotope measuring converters (RMC).

S1 is the cross section of the pipeline, controlled by the first RMC;
S2 is the cross section of the pipeline, controlled by the second RMC;
L is the distance between sections S1 and S2.

6. Results of the research

As the gas bubbles are missing in the flow (the flow is single-phase and uniform), the output signals from RMC1 and RMC2 are the same. If there is a free gas bubble in the uniform medium flow, the bubble with the flow will be passing section S1 first, and then section S2; therefore, the output signals from RMC1 and RMC2 will then be like those presented in Figures 4 a, b. The bubble speed in Figure 4 a is twice as fast as that in Figure 4 b.

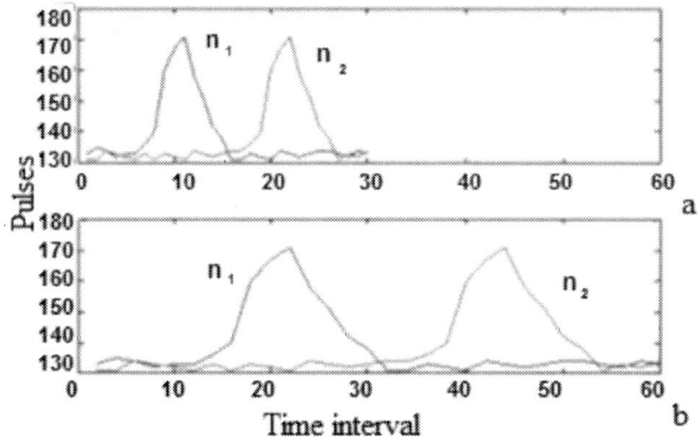

Figure 4. Output signals from RMC1 and RMC2 as the bubble is passing S1 and S2 sections.

Figures 4 a, b demonstrate that the RMC output signals comprise impulses n1 and n2 as the bubble is passing nearby. Impulse n2 comes with the delay after n1, and delay time τ is equal to L/V_b , where V_b is the speed of the bubble. Therefore, by measuring delay time τ of impulse n2 relatively n1, one could calculate the gas bubbles motion speed.

Since a gas bubble moves together with the liquid flow, the bubble motion speed is an indicator of the local (i.e. at the point where the bubble is being at the moment) speed of the liquid flow. In the real oil and gas mixture, not a single bubble moves, but rather a cluster of bubbles with different sizes does [11]. Therefore, measuring the output signal from RMC would give a more complicated result. RMC1 output signal fluctuations would be repeated by the RMC2 output signal, delayed by time τ. Time τ would depend on the speed values of all bubbles in the flow (i.e. located in various points of its cross section), i.e. on the average speed of gas V_g.

Inside the oil and gas mixtured flow, large bubbles of free gas mainly move through the central area of the flow, where the local speed is higher than that near pipeline walls. Smaller bubbles are distributed more or less evenly throughout the flow cross section, with their speed characterizing the average speed of the liquid flow. If fluctuations caused by the smaller bubbles were extracted from the RMC output signal fluctuation, then the time delay between RMC2 and RMC1 output fluctuations would characterize the bubbles motion speed thus revealing the average speed of the liquid flow itself [12].

If signals have random nature, the delay time between one signal and the other one is usually measured with the cross-correlation method. Changing the delay time of RMC 2 relatively RMC1 also changes the timescale (width) of RMC1,2 signal impulses (Figure 4). Higher speed compresses the RMC signals along the time axis, whereas the lower speed stretches them. Therefore, provided certain restrictions are imposed on RMC output signal properties, the time delay between RMC2 and RMC1 signals may be substituted by measuring the timescale (width) of the RMC 1-τ. Usually, the timescale (width) of the output signal is measured using the autocorrelation method (ACM) [10].

The speed of liquid and gas is calculated in accordance with the following expressions:

$$V_{lq} = L_{lq}/\tau_{lq} + V_{lq0}, \tag{3}$$

$$V_g = L_g/\tau_g + V_{g0}, \tag{4}$$

where L_g, L_{lq}, V_{lq0}, V_{g0} are calibration factors;

τ_{lq} is the width of the autocorrelation function, fluctuations, caused by "frozen-in" bubbles;

τ_g is the width of the autocorrelation function, fluctuations, caused by all bubbles within the flow.

7. Conclusion

So, the necessary multiplicity of measurements of the target value (0.2 s or 0.02 s), improvement of calibrated characteristics by any informative stream parameter that can be marked by registering the change in stream density correlated with the measuring value are performed in statistical measurements with the help of the radioisotope y-method. Even if the distinction of the oil density value is insignificant compared to the total stream, the controller switches calibrated characteristics to the matching stream component.

Taking all the aforesaid into consideration, we draw the following conclusion: a random value in this instrument acts as a metrological device, viz. switching of the measuring instrument to the required mode and permanent automatic updating of calibrated characteristics throughout a dynamic range.

References

[1] Voytyuk I N and Kopteva A V 2015 *Int. Conf. on Mech. Engineering, Automation and Control Syst.* (Tomsk) vol 1 (Tomsk Polytechnic University: IEEE Xplore) p 1-4
[2] Gazin D I and Kratirov V A 2003 *Microprocessor Measuring Instruments Ed. III* (S.Petersburg: Nestor, SPGPU) pp 48-54
[3] Ibragfimov G Z 2005 *Technique and Technology of Oil and Gas Production* (Moscow: MGOU)
[4] Luai M, Shaahid S M, Lukman O and Al-Sarkhi A 2014 *The Sci. World J.* 20-31
[5] Yaryshev G M, Yaryshev Yu G and Gorchakov V G 2009 *Exposition Oil and Gas* 20–21
[6] Lishchuk A N 2013 *Oil Industry* 3 1–3
[7] Kopteva A V and Voytyuk I N 2015 *Int. Conf. on Mech. Engineering, Automation and Control Syst.* (Tomsk) vol 1 (Tomsk Polytechnic University: IEEE Xplore) p 1-4
[8] Ochoa B, Kruspe T and Goodbread J 2014 *IEEE Int. Conf. Sensors and Measuring Syst.* pp 1-6
[9] Barak A M 2015 *Oil and Gas* 29–32
[10] Lur'e M V 2003 *Mathematical modeling of oil pipeline transport processes of oil and gas* (Moscow: Oil and gas) p 235
[11] Lishchuk A N 2013 *Oil Industry* 1–3
[12] Bujdosó E 1996 *J. of Radioanalytical and Nuclear Chemistry* **1** 557-577

AC motor diagnostics system based on complex parametric analysis

N A Korolev and S V Solovev

Saint-Petersburg mining university, 2, 21st line, Vasilevskiy Island, St Petersburg, 199106, Russia

E-mail: korolev_na_spmi@mail.ru

Abstract. The article deals with the principle of evaluation of technical condition, based on a comprehensive analysis of the motor parameters which is a main unit in mechanical engineering. Diagnostics system and residential life assessment of electromechanical equipment is presented based on the AC engine and algorithms of its work. The important challenge of diagnostics remains the well-timed faults detection and maintenance and repair organization. The solution of such challenge remains accuracy and reliability of diagnostic systems.

1. Introduction

There are more than 75% of malfunctions presented by machine unit faults on the heavy industry facilities. Consequently the level of its reliability and safety mostly is defined by technical state. Diagnostics of technical devices including electromechanical equipment is an area of technical knowledges that covers theory, methods and means of electric equipment technical state determination in its operation conditions. [1-3]. Diagnostics as an area of science and it practical application located between different sections of science. First of all, this applies to physics and chemistry in terms of behavior and change of materials properties of different kind and the processes that are occurring due to the different factors in them. This also applies to mathematics in terms of experimental data analysis and statistical process control and software to solve complex problems on a computer. Electromechanical equipment exposed to a significant risk of defects and malfunctions, and has a high accident rate. Methods of prevention and detection of faults that are applied today [3] allow only reduce failures in some degree.

The main tasks of diagnostics are [1]:

- Determination of electrical equipment technical state in changing operating exposes conditions,
- Detection of type and degree of defect danger
- Forecasting residual resource and service life.

The decision of which type of diagnosis need to be used, is based on its feasibility study. Such decision is a purely consumer. On the one hand it seeks to objectively determine the condition of the equipment, using a range of diagnostic parameters and their functional connections. On the other hand it is revealed the need for limitation of material resources. So the economy can significantly narrow diagnostic tasks that may affect the objective determination of equipment condition.

IOP Conf. Series: Materials Science and Engineering **177** (2017) 012007 doi:10.1088/1757-899X/177/1/012007

2. Diagnostic methods

The most widely spread type of electrical machines are AC motors. The use of induction motors (IM) is constantly growing and, according to the forecast of experts, their share in electrical drive area may reach 85-90% [4-6]. Diagnostics of IM state and assessment of its vital resources is an important practical task.

Well-timed defect detection at earlier stages of their development and violation of individual machine components operation may prevent emergency situation that can lead to negative consequences [7,8].

Currently, diagnosis of induction motor is based on the following methods:

- vibration analysis of individual elements of the machine;
- current and voltage spectral analysis;
- running machine vibrations acoustic analysis;
- air gap magnetic flux analysis;
- secondary magnetic fields analysis;
- separate parts of machine temperature analysis;
- analysis of the iron content in oil;
- insulation state analysis
- signature analysis

The most widespread received methods based on analysis of vibration parameters in different machine spots, electrical parameters of running equipment (current, voltage, consumed power) and thermosvision analysis. However, each method has its own disadvantages that can't let determine the defects and the reasons which caused them [8]. Also due to the complexity of diagnostics that are caused by difficulties of operation conditions, it is necessary to use indirect parameters if it cannot be define by some reasons. Therefore diagnostic system implementation bases on the complex analysis, will improve the accuracy by evaluating the technical condition of AC machines. Since these two methods can diagnose the greatest number of defects in the AC motor[9].

3. Result and discussion

Currently there is no need to prove the usefulness of diagnostic systems, but doesn't solve problems in this area remains reliability, accuracy and assessment quality of the state and residual life. This happens because of the lack of information about the objects (the amount and timing of ongoing repairs and maintenance time and operating modes, et al.) and erroneous conclusions, put forward at the lack of information.

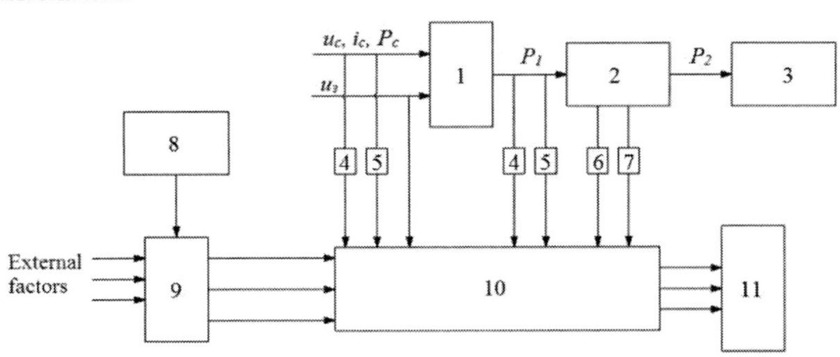

Figure 1. Schematic structure of electromechanical device with induction motor technical condition 1 – static power converter; 2 – electric motor; 3 – executive agency;4 –current sensors; 5 – voltage sensors; 6 – vibration sensors; 7 –temperature sensor; 8 – process automation system; 9 – data preprocessing unit; 10 – data acquisition board; 11 – personal computer; u_c, i_c – grid current and voltage; P_c – grid power; P_1 – power supplied to the motor; P_2 – power transferred to an executive body; $u_з$ – converter control signal.

IOP Conf. Series: Materials Science and Engineering **177** (2017) 012007 doi:10.1088/1757-899X/177/1/012007

Modern automation systems allows electromechanical equipment be equipped by different types of sensors that can allow to determine electrical machine parameters without its stop and installation (is our case this is electrical, vibrating, temperature and indirect parameters). By these parameters it is possible to estimate the actual state of the AC motor. On the basis of a comprehensive analysis of the diagnosis system was developed (Figure 1)

Three-phase voltage and current signals recording from converter input and motor input carried by current sensor 4 and voltage sensor 5. Also the signals from vibration sensors 6 placed on the frame, temperature sensor 7 and converter reference signal are recording. The system also takes into account the process parameters and indirect factors (ambient temperature, humidity, human factor) form the basis of circumstantial evidence in the data preprocessing unit, which complements the main. All received data goes on data acquisition board, where the man database of the motor parameters is forming with further software processing on a personal computer. Each parameter xi, depending on the area to which it refers (electrical, vibration, temperature, indirect) is determined by the relative weight value of X_1, X_2, X_3, ...X_n, showing the levels of development of the defect (Figure 2).

Depending on parameters value they is need to be divided on following state values:

a) passport (reference), parameters after commissioning works;

b) normal, when parameters stay within the permissible range;

c) pre-alarm, when the value of at least one of the parameters exceeds the permissible level, and the other area closer to a threshold value and there is no violations of technical and technological capabilities. In this case it is possible to return to normal by taking precautionary measures (load changes, reducing productivity, etc.);

d) emergency, when inflated all or almost all indicators that separate normal and emergency conditions, and the suspension of production becomes inevitable.

	Electrical parameters																
Per unik level of parameters / Defects	Winding insulation damage on shell	Phase-to-phase insulation damage	Winding short circuit	Stator winding defect	Stator winding short circuit	Rotor winding defect	Bearing damage	Rotor damage	Rotor core flux guive damage	Stator core flux guive damage	Dinamic eccentricity	Static eccentricity	Contact connection defect	Rotor masses disbalance	Mechanical weakenings	Reducer defect	Shafts misalignment
Passport	0,3	0,09	0,1	0,1	0,08	0,09	0,1	0,1	0,14	0,1	0,09	0,1	0,1	0,12	0,1	0,06	0,04
Normal state	0,52	0,1	0,12	0,1	0,08	0,1	0,1	0,1	0,2	0,12	0,1	0,1	0,1	0,16	0,12	0,04	0
Emergency state	0,6	0,18	0,2	0,2	0,16	0,18	0,2	0,2	0,28	0,2	0,18	0,2	0,2	0,24	0,2	0,12	0,08
Emergency state	0,68	0,26	0,28	0,3	0,24	0,26	0,3	0,3	0,36	0,28	0,26	0,3	0,3	0,32	0,28	0,2	0,16

Figure 2. Example of relative weight levels table upon electrical, vibration, temperature and indirect parameters indications.

Summary relative parameter X determined for each state (passport, normal, pre-emergency and emergency), and each defect (1), and is a qualitative indicator of the technical condition of the electric motor.

$$X = X_1 + X_2 + X_3 + ... + X_n \quad (1)$$

Technical diagnostics system is a collection of objects and tools needed to carry out the diagnosis by the rules established in the normative and technical documentation.

Block diagram of the diagnostics system is multilayered and can be represented in the form of its three functional units (subsystems), figure3.

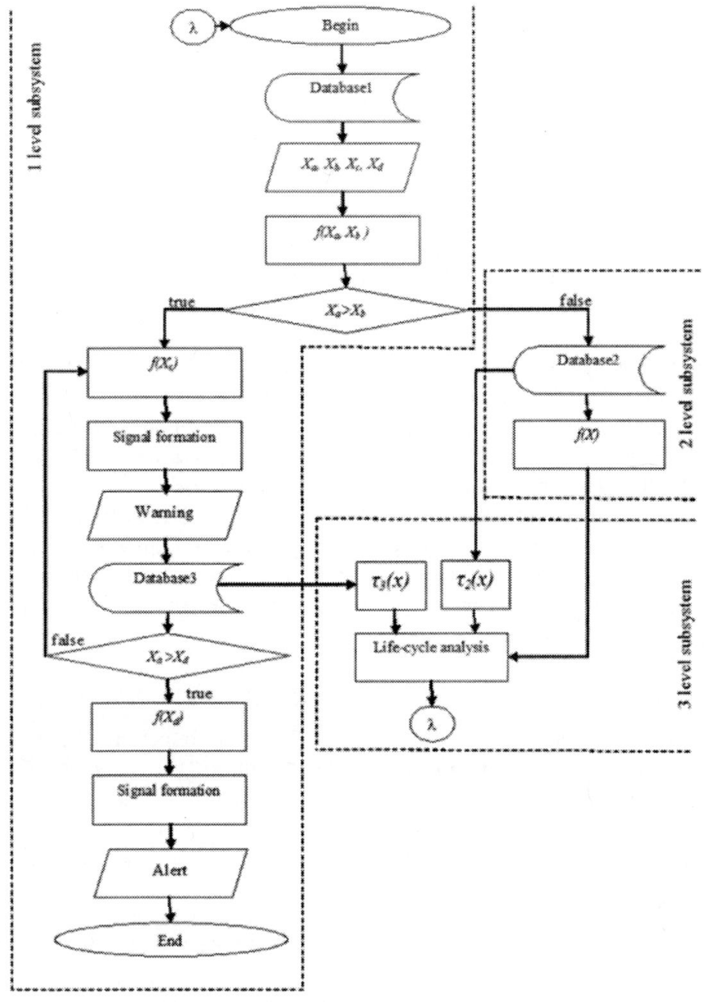

Figure 3. Block diagram of the diagnostic system

Subsystem of 1[st]level, provides:

- storage of informative parameters;
- measuring of diagnostic parameters;
- processing of measured diagnostic parameters by representing them in normalized form and transmitting the results of measurement in processed form to database
- determination of compliance with the values of measured parameters normalized values and the formation of conclusions about the technical condition of the object;
- alert signal formation in excess of the measured parameters
- receiving command signals to change the sensor sensitivity, duration and frequency measurement and synchronization;

In 1st level subsystem retrospective database (Figure 2) is presented which is reflecting a probability value of diagnosed parameters characterizing the degree of influence of the parameter on the technical condition and trends.

Subsystem of 2nd level provides:

- determination the type and level of developing defects;
- calculation of derivatives and dynamic diagnostic parameters;
- defining the retrospective diagnostic parameters;
- defining danger degree of developing parameters;
- subsystem algorithm change by top-level command;
- transmitting calculation results and the analysis to the database;
- forming the signal to change the operating mode of the object;

The evaluation of actual technical state is carried as it follows by (1).

Subsystem of the 3rd level provides an assessment of residual life of the object.

Based on the main tasks of technical diagnostics, primary act should be to identify the type of technical condition. When determine whether a fault and defect, next step is aimed on the determination of type of effect and the danger either to determine reasons of malfunction.

The object can be formed and develop explicit and implicit types of defects, but only obvious defects can be diagnosed. The implicit defect category refers to defect which cannot be determine due to the lack of method and means of detection.

Definition of the technical condition of parts and pieces of equipment and all kinds of disturbances in their functioning occurs with the use of diagnostic parameters. Diagnostic parameters are presented as a coefficient K, which is taking into account the step change in the boundaries of the parameter over time (Figure 4) and let forecasted parameter value can be estimated. Consequently, the amount of residual life of the electric drive τdefined as:

$$\tau = K_1 \cdot X_1 + K_2 \cdot X_2 + K_3 \cdot X_3 + ... + K_n \cdot X_n (2)$$

Where K_i – coefficient which is taking into account borders statement of i-th parameter with taking into account the defects detected at time t, and depending on the passport (reference) normal, pre-emergency and emergency conditions.

The databases of technical parameters of AC drive and dynamics of changes over time, the actual technical condition assessment and predictive value of the residual life will be the results of diagnostic system based on parametric analysis.

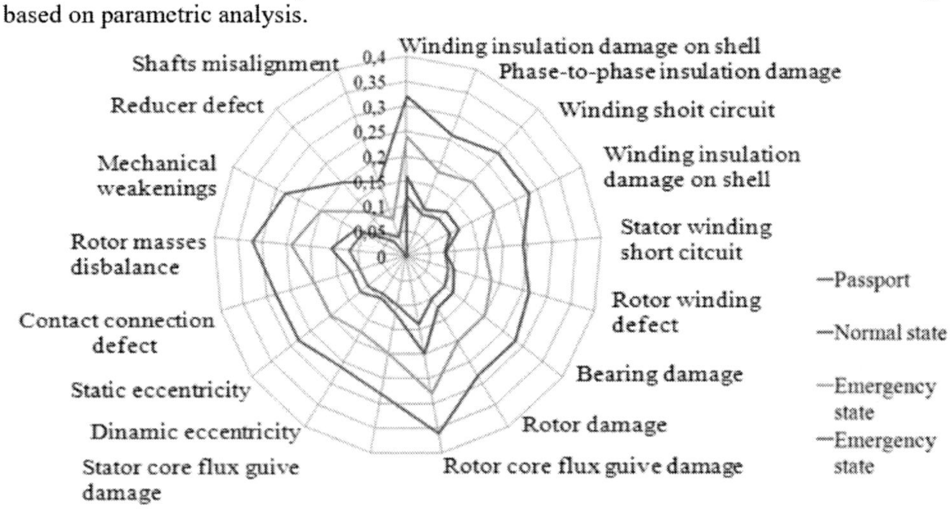

Figure 4. Electrical drive technical condition diagram

IOP Conf. Series: Materials Science and Engineering **177** (2017) 012007 doi:10.1088/1757-899X/177/1/012007

4. Economic feasibility.

Substantiation of economic efficiency is performed in the planning of monitoring and diagnostics of technical condition of electrical equipment. The basis of cost-effectiveness study is a technical and economic assessment in two variants of object exploitation: equipped with a system for monitoring and evaluation of residual life without additional equipment. The main indicators for determining the feasibility of the technical condition of the monitoring system and the reduction of production costs, are as follows:

- Forecasted derating of failure risk;
- lowering the cost of production due to the economic effect for maintenance

The effectiveness of implementation of monitoring and diagnostic system is evaluated against the value of reducing the total financial damage from possible accidents for the predicted time operation of the facility equipped with the monitoring system, to the value of its implementation and maintenance costs over the same period (figure 5).If the cost efficiency is not sufficient, then the re-development of the monitoring system to achieve acceptable performance for economic enterprise, provided that the security level is not reduced object, characterized by an indicator of risk.

The main costs in the implementation of the monitoring system:

- for purchasing and installation of the system on site;
- for the organization of an information control center to work with the data;
- for purchasing the appropriate computer equipment to the information center;

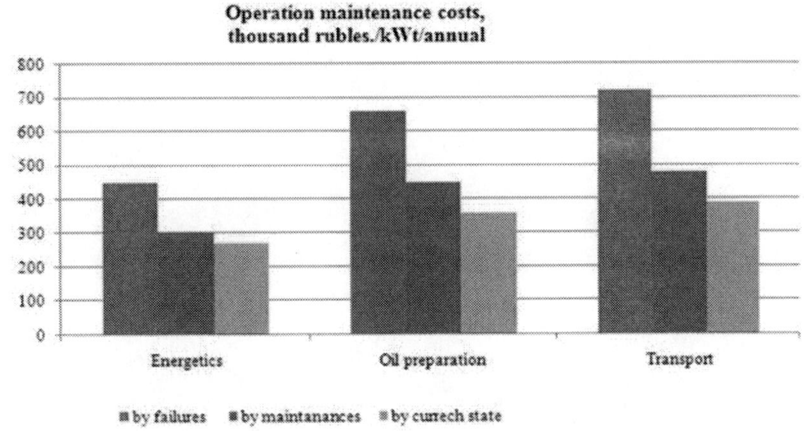

Figure 5. Specific operational costs for industries of basic equipment

5. Conclusion

The effectiveness of diagnostic systems depends on its equipment with devices for control of direct and indirect factors of loss of the resource that leads to the failure, and by the quality of the system, providing a high percentage of defects and damage detection and the keeping demanded operation efficiency.

The transition to a complex diagnostic system of electromechanical equipment with considering external factors will improve accuracy of the technical condition and residual life assessment, by conclusion of which the repairs and maintenance can be well-to-scheduled. Also it is economically feasible to extend the life of equipment.

Implementation of diagnostic system sand assessment of residual life of AC motor that leading today in electrical drive area of use, reducing enterprise costs of maintenance and repair, emergency response and operation of pre-emergency equipment.

IOP Conf. Series: Materials Science and Engineering **177** (2017) 012007 doi:10.1088/1757-899X/177/1/012007

References

[1] Kostukov V N 1985 *Proc. of Hydraulic and Building Control Systems, Traction and Road Machines* (Ufa: OmPI)

[2] Malov E A, Bronfin I B, Dolgopyatov V N, Mikerin B I, Kostukov V N and Boychenko S N 1994 *Safety in Industry* **8** 19

[3] Maks J 1983 *Methods and approaches for signals processing on physical measurement* (Moscow: Mir)

[4] Birger I A 1978 *Mechanical engineering* (Moscow)

[5] Zhukovskiy U L and Koteleva N I 2014 *Modern Scientific Res. and Innov.* **5** 1

[6] Shuhgalter M L 2009 *Economics and Life* **26** 113

[7] Mironova I S and Bashirov M G 2010 *Proc. of the Scientific Conf. Environmental Problems of Oil Production* (Ufa: Oil and gas business) p 506

[8] Kozyaruk A E, Krivenko A V, Zhukovskiy U L, Korzhev A A, Baburin S V, and Cheremushkina M S 2013 *Maintenance of machines' electrical equipment with an evaluation of residual lifetime* (St. Petersburg: Mining university)

[9] Kozyaruk A E and Zhukovskiy U L 2014 *Mining Machinery and El. Mech.* **10** 8

Development and program implementation of elements for identification of the electromagnet condition for movable element position control

R I Leukhin, D V Shaykhutdinov, K M Shirokov, N D Narakidze and A S Vlasov

Platov South Russia State Polytechnic University (NPI), 132, Prosveschenija St., Novocherkassk, 346400, Russia

E-mail: leukhin@npi-tu.ru

Abstract. Developing the experimental design of new electromagnetic constructions types in engineering industry enterprises requires solutions of two major problems: regulator's parameters setup and comprehensive testing of electromagnets. A weber-ampere characteristic as a data source for electromagnet condition identification was selected. Present article focuses on development and implementation of the software for electromagnetic drive control system based on the weber-ampere characteristic measuring. The software for weber-ampere characteristic data processing based on artificial neural network is developed. Results of the design have been integrated into the program code in LabVIEW environment. The license package of LabVIEW graphic programming was used. The hardware is chosen and possibility of its use for control system implementation was proved. The trained artificial neural network defines electromagnetic drive effector position with minimal error. Developed system allows to control the electromagnetic drive powered by the voltage source, the current source and hybrid sources.

1. Introduction

Electromagnets, along with transformers, are the most common representatives of electromagnetic designs. Proportional electromagnets are the most complex system [1, 2]. This type of electromagnetic drives requires systems that can control parameters of electromagnets elements movement. In experimental development of new products, it is important to setup parameters of regulators and make complex testing of electromagnets. Also, high reliability requirements are denoted for this class of devices when it's used as a part of increased responsibility systems. Complexity and continuous improvement of the magnetic systems now demand a new high-precision approaches to control of the movable elements position. Automated systems controlling production of the intelligent electric drives are required for implementation of an effective approach. One of the basic elements of production systems is the stand for diagnostics and regulator parameters setup. When developing an experimental production of new proportional electromagnets of direct current, the specified stand can be presented in the form of the automated system for researches and complex tests. These operations are intended to improve productivity of optimization processes of electromagnets designs, including the use of new materials or known materials in limit operating modes.

From the consumer's point of view, electromagnet main characteristics are mass-dimensional indicators and energy parameters. However, using electromagnets must also be considered in determining their particular performance. The performance of the electromagnet in creating a traction force value estimated by the time constant of a current rise in the winding. Usually, dynamic properties of the electromagnet are characterized by time constant τ_M, which is calculated by the formula:

$$\tau_{\mathrm{M}} = \frac{L}{R},$$

where L – inductance of the winding, R – resistance of the solenoid coil.

2. Materials and Methods

To improve the performance, the force of an electromagnet with a negative feedback current in the electromagnet winding is used. In this case, the source of the power control system is converted to the type 'voltage source', referring to the 'current source' or 'hybrid source' that requires further analysis and consideration in the design.

When boosting the electromagnet coil connected to power supply voltage U and current I, it appears that the current-loop can reach steady state value I_S, which is defined as

$$I_{\mathrm{S}} = \frac{U}{R}.$$

Current feedback limits the amount of current I at the level of the nominal value of I_N, corresponding to a given input value.

The main characteristics of the power supply of the electromagnets control systems include: maximum voltage U_M, maximum current I_M and maximum power P_A, which is calculated according to the formula:

$$P_{\mathrm{A}} = U_{\mathrm{M}} I_{\mathrm{M}}.$$

If I_S is estimated at a fraction of I_M by using coefficient $k_A < 1$, then we have:

$$I_{\mathrm{S}} = k_{\mathrm{A}} I_M.$$

In this case, the following conclusion can be done: for the case when the current source or a hybrid power source was used, the time constant is defined by the expression:

$$T_{\mathrm{M}} = \frac{\tau_M I_N}{I_S},$$

and it also can be expressed by the formula:

$$T_{\mathrm{M}} = L\frac{I_N}{U_M}.$$

Consequently, we will have:

$$T_{\mathrm{M}} = \frac{L I_N^2}{k_A U_M I_M} = \frac{2 F_N \delta}{k_A P_A},$$

where δ – the air gap between the armature and the yoke of the electromagnet; F_N - traction, determined by the formula:

$$F_{\mathrm{N}} = \frac{B_N^2 S_M}{\mu_0} = \frac{L I_N^2}{2\delta},$$

where S_M – the square of the armature, $\mu_0 = 4\pi 10^{-7}$ – the magnetic constant, B_N – induction in the gap defined by the formula:

$$B_{\mathrm{N}} = \mu_0 I_N \frac{w}{2\delta},$$

where w – the number of turns in the electromagnet winding.

The expression for the T_M time constant corresponding to the current source indicates that with chosen values of k_A and δ, electromagnet performance depends on the ratio of F_N and P_A. Therefore, for different basic designs of electromagnets, the problem of speed increasing can be solved by varying the power supply.

With high complexity of the control object (for example, a current source having a power limit or configuration of the magnetic system of the electromagnet has a high geometrical complexity) or in the absence of complete a priori information about the object it becomes necessary to apply intellectual technologies of control [3-6]. Now, a very large number of publications are devoted to intellectual technologies that confirms relevance and popularity of these control systems.

The proposed approach is based on the following assumptions. The position of the electromagnet movable element x defines a degree of magnetic circuit closure d. The degree of of magnetic circuit closure d determines the level of magnetic flux F at a preset value of magnetopropellent Iw. The level of a magnetic flux has direct interrelation with the flux linkage of the winding of magnetic conductor ψ. Thus, there is physical interrelation between the weber-ampere characteristic of electromagnet $\psi(Iw)$ and the position of its movable element, $x: x \rightarrow d \rightarrow F(Iw) \rightarrow \psi(Iw)$. Based on this assumptions, the automatic control system for rather high complexity of the controlled object in the absence of rather complete information can be developed (figure 1).

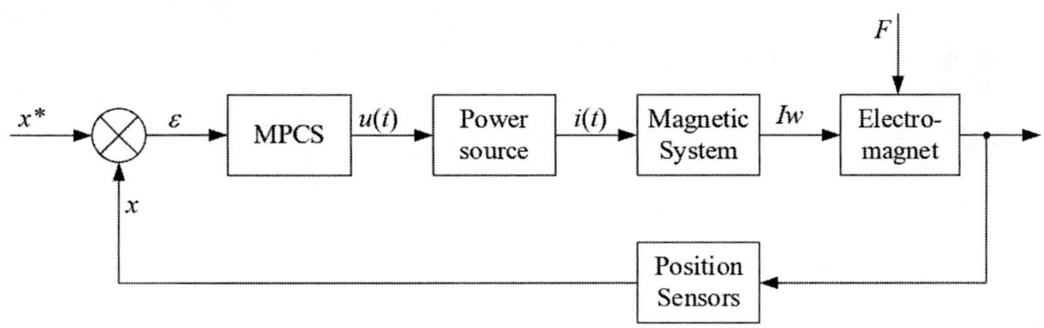

Figure 1. The block diagram of the automatic control system of the electromagnet.

Parts of the system (figure 1) are: the microprocessor control system (MPCS), the managed power source, the magnetic system, the position sensor, the electromagnet.

The purpose of the work is to develop software for identification of the electromagnet condition (Position Sensor) and implement it to system of movable element position control. Herein the weber-ampere characteristic has to be used as information source [7]. Researches were conducted with the use of elements of the theory of magnetic field, the theory of automatic control, mathematical modeling with the use of the licensed package of the programs '*Matlab, LabVIEW*'.

3. Results and Discussion

Automatic control system represent as the hardware-software complex providing continuous and effective control of complex technical systems or the proceeding processes in real time.

Generally, for creation of the automatic system for electromagnetic drive control, the aim to provide high-speed performance, a movable part of the drive definition accuracy with no need of large-size sensors using is pursued.

In present work, software for the control system of the electromagnetic drive movable element is developed in the environment of graphic programming of *LabVIEW* [8]. The software includes the artificial neural network developed in *MathScript* component of *LabVIEW*. *MathScript* uses the *MATLAB* library for implementation and training of the artificial neural network. The Weber-ampere characteristic is input data for the artificial neural network (figure 2a). The first entrance is values of current (i), the second entrance is a values of flux linkage (ψ). Positions of the electromagnetic drive movable element (figure 2b) received by laser sensor were used as the training data.

The trained artificial neural network [9] defines position (x) of the movable element by the weber-ampere characteristic processing. It gives the opportunity to exclude the position sensor from the automatic control system that influences its dimensions, weight and cost. The artificial neural network after training needs to be integrated into the control system hardware. The hardware component with its own software and a possibility of integration of applied programs for autonomous system is necessary. The possibility of connection analog input and output signals, digital outputs for current source effective control is also necessary [10]. Availability of constant memory or connection of peripheral equipment as information storage with possibility of data accumulation for artificial neural network training is necessary.

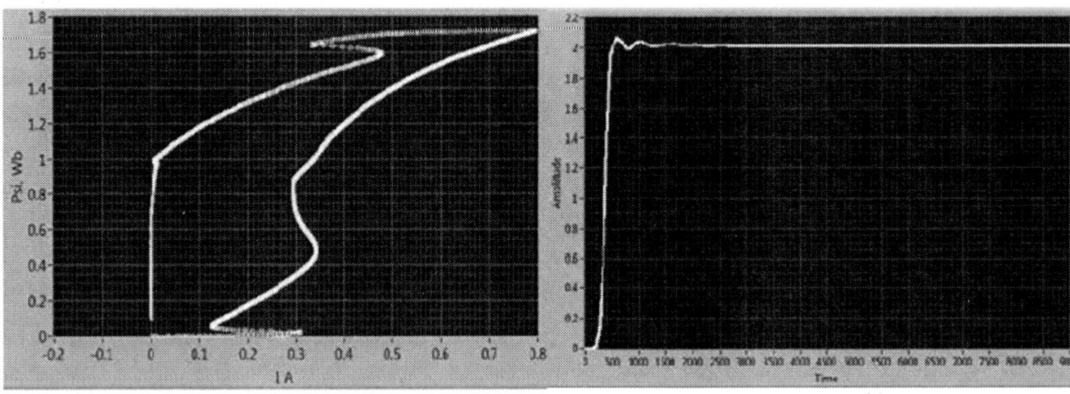

a) b)

Figure 2. The weber-ampere characteristic (a) and the schedule of position of the movable element (b) of the operating cycle of the electromagnetic drive.

The industrial controler 'National Instruments SBRIO-9636' meets all necessary requirements. The chosen controler has its own software RAM size of 512 MB, USB storage devices connection possibility, 16 analog single inputs or 8 differential, 4 analog outputs. 'National Instruments SBRIO-9636' has small dimensions in comparison with analogs that allows one to get a compact size of the control system.

As an object of control the direct current electromagnetic drive was used. The dynamic weber-ampere characteristic was measured, and also position *(x)* of the movable part by means of the laser sensor in the course of operation was defined.

As a basis of the control system, the three-layered artificial neural network with the return distribution of the mistake has been chosen. For its training data on the position of the movable part of the electromagnetic drive were used. The result of this artificial neural network depends on the number of the data which are used for training. For minimal error achievement, a training network with 1000 epochs was used. Epoch is one of all examples of the training data with simultaneous correction of network weights (for specific examples, depending on network solving correctness). When an artificial network was trained, a little number (hundreds, or even tens) of epochs can be demanded. During the epoch, network scales can be corrected not on every training example. If the artificial network was already trained to solve correctly this specific example, it is possible to cease to demand continuous "improvements" of quality of the solution of this example (as these improvements will already be insignificant) and to allow artificial network to direct its attention to other examples which for the present are solved with insufficient accuracy. The result of training of artificial neural network is shown in figure 3.

Regarding the implementation of control system for electromagnetic drive it should be noted that in it the PID-law of control on current (*I*) is implemented. Into analog channel AI0 of the SbRIO-9636 controller the value of current is transmitted and into the channel AI1 the value of the electromagnetic drive windings voltage is transmitted. The output signal from the industrial controller SbRIO-9636 is depending on the calculated value of electromagnet movable element position. This output is the control signal for the current source.

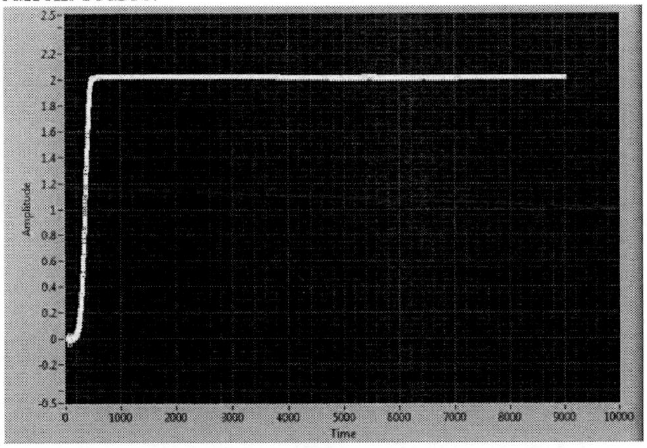

Figure 3. Result of work of artificial neural network.

4. Conclusion

After training, artificial neural network defines the position of electromagnetic drive movable element by its weber-ampere characteristic data. But thus for each separate electromagnetic drive, it is necessary to train new neural network for its functioning without increase in error. Position that received by using the trained artificial neural network and real positions differ less than for 5 %. It confirms that developing control system based on artificial neural network is perspective task as the accuracy of movable element positioning aims at 95 % with no need to use the special position sensor.

Acknowledgments

The results obtained with the support of the grant RFBR № 15-38-20652 'Development of the theory of predictive methods for sensorless control and diagnosis of electric drives' using equipment center for collective use 'Diagnosis and energy-efficient electrical equipment' (NPI).

References

[1] Leukhin R I, Shaykhutdinov D V and Leukhin V I 2016 *Proc. Eng.* **150** 347-353
[2] Shaykhutdinov D, Lankin A, Narakidze N, Grechikhin V, Shirokov K and Gorbatenko N 2015 *Research J. of Appl. Sci.* **10(10)** 555-557
[3] Zolfaghari M, Taher S A and Munuz D V 2016 *Ain Shams Eng. J.* **7(2)** 729-740
[4] Zohoori A, Vahedi A, Meo S and Sorrentino V 2016 *J. of Intellig. and Fuzzy Syst.* **30(1)** 159-169
[5] Kiran Kumar B, Siva Reddy Y V, Vijaya Kumar M 2015 *Int. J. of Control Theory and Appl.* **8(1)** 71-76
[6] Choi J-S, Ko J-S, Chung D-H 2006 *2006 SICE-ICASE Int. Joint Conf.* 690-695
[7] Gorbatenko N I, Lankin A M and Lankin M V 2016 *Proc. Eng.* **150** 1027-1031
[8] Meng F, Zhang H, Cao D and Chen H 2016 *IEEE/ASME Transactions on Mechatronics* **21(3)** 1742-53
[9] Grechihin V V, Lankin M V, Lankin A M 2015 *Proc. Eng.* **129** 793-798
[10] Jiang J and Tian J 2015 *J. of Chongqing Univ.* **38(4)** 75-79

Combined finish machining of nonrigid parts by the brush electrode

V P Smolentsev, O N Kirillov, O N Fedonin and A Yu Ryazantsev

Voronezh State Technical University, 14 Moscow Ave., 394026, Voronezh, Russia

E-mail: ryazantsev86@mail.ru

Annotation. The capacities of extensive use of technology and a combined machining tool by an unshaped brush electrode are discovered. This type of machining occurs during dimensional finish formation of composite parts of small rigidity (including the parts made of difficult-to-machine materials). This process takes place provided that there is required pressure of wire beams in the brush electrode and a combination of electrical modes of the process with tool parameters. The ways of maintaining the process stability, increasing technological parameters, which allows extending the application field of universal high-productive finish machining by an electrode, are presented. This problem has been solved for the first time, and its results make it possible to create items of new generations, which is actual for machine building

1. Introduction

With the complication of the shape of constituent elements of modern high technology production in machine building along with the increase of the unit load on the construction under the influence of operating loads and hereditary phenomena, we can observe the changes in the part shape. In addition, errors, which are capable of intervening failures in the whole article, arise. This is particularly apparent during production of parts for aircrafts, where openwork parts with small rigidity are applied to reduce the weight of the construction. New tools that have been recently produced with universal working section, which is unshaped and changeable during the machining process, allowed performing finish machining of nonrigid parts having a composite profile by a single tool. The metal brush electrode [1–3] is the tool, which is successfully used for dimensional machining of composite surfaces with any rigidity of a semi-finished product. The creation of the process control system, taking into account rigidity [2, 3] of the machined part section, has been required for the subsequent use of the promising combined technological process, which will be viewed further in this article.

2. The mechanism of interaction of the brush electrode with a machined object of limited rigidity

Nonrigid parts in respect to machining of metal parts by the brush electrode are parts that contain sections, which under the influence of tool pressure force and discharge activity, deviate by the value that exceeds the available dimension change of the finished part and the dimension of brush electrode pressure for finish machining [3, 4].

The scheme of interaction of the brush electrode with the machined nonrigid part section is presented in Figure 1.

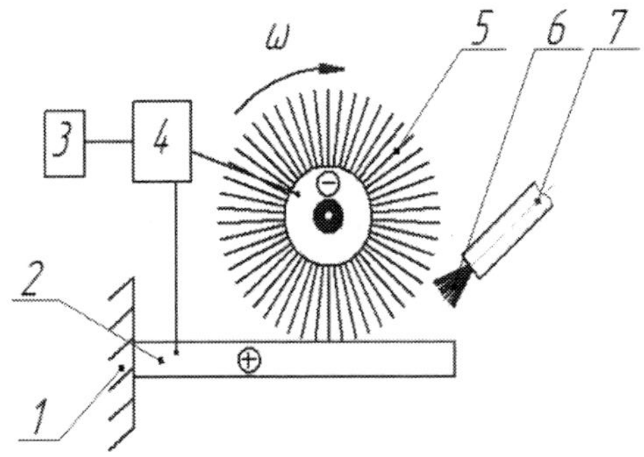

Figure 1. The scheme of combined machining of a nonrigid part by the brush electrode: 1 – the base surface, 2 – the nonrigid part element, 3 – the control unit of mode parameters, 4 – the nozzle for supply of working fluid, 5 – the brush electrode, 6 – the weak electrolyte based on neutral salts, 7 – the source of low voltage current.

The mechanism of finished machining of nonrigid parts [3] and calculation methodology of the rush electrode [5, 6] are viewed. It is necessary to set sections with minimum rigidity practically in every item, for which the tool should be created. Thus, the relations among the properties of nonrigid machined objects, parameters of brush electrode and modes of machining are formalized.

A key parameter in the brush electrode is a material and a wire diameter combined in beams. It is known that for finished machining, it is necessary to apply brass, and the wire diameter, used in the case under consideration during high velocity machining, amounts to 0.1 up to 0.3 mm.

Machining by the brush electrode includes several types of action, such as:

1. The power from discharge pulse at the moment of beam coming to the machining area (Figure 2).

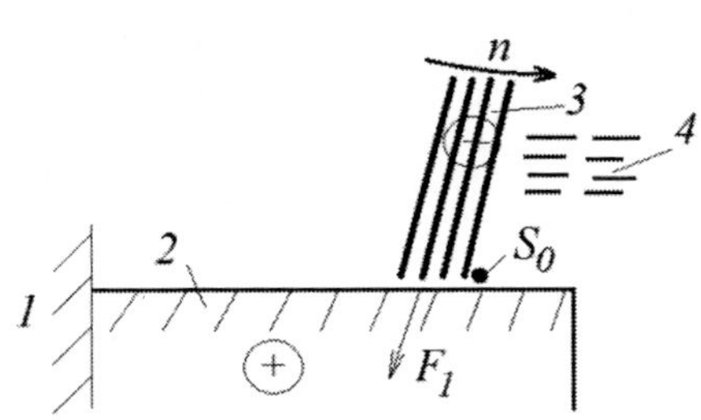

Figure 2. The start machining process by the wire beam in working liquid 4: 1 – base, S_0 – the tolerance between electrodes at the moment of wire 3 running on semi-finished product 2 and the start of discharge, F_1 – the power arising at the moment of discharge.

The limitation of this stage is a beam size and a step between beams, where it is necessary that the first beam leaves the machining area completely before the next beam contacts a semi-finished product. In this case, the cross-sectional size (diameter) of the beam should be minimum.

2. Cleaning the machining area by wire ends (Figure 3), where the contact is provided by approaching the wire ends to a semi-finished product by such value as minimum guaranteed pressure Pr_G. This stage allows preparing the surface for the anode removal of the tolerance after the beam has left the machining area.

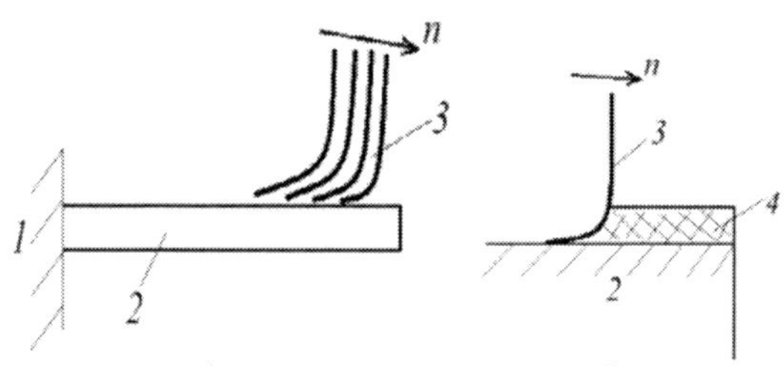

Figure 3. Cleaning of console section 2 of the semi-finished product surface, fixed on base 1, by the mechanical influence by wire ends 3: a – the scheme of cleaning, b – removal of pollutants,4 – oxides, slivers and other materials on the surface of the machined part.

Technological recommendations on the choice of pressure (not less than Pr_G, but not more than one contact by the side surface of the curved wire of the machined surface). As a rule, the pressure for the calculated tool is $Pr_G = 0.2 \div 0.6$ mm.

3. Tolerance Z (Figure 4) is removed owing to material anode dissolution

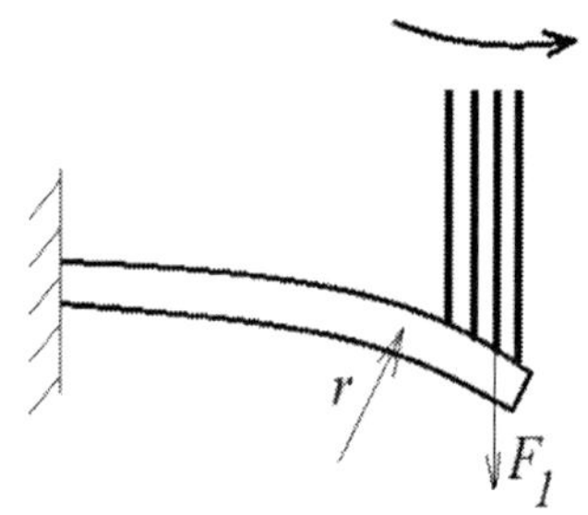

Figure 4. The scheme of nonrigid part machining: r – the radius of a nonrigid part bend owing to pulse actions with power F_1.

The magnitude of the bend depends on the installation scheme of the section of a semi-finished product: in the presence of support, it will be a maximum deflection when the tool is located in the middle of the section ($l/2$).

The tolerance, removable per single revolution of the brush electrode (Z_0), can be calculated according to the following formula:

$$Z_0 = k_0 Z_1 n_0,$$

where k_0 – the coefficient taking into account the portion of tolerance removal due to anode dissolution at the moment of the beam approach to a semi-finished product (k_0=1.5÷1.6); Z_1 – the material removal due to electroerosion tolerance removal by a single beam; n_0 – the quantity of beams in the brush electrode.

A nonrigid semi-finished product (or its sections) (Figure 4) can initiate a bend of the wire and the whole beam with radius r under the influence of discharge power, which can exceed the pressure value, initiate autovibrations and break the machining process by the brush electrode.

3. Designing of the technological process

Combined machining by an unshaped brush electrode represents a combination of three processes: electroerosion, mechanical and anode dissolution of the metal [2, 8]. Processes influence each other mutually, which is evaluated by the coefficient of mutual influence. During designing of the technological process of finished combined machining by the brush electrode, it is necessary to combine and to enhance the advantages of influence constituents: high accuracy of electroerosion machining, high production and a low height of microroughnesses of electrochemical machining, intensification of the process by mechanical action of brush electrode beams and a high surface quality, determined by geometrical, physical and mechanical parameters of the surface layer. Owing to the mechanical action of the beams of the tool working part, there is an opportunity of removal of nonmetallic inclusions and oxides from the machined surface, which allows intensifying the machining process and enhancing technological capabilities.

The technology of high-velocity machining by the brush electrode includes several independent stages.

1. At the first stage, the type and the sizes of the electrode tool [5, 6] are chosen, at which it is necessary to consider:

- rigidity, elasticity, electrotechnical properties and other parameters of the brush electrode, which allow finding the value of beams pressure to a semi-finished product at any tolerance value;
- the packing density of brush electrode beams must provide the movement trajectory of electrodes for the completion of the pulse after leaving the machining area by the side of the tool;
- justification of the choice of the brush electrode wire material: the copper and brass wire of small diameter is chosen during finish machining of current conducting items of composite geometrical shape with limited rigidity for production of the tool working part.

2. The choice of the working medium. Parameters of the surface layer, productivity, accuracy depend on properties of the working medium during high-velocity combined machining. At this, anode dissolution increases the quality of the surface layer and the velocity of metal removal from the surface of a semi-finished product. In works [7, 8 and others], a possibility of using low conducting lubricating-cooling liquids as working media, sometimes with additives (usually up to 3%) of neutral salts (for example NaCl), is presented. It allowed establishing the maximum rate of tool rotation, which is necessary during the work with high peripheral velocities of the brush electrode.

3. Special or modernized for machining brush electrode machines [9] that permanently provide a set value of beams pressing to the spot of part machining are required for high-velocity machining.

4. Electric field strength for high-velocity dimensional machining is chosen by taking into account the required roughness and wear of the brush electrode.

Optimum machining modes of nonrigid parts are given in the table below.

Table 1. Modes of high-velocity machining by the brush electrode

Parameter name	Value
Voltage, V	4–6
Size of brush electrode pressing to semi-finished product, mm	0.1–0.3
Brush electrode annular velocity, m/s	More than 30
Electrode tool diameter, mm (is selected taking into account the equipment technological characteristics)	100–300
Wire diameter in beams *, mm	0.1–0.3

5. The time of high-velocity finish machining by the brush electrode depends on sizes and configurations of a machined item, predetermined technological parameters and machining modes.

4. Conclusion

The control mechanism of the contact process of finish machining and brush electrode calculation methodology for machining nonrigid composite shape parts are developed. They are developed as a technological system along with supporting the position of the machining section relatively the tool and stabilization of the pressure value. This allowed machining nonrigid composite shape parts with high accuracy and quality of the surface per one brush electrode transfer, which increases technological capacities of the process.

References

[1] V P Smolentsev, O N Kirillov and E V Smolentsev, RF Patent No. 2,241,582 (2 march 2003)
[2] Pisarev A V 2002 *Russian J. Nontraditional Machining Methods* 22–31
[3] Ryazantsev A Y 2015 *Bulletin of Voronezh State Techn. Univ.* **11**(5) 8–13
[4] Kirillov O N 2010 *Fund. and Appl. Probl. of Equipment and Technology Eagle* **31-34** 156
[5] Ryazantsev A Y and Kirillov O N 2015 *Bulletin of Voronezh State Techn. Univ.* **11**(5) 4–8
[6] Kirillov O N 2009 *Bulletin of Voronezh State Techn. Univ.* **5**(11) 39–42
[7] Kirillov O N 2010 *Voronezh State Techn. Univ.* **110** 254
[8] Kirillov O N 2009 *News of Orel State Technical Univ., Edition 'Basic and Appl. Probl. of Eng. and Techn.'* 2009 **560** 66–71
[9] Kirillov O N 2010 *Bulletin of Voronezh State Techn. Univ.* **6**(1) 20–22

Intelligent fuzzy-neural pattern generation and control of a quadrupedal bionic inspection robot

D Sayfeddine and A G Bulgakov

South Russian State Polytechnic University, 132, Prosveshenie str., Novocherkassk, Russia

E-mail: daher@live.ru

Abstract. This paper represents a case study on 'single leg single step' pattern generation and control of quadrupedal bionic robot movement using intelligent fuzzy-neural approaches. The aim is to set up a flip-flop mechanical configuration allowing the robot to move one step forward. The same algorithm can be integrated to develop a full trajectory pattern as an interconnected task of global path planning for autonomous quadrupedal robots.

1. Introduction

The tendency of designing robotic systems based on biomimetics has effectively started with the Arab polymath, Al-Jazari, (1136-1206) with his invention of the musical robotic band working based on hydraulic switching. Since then, engineers continued developing bionic robots and control systems by 'inhaling' the massive designs prototypes in the nature.

Robots in construction are adopted not only to automate the construction process, but also to carry out quality management tasks. For instance, flying drones can inspect high-rise structures or can serve to execute photogrammetry. Telerobotic solutions are used to inspect shafts, air conditioning systems, and manholes. Climbing robots are used to inspect bridges, towers and hidden difficult corners of the building. The success of implementing intelligent solutions to carry out the inspection tasks strictly depends on the performance of the robotic control system, the stability and reliability of the design mechanism and the inspected site condition. The latter can create major obstacle for the robot to move. For instance, with the vertical take-off and landing flying robots such as quadrotors, stability tends to minimize when the altitude increases. On the other hand, wet fields paralyze the robot pedal mechanisms.

Recurring to the control methodology of the robot, we can state that synchronization of different end-effector mechanisms is essential to start the movement and thus to track the desired trajectory. For example, a rotorcraft, UAV, has to be controlled so that to produce equivalent rotational speed of a chosen rotors combination in order to shift between four flight regimes: hover, roll, pitch and yaw. For walking robot, legs located on different body axis sides have to work based on a flip-flop regime in order to shift the center of gravity of the robot and to make it move from its current position. These regimes are called patterns. In our paper, we discuss a quadrupedal robot, performing inspection of glass facades. The study will be limited to a mathematical model of the locomotion end effectors, global path planning control based on prototype patterns. The results are simulation graphs depicting the performance of tracking the generated pattern trajectory.

2. Walking as a technical bionic task

In the engineering concept, movement is realized via mechanisms, which generate lifts forces and torques. The end effector, which interacts with the surrounding, can be tires, chains, and propellers etc. For walking mechanisms, the concept is different- a "foot" is touching the ground. The walking mechanism is minimally divided into three mechanical parts. It can reach higher grades of complexity depending on the manoeuvres, which are required to be accomplished and degrees of movement freedom. While walking, the end effector exerts forces parallel to the main movement axis. As a result, the mechanism will act as a pendulum. The frequency of harmonic oscillation of the mechanism depends on the walking speed.

Although it can be divided into smaller functional components, the walking process is presented as a single input-single output system. It is not wise to adopt a multiple systems approach as it is not optimal to use from the point of view of dynamics, feedback control and signals analysis. In the wake of the above-mentioned, it is not so important to study how a separate component works, but how it can integrate with other components and how it is going to influence the overall performance of the consolidated system. This technical task is topical from the point of view of robotics, and it can be categorized under swarm control, where swarm designates the combination of different mechanisms working in the adaptive synchronized regime in order to achieve a global control task.

Based on bionic researches, the walking process is categorized as anticipatory feed-forward (proactive) regulation. It is the case of the most repetitive behaviour reacting to external stimuli. More complex walking systems comprise more complex stimulation generation and coordination control. This rhythmic dynamics is known also as central pattern control (figure 1).

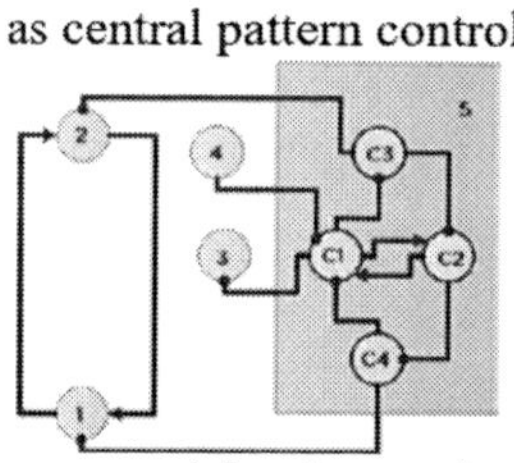

Figure 1. The structure of central pattern control for the quadrupedal robot.

In figure 1, the following parameters are depicted: neurons with index 1-3 are motor neurons, where neurons 1 and 2 are used to steer right and left, neuron 4 is sensory, and C1 to C4 are intermediate neurons. Arrows represent an excitatory connection and the point arrows are for inhibitory connection. As it can be seen, it exist inhibitory connection between C1 and C3 thus allowing the oscillation pattern to be generated. The synchronization between legs movement known as periodic gait requires rules and timing control, otherwise inhibited legs will act as 'car-break'. Consequently, the smoothness of the walking process is not achieved and a possible damage to the mechanical parts is forecasted. The general combination of possible gaits for a mobile robot is calculated as follows:

$$N_G=(2n-1)! \qquad (1)$$

where N_G – the possible number of gaits for a given robot and n – the number of legs of the robots; in our case, $n = 4$.

3. Control of a bionic leg

The most optimal control strategy from the stability perspective is to have the local controller for each leg separately and then to generalize the flip-flop rules according to periodic gaits. This cyclic coordination is necessary in order to obtain smooth and natural oscillation of the mechanical part. This will reduce the energy consumption, as the oscillation will be obtained with no power prerequisites. The other benefit of this approach is to minimize the wear and tear frequency of the mechanical parts as there is no need for any breaks and break releasing mechanisms.

Beside synchronization and periodic gait control, posture control is essential for safe movement. By posture control, we understand the ability of the control system to maintain the centre of mass of the leg in a controllable stable diapason. From the design point of view, the centre of mass geometric point should be within the geometric focus point and the body, thus allowing the robot for better manoeuvrability in case of crosswind. This is essential when having the robot moving at high altitudes.

As a consequence, the global control task for the walking robot includes the following subtasks:

a. Stance control is associated with the kinematic configuration of the legs;
b. Posture control is responsible for centering of the mass position and mostly studied with dynamic analysis;
c. Swing control overlooks the frequency synchronization between different mechanisms in accordance with the first two approaches.

3.1. Literature review

The most optimal control aspect of walking robots cannot be perfectly liaised with natural processes, as the latter is extremely difficult to copy and far more complex than the task of reverse engineering. In addition, the natural biological tissues react somehow differently to exerted forces and applied moments. Nevertheless, different studies were oriented to incorporate the biological concept to the control system of the walking robots. For instance, [3] suggested decentralized control for a top-to-down hierarchy system. The behaviour of the mechanical components of the walking system was studied in depth. In [6], the author studied the topic from the point of view of a central pattern generator. Although the single leg control approach was adopted, they have analysed further the implementation of an oscillatory pacemaker regulating the period of activation of each leg.

Regarding the stance control, the literature analysis reflects the complexity of the task, as it requires real-time monitoring, reading of angular and position sensors and computation of regulating voltage. For a small mobile robot, the on-board microcomputer is not optimal when the stance problem is targeted. This will cause a slow walking process, and it can be treated as a local minimum task as the robot evaluates all possible outcomes of the control algorithm. Taking into consideration that robots are moving in unknown and risky environment, the average of data to be analysed is tremendously huge for on-board electronics.

Pertaining the swing control, the task is far simpler than the previous task as it is mechanically detached from the axis-body of the robot. In [3,7], the authors suggested two-layer designs of the feed-forward neural network with six inputs and two outputs representing the angular velocity of each mechanical joint.

As a conclusion of the literature review and, particularly, of [1-5], the three control subtasks were re-identified. The global control problem of the walking mechanism encompasses the following tasks:

a. Centre of the mass position and speed control;
b. Gait control according to the central pattern generator;
c. Single leg control including the stance, swing and obstacle avoidance;
d. Direction control and steering of each leg with reference to the global trajectory;
e. Control of internal components affecting the excitation and inhibition of a chosen leg.

3.2. Quadrupedal Robot Model

Modelling and control of legged robots is a topical technical problem. Many researchers from different scientific backgrounds combine their efforts on a regular basis in order to obtain an adequate mathematical model. As it was cited earlier, the simplest approach is to adopt decentralized control for each modelled leg of the robot taking into consideration a central pattern generator. This can be explained as follows: to walk, human being defines his orientation based on the need (desired trajectory), communicates and formulates this need into pulses (central generator), transfers the message to the local control post (decentralized controller), which imposes certain behaviour on the end-effector. The same idea is applicable in case of the bionic robot. The structural diagram of the decentralized control task with reference to the central pattern generator is depicted in figure 2. As it can be noted, the desired moment for each leg is obtained from the pattern registrar as desired torque

τ_d, the output is the torque applied to load τ_l, where the latter is being modulated via the motor rotational velocity as indicated in the following equation:

$$\omega_M = \tau_l\left(\frac{1}{J_l p} + \frac{p}{K_{st}}\right), \quad (2)$$

where ω_M – the motor angular velocity; J_l – the load inertia; p – the Laplace operator and K_{st} – the stiffness coefficient.

On the other hand, the torque exerted by a spring caused by angular deflection is computed as follows:

$$\tau_l = -K_{st}(\theta_l - \theta_M), \quad (3)$$

where $(\theta_l - \theta_M)$ represents the deflection between the motor and the load.

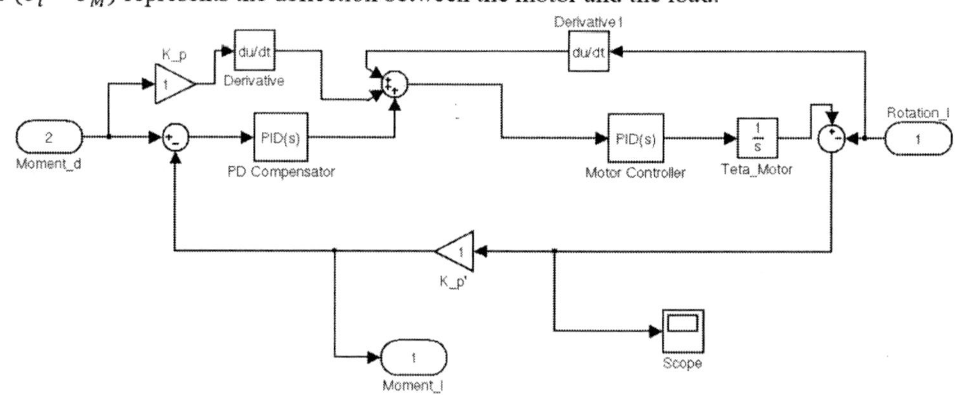

Figure 2. The simulink model of decentralized leg control.

It is easy to find the motor torque using the motor angular velocity, which in its turn can be found from the angular position. Therefore, it is possible to write the following equation:

$$\omega_M = \frac{\tau_l}{K_{st}p} + \theta_l p; \quad (4)$$

As it can be seen from (4), the term of inertia in rotational moment J_l does not appear in the right-hand side of the equation. Taking figure 3 as a reference point, it can be noted that equation (4) is being compensated using the feed-forward approach, where the PD transfer function represents the uncertainty compensation, which is added to the mechanical interference and fed to the motor controller as a single input equal to desired rotational velocity of the motor. In the wake of the above-mentioned, it is obvious that the position of the leg (stance control) is not crucial for the swing mechanism, thus the control task is formulated as a velocity-based approach. Randomly, we have chosen the numerical values of the parameters highlighted in the aforementioned equations. The task is to achieve

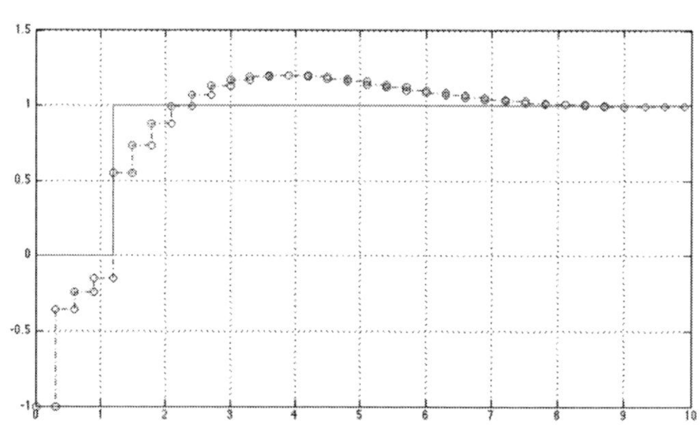

Figure 3. Simulation results of the decentralized leg control: *x-axis: time, [s], y-axis: torque [Nm]*

3.3. Fuzzy Logic Pattern Generator

Let us study the gait of a quadrupedal animal and try to implement the same pattern in the mobile robot. Figure 4 illustrates the pattern combination of each leg. We define eight working elements as each leg consists of two mobile sections: the upper part (U) rotating with reference to the fixed body axis and the bottom part (B) rotating with reference to the upper part. As we can see from figure 5, the three cinematic snapshots have a single result as an output: moving the right front leg (RU2, RB2) one step forward6 and, consequently, the leg (RU1, RB1) is also moving. To achieve that, the decentralized control system for each leg should follow pattern rules, which are summarized in table 1 for 'single leg single step' movement.

Figure 4. 'Single leg single step' movement representation

Table 1. The central pattern for single leg movement

Leg Section	Phase 1	Phase 2	Phase 3
LU1	1	1	1
LB1	1	1	0
RU1	0	0	1
RB1	0	0	1
LU2	0	0	0
LB2	0	0	0
RU2	1	1	1
RB2	1	1	1

The coloured filling in table 1 represents the activation of the end effector mechanism. For instance, in the first phase, the upper and bottom sections of the left leg (1) and right leg (2) are activated, whereas the right leg (1) and the left leg (2) are inhibited. The same can be represented in a binary form in order set up linguistic rules for the fuzzy logic pattern generator, the design of which incorporates the following steps:

a. Eight inputs with a two membership function representing the gait of each leg section;
b. Four outputs representing in detail the activation and inhibition status of each leg, serving as an input for the swing control.
c. Fuzzy rules representing the possible linguistic combination

The functional block diagram of the swing control system with reference to the fuzzy logic pattern generator can be consolidated as shown in figure 5.

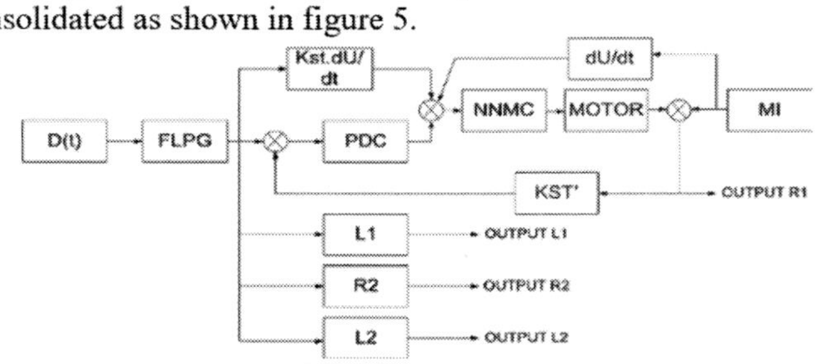

Figure 5. The consolidated swing control functional block diagram

D(t) – desired trajectory; FLPG- Fuzzy Logic Pattern Generator; PDC – PD Compensator; NNMC – Neural Network Motor Controller; MI – Mechanical Interference; K_{st} – Stiffness coefficient

3.3. Neural network motor controller

In this paragraph, we discuss the optimization of the motor controller in order to avoid overshoots in velocity. This is necessary as it will allow for better manoeuvrability of the robot and obstacle avoidance as well as minimizing the mechanical interference. The implementation of the neural network is straight forward. Data collected from the PID motor controller will serve to evaluate and train the nonlinear autoregressive neural network with an exogenous input taking into consideration the desired output and the mechanical effect on the leg section. Figures 6a and 6b illustrate the simulation results of the control approach.

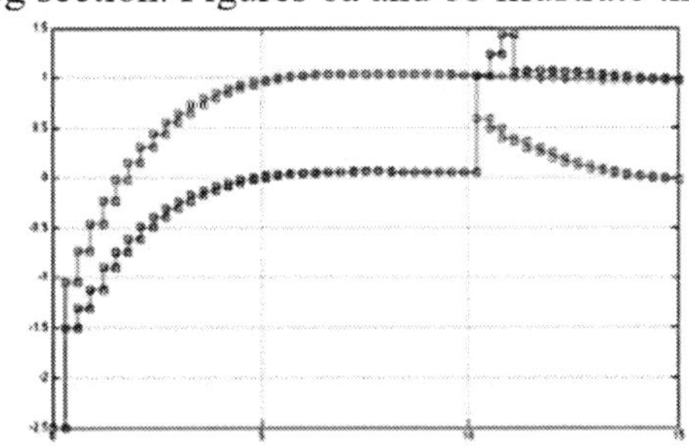

a

Figure 6a. Control of legs L1 and R1 according to FLP: *x-axis: time, [s], y-axis: torque [Nm]*

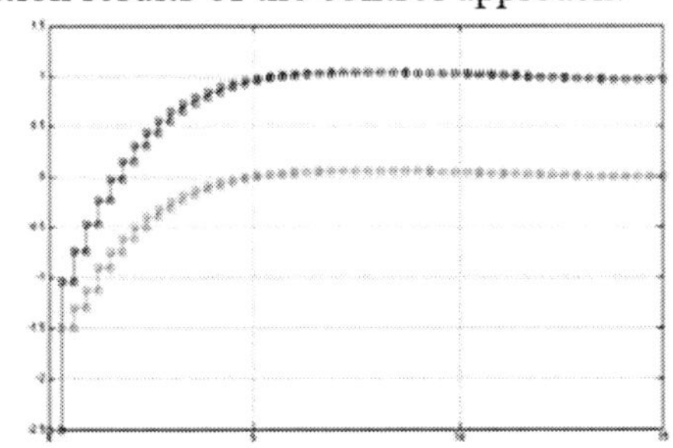

b

Figure 6b. Control of legs L2 and R2 according to FLP: *x-axis: time, [s], y-axis: torque [Nm]*

4. Conclusion

The results shown in figures 6a and 6b can be understood as follows: according to the generated pattern developed by the fuzzy logic patter generator, a combination of end-effectors movements of the quadrupedal robot are assigned as a control task. The monitoring and regulation of the end-effector outputs is managed by the neural network controller. As we can see, the leg systems successfully switch between excitation and inhibition states of leg sections allowing legs R1 and R2 to move one step forward in three phases. It is noticed as well the absence of overshooting.

On the other hand, we can clearly see that the process time has increased from 1.5 s to 3.5 s. This fact is due to the time needed for a fuzzy logic pattern generator to analyze the posture and stance control feedback. Accordingly, the neural network response time is considered in the slippage time. This proves the hypothesis stated earlier about computation time and capabilities of the robot autopilot. Taking into consideration random coefficient parameters, we can clearly see that moving the R1 and R2 legs consumed 15 s in total. This crossed distance estimated is proportional to the specifications of the end-effector (length, mass etc...). The response time can be slightly improved when optimizing the stiffness coefficient of the articulations.

References

[1] Cruse H, Kindermann T et al. 1998 *Neural Networks* **11** 1435-47

[2] Martyna Grabowska et al *J. of Experimental Biology* **215** 4255-66

[3] Cruse, H., Dean, J., Durr et al 2000 *Symposium on Unmanned Vehicles for Aerial, Ground and Naval Military Operations*

[4] Bucher, D., Akay, T., DiCaprio et al 2003 *J. of Neuroph.* **89** 1245-1255

[5] Cruse, H., Dean, J., et al 2002 *A Decentralized, Biologically Based Network for Autonomous Control of Walking In Neurotechnology for Biomimetic Robots* (MIT Press Cambridge: USA) p 383-400

[6] Delcomyn F & Nelson M 2000 *Robot & Auton Syst* **30** 5-15

[7] Quinn R, Nelson D et al 2002 *Towards the Development of Agile and Mission Capable Legged Robots In Neurotechnology for Biomimetic Robots* (MIT Press, Cambridge: USA) p 401-418

The digital algorithm for fast detecting and identifying the asymmetry of voltages in three-phase electric grids of mechanical engineering facilities

O B Shonin, S B Kryltcov and N G Novozhilov

Department of Electrical Engineering, Saint-Petersburg Mining University, 21st Line, 2, St Petersburg, 199106, Russia

E-mail: ninosh_eltech@mail.ru, kryltcov@outlook.com, nikita_novozhilov@mail.ru

Abstract. The paper considers a new technique for the fast method of extracting symmetrical components of unbalanced voltages caused by the faults in electric grids of mechanical engineering facilities. The proposed approach is based on the iterative algorithm that checks if the set of at least three voltage discrete measurements belongs to a specific ellipse trajectory of the voltage space vector. Using classification of unbalanced faults in the grid and results of decomposing the voltages into symmetrical components, the algorithm is capable to discriminate between one-phase, two-phase and three-phase voltage sags. The paper concludes that results of simulation in Simulink environment have proved the correctness of the proposed algorithm for detecting and identifying the unbalanced voltage sags in the electrical grid under condition that it is free from high order harmonics.

1. Introduction

Unbalanced grid conditions often emerge because of asymmetrical voltage sags due to faults in transmission lines and in power transformers. The problem of voltage sags is the most acute for electric grids of mechanical engineering facilities with equipment of several megawatts such as arc furnace, cold/hot roll mills, blast furnace etc., which are sensitive to the electrical energy quality. Voltage sags may lead to disconnection of such equipment from the grid, interruption of the production process and consequent severe economic losses for the enterprise.

Fast and accurate identification of unbalanced grid conditions provide benefits for a number of power equipment. For the variable frequency drive, it can improve its ride-through capability via timely reconfiguration of the control system for implementing an auxiliary mode of operation in order to maintain the controllability of the drive. For protection devices of transmission systems, it may lead to faster isolation of faulted grid segments, therefore limiting voltage sag consequences for other parts of the power system. As to compensation devices in distribution networks, it may provide their faster respond to the advent of asymmetry and, therefore, proper and timely correction of the disturbances via putting into operation additional devices to secure nominal voltage for end-user devices.

The unbalanced state of the grid is usually described in terms of Fortescue's symmetrical components theory [1], which claims that any unbalanced three-phase system in a steady state can be represented as three balanced three-phase systems called sequences. Parameters of phasors of symmetrical components normally are estimated by applying the Fast Fourier Transform (FFT) to a set of measurements. Extracted harmonics of each phase are analyzed to determine magnitudes and angles of positive and negative sequences [2].

An index of asymmetry for the unbalanced three-phase system usually is estimated through α and β components of the voltage space vector by extracting basic and second harmonics with consequent calculation of the ratio of their magnitudes [3]. The performance speed of algorithms based on FFT is limited by the necessity to have a period of being processed signal in order to obtain the correct estimation of the phasors parameters - otherwise application of FFT to a part of sinusoid leads to the spectrum leakage and false computations [4].

In case of dynamical signals, Lyon's theory of generalized symmetrical components is used [5]. This theory allows transforming the dynamic phasors to time domain signals by putting into correspondence the phase shift between phasors to a respective time delay. Delayed signals are possible to obtain via direct accumulation of measurement samplings [6] or by digital filters, which provide additional benefits such as attenuating the harmonics and adaptive tuning. The typical examples of these circuits are second-order low-pass filters [7] and dual second order generalized integrators [8]. However, the performance speed of such circuits is bounded by the necessity to implement time delay of quarter of a period.

Frontier industrial controllers have processors that provide calculating sophisticated math functions during a cycle of computation. For real-time control systems in this case, the basic purpose is developing algorithms, which effectiveness mainly depends on computational performance of the controller instead of amount of data collected. The aim of the paper is to present this algorithm applied to fast and reliable detection of the grid voltage parameters under steady state unbalanced conditions.

2. Theoretical basis

Fault events in the power systems affect both magnitude and phase angle of voltages. During the sags, a voltage magnitude goes down to less than 90% of nominal value. Changes in phase angles are in the form of a phase angle jump, which is defined as a shift of zero crossing of instantaneous voltage in comparison with pre-event observations. Depending on different drops of RMS voltages and a phase angle between phases, the sags are classified into seven categories according to the ABC classification [9]. It is useful to find relations between these categories of sags and parameters of a rotating space vector, representing a three-phase system in different frames.

According to the Clark transform [1], coordinates of such vector in a stationary $\alpha\beta$ plane can be obtained from instantaneous voltage measurements in abc system as follows:

$$\begin{bmatrix} v_\alpha \\ v_\beta \end{bmatrix} = \frac{2}{3} \cdot \begin{bmatrix} 1 & -1/2 & -1/2 \\ 0 & \sqrt{3}/2 & -\sqrt{3}/2 \end{bmatrix} \cdot \begin{bmatrix} v_a \\ v_b \\ v_c \end{bmatrix} \tag{1}$$

Using complex vector notation, we can write down:

$$\vec{v} = v_\alpha + jv_\beta \tag{2}$$

According to the Fortesque's theory [1], each voltage vector of the unbalanced system can be represented as sum of positive and negative sequence components:

$$\vec{V}_a = \vec{V}_a^{\,+} + \vec{V}_a^{\,-}; \quad \vec{V}_b = \vec{V}_b^{\,+} + \vec{V}_b^{\,-}; \quad \vec{V}_c = \vec{V}_c^{\,+} + \vec{V}_c^{\,-} \tag{3}$$

Combining equations (1), (2) and (3) gives representation of space vectors in terms of positive and negative sequences:

$$\vec{v} = \vec{v}^{\,+} + \vec{v}^{\,-} \tag{4}$$

Vectors $\vec{v}^{\,+}$ and $\vec{v}^{\,-}$ rotate in opposite directions. For unbalanced systems, the vector has an elliptical locus with parameters a, b, θ indicated in Figure 1.

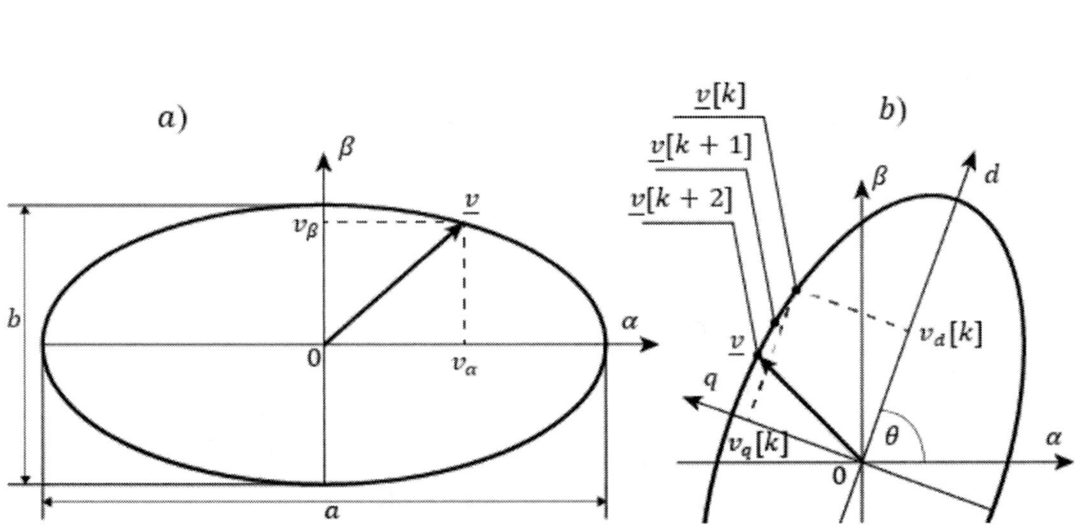

Figure 1. Ellipse locus of a grid voltage vector during unbalanced operation in: $\alpha\beta$ coordinates – a) and dq coordinates – b)

The ellipse equation is as follows:

$$v_\alpha{}^2 \cdot a^{-2} + v_\beta{}^2 \cdot b^{-2} = 1 \quad (5)$$

Here *a, b* – major and minor axes of ellipse.

According to equation (4), these quantities are the maximum and minimum sum of magnitudes of positive and negative components:

$$a = V^+ + V^- ; \quad b = \left| V^+ - V^- \right| \quad (6)$$

There are three types of voltage sags according to the classification based on symmetrical components. During one-phase voltage sags only axe *a* remains unaffected while *b* decreases. Two-phase sags lead to unequal decrease in both axes a and b. Three-phase voltage sags lead to an equal decrease in axes *a* and *b*.

3. Proposed algorithm

The proposed algorithm is based on verification of belonging of at least 3 measured samplings to one of the ellipse-like trajectories of the rotating space vector. For simplicity, the proposed algorithm will be discussed in detail for the set of two of samplings.

In this case, equation (5) goes to the system of two equations in respect to two variables:

$$v_\alpha[k]^2 \cdot a^{-2} + v_\beta[k]^2 \cdot b^{-2} = 1; \qquad v_\alpha[k+1]^2 \cdot a^{-2} + v_\beta[k+1]^2 \cdot b^{-2} = 1 \quad (7)$$

Solution of system (7) for axes a and b can be written as follows:

$$a = \left(\left| v_\alpha{}^2[k+1] \cdot v_\beta[k]^2 - v_\alpha{}^2[k] \cdot v_\beta{}^2[k+1] \right| \right)^{\frac{1}{2}} \cdot \left(\left| v_\beta[k]^2 - v_\beta{}^2[k+1] \right| \right)^{-\frac{1}{2}};$$

$$b = \left(\left| v_\alpha{}^2[k] \cdot v_\beta[k+1]^2 - v_\alpha{}^2[k+1] \cdot v_\beta{}^2[k] \right| \right)^{\frac{1}{2}} \cdot \left(\left| v_\alpha[k]^2 - v_\alpha{}^2[k+1] \right| \right)^{-\frac{1}{2}} \quad (1)$$

Equation (8) is valid for the case axe a of the ellipse is aligned along with the alpha-axis. With asymmetrical faults, the ellipse-like trajectory may be rotated at arbitrary angle θ depending on the parameters of the unbalanced system. Therefore, all projections should be also rotated at this angle, which may be implemented by using the synchronous reference frame (Figure 1b). Coordinates of a vector in the orthogonal reference frame rotated at angle θ are calculated via the Park transform [10]:

$$v_d = v_\alpha \cdot \cos\theta + v_\beta \cdot \sin\theta; \qquad v_q = v_\beta \cdot \cos\theta - v_\alpha \cdot \sin\theta \quad (9)$$

In a new coordinates system, the ellipse axes a and b are defined as follows; (2)

$$a=\left(\left|v_d^{\,2}[k+1]\cdot v_q[k]^2-v_d^{\,2}[k]\cdot v_q^{\,2}[k+1]\right|\right)^{\frac{1}{2}}\cdot\left(\left|v_q[k]^2-v_q^{\,2}[k+1]\right|\right)^{-\frac{1}{2}};$$
$$b=\left(\left|v_d^{\,2}[k]\cdot v_q[k+1]^2-v_d^{\,2}[k+1]\cdot v_q^{\,2}[k]\right|\right)^{\frac{1}{2}}\cdot\left(\left|v_d[k]^2-v_d^{\,2}[k+1]\right|\right)^{-\frac{1}{2}} \quad (10)$$

Solving equation (5) for the third sampling under condition that axes a and b are found from equations (10), we obtain formula for assessment of deviation E_θ of sampling $\underline{v}[k+2]$ from the ellipse trajectory:

$$E_\theta=\left(v_d^{\,2}[k+2]\right)\cdot a^{-2}+v_q^{\,2}[k+2]\cdot b^{-2}-1 \quad (3)$$

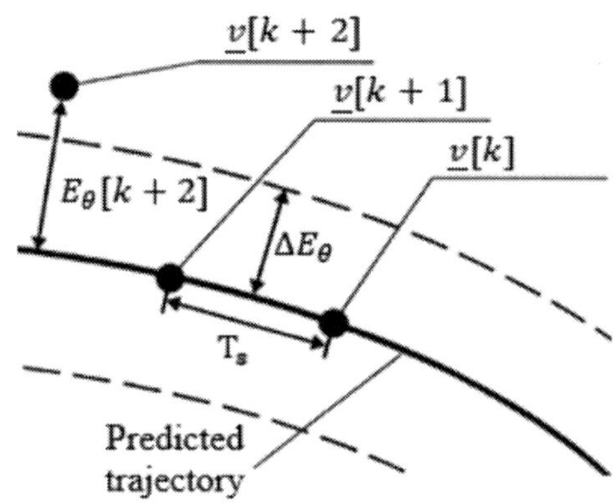

Figure 2. Possible placement of samplings in respect to the ellipse trajectory

One of the easiest ways to estimate the angle θ is to solve equation (11) for a predefined set of angles in the range: $\theta=0-90°$. The quantization of angles affects accuracy of the algorithm and is dependent on the performance of the controller.

As the next step, we choose the solution with minimal absolute value of E_θ and compare it with the predefined threshold ΔE_θ, defining the tolerance curve (Figure 2). If the solution error is greater than tolerance value ΔE_θ, then we assume that the transients exists and therefore wrong extraction of symmetrical components takes place. Otherwise, we detect the unbalanced steady-state grid operation with positive and negative sequence components, which magnitudes can be calculated by substitution of estimated values of axes a and b to equation (6):

$$V^-=|a-b|\cdot 2^{-1};\qquad V^+=a-V^- \quad (4)$$

The balanced system has the circle trajectory of the voltage vector and, therefore, will be represented as a special case of the ellipse with $a=b$. The type of unbalanced voltage sag can be determined by deviation of quantity a form the pre-event value.

The proposed algorithm is valid for the case of ideal asymmetry of the system without distortions of the voltage waveforms. In case of distortions, the algorithm should be modified for the instance by extension of measurements and filtering the signals.

4. Results of simulation

Simulation has been carried out in the Simulink environment with assistance of a processing unit built up in accordance with the proposed algorithm and controlled three-phase voltage source with the predetermined type of the voltage asymmetry in accordance with the symmetrical components classification. The controller is assumed to be fast enough to compute the whole cycle of operations between measurements at the sampling frequency of 10kHz. A step of discrete variation of angle θ was adopted as one degree.

Results of the simulation in per units are shown in Figure 4.

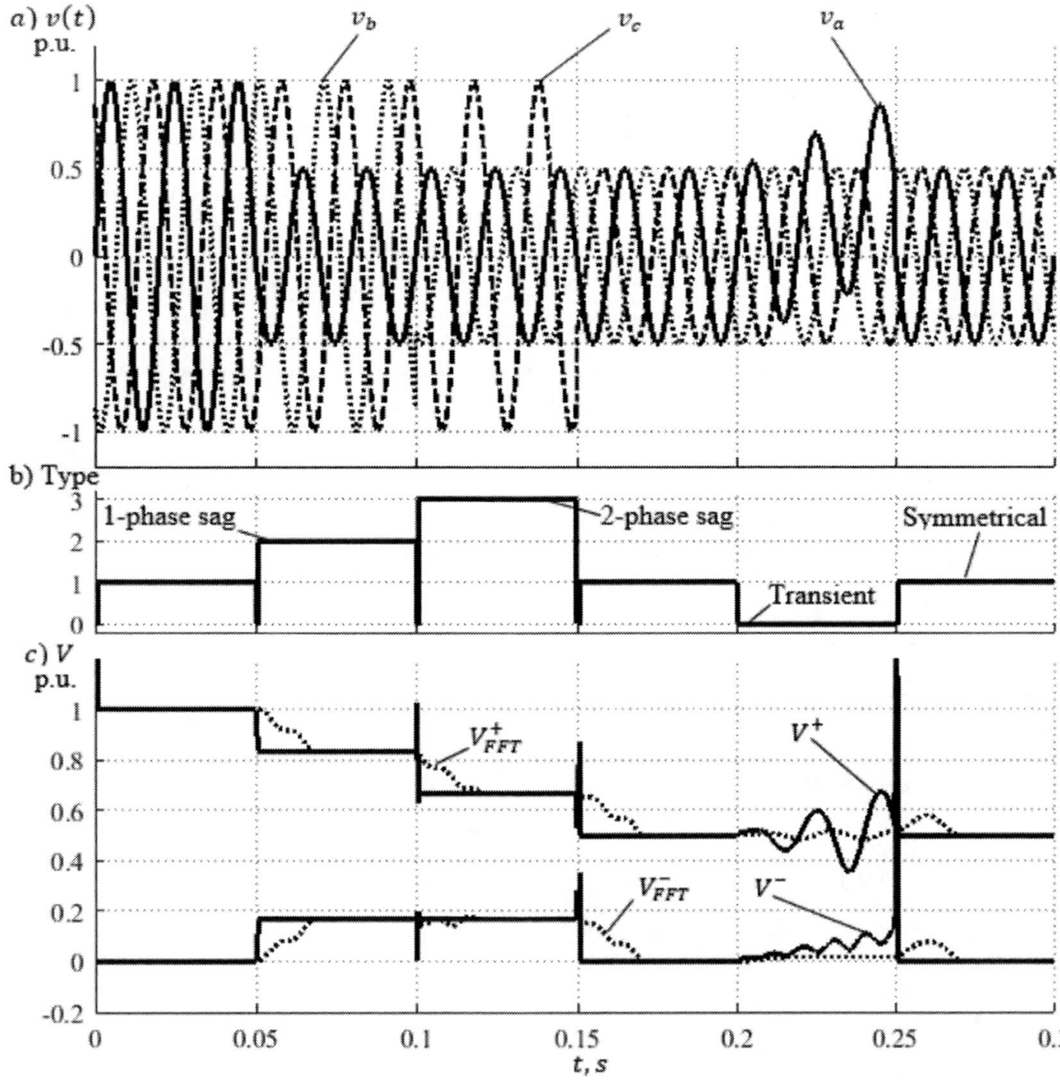

Figure 3. Results of the proposed algorithm study under different conditions. Top-to-down: phase voltages waveform – a); fault types identified by the algorithm: transients, symmetrical state, one-phase sag, two-phase sag – b); positive V^+ and negative V^- sequence magnitudes extracted by the proposed algorithm in comparison with magnitudes $V_{FFT}{}^+$, $V_{FFT}{}^-$ obtained with the FFT-based algorithm – c).

Figure 4a demonstrates the modelled voltage waveform during asymmetrical operation of the system. Figure 4b shows the type of asymmetry detected by the algorithm. Figures 4c-d show positive and negative sequence components extracted by the algorithm and for comparison the same components extracted by FFT-based algorithm are shown as well. The electrical grid operates in a normal mode in the interval of time $0 \le 0 \le 0.05s$.

During the one-phase sag at time t=0.05 – 0.1s the voltage in phase A decreases to 50% of nominal value. It can be seen that proposed algorithm is possible to detect one-phase sags and extract the magnitudes of positive and negative sequences nearly instantly. The same is fair for two-phase voltage sags taking place at time t=0.1 –0.15s. It is evident that the algorithm distinguishes between one-phase and two-phase faults. During a three-phase voltage sag, at time t=0.15 – 0.2s, the algorithm detects the circle form of the voltage vector trajectory and classifies the system as a symmetrical one.

To study a respond of the algorithm to transients, the voltage in phase A was changing at the span of t=0.2 – 0.25s. Because the minimal error of evaluating the ellipse parameters proved to be higher than tolerance threshold, the algorithm detects transients, and therefore identification of positive and negative components is not possible.

5. Conclusions

Paper presents a novel digital algorithm that may be effectively used for fast extraction of the positive and negative sequences from asymmetrical voltage systems. In comparison with other similar algorithms, no spectrum analysis or filtering is needed. The proposed algorithm iteratively solves a set of equations to align the synchronous reference frame along the major or minor axes of the elliptical locus formed by a grid voltage space vector. To detect asymmetrical conditions in the system, at least 3 measured samplings are required.

In the presence of additional harmonics in the system, the algorithm is to be modified by the increase in the number of samplings to calculate the deviation error or by using filtered signals.

The algorithm performance has been validated by computer simulation of processing unbalanced signals of different origins in order to detect and identify one-phase, two-phase and three-phase voltage sags on the basis of symmetrical components classification. Results of simulation have revealed nearly instantaneous extraction of positive and negative sequences and clear detection of every type of faults.

References

[1] Marx S, Bender D 2016 *33th Annual HANDS-ON relay school* (Washington: USA) p 76
[2] Tenti P, Willems J, Mattavelli P and Tedeschi E 2007 *Electrical Power Quality and Utilisation J.* **13(1)** 9-15
[3] Paap G C 2000 *IEEE Transactions on Power Systems* **15(2)** 522-28
[4] Lyon D 2009 *J. of Object Technology* **8(7)** 23-34
[5] Leva S 2009 *J. of Electromagnetic Analysis and Applications* 205-13
[6] Haque M T, Hosseini S H A 2001 *The 2nd Intl. Conf. on Electrical and Electronics Eng.* (Bursa:Turkey) p 152
[7] Robles E, Ceballos S, Pou J, Martın J L, Zaragoza J and Ibanez P 2010 *IEEE Transactions on Power Electronics* **25(10)** 2552-63
[8] Luna A, Rocabert J, Candela I, Rodriguez P, Teodorescu R and Blaabjerg F 2012 *In 2012 IEEE Energy Conversion Congress and Exposition (ECCE)* 2769
[9] Kamarudin M S, Zulkifli S A, Sulaiman E and Mokhtar N H 2008 *The 2nd Int. Power Eng. and Optimization Conf.* (Selangor: MALAYSIA) p 267
[10] Popescu M 2000 *Induction Motor Modelling for Vector Control Purposes* (Espoo: Helsinki University of Technology)

On potentiality of brake rigging unification for freight cars with separate braking

V V Sinitsyn[1] **and V V Kobishchanov**[2]

[1] Management company 'Bryansk engineering plant', 26, Ulyanov St., Bryansk, 241015, Russia
[2] Bryansk State Technical University, 7, 50-letiya Oktyabrya bul., Bryansk, 241035, Russia

E-mail:universalmasch@yandex.ru

Abstract. A possibility of unified rigging use with a brake cylinder (BC) in bogie areas or on a bogie in a freight car is defined. Two ranges of transmission ratios were obtained for freight cars with pressure control in cylinders with the aid of the automated mode. To change transmission ratios, two openings are supposed to be used in rigging.

1. Introduction

One of the ways of brake system updating in domestic freight eight-wheel cars consists in the location of the mechanical unit on a bogie.

This direction is most developed abroad. We should mention such characteristic examples of a brake system used on the bogie as Motion Control manufactured by Amsted Rail Co. Ltd. (USA),a brake system developed by Cardwell Westinghouse (incorporated into Wabtec Co.) for Barber bogie manufactured by Wabtec Co. (USA), and Knorr-Bremse (Germany) for European bogie Y 25.

According to [1-3], systems with the location of an actuator component of a brake on bogies allows simplifying a brake system and with further substitution of existing arrangements. Furthermore, when arranging the system on bogies, an undercar area becomes almost absolutely free from brake equipment.

It becomes particularly important during designing railroad cars for increased axial loads. At present, this drive is used on a closed-bottom gondola car of model 12-9828 with a lowered floor level and the axial load of 27 t manufactured by Roslav 1 Railway-Carriage Repair Works Co. [4]. In comparison with analogs this car has an increased carrying capacity at the expense of floor lowering in the central area ensuring additional body space. Here, the original design of a brake mechanical unit with a separate bogie drive is realized. A rigging structure includes a brake cylinder of model TCR-10-85 (a brake cylinder with an integrated regulator of a rod outlet manufactured by Trans –pneumatics Co.) installed on a carriage under frame close to a bogie. A cylinder conveys an effort to a bogie rigging by means of levers fixed on a horizontal shaft. A shaft is installed on a carriage under frame through bushings with elastomeric spacers.

The experience of this work is significant from the point of view of the particularly compact brake equipment use in a freight car. In this connection, the application of just cylinders TCR in an actuator unit of the brake of freight eight-wheel cars with drives located in bogie areas is promising.

A possibility of such rigging use for a maximum amount of car types is limited to a high extent by a certain scatter of transmission ratios in riggings used in domestic railroad cars.

It is necessary to determine a possibility of a unified rigging (with cylinder TCR) use in railroad cars with pressure control in cylinders with the aid of an automated mode. A similar problem should be solved also for railroad cars having no possibility to adjust pressure in cylinders with the aid of an automated mode.

At the same time, it is necessary to consider a possibility of the location of a brake system with cylinder TCR both in bogie areas and on a bogie for a maximum possible number of railroad types. In addition, we shall define limitations on the correlations of a lightweight and carrying capacity for freight cars operated in the RF, within the limits of which it is possible to use a unified brake system. It is supposed to use a cylinder in a brake system with its location on a bogie, as a system least dependent on design peculiarities of railroad cars. The cylinder of the TCR-10-40 type is chosen as one of the most possible variants of brake cylinders.

In accordance with a routine calculation [5], we will define the ranges of transmission ratios of the rigging with cylinder TCR-10-40 TR 24.05.382, within the limits of which, it is possible to use a brake system, the actuator of which is located on the bogie of type 18-100 or corresponding to it with the axial load of 25 tons (trucks types 2 and 3, GOST 9246-2013).

According to [5], for railroad cars with separate braking and compact cylinders, there is no standard foreseen for a rod outlet to regulate a system on cast-iron shoes. Kinetic friction modeling is considered thoroughly in [6]. In this connection, we will consider a brake system equipped only with composite shoes.

For freight cars having a gross weight of 100 t and a light weight in accordance with table 2.2 [5] (corresponding table 10 [7] has a lower upper limit of pressure taking into account skid),let us determine transmission ratios. At the same time, we find an inferior limit with the lightweight of up to 27 t in accordance with the formulae of the standard computation [5]. Now, we will define a minimum possible transmission ratio (during computation of brake effectiveness) for a fright car with a gross weight of 100 t (25 tons per axle). This transmission ratio will be a maximum during computation taking into account the skidding for a fright car with a minimum possible lightweight. In the same way, we will define transmission ratios for freight cars with a minimum possible light weight having a gross weight of 94 t (23,5tnfper axle) and a lightweight in accordance with Table 2.2 [5].

2. Results and discussions

We will determine a minimum possible light weight.

Let us determine $K_{r\,min}$ under the complete freight car load reasoning from the relation of:

$$[\delta_r] = \frac{mK_{r\,min}}{Q}, \tag{1}$$

where Q –the gross weight of a freight car (100 t); $[\delta_r] = 0.14$.

$$K_{r\,min} = 1.75\,\text{tf}.$$

Let us determine $K_{D\,min}$ – the minimum possible force on a shoe in the system with composition brake shoes under the maximum load of a freight car. For that, we will solve the following equation:

$$K_{r\,min} = \frac{1.22K_D(K_D + 20)}{4K_D + 20}; \tag{2}$$

$$K_{D\,min} = 1.79\,\text{tf}.$$

Let us determine n taking into account the pressure for the computation of effectiveness under maximum load. In addition, we will take into account the absence of spring compression force of an automatic regulator, which leads to the rod (F_2) in the system with TCR (BC).

$$K_D = \frac{((\pi d_c^2 p_c \eta_c / 4) - F_1) n \eta_p}{1000m}; \quad (3)$$

$$n = 5.67.$$

For an empty car, taking into account the obtained n and pressure in cylinders for an empty car during the computation for a skid through formulae (3) and (2), we will find $K_{D min\ skid} = 0.854$ tf and $K_{r min\ skid} = 0.928$ tf.

At the same time,

$$\delta_r = \frac{mK_{r\,min\,skid}}{T}.$$

In accordance with formula $\delta_r \varphi_{kr} \le [\psi_r]$, we will obtain the equations set:

$$[\psi_r] = \frac{mK_{r\,min}\varphi_{kr}}{T}; \quad (4)$$

$$[\psi_r] = \psi_{(q0)}\psi_{(V)},$$

where $\psi_{(q0)} = 0.17 - 0.0015(q_0 - 5)$; $q_0 = T/4$ – load on a wheel pair (axle). At that, it should be taken into account that $q_0 \ge 5$ tf, that is, a car cannot have light weight less than 20 t.

For its solution, we find:

$$\varphi_{kr20} = 0.322;\ \varphi_{kr100} = 0.257;\ \varphi_{kr120} = 0.249;$$

$$\psi_{(20)} = 0.7829;\ \psi_{(100)} = 0.5638;\ \psi_{(120)} = 0.5447$$

Let us solve the equations set (4) and choose the highest values:

$$[\psi_r]_{120} = 0.0947;\ T_{120} = 19.519.$$

In accordance with GOST 9246-2013, the minimum lightweight is 21 tons for trucks types 2 and 3.Let us accept minimum packaging as 21 t.

According to Table 1, the transmission ratios of 5.7/6.3 in the light weight range of 21-27 t are within the limits of 22.5-27 t. Thus, we obtain the transmission ratio of 5.7/5.9 under the light weight of 21-22.5 t. For the rest of cars with the light weight from 22.5 t, the range of transmission ratios is 5.7/6.3.

The inferior limit with the lightweight of up to 27 t of freight cars with composition shoes and gross weight of 94 t, determined according to the mentioned method, will also make 21 t (Table 2).

Table 1. The ranges of transmission ratios for the rigging with cylinder TCR-10-40, under standard pressure control in cylinders for cars with a gross weight of 100 t.

Lightweight, 21-27t; P=0.13/0.16 MPa		Lightweight, 27-32 t; P=0.15/0.19 MPa		Lightweight, 32-36 t; P=0.18/0.22 MPa		Lightweight, 36-45 t; P=0.21/0.24 MPa	
Lower limit	Upper limit	Lower limit	Upper limit	Lower limit	Upper limit	Lower limit	Upper limit
5.7/5.9	5.8/7.8	5.7/6.3	5.8/7.6	5.7/6.4	5.7/7.3	5.7/6.55	5.7/8.45

In the numerator there is a value for brake effectiveness, in the denominator – for skidding.

Table 2. The ranges of transmission ratios for the rigging with cylinder TCR–10-40 under standard pressure control in cylinders for cars with a gross weight of 94 t.

Lightweight, 21-27 t; P=0.13/0.16 MPa		Lightweight, 27-32 t; P=0.15/0.19 MPa		Lightweight, 32-36 f; P=0.18/0.22 MPa		Lightweight, 36-45 t; P=0.21/0.24 MPa	
Lower limit	Upper limit	Lower limit	Upper limit	Lower limit	Upper limit	Lower limit	Upper limit
5.25/5.95	5.8/7.8	5.25/6.3	5.8/7.6	5.25/6.4	5.3/7.3	5.25/6.6	5.65/8.45

According to Table 2, the transmission ratios of 5.8/6.3 in the range of the light weight of 21-27 t are within the limits of 22.5-27 t. Therefore, we will obtain the range of transmission ratios of 5.25/5.95 under the light weight of 21-22.5 t. For the rest of cars with a light weight from 22.5 t, the range of transmission ratios is 5.8/6.3.

In the numerator, there is a value for brake effectiveness, in the denominator – for a skidding. In the general case, for cars having a maximum lightweight within the range of 21-45 t and a gross weight of 94 or 100 t, we obtain two ranges of transmission ratios: 5.7/5.9 for cars with the light weight of 21-22.5 t and 5.8/6.3 for the rest with the light weight of 22.5-45 t.

Table 3. The ranges of transmission ratios for the rigging with cylinder TCR-10-40, under the standard control of pressure in cylinders for cars with the gross weight of 94-100 t

Lightweight, 21-27 t; P=0.13/0.16 MPa		Lightweight, 27-32 t; P=0.15/0.19 MPa		Lightweight, 32-36 t; P=0.18/0.22 MPa		Lightweight, 36-45 t; P=0.21/0.24 MPa	
Lower limit 21-22.5	Upper limit 22.5-27	Lower limit	Upper limit	Lower limit	Upper limit	Lower limit	Upper limit
5.7/5.9	5.8/6.3	5.8/6.3	5.8/6.3	5.8/6.3	5.8/6.3	5.8/6.3	5,8/6.3

In the numerator, there is a value of brake effectiveness, in the denominator – for a skid. Now, we will determine the limits of possible use of pressure automatic control in cylinders. The possibility of the automatic mode use in cars could be determined in accordance with the diagram [8] taking into account standard adjustments of pressure [7]. In Figure 1, there are values of minimum possible deflections of a car suspension under a paying load, that is, at lesser deflections, the value of pressure will be less than 0.3 MPa. It should be emphasized that on the basis of the diagram mentioned above under the pressure lower than 0.3 MPa, we will obtain sufficient brake effectiveness without the skid. But at present, this problem in the regulation documentation is not shown. This circumstance may be one of the reasons for the use of a single-mode brake (along with the necessity of an automatic mode (automatic modes) installation and supporting beams and their maintenance).

Cars having lightweight up to 27 t must have the deflection of the spring suspension not less than 35mm. Those cars having light weight from 27 to 32 t – not less than 27.3 mm. And the cars having lightweight from 32 to 36 t – not less than 21.6 mm, and those having lightweight from 36 to 45 t – not less than 17.5 mm.

Accordingly, the minimum carrying capacity of cars on the bogies of type 18-100 with the flexibility of a spring set equal to $\lambda = 1.13$ mm/t funder light weight of up to 27 t should be not less than 61.9 t. Under the light weight from 27 to 32 t – not less than 48.3 t. With the lightweight from 32 up to 36 t – not less than 38.2 t. Under the light weight from 36 to 45 t – not less than 31 t.

Practice shows that these cars are operated both with an automatic mode and without one [9]. Under the complete load, the pressure of an automatic mode may be less than 0.3 MPa. In particular, freight cars for the motor car transportation possess this property. It is assumed, according to an agreement with a customer "RRW" Co. to be limited with modes switched manually for cars loading, the complete rated loading of which does not exceed 70% of light weight [5]. The computations of a brake for some types of cars in this case show a possibility of skid under a loaded state of a car under a complete rated load. In this connection, the advantages of a single-mode brake are evident.

The use of a single-mode brake has also limitations. According to the computations carried out, one limitation is a ratio of a light weight to a carrying capacity equal to ~0.8 (carrying capacity should not exceed 125% of a light weight). With the carrying capacity less or equal to this ratio, it is possible to obtain a stable transmission ratio meeting the normative requirements. With larger carrying capacity, at the minimum possible transmission ratio of rigging, we obtain skid possibility.

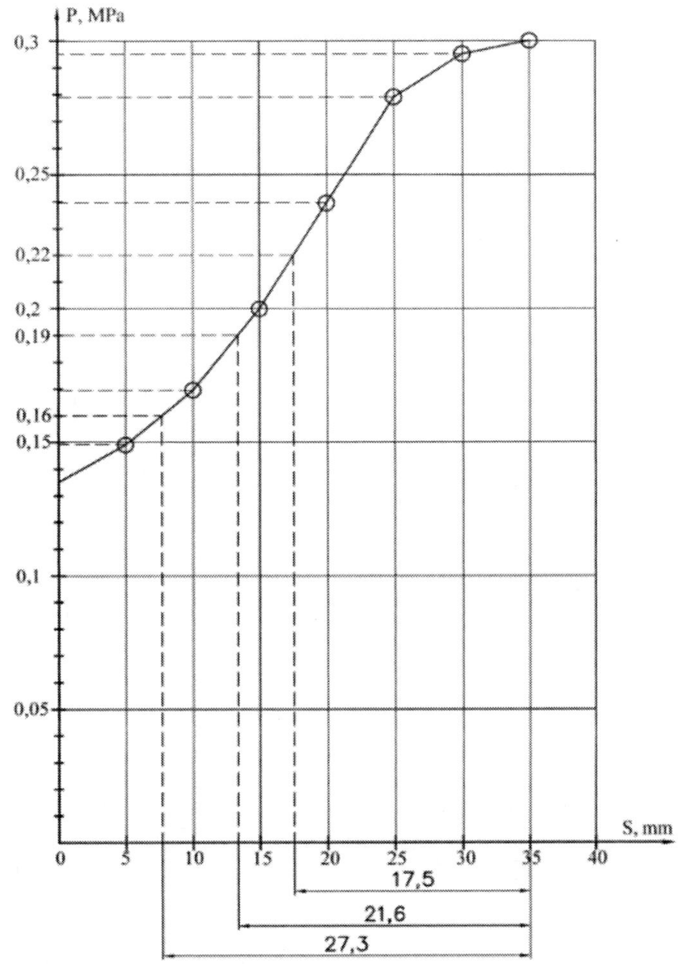

Figure 1. The dependence of the output pressure of automatic mode 265A-4 upon a deflection of a car spring unit [8] and the value of minimum possible deflections at pressure adjustment depending on the lightweight.

Now, we will determine ranges of transmission ratios for a single-mode brake. According to [10], air distributors must be switched on in a medium mode in cars having a board 'single-mode' on a car body supplied with composite shoes.

The minimum carrying capacity may be equal to zero. The maximum carrying capacity of such cars, according to the dependences obtained within the limits of adjustable pressures for such cars on the bogies of type 18-100 under the light weight, is up to 27 t – 61.9 t (229 % of the light weight). Under the lightweight from 27 to 32 t – the maximum carrying capacity is 48.3 t (151 % of the light weight). Under the lightweight from 32 to 36 t – the maximum carrying capacity is 38.2 t. Under the lightweight from 36 to 45 t – the maximum carrying capacity is 31 t (69 % of the lightweight).

The analysis of transmission ratios for the maximum ranges of the paying load spread under the pressure corresponding to the average mode of an air distributor shows the following: under the light weight of 36-45 t (carrying capacity is within limits from 0 up to 31 t), we can obtain transmission ratios of 4.25/4.35, under the light weight of 32-36 t, (carrying capacity is within the limits from 0 up to 38.2 t), we obtain two ranges of transmission ratios — 3.75/3.8, within the limits of the lightweight of 32-33 t, and 3.9/3.95 — within the limits of the lightweight of 33-36 t.

Under the lightweight of 27-36 t (carrying capacity is within the limits from 0 up to 48.3 t), and also under the light weight from 21 up to 27 t (carrying capacity is within the limits from 0 up to 61.2 t) of a stable transmission ratio in the range from minimum to maximum carrying capacity, a stable transmission ratio is impossible to obtain. In this connection, for these cars, a level of carrying capacity must be considerably lower. Besides, in this case,a considerable number of transmission ratios depending on individual peculiarities of every car will be obtained.

3. Conclusion

Thus, we have obtained a possibility of using a unified rigging (with cylinder TCR) for cars with pressure adjustment in cylinders with the aid of an automatic mode. At the same time,two openings will be used for the changes of transmission ratios in the rigging.

It should be pointed out that cylinder TCR-10-40 at the obtained transmission ratios will not allow using standard gaps between shoes and wheelsto a sufficient extent. In this connection, they can be used when coupled for one bogie, or a cylinder with a large rod travel has to be used.

For cars without a possibility of pressure adjustment in cylinders with the aid of an automatic mode, we have a considerable number of transmission ratios depending upon individual characteristics (a lightweight and carrying capacity) of cars. This circumstance does not allow using a unified rigging (including that with cylinder TCR).

References

[1] Nikitin G B, Kazarinov A V and Nazarov I V 2008 *Russian J. Railroad Transport* **4** 62-64
[2] Bityutsky A A and Klushantsev D E 2011 *Russian J. Science and Transport Equipment* **2** 8-15
[3] Babayev A M 2010 *Russian J. Heavy Machinery* **3** 37-39
[4] Batenkov V A 2010 *Eurasia Vesti* **5** 12-13
[5] *Technical Requirements to the Brake Equipment of Freight Cars Constructed at the Plants of the RF* 1996 (MPS of the Russian Federation) p 77
[6] Goryacheva I G andMakhovskaya Yu Yu 2010 *Russian J. Mech. Solids* **45 (3)** 390-398
[7] *General Manual on the Repair of the Brake Equipment of Cars 732-TSV-TSL* 2011 (PKB TSV JSC 'RZD') p 198
[8] Sipyagin E S 2009 *Russian J. Cars and Car Economy* **2** 32-34
[9] Sinitsyn V V and Gerasimov K V 2013 *Russian J. Heavy Industry* **9** 26-27
[10] *Rules of Maintenance of Brake Equipment and Brake Control of Rolling Stock Approved by the Railroad Transport Council of the CIS States* 2014 p 181

IOP Conf. Series: Materials Science and Engineering 177 (2017) 012013 doi:10.1088/1757-899X/177/1/012013

Operational control and diagnostics of the equipment by the parameters of the electric drive power supply by the example of deep well pump units

D I Shishlyannikov[1] and M A Vasilyeva[2]

[1] Perm National Research Polytechnic University, 29, Komsomolsky Ave., Perm, 614990, Russia
[2] Sankt-Petersburg Mining University, 2, 21-st Line, Vasilyevskiy Island, St Petersburg, 199106, Russia

E-mail: saturn.sun@mail.ru

Abstract. Currently, the vast majority of Russian oil wells use oil well pumping units (OPU), equipped with a mechanical drive – a balancing pumping unit. The significant lifetime of balancing pumping units used in oil fields of Russia amounts to full or close to full resource generation and to a high accident rate of the mechanical drive rod of OPU. Reduced material costs for operation and maintenance of OPU may be provided by the rejection of the system of preventive maintenance and service during the transition to the actual technical condition. The article analyzes the statistics of the most frequent crash balancing pumping units of OPU and their causes. The prospect of applying the method of instrument control parameters is proved, and the technical condition of the OPU is assessed based on the analysis of the magnitude and the nature of changes in load drive motors, determined by the recording of the instantaneous values of power consumption. The authors consider the construction and operation of the programmable controller 'AKD-SK' manufactured by JSC R&D company 'ROS' (Perm). Fundamentals of vattmetrogramm analysis of OPU drive motors are stated, and the character of displaying key defects of submersible pumps and balanced beam unit is described.

1. Introduction

For oil-producing enterprises engaged in well operation using deep well pumping units (DWPU), the tasks of enhancing the efficiency of using DWPU by means of reducing material costs for the maintenance of their working capacity, reducing the number of crashes and by the transition from a planned preventive repair system of pumping equipment to servicing according to the actual technical condition remain relevant

Solution of these problems is possible through the development and widespread use of methods and tools of instrument control, implementing a continuous monitoring of DWPU operating parameters. They allow quantifying the loading drive and to monitor the intensity of the flow of the degradation processes in the DWPU subassemblies which are most subject to wear [1, 2].

Currently, in the Russian and CIS countries oil fields, DWPU with a mechanical drive – balancing beam pumping unit — are the most common. These units represent a conservative set of equipment, the construction of which has not been fundamentally changed over the last 40 years. The main advantages of DWPU with a mechanical drive can be related to long periods of operation, simplicity of the construction and maintenance, as well as to the use of rod pumps at marginal wells under complicated geological conditions (high water content, significant content of abrasive, sulfur, paraffin reservoir fluid, etc.). However, significant dynamic components and variability of external loads, exposure of the beam-pumping unit (BPU) to the adverse effects of the atmosphere, installation and adjustment errors, improper maintenance and operation of the BPU beyond the normative periods stipulate a high accident rate of DWPU, which leads to an increase of operating costs and a reduction of profitability of an oil extraction process [3].

2. Materials and methods

According to data provided by the JSC 'Lukoil-Perm' company (Table 1), the most frequent emergency failures of BPU balancers of the DWPU include: destruction and withdrawal of the finger out of the crank (41%), depreciation and destruction of the low-speed shaft of the gearbox (20%), rapture of the connecting rod (17, %). Among the main reasons stipulating the occurrence of unexpected failures are metal fatigue, exceeding normative load, low level of service of DWPU.

Table 1. Statistics of emergency failures of balancers according to data of JSC 'Lukoil-Perm'

Emergency failure	The proportion of the total number of failures, %
Getaway of the crank from the gear box driven shaft	3
The output pin from the crank	8
The destruction of the crank pin	33
Breakage of the balancer	3
Breakage of the traverse	5
Breakage of the connecting rod	17
The destruction of the traverse support	3
Breakage of the head rocker	8
The destruction of the gear (low speed shaft breakage)	17
The destruction of the gearbox gears	3

The most common causes of failure of the deep well pumps are the parted rods, defects of receiving and discharge valves, corrosion and abrasive damage of work surfaces.

It should be noted that one of the key factors that have a significant impact on the operating time of DWPU is the balanced state of BPU, determining the level of dynamic loads on the machine components and the amount of specific power inputs for the rise of the reservoir fluid. Today, the following methods are most commonly used for balancing mechanical BPU:

– weight balance – the oldest and simplest method, implemented by using a set of heavy plates, mounted on the rear end of the rocker BPU;

– rotor balancing – the loads in the form of counter-balances are placed on heavy cranks, the mass of which is also involved in the process of balancing of the equipment;

– combined, rotary-balancing counterbalance – the most effective, is used in most of the BPU domestic and foreign production.

The main purpose of balancing BPU devices is accumulation of potential energy when the rods move downwards and output of the energy when the rods move upwards. Supplied potential energy is converted into work, which, together with the work done by the drive motor, is spent on the movement of the suspension point of the column of rods upwards and lifting oily liquid from the well.

Currently, at oil fields of Russian, BPU balancing is generally performed using a current clamp, controlling the amount of current in the stator windings of the asynchronous motor. Because the drive motors are usually underused in the BPU operation, the motor windings are dominated by the reactive power components that are not related to the load, resulting in low efficiency of balancing the BPU under existing currents. According to the contracting organizations carrying out technical service of DWPU, about 65% of the existing BPU park are incorrectly balanced at Russian oil fields.

Difficult operating conditions of oilfield equipment, a multifactorial process of interaction of elements DWPU among themselves and with the external environment, as well as price restrictions (profitability, product competitiveness in the market conditions), necessitate a technically simple and reliable solutions in the design of automatic parameter control systems work and evaluation of technical state pumping systems. One of the most promising ways of instrument control and diagnostics of DWPU is to analyze the magnitude and the nature of changes in the external load, determined by measuring the instantaneous values of currents, voltages, and to calculate the power consumed by the motor of BPU [4-12].

3. Research and discussion

Employees of JSC R&D 'ROS' company (Perm) developed and commercially produced programmable controller 'AKD-SK', designed for continuous monitoring of machine-rocking electric parameters. The complex is used for registration vattmetrogramm on energized equipment without prior disconnection that allows us to diagnose and monitor the balance of the BPU, and send the results to the operator of the oil field network.

The structure of the 'AKD-SK' equipment complex (Figure 1) includes a controller with light diodes to indicate, a clamp meter and a timer of a magnetic type. The complex is directly mounted in the control station of BPU, has the external output RS 482 connector (Figure 2) for connecting of an operator of the oil field to a network and the additional option of wireless data transfer from the controller to the network, laptop or tablet [13].

Figure 1. The external view of the programmable controller 'AKD-SK'.

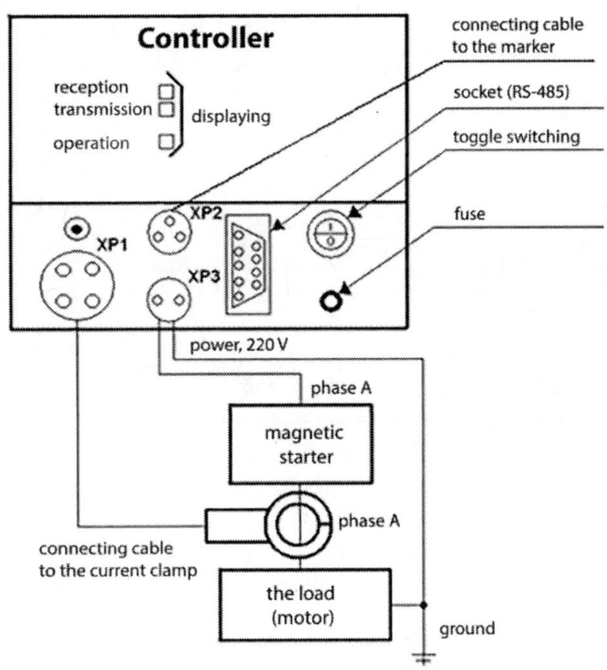

Figure 2. The structural scheme of the programmable controller 'AKD-SK'.

For the power supply of the complex and registration of instantaneous voltage values consumed by the BPU engine drive, the power cable connecting the controller via connector XP3 is put on one of the input phases (e.g., phase A) of the magnetic actuator and the ground bus. The clip-on ammeters are set on one of the phase wires coming from the output of the magnetic contactor and connected to the controller via connector XP1. Fixing the lower (upper) position of the balance head of BPU is carried out by using a timer of the magnetic type, which is mounted on the frame of BPU near the output shaft of the gearbox and is connected to the controller via connecting cable connector XP2. At the same time on the gearbox output shaft, the magnet was mounted in a position corresponding to the lower (upper) position of the head of the balance of the SC. The obtained data are stored in the controller's volatile memory and are transmitted on request to the operator of a network of an oil field. Visualization and vattmetrogramm treatment (Figure 3) is carried out using the specialized software installed on personal computers (and / or notebooks) of the technical services staff of the oilfield.

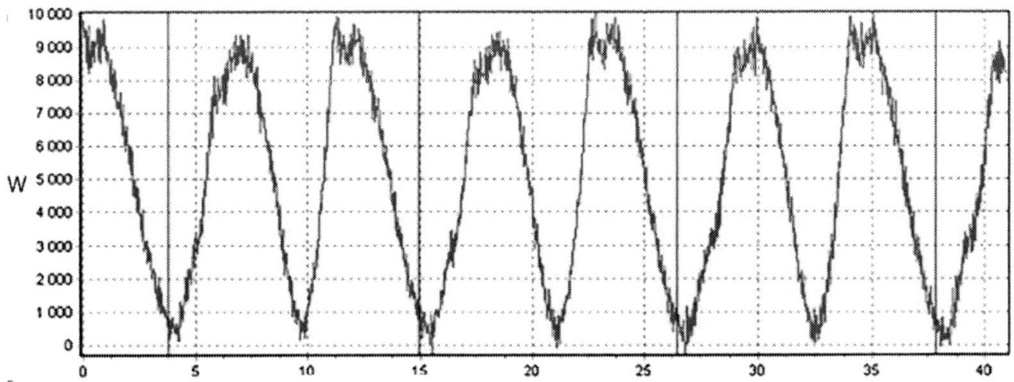

Figure. 3. Vattmetrogramms of fully balanced BPU, good technical condition.

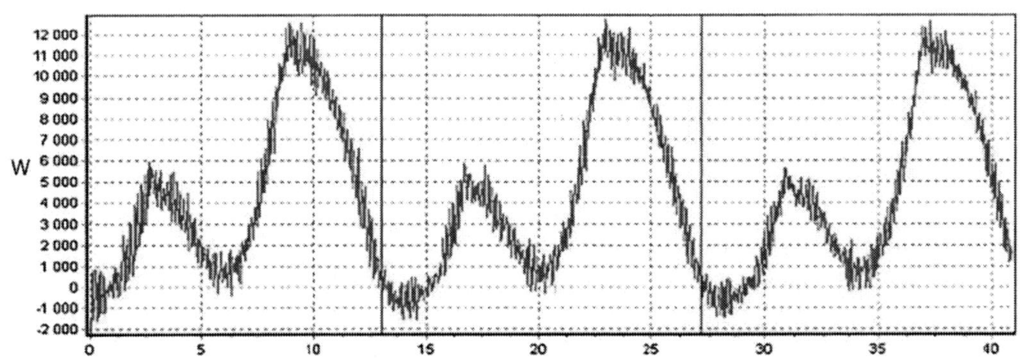

Figure. 4. Vattmetrogramms of unbalanced BPU (small load).

Vattmetrogramma of the fully balanced and serviceable BPU (Figure 3) for each full stroke of the downhole pump rod has two half-cycles with distinct peaks corresponding to the horizontal position of the crank. According to the established standards, the difference values of the maximum values of power consumed by the drive motor of the BPU when lowering and raising the rod well pump should not exceed 10%. An insufficient magnitude of the anti-moment generated by crank loads when lowering the rod string down of the hole pump during an unbalanced BPU (Figure 4) determines a transition of the drive motor in the generator mode of operation, and the working stroke of the pump shaft is accompanied by increased loads on the gearbox and the BPU engine. The consequence of these processes is an increase of the specific energy consumption during the rise of the formation fluid, high dynamics and exceeding the loads regulatory values in the BPU elements.

Analysis of vattmetrogramms allows us to predict the most common defects of deep well drilling pumps. Diagnostic features of breakage of the OPU rod string is an increase of the peak active power values consumed by the motor when lowering the head rocker (the rod string weight does not compensate an anti-torque produced by the weights crank). At that, the half-period of the power increase is absent from the vattmetrogramme when the rocker is lifted, which is caused by the absence of an external load exerted when lifting the reservoir fluid column of the well. Similarly, the defects of the discharge valve of the submersible pump are shown. They are a substantial reduction in the power consumed by the motor when the piston rises due to leaking of a greater part of formation fluid through the faulty discharge piston valve back into the pump cylinder, and, as a consequence, a significant decrease in the load suspension point of the column bars. The unevenness of the load growth while lowering the rod string, an increase of dynamic components of a force on the drive of BPU indicate a malfunction of the valve of the receiving borehole pump.

Spectral analysis of the active power signal (Figure3) obtained by the programmable controller 'AKD-SK' allows revealing frequency components from 0 to 30 Hz characterizing the kinematic chain fluctuations 'borehole pump – pumping unit – gearbox – drive motor'. Defects in working units and mechanical gears of OPU determine the occurrence of variable loads, which causes the appearance of new spectral components. Periodic measurement values in the power spectrum, characterizing the specific defects in the drive motor and the mechanical transmission, allows carrying out the assessment of technical condition OPU in the most simplest mode and, if necessary, to carry out repairs exposure to prevent crashes.

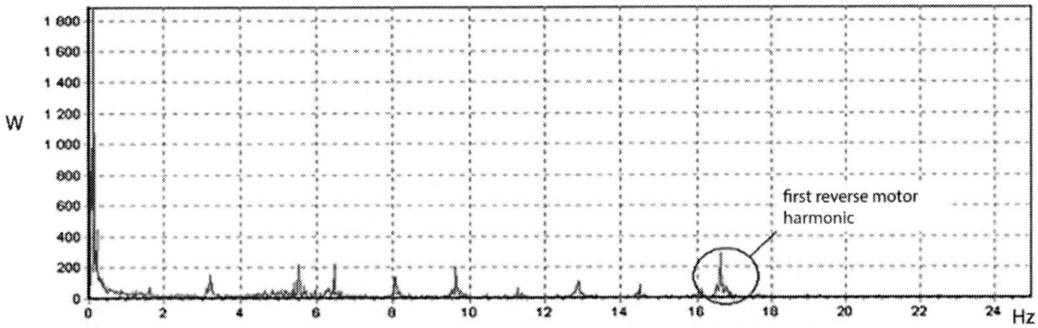

Figure 5. Spectra vattmetrogramm: a) in the presence of a defect motor pumping unit.

Through spectral analysis of records of power consumption in induction motors (Figure 5), we can find defective electric parts of the rotor including the rotor winding breaks and short circuit between laminations; electrical parts of the stator including breakages and electrical asymmetry of power winding, short circuit between laminations; static and rotating eccentricities; bearing defects leading to fluctuations of the air gap forms. By the nature of changes in the spectrum of the signal of power consumption, we can diagnose fault gears, the fits of gears on the shaft, misalignment of driven shafts and their supports of rotation, defects of V-belt transmission (Figure 5).

The disadvantages of the method of estimation of the technical condition of DWPU relatively its power consumption refer to the difficulty of identifying a number of defects at the initial stage of development. First, these are the failures of rolling bearings, cranking, deterioration and destruction of crank pins. The origin and development of these defects are accompanied by a change in the diagnostic signals of the spectra in the higher frequency range [4, 6, 14, 15].

The increase in informativeness and accuracy of diagnosis is possible through the implementation of multi-parameter technical monitoring (acoustic emission, vibration diagnostics), implemented as an option controller 'AKD-SK'. For example, setting wireless acoustic emission and/or vibration sensors on crank pins will prevent more than 40% of BPU sudden failures [12, 13].

4. Conclusion

Thus, the continuous recording of signals of active power consumed by drive motors SHSNU allows us to carry out the control over the operational loading of nodes of the pumping unit in the most simple and reliable way, to qualitatively perform the balancing of BPU resulting in reduction of the specific energy consumption during the rise of the reservoir fluid and reduction of dynamic loads on OPU details. Analysis of relevant information on the magnitude and nature of the change of electric drives loading of BPU enables assessment of the technical condition, complete and residual life of OPU elements.

The results of studies, presented in the article, show the prospects of development tools for diagnosis, carrying out the control values and the changing of the nature of power consumed by the drive motors. The proposed technical solutions and methodological bases can also be successfully used for diagnostics and operational control of mining machines and technological equipment of mining enterprises.

References

[1] Sugak EV, Vasilenko NV, Nazarov G G and others 2001 *The reliability of technical systems* (Krasnojarsk: Rasko)
[2] Molchanov A G 2014 *Drilling and Oil* **2** 3-9
[3] Sungatullin A A, Norkina O A and Sadykova R R 2011 *Proc. of the Int. Sci. and Techn. Conf.* (Tjumen) vol 4 (Tjumen: TjumGNGU) p 251-254
[4] Barkov A V, Barkova N A, Borisov A A 2012 *Methods of diagnosing mechanisms by electric current consumption* (Saint-Petersburg: Sevzapuchcentr)
[5] Sidorov V A, Kravchenko V M, Sedush V Ja and others 2003 *Technical diagnosis of mechanical equipment* (Doneck: Novyj mir)
[6] Asonov S A, Gabov V V, Ivanov S L, Trifanov M G, Chekmasov N V and Shishljannikov D I 2015 *Bull. of Perm National Res. Polytechnic Univ. Geology. Oil and Gas and Mining* (Perm: Perm National Research Polytechnic UniversityPress) p 62–71
[7] Sidel'nikov L G and Afanas'ev D O 2013 *Bull. of Perm National Res. Polytechnic Univ. Geology. Oil and Gas and Mining* (Perm: Perm National Research Polytechnic University Press) p 127-137
[8] Gur'ev P A and Nuss S V 2006 *Bull. of Perm National Res. Polytechnic Univ. Geology. Oil and Gas and Mining* (Perm: Perm National Research Polytechnic UniversityPress) p 165-175
[9] Thorsen V and Dalva M 1997 *University of Cambridge* (Cambridge: University of CambridgePress) p 1-13
[10] Randy R S, Thomas G H, Kamrah F and Rodert G B 1995 *IEEE Transactions on Industry Applications* **4**(6) 52-59
[11] William T T and Fenger M 2001 *IEEE Industry Application Magazine* **7** 23-29
[12] Shishljannikov D I, Trifanov M G, Romanov V A, Ivanov S L and Asonov S A 2015 Mining *Inform. and Analytical Bull.* (Moscow: Moscow State mining University Press) p 227-233
[13] Passport 'Stationary programmable controller 'AKD-SK' (Perm: R&D company 'ROS')
[14] Petukhov V 2008 *Electrical Eng. News* **22**(6) 33-37
[15] Kuptsov V V, Gorzunov A S and Sarvarov A S 2009 *Bull. of South Ural State Univ.* (Ekaterinburg: South Ural State University Press) p 123-129

Automated system for definition of life-cycle resources of electromechanical equipment

Y Zhukovskiy and N Koteleva

Mining University, 2, 21st Line, St Petersburg, 199106, Russia

Email: spmi_energo@mail.ru

Abstract. The frequency of maintenance of electromechanical equipment depends on the plant, which uses and runs this equipment. Very often the maintenance frequency is poorly correlated with the actual state of the electromechanical equipment. Furthermore, traditional methods of diagnosis sometimes cannot work without stopping the process (for example, for equipment located in hard to reach places) and so the maintenance costs are increased. This problem can be solved using the indirect methods of diagnosing of the electromechanical equipment. The indirect methods often use the parameters in the real time and seldom use the parameters of traditional diagnostic methods for determination of the resource of electromechanical equipment. This article is dedicated to developing the structure of a special automated control system. This system must use the big flow of the information about the direct and indirect parameters of the equipment state from plants from different areas of industry and factories which produce the electromechanical equipment.

1. Introduction

Technological processes safety is a primary task of plants in the mining industry. The whole electromechanical equipment used in the plant, and especially electromechanical equipment used in hazardous areas, plays the most important role in ensuring process safety. The different methods of diagnosis allow one to increase the reliability of the electromechanical equipment. Using the methods of determination of the resource data of the life cycle of electromechanical equipment, it is possible to predict the development of defects and, thus, prevent accidents at work, directly related to the incorrect operation of the equipment. To ensure the reliable operation of electromechanical equipment in the mining industry enterprises, traditional methods of defect detection are predominantly used. They are based on periodic maintenance and inspection of equipment. Periodic maintenance provides information to the system from time to time. So the diagnostic system has information insufficiency. Despite the high automation and the presence of specific systems, such as ERP and MES systems, the most modern complex system of diagnostic does not solve this problem. The first reason, these systems do not integrate and do not have links with each other. The second reason, these systems do not have special modules which help to create knowledge which parameters influence the defect of equipment. So, for improving the diagnostic methods, we must firstly integrate different systems and, secondly, add special modules. Developing a complex integrated system is necessary for solving this problem.

2. A traditional structure of an automated system for definition of life-cycle resources of electromechanical equipment

The problems of reliable operation of electromechanical equipment and the waste of energy could be solved using the special automated system for definition of life-cycle resources of electromechanical equipment. This system must contain not only usual modules of safety and control, it must contain the special predict module.

The determination of electromechanical equipment' resources must be realized in three steps. At the first step, the main parameters must be monitored. The main parameters are the parameters that could characterize the basic state of electromechanical equipment. At the second step, the parts of electromechanical equipment must be diagnosed using the different methods. At the third step, the special knowledge must be produced and used. This knowledge shows the influence of different defects on every part of electromechanical equipment and the life-cycle resources of electromechanical equipment. The knowledge helps to make decisions that will influence the life-cycle resources of electromechanical equipment.

A traditional approach that helps to solve this task bases on the automated system for definition of life-cycle resources of electromechanical equipment. This system gives the possibilities for collecting, aggregating and analyzing of data. It has the traditional structure that is shown in figure 1.

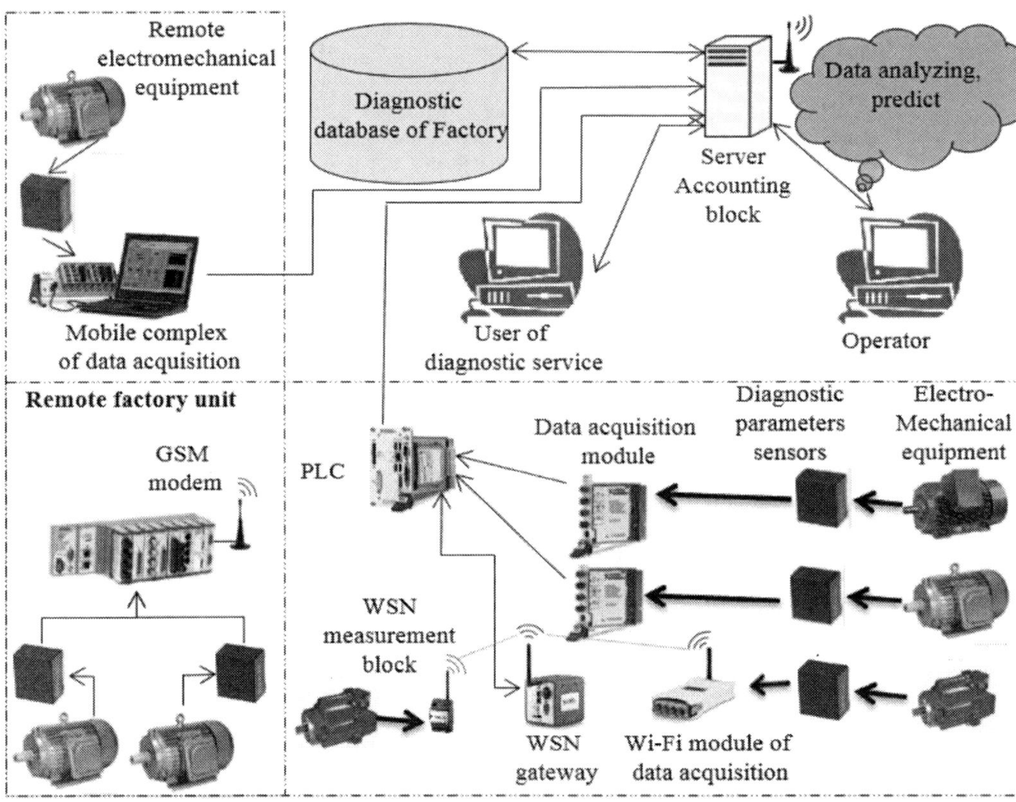

Figure 1. The traditional structure of automated system for definition of life-cycle resources of electromechanical equipment.

According to the figure, a plant has a special diagnostic parameters database. This database contains the information about the state of electromechanical equipment, including the remote equipment. This information is collected, aggregated and analyzed. The data are aggregated inside the database and could be printed out for the specialist using the special interface in different forms. This report helps do an analysis for the specialists. The tradition automated system has a special function – the indirect parameter calculation. These parameters are calculated according to technological processes data, data of an electromechanical equipment state, diagnostic data and other information that is stored in the database. When the operator has this full information, he can chose the right decision that helps to increase the duration of life-cycle of electromechanical equipment. Often, however, the decision is made on the poor information basis. This information is poor because it is collected according to the instructions of repair. So this dynamic analysis is not carried out. However dynamic analysis has a significant advantage – can detect a fault inherent on the operating mode. In the system, the decisions that affect the continued cycle of electromechanical equipment (repair, replacement parts and the like) are traditionally only fixed, but do not evaluate their influence to the continued functioning of the quality of electromechanical equipment.

Thus, the specialists choose a certain way based on their own knowledge and technological intuition or relying on experience of engineers of technical support of the manufacturer. This method is not always effective and sometimes is of low effectiveness because it uses some experimental averaged data of industrial experience about special facts which can affect the work of the electromechanical equipment. So, the information inside this database usually is not complete and useful. The typical plant cannot solve this problem alone because it uses its own experience but sometimes the reasons of fault of equipment are very different. This problem could be solved using the global information that are collected by different plants which use this equipment and by factories which produce it. So the automated control system must be modified.

3. An advanced structure of the automated system for definition of life-cycle resources of electromechanical equipment

The traditional approach to evaluating the electromechanical equipment life cycle has a number of drawbacks, the main of which is the localization of data on the equipment, the presence of incomplete information about defects and the reasons that influenced their development, often within a single enterprise using this electromechanical equipment and in very rare cases, as part of a partnership the interaction between the company and the manufacturer. This disadvantage can solve the globalization of the electromechanical equipment data. This globalization is possible only with the transition to the new structure of the system (Figure 2).

The functions of the advanced system are realized in a special services form. This services are provided using the Industrial Internet technology. So the database has a two different forms – the private database, which is used in the plant or factory in the private area, and the global database which is used in the public area. Each private database can be aggregated into a single global information system. It is a special intelligent system which must be self-learning and smart. The "smart" means what the system can be adapted under different conditions to aggregate data of different private databases, to produce and to collect the special knowledge during the life-cycle of electromechanical equipment.

This system must have the possibilities to increase the functionality; therefore, the system must have special capabilities:

- it must be integrated, i.e. the ability to be implemented into existing systems,
- it must be self-learning, i.e., the ability to adjust the tuning coefficients during operation;
- it must be adapted, i.e. the ability to maintain their functionality under varying operating conditions;
- it must be self-organized, i.e., preservation of health in terms of commissioning of new loops and variables.

This system can work in different scenarios. To date, the most optimal of which are the following:

- equipment state monitoring;
- a defect detection;
- simulation of the future work of the electromechanical equipment with the identified defect;

- the calculation of the residual life of the equipment;
- development of equipment repair plans.

The first recognition stage - the recognition of the state of equipment defects, without interference in its mode of operation by periodic or continuous observation, measurement - called as the functional diagnosis. In the case of the first stage of monitoring signs of defects, a deeper analysis of the data using the simulation condition of electromechanical equipment under the various load conditions is held.

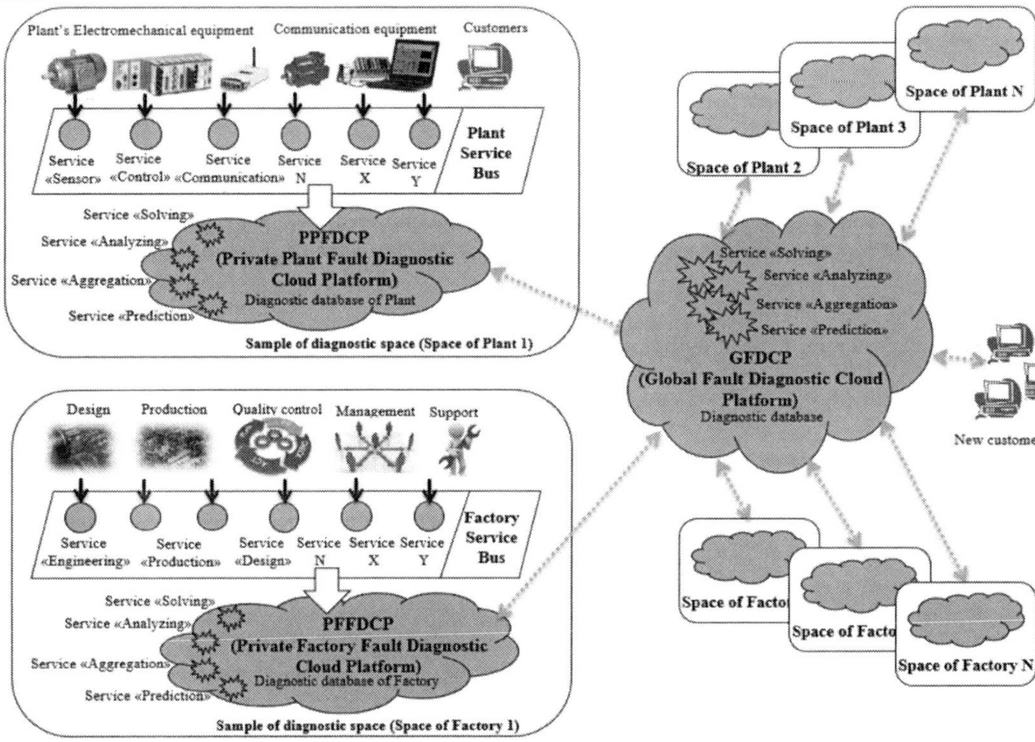

Figure 2. An advanced structure of the automated system for definition of life-cycle resources of electromechanical equipment.

The various subsystems must be implemented for realization of a global automation system, including:

- the subsystem information collection;
- the information processing subsystem;
- the subsystem of monitoring of resources consumption;
- the subsystem modeling;
- the diagnostics subsystem;
- the subsystem forecasting residual resource;
- the management and control subsystem.

The created subsystem must cover all workflows from sensors and transmitters, finishing controlled actuators.

The global automated system will improve the accuracy and quality of the assessment of the state and remaining life of the electromechanical equipment, taking into account the quality of the main supply and operating conditions that can set the timing of repairs more accurately and justify the excess period of operation of the unit. In addition, the system will increase the efficiency of the electromechanical equipment fault detection at an early stage in the process of operation, increase productivity, increase the accuracy of diagnosis and provide forecasting technical condition of electromechanical equipment.

4. Conclusions

Creating a global advanced diagnostics systems and assessing the residual life of the electromechanical equipment will integrate industry enterprises into a single diagnostic network, will organize the collection and processing of statistical data for more accurate operation of the system and will allow remote users to calculate the state and life-cycle resource of electromechanical equipment according to their actual state. It will lead to lower costs maintenance and repair, and will reveal the electromechanical equipment with high power consumption. The system will combine the expertise of a commercial plant using this electro-mechanical equipment and the experience of industrial enterprises producing electromechanical equipment. This alliance will provide advanced knowledge of the nature of defects, reduce operating costs and improve the energy efficiency of production and operation of electromechanical equipment.

References

[1] De Aguiar, E P Nogueira, F M D A, Amaral R P F, Fabri D F, Rossignoli S C D A, Ferreira J G and Ribeiro M V 2016 *Neural Computing and Applications* **27**(5) 1175-89
[2] Lili G, Kebao X and Zhaoxue L 2015 *Open Electrical and Electronic Eng. J.* **9(1)** 166-169
[3] Pandya H J, Park K, Chen W, Goodell L A, Foran D J and Desai J P 2016 *IEEE Transactions on Biomedical Eng.* **63**(7) 1347-53
[4] Dumont P T 2015 *Proc. Annual Reliability and Maintainability Symp.* USA (Florida: IEEE)
[5] Leonidovich Z Y and Urievich V B 2015 *Int. J. of Appl. Eng. Res.* **10**(20) 41150-55
[6] Mokhtar M, Edge JC and Mills A R 2014 *Proc. of the Annual Conf. of the Prognostics and Health Management Society* USA (Fort Worth: Prognostics and Health Management Society) p 367-377
[7] Aranda Carmona A J 2012 *44th Int. Conf. on Large High Voltage Electric Systems* France (Paris: CIGRE) pp 18-27
[8] Xu W, Appel J and Chae J 2012 *J. of Microelectromechanical System* **21**(2) 302-307

The application of mathematical modeling for the development of devices as an example of viscous fluid purification from magnetic impurity

V A Zhukov[1], **E P Masyutkin**[2] **and B A Avdeyev**[2]

[1] Admiral Makarov State University of Maritime and Inland Shipping, 5/7, Dvinskaya str., St Petersburg, 198035, Russia
[2] Kerch State Maritime Technological University, 82, Ordzonikidze str., Kerch, 298309, Russia

E-mail: dirigeant@mail.ru

Abstract. The mathematical model of a coagulation phenomenon and a coagulation process of ferromagnetic particles under the influence of a constant magnetic field are given in the article. The principles that need to be followed during mathematical modeling are stated.

1. Introduction

Methods of technologies and hardware with application of classical modeling techniques are well-known. They are logical, geometrical, physical and mathematical ones, the basis of which is presented by the fundamental laws of natural sciences.

All the above mentioned facts, in conjunction with a qualified use of contemporary computing machinery, allow us to introduce leading technologies and corresponding equipment into various branches of human activity: power engineering, metallurgy, mechanical engineering, chemistry, transport, etc.

Colligating and analyzing the existing modelling techniques, facilities of computing machinery as well as training of specialists and a highly qualified staff in higher education establishments, it is possible to establish the fact that the use of previous experience in the area of equipment modelling, mechanism, and manufacturing processes is still not enough.

2. Statement of the problem

Intuition and the downsized versions of physical and mathematical modeling techniques are used now in most cases for solution of scientific and technical tasks of hardware creation that, in our opinion, significantly affects the quality of the equipment being developed. Complexity of machines, mechanisms, devices is well-known: the significant means and amount of time are spent for their creation, and manufacturing lead time and implementation last for many years. Therefore, another approach to creation of technical means is necessary, namely – engaging the modern complex modeling techniques, based on comprehension of substance processes.

The main component of such methods of hardware development is the mathematical model created on the basis of classical laws of natural sciences and iteratively tested regularities. As a rule, the mathematical models have to be formulated in the form of differential equations that allows investigating various modes of a prototype system, including transient processes.

The objective of work is a demonstration of use of mathematical modeling to coagulation process of particles in magnetic field.

3. Materials and methods

Mathematical modeling of the physical phenomenon, created on basic physical laws of natural sciences was used as a method. The process of the numerical solution was conducted with use of a MathCAD package. Input data and initial data are represented in the article. The pictures of behavior of globular steel particles in magnetic field were taken as experimental verification of coagulation. The particles were placed into the transparent flask filled with glycerin. Field strength changed by means of increasing of current in an electromagnet as follows: H_1=0.8·10^4 A/m; H_2=1.6·10^4 A/m; H_3=2.4·10^4 A/m; H_4=3.2·10^4 A/m; H_5=4.0·10^4 A/m.

4. Results and discussion

The authors' experience gained in mathematical modeling of processes and devices of various applications [1-3] is useful if fragments of hardware creation for granular and dispersive media (cleaning of magnetic impurity) are given as an example.

The analysis of theory bases, calculation methods and design of purifiers which have different operating principles, designed for extraction of mechanical impurities from viscous fluids, showed prospects of application of force field of electrical nature which raise a rate of gases purification and liquids purification [4].

On the basis of the above-stated information, during modernization of technologies and devices for purifying medium, it is necessary to observe the following principles:

- a maximum application of the positive characteristics of the existing designs (high performance of work, manufacturability, reliability and simplicity of operation);
- an application of auxiliary energy in the form of force field of electrical nature of the apparatus being designed, should be carried out so that the influence upon particles promoted increase in effectiveness of their extraction from the medium violating the flow patterns insignificantly. Purifiers with the field (electric, magnetic, ultrasonic or combined) can be divided into coagulators, sumps, filters, inertial devices and separators, in which particles of impurity interact with each other under certain conditions, forming floccules, and with the corresponding electrodes (poles or other settling surfaces, which finally leads to an increase of purification efficiency and reliability of mechanisms) [5-6]. These devices are shown in figure 1.

The basic forces influencing ferromagnetic particles are attractive forces (1), the magnetic force (2), Coulomb force for magnetic particles (3), the centrifugal force (4) and the drag force (5):

$$F_G = m \cdot g = (\rho_p - \rho) \cdot V \cdot g \tag{1}$$

$$F_M = \mu_0 \cdot \chi \cdot V \cdot H \cdot \mathrm{grad}\, H \tag{2}$$

$$F_F = \frac{4 \cdot \pi \cdot M_1 \cdot M_2}{\mu_0 \cdot \mu \cdot r^2} \tag{3}$$

$$F_C = \frac{m \cdot U^2}{R} \tag{4}$$

$$F_D = -3 \cdot \pi \cdot d \cdot \eta \cdot U \tag{5}$$

where m – the mass of the body, kg; g = 9.81 – free-fall acceleration, m/s^2; ρ_p – the density of the body, kg /m^3; ρ – the density of the medium kg/m^3; V – the volume of the body, м^3; μ_0 – the magnetic constant, equal to 4·π·10^{-7}, H/m; χ – the magnetic perception of the body, which is dimensionless; M_1, M_2 – magnetic forces of the particle, m^{2}kg/s^2 A; μ – magnetic permeability, H/m; H – the magnetic field strength, A/m; r – the interparticle distance, m; U – the velocity of the body, m/s; R – the radius along which the particle is moving, m; d – the diameter of the particle, m.

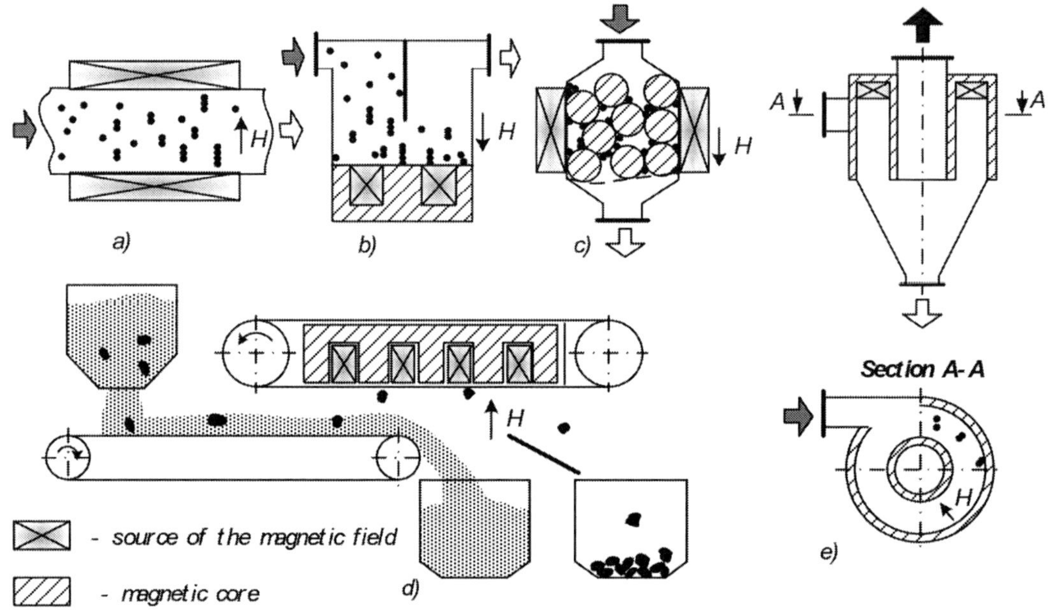

Figure 1. Schemes of electromagnetic cleaning devices:
a) coagulator, b) sump; c) filter; d) separator; e) hydrocyclone

The coagulation process is presented in figure 2.

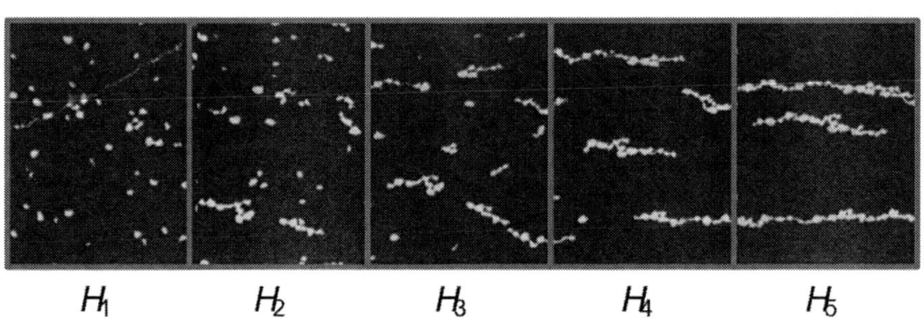

Figure 2. The coagulation process in the magnetic field ($H_1<H_2<H_3<H_4<H_5$).

Effectiveness of extraction of mechanical impurities from the granular medium and viscous fluid depends on:

- properties of particles – density ρ, size d, form ψ, their concentration C, permittivity ε, magnetic susceptibility χ, energy absorption rate α, etc.;
- key parameters of the granular medium and viscous fluid – density ρ_p, viscosity η and hydrodynamics of the flow;
- characteristics of the force field – strength H (E), speed of its change - grad H (grad E), intensity J and frequency f of a source of ultrasonic vibrations.

Mathematical modeling of dynamics of particles in clearing devices with force field of electric nature was carried out with the help of the following algorithm:

-setting up of mathematical models;

- computational investigation;

- processing and the analysis of results of computational investigation - physical (experimental) simulation for the purpose of specification of the coefficients connecting parameters of particles, the medium and the field;
- correction of initial mathematical models taking into account the results of physical modeling;
- computational reinvestigation or multiple computational investigations of the corrected mathematical model;
- development of the computational methods suitable for designing the defined (desired) construction;
- experimental-industrial control and commercialization.

Setting up mathematical models was carried out with use of the well-known laws and regularities of solid mechanics, liquid mechanics and gas mechanics, electrodynamic mechanics. Other components of the algorithm, in our opinion, require no explanation. We will consider a problem of coagulation of ferromagnetic particles in the magnetic field as an example. Let us assume that the magnetic field in the area of particles arrangement is homogeneous (dH/dt=0), as shown in figure 3 a. The effect of forces on them coincides with the direction of a vector of field strength H. The force of magnetic concentration is equal to zero. Without gravity and buoyancy force, particles will be affected by forces of coagulation F_F and resistance of medium F_D. It is possible to write down the equations in a differential form according to the second law of dynamics for each of the two particles having mass m:

$$\begin{cases} -m_1\dfrac{d^2y_1}{dt^2} = -3\cdot\pi\cdot d_1\cdot\eta\cdot U_1 + \dfrac{\pi^3\cdot\mu_0}{9}\cdot\dfrac{d_1^2\cdot d_2^2}{(y_1+y_2)^2}\cdot\chi^2\cdot H^2 \\ -m_2\dfrac{d^2y_2}{dt^2} = -3\cdot\pi\cdot d_2\cdot\eta\cdot U_2 - \dfrac{\pi^3\cdot\mu_0}{9}\cdot\dfrac{d_1^2\cdot d_2^2}{(y_1+y_2)^2}\cdot\chi^2\cdot H^2 \end{cases} \quad (6)$$

Initial conditions:

$$t = 0; t(0) = 0;\ \frac{dy_1}{dt}(0) = U_{10};\ \frac{dy_2}{dt}(0) = U_{20};\ y_1(0) = y_{10};\ y_2(0) = y_{20} \quad (7)$$

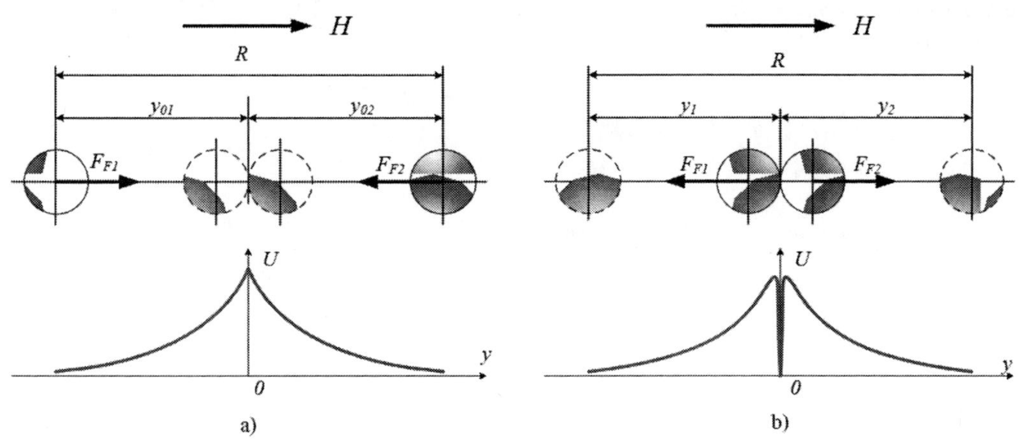

Figure 3. Coagulation of particles in the magnetic field: a) direct problem; b) inverse problem

The system of equations (6) with initial conditions (7) is a mathematical model of the process of magnetic coagulation of two particles, which sizes are close to each other in the viscous fluid. Solution (6) allows finding conditions of formation of floccules, if the values of parameters of particles are known, as well as basic data for computation of the magnetic field.

However, as it appears from equation (3), the effect of force F_F is infinite in space, i.e. particles, being in the field, always experience the influence of each other and whereupon they move having, at each instant of time, defined values of path y_1 and y_2 and speed of U_1 and U_2. Therefore, it is impossible to solve the system of equations (6) even by means of a computational method if initial conditions are unknown (7).

For the solution of the system of equations (6), we will consider two ferromagnetic particles which are affected by forces of F_D and F_F, as shown in figure 3 b. In this case, the physical meaning of the effect of forces is as follows. Floccules consisting of two particles collapses under the influence of F_F. Force F_D prevents this. At the start of a motion path at t = 0, F_F has the maximal value, F_D is equal to zero, therefore the motion of particles to the opposite sides is accelerated. F_F decreases with an increase of the distance between particles, while the force of F_D increases at the expense of speed increase U. At instant $t = t_1$, forces F_F and F_D converge to a maximum. F_F and F_D converge to zero while further motion of particles and forces F_F, F_D can be neglected at time $t = t_\kappa$ on account of their smallness. Thus, during time $0...t_\kappa$, particles will diverge from distance R. At the same time, dynamics of particles, characterized by the same differential equations (only symbols in front of members change), will be described. The equations, more suitable for computational investigation, will be of the following kind:

$$\begin{cases} \dfrac{d^2y_1}{dt^2} = \dfrac{18\cdot\eta_c}{d_1^2\cdot\rho}\cdot\dfrac{dy_1}{dt} - \dfrac{2\cdot\pi^2\cdot\mu_0\cdot d_2^2}{3\cdot d_1\cdot\rho}\cdot\dfrac{\chi^2\cdot H^2}{(y_1+y_2)^2}; \\ \dfrac{d^2y_2}{dt^2} = \dfrac{18\cdot\eta_c}{d_2^2\cdot\rho}\cdot\dfrac{dy_2}{dt} + \dfrac{2\cdot\pi^2\cdot\mu_0\cdot d_1^2}{3\cdot d_2\cdot\rho}\cdot\dfrac{\chi^2\cdot H^2}{(y_1+y_2)^2}; \end{cases} \tag{8}$$

Initial conditions: t=0; $t(0)$=0; $y_1(0)=y_2(0)=d/2$; $U_1(0)=U_2(0)=0$,

Equations (8) are described in [7] in more detail.

The system of equations (8) allows one to define the main dynamic characteristics of coagulation of particles: speed $U=f(t,R,d,\eta,\chi,H)$ and the distance at which they interact $R=f(t,U)$.

The results of a numerical integration of the system of equations (8), representing a coagulation of particles in the magnetic field are adduced in figure 4: variation of distance y_1 and y_2 between them, and their speeds u_1 and u_2 depending upon time: $d=5\cdot10^{-6}$ m; $H=4\cdot10^4$ A/m; $\chi=29.25\cdot10^{-3}$; $\rho=7.8\cdot10^3$ kg/m3; $\eta=1.85\cdot10^{-3}$ Pa·s.

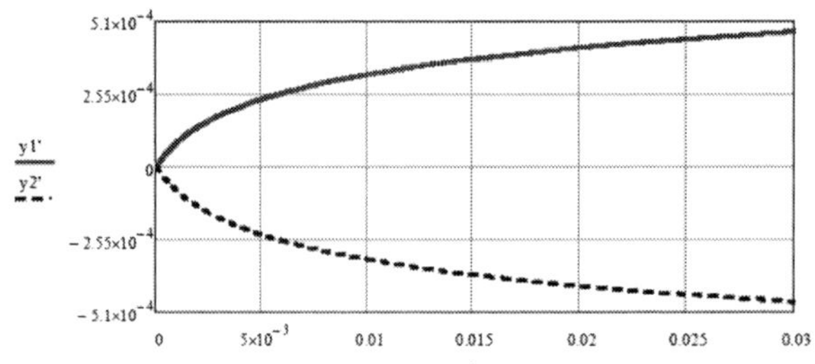

a

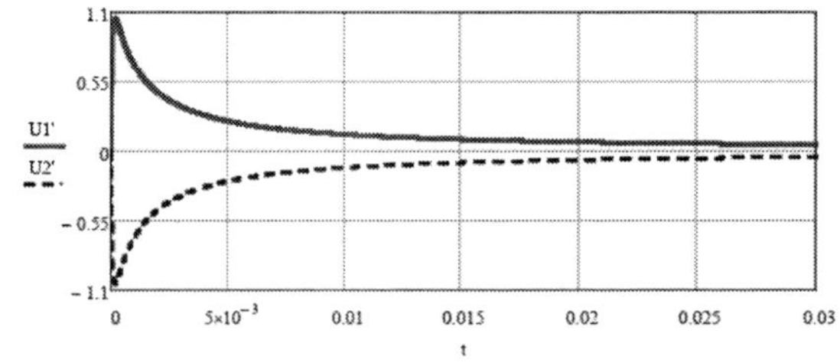

b

Figure 4. Paths (a) and speed (b) of particles.

Varying the parameters of particles the parameters of medium and the parameters of field iteratively from solution to solution, the optimal conditions of coagulation of particles of cleaning units (purifiers) are identified.

The data obtained by the method of computational investigation of the inverse problem of coagulation of particles were checked experimentally with the help of a primal problem. The maximal speed, observed at the time of particle collision, differed by no more than 5% of the maximal speed obtained as a result of the solution of the system of equations (8). The computational solution of a direct and inverse problem of coagulation is given in figure 5. Thus, the possibility of using the system of differential equations (8) is decisive.

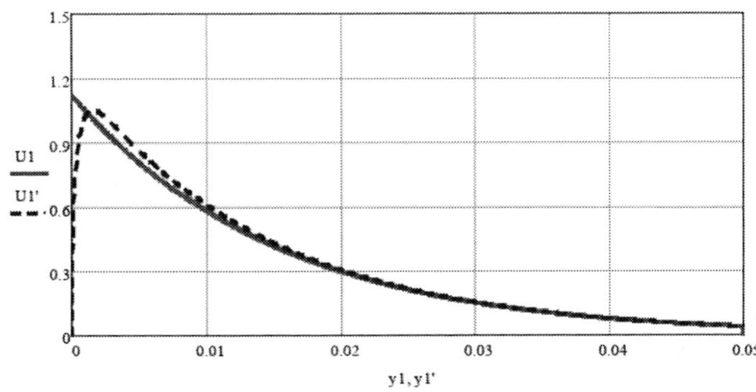

Figure 5. Dependence $U=f(y)$ of the direct and inverse problems of coagulation.

The differential equations for the electric field or ultrasound are being composed and investigated numerically in a similar manner [8-11]. We hope that the above-mentioned examples and the example in mathematical modeling of the phenomena and processes when imposing fields onto dispersive media will be useful for scientists, graduate students and students.

4. Conclusion

Mathematical modeling plays an important role in development and designing of various devices and mechanisms. The inverse problem of coagulation of particles in the magnetic field was considered as an example of successful utilization of modeling in engineering, owing to which, the most optimal conditions of coagulation of particles in purifiers are identified which is impossible for a direct problem.

For example, so that the particles of 5 microns in diameter could interact with each other at the field strength of $4*10^4$ A/m, their concentration should not be less than 0.5 g/m^3.

References

[1] Avdeyev B A, Masyutkin E P and Prosvirnin V I 2014 *Univ. News. North-Caucasian Region. Tech. Sc. Ser.* **4(179)** 13

[2] Avdeyev B A, Masyutkin E P and Prosvirnin V I 2015 *Bulletin of Admiral Makarov State University of Maritime and Inland Shipping* **2 (30)** 127

[3] Avdeyev B A 2016 *Improving the efficiency of the engine oil cleaning in ship diesel engines through the use of magnetic hydrocyclones 'Monograph'* (Ulyanovsk: zebra)

[4] Zoua Q C , Jieb J C, Sun J L [and others] 2015 *Separation and Purification Technology* **142** 101

[5] Bulyzhev E M and Menshov E N 2014 *Proc. of the Samara Sc. Center of the Rus. Ac. of Sc.* **16** 247

[6] Chernyi S 2016 *Oil Industry* **2** 106-110

[7] Aleksandrov E E, Kravec I A, Lysikov E N and others 2006 Increasing the service life of the technical systems through the use of electric and magnetic fields (National Technical University 'Kharkiv Polytechnic Institute)

[8] Ebner A D, Ritter J A and Ploehn H J 1997 *Separation and Purification Technology* **100** (3) 199
[9] Chernyi S and Zhilenkov A 2015 *J. of Transport and Telecommunication* **16** (1) 73
[10] Chernyi S 2016 *J. of Transport and Telecommunication,* **17** (3) 202
[11] Sinha S, Bae K I and Das S 2016 *Colloids and Surfaces A: Physicochemical and Engineering Aspects* **489** 216

Analysis of noisy signal restoration quality with Lanczos filter

A S Belyaev, I A Tutov and S A Maykov

Tomsk Polytechnic University, 30, Lenina ave., Tomsk, 634050, Russia

E-mail: belyaewas@mail.ru

Abstract. The differential component of the feedback is used to enhance dynamics of controlled mechanisms in control systems. However, the use of the differential component increases the noise interference, resulting in the need to use different filtering methods. This paper describes the research of a differentiated signal input in a differential Lanczos filter. The influence of the filter order on the integrated square error of the filtered signal is defined. The impact of signal dispersion and a sampling interval on the integrated square error of the filtered signal is identified. Moreover, the most convenient parameter values for the Lanczos filter are determined. The paper also includes the comparison of the exponential moving average filter and the Lanczos filter with selected values of noise dispersion and a sampling interval. The research shows the possibility to apply the Lanczos filter for processing of the differentiated signals.

1. Introduction

Control systems are used in many different fields of mechanical engineering such as space, aviation, shipbuilding and other industries. For enhancing dynamics of controlled mechanisms, a differential component is used to implement real-time control systems with high control quality; however, the use of the differential component increases the noise interference, resulting in the need to use filtering methods [1-5]. There are 7 basic types of filtering:

1) Methods based on substitution of infinitely small increments of differentiated signal *ds* and time *dt* with their finite values Δs and Δt;
2) Methods based on the use of different interpolating polynomials.
3) Methods based on the use of splines.
4) Methods based on the use of various approximating functions.
5) Methods based on integral equations.
6) Methods based on Fourier and Laplace transforms.
7) Methods based on wavelet transform [6]

However, algorithmic implementation of these methods requires high computational power and increases signal processing time, which limits their usage in real-time systems. It forces specialists to find a compromise between the quality of the filter and the amount of processing time. A good combination of implementation simplicity, low computational cost and relatively high accuracy lead to widespread occurrence of the exponential smoothing method.

The paper [7] describes a study of a standard filter based on the method of an exponential moving average. However, this filter does not differentiate the signal; therefore, it is rarely used for such signals. Differentiation filters are used most often; however, these filters do not typically handle noise suppression well. Combining good noise suppression and differentiation, the Lanczos filter is widely used in filtering radio signals.

2. Methods

The choice of the structure and parameters of the differentiating filter will be implemented in the Lanczos filter class [8], the general equation of which is as follows:

$$y\left[nT\right]=\frac{3}{TN(N+1)(2N+1)}\sum_{k=0}^{2N}(N-k)x\left[(n-k)T\right] \qquad (1)$$

where N – filter order.

This filtration method can be implemented in different application [9, 10]. For the processing of a real analog signal, the signal received by the sensor must be converted by means of an analog-to-digital converter (ADC). Because this filter differentiates the original signal, it should be integrated, and then noise with a certain value of dispersion should be added. At the next stage, the signal is converted into a digital value by means of the ADC with a certain quantization level defined by the resolution of the ADC.

In order to define control quality indicators, it is necessary to compare the transformed signal and the source. The comparison method used is the integral square error (ISE), which is defined as:

$$err=\sum_{i=1}^{n}(S_i-S_i^{'})^2\,, \qquad (2)$$

where S_i is the value of the filtered signal at the i-th iteration, $S_i^{'}$ is the true signal value at the i-th iteration. Based on this model, the transformed signal will be affected by the following parameters:

1) Noise dispersion - σ;
2) ADC quantization level - Δ;
3) Filter order.

But how changes in these parameters will affect the transformed signal.

3. Filter analysis

For the research of these dependencies, a model was implemented in the MATLAB&Simulink environment. As the test signal for the analysis of the ESS, a signal with the following basic levels is used: a negative level (range 0-100), step transition, positive transition (100-300), dynamic transition (300-380), and a zero level.

In order to analyse the relation between the ISE and the filter order, the order was changed in the range from the first to the twentieth. In so doing, the quantization level was set as 0.5, and the noise dispersion values were made equal to the following five values $\{0.01; 0.03; 0.05; 0.07; 0.09\}$.

Figure 1 shows that when the filter order is increased, the integral square error first decreases significantly, and then begins to increase; also, an increase in noise dispersion causes an increase in the ISE, i.e. there is a directly-proportional relation between these parameters.

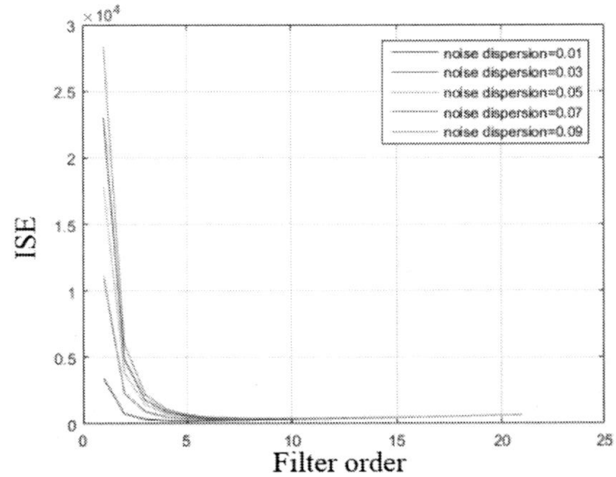

Figure 1. Dependence of the integral square error on the filter order at different noise dispersion values.

The diagrams show that an increase in the dispersion value greatly increases the integral error in the region of filter order values of 1-5. In the region of values over 6, the effect of the quantization level on the error is negligibly small. However, under real-life conditions, noise dispersion is not changed; therefore, it is necessary to determine the effect of the quantization level on the integral error.

We will make an evaluation of the effect of the quantization level and the filter order on the integral component. The quantization level for this value was equal to the following three values {0.01; 0.5; 1} while dispersion was equal to 0.5. The result is shown in Figure 7. Figure 2 shows that an increase in the quantization interval leads to an increase in the ISE, i.e. there is a directly-proportional relationship between these parameters.

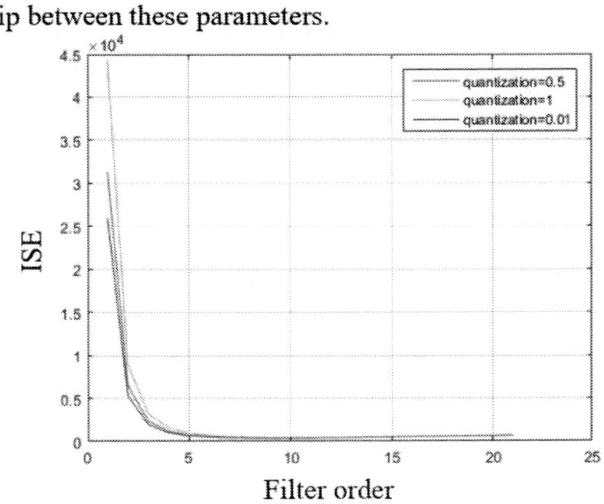

Figure 2. Dependence of the integral square error on the filter order at different quantization level values.

The diagrams show that an increase in the quantization level significantly increases the integral error in the region of filter order values of 1-4. In the region of values over 5, the effect of the quantization level on the error is negligibly small.

Let us check the results for multiple values of the quantization level [0.01 1] and dispersion [0.01 1] at 0.01 intervals. In order to achieve this, two experiments were carried out; in the first one, the dispersion was equal to 0.5 and maintained as constant, and the filter order was changed [1 22], figure

3a. On the contrary, the quantization level value was equal to 0.01, and the filter order and the dispersion value were changed, figure 3b.

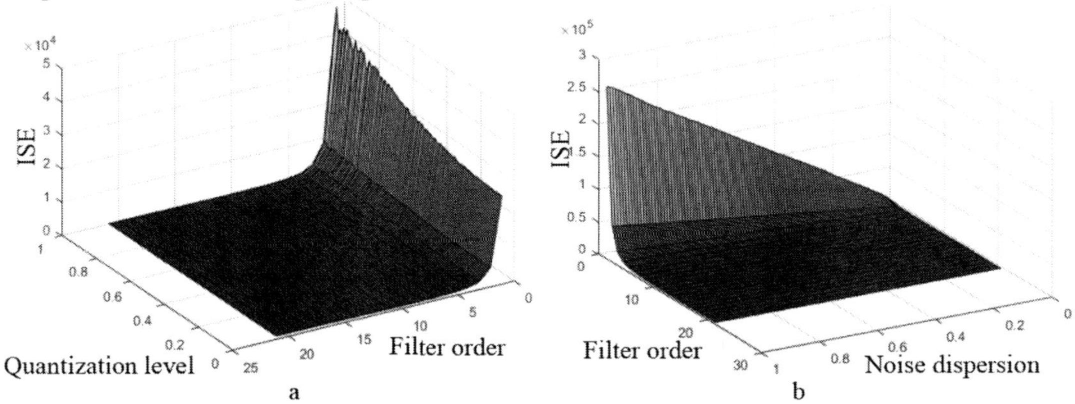

Figure 3. Dependence of the integral error on the filter order: a – quantization level; b – noise dispersion

The diagrams only confirm the results obtained earlier.

Now we will analyze the opportunity to compensate effects of dispersion and the quantization level in relation to the ISE. To achieve this, a dependence between noise dispersion and the ISE was found for the best filter order (figure 4a) and the relationship between the quantization level and the ISE, as well as for the best filter order values was established (figure 4b).

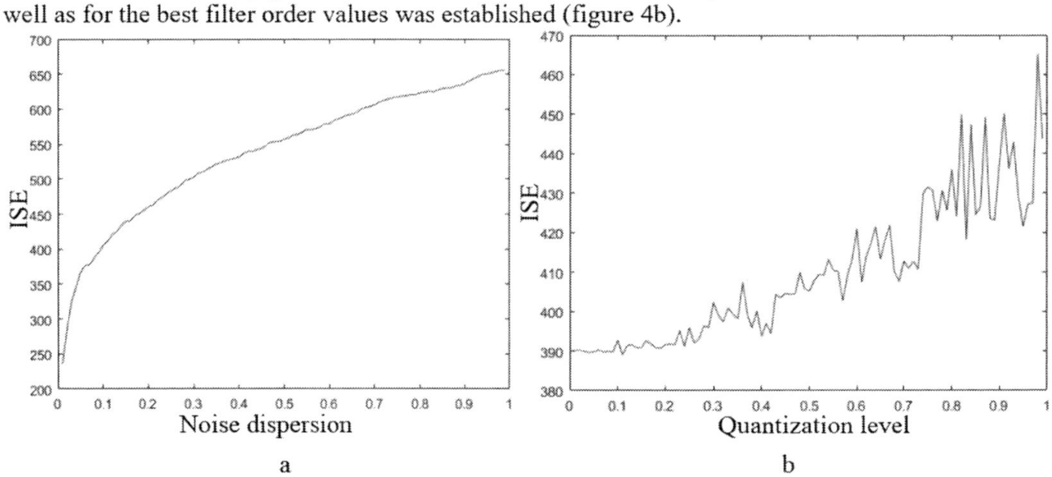

Figure 4. Dependence of the ISE (with the best filter order settings) on a – noise dispersion; b – quantization level.

An increase in the filter order reduces the integral square error only to a certain value, after which the ISE gradually increases.

The result of the research carried out is as follows:

- Noise dispersion has a significant effect on the ISE at filter order values under 6; at higher values, effects of noise dispersion are compensated.
- The quantization level has a significant effect on the ISE at filter order values under 4; at higher values, effects of the quantization level are compensated.

- As shown in the diagrams, it is not possible to compensate the effect of noise dispersion and the quantization level on the ISE completely; however, the increase in the error is not very significant.

4. Comparison of the results between Lanczos filtering and exponential moving average methods

As described in [7], the ISE has a directly-proportional dependence on dispersion and the quantization interval, as well as the Lanczos filter. Three experiments were conducted to compare the filters: with dispersion equal to 0.01 and the quantization interval equal to 0.01 (figure 5a); with dispersion equal to 0.5 and the quantization interval — to 0.5 (figure 5b); with dispersion equal to 1 and the quantization interval — to 1 (figure 5c). The diagrams showing the output signals of the filters are below.

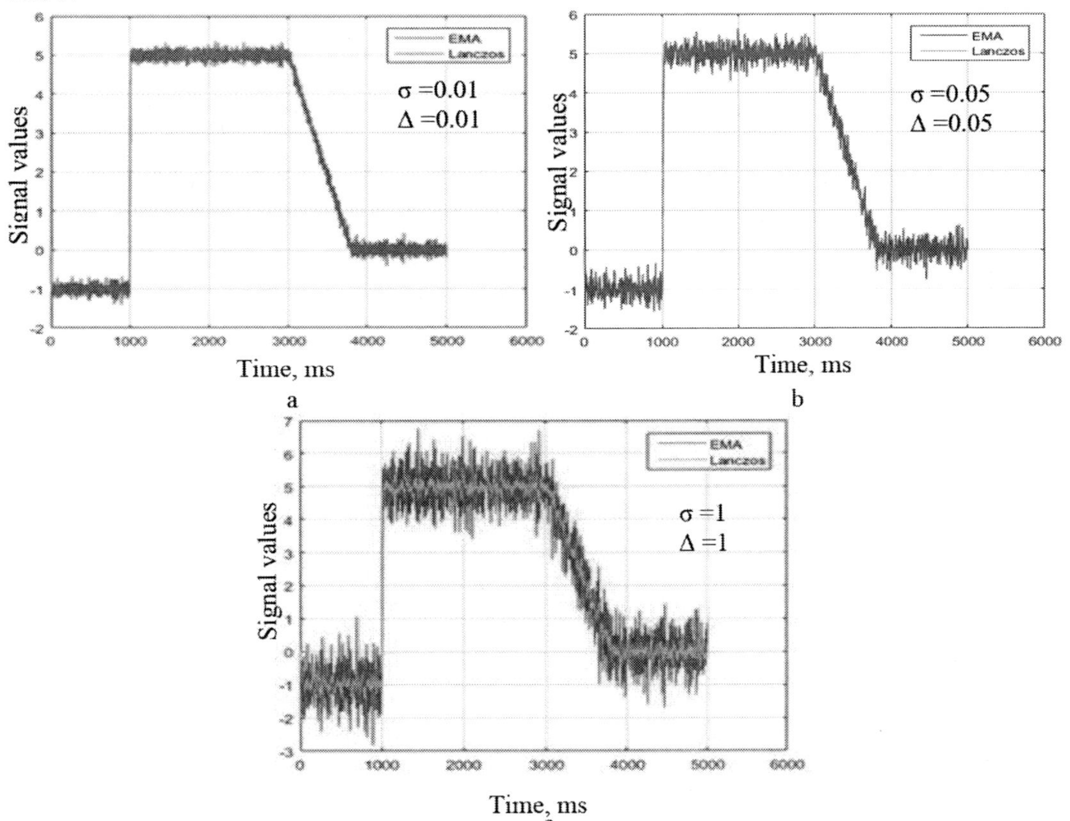

Figure 5. Filtered signals with dispersion and the quantization level equal to: a – σ = 0.01, Δ=0.01; b – σ =0.05, Δ=0.05; c – σ =1, Δ=1.

According to the diagrams of transformed signals, it is evident that with the values of dispersion and the quantization interval equal to 0.01, the filters work with close mean-square error values; however, when increasing the values of dispersion and the quantization interval, the Lanczos filter performs better with a lower integral mean-square error.

5. Conclusion:

1. The use of the Lanczos filter for filtering digital signals is more appropriate for systems with higher noise dispersion.

2. The use of the Lanczos filter of the 6th or 7th order allows one to virtually compensate effects of noise dispersion and the quantization level on the ISE.

References

[1] Kim J and Kim T 2009 *Proc. 6th Int. Conf. on Computer Graphics, Imaging and Visualization: New Advances and Trends* (Tianjin: IEEE) pp 367-74
[2] Long H, Qu Z, Fan X, and Liu S 2012 *Proc. 10th World Congress on Intelligent Control and Automation* (Beijing, IEEE) pp 4315-9
[3] Yudan H, Zhihua C, Wenyin G, and Qin L 2014 *IEEE Congress on Evolutionary Computation* (Beijing, IEEE) pp 2893-900
[4] Shelomentcev E and Alexandrova T 2014 *IOP Conf. Series: Mat. Sci. and Eng.* (Tomsk) vol 66 (IOP Publishing) 012027
[5] Pandey C and Nemade S 2014 *Int. IEEE India Conf.* (Pune, IEEE) pp 1-6
[6] Lycashin Y 2003 *Adaptivie metodi kratkosrochnogo prognozirovaniy vremennih rydov* (Moscow: Finance and statistica)
[7] Belyaev A, Tutov I and Butuzov D 2016 *Int. Siberian Conf. on Control and Communications* (Moscow) *Electronic Materials* (IEEE) 2380-6516
[8] Hamming R 1977 *Digital filters* (Englewood Cliffs, NJ: Prentice-Hall)
[9] Jonathan Taquet and Claude Labit 2010 *Multimedia Signal Proc. IEEE Int.Workshop* (ST. Malo, IEEE) pp 122-7
[10] Amina K, Nejmeddine B, Nouri M, Amine S and Mohamed A 2014 *1st Int. Conf. on Adv. Techn. for Signal and Image Proc.* (Sousse-Tunisia, IEEE) pp 148-52

Simulation model of load balancing in distributed computing systems

I A Botygin, V N Popov and S G Frolov

Tomsk Polytechnic University, 30, Lenina ave., Tomsk, 634050, Russia

E-mail: bia@tpu.ru

Abstract. The availability of high-performance computing, high speed data transfer over the network and widespread of software for the design and pre-production in mechanical engineering have led to the fact that at the present time the large industrial enterprises and small engineering companies implement complex computer systems for efficient solutions of production and management tasks. Such computer systems are generally built on the basis of distributed heterogeneous computer systems. The analytical problems solved by such systems are the key models of research, but the system-wide problems of efficient distribution (balancing) of the computational load and accommodation input, intermediate and output databases are no less important. The main tasks of this balancing system are load and condition monitoring of compute nodes, and the selection of a node for transition of the user's request in accordance with a predetermined algorithm. The load balancing is one of the most used methods of increasing productivity of distributed computing systems through the optimal allocation of tasks between the computer system nodes. Therefore, the development of methods and algorithms for computing optimal scheduling in a distributed system, dynamically changing its infrastructure, is an important task.

1. Introduction

Currently, the design and production preparation in mechanical engineering implemented a wide range of engineering calculations (strength, hydrodynamic, thermal, etc.). These tasks require a much more machine resources than the systems of design or production preparation. Modern systems of automation engineering calculations, depending on the scale of the task, are carried out parallel calculations using different methods and structures of distributed computing. One of the major problems in distributed computing is scheduling of loading of nodes in the system.

Today, there are many methods of the load balancing: Round Robin, Weighted Round Robin, Least connections, Weighted least connections, and others. The load balancing is also divided into static, semi-dynamic and dynamic. Static balancing is performed prior to the start of the task, semi-dynamic load balancing requires that the distribution of tasks carried out at the stage of initialization and dynamic balancing is performed in the course of computation. If we talk about how to interact the load balancing in a computer system, the balancing is divided into central, where the special compute node (manager) distributes the modules of a task between calculators, and distributed, in which status data is exchanged between all nodes in a distributed computing system by using a special algorithm [1–14].

The article describes the method of computational load balancing in distributed automation systems which are focused on multi-agent and multi-threaded data processing. A scheme for the control of request processing from the terminal devices is proposed. This scheme provides an effective dynamic horizontal scaling of computing power at high loads. A GPSS-model and the results of model experiments for developed algorithm load dispatching are represented. This model and results show the effectiveness of the algorithm even with a significant increase in the number of connected nodes and with an increase in architecture of distributed computing system.

2. The architecture of a distributed computing system

Modeling of functioning of load balancing algorithm was investigated in the functional structure of a distributed computing system shown in Figure 1.

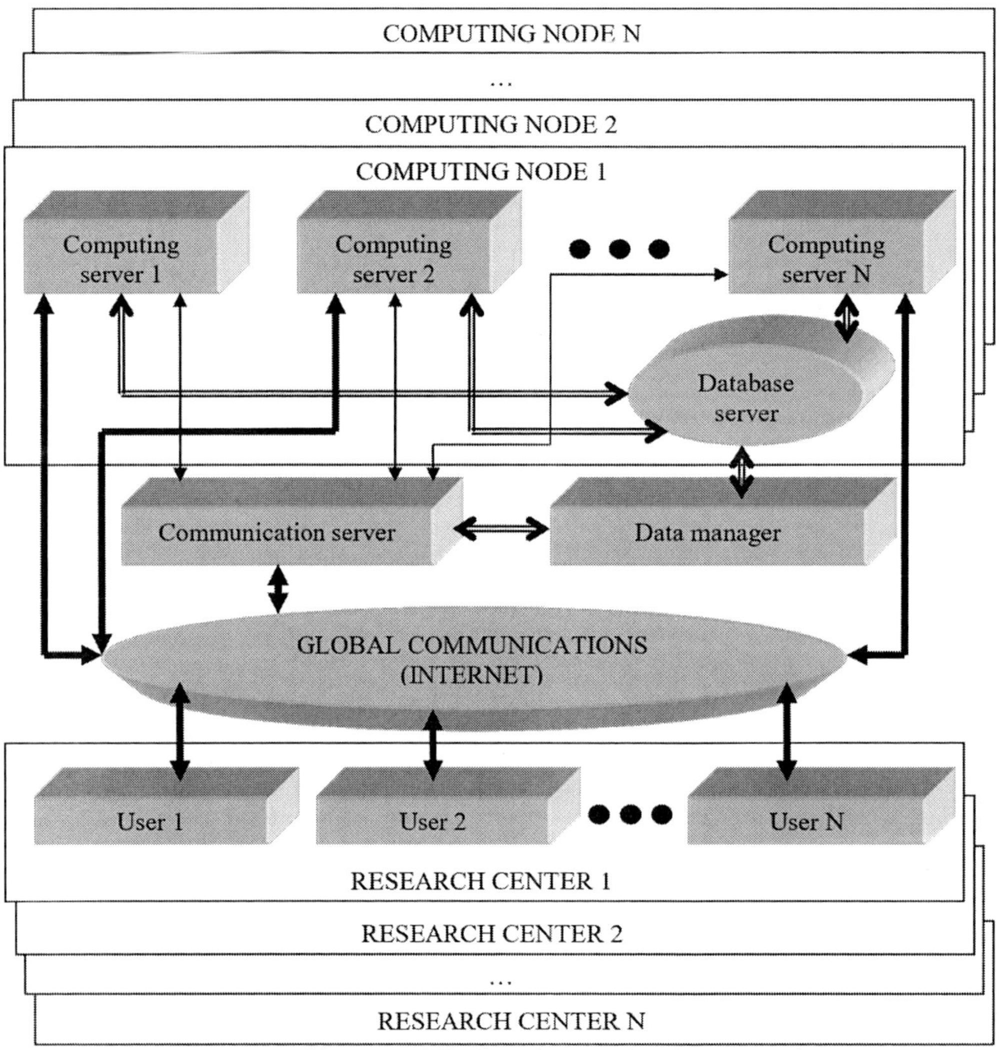

Figure 1. Architecture of distributed computing data processing system.

A feature of this system is the ability to increase performance using horizontal scaling computing nodes and resources. The system architecture is based on the components used for the construction of modern tools of distributed computing (grid-systems). But the internal structure of the complex, the contents of functional components (middleware) protocols and their interaction are original. Architecture of this complex virtualizes three basic technical resources on which the high-performance data center (computer systems, data storage systems and global communication) is constructed, and then collects them into a single virtual computer to provide users of this center with its resources as services.

Computing servers, a communication server and a database server are allocated in the structure of distributed computing data processing system. Planning and dispatching of data processing is entrusted to the special communication server. The interaction of all users with this complex is carried out only through the communication server. The main task of the communication server is to provide optimal loading of the available computing servers and to provide users with real-time mode (online). In this situation, the user does not know what particular network node performs his task. The user just consumes a certain number of virtual processing power capacity available in the network. Mathematical data processing is performed using of the computing servers.

A scheme of the developed algorithm for selecting of the computing server by means of the communication server is shown in Figure 2.

The algorithm uses information that is stored in the profiles (meta descriptions) of computing servers (a table of processors, a table of compliance of tasks and a table of tasks).

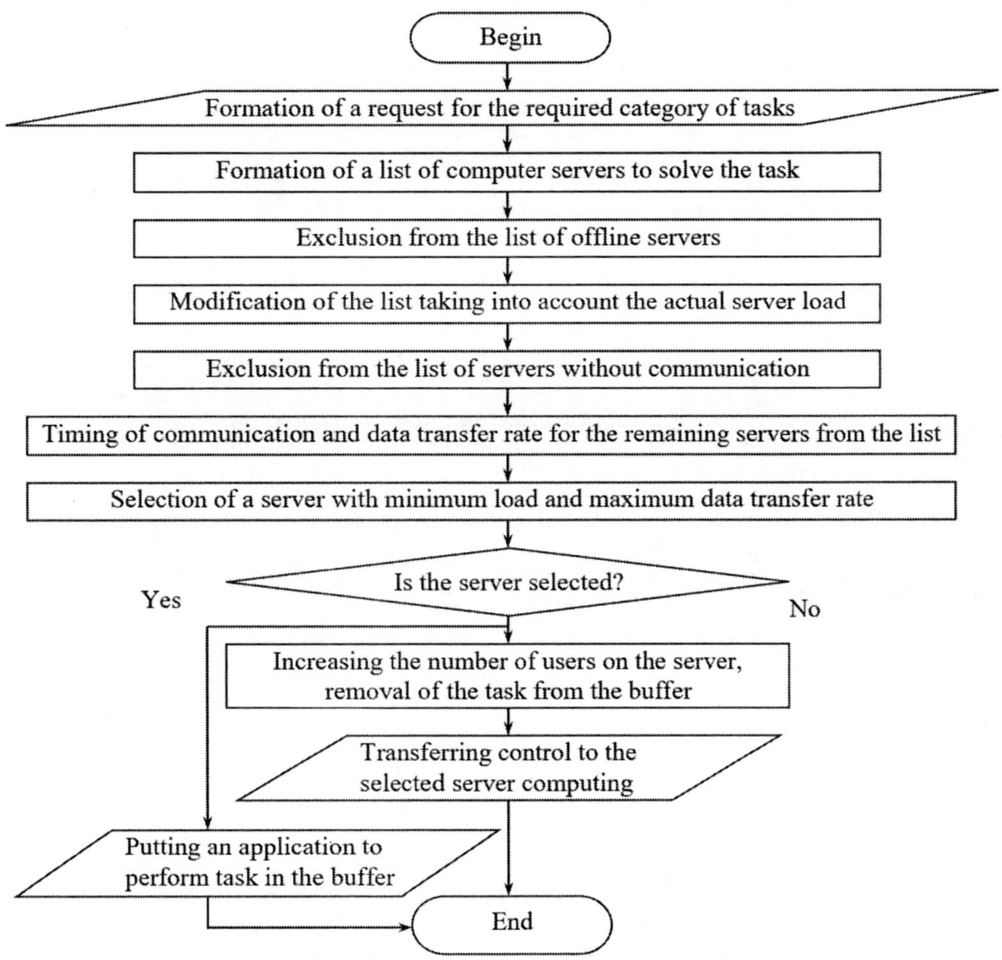

Figure 2. Algorithm of functioning of communication server

The main steps of the algorithm:

1. Creating a list of computer servers to solve a task.
2. Servers which are offline are deleted from the list.

3. Servers which are busy are deleted from the list.
4. Servers which have no connection by technical reasons are deleted from the list.
5. Determination of the connection time and data transfer speed of the servers.
6. Determination of the server with the most free computing power.
7. If there are several servers with the same processing power, then the server that has a greater data transfer rate is selected.
8. The value of the field 'Number of users' is increased by one.
9. In the absence of servers with free computing capacity, an application for the task execution is placed in the buffer of the communication server.
10. Control is transferred to the selected computing server.

The selection criterion is to find a server which has a minimum load and a maximum speed of data exchange.

3. GPSS-model for assessment of the effectiveness of the load balancing algorithm

A simulation model in the GPSS World language was developed for assessment of the effectiveness of a distributed computing system.

Figure 3 shows an enlarged model of the load balancing algorithm.

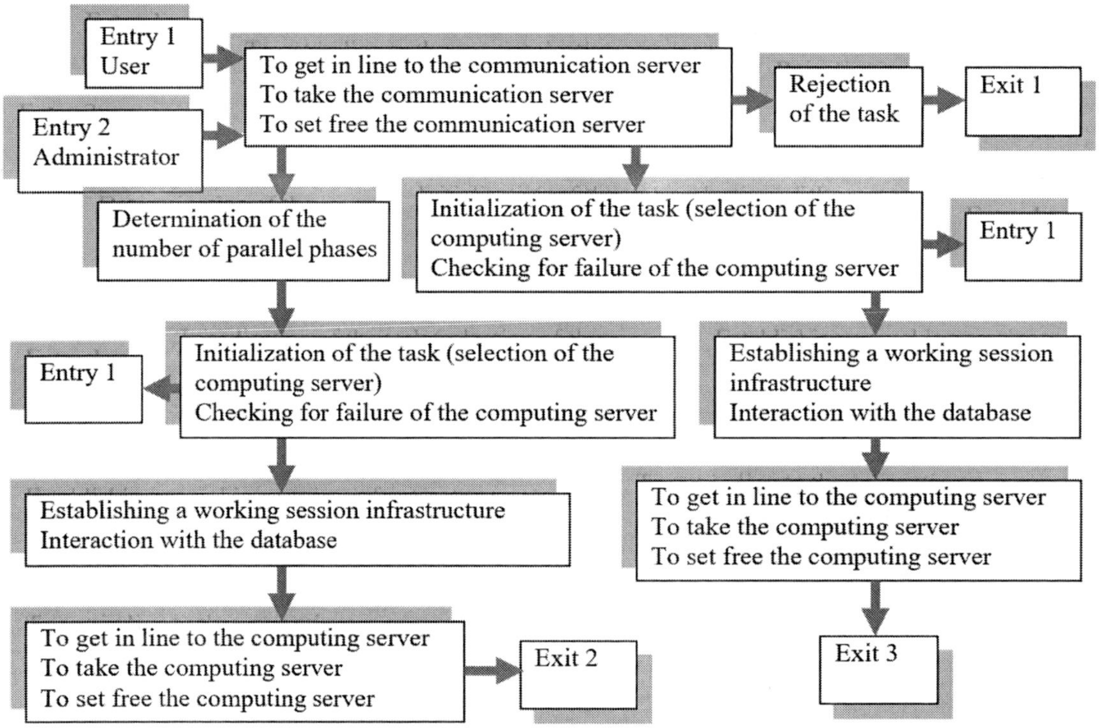

Figure 3. Gpss model for assessment of effectiveness of load balancing algorithm.

The main objects of the modeled system are flows of input tasks, computing segments, computing servers of segments, the communication server, database servers, the database manager. Calculation of the moments of task occurrences (a random value associated with the time interval between occurrences of two neighbor tasks) is performed according to the normal distribution law. Service intervals are also a random value which is distributed exponentially.

Variable parameters of the model are: the number of computing servers, maximum time of task processing, the number of task types, average interval between tasks, the proportion of parallel tasks, the proportion of rejected tasks, the average time of task processing by the communication server, deviation from the average task processing time of the communication server, failure probability of a computing server, recovery probability of a computing server, the recovery average time of a computing server, deviation from the recovery average time of a computing server, the number of communication servers, the maximum time of administrator task processing. Typical tasks in this model are the tasks of collection, storage and processing of meteorological data for the problems of analysis and forecasting of climatic processes [15, 16].

The authors conducted a study of dependency of some criteria (the total number of tasks, the number of rejected tasks, the number of tasks without parallelism and with parallelism, the number of emergency situations with the recovery and without the recovery (for tasks without parallelism and with parallelism), the number of tasks in the queue) from the following parameters: the number of computing servers, maximum time of task processing, the number of task types, the interval between tasks, the number of communication servers (Figure 4).

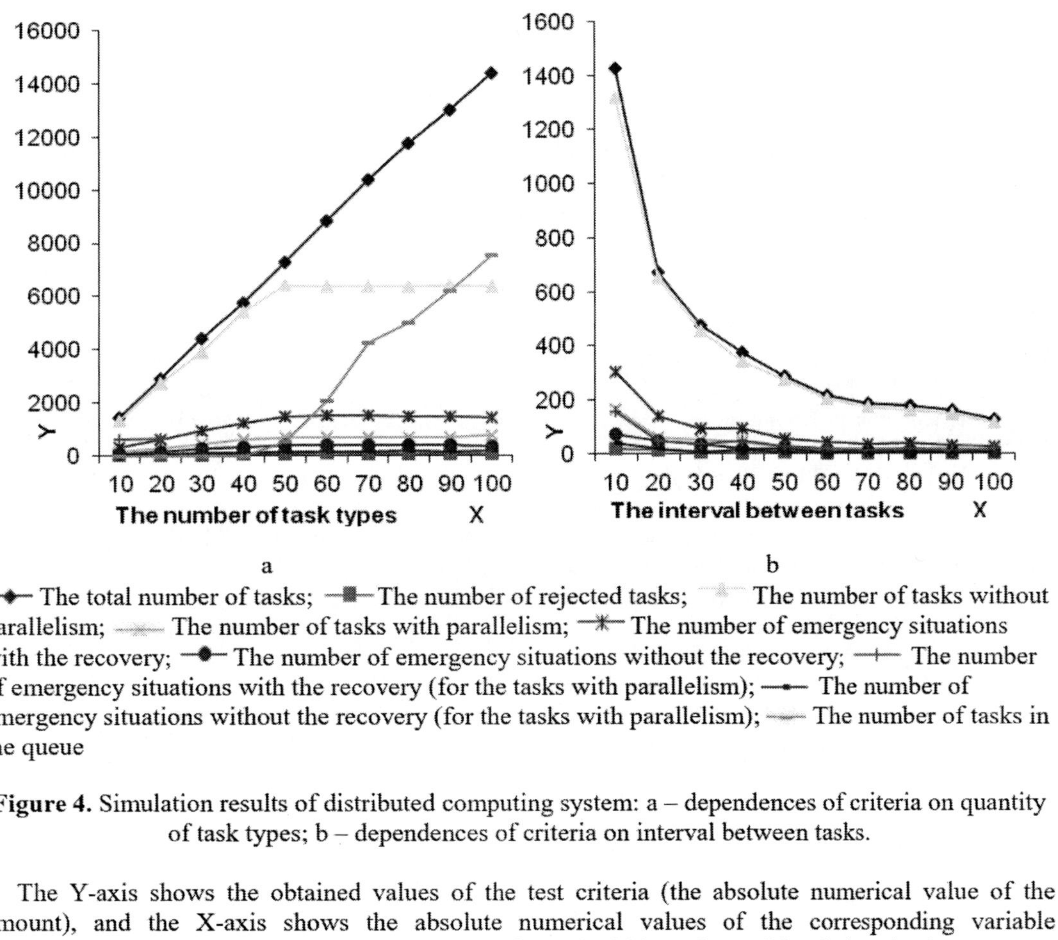

Figure 4. Simulation results of distributed computing system: a – dependences of criteria on quantity of task types; b – dependences of criteria on interval between tasks.

The Y-axis shows the obtained values of the test criteria (the absolute numerical value of the amount), and the X-axis shows the absolute numerical values of the corresponding variable parameters. In all experiments, simulation was performed within 24 hours (simulation time).

The following performance parameters of computing servers were chosen in the model: the number (over 100), the maximum time of task processing (up to 100 units), the number of communication servers (up to 10). This combination of parameters has essentially no effect on the studied criteria. The number of task types significantly affect the number of tasks performed in the system and the communication server queue.

Changing the range of task income greatly affects the studied criteria especially in the early stages of its increase.

Simulation experiments of the functional structure of a distributed computing system have shown its efficiency even with a significant increase in the number of connected nodes and scale of architecture. The proposed model parameters and static criteria for evaluating its operation allowed comprehensively and independently evaluate the quality of the developed model.

4. Conclusion

The question of the load planning in a distributed computing system should be addressed at an early stage of development of any service. Initially, the problem of inadequate performance of a distributed computing system in connection with increasing loads can be solved by building up the capacity of its servers but the effective method is the optimization of the algorithms and program codes.

References

[1] Kshemkalyani A D and Singhal M 2008 *Distributed Computing: Principles, Algorithms, and Systems* (Cambridge University Press)

[2] Md. Firoj Ali and Rafiqul Zaman Khan 2012 The study on load balancingstrategies in distributed computing system *Int. J. of Comp. Science and Engineering Survey* **3** 19-30

[3] Jain P and Gupta D 2009 An Algorithm for Dynamic Load Balancing in Distributed Systems with Multiple Supporting Nodes (Academy Publisher) pp 232-236

[4] Chhabra A, Singh G, Waraich S S, Sidhu B and Kumar G 2006 *Word Academy of Sci., Eng. and Techn.* 39-42

[5] Mesheryakov R, Moiseev A, Demin A, Dorofeev V and Sorokin V 2016 *Key Eng. Mat.* **685** 943-947

[6] Sanchez-Rodriguez D, Macías E M and Suarez A 2003 *Int. Conf. on Parallel and Distributed Processing Techniques and Applications* (Las Vegas, Nevada, USA) vol 1 (Providence: CSREA Press) pp 473-479

[7] Semchedine F, Bouallouche-Medjkoune L and Aissani D 2011 *J. of Network and Comp. Appl.* **34** 1123-30

[8] Sherstnev V S 2005 *The 9th Russian-Korean Int. Symp. on Sci. and Techn.* (Novosibirsk) vol 1 (Novosibirsk: NSTU Press) pp 696-700

[9] Shubhinder Kaur and Gurpreet Kaur 2015 *Int. J. of Comp. Applications* **125** vol 9 pp 25-28

[10] Imas O, Kaminskaya V and Sherstneva A 2015 *Proc. of 2015 Int. Conf. on Interactive Collaborative Learning* (Florence, Italy) vol 1 (IEEE Xplore) pp 511-514

[11] Shilpa Gambhir and Er Sonia Goyal 2014 *Int. J. of Adv. Res. in Electronics and Communication Eng.* **3** 659-663

[12] Fouzi Semchedinel, Louiza Bouallouche-Medjkoune, Leila Bennacer, Naim Aber and Djamil Aissani 2012 *Wireless Personal Communications* **67** 105-112

[13] Shubhinder Kaur and Gurpreet Kaur 2015 *Int. J. of Comp. Applications* **121** 45-47

[14] Rafiqul Z Khan and Md F Ali 2014 *Int. J. of Information Technol. and Comp. Sci.* **6** 65-71

[15] Popov V N, Botygin I A and Koshelev N V 2016 *Key Eng. Mat.* **685** 925-929

[16] Botygin I A, Popov V N, Tartakovsky V A and Sherstnev V S 2015 *Proc. SPIE 21st Int. Symp. Atmospheric and Ocean Optics: Atmospheric Physics* (Tomsk) vol 9680 (SPIE Digital Library) pp 1-4

IOP Conf. Series: Materials Science and Engineering **177** (2017) 012018 doi:10.1088/1757-899X/177/1/012018

Feasibility of the Davidenkov method for investigation of hoop residual stresses in cold expanded cylinders

V F Skvortsov, A Yu Arlyapov, A O Boznak, I I Ogol and A B Kim

Tomsk Polytechnic University, 30 Lenin Ave., Tomsk, 634050, Russia

E-mail: arlyapov@tpu.ru, aleksey@tpu.ru, iio@tpu.ru, bogowhich@tpu.ru

Abstract. Cold expansion of holes in hollow cylinders is accompanied by raise of residual stresses in the bulk of the cylinder, the largest of which are hoop residual stresses. Since hoop residual stresses affect critical performance characteristics, assessment of their nature and magnitude is one of the most significant engineering tasks. The most wide spread mechanical methods to define magnitude of hoop residual stresses are G Sachs and N N Davidenkov methods. The paper presents result of experimental studies by Sachs and Davidenkov methods of hoop residual stresses generated in hollow cylinders made of carbon steel AISI 1050 after cold expansion. It is shown that the shape of curves of hoop residual stresses calculated by the mentioned methods, depending on the hoop residual strain generated on the outer surface of the cylinder, can differ (strain is elastic) or coincide (strain is elastic-plastic). It was established that the absolute values of the stresses defined by the mentioned methods in the area adjacent to the hole differ by an average of 15%, and in the area adjacent to the outer surface differ by an average of 20%.

1. Introduction

Cold expansion is one of the most effective methods for finishing and hardening of holes in hollow cylinders [1-6]. However, with the ratio of cylinder external diameter to hole diameter $D/d \geq 3$, cold expansion forms significant residual stresses in entire volume of a cylinder [1, 2, 7]. At the same time, the largest in absolute value stresses are the hoop residual stresses, which define the most important performance characteristics of components, such as fatigue strength [8-10], wear resistance [3, 7], dimensional stability [11, 12], etc. In this regard, determination of hoop residual stresses generated in a cold expanded cylinder is an important engineering problem.

Review has shown [13-16] that most of the existing methods for determination of residual stresses in cylinders are mechanical methods (the use of which is usually accompanied by destruction of the product). The most widespread mechanical methods include methods of G Sachs [17] and N N Davidenkov [18]. Sachs method consists in successive removal of layers of metal from the internal or external surface of the samples and measurement of hoop and axial strains on the outer or inner radius, respectively [14]. The method helps to define hoop, radial and axial residual stresses simultaneously. It is noted [13] that this method is very labor-intensive, time-consuming, and places high demands on the means of measurement due to the small quantities of strains to be measured. In cases only hoop residual stresses have to be determined, it is reasonable to apply Davidenkov method in order to save time and make it less labour intensive. This method implies cutting the cylinder along its axis and measuring the change in the median diameter without removal of metal from the internal surface of the cylinder [13, 14]. In [13], it is noted that the magnitude of hoop residual stresses determined with this method (by cutting the cylinder) is approximate, since its calculation is made on the assumption that the stresses are distributed in the cylinder wall hyperbolically as in bending of a curved beam. In this regard, it is reasonable to expect that the stress distribution may be different and the results obtained will have an error, but the magnitude of this error is obscure.

The purpose of this paper is to investigate experimentally hoop residual stresses in the cold expanded hollow cylinders using methods of G Sachs and N N Davidenkov (reduced to cutting the cylinder along its axis) and to compare the results.

2. Methods of research

Experiments were carried out on samples (figure 1) made of steel AISI 1050 with a hole diameter of $d = 5$ mm. External diameter D of the samples was 15 mm ($D/d = 3$) and 25 mm ($D/d = 5$). Length L of the sample was equal to 10 and 40 mm. Samples were machined on a lathe 'DMG Mori CTX 310 ecoline'; holes were drilled and then reamed with hand reamers. The total number of samples was 72. Three samples were used in each experiment.

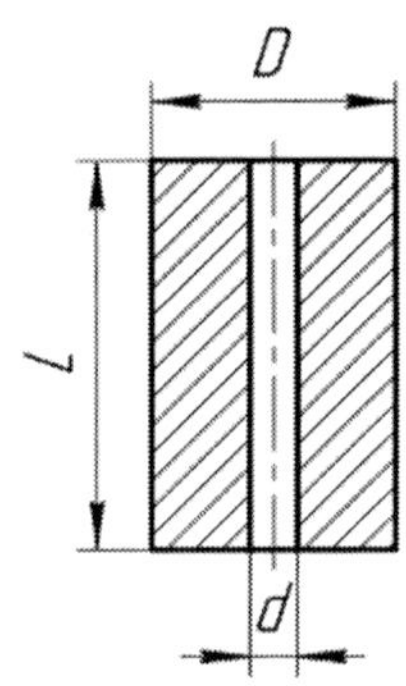

Figure 1. Sample

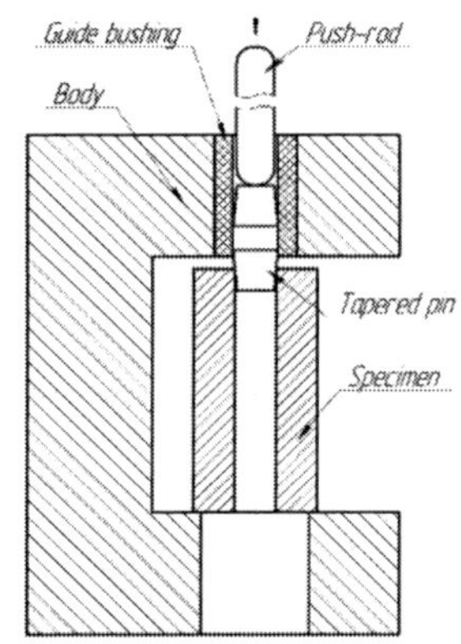

Figure 2. Direct cold expansion

Direct cold expansion was used in this study (figure 2). To achieve cold expansion, a tapered pin was forced through the hole. Interference was 0.9%, 3.4% and 7.1%. It was defined by the following equation:

$$Interference\ \% = \frac{d_p - d}{d} \times 100,$$

where d_p represents the diameter of the cylindrical part of the pin, and d represents the specimen hole diameter.

The tapered pin (figure 2) used for cold expansion was made of WC-8%Co (92% of tungsten carbide and 8% of cobalt) with a cones angle equal to 6° and a width of the cylindrical margin equal to 3 mm. Oil fluid MR-7 was used as a lubricant. Tests were performed on the universal testing machine 'IR5082-500'. The cold expansion speed was equal to 0.008 m/s. The sample was held in the device shown in figure 2.

According to Davidenkov method, hoop residual stresses σ_θ were determined by the change in the median diameter of the sample (a hollow cylinder) after cutting it along the axis (figure 3). The calculation was performed using formula [13], which in our notation (figure 4) is as follows:

$$\sigma_\theta(b) = \frac{1}{1-\nu^2} \cdot \frac{E}{D_C - 2\gamma} \cdot \left(\frac{b - \frac{t}{2} + \gamma}{b - \frac{t}{2} + \frac{D_C}{2}} \right) \cdot \Delta D_C, \quad (1)$$

where E is the Young's modulus of the sample material, MPa; ν is the Poisson's ratio of the sample material; D_C is the diameter of the median surface of the sample, mm; γ is the distance from the neutral axis of the sample wall section to the sample center of gravity, mm; t is the specimen wall thickness, mm; b is the distance from the given section to the internal surface of the sample, mm; ΔD_C is the increment of the median surface diameter of the sample after cutting it along the axis, mm.

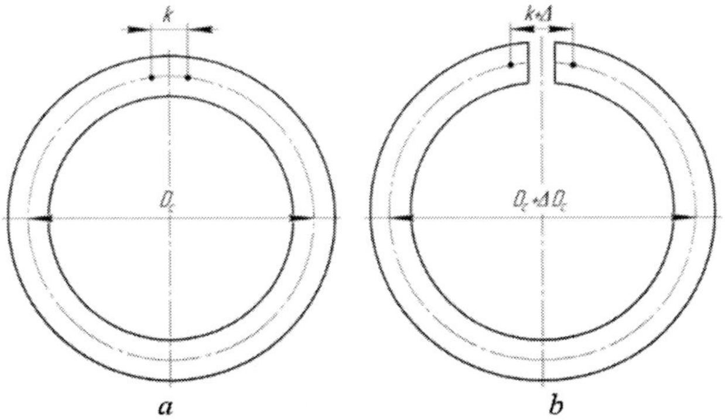

Figure 3. Change in the median diameter of the sample: a – before cutting; b – after cutting

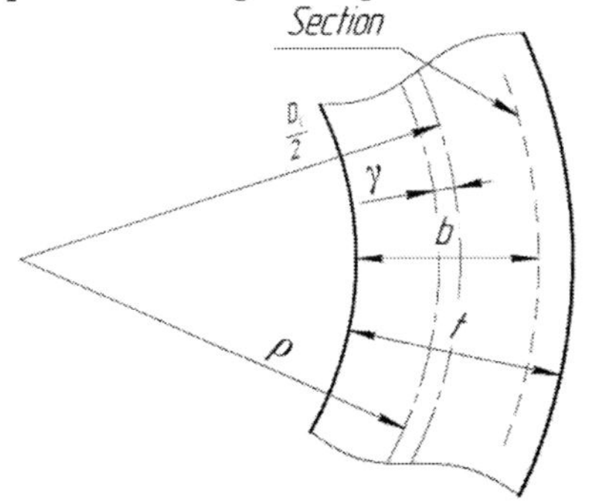

Figure 4. Coordinates of the sample section

Distance γ was found by the formula [13]:

$$\gamma = R_C - \rho = \frac{D_C}{2} - \frac{t}{\ln \dfrac{D}{D-2t}}, \quad (2)$$

where R_C is the radius of the sample median surface, mm; ρ is the radius of the sample neutral layer, mm; D is the external diameter of the sample, mm (considered to be the same after cutting). The ΔD_C value was determined by increment Δ of distance k between the indentations of the conical indenter left on the median circle of the sample end, using ratio $\Delta D_C = \Delta/\pi$. The specified distance was measured before and after the cutting using the microscope 'UIM 21' (the distance before cutting was about 4 mm). Cutting was performed on a CNC wire-EDM machine, model DK 7725.

Hoop residual stresses were calculated by Sachs method with the use of I A Birger formula [14], which in our notation is as follows:

$$\sigma_\theta(r) = \frac{E}{1-\nu^2} \cdot \left[\frac{R^2 - r^2}{2r} \cdot \left(\frac{d\varepsilon_\theta}{dr}(r) + \nu \frac{d\varepsilon_z}{dr}(r) \right) - \frac{R^2 + r^2}{2r^2} \cdot \left(\varepsilon_\theta(r) + \nu\varepsilon_z(r) \right) \right], \quad (3)$$

where R is the radius of the external surface of the sample, mm; r is the radius of the internal surface of the sample (cutting radius), mm; $\varepsilon_\theta(r)$ is the hoop strain at the outer radius as a result of cutting internal surface to radius r; $\varepsilon_z(r)$ is the axial strain at the outer radius as a result of cutting the internal surface to radius r. Residual stresses were calculated using the experimental technique described in [19, 20].

Young's modulus $E = 210$ GPa and Poisson's ratio $\nu = 0.3$ were used in calculations.

3. Results and discussion

Distribution of hoop residual stresses σ_θ along radius r of the cold expanded samples (stress curves) is shown in figure 5 ($D = 15$ mm) and figure 6 ($D = 25$ mm). We can see that both methods show compressive hoop residual stresses in the zone adjacent to the hole.

Samples with $D = 15$ mm cold expanded with 0.9% interference (figure 5, a and d), which leads to elastic hoop residual strain $\Delta D/D$ at the external surface equal to 0.04% for the samples with the length of 40 mm and 0.03% for the samples with the length of 10 mm, show different forms of the curves obtained by different methods. With the increase in the interference to 3.4% (figure 5, b and e), hoop residual strain becomes elastic-plastic (for the length of 40 mm, it is equal to 0.33%; for the length of 10 mm, it is 0.29%), and the forms of curves obtained by different methods coincide. With further increase in the interference to 7.1% (figure 5, c and f), hoop residual strain continues to increase (to

0.79% for 40 mm long samples, to 0.74% for 10 mm long samples), and the forms of curves obtained by different methods also coincide.

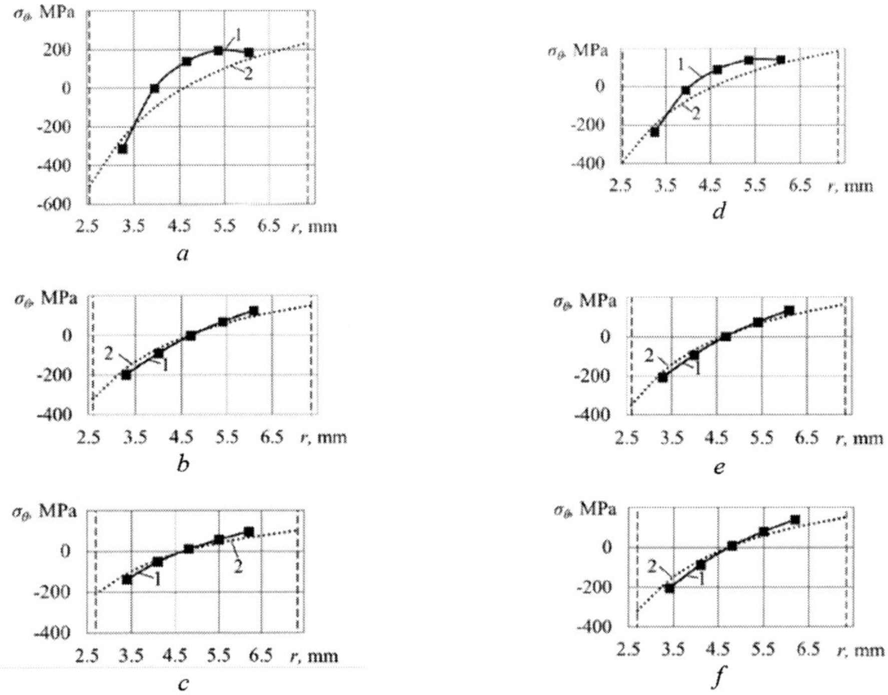

Figure 5. Curves for hoop residual stresses obtained by Sachs method (*1*) and by Davidenkov method (*2*), in the cold expanded samples with $D = 15$ mm:
a – Interference = 0.9%, $L = 40$ mm; *b* – Interference = 3.4%, $L = 40$ mm;
c – Interference = 7.1%, $L = 40$ mm; *d* – Interference = 0.9%, $L = 10$ mm;
e – Interference = 3.4%, $L = 10$ mm; f – Interference = 7.1%, $L = 10$ mm.
Here and below, dashed lines show internal and external surfaces of the samples

Only in one of the experiments on samples having $D = 25$ mm (interference is 7.1% and sample length is 40 mm), hoop residual strain on the external surface was elastoplastic (0.24%), and the form of curves of hoop residual stresses was the same for both methods (figure 6, *c*). In other experiments, hoop residual strain was elastic ($\Delta D/D < 0.2\%$). Particularly, when the interference is 0.9% (figure 6, *a* and *d*), the strain is 0.05% for 40 mm long samples, and it is equal to 0.03% for 10 mm long samples. With the increase in the interference to 3.4% (figure 6, *b* and *e*), hoop residual strain increases to 0.12% for 40 mm long samples, and it increases to 0.08% for 10 mm long samples. When interference is 7.1% and the sample length is 10 mm (figure 6, *f*), hoop residual strain is 0.18%. Curves obtained by different methods for samples with $\Delta D/D < 0.2\%$ are different (figure 6, *a*, *b*, *d*, *e* and *f*). It is obvious that with increasing hoop residual strain, the difference in the form of curves becomes less significant. This is probably due to the fact that distribution of hoop residual stresses in the wall of the sample becomes close to the hyperbolic law under the whole plastic deformation ($\Delta D/D > 0.2\%$).

The absolute value of hoop residual stresses in the area adjacent to the hole (at a distance of 0.7 mm from the surface) for the samples of both diameters determined by Davidenkov method differs from the value defined by Sachs method on the average by 15% (less than 50 MPa). In the region adjacent to the external surface (at a distance of about 1 mm from the surface for the samples with

$D = 15$ mm and at a distance of about 3 mm for the samples with $D = 25$ mm), these stresses differ on the average by 20% (less than 40 MPa).

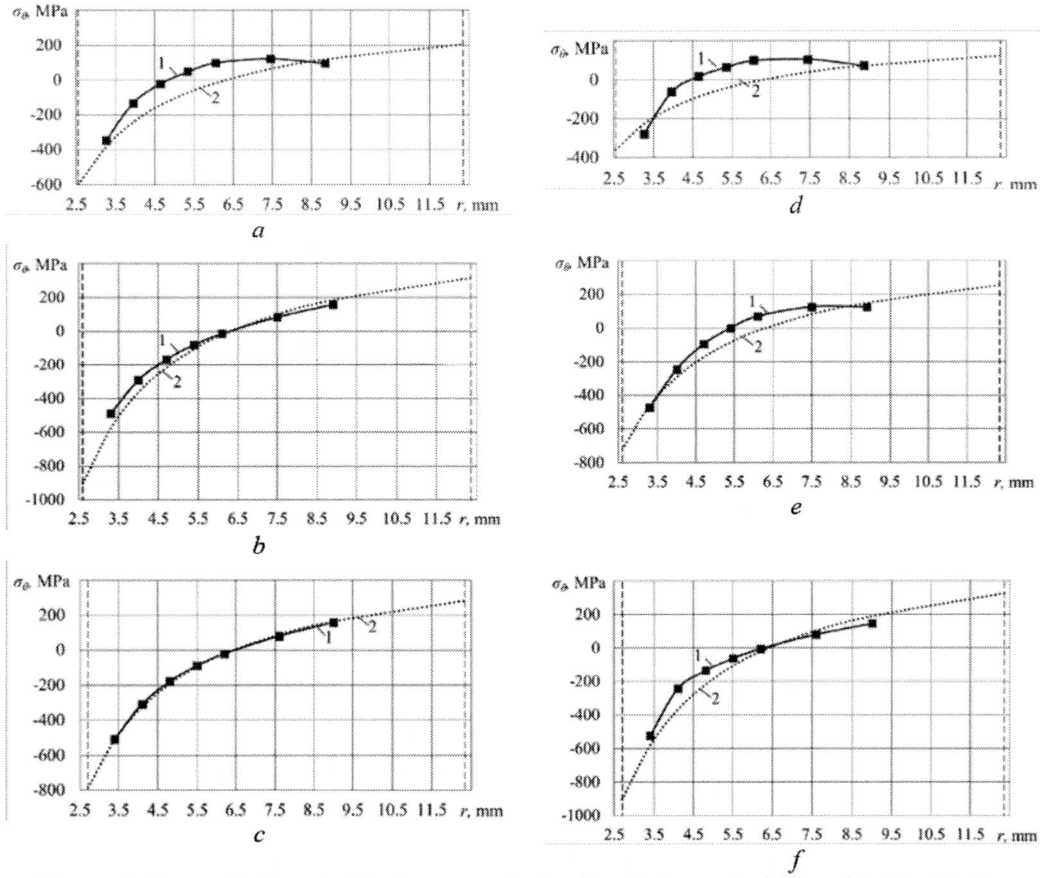

Figure 6. Curves for hoop residual stresses obtained by Sachs method (*1*) and by Davidenkov method (*2*), in the cold expanded samples with $D = 25$ mm:
a – Interference = 0.9%, $L = 40$ mm; *b* – Interference = 3.4%, $L = 40$ mm;
c – Interference = 7.1%, $L = 40$ mm; *d* – Interference = 0.9%, $L = 10$ mm;
e – Interference = 3.4%, $L = 10$ mm; *f* – Interference = 7.1%, $L = 10$ mm

4. Conclusions

1. The form of curves for hoop residual stresses generated during cold expansion of hollow cylinders made of AISI 1050 (the hole diameter is 5 mm, the outer diameter is 15 mm and 25 mm, the length is 10 mm and 40 mm, the interferences from 0.9% to 7.1%) obtained under elastic-plastic hoop residual strain on the external surface of the cylinder ($\Delta D/D > 0.2\%$) by the method of Davidenkov (reduced to cutting of the cylinder along the axis) coincides with the form of curves obtained by Sachs method. When $\Delta D/D < 0.2\%$, the form of the curves obtained by these methods differs, and the smaller the value of $\Delta D/D$ is, the greater the difference is.
2. The nature of the hoop residual stresses (their sign) defined by different methods is the same. Thus, the absolute value of the stresses in the region adjacent to the hole (at the distance of 0.7 mm from the surface) differs on the average by 15% (not more than 50 MPa), and in the region adjacent to the external surface (at a distance of about 1 mm from the surface for the

samples with $D = 15$ mm and at a distance of about 3 mm for the samples with $D = 25$ mm) these stresses differ on the average by 20% (not more than 40 MPa).

3. The method of N N Davidenkov, reduced to cutting of the cylinder along its axis, without further incremental removal of metal layers from the internal surface, is a simple and effective method to determine hoop residual stresses with the required accuracy and can be used instead of the method of G Sachs.

Acknowledgments
The authors gratefully acknowledge the guidance and role of Professor Vladimir Skvortsov during the conceptual stage of this study. The authors also wish to thank Lyudmila Petrova for invaluable metrological support.

References

[1] Monchenko V P 1980 *Effective technology of hollow cylinders production* (Moscow: Mashinostroenie)
[2] Proskuryakov Yu G 1961 *Holes cold expansion* (Sverdlovsk: Mashgis)
[3] Rosenberg A M and Rosenberg O A 1990 *Mechanics of plastic deformation in processes of cutting and deforming broaching* (Kiev: Naukova dumka)
[4] Schneider Yu G 1998 *Finishing forming processes: Handbook* (Saint Petersburg: Politechnika)
[5] Fu Y, Ge E, Su H, Xu J and Li R 2015 *Chin. J. of Aeronautics* **28(4)** 961-73
[6] Skvortsov V F, Arlyapov A Yu, Okhotin I S and Kim A B 2012 *The 7th Int. Forum on Strategic Technology (Tomsk)* vol 1 (Tomsk: TPU Publisher) p 1
[7] Rosenberg A M Rosenberg O A, Gritsenko E I and Posvyatenko E K 1977 *Surface finish after deforming broaching* (Kiev: Naukova dumka)
[8] Brondz L D 1986 *Technology and support of airplane age* (Moscow: Mashinostroenie)
[9] Chakherlou T N, Alvandi-Tabrizi Y and Kiani A 2011 *Int. J. of Fatigue* **33(6)** 800-10
[10] De Matos P F P, McEvily A J, Moreira P M G P and de Castro P M S T 2007 *Int. J. of Fatigue* **29(3)** 575-86
[11] Proskuryakov Yu G, Romanov V N and Isayev A N 1984 *Three-Dimensional holes cold expansion* (Moscow: Mashinostroenie)
[12] Zhuang W Z and Halford G R 2001 *Int. J. of Fatigue* **23** 31-37
[13] Babichev M A 1955 *Methods of residual stresses determination in machine parts* (Moscow: AN SSSR)
[14] Birger I A 1963 *Residual stresses* (Moscow: Mashgiz)
[15] Kobrin M M and Dehtiar L I 1965 *Determination of residual stresses in cylindrical machine parts* (Moscow: Mashinostroenie)
[16] Rossini N S, Dassisti M, Benyounis K Y and Olabi A G 2012 *Mater. & Design* **35** 572-88
[17] Sachs G 1927 *Z. Metallkunde* **19(9)** 352-7
[18] Davidenkov N N 1950 *Zavodskaya laboratoriya* **16(2)** 188
[19] Skvortsov V F, Arlyapov A Yu, Boznak A O and Kim A B 2014 *Int. Conf. on Mech. Eng., Automation and Control Systems* (Tomsk) vol 1 (Tomsk: TPU Publisher) p 1
[20] Skvortsov V F, Boznak A O, Kim A B and Arlyapov A Yu 2015 *Inst. of Phys. Conf. Ser.: Mater. Sci. and Eng.* (Yurga) vol 91 (Tomsk: TPU Publisher) p 12

The algorithm of crack and crack tip coordinates detection in optical images during fatigue test

S V Panin[1,2], V O Chemezov [1], P S Lyubutin[1,2] and V V Titkov[1]

[1] Institute of Strength Physics and Materials Science of Siberian Branch of the Russian Academy of Sciences, 2/4, Akademicheskii ave., Tomsk, 634055, Russia
[2] National Research Tomsk Polytechnic University, 30, Lenina ave., Tomsk, 634050, Russia

E-mail: vpointc@rambler.ru

Abstract. An algorithm of crack detection during fatigue testing of materials, designed to automate the process of cyclic loading and tracking the crack tip, is proposed and tested. The ultimate goal of the study is aimed at controlling the displacements of the optical system with regard to the specimen under fatigue loading to ensure observation of the 'area of interest'. It is shown that the image region that contains the crack may be detected and positioned with an average error of 1.93%. In terms of determining the crack tip position, the algorithm provides the accuracy of its localization with the average error value of 56 pixels.

1. Introduction

When studying the fatigue fracture of solids, the problem of estimating crack growth parameters (length, growth rate, crack opening displacement, etc.) is one of the most important. Thus, the tracking of the fatigue crack tip is an actual engineering problem. The issues of fatigue failure, as well as the problems of mutual influence of the damage accumulation process and stress-strain state evolution in the vicinity of the crack tip, have been constantly attracting the attention of researchers from the ancient times and are of fundamental and practical importance today [1]. For example, in [2] the basic regularities of deformation and damage accumulation processes are summarized and mathematical models to describe this process development are presented.

Currently, a contactless strain measurement optical system is widely used, since it enables quantitative analysis of the fatigue crack growth characteristics and recording of strain distribution patterns in the vicinity of the crack tip. The commercial systems, like VIC 3D, Strain Master [3, 4] and others first of all in their stereoscopic realization have become very popular for appreciable aiding the experimental mechanics. These systems comprise a pair of TV cameras operating in a macro shooting mode. They are usually aimed at obtaining the macroscale pattern of strain distribution on the surface of the specimen under loading. At the same time, from the point of view of physical mesomechanics and fracture mechanics, the studies involving the crack with the length not exceeding tens of microns are of particular interest. In this case, the crack growth observation at macroscale is impossible because the spatial resolution does not resolve the crack as well as the processes developing at its tip.

To solve this problem, optical microscopes with a photo or video camera attached are employed. However, when using high magnification of the optical microscope with the cyclic loading, a fatigue crack tip will constantly go outside the observation area. In this concern, it is necessary to design a

IOP Conf. Series: Materials Science and Engineering **177** (2017) 012019 doi:10.1088/1757-899X/177/1/012019

hardware/software setup (systems) that allows controlling the mechanical scanner, which displaces the optical system (comprising a microscope and video camera) in relation to the sample surface where the crack tip is to be detected and located. In this paper, the problem of image analysis algorithm development was solved in order to ensure fast crack detection by a series of images for further decision-making on the magnitude and direction of the displacement of the optical tracking system.

The literature reports on several different methods available for crack detection including image-processing methods (wavelet and Fourier transforms, Canny filters, Sobel filters, etc.) [5], percolation-based techniques [6–8], impact echo method [9], PCA-based techniques [10], etc. Automated crack inspection systems were developed in adjacent engineering areas [11], where the Laplacian of Gaussian algorithms have been implemented. Stereovision techniques have been used in automated distress survey systems [12]. Another crack detection scheme makes use of maximum a posteriori (map) classifier, which relies on a set of geometric characteristics of the segmented binary regions [13].

2. Crack detection

The crack detection algorithm proposed in this paper consists of two approaches: 1) analysis of the image pairs (series) based on data of objects motion; 2) analysis of individual images based on the contour detection. Let us consider each of them in detail.

2.1. The analysis of image pairs (series) based on data of objects motion.

When analyzing an image series (or video sequence) that illustrates the crack growth, key information can be extracted from the analyzed motion of image fragments. In this case, the surface texture is of low informativeness, since it mostly allows improving the reliability (accuracy and noise robustness) during displacement determination. For further analysis, the data on the optical flow resulted from the discontinuity (crack) growth are used. Thus, the object of the analysis is the pattern of the optical flow (Figure 1a), where a discontinuity (abrupt variation in displacements at the edges of the opening crack) has to indicate the presence of the crack.

However, the opening displacement values for short cracks can make only a few pixels; in this case, for the purpose of additional contrasting (edge enhancement), the optical flow differentiation procedure is applied (Figure 1b). Its implementation is realized with the help of Sobel operator for a window size of 3x3 [14]. Remarkably, the use of the first approach results in a number of errors due to inherent errors of optical flow determination algorithms. For this reason, the first approach to crack detection can be used for the initial (coarse) detection of the image region where a crack might present (Figure 1c).

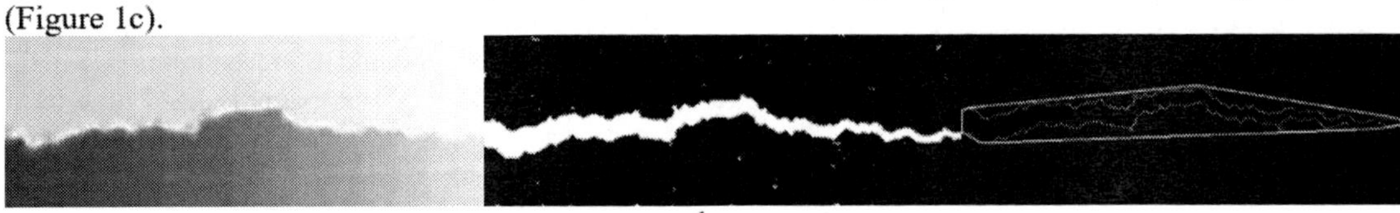

a b c

Figure 1. Crack detection by the analysis of the image series based on object motion data: (a) the optical flow pattern; (b) differentiation of the optical flow; (c) selecting the region of probable crack location

2.2. The analysis of individual images based on contour detection.

During the processing of single images (without considering other images in the series or a video stream), the information about the surface texture is analyzed (Figure 2a). In the process of crack tracking, two key tasks are to be solved: 1) detection of a crack in the image and 2) determination of its location. It is implemented through image contour analysis, which is based on the decomposition of the 'scene' into individual objects and the background (Figure 2b) (segmentation). The use of the contour analysis makes it possible to allocate small crack-like objects (Figure 2c). During the segmentation, a large number of objects are detected, while only one or a few of them are the cracks indeed.

It is a priori known that a crack appears as the darkest object in an image. In order to reduce computational costs and to increase the robustness of the crack detection, the developed algorithm employs binarization that makes it possible to reduce the amount of objects under analysis. In doing so, the binarization algorithm is based on histogram analysis with the determination of the (global) brightness threshold value. It is related to the fact that the threshold value mostly corresponds to low brightness levels. Subsequent processing of the resulting black-and-white image is related to the use of the 'morphological closing' operation. It is aimed at removing small dark fragments in the image that correspond to surface texture elements, but do not belong to the crack(s). Then, the resulting black and white image is subjected to contour analysis in order to detect the object(s), whose geometrical characteristics in the maximum value correspond to those of the crack.

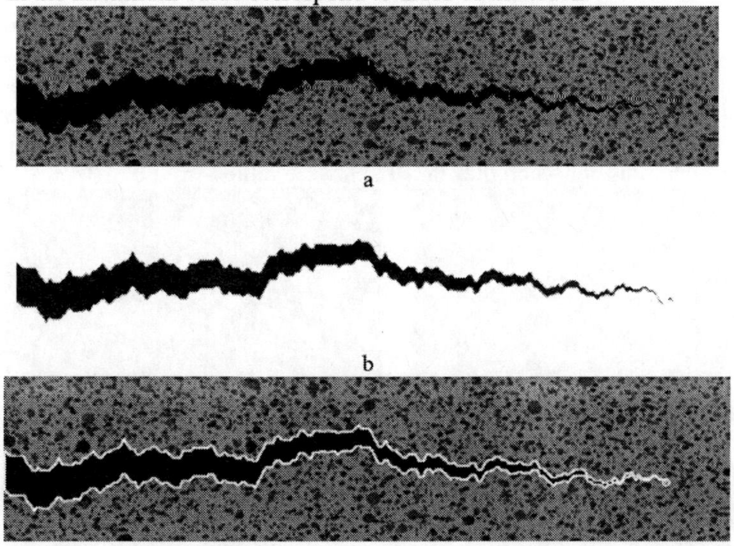

Figure 2. Crack detection by the analysis of individual images based on the contour detection: (a) input image; (b) binarized image used for segmentation; (c) detected crack

2.3. Crack tip detection.

A region in the image corresponding to the crack tip is usually of low contrast. Hence, the procedure of determining its exact coordinates is a separate challenge. It is proposed to construct the skeleton [15] of crack-like objects with subsequent assessment of the crack opening value towards the axis of the skeleton. For the crack skeleton construction, the shortest distance from the point inside the crack towards its boundaries (faces) is evaluated (Figure 3). It is based on the fact that the crack skeleton represents a curve that combines the points with the maximum values of the estimated distance.

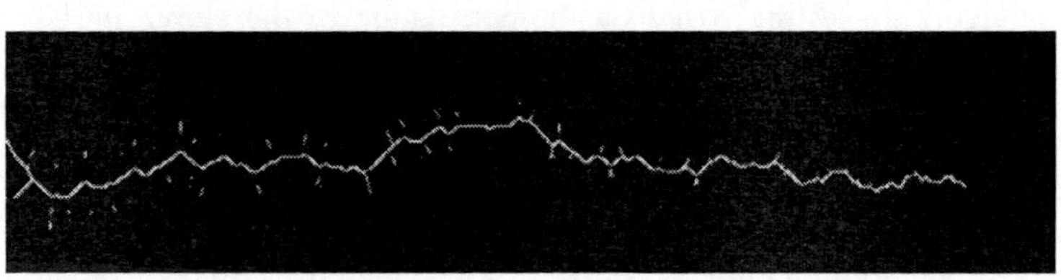

Figure 3. The crack skeleton.

IOP Conf. Series: Materials Science and Engineering **177** (2017) 012019 doi:10.1088/1757-899X/177/1/012019

3. Testing

The developed algorithm was tested with the use of a model image series. The efficiency of the algorithm was first evaluated in the absence of external perturbations (noise, geometric distortion, etc.). The model image series was formed as a textured surface (simulating a sprayed speckle) with software-assisted superimposition of a crack (Figure 4 a, b). The construction of the model images was made in several stages: 1) synthesis of the material surface model (speckle image formation); 2) setting the crack parameters (length, opening, shape of edges that imitate its branching); 3) material model variation (image distorting) with respect to the change in crack parameters.

In addition, in order to determine the robustness of the proposed algorithm, the model images were distorted by blurring and noising (to imitate the distortion specific for image acquisition in laboratory experiments). Also, crack detection was carried out when processing experimental images (Figure 4 c, d) made under cyclic tension of aluminum alloy 2024.

To assess the effectiveness of the crack detection algorithm, two measures were introduced: 1) the difference between the coordinates of the crack tip in the model image (a known value) and those determined with the use of the algorithm (see Section 2); 2) the difference between the crack area (the principle of the calculation is given in Sec. 2) detected using the proposed algorithm and that known from the model image. The correctness level of the proposed algorithm functioning might be evaluated through calculating each indicator when they take the lowest value.

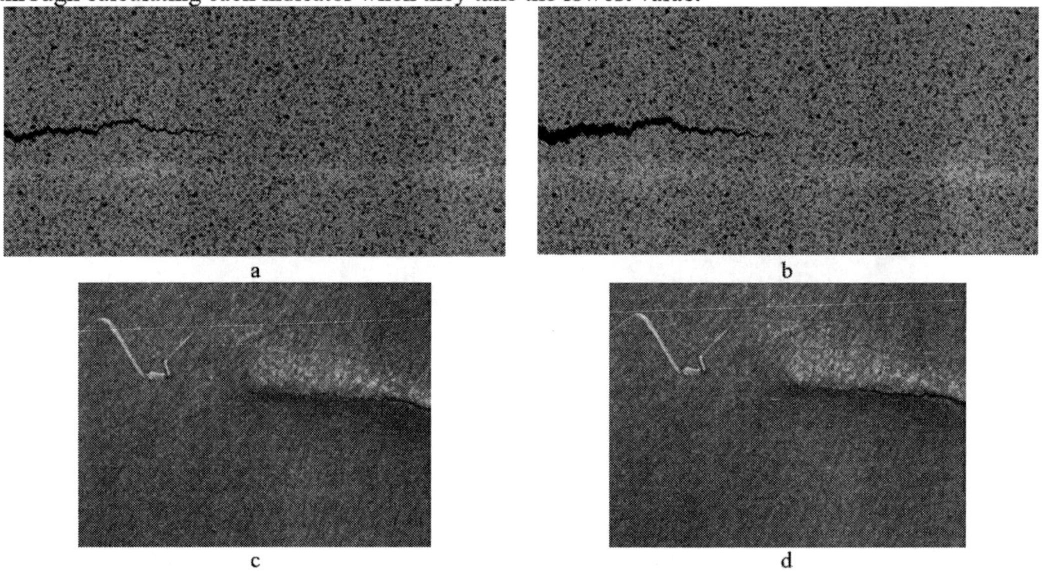

Figure 4. Model (a, b) and experimental (c, d) images illustrating the crack growth.

4. Results and discussion

The results of the algorithm testing based on the calculation of the above introduced measures are shown in Figure 5. At first, the comparison of the area of the algorithm-detected crack against the model value was performed. Then, the detected crack tip coordinates were compared against known (model) ones. Figure 5 also illustrates the results of crack detection when various levels of noise and blur were applied to the model images.

The results of model image processing (including blurred and noisy ones) have shown that the algorithm is capable of determining the coordinates of the crack tip with the average error of 56 pixels. During the crack area determination, the algorithm ensures the accuracy to locate its position (including edges) with an average error of about 1.93%.

During testing 8 pairs of model images were generated, thus a statistical characteristic could be obtained. Figure 5 shows the standard deviation of the crack area detection error equal to 0.21% and the standard deviation of the crack tip coordinates detection error equal to 15 pixels from its average value.

The analysis of the data presented in Figure 5 shows that the proposed algorithm gains a maximum error of crack allocation equal to 5.7% when noisy images (s = 45) are processed, while the error increases sharply under their blurring and reaches 40% when the blur filter with $\sigma = 4.5$ is employed.

In the case of crack tip positioning, the trend of an error increase persists for both noisy and blurry images. However, the model images have the size of 2000×1000 pixels (which corresponds to their diagonal length of 2236 pixels), while the maximum error of coordinate determination is about 350 pixels. It makes only 15% of the diagonal length of the entire image. Therefore, this error value would not significantly affect the accuracy of the optical system positioning during the tracking of a fatigue crack.

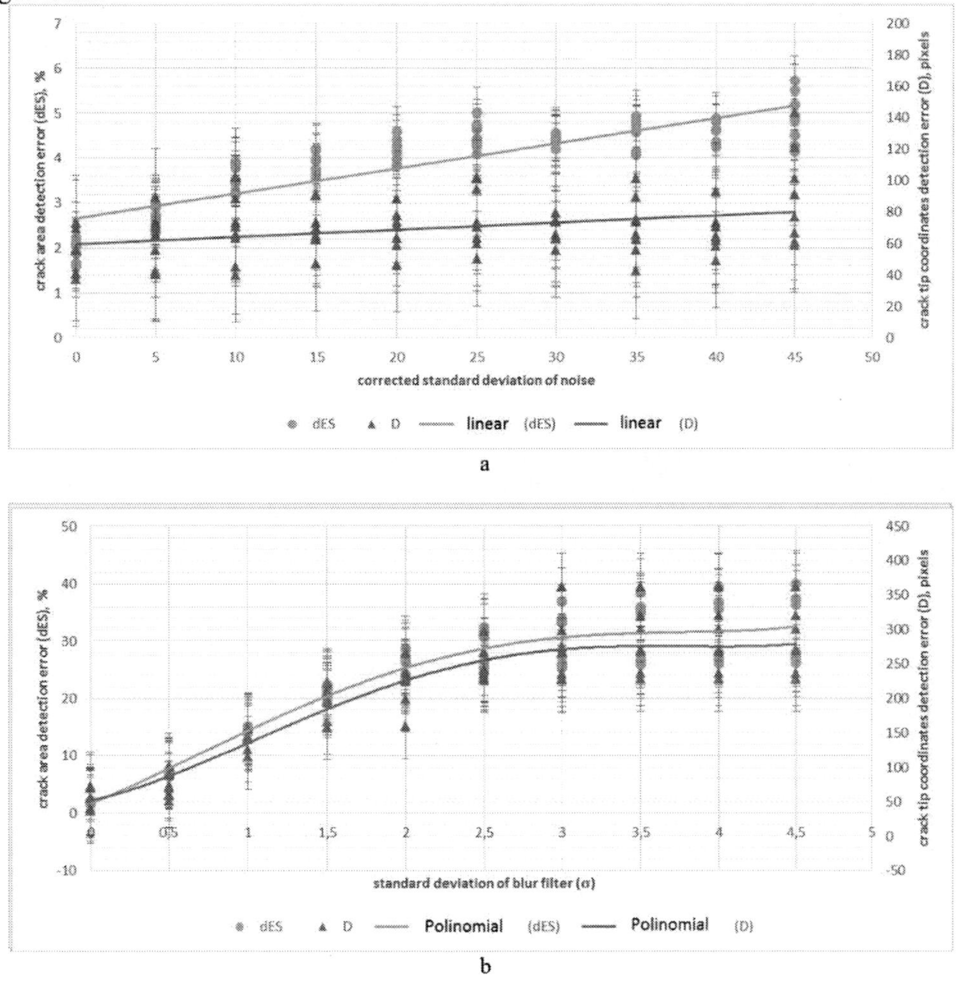

Figure 5. Plots of the crack and crack tip detection under noise (a) and blur (b).

The result of crack detection in the experimental images is shown in Figure 6. It is seen that the contour delineates precisely the crack edges. Thus, the possibility of applying the proposed algorithm to solve the problems of crack detection in optical images was proven for both model and experimental images.

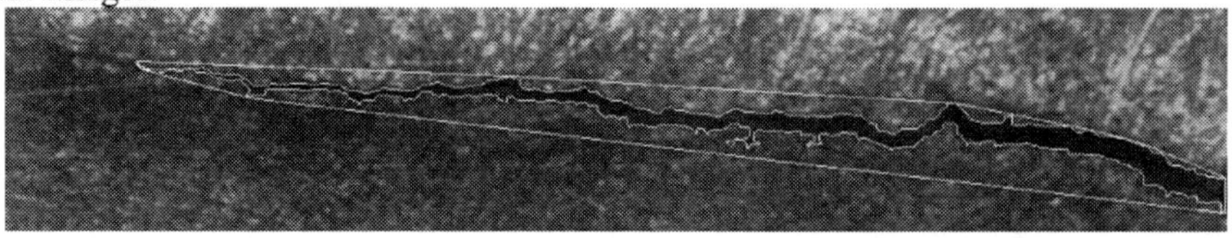

Figure 6. A crack detected in the experimental image.

5. Conclusion

The algorithm for automatic crack and crack tip detection in optical images under fatigue testing was developed and tested. The results on model image processing (including blurred and noisy ones) have shown that the algorithm is capable of determining the coordinates of the crack tip with the average error of 56 pixels. During the crack area determination, the algorithm ensures the accuracy to locate its position (including edges) with an average error of about 1.93%. The maximum error of crack area location is 5.7% for noise content, but it rapidly increases for blurred one and is equal to 40%. Accurate performance of the developed algorithm under processing of experimental images was proven. The developed algorithm will be used to solve problems of fatigue crack tracking automation when large magnification of the optical system is used.

Acknowledgments

The work was conducted as a part of the program of fundamental researches of the State Academies of Sciences (PFR SAS) for 2013-2020 No. 23.1.3, as well as RFBR grant No. 16-38-00526, and the grant of the President of the RF No. SP-1529.2015.5. Authors acknowledge A V Eremin for provided optical images and participating in results discussion.

References

[1] Stepanova L V 2011 *Vestnik of Samara State Univ. Natural Science Series* **2(83)** 105–15
[2] Volkov I A and Korotkih U G 2008 *Equations of state viscous elastoplastic media with damage* (Moscow: Fizmatlit) p 424
[3] Correlated Solutions Inc. 2016 The VIC-3D System, [Online]. Available: http://www.correlatedsolutions.com/vic-3d/.
[4] LaVision 2016 StrainMaster [Online]. Available: http://www.lavision.de/en/products/strainmaster/strainmaster-dic.php.
[5] Abdel-Qader I, Abudayyeh O and Kelly M E 2003 *J. of Comp. in Civil Eng.* **17(3)** 255-63
[6] Yamaguchi T, Suzuki K, Hartono P and Hashimoto S 2008 *Proc. of IEEE Conf. on Industrial Electr. and Appl. 2008* (Singapore: IEEE) vol 3 pp 1875-1880
[7] Yamaguchi T and Hashimoto S 2006 *Transactions on Information and Systems* (Oxford: Oxford University Press) vol E89-D(7) pp 2044-52
[8] Wang P and Huang H 2010 *The 3rd Intern. Cong. on Im. and Sig. Proc.* **1** 2530-2533
[9] Hlava Z 2009 *Proc. of the Ultragarsas (Ultrasound)* (Kaunas: University of Technology) vol 64(2) pp 12-6
[10] Abdel-Qader I, Pashaie-Rad S, Abudayyeh O and Yehia S 2006 *Adv. in Eng. Soft.* **37(12)** 771-8
[11] Salim Lim R, Manh La H, Shan Z and Sheng W 2011 *IEEE Intern. Conf. on Rob. and Autom.* (Shanghai: IEEE) 6288-93
[12] Kelvin C Wang P and Weiguo G 2007 Automated Real-Time Pavement Crack Detection and Classification (Arkansas: *Transportisation Research Board*) pp 33
[13] Oliveira H and Correia P L 2010 *The 18th Europ. Sig. Proc. Conf.* 274-78
[14] Gonzalez R C and Woods R E 2002 *Digital image processing* (Upper Saddle River: Prentice Hall) 793
[15] Glantz R and Kropatsch W G 2001 *Lect. Notes in Comp. Sc.* **2059** 186-195

Optimization of the coherence function estimation for multi-core central processing unit

A G Cheremnov[1], **V A Faerman**[2] **and V S Avramchuk**[3]

Tomsk Polytechnic University, Tomsk, Russian Federation

E-mail: [1]agc1@tpu.ru, [2]vaf@tpu.ru, [3]avs@tpu.ru

Abstract. The paper considers use of parallel processing on multi-core central processing unit for optimization of the coherence function evaluation arising in digital signal processing. Coherence function along with other methods of spectral analysis is commonly used for vibration diagnosis of rotating machinery and its particular nodes. An algorithm is given for the function evaluation for signals represented with digital samples. The algorithm is analyzed for its software implementation and computational problems. Optimization measures are described, including algorithmic, architecture and compiler optimization, their results are assessed for multi-core processors from different manufacturers. Thus, speeding-up of the parallel execution with respect to sequential execution was studied and results are presented for Intel Core i7-4720HQ и AMD FX-9590 processors. The results show comparatively high efficiency of the optimization measures taken. In particular, acceleration indicators and average CPU utilization have been significantly improved, showing high degree of parallelism of the constructed calculating functions. The developed software underwent state registration and will be used as a part of a software and hardware solution for rotating machinery fault diagnosis and pipeline leak location with acoustic correlation method.

1. Introduction

Solution of the number modern mechanical engineering problems requires applying of information technologies. Despite the fact that the most of these tasks are simple in terms of computations, some of them (as real-time control, prediction of technological process parameters or automated fault diagnosis) still require a huge amount of computation.

Continuous increase in volume of processed information and development of new signal processing methods requiring more computational resources lead to development of computation technology and facilitate appearance of new approaches to efficient use of computer hardware.

Primary resources of modern personal computers (PCs) are their central processing units (CPUs) whose performance is largely determined by their clock rate [1, 2]. Due to that until recently the main way to increase performance was connected to increase of the clock rate. At the same time, raised clock rate causes increases in both power consumption and heat emission inside the computer. The latter was de-emphasized by node miniaturization in the production of processors. Currently, however, further miniaturization is deemed problematic due to principal limitations, thus raising performance by increasing the number of cores becomes ever more important [2].

Introduction of parallel processing into the modern CPUs architecture creates several requirements to computation structuring: scalability, portability and adaptivity [3]. Only compliance with conditions of parallel decomposition and agglomeration leads to full use of the processing power and brings performance close to optimal values. Provision of the latter requires application of different approaches towards optimization and some specific technologies, such as Open MP, Intel TBB, Intel Click Plus and others [4].

This paper states and solves an optimization problem in coherence function evaluation with the aim to increase the computation performance on multi-core CPUs. This function is important in digital signal processing (DSP) and is commonly used for the spectral analysis of rotating machinery vibration signals and also for pipeline leak location with acoustic correlation method.Performance characteristics obtained as a result of the optimization are shown below.

The results are of significant practicality and will find their application in creation of a software and hardware solution for the technical condition assessment of mechanical engineering equipment, fault diagnosis of vehicles engines and transimission gearboxes. Necessity for effective use of hardware is determined by both significant amount of calculations and limited potential of hardware used in the project.

2. Coherence function in signal processing

The coherence function may be seen in Digital Signal Processing as an analogue of a mutual correlation function over frequency domain, reflecting a degree of linear dependence between harmonic components of the signals being analyzed [5]. Mathematical definition of the coherence function is a ratio of mutual spectrum of the signals and square root from product of their energy spectra [5]:

$$\gamma(f_k) = \frac{\left|E(P_{AB}(f_k))\right|}{\left(E(P_{AA}(f_k) \cdot E(P_{BB}(f_k)))\right)^{0.5}}, \qquad (1)$$

where $E(\cdot)$ is an averaging operator for the ensemble over non-overlapping epochs; $P_{AA}(k)$ $P_{BB}(k)$ are energy spectra represented by digital real signal samples $s_A(t)$ and $s_B(t)$; $P_{AB}(k)$ is a mutual energy spectrum of the signals. The argument of the coherence function takes the values

$$f_k = k \cdot \frac{f_d}{2n}, \qquad (2)$$

where f_d is signal sampling frequency $s_A(t), s_B(t)$; $k = 0,1,\ldots, n$; $n+1$ is a number of spectral samples.

Energy spectra and the mutual spectrum may be obtained through the following calculation:

$$P_{AA}(f_k) = F_D^*(s_A(t_i)) \cdot F_D(s_A(t_i)) = \left|F_D(s_A(t_i))\right|^2, \qquad (3)$$

$$P_{BB}(f_k) = F_D^*(s_B(t_i)) \cdot F_B(s_B(t_i)) = \left|F_D(s_B(t_i))\right|^2, \qquad (4)$$

$$P_{AB}(f_k) = F_D^*(s_A(t_i)) \cdot F_D(s_B(t_i)), \qquad (5)$$

where F_D is direct discrete-time Short-Time Fourier Transform (STFT); F_D^* is a complex conjugate of the STFT results. The studied signals are defined over a time interval

$$t_i = i \cdot \frac{1}{f_d}. \qquad (6)$$

The number of samples is determined by chosen width of a STFT chunk frame.

The coherence function is an important tool for signal processing allowing justified determination of a frequency domain where a stationary determined signal appears that carries information, [5] and then apply necessary digital bandpass and barrier filters [6].

3. Optimization of the coherence function evaluation

From the programming point of view the computational procedure may be represented as a main loop which is executed a preset number of times. At that, each iteration consists of data sampling to form the STFT chunk frames, calculation of the signal spectra and their step-by-step accumulation as per (1). Final value of the coherence function is determined at the loop termination by substituting obtained values into the equation (1).

Application of the coherence function to solution of practical tasks in signal processing, particularly that of leak location in pipelines with an acoustic emission method assumes treatment of large volumes of data. The latter imposes significant constraints on the software implementation for the project, making it more expensive, particularly if there is a necessity to provide an operating mode close to real-time. Due to that parallel processing is a preferred way to arrange the computations.

As was mentioned before, use of the coherence functions requires significant number of STFT evaluations. The Cooley-Tukey algorithm [7] for Fast Fourier Transform with time decimation over a fixed base of 2 was selected as an algorithm to calculate STFT to increase computation performance speed. Choice of the algorithm is determined by its ease of implementation, clarity and high degree of internal parallelism [8].

Additional increase in STFT performance may be achieved by preliminary treatment of data, namely by its bit-reversal (change of bit order in the number's binary representation to the opposite) [7, 8]. In this case only array elements' indexes are subjected to the transformation. As a result, an order of the input data is changed, while the numerical values are the same.

Parallelizing was performed with OpenMP and Intel Threading Building Blocks (TBB) technologies. Choice of OpenMP was determined by spacial localization of the data, while tools from the Intel TBB library were used to arrange recursive traversal of the 'butterfly' graph during the STFT evaluation with time decimation following the Cooley-Tukey method, because this approach suggests parallelism of tasks, rather than that of data. Built-in graphic subsystem is used as an additional co-processor to execute general calculations.

Despite significant increase in calculation performance due to the above mentioned measures the software implementation had been far from ideal. Further runtime optimization was attained with scalar optimization methods [7]. The scalar optimization includes constant folding, constant propagation and copy propagation. Constant folding is a process of preliminary calculation of constants during compilation of the application; constant propagation is substitution of previously calculated constants into equations; copy propagation is a process to substitute variables with their respective values [7].

To reduce time burden for resource distribution during parallel calculations, a manual loading of data into CPU cache from the RAM is implemented [9]. This is achieved with the data prefetching function which fetches into the into the CPU cache a 64 bytes-long line of data starting from a given address .

Practical effect manifests as reduced data access time for the soon-requested data. The reason for this is that such data are likely already in the CPU cache as a result of this manipulations.

The final optimization measure is making the mathematical core with the optimizing Intel C++ compiler.

4. Assessment of optimization and speed

In accordance with the algorithm outlined above a software solution has been developed for experimental assessment of the coherence function estimation speed. Studies were performed using two CPUs: Intel Core i7-4720HQ and AMD FX-9590. Total number of signal samples $\boldsymbol{N}$ varied from 524288 to 4194304 samples. The time to calculate the coherence function was measured one thousand times.

Because there is a multitude of factors (including pseudoparallel operation of the OS core process scheduler, interrupt processing) that can slower but not accelerate calculations [10], the shortest time was held as the execution time to increase fidelity of the study. Time measurements were carried out with a special function tbb::tick_count() in the Intel TBB library that takes into account possible changes in the processor speed during the calculations.

As an example there are results of speed assessment for calculations with Intel Core i7-4720 HQ given in Table 1.

Table 1. Execution time assessment for Intel Core i7-4720HQ CPU

Chunk frame width	Number of signal samples (N)	Sequential execution time	Parallel execution time
32768	524288	0.1801	0.0622
	1048576	0.3603	0.1241
	2097152	0.7413	0.2541
	4194304	1.4673	0.5057
65536	524288	0.1927	0.0643
	1048576	0.3875	0.1307
	2097152	0.7815	0.2586
	4194304	1.5604	0.5206
131072	524288	0.2089	0.0692
	1048576	0.4061	0.1382
	2097152	0.8245	0.2725
	4194304	1.6316	0.5195

Dependences of the acceleration parameter [8] of the sample size for different chunk size are given in Figures 1 and 2.

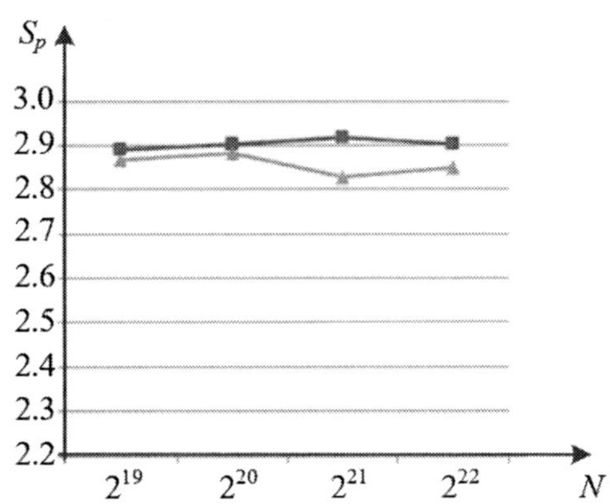

Figure 1. Dependence of acceleration on the sample size for the chunk frame size of 32768 (green – AMD FX-9590, red – Intel i7-4720 HQ).

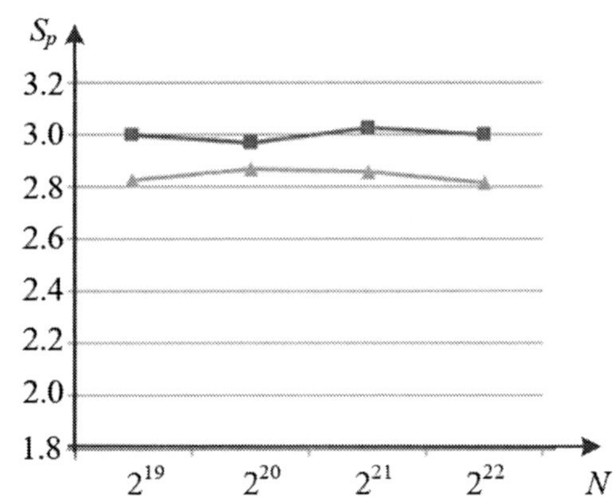

Figure 2. Dependence of acceleration on the sample size for the chunk frame size of 65536 (green – AMD FX-9590, red – Intel i7-4720 HQ).

Acceleration values are insignificantly reduced with the change of the initial sample, thus indicating good scalability of the software solution.

A study of load balancing was carried out during the calculations with the aim to determine a degree of load distribution. This study showed that the CPU utilization curve (when all cores are utilized) is almost symmetric with respect to the central value, thus indicating that there is a uniform distribution of load between the cores. However, from Figure 3 we may see, that only about 60% of the potential capability of the CPU is utilized, which we deem insufficient. Further studies with Intel

Parallel Studio debugger and IDA Pro disassembler helped to find the cause, which was sequential operation of the C++ memory allocation tools. Use of the Scalable_malloc function from the Intel TBB library allowed to increase performance by parallelizing memory allocation. At that, average CPU utilization reached 85%, which is reflected in Figure 4.

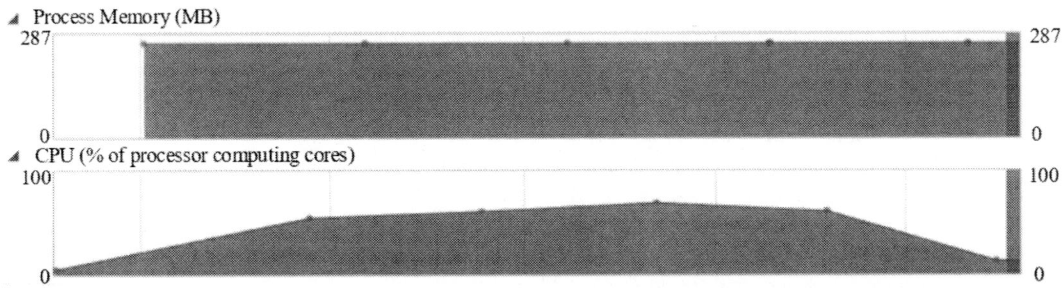

Figure 3. Memory usage and CPU utilization as functions of time, before optimization of memory allocation.

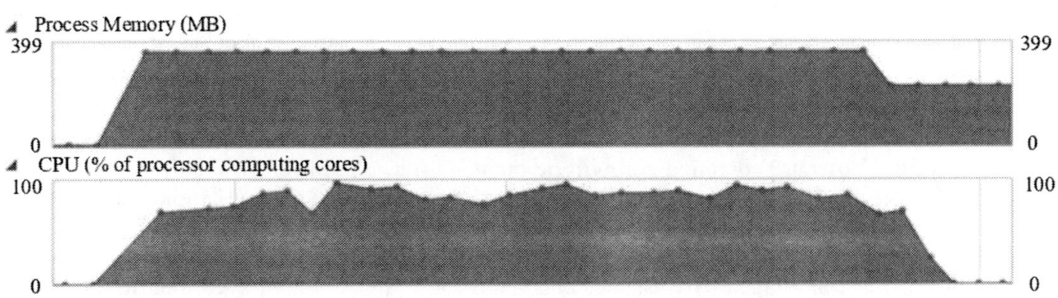

Figure 4. Memory usage and CPU utilization as functions of time, after optimization of memory allocation.

Thus, use of parallel memory allocator allowed to raise CPU utilization and by that increase performance.

For comparison indicators of both optimized and non-optimized versions of the coherence function estimation on the Intel Core i7-4720HQ CPU are given in Figures 5 and 6.

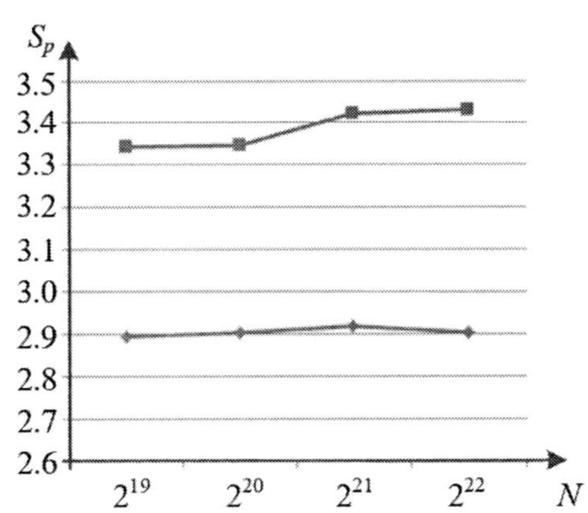

Figure 5. Acceleration as a function of sample size for the chunk frame size of 32768 (blue –

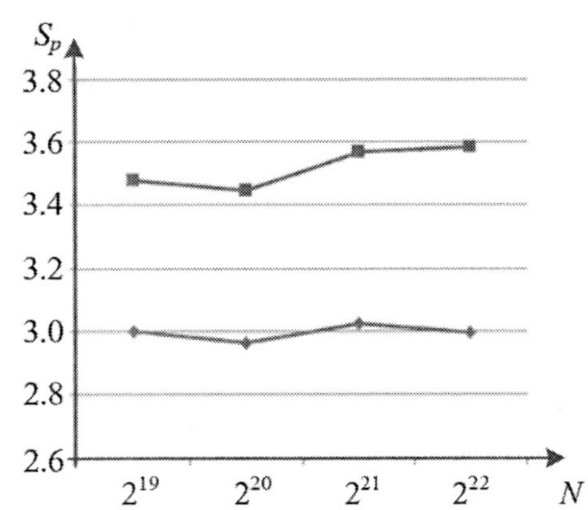

Figure 6. Acceleration as a function of sample size for the chunk frame size of 65536 (blue –

AMD FX-9590, red – Intel i7-4720 HQ). AMD FX-9590, red – Intel i7-4720 HQ).

5. Conclusion

This paper outlines a solution of a calculation optimization task for calculation of the coherence function on multi-core CPUs. This task is of great applied significance because the coherence function is widely used in solving different tasks of digital signal processing, for example in pipeline leak location with correlation acoustic method.

Introduction of parallelism is deemed the most approachable way to raise hardware resources utilization without significant additional costs. In this case a PC with a modern multi-core CPU serves as a computational platform.

Compiler, architecture and algorithmic optimization methods were employed to increase parallelizing performance during the coherence function estimation. The obtained data allowed for development of specialized software that provides automated selection of efficient division of the task into subtasks for parallel data processing as well as for selection between sequential and parallel methods of data treatment depending on the sample size.

The performed analysis of the coherence function parallelizing attests to efficiency of its software implementation. In particular, manual data prefetching into the cache line of the CPU significantly increases performance because there is no need for information exchange with the RAM during the calculations.

Further increase of performance may be attained by delegating the most resource-consuming part of the task, i.e., fast Fourier transform to graphics processors using such technologies as CUDA or OpenCL [11].

The software that has been developed during this study underwent state registration with the authorities of the Russian Federation and will be included into a hardware and software solution for rotating machinery fault diagnosis and pipeline leak location with acoustic correlation method.

Acknowledgements

The reported study was funded by Russian Foundation for Basic Research according to the research project No. 16-37-00049 mol_a.

Reference

[1] Blake G, Dreslinski R G and Mudge T 2009 *IEEE Signal Proc. Mag.* **26(6)** 26-37
[2] Herlihy M 2007 *Lect. Notes Comput. Sc.* **4855** 1-8
[3] Shukla S K, Murthy C N S and Chande P K 2015 *Adv. in Intelligent Syst. and Comp.* **1089** 537–545
[4] Kim H and Bond R 2009 *IEEE Signal Proc. Mag.* **26(6)** 1-8
[5] Carter C G 1987 *Proc. of IEEE* **75(2)** 236-255
[6] Ionel R, Ionel S and Ignea A 2010 *Proc. of Int. Joint Conf. on Computational Cybernetics and Technical Informatics (Timisoara)* (IEEE: USA) 5491234
[7] Avramchuk V S, Luneva E E and Cheremnov A G 2014 *Proc. of Int. Conf. on Mechanical Engineering, Automation and Control Systems* (IEEE: USA) 6986858
[8] Gupta A and Kumar V 1993 *IEEE Parall. Distr.* **4(8)** 922-932
[9] Avramchuka V V, Luneva E E and Cheremnov A G 2014 *Adv. Mat. Res.* **1040** 969-974
[10] Manikandan N and Subha S 2016 *Int. J. of Pharmacy and Techn.* **8** 3916-27
[11] Skirnevskiy I and Korovin A 2015 *Key Eng. Mat.* 685857-862

Electronic load as part of the test complex of the power processing unit of electric and plasma propulsion

S V Chubov and A I Soldatov

Tomsk Polytechnic University, 30, Lenina Ave., Tomsk, 634050, Russia

E-mail: Chubov@tpu.ru

Abstract. This article provides the advantages and technical solutions for the use of electronic loads as part of a testing complex of power and management systems of electric and plasma propulsion of three types. The paper shows the parameters that were applied to select the electronic loads and describes their functionality.

1. Introduction

The electric propulsion system based on the stationary plasma thruster (SPT) is currently widely used for the correction of the orbit of geostationary satellites. Apower processing unit (PPU) is one of the main propulsion units, significantly affecting its output parameters and characteristics. The quality of testing PPU during ground tests largely depends on the extent to which the actual load simulations correspond to the real SPT [1 - 10]. The simplest equivalent of the SPT supply chain is resistive load, but it has several drawbacks when used under high power. The electronic intelligent load is a modern alternative solution of the classical embodiment of resistive load. Further researches using electronic loads are conducted to create and improve the PPU performance.

2. Materials and methods

The load of PPU is a chain of electric power supply SPT shown in figure 1, where A – anode; HC, HC(p) – primary and backup heaters of cathodes; IE, IE(r) – primary and backup ignition electrodes; MC – the magnetic coil (MC) which may have a series connection with the anode circuit without using additional power supply MC; FR, FR (p) – main and reserve flow regulators. In addition, the PPU load deals with an engine valve (EV), which is a part of the propulsion system (not shown in the image).

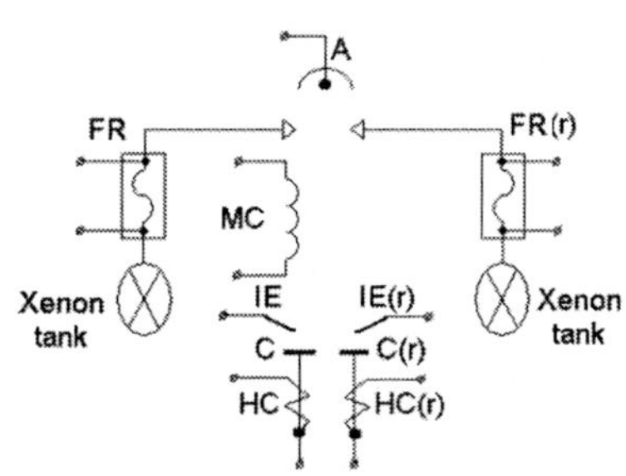

Figure 1. The structural schemeof the SPT power supplies circuit.

IOP Conf. Series: Materials Science and Engineering **177** (2017) 012021 doi:10.1088/1757-899X/177/1/012021

When testing PPU, the load is installed equivalently to the resistance circuits taking into account their inherent electrical parameters change during the operation within certain limits. Table 1 shows the electrical circuit of three different settings of engine types.

Table 1. The electrical parameters of the SPT power elements

Circuit	Parameter	Mode	Engine SPT-140D	Engine KM-75	Engine M-100
A	U_A, V	Idling	≤350	≤900	≤350
		Operating	300^{+15}_{-30}	810^{+25}_{-35}	300^{+15}_{-17}
	I_A, A	Pк	4.5±0.5	1.8±0.2	1.5±0.2
		Nominal	15±0.3	2.8±0.1	4.5±0.1
		STZA	22±1	3.7±0.3	7.0±0.7
	R_A, Om		11 – 79	193 – 522	36 – 243
	Type of loading (amount)		BK8522 (3 pcs.)	AKIP-1350 (1 pc.)	BK8522 (2 pcs.)
HC	I_{HC}, A	t=180 μs	17±0.5	8.5±0.25	12±0.3
	R_{HC},Om		0.37 – 1.0	0.55 – 1.55	0.35 –0.4
	Type of loading (amount)		BK8510 (1 pc.)	BK8500 (2 pcs.)	BK8500 (2 pcs.)
FR	I_{FR}, A	Standby	1.5 – 1.75	1.5±0.5	1.5±0.5 (3.6 – 4)
		Nominal	0 – 4	0 – 4	0 – 4
	R_{FR},Om		0.24 – 0.45	0.5 – 0.8	0.2 – 0.4
	Type of loading			Resistive	
MC	I_{MC}, A		5.25±0.25	(6–10)±0.25	–
	R_{MC}, Om		1.1 – 3.2	0.85 – 1.65	–
	Type of loading (amount)		BK8500 (1 pc.)	BK8500 (1 pc.)	–
IE	U_{IE}, V	Before ignition		320±30	
		After ignition		20–40	
	R_{IE}, Om	Before ignition	3±0.5	10±1	10000±1000
		After ignition	5±0.5	20±10	5±0.5
	Type of loading			Resistive	
EV	U_{EV} /I_{EV}	Switching on	25–34V	0.74±0.06 A	22.3–32.7
		Maintenance	10.5±2.5 V	0.26±0.3 A	8–13
	R_{EV}, Om		210±5	8.5±0.3	210±5
	Type of loading			Active-inductive	

Two variants of the load can be considered as the equivalent circuits of the real engine power: resistive and electronic loads.

Resistive load, but it has several drawbacks when used under high power:

- allocation of a large amount of heat (Figure 2), which requires the use of additional forced cooling units both as a part of test equipment and indoors;
- great weight and size parameters, which have a negative impact on thetransportation of the test set from a manufacturing organization to a customer organization;

- the inability to adjust the load resistance automatically;
- load resistance change due to the presence of the temperature coefficient;
- an inductive load component.

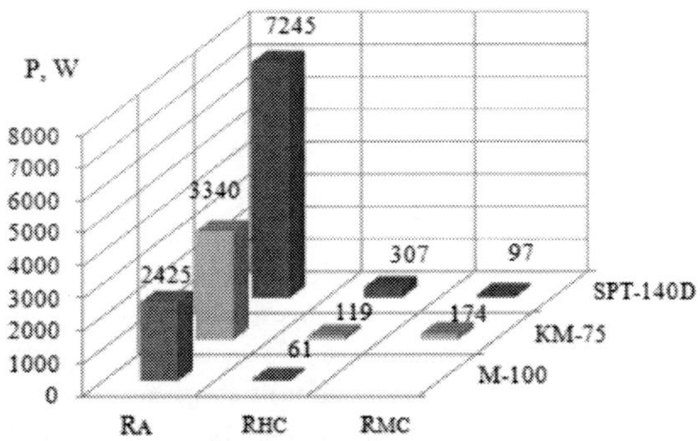

Figure 2. The load circuit power equivalent to various circuits engines (SPT-100D, KM-75, M-100).

Nowadays, there are many manufacturers, which produce electronic loads. The selection of these electronic loads to be used as a part of the PPU testing system is conducted according to the following specifications:

- the maintenance of required voltage, current, power, and a wide range of their control;
- the ability to control and manage the electrical parameters both via a digital display located on the front panel and via computer interfaces in order to integrate into an automated test equipment system;
- the high accuracy rates.

BK85hh (B&K Precision Corp.) and AKIP (PRODIGIT ELECTRONICS CO., LTD) machinery satisfy by the above-mentioned requirements. To solve the current issue, we can outline the most suitable models: BK-8500, BK-8510, BK-8522, AKIP-1350.For example, loads BK-8510 and BK-8522 (Figure 3) serve to simulate the engine SPT-140D elements on the test bench.

Figure 3. A test bench with electronic loads.

3. Results and discussion

Similar devices can be used as a complex and separately in various modes during the configuration, debugging, and testing of individual components and modules PPU. When checking the voltage sources, it is convenient to use the constant current mode (CC) whenthe setpoint current flows through the electronic load. This current is maintained when the value of the input voltage changes. In a constant voltage mode (CV), the power sources are tested to determine their ability to limit the current. A constant resistance mode (CR) is needed for testing the power sources when determining the maximum possible outputted current values.

The check of PPU modes on the stand with electronic loads is based on the SPT work principle. The first stage is a work preparation, which includes the following stages: a cathode choice, power supply to the cathode heating circuit within 180 seconds, and standby mode of the flow regulator is provided. The flow regulator regulates the working substance flow supplied to the anode path from its storage unit. The HC circuit imitates the electronic load in a constant resistance (CR) mode (Figure 4). The flow regulator (FR) electronic circuit is a resistive one. The HC circuit resistance varies within a predetermined range (due to the electronic load functionality, it may vary automatically). At this point, the stabilization of I_{HC} heating current is controlled. Then the anode voltage of idling U_A is introduced to the process. The anodic load operates in a measurement mode. Ignition voltage is supplied to the ignition electrode with respect to a selected cathode in the form of pulses of positive polarity. After performing these manipulations, a real engine should start, and on the stand the inclusionof the plate load in a constant resistance mode is an imitation of the engine start. The discharge current I_A will appear in the A chain (between the anode and the cathode). The change of e-load resistance achieves the discharge current set point (approximately 0.3 times greater than the nominal value). Then, the PPU gives a 'P_K' command, which disables the cathode heat and transfers the flow regulator into the control mode depending on the discharge current. When I_A rises over the nominal limit, the current flow regulator also increases to a maximum value. This leads to the discharge current decrease. When the discharge current falls below the nominal value, the process is reversed. That is, flow regulator current is reduced to zero, and the discharge current increases. A significant increase of the nominal discharge current value results in PPU swiching off the anode voltage and activating the overcurrent protection signal 'STZA'. This phenomenon is accompanied by abnormal processes in the engine.

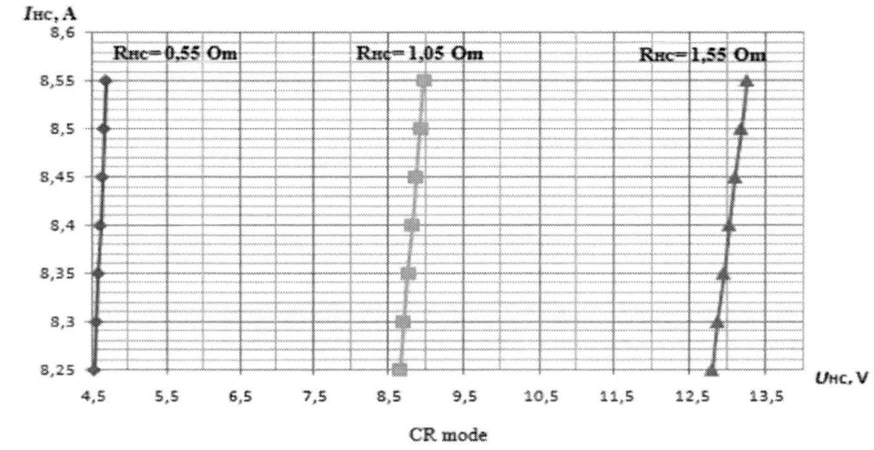

Figure4. TheHC circuit (Engine KM-75) imitatesthe electronic load in theoperation CR mode.

It is important to note the following advantages of electronic loads BK85hh (B&K Precision Corp.):

- the possibility of a remote load control via serial interfaces (RS-232, USB) (Figure 5) using the following steps: the selection of a constant voltage mode, constant current, constant power, and constant resistance imitation, as well as the load imitationvarying dynamically over time;

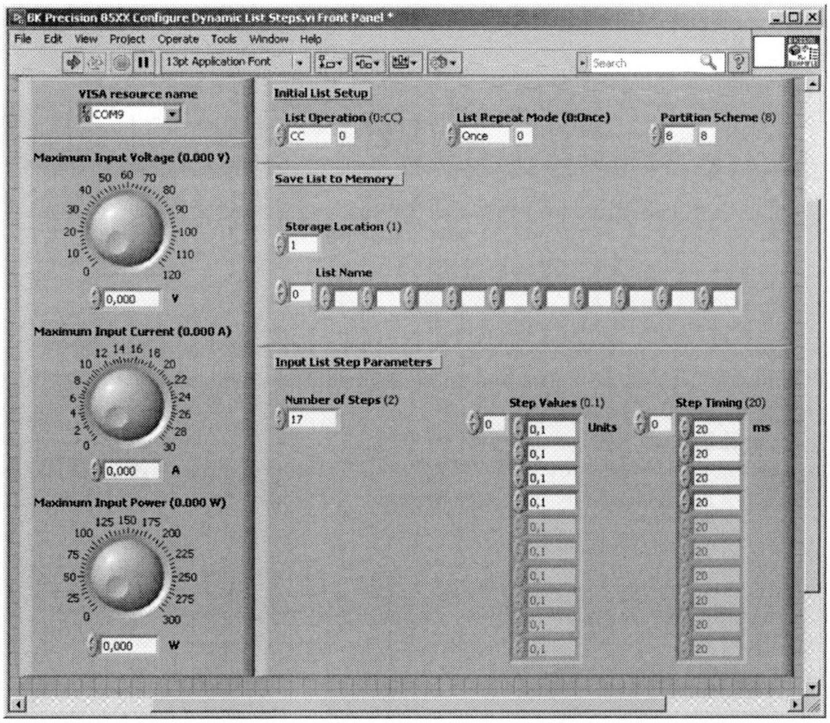

Figure 5. The software user's interface in a BK Precision 85xx LabView software.

- load shorting imitation;
- the connected source protection function which prevents excessive current levels, voltage, power, and reverse polarity;
- implementation of successive tests with multiple transitions states.

Electronic load AKIP -1350 has the following special features:

- the ability to operate up to 5 kW power at a maximum voltage of 1000
- simultaneous displaying of current, voltage, and power on a large LCD display
- five operating modes: constant voltage, constant current, constant resistance, constant power, a dynamic mode with a variable speed load growth (from 50ms to 10s).

When using the machinery of two manufacturers, one has the opportunity to install discrete input parameters.

4. Conclusion

When testing, the complex of electronic loads with a wide range of service functions, high accuracy performance, and serial interfaces allowthe qualitative solution of the problems related to debugging, testing and PPU testing in an automatic regime using computer control. Thus, we can save the results on the computer and the subsequent analysis of emerging faults, the possibility of long-term continuous testing without the operator. In addition, it is possible to save results and the formation of the test report, a configurable test scenario, dynamic load changes from a minimum to a maximum and vice versa at different speeds.

References

[1] Mikhailov M V and Moshnyakov A A 2015 *Proc. XIX Int. Scient. Conf. Reshetnev Readings (Krasnoyarsk)* Vol 1 (Krasnoyarsk: Siberian State Aerospace University Press) pp 165-167

[2] Chernyshev A I, Shinyakov Y A, Galaiko V N, Volkov M P, Gordeev K G and Goroshkov I N The energy supply method of power electric plasma propulsion and its implementation machinery. Patent RF, no. 2220322, 2001
[3] Katasonov N M Electric Plasma Propulsion Starting and Supply System. Patent RF, no. 2162623, 1999
[4] Mikhailov MV, Katasonov NM and Chernishev A I 1998 *Proc. of KORUS'98 the 2-nd Int. Symp.* Vol 1 (Tomsk: TPU press) 528 p
[5] Mikhailov MV, Katasonov N M and Chernishev A I 1999 *Proc. of the 3-rd Int. Symp. on Science and Technology (Novosibirsk)* Vol 2 (Novosibirsk: Novosibirsk State Technical University Press) 771 p
[6] Katasonov N M Electric plasma propulsion simulator Patent RF No. 2395716, 2009
[7] Mikhailov M V and Moshnyakov A A 2016 *Web of Conferences* 48 01007
[8] Lazurenko A, Vial V, Bouchoule A, Skrylnikov A, Kozlov V and Kim V 2006 *J. of Propulsion and Power* 22(1) 38-47
[9] Mukund R P 2005 Spacecraft power systems (Florida: CRC Press)
[10] Lesnevskii V A, Mikhailov M V, Hodnenko V P and Khromov A V 2012 *Electromechanical matters VNIEM Studies* Vol 126 (Moscow: JC VNIEM Corporation) pp 25-28

Computer-assisted design of flux-cored wires

Yu N Dubtsov, I V Zorin, G N Sokolov, A A Antonov, A A Artem'ev and V I Lysak

Volgograd State Technical University, 28, Lenina ave., Volgograd, 400005, Russia

E-mail: dubcow@mail.ru

Abstract. The algorithm and description of the AlMe-WireLaB software for the computer-assisted design of flux-cored wires are introduced. The software functionality is illustrated with the selection of the components for the flux-cored wire, ensuring the acquisition of the deposited metal of the Fe-Cr-C-Mo-Ni-Ti-B system. It is demonstrated that the developed software enables the technologically reliable flux-cored wire to be designed for surfacing, resulting in a metal of an ordered composition.

1. Introduction

Due to their high workability in surfacing processes, flux-cored wires are the parent material when it is required to surface the alloy with special properties [1-3]. The Fe-Cr-C-Mo-Ni-Ti-B alloying system is of interest as the basis for the development of abrasion-resistant alloys with a thermoresistant structure at temperatures of up to 550 °C [4, 5]. Yet such a structure can be acquired with optimal boron and carbon alloying on the one hand responsible for solid phase formation, and, on the other hand, for exceessive increases in the metal's brittleness [6, 7].

To obtain this optimum, the components' content and the filling factor of the flux-cored wire are required to be recalculated many times. To automatise these operations, special software is required to account for specific mass (g/mm^3) of an element, its content in the metal band or powder and the alloy transfer effervescive. Thus, our objective was to develop the software enabling flux-cored wire design to acquire the ordered composition of the deposited metal with the best working properties.

2. Materials and methods

With the the developed AlMe-WireLaB software, the flux-cored wire was designed to ensure the following deposited metal composition, % wt.: 3–3.2 C; 12.5–13.0 Cr; 1.8–2.1 Mo; 1.0–1.4 Ni; 1.2–1.5 Ti; 0.4–0.5 B; and Fe – the remainder. The coating of the flux-cored wire of 3.0 mm in diameter was formed from steel strap Sv-08kp, which is 12 mm wide and 0.5 mm thick [8]. The core of the designed flux-cored wire consisted of aluminothermic chrome, ferromolybdenum, sodium silicofluoride, commercial nickel and silver graphite powders as well as the titanium diboride powder.

Arc surfacing on the steel of grade 20 with the designed flux-cored wire was performed in the shielding gas (argon) with direct current of the reversed polarity. The weld fabrication characteristics of the wires were controled by recording oscillograms of the welding current and arc voltage during the reference samples' surfacing with the computer-assisted control method and the analog-to-digital converter 'LA-20USB'. The chemical composition of the deposited metal was specified with the spark optical emmission spectrometer 'PMI MASTER Pro'. The deposited metal structure was studied with electron microscopy (scanning dual beam electron-ion microscope FEI Versa 3D).

3. Flux-cored wire calculation methods

The aim of these calculation methods was to specify the flux-cored wire composition that, at a certain filling factor of the wire, will ensure the ordered composition of the deposited metal. To calculate the composition of the flux-cored wire core, the following data are required:

- the composition of the deposited metal;
- the rated chemical composition of the wire specified as a function of the deposited metal composition, adjusted for alloy oxidation during surfacing and for the base metal share in the deposited metal;
- the percentage ratio of the core and coating masses in the flux-cored wire (filling factor);
- the chemical composition of the materials used to produce the wire (viz. metal band, metal powders, ferroalloys, minerals, and other components).

The calculation is performed on the basis of required wire mass M_{fcw} (kg) and diameter D_{fcw} (mm). To attain the required content of the elements in the deposited metal, a somewhat larger amount of alloying powders should be added to the wire content in accordance with their alloy transfer effervescives.

The alloy transfer effervescives from the wire into the deposited metal accounting for the losses in the weld fire point due to oxidation, vaporation, transition into the slag etc., are chosen with an approximation, using known data (Table 1); then they are experimentally ascertained and corrected [9].

Table 1. Experimental values of alloy transfer effervescives from the wire into the deposited metal during arc welding with a consumable electrode in the shielding gas (argon)

	Alloying elements				
	C	Cr	Mo	Ti	Ni
Alloy transfer effervescive	0.86	0.76	0.7	0.3	0.7

When calculating, it is necessary to consider that the alloy transfer effervescives depend on the surfacing procedure and mode, mostly on the welding current rate and arc voltage, the component's content in the wire, the presence of the deoxidizing agents, powders' granulation in the core of the flux-cored wire, and the alloying system of the deposited metal [10].

The flux-cored wire content is calculated for 100 g of the wire. The required amount of the component via which the alloy is added can be calculated as:

$$p = \frac{[X]_e \cdot 100}{b}, \tag{1}$$

where p is the mass of the flux-cored wire core component, g; $[X]_e$ is the required amount of the calculated alloy in the wire core, g; and b is the content of the considered alloy in the added component, % wt.

As long as accompanying alloys can be added to any core component except for pure powders of metals, graphite and chemical compounds, their amount is accounted for and calculated as:

$$d = \frac{P \cdot c}{100}, \tag{2}$$

where d is the accompanying alloy mass, g; c is the content of the accompanying alloy in the added component, % wt; and P is the mass of the added component, g.

It should be borne in mind that the accompanying alloys are also added to the metal band. If the total amount of impurities exceeds the prohibitive limit, the material adding excessive impurities should be substituted by the component with their minimal content or by the pure metal powder.

Next, the preliminary calculation of the filling factor of the flux-cored wire (K_p^r) is made:

$$K_p^r = \sum_{i=1}^{n} m_{kom}, \quad (3)$$

where m_{kom} is the mass fraction of all the core components with an allowance for the accompanying impurities.

The geometrical parameters (i.e. thickness and width) of the band for the coating of the experimental flux-cored wire are chosen according to the data in Table 2.

Next, the powder mixture is prepared (50–100 g) and the wire is produced by 6x or 4x drawing-down. Then, a 100 mm piece is cut from the middle part of the wire and weighed with at least a three-place balance, and then the core is removed and the coating mass is specified.

Table 2. Filling factors of the flux-cored wire provided by different bands

Wire diameter, mm	3.6			3.0			2.5		
Band width, mm	15			12			10		
Band thickness, mm	0.8	0.5	0.4	0.8	0.5	0.4	0.8	0.5	0.4
Filling factor, K_3	0.28–0.3	0.38–0.4	0.52–0.55	0.22–0.25	0.3–0.33	0.4–0.42	0.18–0.2	0.28–0.3	0.33–0.38

Next, core mass M_s^e is calculated by formula 3.15:

$$M_s^e = M_p^e - M_o^e, \quad (4)$$

where M_p^e is the mass of the flux-cored wire piece; and M_o^e is the mass of the flux-cored wire coating.

With the values obtained, the actual filling factor (K_p) is then calculated:

$$K_p = \frac{M_s^e}{M_p^e} \quad (5)$$

If $K_p > K_p^r$, then they are evened out with neutral additives (for ferrous alloys – iron powder). To eliminate cavitation, 1.5–2 % of sodium silicofluoride ($Na\text{-}SiF_6$) is added to the mixture consisting mainly of ferrous alloys [11].

If $K_p < K_p^r$, the calculation is refined. For this purpose, first, the specific mass of the cogged core (ρ_s) is calculated:

$$\rho_s = \frac{M_s^e}{\frac{\pi}{4}(d-2s)^2 \cdot L}, \quad (6)$$

where d is the FCW diameter; s is the thickness of the flux-cored wire coating; and L is the length of the flux-cored wire piece (i.e. 100 mm).

With known ρ_s, it is possible to calculate precisely the coating thickness (s) required to provide for the needed filling factor (K_p^r):

$$s = \frac{d}{2}\left(1 - \frac{\sqrt{K_p}}{K_p}\right), \tag{7}$$

where factor K_r, dependent on the physical properties of the flux-cored wire components, is calculated as:

$$K_r = \frac{\rho_s}{\rho_o}\left(\frac{1}{K_p^r} - 1\right) + 1, \tag{8}$$

where ρ_o is the specific mass of the coating metal.

The coating thickness (s) should correspond to the nomenclature value of the manufactured metal bands. For this purpose, value s is rounded upwards; this calculation refines filling factor K_p^r:

$$K_p^r = \frac{\frac{\rho_o}{\rho_s}}{\frac{d^2}{4(d \cdot s - s^2)} - 1} + 1 \tag{9}$$

The band width (h) is calculated as:

$$h = 1.06 - \pi(D - s), \tag{10}$$

where D is the drawing die diameter.

4. Description of the AlMe-WireLaB software

Development of the flux-cored wire ensuring the required chemical composition of the deposited metal requires reiterated calculations of the components' content and geometrical parameters of the wire. To automate this calculation, the AlMe-WireLaB software was developed. Its algorithm is introduced in Figure 1.

The main interface window of the program shows the saved calculations list – block 1, which can be reviewed, edited, printed – block 7, or deleted. The design of the flux-cored wire starts with the button 'add'. In the opened window, one can specify:

- name (at least four and no more than 200 characters);
- calculation type – 'flux-cored wire';
- flux-cored wire diameter in mm;
- flux-cored wire mass;
- mass units (kg or g);
- surfacing method.

To work with the created calculation, one should select it from the list. To enter the required composition of the deposited metal, select the required tab in the 'Database menu':

- 'Elements database' – contains the name and the specific mass (g/mm^3) of the element from the periodic system;
- 'Material database' –contains the compositions of the components used to produce the flux-cored wire.

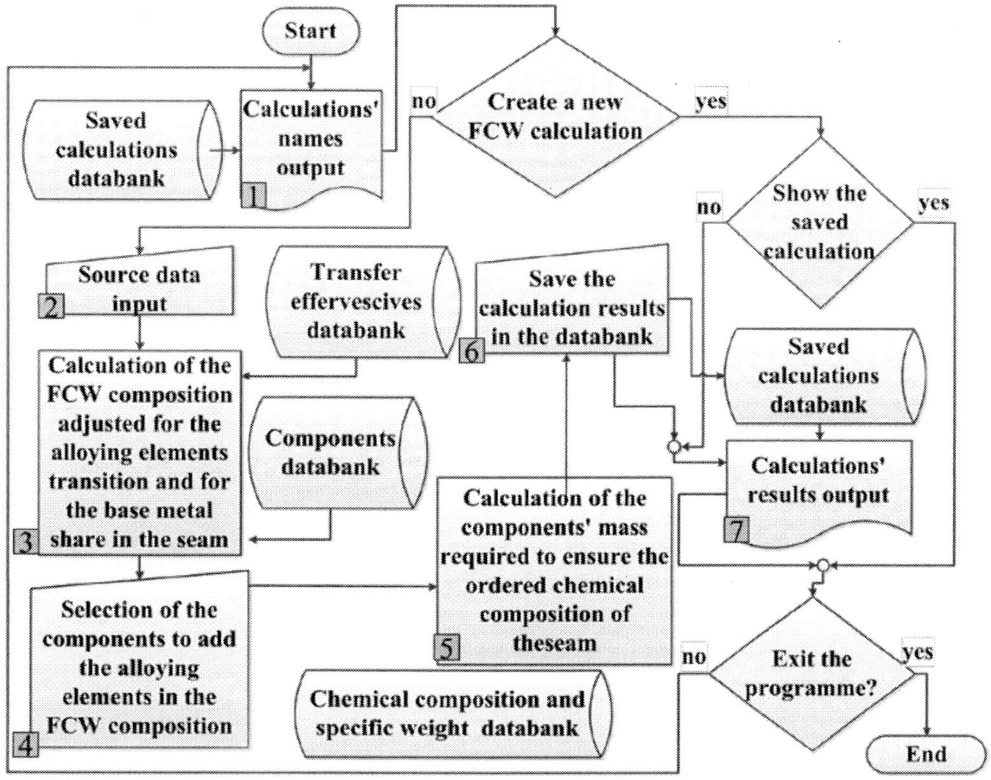

Figure 1. The control-flow chart for the operation of the software for flux-cored wire design.

If we know only the elemental composition, let us select the first database; if we know the mass fraction of the component added into the flux-cored wire, let us select the second database. To add a component to the window 'source data' – (algorithm block 2), one should point the cursor to its name and click the right mouse button – the line will be highlighted. Then the form 'material parameters' should be filled in with the component's mass fraction and the button 'add component' should be pressed. If a component should be deleted from the table 'source data', one should select the line with its name and press the button 'delete' above. When the sorce data list is formed, press the button 'introduce $k_п$'. In block 3, the composition is recalculated taking into account transition coefficients of the alloying elements, which values are taken from the software database (Figure 2, a).

The flux-cored wire composition is configured by choosing source materials as per GOST standards (algorithm block 4).

For this purpose, in the tab 'source data', let us select the line with the alloy name and press the button 'suggest' (Figure 2, a). In the opened tab, we choose the mass fractions of the impurities that will be added to the wire together with the component. After the component is chosen from the list, we should click on the line 'total' of this component and press the button 'add' (Figure 2, b, c). After this, the selected component will replace the alloy in the source data table, and the flux-cored wire composition will be recalculated with provision for the impurities occuring in the wire together with the component. To specify the composition of the band from which the flux-cored wire coating will be formed, we should select it from the list 'material database' and press the button 'add component'; the mass fraction, however, is not specified. After all, the alloys are replaced with the components and the coating material specification, we press the button 'calculate'. Next, in block 5, the automated calculation is performed. The program opens the tab 'calculated', where the result of the performed calculation (algorithm block 6) is shown.

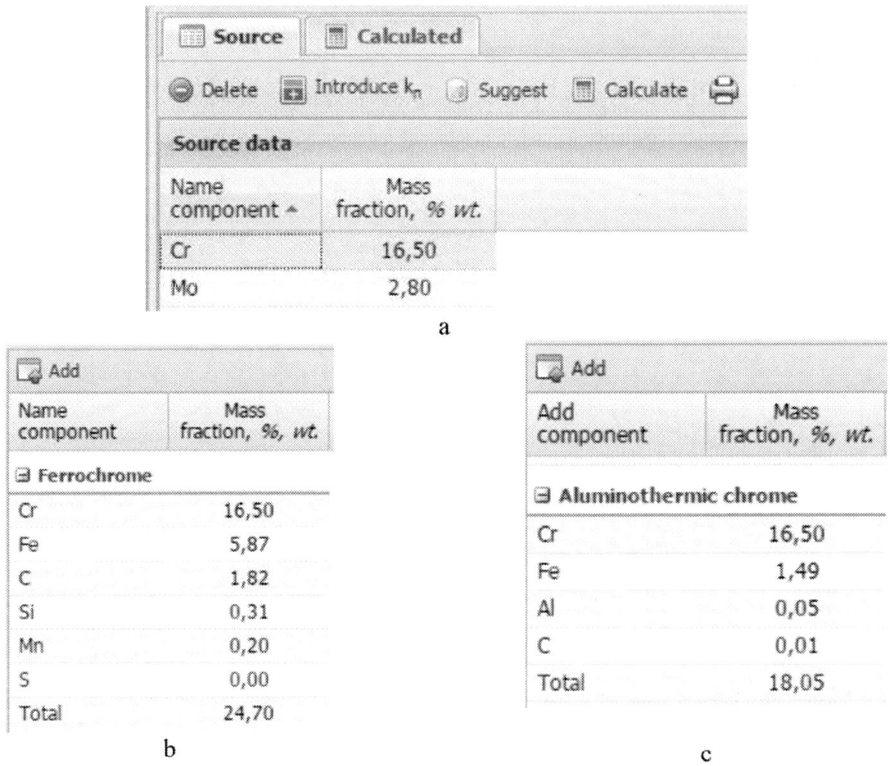

Figure 2. Alloy selection (a) and variants of components: ferrochrome (b) and aluminothermic chrome (c) with which it can be added to the flux-cored wire.

To compare the component composition of the calculated flux-cored wire with the source, we press the button 'chemical composition'. The opened tab (Figure 3) shows the calculated and the ordered chemical composition and their divergencies from each other. The chemical elements specified in the source data are highlighted in green to separate them from the impurities.

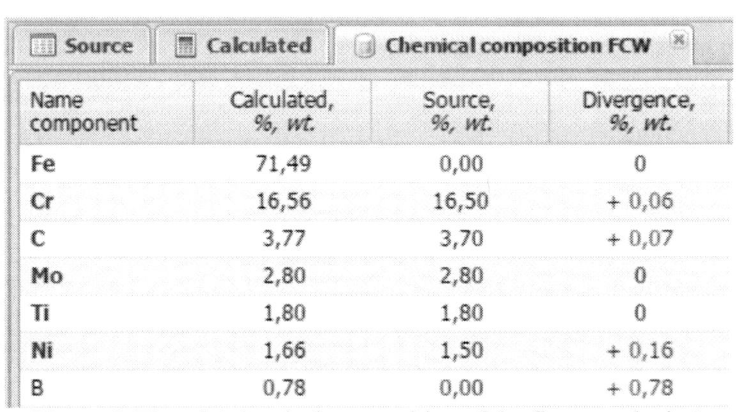

Figure 3. The tab 'chemical composition of the flux-cored wire'.

To reduce the impurity content in the flux-core wire, it is required to replace the components excessively adding them with a component with minimal impurity content or to add chemically pure metal powder. To this end, this component is visually highlighted in the tab 'calculated'; then we pass to the tab 'source', we highlight the line in the table with the component and press the button 'delete'. Next, with the previously described algorithm of replacement, for example, 'Cr' — with aluminothermic chrome (Figure 2, a, c), and the remaining carbon – with 'Silver graphite POL-2,' we press the button 'calculate'. Now, there should be no divergencies from the ordered composition (Figure 4). However, there may still be impurities (for example, Mn), added by the band. To reduce them, if it is possible, the iron powder is added to the filler, and the band thickness is reduced.

Source | Calculated | Chemical composition FCW

Name component	Calculated, %, wt.	Source, %, wt.	Divergence, %, wt.
Cr	16,50	16,50	0
Fe	72,22	0,00	0
Mo	2,80	2,80	0
Ni	1,50	1,50	0
Ti	1,80	1,80	0
C	3,70	3,70	0
B	0,78	0,00	+ 0,78
Mn	0,34	0,00	+ 0,34

Figure 4. The tab 'chemical composition of the flux-cored wire after conversions'.

Next, according to the data in Table 2, the geometric parameters (i.e. thickness and width) of the band are chosen to produce the coating of the experimental flux-cored wire, the powder mixture is prepared (50–100 g), and the flux-cored wire is produced by 6x or 4x drawing-down. Then, with Equation (6), value ρ_c is specified and entered in the program via form q, opened with the button 'value ρ_c'. The calculation results are shown in the tab 'source' (Figure 5) in line with the ordered band parameters. To print out the calculation results in a convenient form, we press the respective button – block 7.

Source data

Name component	Position	Form	Mass fraction, %, wt.	Diameter, mm	Width, mm	Thickness, mm
Steel strap Sv-08kp (GOST 9045-93)	outer shell	strap	65,00	3	12	0,5
Aluminothermic chrome	inside	powder	17,98	-	-	-
Iron powder	inside	powder	4,76	-	-	-
Silver graphite powder	inside	powder	3,61	-	-	-
Titanium diboride powder	inside	powder	2,66	-	-	-
Ferromolybdenum	inside	powder	4,64	-	-	-
Electrolytic nickel	inside	powder	1,34	-	-	-

Figure 5. The calculation results.

The administration mode enables one to extend and to edit the database of materials and their properties as well as that of the alloy transfer effervescives.

5. Confidence Rating

According to the results of the calculation, 10 m of the flux-cored wire were produced. The filling factor estimation performed for several sections of the produced wire showed good convergence (i.e. no more than a 5% error) of the empirical and calculated values.

The acquired oscillograms of the welding current and arc voltage during surfacing indicate a stable drop transfer with a drop detachment frequency of 0.126 s, which means uniform melting of the flux-cored wire components.

The deposited metal structure (Figure 6) represents a carboboride eutectic composed of a solid solution of iron, iron-chrome carboborides, and titanium molybdenum carbides.

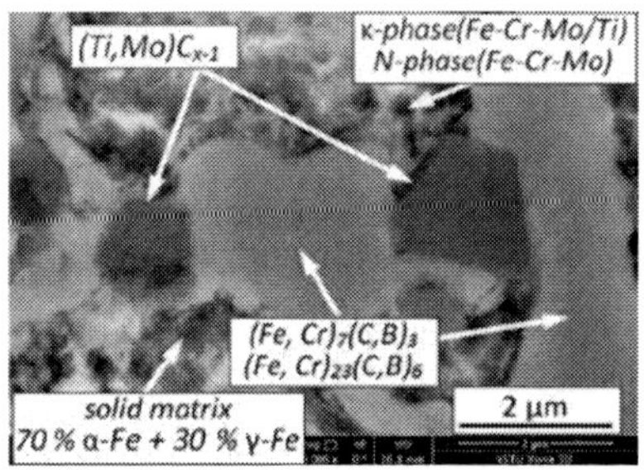

a

$(Fe, Cr)_7(C, B)_3$ $(Fe, Cr)_{23}(C, B)_6$	
Element	Wt. %
C	6.2
Mo	1.67
Ti	0.59
Cr	28.57
Mn	1.73
Ni	0.28
Fe	the remainder

solid matrix	
Element	Wt. %
C	2.86
Mo	0.26
Ti	0.46
Cr	6.61
Mn	1.14
Ni	1.38
Fe	the remainder

b

Figure 6. Deposited metal microstructure (a) and chemical composition (b) in the segments of solid solutions and carbides.

The spark optical emission spectrometer proved the ordered composition of the deposited metal.

6. Conclusions

Due to the introduction of the alloy transfer effervescives and the specific mass of the cogged core, the suggested method enables one to select, with maximum precision, the width and thickness of the band for the flux-cored wire coating in order to ensure the required filling factor.

The developed AlMe-WireLaB software, containing the wide appendable database of the alloy transfer effervescives of the core powder components, as well as a simple interface, facilitates the flux-cored wire design methods.

The functionality and reliability of the calculation methods and the software were proven by the fact that the arc surfacing with the designed flux-cored wire resulted in the formation of the deposited metal of the Fe-Cr-C-Mo-Ni-Ti-B system with the ordered composition.

Acknowledgments

The reported study was funded by the Russian Foundation of Fundamental Research (RFBR 16-08-01276_a; 16-38-00764_mol_a) and a grant from the President of the Russian Federation, No. MK-4713.2016.8.

References

[1] Berns H, Fischer A 1986 *Wear* **112** 163–180
[2] Gualco A, Marini C, Svoboda H, Surian E 2015 *Procedia Mat. Sci.* **8** 934–943
[3] Coronado J J, Holman C F, Go′mez A L 2009 *Tribol. Int.* **42** 745–749
[4] Badisch E, Katsich C, Winkelmann H, Franek F, Roy M 2010 *Tribol. Int.* **43** 1234–1244
[5] Ke Yang, Shengfu Yu, Yingbin Li, Chenglin Li 2008 *Appl. Surf. Sci.* **254** 5023–5027
[6] Grinberg N A, Mongait I A 1978 *Met. Sci. Heat Treat.* **20(12)** 1016–1019
[7] Eremin E N, Losev A S 2016 *Weld. Int.* **30(1)** 48-54
[8] GOST RF 9045-93, Steel strap of the type Sv-08kp, Moscow, 1997
[9] Dubtsov Yu N, Zorin I V, Sokolov G N, Artem'ev A A and Lysak V I 2015 *Key Engineer. Mater.* **685** 495–499
[10] Pokhodnya I K, Suptel A M, Shlepakov V N 1972 *Welding flux cored wire* (Kiev: Naukova dumka)
[11] Sokolov G N 2001 *Weld. Int.* **15(4)** 309–311

The comparative analysis of different computations methods of strength of materials by the example of calculations of the axle beam

S I Evtushenko, I A Petrov, M N Shutova and A S Alekseeva

Platov State Polytechnic University (NPI), 132, Prosvescheniya str., Novocherkassk, 346428, Russia

E-mail: evtushenko_s@novoch.ru

Abstract. The paper presents data of calculation the main characteristics of resilience by different ways. The basic data for the article were the calculation of the guiding axle beam of the vehicle. The calculation was performed by the analytic method and it was necessary to re-check strength of materials by any other method for reliability of the carried-out work. The finite element method was chosen as the competing option.

1. Introduction

The main goal of research is evaluation of using a numerical method (finite element method) compared with the analytic method of stress strain analysis of the guiding axle beam.

As a rule, to produce new parts and to improve old parts, before starting batch production, their load capacity ratings are carried out. Mechanical parts' main characteristics, which identify the suitability or unsuitability of constructions, are yield factor and fatigue safety factor.

At first, the analytic method was calculated and then the finite element method was used for the comparative analysis.

It should be noted that the dual motion mode was produced for this construction:

1. Operating mode;
2. Transport mode.

Efforts and safety margin were determined for four cases:

1. Vehicle skidding (maximum transverse force, zero longitudinal force);
2. Motion on barriers (maximum vertical force, zero transverse and longitudinal forces);
3. Hitting on obstruction (maximum longitudinal force);
4. Motion excluding dynamics (vertical force is vehicle weight, longitudinal force is the rolling resistance).

Both motion modes were calculated for each case, i.e. there are 8 calculations.

There is a diagram of the axle in Figure 1. Gravity load is applied at points A and B depending on the mode and loading case. In general, the load is presented by concentrated force of X, Y and Z directions. The stress was calculated in main sections A-A, B-B, C-C and D-D. It was found in the course of the solution that the minimal yield factor was in section D-D, which is shown in Figure 2.

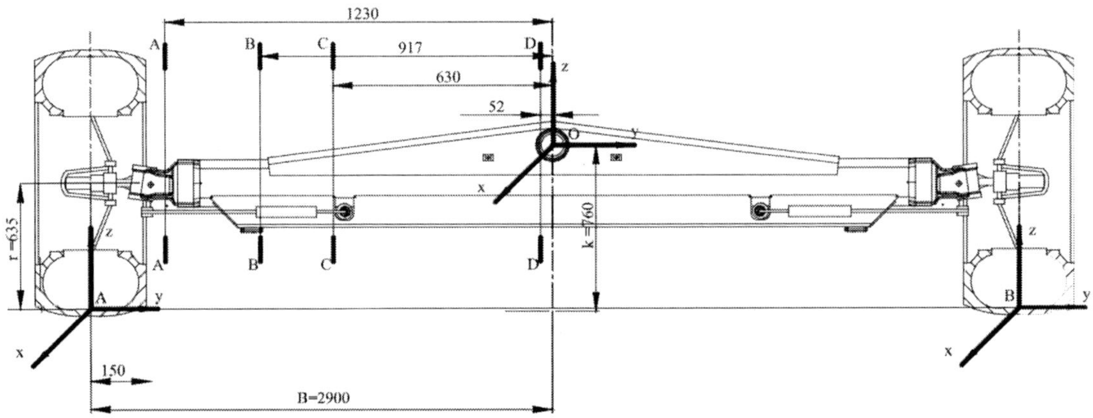

Figure 1. The analytical model and the sections' layout.

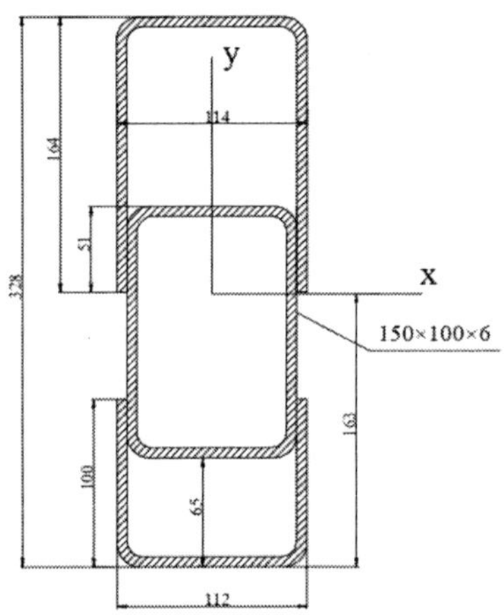

Figure 2. Beam cross section D-D.

2. The analytical method

The analytical method means that point moment equations are recorded, moments in horizontal and vertical planes are calculated, horizontal and vertical forces and von Mises equivalent stress are estimated for each section, and relies on this data the yield factor can be found:

$$n_y = \frac{\sigma_y}{\sigma} \geq [n], \tag{1}$$

σ_y – yield stress of the material, which is 295 MPa for steel 20,

$[n]$ – minimal yield factor, which is 1.5.

As the Table shows, the factor of safety is lower than permissible values (which is 1.5) when the vehicle is running over barriers or insurmountable obstacle.

Fatigue safety factor is determined by the following formula:

$$n = \frac{\sigma_{-1}}{k_\sigma \cdot \sigma_a + \psi_\sigma \cdot \sigma_m} \leq 2.0, \tag{2}$$

where $\sigma_{-1} \approx 0.4 \cdot \sigma_v$ = 168 MPa; σ_v = 420 MPa (steel 20); k_σ – fatigue notch factor;

$\sigma_a = \sigma(k_D - 1)$ – the amplitude of stress; ψ_σ – stress ratio, 0.1; k_D – dynamic factor, 1.2.

The results of calculations are shown in Table 1.

Table 1. The factor of safety against yielding of the axle beam.

Beam sections	Motion mode	Load case			
		1	2	3	4
A-A	Operating mode	1.69	1.73	1.13	2.70
	Transport mode	2.07	1.98	1.22	3.66
B-B	Operating mode	2.16	1.34	0.82	2.06
	Transport mode	2.78	1.53	0.88	2.82
C-C	Operating mode	2.22	1.16	0.71	1.78
	Transport mode	1.90	1.73	0.75	2.43
D-D	Operating mode	2.12	1.10	0.60	1.97
	Transport mode	2.70	1.65	0.64	2.27

The result of calculation for section C-C is:

$$n = \frac{168}{1 \cdot 27.5 + 0.1 \cdot 149.5} = 4.07.$$

The result of calculation for section D-D is:

$$n = \frac{168}{1.5 \cdot 29.85 + 0.1 \cdot 149.24} = 2.81.$$

3. Finite element method

The computation is also made be means of finite element analysis using appropriate software. The finite-element model consisting of 125400 is developed for the calculation. The model is shown in Figure 3. The actual loads are transmitted through the concentrated forces, bending moments and surface pressure.

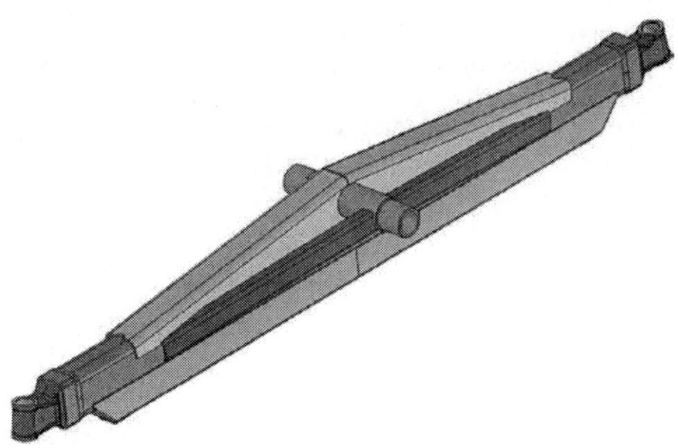

Figure 3. The finite-element model of the guiding axle beam.

Workloads obtained from the analytic method are applied to the same points. The result of calculation is Von Mises equivalent stress pattern on the surface of the construction and inside it.

For example, the results of some modes are shown in Figures 4 and 5.

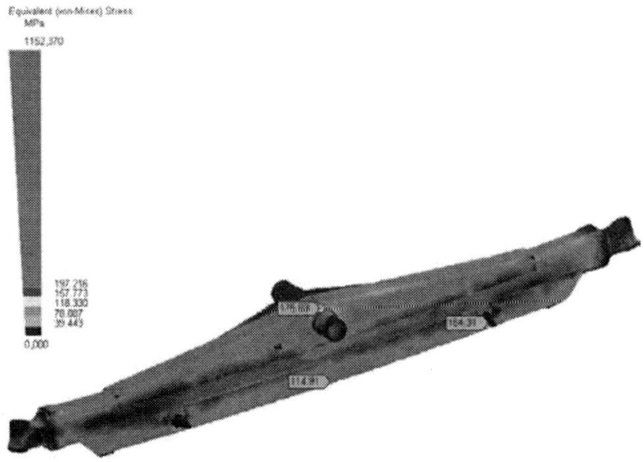

Figure 4. Von Mises equivalent stresses of the transport mode in case of motion on barriers, MPa

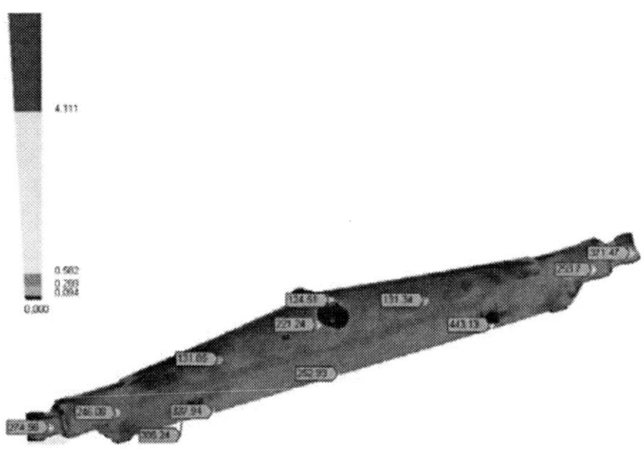

Figure 5. Von Mises equivalent stresses of the operating mode in case of hitting the insurmountable obstacle, MPa.

As we can see from figures, there are isolated local areas with stresses exceeding permissible values. The results of the finite element analysis are shown in Table 2. Furthermore, maximum values of the stresses are set for each section. For example, the stress for section C-C is 164.31 MPa, the stress for section D-D is 175.84 MPa.

Table 2. The factor of safety against yielding

Beam sections	Motion mode	Load case			
		1	2	3	4
A-A	Operating mode	1.75	1.76	**1.08**	2.75
	Transport mode	2.21	1.96	1.24	3.60
B-B	Operating mode	2.23	1.32	**0.85**	2.12
	Transport mode	2.77	**1.38**	0.87	2.85
C-C	Operating mode	2.24	1.24	**0.67**	1.85
	Transport mode	1.94	**1.79**	0.80	2.48
D-D	Operating mode	2.14	1.13	**0.65**	2.02
	Transport mode	2.74	**1.68**	0.68	2.32

As Table 2 shows, the difference in the magnitude of the factor of safety against yielding is slight. In general, miscalculation of n by different ways can be written as:

$$\Delta n = \frac{\sum\left(n_{sek} - n_{f.e.}\right)}{\sum n_{sek}} \quad (3)$$

Where n_{sek} – yield factor, calculated by the analytic method; $n_{f.e.}$ – yield factor, calculated by the finite element method; $\Delta n = 0.041$ or 4.1%, which is lower than permissible 5%.

Yield factor is calculated using the same formula (2), but using other way of finding the components.

The computation is made for the longest work mode, i.e. for the motion excluding dynamics. Meanwhile, the maximum principal stresses are measured in the operating mode and in the transport mode at the same point (at 12 mm from the section D-D).

The following components of the formula (2) are changed:

- $n = \frac{\sigma_{-1}}{k_\sigma \cdot \sigma_a + \psi_\sigma \cdot \sigma_m} \le 2.0$;
- $\sigma_a = \frac{\sigma_{op} - \sigma_{tr}}{2}$ – the stress amplitude;
- σ_{op} – stresses of the operating mode, Figure 6;
- σ_{tr} – stresses of the transport mode, Figure 7.
- Mean values of stress amplitude, $\sigma_m = \frac{\sigma_{op} - \sigma_{tr}}{2}$.
- ψ_σ – the stress ratio, which is 0.3 for FEA.

$$n = \frac{168}{1 \cdot 11.7 + 0.3 \cdot 116.12} = 3.20.$$

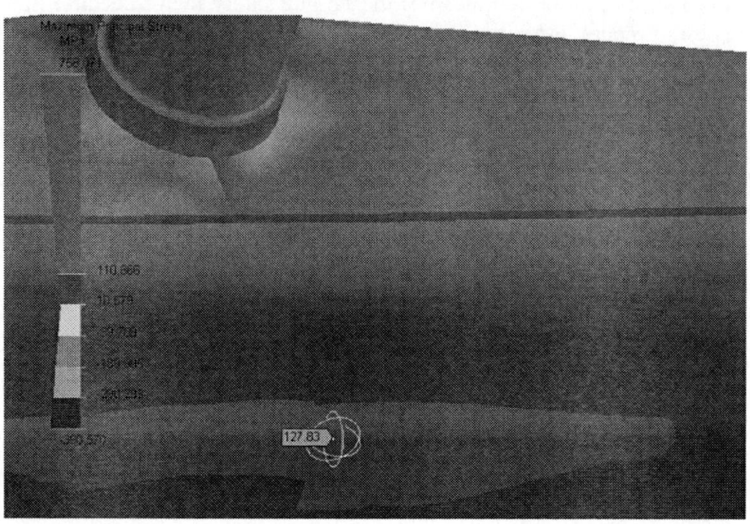

Figure 6. Maximum principal stresses in the operating mode with the motion excluding dynamics.

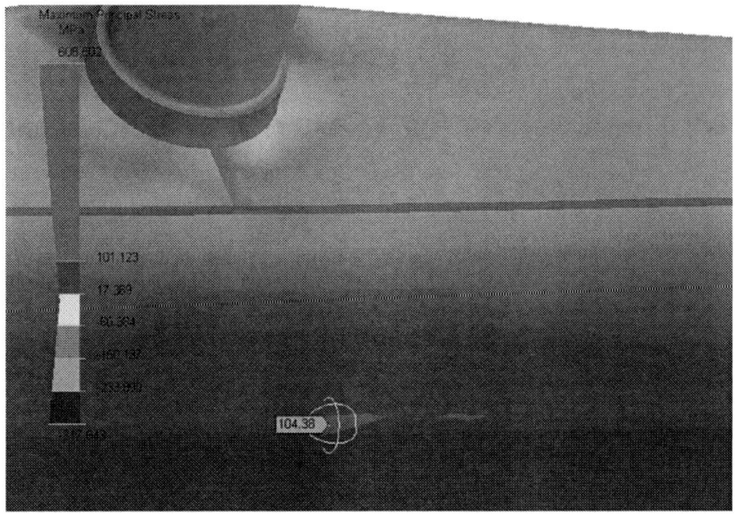

Figure 7. Maximum principal stresses in the transport mode with the motion excluding dynamics.

4. Conclusion

Consequently, the quintessence of researching is captured in the following key observations:

The difference between the values of the factors is insignificant, which indicates the correctness of calculations with the use of both methods.

Finite-element analysis is more vivid and it allows one:

- to perform a visual check of locations and sizes of the areas with critical stresses;
- to find stress raisers;
- to make a 3D model;
- to save time of calculations.

The finite element gives a close definition of the fatigue safety factor because this method uses the real loads for the design model instead of using the reference factor, which can be changed depending on the engineer's decision.

References

[1] Evtushenko S I, Shutova M N, Shkurakov L V, Petrov I A and Fedorchuk V E 2008 *The expertise of strength calculation of the guiding axle beam* Int. Conf. on Engineering (Rostov-on-Don: Rostov State University of Civil Engineering) pp 56 – 58

[2] Perelmuter A V and Slivker V I 2011 *Design models of structures and possibility of their analysis* vol 4 (Moscow: SCAD SOFT)

[3] Perelmuter A V and Slivker V I 2010 *Stability of equilibrium of structures and related problems* vol 1 (Moscow: SCAD SOFT)

[4] John R Taylor 1982 *An Introduction To Error Analysis* vol 2 (California: Mill Valley/University Science Books)

[5] Gaydzhurov P P and Isakova E R 2011 *News of Higher Educational Institutions of North Caucasus region* **3** 26-29

[6] Gaydzhurov P P and Isakova E R 2011 *News of Higher Educational Institutions of North Caucasus region* **4** 7-13

[7] Kindmann R and Kraus M 2007 *Finite element method in steel constructions*

Evaluation of the vehicle state with vibration-based diagnostics methods

V E Gai, I V Polyakov, M S Krasheninnikov, A A Koshurina and R A Dorofeev

Nizhny Novgorod State Technical University named after R E Alekseev, 24, Minin str., Nizhny Novgorod, 603950, Russia

E-mail: iamuser@inbox.ru

Abstract. Timely detection of a trouble in the mechanisms work is a guarantee of the stable operation of the entire machine complex. It allows minimizing unexpected losses, and avoiding any injuries inflicted on working people. The solution of the problem is the most important for vehicles and machines, working in remote areas of the infrastructure. All-terrain vehicles can be referred to such type of transport. The potential object of application of the described methodology is the multipurpose rotary-screw amphibious vehicle for rescue; reconnaissance; transport and technological operations. At the present time, there is no information on the use of these kinds of systems in ground-based vehicles. The present paper is devoted to the state estimation of a mechanism based on the analysis of vibration signals produced by the mechanism, in particular, the vibration signals of rolling bearings. The theory of active perception was used for the solution of the problem of the state estimation.

1. Introduction

Continuous monitoring of the equipment condition is essential for the failure risk reduction of equipment [1]. There are several approaches for the solution of this problem. One of the solutions is the usage of methods of vibration-based diagnostics. It allows evaluating the technical condition of the research object. The diagnosis is performed in operational conditions without mechanism dismantling. The diagnosis is based on the analysis of the characteristics of vibro-acoustic processes accompanying the work of the mechanism. The vibro-acoustic diagnostic techniques, in a short time, allow identifying the problem and detecting the emerging defects at the early stages.

The solution method for the task of vibration-based diagnostics can be decomposed into three stages: a preliminary signal processing, calculation of signs, and decision-making. In the most general case, we can assume that the analysis of vibroacoustic nonstationary signals is performed.

Let us consider the disadvantages of known approaches to the solution the problem of vibration diagnostics, considering the decomposition of the problem into three stages [2, 3].

The preprocessing step. This step usually consists in filtering the vibration signal. Considering the fact that the recognition is achieved under conditions of a priori uncertainty, which means that there are no information about a hindrance, the selection of the appropriate filter is difficult. The well-known works pointed that the dividing of the signal into segments does not solve the problem of the choice of the signal segment length. However, this division is necessary in the analysis of non-stationary signals.

The step of a system signs forming. The problem of formation of the system of the signs is associated with the selection of the final set of features to ensure the uniqueness of the solution of the classification problem at the recognition stage and meets the requirements of necessity and sufficiency. The step of choice of the system of the signs is necessary for reduction of the dimensionality of the input description.

The decision step. The decision step consists in the comparison of the sign descriptions of the analyzed vibration with the existing benchmark. It is assumed that a compact set of points in the signs system complies with the benchmark. However, hindrances, structural changes of the same class representative, lead to overlapping classes. Therefore, the problem of the decision-making closes on the problem of the signs system forming, which allows generating the benchmark having a compact representation.

The purpose of the work is the development of new approach for further solution of the task of vibration-based diagnostics of vehicles. The diagnostics is based on the theory of active perception, which allows solving issues, connected with implementation phases of preliminary signal processing and the signs system formation.

2. The proposed approach to the implementation of the vibration-based diagnostics system

The proposed approach to the solution of the vibration diagnostics problem of a rolling bearing is based on the using of machine learning techniques and active perception theory [3]. Considering the vibration-based diagnostics system as a pattern recognition system, it is possible to distinguish three stages of information processing: pre-treatment; calculation of signs and decision-making.

Pretreatment means the performance of the Q-transformation, which consists in the application of the addition operation of the original signal segments:

$$g(t)=\sum_{k=(t-1)L+1}^{t\cdot L} f(k), t=\overline{1,N},$$

where L – the number of readings, within the segment, N – the number of signal segments, g – the result of application of the Q-transformation to signal f, $f(k)$ – k –reading of signal f, $g(t)$ – t-reading of signal g.

Formation of the feature description of the original signal is the application of the set of Walsh filters of the Hartmut system to signal g:

$$\mu(k,c(t))=\sum_{i=0}^{M-1} F_k(i)\cdot g(((t-1)\cdot M+1):(t\cdot M)),$$

where $\mu(k,c(t))$ is the result of application of the sets of Walsh filters of the Hartmut system to signal g, $k=\overline{0,M-1}$, $t=\overline{0,|c|-1}$, $c=\{1,P,2\cdot P,3\cdot P,...,N-T\cdot P\}$ is a set of the offset values for signal g, $|c|$ – is a cardinal number of set c, P - is an offset value for signal g ($1\le P\le M$), M – is the number of the used filters. Thus, the feature description of the signal represents the size of matrix $M\times|c|$ and each line of the feature description is a result of the U-transformation the signal segment.

The consequent application of the Q-transformation and filter systems to the signal implements the U-transformation, which is basic in the theory of active perception.

The U-transformation has the lowest possible computational complexity, since for its realization only simple operations are used – addition and subtraction. Standard transformations require the implementation of convolution, and the level of weighting coefficients – the operation of arithmetic multiplication.

Let each filter $F_i \in \{F_i\} \equiv \boldsymbol{F}$ correspond to coordinate-defined binary operator $V_i \in \{V_i\} \equiv V$; Then, components $\mu_i \neq 0$ of vector μ is permissible to put in line with operator Vi or $\overline{V_i}$ depending on the component sign. As a result, vector μ is associated with a subset of operators $\{V_i\}$, with a design similar to the filter, but having a different meaning of the vector elements ($+1 \leftrightarrow 1$; $-1 \leftrightarrow 0$). By specifying set-theoretic operations of multiplication and addition on set $\{V_i\}$, we have the algebra of the signal description of one-dimensional Boolean functions. Taking into account the reversals, all in all, there are 15 operators that can be used in the formation of feature descriptions, as the V_0 operator receives only the direct value.

On the set of operators, the algebra group is formed (the stage of synthesis) of the analyzed signal:

1) The family of algebraic structures (called the complete groups) $\{P_{ni}\}$, where $P_{ni} = \{V_i, V_j, V_k\}$ with the power of 35;

2) The family of algebraic structures (called the closed groups) $\{P_{si}\}$, where $P_{si} = \{V_i, V_j, V_k, V_r\}$ with the power of 105, where each group is formed of a pair of complete groups connected in a certain way.

Among the complete groups, we can allocate the complete groups for the operations of addition and the operation of multiplication, among the closed groups - closed groups and closed sets.

The set of operators describing the signal is formed with using of the spectral representation of signal μ. Then, the set of complete and closed groups is formed as well:

$V = GV[\mu]$, $P_{na} = GP_{na}[\mu, V]$, $P_{nm} = GP_{nm}[\mu, V]$, $P_s = GP_s[\mu, V, P_{na}, P_{nm}]$, $P_c = GP_c[\mu, V, P_{na}, P_{nm}]$,

where GV – is an operator for calculation of feature descriptions V using spectral representation of the signal on the basis of operators: GP_{na} (GP_{nm}) – on the basis of complete groups for the operations of addition P_{na} (multiplication, P_{nm}), GP_c (GP_s) – on the basis of closed groups P_s (closed sets P_c).

A model of signs descriptions, which takes into account the connection between adjacent segments of the signal, is proposed. In this case, the calculation method of the signs descriptions consists in the forming of the matrix of transition probabilities between neighboring segments descriptions.

The signs systems based on the matrices of transition probabilities are proposed. The following variants of the systems are considered:

1) The signs system describing the probabilities of transitions between the values of the operators (the operator can take three possible values: direct, inverse and zero). The values are calculated from the adjacent signal segments, without considering the connections between the various operators, the dimension of the feature space – 3 × 3 × 15;

2) The signs system describing the probabilities of transitions between the values of the operators. The values are calculated on neighbour segments of the signal, taking into account the connection between operators, the dimension of the feature space – 45 × 45.

3) The signs system of describing the probabilities of transitions between segments of the descriptions is provided in the form of complete groups, the dimension of the feature space – 140 × 140; in case of using the full group, it is allowed using only a few groups of maximum weight;

4) The signs system describing the probabilities of transitions between the descriptions of the segments provided in the form of closed groups, the dimension of the feature space – 840 × 840; with using closed groups, it is allowed using only a few groups of maximum weight.

The formation algorithm of signs system based on the complete groups (closed groups, $Desc_i$ – description of the i -th segment signal in the form of complete or closed groups):

$$\forall i = \overline{1, N}$$

$$\forall k = \overline{1, |Desc_i|}$$

$$\forall l = \overline{1, |Desc_{i+1}|}$$

$$MPG_{Desc_i}(Desc_i(k), Desc_{i+1}(l)) = MPG_{Desc_i}(Desc_i(k), Desc_{i+1}(l)) + 1.$$

The following notations will be used: PP_{na} (PP_{nm}) – the matrix of transition probabilities between the full groups (multiplication), PP_c – the matrix of transition probabilities between the closed groups.

The formation algorithm of the PVI signs system based on operators without considering connections (V_i – a description of the i-th signal segment as operators):

The formation algorithm of the PVD signs system by operators, considering the connections between operators:

$$
\begin{array}{l}
\forall i = \overline{1,N} \\
\quad \forall k = \overline{1,15} \\
\quad\quad if\ V_i(k) == -1 \\
\quad\quad\quad p = 1 \\
\quad\quad if\ V_i(k) == 0 \\
\quad\quad\quad p = 2 \\
\quad\quad if\ V_i(k) == +1 \\
\quad\quad\quad p = 3 \\
\quad\quad p = p + (k-1)\cdot 3 \\
\quad\quad if\ V_{i+1}(k) == -1 \\
\quad\quad\quad q = 1 \\
\quad\quad if\ V_{i+1}(k) == 0 \\
\quad\quad\quad q = 2 \\
\quad\quad if\ V_{i+1}(k) == +1 \\
\quad\quad\quad q = 3 \\
\quad\quad q = q + (k-1)\cdot 3 \\
\quad\quad PVI_{Desc,k}(p,q) = PVI_{Desc,k}(p,q).
\end{array}
$$

$$
\begin{array}{l}
\forall i = \overline{1,N} \\
\quad \forall k = \overline{1,15} \\
\quad\quad if\ V_i(k) == -1 \\
\quad\quad\quad p = 1 \\
\quad\quad if\ V_i(k) == 0 \\
\quad\quad\quad p = 2 \\
\quad\quad if\ V_i(k) == +1 \\
\quad\quad\quad p = 3 \\
\quad\quad p = p + (k-1)\cdot 3 \\
\quad\quad \forall l = \overline{1,15} \\
\quad\quad\quad if\ V_{i+1}(k) == -1 \\
\quad\quad\quad\quad q = 1 \\
\quad\quad\quad if\ V_{i+1}(k) == 0 \\
\quad\quad\quad\quad q = 2 \\
\quad\quad\quad if\ V_{i+1}(k) == +1 \\
\quad\quad\quad\quad q = 3 \\
\quad\quad\quad q = q + (k-1)\cdot 3 \\
\quad\quad PVD_{Desc}(p,q) = PVD_{Desc}(p,q).
\end{array}
$$

The matrices of transition probabilities between descriptions of the segments can be used as the signs systems. At the same time, it is possible to take into account the transition probabilities between many segments and not only between neighbor segments (*i* and (*i+1*)). For the i -th segment, we can consider not only (*i+1*), (*i+2*) and further segments, but also (*i – 1*), (*i – 2*) segments, i.e. not only the 'next' segments can be considered, but 'previous' segments can be considered as well.

3. The computational experiment

3.1. The experiment description

The computational experiment was carried out based on the database of vibration signals, described in [7]. The database contains records of vibration signals recorded by defects of rolling bearings (on the outer and inner track roller body) and without defects. The defects dimensions constitute 0.007, 0.014, 0.021 and 0.028 inches in diameter. In the twenty-one, the bearing condition can be classified in the database: 1) normal; 2-5) rolling element defect (defect sizes: 0.007, 0.014, 0.021, 0.028); 6-9) defect in the inner tracks (defect sizes: 0.007, 0.014, 0.021, 0.028); 10-21) outer race defect (defect positions: centr, orth, opp; defect sizes: 0.007, 0.014, 0.021, 0.028).

The computational experiment was directed to study of the accuracy of determining the condition of the bearing depending on the features of the system. Solution of the problem of classification is performed based on the method of the support vector machine (SVM) and the k-nearest neighbor (KNN).

3.2. Results

The tests results of existing vibration-based diagnostics methods of a rolling bearing are shown in Table 1.

Table 1. The tests results of existing vibration-based diagnostics methods

Source	Signs	Signs amount	Classification method	The amount of the classification conditions	Accuracy of recognition (R, в %)
[5]	Mel frequency cepstral coefficients	16	Gaussian mixture model	4	94
[6]	Statistical signs	26	The method of support vector machine	21	92.88
[6]	Spectrum of the complex envelope	72	The method of support vector machine	21	85.47
[7]	Properties coefficients of wavelet decomposition	5	Radial basis function network	4	72.1
[7]	Properties coefficients of wavelet decomposition	5	Multilayer perceptron	4	100

3.3. The study of the proposed approach
Table 2 shows the results of the accuracy of the rolling bearing condition evaluation. Various systems and signs of classifiers were used during the evaluation.

Table 2. The results of the computational experiment

Classificator / Signs system	PVI	PVD	PP_{na}	PP_{nm}	PP_s
KNN	85	95	95	97	-
SVM	81	91	91	97	-

Conclusions of the experiment results:

1) using the system of signs PPnm and classifier KNN provides the best results;

2) the testing of the system of signs PPs was not performed successfully because of lack of RAM;

3) the accuracy of the achieved results is comparable with the known ones, and in some cases exceeds them.

4. Conclusion
The vibration-based diagnostics methods of a rolling bearing were considered in this paper. The solution was performed from the standpoint of the theory of active perception. The investigations of the proposed method showed the ability to increase, as compared to known approaches, the accuracy of the estimate of the rolling bearing condition. The results are a significant advance in the field of forecasting and monitoring the state of the diagnosed object. Timely assessment of critical components and mechanisms is an appropriate stage of development and improvement of various all-terrain vehicles.

Acknowledgements
This work was carried out at the NSTU named after R.E. Alekseev, with financial support from the government in the face of the Russian Ministry of Education under the Federal Program "Research and

development on priority directions of the scientific-technological complex of Russia for 2014-2020", the unique identifier of the project: RFMEFI57714X0105.

References

[1] Krasheninnikov M, Kulashov A, Shapkin V and Koshurina A 2013 *J. Lecture Notes in Electrical Engineering* **195** 477
[2] Jing W, Liu H and Lu C. 2012 *Proc. Int. Conf. on Future Information Technology and Management Sci. & Eng. Lecture Notes in Information Technology* (Hong Kong) p 128
[3] Gai V 2014 *Proc. Int. Conf. on Mechanical Eng., Automation and Control Systems*
[4] Utrobin V A 2004 *J. Adv. in Phys. Sci.* **47** 1017
[5] Nelwamondo F V, Marwala T and Mahola U 2006 *Int. J. of Innovative Computing, Information and Control* 1281
[6] Boldt F, Rauber T W and Varejão F M 2013 *Proc. Int. Conf on Inteligência Artificial e Computacional* (São Carlos, SP, Brazil) p 213
[7] Al-Raheem K F and Abdul-Karem W 2010 *Int. J. of Eng., Sci. and Techn.* 278

Designing a combined casting mold for manufacture of a gasoline centrifugal pump body using CAD/CAM-systems

N E Galin[1], I I Ogol[1], Yu B Chervach[1], V Kh Dammer[2] and Jia Hong Ru[1]

[1] Tomsk Polytechnic University, 30, Lenina ave., Tomsk, 634050, Russia
[2] Scientific-productional Centre ‘Polus’, 56c, Kirova ave., Tomsk, 634050, Russia

E-mail: negalin@tpu.ru

Abstract. The present paper examines designing of a combined casting mold for manufacture of a gasoline centrifugal pump body. The paper offers technological solutions for obtaining high quality castings at the testing stage of the finished mold. The paper is intended for practical use and prepared by order of JSC ‘Tomsk Electrical Engineering Plant’ using software and equipment of the department ‘Technologies of Computer-Aided Machinery Manufacturing’ of the Tomsk Polytechnic University (TPU) under the economic contract within state import substitution program. In preparing the paper, CAD/CAM-systems KOMPAS-3D and PowerMILL were used. In 2015, the designed casting mold was introduced into the production process at JSC ‘Tomsk Electrical Engineering Plant’.

1. Introduction

Method of casting into long-life metallic forms (chill molds) is applied for producing more than one third of the total amount of products from nonferrous metals and alloys that become essential components for machinery industry [1]. The variant of chill casting ensures higher density and better mechanical properties of metal, higher accuracy and quality of a cast surface, and higher operating rate as compared to sand casting [2].

This method is used in series and large-scale manufacturing [3] and involves designing of a casting mold for product manufacturing. At present, high manufacturability and competitive ability of this production method are provided due to using of CAD/CAM-systems and NC machine tools especially if we talk about designing products of a complicated form.

According to the technical task of JSC ‘Tomsk Electrical Engineering Plant’, it was required to design a long-life metallic casting mold for manufacturing a body of a gasoline centrifugal pump (GCP) from aluminum grade Al-6%Si-0.2%Mg [4] meeting all the constructive and process requirements. This type pump is used in a fuel system of military machinery and is intended for providing light fuel (gasoline) operation of a multifuel diesel engine. It should be noted that previously these products have been manufactured only by sand casting.

2. Task setting

A 3D-model of a gasoline centrifugal pump body (Figure 1 a) was designed using CAD-system KOMPAS-3D based on the drawing submitted by the customer.

After analysis of the body design that showed that the product has a complicated form requiring nonstandard solutions, the following tasks were formulated:

1. Providing a mold design with the least number of jointing planes.
2. Providing mold filling when producing thin-wall castings with many undercuts.
3. Creating favorable conditions for air venting from casting volume.
4. Reducing labour intensity of mold assembly/disassembly.
5. Reducing costs for manufacture of a chill mold.
6. Minimizing rejection of castings.

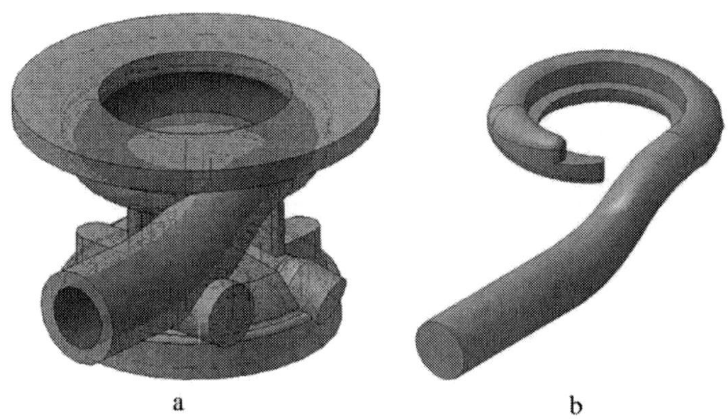

Figure 1. a – 3D-model of a gasoline centrifugal pump body, b – sand core.

3. Designing a combined casting mold

In most cases, a chill mold has 1-2 jointing planes [5]. Due to complexity of the casting configuration, the casting mold designed shall have a combined (curved) jointing plane, and shall consist of four main elements (Figure 2 a). Arrows on the figure indicate parting directions.

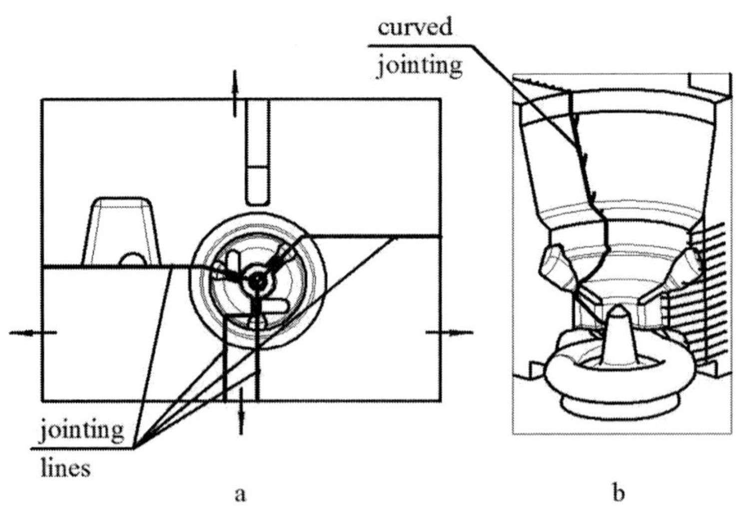

Figure 2. a – Scheme of location of jointing lines and parting directions, b – curved jointing.

A problem of formation of a complex nozzle channel was solved by using a sand core [6] – 'volute' (Figure 1 b), for which purpose a sand box was manufactured.

A curved jointing (Figure 2 b) that allowed maintaining the body original form as full as possible, was made between nozzle-forming parts. The body inner cavity and lug (ridge) were formed by a core rod installed on the lid. Metal was fed from the bottom, and a dashpot was made in the lower plate. Air was vented through the lid. The designed casting mold is shown on Figure 3.

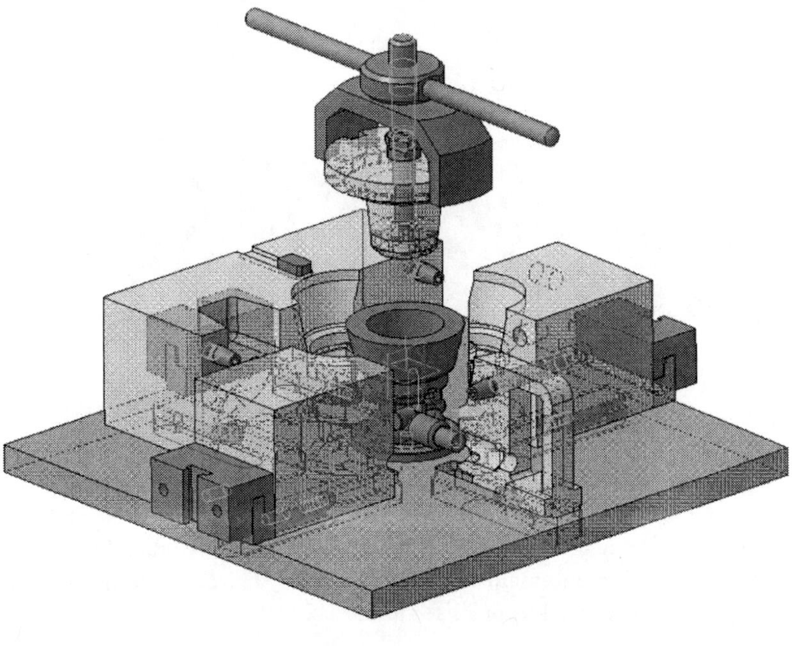

Figure 3. Combined casting mold.

For reducing manufacturing costs and time of the project implementation, it was decided to manufacture the mold from aluminum grade Al-0.5%Si-1.5%Mg [7] with coating of the working surfaces with refractory ceramic. Technique of applying refractory coatings is widely used in product manufacturing by casting method [8, 9]. In our case, it allowed preventing adhesive bonding of the casting and the mold.

Shape-generating components of the casting mold are products of a complicated form that is why it is practically impossible to manufacture them without Numerical Control (NC) tools. For preparation of a control program for a NC machine tool, CAM-system PowerMILL was applied. For manufacture of the components, the machine tool DMU 50 was used.

4. Testing the mold

After the first tests, rejection amounted to 45%. In addition to a severe shrinkage defect 1 in the casting central part, places of 'misruns' 2 in the pump lower part and on lug bosses 3, and pores 4 on the nozzle were detected on some specimens (Figure 4).

Despite the fact that rejection 45% for chill casting can be considered acceptable, we were not satisfied with this result. For preventing shrinkage in the area 1, it was decided to install an additional central metallic core.

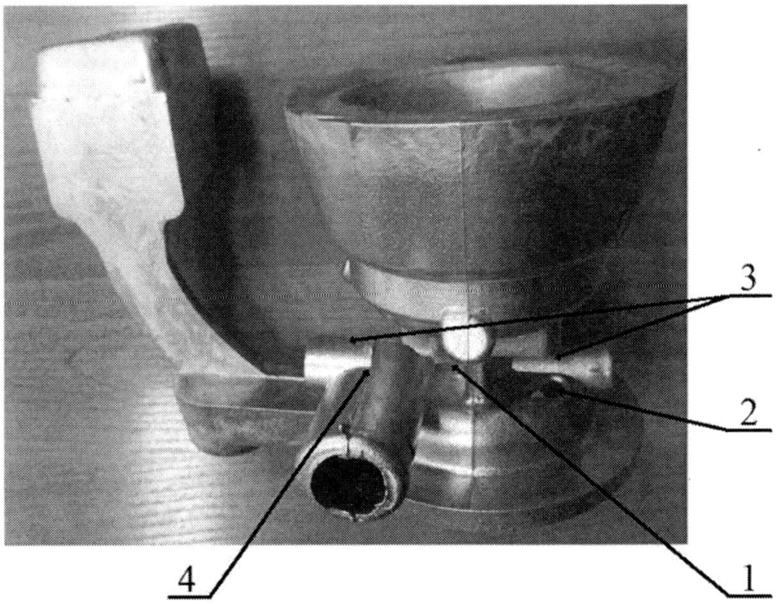

Figure 4. Casting defects: 1 – defect shrinkage, 2 and 3 – places of 'misruns', 4 – pores.

For preventing shrinkage [10] on lug bosses, copper coolers with a snakelike slot (Figure 5 a) contributing to both air cooling and venting, were built into the casting mold parts. The problem of pore formation was solved by making additional air venting channels (figure 5 b) on all the mold joint surfaces. During further testing, the mold initial temperature [11, 12] was increased from 300 °C to 400 °C, and casting time was reduced from 20 sec to 10 sec.

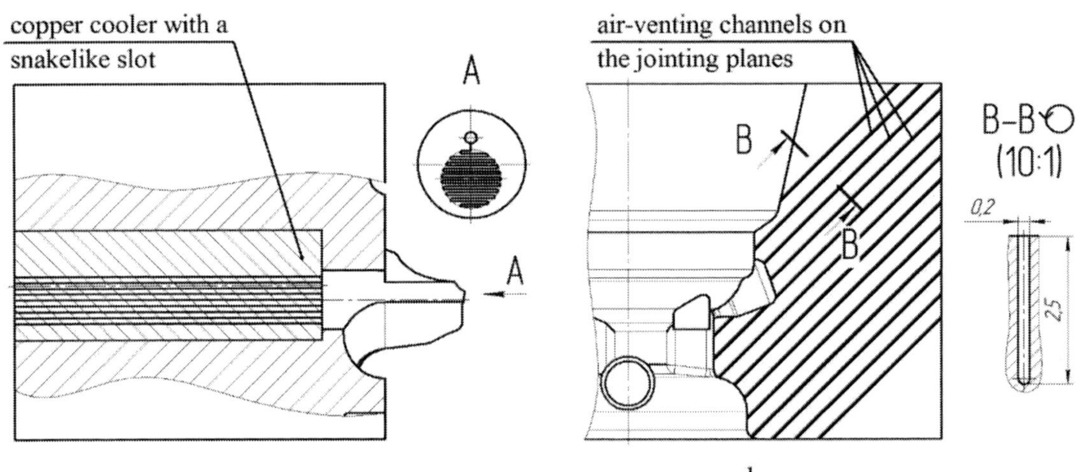

Figure 5. Methods of defect prevention.

5. Conclusion

The solutions implemented increased the amount of acceptable castings to 80%. At present, practical work on selecting casting modes is being conducted for further rejection reduction. Theoretical work on calculating temperature patterns, solution phase field, and velocity and pressure fields, on calculating segregation (chemical inhomogeneity) and defects, and on calculating stresses and deformations, is also being planned.

Within the project implementation, a computer-aided installation for casting a gasoline centrifugal pump body was constructed and introduced into production at JSC 'Tomsk Electrical Engineering Plant'.

References

[1] Burakou S L, Veinik A I and Dubinin N P 1980 *Chill casting* (Moscow: Engineering)
[2] Galdin N M 1989 *Color Cast Guide* (Moscow: Engineering)
[3] Orlov N D and Mironov V M 1971 *Directory caster* (Moscow: Engineering)
[4] GOST RF 2685-75, Aluminum alloy. The mechanical properties of aluminum casting alloys, Moscow, 1975
[5] Petrychenko A M 1967 *Theory and chill casting technology* (Kiev: Tehnika)
[6] Campbell J 2003 *Castings.* (Burlington: Butterworth-Heinemann)
[7] GOST 4784-97, Aluminium and aluminum wrought alloys. Stamps, Moscow, 1997
[8] Sadokha M A 2011 *Casting and Metallurgy* **62(3)** 58-60
[9] Mitterer C, Holler F, Üstel F and Heim D 2000 *Surface and coatings Technology* **125(3)** 233-9
[10] Couper M J, Neeson A E and Griffiths J R 1990 *Fatigue & Fract. of Eng. Mat. & Str.* **3(3)** 213
[11] Meena A and Mansori M El 2012 *J. of Mat. Proc. Techn.* **11(11)** 2484-95
[12] Chernyshev E A and Evlampiev A A 2013 *Int. J. of Exper. Educ.* **11(1)** 81-4

The analysis of the influence of the material antifrictional layer frictional properties on the parameters of the spherical bearing contact zone

A A Kamenskikh and N A Trufanov

Perm National Research Polytechnic University, 29, Komsomolsky Ave., Perm, 614000, Russia

E-mail: anna_kamenskih@mail.ru

Abstract. The paper presents data on the influence of the frictional properties of a material antifrictional layer on the parameters of the spherical bearing contact zone. The dependences of the friction coefficient from the load were obtained as a result of the study. Series of numerical experiments were conducted to investigate the frictional properties of a materials contact pair in the work. Regularities of the relative contact pressure and relative contact tangential stress were obtained for seven variants of the load-friction coefficient for the spherical bearing with a layer of modified fluoroplastic. The study puts emphasis on the fact that that adhesion area of the contact surface is reduced and the load is increased taking into account the fact that the friction properties of the layer has been fixed in the study.

1. Introduction

The contact units with antifrictional coatings and layers are widely applied in engineering, construction, medicine and other industries. These designs are costly. It is difficult to repair them. High requirements with respect to durability, reliability and service life are characteristic of them. Such contact units are working within the contact mechanics. The stress-strain state of units has a complex three-dimensional character, and the design may include more than one sliding surface. The bearing bridges belong to these constructions [1, 2]. The modern antifrictional polymer materials are used in these designs as coating and layers materials. Modified fluoroplastic refers to these materials. A special topic of research is determined physical and mechanical, frictional and performance properties of the antifrictional interlayer materials. Research of the influence of material properties on contact area parameters within the contact mechanics is also a separate scientific topic.

It is worth mentioning that the fundamentals of contact mechanics were found by several scientists. They considered a one-way contact or contact bodies of canonical form [3]. Works on the study of contact interaction of the bodies with coatings and layers are also exist [4, 5]. The bodies of the canonical form, space and half-space are considered in these works. The analytical and semi-analytical solutions methods are mainly used in these studies. Effective numerical approaches to solving contact problems are also exist, these include the finite element method and the boundary element method [6]. These methods are applicable to the implementation of contact interaction of the bodies with coatings and layers. There is not enough information about the complex mathematical description of the contact status for bodies with coatings and layers having multiple sliding surfaces and a complex spatial configuration, and numerical methods require highly skilled researchers and multiprocessor computers.

The contact interaction of spherical bearing elements through the antifriction layer is made within the study. The influence of mechanical and frictional properties of material layer on the parameters of the contact zone is considered in the work.

2. Mathematical statement, materials and methods

The general mathematical problem statement of contact between two elastic bodies through the antifriction layer is made in the work (Figure 1). Deformation plasticity theory is selected to describe the behavior of the material layer [2].

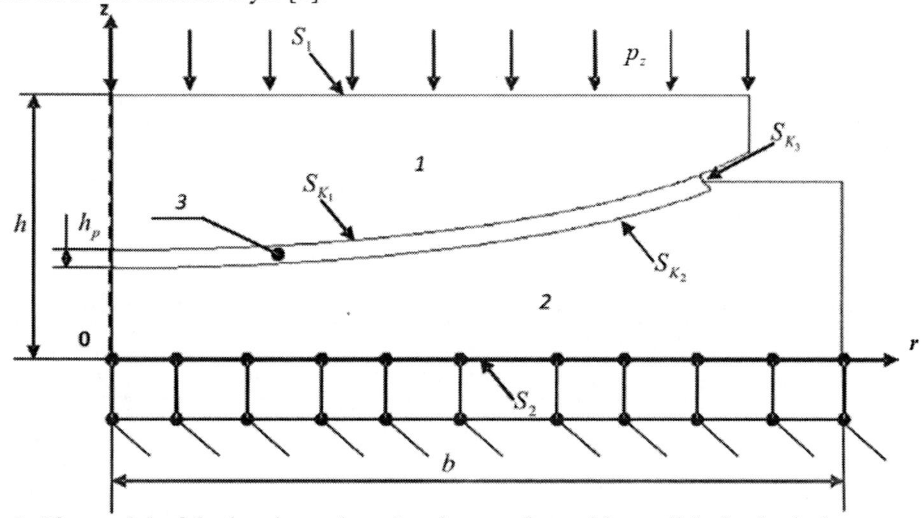

Figure 1. The model of the bearing, where 1 – the top plate with a polished spherical segment; 2 – the bottom plate with a spherical cut; 3 – the interlayer from antifriction material.

The contact boundary conditions applied to the surface are $S_K = S_{K_1} \cup S_{K_2} \cup S_{K_3}$. Wherein the two bodies are in contact under S_K and which have conditioned numbers 1 and 2. The types of contact interaction that are implemented in the problem are given below:

– sliding with friction (friction of rest): $\vec{u}^1 = \vec{u}^2$, $\sigma_n^1 = \sigma_n^2$, $\sigma_{n\tau_1}^1 = \sigma_{n\tau_1}^2$, $\sigma_{n\tau_2}^1 = \sigma_{n\tau_2}^2$, wherein $\sigma_n < 0$, $|\sigma_{n\tau}| < q(\sigma_n)|\sigma_n|$;

– sliding with friction (sliding friction): $u_n^1 = u_n^2$, $u_{\tau_1}^1 \neq u_{\tau_1}^2$, $u_{\tau_2}^1 \neq u_{\tau_2}^2$, $\sigma_n^1 = \sigma_n^2$, $\sigma_{n\tau_1}^1 = \sigma_{n\tau_1}^2$, $\sigma_{n\tau_2}^1 = \sigma_{n\tau_2}^2$, wherein $\sigma_n < 0$, $|\sigma_{n\tau}| = q(\sigma_n)|\sigma_n|$;

– noncontact: $|u_n^1 - u_n^2| \geq 0$, $\sigma_{n\tau_1} = \sigma_{n\tau_2} = \sigma_n = 0$;

– adhesion: $\vec{u}^1 = \vec{u}^2$, $\sigma_n^1 = \sigma_n^2$, $\sigma_{n\tau_1}^1 = \sigma_{n\tau_1}^2$, $\sigma_{n\tau_2}^1 = \sigma_{n\tau_2}^2$,

where $q(\sigma_n)$ – the friction coefficient, τ_1, τ_2 – axes designations which lie in a plane tangent to the contact surface, u_n – displacement along a normal to a corresponding contact edge, u_{τ_1}, u_{τ_2} – displacement in a tangential plane, σ_n – stress along the normal to the contact boundary, $\sigma_{n\tau_1}$, $\sigma_{n\tau_2}$ – tangential stresses at the contact boundary, $\sigma_{n\tau}$ – the value of vector tangential contact stresses.

The mathematical formulation is complemented by kinematic boundary conditions on surface S_2: $u_z = 0$, $\sigma_{rz} = 0$, $r \in S_2$, and also by static and kinematic boundary conditions on surface S_1:

$\int_{S_1} p_z dS_1 = -Q_z$, $u_z(r,h) = U = const$, $\sigma_{rz} = 0$, $r \in S_1$, where Q_z – the vertical force applied to S_1, U – the unknown quantity and the remaining outer surfaces are free of load.

The defining relations of the model of modified fluoroplastic were verified in the framework of a numerical experiment. The numerical experiment was implemented as a deformation of cylindrical samples by the plates of a press in the framework of contact mechanics. It was found that the numerical calculation with the selected defining relations provides a good quantitative agreement between the experimental results obtained at the Continuum Mechanics Institute. Determination of the frictional properties of antifriction polymeric materials is the next stage of the study. Special equipment has been developed and a multi-stage experimental program was made by professor of Continuum Mechanics Institute AA Adamov for conducting the experiment to determine frictional properties of the layer material. The dependences of the friction coefficient on the specific pressure on the sample were obtained from the experimentes for two variants of the contact interaction: without lubrication and with lubricant. The numerical experiment program has been selected for two variants of contact interaction for analysis of the friction coefficient influence on the contact parameters in the design of the spherical bearing. The experimental results are presented in table 1 for a modified fluoroplastic.

Table 1. The numerical experiment, the ratio of pressure and the friction coefficient

№	Contact without lubrication		Contact with lubrication	
	Friction coefficient	Pressure, MPa	Friction coefficient	Pressure, MPa
1	0.037	4	0.022	3
2	0.033	9	0.02	5
3	0.029	12	0.016	10
4	0.025	15	0.012	15
5	0.021	25	0.01	25
6	0.017	35	0.009	35
7	0.015	45	0.007	40

The solution is implemented by using the theory of elastic-plastic deformation in the software 'ANSYS'. The convergence of the numerical solutions has been considered in [2] under deformation of the spherical bearing with the antifrictional layer made of the modified fluoroplastic with a constant friction coefficient. It is established that eight elements throughout the thickness of a layer is enough to ensure quality solutions in axisymmetric formulation, this corresponds to a finite element model with two hundred thirty five thousand nodal unknowns. The implementation of the contact boundary conditions for the tangential contact was checked, a displacement and deformation path in the main space of deviatoric strain values are considered in the design.

3. Results and Discussion

Series of numerical experiments were realized to study the unfluence of the impact of antifrictional properties of the friction material layer on the contact zone parameters. The field distribution relatively contact pressure (Figure 2a and 2b) and relatively tangential contact stress (Figure 2 c and 2d) at contact without lubricant and with lubricant were obtained by solving the problem. They are shown on contact surface S_{K_1}, at which the rotation of a spherical segment of the top plate was realized. The friction coefficient is decreased when the load acting on the contact unit is increased.

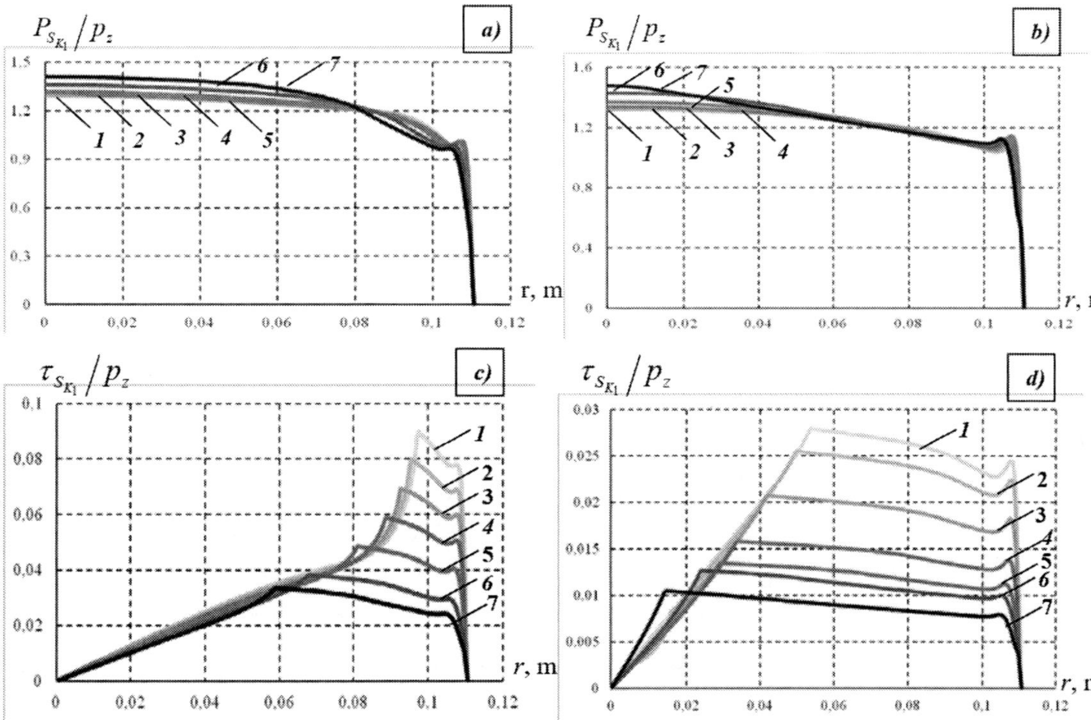

Figure 2. The relative contact pressure (a, b) and the relative contact tangential stress (c, d) on contact surface S_{K_1} : (a, c) contact without lubrication and (b, d) contact with lubrication, lines 1-7 are the result of numerical experiment № 1-7, respectively.

The level of the maximum relative contact pressure depends only slightly on the load and the friction coefficient. The relative contact tangential stress is the most sensitive to changes of the friction coefficient. The tangential stress level is decreased as the load grows. And the distribution of the relative contact tangential stress is becoming more uniform. It is more pronounced by contact with lubricant. The value of the relative tangential stress increases in the trapping zone. Its maximum value is observed in the transition contact zone from the adhesion state to the sliding state.

The dependences of maximum relative contact pressure and maximum relative tangential contact stress on the load are shown in Figure 3.

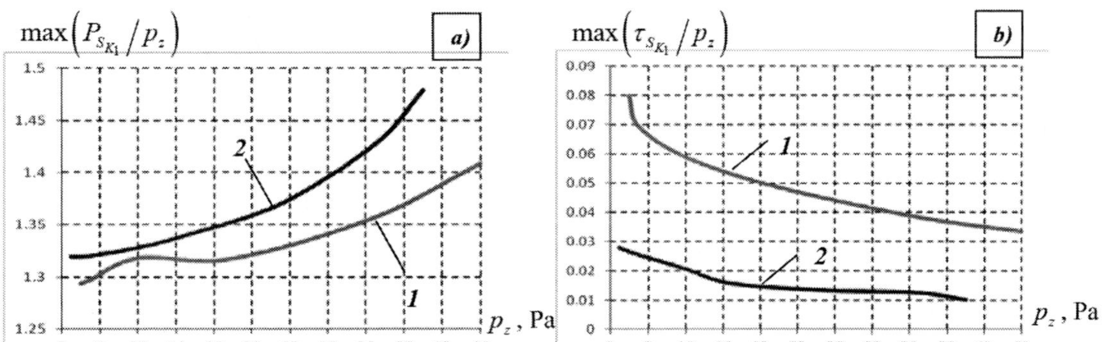

Figure 3. The dependence of maximum relative contact pressure (a) and maximum relative contact tangential stress (b) of p_z : 1 – without lubrication, 2 – with lubrication.

The dependences of maximum relative contact pressure and maximum relative tangential stress of the load were set in the framework of a series of numerical experiments:

- maximum contact pressure is increased with the increasing load and the decreasing friction coefficient in all variants of contact interaction, the contact pressure is greater in the case of contact with lubricant;
- the antifrictional properties of material are observed in the case of relative contact tangential stress: its maximum value decreases as the load grows;
- the maximum value of the relative contact tangential stress is much less in contact with lubricant at an identical load level.

The decreasing adhesion area of the contact surface has been registered, as shown in Figure 4.

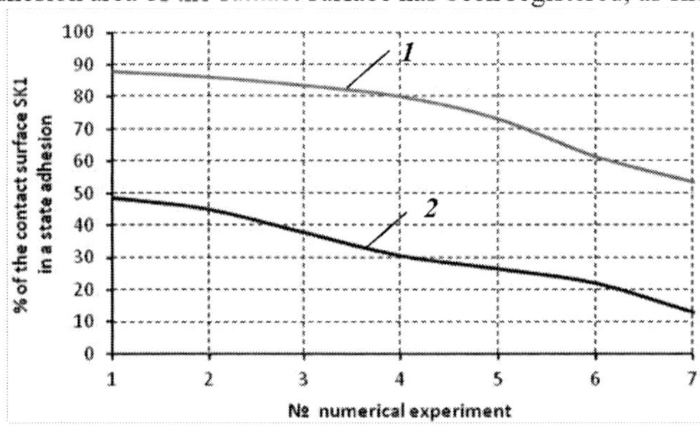

Figure 4. The percentage of the contact surface area being in a state of adhesion: 1 – without lubrication, 2 – with lubrication.

The tendency towards the reduction of the adhesion area of the contact surface has been registered. The friction coefficients are very different for models without lubricant and with lubricant at the same load level, it influences the adhesion contact area:

- about ninety percent of the contact surface is in the state of adhesion if the bearing of minimum load is deformed by the contact without lubrication, and less than fifty percent of the contact surface is in the state of adhesion if the bearing of the minimum load is deformed by the contact with lubrication;
- about fifty percent of the contact surface is in the state of adhesion if the bearing of the maximum load is deformed by the contact without lubrication, and less than fifteen percent of the contact surface is in the state of adhesion if the bearing of the maximum load is deformed by the contact with lubrication.

4. Conclusion

Some results of solving the task can be formulated as the conclusions:

1. The mathematical problem statement of the contact interaction of two elastic bodies with antifrictional layer was made in the work.
2. The defining relations behavior model of the modified fluoroplastic was verified in the framework of a numerical experiment. The good qualitative and quantitative agreement between the results of experiments and numerical experiments was established.
3. The problem of the contact interaction of elements of spherical bearing with antifrictional polymer layer in axisymmetric and three-dimensional productions was performed at a constant friction coefficient equal to 0.04: convergence of the numerical solution was installed, distribution of contact zones was obtained, nature distribution of contact pressure and contact tangential stress was set.
4. Dependences of the friction coefficient on specific pressure were obtained for the modified fluoroplastic for the contact without lubricant and with lubricant. The numerical experiment program was based on the results of the field experiment.

5. The comparative analysis of the deformation behavior in the contact assembly with the antifrictional layer from the modified fluoroplastic with different ratios of the load and the friction coefficient was made.
6. Dependences of the maximal values of relative contact pressure and relative contact tangential stress on the load acting on a spherical bearing were set (its own friction coefficient corresponds to each load).
7. The tendency towards reducing the adhesion area was set when carrying out numerical experiments: the area of adhesion for contact without lubrication is by 35-40% less than for contact with lubricant at the same load level.

Acknowledgments
The reported study was funded by the RFBR, according to research project No. 16-38-00390 mol_a.

References
[1] Xing J and Diao H 2012 *Procedia Earth and Planetary Science* **5** 358-363
[2] Kamenskih A A and Trufanov N A 2015 *Friction and Wear.* **36**(2) 170-176
[3] Dag S, Guler M A, Yildirim B and Ozatag A C 2013 *Acta Mech.* **224** 1773-89
[4] Chidlow S J, Teodorescu M and Vaughan N D 2011 *Int. J. of Solids and Struct.* **48** 3243-56
[5] Goryacheva I G and Torskaya E V 2012 *Wear.* **286** (11-12) 1417-422
[6] Rogovoy A and Ivanov B 1997 *Computers & Structures* **62**(1) 133-139

A fuzzy set approach for reliability calculation of valve controlling electric actuators

D P Karmachev, A A Yefremov and E E Luneva

Tomsk Polytechnic University, 30, Lenina ave., Tomsk, 634050, Russia

E-mail: karmachevd@mail.ru

Abstract. The oil and gas equipment and electric actuators in particular frequently perform in various operational modes and under dynamic environmental conditions. These factors affect equipment reliability measures in a vague, uncertain way. To eliminate the ambiguity, reliability model parameters could be defined as fuzzy numbers. We suggest a technique that allows constructing fundamental fuzzy-valued performance reliability measures based on an analysis of electric actuators failure data in accordance with the amount of work, completed before the failure, instead of failure time. Also, this paper provides a computation example of fuzzy-valued reliability and hazard rate functions, assuming Kumaraswamy complementary Weibull geometric distribution as a lifetime (reliability) model for electric actuators.

1. Introduction

Reliability is the essential property of technical system or equipment. The key aspect of the management and control for any kind of industrial facility is reliability-centered maintenance (RCM) methodology [1-3]. It demands accurate definition of system reliability measures for decision making process concerning equipment maintenance [3, 4].

Failure data analysis allows selecting adequate reliability model in accordance with feasible lifetime distribution and estimating reliability model parameters [5]. In this context, times to failure of identical items are considered as random numbers drawn from the parent population. However, during the period of operation items (including valve controlling electric actuators) perform in various operation modes, subjected to the transient load under diverse environmental conditions [6]. In this regard, the reliability estimation of an electric actuator could be based on either random times or number of OPEN/CLOSE cycles to failure.

Considering the fact of ambiguity in operational conditions of electric actuators, it seems appropriate to apply reliability models with fuzzy parameters [7-10]. With this background, in order to determine joint influence of operating time and amount of work on reliability of an electric actuator, in this study the number of cycles to failure is regarded as a fuzzy value, whilst the reliability itself is considered to be a function of time.

2. Basic reliability measures

As a part of the study we suggest a technique for obtaining 3D-plots of fuzzy-valued functions of reliability and hazard rate [7, 9].

Reliability or survival function is the essential reliability measure defined as the probability that a failure doesn't occur before specified time t:

$$R(\Theta,t)=\Pr\{T>t\mid\Theta\}=1-F(\Theta,t),\quad t\geq 0.$$

Here T is a random time to failure and $F(\Theta,t)$ is the cumulative distribution function (cdf) with parameter vector $\mathbf{\Theta}$.

Hazard rate function (hrf) is a reliability measure which allows, among other things, defining system lifecycle phases according to the shape of hrf given by

$$\lambda(\Theta,t)=\frac{F'(\Theta,t)}{1-F(\Theta,t)}. \quad (1)$$

Generally, hrf has a bathtub-shaped plot depicted in figure 1.

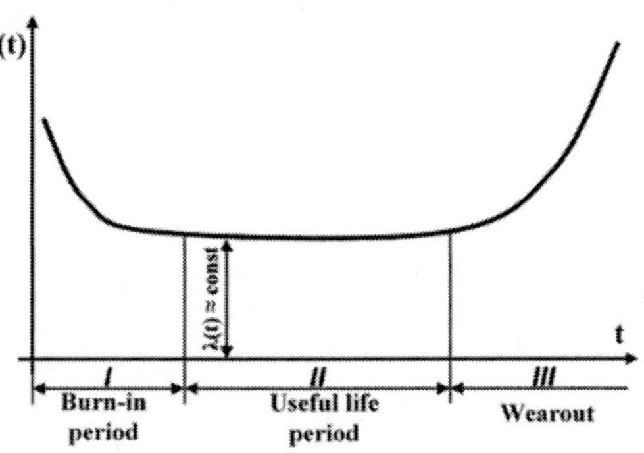

Figure 1. Bathtub-shaped hazard rate function.

Burn-in period is specified by the failures of substandard equipment and by the failures caused by design flaws, construction and start-up operation errors. Useful life period is characterized by the lowest and nearly constant values of the hazard rate function. Finally, the third period represents failures, occurred due to wear-out or ageing [1-4].

3. Failure Data Preparation and Results

Electric actuator failure data is represented as failure times and amount of work completed before the failure, i.e. number of OPEN/CLOSE cycles to failure. Failure data on 286 electric actuators, received for the study, were grouped into 7 equally sized bins. Since the number of cycles to failure varies between 0 and 17500, the width of each bin is 2500.

Figure 2 shows the histogram of number of cycles, provided that each bin is assigned with normalized membership value, proportional to its height. Then, times of failures, associated with i^{th} bin, should be analyzed to estimate parameters of the lifetime model [11].

For this study Kumaraswamy complementary Weibull geometric (Kw-CWG) distribution, introduced in [12], was selected as a reliability model. This distribution has five parameters, providing opportunities for a fine tuning the shape of reliability function which is essential in parameter estimation. Moreover, the reliability model based on the Kw-CWG distribution is very flexible and contains 23 sub-models as special cases [12], including Weibull distribution, widely used in reliability engineering and survival analysis [13]. It is possible to obtain constant, increasing, decreasing, bathtub and unimodal shape of hrf by changing the set of distribution parameter values. According to Kw-CWG model the reliability function is determined as

$$R(t)=\left\{1-\alpha^{a}\left(1-\exp\left[-(\gamma t)^{\beta}\right]\right)^{a}\left(\alpha+(1-\alpha)\exp\left[-(\gamma t)^{\beta}\right]\right)^{-a}\right\}^{b}, \quad (2)$$

where $\gamma > 0$ is a scale parameter and $\alpha > 0, \beta > 0, a > 0, b > 0$ are shape parameters.

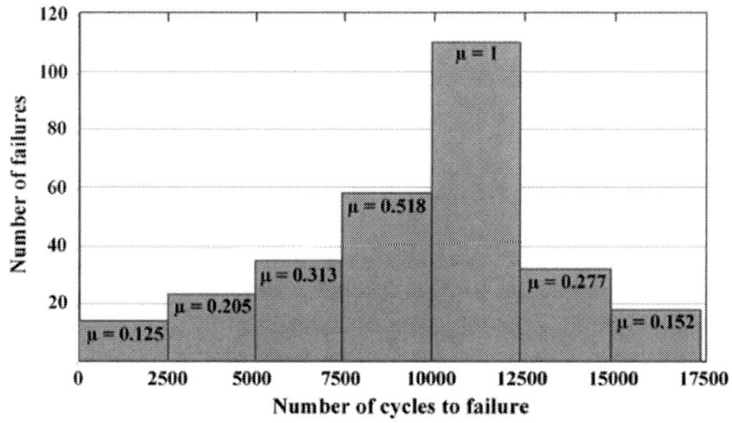

Figure 2. Histogram of number of OPEN/CLOSE cycles to failure.

The hrf of Kw-CWG reliability model is expressed as

$$\lambda(t)=\frac{\alpha^a\beta\gamma ab(\gamma t)^{\beta-1}\exp\left[-(\gamma t)^\beta\right]}{\left\{\alpha+(1-\alpha)\exp\left[-(\gamma t)^\beta\right]\right\}^{a+1}}\times\left\{1-\left\{\frac{\alpha-\alpha\exp\left[-(\gamma t)^\beta\right]}{\alpha+(1-\alpha)\exp\left[-(\gamma t)^\beta\right]}\right\}^a\right\}^{-1}. \quad (3)$$

We assume that electric actuators operate under similar conditions if their respective numbers of cycles to failure fall into the same bin of the histogram (figure 2). Proceeding from this assumption, seven reliability functions were obtained, each corresponding to the relevant bin (figure 3).

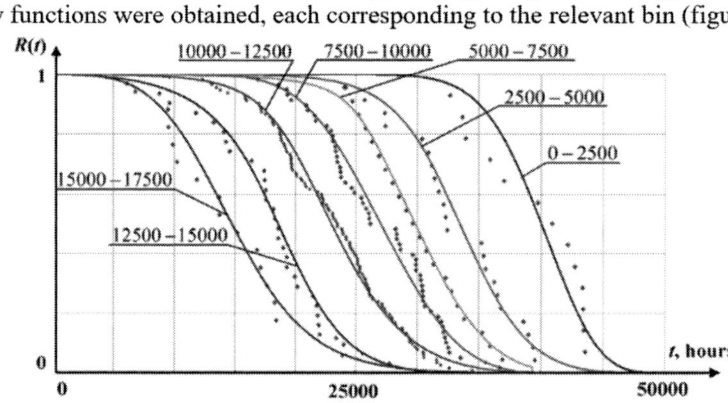

Figure 3. Reliability functions for the different bins.

Point estimations of parameters $\hat{\mathbf{\Theta}}_i$ for the reliability functions, obtained by maximum likelihood estimation [11-13], are accumulated in table 1.

Since each bin of the histogram is attributed with the membership value, we plot these reliability functions in 3D space, assigning each one with the corresponding value along Z-axis.

Table 1. Estimated parameters for Kw-CWG distribution

	Number of bin						
Parameter	1	2	3	4	5	6	7
α	1.28	0.16	0.03	0.72	0.06	0.09	0.18
β	10.59	4.94	5.19	5.27	2.93	2.71	1.52
γ	$2.63\cdot10^{-5}$	$3.95\cdot10^{-5}$	$5.18\cdot10^{-5}$	$5.25\cdot10^{-5}$	$7.77\cdot10^{-5}$	$8.57\cdot10^{-5}$	$1.4\cdot10^{-4}$
a	1.76	1.33	0.59	1.82	0.99	0.79	1.79
b	0.49	0.35	0.10	0.13	0.25	0.43	0.68

The result is a fuzzy-valued reliability function $\tilde{R}(\Theta,t)$ presented itself as a 3D surface (figures 4, 5).

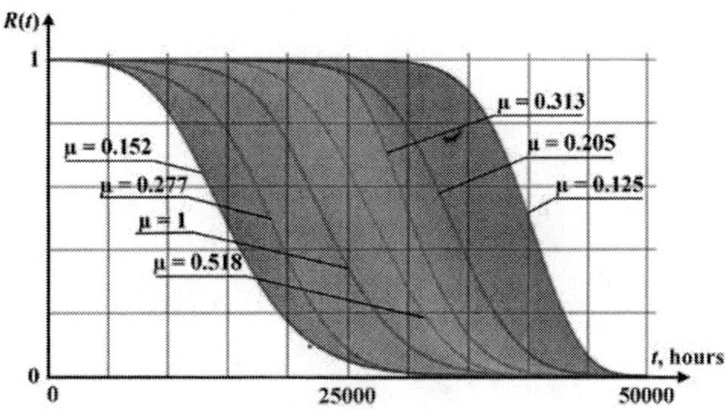

Figure 4. Plane view of fuzzy-valued reliability function of electric actuator.

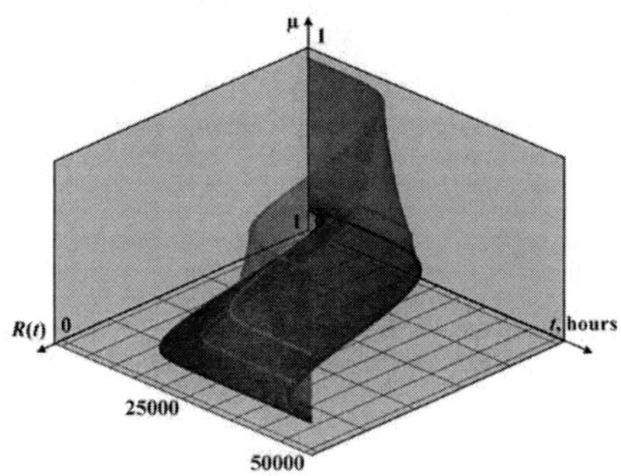

Figure 5. Isometric view of fuzzy-valued reliability function of electric actuator.

Fuzzy-valued hrf (figures 6, 7) was obtained by a similar way, provided that estimated parameters were substituted into equation (3).

Figure 6 shows the absence of decreasing hazard rate segment for electric actuators, indicating that there were no errors committed during construction and start-up operations and that all flawed items were rejected earlier.

IOP Conf. Series: Materials Science and Engineering **177** (2017) 012027 doi:10.1088/1757-899X/177/1/012027

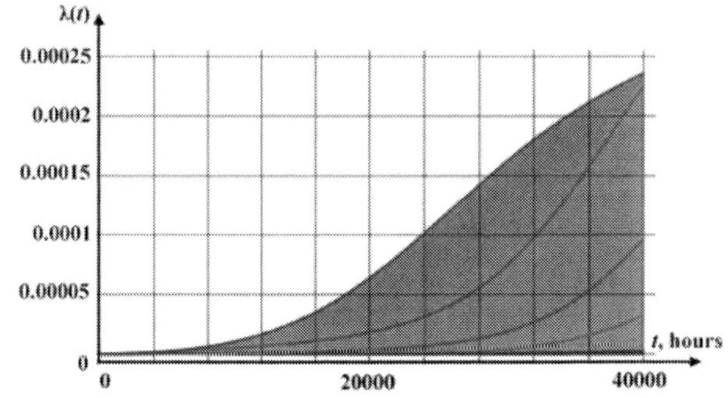

Figure 6. Plane view of fuzzy-valued hrf of electric actuator.

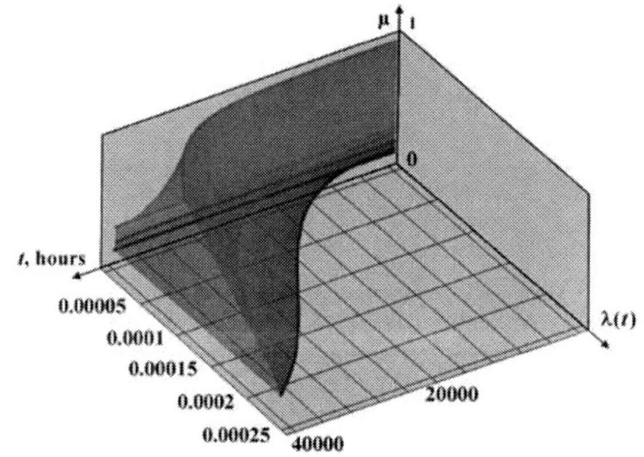

Figure 7. Isometric view of fuzzy-valued hrf of electric actuator

Useful results can be obtained by cutting the surface plot of the fuzzy-valued reliability function with planes, perpendicular to the coordinate axes. For example, if the cutting plane is defined as $t = t^*$, the trace of the surface can be interpreted as a fuzzy value of reliability for a mission time t^*. Similarly, cutting $\tilde{R}(\Theta, t)$ with a plane $R = \gamma$ gives us a fuzzy value of $100(1-\gamma)th$ percentile life [1, 7]. Figure 8 shows the examples of such cuts for the electric actuators under study.

These results can be defuzzified by Centre of Gravity (or any other) method [14, 15] in order to obtain the crisp values of reliability and percentile life.

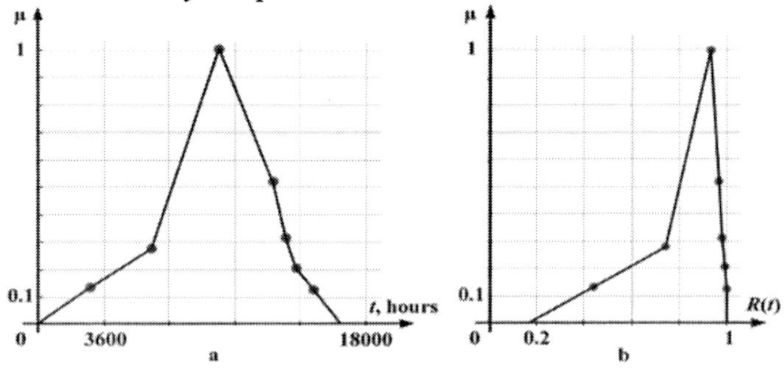

Figure 8. Traces of fuzzy-valued reliability function for electric actuators: a – fuzzy percentile life; b – fuzzy reliability.

4. Conclusion

In this paper the approach for determining of fuzzy reliability measures was suggested. Within this framework, fuzzy-valued reliability function of an electric actuator was obtained, as well as its hazard rate function. These measures can be applied along with material approach of reliability assessment, based on the analysis of electrophysical effects and mechanical friction and wear-out processes, in order to increase fault tolerance of electric actuators.

Informative and examined in detail performance reliability measures are essential for RCM analysis of technical systems. It is necessary to identify the causes of equipment failures and to acquire adequate reliability models in order to enhance the efficiency of preventive maintenance management program. In particular, fuzzy-valued reliability measures of an electric actuator, acquired in this paper, facilitate the integration of time and work completed before the failure and, hence, the assessment of their joint influence on system performance.

References

[1] Stapelberg R F 2009 *Handbook of Reliability, Availability, Maintainability and Safety in Engineering Design* (London: Springer-Verlag London)
[2] Manzini R, Regattieri A, Pham H and Ferrari E 2010 *Maintenance for Industrial Systems* (London: Springer-Verlag London)
[3] Yang G 2007 *Life Cycle Reliability Engineering* (Hoboken, NJ: John Wiley & Sons, Inc.)
[4] Nakagawa T 2008 *Advanced Reliability Models and Maintenance Policies* (London: Springer-Verlag London)
[5] Tang Y, Zou Z, Jing J, Zhang Z and Xie C 2015 *J. Nat. Gas Sci. Engineering* **26** pp 1050-58
[6] Tsukao S, Sato T, Hiramoto K, Kida J, Toyoda M and Ito H 2008 *42nd Int. Conf. on Large High Voltage Electr. Syst. CIGRE 2008* (Paris) A3-108
[7] Yefremov A A 2014 *Proc. Int. Conf. on Mechanical Engineering, Automation and Control Systems MEACS 2014* (Tomsk) 6986900
[8] Luneva E E, Banokin P I, Yefremov A A and Tiropanis T 2016 *Key Eng. Mat.* **685** 847-51
[9] Sedova N A, Sedov V A and Glushkov S V 2016 *Vibroengineering Procedia* **8** 506-511
[10] Sedov V A, Sedova N A and Glushkov S V 2016 *Vibroengineering Procedia* **8** 453-458
[11] Lefebvre M 2006 *Applied Probability and Statistics* (New York: Springer Science+Business Media LLC)
[12] Afify A Z, Cordeiro G M, Butt N S, Ortega E M M and Suzuki A K 2016 A new lifetime model with variable shapes for the hazard rate *Brazil. J. Prob. Stat.* In press
[13] Birolini A 2014 *Reliability Engineering* (Berlin: Springer-Verlag Berlin Heidelberg)
[14] Siler W and Buckley J J 2005 *Fuzzy Expert Systems and Fuzzy Reasoning* (Hoboken, NJ: John Wiley & Sons, Inc.)
[15] Buckley J J 2005 *Simulating Fuzzy Systems* (Berlin: Springer-Verlag Berlin Heidelberg)

A simple algorithm for distance estimation without radar and stereo vision based on the bionic principle of bee eyes

A A Khamukhin

Institute of Cybernetics, Tomsk Polytechnic University, 30, Lenin ave., Tomsk, 634050, Russia

Email: aaxtpu@tpu.ru

Abstract. Simple navigation algorithms are needed for small autonomous unmanned aerial vehicles (UAVs). These algorithms can be implemented in a small microprocessor with low power consumption. This will help to reduce the weight of the UAVs computing equipment and to increase the flight range. The proposed algorithm uses only the number of opaque channels (ommatidia in bees) through which a target can be seen by moving an observer from location 1 to 2 toward the target. The distance estimation is given relative to the distance between locations 1 and 2. The simple scheme of an appositional compound eye to develop calculation formula is proposed. The distance estimation error analysis shows that it decreases with an increase of the total number of opaque channels to a certain limit. An acceptable error of about 2 % is achieved with the angle of view from 3 to 10° when the total number of opaque channels is 21600.

1. Introduction

Simple navigation algorithms are needed for small autonomous unmanned aerial vehicles (UAVs) used, for example, for an intelligence service. These algorithms can be implemented in a small microprocessor with low power consumption. This will help to reduce the weight of the UAVs computing equipment and to increase the flight range.

Satellite navigation systems (GPS, GLONASS, etc.) are highly vulnerable to various classes of interference. In addition, they do not solve the problems of the local navigation (collision avoidance, detection of the targets, autonomous takeoff and landing, etc.).

Radars are the most often used in the distance and speed measurements, but radar's radiation is easy to determine, which is unacceptable in secret flying. Stereo vision requires a precise installation of cameras and complex computer image processing.

Bees and other flying insects have extraordinary navigation capabilities without radars and stereo vision, despite their tiny brains. Scientists try recreating a similar light-weight imaging system, which could change the way to build mobile robots and small flying vehicles.

Researchers from the University of Bielefeld (Germany) designed a small catadioptric system with a field of view of 280 degrees, simulating the eyes of a bee [1]. However, this system does not divide the optical beam into parts, as it occurs in the eye of a bee, and related to catadioptric systems, which contain elements reflecting and refracting light; in other words, they are composed of mirrors and lenses. Therefore, the processing of the optical flow requires a sufficiently powerful microprocessor (Core 2 Duo T7500 with 2.2 GHz [1]).

Fundamental review [2] describes the principles of visual guidance of honeybees. The experiments described in this review show that bees stabilize flight, negotiate narrow passages, and orchestrate smooth landings by using what seem to be a series of simple, low-level visual reflexes. But they do not tell us whether flying bees 'see' the world in three dimensions in the way we do. It was shown only that the bees were able to distinguish flowers at different heights. 'Under the experimental conditions, the only cue that a bee could have used to gauge the height of each flower would be the speed of the flower's image as a bee flew over it: the higher the flower, the faster the motion of its image.' [2].

Additional experiments have shown how it is possible to change trajectories of bee landing when the disk displayed a four-arm spiral pattern that was either stationary or rotating at speeds of 0.5 or 1 rotation per second, either in the counterclockwise or the clockwise direction, so as to increase or decrease the apparent rate of expansion, respectively [3].

Australian scientists create a guidance system for achieving automatic landing of a fixed-wing aircraft in the unstructured outdoor terrain, using onboard video cameras. The system uses optic flow information for sensing and controlling the height above the ground. At low heights, when the optic flow is unreliable, stereo information is used to guide the descent, close to touchdown [4]. The main advantages of such systems are that they do not emit active radiation and do not rely on any external information such as a global positioning system or an instrument landing system.

An onboard monocular vision system for autonomous takeoff, hovering and landing uses the letter "H" surrounded by a circle with known sizes [5]. A vision algorithm for robust and real-time landing pad recognition is implemented. But we are interested in the distance estimation of unknown size objects.

Yet another video method for movement measurements is an analysis of the images sequence received by a single camera [6]. Police use this method for vehicle velocity measurements on the roads. But it needs a powerful computer that is caused by a complex image processing [6]. Thus, this method is not applicable for small UAVs.

Thus, there are experiments showing that bees (and similar flying insects) are able to determine the distance to the target object of unknown size using the optical flow. However, published algorithms are based on determining the distance using stereo vision, optic flow analysis frame by frame or using geometric shapes with known parameters. Algorithms based on stereo vision or optic flow analysis are very complex to implement in a small microprocessor. Algorithms based on using geometric shapes with known parameters are unacceptable for unknown size objects.

2. Research objective

The aim of this study is to create a simple algorithm for distance estimation to unknown size objects, using some properties of a bee eye. We assume that the main bee eyes feature is the segmentation of the entire image to the pixels via opaque channels named as ommatidia (Figure 1). Such division allows a preliminary signal processing that does not require the participation of a small bee brain, which is often used in engineering: see, e.g. [7].

We will also assume an equable motion toward the target object. It should be noted that this is not a complete list of special properties of bee eyes, but in this case, it is enough to create a simple algorithm for the distance estimation.

3. Research methodology

The research method is simply based on the analysis of the sensor in motion similar to the compound eye of flying insects (Figure 1).

'Apposition compound eyes are made up of ommatidia. In conventional apposition eyes, the receptive rod (rhabdom) acts as a detector that measures the average brightness of a small region of space, typically about 1° across. The overall erect image seen by the animal is the mosaic formed by these adjacent fields of view.' [8].

Each pair maximally accepts only the normally impinging rays.

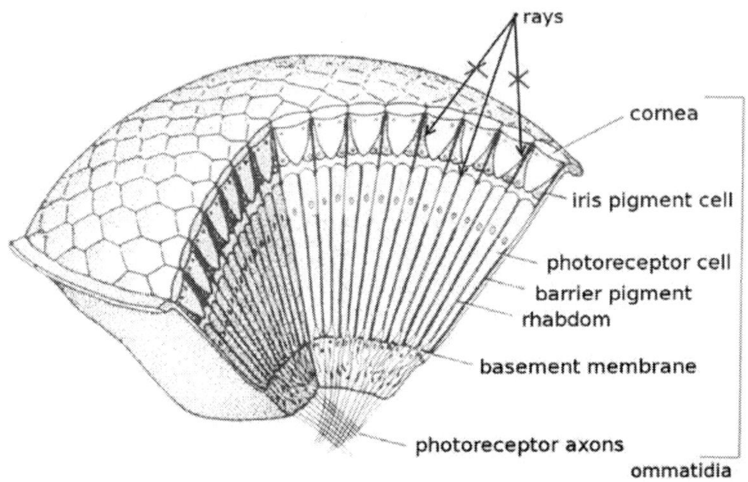

Figure 1. The schematic of a typical compound eye, inside and out. Modified from [7] Figure 7.3.

There are more complex superposition eyes of flying insects. Normally impinging rays are perceived as not only by one of the ommatidia, but also be those that are nearby. Resolution can be as good as in an apposition eye with similar-sized facets, and the sensitivity is usually much greater than in an apposition eye of the same size [8]. In this study, we will not consider such superposition eyes.

We used a simple two-dimensional model of the compound eye to keep simple analytical formulas. Further, they can be easily converted into a three-dimensional model.

We also did not consider the problem of selection of the target object in the landscape, which can be solved by using a two-stage continuous wavelet transform [9] and by artificial neural networks [10].

4. The research scheme and results

The simple two-dimensional research scheme for distance estimation by a compound eye is presented in Figure 2.

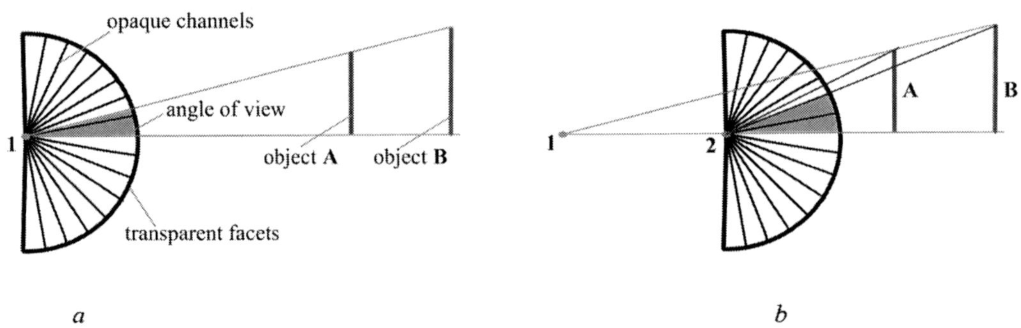

Figure 2. The simple two-dimensional research scheme for distance estimation by an observer with a compound eye: *a* – the eye center is in location 1 where the angles of view for objects A and B are the same; *b* – the eye center shifted to location 2, where the angles of view for objects A and B are various.

This research scheme demonstrates that two objects with different heights at various distances look the same, because they have the same angle of view (Figure 2a). They can only be discerned by an observer moving from location 1 to location 2 (Figure 2b). Such research scheme makes it easy to calculate the distance and height of objects A and B relative to the distance traveled by an observer from location 1 to 2 (d_{12}):

$$\frac{h_B}{d_{1B}} = tg\alpha,$$
$$\frac{h_B}{d_{1B} - d_{12}} = tg\beta,$$
$$d_{1B} = \frac{tg\beta}{tg\beta - tg\alpha} d_{12},$$
$$h_B = \frac{tg\alpha \cdot tg\beta}{tg\beta - tg\alpha} d_{12}, \qquad (1)$$

where d_{1B} is the distance between location 1 and object B, h_B is the height of object B, α is the angle of view of object B from location 1, β is the angle of view of object B from location 2.

The distance between location 1 and object A and the height of object A can be calculated similarly by equations (1).

Equations (1) are based on the measurement of angles of view of objects. But they are very intricate. Bees' eyes are probably able to assess the relative number of receptors that are hit by the rays of the target received through ommatidia of a compound eye. The device, similar to a compound eye, can calculate the relative amount of opaque channels through which the beams fall on the target. This relative amount will correspond to the angle of view with some error. It is obvious that it can be used instead of the tangent in equations (1). The tangent of an angle is equal to this angle at low angles with an error of about 2 % (to 15°) and with an error of about 10 % (to 30°). This error is quite acceptable in the assessment of long distances in order to avoid collisions. Then, equation (1) can be rewritten as follows:

$$d^*_{1B} \approx \frac{N_{2B}}{N_{2B} - N_{1B}} d_{12},$$
$$h^*_{1B} \approx \frac{N_{1B} \cdot N_{2B}}{N_{2B} - N_{1B}} d_{12}, \qquad (2)$$

where N_{1B} and N_{2B} are the relative amount of opaque channels through which the beams fall on the target from location 1 and 2, respectively.

5. Error estimation

The first source of error caused by the replacement of the tangent of the angle by the angle value is described above.

The second source of error is a resolution error. It is caused by a limit of the number of opaque channels. The number of ommatidia in various flying insects varies from 5 to 28 thousand pcs. It is limited by the diffraction of light. Diffraction effects are generally most pronounced for a wave which wavelength is roughly comparable to the dimensions of the diffracting object or slit. The device, similar to a compound eye, can be made larger than the eyes of a bee. It will decrease diffraction effects.

Therefore, error estimation of the formulas (2) depending on the number of channels in this device is very interesting. Error δ was calculated as:

$$\delta = \frac{d_{1B} - d^*_{1B}}{d_{1B}} 100\%. \qquad (3)$$

Formula (2) takes into account both sources of the error noted above.

Figure 3 shows the relationship between the calculated distance error and the angle of view of the target object for the different number of opaque channels through which the beams fall from the target.

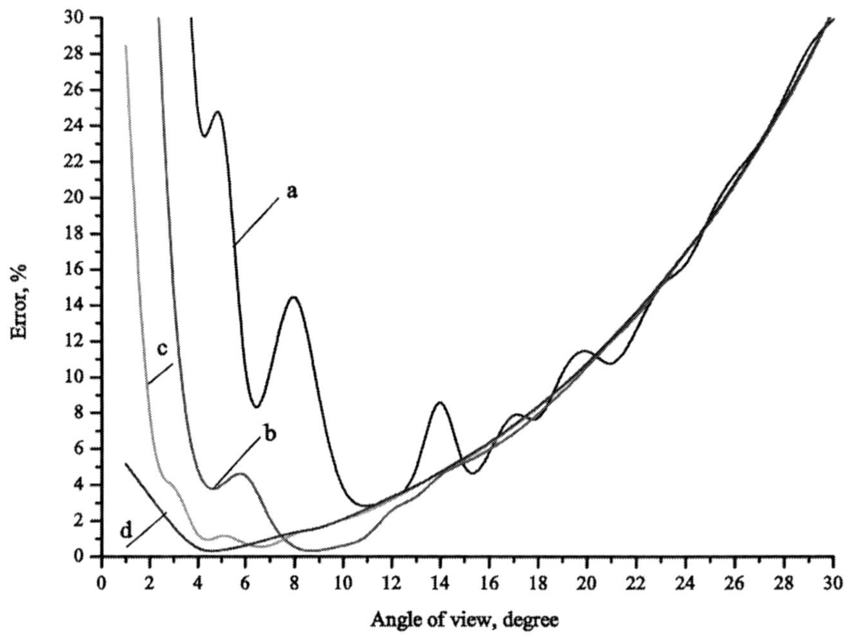

Figure 3. Relationship between calculated distance error and the angle of view of target object for the research scheme of a compound eye: *a*, *b*, *c*, *d* are the curves for the number of opaque channels 720, 4320, 21600, 43200, respectively.

We can observe in Figure 3 that the distance error using the compound eye depends on the angle of view of the target object and the total number of opaque channels. Increasing the total number of opaque channels from 720 to 21600 allows reducing the error very significant.

The error of about 2 % and lower is achieved when the angle of view is from 3 to 10°, and the total number of opaque channels is 21600. It corresponds to 1' of the angle resolution. The further increase of the total number of opaque channels does not reduce the error significantly except for tiny angles of view.

So the total number of opaque channels and the angle of view are very important characteristics for designing of this device.

6. Discussion

Which error is acceptable for the distance estimation using this algorithm?

Figure 3 shows that the distance estimation error is very high at near and far distances, when we see the target at a viewing angle of less than 2° or above 15°.

Obviously, the accuracy of 10 % or even 2 % is unacceptable for precise landing. It is known that bees have three simple eyes in addition to a pair of compound eyes. Perhaps a bee uses them for short-range navigation, when the compound eyes are working with a serious error. Australian scientists create their system for automatic landing of a fixed-wing with a similar decision. At low heights, when the optic flow is unreliable, stereo information is used to guide the descent close to touchdown [4].

However, such type of an error, from 2 to 10 %, is acceptable for avoiding collisions while moving. To avoid collision, we do not need to measure the distance to the obstacle with high accuracy. Therefore, the proposed algorithm is quite suitable for the collision avoidance function for UAVs. Perhaps, there are still applications for this algorithm, which we do not know yet.

7. Conclusions

The simple algorithm for distance estimation without radar and stereo vision based on the bionic principle of bee eyes is created.

The simple two-dimensional research scheme for distance estimation by a compound eye is presented.

The formula for calculating the distance to an object of the unknown size, based on the calculation of the relative amount of opaque channels through which a target can be seen by an observer moving from location 1 to 2 toward a target, has been presented.

Analysis of the distance estimation error for the proposed algorithm has been executed. It was shown that the error decreases with an increase of the total number of opaque channels to a certain limit.

It is shown that the minimal error lower 2 % is achieved at the angle of view of a target from 3 to 10° when the total number of opaque channels is 21600. However, the acceptable error from 2 to 10 % is achieved at the angle of view from 2 to 20°.

References

[1] Sturzl W, Boeddeker N, Dittmar L and Egelhaaf M 2010 Mimicking honeybee eyes with a 280^0 field of view catadioptric imaging system *Bioinspiration & Biomimetics* vol 5 no 3 (Bristol: IOP Publishing Ltd)

[2] Srinivasan M V 2011 Honeybees as a Model for the Study of Visually Guided Flight, Navigation, and Biologically Inspired Robotics *Physiological Reviews* vol 91 no 2 (Rockville Pike: The American Physiological Society) pp 413–460

[3] Bairda E, Boeddeker N, Ibbotson M R and Srinivasan M V 2013 A universal strategy for visually guided landing *PNAS* vol 110 no 46 (Redwood City: HighWire Press) pp 1–6

[4] Thurrowgood S, Moore R J D, Soccol D, Knight M and Srinivasan M V 2014 A Biologically Inspired, Vision-based Guidance System for Automatic Landing of a Fixed-wing Aircraft *J. of Field Robotics* vol 31 iss 4(Hoboken: Wiley) pp 699–727

[5] Yang S, Scherer S A and Zell A 2013 An Onboard Monocular Vision System for Autonomous Takeoff, Hovering and Landing of a Micro Aerial Vehicle *J. of Intelligent & Robotic Systems* vol 69 iss 1 (Berlin: Springer) pp 499–515

[6] Roach J W and Aggarwal J K 1980 Determining the movement of objects from a sequence of images *IEEE Transactions on Pattern Analysis and Machine Intelligence* volume PAMI-2 iss 6 (New York: IEEE) pp 554–562

[7] Khamukhin A A and Khamukhin Al A 2014 Preprocessing of Coefficients for Reusable Continuous Wavelet Transform *Advanced Materials Research* vol 1040 (Pfaffikon: Trans Tech Publications) pp 975–979

[8] Land M F and Nilsson D E 2012 Animal Eyes (Oxford: Oxford University Press)

[9] Khamukhin A A, Perminov R I and Yagunov T A 2014 Numerical simulation of hydroacoustic noise signals detected by a two stages wavelet transform *Proc. Int. Conf. on Mechanical Eng., Automation and Control Systems (Tomsk)* (New York: IEEE) pp 1–4

[10] Schmidhuber J 2015 Deep learning in neural networks: An overview *Neural Networks* vol 61 (Amsterdam: Elsevier) pp 85–117

Automation of data processing and calculation of retention parameters and thermodynamic data for gas chromatography

A I Makarycheva[1] **and V A Faerman**[2]

[1] National research Tomsk State University, 36, Lenina ave., Tomsk, 634050, Russia
[2] National research Tomsk Polytechnic University, 30, Lenina ave., Tomsk, 634050, Russia

E-mail: sandra_tsu@mail.ru

Abstract. The analyses of automation patterns is performed and the programming solution for the automation of data processing of the chromatographic data and their further information storage with a help of a software package, Mathcad and MS Excel spreadsheets, is developed. The offered approach concedes the ability of data processing algorithm modification and does not require any programming experts participation. The approach provides making a measurement of the given time and retention volumes, specific retention volumes, a measurement of differential molar free adsorption energy, and a measurement of partial molar solution enthalpies and isosteric heats of adsorption. The developed solution is focused on the appliance in a small research group and is tested on the series of some new gas chromatography sorbents. More than 20 analytes were submitted to calculation of retention parameters and thermodynamic sorption quantities. The received data are provided in the form accessible to comparative analysis, and they are able to find sorbing agents with the most profitable properties to solve some concrete analytic issues.

1. Introduction

Despite the considerable progress in the hardware implementation of gas chromatography analyses, the software solutions in the field of chromatography processing data are slightly limited in appliance nowadays [1]. Many kinds of both foreign and native software solutions for gas chromatography data processing are known ('Chemstation', 'Multichrom', 'Kristall' etc.).

However, most of them do not correspond to the needs of investigators in full. This happens due to the variety of real analytic issues and their specificity of decision calculations. In particular, the main issue of such kind of software solution is the automation of performance of the quantitative chromatographic analysis, while for the characteristic of new chromatographic materials, the specific algorithm series for the processing of analytes retention parameters is needed, and the calculation of the range of volumes on their basis is required. Thus, the sorbent`s polarity and selectivity investigation is complicated by the existence of various approaches of these concepts definition [2-4]. In this regard, the investigator faces the need to modify methods of experimental data processing, and hence the necessity of programming modification in calculation methods is clear.

2. Chromatographic data operations

2.1. Chromatographic data acquisition

The investigation of chromatographic sorbents` properties was conducted on the gas chromatograph 'MAESTRO 7820' (Agilent Technologies). The device is equipped with a flame-ionization detector and an injection port for packed columns with an electronic gas flow/pressure control. The data acquisition was performed at temperatures of 150, 170, 200°C and with a carrier gas speed (helium) of 30 ml\min. Detector's signals were processed with a help of software solution ChemStation. The further experimental data processing was conducted in compliance with the given method.

2.1. Experiential data processing

The experiential data processing was conducted in compliance with the given method. The adjusted retention time (t'_R) is determined from the formula [11]:

$$t'_R = t_R - t_M ,$$

where t_R - retention time; t_M – time of inert compound being in the chromatographic system, min, in the function of which methane was used.

The adjusted retention volume (corrected retention volume V_r' [11]:

$$V'_R = (t_R - t_M)F_C = V_R - V_M ,$$

where F_C – volume velocity of a carrier gas, ml/min.

According to the next formula, the Kovach's retention indexes were calculated and then they were used as the relative retention parameters, which do not depend on little accidental fluctuations of chromatographic analysis parameters [11]:

$$I = 100\frac{\log t'_{R(x)} - \log t'_{R(n)}}{\log t'_{R(n+1)} - \log t'_{R(n)}} + 100n ,$$

where $t'_{R(x)}, t'_{R(n)}, t'_{R(n+1)}$ – the adjusted times of test sorbate retention and n-alkanes with carbon number n and n+1, eluting before and after the investigated sorbate.

The values of differential molar free adsorption energy of test compounds were calculated using the given formula [9]:

$$\Delta G_i = -2.3RTb ,$$

$$\Delta G_i = -2.3RT\left[\frac{I - 100n}{100}b + \log\frac{V_{gn}T}{273}\right],$$

where R = 8.314 J/mol·K; T – the column`s temperature, K; $I$ – the compound retention index at column`s temperature T; n – the carbon number value of one of the referent n-alkanes; $b = \log t'_{R(n+1)} = \log t'_{R(n)}$; t'_R – the adjusted retention time; V_{gn} – the specific retention volume of n-alkane with a carbon number value equal to n.

The values of specific retention volumes are determined from the formula [11]:

$$V_g' = \frac{3}{2}\frac{(t_R - t_M)F_C \cdot 273.15}{W_S T_C} \cdot \frac{(p_i/p_0)^2 - 1}{(p_i/p_0)^3 - 1},$$

where F_C – the volume velocity of a carrier gas ml/min; W_S – the sorbent`s mass, g; $T_C$ – the column`s temperature, °C; ρ_0 – the barometric pressure, Pa; ρ_i – the pressure in the column inlet, Pa.

Also, the calculation of thermodynamic adsorption`s characteristics is executed on the ground of the specific retention volume. The values of partial molar solution enthalpy are:

$$\Delta H = 2.306R\frac{\log V_{g(1)} - \log V_{g(2)}}{T_1^{-1} - T_2^{-1}},$$

where ΔH – partial molar solution enthalpy, R – universal gas constant; $V_{g(1)}$ and $V_{g(2)}$ – specific retention volumes at a temperature from columns T_1, T_2.

Isosteric heat of adsorption Q:

$$Q = \frac{\ln(V_{g(1)} / T_1) - \ln(V_{g(2)} / T_2)}{T_1^{-1} - T_2^{-1}}.$$

As noted previously, all calculated parameters are archived and used in further analyses.

3. Automation of chromatographic data processing

From an organizational point of view, the simplest but the least efficient approach of the operations mentioned above is the hand calculation with a help of a scientific calculator. The drawbacks of this approach are clear. Firstly, the analysis of large data volumes requires considerable time expenditures. Secondly, the error possibility is increasing with the complexity degree of applied mathematical methods of data processing. The detection and elimination of them may require additional efforts [5]. Thirdly, the raw data as well as the results of their analysis are kept without strict binding to a define format. That fact complicates their further appliance and perception by other members of collective. All complications mentioned above take place in the processing of chromatographic parameters. Within these frameworks, it is necessary to automate this process.

3.1. Methods of data processing automation

Despite the fact that the most obvious decision is to use some special software, which is oriented to data processing of physical and chemical investigations, their appliance appears to be still limited. Thus, one can talk about the 'flexibility' as one of the key requirements for software solutions. That limits the appliance of the proprietary products with a closed source.

As an example of an alternative decision, the development of special software solution can be considered. This solution can be generated using one of the object-oriented languages of high-level software engineering (such as C#, Visual Pascal, C++ etc.) and meet all requirements to the full extent. However, the imperfection of this approach is the necessity to engage the group of qualified software engineers for the development and maintaining of the software programme. In addition, it requires a particular specialist in the field of system analysis for task formalization and requirement list preparation. It is worth mentioning that as a rule, small research groups do not have enough financial assistance to engage some outside experts.

In view of this fact under the conditions of tight lack of available resources, the software solutions for data processing and storage with increasing frequency are received without engaging of professional software engineers [5]. These solutions are developed on the ground of available office applications and mathematical packages, which are common in the academic environment.

3.2. Methods of data processing automation

Thus, for example, the use of MS Excel spreadsheets is the common decision for chemical data processing [5]. On the one hand, the widespread use of this software solution can be explained with the mass availability of the MS Office package and MS Excel as a part of it. On the other hand, there is a majority of investigators, who are not IT professionals and have enough skills to use a spreadsheet.

The advantages of the MS Excel use are a good structure and visibility of experimental data and results of its processing. Moreover, relatively simple spreadsheets allow performing the organization of big similar data processing in accordance with simple algorithms, which require the sequence of arithmetical operations and logic condition tests only [5]. Nevertheless, the limitation of the given value set and their target in processing tabular data complicates the realization of more difficult algorithms that can process the results of investigation. Among these algorithms are, for example, value extrapolations of some data array, regression's modeling, solution of differential equations, Fourier transformations and other algorithmically difficult mathematical methods. To wide extent, the reason of the given drawback is in the impossibility of the direct realization in the field of spreadsheets of cycles and recursions [5]. Despite the fact that the realization of these algorithmic constructions in MS Excel is possible with a help of special software, which is written with the use of Visual Basic for the Applications language (VBA) [5], this process demands the involvement of professionals.

With the point of view of these algorithms realization, the mathematical packages have more potential. Their advantage is their combination of a large set of various mathematical functions with basic programming software, rather simple to be used by someone who is a non-programmer [6]. In general, mathematical packages Mathcad and Matlab are most frequently used for data processing and for scientific work automation, especially in the academic environment. Generally, the given software solutions have similar abilities (in relation to the solution of the generalized problem of processing of data, which were received during chemical experiments). However, they also have their own specificity. Therewith, Mathcad, having a visually oriented input language, is easier in mastering and seems to be a more preferred solution [6]. Besides, the lower academic license cost can also be its merit.

However, with the use of the mathematical package directly for data processing (and for their visualization), it is preferable to save the visibility of numerical data presentation. This is possible due to the appliance of spreadsheets. In this regard, the conceptual model of the solution, which is shown in Figure 1, was accepted.

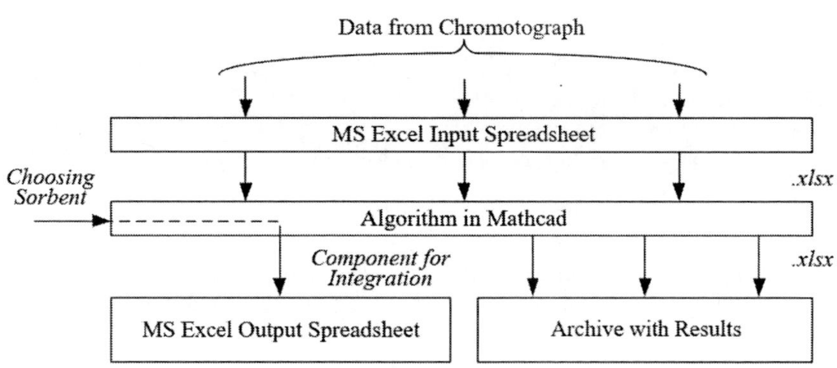

Figure 1. The conceptual diagram of the software solution.

4. Testing of software solution

In accordance with the conceptual diagram in Figure 1, the simple software solution was developed. It solves the problem of chromatographic data storage and efficient processing, as well as it meets all requirements mentioned above.

For the software solution test, the retention times of test compounds and the parameters of chromatographic analyses for the range of sorbents were processed as raw experimental data. Among them were silica gel of Silchrom 80 and the sorbents on its basis, which were adsorptively modified with 8-hydroxyquinoline and 1-phenylazo-2-naphtolate complexes of cuprum (II), cobalt (II) and nickel (II) [9-10]. For the appraisal of chromatographic polarity and selectivity of investigated chromatographic materials, the organic compounds of various classes were used: alkanes, alkenes, alcohols, aldehydes and ketones, nitro and aromatic compounds.

4.1. Input data setting

The retention times of test compounds, which were aggregated in the MS Excel file, are used as an input of the algorithm realized in Mathcad in the form of a two-dimensional array. It was the input data. In this connection, the time retention value, corresponding to the specific test substance, and experiments` conditions (corresponding to the column`s temperature and to the sorbent) were included in the cell. This cell was located at the intersection of the test compound line and the test condition column. A section of the spreadsheet with input data is displayed in Figure 2.

	A	C	D	E
1	Name	C80 4 Ni(SudanI)2	C80 4 Ni(SudanI)2	C80 4 Ni(SudanI)2
2	Date	April 2016	April 2016	April 2016
3	Temperature	150	170	200
11	hexane	0.487	0.378	0.313
12	heptane	0.628	0.482	0.368
13	octane	0.962	0.699	0.468
14	nonane	1.553	0.948	0.603
15	decane			
16	undecane			
21	ch2cl2	0.357	0.348	0.27
22	chcl3	0.453	0.396	0.312
23	ccl4	0.48	0.411	0.32
24	1,2-dichloroethane	0.6	0.488	0.352
25	nitropropane	2.646	1.483	0.784
26	heptene-1	0.747	0.541	0.399
27	benzene	0.599	0.473	0.349
28	toluene	0.966	0.7	0.459

Figure 2. Part of Excel spreadsheet with input data.

Due to the fact that every column of input data, brought to the programme, represents the set of experiential results with various test substances, produced in the conditions of the fixed column`s temperature and with the single sorbent, its processing was fulfilled separately from other data. These results should be recorded separately too.

Thus, for the choice of analysis data, the standard element of control ListBox was realized in the Mathcad with a help of VBScript (Visual Basic Scripting Edition). It allows choosing the only option from the given set, in which the quantity of elements was defined by the number of columns in the input array.

4.2. Presentation of processing results

With foundation of retention time values, which a chosen array contains, the calculations of target values for each of the used test compound were executed. The results of calculations were shown in the MS Excel form with the use of a Mathcad spreadsheet integrated component. The results of experimental data processing are included in table 1.

Thus, using the example of Silochrom C80, which was modified with 1-phenylaso-2-naphtholates of metals, the appraisal of chromatographic polarity with the use of software calculated values ΔG, ΔH, Q was realized. These data are convenient to compare with characteristics of other sorbents in order to choose the best one for separation of a definite class of organic compounds.

Table 1. Fragment of table with calculated chromatographic parameters and thermodynamic data for Silochrom C80 modified with 1-phenylaso-2-naphtholate of nickel (II)

Agents	t_R	t'_R	V'_R	V_g	I	ΔG	ΔH	Q
Benzene	0.60	0.35	10.47	1.69	675.00	3.31	37.87	4.98
Etanol	1.37	1.12	33.63	5.42	878.00	7.52	98.54	12.27
Butanon-2	3.91	3.66	109.77	17.68	1084.00	12.15	55.76	7.13
Nitropropan	2.65	2.40	71.88	11.58	1010.00	104.65	55.15	7.06
Pyridin	20.05	19.80	594.03	95.70	1378.00	18.89	82.06	10.29

5. Conclusion

With a use of software package Mathcad, combining with the MS Excel spreadsheets` abilities, the method of chromatographic information processing, which allows handling the big data volume, was developed. The big data volumes show the information about the date and the investigation`s objects (analytes` nature and sorbate materials, which were used), physical properties of a chromatographic process (column`s temperature, carrier gas speed, pressure in a column inlet etc.). Besides, they show the direct results of chromatographic analyses (times of analytes retention). The received method allows processing chromatographic data quickly and with satisfactory accuracy, giving the results of retention parameters` calculation and thermodynamic quantities of analytes` sorption in the format, which is convenient for the further interpretation.

Based on the data, obtained from approbation of the introduced solution, it is shown that the modification of Silochrom leads to a decrease of investigated sorbent`s polarities in relation to electron-donating molecules. The ΔG volumes are minimal for aromatic compounds. It is connected with their lower ability to create π-complexes in comparison with other kinds of molecular interactions. Molar adsorption energy of the methylene unit increases in reference to the unmodified silica gel, which is reflected in the increasing of n-alkanes elution time.

The realized software solution allows a chromatographist to make modifications and additions to the data processing algorithm on their own. It widens the limits of its applicability for calculation of various values on the basis of chromatographic retention parameters for the purpose of the detailed survey and the characteristic of new sorption materials.

Reference

[1] Miller J M 1988 *Chromatography: Contrasts and Concepts* (New York: Willey)
[2] Abraham M H, Poole C F and Poole S K 1999 *J. Chromotograph. A* **842** 79-114
[3] Schurg V and Juza M 1997 *J. Chromotograph. A* **757(1)** 119-35
[4] Zhang Y and Carr P W 2009 *J. Chromotograph. A* **1216(39)** 6685-6694
[5] Tanganova BB and Krupennikova V E 2011 *Int. J. of Appl. and Basic Res.* **5** 35-38
[6] De Levie R 2004 *How to use excel in analytical chemistry and in general scientific data analysis* (Cambridge: Cambridge University Press)
[7] Maxfield B 2006 *Engineering with Mathcad: using Mathcad to create and organize your engineering calculations* (Oxford: Butterworth-Heinemann)
[8] Zielenski T J 1998 *J. Chem. Educ.* **75(9)** 1189-1191
[9] Makarycheva A I and Slizhov Yu G 2016 *Key. Eng. Mat.* **670** 246-251
[10] Makarycheva A I and Slizhov Yu G 2014 *Adv.Mat.Res.* **1040** 405-409
[11] McNair H M and Miller J M 2011 Basic gas chromatography (New Jersey: Willey)
[12] Sweredoski M J, Smith G T, Kalli A, Graham R L J and Hess S 2011 *J. of Biomolecular Techniques* **22(4)** 122-126

Scheduling based on a dynamic resource connection

A E Nagiyev[1], I A Botygin[1], A I Shersntneva[1] and P A Konyaev[2]

[1] Tomsk Polytechnic University, 30, Lenina Ave., Tomsk, 634050, Russia
[2] V E Zuev Institute of Atmospheric Optics Siberian Branch of Russian Academy of Sciences, 1, Academician Zuev Sq., Tomsk, 634055, Russia

E-mail: andrew_nagiev09@mail.ru

Abstract. The practical using of distributed computing systems associated with many problems, including troubles with the organization of an effective interaction between the agents located at the nodes of the system, with the specific configuration of each node of the system to perform a certain task, with the effective distribution of the available information and computational resources of the system, with the control of multithreading which implements the logic of solving research problems and so on. The article describes the method of computing load balancing in distributed automatic systems, focused on the multi-agency and multi-threaded data processing. The scheme of the control of processing requests from the terminal devices, providing the effective dynamic scaling of computing power under peak load is offered. The results of the model experiments research of the developed load scheduling algorithm are set out. These results show the effectiveness of the algorithm even with a significant expansion in the number of connected nodes and zoom in the architecture distributed computing system.

1. Introduction

At present, distributed computing systems (DCS) are increasingly intensely [1]. Parallel processing using a large number of computing nodes is a distinctive feature of the distributed computing systems. The advantage of this approach is the possibility of a rapid increase in productivity by the horizontal scaling of computing nodes and resources.

At the moment, there are three types of distributed systems [2]:

- A cluster —a certain number of computing nodes, combined with the LAN. Resources are used by only one working group [3]. In the DCS, clustering is performed only at the software level.
- An enterprise-class computing system is a computer system. The resources of this system are used to fulfil the tasks of several working groups.
- A Grid System — a system that combines a large number of geographically dispersed computing nodes. This system is intended for simultaneous processing of a large number of working groups.

2. DCS and Load Balancing

2.1. Problems of the DCS

There are some problems, which are associated with the development of the DCS:

- The inability to define the global time for the entire distributed computing system. Each node of the system has its own astronomical time. Sometimes it creates difficulties connected with synchronization of the tasks on all of the nodes.

- Communication between nodes is carried out not instantaneously and with significant delays. This requires additional steps for the development of the software and complicates the development of distributed computing systems.
- Communications are insecure that can lead to a loss of data during transmission between the nodes of the system.
- The complexity of the software configuration to complete the tasks on the system due to the heterogeneity of platforms on which the nodes of the distributed computing system are deployed.
- The complexity of ensuring an effective load distribution on the nodes of the DCS, etc.

Thus, the system development is hampered by the fact that computing nodes can have different productivity; communication lines have different bandwidths; tasks may require different processing power and others.

2.2. Description of the Load Balancing

According to the authors, the main attention should be paid to the distribution of load between the nodes (load balancing) in the development of distributed computing systems. Load Balancing (LB) is used to optimize the fulfilment of the distributed computing. Load Balancing allows one to distribute the computational load on the existing nodes in the system effectively. Software, using a predetermined algorithm of the LB, fulfils decomposition of the task, splitting it into modules after transferring the tasks to fulfil in the DCS. These modules have transmitted to the nodes in the system for further processing.

Load balancing is divided into static, dynamic and semi-dynamic [4-7]. The static balancing is fulfilled before the beginning of a task. Semi-dynamic LB assumes that the distribution of tasks is carried out at the stage of initialization. The dynamic load balancing task allocation takes place during the computation. If we imply the way LB interacts in a computer system, the balancing is divided into centralized and distributed. In the centralized LB, a special computing node (the so-called controller) distributes modules tasks between calculators. In the distributed LB using a special algorithm, the data of the status are exchanged between all nodes in the system.

Also, balancing is divided into universal and specialized (designed to fit the specified algorithm in a computing task). [8] At the moment, there are many techniques and load balancing algorithms. Among them are: Round Robin, Weighted Round Robin, Least connections, Weighted least connections [9-11].

3. The Method of a Dynamic Connection of Resources

In this paper, another method of load balancing in distributed computing systems was developed - the method of a dynamic connection of resources (DCR) as a result of the program experiments. Special program experiments were conducted to assess the degree of efficiency of the developed method for horizontal scaling calculators and increasing load.

During the program experiment, at first, the effectiveness of the classic server solution without load balancing was evaluated. Then carrying capacity of this system was compared with the carrying capacity of another system, which interaction of the elements was performed by a method of a dynamic connection of resources.

As it was expected during the program experiment, the classical server solution, which consists of a control server and terminals requests, proved to be ineffective at high loads. Namely, the program experiment conducted by the computer-based system, which consists of a management server and dynamically connected terminals requests from simulators, showed that increasing the number of incoming requests for processing it was failure of the entire system.

In the experiment, on the basis of the computer system using a quad-core processor and simulation of requests in separate threads by the law of uniform distribution at an interval of 10 ms, failure

IOP Conf. Series: Materials Science and Engineering 177 (2017) 012030 doi:10.1088/1757-899X/177/1/012030

occurred in the processing of 850 requests. From this it can be concluded that with the increase in frequency, the request generation system productivity decreases. The server does not have time to accept, process and send back the results. Therefore, to increase the effectiveness of tasks (performance, etc.), the method of dynamic connection of resources is used. This approach allows one to dynamically connect additional compute nodes, through which parallel computation of various tasks of the management server is provided. This provides server unloading.

The program experiment was conducted for a practical demonstration of the proposed method. This experiment is described below. The functional structure of a distributed system which is shown in figure 1 is used for the experiment. As can be seen, the main changes have happened to the management server, which was replaced by the main control center (MCC). The system infrastructure has changed. The main control center and the dynamic link load controllers have been added; they will be connected to the additional control centers (ACC). The generator of requests remained unchanged.

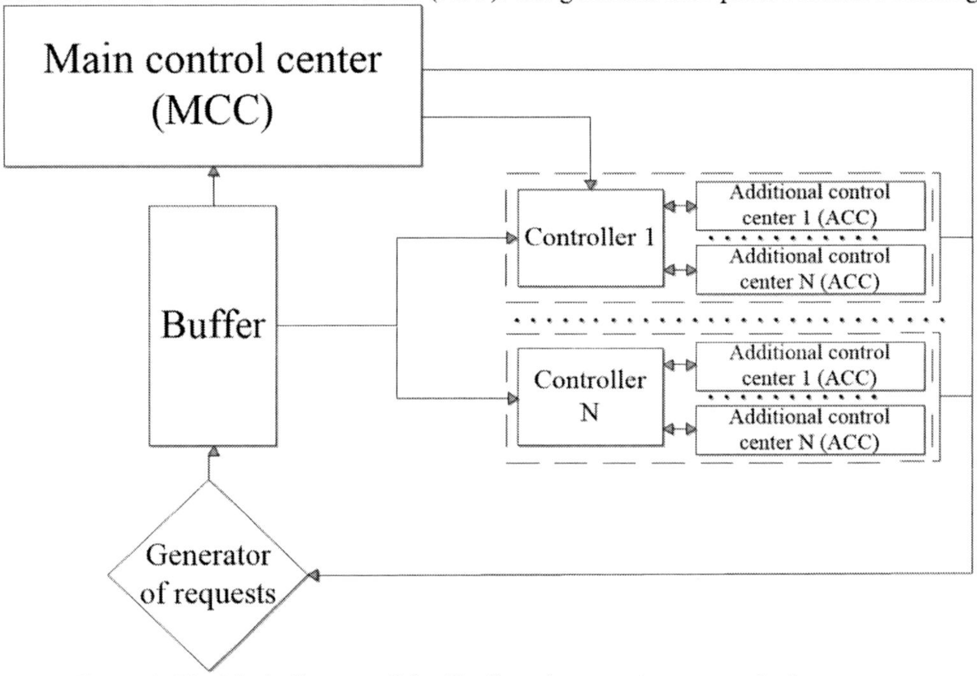

Figure 1. The block diagram of the distributed computing system in the program experiment.

Unlike the previous program experiment, the connection address for each request terminal is recorded in a special buffer of the MCC, which removes this connection after the request is processed and the task which has been set before is solved (Figure 2).

With the increase in the number of incoming requests, MCC does not have time to promptly process all incoming requests. It becomes a 'bottleneck' of the system, which causes the accumulation requests in the buffer. The controller of load balancing starts operating when a certain number of requests which are waiting to be processed in the MCC are reached (Figure 3). The controller starts to run additional control centers. Their number is limited by the rate of incoming requests and the presence of additional computing powers in the system.

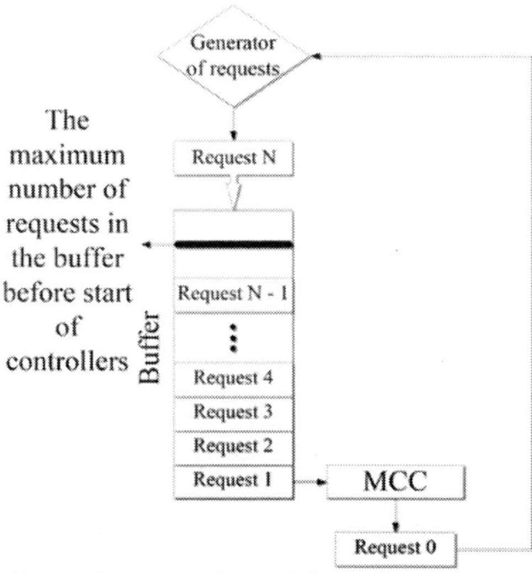

Figure 2. Processing of incoming requests by MCC without additional control centers.

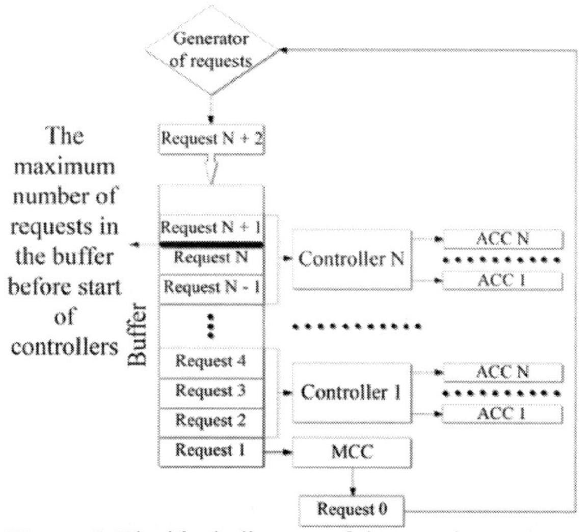

Figure 3. The block diagram of the sample requests which controllers process from the buffer and send them to ACC.

Each of the controllers runs in its own separate thread. The number of controllers is limited only by processing power of the distributed system infrastructure. The controller selects a certain number of requests from the buffer and sends them to the ACC, which it launched in separate threads. ACC is switched off only after processing requests, which were received from the controller.

4. Results of Experiment

As a result of experiments, we can say that the time of terminals' connection and the processing time of requests to MCC is reduced by 57% as compared to the classic version. During the same time, 58.5% more requests from the terminals connect to the MCC than those during working of the main control center without ACC.

During the connection of the second ACC, processing speed has increased by 13% relative to a system with single ACC. During the connection of the third ACC, processing speed has also increased by 24% relative to the system, which employs two additional control centers. The experimental results are shown in graph figure 4.

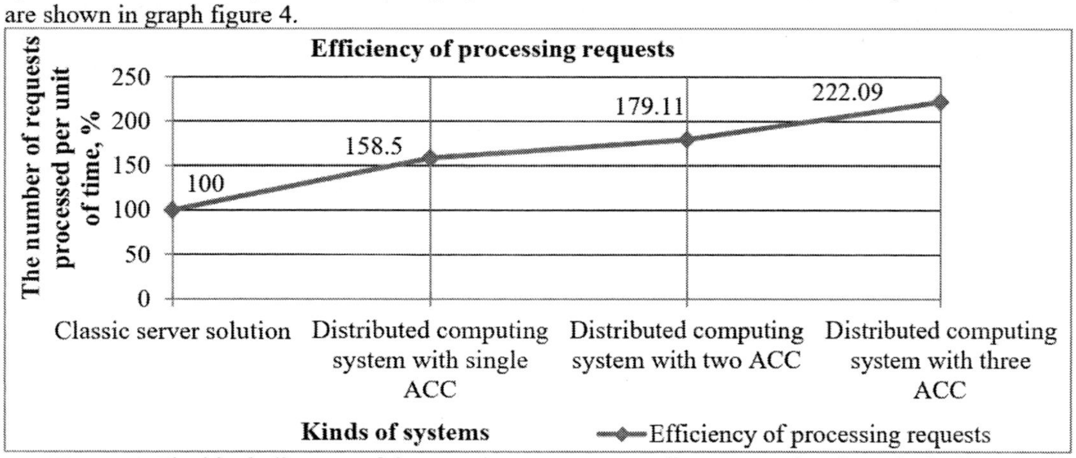

Figure 4. The block diagram of the sample requests controllers processing from the buffer and sending them to ACC.

Experiments were carried out by connecting different numbers of requests terminals (from 10 to 10 000) with uniform distribution in the interval from 0 to 100 arbitrary time units.

If considering CPU usages used in the program experiment qualitatively, the connection of even 250 terminals in substantially separate threads charge CPU. During the connection of 10000 applications for processing, we can clearly see that each of the 4-CPU cores is loaded evenly. In figures 5 and 6, we can see the difference between the load of CPU cores without processing requests from the simulator and with processing the large number of requests according to the method, which was described above.

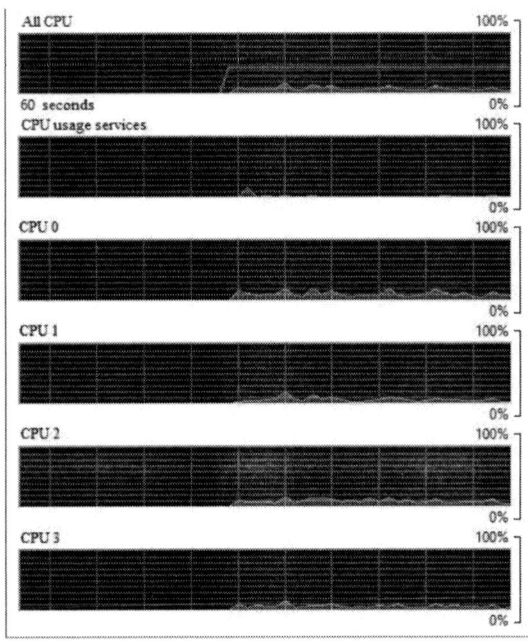

Figure 5. The block diagram of the sample requests controllers processing from the buffer and sending them to ACC.

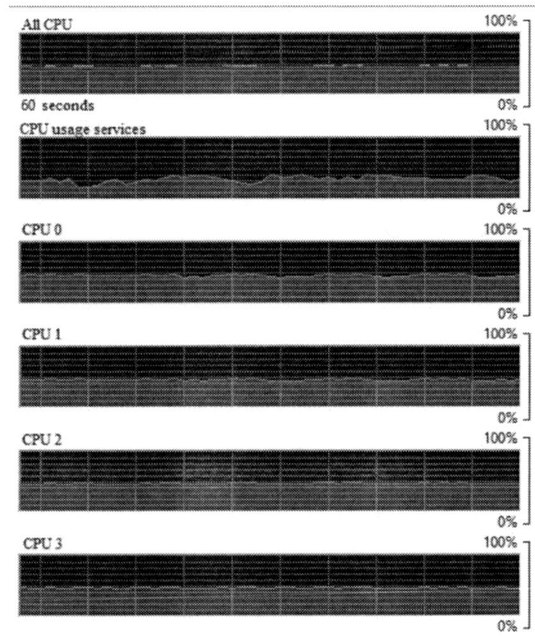

Figure 6. The block diagram of the sample requests controllers processing from the buffer and sending them to ACC.

As is seen from figures 5 and 6, there is a uniform distribution on the computation of the CPU cores based on pseudorandom numbers.

5. Conclusion

The generator of requests, which operates according to the law of uniform distribution and which plays a role of a terminal simulator, experimentally demonstrated performance, efficiency and flexibility of the proposed system, despite the fact that the experiment took place only on one computer with the quad-core processor. Due to this, the ensuring scale of the computing power occurs, which is limited only by the physical abilities of connected additional nodes.

The experiment, which is described above, has demonstrated the effectiveness of the method of dynamic connection of resources. Through modeling of this system, we can state that the considered method for dynamic connection of resources remains effective with a significant increase of the connection of computational nodes.

References

[1] Kshemkalyani A D and Singhal M 2008 *Distributed Computing: Principles, Algorithms, and Systems* (Cambridge, UK: Cambridge University Press)

[2] Tanenbaum A S and Van Steen 2007 *Distributed Systems: Principles and Paradigms* (*2nd*) (Upper Saddle River, NJ, USA: Pearson Education)
[3] Carlos A Varela and Gul Agha 2013 *Programming Distributed Computing Systems: A Foundational Approach Hardcover* (Cambridge, USA: MIT Press)
[4] Load balancing in distributed systems *Intuit* URL: http://www.intuit.ru/studies/courses/1146/238/lecture/6153
[5] Bershadskiy A M, Kurilov L S, Finogeev A G 2009 *Proc. of the Univ. Volga Region* vol 4 (Penza, Russia: PSU Publishing house) pp 38–48
[6] Qiu Y W and Hwang J I G 2016 *IEEE 14th Intl Conf. on Dependable, Autonomic and Secure Computing* (Piscataway, NJ, USA: IEEE Press) pp 565-571
[7] Sommer M, Klink M, Tomforde S and Hähner J 2016 *IEEE Int. Conf. on Autonomic Computing* (Piscataway, NJ, USA: IEEE Press) pp 300-307
[8] Sherstnyov V S, Sherstnyova A I, Botygin I A, Kustov D A 2016 *Key Eng. Mat.* **685** 867-871
[9] Sherstnev V S 2005 *The 9th Russian-Korean Int. Symp. on Science and Technology* vol 1 (Piscataway, NJ, USA: IEEE Press) pp 696-700
[10] Botygin I A, Nagiyev A E 2015 *Int. Conf. on Mechanical Engineering, Automation and Control Systems* (Piscataway, NJ, USA: IEEE Press) pp 1-4
[11] Bal H E, Kaashoek M F and Tanenbaum A S 1992 *IEEE Transactions on Software Engineering* vol 18 (Piscataway, NJ, USA: IEEE Press) pp 190-205

IOP Conf. Series: Materials Science and Engineering **177** (2017) 012031 doi:10.1088/1757-899X/177/1/012031

Models of neural networks with fuzzy activation functions

A T Nguyen[1] and A M Korikov[1,2]

[1] Tomsk Polytechnic University, 30, Lenina ave., Tomsk, 634050, Russia
[2] Tomsk State University of Control Systems and Radioelectronics, 40, Lenina ave., Tomsk, 634050, Russia

E-mail: nguyenanhtu@tpu.ru

Abstract. This paper investigates the application of a new form of neuron activation functions that are based on the fuzzy membership functions derived from the theory of fuzzy systems. On the basis of the results regarding neuron models with fuzzy activation functions, we created the models of fuzzy-neural networks. These fuzzy-neural network models differ from conventional networks that employ the fuzzy inference systems using the methods of neural networks. While conventional fuzzy-neural networks belong to the first type, fuzzy-neural networks proposed here are defined as the second-type models. The simulation results show that the proposed second-type model can successfully solve the problem of the property prediction for time – dependent signals. Neural networks with fuzzy impulse activation functions can be widely applied in many fields of science, technology and mechanical engineering to solve the problems of classification, prediction, approximation, etc.

1. Introduction

Artificial neural networks (NNs) are a computational approach that was developed based on the way the brain solves problems. To model biological neural systems, the elements of artificial NNs (neuron model) imitate the properties and functioning of biological neurons [1]. In recent years, we have seen increasing interest in NNs, as well as their successful applications in various fields, such as business, medicine, science, mechanical engineering, geology, etc. NNs have also been used to solve the problems of forecasting, classification, management, etc. In this paper, we propose the new model of fuzzy-neural networks (FNNs), in which the fuzzy membership functions (MFs) are used as activation functions.

2. Modeling of membership functions as the activation functions

In previous studies, the activation functions of NNs usually take the forms of a unit step function, a linear threshold or sigmoid function [2]. Moreover, impulse and Gaussian distribution functions [3] can be used for activation functions in an effort to simplify the structures of NNs. The results of the above-mentioned studies have revealed the complexity of activation function selections for the development of NNs. To this end, it is necessary to study the validity for a new model that utilizes MFs from the fuzzy system theory [4, 5] as NN activation functions.

According to literature [6], the fuzzy number is a convex, normalized fuzzy set which membership function is at least segmentally continuous and has the functional value at precisely one of the x values. This point x is referred to as the mean value of the fuzzy number.

The support of a fuzzy set is a crisp set and is defined as follows:

$$\operatorname{supp}(\tilde{A}) = [S_L, S_R] : \mu_A(x) > 0, \forall x \in [S_L, S_R].$$

Similarly, the core (kernel) of a fuzzy set is a crisp set:

$$\ker(\tilde{A}) = [K_L, K_R] : \mu_A(x) = 1, \forall x \in [K_L, K_R].$$

Thus, a fuzzy number is defined by four specific points:

$$\tilde{A} = \langle S_L, K_L, K_R, S_R \rangle. \quad (1)$$

However, the fuzzy number will be only completely determined by equation (1) if we know the type of MF. When creating a computer application, we use the MF LR-type, which is defined by the following expression [7]:

$$\mu_A(x) = \begin{cases} f_L(x), & x \in [S_L, K_L); \\ 1, & x \in [K_L, K_R]; \\ f_R(x), & x \in (K_R, S_R]; \\ 0, & x \notin [S_L, S_R]. \end{cases} \quad (2)$$

Equation (2) offers a large variety of MF selections, among which the triangular, trapezoidal and piecewise continuous polynomials are the most commonly-used.

In this paper, we propose a new form of MFs using triangular fuzzy number $N = \langle A, B, C \rangle$:

$$\mu(x) = f_L(x) \cdot H(x - A) \cdot H(B - x) + f_R(x) \cdot H(x - B) \cdot H(C - x), \quad (3)$$

where $f_L(x)$, $f_R(x)$ respectively denote the left and right parts of the MF, which is given by a second-order polynomial, $H(x)$ is the Heaviside unit function [4].

If $f_L(x)$, $f_R(x)$ are second-order polynomials and $f_L'(x) = 0$, $f_R'(x) = 0$ at the characteristic points of the fuzzy values, the MF given by equation (2) could be modified by one of the four following expressions:

$$\begin{cases} f_L'(A) = 0; \\ f_R'(C) = 0. \end{cases} (4); \quad \begin{cases} f_L'(B) = 0; \\ f_R'(B) = 0. \end{cases} (5); \quad \begin{cases} f_L'(A) = 0; \\ f_R'(B) = 0. \end{cases} (6); \quad \begin{cases} f_L'(B) = 0; \\ f_R'(C) = 0. \end{cases} (7)$$

The shapes of MFs using triangular fuzzy numbers based on the conditions given by equation (4) - equation (7) are shown in figure 1

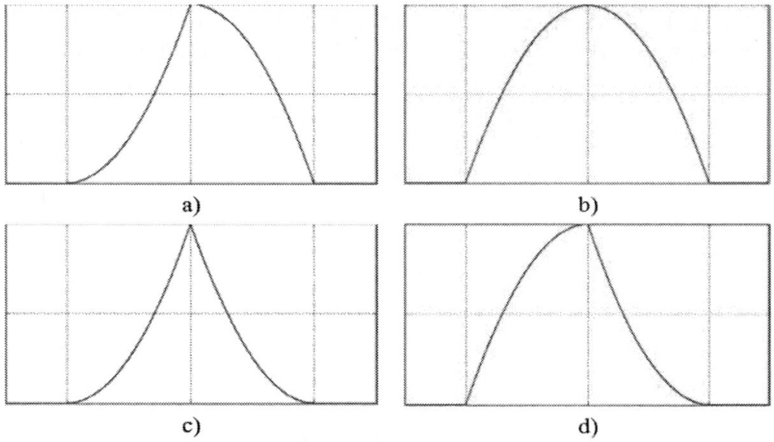

Figure 1. MFs with additional conditions: a) Conditions by equation (4); b) Conditions by equation (5); c) Conditions by equation (6); d) Condition by equation (7).

In next sections, we will investigate the feasibility of fuzzy neuron models and FNNs using these MFs as activation functions.

3. Models of fuzzy neurons

According to [8], a FNN is a clear neural network with a feed-forward signal based on a multi-layer architecture using AND – OR neurons

An AND–neuron is a neuron, in which the multiplication operation of weight w and input x is modeled by conorm $S(w,x)$, and the addition operation of weights is expressed by norm $T(w,x)$.

An OR–neuron is a neuron, in which the multiplication operation of weight w and input x is modelled by norm $T(w,x)$, and the addition operation of weights is given by conorm $S(w,x)$.

The concepts of T–norm and S–conorm were discussed in [8, 9]. T–norm and S–conorm are functions with special properties. They are the actual functions of two variables that are defined in the interval of [0, 1]×[0, 1], and the values of these functions are in the range of [0, 1]. The models of AND–fuzzy neurons and OR–fuzzy neurons are shown in figure 2.

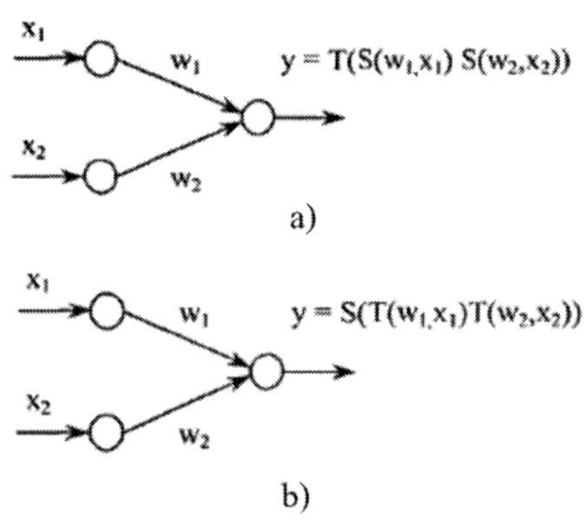

Figure 2. Models of fuzzy neurons:
a) AND–neuron; b) OR–neuron

The FNN, which is mentioned above, is termed the first-type FNN. The present paper proposes the second-type FNN model, whose activation functions have the forms of the MFs given by figure 1.

We can observe that in the first-type FNN model, the fuzzy relationships between neurons are similar to those between the elements of NNs, i.e., the first-type FNNs utilize fuzzy inference systems by NN methods. For the second-type FNN model, fuzziness is an attribute of neurons. While developing the structural representations for FNNs, it is found that it is necessary to study the third type of FNN, which is the combination of the first and second types.

4. Application of second-type FNNs to determine the fundamental frequency of signals with white noise

For many technical applications, it is necessary to identify several important characteristics, such as the frequency and the phase of a fundamental signal that can be interfered by some types of noise. This noise is usually modeled by white noise, which has a constant spectral power density distributed over the whole frequency domain.

In this paper, we investigate the feasibility of the second-type model of FNNs for the problem of determining the fundamental frequency of a time-dependent signal with noise. The methods of so-called 'theory of experiment' [10] are used to process the data of the experiment.

A signal used in this study is given as the combination of a sinusoidal signal and a white noise function:

$$y = a_1 \times \sin(2\pi ft) + a_2 \times r \tag{8}$$

where a_1 is an amplitude of the fundamental sinusoidal signal, a_2 is a constant related to the magnitude of the noise, r denotes a random function, which values are within the interval of $[0,1]$, f is the frequency of the fundamental sinusoidal signal.

The duration from 0 to 0.005 s is studied with 500 time steps, and fundamental frequency f is set between 1 and 2 Hz. Therefore, a set of discrete values of function $y(t)$ corresponding to the 500 time steps is used as an input to our FNN. A second-type FNN is applied to predict frequency f. Similar problems may arise in many applications of FNN, especially in the mechanical engineering.

A multilayer unidirectional network is used to create our FNN. The network consists of an input layer, a hidden layer and an output layer. The hidden layer comprises 10 neurons with fuzzy activation functions of types a) or d). A linear activation function is employed for the output layer. The program selects 70% of the input values for the training process; while the rest 30% are used for the validation process. The training process is based on the Levenberg-Marquardt algorithm.

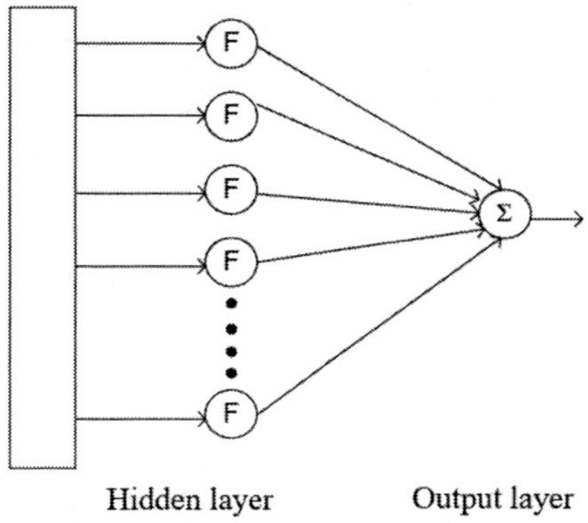

Figure 3. Network structure

Table 1. Test results

Fundamental frequency f (Hz)	Test results with MF of type a)	Test results with MF of type d)
1	1.030	1.165
1.2	1.379	1.332
1.4	1.516	1.373
1.6	1.606	1.678
1.8	1.739	1.758
2	2.012	2.046

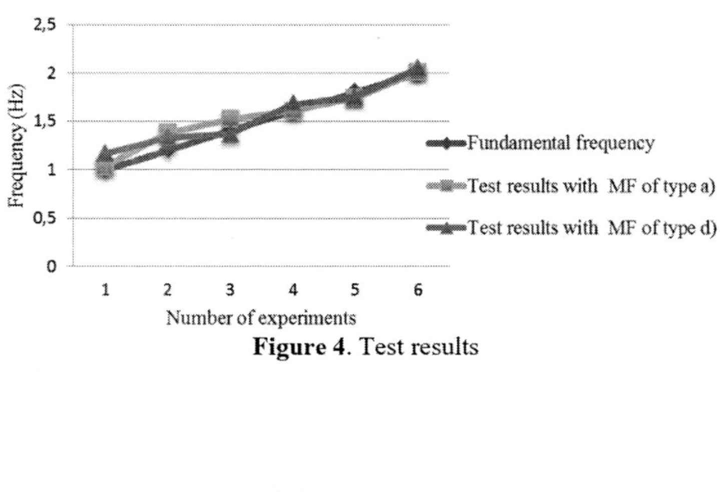

Figure 4. Test results

IOP Conf. Series: Materials Science and Engineering **177** (2017) 012031 doi:10.1088/1757-899X/177/1/012031

5. Conclusions

From the test results it follows that the model of second-type fuzzy neurons and fuzzy neural networks can successfully solve the problems of predicting the properties of time–dependent signals. Therefore, neural networks with fuzzy impulse activation functions may be widely used in many fields of science, technology and mechanical engineering for some problems regarding classification, prediction, approximation, etc. However, these models require significant computing resources while dealing with the time series prediction problems. According to Moore's law [11], the number of elements of a single chip doubles every 18 months. Hence, with the fast development of state-of-the-art computers, this problem can be handled soon. For the next study, we will focus on solving the problem regarding computational cost.

References

[1] Wasserman F 1989 *Neurocomputing technics: the theory and practice* (New York: Van Nostrand Reinhold)
[2] Galushkin A I 2010 *Neural Networks: basic theory* (Moscow: Hotline – Telecom)
[3] Aksenov S V and Novoseltsev V B 2006 *The organization and the use of neural networks (methods and technologies)* (Tomsk: Publishing house of the YTL) 128 p
[4] Yefremov A A 2015 *J. Proc. of Tomsk State Univ. of Control Syst. and Radioelectr.* **2 (36)** 136
[5] Yefremov A A 2013 *Youth and Modern Information Technology* M **75** pp 129–131
[6] Beer M and Moller B 2004 *Fuzzy Randomness: Uncertainty in Civil Engineering and Computational Mechanics* (New York: Springer – Verlag)
[7] Yefremov A A and Korikov A M 2011 *Siberian Journal of Science* **1(1)** 340–343
[8] Yarushkina N G 2004 *Lectures on Neuroinformatics* (Moscow: Moscow Engineering Physics Institute) part 1 p 151
[9] Gupta M M and Qi J 1991 *Fuzzy Sets and Systems* **40** 431–450
[10] Korikov A M 2015 *J. Proc. of Tomsk State Univ. of Control Syst. and Radioelectr.* **2(36)** 148
[11] Thackray A, Brock D and Jones R 2015 *Moore's Law: The Life of Gordon Moore, Silicon Valley's Quiet Revolutionary* (New York: Basic Books)

Research of thermal deformation of a kinematic wave reducerwith a modified tooth profile during the work in low temperature conditions

D L Stepanova and P A Nikulin

Tomsk Polytechnic University, 30, Lenina Ave., Tomsk, 634050, Russia

E-mail: pashanik91@gmail.com

Abstract. In the conditions of the Extreme North working resource of mechanicaltools and machineelements is reduced because of bad weather conditions in this region.At a low temperature materials are exposed to deformation which is capable to break operability of the mechanism.In connection with the high requirements to the accuracy of a kinematic wave reducer, it is necessary to conduct a research for the purpose of comparison of value of thermal deformation and the appointed admission on a reducer detail.If value of thermal deformation is more admission, then it can lead to jamming of the mechanism.The research was conducted for a collected reducer and separately for not loaded driver gear.

1. Introduction

A simplified diagram of a modified gear tooth profile is illustrated in Figure 1. This profile of teeth driven gear provides a possibility to create the transmission, which has a difference between the driver and the driven gear in one tooth. It is a feature of a reducer with the modified shape of the tooth. The use of a similar profile and a difference of teeth of a driver and driven gear provide the big area of contact, in comparison with evolvent gearing. Such difference allows several teeth to perceive a high value of ability of loading.

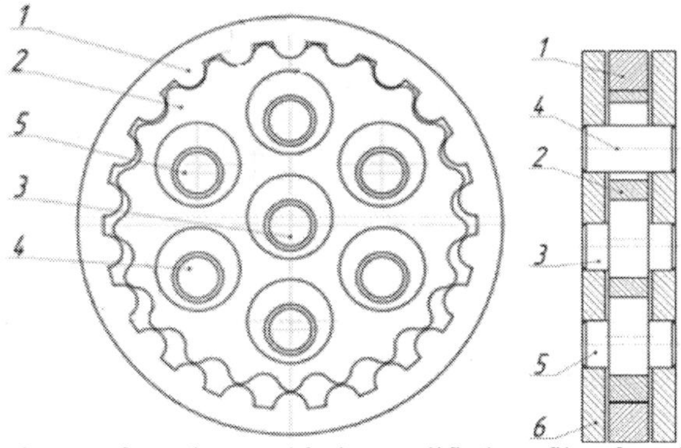

Figure 1. The simplified scheme of a reducer with the modified profile of tooth: 1 – Driven gear; 2 – Driver gear; 3 – Eccentric shaft; 4 – Hub; 5 – Fingers; 6 – Mobile wheel.

Analysis of the details of a reducer revealed thatthe mostdeformable detail is the drivergear.Deformation of thedrivergear can lead to jamming of a reducer. Therefore, we will conduct a research of values of possible deformation for driver gear with the purpose to exclude a possibility of jamming of a reducer under the influence of temperature changes.

2. Materials and methods of a research.

Driver gear made of steel 40X [11]. Let's define deformations of a gear wheel as separate detail. The driver gear is established on 4 bearings according to the kinematic scheme of a reducer. The research was conducted of using the SolidWorks Simulation program at a temperature of driver gear of –50 °C.

Results of thermal deformation are visible in figure 2. The figure shows that the maximum deformation occurs at places with the smallest amount of material. In these points, the value of resultant movement makes 0.02 mm (figure 2 a), and value of equivalent deformation makes 0.0014 N/m^2 (figure 2 b).

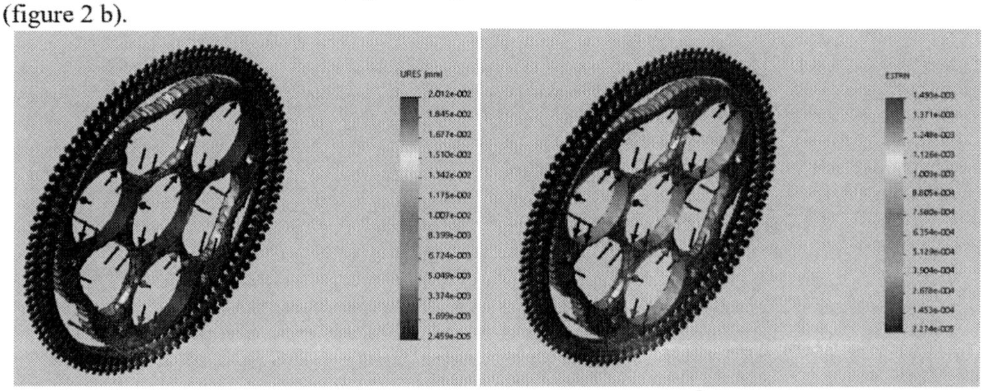

a b

Figure 2. Results of thermal deformation: a – value of resultant movement; b – value of tension.

3. Thermal deformation on assembly of a reducer with temperature in a zone contact of teeth of a driver and driven gear +90°C and on all parts of a reducer at a temperature of -50 °C.

All parts of a reducer are made from steel 40X [11]. As a result of the analysis diagrams of resultant movements and tension are received.

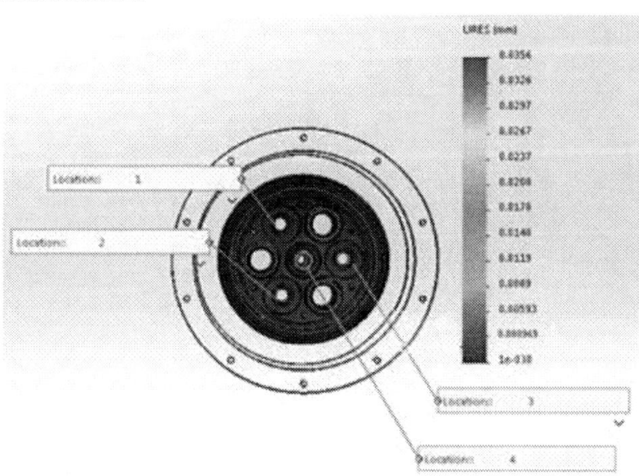

Figure 3. Resultant movement on eccentric and input shaft: Location 1 – 0.00694 mm; Location 2 – 0.00829 mm; Location 3 – 0.00677 mm; Location 4 – 0.00855 mm.

The greatest value of resultant movement on eccentric makes 0.00855mm (figure 3). This value is less than value of a gap in bearings. The greatest value of tension makes 4.966 $\cdot 10^7$ N/m^2 (figure 4), this value of tension doesn't exceed the yield strength of steel 40X σ_t =2.068 $\cdot 10^8$ N/m^2, it follows from this that durability of a design is provided.

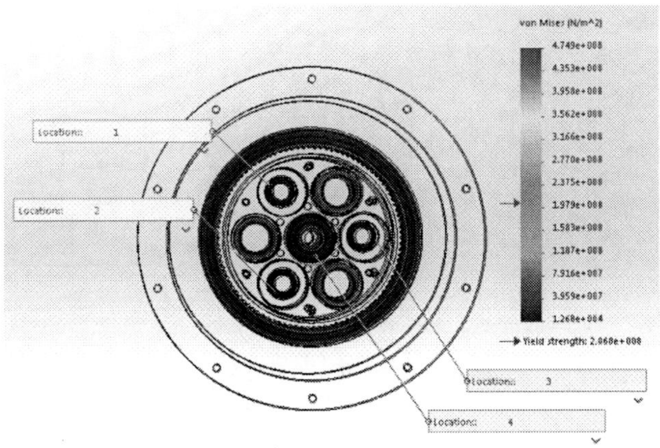

Figure 4. Value of tension in eccentric: Location 1 – 4.506 $\cdot 10^5$ N/m^2; Location 2 – 2.024$\cdot 10^5$ N/m^2; Location 3 – 6.232$\cdot 10^5$ N/m^2; Location 4 – 4.966$\cdot 10^7$ N/m^2.

4. Thermal deformation on assembly of a reducer with temperature in a zone contact of teeth of a gear wheel and a wheel + 90°C and on all parts of a reducer at a temperature of +50 °C.

All parts of a reducer are made from steel 40X [11]. As a result of the analysis diagrams of resultant movements and tension are received.

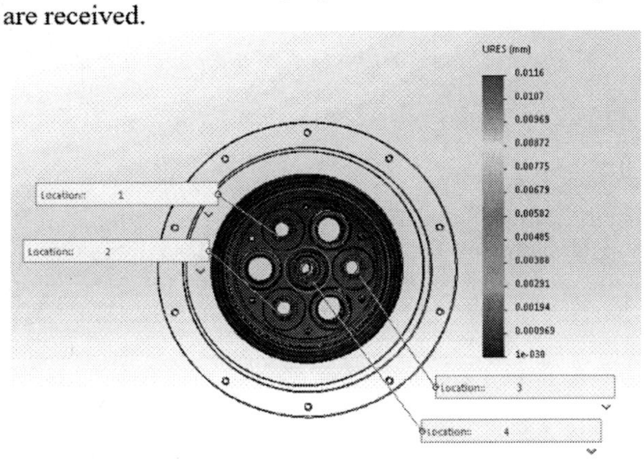

Figure 5. Resultant movement on eccentric and input shaft: Location 1 – 0.00219 mm; Location 2 – 0.00293 mm; Location 3 – 0.0024 mm; Location 4 – 0.00298 mm.

The greatest value of resultant movement on eccentric makes 0.00298mm (figure 5). This value is less than value of a gap in bearings. The greatest value of tension makes 1.865 $\cdot 10^7$ N/m^2 (figure 6), this value of tension doesn't exceed the yield strength of steel 40X σ_t =2.068 $\cdot 10^8$N/m^2, it follows from this that durability of a design is provided.

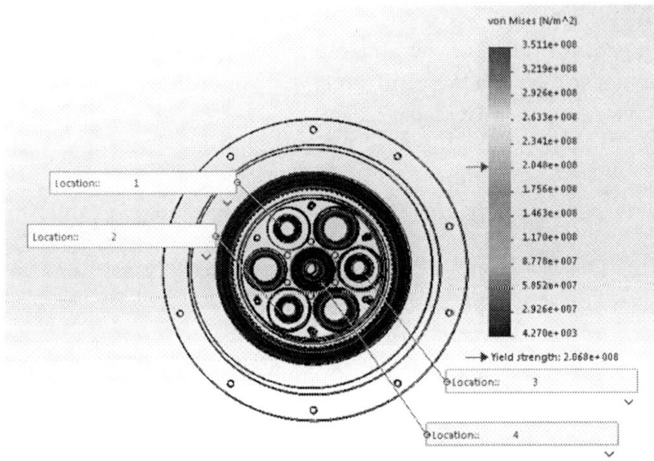

Figure 6. Value of tension in eccentric: Location 1 – $2.001 \cdot 10^5$ N/m^2; Location 2 – $1.004 \cdot 10^5$ N/m^2; Location 3 – $2.517 \cdot 10^5$ N/m^2; Location 4 – $1.865 \cdot 10^7$ N/m^2.

5. Conclusion

The mechanism has difficult geometry and great demands are placed on production accuracy. The thermal analysis has revealed that, the maximum resultant movement makes, at a temperature of influence of –50 °C on all details and +90 °C in a gearing zone of driver and driven gear makes 0.0225 mm. This value is less than value of the lower limit admission which makes 0,025 mm therefore this influence won't bring to get jammed a reducer.

References

[1] Marisova A F 2010 *Epicyclic gearing. Calculation and design* (Rostov-on-Don, Don State Thechnical University)
[2] Dunaev P F, Lelikov O P 2008 *Construction components and machine parts* (Publishing center Academy)
[3] Fishenko V N 2016 *Handbook for constructor. Book 2. Design mechanism and parts.* (Publishing center Infra-Engineering)
[4] Kudravchev I V, Halimanovich I V, Lekanov A V, Masanov A G, Silchenko P N, Novikov E S 2013 *The 14 Russian Conf. Young Sci. on Math.Modeling and Information Techn.* **1** 132-134
[5] Semenov Y A Semenova N S, 2015 *Theory of Mechanisms and Machines* (Saint Petersburg, Polytechnic University)
[6] Stepanova D L 2014 *Modern Thechnics and Thecnonogy* **1** 201-202
[7] Stepanova D L 2014 *Modern Thechnics and Thecnonogy* **1** 283-284
[8] Luminarskiy I E 2014 *Bullatin of University. Mechanical engineering* **3(648)** 8-15
[9] Timofeev G, Egorova O, Samailova M 2015 *The 14 IFoMM World Congress*
[10] Tarabarin V B, Tarabarina Z I *Bull. of Bauman Moscow State Techn. Univ. Mech. Eng.* **5** 117
[11] GOST RF 7370-98, Railway arrow of the types P75, P65 and P50, Moscow, 1999

The method of optimization of neuro-based concurrent operations in neurocomputers

V A Romanchuk

Ryazan State University named for S Yesenin, 46, Svobody ave., Ryazan, 390000, Russia

E-mail: v.a.romanchuk@rsu.edu.ru

Abstract. The article deals with the task of optimization of neuro-based concurrent operations to be implemented in neurocomputers. We define mathematical tools of this optimization that can employ the set-theoretic approach towards such concepts as task, operation, and microcommand. We consider segmentation and parallelization of operations as methods to use, depending on precedence relations among operations that constitute these segments. The task solution of optimization of neuro-based concurrent operations in neurocomputers can be applied to a whole class of neurocomputers, regardless of the manufacturer, the model or the product line, since we only address the general properties and principles of the neurocomputer operation. We select criteria and define methods of evaluating the effectiveness of parallelization of concurrent operations, when they are implemented in neurocomputers. We describe our empiric research in the form of a software system that automatically optimizes neuro-based concurrent operations in neurocomputers on the NP Studio platform.

1. Introduction

For a long time, neurocomputers were considered effective for solving so-called non-formalized and low-formalized tasks in situations when this requires an algorithm of task-solving training with authentic empiric data. The first type of these tasks was connected with approximation of specific functions to a discreet set of values, i.e., the task of image recognition [1]. At present, neurocomputers are widely used in designing data processing systems and in control of complex objects (including manufacture of machinery).

By the present time, another class of tasks has been added, those which do not always require training and empiric data, but are easy to represent in a neuro-network logic basis. Among them, we see tasks that display expressed natural parallelism in processing of signals, images, et al. We find the proof of better prospects for neurocomputers than for other architectural solutions in the fact that recently, the range of general mathematical tasks solved on a neuro-computer basis has increased. We can also list such tasks as solving large-size linear and non-linear algebraic equations and inequations; systems of non-linear differential equations; partial-derivative equations; optimization procedures, and other tasks [2, 3].

Yet currently, there is an issue with the lack of mathematical tools and optimization techniques for neuro-networking tasks for neurocomputer solutions, which renders this task solving ineffective. The results of this research have been tested during development of algorithms and software for a specialized computing device based on neuroprocessors, whose function is to automatically control electromechanic system modules (illustrated in a hexapode device) in a near-real time mode.

2. Mathematical tools of optimization of neuro-based concurrent operations in neurocomputers

Let us suppose that $Z^{(j)}$ is a certainneuro-networking task in a neuroprocessor platform implementation. Neuro-network task $Z^{(j)}$ can be set as a tuple ofparameters and specifications:

1. A set of artificial neuron network (ANN) inputs $X = \{x_1,....x_i,...,x_{xn}\}$, where each x_i input is defined by its digit capacity X_{R_i} andby the X_{N_i} type.

2. A set of artificial neuron network inputs $Y = \{y_1,....y_i,...,y_{yn}\}$, where each y_i output is defined by its digit capacity Y_{R_i} and by Y_{N_i} types.

3. A set of neurons $N = \{n_1,....n_i,...,n_{nc}\}$, where each requires emulation for solving the $Z^{(j)}$ task.

4. A set of neuron associations defines the dependencies among the artificial network neurons: $N_E = \{N_{11}, N_{12},..., N_{1nc},..., N_{ncnc}\}$,which can be represented as a connection matrix of neurons, inputs and outputs of the artificial neuron network.

$N_E \to MNet = [MNet_{ij}]$ with dimensions $(nc + xn + yn) \times (nc + xn + yn)$.

This matrix shows an artificial neuron network topology, the number of ANN layers, availability or the absence of bypass connections, transfer functions of neurons.

We introduce the notion of precedence, $\prec$, set as $\forall n_i, n_j \in N, n_i \prec n_j$, where, for its input, the n_j neuron employs the output of the n_i neuron, that is why, the emulation of the n_j neuron can run without preliminary emulation of the n_i neuron.

Each element of the matrix can take these values: $MNet_{ij} = \{'-', 'n', 'y'\}$:

$'n'$ ('no') – no precedence relationship $\neg(n_i \prec n_j)$ between subprogrammes n_i and n_j ($j > i$);

$'y'$ ('yes') - precedence relationship $(n_i \prec n_j)$; $j > i$; between subprogrammes n_i and n_j ;

$'-'$ - precedence relationship is impossible between neurons.

As an illustration, we can consider the matrix of associations between six neurons $N = \{n_1, n_2, n_3, n_4\}$, two inputs $X = \{x_1, x_2\}$ and single output $Y = \{y_1\}$, the matrix having a perceptron topology (Figure 1).

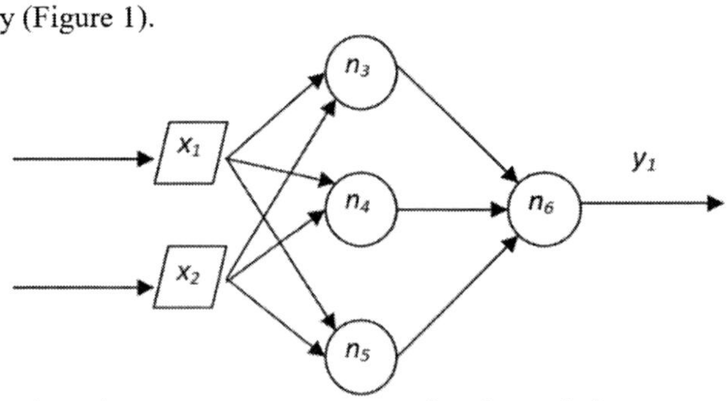

Figure 1. The illustration of a perceptron to set a matrix of associations.

$$MNet = \begin{bmatrix} - & x_1 & x_2 & n_1 & n_2 & n_3 & n_4 & y_1 \\ x_1 & - & n & y & y & y & n & n \\ x_2 & n & - & y & y & y & n & n \\ n_1 & n & n & - & n & n & y & n \\ n_2 & n & n & n & - & n & y & n \\ n_3 & n & n & n & n & - & y & n \\ n_4 & n & n & n & n & n & - & y \\ y_1 & n & n & n & n & n & n & - \end{bmatrix}$$

In the *MNet* matrix, all elements below the main diagonal line equal $'n'$, as a perceptron does not have back links.

To solve the $Z^{(j)}$ task, all neurons require emulation in accordance with the relationships of precedence between neurons; then the $Z^{(j)}$ task can be represented as a tuple ofneuron emulation operations like this: $y = f(g) = f(\sum_{i=1}^{n} a_i x_i + a_0)$.

We introduce the set of operations $O = \{O_1, O_2, ..., O_i, ..., O_{nc}\}$. Then, neuro-network task $Z^{(j)}$ is a tuple of operation $O_1, O_2, ..., O_m, ..., O_{nc}$ with a length of $nc = |Z^{(j)}|; j = \overline{1, N}$, i.e.: $Z^{(j)} = < O_1, O_2, ..., O_m, ..., O_{nc} >; j = \overline{1, N}$.

Thus, the number of operations $O_1, O_2, ..., O_m, ..., O_{nc}$ corresponds to the number of emulated neurons of the ANN, as a certain O_l operation represents a mathematical model of a formal neuron: $O_l = f(\sum_{m=1}^{n} a_m x_m + a_0)$.

For certain operations O_i and O_j, the precedence relationship $\prec$ is also applicable, and it is set in this way: $\forall O_i, O_j \in O, O_i \prec O_j$: for input, the O_j operation employs the output data of the O_i operation, so the O_j operationcannot run unless the O_i operation has been performed.

The O_m operation can be regarded as a set of autonomous operations, based on the associations and functional units of a neurocomputer that implement the set: $O_m = \{O_{m.1}, O_{m.2}, ..., O_{m.NP}\}$,where NP is the number ofpossible functional stages of neuron emulation operations in the neuroprocessor.

In general, the number of operations during the neuron emulation is as follows: $NP = xn + (xn - 1) + 1$, where xn is the number of neuron inputs, depending on the ANN.

Here, the first term defines the u_m operation that multiplies the x_m input value by the a_m link weight. The second term defines the s_m operation of adding the results of the multiplication product. The third term is the operation of activating the a_m neuron.

Precedence relationship for specific functional modules of a single neuronal emulation operation can be represented as an *MNeuron* matrix with dimensions $(2 * xn + (xn - 1) + 2) \times (2 * xn + (xn - 1) + 2)$. Then, the ANN represented in Figure 1 can be represented as a set of concurrent operations (Figure 2).

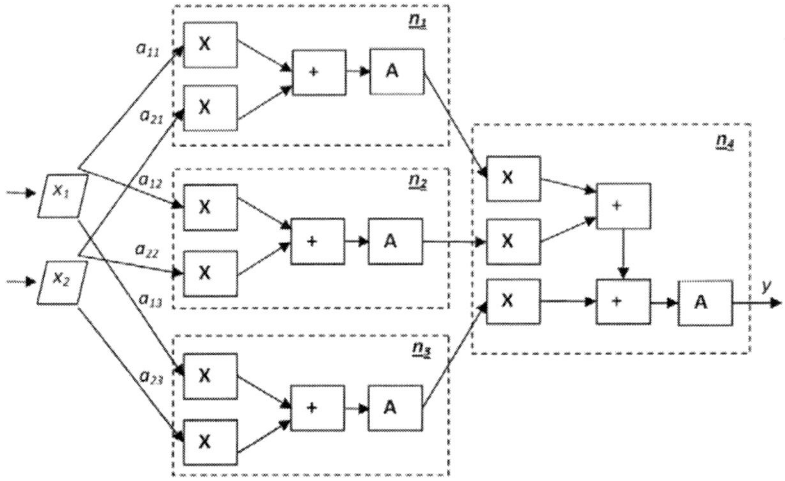

Figure 2. The illustration of a perceptron represented as a tuple of operations.

Then, for a specific $Z^{(j)}$ task, the *MNetA* matrix, describing all precedence relationships between operations can be defined as a combination of *MNeuron* and *MNet* matrices, where each element of the $MNet_{ij}$ matrix can be defined as an γ image, i.e., as an *MNeuron* matrix:

$\gamma : MNet_{ij} \rightarrow MNeuron$.

The precedence relationships of the *MNet* matrix are preserved as those associated with input and output values of the *MNeuron* matrix.The dimensions of the *MNetA* matrix are as follows: $(nc + xn + yn) * (2 * xn + (xn - 1) + 2) \times (nc + xn + yn) * (2 * xn + (xn - 1) + 2)$.

At the stage of software development, each operation O_i orset of operations $O_m = \{O_{m.1}, O_{m.2}, ..., O_{m,NP}\}$ is assigned a tuple ofneuroprocessor microcommands in its internal coding $< MK_1^i, MK_2^i, ..., MK_K^i >$; where K is the minimum number of microcommands necessary for implementation of a certain O_i operation. Most commonly, since a neurocomputer is a highly concurrent device, a single microcommand may correspond to a tuple of operations of neuron emulation: during a single cycle of a neuroprocessor, a set of neurons can be emulated. Then, several variants are possible: $O_i \rightarrow < MK_1^i, MK_2^i, ..., MK_K^i >; K <> 0$ or $O_m = \{O_{m.1}, O_{m.2}, ..., O_{m,NP}\} \rightarrow < MK_1^i, MK_2^i, ..., MK_K^i >; K <> 0$ or $O_i \rightarrow MK_i ; K = 1$ or $O_m = \{O_{m.1}, O_{m.2}, ..., O_{m,NP}\} \rightarrow MK_m; K = 1$.

The task of optimization consists in defining operation tuples that correspond to a microcommand, in order to maximize the number of neuron emulation operations in a specific microcommand: $\forall MK_p \in MK : MK_p \rightarrow < O_1, O_2, ..., O_i, ..., O_R >: R \rightarrow \max$.

Neurocomputers can be divided into two groups, depending on the neuron emulation methods [4]:

1. Neurocomputers that preform neuron emulation microoperations at the hardware level without assigning apecific functional units to specific concurrent operations (addition, multiplication, activation).

2. Neurocomputers that perform neuron emulation microoperation sat the hardware level stage after the stage, assigning specific functional units to specific concurrent operations (addition, multiplication, activation).

In the first case, since it is impossible to assign special functional units, we employ the *MNet* matrix as a method of data input. In the second case, it is more effective to use the *MNetA* matrix, as this yields better response time, due to reassignment of tuples of concurrent operations.

The goal of the stage of reassignment of concurrent operations $Z^{(j)} =< O_1, O_2,..., O_m,..., O_{nc} >; j = \overline{1,N}$ and their rearrangement in p operation tuples is: $< O_1, O_2,..., O_{nc} >\to< O_{11}, O_{12},..., O_{1nc} >,...,< O_{p1}, O_{p2},..., O_{pnc} >$ for their further representation in neuron emulation microcommand(s): $< O_{11}, O_{12},..., O_{1nc} >,...,< O_{p1}, O_{p2},..., O_{pnc} >\to< MK^i_1, MK^i_2,..., MK^i_K >$.

We introduce the notion of a segment of operations, SO, i.e., a tuple of O operations, without precedence relationships between them: $SO_i =< O_{i1}, O_{i2},..., O_{inc} >; O_{i1} \neg\prec O_{i2} \neg\prec ... \neg\prec O_{inc}$.

Then the whole tuple of operations that implement the $Z^{(j)}$ task can be divided into a tuple ofsegments: $< O_1, O_2,..., O_m,..., O_{nc} >\to< SO_1, SO_2,..., SP_p >$, where $SO_1 \prec SO_2 \prec ... \prec SP_p$.

Segmentation consists of the following stages:

1. Selection of a matrix of precedence relationships, depending on the neuroprocessor type. In the matrix for solving this task, we consider only the area of the neuroprocessor's functional units, restricted to the precedence relationships of input and output data.

2. For each value in column $MNet_i = \{n_1, n_2,..., n_{nc}\}$ of the selected area of the $MNet$ ($MNetA$) matrix, the following conditions are set:

- If for n_i, all the values in a column equal $'n'$, the O_i operation(or a tuple of operations), corresponding to the n_i neuron emulation, can be assigned to any SO_j segment.

- All $n_i, i = \overline{1,nc}$ having the same distribution of $'y'$ valuesin matrix areas of the selected area are assigned to a specific SO_j segment.

- If a back connection is necessary for solving a task, and if this connection has to be a precedence relationship between the last and the first neurons in the ANN, the $'y'$ value in the left-hand bottom corner of matrix $MNet_i = \{n_1, n_2,..., n_{nc}\}$ will not be considered.

3. With a set of segments $\{SO_1, SO_2,..., SP_p\}$, the precedence relationships can be defined in the following way:

- if the lowest value $'y'$ of segment operations SO_i has a position in a matrix which is higher than the position of the lowestvalue $'y'$ in the operations of segment SO_j, then $SO_j \prec SO_i$.

With a tuple of segments that are connected by precedence relationships, in order to program and represent operations as a micrommand set of the processor, it is necessary to divide the segments, in order to select a tuple of operations performed in a single cycle, in the following way:

– if the number of operations $|SO_i|$ in segment SO_i is higher than MAC, it is necessary to split the segment in two, where precedence relationships between the segments will not be noted: $|SO_i| > MAC : SO_i = SO_{i1} \cup SO_{i2}$.

As a result, we have a tuple of segments, each of them is represented as a microcommand (or a tuple of microcommands) with a maximum concurrence level of performing the operation set within a cycle.

Further optimization will be possible with semantic analysis of the $Z^{(j)}$ task.

Then it is possible to speak about effective parallellization of operations, which will be expressed through K_{MAC} and $K_{MAC\,i}$ quotients in each processor cycle.

For the first type of neurocomputers, the effectiveness of parallellization can be calculated in the following way: $K_{MAC\,i} = \frac{M_i}{MAC}$, where M is the number of neurons emulated on a neurocomputer within a cycle, and MAC is the maximum number of neurons emulated on the specific neurocomputer within its cycle.

For the second type of neurocomputers: $K_{MAC\,i} = \frac{M_{Oi}}{MAC * NP}$, where M_O is the number of microoperations performed (addition, multiplication, activation) within a cycle, NP is the number of microoperations that constitute neuron emulation operation $O_m = \{O_{m.1}, O_{m.2}, \ldots, O_{m,NP}\}$.

3. Empirical research

For convenience of processing various inputs, we have developed an application that automatically optimizes neuro-based concurrent operations in neurocomputers in the offered subsystem, 'Analysis and optimizations of neurocomputers' within our software system 'NP Studio' [6-10].

4. Conclusion

The theoretic and empirical research that suggest methods of solving operations optimization, when these are performed on neurocomputers, enable us to increase the speed, due to employing all the functional units of a neurocomputer for several classes of tasks, easily represented in a neuro-network logic basis. These are tasks that show pronounced natural parallelism in signal processing, image processing, et al.

References

[1] Pande S, Morgan F, Cawley S, Bruintjes T, Smit G, McGinley D, Carrillo S, Harkin J and McDaid L 2013 *Neural Process Letters* **38** 131-153
[2] Izeboudjen N, Larbes C and Farah A 2014 *Artifical Intelligence Review* **41** 491-534
[3] Kolinummo P, PASI Pulkkinen P, Hämäläinen T and Saarinen J 2000 *Neural Processing Letters* **12** 171-82
[4] Ruchkin V, Romanchuk V and Sulitsa R 2013 *Proc. of the 2nd Mediterranean Conf. on Embedded Computing* (Montenegro: Budva) 58-62
[5] Ruchkin V, Fulin V, Kostrov B, Romanchuk V and Ruchkina E R 2015 *Proc. of the 4nd Mediterranean Conf. on Embedded Computing* (Montenegro: Budva) 45-50
[6] Avetisyan A, Grushin D, Ryzhov A 2002 *Proc. of the Institute for System Programming of Russian Academy of Sciences* 140-147
[7] Ivutin A, Larkin E 2014 *Proc. of the 3rd Mediterranean Conf. on Embedded Computing* (Montenegro: Budva) 236-239
[8] Diken E, Jordans R, Corvino R, Jozwiak L, Corporaal H and Chies F 2014 *Microprocessors and Microsystems* **38** 947-959
[9] Jordans R, Jozwiak L, Corporaal H 2014 *Proc. of the 3rd Mediterranean Conf. on Embedded Computing* (Montenegro: Budva) 32-35
[10] Bekhtin Y, Gurov V, Guryeva M 2014 *Proc. of the 3rd Mediterranean Conf. on Embedded Computing* (Montenegro: Budva) 252-255

IOP Conf. Series: Materials Science and Engineering **177** (2017) 012034 doi:10.1088/1757-899X/177/1/012034

The concept and the principle of the diagnostic observability of the object in problems of monitoring and non-destructive testing

A V Sedov[1,2]

[1] Southern Scientific Centre of the Russian Academy of Sciences, 41, Chehova str., Rostov-on-Don, 344006, Russia
[2] Platov South-Russian State Polytechnic University (NPI), 132, Prosveschenia str., Novocherkassk, Rostov Region, 346428, Russia

E-mail: Sedov_A.V@mail.ru

Abstract. The design of non-destructive testing systems relates to the solution of the problem of analysis of observability. It is the assessment of the opportunities of the set of measured parameters to diagnose the object, to identify emerging defects and the specific modes of operation. The term 'observability' in control theory does not correspond to the tasks of diagnosis and requires the model of the object's state. The proposed concept of 'diagnostic observability' allows to use this term in tasks which aren't related to the control. It doesn't require the model of the object's state, and realizes the estimation on the array of measurements performed for various diagnosed states of the object. The set of diagnostic state variables is determined as coordinates of diagnostic orthogonal space. The proposed principle of diagnostic observability is determined by the method of construction of the space of the diagnostics and by the analysis of the images of states of object. It can become a new basis for building an effective universal adaptive systems of non-destructive testing of complex technical and other objects.

1. Introduction

In control theory [1], the term 'observability' is defined for the object, described in the state space:

$$\begin{cases} \dot{x} = \boldsymbol{f}(x,u,t), \; x \in R^k, \boldsymbol{u} \in R^r; \\ y = \boldsymbol{F}(x,u,t), \; y \in R^\rho, \end{cases} \qquad (1)$$

where x – the vector of state variables, $\boldsymbol{u}$ – the vector of control response, y – the vector of output measured values. The first equation in (1) – the equation of state of the object, the second – the equation of observation of the object. The system (1) is *completely observed* [1], if there is t_1, $t < t_1 < \infty$, and measured values $y(\tau)$ and $u(\tau)$ on interval $t \le \tau \le t_1$ can determine state $x(t)$.

The system (1) for the linear steady-state object is written as follows:

$$\begin{cases} \dot{x} = \boldsymbol{A}x + \boldsymbol{B}u, \; x \in R^k, \boldsymbol{u} \in R^r; \\ y = \boldsymbol{C}x + \boldsymbol{D}u, \; y \in R^\rho, \end{cases} \qquad (2)$$

where $\boldsymbol{A}$, $\boldsymbol{B}$, $\boldsymbol{C}$, $\boldsymbol{D}$ - matrices of appropriate dimensions. The concept of the matrix of observability is introduced [1]:

$$H = \left[C^T \quad A^T C^T \quad \dots \quad \left(A^T\right)^{k-1} C^T \right]. \tag{3}$$

In this case, the object (2) is considered to be *completely observable* if and only if the rank of the matrix of observability is k: rank $H = k$.

The equation of state and the equation of observation of the object are necessary for the assessment of the observability in the existing theory. The production of these equations is possible by structural and parametric identification, which in turn is a complex task and depends on the identifiability of the object [1].

The above-stated traditional definition of observability is somewhat different characteristic than it is required in the diagnosis and non-destructive testing. For example, the matrix (3) characterizes the observability of the state variables only in normal operating conditions and in case in (2) it can be assumed that all matrices A, B, C, D are not changing. In case of non-destructive testing, the observability should be evaluated by other state variables that characterize the appearance or presence of defects, faults in the object. The model (2) is changed because of the defects and faults. In this case, matrices A, B, C, D of the model (2) can not be considered as constant. The application of the matrix of observability (3) is impossible in this case.

Two principles of probing, passive and active, are realized in the non-destructive testing. The detection of moments, which are the most suitable for diagnostic object by y, occurs in the passive monitoring of parameters of operating complex object; special control actions u are not formed. With the implementation of the *active probing* there is a formation of known control action u and a measurement of the output value y, as the response to this action.

In both cases of the probing, the component of control action u in the model of a linear steady-state object (2) can be replaced by constants ξ_d, ζ_d as follows:

$$\begin{cases} \dot{x}_d = A_d x_d + \xi_d, \; x_d \in R^m; \\ y = C_d x_d + \zeta_d, \; y \in R^p, \end{cases} \tag{3}$$

where x_d - the vector of other diagnostic state variables, y - the vector of output measured values.

It is important when the output measured value is scalar response function $y(t)$ measured in interval $[t, t + \Delta t]$ in the context of diagnosis and non-destructive testing. In this case, the equation of observation (3) can be written as follows:

$$y = C_d x_d + \zeta_d, \; y \in R. \tag{4}$$

The production of the equation of state (3) is not required in the diagnosis. The evaluation of the value of vector x_d of *diagnostic of state variables* is only performed using equation (4). The space of value x_d will be called *diagnostic states space* $\mathbf{D} = \{x_d\}$.

2. The methods for evaluating the diagnostic states space

The conctruction of diagnostic space $\mathbf{D}$ is based on the mathematical techniques of reduction of measured signals y in (4), in particular, signals of a response to a probing action. Under the term 'reduction' of measurement y we understand the decrease of the dimension of space of measurement $\mathbf{V} = \{y\}$ with the identification of generalized parameters – diagnostic state variables x_d. Variables x_d must retain a significant variability of measurements y during the transition from one diagnosed state to another. It is important to have a stable unique dependence between the change of the diagnostic state variables x_d and the diagnosed change in object's state. The presence of significant unique dependence corresponds to the *diagnostic observability of object*.

Mathematical reduction of dimension of space $\mathbf{V}=\{y\}$ can be carried out by using different approaches: the traditional spectral analysis [2-4]; the correlation analysis of signals [4-7]; the artificial neural networks [6,8]; the statistical approaches and models [8,9]; the wavelet transform signals [3-5,10]; the decomposition modeling method [7] and others.

Analysis showed that the use of the decomposition modeling method is superior to others. The optimal orthogonal decomposition of signals y on the adaptively adjusted basis based on training sample $\{y\}$ is realized for the construction of diagnostic states space $\mathbf{D}$. Thus, obtained space $\mathbf{D}$ allows one to analyze signals y for signs of defects more effectively. This analysis can be used as a criterion of the quality of measurements y. It requires replacement of the measured signal and measurement principles in case of low-quality detection of defects in the needed group.

This approach has been tested and has shown the ability of effective using in different diagnostic systems: the non-destructive testing of mechanical structures and mechanisms with wave low-frequency probing effects [11]; the determining of corrosive destruction [5]; the electrochemical systems of diagnosis and control of metal alloys [7]; the on-line diagnostics of the autonomous power sources [7]; the diagnostic of modes of operation of large power systems [7]; the analysis of the mechanical loading of items bearing structures [11]. Its more expanded use is possible for detection of damaged blades of wind turbines [8]; for non-destructive testing of materials in engineering [6]; for detection of foreign bodies and its classification with ultrasound using [10]; for fault location with using of environment vibration [2] and others.

3. The principle of reduction in the construction of a diagnostic state space on the basis of the decomposition method of modeling

For simplicity, we consider the one-dimensional case. The output measured value is scalar response function $y(t)=f(t)$, $t\in[0,T_{н}]$ to probing effect *F*. Time interval $T_{н}$ must be sufficient for stable identification of emerging diagnosed conditions. The shock action generating surface waves acts as the probing one of low-frequency non-invasive diagnosis of metal structures.

The learning sample is formed from separate functions $f_i\ (t)$ obtained for different types of defects in the construction, but in case of one type of non-destructive probing effects *F*. Each function $f_i\ (t)$ can be considered as the vector of real values $f_i=[f_{i1},f_{i2},f_{i3},\dots f_{iN}]^T\in\mathbf{R}^N$ of time changing of signal response. Sampling increment Δt or the number of counts N per observation interval is selected according to the sampling theorem [1].

The number of vectors f_i in learning sample $\{f_1,f_2,f_3,\dots,f_n\}\in\mathbf{R}^{N\times n}$ is determined by the number of defects l recognizable in construction $n=l\cdot b$, where b - number of vectors f_i corresponding to one type of the defect. Initially, a set of vectors $\{f_1,f_2,f_3,\dots,f_n\}$ is linearly dependent. It is explained by excess dimension N of vectors f_i, proximity of shapes of graphs f_i, and it is confirmed by the value of rank of matrix $\mathbf{f}=[f_1,f_2,f_3,\dots,f_n]^T\in\mathbf{R}^{n\times N}$ for which it is true that $\mathrm{rank}(\mathbf{f})<<N$.

Construction of diagnostic state space $\mathbf{D}=\{x_d\}$ is finding orthogonal transformation $\Xi\in\mathbf{R}^{N\times m}$ of matrix $\mathbf{f}\in\mathbf{R}^{n\times N}$ into matrix $\mathbf{A}\in\mathbf{R}^{n\times m}$ of form $\mathbf{A}=\Xi\mathbf{f}$, excluding excess dimension $\mathbf{f}$ associated with uninformative, often random variations in graphs f_i. This indirectly follows from equation (4). Thus, matrix $\Xi=[\xi_1,\xi_2,\xi_3,\dots,\xi_m]$ defines this linear subspace in $\mathbf{R}^N$ for which it is true that a set of possible linear combinations of its vectors is also a linear space or a linear span:

$$\mathrm{span}\{\xi_1,\xi_2,\xi_3,\dots,\xi_m\}=\left\{\sum_{j=1}^{m}\beta_j\xi_j:\beta_j\in\mathbf{R}\right\}.$$

In this case, vectors $\xi_1, \xi_2, \xi_3, \ldots, \xi_m$ form an orthonormal basis in $\mathbf{R}^N$, and formula $\Xi^T \Xi = I_m$ is true for matrix Ξ. Conventionally, matrix Ξ can be considered as a matrix of orthogonal compression of linear space $\mathbf{R}^N$ into space $\mathbf{R}^m$. Vectors of responses $f_i \in \mathbf{R}^N$ are converted into images $A_i \in \mathbf{R}^m$, $m << N$ with the help of Ξ, and herewith matrix response $\mathbf{f}$ is converted into the matrix of images $\mathbf{A} = [A_1, A_2, A_3, \ldots A_n]^T \in \mathbf{R}^{n \times m}$. In general, the orthogonal decomposition of original vectors $f_i \in \mathbf{R}^N$ on the basis of Ξ can be expressed in the form of $f_i = \Xi A_i + A_0$, where A_0 - the constant component of the transformation. At the same time, it is true that $x_{di} = A_i$ and $\mathbf{D} = \mathbf{A}$.

At the same time, orthonormal basis $\xi_1, \xi_2, \xi_3, \ldots, \xi_m$ is adaptively configurable, trained and dependent on recognizable sample $\{f_1, f_2, f_3, \ldots, f_n\} \in \mathbf{R}^{N \times n}$. The complex of optimization problems is solved in determining of the basis, in particular:

1) The best reproducibility is $\left\| \mathbf{f} - \mathbf{A}\Xi^T - \mathbf{A}_0 \right\|_2 \to \min$, where $\mathbf{A}_0 \in \mathbf{R}^{n \times N}$ - the matrix of constant components of conversion consisting of vectors A_0.
2) Property of basis orthonormality: $\left\| \Xi^T \Xi - I_m \right\|_2 \to \min$.
3) The property is the best discernibility: $d^2(\mathbf{A}) = \dfrac{1}{m^2 - m} \sum_{\substack{i,j=1 \\ i \neq j}}^{m} \left\| A_i - A_j \right\|_2 \to \max$.

The implementation of all these tasks forming the space of diagnostic state variables **D** is observed when the decomposition method of modeling [7] is used for this purpose.

4. Cluster and functional interpretation of diagnostic observability

The diagnostic observability can be interpreted with the using of images $A_i = (a_{i1}, a_{i2}, \ldots, a_{im}) \in \mathbf{R}^m$ of response function $f_i(t)$ from the learning sample of measurements $\mathbf{f}$, which includes all states of the object that are necessary to diagnose. The images are located in diagnostic state space **D**. Typically, this space is low-dimensional, $m << N$. In the simplest case, this space is two-dimensional $A_i = (a_{i1}, a_{i2}) \in \mathbf{R}^2$ or space, which can be analyzed as a set of two-dimensional spaces - the projections. In space **D,** the boundaries of clusters C_j, $j = \overline{1,s}$ of images corresponding to each of the individual diagnosed object's state are defined. Figure 1 shows four clusters for the seven possible diagnosed object's states. In this two-dimensional projection on the plane of the first two diagnostic state variables a_{i1}, a_{i2} only four cluster object's states $(C_1, C_2), C_3, (C_4, C_5, C_6), C_7$ are clearly distinguished. Five states (C_1, C_2) and (C_4, C_5, C_6) form only two discernible clusters, so state variables a_{i1}, a_{i2} are not enough for separate diagnosis of all the seven existing conditions. In this case, diagnostic observability is *incomplete* - four from seven. It is necessary to use a greater number of state variables $a_{i1}, a_{i2}, \ldots, a_{im}$ and to consider other two-dimensional projection of images on orthogonal planes for discernibility of all seven states.

Perhaps other projections allow one to exclude indistinguishability of these states. A decomposition modeling method arranges state variables $a_{i1}, a_{i2}, \ldots, a_{im}$ in the order of their diagnostic informativeness, which is very convenient in the analysis of the diagnostic observability. At the same time, diagnostic informativeness is understood as further variation $S_N^2(a_j)$ of variable a_j in diagnostic state space $\mathbf{D}$:

$$S_N^2(a_j) = \frac{1}{n} \sum_{i=1}^{n} \left(a_{ij} - \frac{1}{n} \sum_{i=1}^{n} a_{ij} \right)^2 .$$

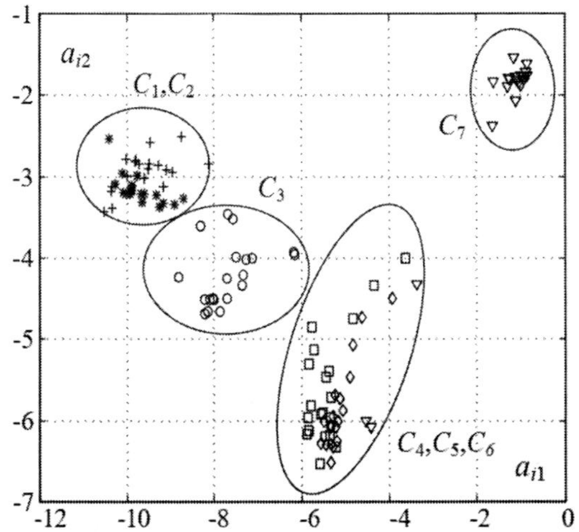

Figure 1. Clusters of the seven states of the object in the diagnostic state space.

If the increase of the dimension of the diagnostic space by the number of state variables $a_{i1}, a_{i2}, \ldots, a_{im}$ does not lead to a complete diagnostic observability, therefore, it is necessary to change the set of measured signals y. The selected set does not meet the principle of diagnostic observability. All of the above-mentioned determines the diagnostic of observability in the *cluster interpretation*.

Besides the visual presentation of diagnostic observability in a cluster interpretation, quantitative estimation η is also possible. The well-known clustering quality criteria [12,13] can be used as a quantitative assessment. Specifically, for each state (cluster) ratio of scatter images D_{C_j} in the cluster to minimum distance R_{ij} from the cluster to the nearest one in the diagnostic space may be estimated: $\eta_j = D_{C_j} / \min_{v} R_{jv}$.

Functional interpretation of diagnostic observability allows us to evaluate how the changing of the value of one or more diagnosed parameters P_c (such as size, location, shape of the defect or otherwise) can be described as functional dependence $\mathbf{F}$ from the changing of the values of state variables – coordinate of images $A_i \in \mathbf{R}^m$ of the learning sample in diagnostic state space $\mathbf{D}$:

$$P_{ci} = \mathbf{F}(a_{i1}, a_{i2}, \ldots, a_{im}).$$

If the construction of this regressive functional dependence is possible, then an assessment of its adequacy can be regarded as quantitative characteristic η of diagnostic observability in the *functional interpretation*.

Finally, the term diagnostic observability is an assessment of the possibility of an available set of measured parameters y by their number and the type which detects defects adequately and accurately, diagnoses and identifies the special emergency mode operation of the object.

5. Conclusion

The customizable orthonormal diagnostic state space for diagnostic systems built with the use of the decomposition simulation method allows us to evaluate the observability of a set of measured data in terms of completeness of diagnostics of the object.

The concept and the principle of the diagnostic observability of the object by the set of measures allowing one to formalize the evaluation of the completeness of the diagnosis of the object in the cluster and functional interpretations have been suggested.

Acknowledgments
The reported study was funded by RFBR, according to research project No. 15-08-08352-a and the Russian Scientific Foundation, project No. 14-19-01676.

References

[1] Krasovskij A A 1987 *Guide to the Automatic Control Theory* (Moscow: Nauka) p 712
[2] Tondreau G and Deraemaeker A 2013 *Mechanical Systems and Signal Proc.* **39(1-2)** 162
[3] Mallat S A 1999 *Wavelet Tour of Signal Processing* (New York: Academic Press) p 851
[4] Gonzales R C and Woods E W 2002 *Digital Image Processing* (New Jersey: Prentice Hall Publishers) p 344
[5] Farhidzadeh A and Salamone S 2015 *Ultrasonics* **57** 198-208
[6] Salazar A, Vergara L and Llinares R 2010 *Mechanical Systems and Signal Processing* **24(6)** 1870-86
[7] Sedov A V 2010 *Modelling Objects with Discrete-Distributed Parameters: Decomposition Approach* (Moscow: Nauka) p 438
[8] Dervilis N, Choi M and Taylor S G 2014 *J. of Sound and Vibration* **333(6)** 1833-50
[9] Cogranne R and Retraint F 2014 *Signal Proc. B* **96** 173-89
[10] Patrick P C and Otman A B 2006 *Ultrasonics* **45(1-4)** 1-14
[11] Bocharova O V, Sedov A V, Andzhikovich I E and Kalinchuk V V 2016 *Russian J. of Non-Destructive Testing* **52(7)** 377-82
[12] Webb A R 2002 *Statistical Pattern Recognition* (New York: John Wiley & Sons) p 496
[13] Fukunaga K 1990 *Introduction to Statistical Pattern Recognition* (London: Academic Press) p 390

Designing of routing algorithms in autonomous distributed data transmission system for mobile computing devices with 'WiFi-Direct' technology

I A Nikitin, V S Sherstnev, A I Sherstneva and I A Botygin

Tomsk Polytechnic University, 30, Lenina Ave., Tomsk, 634050, Russia

E-mail: vss@tpu.ru

Abstract. The results of the research of existent routing protocols in wireless networks and their main features are discussed in the paper. Basing on the protocol data, the routing protocols in wireless networks, including search routing algorithms and phone directory exchange algorithms, are designed with the 'WiFi-Direct' technology. Algorithms without IP-protocol were designed, and that enabled one to increase the efficiency of the algorithms while working only with the MAC-addresses of the devices. The developed algorithms are expected to be used in the mobile software engineering with the Android platform taken as base. Easier algorithms and formats of the well-known route protocols, rejection of the IP-protocols enables to use the developed protocols on more primitive mobile devices. Implementation of the protocols to the engineering industry enables to create data transmission networks among working places and mobile robots without any access points.

1. Introduction

Nowadays everybody uses mobile devices for communication, data transmission, audio and video file playing. The users` abilities are limited by the necessity of connecting to the Internet access point. In this view, the topic of creating a possibility to transmit information among users to a long distance without the network connection is getting more and more relevant with time.

'WiFi-Direct' technology enables two and more users to communicate through P2P-connection [1]. However, this technology had been used very seldom for constructing autonomous distributed data transmission system for data transmission of mobile computing devices [2-6]. The use of such techniques in engineering stimulates enlargement of data transmission network for mobile and static robots without access points. Autonomous system enables to build up the route between two users through intermediated nodes with 'WiFi-Direct' technology if direct connection is impossible. The development of the system consists of two principal phases divided into subtasks.

1. Designing of algorithms:
   A) Analysis of existent routing protocols and the comparative analysis
   B) Designing of the route request algorithms and the request structure
   C) Description of the structure of storing the routing cache in nodes
   D) Designing of data transmission algorithms
2. Software engineering
   A) Designing of human-system interface

B) Creation of Use Case diagram
C) Program realization of designed route algorithms
D) Program realization of data transmission algorithms

The paper is a research of the abilities of using protocols for the system realization and the comparative analysis, the description of the structure of storing the routing cache, as well as the design of the route request algorithms and the request structure, basing on the optimal for the system parameters of existence protocols.

2. Analytical review of the wireless route protocols

The wireless route protocols are divided to proactive and reactive. Proactive protocols build up a few tables with complete information about the system and never start data transmission before getting exhaustive information about the network [7].

One of the modern proactive protocols is OLSR (Optimized Link-State Routing) [8], which is based on gathering and dissemination the ordering information inside the network. As a result of the findings each node is able to build up its own network graph, where the points of graph are nodes and the ribs are the lines of communication. With this information every node can calculate 'optimal' route to any node of the network.

Reactive protocols form the route at the request [9]. Data transmission starts just after disclosure of the way between the source and the destination node. One of the first reactive route protocols in wireless networks are protocols DSR (Dynamic Source Routing Protocol) and AODV (Ad hoc On-Demand Distance Vector) [10-11].

Protocol DSR accumulates information about the route not in the routing table, but in the request packet. The main protocol mechanisms are building up and serving the route. Mechanisms are used together to identify and maintain the route. After the first identification the packets are sent to all possible directions and the information about the past node is attached to the title. To achieve the goal, the packet must contain a completely formed route between the adjusted nodes. After the repeated reception of the first packet the node deletes the packet.

Protocol AODV builds up the route tables on the each node of the network to minimize the information translation time between nodes and finds the route ways independently from the use of the routes. The first step is building up the route tables on the each node. The table holds the information about the shortest way (metric) to the destination node through the neighbour nodes, which exchange metrics until it stops changing. Only then it's possible to pass the information to the optimal route. A comparison of the advantages and disadvantages of different protocols is shown in table 1.

Table 1. A comparison of the different protocols

OLSR	AODV
Proactive protocol	Reactive protocol
Stores new data about the whole network	Stores data about the active routes which is the reason for the less ordering information processing
It takes less time to find the new route as it is displayed in the route table which always updates	More time is spent to find the new route as there is information only about the active routes

Information in the table 1 shows that reactive protocols are more useful for the mobile devices network. Comparative analysis of the main characteristics of reactive protocols is shown in table 2.

Table 2 shows that protocol DSR is the most appropriate route protocol for network with 'WiFi-Direct'. However, protocol cannot be used for networks P2P made with WiFi-Direct. That is why, on its base, route algorithms were projected.

Table 2. Comparative analysis of the main characteristics of reactive protocols

Name	Plural routes	Method of calculating the route	Storing the route	Strategy
AODV	–	The new shortest route	The route table	Complete or partial rewriting of the routes
DSR	+	The shortest or the next in the route table	Route cache	Complete rewriting of the routes

3. Designing of the route search algorithms and forming the clients list

The algorithm updates the route cache regularly. The packet should be informed about the route direction. Thus, information about the route is mentioned in the packet to reach the destination node. Algorithm contains two major mechanisms of the work: the route request mechanism and the route serving mechanism. When opening the route, we can see two possible messages: a route request (RREQ) and a reply to the request (RREP). When the node is about to send a message to a certain destination, it transmits RREQ packet in network. Neighbour node in the broadcasting range gets RREQ-message and adds it to its address, then retransmits it in the network. If PREQ-message does not reach the necessary node, the route is deleted.

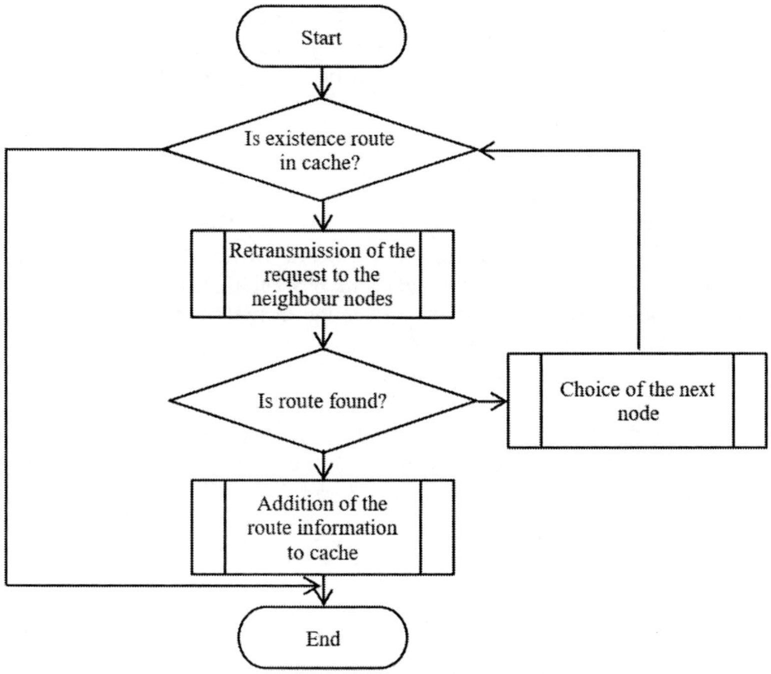

Figure 1. Diagram of route requests.

Every node supports its own route cache, which is stored in memory for the uncovered route. The node will check its route cache for the required destination node until the PREQ-message retransmission. Supporting the route cache on the each node of the network, overhead expenses of the storage, generated while revealing the route, are decreased. If the route is in the routing cache of the intermediate node, it will not retransmit PREQ for the whole network. PREQ-message will be transmitted to the destination node straight away. The first message having reached the destination contains complete information about the route. This node will send PREP packet back to the source, as this route is considered as the shortest one. The source henceforth owns complete information about the route in cache and is able to start the packet transmission. The next function is the packet service. Revealing the route requires two types of messages: message about an error (RERR) and the certification message (ACK). ACK sends a message to the source if the destination node is reached. All intermediate nodes acquire the information about the design route as well. If any errors in the route revealing take place, the source-node gets the message about an error (PRER). It happens if the source-node gets no ACK-message.

In such a way, the source acquires the PERR packet to send the route request again. Having got the RRER message the nodes delete information about the route. The flow diagram of the route request is represented in figure 1.

Fulfilling the routing cache enables to choose the existence optimal route to the destination node or to rewrite it if the previous data is incorrect. If the route isn`t revealed, the request (PREQ) is retransmitted to the neighbour nodes. For the every next node the operation is repeated. After revealing the route the result is written to the routing cache of the source node and the destination node, as well as to all the intermediate nodes.

The format of the PREQ-request message consists of the following fields: UD (16 byte), Source-node (16 byte), Intermediate nodes (16*N byte), Destination node (16 byte), Hops (8 byte), Time-out (8 byte).

In the Unique Identifier (UD) the unique descriptor of the route request is written to avoid making one and the same requests and their rewriting. The appearance of the bails is fixed by the node declining the packet, if the record with UD is already present in cache. The Hops field enables the node to limit the number of intermediate nodes between the source and the destination. With every next intermediate node the Hops value decreases to zero, then the route revealing will be stopped. Time-Out field shows the time left for the route, after which the route will be deleted. It enables to avoid an endless search of the route and to decrease a chance of a fail in data transmission on account of the revealed route being not actual. MAC-address of the source is written in the field "Source-node". MAC-address of the destination is written in the field "Destination-node".

All MAC-addresses of the intermediate nodes are a writer in the field "Intermediate-nodes" in series. N –number of the intermediate nodes can`t be more than 128. PREQ-request is always enlarged by the intermediate nodes until the destination node is reached. Having achieved the destination node, it sends back PREP-reply, which contains complete information about the route and the similar fields.

To avoid a creation of a number of the route requests, which could get less actual with the time, every nodes stores life-cycles of all sent requests in a table. The routing cache of each node includes the next 5 fields: Field 1 – UD of the request (got when initiation of the route request); 'Main MAC' (stores its own MAC-address of the source-node); Field 3 – Stores list of MAC-address of the intermediate nodes); Field 4 – 'Time-out-field' (stores the left time of the route existing); Field 5 – MAC-address of the destination node.

4. Conclusion

A comparative analysis of the types and kinds was hold after the research of the existing algorithms, peculiarities of each of them were revealed. One of them was taken as the base and adopted to our system.

In the progress of work the next results were achieved:

A) Existing routing protocols were analysed and their comparative analysis was carried out;

B) The route request algorithms and the request structure were designed;

C) The structure of storing the routing cache in nodes is described.

The analysis of algorithms has shown the unique specialities of each of them and decide on situations, where these algorithms are the most effective. Algorithms without IP-protocol were designed, and that enabled an increase of the efficiency of the algorithms while working only with the MAC-addresses of the devices. The route cache storing only the necessary information is described.

References

[1] WiFI Alliance Discover WiFi *(Electronic Materials)* http://www.wi-fi.org/discover-wi-fi/wi-fi-direct

[2] Corson S and Macker J 1999 Mobile Ad hoc Networking (MANET): Routing protocol performance issues and evaluation considerations *(Electronic Materials)* https://www.ietf.org/rfc/rfc2501.txt

[3] Divecha B, Abraham A, Grosan C and Sanyal S 2007 *J. of Digital Information Manag.* **5** 19-24

[4] Kustov D, Sherstneva A and Botygin I 2016 *IOP Conf. Series: Mat. Sci. and Eng.* (IOP Publising) 012037

[5] Imas O, Kaminskaya V and Sherstneva A 2015 *Proc. of 2015 Int. Conf. on Interactive Collaborative Learning* (Florence) pp 511-514

[6] Botygin I, Popov V, Tartakovsky V and Sherstnev V 2015 *Proc. of SPIE - The Int. Society for Optical Eng.* 96800J

[7] Reddy K, Raju V and Venkatadri N 2012 *Int. J. of Video&Image Proc. and Network Security IJVIPNS-IJENS (Electronic Materials* vol 12 no 5) http://www.ijens.org/Vol_12_I_05/122005-8484-IJVIPNS-IJENS.pdf

[8] Clausen T and Jacquet P 2003 Optimized link state routing protocol (OLSR) *(Electronic Materials)* https://tools.ietf.org/html/rfc3626

[9] Johnson D, Hu Y and Maltz D 2007 The dynamic source routing protocol (DSR) *(Electronic Materials)* https://tools.ietf.org/html/rfc4728#section-6.2

[10] Perkins C, Belding-Royer E and Das S 2003 Ad hoc on-demand distance vector (AODV) Routing *(Electronic Materials)* https://tools.ietf.org/html/rfc3561

[11] Kaur R, Singh P, Gaba G and Pasricha R 2015 *Indian J. of Science and Technology* IPL0626

A modern approach to storing of 3D geometry of objects in machine engineering industry

E A Sokolova, G A Aslanov and A A Sokolov

North Caucasus Mining and Metallurgic Institute (State Technical University), 44, Nikolaev str., Vladikavkaz, 362021, Russia

E-mail: katya_sea@mail.ru, gevork.aga@yandex.ru, asklv@mail.ru

Abstract. 3D graphics is a kind of computer graphics which has absorbed a lot from the vector and raster computer graphics. It is used in interior design projects, architectural projects, advertising, while creating educational computer programs, movies, visual images of parts and products in engineering, etc. 3D computer graphics allows one to create 3D scenes along with simulation of light conditions and setting up standpoints.

1. Introduction

To study the methods and means of composition, such as reproduction of a space, an environment, light-and-shade, laws of linear, air and color perspectives, the advantages of this kind of computer graphics over vector and raster graphics are quite obvious in this case. In 3D graphics, images (or characters) are modeled and moved around the virtual space, in a natural environment or in an interior, and their animation allows one to see an object from any standpoint, to move it in an artificially created environment and space, accompanying it with special effects [1-5].

3D computer graphics, as well as vector graphics, is object-oriented, which allows changing not only all the elements of a 3D scene, but also each object individually. This kind of computer graphics has great potential when supporting technical drawing. By means of graphic editors of 3D computer graphics, such as Autodesk 3D Studio, we can perform visualization of engineering details and products, and perform modelling of buildings and architectural constructions studied in the relevant section of architectural drafting. Besides, these sections of descriptive geometry as prospect, perspective and orthogonal views can receive graphic support as the principles of building 3D images are partly borrowed from them.

2. Materials and methods

For arts and crafts, 3D computer graphics enables prototyping of future products reproducing the structure and the texture of the materials of which these products will be made. The possibility to see the model of the product from any point of view before its production in the material allows one to change and correct its form or proportions, which can be no longer possible once the work was started (e.g., jewelry, ironwork, etc.). In the same way, 3D graphics can be used while creating sculptures, design, artistic graphics, etc. 3D animation and special effects are also created by means of 3D graphics. Creating video tutorials for training programs can be the main way to use these features of 3D computer graphics [1].

Figure 1. The example of a 3D model of an engine.

Today, there are a lot of various file formats allowing one to store 3D geometry of objects. Let us analyze the three main types, their advantages and disadvantages [7-12,15]. One of the most popular formats is .OBJ. The .OBJ format is a match-all format for storing 3D objects. Files in the .OBJ format can be opened in all modern 3D editors and viewers, such as NewTek LightWave 3D, Adobe Photoshop CC 2015, Autodesk Maya 2016, DAZ 3D DAZ Studio 4, IMSI TurboCAD Pro 2016, Blender, MeshLab, MAXON Cinema 4D, Smith Micro Poser 11, etc.

3. The study of the structure of the files which contain 3D object geometry

Usually, the OBJ file format is a simple data-format that represents 3D geometry alone - namely, the position of each vertex, the UV position of each texture coordinate vertex, vertex normals, and the faces that make each polygon defined as a list of vertices [13,14]. This format is very convenient due to its simplicity, e.g. it is easy to read, and due to its transparent structure, it is easy to understand which geometry elements are described.

Basic elements of an .OBJ file:

- # is a comment (like // in C#)
- usemtl and mtlib describe the material library of the model.
- v is a vertex.
- vt stands for a vertex texture coordinate.
- vn is a vertex normal.
- f is a face (surface).

Another popular format is .3DS. A .3DS file consists of chunks, each of which contains some useful data and, possibly, sub-chunks. Most chunks contain either data or sub-chunks, although there are complex chunks. The general format of each chunk looks as follows:

Table 1. The total format of each unit

Offset	Length	Data
0	2	Chunk id, chunk_id
2	3	Chunk length, chunk_len
4	chunk_len	data or sub-chunks (depending on chunk_id)

The list of the main .3DS-chunks:

- CHUNK_MAIN = 0x4D4D; // [-] main chunk
- CHUNK_OBJMESH = 0x3D3D; // [-] various objects
- CHUNK_OBJBLOCK = 0x4000; // [+] object block
- CHUNK_TRIMESH = 0x4100; // [-] trimesh-objects
- CHUNK_VERTLIST = 0x4110; // [+] vertices list
- CHUNK_FACELIST = 0x4120; // [+] faces descroption
- CHUNK_FACEMAT = 0x4130; // [+] face materials
- CHUNK_MAPLIST = 0x4140; // [+] mapping texture coordinates
- CHUNK_TRMATRIX = 0x4160; // [+] translation matrix
- CHUNK_CAMERA = 0x4700; // [+] camera
- CHUNK_MATERIAL = 0xAFFF; // [-] material
- CHUNK_MATNAME = 0xA000; // [+] material name
- CHUNK_TEXTURE = 0xA200; // [-] material texture
- CHUNK_MAPFILE = 0xA300; // [+] mapping texture file name
- CHUNK_KEYFRAMER = 0xB000; // [-] animation information
- CHUNK_TRACKINFO = 0xB002; // [-] object's performance
- CHUNK_TRACKOBJNAME = 0xB010; // [+] object's name
- CHUNK_TRACKPIVOT = 0xB013; // [+] object's pivot point
- CHUNK_TRACKPOS = 0xB020; // [+] object's track
- CHUNK_TRACKROTATE = 0xB021; // [+] object's rotation track
- CHUNK_TRACKCAMERA = 0xB003; // [-] camera track
- CHUNK_TRACKFOV = 0xB023; // [+] FOV camera track
- CHUNK_TRACKROLL = 0xB024; // [+] roll camera track
- CHUNK_TRACKCAMTGT = 0xB004; // [-] camera target track

The file represents a binary set of data, and in order to read it, one must be sawy to work with such kinds of files [3].

The last format that we are going to discuss is the .ASE format. Like .OBJ file, .ASE files are also presented as a text file, which facilitates the work with this kind of format.

- *MATERIAL_COUNT – number of textures used in a given 3D-scene.
- *MATERIAL – start of a material description block. It is followed by the number of the material.
- *BITMAP – the given material texture address.
- *MATERIAL_REF – reference to the number of the material assigned for the given object.
- Now let us study the description of geometry and texture coordinates.
- *GEOMOBJECT – refers to a new geometrical sub-object in the 3D scene loaded from the file.
- *MESH_NUMVERTEX – refers to the number of vertices that describe the geometry of the given 3D object (sub-object).
- *MESH_NUMFACES – refers to the number of polygons in the given geometrical sub-object.

- *MESH_NUMTVERTEX, *MESH_NUMTVFACES – describes texture coordinates and their bound to polygons.
- *MESH_VERTEX, *MESH_FACE, *MESH_TVERT, *MESH_TFACE – straight description of vertices and polygons [4].

4. Results and discussion

In the present study, to solve the given problem of developing a method of compression of digital scans lossless, a model of representation of spatial objects using an ordered one-dimensional representation of the point cloud of a 3D object is developed, where S – the point cloud, G – the curve that specifies the order of the points, R(G) – a scan function corresponding to the order of bypass of points G. The following functional description is a model-based representation of spatial objects using an ordered one-dimensional view.

Let us introduce the following notation:

O – 3D object,

b – bit pixel code,

M – memory in bytes

T – code table,

Tb(i) – the position of the code bits

B – block consisting of 3-bit items merger

x,y,z – a 3-bit header of the code in the block

Ts – table of offsets

i,j,k – codes, additional bits, the distance in the offset table.

For each three-dimensional object, we must specify the set of nodes and their connections with each other.

Each pixel is represented by a bit code. Then each of the codes must be placed in byte memory, while fulfilling the condition when each subsequent bit code is greater than the previous one.

The position of a bit code is being constantly checked when entering it to the storage.

Then, the resulting sequence is entered to the block, and, simultaneously with it, the information is analysed by constructing a bit tree, and on the basis of it — of the table of offsets.

All the above-mentioned can be combined into a mathematical model:

$$\begin{cases} \forall O: \\ T \in M(b)T, \text{ where} \\ \forall\, i \in (1,m): Tb_{i-1} < Tb_i \\ B = b_x \cup b_y \cup b_z \\ \forall\, Ts(i,j,k) \equiv Tb \\ I(k) = \sum_{j=1}^{n} j q_j\,(k) \\ x_0 = \min x_i, y_0 = \min y_i, z_0 = \min z_i \end{cases} \quad (1)$$

Quantitative characteristics of products:

– the mathematical model which will allow processing images of 3D formats (*.obj*.3ds, *.ase)

– the operation time of the mathematical model depends on the quantity of processed images (directly proportional to the size of the uploaded file)

– the created method of lossless compression for representation of objet d'art obtained in the 3D format will allow us to get images without artefacts with a probability of 0.87.

A number of experiments were conducted, on the basis of which the graph of dependence of running time of the mathematical model on the size of the file was constructed (figure 2).

The appearance of artefacts depends on the number of recovered points on the object. Figure 3 depicts a histogram that shows the percentage of the recovered feature points and the probability of occurrence of artefacts. The proposed software allows viewing all 3 of the above mentioned formats. The developed software can be useful in the machine engineering industry. Due to the fact that the application is not overloaded with unnecessary functions, it can quickly display the necessary drawing on all its sides. The digital model simulates the final product and gives engineers the ability to visualize, optimize and control the result before creating a physical prototype.

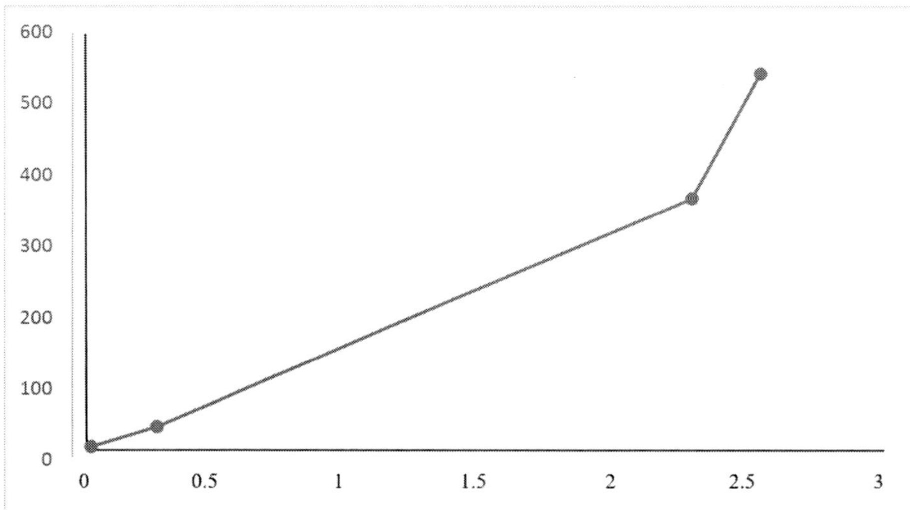

Figure 2. The graph of dependence of running time of the mathematical model on the size of the processed images.

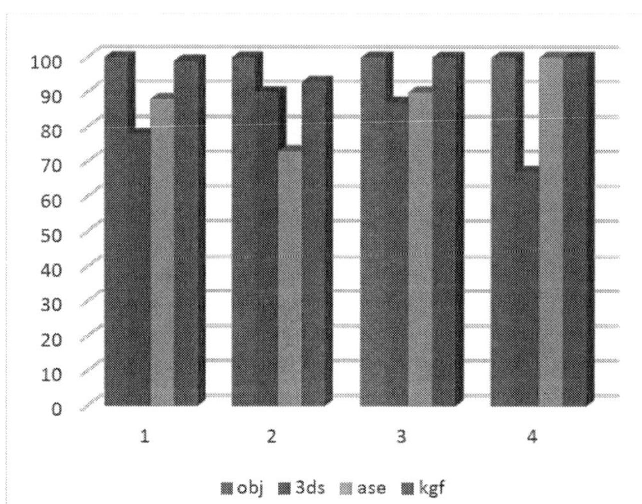

Figure 3. Probability of artefacts occurrence

5. Conclusion

The application's main function is to work with the new format for storing 3D geometry, which is based on the basic principles of the .OBJ format.

The advantage of the new file formats is that they occupy much less storage space than .OBJ files do. According to our studies and experiments, the same object described by the new format file is about 28 times smaller than the initial one. This is due to the fact that this format uses a modern algorithm for lossless compression, which allows us to achieve this result. This feature will be very useful for the majority of users and machine engineers in particular.

Acknowledgments
This article was prepared by the results of the research carried out with the financial support of the Russian Ministry of Education in the framework of the basic part of the state task (project No. 3943 on the theme 'Development of special mathematical software for the new structure of information analysis and the processing system for man-made cycles of enterprises of the mining and metallurgical complex').

References
[1] Zhi-liang Zhu, Wei Zhang, Kwok-wo Wong and Hai Yu 2011 *Inf. Sci.* **6** 1171-86
[2] Wei Zhang, Kwok-wo Wong, Hai Yu and Zhi-liang Zhu 2013 *Commun. Nonlinear Sci. Numer. Simul.* **3** 584-600
[3] Xingyuan Wang, Lin Teng and Xue Qin 2012 *Signal Process* **4** 1101-08
[4] Yicong Zhou, Weijia Cao, and Philip Chen 2014 *Signal Process* **8** 197-207
[5] Kumaritov A M, Sokolova E A and Sokolov A A 2016 *Mining J.* **2** 94-96
[6] Murillo-Escobar M A, Cruz-Hernández C, Abundiz-Pérez F 2015 López-Gutiérrez R M and Acosta Del Campo O R *Signal Process* **109** 119-131
[7] Zhongyun Hua, Yicong Zhou, Chi-Man Pun C L and Chen Philip 2016 *Inf. Sci.* **10** 80-94
[8] Yingqian Zhang and Xingyuan Wang 2014 *Inf. Sci.* **20** 329-351
[9] Wei Zhang, Kwok-wo Wong, Hai Yu and Zhi-liang Zhu 2013 *Commun. Nonlinear Sci. Numer. Simul.* 8 2066-80
[10] Leo Yu Zhang, Xiaobo Hu, Yuansheng Liu, Kwok-Wo Wong and Jie Gan 2014 *Commun. Nonlinear Sci. Numer. Simul.* **10** 3653-59
[11] Hongjun Liu and Xingyuan Wang 2011 *Opt.Commun.* **16-17** 3895-3903
[12] Shannon C E, J. 28 1949 *Bell Syst. Tech.* 656-715
[13] Chen L, Jingyu Liu, Jisen Wen, Xiong Gao, Haidan Mao, Xiaoyan Shi and Qingling Qu 2015 *Opt.Laser Technol.* **6** 80-86
[14] Yicong Zhou, Long Bao and Philip Chen C L 2014 *Signal Process* **97** 172-182
[15] Ping Ping, Feng Xu and Zhi-jian Wang 2014 *Signal Process* **105** 419-429

Development of an algorithm for automated enhancement of digital prototypes in machine engineering

E A Sokolova and G A Dzhioev

North Caucasian Institute of Mining and Metallurgy (State Technological University), 44, Nikolaeva str., Vladikavkaz, 362023, Russia

E-mail: katya_sea@mail.ru, kel.tas204@gmail.com

Abstract. The paper deals with the problem of processing digital prototypes in machine engineering with the use of modern approaches to computer vision, methods of taxonomy (a section of the decision theory), automation of manual retouching techniques. Upon further study of the problem, different taxonomic methods have been considered, among which the reference method has been chosen as the most appropriate for automated search of defective areas of the prototype. As a result, the algorithm for automated enhancement of digital prototypes of the digital image has been developed, using modern information technologies.

1. Introduction

A distinctive feature of the machines developed now is the computerization of all stages of design, production and operation in the context of the life cycle of products [1-5]. The fact that information systems tend to capture both basic and related aspects of the machine engineering brings about the need to create computer techniques to develop new materials and integrate them into a common data stream.

Lack of verified behavioural models for the used metallic materials and advanced composite materials in technological processes and loaded structures impairs the accuracy of computer modelling in machine engineering.

The process of product development in the machine engineering industry is currently divided into several stages, each of which has its own technical problems:

- At the stage of conceptual design, technical designers and engineers often use paper-based or digital formats that are incompatible with the digital data used at the construction stage. The lack of digital data, compatible formats and automation is a characteristic feature that distinguishes this stage from the construction and production stages. These conceptual design data has to be re-created later in a digitalised form, which leads to the waste of time and resources.
- At the construction stage, mechanic and electrical engineers use different systems and formats; lack of automation does not allow reacting quickly to changes in production requirements for products. Another problem at the construction stage is that in a typical 3D CAD aimed at work with geometric entities, it is difficult to create and use digital prototypes for testing and optimization of products before their actual implementation. So there arises a need to build complex and costly physical prototypes.

- The production stage is isolated from all digital processes: conceptual design, mechanical and electrical components design. Background information is provided in an analog form, i.e. as drawings. The result is a high need for physical prototypes, which adversely affects the performance and innovation aspects of the process.

2. Problem statement

Image editing is a change in the original image by means of classic or digital methods. It may also be defined by the term 'improvement' [5-10,18]. The main task of improvement is to fix any defects of the digital material. Its purpose is to bring the image to perfection, but at the same time to preserve its character and nature, preparing the frame for further processing.

Digital prototype processing takes longer than its creating, and most professionals spend a lot of time on performing this routine yet absolutely essential function. To solve this problem is the main purpose of this study, which is supposed to simplify work with digital prototypes.

The purpose of this work is to develop a fundamentally new technology (algorithm) of digital prototype defect fixing that will be applied automatically with decision-making techniques, which have not been used in this field before, combined with modern image and object recognition approaches.

To develop the enhancement algorithm, the following tasks have been performed:

1. Review of a prototype research methods and its definition has been made.
2. Research in the field of conservation of 3D images has been carried out.
3. Research on manual approaches to digital prototype defect fixing in machine engineering has been carried out.
4. A defect search algorithm based on the reference method has been developed.

3. Materials and methods

As a solution of the problem there has been developed an algorithm that implements [6, 8-11]:

- **Automated face detection in an image by Viola-Jones.**

Viola-Jones is one of the best in the 'recognition efficiency/operation speed' ratio. Also, this detector has an extremely low probability of false detection of the target object. The algorithm also operates well even at a sharp angle, for instance of about 30 degrees. At an angle of inclination greater than 30 degrees, the percentage of detections slumps drastically.

- **Color-based segmentation of the prototype:**

The object detection algorithm suggested by P. Viola and M. Jones, in some cases, commits errors of the second kind (false response) on the objects that are significantly different from the background in color. To reduce the error, the stage of scanning candidate areas for color matching was added to the detection algorithm. Thus, in order to modify the object detection algorithm, the following approach was applied:

Analyzing a sample of 1000 digital prototypes containing machine engineering objects, we have found out that the color of the object without a background can be described in the RGB color space by the following relation. A pixel (R, G, B) refers to the skin if all the following conditions are met:

- $R > 100$; $G > 40$; $B > 20$;
- $\max\{R, G, B\} - \min\{R, G, B\} > 15$;
- $(R - G > 15)$; $(R > G)$; $(R > B)$,

where R, G, B are respectively the values of the red, green and blue components of the pixel brightness. Using this relation, each pixel of the image is checked on belonging to a specific color range. All pixels that fall in this range will determine the areas of the object, which will then be tested for defects.

- **Detection of defective areas by the reference method**

The reference method is a variant of the taxonomy with possibilities of its effective usage in the decision-making technology and the guaranteed receipt of Pareto-optimal solutions with regard to multi-criteria problems [6-9].

After defining the object's areas, we obtain an object (pixel) vector meeting the conditions where each object describes the brightness of the color component: red, green, blue.

The following is required to detect imperfections:

1 To normalize initial pixel array;
2 To determine the reference object;
3 To calculate the distance of all objects from the reference one;
4 To rank each object (pixel) with regard to the reference object;
5 To take the most distant pixels with a certain inaccuracy (Eps) as defected areas.

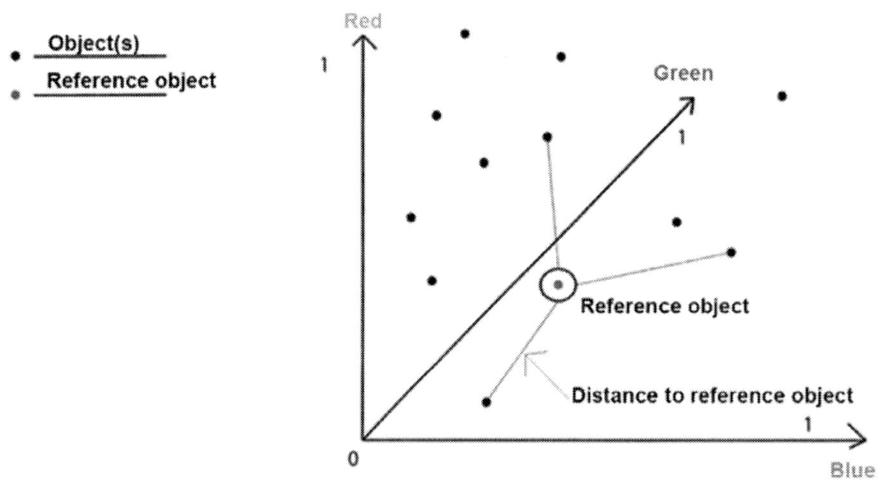

Figure 1. The distance from objects to the reference object as the main criteria of defected areas.

- **Formal statement of the problem:**

$$\begin{cases} f=\sum_{k=1}^{n} G(R_k) \rightarrow \max; \\ \forall: R_k=\sqrt{\sum_{j=1}^{3} \left(X_{k_j}-M_j\right)^2}; \\ \forall: G(R_k)=1: \ R_k \geq Eps; \\ \forall: G(R_k)=0: \ R_k<Eps; \\ Eps=Max(R_1,R_2,\ldots,R_k)*0.2; \end{cases} \quad (1)$$

where:

$k=1,2,...,n;$ – the index of the object being examined;

n – the number of objects (pixels);

X_{k_j} – the value assumed by a j-characteristic property of a k-object;

j=1,2,3; – the index of the corresponding characteristic of the pixel

R_k- the distance from a k-object to the reference object;

$$\bar{M}=\{M_1,\ldots,M_j\} \text{ – Reference object;}$$

$$\begin{cases} M_1=\dfrac{\sum_{k=1}^{n} X_{k_1}}{n}; \\ M_2=\dfrac{\sum_{k=1}^{n} X_{k_2}}{n}; \\ M_3=\dfrac{\sum_{k=1}^{n} X_{k_3}}{n}; \end{cases}$$

where:

$k=1,2,\ldots,n$;-the index of the object being examined;

n-the number of objects (pixels);

X_{k_j}-the value assumed by a j-the characteristic property of a k-object;

$j=1,2,3; \ldots$-the index of the corresponding characteristic of the pixel

- **Elimination of defective areas using frequency decomposition:**

Decomposition into spatial frequencies is a method allowing spreading small and large image details into separate layers. One layer will contain the information only about the details (high frequencies), the other layer will contain information about the color and tonal transitions (low frequencies). This makes it easy to correct the shape without affecting the texture of the surface or to correct the texture without affecting the form [4].

4. Results and discussion

The algorithm developed by us allows for automated enhancement of the digital prototypes quality. For a clearer description of the algorithm of automated skin retouch, there is a figure showing the result of the downloaded image processing by a software package (Figure 2).

Figure 2. An example of image processing by the algorithm.

Optimization of the quality of defect repairing is provided by the use of decision-making methods together with advanced computer vision approaches. Using image frequency decomposition and color-based segmentation allows one not to produce the object and background distortion in the image. With further development, the enhancement algorithm can be implemented in the modern software.

For conducting experiments, we used a PC with the following characteristics:

The processor:

The manufacturer of the processor is Intel Company

The processor type is Core i7 with a frequency of 2.5 GHz

The number of processor cores is 4
Maximal bit-timing frequency is 3.7 GHz
Cache memory is 6 MB
RAM:
Random access memory (RAM) is 16 GB
Memory type is ddr3l
Memory frequency is 1600 MHz
Video card:
The video card manufacturer is AMD
Graphic card is Radeon R9 2GB M370X
Hard drive:
Hard disk-drive (SSD) is 512 GB

The main characteristics of the software product is the quality of elimination of defects of the developed algorithm on the loaded image. The experiment showed that the optimal size of the convolution matrix is equal to the width and height, in the area of which the face is defined, where the width and the length are respectively divided into the range of integers, approximately from 64 to 200. The result of dividing the width and the length is an integer < 3.

5. Conclusion

Nowadays, experts are paying a lot of attention to the problem of image processing. However, the algorithm and the solution do not exhaust all varieties of existing problems in the field of digital processing of visual information in the machine engineering industry. It illustrates the relevance of the research and development that we conduct.

References

[1] Kumaritov A M, Sokolova E A and Sokolov A A 2016 *Mining J.* **2** 94-96
[2] Murillo-Escobar M A, Cruz-Hernández C, Abundiz-Pérez F 2015 López-Gutiérrez R M and Acosta Del Campo O R *Signal Process* **109** 119-131
[3] Zhongyun Hua, Yicong Zhou, Chi-Man Pun C L and Chen Philip 2016 *Inf. Sci* **10** 80-94
[4] Yingqian Zhang and Xingyuan Wang 2014 *Inf. Sci* **20** 329-351
[5] Wei Zhang, Kwok-wo Wong, Hai Yu and Zhi-liang Zhu 2013 *Commun. Nonlinear Sci. Numer. Simul.* 8 2066-80
[6] Leo Yu Zhang, Xiaobo Hu, Yuansheng Liu, Kwok-Wo Wong and Jie Gan 2014 *Commun. Nonlinear Sci. Numer. Simul.* **10** 3653-59
[7] Puzin V S Khoroshev A S Tchoutchkin D A Shevchenko E V Batishchev D V Gummel A A, 2014 *Life Sci J* **11** 141
[8] Hongjun Liu and Xingyuan Wang 2011 *Opt.Commun.* **16-17** 3895-3903
[9] Shannon C E, J. 28 1949 *Bell Syst. Tech.* 656-715

Energy efficiency improvement of electric drive of cold pilgering mill

D A Sychev, N V Savosteenko and A A Gryzlov

RiK-Energo Ltd, 15, Tarasova str., Chelyabinsk, 454048, Russia

E-mail: m9191236713@gmail.com

Abstract. Different ways to improve the energy efficiency of electric drives of cold pilgering mill are considered. Mathematical modeling methods represent studies according to which it is possible to evaluate quantitatively the energy savings. For example, cold pilgering mill 450 shows the mean square of the armature current of the main drive motor related to critical frequency of the speed loop. The possibility of energy saving by previous field weakening of the motor before the operating cycle of rolling is considered. The optimal energy saving points of supply and termination of the pulse are determined by the field weakening. The correlation of the parameters of the dynamic units and the change of the electric drive work schedule provides the greatest loss reduction in the main drive of the cold pilgering mill stand based on the conditions. Activities aimed at improving energy efficiency of electric drives of mills of this group are reviewed, which reduces the electric energy consumption for the cycle rolling by 20-25 %.

1. Introduction

The main drive of a stand cold pilgering mills seamless pipe has an irregular periodic load curve of the operating device [1]. This fact leads to an overstatement of the installed capacity of power facilities. Electric drives of the mills of the reporting group realized on the basis of DC drives, in which a significant portion of the electrical losses is a part of the variable losses in the motor. The question of the modernization relevance, taking into account the significant productivity and capacity of power facilities, is dictated by the fact that most of the mills are working with old control schemes. Due to the updating of the electric drive elements, methods and control systems, it becomes possible to estimate the resources of energy saving in the electric drive for these types of mills [2].

2. Statement of energy saving problem

The mills of the considered groups of the main drives of the stands, implemented on the basis of the DC drives, are designed by the principle of minor regulation. Among related studies aimed at improving the efficiency of pilgering mills, it is possible to note the work, based on the decrease of the rate of the rectified EMF thyristor converter due to the automatic correction of the EMF of the motor when reducing the consumption of reactive power because of the main voltage deviation [3]. The object in this case was made by the broadband hot rolling mill.

Historically, that question of saving energy and reducing costs on power facilities was solved in the installed flywheel of open-loop electric drive systems. In [4], the methodology of the model calculations and the choice of a flywheel of the electric drives of the rolling mills of different types, having periodic load, are considred. There are, however, some enterprises manufacturing the electric drive of the cold pilgering mills of seamless pipes that do not have flywheels installed in the kinematic chain of the motor [5].

It is possible to bring an artificial softening of the mechanical characteristics in the electric drives of the powerful reversible hot rolling mill as another variant to solve this problem, where the system generator-motor was used, and the wound rotor induction motor (aggregates of Ilgner) was used to rotate the generator. This provides a reduction of the pulse load on the supply network in the moments when metal in the rolls [6].

Rolling on the cold pilgering mill is periodic; it provides irregular load torque for the electric drive during the cycle [7]. The process that requires the supply and rotates the workpiece means a lack of metal in the rolls, so the rolling tool is made so that the bite is formed between the rollers in the initial (I) and ending (II) positions during rolling [8].

From the point of view of energy saving, especially taking into account the achievements of power electronics [9], computer engineering, systems and ways of control, it becomes possible to estimate the resources of energy saving in the electric drive for these types of mills [10]. The energy efficiency of electric drives of mills for the cold pilgering mill are proposed to solve the reduced RMS current in the armature chain that is placed in the existing drive. Thus, it is possible to soften the mechanical characteristics of the closed loop system of the electric drive by preliminary field weakening of the motor because it increase the rated no-load speed before rolling [11].

3. Softening mechanical characteristics of the drive closed system

Mechanical characteristics of the drive closed system can be softened. The idea of the first method is to withdraw the part of the energy spent on the deformation of the metal from the rotating mass of the electric drive, thereby to increase the evenness of armature current during the rolling cycle and to reduce the magnitude of the losses [12].

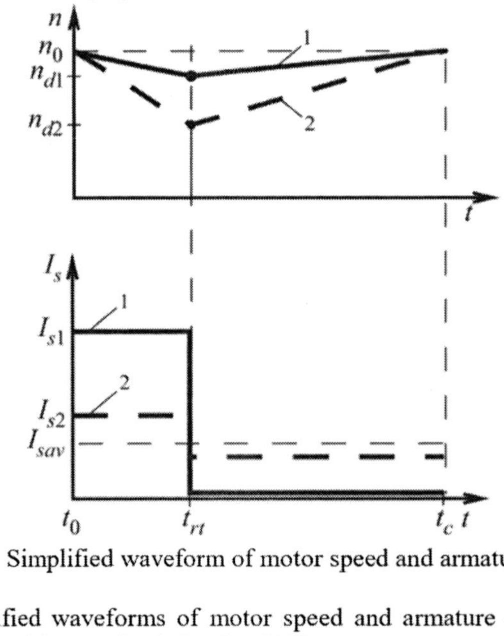

Figure 1. Simplified waveform of motor speed and armature current.

Figure 1 shows simplified waveforms of motor speed and armature current. The energy of the rotating mass of the electric drive can be defined as [13]:

$$W_{max} = J \cdot (n_o^2 - n_d^2)/2 \tag{1}$$

where W_{max} – rotating mass energy; n_o – the ideal no-load speed of the engine; n_d – value of dynamic decreased velocity; J – the moment of inertia of the electric drive. The energy savings will be more significant than most of the energy of W_{rol} required for rolling, which will be covered due to the energy of the rotating mass of the electric drive. Simplified waveforms (Figure 1, (1)) of current and velocity correspond to the original case mechanical characteristics of the closed system of the drive. Softening this feature allows reducing the RMS armature current to the average value (Figure 1 (2)), thereby reducing the amount of electric losses in the armature chain [14].

The estimation of reducing RMS current was carried out using the mathematical model of 'electric drive – pilgering mill' cold pilgering mill 450 of seamless tubes. This model included the main electric drive of the pilgering mill, which is the scheme of slave control [15]; the mechanical part, taking into account the mechanical transmission in the system; block cast torques rolling along with the motor shaft. References data of motor [16] are:

$$U_{ref} = 620\ V,\ I_{ref} = 2340\ A,\ M_{ref} = 310.2\ kN \cdot m \quad (2)$$

$$\omega_{ref} = 4.2\ rad/s,\ J = 19875\ kg \cdot m^2 \quad (3)$$

Softening mechanical characteristics is produced because of the decrease in the gain of proportional speed control K_{rs}. Figure 2 shows the ratio of RMS armature current to the speed control loop critical frequency. If we choose critical frequency of 10 rad/s, the RMS current value is reduced by 12 %. The real decrease in the RMS current (up to 50 %) is achieved only when [17].

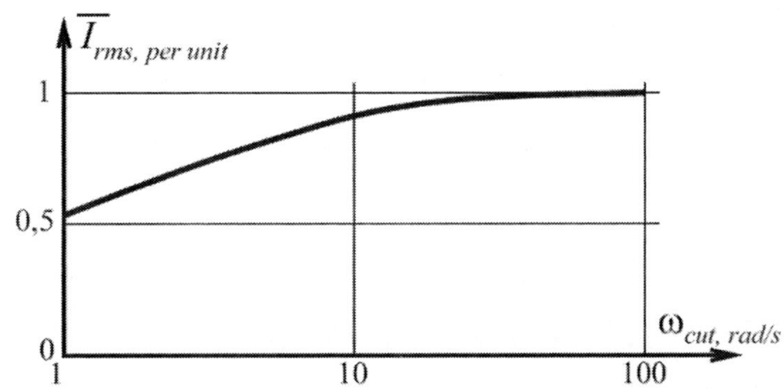

Figure 2. Dependence of RMS armature current of critical frequency speed control loop.

Low critical frequency and significant drawdown speed cannot be tolerated in real systems. As a result, the energy savings obtained per unit of output due to lower productivity of the mill are insignificant. This fact is connected with the cold pilgering mill 450 due to the fact that rolling time t_{rt} is relatively large, about 40 % of total cycle t_{tc}. The mass energy of the drive will be better if the rolling time is much less than the time of the entire cycle [18].

4. Preliminary field weakening before impulse stroke

The idea of the second method for energy saving is to use high-speed differential Δn_d at the stage of the imposition of load torque, to prevent a significant reduction in the average rolling speed. This is possible if at the stage of cycle, when the metal is not in the rolls, we increase the speed of the drive considerably above the ideal no-load speed of the drive at the rated voltage in the armature, i.e. at the cost of preliminary field weakening at the beginning of the impulse stroke [19].

Before each rolling cycle, the drive speed is artificially increased to n_{01} with weakening motor field, thereby achieving an increase of the RMS of speed per cycle. In the article, by the example of cold pilgering mill 450, we will consider the effect of the degree of flux attenuation of drive Φ (initial overspeed n_{01}) relatively the RMS current of the armature and the output of the mill [20]. At the same time, the speed limit should be set, based on the conditions of limitation impulse loads in the mechanical equipment for the metal pickup. Because the field weakening process occurs due to dynamic, it is also important to choose the time of pulsing and pulse stop during the field weakening (Figure 3). The problem unfolded in the function of the angle of rotation of the rolls, since all state variables are easier and clearer considered by this parameter. Figure 4 shows the relation of armature current RMS to the rotation angle of rolls α_{pre} corresponding to the start pulse to supply field weakening. The greatest effect for energy saving of cold pilgering mill 450 is obtained by the flow of drive $F = 78$ % of the nominal rate, $\alpha_{pre} = 0.75$ *rad*, and was about 25 %.

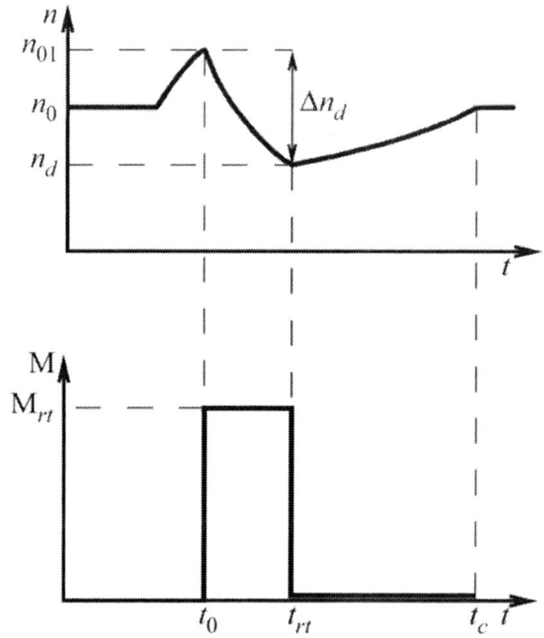

Figure 3. Preliminary field weakening.

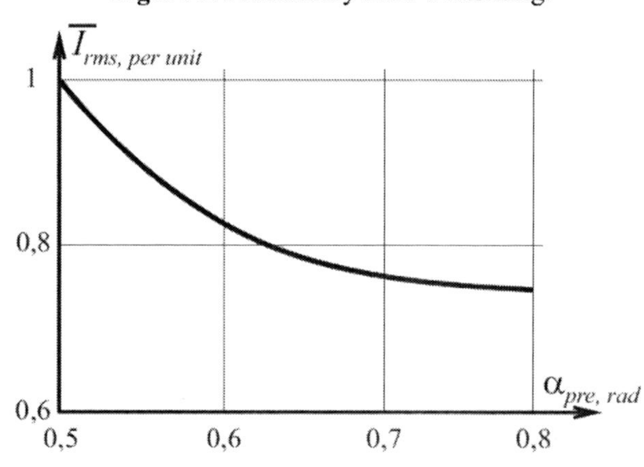

Figure 4. Dependence of RMS current of armature on angle of rotation of roll; pulse corresponds to the start of field weakening.

5. Conclusion

This article discusses the ways to improve the energy efficiency of the electric drive of cold pilgering mills. As a result of solving the problem, we have identified conditions that allow justifying the ratio of the dynamic chains and making changes in the work schedule of the drive, which provides the most effective reduction of losses in the drive of the stand roll cold pilgering mill.

References

[1] Kazantsev V P and Dadenkov D A 2015 *Russian Electrical Eng.* **86**(6) 344-349
[2] Nos O V and Kharitonov S A 2015 *Russian Electrical Eng.* **86**(2) 72-78
[3] Kruglikov O V 2015 *Russian Electrical Eng.* **86**(3) 118-124
[4] Osipov O I 2015 *Russian Electrical Eng.* **86**(1) 5-8
[5] Karandaev A S, Kornilov G P, Khramshin T R and Khramshin V R 2015 *Russian Electrical Eng.* **86**(4) 201-207
[6] Gorozhankin A N, Grigor'ev M A, Zhuravlev A M and Sychev D A 2015 *Russian Electrical Eng.* **86**(12) 697-699
[7] Grigor'ev M A 2015 *Russian Electrical Eng.* **86**(12) 694-696
[8] Grigor'ev M A, Naumovich N I and Belousov E V 2015 *Russian Electrical Eng.* **86**(12) 731-734
[9] Savos'kin A N, Kulinich Y M and Garbuzov I I 2015 *Russian Electrical Eng.* **86**(9) 540-547
[10] Feoktistov V P, In'kov Y M and Tretinnikov O V 2015 *Russian Electrical Eng.* **86**(9) 514-518
[11] Grigor'ev M A, Sychev D A, Zhuravlev A M, Khayatov E S and Savosteenko N V 2015 *Russian Electrical Eng.* **86**(12) 728-730
[12] Funk T A, Saprunova N M, Belousov E V and Zhuravlev A M 2015 *Russian Electrical Eng.* **86**(12) 716-718
[13] Grigoryev M A 2014 *Russian Electrical Eng.* **85**(10) 601-603
[14] Pavlenko A V, Vasyukov I V, Puzin V S, Grinchenkov V P and Bol'shenko A V 2015 *Russian Electrical Eng.* **86**(8) 453-458
[15] Grigoryev M A, Kinas S I 2014 *Russian Electrical Eng.* **85**(10) 645-648
[16] Antonov B M, Baranov N N and Kryukov K V *Russian Electrical Eng.* **86**(7) 385-390
[17] Romodin A V and Kuznetsov M I 2015 *Russian Electrical Eng.* **86**(6) 339-343
[18] Korshunov A I 2015 *Russian Electrical Eng.* **86**(4) 187-193
[19] Ladygin A N, Bogachenko D D and Kholin V V 2015 *Russian Electrical Eng.* **86**(1) 14-17
[20] Bychkov M G, Kuznetsova V N, Vasyukov S A and Krasovsky A B 2015 *Russian Electrical Eng.* **86**(1) 22-28

Application of Lucas–Kanade algorithm with weight coefficient bilateral filtration for the digital image correlation method

V V Titkov[2], S V Panin[1,2], P S Lyubutin[1,2], V O Chemezov[2] and A V Eremin[1,2]

[1] Tomsk Polytechnic University, 30, Lenina ave., Tomsk, 634050, Russia
[2] Institute of Strength Physics and Material Science of Siberian Branch Russian Academy of Sciences, 2/4, Akademicheskii ave., Tomsk, 634055, Russia

E-mail: svp@ispms.tsc.ru

Abstract. Application of bilateral filer weight coefficients for computing the weight function in Lucas–Kanade algorithm for optical flow determination is proposed. The obtained results based on processing different image types demonstrate the error reduction in optical flow determination. The approach to weight function parameters selection for bilateral filter is proposed and investigated.

1. Introduction

A digital image correlation (DIC) technique being currently widely spread both in research and industry belongs to optical methods for strain evaluation and makes it possible to study deformation processes and fracture of heterogeneous materials (metals, alloys, ceramic, polymers etc.), and moreover to evaluate the structural integrity of loaded machine components and structure elements [1, 2]. The DIC technique ensures non-contact numerical characterization of the processes which develop on the surface of loaded materials with spatial resolution of several micrometers. Basic principles of strain estimation provided by the DIC are i) construction of displacement vectors field and ii) computation of displacement vectors derivatives (calculation of deformation components).

Traditionally, optical flow is determined within two stages which are characterized by pixel and subpixel computation accuracy, correspondingly. A classical algorithm for subpixel accuracy optical flow determination is Lucas–Kanade one [3]. Computation of optical flow (V_x, V_y) within a window is carried out by Lucas–Kanade algorithm when solving the system of equations as follows:

$$\begin{bmatrix} V_x \\ V_y \end{bmatrix} = \begin{bmatrix} \sum_i I_x(q_i)^2 & \sum_i I_x(q_i)I_y(q_i) \\ \sum_i I_x(q_i)I_y(q_i) & \sum_i I_y(q_i)^2 \end{bmatrix}^{-1} \begin{bmatrix} -\sum_i I_x(q_i)I_t(q_i) \\ -\sum_i I_y(q_i)I_t(q_i) \end{bmatrix}$$

where $q_1, q_2, \ldots, q_n$ – pixels within the window; $I_x(q_i)$, $I_x(q_i)$, $I_x(q_i)$ – partial derivatives $I_x(q_i)$ with respect to coordinates x, y and time t, calculating in point q_i.

In doing so, all pixels within the window exert the same influence onto the final results. Weight coefficients (weighted solution) are used to take into account the distance from the pixel where optical flow is determined for all other pixels in the window:

$$\begin{bmatrix} V_x \\ V_y \end{bmatrix} = \begin{bmatrix} \sum_i w_i I_x(q_i)^2 & \sum_i w_i I_x(q_i) I_y(q_i) \\ \sum_i w_i I_x(q_i) I_y(q_i) & \sum_i w_i I_y(q_i)^2 \end{bmatrix}^{-1} \begin{bmatrix} -\sum_i w_i I_x(q_i) I_t(q_i) \\ -\sum_i w_i I_y(q_i) I_t(q_i) \end{bmatrix}$$

where $w_1, w_2, \ldots, w_n$ – weight coefficients.

Traditionally, normal distribution of the distances between pixel q_i and the window center is used in the capacity of weight coefficients w_i. Application of cross-bilateral filtration coefficients [4], which are computed over the input (initial) image, is investigated in the present paper.

2. Algorithm and test method description

In order to calculate bilateral filter coefficients both the described distance from pixel q_i to the window center as well as the information about pixels brightness (difference between pixel brightness which is used for optical flow determination and the rest pixels within the window) should be considered. When bilateral filtration coefficients are utilized, the weight function has the following form:

$$w_i = w_{c_i} w_{d_i},$$

where $w_c = exp(-\frac{(I_t(\bar{i}) - I_t(\bar{i}+\Delta\bar{i}))^2}{2\sigma_c^2})$ – weight function for color data,

$w_d = exp(-\frac{(\Delta\bar{i})^2}{2\sigma_d^2})$ – weight function for distance data,

I_t – image at the initial time instance,

I_{t+1} – image at the current (next) time instance,

$I(\bar{i})$ – pixel brightness in point $\bar{i} = (i, j)$.

Figure 1, b demonstrates the example of weight coefficient distribution, which is computed for the central pixel of the image fragment presented in figure 1, a. It is seen that for this pixel, the highest weights belong to the ones with the approximately similar brightness level as well as pixels at its closest vicinity.

a b

Figure 1. Image region (a) and the corresponding distribution of weight coefficients (b).

The equation for computing weight functions with considering their coefficient values for both images in analyzed pair is proposed:

$$w = w_{c_t} w_{d_t} w_{c_{t+1}} w_{d_{t+1}},$$

where w_{c_t}, w_{d_t} – weight functions at the initial time instant (for initial image);

w_{c_t}, w_{d_t} – weight functions in current (next) time instant (for current frame).

In the present paper, the three-dimensional recursive search (3DRS) algorithm was used as a basic algorithm for optical flow determination with pixel accuracy. The 3DRS method is widely employed for video data processing as far as it allows reducing the time for the displacement vector field construction [5, 6].

Optical flow computation with pixel accuracy in the three-dimensional recursive search (3DRS) algorithm is implemented through the searching procedure with similarity metric minimum. The sum of absolute differences (SAD) of image patch brightness in current and initial images is utilized in the capacity of this metric, whereas Lucas–Kanade algorithm is used at the second stage for a more precise displacement determination.

The efficiency of the weight coefficient application was studied with the use of test images (in gray scale) presented in open public database Middlebury benchmark [7, 8]. Computation of the vector field was conducted for 8 image pairs with known (model) information on the optical flow – Dimetrodon, Grove2, Grove3, Hydrangea, RubberWhale (figures 2, a, c), Urban2, Urban3, Venus.

As a metric for the efficiency evaluation of optical flow determination, the Average Endpoint Error (*AEE*) [8] was taken:

$$AEE=\frac{1}{h\cdot w}\sum_{y=1}^{h}\sum_{x=1}^{w}\sqrt{(u-u_{GT})^2+(v-v_{GT})^2}$$

where h, w – image height and width, u, v – vector components (displacements) of the calculated optical flow, u_{GT}, v_{GT} – vector components (displacements) of the modeled optical flow.

In order to evaluate the efficiency of the proposed approach at the objects (fragment) edges, their detection was carried out by the widely spread Canny algorithm (Canny Edge Detector) [9]. In doing so, the program codes were taken from the Computer Vision public library OpenCV [10]. For this purpose the mask of object edges was formed (figure 2, b) which then was employed at the *AEE* computation. The influence of the weight coefficients on the general “quality” (accuracy and robustness) of the vector field construction was performed through calculation of the *AEE* both for entire (full) images and for ones with excluded edges.

3. Testing results

Using subpixel computation/refinement of optical flow by Lucas–Kanade algorithm (LK) in comparison with pixel accuracy, one reduces the total *AEE* (table 1-3) for 8 pair of images for:
- 1.174 (19.7%) at full image analysis;
- 1.115 (15.2%) at analysis of edges in the images;
- 1.218 (22.7%) at image analysis excluding edges.

Employment of weight coefficients in Lucas–Kanade algorithm (wLK) results in a subsequent error reduction for determination of optical flow (table 1-3) for:
- 1.414 (23.7%) at full image analysis;
- 1.282 (17.5%) at analysis of edges in the images;
- 1.491 (27.8%) at image analysis excluding edges.

The error is reduced by 4%, 2.3% and 5.1%, correspondingly, in comparison with the case where weight coefficients of the bilateral filter were not used.

Application of Lucas–Kanade algorithm with a glance of weight function coefficient values for both images in the analyzed pair (wwLK) give a rise to the *AEE* reduction as well:
- 1.524 (25.5%) at full image analysis;
- 1.410 (19.2%) at analysis of edges in the images;
- 1.569 (29.2%) at image analysis excluding edges.

Thus, *AEE* was decreased by 1.8%, 1.7% and 1.4%, that is higher than in the case when the wLK is employed.

Table 1. Average endpoint error of optical flow determination for full image analysis.

Algorithm	Pair of test images								Total
	Dimetr.	Grove2	Grove3	Hydr.	RubbWh.	Urban2	Urban2	Venus	*AEE*
basic	0.441	0.475	0.789	0.353	0.327	1.028	2.088	0.466	5.968

LK	0.243	0.264	0.646	0.271	0.206	0.789	1.918	0.457	4.794
wLK	**0.225**	0.206	0.622	0.231	0.162	**0.748**	1.946	0.414	4.554
wwLK	0.246	**0.190**	**0.567**	**0.220**	**0.146**	0.761	**1.917**	**0.397**	**4.444**

Table 2. Average endpoint error of optical flow determination for analysis of image edges only.

Algorithm	Pair of test images								Total
	Dimetr.	Grove2	Grove3	Hydr.	RubbWh.	Urban2	Urban2	Venus	*AEE*
basic	0.410	0.516	0.945	0.672	0.427	1.525	2.292	0.552	7.337
LK	0.157	0.320	0.826	0.545	0.407	**1.293**	**2.148**	0.526	6.222
wLK	**0.151**	0.274	0.808	0.514	0.330	1.305	2.185	0.487	6.055
wwLK	0.189	**0.245**	**0.743**	**0.485**	**0.277**	1.349	2.170	**0.473**	**5.927**

Table 3. Average endpoint error of optical flow determination for analysis of image without considering the edges.

Algorithm	Pair of test images								Total
	Dimetr.	Grove2	Grove3	Hydr.	RubbWh.	Urban2	Urban2	Venus	*AEE*
basic	0.448	0.437	0.550	0.228	0.301	0.939	2.044	0.419	5.367
LK	0.263	0.212	0.371	0.164	0.153	0.670	1.868	0.418	4.149
wLK	**0.242**	0.142	0.337	0.121	0.119	**0.649**	1.894	0.372	3.876
wwLK	0.260	**0.139**	**0.297**	**0.118**	**0.111**	0.656	**1.862**	**0.354**	**3.798**

Figures 2 d–i demonstrate the examples of optical flows and *AEE* fields computed for the RubberWhale image pair.

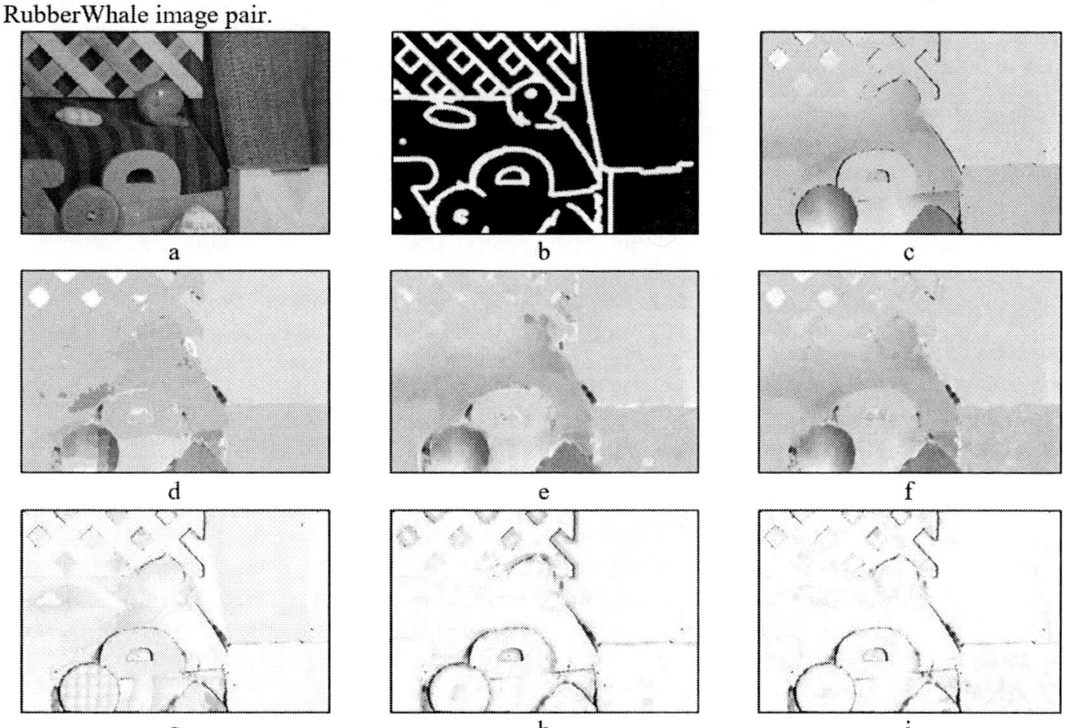

Figure 2. Test image of: RubberWhale (a), edge mask (b), modeled optical flow (c). Optical flow (d-f) and *AEE* field (g-i) for the algorithms: d, g – basic; e, h – LK; f, i – wLK.

It is seen that pixel computation accuracy of optical flow provides gradient changes of the *AEE* field, while these changes are less significant when the LK is applied (figure 2, h). However, utilization of the LK leads to increasing of errors at object edges, but these errors are lower when the bilateral filter with weight coefficients is applied (figure 2, i). Thereby, application of the bilateral filter with weight coefficients allows using optical flow pixel computation accuracy by the Lucas–Kanade algorithm without increasing the error at the object edges.

Detailed analysis of the results has shown that application of weighted similarity metric reduces error for full image analysis. In doing so, for the Urban3 image pair, the minimum changes is achieved at – 0.5%, while for the RubberWhale image pair the maximum improvement has been attained at – 29.1%. Thereby, application of weight coefficients in Lucas–Kanade algorithm gives rise to a significant decrease of the optical flow determination error for the most of analyzed image pairs.

4. Conclusion

The method for computation the bilateral filter coefficients for two images of the pair under analysis is proposed, that is, based on the product of bilateral filter coefficients of both images. This approach results in error reduction by more than 1.4% when applying similarity metric computation with considering pixels at the first and the second images, which have brightness similar to the pixel located in the center of the window.

Investigation of optical flow determination with subpixel accuracy has shown the possibility of *AEE* decreasing by more than 29% by employing the weight coefficients. Thereby, the efficiency of Lucas–Kanade algorithm was increased by introduction of bilateral filter coefficients for the standard weight function.

Acknowledgments

This work was supported by the Fundamental Research Program for the National Academy of Sciences (FRP NAS) in 2013–2020 on theme 23.1.3 'Scientific basic foundation for diagnostic of state prior to failure, estimation of mechanical state and monitoring of multiscale structurally heterogeneous matters', grant of the Russian Foundation for Basic Research (RFBR) No. 16-38-00526 and grant of the Russian Federation president No. SP-1529.2015.5.

References

[1] Sutton M A, Cheng M Q, Peters W H 1986 *Image and Vision Computing* **4(3)** 143-151

[2] Sutton M A, Orteu J-J, et al 2009 Image correlation for shape, motion and deformation measurements (New York) p 321

[3] Lucas B D and Kanade T 1981 *Proc. Defense Advanced Research Projects Agency Image Understanding Workshop* vol 2 (San Francisco: Morgan Kaufmann Publishers Inc.) pp 674–679

[4] Yoon K-J and Kweon I S 2006 *Institute of Electrical and Electronics Engineers Transactions on Pattern Analysis and Machine Intelligence* vol 28 (Washington: Computer Society) chapter 4 pp 650-656

[5] Braspenning R A and de Haan G 2004 *Proc. of the Society of Photo-Optical Instrumentation Engineers* vol 5308 (Netherlands: Philips Research Labs) pp 396-407

[6] Panin S V, Titkov V V and Lyubutin P S 2015 *Optoelectr., Instr. and Data Proc.* **51(2)** 124-133

[7] Fortun D, Bouthemy P, Kervrann C 2015 *Computer Vision and Image Understanding* **134** 1-21

[8] Horn B, Schunck B 1981 *Artificial Intelligence* **16** 185-203

[9] Canny J 1986 *Institute of Electrical and Electronics Engineers Transactions on Pattern Analysis and Machine Intelligence* vol 8 (Washington: Computer Society) chapter 6 pp 679–698

[10] Itseez OpenCV 2016 [Online] Available: http://www.opencv.org

Design and optimization of the micro-engine turbine rotor manufacturing using the rapid prototyping technology

R A Vdovin and V G Smelov

Samara National Research University, 34, Moskovskoe Hwy. Samara, 443086, Russia

E-mail: vdovin.ssau@gmail.com

Abstract. This work describes the experience in manufacturing the turbine rotor for the micro-engine. It demonstrates the design principles for the complex investment casting process combining the use of the ProCast software and the rapid prototyping techniques. At the virtual modelling stage, in addition to optimized process parameters, the casting structure was improved to obtain the defect-free section. The real production stage allowed demonstrating the performance and fitness of rapid prototyping techniques for the manufacture of geometrically-complex engine-building parts.

1. Introduction

In the modern aviation background results in a new class of aviation equipment – unmanned aerial vehicles (UAV), which combine the experience of introducing the scientific research results and achievements in the production process. The analysis of the manufactured UAV showed that their outstanding feature is the high speed at comparatively small engine dimensions. Consequently, there is a need in manufacturing of the brand-new micro gas turbine engines (GTE). As we know from the GTE theory, the turbine is the most critical component of the engine. This is connected with the high continuously growing level of temperatures and stresses occurring in the turbine blades as well as with the necessity to considerably extend their service life. As a rule, turbine blades are manufactured of heat-proof nickel alloys (HA) by the directional crystallization casting due to the fact that the HA-category high-melting point alloys are almost non-machinable materials [1-5].

The micro GTE turbine rotor (Figure 1) shall also comply with the raised heat-resistance and high-temperature stability as well as corrosion resistance requirements. For this reason, when manufacturing the hot component of GTE, HA-category high-temperature nickel alloys with the melting temperatures of 1550-1600 °C are used.

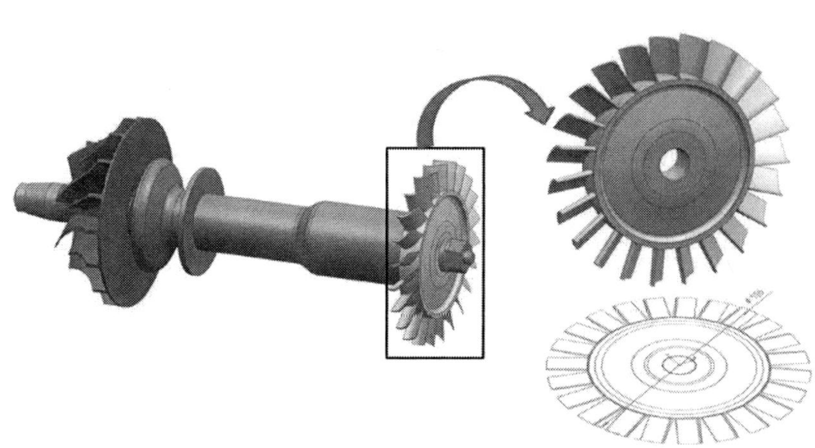

Figure 1. The 3D model of the turbine wheel.

2. The experimental research of the casting process

The traditional manufacturing process of the micro GTE turbine rotor involves casting with directed crystallization in batch pusher furnaces. The whole manufacturing process consists of several stages including: producing metal moulds, creating wax models, manufacturing ceramic shell moulds and making the metal casts. In work [6], a similar manufacturing process for the turbine rotor for Drone's micro engines is considered, however the specific feature of this work is the introduction of a temporary outer ring in the design; the ring thickness is h=1 mm. From the technical point of view, its expediency is explained by the additional metal feed of the blade airfoil end zone due to the considerable airfoil thickness; by the additional structural rigidity – due to the absence of transversal movement of these zones during coating and by the absence of longitudinal movement due to the hampered volume shrinkage, also the ring serves as the process support when coating the model (Figure 2).

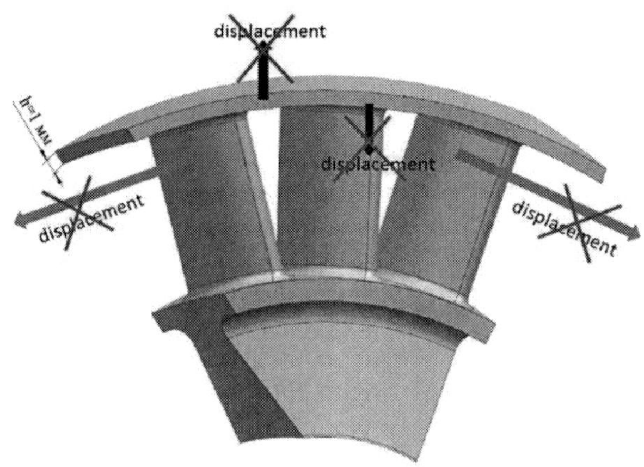

Figure 2. The 3D model of the turbine rotor with the outer rim.

The experimental pouring of the improved turbine rotor structure was carried out within the research work. However, the obtained casts indicated in Figure 3 had defects represented by misruns and unshaped zones within end sections. The casting process analysis showed that the most apparent and objective reason of defects was the inefficient outer rim structure, namely inability to access the most remote zones of the central feeder due to its small flow area (S=9 mm^2). Therefore, the exploratory optimization studies shall be carried out to select the most optimal structure of the outer rim.

Figure 3. Casting defects of the turbine rotor.

Considering the fact that the directed-crystallization casting technology is a costly process, it is expedient to use the computation modelling systems with performing the experimental studies on developing the casting process parameters on a PC screen. Using the computer modelling system will considerably reduce the cycle time, production costs, and, undoubtedly, improve the quality of the finished part [7-11].

3. Results and discussion

3.1 The results of numerical simulation

Computer modelling was performed using the ProCast software, using the finite element method (FEM). The adequacy of software employed and the convergence of results obtained have been tested many times by the authors and described in works [12-14].

Figure 4 shows the computer modelling results for the turbine rotor casting process of the micro GTE featuring a 1 mm thick outer ring in its design.

By iteration modelling, the optimal process parameters were selected:

Pouring temperature – 1530 °C;

Temperature heating of a ceramic mould – 1300 °C;

Pouring time – 4 s;

Heat transfer rate – 2000 $W/m^{(2\text{-}K)}$;

Pouring rate 130 mm/s.

It should be noted that during modelling, the cast material, namely heat-resistant nickel alloy JS-30 VI, as well as the ceramic shell material – electrocorundum (Al_2O_3) – remained the same.

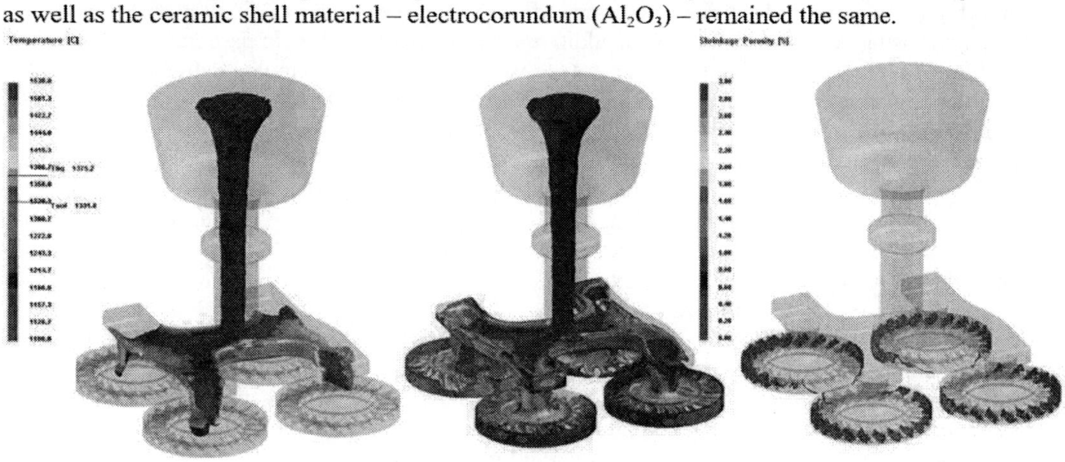

a *b*

Figure 4. Computer modelling results: *a* – mould filling with metal before the crystallization commencement; *b* – formation of shrinkage porosity 2-3%.

However, according to the virtual process calculations, the porosity of about 3-5%, local misruns and unshaped blade airfoil areas were observed within the end section of blades (which is essential) as well as within the outer ring area. In this connection, it was decided to double the thickness of the outer ring in order to ensure the additional metal supply. Thus, the outer rim acted as a top which subsequently had to be removed during machining.

The results of the virtual experiment with the increased outer ring thickness are given in Figure 5. The analysis of modelling results shows that the defects of blades represented by misruns and unshaped airfoil zones were managed to be compensated by increasing the outer rim thickness. The resulting porosity is less than 0.3% and is concentrated in the outer ring (top), which is cut off in the course of machining. Therefore, by optimizing the structure and casting process parameter, the stage of full-scale experiment on turbine rotor casting manufacture may be proceeded to.

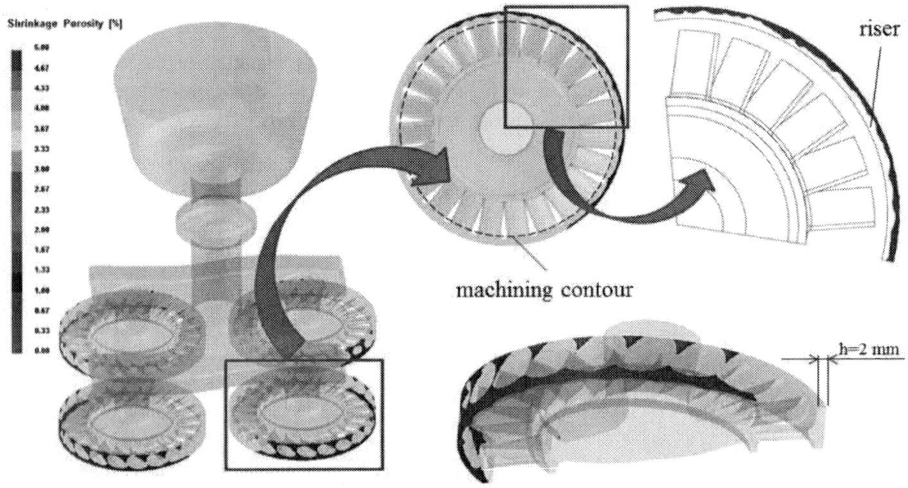

Figure 5. The results for porosity occurring in the outer ring.

3.2 Design manufacturing technology of the turbine rotor

Manufacturing the turbine wheel prototype

The first step in the micro GTE turbine rotor manufacturing process was building the part prototype (a master model) using its volume model with additive technologies [15, 16]. In particular, a success was achieved in manufacturing of the turbine prototype using the PolyJet technology (figure 6 a) with the reproduction precision of 0.02 mm along the X and Y axes and 0.016 mm along the Z axis.

Manufacturing silicone accessory

An elastic silicone mould (Figure 6 b) was manufactured using the built turbine wheel prototype serving for the subsequent pouring of wax and receiving the wax turbine models. An advantage of using the elastic form is its flexibility which, due to the complex structure of the turbine wheel, makes the process of creating the production accessories easier and faster. In addition, silicone is ideal to reproduce the geometry and texture of a part.

Manufacturing the turbine rotor wax models

The turbine rotor wax models were manufactured by the Paramelt investment compound pouring into the prepared silicone mould (Figure 6 c). Pouring with melted wax of the mould cavity was carried out at the temperature of 90 °C. The components of pouring gate and gating system were also manufactured of the wax investment compound. Using heating elements and a soldering station, wax models were assembled in model blocks as shown in Figures 4 and 5.

IOP Conf. Series: Materials Science and Engineering **177** (2017) 012040 doi:10.1088/1757-899X/177/1/012040

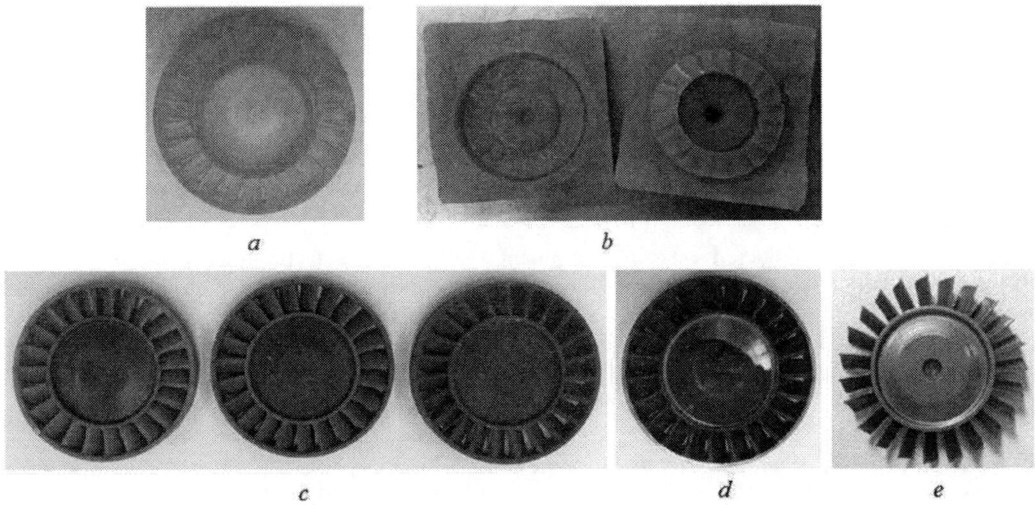

Figure 6. The technology of manufacturing the turbine rotor: a – master model; b – silicone form; c – wax models; d – casts; e –turbine rotor after machining

Manufacturing ceramic moulds and producing casts

The technology for manufacturing ceramic shell moulds consists in coating of moistened casting models with quartz sand and the subsequent drying of each layer [17]. The suspension containing the hydrolized ethyl silicate solution as a binding component was utilized for the first two layers of the mould; the suspension containing liquid glass was employed for the subsequent layers. Silicon dioxide with the grain size of 100 μm was used as a refractory filler of the suspension and a refractory coating material. After the application of the initial two layers using ethyl silicate as a binding hydrolized solution, the alumo-boron phosphate concentrate with the density of 1.25-1.30 g/cm^3 was applied as a separation layer. The nickel-based JS-30 VI heat-resistant alloy was melted and poured into a ceramic mould using a pusher continuous furnace implementing the directed-crystallization pouring method. The results of the experimental pouring of the micro GTE turbine rotor are given in figure *6d*.

4. Conclusion

The GTE turbine rotor represents the most critical and highly loaded part stipulating the service life and operational reliability of the engine. The performance of the turbine rotor in general, and of its blades in particular, is conditioned not only by the strength characteristics of materials employed and the geometric shape, but, in many respects, by the developing manufacturing methods. Development of the turbine rotor manufacturing technology is connected with the application of experience accumulated in machine building to ensure machining of a relatively high amount of parts with the developed crooked spacial surfaces, including with the use of rapid prototyping methods.

The existing manufacturing process applied for similar category of parts with the developed crooked spacial surfaces is based on the trail-and-error method. Frequently, in practice, several manufacturing cycles are performed in order to achieve the required quality level. This multioptional pattern always results in considerable time and material losses.

Therefore, implementing a complex process combining the computer experiment on optimizing the casting process parameters and its bringing to a perfect state as well as the rapid propotyping method allow reducing costs for the manufacture of parts considerably. The use of the computer modelling system for casting processes allows solving the issue related to the casting defect forecasting in a most complex manner. Using this system allows learning all the processes and phenomena characterizing the cast during crystallization in a rather profound way, understanding the reasons of defects, evaluating the efficiency of the process employed. When the rapid prototyping techniques are employed as the most flexible blank production technology, the initial casts of parts may be manufactured within the shortest term (2-3 weeks). As to the traditional technologies, the manufacture of metal die moulds takes from 3 to 6 months.

Owing to this combined process employed for producing the micro GTE turbine rotor, timeframes required for designing its manufacturing technologies were reduced along with a considerable decrease of the production costs.

Acknowledgments
These studies were conducted on the equipment of the CAM technology common use center (RFMEFI59314X0003). This work was supported by the Ministry of education and science of the Russian Federation in the framework of the implementation of the Program 'Research and development on priority directions of the scientific-technological complex of Russia for 2014-2020'.

References
[1] Takeshi N, Takeshi O, Kuniyuki I, Ken-ichi S and Massato I 2014 *Eng. Review* **47** 29-32
[2] Power D C 1995 *Platinum Metals Review* **39** 117-126
[3] Gell M, Duhl D N and Giamei A F 1980 *Superalloys* **2** 205-214
[4] Rzyankina E Szeliga D Mahomed N and Nowotnik A 2013 *Appl. Mech. and Mat.* **372** 54-61
[5] Vdovin R A, Smelov V G, Bolotov M A and Pronichev N D 2016 *IOP Conf. Series* (Yurga) vol 142 (IOP Publishing) pp 1-8
[6] Roskosz S, Szeliga D, Cydan R and Rokicki P 2016 *Trans Tech Publications* **246** 189-192
[7] Tarasevich N I, Korniyets I V, Tarasevich I N and Dudchenko A V 2010 *Metal and Molding of Ukraine* **5** 20-25
[8] Szeliga D, Kubaiak K and Jarczyk G 2013 *Int. J. of Metalcasting* **7** 17-23
[9] Kermanpur A, Eskandari M, Pumohamad H, Soltani M A and Shaten R 2010 *Mat. and Design* **31** 1096-1104
[10] Bikmeev A T, Ivanov S V and Perepelitsa S I 2012 *Vestnik UGATU* **16** 38-42
[11] Dai H J, Souza N D and Dong H B 2011 *Metallurgical and Mat. Transactions.* A **42** 3430-38
[12] Vdovin R A 2013 *Bulletin of Samara Scientific Center of Russian Academy of Sci.* **15** 612-619
[13] Vdovin R A and Smelov V G 2014 *Int. J. of Eng. and Techn.* **6** 2269-75
[14] Vdovin R A and Smelov V G 2016 *IOP Conf. Series* (Tomsk) vol 156 (IOP Publishing) pp 1-7
[15] Agapovichev A V, Balaykin A V and Smelov V G 2015 *Modern Appl. Sci.* **9** 335-343
[16] Agapovichev A V, Balaykin A V, Smelov V G and Agapovichev A V 2015 *Modern Appl. Sci.* **9** 151-159
[17] Kuchin P S and Maltsev H 2008 *Molding Production* **10** 37-39

Anchoring of development workings in a zone of influence of mining in case of the level anchoring system

V F Demin[1], O B Fofanov[2], T V Demina[2] and V V Yavorskiy[3]

[1] Karaganda State Technical University, 56, Mira bul., Karaganda, 100027, Kazakhstan
[2] Tomsk Polytechnic University, 30, Lenina ave., Tomsk, 634050, Russia
[3] Karaganda State Industrial University, 30, Republic ave., Temirtau, 101400, Kazakhstan

E-mail: yavorskiy-v-v@mail.ru

Abstract. Regularities of the change of the stress-strain state of coal containing rock masses, depending on mining-geological factors, were revealed. These factors allow establishing rational parameters of anchoring of wall rocks to enhance the stability of development workings. Specific conditions of the deflected mode, displays of rock pressure, terms of maintenance depending on technological parameters are investigated. Researches allowed determining the degree of their development influence on the efficiency of application of the anchoring of the hollow making and will allow a reasonable application of anchoring certificates, provide stability of the rocks mining and reduce expenses on its realization and maintenance.

1. Introduction
The increase of the underground coal mining is possible only with high-performance technology and maintenance development workings, providing an increase of the volume of mining and preparatory works. An important task is the determination of the deflected mode of the massif taking into account the factors influencing the stability of the borders of the mine workings. Setting of parameters of rock massif deformation around development workings with the anchorage is an important task of the mining industry [1-7].

2. The influence of controllability of the roof rocks (with the increase of the layer of free caving mudstone) with a trapezoidal shape of a roadway cross-section
The parameters of the design diagram are: the stratum dip angle is 15^0, the stratum depth is 3.8 m; the mining depth is 400 m; the mine working cross section is 1.5 m^2; the achor length is 3 m with the anchor diameter of 0.022 m.

Figure 1 presents the distribution of the longitudinal stresses in the 7.5 m layer of claystone along the border of the mine working. Analysis of stress distribution shows that the areas of unstable rocks appear around the mine working. To a greater extent, it concerns the roof and the soil of the mine working, as well as its sides in the area of the lower part of lateral faces of the mine working boundaries. The maximal value of normal stresses occurs in the anchor located on the roof of the mine working in the rightmost anchor in the place of its attachment. The maximal value of the longitudinal stress arises in the anchor located on the right lateral surface of the mine working (the first one from the bottom).

In order to reduce costs while maintaining the mine workings in the mines of Karaganda basin, the anchor support is used, which quickly takes the load and prevents the stratification of the surrounding massif and the increase of the sizes of the areas of inelastic deformations, which results in the reduction of the support load. The rock, 'stitched' by anchors, acts as a support and helps to reduce rocks displacement along the border of the mine workings cross section.

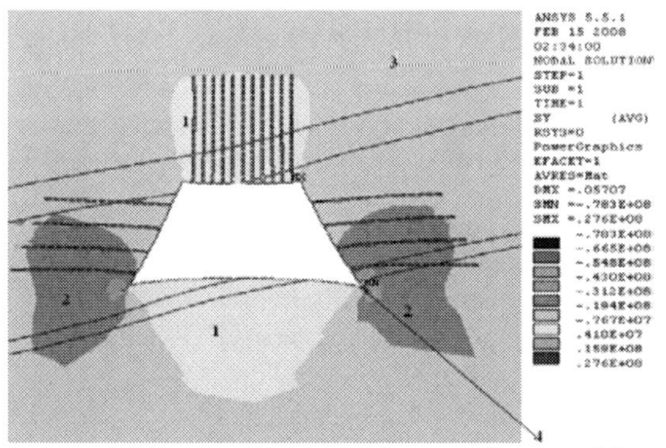

1 – very unstable area; 2 – unstable area; 3 – unstable area;
4 – area of average resistance

Figure 1. Distribution of areas of equal longitudinal stresses in the massif.

Rock pressure manifestations with the determination of the degree of technological factors influence are detected by the finite element method with program-methodical complex ANSYS. This method is used for the conditions of the manufacturing scheme of treatment works with reciprocal-flow airing on the K_7 stratum of the mine named after Kostenko (MD JSC 'ArcelorMittal Temirtau') with the lava 200 m long using the anchor support of mine workings. The change in stresses of the rocks massif is determined depending on the inclination angle of the anchors in the roof with the anchor length and the anchor diameter equal to 2.4 m and 0.022 m, respectively, in the mine working having a 17.5 m^2 section at a 650 m depth.

By changing the inclination angle of the anchors in the roof of the mine working, the normal (along axis *Y*) and shear stresses are changed as well. Let us present the stress distributions by the example of shear stresses τ_{xy} in the rock massif corresponding to the inclination angles of the anchors in both directions from the axis of the mine working relative to horizontal plane $\alpha = \beta = 45°$ (figure 2, a), $\alpha = \beta = 75°$ (figure 2, b), $\alpha = \beta = 90°$ (figure 2, c). Table 1 presents the values of maximal normal and shearing stresses using the model of maximal displacement.

Table 1. Values of maximal normal and shearing stresses

α, β (degree)	σ_x (MPa)	σ_y (MPa)	τ_{xy} (MPa)	u_x (m)	u_y (m)
30	93.6	19.7	54.1	0.0028	0.1382
45	93.4	28.2	46.5	0.0028	0.1382
60	93.4	26.1	30.6	0.0028	0.1382
75	93.4	9.05	29.1	0.0028	0.1382
90	93.4	5.73	29.1	0.0028	0.1382

It has become clear from the undertaken study that the optimal minimal angle of anchors location in the roof are the angles of their installation relative to the horizontal plane equal to $\alpha = \beta = 75°$. At that, minimal normal stresses are provided from the direction of the roof rocks (2 - 3 times), and lengthy anchors (cable ones) do not fall into the failure zone of the rocks belonging to the immediate and main roof. The dependences of maximal normal stresses along axis *Y* are determined depending on the inclination angle of the anchor relative to the horizontal plane:

$$\sigma_{max}^{y}(\alpha) = 2.7 \cdot 10^{-5} \cdot \alpha^4 - 5.9 \cdot 10^{-3} \cdot \alpha^3 + 0.4 \cdot \alpha^2 - 13.1 \cdot \alpha + 153.5$$

In accordance with the undertaken studies, the certificates of anchors, intended for development workings in the area of influence of mine workings in case of a three-level anchoring system, were designed: cable anchors – 5 m long, steel-polymeric anchors – 2.4 m long, and lateral plastic anchors – 1.8 m long. Such kind of system allowed stabilizing vertical and lateral deformations of lateral rocks. These certificates are realized when mining lava of stratum K_{12} under conditions of the mine 'Saranskaya' in Karaganda coal basin (figure 3).

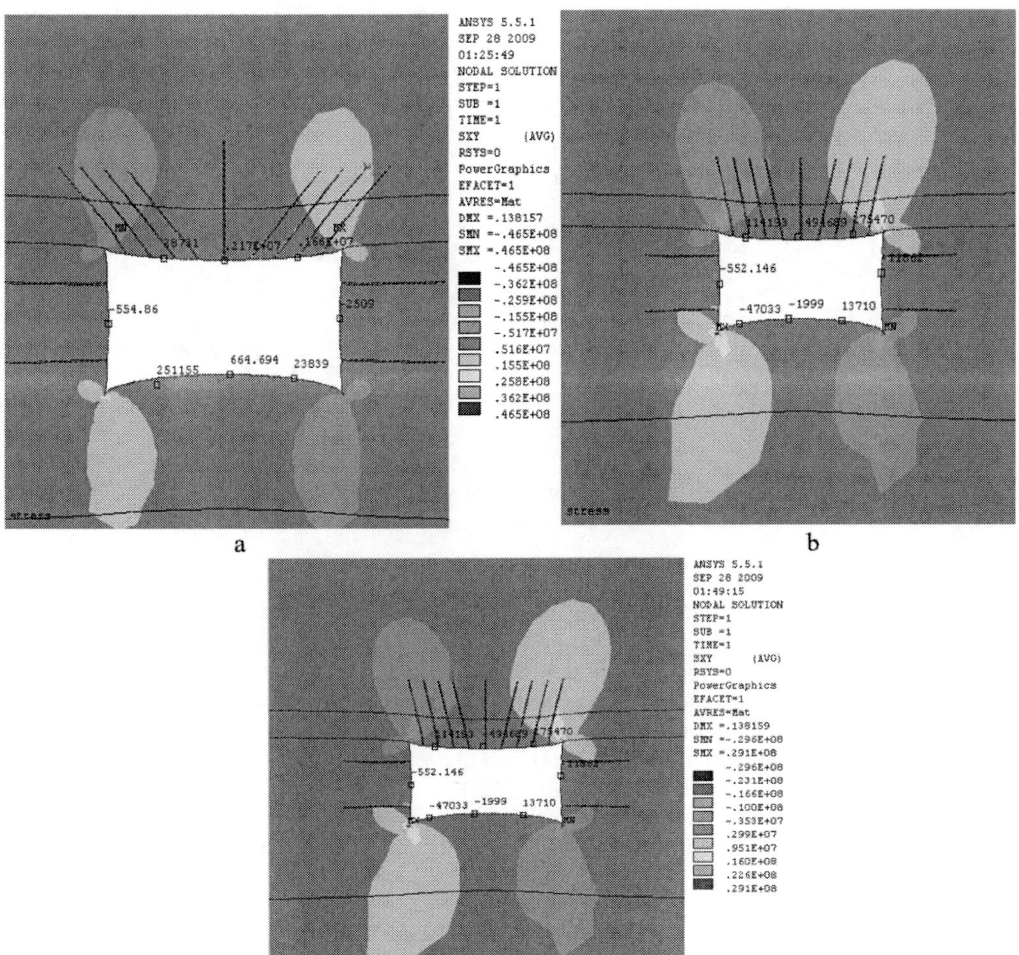

Figure 2. The distribution of shearing stresses τ_{xy}.

3. Anchoring of development workings

In connection with the identification of fractures in the upper and lower layers of the roof rocks and their subsidence over the area of linking of lava to the belt roadway of stratum K_{12}, rope anchors 5 m long were installed at a distance of 1.5 m between them and a number of anchors. In total, at the mine working site, 100 rope anchors were installed from PC 119 to PC 112. Further observations of the state of the roof rocks with the cable anchors have established that grabs deflection in these zones is much smaller (0.3-0.4 m delamination) and allows reducing repair costs and maintenance of workings.

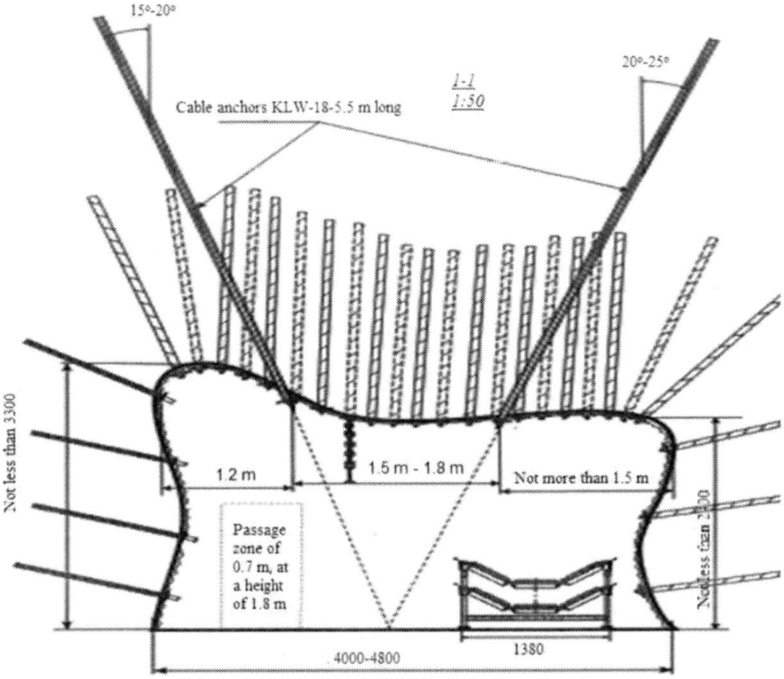

Figure 3. Development workings anchoring in the zone of influence of mining in case of a three-level anchoring system.

In future, when mining an extraction pillar in stratum K_7, a progressive certificate of linking of the lava to the belt entry has been introduced. Anchoring of the fractured zone between the linear sections of the powered roof support and the road support with the help of steel-polymeric and rope anchors will allow consolidation of the rock in this region, which in its turn will provide the stability of rocks in the roof and reduce the labor expenditures required for retimbering of the fractured zone (figure 4). At a distance of 10-15 m ahead of the lava, 1-2 'threads' of the summer made of special section SVP-22(27), each 2.0-2.5 m long, will be hung up simultaneously with installation of hydraulic props GVKU under these threads with one meter spacing [8-10].

The discovered regularities of the change of the stress-strain state of coal containing rock massifs depending on mining and geological factors allow establishing rational parameters of adjoining rocks anchoring in specific external environment to enhance the stability of development workings and their linking to lavas.

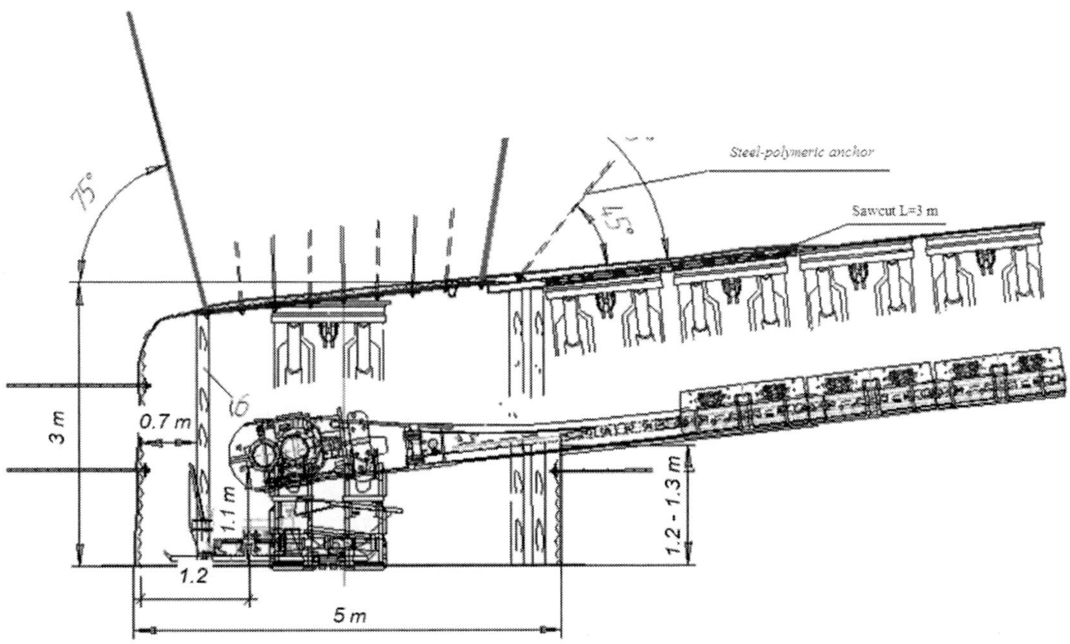

Figure 4. Anchoring of the fractured zone between the linear sections of the powered roof support and the road support.

4. Conclusion

Application of the anchoring support allows us to considerably increase labor productivity and its rate (by 30-40 %), to reduce expenses on mine workings (by 50-60 %), to lower expenditure and to reduce the cost of dunnage, repair costs and mine workings maintenance, to improve the condition of the mine workings, to approximate the cross-sectional area of the mine working in clear to the area of advance and to tighten the safety measures of mine workings.

The empirical dependences of changing of maximal normal stresses, depending on the change in the anchor length as well as on the anchor inclination angle for generation of the rectangular cross section, are obtained.

A comparative evaluation of the conducted researches and testings in the plant conditions has been made. It has shown a satisfactory convergence of the considered parameters of the deflected mode of rocks massifs.

References

[1] Molinda G M and Dennis R D 2007 *Analysis of roof bolt systems Christopher Mark* Mining Engineer NIOSH (Pittsburgh: Research Laboratory Pittsburgh, Pennsylvania USA) pp 34-56

[2] Laubscher D H and Jakubec J J 2000 *Rock Mass Classification System for Jointed Rock Masses* (London: The IRMR/MRMR) pp 40-45

[3] Hudson J A and Harrison J P 1997 *Engineering Rock Mechanics* An Introduction to the principles (London: Chapman and Hall)

[4] Kowaltchuk A B 2003 *Neues Konzept dei der Projektierung und Herstellung der Strebausrüngungen* (Gluckauf) vol 1 (Zittau: Verlag Deutchland) p 20-23

[5] Demtschenko A I and Kulassek M 2002 *Ingenieurgrundlage der stabile Bewetterung der Hochleistungsbetriebe unter Bedingungen des Flötzes, des Bergwerks Krasnoarmejskaja – Zapadnaja* (Gluckauf) vol 1 (Zittau: Verlag Deutchland) p 20-23

[6] Daniarov N and Mozer D and Greeb I 2007 *Zittauer Seminar zur energiewirtschaftlichen Situation in den Laendern Mittels und Osteuropas* (Zittau: Verlag Deutschland) p 347-355

[7] Brady B H and Brown E T 2009 *Rock Mechanics for underground mining* (Zittau: Verlag Deutschland) p 628

[8] Demin V F and Yavorskiy V V and Demina T V 2015 *Study of rock pressure around the workings of given principal stresses* (Moscow: Moscow State University Press) pp 11-14

[9] Demin V F and Nemova N A and Demina T V and Karataev A D 2015 *Scientific-Technical J. National Mining Univ.* **4**(148) 35-38

[10] Demin V F and Yavorskiy V V and Demina T V 2015 *J. Adv. in Current Natural Sci.* **12** 95-99

Deflected mode of marginal rock massif around mine working boundaries depending on anchoring parameters

V F Demin[1], O B Fofanov[2], T V Demina[2] and V V Yavorskiy[3]

[1] Karaganda State Technical University, 56, Mira bul., Karaganda, 100027, Kazakhstan
[2] Tomsk Polytechnic University, 30, Lenina ave., Tomsk, 634050, Russia
[3] Karaganda State Industrial University, 30, Republic ave., Temirtau, 101400, Kazakhstan

E-mail: yavorskiy-v-v@mail.ru

Abstract. The deflected mode, rock pressure manifestations, conditions of maintenance of mine workings depending on geological and technological parameters have been studied. Researches have allowed establishing the degree of these parameters impact on the effectiveness of anchoring of mine workings. The paper researches such aspects of the problem as rock pressures, terms of maintenance of mine workings, technological parameters and others. The authors consider the expediency of the introduction of anchor certificates which will allow for stability of the rock mining and a reduction of the expenses on realization and maintenance of mine workings.

1. Introduction

The maintenance and the increase of the volumes of underground coal mining are possible only at the presence of highly efficient technology of realization and maintenance of development workings, which provide an increase of the volume of mining and preparatory work. The aim of the research is the creation of technology of intensive and safe mining based on identified patterns of behavior of adjacent rocks, optimization of the parameters of flow charts of preparatory work, providing the increase in the functioning efficiency of underground mining practice [1-7].

2. Ensuring the stability of mine workings

When operating mines with the increasing depth of mine working, one of the problems demanding solution is ensuring the stability of mine workings. To maintain the mine workings in the mines of Karaganda coal basin, metal yielding supports (MYS) of the arch type and the anchoring technology in the limited volume (in a pure form up to 22 %) are applied. The expenditures on realization and anchoring of 1 m of mine working amount to 13-18 thousand rubles, the consumption of the metal-roll is 0.3-1.0 t. At that, the maintenance cost is not less than 10-15 % of the mining costs. With the approved systems of mine working, it is required to perform 5.0 to 5.5 km of mine workings per 1 million tons of coal, which requires considerable expenses for the preparation of excavation sites.

In-seam workings are most subjected to the influence of the rock pressure. The loss of their cross-sectional area reaches up to 60-70 %. This leads to the fact that 20% of mine workings are annually repaired and retimbered. The share of the expenses on carrying out, anchoring and maintenance of mine workings reaches up to 15-20 % of the coal production cost. More than 10 % of the underground workers are engaged in repair of mine workings.

The conditions of maintaining the mine workings with different types of anchoring in the zone of influence of the stope were studied by the example of intermediate belt entry 49k_{10}-z of lava in the mine named after Kostenko, Karaganda coal basin. The working-bed height of a k_{10} stratum in the West wing of the mine is 3.7 m to 4.0 m. The immediate roof varies along the strike from 3 m to 7 m and is represented by mudstone. The main roof is composed of loosely crumbling sandstones with a thickness of 24-32 m. The maximum value of blowing of soil rocks after two years of maintenance of mine working amounted to 0.35-0.8 m. To ensure the necessary cross-section ahead of the lava at a distance of 50-80 m, the ripping of the soil rocks of the entry was carried to a depth of 0.5 m to 0.8 m.

Figure 1 shows the status of the studied mine working in the area influenced by the stope. The most favorable conditions of the maintenance were provided in the area of intermediate belt entry 49k_{10}-z with a length of 50 m of a semi-arched form, anchored by the composite support (anchors in combination with the MYS) with a density of 1.33 frame/RM. m (figure 1 a).

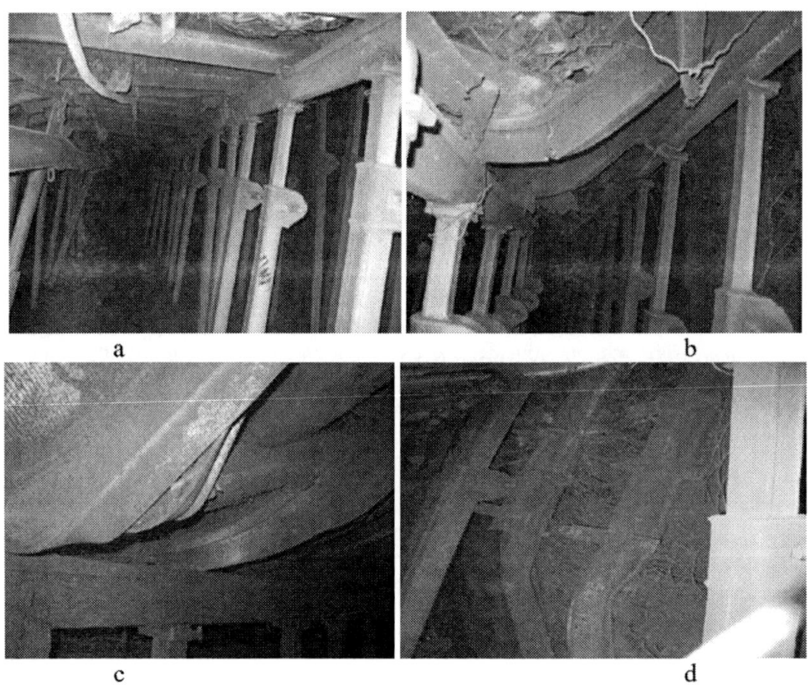

a – composite support; b – deformation of the roof bar;
c – impulse of the roof bar; d – deformation of the props

Figure 1. The condition of intermediate belt entry 49$_{K10}$-z of the mine named after Kostenko of Karaganda coal basin in the zone of influence of stopes.

For this area of mine working, the following changes of deformations in the support condition are typical: for the roof bar and its impulse (figure 1 b, c) along the borders of summers – 60 %; for the composite props (figure 1 d) in a vertical plane – 1.5 %; the deviation of friction props from the vertical position, preliminary along the first summer from the stope – 70 %.

3. The impact of stope

In this regard, the study of features of deformation of the rock massif around development workings with anchoring at various incidence angles of the stratum and depth of anchorage, justification of the installation of the anchor support and the definition of the rational sphere of its use are an important task of the mining industry.

In the course of studies, the authors determined the deflected mode (DM) around the mine working: of the roof, soil and sides; the strain rate, the area of delamination (cracking); stresses (compressive, tensile and shear), shifting of the service life (in progress). Relatively favorable, average and difficult conditions of mine workings have been considered.

To determine the conditional zones of inelastic deformations, the program, which allows determining the deflected mode in the considered point of the technogeneous space, and then setting the durability of an object (the time to failure) and evaluating the stability of bed outcrops for subsequent adoption of technological measures. For the geomechanical interpretation of the modeling results, belt entry 64_{K10}-z with the 16.2 m^2 cross-section of stratum K_{10} of the mine 'Abaiskaya' in Karaganda coal basin, traversed at a depth of 630-640 m (figure 2).

In the present research, the analytical modeling is performed using the numerical finite element method. The simulation is performed for the conditions of the strata conveyor working of stratum K_{10} of the mine 'Abaiskaya' in Karaganda coal basin at a mine working depth of 400 m and a bed thickness of 3.8 m. The deflected mode of the massif around the current mining working is considered. The problem is reduced to the flat working. The solution is carried out in the elastic formulation due to the relatively short time of deformation of rocks in the area of the development face during its development. In contrast to the known approaches, the sizes of the zones of deformation propagation with the analysis of their parameters are specified.

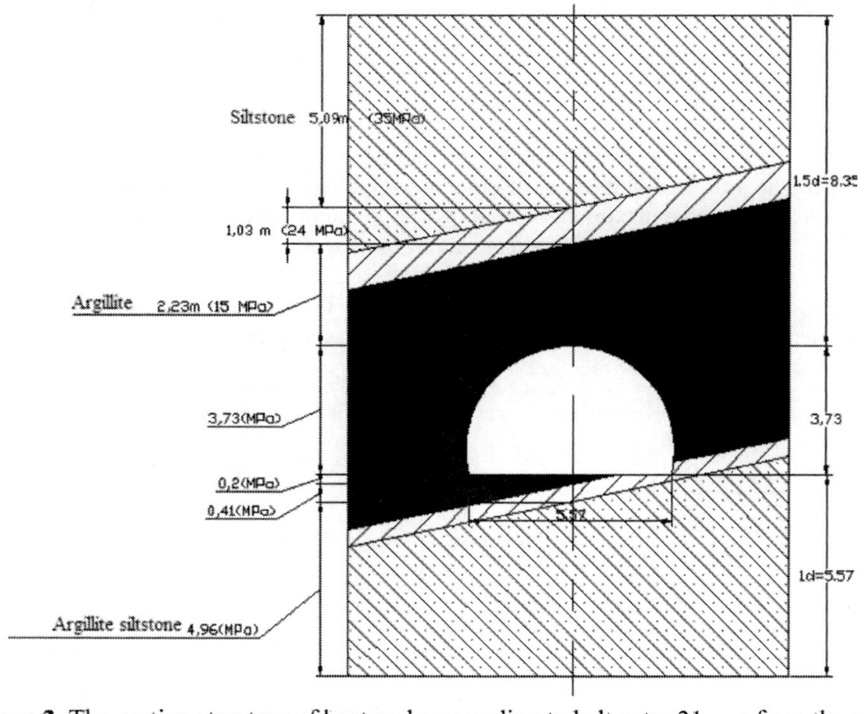

Figure 2. The section structure of host rocks according to belt entry 21_{K12}-s from the mine 'Abaiskaya' in Karaganda coal basin.

The study produced mathematical models using ANSYS software allowing us to determine the influence of geological factors on the operating conditions of the powered roof support of mine workings.

IOP Conf. Series: Materials Science and Engineering **177** (2017) 012042 doi:10.1088/1757-899X/177/1/012042

The ANSYS model was built to the massif of enclosing rocks, corresponding to the conditions of occurrence of stratum K_{10}.

At the first stage, we studied the influence of the shape of the cross section of the mine working and the incidence angle of the coal seam on the value of the occurring maximum stresses in the rock massif when mounting the working with the anchor support.

In case of a vaulted (arched) shape of the cross section of extraction production, normal stresses (σ_y) grow with the increase of the incidence angle (α) of the stratum from 10° to 40° according to the exponential function in the range from 10 to 13.5 MPa (figure 3 a). Longitudinal stresses (σ_x) are increased at angle α from 10° to 20° in the range of 63.2 to 64.1 MPa, and then the influence of the incidence angle is not seen (figure 3 b). Shear stresses (τ_{xy}) decrease from 50 to 33 MPa proportionally in the range of incidence angles α = 10° – 30° of the stratum, and at α = 30° – 40° – increase from 33 to 37 MPa (figure 3 c).

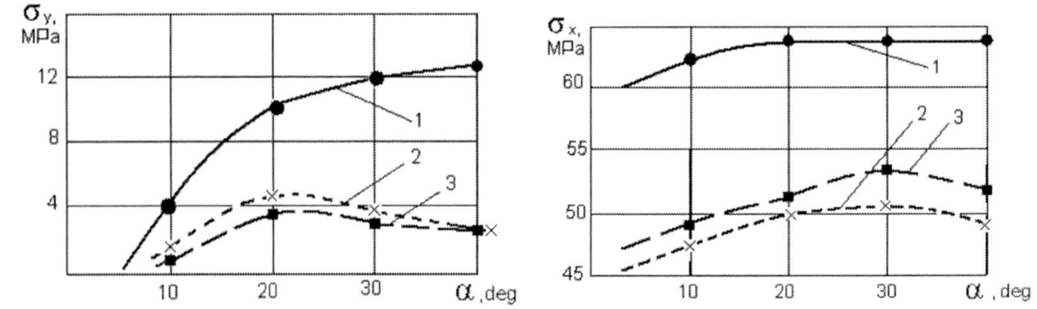

a b

1 – arched; 2 – polygonal; 3 – rectangular

Figure 3. The influence of the type of the cross-sectional shape and the incidence angle of the stratum on the value of the maximum normal (a) and longitudinal (b) stresses in the rock massif in case of anchoring of mine workings.

Moreover, large stresses occur along the coal seam rise. The distribution of stresses in the zones of country wall rocks surrounding the mine working is presented in figures 4 a and 4 b.

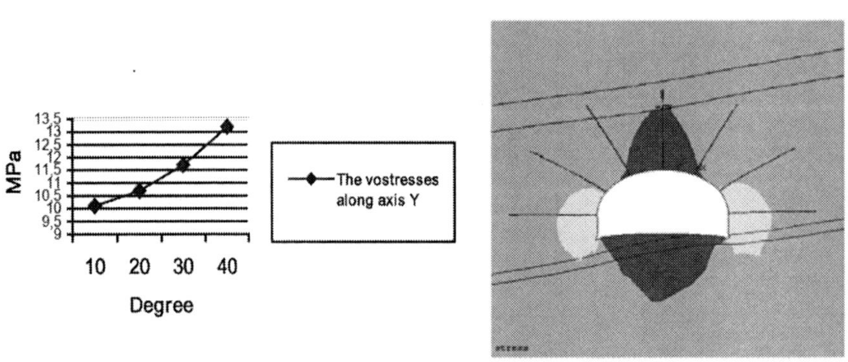

a b

a and b – the nature of changes and the plot (when $\alpha = 10^\circ$) of maximum normal stresses

Figure 4. The change and distribution of the maximum stresses in the zones surrounding mine working.

For a rectangular shape of the cross-section of excavation mine working, maximum normal stresses σ_y grow at $\alpha = 10^\circ - 20^\circ$ from 1.2 to 3.5 MPa, and then slightly decrease when $\alpha = 20^\circ - 40^\circ$ from 3.5 to 3.0 MPa. Longitudinal stresses (σ_x) increase from 49 to 53.4 MPa when $\alpha = 10^\circ - 30^\circ$, and then sharply decrease to 52 MPa during angle shifting $\alpha = 40^\circ$. Shear stresses (τ_{xy}) grow according to the mild exponential function from 18 to 38 MPa at angle change $\alpha = 10^\circ - 40^\circ$. Stress values in the areas surrounding the mine working, are presented in figures 5 a, b.

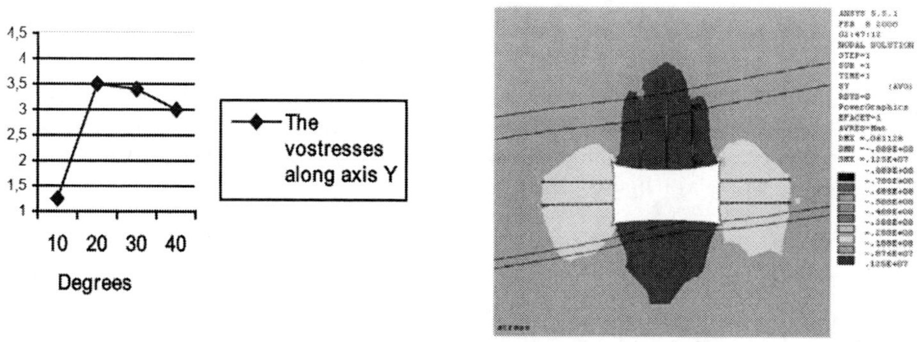

a b

a and b – the nature of changes and the plot (when $\alpha = 10^\circ$) of maximum normal stress

Figure 5. The maximum stresses in the areas surrounding the mine working of a rectangular cross-sectional shape.

In case of a polygonal shape of the cross section of mine working, the trends of the change in the deflected mode roughly resemble the nature of the change of dependencies in case of a rectangular shape of the cross-section mine working. There are stresses σ_y that are 1.5 times greater in magnitude, and, on the contrary, stresses σ_x that are less by 2-3 MPa, and τ_{xy} that are 1.5 – 2.0 times greater. Change and distribution of stresses in the areas surrounding the mine working are presented in figures 6 a and 6 b.

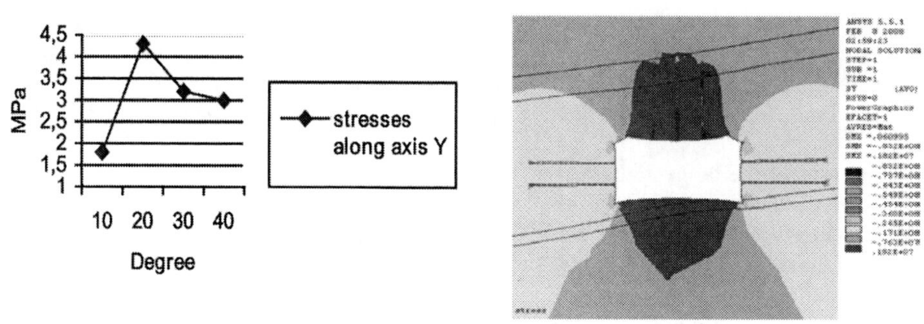

a b

a and b – the nature of changes and the plot (when $\alpha = 10^\circ$) of maximum normal stress

Figure 6. The change and distribution of the maximum stresses in the areas surrounding the mine working.

The conducted research allows making a conclusion on the preference of applying a rectangular shape of the cross-section of excavation workings with the anchoring of country rocks for conditions of stratum K_{10} exploitation of the mine 'Abaiskaya', MD JSC 'ArcelorMittal Temirtau' [8-10].

The deflection mode of country rocks depending on the layer thickness of free caving rocks in case of different lengths of anchoring is studied. The studies were performed by the example of the mine working of the trapezoidal cross-sectional shape with following parameters of the design diagram: the incidence angle of the stratum is 15°, the stratum thickness is 3.8 m; the dredging depth is 400 m; the cross section of the mine working is 15.5 m^2; the anchor diameter is 0.022 m.

The nature of change and distribution of stresses in the roof, soil and sides of the mine working. When the value of the layer of free caving rocks is from 1.03 m to 6.0 m and the length of the anchor is from 2.4 m to 5.0 m, the following changes of stresses around the mine working take place. The maximum and minimum normal stresses grow in a proportional linear dependence in case of the increase of the anchor length (from 1.5 to 6 m) and of the layer thickness of free caving rocks, for example, composed of claystone, (from 1 to 6 m) (figure 7 a). The changes of stresses in the considered range in the longitudinal plane with the increase of the anchor length and the thickness growth of the free caving rocks layer have the following tendencies: stretching stresses are reduced, and compressive ones have a skip when the anchor length is 3.0 – 3.5 m and are generally in the narrow range (42-48 MPa) – figure 7 b.

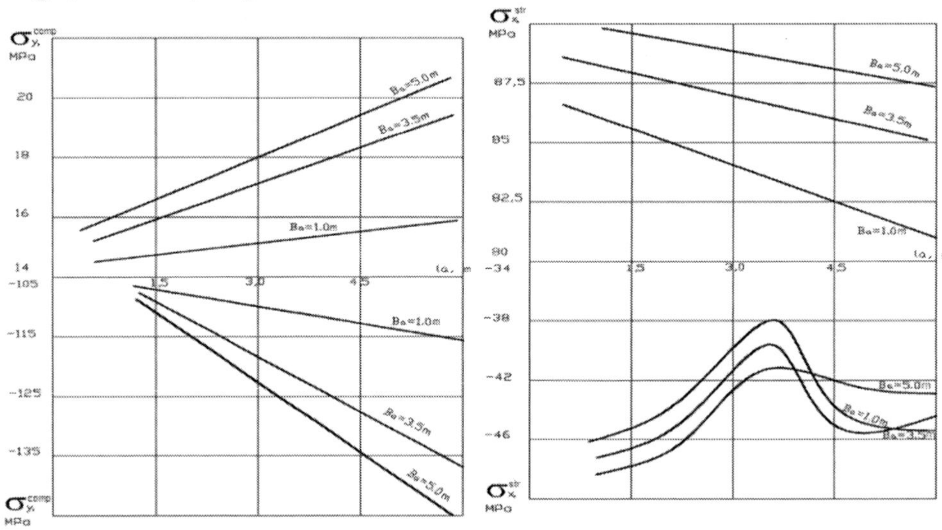

Figure 7. The change of the normal and lateral stresses in the marginal rocks of the development working, depending on the anchoring length and the layer thickness of the failed rocks.

The patterns of the tangential stresses change are presented in figure 5. They tend to rise when the mudstone layer thickness is 5 m and the claystone layer thickness is 1.0 to 3.5 m. They increase when the anchor length changes from 1.5 to 3.0 (3.5) m, and then decrease. At that, the increase in the diameter of blast holes (up to 0.05 m) has a negative impact on emerging stresses and leads to their twofold growth over the entire range.

The conducted research of the deflected mode of country rocks depending on the layer thickness of free caving rocks when the anchoring length varies allowed us to establish the following behavior of the lateral rocks by zones of their location (figure 8 a, b).

IOP Conf. Series: Materials Science and Engineering 177 (2017) 012042 doi:10.1088/1757-899X/177/1/012042

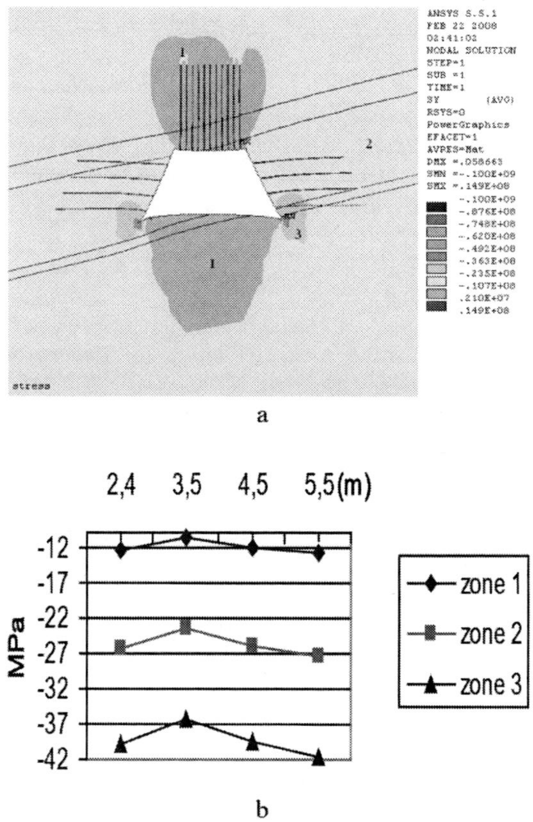

Claystone layer is 1 m, anchor length is l=3.5 m

Figure 8. The distribution plot (a) and dependence of the change of normal stresses (b) in country rocks on the depth of their anchoring.

4. Conclusion

Analysis of stress distribution shows that the areas of unstable rocks appear around the mine working. To a greater extent, this applies to the roof and soil of mine working, as well as its sides in the lower part of the lateral sides of the mine working boundaries. The maximum value of the normal stresses occurs in the anchor located on the roof of the mine working in the rightmost anchor in a place of its anchoring. The maximum value of the longitudinal stress arises in the anchor located on the right lateral surface of the mine working (the first one from the bottom).

The established regularities of the change of the deflected mode of the coal in rock massifs (displacements, stresses, zones of cracking), depending on the basic geological and mining factors, will allow establishing the optimal parameters of the anchoring in order to increase the stability of development mine workings under specific conditions, which will allow one to develop new and improve the existing technologies of the efficient and safe anchoring of the marginal rocks, when conducting mine workings on steep and inclined coal seams, which will be adaptive to changing geological and mining conditions.

References

[1] Molinda G M and Dennis R D 2007 *Analysis of roof bolt systems Christopher Mark* Mining Engineer NIOSH (Pittsburgh: Research Laboratory Pittsburgh, Pennsylvania USA) pp 34-56

[2] Laubscher D H and Jakubec J J 2000 *Rock Mass Classification System for Jointed Rock Masses* (London: The IRMR/MRMR) pp 40-45
[3] Hudson J A and Harrison J P 1997 *Engineering Rock Mechanics* An Introduction to the principles (London: Chapman and Hall)
[4] Kowaltchuk A B 2003 *Neues Konzept dei der Projektierung und Herstellung der Strebausrüngungen* (Gluckauf) vol 1 (Zittau: Verlag Deutchland) p 20-23
[5] Demtschenko A I and Kulassek M 2002 *Ingenieurgrundlage der stabile Bewetterung der Hochleistungsbetriebe unter Bedingungen des Flötzes, des Bergwerks Krasnoarmejskaja – Zapadnaja* (Gluckauf) vol 1 (Zittau: Verlag Deutchland) p 20-23
[6] Daniarov N and Mozer D and Greeb I 2007 *Zittauer Seminar zur energiewirtschaftlichen Situation in den Laendern Mittels und Osteuropas* (Zittau: Verlag Deutschland) p 347-355
[7] Brady B H and Brown E T 2009 *Rock Mechanics for underground mining* (Zittau: Verlag Deutschland) p 628
[8] Demin V F and Yavorskiy V V and Demina T V 2015 *Study of rock pressure around the workings of given principal stresses* (Moscow: Moscow State University Press) pp 11-14
[9] Demin V F and Nemova N A and Demina T V and Karataev A D 2015 *Scientific-Technical J. National Mining Univ.* **4**(148) 35-38
[10] Demin V F and Yavorskiy V V and Demina T V 2015 *J. Adv. in Current Natural Sci.* **12** 95-99

A quality evaluation of stabilization of rotation frequency of gas-diesel engines when using an adaptive automatic control system

A A Zhilenkov[1] **and A A Efremov**[2]

[1] Department of Control Systems and Informatics, Saint Petersburg National Research University of Information Technologies, Mechanics and Optics, 49,Kronverkskiy ave., St Petersburg, 197101, Russia
[2] Department of System Analysis and Control, Peter the Great Saint-Petersburg Polytechnic University, 29, Polytekhnicheskaya str., St Petersburg, 195251, Russia

E mail: zhilenkovanton@gmail.com

Abstract. A possibility of quality improvement of stabilization of rotation frequency of the gas-diesels used as prime mover of generator set in the multigenerator units working for abruptly variable load of large power is considered. An evaluation is made on the condition of fuzzy controller use developed and described by the authors in a number of articles. An evaluation has shown that theoretically, the revolution range of a gas-diesel engine may be reduced 25-30 times in case of optimal settings of the controller in the whole power range. The results of modelling showing a considerable quality improvement of transient processes in the investigated system during a sharp change of loading are presented in this article.

1. Introduction

An introduction of gas-diesel engines as prime movers of generating sets is one of the most perspective trends of multigenerating autonomous grids of a number of transport objects and an industry. First of all, it concerns self-contained drilling rigs where associated gas can be used as fuel of gas-diesel engines that increases their efficiency, profitability, reliability and reduces the number of hazardous emissions. To date, gas-diesel engines have been used as primary engines only in one-element generating sets because of a low revolution stability. Electricity supply of powerful consumers of a drilling platform requires the use of the multigenerating systems providing a possibility of operation of several generating sets in parallel. Such kind of mode sets stringent requirements for parameters of the electrical energy produced by generators, first of all –for frequency stability of the generated current and the voltage level. The first condition directly depends on ensuring stability of a shaft rotation frequency of a gas-diesel primary engine of the generating set (GS).

The reasons of low stability of the gas-diesel engine revealed on the basis of the analysis of specialized literature and our own full-scale studies have been described by authors in a number of works [1-4]. A considerable change of calorific values of gas under various conditions and a lack of account of an effect of turbocharging during synthesis of the regulator of gas-diesel engine revolutions is observed. Let us refer to the main of them.

2. The improved dynamic model of GDGS

Authors have offered a specified model of GDGS describing the generating set both in a statics, and in dynamics for a research of dynamic characteristics. Authors have shown [4] that the dynamic model of the gas-diesel engine owing to an effect of disturbances on loading cannot be adequately described by the mathematical models offered in the literature which are not taken in to account during turbo-charging. An offered chain diagram, describing GDGS, is shown in figure1.

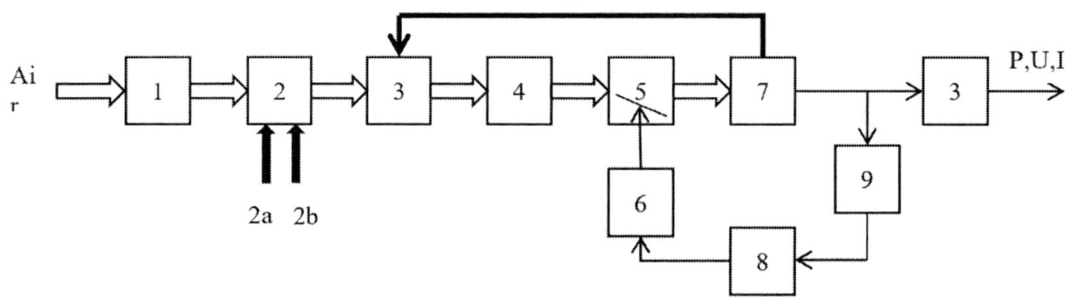

Figure1.A chain diagram of the gas-diesel generating set

1 – air filter; 2 – mixer; 2a – main piping of gas; 2b – gas piping of no-load operation; 3 – turbocharger; 4 – cooler; 5 – shutter; 6 – DC machine; 7 – cylinder block of gas diesel engine; 7a – outlet pipe of exhaust gases; 8 – regulator; 9 – Hall Effect Sensors.

This model is described in detail in [1-4].

3. Adaptive nonlinear control system of GDGS revolutions

The authors have suggested the adaptive regulator which is carrying out automatic identification of GDE parameters and adaptive correction of the PID-regulator settings of GDE [1-4] for the increase in efficiency stabilization systems of rotation frequency of gas-diesel engines (GDE) in the gas-diesel-generator set (GDGS) of the autonomous grid (AG) operating with abruptly variable loads.

The block diagram of the investigated system including the controller (C) with the tuning unit (TU) of parameters of the controller is represented in figure2.

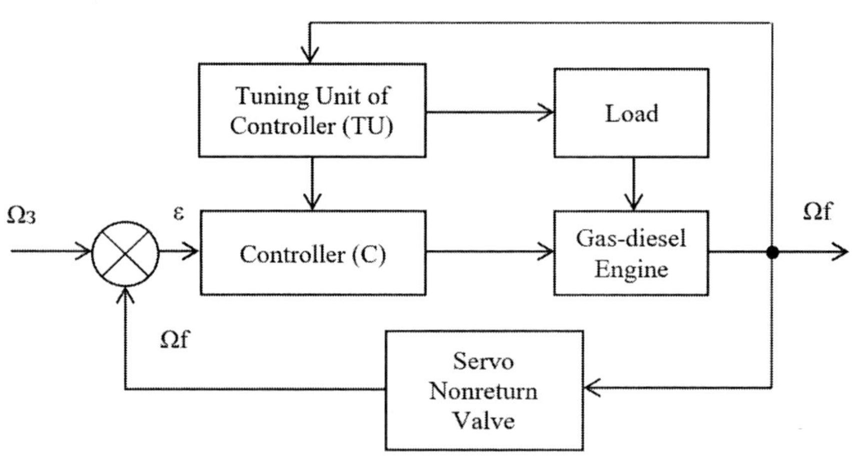

Figure 2. A block diagram of adaptive controller GDE with automatic identification of parameters: Ω_3 – assigned rotation frequency; Ω_f –actual rotation frequency; ε – error signal.

The developed identification algorithm of parameters and settings of the PID-controller uses the controlled load shedding and demanded load of GDGS with a step of 20%. Primary identification is made once during system tuning. Further, the fuzzy controller of parameters of the PID-controller is included.

The existing methodology allows providing the simple and fine PID controller tuning under a particular service condition. Further, the adjusted PID controller is capable of providing satisfactory stabilization of rotation frequency at its insignificant deviations from a specified value. But, as is well known, the quality of transient processes in the system with the PID-controller will be unsatisfactory with high probability during abrupt changes of an operating mode of GDGS or during its transition to another operating mode [4-7]. A tuning of the PID controller parameters is required under the mentioned conditions. The authors suggested a use of the controller on the basis of fuzzy logic for ensuring automatic correction of settings.

A block diagram of the possible implementation variant of the fuzzy controller for adaptive tuning of the PID controller coefficients of GDGS revolutions (during which the regular controller supplied with the GDE can also be used) is shown in figure 3.

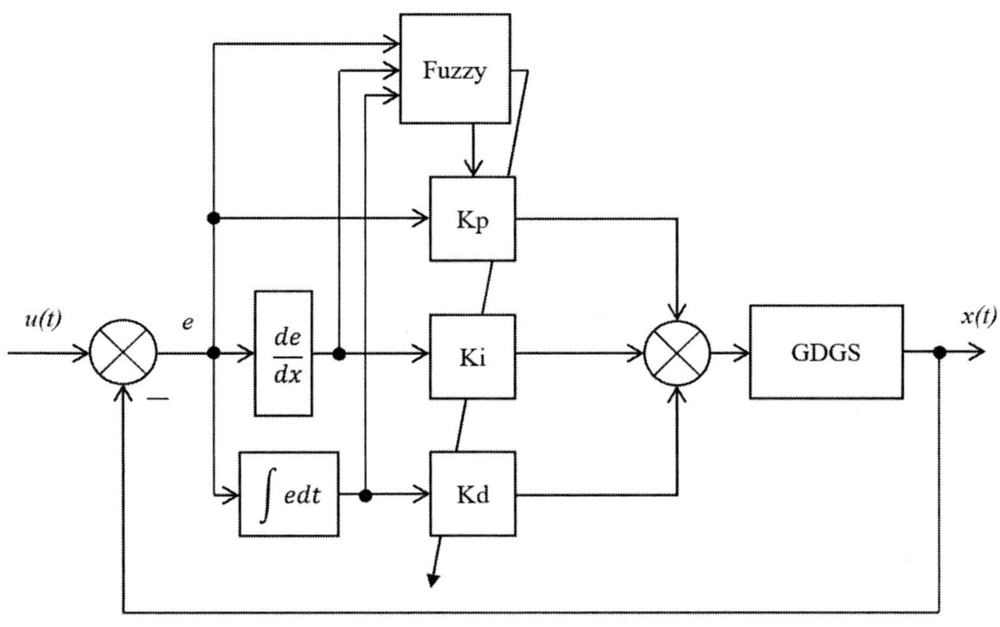

Figure 3. The use of the fuzzy controller of PID-regulator settings of rotation frequency of GDGS.

Theoretically, incorporation of automatic correction of controller settings of rotation frequency of GDGS allows expanding the range of stable operation of the latter up to emergency operation.

4. Evaluation of theoretical efficiency of stabilization of GDE revolutions during adaptive control

As it has been found in [2], really functioning GDGS can be considered as the perfect machine which is effected by the normal-mode interference with a certain [4-10] physical nature and parameters. It also may be assumed that the suggested control algorithm [2] will allow lifting significantly the stability of revolutions of GDE in all power ranges of load.

The block diagram of the GDE is represented in figure5. The normal-mode interference in the form of the destabilizing function is summed up with an output signal – the rotation speed of a GDE shaft. Let us present the description of blocks of the scheme in figure4:

$$\left.\begin{aligned} W_{\text{PID}} &= \frac{T_{\text{p}}T_g p^2 + K_{\text{p}}T_{\text{i}}p + 1}{T_{\text{i}}p}; W_y = \frac{K_y}{T_y p + 1}; \\ W_{\text{D}} &= \frac{(T_{TH}p + 1)}{\dfrac{T_{TH}}{1 - K_d K_{TH}}p + 1} \cdot \frac{K_d}{1 - K_d K_{TH}}; W_\omega = \frac{1/D}{\dfrac{T_e}{D}p + 1} \end{aligned}\right\} \quad (1)$$

Assuming $W_K = W_{PID} \cdot W_{TH} \cdot W_{\text{d}} \cdot W_\omega$,let us calculate the interference level at the output of closed loop system $f_{\omega з}$:

$$f_{\omega z} = \frac{f_\omega}{1 + W_K} = f_\omega \cdot \frac{1}{1 + W_K} = f_\omega \cdot W_Z \cdot$$

Interference power spectral density at the output of the closed loop

$$S_{\omega z}(\omega) = S_\omega \cdot |W_z(j\omega)|^2 = S_\omega \cdot \left| \frac{1}{1 + W_K(j\omega)} \right|^2 .$$

From the given formula, taking into account a character of linear amplitude frequency characteristic (LAFC) of the stable system for the area of effective interference cancellation, where $|W_K| >> 1$:

$$S_{\omega з}(\omega) \approx S_\omega \cdot \frac{1}{|W_K(j\omega)|^2} .$$

Let us consider now LAFC of the closed loop system in more detail. Taking into account (1), it is possible to write down:

$$W_K = \frac{T_{\text{i}}T_g p + K_p T_{\text{i}} p + 1}{T_{\text{i}} p} \cdot \frac{K_y}{T_y p + 1} \cdot \frac{K_g \cdot (\text{T}_{TH}\text{p} + 1)}{\dfrac{\text{T}_{TH}}{1\text{-K}_y\text{K}_{TH}} p + 1} \cdot \frac{1/D}{\dfrac{T_e}{D} p + 1} =$$

$$\frac{K_y}{D} \cdot \frac{K_{\text{d}}}{1\text{-K}_{\text{d}}\text{K}_{TH}} \cdot \frac{a_3 p^3 + a_2 p^2 + a_1 p + 1}{(b_3 p^3 + b_2 p^2 + b_1 p + 1).p} \cdot \frac{1}{T_{\text{i}} p},$$

where $a_1 = K_{\text{p}} T_{\text{i}} + T_{TH}$; $a_2 = T_{\text{i}} T_{\text{g}} + K_{\text{p}} T_{\text{i}} T_{TH}$; $a_3 = T_{\text{i}} T_{\text{d}} T_{TH}$;

$$b_1 = T_y + \frac{T_e}{D} + \frac{T_{TH}}{1 - K_{TH} K_d}; \quad b_2 = T_y \frac{T_e}{D} + \frac{T_y T_{TH}}{1 - K_{TH} K_{\text{d}}} + \frac{T_e T_{TH}}{D(1 - K_{TH} K_{\text{d}})}; \quad b_3 = \frac{T_y T_e T_{TH}}{D(1 - K_{TH} \text{Kd})} . .$$

According to the results of calculations of polynomial coefficients $A(a_i, p)$ and $B(b_i, p)$, it has been revealed that they actually compensate each other. At the same time, the error does not exceed in the narrow frequency range of 10 dB.

5. Quality research of rotation frequency control and power control via computer model

The general block diagram of GDGU operating in parallel is represented in figure 4. Active power allocation between GU of APS depends on the control accuracy of GDGU's revolutions.

The APS computer model with GDGU is represented in figure 4. It consists of two GDGUs operating for active and inductive load of different power (0.2 MW and 0.3 MW with the power of each GDGU equal to 0.67 MW). GDGU incorporates the following blocks: a model of the gas-diesel engine, excitation systems of the generator, switchboards, measuring units. Control of settings of regulators is carried out via virtual COM ports from the external fuzzy controller.

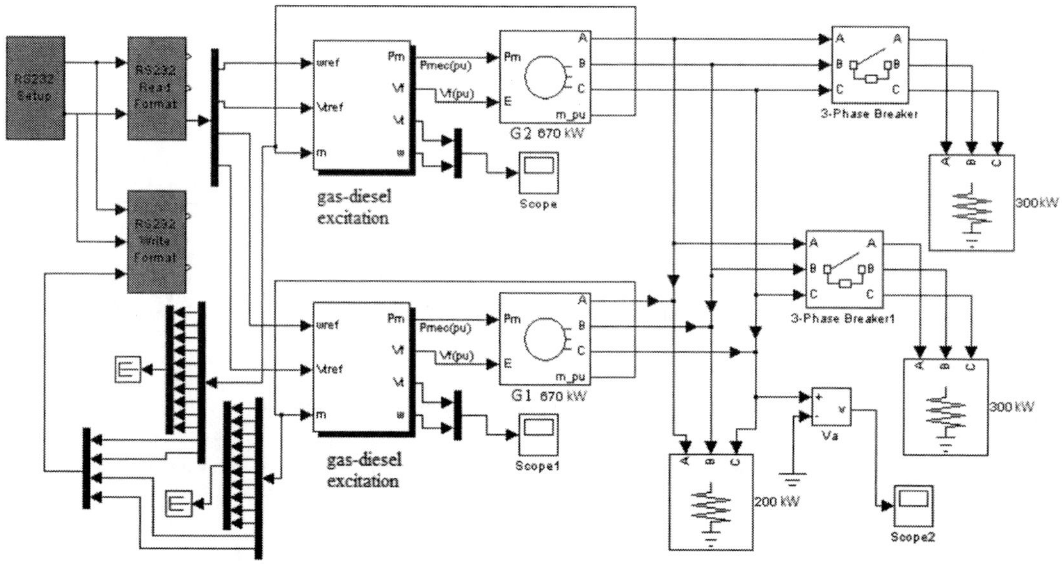

Figure4. The model of APS with two GDGUs operating in parallel.

6. Conclusion

The analysis of results shows that the use of the adaptive regulator of rotating speed of the gas-diesel engine including a fuzzy controller provides reducing of oscillations of GDGU's revolutions theoretically about 25-30 times.

7. Reference

[1] Chernyi S and Zhilenkov A 2015 *Transport and Telecommunication* **16** 73-82

[2] Nyrkov A, Budnik V, Sokolov S and Chernyi S 2016 *IOP Conf. Ser.: Mat. Sci. and Eng.* (Tomsk) vol 142 (IOP Publising) 012121

[3] Nyrkov A, Sokolov S, Zhilenkov A and Chernyi S 2016 *2016 IEEE NW Russia Young Res. in Electrical and Electronic Eng. Conf.* (St Petersburg: IEEE) 636-640

[4] Alekseev N A and Baranov A P 2000 (St Petersburg: Admiral Makarov State Maritime Academy)

[5] Zhilenkov A 2016 *Vibroengineering Procedia* **8** 17-21

[6] Fatuev V A et al 1998 *Reliability of automated Information Systems*: manual (Moscow: Izdat)

[7] Sedov V A, Sedova N A and Glushkov S V *Vibroengineering Procedia* **8** 453-458

[8] Director S and Rocker R 1974 (Moscow: Mir)

[9] Gorbatov V A, Smirnov M I and Khlytchev I S 1991 (Moscow: Energoizdat)

[10] Boriev Z, Nyrkov A, Sokolov S and Chernyi S 2016 *IOP Conf. Ser.: Mater. Sci. Eng.* (Tomsk) vol 124 (IOP Publising) 012006

Implementation of 14 bits floating point numbers of calculating units for neural network hardware development

I V Zoev, A P Beresnev, E A Mytsko and A N Malchukov

Tomsk Polytechnic University, 30, Lenina Ave., Tomsk, 634050, Russia

E-mail: zoev.ivan@yandex.ru

Abstract An important aspect of modern automation is machine learning. Specifically, neural networks are used for environment analysis and decision making based on available data. This article covers the most frequently performed operations on floating-point numbers in artificial neural networks. Also, a selection of the optimum value of the bit to 14-bit floating-point numbers for implementation on FPGAs was submitted based on the modern architecture of integrated circuits. The description of the floating-point multiplication (multiplier) algorithm was presented. In addition, features of the addition (adder) and subtraction (subtractor) operations were described in the article. Furthermore, operations for such variety of neural networks as a convolution network - mathematical comparison of a floating point ('less than' and 'greater than or equal') were presented. In conclusion, the comparison with calculating units of Atlera was made.

1. Introduction

Neural networks are of great importance in the modern world. They are used in problems of prediction, control and classification and in process automation. The scope may be vast, industrial, agricultural equipment, and automatic control of various air and ground equipment. For example, automation of warehouse loaders, agricultural harvester, autopilot in cars.

Current researches are carried out in the direction of increasing the size of neural networks in order to improve accuracy of their work [1]. However increasing the size of network leads to an increase of the perform time. One of the ways to reduce the operating time of the neural network is the use of optimization algorithms. The other way is to experiment with computing architecture of NN. Today, most of the studies are primarily focused at the software level [2-5].

However, if the optimization of calculations is made at the hardware level, it will be possible to achieve better performance in a system of small size. Also, hardware systems require less power than others do. But this task is not as simple as building a network in hardware as this is only one part of the work. For normal operation, the network must be trained. But hardware implementation of learning is not trivial. Since there are many program implementation of learning algorithms, it is simpler to train the network at the program level than to transfer the parameters trained network to hardware.

2. Numbers representation

To transfer the parameters (weights) of the network, firstly, it is necessary to determine in which form they should be stored. Most of the neural networks are working with single precision floating point numbers, rarely with double-precision ones. However, in a hardware implementation, there is a problem the number of logical cells required to implement. And the smaller the bitness of the number, the less logic cells are required for the operation with it. Table 1 shows the number of cells required, depending on the bitness of a number of different accuracies.

Table 1. Dependence of logic cells' numbers on representation of floating point bitness.

Bitness of floating point	Multiplication			Addition		Comparison	
	LUT	REG	Multiplier	LUT	REG	LUT	REG
Fp16	75	27	1 (18x18)	240	29	44	35
Fp32	127	32	1 (27x27)	524	45	82	67
Fp64	307	38	4 (27x27)	1121	107	167	133

The smallest bitness, according to the IEEE 754 standard, is represented with a half floating point [6], which is also supported by Nvidia graphic cards manufacturers, and it can accelerate the process of training neural networks [7].

If we consider the multiplication operation of floating point numbers and compare them with the field-programmable gate array (FPGA) architecture, in which hardware implementation is made, then we can note an interesting feature. For example, Cyclone V has special units for multiplying (DSP – a digital signal processor), which contain the hard multipliers size and number: 3 - 9x9, 2 - 18x18, 1-27x27[8]. In order to save DSP data blocks, it is logical to use variation with the largest number of multipliers. For example, there is only one configuration in Spartan 6 by Xilinx — 18x18, but it can operate at a higher clock frequency [9].

If we go back to the representation of numbers and look at the part of the number that is multiplied (10-bit mantissa + 1 implicit bit), we will find that it is necessary to use an 11x11 multiplier. Therefore, DSP multipliers of 18x18 size will be used to implement this multiplication block. The loss of one unit of the multiplier due to a 2-bit mantissa for hardware implementation of neural networks is not equivalent. Therefore, for the hardware implementation, it has been decided to use a format of a 14-bit floating point from half precision with two bits of the mantissa truncated. This allows using FPGA multipliers optimally, and is compatible with the format of 16 bits.

This transformation affects the calculation accuracy in the neural network, but in the classification task, this problem does not affect the result [10].

3. Multiplication

According to the rules of the IEEE 754 standard operation with floating point numbers, multiplication takes place as follows. The sign of the resulting number occurs from an XOR operation on the signs operands of multiplication. Exponents of operands are added together and the resulting exponent is chosen from the recent multiplication mantissa (simple sum or increment sum). The mantissa is obtained by multiplying, and the result is written into the result number with the truncation up to 9 bits. Representation of the circuit is shown in figure 1.

This scheme does not describe the whole IEEE 754 standard, which may affect the operation of the scheme. Namely, processing of plus/minus infinity cases is lacking. But the neural network almost does not work with these values.

Table 2 shows the value of occupied cells on the FPGA implementation multiplier for the full and non-full IEEE 754 standard.

Table 2. Dependence logic cells of the multiply unit with and without the case of infinity.

	LUT	Different LUT	Multiplier
Fp14 multiplier with infinity	35	2	1 (9x9)
Fp14 multiplier without infinity	33		

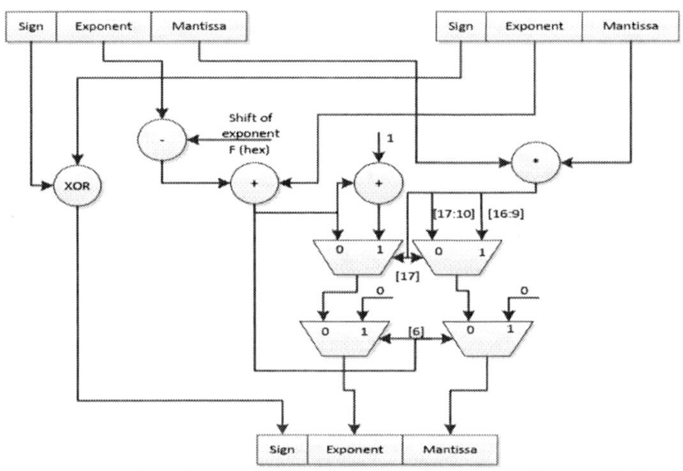

Figure 1. The scheme of the hardware floating point multiplier.

4. Addition and subtraction

Hardware implementation of the addition for floating point numbers is more difficult than multiplication. According to the rules, the first action brings mantissa values to one order.

This requires bringing of the two numbers to a fixed point format. The result of the conversion will be the 40-bits representation. After, addition occurs. The result will be represented in a 41-bit fixed point. Then it is necessary to perform the reverse conversion to the floating point format. The functional diagram is shown in figure 3.

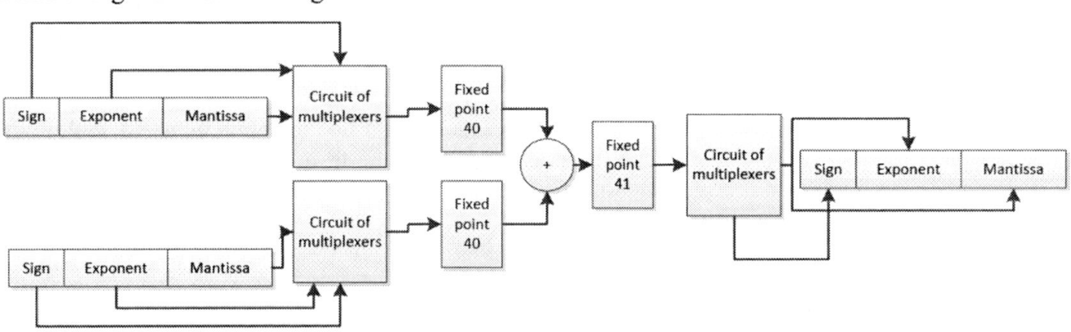

Figure 3. The scheme of the hardware floating point adder.

At the stage of conversion to a fixed point, a sign of operands must be considered and a conversion into two's complement representation for negative numbers has to be made. To implement subtraction, simply change the sign of the second operand.

5. Comparison

Also, other operations are used in neural networks besides the addition and multiplication. For example, a selection of the maximum element is used in convolutional networks. To implement the search of the maximum number, it is necessary to use the adder performing the subtraction. If the result is negative, then the subtrahend is larger than the minuend. If the result is positive, the subtrahend is greater than or equal to the minuend. The scheme is shown in figure 4.

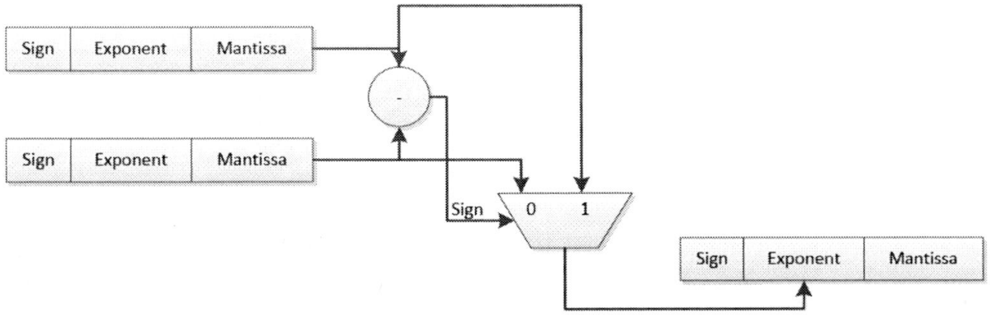

Figure 4. The scheme of the hardware floating point comparator.

6. Result & Conclusion

Implementations of calculating blocks were created using hardware description languages. Altera has a free IP core performing the same functions. However, their modules cannot be configured for the desired parameters in a 14-bit floating point (16-bit representation minimum).

The presented implementations have a form of combinational circuits; therefore, the sync impulse is not required. The delay speed depends on the logical cells. The comparative characteristics tables (tables 3-5) show the difference of the characteristics of the two implementations.

Table 3. Comparison of the multiplier of Altera with logic elements, registers, and performance.

	LUT	REG	Multiplier	F_{max} (MHz)	Latency
Fp16	75	27	1 (18x18)	58	2
Fp14	35	0	1 (9x9)	123	1
(Fp14 inf)	(33)			(137)	

Table 4. Comparison of the adder of Altera with logic elements, registers, and performance.

	LUT	REG	F_{max} (MHz)	Latency
Fp16	240	29	44	1
Fp14	497	0	35	1

Table 5. Comparison of the comparator of Altera with logic elements, registers, and performance.

	LUT	REG	F_{max} (MHz)	Latency
Fp16	44	35	225	1
Fp14	279	0	60	1

Based on table 3, we can conclude that there are better results of performance and less occupied logic cells compared to the multiplier of Altera. However, Altera wins in the search of the maximum element and the adder by the numbers of occupied logic cells. But if the register is considered as a classical D flip-flop, which contains 6 logic cells, the difference between the modules of Altera becomes smaller.

The optimization problem of these two calculators and the reduction of occupied LUT and/or the increase of the performance are still relevant and give grounds to continue the research of these functional units.

Despite drawbacks, the presented implementation provides the basis for the creation of hardware development of neural networks with software learning. Classic, recurrent and convolution neural networks can be created with these implementations. And the implementation on FPGA provides parallel work of neurons [9].

References

[1] Simonyan K , Zisserman A 2015 *Int. Conf. on Learning Representations* (San Diego: Corenell University Library) p 1

[2] Jia Y, Shelhamer E, Donahue J, Karayev S, Long J, Girshick R, Guadarrama S, Darrell T 2014 *ACM Conf. on Multimedia* (Orlando: Association for Computing Machiner) p 675

[3] Cireşan D, Meier U, Masci J, Schmidhuber J 2012 *J. Neural Networks* **32** 333-338

[4] Dong Y, Eversole A, Seltzer M L, Kaisheng Y 2015 *An Introduction to Computational Networks and the Computational Network Toolkit* Tech. Rep. MSR-TR-2014-112

[5] Gu J, Liu Y, Gao Y, Zhu M 2016 *4th Int. Workshop on OpenCL* (Vienna: Association for Computing Machinery) pp 1-12

[6] IEEE Computer Society 2008 *IEEE Standard for Floating-Point Arithmetic* (New York: IEEE)

[7] Dixon P R, Oonishi T, Furui S 2009 *IEEE Int. Conf. on Acoustics, Speech and Signal Proc.* (Taipei: IEEE) p 4321

[8] Pozniak K T, Czarski T, Romaniuk R S 2004 *The Int. Society for Optical Eng.* (Warsaw) p 1

[9] Omondi A R, Rajapakse J C 2006 *FPGA Implementations of Neural Networks* (Netherlands: Springer)

[10] Gupta S, Agrawal A, Gopalakrishnan K, Narayanan P 2015 *32nd Int. Conf. on Machine Learning* (Lille: International Machine Learning Society) p 1

Information and diagnostic tools of objective control as means to improve performance of mining machines

I E Zvonarev[1] and D I Shishlyannikov[2]

[1] Mining University, 2, 21st line, St Petersburg, 199106, Russia
[2] Perm National Research Polytechnic University, 29, Komsomolsky ave., Perm, 614990, Russia

E-mail: ZVano@mail.ru

Abstract. The paper justifies the relevance of developing and implementing automated onboard systems for operation data and maintenance recording in heading-and-winning machines. The analysis of advantages and disadvantages of existing automated onboard systems for operation data and maintenance recording in heading-and-winning machines for potassium mines are presented. The basic technical requirements for the design, operating algorithms and functions of recording systems of mining machines for potassium mines are formulated. A method of controlling operating parameters is presented; the concept of the onboard automated recording system for the Ural heading-and-winning machine is outlined. The results of experimental studies of variations in loading of the Ural-20R miner's operating member drives, using the VATUR portable measuring complex, are given. It is proved that existing means of objective control of operating parameters of the URAL-20R heading-and-winning machine do not assure its optimal operation. The authors present a technique of analyzing the data provided by parameter recorders that allow increasing efficiency of mechanical complexes by determining numerical values characterizing the technical and technological level of potassium ore production organization. The efficiency assessment criteria for engineering and maintenance departments of mining enterprises are advanced. A technology of continuous automated monitoring of potassium mine's outburst hazard is described.

1. Introduction

The objectives of increasing production volumes and unit performance of shaft-sinking clearing combines, optimizing the costs of maintenance and repair while reducing the number of emergency failures remain relevant for the enterprises engaged in the underground extraction of potash ore. Currently, there is no systematic approach to address these challenges, making it difficult to implement measures to improve the reliability of mining equipment and the quality of its technical service.

Experience of leading machine-building and mining enterprises shows that increased equipment performance can be achieved through the application of measuring software and recording complexes for continuous control of parameters of the work drives for the winning machines. The implementation of these systems, as well as the rationale for the methods of collecting, processing and further use of the information received, is a complex scientific and technical task that allows us to improve the arrangement of company's engineering and support services, to improve the technological processes of production and transportation of fossil, to generate proposals for equipment enhancement, to improve the safety of mining [1].

Currently, domestic shaft-sinking clearing combines Ural equipped with a planetary disk type and auger type actuators and trucks are widely used at the Russian potash mines. Combine actuators are sets of independent mechanical transmission with three-phase asynchronous motors.

Ural combines are successfully used in domestic potash mines for over 30 years and are generally characterized by high reliability. At the same time, the winning machines do not have the effective on-board diagnostic systems, automatic load limit, the regime parameters control and operating time counter, which causes an increase in the probability of machine crashes, in the majority of cases by the fault of the operator. The lack of the on-board monitoring system does not allow one to make a correct assessment of the operational loading of the machine components, to adequately determine the remaining service life of winning machine's parts and to carry out repairs based on the impact of the actual technical condition in a timely manner.

2. Purpose of research

Challenging operating environment for mining machines, complex processes of interaction of machine elements among themselves and with environment, as well as price restraints (profitability, competitiveness of products in the market), necessitated the search for technically simple and reliable solutions when designing automatic systems for control and monitoring the technical condition of mining machines. In our view, one of the most promising ways to control the parameters of work and technical condition of shaft-sinking clearing combines is an analysis of the intensity and nature of external loads determined through measurement of the instantaneous values of currents, voltages and power consumed by electric actuators [2–4].

3. Idea of technical solution

Employees of the Department of Mining and Oil-Field Machines at Perm National Research Polytechnic University, together with the specialists of 'Regional rope centre' LLC (Perm), developed a prototype of software and recording complex VATUR, which provides measurement, recording and preserving of the fundamental parameters of shaft-sinking clearing combines' actuators [5].

The structure of recording complex VATUR (Figure 1) includes a processing unit, power supply and switching, current clamp, voltage sensors and displacement. The processing unit consists of a single-board computer and an analog-to-digital converter (ADC). Current clamp provides transformation of input current into an output voltage by ratio 1 : 1. Voltage sensors convert the input voltage in the ratio of 1 : 3. These converted signals are supplied via the ADC directly into the computer. Distometer is made on the basis of incremental turning angle encoder PUF-6 with 360 strokes and vice versa. The distometer is mounted on the side ski of the combine from the direction of the magnetic station. After the measurement, the data are stored on a removable hard disk and then transferred to the office computer.

Software and recording complex VATUR carries out 100 measurements within one 20 ms period of supply mains. The values of effective power I_{eff}, voltage U_{eff}, active P_{a} and full power S are determined from the measured instantaneous values of current Ii and voltage U_{i}:

$$I_{\text{eff}} = \left(N^{-1} \sum_{1}^{N} I_i^2 \right)^{1/2}, U_{\text{eff}} = \left(N^{-1} \sum_{1}^{N} U_i^2 \right)^{1/2}, P_a = 3N^{-1} \sum_{1}^{N} U_i I_i, S = 3U_{\text{eff}} I_{\text{eff}}, \quad (1)$$

where N is the number of measuring points over a fixed time interval, $N = 100$; time interval is 20 ms.

The supply voltage and the load on the combine engines are symmetrical, so active power measurement in three-phase circuit is made by the active power meter with artificial zero. Visualization and data processing is carried out by means of a specially developed software Vatur-of (Figure 2).

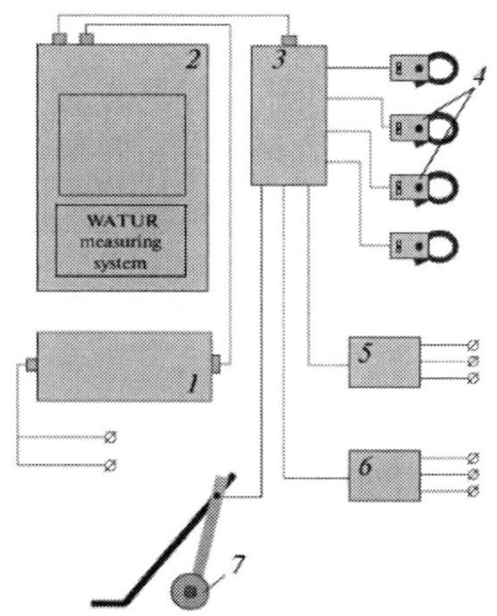

Figure 1. The block diagram of the VATUR measuring system: 1 – power unit, 2 – processing unit, 3 – switching unit, 4 – current sensors (clamp meters), 5 and 6 – voltage sensors, 7 – motion sensor.

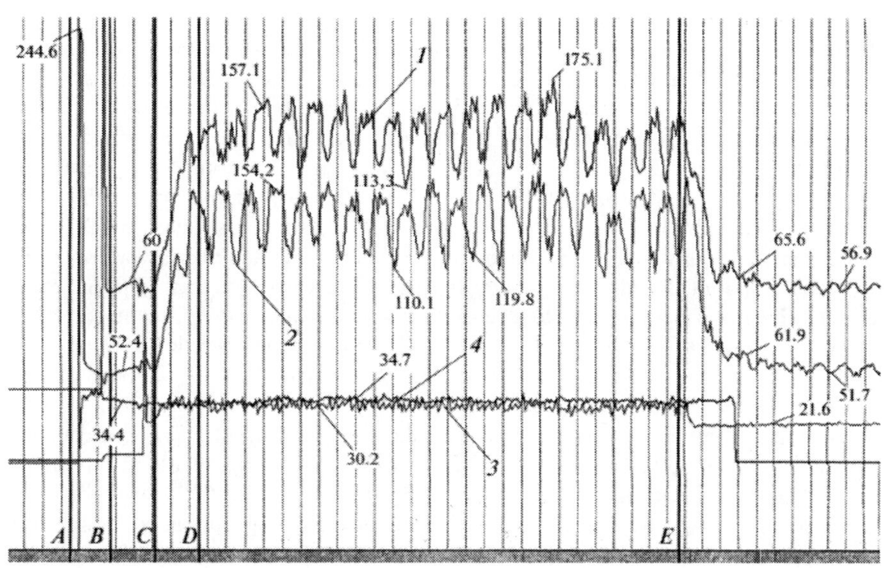

Figure 2. Diagrams of active three-phase powers of electric motors of the Ural-20R machine at an average speed of 10.5 m/hour: 1 and 2 – the drives providing the relative motion of cutter wheels; 3 – fender drive; 4 – conveyer drive. Grid line pitch – 5 s.

On the above-mentioned chart of change of active power of the electric drives for cutter wheel relative motion, there can be identified launch areas (A-B), idle move (B-C), cutting (C-D) and steady state operation (D-E). Software and recording complex VATUR allows one to create and store of data arrays containing information about the long periods of work of the shaft-sinking clearing combine [5]. Wattmetrogram analysis using methods of mathematical statistics allows one to determine the following.

1. Operation specifications of the combine: turn on-turn off frequency, productive work time and downtime. These parameters characterize the rhythm of technological processes of minerals extraction and transport in the cleaning chamber, allow us to estimate the level of work arrangement and the degree of equipment utilize capacity, to identify negative trends and factors affecting the performance of the mechanized complex.

2. The frequency of occurrence and duration of excessive loads, that characterizes the expertise and industrial practices that are shown by operators of mining machinery. In case of emergency the wattmetrogram is a document that allows us to identify the causes of the accident and to judge the actions of the staff. The automatic load control and combine management systems can operate on the basis of relevant information about the operation of actuators.

3. The average and peak values of the loads affecting the transmission parts. The average values of the active power and its fluctuations characterized by a standard deviation allow a qualitative estimation of the dynamic component of the external load, precision of manufacturing and assembly of transmission elements, the intensity of degradation processes in combine actuators. Information on the nature and magnitude of change in actuator load is delivered to the manufacturer for further design calculations, changes in the machine design at the design stage, the modernization of the combines in operation.

4. Technical performance and service hours of the combine. The findings correlated with the costs of maintenance and repairs determine the cost of vehicle-hour of combine availability and objectively characterize the efficiency of the enterprise maintenance services. The increase in the cost of vehicle-hour of availability of the particular mining machine compared with the rest of the combine in operation at the mine can reasonably show the necessity of its decommission [1].

5. The total and remaining transmission elements lifetime. It is known that the operating time of the mining combine is determined by the strength properties of machine elements, as well as the size and nature of the actual operating loads. Total and residual lifetime of the uniform transmission can be determined with a high degree of probability on the basis of continuous recording of the active power of the engine (its operational load) in a turnaround time according to the procedures detailed in [2-3, 6-7].

A generic parameter that allows us to assess the efficiency of mining machine is the value of specific energy consumption for the process of destruction of potash mass by the combine cutters, which characterizes the physical and mechanical properties of potash mass, work operating parameters and the combine technical condition. The value of specific energy consumption of the process of destruction of the mass by the actuators' cutters was calculated for plots of steady state combine operation according to formula $H_w = P_a(V_p F_d \gamma)^{-1}$, where H_w is the specific energy consumption of the cutting process, kWh/t; F_d is the face area treated by the actuators, m^2; V_p is the speed of the combine at the face, m/h; γ is the ore density in the mass, t/m^3.

Analysis of the experimental results (Figure 3) showed the ability to identify accurately the change of specific energy consumption of the process of mass destruction by actuators' cutters of the combine. The methods and equipment make it possible to register a decline in the energy intensity of the mass destruction process caused by gas-dynamic phenomena, to carry out the identification of outburst zones, thus ensuring the safe conduct of mining operations [8].

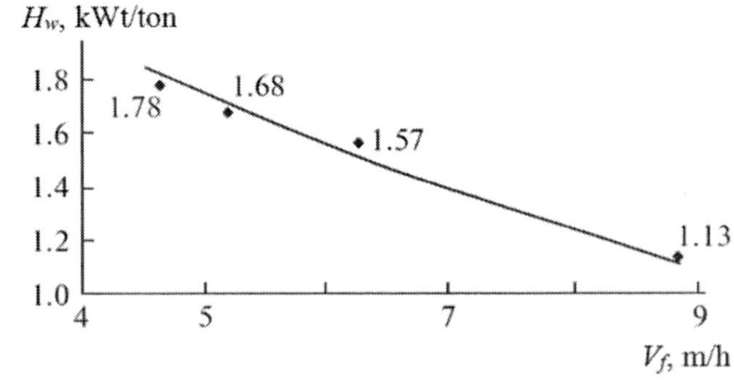

Figure 3. Dependence of change in specific energy costs on the velocity of the Ural-20R machine feed.

Mathematical processing of change functions of the instantaneous values of currents consumed by combine electric motors was carried out using spectral analysis techniques. According the Nyquist theorem, the analogue signal spectrum can be uniquely represented by its discrete samples taken at intervals of $\tau = 0.5f^{1}$, where τ is the sampling interval of the analog signal function by time, s; f is the frequency of the analog signal range limit, Hz.

Spectral analysis of analog current signals (Figure 4) obtained by the measurement software and recording complex VATUR allows to reveal frequency components from 0 to 1000 Hz, characterizing the fluctuations of kinematic chain actuator – gearbox – drive motor. Defects in working units and mechanical gears cause the occurrence of variable loads, which causes the appearance of new spectral components. Periodic measurement of values in the current spectrum that characterizes the specific defects in the drive motor and the manual transmission, allows us to carry out the assessment of technical condition of the drive mining machinery and, if necessary, to carry out repairs to prevent crashes.

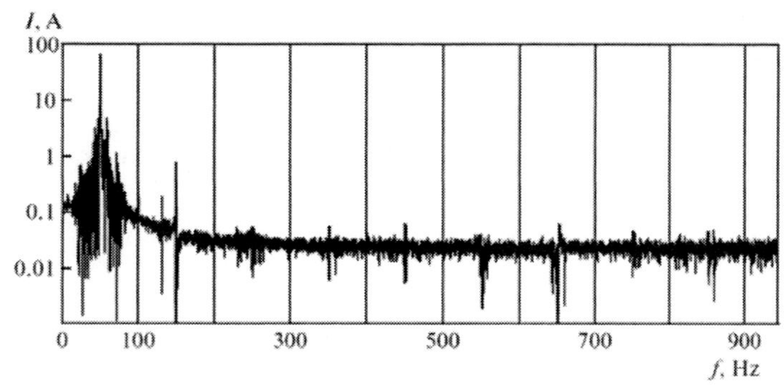

Figure 4. The spectrum of the current of the drive induction motion relative to the motion of the cutter wheels of the Ural-20R machine (logarithmic scale of amplitude axes).

4. Conclusion

By analyzing the consumed current records in asynchronous motors, the defective electric part of the rotor may be found, including the rotor winding breaks and the short circuit of the active core plates; the electrical parts of the stator, including breaks and electrical asymmetry of power winding, the short circuit of the active core plates; static and rotating eccentricities; bearing defects, leading to fluctuations in the air gap form. In addition, an important diagnostic sign is the value and duration of the occurrence of starting current. In the actuators' transmission the fault gears, the landing gear on the shaft, the misalignment of slave shafts and their rotation bearings, the misalignment of the driven shafts arrayed in a line can be diagnosed by the nature of the current signal changes.

The disadvantages of the method of evaluation of the technical condition of transmissions by current consumption relates the difficulty of identifying a number of defects at the initial stage of development. Primarily, these are the defects of rolling bearings, the identification of which is carried out indirectly, for example, by changes in the characteristics of gearing [9]. The increased information content and accuracy of diagnosis is possible through the implementation of periodic multi-parameter monitoring (acoustic emission, vibration diagnostics) of bearing assemblies.

Continuous recording of capacity consumed by combine electric motors allows the simplest and the most reliable monitoring of the operational load of mining machine's actuator assemblies, carrying out an assessment of technical condition, complete and remaining life of the transmission components [10]; determining the operating parameters, technical performance and the combine error-free running time, improving the safety of mining operations. The combine automatic control systems may effectively operate on the basis of the current information about the amount and nature of actuator load.

References

[1] Andreeva L I 2004 *The way to form technical service for mining equipment at coal mining plant* (Yekaterinburg: UGGU) p 20

[2] Ivanov S L 1999 *Increase the resource of transmissions of mining machines based on the assessment of energy loading of their elements* (Sankt-Petersburg: Mining Institute) p 92

[3] Kolomiitsev M D 1988 *Operation of mining machines and automated systems* (Leningrad: Mining Institute) p 96

[4] Krasnikov Yu D, Solod S V and Khazanov Kh I 1989 *Increased mining excavation machinery reliability* (Moscow: Nedra) p 215

[5] Trifanov G D, Knyazev A A, Chekmasov N V and Shishlyannikov D I 2013 *Mining Equipment and Electromechanics* **2** 41–44

[6] Zvonarev I E, Ivanov S L, Fokin A S and Semenov M A 2012 *Tech. sci. Siberian Consultant Association* 53-62

[7] Kolomiitsev M D 1987 *Notes of Leningrad Mining Institute* **117** 69-76

[8] Laptev B V 1994 *Prevention of gas-dynamic phenomena in the potash mines* (Moscow: Nedra)

[9] Barkov A V, Barkova N A and Borisov A A 2012 *Methods of diagnosing mechanisms for electric current consumption* (Sankt-Petersburg: Severo-zapadnyy uchebnyy tsentr) p 67

[10] Ivanov S L 2013 *Mining Equipment and Electromechanics* **11** 28-33

Forming system of strategic innovation management at high-tech engineering enterprises

O T Ergunova[1], **V G Lizunkov**[2], **E Yu Malushko**[3], **V I Marchuk**[2], **A Yu Ignatenko**[2]

[1] The Ural State University of Economics, 62, 8th of March Str., 620219, Yekaterinburg, Russia
[2] Yurga Institute of Technology, 26, Leningradskaya Str., Yurga, Kemerovo Region, 652055, Russia
[3] Volgograd State University, 100, Universitetskiy Ave., Volgograd, 400062, Russia

E-mail: vladeslave@rambler.ru

Abstract. The article considers the processes of forming the strategic system of innovative activity management at the enterprises of a high-tech mechanical engineering complex (MEC) that are traditionally decisive in shaping the economic base of Russia. The authors proposed a method of designing a strategic system of innovative activity management at the MEC's enterprises of the region which is based on the consideration of means and opportunities of the enterprise to implement intellectual activity. The proposed methodology and the empirical results constituted a ground for developing a system of strategic innovation management when performing the state-guaranteed order at the hi-tech MEC enterprises. Its implementation will help to reduce the level of uncertainty throughout the entire life cycle of an innovative activity product.

1. Introduction

In the context of the country's transition to innovation-oriented post-industrial economy with a significant intensification of state innovation support, we witness a continuous decline of innovative competitiveness and efficiency of the high-tech mechanical engineering enterprises due to the low level of their strategic management efficiency. However, under conditions of high rates of the world's scientific and technological development, stimulation of innovative activity in mechanical engineering is not only important for the competitiveness of the national engineering production, but also for the development of related industries utilizing mechanical engineering products. We develop a methodology taking into account the resources application potential to improve the degree of implementation of innovations into the high-tech enterprises of the machine-building complex (MEC). The conducted systemic analysis of the strategic innovation management at the one of MEC enterprises leads to the conclusion that processes of strategic innovation management in such enterprises are quite random and insecure.

The developed methodology is based on determining a horizontal resource potential of strategic innovation management and allows one to determine the potential of resources application at each stage of the product lifecycle [1]. Comparison of actual and theoretical values of the horizontal resource potential will allow us to adjust the distribution of each resource within the stage and thus to optimize the implementation of the given tasks at a particular stage [2]. On the basis of the proposed methodology and the empirical results, we developed a scheme of strategic management of innovative activity. It is supposed that its implementation will help to reduce the level of uncertainty throughout the entire life cycle of product innovation when performing the state-guaranteed order at MEC enterprises.

2. Materials and methods

The methodological base of the research is represented by the method of analysing financial results of enterprises to assess their sustainable development [3], methodological apparatus of rating estimation of their financial and economic state [3], methods of evaluating the effectiveness of the innovative projects implementation at the enterprises [4], methods of evaluating and analysing competitiveness and sustainable development of mechanical engineering enterprises [5].

The study was conducted in two directions. The first direction is devoted to the investigation of the functional elements of the integrated methodology for calculating the resource potential of the enterprise. The second direction is related to obtaining specific results defining a set of tools and capabilities of the enterprise in implementing intellectual property, and it will be based on the results obtained in the first study [6]. After analyzing and synthesizing ways to optimize the existing system of strategic innovation management, we analyzed the horizontal distribution of resources within each stage of the product life cycle, the vertical distribution of each kind of resources at all stages. Then the results were summed up into a single model to develop recommendations how to optimize the existing system. Expert assessment was conducted through the survey of senior and middle managers at one of the high-tech mechanical engineering enterprises of the Siberian Federal District (SFD). The results were processed by the method of mathematical statistics.

3. Situation Analysis in the industry

The most active sphere of implementing innovations in mechanical engineering according to the number of companies investing in innovation is a sub-sector of producing electronic and optical equipment: new products or technologies are developed at about 30% of enterprises (figure 1). The same situation is observed if only technological innovations are taken into account (figure 2).

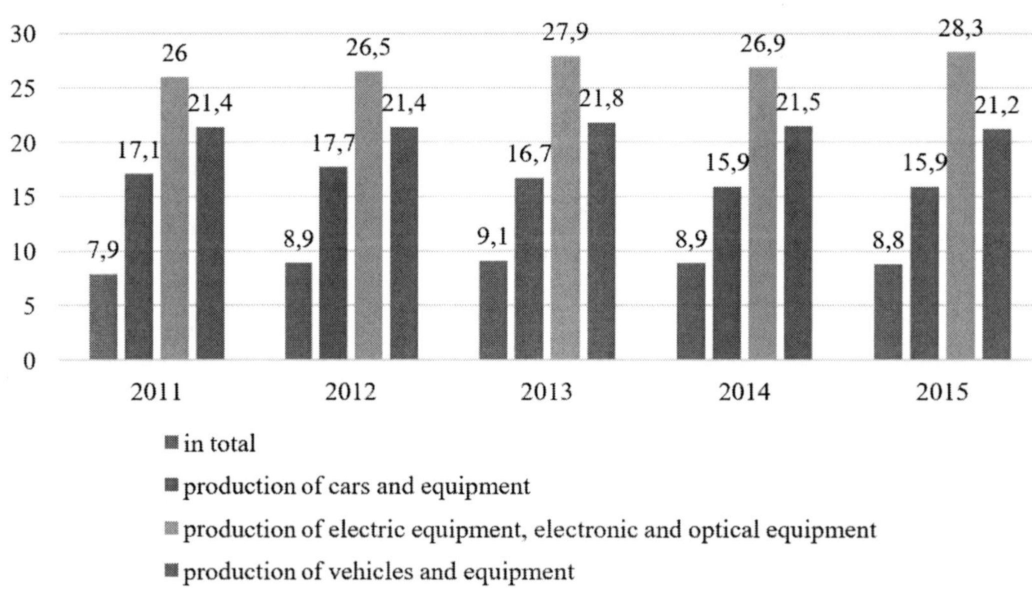

Figure 1. The share of mechanical engineering enterprises engaged in innovation in 2011-2015, as

compared to the data for the whole Russian economy [7]

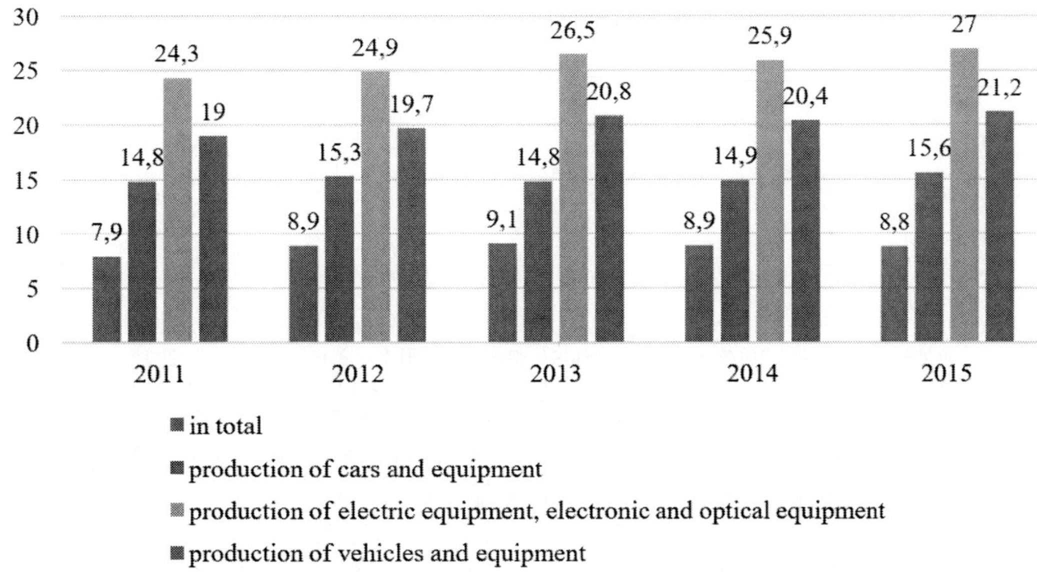

Figure 2. The share of mechanical engineering enterprises engaged in technological innovation in 2011-2015, as compared to the data for the whole Russian economy [8]

The extent of financial investments aimed at promoting technological innovations shows a stable upward trend though it was a slight decrease in 2015. In 2015, these expenses amounted to 735.8 billion rubles that is lower by 11% in fixed prices as compaired to 2014. About two thirds of the technological innovation costs belong to the most economically successful industries, such as extractive industries (17.1%), automotive industry (7.9%), production and distribution of electricity, gas and water (6.3%), metallurgy (5.9%), aircraft and spacecraft manufacturing (5.4%). Expenditures on technological innovation are mainly related to the introduction of process innovations aimed at enhancing the effectiveness of production processes: their share exceeds 64% of the total investments. The costs of pioneering product innovations implying a significant breakthrough in product manufacturing are 1.8 times lower (35.9%). It is compared favorably with high-tech and separate medium-tech sectors (chemical industry, automotive industry, machinery and equipment production) where the share of product innovation costs is about three-quarters of total expenditures on innovation. The intensity of technological innovation expenditures in industrial manufacturing amounted to 1.8% in 2015 (versus 2.1% in 2014). In high-tech sectors, the figure is three times higher (5.9%), and, in some industries, it reaches maximum values comparable to global counterparts (for example, in medical equipment and appliances manufacturing - 6.6%).

Figure 3 demonstrates the comparison of indices characterizing the dynamics of the turnover, the costs of technological innovation and manufacturing innovative products, and proves the absense of a clear dependence between these indicators.

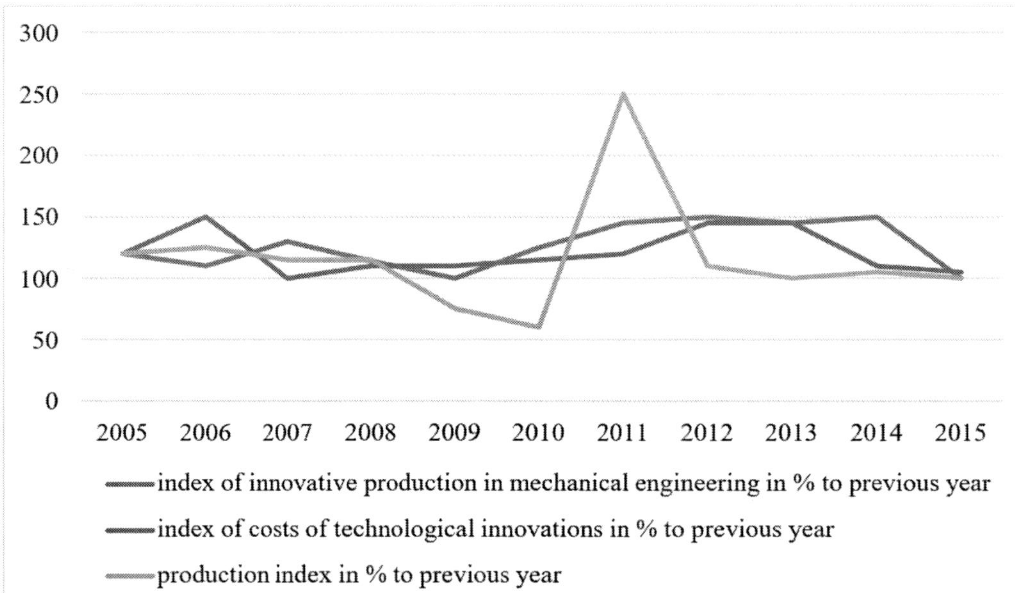

Figure 3. Comparison of indices characterizing the production and innovation of engineering enterprises in 2005-2015

At the same time, fluctuations of technological innovation costs are smoother and have lower amplitude as compared to other indicators. This can be explained by the fact that the investment programs for the machine-building enterprises are medium and long term while the dynamics of production volumes both of innovative products and output as a whole is more susceptible to the economic crisis. In addition, strategic management at the organizational level, including innovation activities, at the high-tech mechanical engineering enterprises is relatively new in Russia as it was previously common to carry out only operational management for individual projects.

It should be noticed that machine-building enterprises are first increasing innovative products volume when exiting the crisis (after the crisis of 2008, the innovative products volume increased in 2012 while total output reduced). This brings us to the conclusion that innovative products manufacturing contributes greatly to the anti-crisis measures. Thus the research of the impact of strategic management methods in the innovative sphere of high-tech machinery enterprises, of their further development in order to enhance economic efficiency proves to be highly relevant.

4. Study procedures

On the basis of product life cycle stages and existing traditional resource types, the study encouraged the development of methods to measure the horizontal resource potential of the strategic innovation management of innovative at the n-th stage [9]. The methodology allows defining the resource application potential at each stage of the product life cycle, as well as to redistribute and manage resources most efficiently within each stage. Thus regarding the already conducted study [10], the horizontal potential of applying resources for the strategic management of innovative activity at the n-th stage of the product life cycle can be calculated as additive components with weight coefficients:

$$P_n = k1K_n + k2L_n + k3M_n + k4I_n, \quad (1)$$

where K_n is financial resources at the n-th stage of knowledge generation, L_n - labour force at the n-th stage, M_n - material resources at the n-th stage, I_n - information resources at the n-th stage.

In order to assess specific types of resources, it is necessary to determine:

1) the payroll normalized to the number of employees participating in performing a specific stage of the product life cycle;

2) the share of the human resources at every stage of the product life cycle equated to the share of employees who are in charge of performing this stage;

3) distribution of material resources based on a cost component of a technical upgrading plan for each stage of the product life cycle, but a cost component is standardized throughout the technical upgrade plan;

4) the share of information resources in terms of information and computer support at every stage of the product life cycle.

Comparing actual and theoretical values will allow one to adjust the distribution of each resource within the stage and thus to optimize the implementation of the tasks at a particular stage [11].

In order to assess the real resource potential of strategic innovation management, we used data received from the industrial activity report of one large state-owned MEC enterprise in the SFD in 2016. According to life cycle stages of the product, the actual data on the use of resources were measured at each stage and then compared with the ideal values. To obtain ideal values of resource allocation, senior and middle managers were interviewed at the enterprise [12]. The actual measurement data are accurate as every stage of the product life cycle is carried out by different departments of the enterprise [13]. Then the received percentages of financial, human, material and information resources are normalized for each phase in order to obtain the allocation of resources at every stage of the product life cycle.

In support of the proposed methodology for MEC enterprises in the region, a schematic diagram is given to visually represent the strategic management of innovative activity when performing the state-guaranteed order. The proposed scheme of strategic management of innovative activity helps to increase the efficiency of each stage. As a result of implementating the scheme of strategic innovation management, the commissioning dates are reduced and the number of warranty repairs decreases. In general, the implementation of the given scheme at MEC enterprises leads to decreasing the level of uncertainty throughout the product life cycle and consequently to reducing the deadlines and increasing the quality of executed works which inevitably results in increasing the effectiveness of the state-guaranteed order.

5. The results of the empirical research

On the basis of the above mentioned methodologies, we have studied the system of generating knowledge at the MEC enterprise and obtained ideal and empirical values of the horizontal distribution of resources throughout the product life cycle (table1).

Table 1. Ideal (I) and empirical (E) elements of the contribution value of the resource potential of innovation (%)

Stages of the product lifecycle	resource allocation									
	labor		material		financial		information		total	
	I	E	I	E	I	E	I	E	I	E
marketing research	45	20	15	25	30	15	10	40	100	100
simulation	50	23	20	26	20	11	10	40	100	100
software	40	20	20	16	30	32	10	32	100	100
engineering	40	21	15	21	35	5	10	53	100	100
technology development	25	28	25	31	10	27	40	14	100	100
manufacturing	25	28	25	31	10	27	40	14	100	100
test	15	28	40	27	20	22	25	23	100	100

It should be noted that the ideal allocation of resources, according to respondents, is optimal for functioning of MEC companies and does not depend on a specific state-guaranteed order. Comparing empirical data and weighting coefficients of the theoretical horizontal capacity of using resources shows significant discrepancies in actual and theoretical values of used resources persentage at each stage of the product life cycle [14]. The stage of research and marketing analysis shows a significant redistribution of resources towards increasing the share of information and financial resources. The second stage of the product life cycle - the stage of technical specifications preparation, research, documentation and modeling – indicates a significant increase in the proportion of material and financial resources in comparison with theoretical estimates, but a significant decrease in the use of labor resources [15]. The analysis of the third stage of the product life cycle - the stage of developing electric circuits and software – demonstrates redistribution of resources so that the material and information ones underwent an increase in the amount, but the use of the labor force decreased twice as compared with the theoretical model. The results of the fourth stage - the stage of developing the design documentation – reveals redistribution of resources towards the material ones. At the same time, there was a noticeable decline in actual levels of labor and information resources application in comparison with theoretical model estimates. The analysis of the fifth stage of the product life cycle - the stage of developing technical documentation – shows that the resources are redistributed again and this time financial and information resources experience a significant increase. Obviously, these processes reflect the decline in the use of labor and material resources at this stage of the product life cycle. At the sixth stage of the product life cycle - the stage of manufacturing products – the increase in employing labor and financial resources comes with a decrease in using information resources in comparison with theoretical estimates. When analyzing the seventh stage of the product lifecycle - stage of product testing - it is visible that resources are redistributed towards the use of labor forces to compare with theoretical estimates. But this process is accompanied with the parallel decline in applying financial resources.

Comparison of theoretical (potential) and actual estimates provides data on development reserves for strategic innovation management when a real production process is organized. Thus the analysis of empirical data makes it possible to define the more likely way of optimizing the use of each resource apart and at every stage of the product life cycle.

6. Conclusion

Developing the integrated methodology of strategic management of innovative activity at the MEC enterprises in the region as a set of enterprises assets and capabilities when implementing intellectual property allowed obtaining the following theoretical and practical results. Firstly, the proposed theoretical methodology allows evaluating the capabilities of the enterprise in implementing intellectual activity. Secondly, a scheme of strategic management of innovative activity is developed to effectively perform the state-guaranteed order; its implementation contributes to a reduction of the degree of uncertainty throughout the entire product life cycle and, as a result, to a shortening of the implementation period and to improvement of the quality of all performed work. It inevitably leads to an increase in the effectiveness of performing the state-guaranteed order. Thirdly, the empirical research becomes the blueprint for determining the real proportion of allocated resources at every stage of the product life cycle at the high-tech engineering enterprises. The proposed methodology accounts for the potential use of resources to improve the effectiveness of the strategic management of innovative activity at the high-tech engineering enterprises as an engine of innovation in the region is able to improve efficiency of production at MEC enterprises and thus to increase the level of innovative development of the given region.

7. Acknowledgments

The reported study was funded by the Russian Humanitarian Science Foundation (№ 16-33-00010).

References

[1] Popov E V, Vlasov M V, Shishkina A Yu 2014 Priorities of knowledge generation at the state enterprises. *National interests: priorities and safety* **26** 2-11

[2] Padalkin V Y, Vorontsov E I 2012 Using the methods of the analysis of financial results for the assessment of sustainable development of enterprises Economics MSK *Theory and radio equipment* **4** 119-123

[3] Batkovsky A M, Trofimets V J, Trofimets E N 2014 Development of the methodical device-rated financial and economic state enterprises of the military-industrial complex *Finances and Credit* **48 (624)** 34-45

[4] Shashkarova M V 2015 Evaluating the effectiveness of the implementation of innovative projects at the enterprises oboronnopromyshlennogo complex *Innovations and investments* **5** 10-13

[5] Nesteruk D N, Momot M V 2014 Information Technology of Estimation and Forecasting Innovative Activity Based on Distributed Data Input. *Applied Mechanics and Materials* **682** 579-585

[6] Goldman A A, Nikiforov I I, Ivanova R P 2015 Priorities of Higher Social Education at Mirny Polytechnic Institute of NEFU. *Issues on Science and Education* **2-2** 333

[7] Lizunkov V G, Minin M G, Malushko E Y, Medvedev V E 2016 Developing economic and managerial competencies of bachelors in mechanical engineering *SHS Web of Conferences* **28** UNSP 01063

[8] Prokopenko S A 2005 The renewed technical university - for the modern coal industry of kuzbass. *Ugol'.* **12** 48-51

[9] Stepanova N V, Razumakov A A 2013 *The 8 international forum on strategic technologies (IFOST 2013)* **1** 240–242

[10] Lizunkov V G, Malushko E Y, Maletina O A, Tsybaneva V A 2016 Use of virtual learning system for educating students with disabilities and special needs *3nd international multidisciplinary scientific conference on social sciences & arts SGEM 2016: conference proceedings* **1(1)** 481-487 doi: 10.5593/SGEMSOCIAL2016/B11/S03.062

[11] Morozova M V, Kust T S, Sokolova E Y, Osipova S I, Gafurova N V, Vaganova V I, Ovchinnikov VA, Agavelyan R O 2016 The management of youth employment in a lifelong engineering education system. *International Review of Management and Marketing* **6(2)** 247-252

[12] Tashchiyan G O, Suzdalova M A, Grichin S V, Goriaynova E S, Kaz M S 2016 A system of evaluation of engineering solutions competitiveness of a company based on its image *IOP Conference Series: Materials Science and Engineering* **142(1)** 012103 doi:10.1088/1757-899X/142/1/012103

[13] Zimin A V 2015 RIVS Science and Production Association: Contemporary strategy and principal areas of activity *Eurasian mining* **1** 3-4

[14] Zyryanov I V, Dvoychenkova G P, Kovalchuk O Y 2014 Scientific Educational Centre of MPTI – the Platform of Higher Education Development in Western Yakutia. *Mining Journal* **1** 31–33

[15] Goldman A A 2015 Petroleum Engineering Education in NEFU, Yakutia. *Journal of Siberian Federal University* **8** 1551–59

New Methods of Three-Dimensional Images Recognition Based on Stochastic Geometry and Functional Analysis

N G Fedotov[1], A V Moiseev[2], A A Syemov[1], V G Lizunkov[3] and A Y Kindaev[2]

[1] Penza State University, 40, Krasnaya Street, Penza, 440017, Russia
[2] Penza State Technological University, 1a/11, Baidukova Passage/Gagarina Street, Penza, 440039, Russia
[3] Yurga Institute of Technology, TPU Affiliate, 26, Leningradskaya Street, Yurga, 652055, Russia

E-mail: vladeslave@rambler.ru

Abstract. A new approach to 3D objects recognition based on modern methods of stochastic geometry and functional analysis is proposed in the paper. A detailed mathematical description of the method developed on the approach is also presented. The 3D trace transform allows creating an invariant description of spatial objects, which better resist distortion and coordinate noise than the one, obtained as a result of the object normalization procedure, does. The ability to control properties of developed features increases intellectual capacities of the 3D trace transform significantly, which can be mentioned as its undeniable advantage. The justification of the proposed theory and mathematical model is a variety of worked out theoretical examples of hypertriplet features that have particular described properties. The paper considers in detail scan techniques of the hypertrace transform and its mathematical model as well as approaches to developing and distinguishing informative features.

1. Introduction

Current progress in science and technology has resulted in an annually growing number of people involved in information processing. One of the most important tasks connected with information systems creation is automating the process of pattern recognition. Ample researches are being conducted to perform the task and among them there are those that help us to try and to understand one of the main abilities of a human brain – the ability to recognize [1-3].

They develop and generate decisive prerequisites for constructing intelligence systems.

It is worth noting that imitating human brains activity is not the only method of creating such systems [4-8]. Technology has its own ways of solving the problem, the ways that do not copy natural abilities of a human being technically, but take into account merits and particular properties of natural visual systems. The paper discusses some of the above-mentioned ways using stochastic geometry and functional analysis.

Here is the description of the method mathematical structure with extensive comments where they are necessary.

2. Techniques of 3D image scanning and choice of scanning element

Let F denote a source model of a 3D image. We define plane $B(\eta,r)=\{x\,|\,x^T\cdot\eta=r\}$ as the tangent to the sphere that has the center at the origin of coordinates and radius r, the plane passes through given point X at distance r from the origin of coordinates with given angles ω and φ, where $\eta = [\cos\omega \cdot \sin\varphi, \sin\omega \cdot \sin\varphi, \cos\varphi]$ is the unit vector in R^3, ω is the angle between axis 0x and the projection of interval OX on plane $0xy$, φ is the angle between axis $0z$ and interval OX (fig. 1).

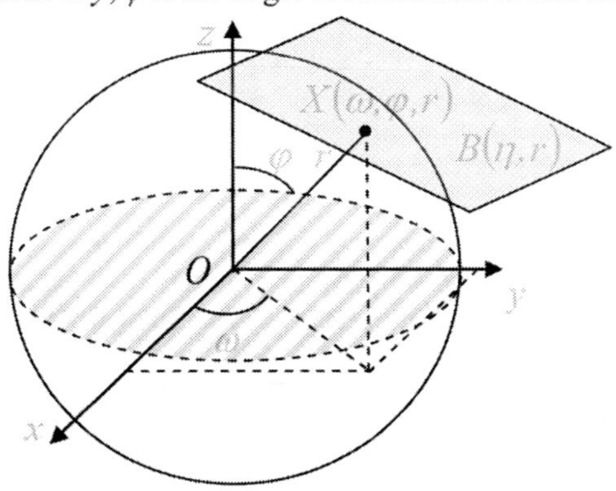

Figure 1. Defining spherical coordinates of the plane

3D images will be scanned by planes. The choice is justified by the following: intersection of a plane with any other extensive geometric primitive is explicit in Higher Mathematics and has strict analytical representation that will be useful for developing the analytic structure of a feature.

Moreover, a grid of parallel planes helps to give invariant description of an object as it provides us with the condition necessary for developing features that are invariant to a group of motions of a 3D image.

One should aim for the situation where all scanning grids of parallel planes at different angles ω and φ of viewing a recognized 3D image will coincide with each other after any its spatial shift or rotation (discrete step disregarded).

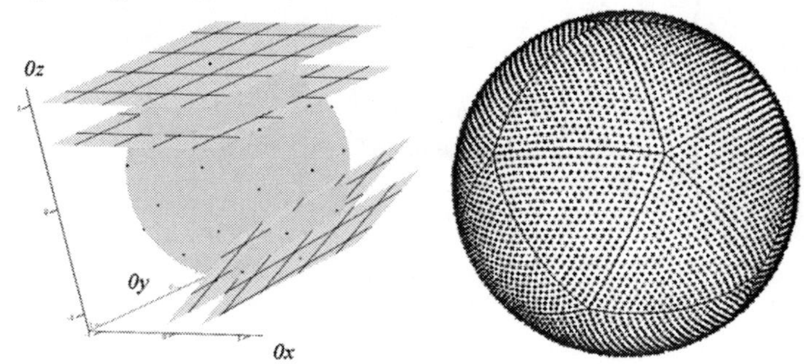

Figure 2. a – a sphere base grid and grids of parallel scanning planes conformed with it; b – an example of a sphere uniform grid

In terms of topology for the continuous case, stereotyped search of all angles ω and φ that identify every scanning grid of parallel planes gives a model of concentric spheres with the center at the origin of coordinates. Every scanning grid of parallel planes on a sphere will be matched to a point that will be a point of tangency with a sphere of the plane that is parallel to planes of the given grid (separately for each pair (ω, φ) of view angles). A set of such points on a sphere will make a grid that will be called a base one (fig. 2a).

It should be noted that a pair of angles (ω, φ) uniquely defines a base grid node that matches up the only tangent plane to the sphere in the point and, therefore, the only scanning grid of parallel planes.

IOP Conf. Series: Materials Science and Engineering **177** (2017) 012047 doi:10.1088/1757-899X/177/1/012047

For a discrete case on a standard global map, one can find a thicker concentration of points near the Pole than near the Equator. So if, while rotating, we superpose the Pole and a point at the Equator, then deviation of points of the source grid and the rotated one will be evident. As every point of a sphere base grid uniquely defines a tilting angle of a planes grid, changing the tilting angle of the planes grid will result in changing the form of sections produced. Thus feature calculation errors will be bigger and the accuracy of a 3D object recognition will be lower.

On the other hand, if, while rotating around the center of the sphere, the base grid coincides with itself, then respective grids of tangent parallel planes will completely coincide with each other and the resulting sections will be the same (will not change their form). Hence the calculated feature value will not change.

Thus one need to create a base grid that will ensure a uniform grid of a sphere in order to diminish errors in convergence of the base grid nodes while rotating it because of discrete step of scanning. Uniform distribution of base grid points on a sphere (fig. 2b) will guarantee the absence of spots with high concentration of base grid nodes on the sphere surface, those that define sections at different viewing angles of the object.

The property of uniform distribution of base grid points on a sphere is the necessary condition of invariance of developed features during rotation.

3. Mathematical model of hypertrace transform

Scanning 3D images is done with a grid of parallel planes having distance Δr between the planes and given angles ω and φ (fig. 3a). Mutual location of 3D image F and every scanning plane $B(\eta(\omega,\varphi),r)$ forms a section (fig. 3b) that is defined by number G being calculated according to a certain rule: *HyperT*: $G = HyperT(F \cap B(\eta(\omega,\varphi),r))$. One can use computations of the perimeter or area of the section, the number of intersection of the plane and the source object, the properties of the neighborhood of the section and so on as such rule. In other words, functional *HyperT* defines the property of a section feature.

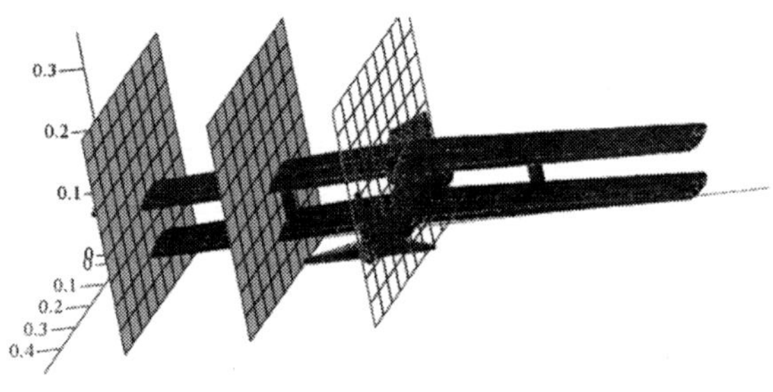

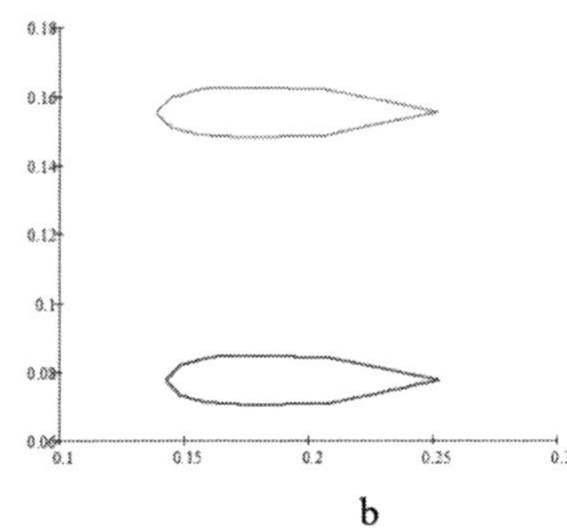

a b

Figure 3. a – Process of 3D object scanning with planes; b – One of the sections

From here on, scanning with a grid of parallel planes is done for a new value of angle $\omega+\Delta\omega$ and $\varphi+\Delta\varphi$ that has received discrete increments $\Delta\omega$ and $\Delta\varphi$, respectively, with the same value Δr between scanning elements of the planes grid. The same earlier chosen rule *HyperT* is used for sections of a new grid of planes $B(\eta(\omega+\Delta\omega,\varphi+\Delta\varphi),r_i)$. It is important to stress that angles are changed not randomly, but in accordance with the construction of the sphere base grid.

A computational result of the *HyperT* functional depends on three parameters of a plane (r, ω, φ). So if any 2D image received after intersecting the source 3D model with a scanning plane, certain informative feature $\Pi(F_{\sec t})$ will be given according to the rule *HyperT*. Then for numerical analysis it is convenient to present the result of the 3D trace-transform as three-dimension hypertrace matrix *3TM*, which has a vertical 0φ axis, a horizontal 0ω axis and a deep-directed $0r$ axis.

For example, every deep-directed row of the matrix contains feature-elements that are calculated using 2D images [9], which resulted from intersecting a source 3D object with scanning planes after direct search of all values of distance variable r with fixed values of angles ω and φ. If plane B does not intersect the 3D image, that is $F \cap B(\eta(\omega,\varphi),r) = \phi$, then the value of a hypertrace functional is assumed to equal zero: $HyperT(F \cap B(\eta(\omega,\varphi),r)) = 0$.

After filling the three-dimension hypertrace matrix, we process its deep-directed rows with the help of the *HyperP* functional, that can be defined, for example, as follows: $Hyper\mathrm{P} = \int G(\omega, \varphi, r)dr$. As a result, the source three-dimension hypertrace matrix evolves into two-dimension matrix *2TM*. Then we process the *2TM* matrix column after column using functional *HyperΩ*, which can be defined, for example, in the following way: $Hyper\Omega = \max_{\varphi} G(\omega,\varphi)$. As a result, we get horizontal row *1TM* (values vector) that has a continuous analog in the form of the 2π-cyclic curve. One can apply the *HyperΘ* functional to the received set of numbers and obtain a certain number – image feature *Res(F)*. The functional can be defined, for example, by the amplitude of the second harmonics of Fourier series.

Hence, the hypertriplet feature of 3D image F has the structure as a combination of four functionals each of which, apart from functional *HyperT*, reduces the matrix *3TM* size by one item if we use them consequently:

$$Res(\mathrm{F}) = Hyper\Theta \circ Hyper\Omega \circ Hyper\mathrm{P} \circ Hyper\mathrm{T}(F_{sect}).$$

Every 2D image received after intersecting the source 3D model with a grid of parallel planes at different viewing angles needs scanning to elicit some relevant features (for example, the perimeter of a section figure). To define features of a two-dimension image of a section, we use 2D trace-transform [10].

The triplet feature of 2D image *Fsect* has the structure as a composition of three functionals each of which, apart from functional T, reduces the matrix *TM* size by one item if we use them consequently [10]:

$$\Pi(\mathrm{F}_{sect}) = Hyper\mathrm{T}(\mathrm{F}_{sect}) = \Theta \circ \mathrm{P} \circ \mathrm{T}(\mathrm{F}_{sect} \cap l(\theta,\rho)).$$

Combining the formulas for *Res(F)* and $\Pi(F_{sect})$, we finally receive the following analytic structure of a feature of a 3D image in the form of many functionals composition:

$$Res(\mathrm{F}) = Hyper\Theta \circ Hyper\Omega \circ Hyper\mathrm{P} \circ Hyper\mathrm{T}(\Theta \circ \mathrm{P} \circ \mathrm{T}(\mathrm{F}_{sect} \cap l(\theta,\rho))).$$

Due to the composite structure of functionals, presenting the analytic structure of features $\Pi(F_{sect})$ and *Res(F)* respectively, it is possible to receive an abundance of features. Moreover, some features have distinct geometric interpretation, which facilitates developing features and increases their discriminative power. Particular qualities of hypertriplet and triplet features allow creating features both sensitive and invariant to group of motions and scaling, which results in better intellectuality and flexibility of the 3D trace-method used for objects recognition.

A more detailed description of a stochastic scanning technique, its merits and influences on a feature development can be found in [10].

4. Conclusion

The method under discussion is rather versatile as the scanning scheme is not stipulated by geometric parameters of a source model. So the proposed technics can deal with objects of any layout or degree of complexity, which increases intellectual capacity of 3D transform.

The described method affords to give an invariant description of a source 3D object under any translation or rotation motions and homothetic transformation. Furthermore, the developed hypertriplet features are strongly resistant to coordinate noise and linear distortions. They also possess distinct geometric interpretation facilitating intellectual capabilities for analysis.

Effective intellectual analysis of hyper-trace transform is stipulated by the ability to control properties of the developed features. Moreover, sensitive features allow receiving a lot of additional information about the object using the same scanning technique (for example, information about the object transformations), which is essential for solving a certain class of practical problems (in robotic technology and the like).

Another intellectual peculiarity of the proposed method is a high level of pre-processing, processing and after-processing of a 3D image using the same scanning technique.

The authors plan to improve the method in such way that it could be applied for analyzing not only binary and monochrome 3D images, but also those having color and texture. Similar results have already been obtained while analyzing 2D images having color and texture in [10]. We expect the intellectual level of hypertrace transforms to be higher after our adding affine-invariant transforms to the theory.

When introducing a small amount of the aluminium oxide powder, the distance between the axes of the second-order dendrites and the average grain size reduces. This implies that a considerable part of the powder particles is effective crystallization centres. When increasing the nanopowder content, the structure starts coarsening relatively that which was obtained using low powder concentrations. The introduction of the large quantity of the modifier leads to its coagulation and reduction of its influence on the structure.

References

[1] Kiy K I 2015 Pattern Recognition and Image Analysis. *Advances in Mathematical Theory and Applications* **25(2)** 338-346

[2] Myasnikov V V 2016 Pattern Recognition and Image Analysis. *Advances in Mathematical Theory and Applications* **26(1)** 22-33

[3] Vasil'ev K K, Dement'ev V E and Andriyanov N A 2016 Pattern Recognition and Image Analysis, *Advances in Mathematical Theory and Applications* **26(1)** 240-247

[4] Song S and Xiao J 2014 *Computer vision* 8694 634-651

[5] Wang C and Huang K-Q 2015 *Journal of Computer Science and Technology* **30(2)** 325-339

[6] Zhang Y, Song S, Tan P and Xiao J 2014 *Computer vision* **8694** 668-686

[7] Andreux M, Rodolà E, Aubry M and Cremers D 2014 *Computer vision* **8928** 299-312

[8] Lmaati E A, Oirrak A E, Kaddioui M N, Ouahman A A and Sadgal M 2010 *International Arab Journal of Information Technology* **7(3)** 264-270

[9] Fedotov N G, Ryndina S V and Semov A A 2014 Pattern Recognition and Image Analysis. *Advances in Mathematical Theory and Applications* **24(4)** 566-574

[10] Fedotov N G 2009 The theory of pattern recognition features based on stochastic geometry and functional analysis, FIZMATLIT **304**

IOP Conf. Series: Materials Science and Engineering **177** (2017) 012048 doi:10.1088/1757-899X/177/1/012048

Hardware-software complex of informing passengers of forecasted route transport arrival at stop

V Yu Pogrebnoy[1], M I Pushkarev[1], A S Fadeev[1]

[1]Tomsk Polytechnic University, 30, Lenina Ave., Tomsk, 634050, Russia

E-mail: pogrebnoyvadim@tpu.ru

Abstract. The paper presents the hardware-software complex of informing the passengers of the forecasted route transport arrival. A client-server architecture of the forecasting information system is represented and an electronic information board prototype is described. The scheme of information transfer and processing, starting with receiving navigating telemetric data from a transport vehicle and up to the time of passenger public transport arrival at the stop, as well as representation of the information on the electronic board is illustrated and described. Methods and algorithms of determination of the transport vehicle current location in the city route network are considered in detail. The description of the proposed forecasting model of transport vehicle arrival time at the stop is given. The obtained result is applied in Tomsk for forecasting and displaying the arrival time information at the stops.

1. Introduction

The rapid development of new technologies makes them an essential element of different fields of activity of modern society beginning from the primary education and ending with the investigation and solution of complex problems in different fields of science.

Thus, for instance, as of today, the first light-emitting diode, proposed in the 1960-s and emitting only incandescence, has a vast colour spectrum and is applied in high-brightness searchlights, car headlamps, flexible self-adhesive ribbons and domestic and industrial lamps.

A separate class of devices is the systems of visualization of the text and graphic information based on light-emitting diode matrices. One of the most widely-spread devices realised on the basis of light-emitting diodes is light-emitting diode information boards. Light-emitting diode boards have found their application in many branches of human life activity:

- advertising light-emitting boards and screens;
- a transport branch: visualization of the transport timetable and electronic boards of arrival;
- a banking sector: rates of exchange and shares;
- establishment of the service sector: display of the information and warnings sent by the Russian Ministry for Emergency Situations, the composition and pollution of the air, the levels of the background radiation, different information on an institute (reception hours of specialists, service cost, news, informers of electronic queues), the information from administrative authorities, dates and times;
- industrial enterprises (notification of the operating personnel).

Thus, light-emitting diode boards are the modern tool of notification of the population and visualisation of information, which can be used in different fields and spheres of human life activity.

In this paper, the application of the light-emitting diode information board at the stops of the public transport, which represents a client side of the automated distributed interactive system intended for informing passengers of the time of passenger public transport arrival at the stop, is considered.

2. Materials and methods

The hardware-software complex of informing the passengers of the passenger traffic consists of:

1) a server side including databases and three basic modules: a module of recording the statistics of traffic, a module of intelligent forecasting and a module of interaction with clients;
2) a client side representing light-emitting diode boards installed at the passenger transport stops or mobile and web-based applications.

The database of the informational system stores the history of movement of transport vehicles (TV), the statistics of transport passage of checkpoints, updated and historical data on the intervals of movement between the checkpoints and on the temporal delays at checkpoints. Storage of the updated data allows representing the current situation on roads, and the storage of historical data — the average statistical information on the traffic in different sections in different periods of time.

The structure of the information system is presented in fig. 1

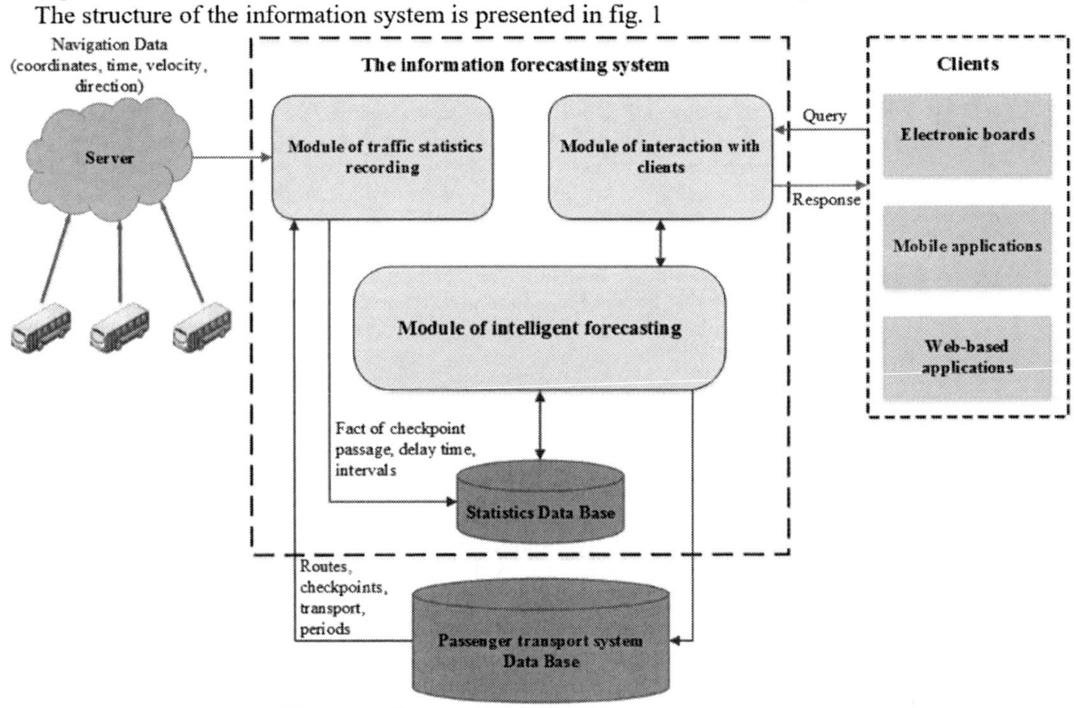

Figure 1. The structure of the forecasting system

The module of recording the traffic statistics performs the function of downloading updated and historical data, obtained with the use of the algorithm of traffic parameters recording, to the databases. At regular intervals, the module receives the latest navigating telemetric data (NTD) of TV and performs the algorithm of recording cyclically. The operational algorithm of this module controls the finite automate of determination of directions in the route, constantly changing TV states in the system. The module records the moments of checkpoints passage by the transport, carries out calculations of updated and average intervals of traffic, as well as updated and average delay periods at checkpoints.

The module of interaction with clients expects the queries from the informers and other systems, receives the necessary data from the module of intelligent forecasting and provides them according to a certain protocol in the required format to the requesting object.

The module of intelligent forecasting analyses the information, stored in the database, in order to forecast the traffic parameters. In the system, the algorithm of forecasting of the time of transport arrival at the object of control by different routes is applied.

3. The model and the algorithm of forecasting

To forecast the time of the transport arrival at the stops, a model based on the historical data has been developed. The model takes into consideration both average values of the parameters of the traffic in separate sections of the route and the latest parameters of the traffic observed on a real-time basis in the same sections of the route. At the same time, the influence of the average values of the parameters has a direct dependence on the TV remoteness from the stop for which it is necessary to forecast the arrival time, and the influence of the latest values has an inverse dependence. Additional storage of the updated values of traffic parameters allows considering the real traffic situation. The algorithm of this model is carried out within the module of intelligent forecasting on a real time basis using the statistics which is accumulated by the module of recording traffic statistics.

A light-emitting diode information board is connected to a information telematic server of the forecasting system of the motor transport enterprise with the use of the Internet. The access to the Internet is realised through the network of cellular communication operators with the use of GSM-protocols. The informational telematic server forms the final information, displayed on the light-emitting diode board, and controls the board operation remotely.

4. The client side

On the basis of the review of the existing analogues of light-emitting diode matrix indicators, a number of drawbacks have been revealed:

1. low reliability of the systems using the built-in computers with operating systems inappropriate for the functioning on a real time basis (Linux);
2. high requirements for the computational capabilities of the systems developed on the basis of the high-level languages, such as Python and C#;
3. substantially bounded functional of devices using standard video processors for advertising light-emitting diode boards;
4. limited temperature conditions and climate designs of the ready-made devices;
5. the necessity of connecting to the Internet and others.

In connection with this, the hardware component of the information board and its control were realised by means of the controller 'Atmega 328p', for which the original software was developed. The developed software (SW) provides the realisation of the following functions:

- installation of the TCP-session intended for the information exchange with the information telematic server;
- formation of queries in accord with the unique protocol to the server of the forecast service;
- decoding of the information, obtained from the server, and formation of virtual pages of the text on its basis;
- alternate displaying of the virtual pages of the text on the light-emitting diode matrix;
- calculation of the arrival time of each tram, trolleybus and bus on the basis of the values of the current time and the information received from the server of the forecast system;
- exclusion of the strings from the readout for those transport vehicles that have already arrived at the current stop.

The original software for the microcontroller 'Atmega 328p', installed in the board, possesses a number of functional peculiarities:

- updated information output about the numbers of routes, destinations and the time of the transport arrival with an opportunity of static displaying of the part of the text at the beginning (a number of the route) and at the end of the string (the time before arrival) and an opportunity of horizontal scrolling of the part of the text in the centre of the string (the name of the last stop);

– information output in the page mode with alternation of all routes, the data of which have been received from the server; automatic formation of pages from the number of strings being in the memory;
– automatic reduction of the value of the time remained before the arrival of a bus, a trolleybus, a tram of each route in the absence of the correcting information from the server;
– automatic exclusion of the displaying of the string, the time value of which has reduced to zero, from the memory and from pages;
– the output of the information on the current time and the temperature in the place of the board installation;
– an opportunity of the use of the entire information field of the board for displaying the coherent text formed on the server side or the creeping lines by the instruction of the server for the possibility to display updated and advertising information;
– an opportunity of changing the speed of scrolling the creeping lines, the speed of changing the information pages, the brightness of diode light emission;
– displaying of the preprogrammed text information in the absence of the opportunity to setup a connection with the server during a certain time interval.

5. The server side

In the sphere of passenger transportation, the application of the systems of automatic locate of TV has become widespread, in particular, of global navigation satellite systems (GLONASS and GPS) [1, 2]. Owing to the use of these systems, there is an opportunity to trace the location of TV in real time and to forecast their further movement, which is a base element of intelligent transport systems (ITS) [3, 4]. The application of NTD, the information on the route combined with intelligent algorithms allows the most accurate forecast of the passage transport arrival time at the required point [3, 5].

The majority of the existing models of forecasting use GPS-statistics, which represents the history of the parameters of TV movement, obtained by the method of the periodic survey of TV, as initial data. These data are interpreted differently in each of the models and are transformed into the input parameters of this model. The output parameters of the model are the forecast times of TV arrival at subsequent route stops [6-11].

6. Designing the forecasting system

The task of the system is to forecast the time of TV arrival at the stop with the use of intelligent models and algorithms of forecasting.

The objects of control of the forecast system are passenger TV: trolleybuses, trams, and buses. The majority of units of the rolling stock of the passenger transport is equipped with navigating-communication equipment GLONASS or GLONASS/GPS, providing determination of the current location and parameters of TV motion.

NTD, obtained with the use of the global navigating systems, do not depend on the network of streets, contain errors, and there is no unique method of coordinates approximation and recording of the parameters of TV motion [3]. A new method of location record and accumulating of the statistics of values of TV motion parameters is proposed. For this purpose, the stops coordinates, the radius or ground, occupied by them, are integrated into the unified system, which allows determining the moment of TV arrival at the stop and the instant of its departure from it. The formed structures of the route in the form of sequence of its stops provide the tracing of the TV movement according to the plan of a specific route.

Let us determine the sought location uniquely by route section s_j with order number j, connecting p_j and p_{j+1} nodes and with distance λ, covered from the beginning of the section. Let us consider three cases of coordinates' location V relatively the perpendiculars drawn through the nodes having coordinates O_j^p (Figure 2a).

Let us call the calculation of λ according to formulae 1 and 2 as approximation of the V point by section s_j:

$$h=\begin{cases}\left|\overrightarrow{O_j^pV}\right|, & \left(\overrightarrow{O_j^pO_{j+1}^p},\overrightarrow{O_j^pV}\right)\le 0\\ \left|\overrightarrow{O_{j+1}^pV}\right|, & \left(\overrightarrow{O_{j+1}^pO_j^p},\overrightarrow{O_{j+1}^pV}\right)\le 0\\ \dfrac{\left|\left[\overrightarrow{O_j^pO_{j+1}^p},\overrightarrow{O_j^pV}\right]\right|}{\left|\overrightarrow{O_j^pV}\right|}, & \text{otherwise}\end{cases},\qquad (1)$$

$$\lambda=\begin{cases}0, & \left(\overrightarrow{O_j^pO_{j+1}^p},\overrightarrow{O_j^pV}\right)\le 0\\ \left|\overrightarrow{O_j^pO_{j+1}^p}\right|, & \left(\overrightarrow{O_{j+1}^pO_j^p},\overrightarrow{O_{j+1}^pV}\right)\le 0\\ \sqrt{\left|\overrightarrow{O_j^pV}\right|^2-h^2}, & \text{otherwise}\end{cases},\qquad (2)$$

where h – remoteness of TV from the section.

Let us determine maximum possible remoteness of TV from route $\tilde{h}$ and function $f_n:(S,V)\mapsto(s_\zeta,\lambda)$, approximating coordinates V by section $s_\zeta=\arg\min\limits_{s_j\in S}(h)$, at that $h>\tilde{h}\Rightarrow\nexists s_\zeta$. Let us call function f_n as approximation of point V by the multitude of sections S. Figure 2b presents a graphic interpretation of approximation of several points by a piecewise-linear diagram presenting some multitude of sections S.

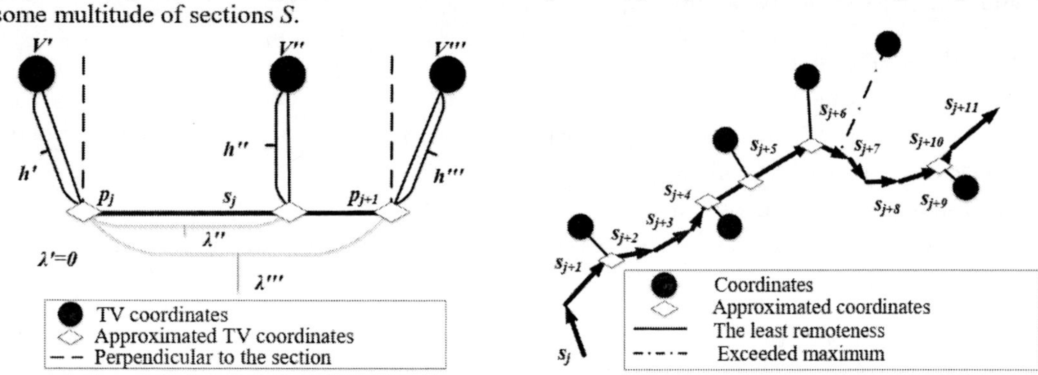

a) b)

Figure 2. Determination of TV location: a – in the section; b – on the route

For determination of multitude S and, correspondingly, for solution of the set problem, a method based on the finite-state machine has been developed [3]. TV can be in ten states: q_0 – location is unknown; q_1 – the route is determined; q_2 – the direction of the route is determined; q_3 – TV turned out to be in the initial node of the section; q_4 – TV turned out to be between the initial and final node; q_5 – TV turned out to be in the final node; q_6 – TV arrived at the initial node; q_7 – TV departed from the initial node; q_8 – TV arrived at the final node; q_9 – TV completed the route.

Transition $q_i \to q_0$ means the absence of NTD or leaving the route. Transitions $q_{i=\overline{3,8}} \to q_{j=\overline{3,5}}$, named as skips, appear in case of passage of nodes and sections. In figure 3, there is a scheme of the finite-state machine without taking into consideration unexpected situations and skips.

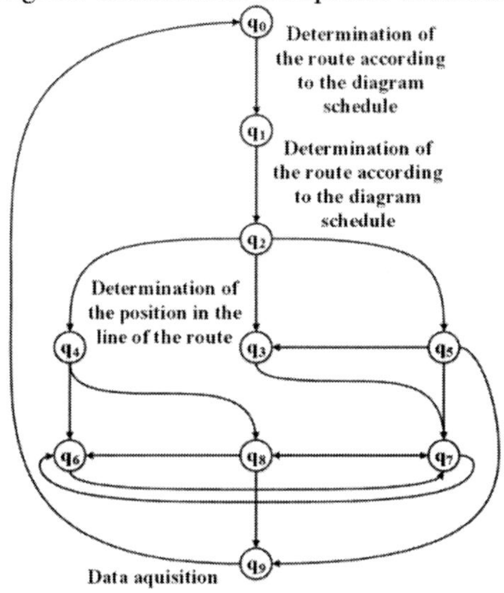

Figure 3. The scheme of the finite-state machine

Transition $q_0 \to q_1$ appears after determination of the route according to the current TV timetable. For transition $q_1 \to q_2$, the method, based on the tracks of TV movement, is constructed in each direction of the route, and based on the choice of the optimum track by the length and a minimal departure from the TV schedule.

7. Parameters recording

Let us determine the moment of the state change as t^{qi}. The interval of movement along the section is calculated during transition $q_7 \to q_8$: $i = t^{q8} - t^{q7}$. The delay time in the node is calculated during transitions $q_6 \to q_7$, $q_8 \to q_7$: $dt = t^{q7} - t^{q6,q8}$.

For determining $t^{q6,q8}$ and t^{q7}, it is necessary to define the location of TV: within the node or outside it. If the node is represented by a circumference with radius ρ, the fulfilment of inequality $\left(x^V - x^O\right)^2 + \left(y^V - y^O\right)^2 \le \rho^2$ is checked; if by a polygon, the method of ray tracing is applied.

Values i and dt are random only in the context of certain conditions: the type and the technical state of TV, the route, the driver's behaviour, the road condition, weather conditions, etc. To decrease the number of criteria limiting the randomness of the specified values, let us distribute the values throughout the sections and nodes of the routes, as well as by the seasons of the year, by weeks and hours.

The obtained random values have normal distribution [4], and there is a possibility of calculation of their statistic parameters:

$$\mathrm{M}_{\eta+1}\left[Z^m\right] = \frac{\eta * \mathrm{M}_\eta\left[Z^m\right] + Z^m_{\eta+1}}{\eta+1}, \tag{3}$$

where Z^m – the random magnitude value of period m; η – the number of the obtained earlier values of random magnitude value Z^m; $M_\eta\left[Z^m\right]$ – mathematical expectation of magnitude Z^m by η values; Z^m_η – the η-th value of random magnitude value Z^m

$$D_{\eta+1}\left[Z^m\right]=\frac{\eta}{\eta+1}\left(D_\eta\left[Z^m\right]+\frac{\left(M_\eta\left[Z^m\right]-Z^m_{\eta+1}\right)^2}{\eta+1}\right), \quad (4)$$

where $D_\eta\left[Z^m\right]$ - the variance of random magnitude value Z^m by η values.

8. Conclusion

On the basis of the problem analysis of the passenger transportation system functioning in Tomsk city, new model and algorithmic solutions of forecasting the transport vehicle time arrival at the stop are proposed in the article. Original methods and algorithms of the transport vehicle current location determination in the city route network, methods of statistics accumulation of transport vehicle traffic parameters on the basis of navigating telemetric data, obtained from global navigation satellite systems (GLONASS and GPS), are considered in detail. The prototype of the electronic information board for public transport stops that allows displaying forecasting data of arrival time of the transport vehicle is proposed. The protocol of interconnection of the forecasting server with electronic information board is represented. Obtained results have been successfully adapted to the intelligent route network of Tomsk city. Electronic information boards, displaying the arrival time of the forecasted transport vehicle on the basis of described methods and algorithms, are set and operate at the public transport stops.

References

[1] Pogrebnoy V Y, Fadeev A S 2013 *Track and passenger transport service* **12** 78-83
[2] Pogrebnoy V Y, Fadeev A S 2013 *Naukovedenie* **6(19)** 100TVN613
[3] Pogrebnoy V Y 2015 *Herald of computer and information technologies. Scientific, technical and production monthly journal* **8** 10-17
[4] Markov N G 2011 *Intelligent navigating telecommunication control systems of mobile objects using cloud computing technologies* (Moscow: Goriachaja linija - Telecom) p 158
[5] Markov N G, Gazizov T T, Leshik Y V et al. 2013 Proceedings of TUSUR **4(30)** 182-187
[6] Jeong R H, 2004 The prediction of bus arrival time using automatic vehicle location systems data p 303
[7] Dihua Sun 2007 *Transportation Research Record: Journal of the Transportation Research Board* **2034** 62-72
[8] Seema S R 2009 *10th National Conference on Technological Trends (NCTT09)* p 193-197
[9] Seyed Mojtaba Tafaghod Sadat Zadeh 2012 *J. of Theoretical and Applied Inf. Techn.* 516-525.
[10] Wei-Hua Lin 2004 Modeling Schedule Recovery Processes in Transit Operations for Bus Arrival Time Prediction *Department of Systems and Industrial Engineering of the University of Arizona* p 23
[11] Wei-Hua Lin. 2004 An experimental study on real time bus arrival time prediction with GPS data *Center for Transportation Research And Department of Civil and Environmental Eng.* p 15

Features of control systems analysis with discrete control devices using mathematical packages

E M Yakovleva[1], V A Faerman[2]

Tomsk Polytechnic University, Tomsk, Russian Federation

E-mail: [1]yakovlevaem@tpu.ru, [2]vaf@tpu.ru

Abstract. The article contains presentation of basic provisions of the theory of automatic pulse control systems as well as methods of analysis of such systems using the mathematical software widespread in the academic environment. The pulse systems under research are considered as analogues systems interacting among themselves, including sensors, amplifiers, controlled objects, and discrete parts. To describe such systems, one uses a mathematical apparatus of difference equations as well as discrete transfer functions. To obtain a transfer function of the open-loop system, being important from the point of view of the analysis of control systems, one uses mathematical packages Mathcad and Matlab. Despite identity of the obtained result, the way of its achievement from the point of view of user's action is various for the specified means. In particular, Matlab uses a structural model of the control system while Mathcad allows only execution of a chain of operator transforms. It is worth noting that distinctions taking place allow considering transformation of signals during interaction of the linear and continuous parts of the control system from different sides. The latter can be used in an educational process for the best assimilation of the course of the control system theory by students.

1. Introduction

Development of the computer technique and hardware of digital processing of signals which has been observed in recent years led to widespread use of digital microprocessor devices as the part of systems of automatic control of technological processors [1]. However, despite advantages of the use of digital devices in the control systems: low cost, flexibility of solvable tasks, possibility of implementation of complex corrective dynamic elements [2], there are complexities in analysis and simulation of a similar class of systems [3].

Specifics of digital control inevitably lead to the fact that a controlling signal on the controller output exists only in separate moments of time, that is, it has an impulse character [4]. The latter leads to appearance of additional restrictions which are superimposed on parameters of the control system and which non-compliance can lead to loss of stability [5]. At the same time, parameters of a control sequence of pulses have a significant effect on control quality.

In this connection, analysis of pulse control systems (PCS) is the most important task of the modern theory of control, and the ability to reproduce it in practice is a necessary competence of experts in the field of automation [6]. Despite the fact that this task is studied in detail and, in general, is solved [7], practical analysis of PCS still represents complexity for students as it requires understanding of transformations taking place in the control circuit as well as application of specialized software [6].

Such application is difficult in the context of work with the mathematical apparatus describing PCS. The classical method of description of the considered class of systems is compilation and solution of difference equations [8] that is in many respects similar to description of the continuous systems by means of differential equations which in many respects are similar to classical methods of solution of differential equations. In spite of the fact that the solution of state equations (both analytical, and numerical ones) does not represent complexity, it cannot find wide practical application [6]. What is explained by the difficulties connected with obtaining difference equations from differential equations.

This work is directed towards selection of basic theoretical data and working methods with mathematical software Matlab and Mathcad widespread in the academic environment, which are necessary and sufficient for formation of practically significant competences of the analysis of elementary PCS during the study. Results of this work can be used, first of all, in an educational process.

2. Pulse control systems

The main difference between PCS and continuous systems is the passage of the signal on a circuit only in certain discrete moments of time. Therefore, the continuous part reacts only to the discrete values of the continuous signal in the moments of quantization $n \cdot T$ [4, 6, 9]. As a result, to describe PCS one uses the mathematical apparatus of lattice functions $x(nT)$ (where n is integer) instead of continuous functions $x(t)$ [11]. At that, if quantization interval T is set, then the lattice function is uniquely determined with function $x(t)$.

The ratio between lattice function x (n) and its difference $\Delta^k x(n)$ defines the difference equation [6]. If this ratio is linear, then the difference equation is called the linear one. It can be presented in a general view as [6]:

$$a_k \cdot \Delta^k(n) + a_{k-1} \cdot \Delta^{k-1}(n) + \cdots + a_o \cdot q(n) = f(n), \tag{1}$$

or, expressing differences directly through the required functions,

$$b_k \cdot q(n+k) + b_{k-1} \cdot q(n+k-1) + \cdots + b_0 \cdot q(n) = f(n), \tag{2}$$

where $f(n)$ is the known lattice function, $q(n)$ is the required lattice function representing the solution of the difference equation. This difference equation, containing $q(n)$ and $q(n+k)$ in an explicit form, is called the difference equation of the kth order.

Systems of the difference equations find wide application for description of PCS and serve as an analog of the systems of the differential equations describing the state of the continuous systems. [8].

The typical and the most distributed in practice task of analysis of PCS is the task of analysis of single-circuit linear PCS which flow diagram is presented in figure 1 [4, 6].

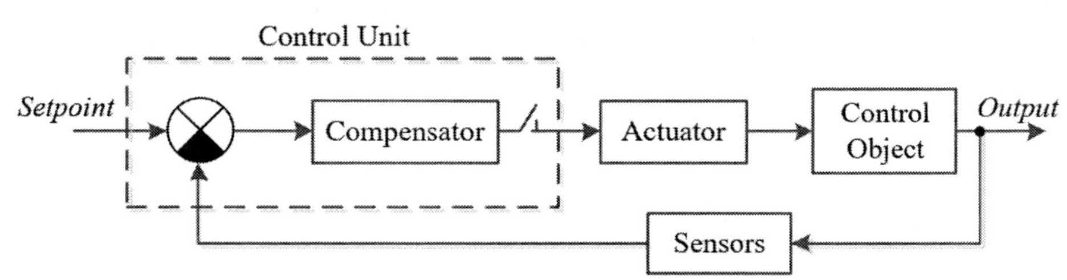

Figure 1. Single-circuit linear pulse system.

Single-circuit PCS can be presented as a combination of digital and continuous subsystems interacting with each other. The continuous part contains a control object as well as intensifying and executive devices [4,6]. The digital part, as a rule, is the control device and integrates the functional elements participating in impulse transformation of a signal.

In the course of transformation of the continuous signal into the discrete one, impulse element executes two operations – quantization on time and pulse modulation [4, 6]. The first one is that the output signal of the impulse element appears in certain discrete, usually equidistant moments $t_i = iT$. As a result [4,6] of impulse modulation, any parameter of the output pulse changes depending on the value of the incoming signal. Depending on the changeable parameter of the pulse [6], one distinguishes different types of modulation: pulse width modulation (PWM), amplitude-pulse modulation and others. The simplest one, from the point of view of implementation, and therefore widespread is PWM.

3. Analysis of impulse control systems

To research linear PCS [4], one uses usually transfer functions (TF) which are a ratio of operator images of output and input signals of the system. As signals exist in PCS in moments of time nT, so they are discrete, TF are called impulse ones and are defined as [4,10]:

$$W(z) = \frac{X(z)}{G(z)}. \quad (3)$$

For lattice functions, it is possible to apply discrete Laplace transform [4, 6]:

$$F(p) = \sum_{i=0}^{\infty} f(i) \cdot e^{-piT}. \quad (4)$$

In case if (4) is converged when $Re(p) < \infty$, original $f(t)$ corresponds to some image $F(p)$. We understand z-transform [1, 4, 6] as an image of the function defined as follows:

$$F(z) = \sum_{i=0}^{\infty} f(i) \cdot z^{-i}. \quad (5)$$

The image exists and corresponds to the original if row (5) converges. Discrete transforms are connected by relation $e^{pT} = z$. The formula of z-transform can be written also for the continuous function in the form of [1, 4, 6]:

$$F(z) = \sum_{i=0}^{\infty} f(t_i) \cdot z^{-i}, \; t_i = iT. \quad (6)$$

According to (5) and (6), transitional function $w(t)$ [4] has to serve as original $W(z)$:

$$W(z) = \sum_{i=0}^{\infty} w(iT) \cdot z^{-i}. \quad (7)$$

3.1. Obtainment of impulse transfer function

Let us consider the single-circuit PCS presented in figure 2 [4, 6]. The impulse element in the system is located after comparing the added device. A continuous part contains strengthening devices and the control object. Correcting devices are connected in the feedback circuit.

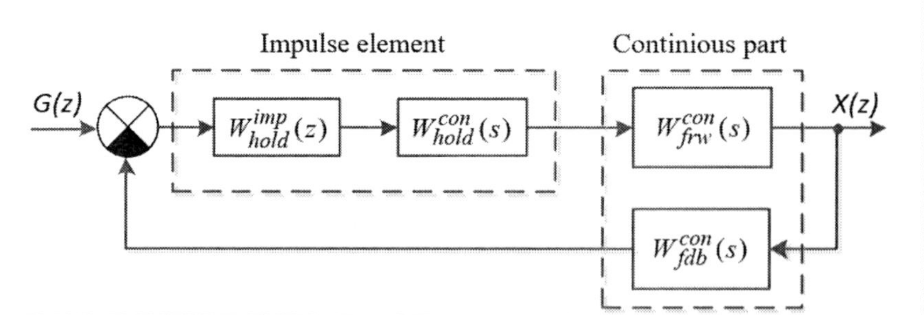

Figure 2. Single-circuit linear impulse system.

To estimate stability of PCS and to determine its quality measures, a TF for the open-loop system should be used. The output signal of continuous system $y(t)$ under zero initial conditions is defined as convolution of the input $x(t)$ signal of a preset value and weight function $w(t)$:

$$y(t) = \int_0^t w(t-\tau)x(\tau)d\tau. \tag{8}$$

The weight function is the original of TF of continuous system $W_{loop}(s)$ so it is connected with this system with inverse Laplace transform. Thus, to pass from continuous Laplace transform to discrete transform, it is enough to do the following actions [6]:

1. To define TF of the continuous part of open system $W_{loop}(s)$, taking into account zero-order hold:

$$W_{loop}(s) = W_{hold}^{con}(s) \cdot W_{frw}^{con}(s) \cdot W_{fdb}^{con}(s). \tag{9}$$

2. With the help of inverse Laplace transform (according to (9)), it is possible to find impulse TF of the open-loop system:

$$w(t) = L_{-1}\{W_{\text{loop}}(s)\}. \tag{10}$$

3. To determine the weight sequence of the system as:

$$w(nT) = w(t)|_{t=Tn}. \tag{11}$$

4. To apply direct discrete Laplace transform:

$$W^0{}_{\text{loop}}(z) = Z\{w(nT)\}. \tag{12}$$

5. To obtain TF of the PCS to multiply the increase of expression (12) on $W_{hold}^{con}(z)$:

$$W_{\text{loop}}(z) = W^0{}_{\text{loop}}(z) \cdot W_{hold}^{imp}(z). \tag{13}$$

3.2. Obtaining transfer function in Matlab

Matlab is the software, being widespread in the academic environment and finding its application when solving automatic control tasks [12,13]. It is explained, on the one hand, by a big set of the modules and instruments focused on the solution of various tasks of the analysis and synthesis of control systems, and on the other hand – by convenience of application of a graphical environment of imitating Simulink modeling for solution of these problems [7].

As example, let us consider the application of Matlab to obtain discrete TF of the linear control system presented in figure 3. Values of parameters of TF of the elements which are parts of the system are given in table 1.

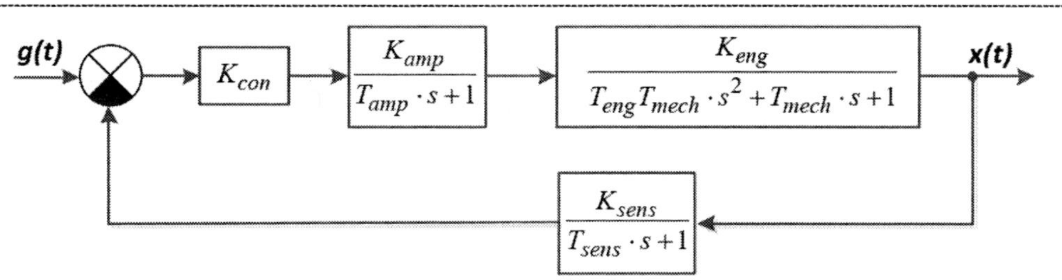

Figure 3. Block diagram of control system.

As it was noted earlier, an essential analysis stage is obtaining of impulse TF of open-loop system $W_{loop}(z)$. For this purpose, one can use the Control System Toolbox module containing a necessary set of functions.

Table 1. Values of parameters of control system.

Parameter	Value
K_{con}	20.00
K_{amp}	18.75
K_{eng}	1.60
K_{sens}	0.02
T_{amp}	0.06
T_{eng}	0.02
T_{mech}	0.30
T_{sens}	0.01
T_{samp}	0.157

The solution of the set problem is reduced to running the following program:

```
>>Kcon=20;Kamp=18.75;Keng=1.6;Ksens=0.02;Tamp=0.06;
  Teng=0.02;Tmech=0.3;Tdc=0.01;
>>  Wcon=tf([Kcon]); Wd=tf([Keng],[Teng*Tmech Tmech 1]);
    Wamp=tf([Kamp],[Tamp 1]); Wsens=tf([Ksens],[Tsens 1]);
>>Wloop=Wcon*Wamp*Weng*Wsens

Transfer function:
                         12
---------------------------------------------------
3.6e-006 s^4 + 0.0006 s^3 + 0.0276 s^2 + 0.37 s + 1

>> Wzloop=c2d(Wloop,0.157)

Transfer function:
   2.665 z^3 + 2.087 z^2 + 0.0386 z + 2.198e-006
--------------------------------------------------------
z^4 - 0.6428 z^3 + 0.042 z^2 - 2.848e-005 z + 4.325e-012

Sampling time: 0.157
```

It is worth noting that when allowing one to work with control systems models at the higher level, Matlab conceals the mathematical aspect of the made transformations, which understanding is essential for students, and for users.

3.3. Obtaining transfer function in Mathcad

To study the mathematical aspect of the problem, the use of Mathcad environment, which allows carrying out symbolical transformations and has a visual graphical interface, seems preferable [14]. The program fragment in Mathcad solving the set problem is presented further.

$K_{con} := 20 \quad K_{amp} := 18.75 \quad K_{eng} := 1.6 \quad K_{sens} := 0.02$

$T_{amp} := 0.06 \quad T_{eng} := 0.02 \quad T_{mech} := 0.3 \quad T_{sens} := 0.01 \quad T_{samp} := 0.157$

$$W_{amp}(s) := \frac{K_{amp}}{T_{amp} \cdot s + 1} \qquad W_{eng}(s) := \frac{K_{eng}}{T_{eng} \cdot T_{mech} \cdot s^2 + T_{mech} \cdot s + 1}$$

$$W_{imp}(z) := \frac{z-1}{z} \quad W_{sens}(s) := \frac{K_{sens}}{T_{sens} \cdot s + 1}$$

$$W_{loop}(s) := W_{amp}(s) \cdot W_{eng}(s) \cdot W_{sens}(s) \quad w(t) := W_{loop}(s) \text{ invlaplace}$$

$$w(n) := w(t) \text{ substitute}, t = T \cdot n \quad W_{tmp}(z) := w(n) \text{ ztrans}$$

$$W_{zloop}(z) := W_{tmp}(z) \cdot W_{imp}(z) \text{ simplify}$$

$$W_{zloop}(z) = \frac{2.665 \cdot z^3 + 2.087 \cdot z^2 + 0.039 \cdot z + 2.202 \cdot 10^{-6}}{z^4 - 0.643 \cdot z^3 + 0.042 \cdot z^2 - 2.848 \cdot 10^{-5} + 4.337 \cdot 10^{-12}}$$

4. Conclusion

The present work provides a summary of some provisions of control theory and of basic control systems of research methods using mathematical packages Mathcad and Matlab, being necessary and sufficient to solve tasks of the analysis of the simplest single-circuit PCS.

Simultaneous application of various software for research allows one to consider interaction of discrete and continuous parts in PCS from different viewpoints and, thus, to form corresponding competences during the study. Besides, the use of special instruments to solve typical tasks of automation promotes acquisition of skills in software automating engineering calculations by students.

Importance of the PCS analysis is caused by wide distribution of digital controllers. The task of the analysis of single-circuit linear PCS considered in the work, in spite of triviality, is important from the practical point of view as typical tasks of digital control of technological objects can be reduced to it.

The materials provided in this work, as well as the method based on parallel presentation of theoretical data, analytical calculations, numerical and structural models were used during composition of the education guidance for the discipline "Automation of design of control systems and means" for the master's degree program of Tomsk Polytechnic University.

References

[1] Xie L B, Shieh L S, Tsai J S H, Wu C Y and Park J 2016 Digital and analog integer delayed modeling and control for multivariable systems with multiple time delays in states, inputs and outputs *Appl. Math. Model.* **40(2)** 1615-1633

[2] Ullah S and Liquat M 2015 *Proc. In. Conf. on Control, Electronics, Renewable Energy and Communications* (IEEE: USA) 7337059

[3] Pushkarev M I, Efimov S V, Gayvoronsky S A and Chenkova A O 2016 Single-loop DC motor control system design with a design of a desired aperiodic degree of stability *IOP C Ser. Mater. S. Eng.* **124(1)** 012026

[4] Chen T and Francis B A 1991 Input-output stability of sampled-data systems *IEEE T Automat. Contr.* **36(1)** 50-58

[5] Liu Xm Marquez H J and Lin Y 2008 Input-to-state stabilization for nonlinear dual-rate sampled data systems via approximate discrete-time model *Automatica* **44(12)** 3157-3161

[6] Faerman V A and Yakovleva E M 2011 Obtaining discrete transfer function of control system using mathcad *Vestnik nauki Sibiri* **1** 287-294

[7] Efimov S V, Zamyatin S V, Gayvoronsky S A and Pushkarev M I 2012 Parametric synthesis of maximum stability degree and specified accuracy of linear automatic control system of PI controller *Mendel* **1** 344

[8] Zhang G, Chen T and Chen X 2007 Performance recovery in digital implementation of analogue systems *SIAM J. Control Optim.* **45(6)** 2207-2223

[9] Hespanha J P, Liberzon D and Teel A R 2008 Lyapunov conditions for input-to-state stability of impulsive systems *Automatica* **44(11)** 2735-2744

[10] Besekerskiy V A and Popov E P 2004 *Control System Theory* (St. Petersburg: Professiya) p. 767

[11] Avramchuk V S and Yakovleva E M 2016 Spectral analysis of narrow band periodic signals *Key Eng. Mat.* **685** 225-229

[12] Hou L, Michel A N and Ye H 1997 Some qualitative properties of sampled-data control systems *IEEE T Automat. Contr.* **42(12)** 1721-1725
[13] Goncharov V and Rudnicki V 2008 Real interpolation method in automatic control systems self-adjustment problem *Systems Science.* **36(3)** 1-19
[14] Fumo N 2014 Using mathcad to simplify uncertainty computations in a laboratory course *Comput. Appl. Eng. Educ.* **23(2)** 250-257

Metal cutting of tooth gears by a method of a nonlinear generating process of a profile

S K Ambrosimov[1] and A V Khandozhko[2]

[1] Lipetsk State Technical University, 30, Moscovskaya ave., Lipetsk, 398600, Russia
[2] Bryansk State Technical University, 7, 50-letiya Oktyabrya bul., Bryansk, 241035, Russia

E-mail: ambsk@mail.ru, avkbstu@mail.ru

Abstract. In this article, the perspective methods of milling of tooth gears by disk cutters with the nonlinear coordination between feeding movements are presented. Besides, the algorithm of the contour calculation of sophisticated reconciled movements of a tool and a gear blank is described.

1. Introduction

Metal cutting of gear wheels of a various profile including an involute one is carried out by special pitch tools: hobbing cutters, gear shape cutters, chasers and other pitch tools. Such surfaces are shaped by two coordinated movements of shaping at a constant ratio of a feeding rotation of a work piece and gear shape cutters, chasers tools, the rotational main movement of a cutter and a feeding rotation of a work piece for hobbing cutters. Thus the shape of a tool generating flat varies at the profile form change of a gear wheel tooth contour, a module, an angle of involute profile. It demands the application of a big spectrum of complicated, expensive and special tools [1, 2]. Therefore, these methods are low efficient in small-batch and unit productions, and the application of tools running based on a method of copying, does not ensure high accuracy and quality of a surface [3-5].

For the purpose of the efficiency increase of the cutter life use, it is necessary to develop a system of control and cutter recovery taking into account the mentioned factors of dispersion. For this, the development of a simulator for the process of adequate cutter life use, taking into consideration the factors emphasized, is necessary [8, 9].

2. Materials and methods

During the development of the simulator of cutter life use, taking into account the dispersion factors mentioned, the following models of tool wear were used: a fan model, an accumulation model, a complex model and a breakout pattern. To solve the problem identified, we used the methods of probability theory and those of mathematical statistics.

3. Metal cutting of gear wheels with three nonlinear movements of forming

Shaping of working surfaces of teeth is carried out by a disk cutter of a trapezoidal (conical) profile, herewith, each working side of a gear tooth profile is executed by the rectilinear corresponding side of a tool profile, and a combined, slot profile of the gear wheel coupling with working sites of a work piece profile is carried out by its cylindrical surface.

Profile processing of each slot is made by three simultaneous forming movements which are coordinated nonlinearly and lie in one plain of shaping and the principal rotation of cutter ω_{zo} (Figure 1). One of them, a uniform rotation of gear wheel ω_{X1}, and the other two, S_Z and S_Y, will agree with rotational ω_{X1} so that during processing of profiles sites, rectilinear generators roll along a work surface. In this manner, the rectilinear generator of a flared section of a cutter should be sequentially tangent to each point of a processed profile. Ratios between separate movements constantly vary depending on the position of the top of the cutting edge and the magnitude of its slippage regarding the cutting surface.

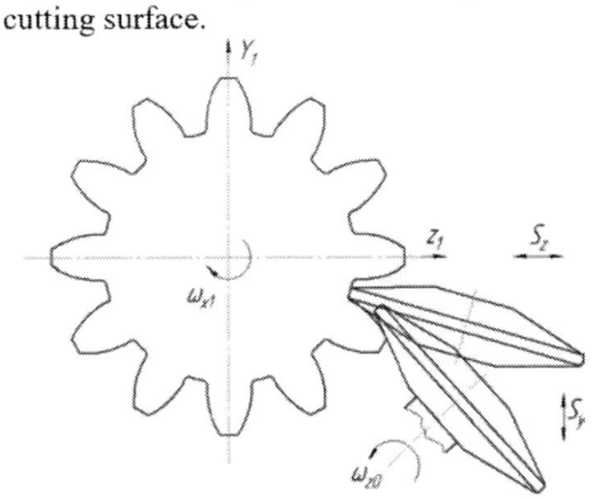

Figure 1. The diagram of a nonlinear rolling of a gear wheel slot with a disk cutter of a trapezoidal profile with three sophisticated and coordinated forming movements.

The designed method of processing of difficult surfaces allows increasing precision owing to profiles generating process of a work piece by rectilinear generators of a tool; besides, it is multipurpose since involute wheels processing of a various module and wheels with non-standard profile angles can be manufactured by cutters of one standard size.

4. Sophisticated trajectory installation and coordinated feedings movements during processing of gear wheels of a non-evolvent profile

The method of a nonlinear rolling is applicable for wheels of a non-evolvent profile, for example, for Novikov's meshing wheels or a cycloidal profile [10].

The main task for the implementation of the described method by any function is establishment of the functional correlation between separate movements of shaping that is reduced to the establishment of tool top migrations on the shaft, $Z_1\ \Delta S_Z$, and $Y_1\ \Delta S_Y$ in step-by-step increments depending on a rotation angle of a work piece φ in the direction of angle feeding ω_{X1}. For this purpose, the cutter should sequentially touch the working surface of the tooth at each point of the profile [9]. That is, the processing side of the tool profile at each point should be tangent to the processed side of the profile of a work piece (Figure 2).

The following designations were used: 1, $1'$ – consecutive positions of the side of a tool profile; 2, $2'$ – consecutive positions of the rolled side of the work piece profile; A_1A' – an initial point of the contact of the tool with the processed surface in two consecutive positions of a work piece; A_2A' – correspondingly subsequently contact point; B_1B' – two consequent positions of the tool top during migration of the tool along shafts Z and Y without rolling; B'' – the position of the tool top in the following engagement point with a work piece during the tool migration simultaneously on shafts Z and Y during rolling.

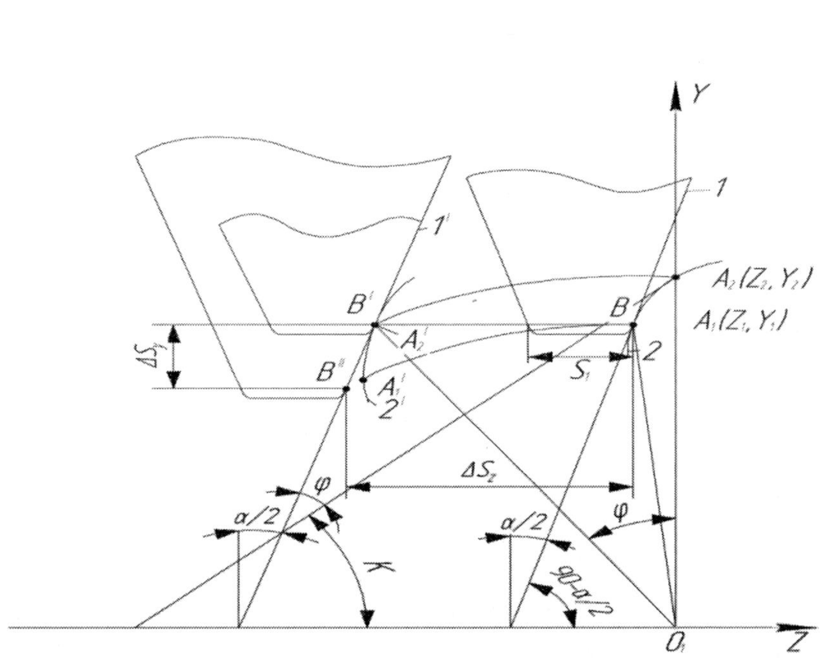

Figure 2. The scheme of the tooth profile processed by the cutter rolling.

In the course of nonlinear rolling, moment positions of tooth profile 1 are defined by an angle of its tilting to shaft O_1Z in coordinate system O_1YZ during certain moments of time or an angle of a conical site of cutting profile $\alpha/2$. Positions of the tool working profile vary in coordinate system O_1YZ under the following law:

$$\varphi = \omega t . \qquad (1)$$

The cutter profile should be tangent to all consecutive positions of the surface profile described by function $y = f(z)$ in the same coordinate system. During the turning movement of the tooth profile to angle φ, the cutter profile will be displaced from point $A_1(z_1, y_1)$ to point $A_2'(z_2', y_2')$; it means during processing with slippage $p = 0$ (the tooth top of the cutter does not change its position in processing relatively the cutting edge), profile top B transfers on shaft Z from point A_1 to point $A_2'(B')$.

For rolling provision without a slippage $(p = 1)$ (the tooth top of the cutter changes its position in processing relating to the cutting edge) the profile top is displaced from a point B to the point B'' so that the length of the profile $B'B''$ was equal to the curvilinear site of the processed profile $A_1'A_2'$ or A_1A_2. Therefore, the initial position of the profile top in shaft direction O_1Y makes interchange ΔS_Y in shaft direction $O_1Z - \Delta S_Z$.

Thus, the problem of calculation of the coordinates of the work piece migration during processing is reduced to the definition of the first derived functions describing the tooth profile, which defines the position of the tangents drawn at an angle to shaft O_1Z, in two positions of functions $y = f(z)$, different from angle of rotation φ. This problem is solved quite easily for the initial profile position at interval A_1A_2, and it is difficult enough for the following position of functions at interval $A_1'A_2'$.

A tangent angle is set with regard to the function of the tooth profile, drawn to some point A_2 (with coordinates z_2, y_2) belonging to the profile in its home position. The tangent analysis is conducted in a similar way with z-coordinate value z_2, there can be some equations, concerning the given function. The tangent angle of inclination at point A_2 is the following:

$$K = arctg\left(\frac{df}{dz} \right)\Big|_{z_2} . \quad (2)$$

A new position of point A_2 is set (point A_2' with coordinates $(z_2' \; y_2')$, in which the tangent will occupy tool profile position 1; in other words, the tangent position turned at angle φ, will occupy a position at the angle to shaft O_1Z:

$$\Delta\varphi = 90^\circ - \alpha/2 - K . \quad (3)$$

As point A_2 (A_2') turns at angle φ towards rotation centre O_1, its new coordinates are recorded with the function of the coordinated transformation:

$$\begin{aligned} z_2' &= -y_2 \cos\varphi + z_2 \cos\varphi \\ y_2' &= y_2 \cos\varphi + z_2 \cos\varphi \end{aligned} \quad (4)$$

Coordinates of point A_1 are set, through which the tangent parallel to the position of tool profile 1 and makes an angle $(90^\circ - \alpha/2)$ with the shaft:

$$\frac{df}{dz}\Big|_{z_1} = tg(90^\circ - \alpha/2). \quad (5)$$

Substituting discovered value z_1 in the equation of the profile function, coordinates of tangent point A_1 are set during profile turning movement at angle φ.

Migration of the tool top is set from point B' to point B without offset of the tooth top concerning cutting edge $(p = 1)$:

$$\begin{aligned} \Delta S_{z1} &= z_1 - z_2', \\ \Delta S_{y1} &= y_1 - y_2' \end{aligned} \quad (6)$$

Migration of the tool top is set from point B_1 to point B'', taking into account pure rolling (curve *AB* can be substituted with the sufficient degree of accuracy by a chord).

With regard to pure rolling $(p = 1)$, the following holds:

$$\begin{aligned} \Delta S_z &= \Delta S_{z_1} + \left| \left[(z_2 - z_1)^2 + (y_2 - y_1)^2 \right]^{1/2} \right| \sin \alpha/2 \\ \Delta S_y &= \Delta S_{y_1} + \left| \left[(z_2 - z_1)^2 + (y_2 - y_1)^2 \right]^{1/2} \right| \cos \alpha/2 \end{aligned} \quad (7)$$

During the rolling action with slippage $(p \neq 1)$, the following is true:

$$\begin{aligned} \Delta S_z &= \Delta S_{z_1} + p \left| \left[(z_2 - z_1)^2 + (y_2 - y_1)^2 \right]^{1/2} \right| \sin \alpha/2 \\ \Delta S_y &= \Delta S_{y_1} + p \left| \left[(z_2 - z_1)^2 + (y_2 - y_1)^2 \right]^{1/2} \right| \cos \alpha/2 \end{aligned} \quad (8)$$

5. Trajectory establishment of sophisticated and coordinated movements of feedings during teeth processing of the evolvent profile

Setting of the functional correlation between separate forming movements for tooth gears of the evolvent profile begins with setting of the profile angle of cutter $\alpha/2$ (Figure 3).

The profile angle is defined by the following formula:

$$\alpha/2 = \eta_o + \gamma_{of} = \frac{ob \cdot 180^\circ}{r_в \cdot \pi} + \gamma_{of}, \quad (9$$

where ob is an arc site on the basic circle of a wheel, which is selected with regard to the possibility of wheel processing of several various modules by a cutter of one standard size and makes $(0.3-0.4)E$, where E is the width of a slot on the arc of basic radius $r_в$:

$$r_в = (mz \cdot \cos\alpha_o)/2 \quad (10)$$

where α_o is a standard or an accepted angle of linkage,

y_{of} is the angle of the evolvent at point f:

$$\gamma_{of} = (tg\alpha_{of} \cdot 180^\circ)/\pi - \alpha_{of}; \quad (11)$$

$$\alpha_{of} = \arccos\ r_в/r_f \quad (12)$$

r_f defines the junction point of a working site of the tooth profile with a slot:

$$r_f = 0.5mz - m. \quad (13)$$

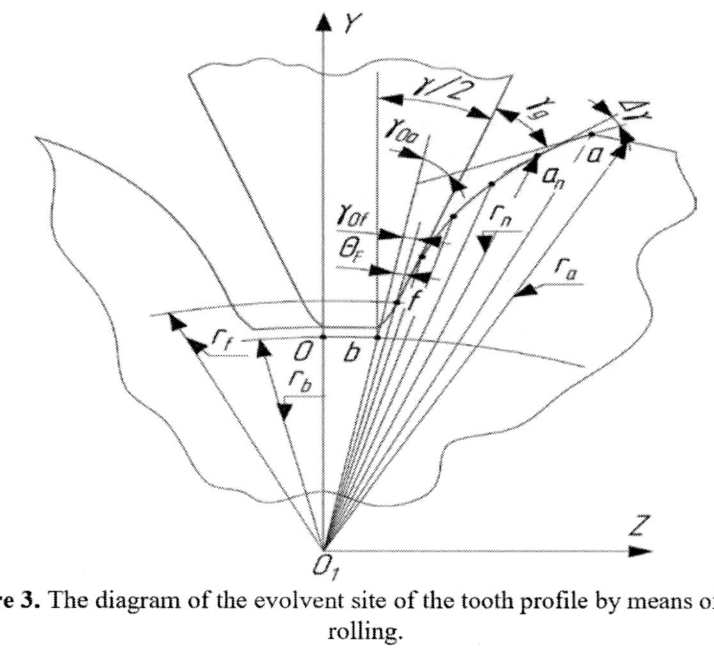

Figure 3. The diagram of the evolvent site of the tooth profile by means of cutter rolling.

Profile processing accuracy depends on the amount of supporting reference points N on the evolvent tooth profile and the amount of teeth. The maximum angle of profile rotation γ_g during processing of a working site of the tooth profile from point a to point f is defined by the following formula:

$$\gamma_g = \alpha_{oa} - \alpha_{of}; \quad (14)$$

$$\alpha_{oa} = \arccos(r_в / r_a), \quad (15)$$

where γ_{oa} is the angle of the evolvent at point f:

$$\gamma_{oa} = (tg\alpha_{oa} \cdot 180^\circ)/\pi - \alpha_{oa}, \quad (16)$$

where r_a is the radius of teeth tops:

$$r_a = 0.5mz + m. \quad (17)$$

An angle of rotation is set between two subsequent reference points $\Delta\gamma$ and the absolute angle of rotation $\boldsymbol{n}$ of the point of tooth profile α_n, for the provision of the condition of the contingence of point α_n (the first point is point α)

$$\Delta\gamma = (\alpha_{oa} - \alpha_{of})/N; \quad (18)$$

$$\alpha_n = \alpha_{oa} - \Delta\gamma n. \quad (19)$$

Radii of all consecutive points of the contingence from α to f are set. For point a_n with radius r_n, the following is true:

$$r_n = r_в / \cos(\alpha_{oa} - \Delta\gamma n). (20)$$

Coordinates of point $\boldsymbol{n}$ of the evolvent profile are determined by formulas

$$y_n = r_n \cos\left(\eta_o + tg\frac{\alpha_n \cdot 180^\circ}{\pi} - \alpha_n\right), \quad (21)$$

$$z_n = r_n \sin\left(\eta_o + tg\frac{\alpha_n \cdot 180^\circ}{\pi} - \alpha_n\right). \quad (22)$$

Since the absolute angle of rotation of each point of profile is $y_n = \varphi$, (Figure 2), new coordinates of points of the profile during the contingence moments are defined by formulas (4) of coordinated transformations:

$$y_n' = y_n \cos\Delta\gamma n + z_n \sin\Delta\gamma n; \quad (23)$$

$$z_n' = -y_n \sin\Delta\gamma n + z_n \cos\Delta\gamma n. \quad (24)$$

Calculation of sophisticated and coordinated tool migrations taking into account profile rolling is made by formulas (6) and (7).

Processing of the right working part of the tooth profile is completed at point f and begins at point a; for the left working part of the profile, it begins at point f_1 and completes at the point.

6. Trajectory establishment of feeding movements during slot processing of tooth gears of the evolvent profile

Slot processing is conducted after processing of the working site of the profile from the tooth top (point $\boldsymbol{a}$ with radius r_a) to the conjugated site at the tooth fillet (point f with radius r_f) (Figure 4).

For this purpose, the motion on shaft Z is set for the cutter on shaft Z at the value of a difference between doubled intercept F and intercept S_1, and the wheel is simultaneously turned at radius r_f at the value of a difference of intercepts E and $2F$ with working feeding which equals the sum of rates of these motions. Such motions ensure the contingence with the opposite side of the tool profile of the working site of the opposite tooth profile at point f_1, that is at the junction point of the working site of the tooth profile with a slot.

Thus, the tool and the work piece occupy a new position (Figure 4, it is shown by a dotted line). From point f_1 to point a_1, the opposite profile of the working site of the tooth is processed by the profile processing algorithm at the site, but in the opposite direction.

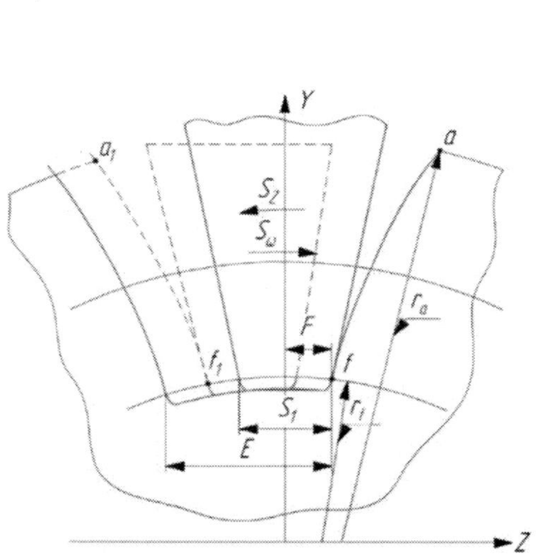

Figure 4. The slot processing diagram of the tooth gear of the evolvent profile from point f to point f_1 with two coordinated feeding movements.

7. Conclusions

Methods of nonlinear rolling of tooth gears have been developed and presented. These methods allow processing of gears with a working surface of various forms of a profile by a disk cutter of a trapezoidal profile. The narrow range of machines with NC allows enhancing the accuracy and productivity of processing under conditions of serial and individual manufacture.

Algorithms of trajectories calculation of sophisticated and coordinated movements of the work piece and the tool for processing of wheels with the various forms of the tooth profile are presented.

References

[1] Kazakyavichyus S M, Stanovskoy V V, Remneva T A, Kuznetsov W M, Bubentchikov A M and Shcherbakov N R 2011 *Russ. Eng. Res.* **31(3)** 197-199
[2] Bubenchikov A M, Kazakavitschyus S M and Shcherbakov N R 2016 *Conf.: Ser. Mater. Sci. Eng.* (Tomsk) **124(1)** (IOP Publishing) 012095
[3] CheGan P, Bolou V, Pozdirca A and Peterlicean A 2013 *Procedia Technology* **12** 420-426
[4] Erdim H, Lazoglu I 2012 *Procedia Eng.* **1** 96-101
[5] Siebrecht T, Kersting P, Birdman D, Odendahl S and Bergmann J 2015 *Procedia CIRP* **37** 188-192
[6] Indakov N, Gordeev Y and Binchurov A 2016 *IOP Conf.: Ser. Mater. Sci. Eng.* (Tomsk) **124(1)** (IOP Publishing) 012150
[7] Gordeev Y I, Abkaryan A K, Binchurov A S and Jasinski V B 2014 *Adv. Mat. Res.* **1040** 13-18
[8] Ambrosimov S, Zhirkov A 2016 *IOP Conf.: Ser. Mater. Sci. Eng.* (Tomsk) **124(1)** (IOP Publishing) 012002
[9] Ambrosimov S, Harlamov G and Cherepenko A 2016 *Procedia Eng.* **150** 1089-97
[10] Shatalov E V, Efremenkov E A and Shibinskiy K G 2016 *IOP Conf.: Ser. Mater. Sci. Eng.* (Tomsk) **124(1)** (IOP Publishing) 012007

Analysis of vibrational load influence upon passengers in trains with a compulsory body tilt

D Ya Antipin[1], V V Kobishchanov[1], V F Lapshin[2], A S Mitrakov[2] and S G Shorokhov[1]

[1] Bryansk state technical university, 7, 50-letiya Oktyabrya bul., Bryansk, 241035, Russia
[2] Ural State University of Railway Transport, 66, Kolmogorova str., Ekaterinburg, 620034, Russia

E-mail: adya24@rambler.ru

Abstract. The procedure for forecasting the vibrational load influence upon passengers of trains of rolling stocks equipped with a system of a compulsory body tilt on railroad curves is offered. The procedure is based on the use of computer simulation methods and application of solid-state models of anthropometrical mannequins. As a result of the carried out investigations, there are substantiated criteria of the comfort level estimate for passengers in the rolling-stock under consideration. The procedure is approved by the example of the promising domestic rolling stock with a compulsory body tilt on railroad curves.

1. Introduction

The tendency of the development of passenger rolling-stock consists in a railway traffic speed increase and decrease of the travel time that will allow increasing the competitiveness in comparison with other types of passenger vehicles. The increase of the railway traffic speed took place mainly on the existing trunk-railways. The majority of trunk-railways are designed without taking into consideration the organization of speed and high-speed running, and the state of groundwork is an obstacle for the railway traffic speed increase of a passenger rolling-stock. The analysis of the world experience in the organization of modern passenger railway communications on existing trunk-railways has shown that the average route speed of railway traffic and a decrease of the travel time is obtained at the expense of application of the rolling stock with systems of the compulsory body tilt. The effect mentioned above is achieved at the expense of the speed increase while passing railroad curves without a safety level and a passenger comfort decrease. It is obtained through the compensation of centrifugal forces affecting the rolling stock and passengers due to the compulsory rolling-stock tilt on the curve inside of railroad curves.

Theoretically, the system of tilt control following comfort criteria described above must ensure such body tilt at which passenger ride comfort would differ little from the comfort experienced in common trains running at lower speeds. It is possible to do having introduced the idea of an ideal tilt at which the angle of a body slope, for instance, at the entry to the curve, must gradually increase in accordance with the increase of values of transverse acceleration and an angle of an external high rail increase. It will give an opportunity to obtain 'ideal' values of the passenger comfort index. Then, comparing the control system with an ideal case, it is possible to define its functioning quality.

Speaking of the efficiency of control system functioning, we should define clearly a basic aim and emphasize requirements reasoning from the procedure presented.

The functional purpose of the system of a tilt consists in ensuring a quick reaction to the changes in a railroad curve and an outer rail height, without triggering the system of a tilt at random imperfections when running along straight track areas. For this purpose, it is necessary to make a compromise between responsiveness to the beginning of easy and random imperfections on straight track areas.

2. Results and discussion

Basic demands made to the control system are:

1. Railroad change monitoring,
2. Computation of an essential angle of a tilt taking into account the assurance of comfort and safety,
3. System fault monitoring,
4. Speed control.

For tilt control systems, there are some different approaches. In general, these approaches could be combined into two different groups: monitoring of the change in the track geometry and a system using geo-location data.

Nullification is one of earlier control systems which tried to reduce a measured acceleration in a steady-state curve to zero. A feedback in this kind of a system is ensured by an accelerator installed in a body (an acceleration measuring device). An evident advantage of this system consisted in the fact that a sensor was located after the second stage of a suspension. This type of location acts as a mechanical filter of track random imperfections. But, a sensor location in a tillable body can cause an interaction between the suspension and dynamic control that will result in problems of stability in the control system. The experience of this control system use has shown that the complete compensation of lateral accelerations is connected with a large speed of a tilt and does not ensure optimum comfort.

The investigations described in [1] have shown that in case of complete compensation of a centrifugal force, the effect of rocking by daylight is observed among passengers because of the non-coordination of information from a vestibular apparatus and an observed visual image. In other words, rocking arose because passengers saw a slope, but did not feel it. At the same time, the speed of turning and jerks increase a sensory conflict.

To reduce this kind of a phenomenon of rocking, it is assumed to limit the level of compensation at the level of 60-75% of the level of complete compensation.

As the development of the 'nullification' strategy became a strategy of 'command control' giving a possibility of partial compensation of accelerations. It is carried out through the introduction of an amplification factor which can change within the limits of 0 – 'the system is switched off' to 1- 'complete compensation of unabsorbed acceleration'. This system is equipped with both an accelerometer and gyroscopes to increase sensitivity and raise a possibility of monitoring the outer rail height.

It should be noted that a similar effect can be obtained in a nullification system of control having introduced a reverse angle in a controller, but this method is more complicated in a fulfillment.

In the command system of control, the sensors have no suspension operation influence upon them as they are located on a bogie. It allows placing beyond a control loop and increasing the stability of the control system. But, because of rigid connections with a bogie, an accelerometer apprehends accelerations not only in curves, but also on random imperfections. It can result in the system triggering on straight areas of the track. It is possible to avoid the system spontaneous triggering through the introduction of Kalman's filter [2]. This filter is intended for the recursive after-estimate of the vector of state of a priori known dynamic system, that is, for the estimate of the current state of the system, one uses current measurements and a previous state of the filter itself. The filter introduction will reduce an effect of track imperfections for satisfactory passing through straight areas of the track, but will result in a delay of the control system triggering on curve sections of the track. To eliminate this drawback, the command control can be supplemented with the principle of the leading triggering used in rolling-stock X2000.

In the control system, formed according to the principle of advance, a signal, received from sensors installed on bogies of head cars, is transmitted in a digital way to the following cars with a corresponding time delay. For this type of the system it is necessary to take into account the speed and the direction of motion.

The drawback of this system is a complex procedure of the fulfillment and the necessity of the information transfer between cars. Besides, in this kind of systems, a head car is usually excluded from the operation of the system, or makes it with delay. The solution of the problem mentioned is the use of the control system based on the geo-positioning similar to that used in trains of the Japanese railroad company 'JR Central'.

In such control systems, an on-board computer is used with preliminarily recorded information of curve sections of the track, including a track curvature, an outer rail height and so on. The system is formed on the essential information transmission from a database to the device of a body shift on reaching an entry point of a curve section of the track by a rolling-stock. The main problem of this type of a control system is the definition of an exact position.

In the places of the beginning of a track curve section, a road mark is installed, approaching to which an on-board computer reads essential information and transmits it to the device of a body shift, at the end of a curve section, the on-board computer receives information about the end of a body tilt from other mark. This system for tilt control ensures considerable smoothness of work in track curve sections and stability in straight ones independently of random imperfections. A drawback consists in high cost, availability of an essential infrastructure and impossibility of the rolling-stock use on other routes without preliminary training. Besides, when using track marks, there is likelihood that the information will not be read from a stationary transmitter. To increase reliability, the tilt control system can be equipped with an odometer reading a car travel. It will allow controlling a route traversed between track marks. A drawback in the use of the odometer consists in error accumulation.

It is possible to eliminate the drawback of the odometer by means of the introduction of the GPS/GLONASS systems of geo-positioning into the system of control. The application of such systems allows defining absolute coordinates of a rolling-stock, which makes it possible to use a partial failure of track marks. Having considered the control systems described in accordance with the criteria of choice, it is possible to draw a conclusion of the usefulness of application of one of two systems functioning either on the principle of advance or on the principle of geo-positioning for a domestic rolling-stock. For existing tracks, it is offered to use a system formed by the principle of a leading triggering and for new tracks — allowing the use of automatic train monitoring – a geo-positioning system supplemented with the GLONASS system.

When using a rolling-stock equipped with the systems of a compulsory body tilt, a significant problem consists in the control of passenger safety and comfort when passing curve sections of the track. The criteria for passenger safety and comfort consist in:

a) The level of non-absorbed accelerations affecting a person.

b) The elimination of rocking effect.

For a motor-car rolling-stock operated on domestic railway lines, the criterion of safety and comfort can be a level of transverse non-absorbed acceleration and functioning at the level of an axle box during the motion along the curve track sections, not compensated by the outer rail height [3] and an index of motion evenness in vertical Wz and horizontal Wy cross directions [2].

In accordance with [3], a maximum admissible value of non-absorbed acceleration makes 0.7 m/sec^2. On the tracks where speed trains and passenger rolling-stocks are operated with updated dynamic characteristics with the permission of RZhD, an admissible value of anp can be increased up to 1.0 m/sec^2. The results of complex psycho-physiological investigations of passengers carried out by VNIIZhT jointly with VNIIZhG have shown that when increasing a non-absorbed acceleration above 0.8 m/sec^2, an increasing degradation of a functional state and availability of vivid signs of a subjective discomfort were observed among the majority of passengers [4]. At the same time, the safety of passengers is ensured up to the level of a non-absorbed acceleration of 1.2 m/sec^2 [4]. In work [5], it is pointed out that when exceeding the limit of 2 m/sec^2 by horizontal accelerations affecting a passenger, sitting in an armchair, he begins to feel discomfort at accelerations of above 3 m/sec^2. For standing passengers, the admissible accelerations are those which do not exceed the limits of 0.7-1 m/sec^2.

In European Standard EN 13803-1 [6], limit values are the regulated non-absorbed quasi-static cross accelerations of a body and the units of a rolling-stock in the range of 1.0-1.5 m/sec^2.

The indices of motion evenness are defined through the dependences recommended by GOST RF 55495-2013 [7]:

- for vibrations acting in a vertical direction:

$$W_z=4.346\tilde{a}_k^3; \quad (1)$$

- for vibrations acting in a horizontal cross direction:

$$W_y=4.676\tilde{a}_k^3, \quad (2)$$

where $\tilde{a}_k$ – the quadratic mean value of corrected vibro-acceleration in the k-th range of motion speed, m/sec^2.

The indices of motion evenness in vertical and horizontal directions must not exceed the value of 3.25.

For the rolling stock equipped with a system of the compulsory body tilt, the level of comfort is defined according to the European procedures described in CEN12299 [8] and materials of Japanese specialists [9].

Within the bounds of Standard EN 12299, let us define the passenger comfort in straight track sections using a coefficient of average comfort, N_{MV} [8], defined through dependence

$$N_{MV}=6\cdot\sqrt{(a_{XP}^{Wd})^2+(a_{YP}^{Wd})^2+(a_{ZP}^{Wd})^2}, \quad (3)$$

where a_{XP}^{Wd} – accelerations affecting a passenger in the longitudinal direction; a_{YP}^{Wd} – accelerations affecting a passenger in the side direction; a_{ZP}^{Wd} – accelerations affecting a passenger in the vertical direction.

This coefficient takes into account accelerations affecting a person in all directions at the level of the floor.

To determine the percentage of passengers feeling discomfort in [8], the following dependence is used:

$$P_{CT}=100\left\{max\left[A\ y_{ls\ max}+B\ y_{ls\ max}\ C;0\right]+\left(D\ \ _{ls\ max}\right)^E\right\}, \quad (4)$$

where $\ddot{y}_{ls}$ – the side acceleration of a body, m/sec^2; $\dddot{y}_{ls}$ – the side acceleration change of a body during 1 sec, m/sec^3; $\dot{\varphi}_{ls}$ – the angular velocity of a body, rad/sec; A, B, C, D, E – constants admitted in accordance with [8].

In accordance with [9], the level of discomfort according to a four-point scale is defined through the following dependence:

$$TC_T=ay+by+c+\dot{d}e, \quad (5)$$

where $\ddot{y}$ – the side acceleration of the body, m/sec^2; $\dddot{y}$ – the side acceleration change of the body during 1 sec, m/sec^3; $\dot{\varphi}$– the angular velocity of the body, rad/sec; $\ddot{\varphi}$– the angular acceleration of the body, rad/sec^2; a, b, c, d, e – constants adopted in accordance with [9].

For the rolling stock equipped with the system of the compulsory body tilt, it is necessary to take into account the effect of rocking. The researches of this effect by European scientists [4] have shown that the most common reason for its arising is a conflict of the human sensor system. For example, during the complete compensation of a centrifugal force, passengers feel the absence of corresponding accelerations in the daytime as the effect of rocking because of non-coordination feeling in a body position and a visual image [8, 10]. Spreading of sickness depends on a number of factors: physical properties, duration of an effect, human susceptibility, the task to be carried out and other factors.

The effect of passengers rocking during the curve track section passing is estimated in accordance with [10], on the basis of the definition of a rocking dose defined through dependence

$$MSDV_z(t)=k_{MSDV}\cdot\sqrt{\int_0^t a_{wf}^2(t)dt}, \quad (6)$$

where a_{wf} – the frequency-weighted vertical acceleration, m/sec^2; k_{MSDV} = 1/3 – the coefficient for mixed population of grown-up men and women incapable of the adaptation.

On the basis of a rocking dose, it is also possible to prognosticate a level of indisposition. The value of 'rate of illness' IR(t) is graduated according to the scale from 0 ('I feel good') up to 3 ('I feel absolutely terrible' and is defined through dependence [9]:

$$IR(t)=\frac{MSDV_z(t)}{50}. \qquad (7)$$

In [11], there is a derived 'neat' dose of rocking ND(t) as a function of time taking into account the rise of the effect of rocking and recovery after it, which is definable through the dependence:

$$ND(t)=C_a\cdot\int_0^t A(\tau)\cdot e^{-C_L\cdot(\tau-t)}dt, \qquad (8)$$

where $A(\tau)$ describes a motion, C_a and C_L – constants.

In the procedures shown, the level of comfort is estimated on the basis of body accelerations of a unit of a rolling stock, taking into account only two passenger positions – 'upright' and 'sitting', whereas on the territory of the country, the revenue service of trains 'Talgo 250' with cars equipped with berths began.

The estimation of safety and the level of passenger comfort in the paper is carried out by methods of the mathematical modeling and computer simulation of rolling stock working with the use of anthropometrical models of mannequins.

a five-car local train equipped with air springs of a central stage of bogie suspension and a system of the compulsory car body tilt on track curve sections with an electric locomotive were analyzed as a test object.

The estimation of accelerations affecting train elements and passenger models is carried out on the basis of a solid-state computer model developed in the environment of the 'Universal Mechanism' software complex (Figure 1).

A model represents a two-level system including subsystems of the first level – a 'head' car, a 'trailer', and a 'powered' car, as well as subsystems of the second level describing functioning chassis and the automatic coupling device. The subsystems of the first level are formed with absolute solids describing inertial characteristics of a car body and connected with special power elements and hinges with the subsystems of the second level – 'a bogie of a trailer' and 'an automatic coupling device'.

During a car body modeling, an actual location of heavy internal and external equipment and passengers is taken into account. Computer models of the 'trailer bogie' and 'motor car bogie' subsystems represent a system of absolute solids connected by power elements and hinges. The modeling of an air spring is carried out within the limits of the GENSYS model described in the digital Matlab/Simulink processor and integrated into a developed computer model with the use of the UM Control module of the 'universal mechanism' complex. On the basis of the data of body acceleration, the angle of slope is computed in the Matlab/Simulink program, a control effect is transferred to the model of an air spring changing the angle of the body tilt.

The modeling of a car motion is considered taking into account track micro-roughness. The formation of random imperfections is carried out on the basis of the spectrum density function, equivalent to a computed roughness in vertical and horizontal directions, defined for speeds of a car motion up to 200 km/h according to the procedure shown in RD 32.68-96.

For the estimation of dynamic influence upon passengers during a train motion, the solid-state models of car bodies were supplemented with computer models of anthropometrical mannequins of the Hybrid III 50th Male Dummy type.

Mannequins are located in different passenger places of a car, as well as in passages, platforms and service rooms in the 'upright', 'sitting' and 'lying' positions (Figure 2).

As a result of multi-version computations carried out, we defined maximum accelerations of axle-box units, car bodies, as well as dummy accelerations taking into account the 'worst' positions and places of location from the point of view of comfort assurance.

The analysis of results obtained has shown that transversal non-absorbed accelerations operating at the level of an axle-box at the motion along the track curve sections at speeds up to 200 km/h, do not exceed the recommended level [3]; the transversal accelerations of a car body correspond to the requirements of EN 13803-1 [6]. Dummy accelerations in a 'sitting' and 'upright' position do not exceed the recommended levels [5].

The estimation of motion evenness indices has shown their correspondence with the requirements of GOST RF 55495-2013 with considerable indices of reserve.

Figure 1. The solid-state computer model of the electric train: 1 – 'head car' subsystem; 2 – 'trailer' subsystem; 3 – 'powered car' subsystem.

Figure 2. The location of solid-state models of anthropometrical mannequins in a car body.

When estimating the comfort level on the basis of procedures recommended in [8-10], both accelerations of a car body and accelerations affecting computer models of mannequins were assumed as initial data.

The analysis of results of the passenger comfort level has shown that the results obtained when we use accelerations of car bodies by 10–45% higher than the results obtained with the use of dummy accelerations as initial data. The largest discrepancies are obtained for a passenger 'upright' position on the platform of a car near an entrance door taken as the position with the lowest level of comfort. The least ones were for the 'sitting' position in the medium row of a car in the armchair by a passage.

3. Conclusions

On the basis of the obtained results, it is possible to draw a conclusion about the correspondence of the design of a local train, driven by an electric locomotive, with the requirements of domestic and foreign regulation documentations and an expediency of the use of computer models of anthropometrical mannequins to specify the estimation of the passenger comfort level in trains equipped with a system of the compulsory car body tilt in track curve sections.

4. References

[1] Forstberg J, Andersson E, Ledin T 1999 *Proceedings* **CO3-4**

[2] Ribeiro M, Kalman and Extended Kalman Filters 2004 *Concept, Derivation and Properties, Institute for Systems and Robotic Av. Rovisco Pais* (Lisboa Portugal) vol 1 1049-001

[3] 2012 *Instruction on Current Maintenance of Railroad Track, Approved by the order of RZD.* 29.12.2012 № 2791 p RZD 274

[4] Smelyansky I V 2008 *Development of Standards for Non-Absorbed Acceleration and Its Increment for the Modern Rolling Stock at the High-Speed Operation: author's thesis for Candidate of Eng. Competition of I.V. Smelyansky* (Moscow) p 24

[5] Shakhunyants G M 1987 *Railroad Track* Moscow 479

[6] *Railway applications - Track - Track alignment design parameters - Track gauges 1 435 mm and wider* Part 1 Plain line German version EN 13803-1:2010

[7] GOST RF 55495-2013 Railcar rolling stock. Requirements for the strength and dynamic qualities Moscow 32

[8] *CEN Railway applications Ride comfort for passengers Measure and evaluation*

EN 12299:2009 *European Committee for Standardization Brussels.*
[9] Persson R 2007 Tilting trains – Description and analysis of the present situation. A literature study VTI report **595A** 81
[10] *ISO 1997 Mechanical vibration and shock – Evaluation of human exposure to the whole body vibrations* Part 1 Generalrequirements ISO 2631-1.2
[11] Kufver B and Förstberg J 1999 *A net dose model for development of nausea. Proc. of the 34th UK Conference on Human Response to Vibration Dunton*

The generalized mathematical model of the failure of the cutting tool

N I Pasko[1], A V Antsev[1], N V Antseva[1] and V P Fyodorov[2]

[1] Tula State University, 92, Lenina ave., Tula, 300012, Russia
[2] Bryansk State Technical University, 7, 50-letiya Octyabrya bul., Bryansk, 241035, Russia

E-mail: a.antsev@yandex.ru

Abstract. We offer a mathematical model which takes into account the following factors: the spread of the cutting properties of the tool, parameters spread of gear blanks and consideration of the factor of a possible fracture of the cutting wedge tool. The reliability function, taking into account the above-mentioned factors, has five parameters for which assessment we propose a method according to our experience. A numerical illustration of the method is shown in the article. We suggest using the model in the optimization mode of the cutting tool preventive measures.

1. Introduction

Efficiency of tooling on metal cutting machines is largely dependent on the reliability (stability) of cutting tools. This efficiency can be increased at the cost of the rational use of the cutting tools life. In addition, by predicting the durability of cutting tools, the cost of tool maintenance of machine-building enterprises can be reduced. These enterprises are characterized by a very wide range of consumable cutting tools [1]. The task of forecasting of cutting tools durability and efficiency improvement of the tool life are complicated by the fact that the cutting tools life (Mean-Time-Between-Failures (MTBF) according to the terminology of the reliability theory) is a random variable that depends on many factors: cutting parameters, cutting properties of tools, the type of cutting, hardness of work pieces, the value of allowances for processing, pre-stress and strain state, vibration, geometric errors of the machine and other factors [2-7].

Cutting tools properties even in one batch are subject to variety. Hardness of work pieces and machining allowances are subject to variety too. All these factors, and the stochastic nature of the process of wear and tear of the cutting tool active part results in the gain of the tool life, and, however, the tool life of even one batch of the cutting tools varies fairly considerably (15-35%) [8].

For the purpose of the efficiency increase of the resource use of the cutting tool, it is necessary to organize the monitoring and replacement system of the cutting tool with account of the specified dispersion factors. This requires the development of a mathematical model of the resource use of the cutting tool, adequately considering the mentioned factors.

2. Materials and methods

When elaborating the mathematical model of the process of the resource use of the cutting tool with account of the specified dispersion factors, the following models of the tool wear have been applied: a fibered model, an accumulation model, a complex model and a fracture model. For the solution to the given task, methods of the probability theory and the mathematical statistics have been used.

3. The generalized mathematical model of failure of the cutting tool

The failure of the cutting tool is usually associated with the deterioration or destruction of the cutting wedge (chipping, sliding fracture, breakage). According to the researchers' data, the proportion of failures because of destruction during the roughing and finishing operation reaches 50% or more. Since the consequences of these failures can be different, while determining the parameters of tools reliability, the type of failure should be considered. The peculiarity of the wear process of the cutting tool also consists in the fact that in case of failure, a wear condition of the tool can be judged by the degree of wear, but it is impossible in case of tool destruction.

If T_1' – the time to failure because of wear and tear, T_2' – the time to failure because of the destruction; the actual time to failure in general (durability) is $T = \min(T_1', T_2')$, and there is a probability of no-failure during operating time t (reliability function), i.e. probability that $T \geq t$ $P(t) = P_1(t) \cdot P_2(t)$, where $P_1(t)$ is the reliability function by wear, and $P_2(t)$ is the reliability function by the destruction of the cutting wedge. It is impossible to evaluate the reliability of private functions $P_1(t)$ and $P_2(t)$ directly by experience, as the wear process and the development process of the destruction take place in parallel and cannot be divided. Sorting out life tests results by the types of failures leads to other conditional functions of reliability $\pi_1(t)$ and $\pi_2(t)$, connected with $P_1(t)$ and $P_2(t)$ by mutually inverse transformations:

$$\pi_i(t) = \frac{1}{p_i}\int_t^{\infty} P(\tau)\cdot\lambda_i(\tau)\,\mathrm{d}\tau; \quad P_i(t) = \exp\left(\int_0^t p_i \frac{\pi_i'(\tau)}{P(\tau)}\mathrm{d}\tau\right), \quad i = 1, 2, \tag{1}$$

and $P(t) = p_1 \cdot \pi_1(t) + p_2 \cdot \pi_2(t)$. Here, p_1 and p_2 are failure shares of the first and second types:

$$p_i = \int_0^{\infty} P(\tau)\cdot\lambda_i(\tau)\,\mathrm{d}\tau,$$

and failure intensity by i factor ($i = 1, 2$) $\lambda_i(t) = -P_i'(t)/P_i(t) = -p_i \cdot \pi_i'(t)/P(t)$.

General distribution density of mean time to failure is $f(t) = -P'(t) = f_1(t)\cdot P_2(t) + f_2(t)\cdot P_1(t)$, where $f_1(t) = -P_1'(t)$, $f_2(t) = -P_2'(t)$ – distribution densities T_1' and T_2', relatively.

Theoretical considerations based on the types of failure modes lead to the following reliability functions for failures related to wear and tear:

$$P_1(t) = \int_0^{\infty} \psi(a)\Phi^*\left(\frac{L - at}{\sigma\sqrt{t}}\right)\mathrm{d}a, \quad \psi(a) = \frac{1}{\sqrt{2\pi}\delta a}\exp\left[-\frac{(\ln a - \ln\hat{a})^2}{2\delta^2}\right]. \tag{2}$$

Here $\psi(a)$ is the density of the logarithmically normal distribution of the average rate of wear by batch tools. $\Phi^*\left(\frac{L - at}{\sigma\sqrt{t}}\right)$ is likelihood that tool wear $Y(t)$ after operating time t at medium intensity a and quadratic wear per life length unit σ does not exceed limit value L, which means it does not fail. $\Phi^*(x) = \int_{-\infty}^{x} \frac{1}{\sqrt{2\pi}}\exp(-x^2/2)\mathrm{d}x$ is the standard normal distribution function, which table is in the reference books.

For failures related to the destruction, we suggest using Weibull's distribution, which has found wide use in failure predicting of technological equipment [9]. When determining this law of durability distribution, we will proceed from this model. Initially, in the body of the cutting wedge, there are many micro-defects, which are gradually developing in the process of cutting because of fluctuations of the cutting force in the fatigue cracks with subsequent destruction of the cutting wedge.

Let us assume that $t_1, t_2, ..., t_N$ is the time of development of i microdefects into a critical size of cracks, leading to the destruction of the tool cutting wedge. The actual wedge breakdown will occur when $T = \min(t_1, t_2, ..., t_N)$.

In such model for T, Weibull's distribution is limited. The distribution density and the reliability function in this case is expressed as follows:

$$f_2(t) = \frac{\beta}{r}\left(\frac{t}{r}\right)^{\beta-1} \exp\left[-\left(\frac{t}{r}\right)^{\beta}\right],\ P_2(t) = \exp\left[-\left(\frac{t}{r}\right)^{\beta}\right], \tag{3}$$

where r and β are distribution parameters.

Mathematical expectation and the coefficient of variation of the lifetime during destruction are expressed through parameters r and β as follows:

$$\bar{T}_2 = r \cdot \Gamma\left(1 + \frac{1}{\beta}\right),\quad K_{T2} = \sqrt{\frac{\Gamma(1+2/\beta)}{\Gamma^2(1+1/\beta} - 1}.$$

Here $\Gamma(x)$ is the gamma-function.

With regard to quotations (1), (2) and (3), we receive

$$P(t) = \frac{1}{\sqrt{2\pi}\delta}\int_0^{\infty}\frac{1}{a}\exp\left[-\frac{(\ln a - \ln \hat{a})^2}{2\delta^2}\right]\cdot \Phi^*\left(\frac{L - at}{\sigma\sqrt{t}}\right)da \cdot \exp\left[-\left(\frac{t}{r}\right)^{\beta}\right],$$

This reliability function has five parameters: $\hat{a}$, σ, δ, r and β.

When $\sigma \to 0$ and $1/r \to 0$, within this range, we get a fibered model [10], whereby the reliability function follows the logarithmically normal law, which means that

$$P(t) = \frac{1}{\sqrt{2\pi}\delta t}\exp\left[-\frac{(\ln t - \ln \hat{T})^2}{2\delta^2}\right],\quad \bar{T} = \frac{L}{\hat{a}}.$$

When $\delta \to 0$ and $1/r \to 0$, within this range, we get an accumulation model according to which the reliability function takes the following form:

$$P(t) \approx \Phi^*\left(\frac{L - \hat{a}}{\sigma\sqrt{t}}\right).$$

When $\hat{a} \to 0$ and $\sigma \to 0$, within this range, we get a real fracture model, according to which $P(t) = P_2(t)$.

When $1/r \to 0$, within this range, we get a complex wear model, according to which $P(t) = P_1(t)$.

4. Parameters assessment of a generic mathematical model of the cutting tool failure

Estimation of parameters $\hat{a}$, σ, δ, r and β by experience is not an easy task. If we assess using the method of maximum likelihood, based on the statistics of operating time to failure t_i, $i = 1, ..., N$, where N – the sample size, then the likelihood function will look like:

$$l(\hat{a}, \sigma, \delta, r, \beta) = \prod_{i=1}^{N} f(t_i), \tag{4}$$

where $f(t) = -P'(t) = f_1(t)P_2(t) + f_2(t)P_1(t)$.

IOP Conf. Series: Materials Science and Engineering **177** (2017) 012052 doi:10.1088/1757-899X/177/1/012052

It is not easy to find parameters $\hat{a}$, σ, δ, r and β that lead to the maximum such as the complex function. A solving method for the system of equations, obtained by equating the partial derivatives of parameters $\hat{a}$, σ, δ, r and β from function (4), seems here too lengthy. It is simpler to look for the function maximum (4) directly, for example by sorting or random search. Although the time of machine calculation increases, the time required for programming and programs debugging reduces significantly.

If we take into account the type of failure during parameters estimation, in other words, proceeding statistics of operating time to failure due to wear $t_{1,1},...,t_{1,N_1}$ and statistics of operating time to failure due to destruction $t_{2,1},...,t_{2,N_2}$ under $N = N_1 + N_2$, parameters $\hat{a}$, σ, δ, r and β can be estimated more accurately by means of inclusion of this additional information. Let us consider this case in more detail, and use the method of moments. In this case the estimates of the mentioned parameters are obtained by solving the following system of five equations:

$$\overline{T}_1^* = \overline{T}_1,\ \overline{T}_2^* = T_2,\ D_1^* = D_1,\ D_2^* = D_2,\ p_1^* = p_1, \tag{5}$$

where $\overline{T}_j^* = \sum_{i=1}^{N_j} t_{j,i}$, $D_j^* = \sum_{i=1}^{N_j} t_{j,i}^2 - (\overline{T}_j^*)^2$, $p_1^* = \dfrac{N_1}{N}$, and

$$\overline{T}_j = \int_0^\infty \pi_j(t)dt,\quad D_j = 2\int_0^\infty \pi_j(t)tdt - (\overline{T}_j)^2,\quad p_1 = \int_0^\infty f_1(t)P_2(t)dt,\ (j = 1, 2).$$

System (5) with respect to $\hat{a}$, σ, δ, r and β is solved numerically. For this purpose, we propose the following algorithm of a random search, consisting of the following basic steps.

1) According to formulas (4), the statistical characteristics of the empiric lifetime choice are calculated.

2) From the a priori consideration, we determine the limits of the possible values of required parameters a', a'', σ', σ'', δ', δ'', r', r'', β' and β''.

3) We generate current parameter values $\hat{a}$, σ, δ, r and β by formulas

$$\hat{a} = \gamma_1(a'' - a'),\ \sigma = \gamma_2(\sigma'' - \sigma'),\ ...,\ \beta = \gamma_5(\beta'' - \beta'),$$

where $\gamma_1,...,\gamma_5$ are pseudo-random numbers distributed uniformly between 0 and 1. These numbers are generated, for example, a Pascal algorithmic language using random functions.

4) By formulas (5) generated by the values of parameters $\hat{a}$, σ, δ, r and β, theoretical characteristics as well as the absolute value of the difference between them are calculated. Let us call them residuals.

$$\Delta_1 = \left|\overline{T}_1 - \overline{T}_1^*\right|/\overline{T}_1,\ \Delta_2 = \left|\overline{T}_2 - \overline{T}_2^*\right|/\overline{T}_2,\ \Delta_3 = \left|\sqrt{D_1} - \sqrt{D_1^*}\right|/\overline{T}_1,\ \Delta_4 = \left|\sqrt{D_2} - \sqrt{D_2^*}\right|/\overline{T}_2^{*p},$$
$$\Delta_5 = \left|p_1 - p_1^*\right|.$$

5) Maximum of the residuals is calculated as $\Delta = \max(\Delta_1,...,\Delta_5)$.

6) If received value Δ is less than the value of the previous iteration, then the value and the corresponding parameters are stored and new parameters values are generated and we return to step 3. Otherwise, just go to step 3. According to the first iteration, Δ is obviously taken as a large value (theoretical infinity).

7) The search process continues for a defined number of search iterations or until the given residual level corresponding to the desired accuracy of parameter estimation is gained. At the end of search iterations, the best values of the required parameters will remain in the computer memory.

5. Illustration of the method of estimating model parameters

We will demonstrate the method of parameter estimation by the example of the source data shown in figure 1.

```
   Source data for the program 'Ocenk_Ob.exe'
   Search ranges of parameter estimates
Average wear intensity per tool, µm/pcs.        1.0 3.0
CV of wear intensity per tool                   0.200 0.400
RMSD of wear per operation time, µm/pcs.        0.1 3.0
Mean time to failure, pcs.                      50.0 200.0
CV of time to failure                           0.100 1.000
Wear limit, mm                                  0.30
Search iterations count                         3000
Failures due to wear count                        28
Failures due to the destruction count             12
    Lifetime choice by wear, pcs.
 158 155 135 107 178 173 118 132 149 162 158 156 153 147
 124 186 163 154 145 104 148 145  87 129 124 162 147 161
    Lifetime choice by the destruction, pcs.
  81 129 154 142 156 180 146 187 100 173 146 131
```

Figure 1. Initial data for evaluating the parameters of reliability function $P(t)$.

The calculation results are shown in figure 2.

The calculation results

Search protocol of optimal parameters

Loop	$\hat{a}$	δ	σ	r	β	Δ
1	0.001	0.204	0.003	82.110	9.062	65.000
2	0.002	0.258	0.001	107.192	7.368	55.552
3	0.001	0.287	0.000	176.472	11.425	22.101
57	0.001	0.253	0.000	156.476	8.425	11.697
198	0.001	0.294	0.002	156.841	9.623	10.857
732	0.001	0.327	0.000	157.794	11.137	10.623
2544	0.001	0.274	0.001	152.051	7.105	9.101

Statistical parameters of the sample

$\overline{T}_1^*=145.0$ $\overline{T}_2^*=143.8$ $K_{T1}^*=0.15$ $K_{T2}^*=0.21$ $\overline{T}^*=144.6$ $K_T^*=0.172$

Durability distribution parameters

$\hat{a}=0.0013$ $\delta=0.274$ $\sigma=0.0008$ r= 152.1 $\beta=7.11$

Residual Δ	9.1
Average wear intensity per tool, µm/pcs.	1.4
CV of wear intensity per tool	0.279
RMSD of wear per operation time, µm/pcs.	0.8
Mean time to failure, pcs.	142.3
CV of time to failure	0.166

Figure 2. The intermediate and final results of the assessment of tools reliability parameters.

The calculation was carried out according to the program, developed in the algorithmic language Pascal. The count of search iterations was 3000, and a better solution was found when the iteration number was 2544. Great accuracy estimates can be obtained by setting a greater number of iterations.

6. Conclusion

Thus, we have proposed a generalized mathematical model of failure of cutting tools, which takes into account the wear and the wear rate of its spread because of tools quality dispersion and the parameters variety, taking into account the fact of the destruction of the instrument. We offer a method for model parameters as well, which is based on our experience. It is assumed that among other factors, the use of the mathematical model allows optimizing the model of cutting tool preventive measures.

Acknowledgments

The research was carried out with the financial support of the Russian Foundation for Basic Research (RFBR) and Tula region administration as a part of research project 15-48-03265 'r_center_a'.

References

[1] Vasin S A, Antsev V Yu and Dolgov D V 2002 *Russia J. Autom. and modern techn.* **2** 3-10
[2] Erenkov O Y, Ivakhnenko A G and Radchenko M V 2013 *Chemical and Petroleum Eng.* **49** 411-417
[3] Kuts V Ivakhnenko A and Khandozhko A 2016 *Proc. of 2015 Int. Conf. on Mech. Eng., Aut. and Control Syst. MEACS 2015* (Institute of Electrical and Electronics Engineers Inc.)
[4] Bannikov A I, Kurchenko A I, Makarova O A, Bannikov A A, Dyatlov N A and Inochkin A S 2016 *Russ. Engin. Res.* **30** 734-735
[5] Petukhov Y E, Domnin P V and Rubets A A 2016 *Russ. Engin. Res.* **36** 309-311
[6] Muratov K R 2016 *Russ. Engin. Res.* **36** 321-323
[7] Chernikov P P and Sharipov B U 2008 *Russ. Engin. Res.* **28** 194-195
[8] Martinov G M and Grigor'ev A S 2013 *Russ. Engin. Res.* **33** 433-437
[9] Pas'ko N I and Antseva N V 2009 *Russ. Engin. Res.* **29** 723-726
[10] Pas'ko N I and Antseva N V 2008 *Russ. Engin. Res.* **28** 692-696

A statistical model of operational impacts on the framework of the bridge crane

V Yu Antsev[1], A S Tolokonnikov[1], A D Gorynin[1] and A A Reutov[2]

[1] Tula State University, 92, Lenina ave., Tula, 300012, Russia

[2] Bryansk State Technical University, 7, 50-letiya Oktyabrya bul., Bryansk, 241035, Russia

E-mail: anzev@yandex.ru

Abstract. The technical regulations of the Customs Union demands implementation of the risk analysis of the bridge cranes operation at their design stage. The statistical model has been developed for performance of random calculations of risks, allowing us to model possible operational influences on the bridge crane metal structure in their various combination. The statistical model is practically actualized in the software product automated calculation of risks of failure occurrence of bridge cranes.

1. Introduction

Bridge cranes of general-purpose are one of the most widespread means of mechanization of cargo handling, transport, warehouse works in various types of productions and warehouse complexes [1]. According to clause 7 of article 4 of the Technical Regulations of the Customs Union TR CU 010/2011, when designing bridge cranes of general-purpose, it is necessary to develop the safety justification including risk analysis of the operation [2-5].

The resource of the bridge crane, its cost and operation safety depend to a greater degree on a metal construction. In its turn, the main and most responsible part of the metal construction of the bridge crane is a framework of the bridge (one or two main rails).

Thus, the designer has a serious task on how to develop such a metal construction of the crane that meets conditions of durability, rigidity, stability, fatigue resistance, possesses an optimum specific quantity of metal and corresponds to the required failure risk level. In the submitted paper, this issue is decided on the basis of statistical modelling of operational influences on the crane metal structure for the purpose of an estimation of failure risks of various variants of metal structures, thrashed out by the designer.

2. Materials and methods

For carrying out the analysis of failure risk of the bridge crane, it is necessary to have statistical data. They can be received in the full-scale tests of the designed crane or on the basis of supervision over similar machines. However, carrying out natural tests and collecting statistical data on the operation under conditions of the operating production are extremely expensive, and, in certain cases, it can be absolutely impracticable. Statistical modeling of the operation of the bridge crane of general-purpose, by means of which it is possible to obtain all necessary data on loading of its metal construction, can be the way out. These data can be transformed into quantitative and qualitative characteristics of reliability and, respectively, risks of failure [6].

To obtain this information, a method of statistical modeling has been used in the paper. Statistical modeling is the type of computer modeling allowing one to consider a random factor of the occurring process and to obtain statistical data on the modelled object.

For this purpose, it is necessary to have basic data about the modelled object. These basic data for the bridge crane of general-purpose are the loading capacity, parameters of driving mechanisms and an operating mode defining performance loadings.

3. Analysis of the failure risk of the bridge crane

Performance loadings for parts of the bridge crane are random variables. The adjustment of the starting and brake equipment as well as the qualification of the crane operator exert an essential influence on the distribution of performance loadings. For definition of the current efforts in parts of the bridge crane design by the method of statistical modeling, it is necessary to present this design in the form of the equations of the movement.

In figure 1, there is a control flow chart of an algorithm of steady-state modeling of loading of the bridge crane of general-purpose which considers dynamic components of loading.

Basic data for steady-state modeling of the operation of the bridge crane of general-purpose are: the crane loading capacity, the flight, the length of the serviced site of the shopfloor, the average height of load lifting, the base of the crane, speeds of executive mechanisms of the crane, the average time of strapping and unstrapping of freight, the weight of components of the crane, the modeling time.

The range of loadings of the bridge crane is set on the basis of an operating mode. The position of the cargo cart on the bridge of the crane during lifting and lowering of freight depends on the arrangement of the serviced equipment in the shopfloor, features of a production cycle, etc.

The analysis of the researches conducted by various authors allows drawing a conclusion that distribution of positions of the cargo cart during lifting and lowering of freight for each half of the bridge is close to symmetric normal distributions regarding one quarter and three quarters of flight.

Expressions for the definition of mathematical expectation a and the error of the mean square of deviation σ of the cargo cart position for the bridge crane of any flight L_k will look as follows:

– for the left half of the bridge:

$$a_l = 0.27 \cdot L_k,\ \sigma_l = 0.12 \cdot L_k;$$

– for the right half of the bridge :

$$a_r = 0.66 \cdot L_k,\ \sigma_r = 0.1 \cdot L_k.$$

Inquiry coordinates for the hoisting-and-transport operation by the crane are x_1, y_1, and unloading place coordinates x_2, y_2 are generated in each cycle of modeling. Thus, the values of these random variables are considered to be distributed according to the normal law.

After passing one cycle of modeling, final coordinates of inquiry are equated to initial ones and the cycle is repeated. The entire procedure will proceed until the counter timer is equal to preset value t_{mod} (modeling time).

We will calculate the time of operations execution using the following formulas:

– movement from an initial point to an inquiry point:

$$t_1 = \max(abs(x_0 - x_1)/v_x, abs(y_0 - y_1)/v_y,$$

where x_0, y_0 – coordinates of the initial position of the crane; v_x – crane movement speed; v_x – load carriage movement speed;

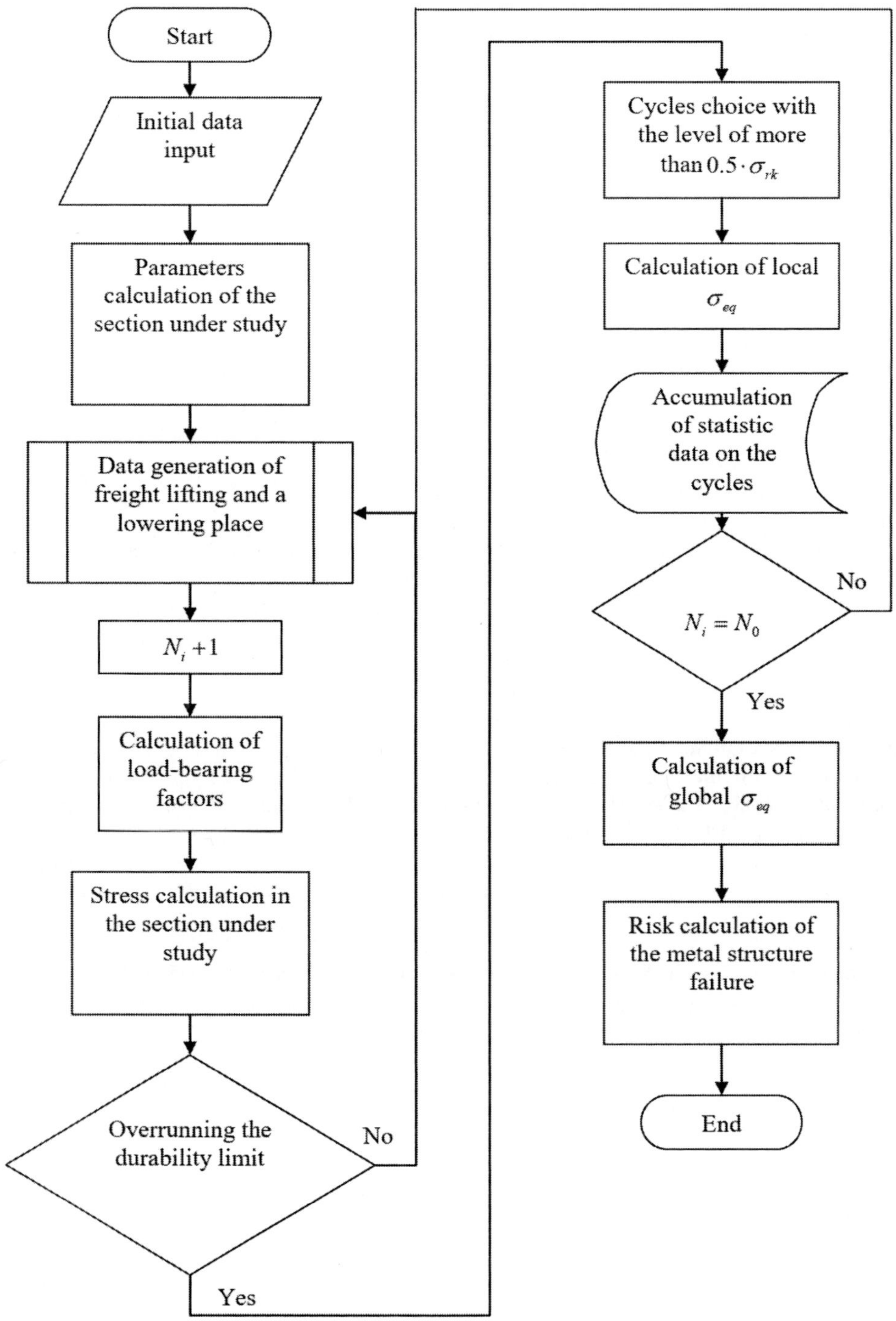

Figure 1. The control flow chart of the algorithm of statistical modeling of loading of the bridge crane.

– movement from an inquiry point to a final point:

$$t_2 = \max(abs(x_1 - x_2)/v_x, abs(y_1 - y_2)/v_y;$$

– lifting and lowering of freight:

$$t_3 = z_{av} / v_l ,$$

where z_{av} – the average height of loads lifting; v_l – speed of lifting and lowering of freight;

– general time of performance of one running cycle is

$$t_c = t_1 + t_2 + 2 \cdot t_3 .$$

The quantity of the passable rail joints for one cycle is

$$\begin{aligned} n_j = & \lfloor abs(x_0/l_r - x_1/l_r) \rfloor + \lfloor abs((x_0 + b_k)/l_r - x_1/l_r) \rfloor \\ & + \lfloor abs(x_1/l_r - x_2/l_r) \rfloor + \lfloor abs((x_1 + b_\kappa)/l_r - x_2/l_r) \rfloor \end{aligned},$$

where l_r – the length of a rail of a crane runway.

This value in combination with the known location of the cargo cart on the crane bridge will allow calculating dynamic loadings when passing rail joints.

Determination of the size of loading in ropes is carried out taking into account the dynamics of the operation of the hoisting mechanism and loads occurring when passing rail joints.

The most dangerous tension can arise because of the possibility of the collapse of the section of the main beam, which can be determined by known dependencies during bending analysis of the main beam [7-10].

4. Conclusion

Thus, the developed statistical model of the crane operation allows one to model operational loading and to obtain data on the operating stresses in a metal construction of the crane bridge for the subsequent calculation of statistical data on the failure risk. This model is used for assessment of the failure risk of cranes with regard to accumulation of elasto-plastic damages in local zones of metal constructions [11]. Authors have enabled the practical realization of the presented statistical model in the software product that allowed automating the calculation of failure risks of bridge cranes.

References

[1] Shadskii G V, Antsev V Yu and Koveshnikov V A 1989 *Soviet Engineering Research* **9 (6)** 66
[2] Popov B E, Levin E A, Kotel'nikov V S and Lipatov A S 2005 *Safety of Labor in the Ind.* **4** 33
[3] Korotkij V A, Simonov D I, Lipatov A S, Duvidovich D I and Molchanov A B 2004 *Safety of Labor in the Ind.* **3** 13
[4] Zaretskij A A, Korotkij A A, Lipatov A S, Mikushevich F E and Fedorov I G 2001 *Safety of Labor in the Ind.* **10** 61
[5] Sokolov S A 2016 *Russ. Engin. Res.* **36** 10
[6] Pas'ko N I, Pushkin N M and Inozemtsev A N 2002 *Automation and Modern Technologies* **5** 7
[7] Kazak S A 1997 *Heavy-duty production* **6** 18
[8] Kazak S A 1994 *Heavy-duty production* **11-12** 6
[9] Lagerev A V, Lagerev I A and Milto A A 2014 *International Review on Modelling and Simulations* **7 (4)** 644
[10] Lagerev A V, Lagerev I A and Milto A A 2015 *International Review on Modelling and Simulations* **8 (2)** 223
[11] Seliverstov G V, Sorokin P A and Tolokonnikov A S 2004 *Heavy-duty production* **1** 14

IOP Conf. Series: Materials Science and Engineering **177** (2017) 012054 doi:10.1088/1757-899X/177/1/012054

Test of the method for calculation of derating of workshop transformers on engineering plants

A Y Shklyarskiy and A I Bardanov

Saint-Petersburg Mining University, 2, line 21, St Petersburg, 199106, Russia

E-mail: andrey.shklyarskiy@gmail.com

Abstract. Today's innovations in engineering production are aimed at the improvement of equipment efficiency in order to reduce losses and to save environment. It is conventional that the adjustable-speed drive improves electrical and mechanical transients of workshop units, but power electronics' devices are current harmonics sources. Presence of high harmonic currents in the workshop's network is a substantial reason for transformer derating. The article discloses the method approbation, the method for calculating additional losses in an idle transformer for non-sine feeding voltage. The approbation was carried out via comparing the calculated data with the experimental data. The article describes the studies method, its strong and weak sides, and it also presents the actual calculation of coefficients for a math model, when calculating losses in a transformer, based on existing experimental data. The conclusion comprises the observation regarding possibility to use the method described to solve practical tasks.

1. Introduction

The operation under nominal load, but with higher harmonic components results in heating certain parts of an electric device, therefore, the entire transformer heated, too. The operation at temperatures above normal accelerates insulation-aging processes for all types of insulation – both organic and nonorganic, resulting further in a decreased lifespan of a transformer [1-3]. A transformer often has to work under bad conditions, with higher harmonics and voltage falls [4, 5]. In connection with this, the operation in a non-sine mode needs the transformer power to be restricted. To do so, we have to evaluate additional losses, emerging in connection with the non-sine currents and voltages, with appropriate accuracy.

Additional losses due to higher harmonics in copper mainly result from the proximity effect and the skin-effect [1-3, 6, 7]. Those losses depend on currents in windings of a transformer and rise when the transformer works for a nonlinear load.

Additional losses in iron comprise losses from re-magnetizing, which are in a proportion with frequency and eddy-current losses, which are in a proportion with the square of the frequency. Such losses appear, when the magnetizing transformer current comprises higher harmonics. It is possible when the steel core of the machine is saturated, or when the feeding voltage is not a sine wave. The latter is particularly bad for power transformers working 24 hours a day, because this increases idle operation losses, which might be substantial [8-10].

A popular method to resolve the problem of additional losses evaluation for a transformer operating in a non-sine mode might be computer simulation of the transformer operation, using the end element method. This method proved to be good [1, 6, 7], but implementing it practically needs much initial data, dedicated software and powerful computers. Therefore, a more preferable approach might be studying simplified analytical models. It is assumed that those models allow predicting idle losses of the power transformer precisely to obtain the derating factor. The model that considers values of voltage harmonics to obtain idle losses of the power transformer is described in the paper. The principal aim of the research is a verification of the model by comparing the calculated losses value with experiments results.

2. Materials and methods

Fuchs and Masoum, in their book 'Power Quality in Power Systems and Electrical Machines', introduce a math model for assessing additional losses in a transformer [3] working idle, the model describes the dependency of additional losses of a transformer operating idle only on feeding voltage harmonics, with known model coefficients. The ratio of additional losses under non-sine voltage to nominal losses in the transformer, operating idle, is to be presented as a function of feeding voltage amplitudes:

$$\frac{W_{total,h\Sigma}}{W_{total,1}} = K_1 \sum_{h=2}^{\infty} \frac{1}{h^k} \left(\frac{V_{ph}}{V_{p1}} \right)^l , \tag{1}$$

where $W_{total,h\Sigma}$ is the additional power losses due to higher harmonics of the feeding voltage, $W_{total,1}$ is the power loss of a transformer working idle in sine mode, V_{ph} is the voltage's h^{th} harmonic amplitude, V_{p1} is the amplitude of the 1st harmonic of voltage in the second case, k, l, K_1 are coefficients of the analytical model.

During the process of making formula (1), the authors made some assumptions, which inevitably affect the accuracy of the model. However, the approach is still acceptable, because using this method significantly simplifies the analysis procedure in connection with the need for lowering the transformer power, as in order to determine additional losses of an idle transformer, one needs to know only coefficients of the model and a voltage harmonics spectrum. Only determining the math model coefficients is complicated. Recommended values for coefficients k and l for a single-phase transformer are given in table 1.

Table 1. Recommended coefficients values

Coefficient	Value
k	0.9 ± 0.3
l	1.75 ± 0.25

3. Results and discussion

To approbate the method, studies data about losses in transformers working idle, obtained by Mohammad Yazdamy-Asrami et al. [7] were used. The experiment that was described in the article was a computer simulation of a single-phase power transformer based on end element method, to assess additional losses due to higher harmonic components of voltage. The applicability of the model was confirmed by matching between the computer simulation results and those obtained via a natural experiment.

The nominal power of the single-phase power transformer is 25 kVA, and the idle losses of the electrical machine is 76 W.

The transformer was studied under three different non-sine voltages, and the harmonic component amplitudes in connection with this are presented in table 2.

Table 2. Spectrums of supplying voltages

Harmonic #	1	2	5	7	9	THD, %	P, Watt
1 case	1	0.2	0.1	0.05	0.01	23	82
2 case	1	0	0.3	0.2	0	36	85
3 case	1	0.3	0.2	0.1	0.05	38	88

The task was formulated to obtain results, which would correspond to those results that were described in article [7], with the help of described math model (1), and to assess the possibility to use the recommended coefficients (table 1) to solve practical tasks. To do so, one has to determine model coefficients, with which the calculation results match the reality. To exclude coefficient K_I from the calculation, functions of ratios between additional losses for different feeding voltages by harmonics amplitudes were studied:

$$\frac{W_h^{case1}}{W_h^{case2}} = \sum_2^{\infty} \frac{1}{h^k}\left(\frac{V_h^{case1}}{V_1}\right)^l \cdot \left(\sum_2^{\infty} \frac{1}{h^k}\left(\frac{V_h^{case2}}{V_1}\right)^l\right)^{-1}, \tag{2}$$

where W_h^{case1} are additional power losses in the first case, W_h^{case2} are additional power losses in the second case, V_h^{case1} is the amplitude of the hth voltage harmonic in the first case, V_h^{case2} is the amplitude of the hth voltage harmonic in the second case.

The right part of equation (2) may be represented as the function of coefficients k and l in 3D space with coordinates k, l, z, where axis z reflects the values of the function:

$$f^{case1,2}(k,l) = \sum_2^{\infty} \frac{1}{h^k}\left(\frac{V_h^{case1}}{V_1}\right)^l \cdot \left(\sum_2^{\infty} \frac{1}{h^k}\left(\frac{V_h^{case2}}{V_1}\right)^l\right)^{-1}. \tag{3}$$

Three experiments were carried out, each resulting in formulation of an equation in accordance with expression (3):

$$\begin{cases} f^{case1,2}(k,l) = \sum_2^{\infty} \frac{1}{h^k}\left(\frac{V_h^{case1}}{V_1}\right)^l \cdot \left(\sum_2^{\infty} \frac{1}{h^k}\left(\frac{V_h^{case2}}{V_1}\right)^l\right)^{-1} \\ f^{case1,3}(k,l) = \sum_2^{\infty} \frac{1}{h^k}\left(\frac{V_h^{case1}}{V_1}\right)^l \cdot \left(\sum_2^{\infty} \frac{1}{h^k}\left(\frac{V_h^{case3}}{V_1}\right)^l\right)^{-1} \\ f^{case2,3}(k,l) = \sum_2^{\infty} \frac{1}{h^k}\left(\frac{V_h^{case2}}{V_1}\right)^l \cdot \left(\sum_2^{\infty} \frac{1}{h^k}\left(\frac{V_h^{case3}}{V_1}\right)^l\right)^{-1} \end{cases}. \tag{4}$$

The dependences of the values of functions $f^{case1,2}(k,l)$, $f^{case1,3}(k,l)$, and $f^{case2,3}(k,l)$ on coefficients k and l, and the plane of true values of ratios are represented by figure 1 (a, b, c).

The solution for each particular equation in the described space is a curve created by crossing between the plane of all possible values of the given function and the plane of its true value. It is clear that the solution for the system of equations (4) shall be surface normal $z = 0$, having common points with all curves that are solutions for the system of equations. The projection of curves, equation system solutions, onto $z = 0$ is represented with figure 1 (d).

We can derive from figure 1 (d) that the system has no solution, for the curves do not cross in one point. However, with acceptable accuracy, we may assume, that the coefficients of the analytical model correspond to some point in surroundings of point *M(0.2, 1.24)*. As it may be seen in this case, the values of coefficients k and l are outside the area, which was specified by the authors of the book (Table 1). Therefore, obtaining satisfactory results using the standard coefficient is not possible.

Coefficient K_l in this example was calculated as the arithmetical mean value of the coefficients obtained for each of the system's equation. Comparison of the calculation results of the analytical model with the results represented in article [5] is presented in table 4.

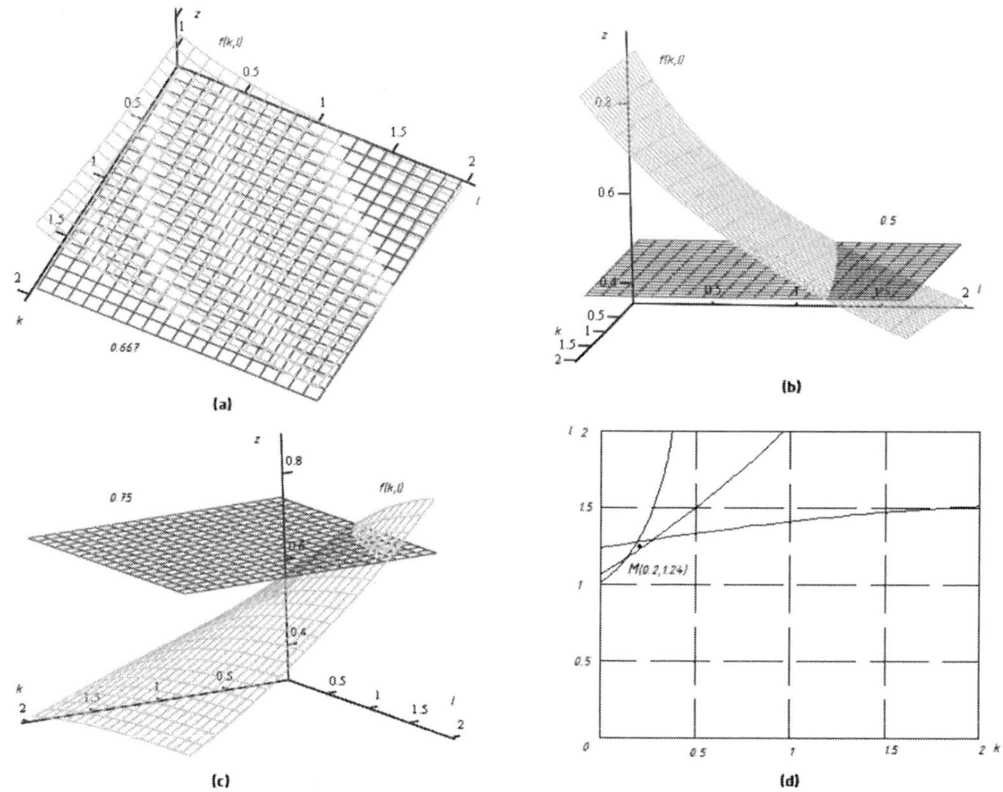

Figure 1. 3D plots of: a – function $f^{case1,2}(k,l)$; b – function $f^{case1,3}(k,l)$; c – function $f^{case2,3}(k,l)$ and d – projections of curves of individual equation solutions onto plane $z = 0$

Table 4. Comparison of calculated and measured results

No. of experiment	Measured value of the transformer idle run additional losses, Watt	Calculated value of the transformer idle run additional losses, Watt	Relative error, %
1	6	6.026	0.43
2	9	9.067	0.74
3	12	11.86	1.15

4. Conclusion

Using analysis, as presented in the article and based on results from table 4, the quantitative correlation was proved to exist between the sum of ratios of voltage amplitude harmonics and additional losses in a transformer.

For a particular transformer, the analytical model coefficients may be selected in a manner when acceptable convergence would be reachable between calculated and experimental data, at least, within small surroundings of changes of the THD voltage, with the harmonics spectrum structure preserved.

It is further worth noticing that additional experimental studies are necessary to make a final conclusion regarding the applicability of the proposed model, which reflects the correlation between losses of a transformer operating idle and the amplitudes of the feeding voltage harmonic components, thus giving a chance to the expanded methodology for adequate determination of coefficients for the studied model to be developed.

References

[1] Faiz J, Ghazizadeh M and Hashem O 2015 *IET Electr. Power Appl.* **9(7)** 486
[2] Sadati S B, Tahani A, Jafari M and Dargahi M 2004 *Iranian J. of Science and Technology* **28(3)** 315
[3] Fuchs E F and Masoum M A S 2008 *Power Quality in Power Systems and Electrical Machines* (New-York: Elsevier Academic Press)
[4] Shklyarskiy Y E, Bunteev Y E and Bardanov A I 2016 *Modern Science and Practice* **2(7)** 34
[5] Abramovich B N, Sychev Y A, Ustinov D A and Shklyarskiy A Y 2012 *Russian J. Industrial Power* **4** 23
[6] Yazdamy-Asrami M, Mirzaie M, Akmal A S and Gholamian S A 2011 *J. Electrical Systems* **7(1)** 12
[7] Yazdamy-Asrami M, Mirzaie M, Akmal A S and Gholamian S A 2013 *Energy* **50** 205
[8] Kefalas T D and Kaladas A G 2010 *IEEE Trans. on Industrial Electronics* **57(1)** 193
[9] Deokar S A and Waghmare L M 2011 *3th Int. Conf. on Electronics Computer Technology* **1** 136
[10] Mohammed O A, Abed N Y and Liu S 2006 *IEEE Trans. Magn* **42(4)** 967

Numerical simulation of the non-Newtonian fluid flow using the indirect boundary element method

M P Bessonova and V A Yakutenok

Tomsk State University, 36, Lenina ave., Tomsk, 634050, Russia

E-mail: bessonova.mp@mail.ru

Abstract. The indirect boundary element method is formulated for a two-dimensional Stokes flow with the moving boundary when gravity force aids the flow. The governing equations of low Reynolds flow are formulated. The numerical technique is described. Two regimes of the fluid flow depending on the Stokes number value were detected: the regime of full filling and the jet flow regime. The comparison of obtained results with data of other authors is presented.

1. Introduction

The process of top-down filling of the vertical channel with non-Newtonian fluid is implemented during the injection molding of polymer melts. This kind of flow has several features. Firstly, the polymeric material flow in the channel is characterized by a moving free surface. Evolution of the free surface can cause formation of gas entrainment in the flow during the filling stage of the mold. This leads to a quality degradation of the finished product. Secondly, it is necessary to take into account the non-linear properties of polymer, i.e. to consider the non-Newtonian fluid flow. The behavior of the non-Newtonian fluid is dependent upon the viscosity variation within the flow domain. This leads to the nonlinear equations in the partial derivatives in contrast to the Stokes equations for the Newtonian fluid. In addition, polymeric materials have high viscosity that allows using the creeping flow model and does not consider the surface tension. The main focus is set on determination of the shape of the free surface which is especially important for manufacturing of polymer products [1-3].

The indirect boundary element method (IBEM) is now a well-established numerical technique which provides an efficient alternative to the prevailing finite difference and finite element methods for the solution of a wide range of engineering problems [4, 5]. Its basic idea consists of transformation of partial differential equations into equivalent boundary integral equations using the fundamental solution. In the present work, IBEM is used to simulate the non-Newtonian fluid flow in the vertical channel when gravity force aids the flow.

2. Governing equations

It is well known that non-Newtonian fluids do not behave as Newtonian fluids and that their viscosity is a nonlinear function of the shear rate. The viscosity of most non-Newtonian fluids is a decreasing function of the shear rate and this is known as shear-thinning behaviour. The non-Newtonian fluid behavior is described by the power law or Ostwald-de-Waele model:

$$\tau_{ij} = 2\eta \dot{e}_{ij}\,, \tag{1}$$

where τ_{ij} – the stress tensor deviator components, $\eta = k\dot{\gamma}^{n-1}$ – the effective viscosity coefficient, k –

power law consistency, n – the power law index, $\dot{\gamma} = \left(2\dot{e}_{ij}\dot{e}_{ji}\right)^{\frac{1}{2}}$ – deformation rate intensity, $\dot{e}_{ij} = 1/2\left(\partial u_i / \partial x_j + \partial u_j / \partial x_i\right)$ – components of the rate-of-strain tensor, u_i – the velocity vector components, x_i – the Cartesian coordinates.

The following scales are adopted for transition to dimensionless variables: for the coordinates – L (a characteristic size of the domain occupied by the fluid), for the velocity – U (characteristic flow velocity), for the pressure – $k\left(U/L\right)^n$. Equation (1) does not change the external veiw, but now $\eta = \dot{\gamma}^{n-1}$ is the dimensionless coefficient of effective viscosity.

The two-dimensional flow of the power law fluid at low Reynolds taking into account the gravity force is described by the dimensionless Stokes equations system:

$$\frac{\partial \sigma_{ij}}{\partial x_j} = 0,\ i,j = 1,2, \tag{2}$$

where $\sigma_{ij} = -p_m\delta_{ij} + \tau_{ij}$ – stress tensor components, $p_m = p + \mathrm{St}x_2$ – modified pressure, p – pressure, $\mathrm{St} = \dfrac{\rho L^{n+1}\cdot(\mathbf{g}\cdot\mathbf{e}_2)}{U^n}$ – Stokes number, $\mathbf{e}_2$ – unit vectors of the x_2 axis, δ_{ij} – the Kroneker delta. Equations (2) are used together with the continuity equation:

$$\frac{\partial u_i}{\partial x_i} = 0. \tag{3}$$

As a result, the flow is determined by two parameters — St, n, and also by the domain configuration if, of course, other forces are not considered in the boundary conditions, for example, the surface tension forces.

The formulation of the problem on filling a vertical channel with the non-Newtonian fluid is given in figure 1.

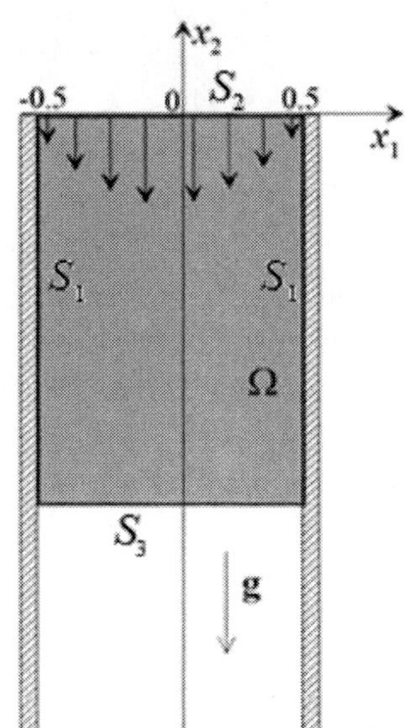

Figure 1. The solution domain $\left(S = S_1 + S_2 + S_3\right)$.

The flow rate of the fluid at the inlet into the channel is considered to be given. The channel width is chosen as a characteristic length, the average flow velocity is characteristic velocity. The boundary conditions at the inlet and on the solid walls are specified by the following relations:

$$u_1 = 0,\ u_2(x_1) = \frac{2n+1}{n+1}\left(1 - |2x_1|^{\frac{n+1}{2}}\right), \tag{4}$$

$$u_i = 0. \tag{5}$$

Dynamic and kinematic boundary conditions are performed on the free surface:

$$t_i = \mathrm{St}_k x_k n_i = t_i^N + \tau_{ij}^{NN} n_j, \tag{6}$$

$$\frac{dx_i}{dt} = u_i, \tag{7}$$

where t_i^N – the linear part of the traction vector, $\tau_{ij}^{NN} = 2(\eta - 1)\dot{e}_{ij}$ – the nonlinear part of the stress tensor. The kinematic condition is written in the Lagrangian form and is used to find the shape of the free boundary. The initial condition is the shape of the free surface at the initial time.

3. Numerical solution

The indirect boundary element method is used for numerical solution. According to the postulates of the IBEM [6, 7], one can write down:

$$\begin{aligned} u_i(x) &= \int_S G_{ij}(x,\xi)\varphi_j(\xi)dS(\xi) + \int_\Omega G_{ij}(x,z)\Psi_j(z)d\Omega(z), \\ t_i^N(x) &= \int_S F_{ij}(x,\xi)\varphi_j(\xi)dS(\xi) + \int_\Omega F_{ij}(x,z)\Psi_j(z)d\Omega(z), \end{aligned} \tag{8}$$

where $\varphi_j(\xi)$ – density of the fictitious sources distributed along boundary S, $\Psi_i = -\partial\tau_{ij}^{NN}/\partial x_j$ – density of the sources distributed about flow domain Ω. Functions G_{ij} and F_{ij} are the singular solutions of Stokes equations [8] and they are determined by formulas

$$G_{ij}(x,\xi) = -\frac{1}{4\pi}\left(\delta_{ij}\ln\frac{1}{r} + \frac{y_i y_j}{r^2}\right),\quad F_{ij}(x,\xi) = \frac{y_i y_j y_k n_k(x)}{\pi r^4}, \tag{9}$$

where $y_i = x_i - \xi_i$, $r = y_i y_i^{1/2}$.

If the values of $t_i^N(x)$, $u_i(x)$ are specified on boundary S, equations (8) make it possible to get the values of unknown boundary forces $\varphi_j(\xi)$ $(\xi \in S)$. But since $\Psi_i(z)$ $(z \in \Omega)$ functions were not known before, one can use an iteration method for this purpose.

For a numerical solution of equations (8), the constant elements and the constant internal cells are applied. The discretized boundary integral equations (8) acquire the following form:

$$\begin{aligned} u_i(x^p) &= \sum_{q=1}^{N}\varphi_j^q \Delta G_{ij}^{pq} + \sum_k\sum_m \Psi_j^{km}\Delta G_{ij}^{pkm}, \\ t_i^N(x^p) &= \sum_{q=1}^{N}\varphi_j^q \Delta F_{ij}^{pq} + \sum_k\sum_m \Psi_j^{km}\Delta F_{ij}^{pkm}, \end{aligned} \tag{10}$$

where $\Delta G_{ij}^{pq} = \int_{\Delta S^q} G_{ij}(x^p,\xi)dS(\xi)$, $\Delta F_{ij}^{pq} = \int_{\Delta S^q} F_{ij}(x^p,\xi)dS(\xi)$, $\Delta G_{ij}^{pkm} = \int_{\Delta\Omega^{km}} G_{ij}(x^p,z)d\Omega(z)$, $\Delta F_{ij}^{pkm} = \int_{\Delta\Omega^{km}} F_{ij}(x^p,z)d\Omega(z)$, x^p is the middle of element p, i.e. node p. The coefficients of the obtained system of linear algebraic equations ΔG_{ij}^{pq} and ΔF_{ij}^{pq} for the case of the constant elements may be computed analytically [9].

To compute domain integrals ΔG_{ij}^{pkm} and ΔF_{ij}^{pkm}, the standard quadrature Gaussian formulas are used without the singularities extraction. During the process of computing, the quadrature formulas with 64 nodes are applied. The system of linear algebraic equations for the Newtonian fluid flow is solved by the Gauss standard method. In the case of non-Newtonian fluid flow, the algebraic equations system becomes nonlinear and is solved by simple iteration method. This method requires values of pseudo forces in each boundary element and each internal collocation point obtained from boundary conditions (for velocities and tractions) and the velocity field, found at the previous iteration. Internal cells sources are estimated numerically within a finite difference approach to velocity derivatives calculation.

4. Results

When filling the channel, the flow pattern of Newtonian and power law fluids is determined by parameters St and n. Figures 2, 3 show the evolution of the free surface shapes of a Newtonian $(n = 1.0)$ and shear-thinning fluids $(n = 0.9$ and $0.8)$.

There are two filling regimes. The change of the filling regimes occurs depending on the value of Stokes number. When $\mathrm{St} = 1$, the fluid continuously flows and fills the channel. The initially flat free surface extends and takes the steady convex shape and moves along the channel. This regime is called the regime of fulfilling.

In figure 4, the comparison of a steady free surface shape, calculated in this paper, with results of [10] for case $n = 1.0$ is presented. We can see that the results are consistent.

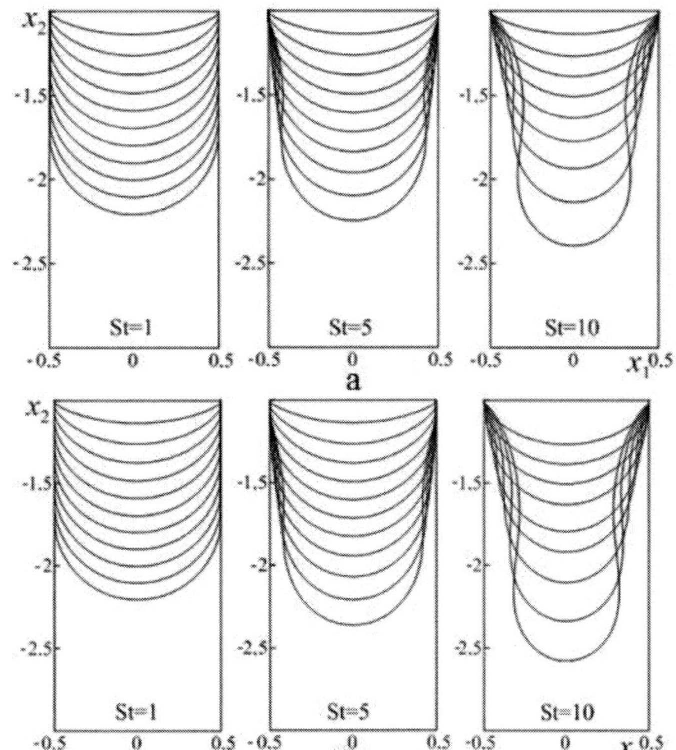

Figure 2. The evolution of the free surface shapes of a hear-thinning fluids: a – $n = 0.8$, $\Delta t = 0.2$, $n = 0.9$, $\Delta t = 0.2$

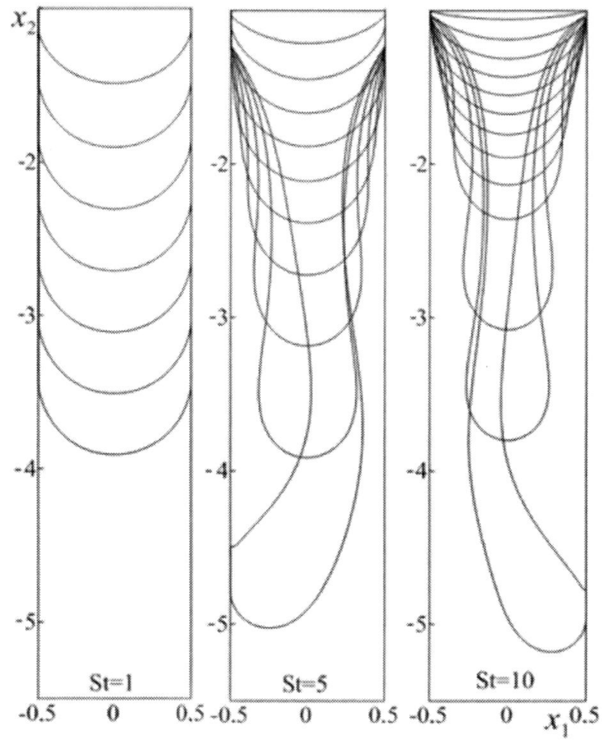

Figure 3. The evolution of the free surface shapes of Newtonian fluid $n = 1.0$, $\Delta t = 0.4$

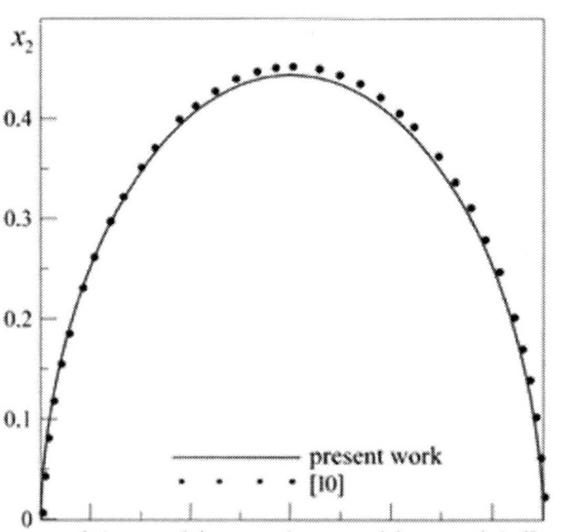

Figure 4. The free surface shapes for $n = 1.0$, St $= 0$.

An increase of the value of Stokes number leads to formation of the jet flow regime. The free surface acquires the drop shape. With time, the minimum transverse size of jet decreases. In addition, we can see that power law index n also influences the free surface shape. With decreasing n, the minimum transverse size increases.

In case $\mathrm{St} = 5$ and 10, the Newtonian fluid flow is characterized by jet deformation under the influence of gravity. In some time point the jet of fluid becomes unstable and flow symmetry is broken (figure 3). This leads to formation of voids which are classified in finished products as defects.

5. Conclusion

The presented results indicate that the indirect boundary element is an efficient means to solve the problems dealing with the non-Newtonian fluids with a free surface. The IBEM combined with the simple iteration method allows modelling the fluid flows at different power law index values. The governing equations were formulated and the numerical simulation method was described. The influence of flow non-linearity and the value of Stokes number on the shape free surface flow were considered. Two regimes of fluid flow were detected: the regime of fulfilling and the jet flow regime.

References

[1] Vos E, Meijer H H. & Peters G M 1991 *Intern. J. of Polymer Processing* **4** 42-50
[2] Zoetelief W F, Peters G M & Meijer H H 1997 *Intern. J. of Polymer Processing* **7** 216
[3] Haagh G V & Van de Vosse F 1998 *Intern. J. for Numerical Methods in Fluids* **28** 1355
[4] Jin X *Polymer Engineering and Science* 1993 **33(19)** 1238–42
[5] Ponomareva M A, Yakutenok V A 2015 *WIT Transactions on Modelling and Simulation* **61** 285
[6] Brebbia C A, Telles J F, Wrobel L C 1984 Boundary Element Techniques. Theory and applications in engineering, Sprinqer-Verlag: Berlin
[7] Banerjee P K & Butterfield R Boundary Element Methods in Engineering Science, McGraf-Hill book company: UK
[8] Ladyzhenskaya O A 1969 The Mathematical Theory of Viscous Incompressible Flow, Gordon & Breach. 2nd ed.: New York
[9] Ponomareva M A, Filina M P, Yakutenok V A 2014 *WIT Transactions on Modelling and Simulation* **56** 289
[10] Mitsoulis E 2010 *AIChE Journal* **56(5)** 1147–62

Numerical simulation of dynamic channel-angular pressing of copper specimens

A S Bodrov[1] and S A Zelepugin[1,2]

[1] National Research Tomsk State University, 36, Lenin ave., Tomsk, 634050, Russia
[2] Tomsk Scientific Center of the Siberian Branch of the Russian Academy of Sciences, 10/3, Akademichesky ave., Tomsk, 634055, Russia

E-mail: alex.bodrov@mail.ru

Abstract. The process of severe plastic deformation of a copper specimen during dynamic channel-angular pressing is numerically investigated in a 3D statement for the dynamic scheme of loading. Computations have been carried out by the finite element method within the framework of the elastic-plastic medium model with allowance for fracture. Loading pressure and initial velocity values are determined to provide the possibility of the process of dynamic channel-angular pressing of copper specimens.

1. Introduction

Bulk nanostructural and ultrafine grained materials are considered to be perspective constructional and functional materials of new generation [1-3]. Investigations of ultrafine-grained metals obtained by severe plastic deformation (SPD) methods have shown that they are characterized by a number of unique properties in comparison with coarse-grained analogues: higher strength combined with high plasticity, low- and high-temperature superplasticity, radiation resistance, etc. The study of such materials properties has theoretical and applied importance.

One of the widely used SPD methods is equal-channel angular pressing (ECAP) [1]. A new SPD method such as dynamic channel-angular pressing (DCAP) is proposed, in which pressing of a specimen through channels is carried out by pulse loading caused by the energy of compressed gases [4, 5]. DCAP allows the specimens of considerably bigger sizes and a smaller number of passes to be used, and the specimen material retains a high plasticity in comparison with the ECAP method.

Experimental studies show that there is a need in the large-scale numerical investigations of DCAP processes to determine the effective parameters of SPD. At present, there are few works devoted to the numerical modeling of DCAP processes [6-8]. This paper presents an investigation of the DCAP process parameters (initial velocity and pressure) for the copper specimen. The SPD process of the bulk copper specimen with the use of DCAP was numerically investigated in the three-dimensional statement for the dynamic scheme of loading. Computations have been carried out by the finite element method within the framework of the elastic-plastic medium model with allowance for fracture [10, 11].

2. Computational approach

The basic numerical model used in our numerical code for solving high-velocity impact problems comprises the set of differential equations of continuum mechanics such as conservation of mass, momentum, energy, and constitutive relationships. The material model includes an equation of state that provides pressure as a function of the mass density and internal energy, a deviatoric elastic constitutive relationship, a yield criterion, and a material failure model.

The model for the damaged medium used in the calculations is characterized by the possibility of microdamages formation. The total volume of the medium, W, consists of undamaged volume W_c of density ρ_c and damaged volume W_f of zero density. The overage density of the total volume is given by $\rho = \rho_c(W_c/W)$. The level of the medium damage is characterized by the specific volume of microdamages $V_f = W_f/(W \cdot \rho)$.

The system of equations describing unsteady adiabatic movements of a compressible medium with allowance for the development of microdamages is [6, 9, 10]

$$\frac{1}{\rho}\frac{\mathrm{d}\rho}{\mathrm{d}t} + \frac{\partial v_i}{\partial x_i} = 0, \tag{1}$$

$$\frac{\mathrm{d}v_i}{\mathrm{d}t} = \frac{1}{\rho}\frac{\partial \sigma_{ij}}{\partial x_j}, \tag{2}$$

$$\rho\frac{\mathrm{d}E}{\mathrm{d}t} = \sigma_{ij}\varepsilon_{ij}. \tag{3}$$

Here, ρ is the density, t is the time, v_i are the components of velocity, $\sigma_{ij} = P\delta_{ij}+S_{ij}$ are the stress tensor components, E is the specific internal energy, ε_{ij} are the components of the strain rate tensor, $P = P_c(\rho/\rho_c)$ is the overage pressure, δ_{ij} is the Kronecker delta, S_{ij} are the components of the stress deviator, P_c is the pressure in the undamaged substance component.

The pressure in the undamaged substance is a function of specific volume, specific internal energy and specific volume of microdamages; and, throughout the range of loading conditions, it is determined by the Mie-Grüneisen equation of state:

$$P_c = \rho_0 a^2\mu + \rho_0 a^2[1-\gamma_0/2+2(b-1)]\mu^2 + \rho_0 a^2[2(1-\gamma_0/2)(b-1)+3(b-1)^2]\mu^3 + \gamma_0\rho_0 E, \tag{4}$$

where $\mu = V_0/(V-V_f)$, γ_0 is the Grüneisen constant; V_0 and V are the initial and current volumes, respectively; a and b are the constants of the Hugoniot shock adiabat, described by relation $u_s=a+bu_p$, where u_s is the shock wave velocity and u_p is the particle velocity behind the shock wave front.

The constitutive relations connect the components of the stress deviator and strain rate tensor and use the Jaumann derivative. The Mises yield criterion is used to describe the plastic flow. A kinetic failure model of the active type for the simulation of fracture in various metals is used for numerical modeling [11]:

$$\frac{\mathrm{d}V_f}{\mathrm{d}t} = \begin{cases} 0, \text{ if } |P_c| \le P^* \text{ or if } (P_c > P^* \text{ and } V_f = 0), \\ -\operatorname{sign}(P_c)\mathrm{K}_f(|P_c| - P^*)(\mathrm{V}_2 + V_f), \text{ if } P_c < -P^* \text{ or if } (P_c > P^* \text{ and } V_f > 0) \end{cases}. \tag{5}$$

Here, $P^*=\mathrm{P_k V_1}/(V_f+\mathrm{V_1})$; $\mathrm{V_1}$, $\mathrm{V_2}$, $\mathrm{P_k}$, $\mathrm{K_f}$ are the experimentally determined constants ($P^*>0$).

Shear modulus G and dynamic yield strength σ were assumed to depend on attained damage level [12]:

$$G = \mathrm{G_0}K_\mathrm{T}\left(1+\frac{\mathrm{c}P}{(1+\mu)^{1/3}}\right)\frac{\mathrm{V_3}}{(V_f+\mathrm{V_3})} \tag{6}$$

$$\sigma = \begin{cases} \sigma_0 K_\mathrm{T}\left(1+\frac{\mathrm{c}P}{(1+\mu)^{1/3}}\right)\left(1-\frac{V_f}{\mathrm{V_4}}\right), \text{ if } V_f \le \mathrm{V_4}, \\ 0, \text{ if } V_f > \mathrm{V_4} \end{cases} \tag{7}$$

$$K_T = \begin{cases} 1, \text{if } T_0 \le T \le T_1, \\ \dfrac{T_m - T}{T_m - T_1}, \text{if } T_1 < T < T_m, \\ 0, \text{if } T \ge T_m \end{cases} \quad (8)$$

where T_m is the melting point of the material, c, V_3, V_4 and T_1 are the experimentally determined constants. In the computations, function $K_T(T)$ was chosen to model the nonthermal character of plastic deformation and the dynamic strength of solids at high strain rates (10^4 s^{-1} or higher).

The interaction of the copper specimen with intersecting channels is considered. The initial (at t=0) and prescribed boundary conditions are introduced on the surfaces in Cartesian coordinates for constitutive equations (1)-(8). The sliding conditions are met on the contact surface between the specimen and the channels walls. The rigid wall boundary conditions are imposed on the channels. Constant load P that simulates the pressure of powder gases is applied to the back surface of the specimen (dynamic loading scheme). The modified finite element method (without the global stiffness matrix) is used for the solution of the formulated problem [13].

3. Computational results

The DCAP process was modeled for the copper specimen with linear dimensions of 16x16x65 mm. The angle of the channels intersection is 90° with an inclined plane making 45° angles with the lateral walls of an external corner. The value of load P and the initial velocity of specimen v_0 are varied (figure 1).

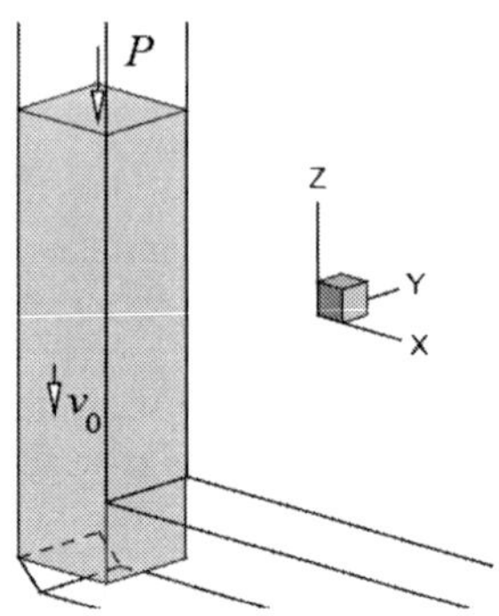

Figure 1. The initial position of the specimen and the geometry of the channels.

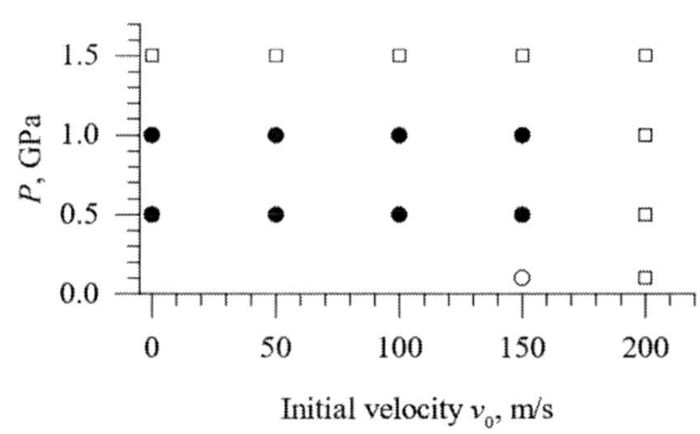

Figure 2. Results of the numerical simulation of the DCAP process: • – successful passage of the specimen through channels, ○ – the specimen is locked, □ – unstable passage or destruction of the specimen.

The analysis of computational results allows us to determine the combination of the initial velocity and load values, which can provide successful passage of the specimen through channels (the black circles in figure 2). The values of load essentially influence the DCAP process (figures 3-4). If the values of load are not high enough, the specimen can be locked in the channels, on the other hand, extremely high values can lead to unstable passage and destruction of the specimen. When the values of load are constant, the increase in the initial velocity leads to elongation of the specimen, and if the load value is insufficient to provide the DCAP process, it can be compensated by increasing the initial velocity. Increasing the load value leads to the fact that the specimen more strongly stretches in the direction of a longitudinal axis (figures 3-4). The speed of the specimen passing through the intersecting channels also increases. As a result, the strain rate increases, the temperature rises inside the specimen. The analysis of the specific energy during shear deformations shows almost full identity of plastic deformation fields in the transverse direction, which testifies the preference of using the square section specimens in DCAP in comparison with the cylindrical ones. At the same time, there is a non-uniformity of deformation occurring along the sample, which is especially seen in the forefront of the specimen and can lead to the necessity of the multiple specimen passes through the channels. It is necessary to notice that the maximum values of specific energy in shear deformations are reached in surface layers of the specimen due to the interaction with the walls of the channels (figure 3b). Melting of the material can take place in these areas. In the upper part of the specimen, the increase of microdamages in the area is observed (figure 4b).

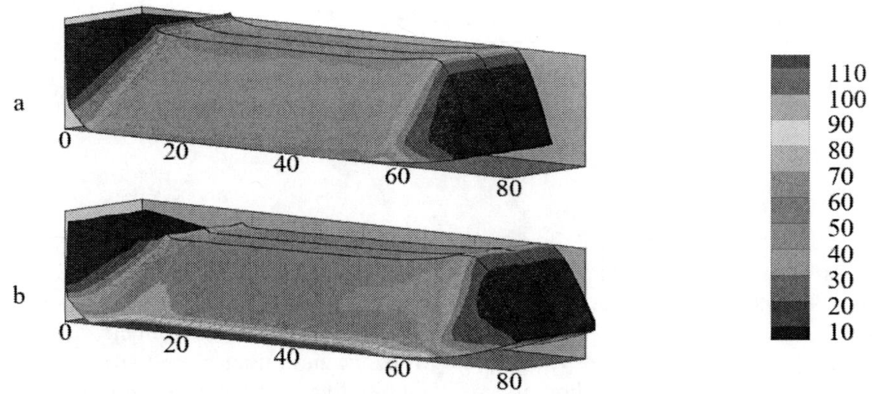

Figure 3. Distribution of the specific energy of shear deformations (kJ/kg) in the specimen: a) P=0.5 GPa, v_0=50 m/s, t=596 μs; b) P=1 GPa, v_0=50 m/s, t=314 μs. The dimensions are in mm.

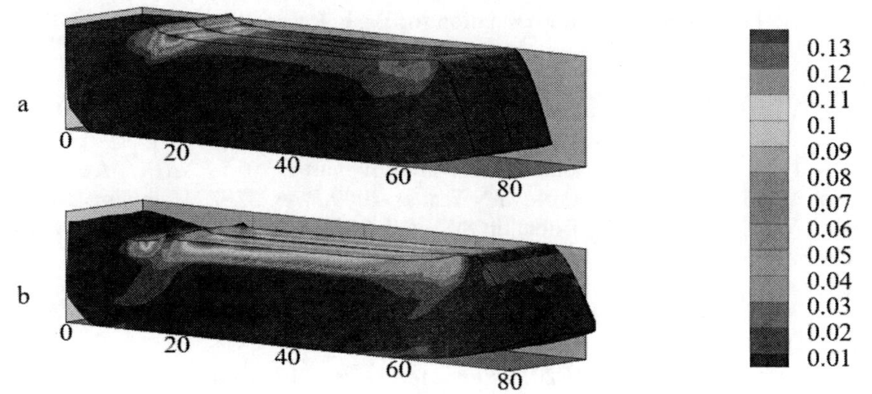

Figure 4. Distribution of the specific volume of microdamages (cm^3/g) in the specimen: a) P=0.5 GPa, v_0=50 m/s, t=596 μs; b) P=1 GPa, v_0=50 m/s, t=314 μs. The dimensions are in mm.

Figure 5 shows the kinetic energy of the specimen for all successful DCAP processes.

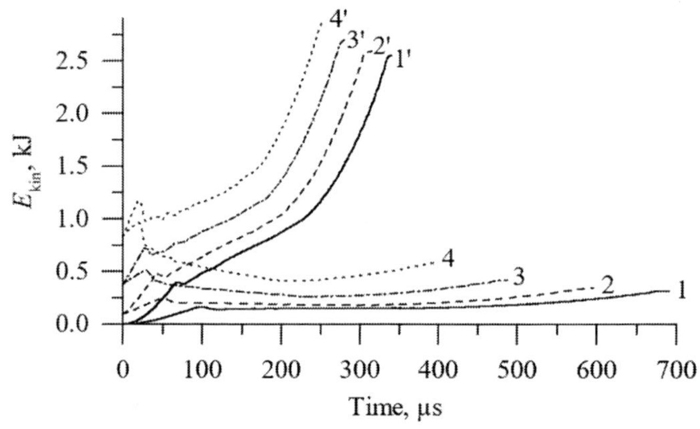

Figure 5. Kinetic energy of the specimen during DCAP for different values of P and v_0:
1 – P=0.5 GPa, v_0=0 m/s,
2 – P=0.5 GPa, v_0=50 m/s,
3 – P=0.5 GPa, v_0=100 m/s,
4 – P=0.5 GPa, v_0=150 m/s,
1' – P=1 GPa, v_0=0 m/s,
2' – P=1 GPa, v_0=50 m/s,
3' – P=1 GPa, v_0=100 m/s,
4' – P=1 GPa, v_0=150 m/s.

4. Conclusion

Numerical investigation is conducted in the three-dimensional statement for the copper specimen deformation processes during DCAP. The diagram of the DCAP process is obtained for the titanium specimen in co-ordinates $v_0 - P$ (initial velocity of the specimen – pressure acting on a rear surface of the specimen). The computational results show that the increase in the velocity of the specimen passing through channels (rate of the specimen deformation) and initial velocity of the sample lead to the lengthening of the sample in a direction of a longitudinal axis, the temperature growth in the sample and the growth of microdamages.

Acknowledgments
The work was funded by the Ministry of Education of the Russian Federation within state task №2014 / 223 (project no. 1943) and the Russian Foundation for Basic Research (project no. 16-43-700774)

References

[1] Valiev R Z, Islamgaliev R K and Alexandrov I V 2000 *Prog. Mater. Sci.* **45**(2) 103
[2] Valiev R 2004 *Nature Mater.* **3**(8) 511
[3] Skryabina N E, Aptukov V N, Romanov P V and Fruchart D 2015 *PNRPU Mech. Bull.* **3** 133
[4] Zeldovich V I, Shorokhov E V, Frolova N Y et al. 2009 *Phys. MET. Metallogr.* **108**(4) 347
[5] Zeldovich V I, Shorokhov E V, Dobatkin S V et al. 2011 *Phys. MET. Metallogr.* **111**(4) 421
[6] Shipachev A N, Il'ina E V and Zelepugin S A 2010 *Deform. Razr. Mater.* **4** 20
[7] Shipachev A N, Zelepugin A S, Il'ina E V and Zelepugin S A 2012 *Deform. Razr. Mater.* **10** 7
[8] Bodrov A S, Olimpieva N V, Zelepugin A S and Zelepugin S A 2016 *Proc. 2015 Int. Conf. MEACS* 7414878
[9] Gerasimov A V and Pashkov S V 2015 *Russ. Aeronaut. Iz.VUZ.* **58**(2) 137
[10] Gerasimov A V, Pashkov S V and Cherepanov R O 2016 *Int. J. Appl. Eng. Res.* **11**(18) 9430
[11] Kanel' G I, Razorenov S V, Utkin A V and Fortov V E 1996 *Shock-Wave Phenomena in Condensed Matter* (Moscow: Yanus-K)
[12] Gorel'skii V A, Zelepugin S A and Sidorov V N 1994 *Int. Appl. Mech.* **30**(3) 193
[13] Johnson G R 2011 *Int. J. Impact Eng.* **38**(6) 456

Mathematical simulation of surface heating during plasma spraying

V I Bogdanovich and M G Giorbelidze

Samara National Research University, 34, Moskovskoe Highway, Samara, 443086, Russian Federation

E-mail: bogdanovich@ssau.ru

Abstract. A mathematical model of temperature distribution over the flat 'coating-substrate' system section during plasma spraying, taking into account a plasma gun travel and coating buildup has been developed. It has been shown that the temperature value in the near-surface layer of the sprayed coating during the plasma gun passage can significantly exceed the temperature values in underlayers.

1. Introduction

The creation of promising aviation gas turbine engines having more advanced technical characteristics is closely connected with the increase of the temperature of the gas flow at the inlet of the turbine, as a result of which increasingly high demands are made for the parts of the hot gas path of the turbine. In view of the fact that the potential of the modern refractories is virtually exhausted, one of the most efficient methods of lowering the temperature loading on the surface of the structural material is application of special coatings, possessing high indices of thermal stability, erosion resistance and providing the effective protection of the parts surface from intercrystalline high-temperature corrosion. Owing to the possibility of combination of high values of physico-mechanical and service properties during formation of the meso-ordered structure, plasma heat-resistant coatings are the best prospect. However, in the process of applying these coatings, the surface of the construction material can be subjected to considerable thermal actions on the part of the plasma jet, as a result of which the problem of temperature distribution in the system 'coating-substrate' is highly relevant [1-10].

2. Statement of the problem of the mathematical model of surface heating

When defining the temperature distribution during plasma spraying, the averaged influence of the heat flux on the system 'coating-substrate' [2] was taken into consideration. In this statement of the problem, the heat obtained by the substrate can be represented as a sum of several constituents:

$$Q = Q_1 + Q_2 + Q_3 + Q_4 + Q_5 - Q_6 - Q_7, \quad (1)$$

where Q_1 is the heat released owing to a conversion of kinetic energy of the particles into thermal energy during the impact on the substrate; Q_2 is the heat obtained by the substrate from the plasma jet; Q_3 is the heat released during crystallization of the molten particles; Q_4 is the heat released during cooling down of the particles; Q_5 is the heat released during exothermic reactions in the sprayed materials; Q_6 is the heat transmitted by the surface owing to convective heat exchange; Q_7 is the heat withdrawn owing to the heat conduction inside the substrate material.

When considering some small volume of the surface of the coating and representing the process of coating in the form of the model with a continuous build-up of the layer, using the heat-balance equation, the boundary condition for the sprayed surface can be represented as:

$$\lambda_2 \frac{\partial T_2}{\partial z} = q(t) - \alpha_2 (T_2 - T_a), \quad q(t) = q_0 f(t), \text{ when } z = \delta(t), \tag{2}$$

where $q(t)$ is the greatest heat flux on the jet axis, α_2 is the heat exchange factor, T_2 is the coating temperature, T_a is the ambient temperature, z is the coordinate denoting the thickness in the 'coating-substrate' system, and $\delta(t)$ is the sprayed-surface coordinate.

The heat of the plasma jet is transmitted to the part through a heating spot with diameter d, which, in general case, differs from the diameter of the spraying spot. Heat flux density q owing to uneven distribution over the heating spot (the amount of heat introduced through the spraying spot surface per time unit) can be described by means of a certain distribution law. As a rule, the density of the thermal energy, introduced by the heated gas and heated particles, is recorded with the use of different laws of distribution. However, when the plasma gun is adjusted in a certain way, so that the maximums of heat flux density coincide, the summation thermal action can be described by a single law of distribution with averaged parameters determined based on the experiments.

Owing to the fact that transmission and transfer of heat in the plasma jet depend on a large number of random factors, it is possible to consider that the density of the heat flux complies with the normal law of distribution. This assumption has found its experimental verification.

From the practical standpoint, an opportunity of considering the periodical influence of the plasma jet on the system 'coating-substrate' when calculating the temperature distribution is of special interest. The distribution of the heat flux of the plasma jet throughout the area of the heating spot in case of high capacity of the heat flux and the significant speed of the plasma gun travel can be recorded by means of the normal law of distribution:

$$q(r) = q_0 \exp(-kr^2), \text{ when } r = (x - Vt)^2 + y^2, \tag{3}$$

where k is the factor of concentration of the jet heat flux, and V is the speed of the plasma gun travel.

Having fixed a spot on the surface of spraying with coordinates $x = y = 0$, let us write the expression for determination of the density of the heat flux during a single passage of the plasma gun through this point:

$$q(r) = q_0 \exp(-kV^2 t^2). \tag{4}$$

In the process of spraying, the plasma gun passes over this point with time interval t_1; the duration of the influence of the plasma jet on this spot is defined by equation $t_2 = d_x / V$, where d_x is the spraying spot diameter. For time t_2, the process of the coating build-up takes place at speed $d\delta / dt$. In this case, the heat flux density has the following view:

$$q(t) = \sum_{n=1}^{N} q_n(t) - \{\eta[t-(n-1)t_1] - \eta[t-(n-1)t_1 - t_2]\}, \tag{5}$$

$$q_n(t) = q_0 \exp\{-kV^2 [t-(n-1)t_1 - 0{,}5t_2]^2\}, \tag{6}$$

where $\eta(t)$ is the asymmetric unit function; index n corresponds to the number of plasma gun passages over this point; and the concentration factor of the heat flux of the jet is related to spraying spot diameter d by relationship $d = 2(\ln 20)^{1/2} k^{1/2}$.

The coating build-up rate and the process of its thickness increase can be presented by expressions:

$$\frac{d\delta}{dt} = V_z(t)\{\eta[t-(n-1)t_1] - \eta[t-(n-1)t_1 - t_2]\}, \tag{7}$$

$$\delta(t)=\begin{cases}(n-1)\delta_0+\int\limits_{(n-1)t_1}^{t} V_z(\tau)d\tau, & t\in\left[(n-1)t_1,(n-1)t_1+t_2\right]\\ n\delta_0, & t\in\left[(n-1)t_1+t_2,nt_1\right],\end{cases} \tag{8}$$

where $\delta_0=\int\limits_0^{t_2} V_z(\tau)d\tau$ is the coating thickness obtained per single plasma gun passage, and $V_z(t)$ is the coating build-up rate per single plasma gun passage over the surface spot under consideration.

3. A mathematical model of surface heating

The mathematical model of the heat problem during spraying the coating onto a substrate surface of a plate-like shape has the following view in case of the averaged thermophysical parameters:

$$\frac{\partial T_2}{\partial t}=a_2^2\frac{\partial^2 T_2}{\partial z^2}, \quad 0<z\le\delta(t); \tag{9}$$

$$\frac{\partial T_1}{\partial t}=a_1^2\frac{\partial^2 T_1}{\partial z^2}, \quad -h<z\le 0; \tag{10}$$

$$\lambda_1\frac{\partial T_2}{\partial z}=q(t)-a_2(T_2-T_c), \quad z=\delta(t); \tag{11}$$

$$\lambda_1\frac{\partial T_1}{\partial z}=a_1(T_1-T_c), \quad z=-h; \tag{12}$$

$$\lambda_1\frac{\partial T_1}{\partial z}=\lambda_2\frac{\partial T_2}{\partial z}, \quad T_1=T_2,\ z=0; \tag{13}$$

$$T_1(z,0)=T_0\ , \tag{14}$$

where T_1 is the substrate temperature, T_0 is the initial temperature of the system, a_i and λ_i are temperature conductivity and heat conductivity factors of the substrate (i=1) and the coating (i=2), respectively.

For the convenience of solving the boundary value problem (9-14), let us introduce dimensionless variables:

$$\xi=\frac{z}{h},\ \zeta=\frac{\delta}{h},\ \zeta_0=\frac{V_z d_n}{hV}, Fo=\frac{a^2 t}{h^2},\ \theta_i=\frac{T_i-T_0}{T_0-T_c}, i=1,2; \tag{15}$$

and parameters:

$$Bi_1=\frac{\alpha_1 h}{\lambda_1};\ Bi_2=\frac{\alpha_2 h}{\lambda_2};\ k_\lambda=\frac{\lambda_1}{\lambda_2};\ k_a=\frac{a_1^2}{a_2^2};\ \beta=kh^2;\ Ki_0=\frac{hq_0}{\lambda_2(T_0-T_c)},$$

$$Fo_1=\frac{a_1^2 t_1}{h^2};\ Fo_2=\frac{a_2^2 t_2}{h^2};\ Pe_z=\frac{hV_z}{a_1^2};\ Pe=\frac{hV}{a_1^2}, \tag{16}$$

where h is the coating thickness.

Relationships (5-8) determining $q(t)$ and $\delta(t)$ in designations (15-16) are transformed as follows:

$$Ki(Fo)=Ki_0\sum_{n=1}^{N} f_n(Fo)\left\{\eta\left[Fo-(n-1)Fo_1\right]-\eta\left[Fo-(n-1)Fo_1-Fo_2\right]\right\}; \tag{17}$$

$$f_n(Fo)=\exp\left\{-\beta Pe^2\left[Fo-(n-1)Fo_1-\frac{1}{2}Fo_2\right]\right\}; \tag{18}$$

$$\zeta(Fo)=\begin{cases}(n-1)\zeta_0+\int\limits_{(n-1)Fo_1}^{Fo} Pe_z(\tau)d\tau, \\ \quad (n-1)Fo_1 \le Fo \le (n-1)Fo_1+Fo_2 \\ n\zeta_0, \quad (n-1)Fo_1+Fo_2 < Fo \le nFo_1 \end{cases} \tag{19}$$

Solving the boundary value problem using the differential series method [3, 4] and neglecting the terms of order of smallness ζ^k, $k \ge 2$, let us write the following conditions on stationary boundary $\zeta = 0$:

$$\left.\frac{\partial\theta_2}{\partial\xi}\right|_{\xi=0} = \Psi(Fo) - \frac{Bi_2}{k_\lambda}\left(1+\theta_1(0,Fo)\right) - \frac{k_a}{k_\lambda}\zeta\left.\frac{\partial\theta_1}{\partial Fo}\right|_{\xi=0}; \tag{20}$$

$$\Psi(Fo) = \frac{Ki}{k_\lambda} - \frac{1}{k_\lambda}\left[Ki - Bi_2\left(1+\theta_1(0,Fo)\right)\right]\frac{Bi_2\zeta}{1+Bi_2\zeta}. \tag{21}$$

The solution of the boundary value problem of thermal conductivity intended for the substrate (having stationary boundaries $\zeta = -1$, $\zeta = 0$) differs from the solution obtained in [3] by the fact that in this approach, owing to supplementary terms of the expression, the heat losses and heating of the lower layers of the coating are taken into consideration. This approach is the most important one when solving the problems with rapidly moving heat sources.

Using a series of the subsequent transformations, including the Laplace transformation by variable Fo, and applying the expansion theorem, let us write the expression for relatively excessive temperature:

$$\theta_1(\xi,Fo) = -1 + 2\sum_{n=1}^{\infty}\frac{A_n\cos\mu_n\xi + B_n\sin\mu_n\xi}{D_n}\exp\left(-\mu_n^2 Fo\right) + \\ + 2\int_0^{Fo}\Psi(Fo-\tau)\sum_{n=1}^{\infty}\frac{\overline{A}_n\cos\mu_n\xi + \overline{B}_n\sin\mu_n\xi}{D_n}\exp\left(-\mu_n^2\tau\right)d\tau \tag{22}$$

Substituting solution (21) in (22) for $\Psi(Fo)$, we will obtain:

$$\theta_1(\xi,Fo) = -1 + 2\sum_{n=1}^{\infty}\frac{A_n\cos\mu_n\xi + B_n\sin\mu_n\xi}{D_n}\exp\left(-\mu_n^2 Fo\right) + 2\int_0^{Fo}\frac{Ki(Fo-\tau)k_\lambda^{-1} + Bi_2Bi_3(Fo-\tau)}{1+Bi_2\zeta(Fo-\tau)} \cdot \\ \cdot\sum_{n=1}^{\infty}\frac{\overline{A}_n\cos\mu_n\xi + \overline{B}_n\sin\mu_n\xi}{D_n}\cdot\exp\left(-\mu_n^2\tau\right)d\tau + 2Bi_3\int_0^{Fo}\theta_2(0,Fo-\tau)\frac{Bi_2\zeta(Fo-\tau)}{1+Bi_2\zeta(Fo-\tau)} \cdot \\ \cdot\sum_{n=1}^{\infty}\frac{\overline{A}_n\cos\mu_n\xi + \overline{B}_n\sin\mu_n\xi}{D_n}\exp\left(-\mu_n^2 Fo\right)d\tau. \tag{23}$$

$$tg\,\mu = -\mu\frac{\overline{k}_a\mu^2 - Bi_1 - Bi_3}{\mu^2\left(1+\overline{k}_aBi_1\right) - Bi_1Bi_3}, \mu > 0, n \ge 1; \tag{24}$$

$$A_n = \mu_n\left(Bi_1 + Bi_3\cos\mu_n\right) + Bi_1Bi_3\sin\mu_n; \tag{25}$$

$$B_n = \mu_n\left(Bi_3\cos\mu_n + Bi_1\overline{k}_a\mu_n\right) + Bi_1Bi_3\left(\cos\mu_n - 1\right); \tag{26}$$

$$\overline{A}_n = \mu_n^2\left(\mu_n\cos\mu_n + Bi_1\sin\mu_n\right); \tag{27}$$

$$\overline{B}_n = \mu_n^2\left(Bi_1\cos\mu_n + \mu_n\sin\mu_n\right); \tag{28}$$

$$D_n = \left[\left(1+Bi_1\overline{k}_a + 2\overline{k}_a\right)\mu_n^3 - Bi_1Bi_3\right]\cos\mu_n - \\ -\left[\overline{k}_a\mu_n - \left(1+Bi_1\overline{k}_a + Bi_1 + Bi_3\right)\mu_n^2 - Bi_1Bi_3\right]\sin\mu_n, \tag{29}$$

where μ is the root of characteristic equation (24); $Bi_3 = Bi_2/k_\lambda$; $\overline{k}_a = k_a/k_\lambda$.

Having inserted $\xi=0$ in (23), we will obtain the Volterra integral equation for determining $\theta_1(0,Fo)$:

$$f(Fo)=\theta_1(0,Fo)=-1+2\sum_{n=1}^{\infty}\frac{A_n}{D_n}\exp\left(-\mu_n^2 Fo\right)+2\int_0^{Fo}\frac{Ki(Fo-\tau)k_\lambda^{-1}+Bi_2Bi_3\zeta(Fo-\tau)}{1+Bi_2\zeta(Fo-\tau)}\cdot$$
$$\cdot\sum_{n=1}^{\infty}\frac{\bar{A}_n}{D_n}\exp\left(-\mu_n^2F_\tau\right)d\tau+2Bi_3\int_0^{Fo}f(Fo-\tau)\cdot\frac{Bi_2\zeta(Fo-\tau)}{1+Bi_2\zeta(Fo-\tau)}\sum_{n=1}^{\infty}\frac{\bar{A}_n}{D_n}\exp\left(-\mu_n^2F_\tau\right)d\tau \quad (30)$$

The Volterra integral equation of the second kind (30) is solved using the method of successive approximations taking into consideration the smallness of the parameter:

$$\left|\frac{Bi_2\zeta(Fo)}{1+Bi_2\zeta(Fo)}\right|<<1.$$

Functions $Ki(Fo)$ and $\zeta(Fo)$ included in (23), (30) are determined by relationships (17-19).

4. The method for calculating the surface heating during plasma spraying

The temperature distribution in the system 'coating-substrate' is realised in several steps:

1) the following parameters of the technological process of the coating spraying are set: $h,\ \alpha_i,\ \lambda_i,\ a_i^2,\ T_0,\ T_c,\ q_0,\ V_z,\ V,\ t_1,\ d,\ n$;

2) using relationships (15-16), the following dimensionless parameters and values are calculated: $Bi,\ k_\lambda,\ k_a,\ \beta, Pe,\ Fo_1,\ \zeta_0,\ \bar{k}_a$;

3) using the method of successive approximations, function $f(Fo)=\theta_1(0,Fo)$ is determined in equation (30), and function $\theta_1(\xi,Fo)$ is determined from expression (23);

4) using ratios (17-19) and (20), in accordance with the developed algorithm, the temperature distribution in the system 'substrate-coating' is calculated for various types of heat sources.

The suggested method is effective when calculating the heat density of the products such as plates during the application and increase of plasma coatings.

5. Analysis of results

The experimental researches of temperature distributions in case of the system 'coating-substrate' were conducted depending on the technological parameters during different durations of the process of spraying on the samples with dimensions 40×40×4 mm. In the centre of these samples, at a depth of 2 mm, there was a chromel-alumel thermocouple with electrodes having a Ø0.1 mm diameter. The temperature data were recorded by means of a 'KSP-4' recorder. Coatings were sprayed on the samples under the following process conditions: current strength is 400 A; orifice gas consumption is the following: $1.7\cdot10^{-4}$ m^3/s (for hydrogen), $8.3\cdot10^{-4}$ m^3/s (for argon); the material consumption is $5\cdot10^{-4}$ kg/s; the linear speed of the plasma gun travel is $11.4\cdot10^{-2}$ m/s. For the purpose of study of the influence of operating practices during spraying on the temperature distribution in the system 'coating-substrate', these parameters were varied in the following limits: current strength is (260-540) A, hydrogen consumption is ($0.8\cdot10^{-4}$-$2.5\cdot10^{-4}$) m^3/s, argon consumption is ($5.6\cdot10^{-4}$-$12.1\cdot10^{-4}$) m^3/s. The coating was sprayed on samples at distances of spraying equal to (0.08-0.22) m, the linear speed of the plasma gun travel varied within $7.3\cdot10^{-3}$ to $21.2\cdot10^{-2}$ m/s.

The results of comparison between the obtained theoretical and experimental data confirm the adequacy and applicability of the suggested mathematical model of surface heating calculation during plasma spraying (Figure 1). Minor discrepancy between the calculated data and experimental ones is connected with the inertia of a thermocouple sensor. The temperature in the increased layer of the coating can be significantly higher than the integral temperature of the system recorded by the thermocouple.

From the analysis of the obtained data (Figure 1), it follows that the adjustment of the heating of the powder material during formation of the coating can be realised both owing to the parameters of the plasma jet [1, 2, 5-10] and by means of controlling the duration of the magnitude of the thermal spike arising during plasma gun passage over the surface area under consideration.

6. Conclusions

A mathematical model of temperature distribution over the flat 'coating-substrate' system section during plasma spraying, taking into account a plasma gun travel and coating buildup has been developed. It has been shown that the temperature value in the near-surface layer of the sprayed coating during the plasma gun passage can significantly exceed the temperature values in underlayers. The heating of the powder material during formation of the coating can be realised both owing to the parameters of the plasma jet and by means of controlling the duration of the magnitude of the thermal spike arising during plasma gun passage over the surface area under consideration.

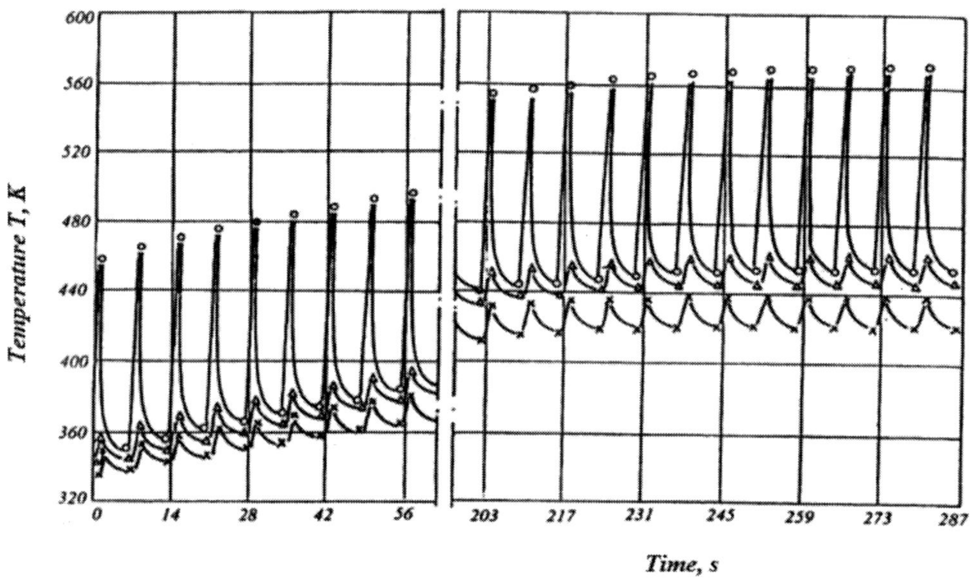

Figure 1. Distribution of the temperature depending on the duration of spraying in the system of the Ni-Co-Cr-Al-Y (o) coating and substrate (Δ, ×) at a depth of $z = -0.5h$: o, Δ – calculation data; × – experimental data.

Acknowledgments

The reported study was funded by the Ministry of Education and Science of the Russian Federation.

References

[1] Barvinok V A 2005 *Plasma in Technology: Reliability and Resource* (Moscow: Nauka i technologii)

[2] Kudinov V V and Ivanov V M 1981 *High-Melting Coating Application by Plasma* (Moscow: Mashinostroyenie)

[3] Barvinok V A and Bogdanovich V I 1982 *Power engineering New York* **20(6)** 101-108

[4] Barvinok V A and Bogdanovich V I 1982 *Power engineering New York* **20(2)** 160-166

[5] Zhang T, Gawne D T and Liu B 2000 *Surface and Coatings Technology* **132(2)** 233-243

[6] Sobolev V V, Guilemany J M, Garmier J C and Calero J A 1994 *Surface and Coatings Technology* **63(3)** 181-187

[7] Isakaev E K, Mordynskii V B, Tyuftyaev A S, Senchenko V N, Chinnov V F and Shcherbakov V V 2013 *Welding International* **27(4)** 304-310
[8] Joshi S V. 1993 *Journal of Thermal Spray Technology* **2** 127-131
[9] Ilichev L L, Gushchin Yu S and Rudakov V I 1991 *Soviet Powder Metallurgy and Metal Ceramics* **30(9)** 789-792
[10] Doltsinis I St, Haller K-U and Handel R 1996 *Structural Eng. and Mechanics* **4(6)** 679-702

Results of the modal analysis of the underground mine hydraulic leg in various modes of its operation

G D Buyalich[1,2,3], K G Buyalich[1] and S V Uvakin[1]

[1] T F Gorbachev Kuzbass State Technical University, 28, Vesennyaya str., Kemerovo, 650000, Russia
[2] Yugra Technological Institute branch of Tomsk Polytechnic University, 26, Leningradskaya str., Yurga, 652055, Russia
[3] Institute of Coal of Siberian Branch of the Russian Academy of Sciences, 10, Leningradskiy ave., Kemerovo, 650060, Russia

E-mail: gdb@kuzstu.ru

Abstract. The article describes the construction of models and the initial conditions for the various modes of legs operation. The article presents the way of setting the initial conditions for the initial thrust with constant pressure and for leg operation in the mode of increased pressure caused by the lowering of the roof. The results of modal analyses conducted for 2 modes of leg operation are shown. It is shown that the form and frequency of natural oscillations of hydraulic legs are determined by their design, pressures of the initial thrust and nominal operating resistance, as well as by the operating modes which differ in schemes of interaction of leg elements with interacting elements of the support and the roof and floor strata. At the same time, changing the operating mode of the legs changes the natural frequency both to the higher and to the lower side, as well as the form of oscillations of certain frequencies. For more accurate calculation of the frequency response of hydraulic legs to the external load of frequencies, it is proposed to take into account not only their design and power properties, but also the modes of their loading.

1. Introduction

In mining the coal seams with hard-to-cave roofs, there are often situations in which the main roof rocks hang as a beam or cantilever plates, and further face advancing leads to their collapse accompanied by the dynamic manifestations of rock pressure [1-3]. According to the results of previous studies [4-6], the amplitude and speed of movement of the roof in such cases depend on the physical and mechanical properties and thickness of the roof rock strata, the pitch of its collapsing, and the type, size and distribution of the support resistance along the width of the near-face area.

The main element of the powered roof supports, preventing the lowering of the roof in the working face, is the hydraulic leg [7-11], failure of which negatively affects the operation of the whole longwall equipment. The total cost of the service of mining equipment for the service life can be 4-8 times the cost of its acquisition.

In this connection, an increase of hydraulic performance is an actual scientific problem. The dynamic loads due to lowering of the roof rocks and acting on the powered roof supports are the main cause of deformation and failure of the elements of hydraulic legs. If the frequencies of the external loads are close to the frequencies of the natural legs oscillation, occurrence of resonance becomes possible, which will lead to their destruction. To detect this condition in the designing of mining machines and their components exposed to dynamic loads, the modal analysis method is applied.

2. Model creation and implementation of the modal analysis

In most cases, the powered support legs have double extension. However, at a first approximation to get the results of modal analysis, the leg model can be made in a simplified way with one extension. This will significantly reduce the number of leg elements, simplify the model and decrease the computation time. As an example, the article presents a model of the leg with the main overall dimensions of the powered support 'M138' (Figure 1).

The model includes the following main elements: a hydraulic cylinder (consisting of cylinder 3, guide sleeve 4, nut 5, piston rod assembly 6), two mounting supports of powered support 2 and 7 simulating the base and the canopy, and two rock blocks 1 and 8 simulating floor and roof, respectively.

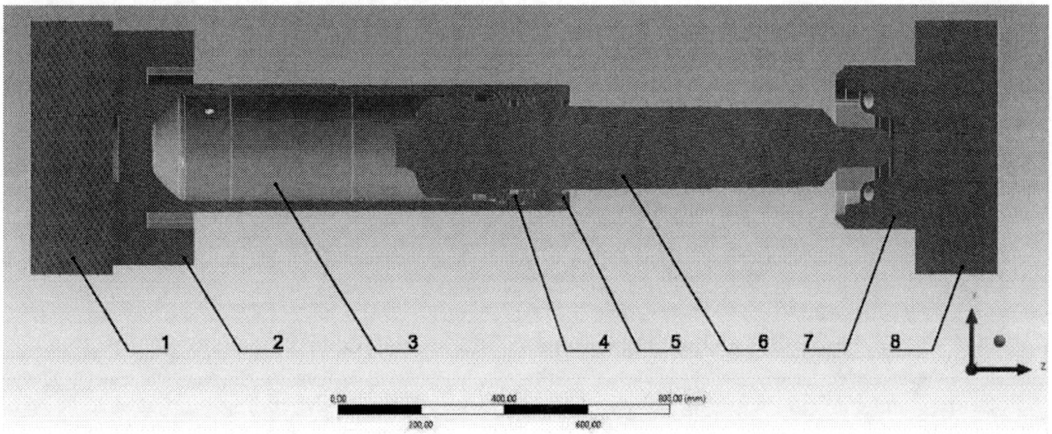

Figure 1. The leg model:
1 – floor block; 2 – lower support; 3 – cylinder; 4 – guide sleeve; 5 – nut; 6 – piston rod assembly; 7 – upper support; 8 – roof block.

All the elements of the model are assigned the density, Young's modulus, Poisson ratio and yield strength of the respective materials. For the hydraulic cylinder, this material is steel 30HGSA, for mounting supports, it is structural steel, and for blocks of the rock – sandstone.

To simulate movable joints in the supports (between the support and the cylinder, between the support and the rod), and movable joints of the leg seal assemblies (between the piston and the cylinder, between the guide sleeve and rod), contact pairs are created.

The principle of operation of the powered supports legs is the same for all, and is as follows. When the hydraulic fluid is fed into the piston chamber, the rod extends resting on the roof, and the pressure of the initial thrust (1st mode) is achieved. Then the pressure in the piston chamber increases due to the movement of the roof rock (2nd mode). When the limit pressure value in the piston cavity corresponding to the pressure setting of the safety valve is achieved, the leg transfers in the mode of constant resistance, i.e. in the operating mode.

To evaluate the effect of the leg operation modes on its natural oscillation frequency, two models are created for two modes of operation. The first model for the first mode corresponds to the leg operation during the initial thrust, wherein the travelling of blocks of roof rocks and floor rocks is prohibited, and constant pressure of 32 MPa is applied to the inside of the piston cavity.

In the second model, the floor block is fixed stationary and the roof block is assigned the travel in the axial direction corresponding to the increment of the pressure from the initial thrust to the working resistance, and the working resistance pressure is applied to the surfaces of the piston cavity of the hydraulic cylinder.

The initial conditions for the second mode of operation are shown in Figure 2.

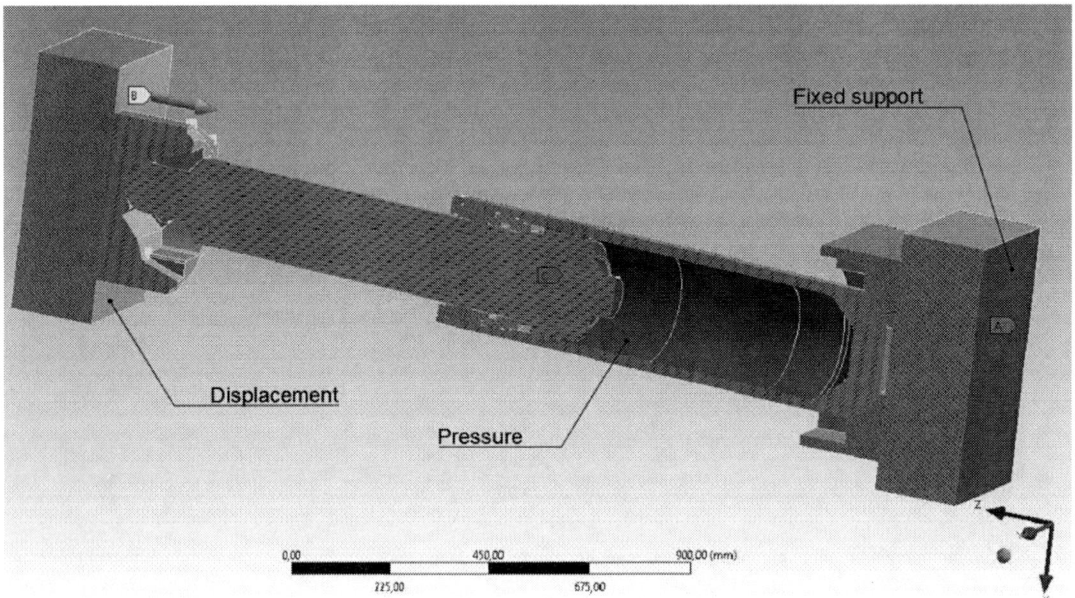

Figure 2. The Initial conditions for the second mode of operation

Pressure change ΔP in the piston cavity of the hydraulic cylinder depending on piston travel ΔL is called the elastic compliance value, the technique of its determination is presented by V.N. Khorin and is determined by the following formula.

$$\Delta P = \frac{\Delta L}{(2c+\beta)L},$$

where ΔL – change in leg extension, m;

L – leg extension, m;

β – the working fluid compressibility factor, m^2/N;

c – the coefficient taking account of the deformations of the working cylinder;

$$c = \frac{\dfrac{R^2+r^2}{R^2-r^2}+\mu}{E},$$

where, R and r – respectively the external and the inner radii of the working cylinder, m;

μ – Poisson ratio of the cylinder material;

E – Young's modulus of the cylinder material, Pa.

In order to solve the tasks of the modal analysis, the first 10 modes were used.

In accordance with the formulas given above, for the 'M138' leg the maximum working pressure in the piston of 42 MPa, corresponding to the nominal operating resistance, corresponds to the displacement of the rod by value ΔL = 5 mm.

Since the leg design is subject to loading corresponding to the initial conditions and the accepted modes of operation, in order to obtain the results of modal analysis the determination of the stress-strain state is performed at first.

3. Results of calculations

The results of the modal analysis presented as the values of natural oscillations and their forms for the first 10 events during the 1st and 2nd modes of the leg operation are shown in Figure 3 and Table 1 which show that in the 1st mode in the lowest modes, the cylinder has multiple frequencies of its own oscillations (except for modes 9 and 10), which is typical for structures with axial symmetry. In the second mode with increased operating resistance, this symmetry is broken.

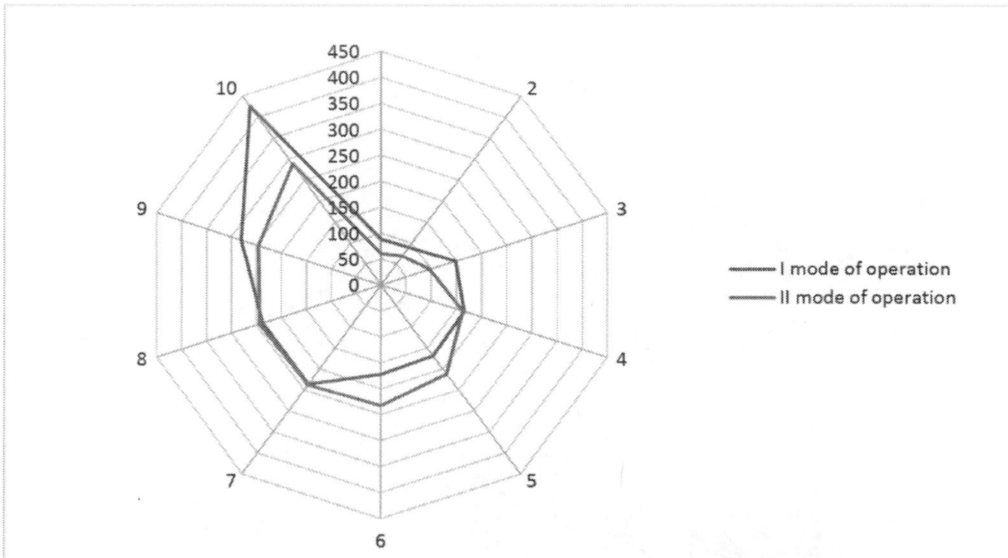

Figure 3. Natural frequencies for the first 10 modes

Table 1. Forms of oscillations of the first 10 modes of legs for various modes of their operation

Mode	Form of oscillation	
	I mode of operation	II mode of operation
1	lateral	lateral
2	lateral	lateral
3	torsional	torsional
4	torsional	lateral
5	lateral	torsional
6	lateral + torsional	lateral + torsional
7	lateral	lateral
8	lateral	lateral
9	torsional	lateral
10	torsional	lateral + torsional

For the modal analysis, the lowest natural frequencies are the most important, as in the event of resonance phenomena, the greatest deviations occur when external vibrations occur in the first (lowest) frequency (mode) of natural oscillations. When the frequency of the external load coincides with higher modes, the intensity of resonance phenomena decreases. Therefore, the properties of the design structure which include both structural features and load values can be characterized using the first frequency. The majority of the finite-element softwares solve the problem of determining the natural vibration frequencies with the best approximation for the lower frequencies and the first frequency does not change, even if the number of elements is increased many times.

This statement is consistent with the mining processes, because the frequencies of external loads from the cutter machine and secondary collapse of the roof are in the order of 100 Hz.

The presented data results of the modal analysis show that the increase of pressure and rod displacement occurring at the transition from the 1st mode to the 2nd lead both to reduction of the natural frequencies (modes 1, 2, 3, 9, 10), and to their increase (modes 5 and 6). In modes 4, 7, 8, the frequencies do not change. While the frequencies of natural oscillations vary on average by ± 10 to 15%. In addition, in modes 4, 5, 9, 10, the form of oscillation changes.

Figures 4 and 5 show the examples of oscillation in mode 4 for the first and second modes of operation.

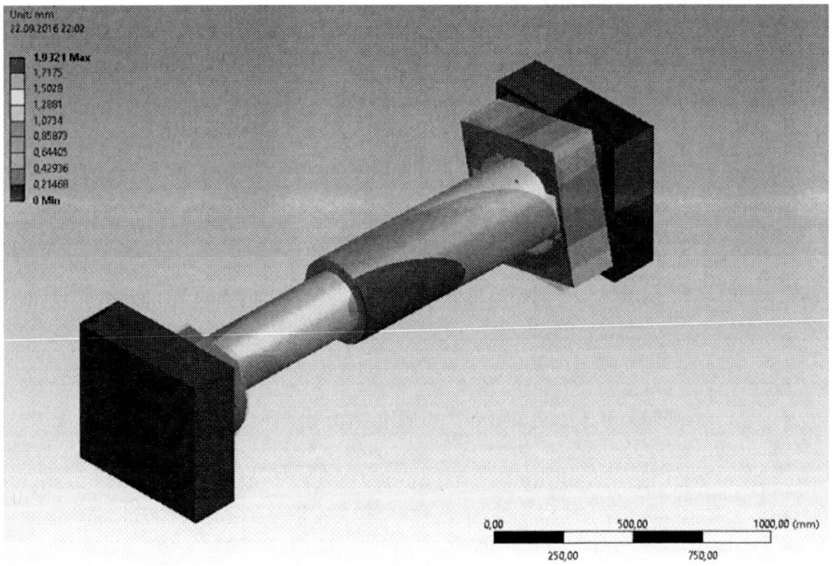

Figure 4. The form of natural oscillations in mode 4 of the 1st mode of the operation (torsial oscillations)

For a better visual display, the results of the modal analysis of the first mode of the operation for the torsional oscillations at a frequency of 165.45 Hz are enlarged 50 times, for the second mode of the operation during the lateral oscillations at a frequency of 155.61 Hz, they are enlarged 5 times.

During torsional oscillations, the maximum displacements occur at the junction of the hydraulic cylinder bottom and the lower support. During transverse oscillations, the maximum displacements occur in the area of the piston location. The amplitude of the transverse oscillations is about 1.7 times higher than torsional oscillations.

Changing of the forms of oscillations during increasing operating resistance indicates the presence of bifurcation points, the location of which is determined when all other things are equated by the working fluid pressure in the piston cavity of the hydraulic leg. This circumstance must also be taken into account when designing the sections of powered supports for specific geological conditions.

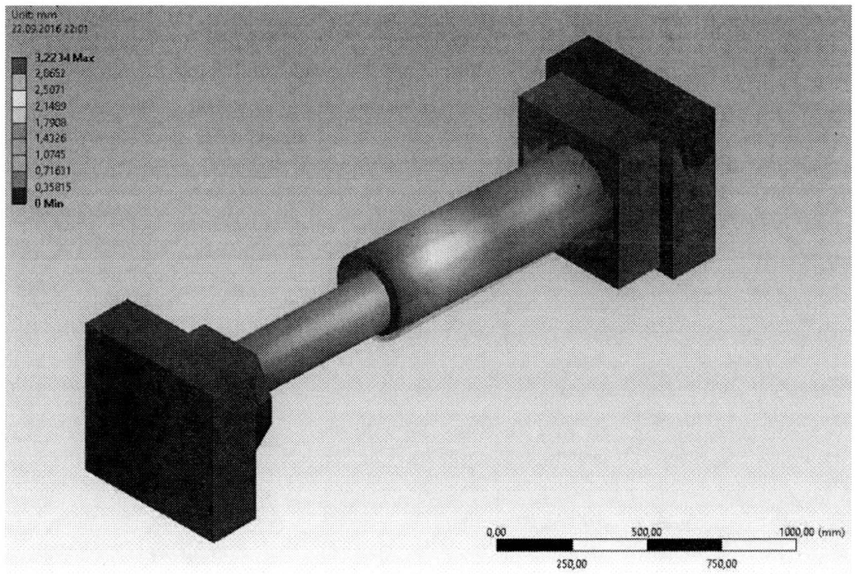

Figure 5. The form of natural oscillations in mode 4 of the 2nd mode of operation (lateral oscillations).

4. Conclusion

The form and frequency of natural oscillations of hydraulic legs are determined by their design, pressures of the initial thrust and nominal operating resistance, as well as by the operating modes which differ in schemes of interaction of leg elements with interacting elements of the support and the roof and floor strata. At the same time, changing the operating mode of the legs changes the natural frequency both to the higher and to the lower side, as well as the form of oscillations of certain frequencies.

For more accurate calculation of the frequency response of hydraulic legs to external load of frequencies, it is proposed to take into account not only their design and power properties, but also the modes of their loading.

References

[1] Klishin V I and Matviets Y V 1993 *J. of Mining Science* **29(2)** 122-127
[2] Buyalich G D, Buyalich K G and Umrikhina V Yu 2016 *IOP Conf. Ser.: Mat. Sci. and Eng.* (Yurga) vol **142** (IOP Publishing) 012120
[3] Klishin S V, Klishin V I and Opruk G Yu 2013 *J. of Mining Sci.* **49(6)** 932-940
[4] Klishin V I, Fokin Yu S, Lekontsev Yu M and Kokoulin D I 2005 *Mining J.* **12** 112-114
[5] Klishin V I 1995 *J. of Mining Sci.* **30(4)** 390-394
[6] Klishin V I and Tarasik T M 2001 *J. of Mining Sci.* **37(1)** 77-84
[7] Buyalich G D, Buyalich K G and Voyevodin V V 2016 *IOP Conf. Ser.: Mat. Sci. and Eng.* (Yurga) vol **127** (IOP Publishing) 012034
[8] Burkov P V, Vorobiev A V and Anuchin A V 2011 *Mining Informational and Analytical Bulletin* **2** 172-183
[9] Chinakhov D A 2011 *Applied Mechanics and Mat.* **52-54** 442-447
[10] Chinakhov D A, Grigorieva E G and Mayorova E I 2016 *IOP Conf. Ser.: Mat. Sci. and Eng.* (Yurga) vol **127** (IOP Publishing) 012013
[11] Chinakhov D A 2013 *Applied Mechanics and Mat.* **762** 717–721

IOP Conf. Series: Materials Science and Engineering **177** (2017) 012059 doi:10.1088/1757-899X/177/1/012059

Inductive sensor to detect metal impurities in non-metallic medium

N A Bykovsky, L N Puchkova and N N Fanakova

Sterlitamak Branch of Ufa State Petroleum Technological University, 2, Octyabrya ave., Sterlitamak, 453118, Bashkortostan, Russia

E-mail: nbikovsky@list.ru

Abstract. The mathematical model for an induction detector intended for detection of metal impurities is examined. The detector consists of three coils. The centre coil serves to induct a magnetic moment in the metal sample, and side coils are used to record this moment during the sample propulsion through the detector. It is shown that at an identical value of the magnetic field induction, created by the induction coil in the unit volume of the sample, the induced magnetic moment is defined by magnetic susceptibility for ferromagnetics, and for nonmagnetic materials – by their electric conductivity.

1. Introduction

In the manufacture of linoleum, lace tablecloths, sticky tapes, adhesive tape, etc., the plasticized mass, consisting of polyvinyl chloride resins, chalk and plasticizers are used. After several preparatory operations, this mass is fed to a 4-roll calender. The practice shows that due to the original substances, a hit of metallic impurities in the plasticized mass takes place [1-10]. This leads, on the one hand, to the reduction of the grade of quality of the finished product, and, on the other hand, – to the damage of calender rolls.

2. The mathematical model

A model of the induction sensor consisting of three coaxial coils to detect metallic impurities was examined. The center (generator) coil is used for the magnetic moment guidance in a metal sample. To do this, the alternating current from an audio generator is conducted through it. Side (receiving) coils are used to register the magnetic moment induced in the metal sample during its movement through the sensor. The interaction of metal particles possessing a magnetic moment with a receiving coil of the sensor is shown in figure 1.

The magnetic flux generated by a metal particle during its movement through the sensor can be determined. It is known [1] that the vector potential of the magnetic field generated by a moving particle is determined by the following formula:

$$\vec{A} = \frac{[\vec{M} \cdot \vec{R}]}{R^3}, \tag{1}$$

where R is the radius – vector from the particle to the point of the magnetic field.

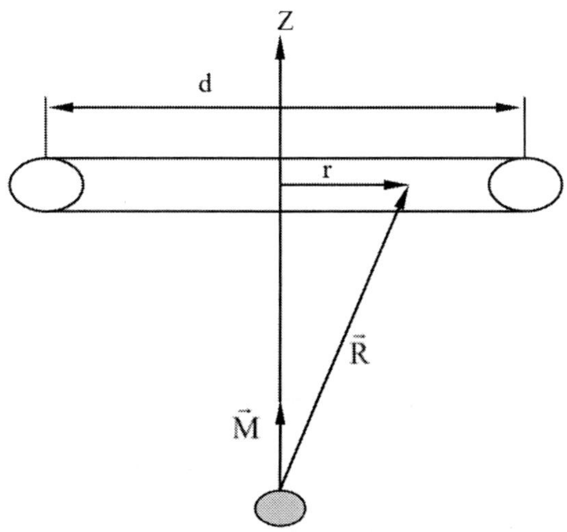

Figure 1. Interaction of metal particles possessing a magnetic moment with a receiving coil of the sensor.

The vector of the magnetic field, generated by a moving particle at the point determined by the radius-vector, is calculated as follows:

$$\vec{B} = rot\vec{A}. \qquad (2)$$

Magnetic flux F, which occurs in the plane of the receiving coil due to the movement of magnetized particles can be determined as follows:

$$F = \int_S BdS = \int_S rotAdS, \qquad (3)$$

where, S is the area of the spool.

Transforming equation (3) in accordance with the Stokes theorem, we get:

$$F = \oint_L \vec{A}d\vec{l} = \oint_L \frac{[\vec{M} \cdot \vec{R}]}{R^3} d\vec{l}. \qquad (4)$$

Transiting into cylindrical coordinates, we get:

$$F = \oint_L \frac{M \cdot \vec{e}_\varphi \cdot d^2 \cdot \vec{e}_\varphi \cdot}{4 \cdot \left(\frac{d^2}{4} + z^2\right)^{3/2}} d\varphi, \qquad (5)$$

where $\vec{e}_\varphi$ is a single vector; d – the diameter of a coil turn; z – the particle coordinate.

After the integration, considering the number of turns n in the receiving coil, equation (5) transforms into:

$$F = \frac{\pi d^2}{2} \cdot \frac{M \cdot n}{\left(\frac{d^2}{4} + z^2\right)^{3/2}}. \qquad (6)$$

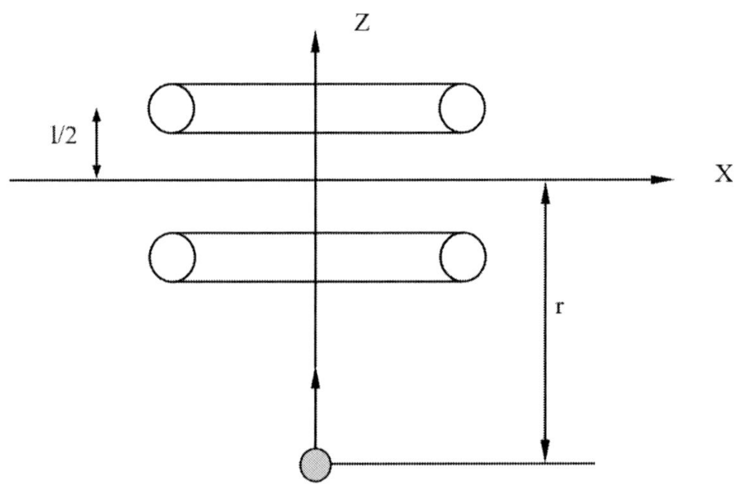

Figure 2. Scheme of receiving coil position in sensor.

As we previously noted, the sensor consists of three coils – two identical receiver coils and one generator coil. The generator coil is located between receiver coils. This sensor design allows obtaining the same electromotive force (EMF) for receiver coils by letting the alternating current through the generator coil. During the opposite connection of the receivier coils, the resulting EMF, induced in them by the generator coil, will be equal to zero. The scheme of the receiving coil position in the sensor is shown in figure 2.

The magnetic flux generated in these coils by a metal particle moving through the sensor, considering its oncoming inclusion, will be determined by the expression:

$$F = \frac{\pi d^2 \cdot n \cdot M}{2} \cdot \left\{ \frac{1}{\left[\frac{d^2}{4} + \left(z - \frac{l}{2} \right)^2 \right]^{3/2}} - \frac{1}{\left[\frac{d^2}{4} + \left(z + \frac{l}{2} \right)^2 \right]^{3/2}} \right\}. \quad (7)$$

Induction EMF e induced by this flow in the receiver coils of the sensor is equal to:

$$e = -\frac{1}{c} \cdot \frac{dF}{dt} \approx -\frac{j\omega}{c} \cdot F, \quad (8)$$

where, c is the speed of light; ω is the cyclic frequency of the current in the generator coil of the sensor.

From equations (7) and (8), it is seen that the EMF generated in the receiving coils varies with the frequency which is equal to the frequency of the current oscillations in the generator coil of the sensor. In this case, the envelope of the peak values for EMF of reception coils during the movement of metallic particles through the sensor is shown in figure 3.

The magnetic moment induced on the metal particle by the magnetic field generated by the generator coil depends on the magnetic properties of this particle.

For ferromagnetic particles, the induced magnetic moment has two components. The first component is determined by the vector of magnetization of the substance, and the second — by induction currents generated in the sample by the alternating magnetic field. It is known [1] that for ferromagnetic materials the first component is bigger than the second one.

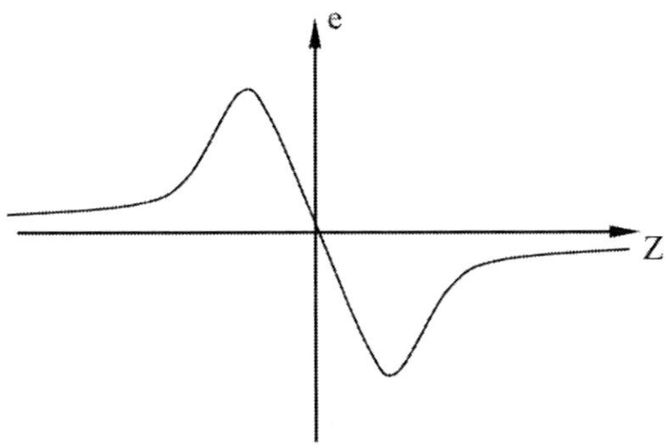

Figure 3. Envelope of the peak values for EMF.

Therefore, with a high degree of accuracy, it can be assumed that the magnetic moment for a ferromagnet is determined by expression

$$M_z = \chi \cdot B_g \cdot V, \tag{9}$$

where χ is the magnetic susceptibility; B_g – the induction magnetic field generated by the generator coil sensor; V – the volume of the particle.

The analysis of equations (7)-(9) shows that in the case of the ferromagnetic material generated in the receiving coils, EMF depends mainly on the frequency of current ω of the generator coil and particle volume V. The dependence of the EMF on speed of the particle movement through the sensor is substantially less than the frequency of the current in the generator coil. Indeed, with the oscillator frequency of 4 kHz, $\omega = 2.5 \cdot 10^4$ 1/s and speed v of a particle passing through the sensor makes an input of order $\frac{dz}{dt} \approx \frac{v}{d} = \frac{2 \cdot 10^2}{8} = 25$ 1/s.

If the metal has paramagnetic or diamagnetic properties ($\mu = 1, \chi = 0$), the magnetization vector is equal to zero. Consequently, the magnetic moment of this sample is created only by induction currents induced therein by the magnetic field of the generating coil.

According to Maxwell's equation [2], the electric field strength can be recorded as follows:

$$rotE = -\frac{1}{c} \cdot \frac{dB}{dt} \approx -\frac{j\omega}{c} \cdot B_g. \tag{10}$$

After transiting to cylindrical coordinates, we will obtain:

$$2\pi \cdot r \cdot E_\varphi = -\frac{j\omega}{c} \cdot B_g \cdot \pi \cdot r^2, \tag{11}$$

where r is the current radius of cylindrical particles.

Consequently, the electric field created by the generator coil of the sensor can be expressed by equation

$$E_\varphi = -\frac{j\omega \cdot r}{2c} \cdot B_g. \tag{12}$$

The density of the induced current in a metallic sample according to Ohm's law will be equal to

$$i_\varphi = \sigma \cdot E_\varphi = -\frac{j\omega \cdot r}{2c} \cdot \sigma \cdot B_g, \tag{13}$$

where σ is the conductivity of the particle.

The magnetic moment generated in a metal sample by inductive currents can be determined by equation

$$\vec{M} = \frac{1}{2c}\int_V \left[\vec{r}\cdot\vec{i}\right]dV. \tag{14}$$

Following the scheme of particle movement through the sensor (figure 1), it is shown that the magnetic moment of the particle is directed along the Z-axis, therefore

$$M_z = \frac{1}{2c}\int_V r\cdot i_\varphi \cdot dV. \tag{15}$$

The equation (13) allows one to obtain

$$M_z = -\frac{j\omega\cdot\sigma}{4c^2}\cdot B_g \int_V r^2 \cdot dV. \tag{16}$$

After integration of equation (16), we have:

$$M_z = -\frac{j\omega\cdot 2\pi\cdot\sigma\cdot B_g\cdot a^5}{15c^2}. \tag{17}$$

where a – the characteristic size of the particle.

Regarding the metal particle as spherical, we will obtain:

$$M_z = -\frac{j\omega\cdot\sigma\cdot B_g}{12.11\cdot\pi^{2/3}\cdot c^2}\cdot V^{5/3}. \tag{18}$$

3. Conclusion

Examining equations (9) and (18), we can conclude the following. At the same value of the magnetic field generated by the generator coil in a single volume of the sample, the induced magnetic moment for ferromagnetic material is determined by the magnetic susceptibility, and paramagnetic and diamagnetic materials depend on the material conductivity. Therefore, for ferromagnetic materials, $M_z \approx \chi$, i.e. it is much greater than one; for para- of diamagnetic materials ($\omega = 4kHz$ $M_z \approx -\frac{j\omega\cdot\sigma}{12.11\cdot\pi^{2/3}\cdot c^2} \approx 0.1$) for copper samples, induced magnetic moments are equal to almost one.

In the case of non-magnetic materials, the induced magnetic moment (see 18) is determined by the particle conductivity and by its volume. Thus, it is evident that the particle volume effect for non-magnetic materials is much greater than that for ferromagnets.

References

[1] Landau L D and Lifshits E M 1982 *Electrodynamics of continuous media* (Moscow: Nauka)
[2] Tamm I E 1989 *Bases of the theory of electricity* (Moscow: Nauka)
[3] Nation J and Jiang W 2016 *Int. J. of Pediatric Otorhinolaryngology* **92** 1-6
[4] Montoya E, Sebastian T, Schultheiss H, Heinrich B, Camley R E and Celinski Z 2016 Magnetization dynamics *Handbook of Surface Science* vol 5 (Elsevier) chapter 3 pp113-167
[5] Dessertine A 2016 *Extractive and Society* **3** 435-441
[6] Rezos M M , Schultz J J, Murdock II R A and Smith S A 2010 *Forensic Sci. Int.* **195** 121-127
[7] Ahmed A S and Kramer G H *Nuclear Instr.s and Methods in Phys. Res. Section A: Accelerators, Spectrometers, Detectors and Associated Equipment* **695** 543-548
[8] Zhigalova O V, Bykovskiy N A and Zakharov A V 2010 *Science Res.* **5** 12-16
[9] Benke V and Brune M, RU Patent No 2,389,172 (20 May 2010)
[10] Mesilov A Yu, RU Patent 2,509,305 (10 March 2014)

Modelling and control algorithms of the cross conveyors line with multiengine variable speed drives

M S Cheremushkina and S V Baburin

Saint-Petersburg Mining University, 2, 21 Line ave., St Petersburg, 199106, Russia

E-mail: cheremushkina-ms@yandex.ru

Abstract. The paper deals with the actual problem of developing the control algorithm that meets the technical requirements of the mine belt conveyors, and enables energy and resource savings taking into account a random sort of traffic. The most effective method of solution of these tasks is the construction of control systems with the use of variable speed drives for asynchronous motors. The authors designed the mathematical model of the system 'variable speed multiengine drive – conveyor – control system of conveyors' that takes into account the dynamic processes occurring in the elements of the transport system, provides an assessment of the energy efficiency of application the developed algorithms, which allows one to reduce the dynamic overload in the belt to 15-20%.

1. Introduction

One of the most productive types of machines for continuous transport are belt conveyors, because they are able to move minerals to considerable distances with minimal operating and energy costs, can be integrated in conveyor lines of the great length and performance, and are also used in the complexes of cyclic-flow technology.

Operating conditions of the conveyor electric drive has a number of specific features that set additional requirements for the choice of a control algorithm: provision of the soft start with restrictions of short-time accelerations; the requirement for creating the starting torque (several times higher than nominal); minimization of the time of starting and braking with no slipping and the restriction of dynamic tension in the belt [1].

2. The proposed model of the cross conveyors line

On the basis of the chosen scheme of conveyors location, in the line with several cross conveyors (Figure 1) in the application 'Simulink MatLab', the model 'conveyor – multiengine electric drive' has been developed. It was created in the form of separate blocks, the inputs of which represent the control and disturbance variables for a given element of the system, and outputs are state variables that are the subject of research or can be used as inputs to other blocks [2].

The object model was provided with the following components:

1.The subsystem 'drive – conveyor' (conv1-conv4) consisting of a mechanical (belt pulley, belt, etc.) and electrical part of the conveyor (multiengine variable speed electric drive).

2.The block which generates the load of the conveyor (loading1, loading2).

3. The block of the control system for the cross conveyors line (Control).

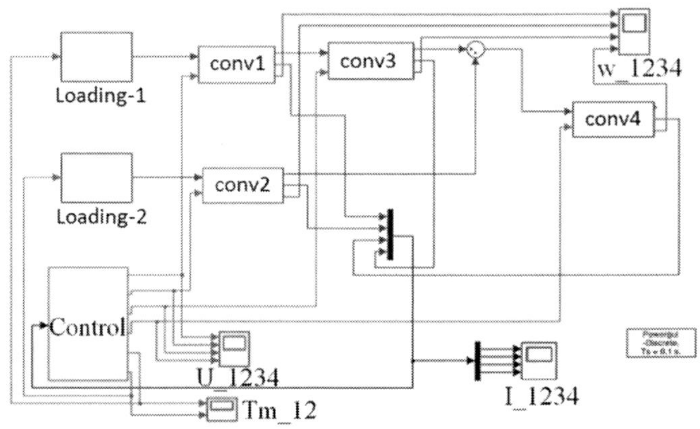

Figure 1. The continuous transport system consisting of four cross conveyors.

The mathematical model of this continuous transport system is a model of four cross conveyors, three of the conveyors are connected in series (conv 1, 3 and 4) and one — parallel (conv 2). In addition to the subsystems, containing models of conveyors (conv 1, 2, 3, 4, Figure 2), are also shown in figure 1: blocks Loading-1,2 (Figure 6) simulating the load and taking into account the random type of traffic; the control block of the transport system (CONTROL) the input of which receives signals (Is) from the conveyors simulating their load, and the outputs are control actions (U1-U4) over the conveyors, and signals Tm1 and Tm2 enter blocks Loading1,2, which allows taking into account the current load on the continuous transport system.

Digital implementation of the continuous transport system was created with the use of the C language in the program 'Matlab'. Using the model of the continuous transport system, it is possible to analyze operating modes of conveyors start-up in a predetermined order with the rated torque up to the rated speed, deceleration, the adjustment of conveyors to the predetermined load, and to obtain functions (monitors w-1234, I_1234, Tm_12) determining the workload of individual conveyors in the system and showing the speed change of the conveyor depending on the load.

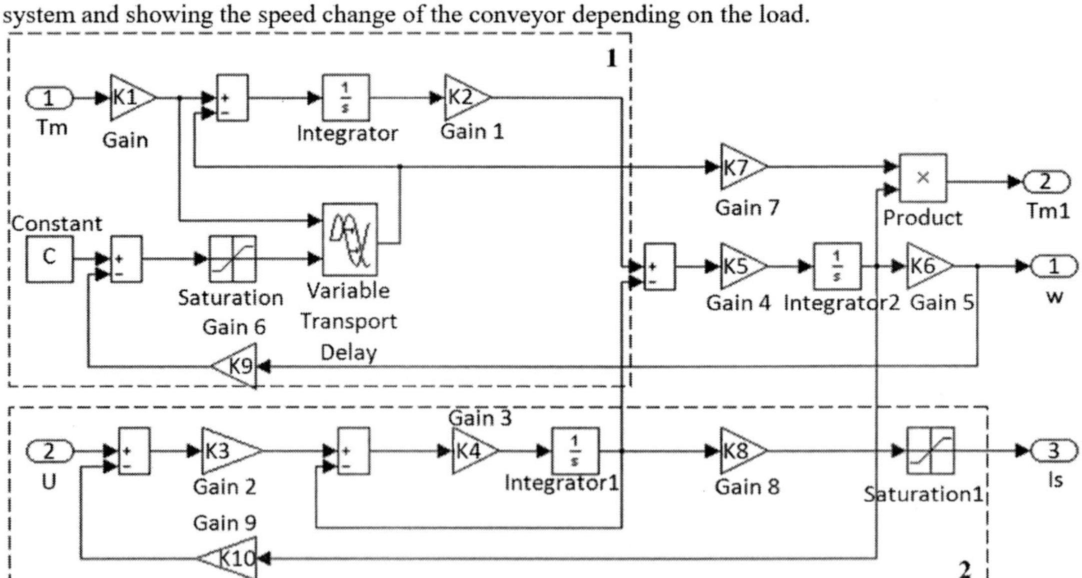

Figure 2. The model of the conveyor in Simulink Matlab.

Unit 1 simulates the variation of the load on the conveyor. The input signals 'Tm' from the block 'Loading' (Figure 1), taking into account the random type of traffic, and a constant value of load 'Constant'. Unit 2 determines the response of the conveyor to input signal U (the control action from the control system of the continuous transport system depending on the conveyors load). The output of the model 'Tm1'is the load supplied to the next belt conveyor (Figure 1, blocks 'conv1-conv4'), 'w' (Figure 2) is the speed of the movement of the belt of the conveyor, which is displayed on the monitor (Figure 1, monitor w1234), Is - stator current that determines the load level of the conveyor (Figure1, monitor I-1234).

To determine the accuracy of the mathematical model of the conveyor drive in the laboratory of the Department of electrical engineering and electromechanics (Mining University), the model of the multiengine variable speed drive for asynchronous electric motors was mounted (Figures 3,4), the parameters of which were incorporated in the computer model.

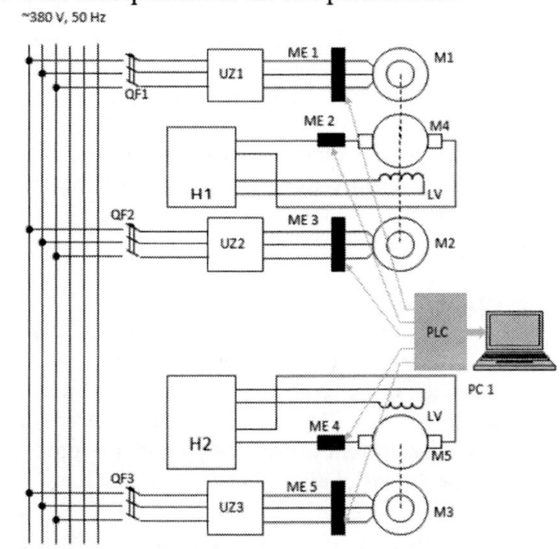

Figure 3. The functional diagram of the laboratory installation.

UZ1, UZ2, UZ3 – variable speed drives; H1, H2 – load (block of resistors); QF1..3 – circuit breakers; M1, M2, M3 – asynchronous motors; M4, M5 – loading machines; LV – excitation winding; ME1..5 – measuring equipment; PLC – controller; PC –personal computer.

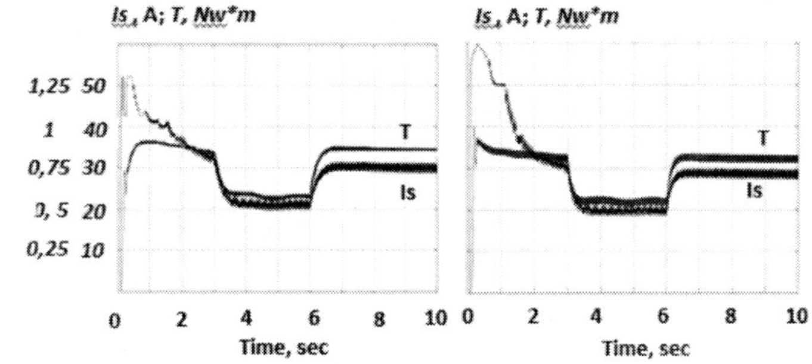

Figure 4. Stator current and torque characteristics (left – experimental, right – simulation).

The comparison of simulation results and experimental data showed that the maximum variation of the estimated and actual characteristics (motor currents and voltages, currents of the load machines in the simulation mode of conveyor acceleration up to full speed and subsequent stop) does not exceed ±11%. This fact allows us to make a conclusion about the sufficient convergence of the results, and consequently, the relevance of the mathematical model to the real object [3].

Figure 5 presents characteristics of motor current and torque when load is applied. The load is set by adjustment of the current of load motors M3 and M4. The comparison of the simulation results showed that the mismatch between the calculated and experimental curves is not more than 10%.

With the presented mathematical model, the system of direct torque control (DTC) for the multiengine asynchronous electric drive of the conveyor was adjusted.

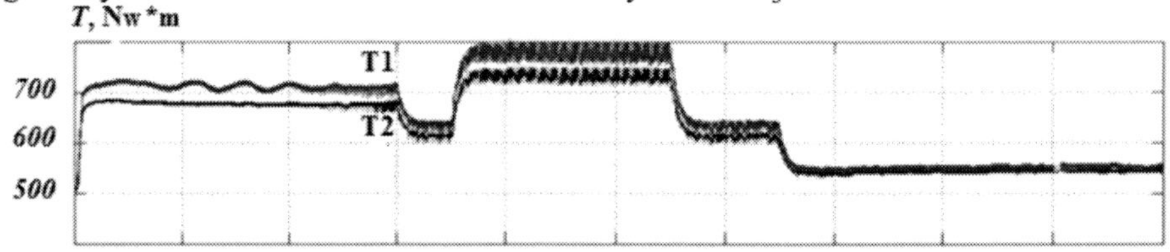

Figure 5. Characteristics of the torque at start-up, load-off and load-on (blue – the existing system of drive control of conveyors; black - direct torque control (DTC)).

Figure 5 presentsthe curves of moment change at start-up under the rated load, load-off and load-on, illustrating the decrease in the dynamic moment (15-20)% compared to the existing control systems for conveyor drives.

3. The designed control algorithm

The reference signals generation algorithm in the control system of the multiengine asynchronous electric drive allows one to take into account the probabilistic type of traffic, to improve the uniformity of load transfer between the drive motors to 90% and to eliminate slippage of the belt with the change of conveyor operating conditions [4]. The algorithm of direct torque control (DTC) which provides maximum performance in the current (torque) loop and a maximum value of the current (up to the rated torque) was used as a local control algorithm for the multiengine electric drive.

One of the main features of the conveyor transport as a technological system is the random incoming of traffic from the interconnected process equipment [5]. The model of load was created in Simulink Matlab (Figure 6), simulating random traffic arriving on the conveyor.

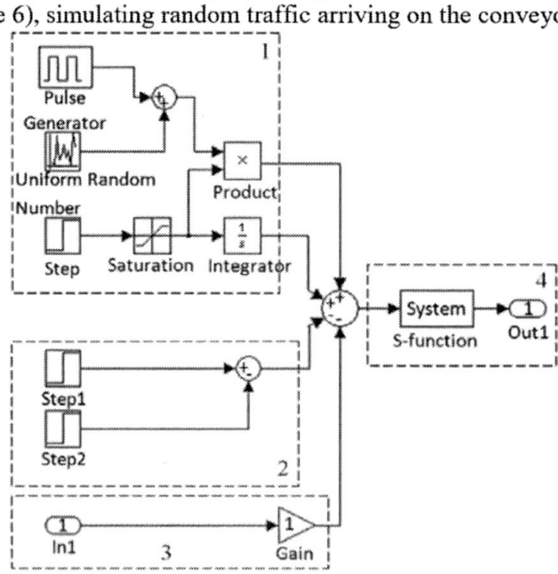

Figure 6. The model of load traffic.

Block 1 simulates the traffic as a discrete sequence of pulses with random duration of incoming load and with random intervals of its absence. Within the duration of impulses, the traffic is described as a continuous random process [6].

Block 2 simulates the load-off mode of the conveyor.

Block 3 simulates the incoming of load from other cross conveyors.

Block 4 is written in the Matlab language s-function that describes the distribution of traffic.

The reference signals generation algorithm in the control system of the multiengined asynchronous electric drive, taking into account the random type of the load, is: the value of load mismatch (actual current of semiconductor converters) is used to adjust the speed of the individual drives [7].

The algorithm of reference signals adjustment is shown in Figure 7.

1. If $T_1 = T_2$, then $K \times (T_1 - T_2) = 0$, a corrective signal is equal to 0.

2. If $T_1 > T_2$, then $K \times (T_1 - T_2) > 0$. The speed controller assignment (PC1) is reduced, and PC2 increases.

3. If $T_1 < T_2$, then $K \times (T_1 - T_2) < 0$. The speed controller assignment (PC1) is increased and PC2 decreases.

Coefficient K regulates the perfomance of torque adjustment [8].

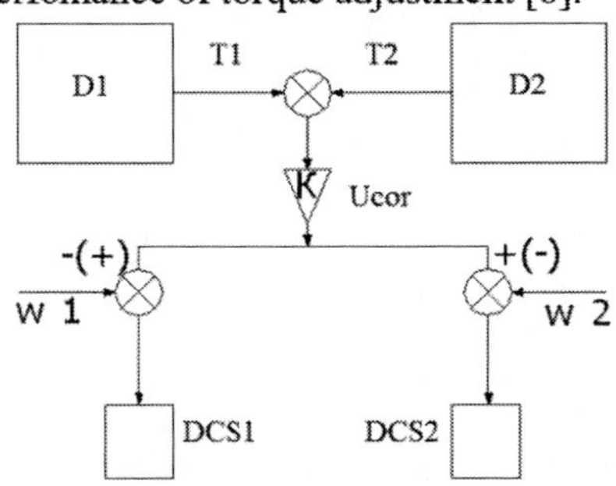

Figure 7. The algorithm of reference signals adjustment (D1,2 – drives, Ucor – correction signal, w1,2 – speed assignment, DCS1,2 – drive control systems).

Figure 8 demonstrates the curves of torque and the speed of the AC drive with the random load from the start-up mode to full speed and subsequent deceleration to a complete stop and stepped load-on. From these curves we can see that the use of the developed algorithm reduces the overshoot and torque oscillations, and therefore speed [9].

Researches of the drive operating modes showed that the implementation of the speed adjustment algorithm of the individual drives has provided increased uniformity of load distribution between drive motors (mismatch is not more than 10%), limiting the overshoot of the engine torque (up to 15%), and therefore, the restriction of dynamic loads on the conveyor drive with the use of the direct torque control algorithm and elimination of the belt slippage when changing the conveyor operating conditions [10].

4. Conclusion

Improving energy and resource savings of conveyor transport requires the installation of the regulated asynchronous electric drive with semiconductor frequency converters and control of operating modes of the cross conveyors line in consideration with the technological requirements and a random type of traffic.

The mathematical model of the electromechanical system of the conveyor – multiengine variable speed drive with implementation in Simulink Matlab allows one to perform researches of the conveyor operating modes taking into account the specifics of the mechanism operation at different control algorithms of electric drives.

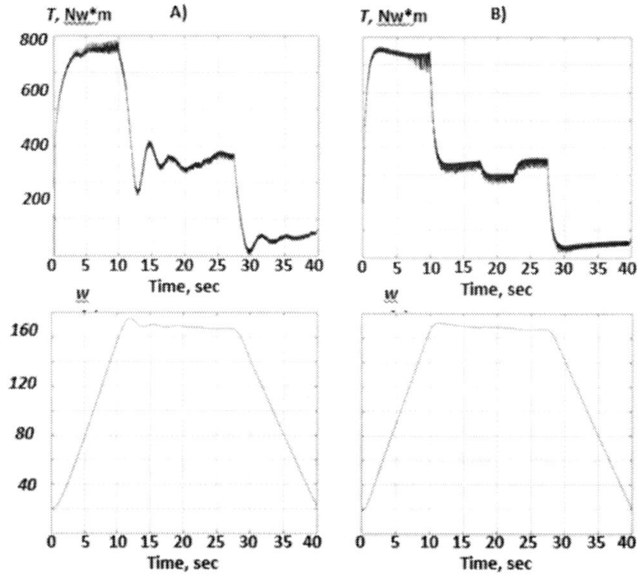

Figure 8. Curves of the drive torque and speed without adjusting the signals (A) and with the developed algorithm (B).

We demonstrated the expediency of applying the algorithm of local digital torque control of the induction motor together with the developed algorithm of reference signals adjustment, providing increased uniformity of load distribution between drive motors and elimination of belt dynamic loads (over-regulation until 15%).

References

[1] Shirong Z, Xiaohua X 2011 *Applied Energy* **88(9)** 3061-3071
[2] Baburin S V, Ustinov D A 2016 *Int. J. of Appl. Eng. Res.* **11(9)** 6402-6406
[3] Jia L, Weijian H, Shirong Z 2015 *J. of Cleaner Prod.* **105** 196-205
[4] Abramovich B N, Sychev Yu A, Ustinov D A and Shkljarskiy A Ya 2014 *Oil Industry* **8** 110-112
[5] Gottlich S, Hoher S, Schindler P, Schleper V, Verl A 2014 *Appl. Math. Mod.* **38(13)** 3295-3313
[6] Bardanov A I and Skamin A N 2015 *Int. Conf. on Mechan. Eng., Autom. and Control Systems (MEACS)* vol **1** (England: IOP Publishing) p 1-3
[7] Molnar V, Fedorko G, Andrejiova M, Grincova A and Michalik P 2016 *Measurement* **94** 364
[8] Chen Z, Wang Z and Li M 2014 *The 40th Annual Conf. of the IEEE Industrial Electronics Society* vol **1** 1041-1047
[9] Jimenez F R, Salamanca J M and Cardenas P F 2014 *IEEET 5th Colombian Workshop Circuits and Systems* **1** 1-10
[10] Kopteva A V 2015 *Exposition Oil & Gas. Russia* **3(42)** 58-61

Interaction of the conical impactor with barriers containing an explosive

A N Ishchenko, M V Habibullin, S A Afanas'eva, A V Chupashev and A I Zykova

Research institute of applied mathematics and mechanics of Tomsk State University, 36, Lenina ave., Tomsk, 634050, Russia

E-mail: ChupashevAV@gmail.com

Abstract. This paper presents the actual problem of the dynamic interaction of a high-speed impactor with a screened explosive. The interaction mathematical model is based within the framework of continuum mechanics considering the mechanism of the shock-wave initiation of detonation in solid explosives. Testing of the method was conducted using experimental data in one dimension. Some problems of interaction of the conical impactor with the explosive protected one-layered and spaced metal barriers are numerically considered.

1. Introduction

In solving a lot of practical problems, including the development of containers for safe storage and transportation of explosives, it is necessary to predict the effects of impact interaction between a solid deformable body and a screened explosives.

The results of experimental and theoretical studies show that the main features of the mechanism of the shock-wave initiation of detonation in solid explosives are determined by their initial inhomogeneity [1-10]. As a result of the shock wave interaction with inhomogeneities due to various mechanical processes (pore collapsing, microclusters on inclusions and pores, grain cracking, friction between the particles, etc.), there ara explosive local nucleation sites or 'hot spots' forming at the spots where exothermic decomposition originates. The necessity of introducing the concept of 'hot spots' based on the fact that volumetric average temperature of the explosive is too low to cause the observed degradation in case of the shock-wave initiation.

The less heated mass of the explosive burns out in the reaction waves propagating from the centers. The development of the explosive transformation process leads to an increase in the pressure behind the initiating shock wave and its amplification. The temperature and the amount of the nucleation site for the reaction igniting surrounding substance increase with the increase of the shock wave. When the shock wave becomes sufficiently strong to react with all of explosives, detonation begins.

The purpose of the research is mathematical modeling to predict the effects of high-speed interaction of the elongated metal conical impactor with the solid explosive screened by a thin plate.

2. The mathematical model of the shock-wave initiation of detonation in solid explosives

According to the model of reaction initiation and the growth [11], the macrokinetics formula includes two parts where the first one describes the ignition process and the second one describes the posterior reaction growth:

$$\frac{dw_1}{dt} = A_1 w_1^x \left(\frac{1}{\delta_1} - \frac{1}{\delta_*} \right)^m + B_1 w_1^x w_2^y p^z , \tag{1}$$

where w_1, $\delta_1 = \frac{\rho_0}{\rho_1}$ – the mass fraction and fractional specific volume of the explosive;

$w_2 = 1 - w_1$ – the mass fraction of detonation products;

δ_*, A_1, B_1, m, z – constants determined by reference to experimental data;

x, y – exponents depended on the combustion geometry.

For describing of explosives (i = 1) and decomposition products (i = 2), the equation of state is applied in the Mie-Grüneisen form, where isentrope is used as a supporting curve with constant Grüneisen parameter γ_0 in the following form:

$$\varepsilon_{si} = \frac{C_i}{R_{1i} \exp(R_{1i}\delta_i)} + \frac{D_i}{R_{2i} \exp(R_{2i}\delta_i)} + \frac{E_i}{\gamma_{0i}\delta_i^{\gamma_{0i}}} - \varepsilon_{0i} , \tag{2}$$

where $\delta_i = \frac{\rho_0}{\rho_i}$;

C_i, D_i, E_i, R_{1i}, R_{2i}, ε_{0i} – empirical constants.

To close the system of equations, describing the movement of reactive mediums as a part of the hydrodynamic model, additivity of volume phases is introduced:

$$\frac{1}{\rho} = \frac{w_1}{\rho_1} + \frac{w_2}{\rho_2} . \tag{3}$$

And the local pressure equilibrium is assumed to be performed in the medium:

$$p = p_1(\rho_1, \varepsilon_1) = p_2(\rho_2, \varepsilon_2) . \tag{4}$$

Additionally, the difference between the sum of two internal energies (where the first one is the explosion products energy defined by isentrope and the second one is explosive energy determined by the Hugoniot adiabatic) and the total internal energy of the elementary volume is believed to be distributed among the components in accordance with the ratio of the first internal energy to the second internal energy. Then:

$$\varepsilon_1 = \varepsilon_{H1} + \frac{\left[\varepsilon - w_1\varepsilon_{H1} - w_2\varepsilon_{s2}\right]\varepsilon_{H1}}{w_1\varepsilon_{H1} + w_2\left|\varepsilon_{s2}\right|} , \quad \varepsilon_2 = \varepsilon_{s2} + \frac{\left[\varepsilon - w_1\varepsilon_{H1} - w_2\varepsilon_{s2}\right]\left|\varepsilon_{s2}\right|}{w_1\varepsilon_{H1} + w_2\left|\varepsilon_{s2}\right|} , \tag{5}$$

where

$$\varepsilon_{H1} = \frac{\left[\dfrac{C_1\left(\delta_1 - \dfrac{\gamma_{01}}{R_{11}}\right)}{\exp(R_{11}\delta_1)} + \dfrac{D_1\left(\delta_1 - \dfrac{\gamma_{01}}{R_{21}}\right)}{\exp(R_{21}\delta_1)} + \gamma_{01}\varepsilon_{01} \right]}{2\delta_1 - \gamma_{01}(1 - \delta_1)} \tag{6}$$

– the explosive internal energy on the Hugoniot adiabatic.

3. Testing of the mathematical model

To verify the model [11] of shock wave initiation of detonation in heterogeneous explosives and to determine the constants of the macrokinetics formula within the one-dimensional deformed state, the calculation for the collision process of the copper plate thickness of 0.76 mm with an open charge explosives PBX-9404 density of 1.844 g/cm^3 at a velocity of 550 m/s was conducted. Results of experimental studies are described in [12]. The calculation was performed using the modified numerical method of large particles [13]. Table 1 shows the parameters of the state equation and the initiation model and reactions progressing in the heterogeneous explosive.

Table 1. The parameters of the state equation, the initiation model and reactions progressing.

Parameters	PBX-9404	Reaction product
C_i, GPa·cm^3/g	10889.37	462.26
γ_{0i}	0.4541	0.38
R_{1i}	9.0	4.6
D_i, GPa·cm^3/g	-141.43	9.77
R_{2i}	4.5	1.3
ε_{0i}, GPa·cm^3/g	0.3	5.53
E_i, GPa·cm^3/g	–	0.6726
A_1, μs^{-1}	20.0	–
B_1, GPa^{-z}/μs	6000	–
z	1.2	–
m	4.0	–
x	0.2222	–
y	0.6667	–
δ_*	0.9822	–

Figure 1 shows a comparison of the calculated pressure profiles in the three sections of the charge (solid lines) and experimental (dashed) pressure profiles. At the initial moment, sensors were apart from the impact surface at distances of 0, 3 and 7 mm. A satisfactory agreement of the data is observed. A decrease of the shock wave amplitude and pressure over the time did not happen as quickly as it did in the experiment.

At different times, Figure 2 shows (microseconds) the calculated pressure profiles and shares of the unreacted explosive depending on the distance by which one can judge the dynamics of the reaction development.

Shock waves with the amplitude of 2.9 GPa, propagating from the moving contact surface, are produced during the collision in the copper impactor and charge of explosive PBX-9404. Explosive decomposition begins directly from the impact surface. The pressure pulse, growing in size and propagating deep into the charge, is generated. This pressure pulse subsequently catches up with the shock wave front and amplifies it. The copper plate shock wave reaching the free surface is reflected in the shock-compressed material as expansion waves. As a result of their action, in spite of the expansion process, the explosive pressure is firstly reduced and then increased (figure 1). At 3.5 μs, the sample pressure reaches the value of the Chapman-Jouquet point, and afterwards, shock wave transition occurs relatively fast with the detonation reaction. The computed pressure in the chemical peak is 54 GPa and in the Chapman-Jouquet point, it is 37 GPa. The detonation wave propagates at a velocity of 8791 m/s, this agrees sufficiently with the experimental data.

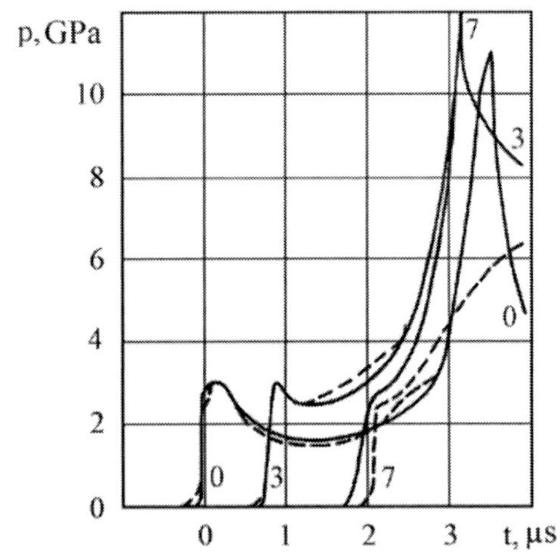

Figure 1. The comparison of the calculated and experimental profiles

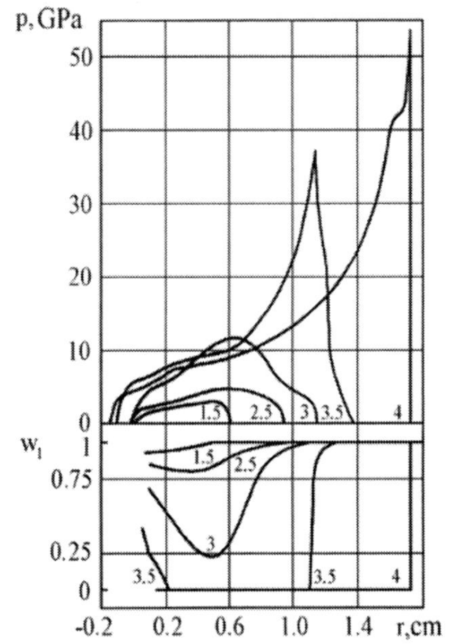

Figure 2. Distribution of pressure and unreacted explosive shares at different times

4. The results of numerical research

Let us consider a numerical solution for some axisymmetric impact problems of the truncated cone (0.12×0.7×8.5 cm) made from composite TNI-95 containing explosive PBX-9404 in design. In all cases, the minimum velocity (accuracy to 100 m/s), when the shock detonation has been initiated for the given target, is taken as the bodies interaction initial rate.

Numerical modeling results are presented in cylindrical coordinates z (cm), r (cm). The right half-plane of the images shows a pressure field (MPa); the left one shows a mass velocity vector field. Here, u_{max} is the longest vector module, t is time. The destroyed material area [14] is marked by the black color in the left half-plate of the images; the reaction zone is marked by red.

Figure 3 shows a collision process with an explosive charge in the AlMg-6 aluminum alloy shell of 0.8 cm thickness with a velocity of 800 m/s.

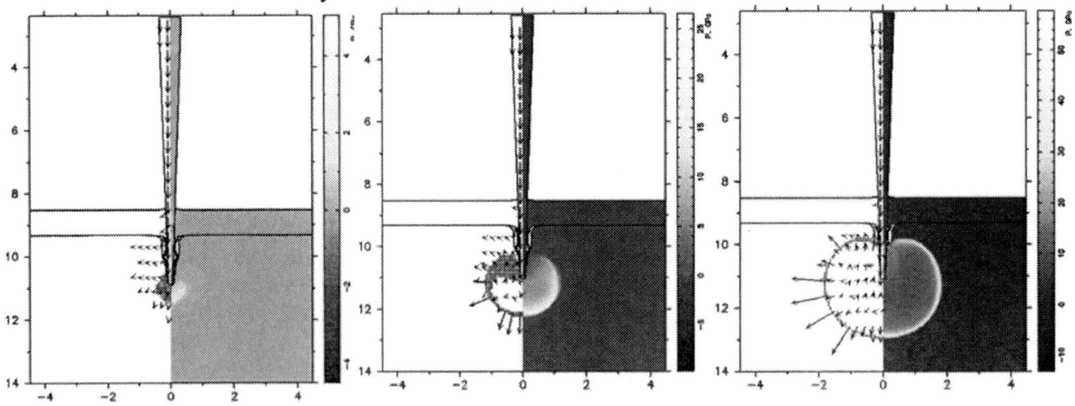

t = *30* μs, u_{max} = 803 m/s t = *32* μs, u_{max} = 1474 m/s t = *33* μs, u_{max} = 2320 m/s

Figure 3. The detonation of explosives in the AlMg6 shell

IOP Conf. Series: Materials Science and Engineering **177** (2017) 012061 doi:10.1088/1757-899X/177/1/012061

The impactor had punctured the shell by the time of 30 μs. And at a certain distance, in the explosive, the impactor initiated the explosive decomposition.

An area occupied by the reaction products had been formed by the time of 32 μs. A detonation complex was formed by 33 μs.

Figure 4 shows the chronogram of shock-wave initiation for detonation of the explosive in the VT1-0 titanium alloy shell of 0.8 cm thickness. A minimum velocity for the barrier blasting is already 1200 m/s.

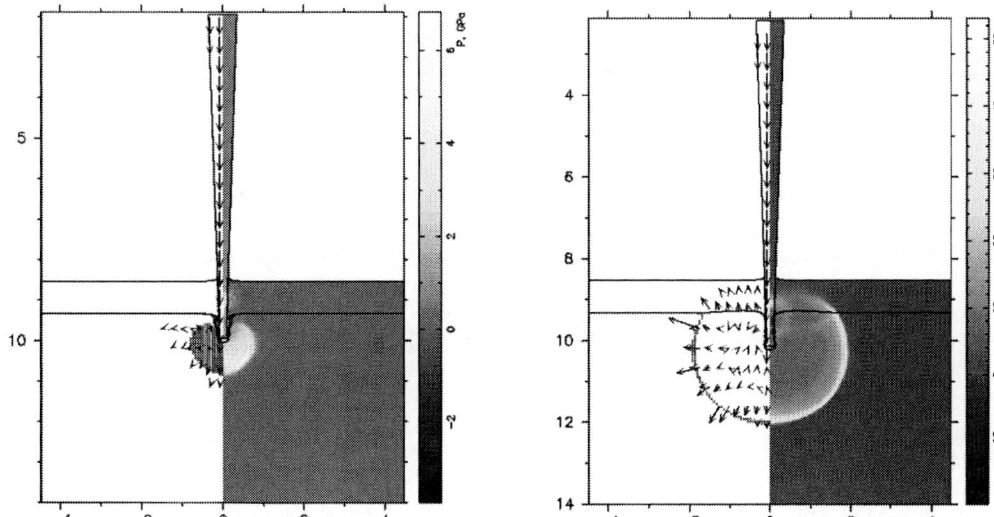

$t = 1$ μs, $u_{max} = 1199$ m/s $t = 18$ μs, $u_{max} = 2349$ m/s

Figure 4. The detonation of explosives in the VT1-0 shell

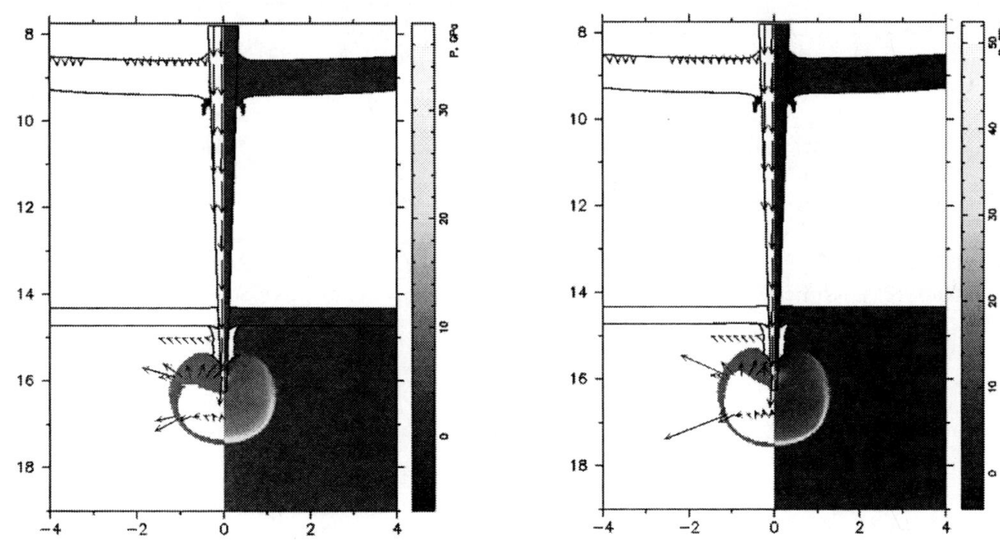

$t = 97.74$ μs, $u_{max} = 889$ m/s $t = 97.8$ μs, $u_{max} = 1747$ m/s

Figure 5. The detonation of screened explosives

Figure 5 depicts a shot at a velocity of 800 m/s in the design, consisting of the AlMg6 aluminum alloy screen of 0.8 cm thickness and the bilayer barrier situated at the distance of 5 cm behind it: the first layer is the AlMg6 aluminum alloy of 0.4 cm thickness, the second one is explosive PBX - 9404.

The cone punctured the screen and shell by 97.74 μs. The explosive pressure value reached the Chapman-Jouquet point, and afterwards the shock wave transition was relatively fast with the detonation reaction (t = 97.8 μs).

5. Conclusion

The interaction mathematical model of the high-speed impactor with explosive protected by the metal barrier in the framework of continuum mechanics, taking into account the mechanism of the shock-wave detonation initiation in solid explosives, is developed.

Numerical modeling of the conical impactor influence on barriers, containing explosive PBX-9404, is conducted in the impact velocities range involving the shock-wave detonation initiation.

We considered the explosives protection by aluminum alloy (AlMg6) and titanium (VT1-0) alloy plates when plates are both adjacent to the explosive surface and separated from the explosive by some distance. Stable initiation of the charge protected by both the AlMg6 aluminum alloy single plate and the two plates spaced by a distance occurs at velocities of about 800 m/s. To initiate the charge in the shell of VT1-0, the considered drummer must have the speed of about 1200 m/s. For initiation of the charge in the VT1-0 titanium alloy shell, the considered impactor must have a velocity of about 1200 m/s.

References

[1] Tarver C M 2012 *J. Energ. Mat.* **30(3)** 220-251
[2] Lee E L, Tarver C M 1980 *Physics of Fluids* **23(12)** 2362-2372
[3] Sheffield S A, Bloomquist D D, Tarver C M 1984 *The J. Chem. Phys.* **80(8)** 3831-44
[4] Wei L, Dong H, Pan H, Hu X, Zhu J 2014 *J. Energ. Mat.* **32(4)** 238-251
[5] Bel'skii B M 2012 *Combustion, Explosion and Shock Waves* **48** 328-334
[6] Lee J, Kuk J, Cho Y, Song S, Lee J W 1997 *Propellants, Explosives, Pyrotechnics* **22** 337-346
[7] Treadway S K, Lloyd A N 2013 *Procedia Engineering* **58** 550-559
[8] Tarver C M, Parker N L, Palmer H G, Hayes B, Erickson L M 1983 *J. Ener. Mat.* **1(3)** 213-250
[9] Gruau C, Picart D, Belmas R, Bouton E, Delmaire-Sizes F, Sabatier J and Trumel H 2009 *Int. J. Impact Eng.* **36(4)** 537-550
[10] Picart D, Ermisse J, Biessy M, Bouton E, Trumel H 2013 *Int. J. Energ. Mat. and Chem. Prop.* **12(6)** 487-509
[11] Green L 1981 *Detonation and Explosives* (Moscow: Mir) pp 107-122
[12] Wakerly J 1981 *Detonation and Explosives* (Moscow: Mir) pp 269-290
[13] Habibulin M V 1997 *Issues of Nuclear Science and Technology* **3** 18-24
[14] Belov N N, Korneev A I, Simonenko V G 1990 *Proc. of Academy of Sciences* **310(5)** 1116-20

The investigation of the influence of thermomechanical treatment of the material of rotary cutter bit toolholders on its hardness

S A Chupin and V I Bolobov

Saint-Petersburg Mining University, 2, 21st Liniya, Saint-Petersburg, 199106, Russia

E-mail: Staseg-88@mail.ru

Abstract. The causes of failure of the tangential rotary cutter bits of the road header during stonedrift in rocks of medium strength are analyzed in the article. It was revealed that the most typical cause of failure of cutter bits is premature wear of the toolholder (body) of the cutter bit. It is well known that the most effective way to improve the wear resistance is to increase hardness. The influence of the thermomechanical treatment of the material of the cutter bit toolholder on its hardness is studied. It was established that the thermomechanical treatment of the cutter bit toolholder material results in the increase of its hardness. It was found that the increase of material hardness is proportional to the increase of material strain intensity during thermomechanical treatment. It was concluded that the use of thermomechanical treatment can lead to the increase of both the hardness and wear resistance of the cutter bit material.

1. Introduction

Currently, the most advanced technique for permanent and service headways during subsurface mining is road headers, while for the mining operation – shearers that break rocks with operating devices of the cutting type.

The effectiveness of rock massif breaking and the productivity of power-loaders are defined by the engineering level of applied cutting tools. The quality of applied cutting tools affects mine costs both directly (via tool cost) and indirectly (via their replacement timing, the reduction of resource elements of transmission, actuators and other power-loader components) [1].

The exploitation of damaged cutter bits causes an increase in power and energy parameters of rock breaking. The above-mentioned reasons cause substantial costs for cutting tool replacement: depending on rock hardness and abrasiveness, they can reach up to 30% of the total cost of the tunneling work [1].

It is conventional that the runout of indexable insert holders is similar to the wearout of cutter bits having a plane face [2, 3]. A finely-divided volume-compressed core built up in the contact zone of the cutting tool and a rock prevents further movement of the cutter bit (Figure 1). Since the cutter bit does not stop moving, and the openness and cleavage of a rock massif are not sufficient for continuous cutter bit movement without crush evacuation, a part of the pressure bulb 'utpours' at a high speed along the cutter bit face towards the free face. Due to high stress of the fragmented rock in the pressure bulb, its 'outpouring' through a narrow gap is accompanied by significant friction on both the undecayed rock and the cutter bit (figure 1, c). In Figure 1, it is shown that most of the time, the hard alloy insert and the bolster of the shoulder of the rotary cutter bit toolholder interacts with the rock.

Thus, rock cutting with the rotary cutter bit causes prompt appearance and destruction of the pressure bulb resulting in tool wear.

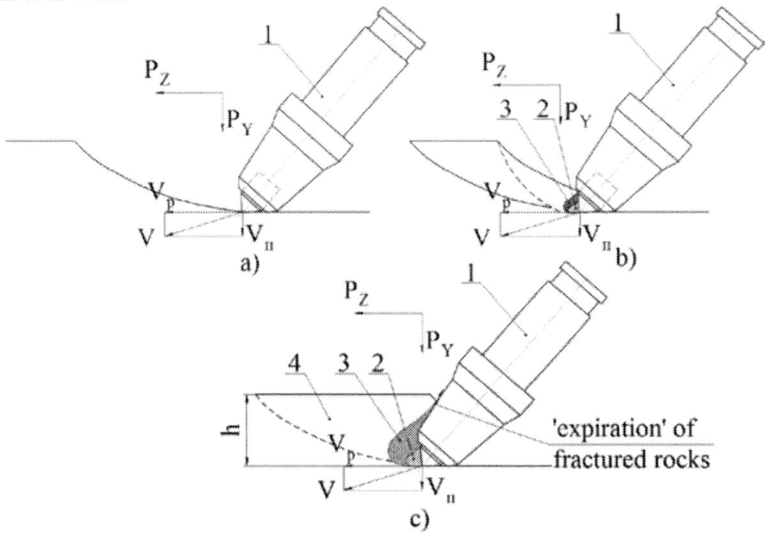

Figure 1. The scheme of cutting the rock massif with the rotary cutter bit at different times during a single big fracture: 1 – rotary cutter bit, 2 – pressure bulb (finely-dispersed fractions of the rock) under the conditions of triaxial compression; 3 – shear zone, 4 – zone of elastic strain of rock massif surrounding areas (big fracture); P_Y – feed force, P_Z – cutting force, V_p – cutting velocity vector, $V_П$ – feed velocity vector.

Most of the tunneling work with the use of mechanical combine systems is realised with the rocks of medium hardness and abrasiveness [4, 5]. Due to the operating experience of roadheaders rotary cutter bits, the main cause for cutter bits malfunction is wearing of the toolholder head accompanied by the breaking out of the hard alloy insert (figure 2). The early wearing of cutter bits in relation to its hard alloy insert leads to incomplete resource consumption of the entire tool. The increased wear resistance of the cutter bit toolholder will lead to the increase of the service life of the rock-destruction tool.

a) b)

Figure 2. The view of rotary cutter bits RSh32-70/16*SK* after their operation: the worn cutter bit a) before the breakage of the hard alloy insert and b) after the breakage of the hard alloy insert [6, 7].

The producers use different methods in order to improve the wear resistance of a rock-destruction tool of mining equipment. A considerable amount of research is done in the area of rational design parameters of the cutter bits, the use of sophisticated and hard-wearing materials or the use of special processing techniques.

The special processing techniques include some kinds of heat treatment, in particular, thermomechanical treatment (TMT) combining steel plastic strain at elevated temperature and heat treatment. The formation of the structure of hardened steel during TMT occurs under conditions of increased density of dislocations caused by technological conditions of the hot strain process [3]. Materials processing using this technique causes a significant increase of strength properties of the processed steel [8 – 11]. It is known that the increase of material hardness leads to the increase of its wear properties [12]. Consequently, one of the effective ways to improve the wear resistance of rotary cutter bit toolholders is to increase their hardness.

This paper presents an investigation of the influence of different values of strain on the hardness of steel of cutter bit toolholders during thermomechanical treatment.

2. Experimental methods and materials

The study of the impact of various modes of TMT on material hardness of the road header rock-destruction tool was studied on the basis of steel 30KhGSA, which is the most commonly used material in the production of tangential rotary cutter bit toolholders. The value of strain intensity ε_p (effective strain) in the samples surface layer is set as a changing parameter of the TMT mode (other parameters are constant).

Cylinder samples with D_0 = 8 mm and are H_0 = 15 mm high (figure 4) were subjected to TMT according to the scheme:

1. Heating in the thermostat to the temperature of 1200 °C and soaking for 10 minutes to establish the process of steel austenitizing;
2. Compression of the sample of height H_0 in line to its axis on the compressed-air hammer at the temperature of 900 °C up to different values of height H_i;
3. Hardening of the sample in oil;
4. Low-temperature tempering at 250 °C.

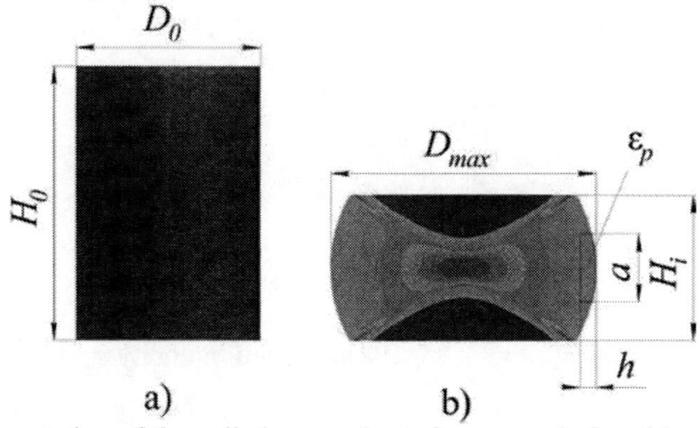

Figure 4. The representation of the cylinder sample strain process before (a) and after (b) uniaxial compression.

The measurements of hardness and strain intensity were carried out on the side surface of the deformed samples in the central (working) part (figure 4 *b*). The plot with a = 3 mm and h = 2 mm was accepted as a working part. Hardness measurements were carried out using universal hardness tester *Zwick ZHU* 187.5.

3. Results and Discussion

The results of hardness measurement of samples made of the 30KhGSA steel, subjected to TMT of various strain intensity are presented in table 1.

IOP Conf. Series: Materials Science and Engineering **177** (2017) 012062 doi:10.1088/1757-899X/177/1/012062

Table 1. Hardness of the working part of samples subjected to thermal and thermomechanical treatment of various strain intensity

№ samples	H_i, mm	Strain intensity ε_p	Hardness,	
			HRC	HB, MPa
1	15	0	44 (35÷45)	4090 (3270÷4210)
2	13	0.17	48	4550
3	12.3	0.22	49	4690
4	11.1	0.32	52	5120
5	9.6	0.44	53	5250
6	8.9	0.49	54	5430
7	6.1	0.69	55	5600

There are also data on hardness of the head part of cutters toolholders in accordance with GOST R 51047-97 'Cutter bits for roadheaders and shearers. General technical specifications', shown in brackets for comparison.

The table shows that the increase of strain intensity leads to a significant increase of hardness in these samples. In this case, maximum strain ε_p = 0.69 conditions of the hardness increase up to 1.2 times compared with the underformed samples.

The curve in figure 5 shows the dependence of hardness variation of the working part of fragments after undergoing TMT, given in terms of HB, on the intensity of plastic strain ε_p, to which this area was subjected before hardening.

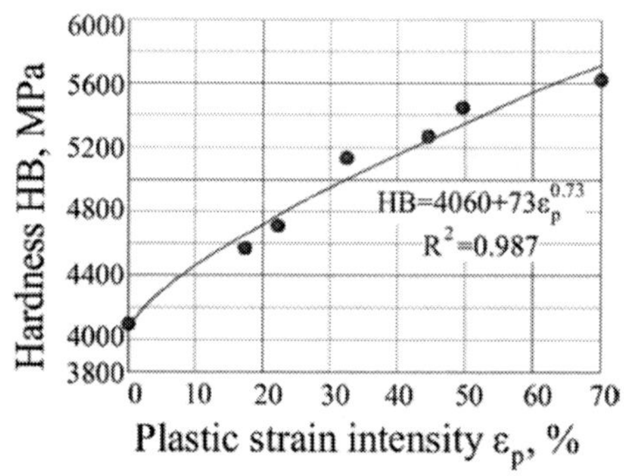

Figure 5. Dependence of the hardness variation of the 30KhGSA steel on the intensity of strain during TMT

As a result of mathematical processing of experimental data, it was found that the variation in hardness depending on strain intensity is accurately described by the equation of the following kind:

$$HB = HB_0 + \Delta HB = HB_0 + A\varepsilon^X,$$

where HB_0 – hardness of the material (steel 30KhGSA) not being subjected to strain before hardening (ε_p = 0); ΔHB – hardness increment due to the strain during TMT; A, x – empirical coefficients, constant for a given material (A =73, x =0.73).

The obtained results of experimental researches were used to create a computer model of the TMT process of rotary cutter bit toolholders TMT using the finite-element method of the software package 'DEFORM-3D' (figure 6).

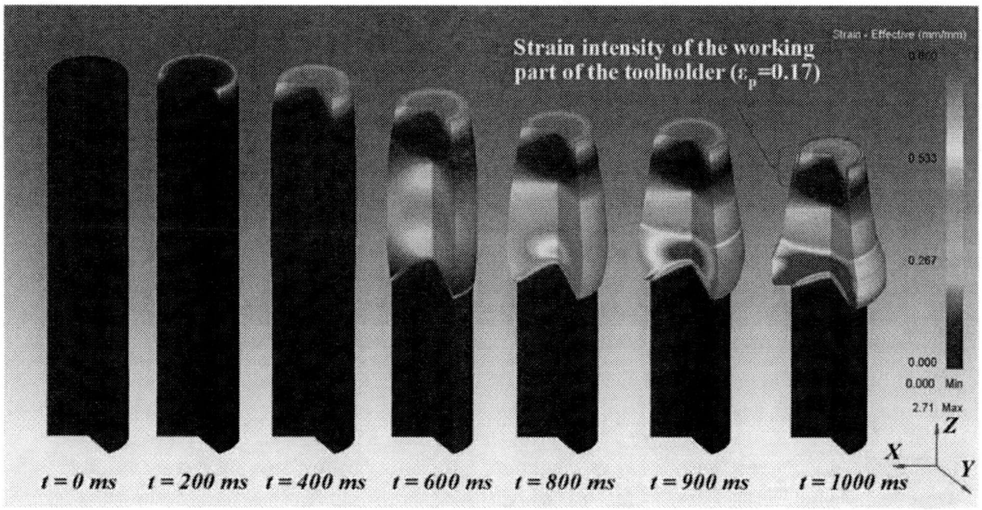

Figure 6. Computer simulation of the change of the initial workpiece shape and the distribution of plastic strain intensity throughout its volume during the manufacturing process of rotary cutter bit toolholder RSh 32-70/16.M1

According to the results of the computer simulation of traditional manufacturing of the shape of rotary cutter bit toolholder RSh 32-70/16.M1, it was established that the use of cylindrical workpieces conditions insufficient strain intensity ε_p = 0.17 for a substantial increase of the toolholder metal hardness as a result of subsequent hardening. The existing technology of cutter bits manufacturing provides a technological stage of forging cooling in the still air during the time period sufficient for the process flow of recrystallization of the deformed metal and removing of the surface hardening of the working part of the toolholder.

The head part of the cutter bit toolholder 'RSh 32-70/16.M1' has the worst wear during cutting of rocks. The change of the direction of the metal flow in the head part of the cutter bit toolholder during forging can increase strain ε_p in this part. As a result of the metal flow investigation, we decided to make a bevel in the head part of the workpiece (Figure 7). As a result of simulation of the metal flow process, it was established that the maximum value of strain ε_p~0.7 is achieved when the bevel length is b = 12...14 mm, and the angle is α = 45...55°.

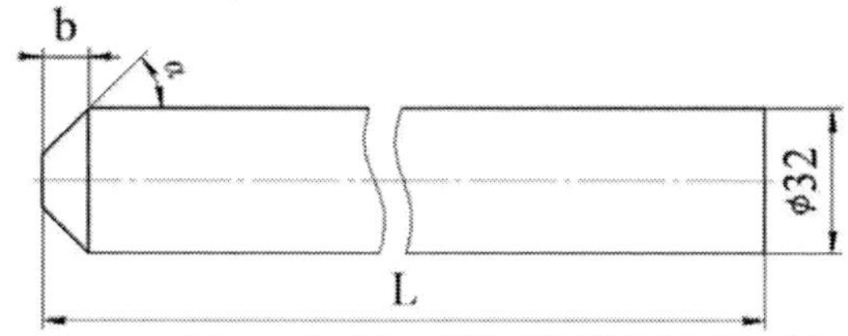

Figure 7. The workpiece of cutter bit toolholder RSh 32-70/16.M1 with the modified shape of the head part

It is recommended that the hard alloy insert fixing should be conducted according to the cold fit technology, used when equipping the rotary cutter bits of foreign manufacture. It should be noted that the proposed technology of cutter bits production does not lead to the increase of the cost of their production [3].

4. Conclusion

1. It was established that both the strain of the cutter bit toolholder steel at elevated temperatures and subsequent hardening increase steel hardness. The strain in the head part of the cutter bit toolholder of about 0.7 and subsequent hardening can increase the hardness up to 1.2 times.

2. The dependence of hardness on strain was determined.

3. It is supposed that such type of the hardness increase will lead to the increase of the wear resistance of this material, but this supposition requires further background studies.

References

[1] Levankovskij L A and Glatman L B 1998 *Rocks and composite materials breakdown by rotary cutter bits: Collected scientific papers* vol 1 (Moscow: Academy of mining science press) pp 81-109
[2] Krapivin M G, Racov I Ja and Sisoev N I 1990 *Mining tools* (Moscow: Nedra)
[3] Chupin S A 2016 *Improving wear resistance of rotary cutter bits of road headers for the mining in medium strength rocks* (Saint Petersburg: Mining University Press)
[4] Thuro K and Plinger R J 1998 *The 8th International IAEG Congress* (Vancouver) vol **2** p 3545-52
[5] Gajewski J and Jonak J 2008 *Methods of artificial intelligence research knives and cutting drums heads* (Warsaw)
[6] Horeshok A A, Mametev L E and Cehin A M 2013 *Manufacturing and exploitation of mining machines breaching tools* (Tomsk: Tomsk polytechnic university press)
[7] Bobrov V L 2012 *Increasing wear resistance of road headers tangential cutter bits by enhancement of heat treatment during their manufacture* (Saint Petersburg: Mining University Press)
[8] Lahtin Ju M and Leonteva V P 1990 *Materials science* (Moscow)
[9] Makarov D N, Komelkov E M, Didenko A P and P'yankov A A 2007 *Metallurgist* **51** 503-504
[10] Kruglova A A, Orlov V V, Khlusova E I and Nemetinov A A 2007 *Metallurgist* **51** 116-120
[11] Naumov A A, Bezobrazov Y A, Kolbasnikov N G and Chernikov E V 2013 *M. Sci. Forum* **762** 62
[12] Bolobov V I and Chupin S A 2015 *Proceedings of the Mining Institute* **216** 44-49

Determination of size properties of the organomineral insulation nanofiller based on the wood matrix

V E Danilov, A M Ayzenshtadt, T A Makhova and M A Frolova

Northern (Arctic) Federal University named after M V Lomonosov, 17, Severnaya Dvina Emb., Arkhangelsk, 634050, Russia

E-mail: v.danilov@narfu.ru

Abstract. This article provides a comparative analysis of the results obtained in determining the particle size of the basalt-based wood-mineral insulation nanofiller using photon correlation spectroscopy in various dispersion media after mechanical milling of a raw material. Aqueous solutions with various concentrations of glycerine were used as diluents. The studies demonstrated that glycerine-based basalt suspensions exhibit sufficient sedimentation stability that produces reliable data on the nanofiller size properties. This is achieved due to such dispersion medium properties as density and permittivity. Furthermore, determinations of the zeta potential of the basalt particle surface in glycerine showed that this dispersion was electrostatically stabilized and thus would be aggregation-stable.

1. Introduction

Economic efficiency of individual processes and production as a whole in various industries greatly depends on the thermal insulation of process equipment. Due to this fact, thermal insulation is widely used, for example, in mechanical engineering, metallurgy, chemical and petrochemical industry, oil refining, pharmaceutical and food industry, etc [1-2]. It is worth noting that the organomineral composite produced on the basis of the wood matrix is currently one of the most efficient insulation materials due to its unique natural properties. Studies [3-5] demonstrated the possibility of producing a thermal insulation material on the basis of the composition consisting of the wood machining waste (sawdust, bark), pre-crushed up to a 1-2 μm size, with its pore volume saturated with basalt nanoparticles. The studies [5] found that the use of basalt nanoparticles as fillers of the wood polymer matrix allowed improving its performance properties. At a filling ratio of around 30-40%, heat resistance, mechanical and heat-conducting properties of a material are improved significantly.

One of the major problems in producing such nanocomposites is provision of interaction between the polymer and the filler. The effect of the nanodispersed filler on the polymer properties strongly depends on the distribution of particles in the polymer matrix. In the initial state, basalt particles may represent agglomerates ranging in size from a few microns to a few nanometers. In order to effectively modify polymer properties, it is necessary to provide dissolution of agglomerates and regular distribution of mineral nanoparticles in the polymer matrix. For this purpose, it is required to obtain reliable data on the modifier size properties at the first stage of composite production.

Determination of the particle size in finely-dispersed systems uses various methods [6, 7]. The most common ones include photon correlation spectroscopy based on the Rayleigh scattering law [8]. In this case, while experimentally implementing this method, it is important to analyze size properties of finely-dispersed systems exhibiting sufficient sedimentation stability in a utilized dispersion medium. Consequently, when determining size properties of finely-dispersed basalt samples, account must be taken of not only the time factor, but also of the dispersion medium properties (density, permittivity) which affect sedimentation stability of analyzed objects.

This study aims to conduct a comparative analysis of results obtained in determining the basalt particle size using photon correlation spectroscopy in various dispersion media after mechanical milling of the raw material.

Sedimentation stability of solid particles in the suspensions analyzed was evaluated by the particle deposition rate (ω_0, m/s) calculated on the basis of Stokes' law by the following expression [9]:

$$\omega_0 = \frac{d^2(\gamma - \gamma_1)g}{18\mu}, \tag{1}$$

where d – a diameter of a deposited particle, m;
γ – density of a deposited particle, kg/m^3;
γ_1 – density of a medium, kg/m^3;
g – acceleration of gravity, m/s^2;
μ – dynamic viscosity of a medium, Pa·s.

2. Materials and methods

Basalt from the Myandukha deposit (Arkhangelsk region, Russia) was selected as the object of the study. This material has high strength, thermophysical and fire-resisting properties, and it is also very susceptible to mechanical dispersion with reproducible determinations of the particle size. In our opinion, aqueous solutions of glycerine are the most optimal choice among various diluents. This is due to the possibility of varying the density value in a water-glycerine system (from 1 to 1.26 g/cm^3) by only changing the component ratio. Control over the quantitative component ratio is realized by the refraction index through the refractometric method. Besides, glycerine is nonvolatile, optically inactive in the visible and ultraviolet spectrum, and highly miscible with water.

Basalt dispersing was carried out by mechanical wet milling in a planetary ball mill 'Retsch PM100', using large grinding bodies of 2 cm in diameter at a rotor speed of 420 rev/min for 30 min. The size and zeta potential of particles was determined with the analyzer 'Delsa Nano Series Zeta Potential' and 'Submicron Particle Size Analyzers (Delsa Nano)' through the dynamic light scattering measuring method. The true density of basalt particles was determined by the pycnometric method. The density of aqueous glycerine solutions was determined by the areometric method. The refraction index was determined by the refractometric method with the XRD unit. Dynamic viscosity was determined with the Brookfield viscometer.

3. Comparative analysis of results obtained in basalt particle size determination

According to the results of true density determination of finely-dispersed basalt test samples, the value of 2700 kg/m^3 was obtained, which was in good agreement with literature data.

Table 1 shows the results obtained in determining the properties of the dispersion medium, size and zeta potential of disperse phase particles.

Figures 1 and 2 show the graph of basalt particle size dependence on the time in utilized dispersion media. Deposition rate values calculated by equation (1) allowed obtaining a graph of dependence of the basalt particles deposition rate on their size (shown in Figure 3).

Table 1. Results of determining the glycerine aqueous solutions properties, size and zeta potential of basalt particles (at 20 °C)

Glycerine % by weight	Density, g/cm^3	Refraction Index	Viscosity, cP	Permittivity [10]	Zeta-potential, mV	Average particle size, nm
0	1.000	1.333	1.005	80.37	-2.25	207.7±56.7
50	1.130	1.398	6.00	65.63	298.34	108.5±28.7
100	1.260	1.474	1410	41.14	601.62	34.7±8.5

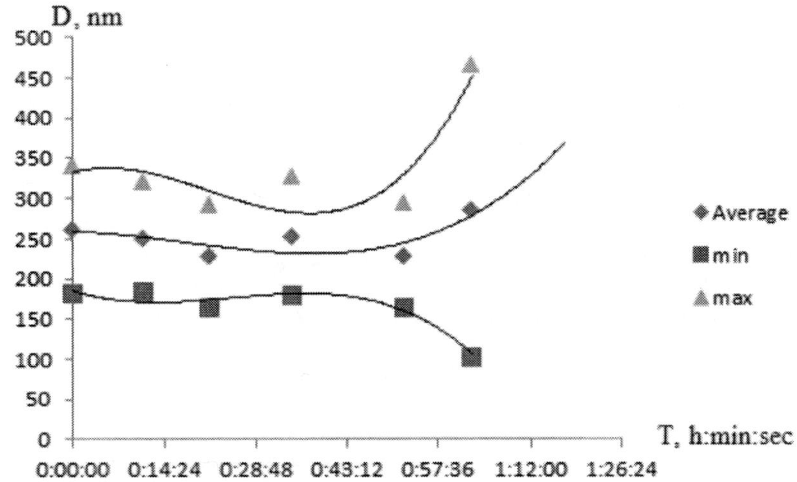

Figure 1. The graph of basalt particle size dependence on time in water

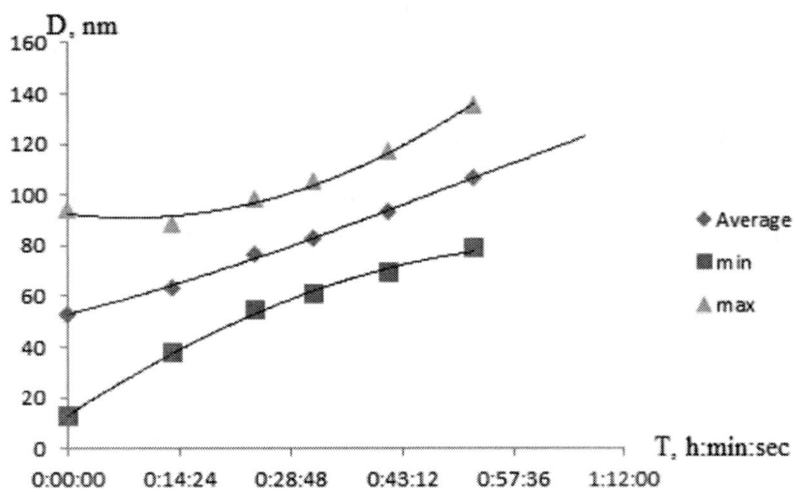

Figure 2. The graph of basalt particle size dependence on time in glycerine

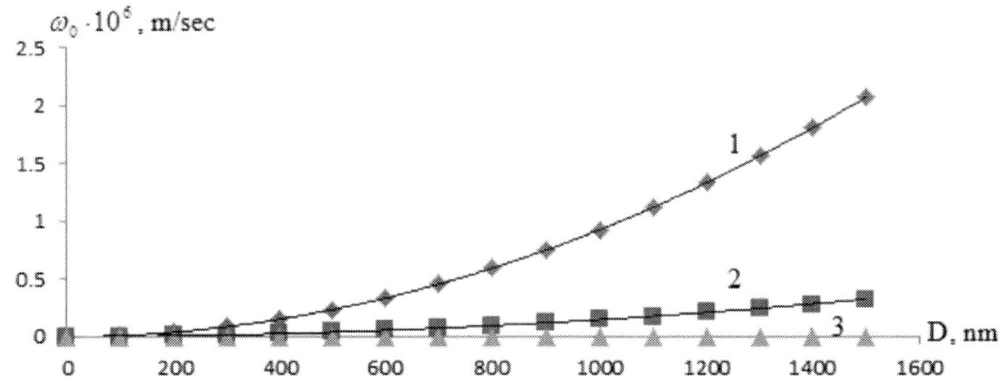

Figure 3. The graph of basalt particles deposition rate dependence on the diameter: 1 - in water; 2 – in the 50% aqueous solution of glycerine; 3 – in glycerine

When comparing the results of studies on the basalt particle size determination in various diluents, the following provisions can be distinguished:

1. Results of determining the basalt particles surface zeta potential in glycerine demonstrated that this dispersion was electrostatically stabilized (ζ>>30 mV), and thus would be aggregation-stable. In contrast, the basalt aqueous suspension is extremely unstable (-30<ζ<30 mV), and, as a consequence, the solid phase tends to coagulate and subsequently sediment in it. Besides, having compared the data from Table 1 and Figures 1 and 2, one can come to a conclusion that with the increase in permittivity which characterizes the polarity of the solvent, the aggregate and sedimentation stability drops significantly.
2. Results of the experimental determination of the size of basalt particles through time (Figure 1, 2) demonstrated that the slow growth of the particle size at a constant rate could be observed in glycerine, while the range of size properties remained virtually the same (a polydispersity index of 0.056). This indicates that by stabilizing particles aggregatively, glycerine not only acts as a diluent, but also as a surface-active agent (surfactant). At the same time, coarsening of particles in water occurs at different rates, whereas the range of size properties increases through time (a polydispersity index of 0.388).
3. Comparison of basalt deposition rates in various diluents (Figure 3) demonstrated that for nano- and especially micro-sized particles, there was a substantial probability of deposition in water in the process of the particle size determination using dynamic light scattering. Consequently, these sedimented particles may be overlooked in the analysis results, which does not provide a complete and accurate information on size properties and fractional distribution in the suspension.

4. Conclusion

After conducted studies of the particle size of the basalt-based wood-mineral insulation nanofiller using photon correlation spectroscopy it is possible to conclude that nanodispersed basalt exhibit sufficient aggregate and sedimentation stability in a glycerine-based dispersion medium.

It is shown that in the aqueous dispersion medium the sedimentation process of basalt particles proceeds intensively. This fact causes significant errors in the assessment of the particle size characteristics of basalt by dynamic light scattering.

It was found that in glycerol nano- and micro-sized particles of basalt electrostatically stabilized and have sufficient aggregate stability for measurements of their sizes. This makes the use of this organic solvent preferred as the diluent.

Acknowledgments
This study was conducted with the support from the Ministry of Education and Science of the Russian Federation within the basic part of the State Task of 2016-2017 (Project No. 3636).

References
[1] Saidal' G I & Kopchenko P V 2007 *Chem Petrol Eng* **43** 159
[2] Skobelkina T N, Yurchenko V G, Bol'shakov V I et al. 2006 *At Energy* **101** 652
[3] Zoltán Pásztory, Ildikó Ronyecz 2013 *Acta Silv. Lign. Hung.* **9** 111-117
[4] Hermawan A, Ohuchi T, Fujimoto N and MuraseY 2009 *J Wood Sci* **55** 74
[5] Danilov V, Ayzenstadt A, Frolova M, Tutygin A 2015 *15th International Multidisciplinary Scientific GeoConference & EXPO SGEM 2015, Nano, bio and green – technologies for a sustainable future*, **1** 227-234
[6] Danilov V E and Ayzenstadt A M 2016 *Nanotechnologies in Construction.* **8**(3) 97-110
[7] Müller H G 1989 *Colloid & Polymer Sci* **267** 1113
[8] Pecora R 2000 *Journal of Nanoparticle Research* **2** 123
[9] Savitskii D P, Dimitryuk T N & Makarov A S 2016 *Solid Fuel Chem.* **50** 39
[10] *Physical Properties of Glycerine and Its Solutions* (New York: Glycerine Producer's Assoc)

Mathematical modeling of the steel ingot teeming and the solidification process

S B Gamanyuk, N A Zyuban, D V Rutskii and S V Palatkin

Volgograd State Technical University, 28, Lenina ave., Volgograd, 400005, Russia

E-mail: rtecmat@vstu.ru

Abstract. The paper presents computer simulation-based findings on the teeming speed effect on the formation of the ingot structure and inner defects. It is reported that for a top-poured ingot, the average teeming speed (1.4-2.0 m/sec) proves to be the best choice as it is accompanied with a decreased number of surface defects and an increased metal yield resulting from a decreased shrinkage cavity.

1. Introduction

In spite of the fact that over the last years there has been a pronounced tendency in the growth of the amount of metal poured with continuous casters, up to 20% of metal is still top poured in ingots in electric furnace shops. Ingot quality is determined by such principal process parameters as teeming speed and melt temperature. The control of metal pouring temperature is more common compared with teeming speed control because there has not been much systematic research of the teeming speed yet.

It is a well-known fact that the amount of surface defects in top-poured ingots depends upon the nature of the stream effluence; the latter, in its turn, is determined by the pouring nozzle shape and metal teeming speed [1-4]. The major effect of the steel teeming process on the metal quality is noted in the works [5-9], which demonstrate a noticeable impact of the teeming rate on ingot macrostructure formation as well as the location, distribution and size of shrinkage defects in the ingot. Somehow, nowadays it is not common to use teeming speed to control the formation of the ingot structure and ingot internal defects.

Computer simulation is now widely commercially used not only to improve product quality, but also to cut down material costs and reduce new product development time [10, 11].

The paper is aimed at the analysis of the effect of steel teeming speed on steel ingot quality using a computer simulation of steel teeming and solidification processes.

2. Materials and methods

A method based on two software applications was developed for the research:

1) The LVMFlow application is designed to study metal convection flows and shrinkage defect location; a controlled volume method. The application makes it possible to visualize the process of mold filling with melt. The application was provided by the MKM Research and Development Shareholders Company Limited (the city of Izhevsk), Version No. 4.4r6 Time, License No. T0027.

2) The 'Crystal' computer simulation application developed to study the metal crystallization process with the finite difference method [12-15].

3. Results and Discussion

The metal teeming speed corresponded to the commercial rates of the metal level rise in the ingot and ranged from 0.7 to 2.5 meters per minute. While performing computer simulation with the LVMFlow application, the speed was altered by varying pouring nozzle diameters from 15 to 65 mm: the low teeming speed (0.7÷1.3 m/min) corresponded to the diameters from 15 to 30 mm; the average speed (1.4÷2.0 m/min) corresponded to the diameters from 30 to 54 mm, and the high speed (2.1÷2.5 m/min) corresponded to 45 up to 65 mm pouring nozzle diameters. In doing so, within each speed range, the effect of speed was studied at 0.1 m/minute intervals.

When analyzing the effect of the teeming speed on the alteration of model ingot crystal zone dimensions and the degree of shrinkage defects in them, the method of finite differences and the 3D modeling option in the LVMFlow application were used.

Since the main criterion for the top-poured ingot quality is the surface condition, it deemed of interest to evaluate the effect of the teeming speed on the spray process. When top pouring an ingot, spraying is the most intensive when liquid steel flow hits the ingot bottom. Consequently, the intensity of spray formation was assessed by the maximum elevation at which liquid metal drops rose from the ingot bottom at the beginning of teeming.

While modeling the teeming process, the following parameters were analyzed: circulation flow distribution throughout the ingot mass and solid phase build-up time when the mold was 50% full and 100% full.

The effect of the teeming speed on the solid phase build-up rate was assessed both with the LVMFlow application and a mathematical calculation:

$$\textit{Solid phase quantity} = 100\,\% - X\,\%,$$

where 100% is the quantity of both liquid and solid phases; $X\,\%$ is the liquid phase quantity.

A comparative analysis of the effect of the teeming rate on the metal spray intensity at the beginning of pouring demonstrated that teeming at a low and a high speed intensifies the metal spray when a liquid steel flow hits the mold bottom; somehow, this phenomenon was not observed when pouring at an average speed. Figure 1 presents simulation results of spray types at the moment when the flow hits the mold bottom at various pouring speeds.

The maximum spray (splashes formed at the height corresponding to ¼ of the mold height ≈ 400 mm) was observed at the speed of 2.5 m/min. When poured at average teeming speeds, the spray was of the order of 70 to 120 mm, while at low teeming speeds, the spray varied from 60 to 150 mm. It is noteworthy that the maximum height of splash from the mold bottom was registered for the teeming speed of 1.1 m/min.

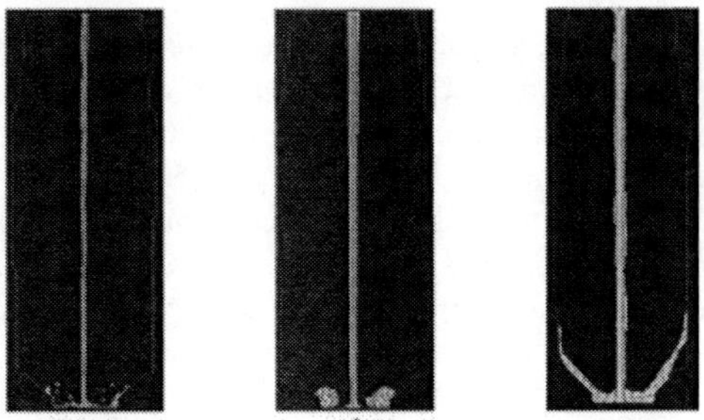

Figure 1. The schematic of spray formation at various teeming speeds: a) drop formation at a low speed; b) a uniform filling of the ingot with the melt poured at an average speed; c) the intense spray when teeming at a high speed.

When teeming at a low speed within the range of 0.7 to 1.0 m/min, a bad structure of the falling stream of metal was noted. The stream was characterized by loss of continuity and a distortion of the flow structure. In addition, the low speed teeming results in heavy cooling of the melt: when the melt is poured into the mold, it loses part of it's overheat. It leads to an intensive oxidation which results in the increased quantity of oxides and increased porosity. Starting from the speed of 1.2 m/min up to 2.0 m/min, no intensive spray was observed due to the formation of a 'protective cushion' of the liquid metal in the mold bottom.

It was demonstrated that after the mold was approximately 50% full, steel flows emerged and started to circulate throughout the metal until the completion of teeming (Figure 2).

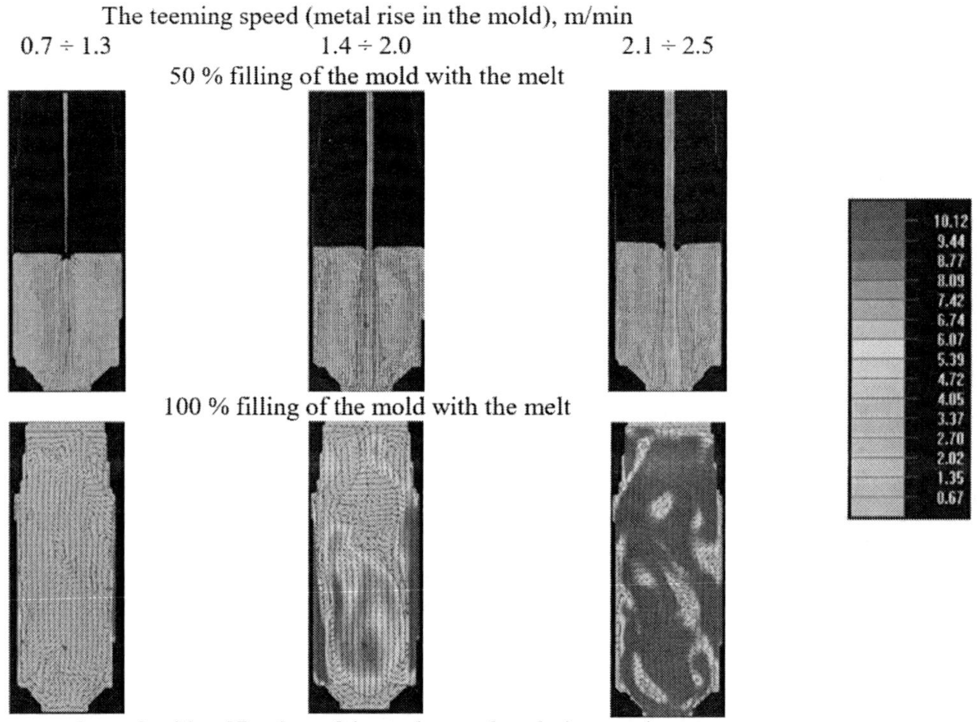

* – color identification of the melt speed variation, cm/s

Figure 2. Liquid steel flow distribution while filling the mold with the melt.

The schematic shows that the intensity of liquid metal circulation flows notably increases with the rise in the teeming speed. When poured at a high teeming speed, the metal flow speed below the poured metal surface decreases approximately twofold when the stream reaches the mold bottom, and the stream diameter increases approximately by a factor of 1.5.

When teeming at an average speed, two classical circulation zones emerge inside the mold: those of rising and falling streams flowing at a speed of about 1.25 cm/s. The rising streams take up nearly 2/3 of the ingot cross-section and evenly rise from the bottom to the middle level. Starting from the middle level and up to the upper level, the flows chaotically mix up which might be caused by termination of teeming (stream effect) and the discard top effect.

The penetration depth of the metal stream while teeming at a low speed (0.7-1.3 m/min) is insignificant because the stream possesses low kinetic energy and cannot penetrate deeply into the melt. The characteristic speed of the circulation contour is the lowest and on the average equal to 0.7 cm/s. A low teeming speed results in a practically laminar flow of the melt, and in this case an upward and a downward flow circulation is observed. Based on speed color identifiers, the average value of the speed in this instance is 0.7 to 3.0 cm/s.

An increase of the teeming speed to 1.2 m/min leads to a more intensive flow circulation, and speed increases on the average to 6.0-8.0 cm/s (throughout the entire ingot). In doing so, the maximum speed is observed in the areas bordering on the mold walls (over 9.0 cm/s) and in the centre of the solidifying ingot within the lower and middle levels.

The study of the teeming speed effect on the intensity of the solid phase growth revealed that within each range of speeds the build-up of the solid phase remains practically the same: the quantity of the solid phase evenly grows within equal time intervals.

Somehow, the most intensive growth of the solid phase was observed in the low teeming speed group: for example, for the solidification time of 15 min, the solid phase quantity of 13 % corresponded to the speed of up to 1.3 m/min; for 8 % of the solid phase, the speed corresponded to the speed of up to 2.0 m/min; for 6 % of the solid phase quantity, the teeming speed was over 2.1 m/min. These figures can be explained by a noticeable cooling action of the mold walls on the relatively small quantity of the melt teemed from small diameter pouring nozzles. Somehow, as the crystallized layer is building up, heat conductivity decreases; it reduces the quantity of the solid phase formed. In this case, the teeming speed does not affect the solidification time because, according to the simulation, the ingots fully crystallize within two hours after the completion of teeming.

A 3D visualization of teeming and crystallization simulation using the LVM Flow application made it possible to locate and specify the typical dimensions of shrinkage defects in 4.5 ton ingots (see Figure 3).

The simulation with metal teeming speeds of up to 1.3 m/min showed the shrinkage cavity penetrated into the chill below the upper level. At average speeds, the shrinkage cavity, as a rule, is located mainly in the hot top metal. The maximum quantity of defects in the axial zone was observed at the teeming speed of above 2.3 m/min.

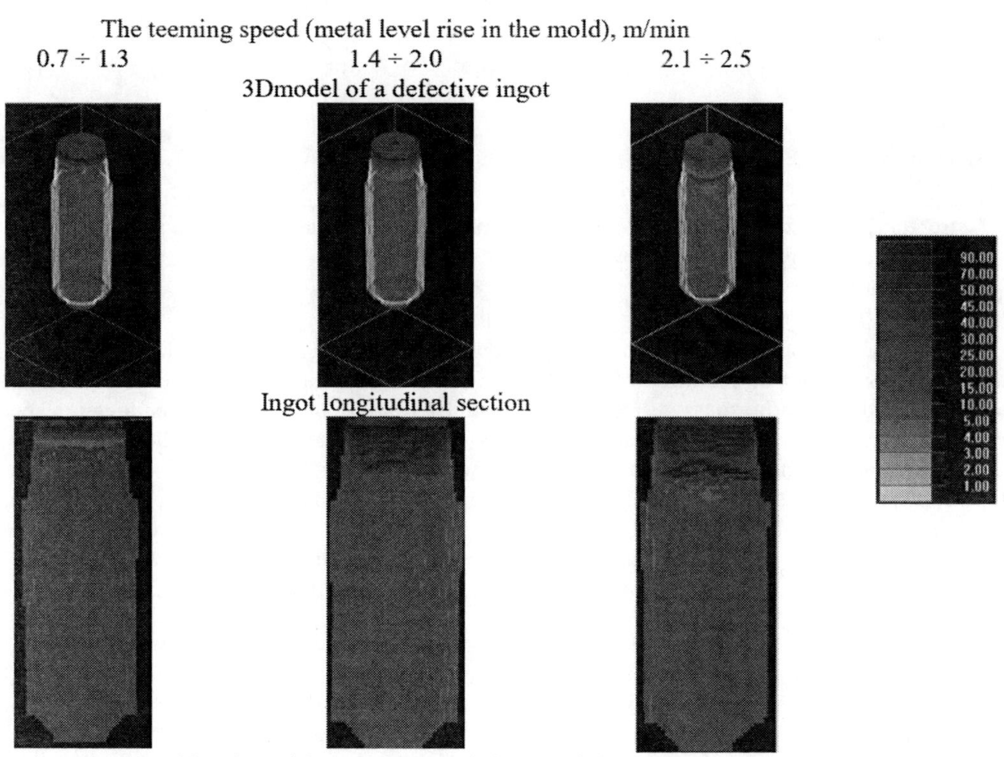

* - color identification of the quantity of shrinkage defects in an ingot

Figure 3. A typical location of the shrinkage cavity inside the chill in a 3D model and in the model axial section.

The metal teeming simulation results were used in the finite difference method to obtain data on the dimensions of typical structural zones for a 4.5 ton ingot; the data are presented in Table 1.

Table 1. The length of structural zones and physical non-heterogeneity zones of model ingots poured at various speeds (finite difference method)

Table 1. The length of structural zones and physical non-heterogeneity zones of model ingots poured at various speeds (finite difference method).

Structural zone type	Zone extension (width) *					
	l, mm			% of the ingot height or diameter		
	teeming speed, m/min					
	0.7 ÷ 1.3	1.4 ÷ 2.0	2.1 ÷ 2.5	0.7 ÷ 1.3	1.4 ÷ 2.0	2.1 ÷ 2.5
Crust zone	14	11	9	2.2	1.74	1.42
Columnar dendrites	81	75	70	12.8	11.5	10.8
Equiaxial dendrites	135	141	137	21.6	22.38	21.3
Sedimentation cone**	593	605	688	29.8	30.5	34.6
Srinkage cavity**	339	309	358	15.1	11.0	18.1
Axial sponginess zone **	369	351	468	18.6	17.7	23.6

* structural zone extension was calculated as a mean value for the three levels
**over the chill height

The logging of the solid phase quantity data during solidification using the LVM Flow application and the measurement of solidified zone extension in solidified ingots using the 'Crystal' application showed that the crust thickness decreases by a factor of 1.5 with an increase in the teeming speed. The columnar crystal zone width decreases by a factor of 1.3 with the increase in the teeming speed, and the sedimentation cone height increases approximately by 1.6 %.

The depth of the shrinkage cavity penetration into the chill as well as the extension of the axial sponginess zone are the shortest at the average teeming speed. It is noteworthy that the volume of the axial sponginess zone reduces to 0.2 %. The computer simulation results showed that the teeming speed alteration effects both the location and dimensions of shrinkage defects.

4. Conclusion

Summing up, the control of the melt teeming speed is an important process parameter which can facilitate the production of ingots of the required structure with a convenient location of shrinkage defects and, thus, can improve the properties of the forgings produced. That is why, ingot quality control based on the properly chosen teeming speed is an important aspect of the ingot production process which does not require costly investment in the production process.

The use of various simulation models of steel teeming and crystallization combined with a metallographic study of the crystallized steel noticeably improves the predictive power of the models defect-wise; these defects can be eliminated in future by choosing optimal regimes of steel teeming.

Acknowledgments
The reported study was funded by RFBR according to the research project №16-08-01029 A and contest SP-2015 (№ SP-4573.2015.1).

References

[1] Skoblo S Ya, Kazachkov E A 1973 *Ingots for large forgings* (Moscow: Metallurgy)
[2] Efimov V A 1976 *Casting and crystallization of steel* (Moscow: Metallurgy)
[3] Zhul'ev S I, Zyuban NA 2001 *Metallurg* **12** 38-39
[4] Basov A V 2011 *Improvement and modernization of steelmaking technology* (Moscow: Moscow State Evening Metallurgical Institute Press)
[5] Zhul'ev S I, Zjuban N A 2003 *Production and quality problems forging ingot* (Volgograd: *Volgograd State Technical University Press*)
[6] Eron'ko S P, Bykovskikh S V 2003 *Casting steel: the Equipment, Technology* (Kiev: Tekhnika)
[7] Eron'ko S P, Bykovskikh S V 1998 *Physical modelling of processes furnace processings and casting steel* (Kiev: Tekhnika)
[8] Nedopekin F V, Belousov V V *Bulletin of Udmurt University: electronic scientific journal Series 4: Physics and Chemistry* **1** 189-199
[9] Palatkin S V, Zjuban N A 2012 *Volgograd State Technical University Volgograd* 179-183
[10] Sapuan S M, Mustapha F, Majid DL, Leman Z, Ariff AHM, Ariffin MKA, Zuhri MYM, Ishak MR and Sahari J 2011 *Key Engineering Materials* **471-472** 601-605
[11] Kermanpura A, Eskandaria M, Purmohamada H, Soltanic MA and Shaterib R 2010 *Materials & Design* **31/3** 1096-1104
[12] Bagmutov V P, Zaharov I N 2006 *Stal'* **9** 23-27
[13] Bagmutov V P, Zaharov I N 2006 *Stal'* **3** 28-33
[14] Reis A, Xu Z, Tol R V and Neto R 2012 *Journal of Manufacturing Processes* **14/1** 1-7
[15] V P Bagmutov, N I Zaharov, N A Zyuban, D V Rutskii, S B Gamanyuk No. 2013613206 (28 March 2013)

IOP Conf. Series: Materials Science and Engineering **177** (2017) 012065 doi:10.1088/1757-899X/177/1/012065

Spherical gearing with intermediate ball elements: parameter ranges with a high contact ratio

M V Gorbenko[1] **and T I Gorbenko**[1,2]

[1] Tomsk Polytechnic University, 30, Lenina ave., Tomsk, 634050, Russia
[2] Tomsk State University, 36, Lenina ave., Tomsk, 634050, Russia

E-mail: gmvski@tpu.ru

Abstract. The paper presents analytical research of the geometry and kinematical parameters of spherical gearing with ball intermediate elements. The main attention is paid to the influence of the offset coefficient on the tooth geometry generation, the contact ratio and the motion transmission angle. Intermediate ball element racetracks on the gear are trochoidal curves on a spherical surface. Two areas for the offset coefficient values providing a high value of the contact ratio – basic trochoid (without offset) and prolate trochoid with abutting racetracks of adjacent ball elements — were revealed. Analysis of the investigated parameters showed that for power transmission, it is preferable to use spherical gearing without an offset, and for kinematic transmission, it is possible to use profiles with a large offset. The present study allows making a rational choice of geometrical parameters depending on the transmission predestination.

1. Introduction

In the modern mechanical engineering, a significant number of gearing types are applied. The development of technology today makes new demands on the reliability, accuracy, stiffness and power efficiency to the transmission devices. This applies both to traditional kinds of transmissions and to newly developed ones. Developing of transmissions with intermediate rolling elements is a challenging trend in the synthesis of gearing because they have high service properties, technological effectiveness as well as low noise, high accuracy parameters and great loading capacity [1-7].

This paper presents an analytical investigation of the offset coefficient influence of spherical gearing with intermediate ball elements [1] on the tooth geometry generation, the contact ratio and the motion transmission angle.

Spherical gearing can be used in transmissions as an alternative to their bevel-gear counterparts, and, thus, calculation of rational gear parameters are important.

2. Spherical gearing geometry

The considered transmission (figure 1) consists of pinion-holder 1 with hemispherical sockets for intermediate bodies – balls (3), gear 2, designed as a sphere, on the surface of which there are tracks for balls of a semicylindrical cross-section, the axial line of the track is a trochoidal curve on a spherical surface [1, 8]. Intermediate elements are standard balls produced by the ball-bearing industry. In order to enable free rotation of ball intermediate bodies (BIB), both radii of nests in the pinion-holder and the radius of channel surfaces are slightly larger than the radius of BIB. To prevent the loss of BIB from nests when leaving the working area, the geometrical closing by the spherical surface of the gear is used.

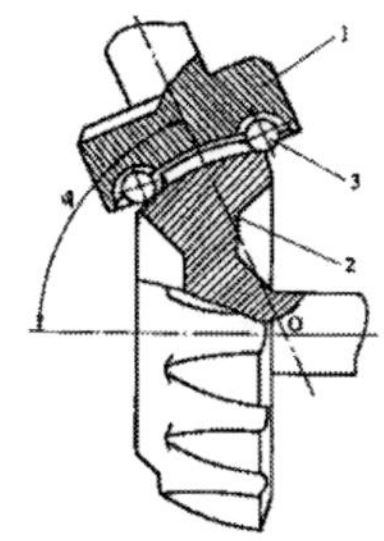

Figure 1. Spherical gearing with intermediate ball elements.
1 – pinion-holder, 2 – gear, 3 – balls.

Let us denote the $\delta=\xi^{(0)}$ kinematical cone angle, $\xi=\delta(x+1)$ – the design cone angle, $x=(\xi-\delta)/\delta$ – the offset factor, φ – the axes intersection angle, φ_1 – the pinion rotation, φ_2 – the gear rotation, U – the gear ratio and R_b. is the rated radius of ball centers concerning the gear sphere center, Z – the number of intermediate ball elements and K is the ball center radius in the pinion-holder in terms of the ball radius.

In the spherical coordinate system, placed on the centre of the gear sphere (R, θ, φ_2), the trajectory of the ball is described by the following system of equations [8]:

$$\begin{cases} \sin\theta = \cos\xi\cos\varphi - \cos\varphi_1\sin\varphi\sin\xi, \\ \varphi_2 = \varphi_1/U, \\ R = R_b. \end{cases}$$

Some calculations are convenient to do in the Cartesian coordinate system, connected with the gear center:

$$\begin{cases} X_{3(s)} = KR_s(\sin\varphi_1\cos\varphi_2 - \cos\varphi_1\cos\varphi\sin\varphi_2 - \mathrm{ctg}\,\xi\sin\varphi\sin\varphi_2), \\ Y_{3(s)} = KR_s(\sin\varphi_1\sin\varphi_2 + \cos\varphi_1\cos\varphi\cos\varphi_2 + \mathrm{ctg}\,\xi\sin\varphi\cos\varphi_2), \\ Z_{3(s)} = KR_s(\mathrm{ctg}\,\xi\cos\varphi - \cos\varphi_1\sin\varphi). \end{cases}$$

The velocity transmission angle at a tangent to the sphere surface plane, perpendicular to the radius-vector of the ball centre, is

$$\mu = \mathrm{arctg}\left(\frac{-\sin\varphi\sin\varphi_1}{\sin\varphi\cos\varphi_1\cos\xi - \sin\xi(U-\cos\varphi)}\right).$$

The tooth form significantly changes depending on the offset coefficient retaining all other parameters. Figure 2 (a–e) represents the tooth form evolution and ball element trajectories for five different offset coefficient values.

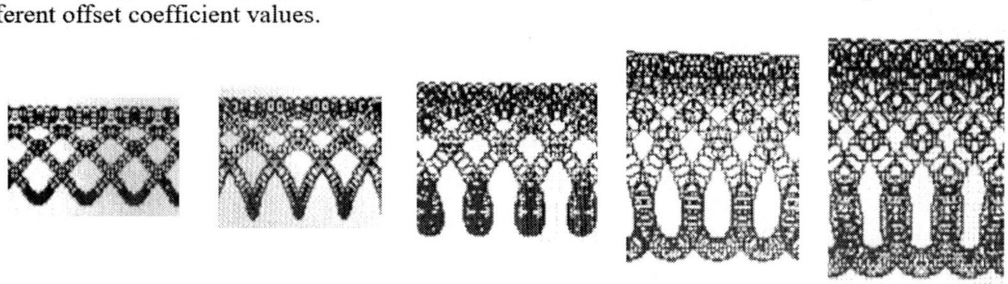

(a) x=-0.2 (b) x=0 (c) x=0.3 (d) x=0.8 (e) x=0.9

Figure 2. The tooth profile evolution vs the offset factor for gearing of U=4, φ=90 , Z=6, K=8 (picture scales are the same)

The angle of contact beginning with the tear-shaped teeth may be found from the equation for self-intersecting of the contact line on the gear:

$$\begin{cases} [\cos\lambda\sin\mu\cos\varphi_{1H2}+\sin\varphi_{1H2}(\cos\lambda\cos\mu\cos\xi+\sin\lambda\sin\xi+K)]\cos\varphi_{2H2} \\ \qquad -[\cos\varphi_{1'2}(\cos\lambda\cos\mu\cos\xi+\sin\lambda\sin\xi+K)-\sin\varphi_{1H2}\cos\lambda\sin\mu]\cos\varphi\sin\varphi_{2H2} \\ \qquad -(\sin\lambda\cos\xi-\cos\lambda\cos\mu\sin\xi+Kctg\xi)\sin\varphi\sin\varphi_{2H2}=0, \\ \varphi_{2H2}=\dfrac{\varphi_{1H2}}{U}. \end{cases}$$

Angle λ depends on the nest radius and the depth of the pinion-holder, the radius and the depth of the ball track and on the clearance between the top land of the teeth and the inner spherical surface of the pinion-holder. For represented gearing without a separator for balls, this angle is about 15–25°, while for schemes with separator [4], it may reach 30-40°, that results in the decreased efficiency. The edge contact pinion-holder nest–ball– semicylindrical track corresponds to $\lambda\approx5$-9°.

Generally, only lower teeth for negative and 0-corrected gears (figure 2a and 2b) are taken into consideration. Analytical research of geometrical features for this gear and derived equation systems allow calculating beginnings (φ_{1H}) and endings (φ_{1K}) of ball engagement with all teeth lines. Borderlines for active meshing are shown in figure 3 and the corresponding contact ratio diagram is in figure 4. Equipotential line (μ=20°) must be considered as borderline for φ_K due to extremely low effectiveness in transmitting motion when μ<20°, even though there is no self-braking up to $\mu\approx7$–9°.

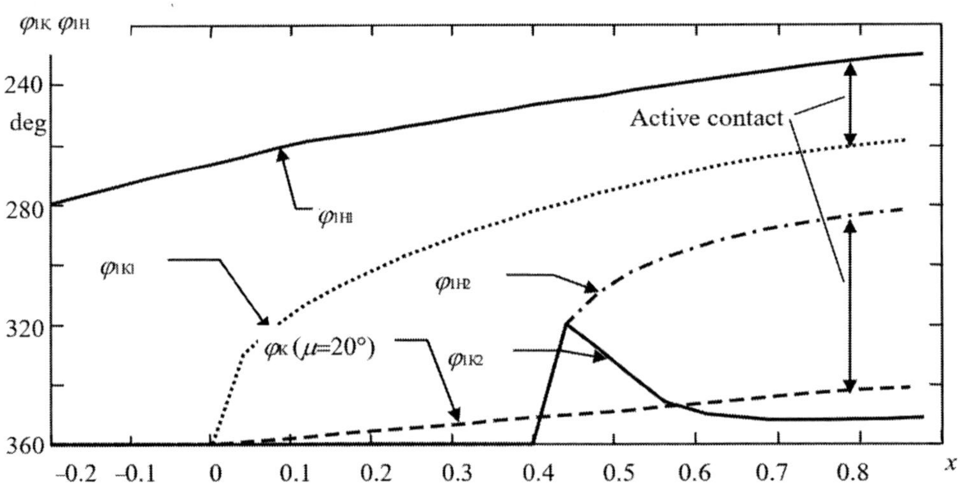

Figure 3. Borders of active meshing in terms of pinion-holder rotation vs the offset factor for transmission: U=2, φ=90°, Z=6, K=5.

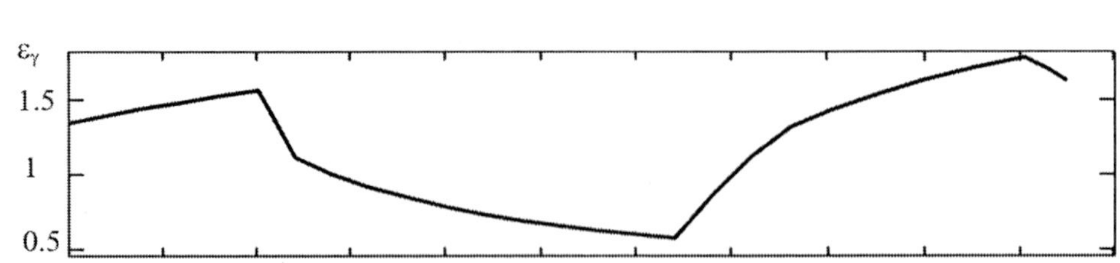

Figure 4. The contact ratio factor vs the factor for transmission: U=2, φ=90°, Z=6, K=5.

The maximum in the contact ratio (figure 4) corresponds to abutting racetracks of adjacent ball elements (figure 2d). The contact ratio drop within the interval of 0<x<0.4 is explained by the first rank tooth undercutting, until a tear-shaped tooth increases and comes into operation. The last drop (figure 4, x>0.8) corresponds to undercutting of the tear-shaped tooth (figure 2e). Lines φ_{1H2} and φ_{1K2} in figure 3 mean the active contact of the tear-shaped tooth.

Calculations of the velocity transmission angle μ through the active part of intermediate body trajectory vs the pinion-holder rotation for the number of offset coefficients are represented in figure 5.

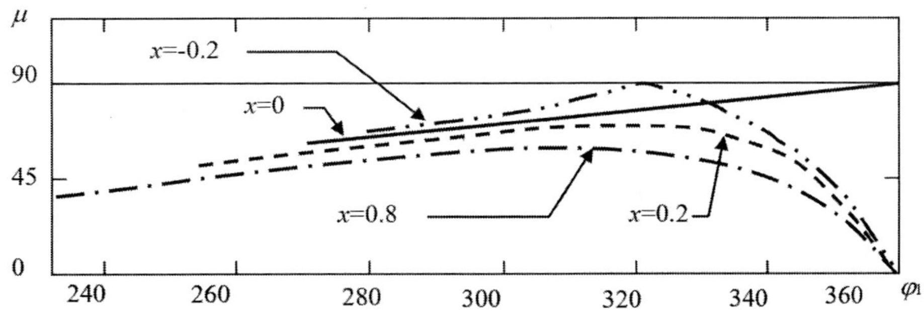

Figure 5. Transmission angle μ vs rotation of the pinion-holder. (U=2, φ=90°, Z=6, K=5).

If μ=90° the radius-vector of contact point drawn from the ball centre and the vector of the ball centre absolute velocity are in one plane. The angle between them is λ. It may be referred as pressure angle. This is the reason why separatorless schemes are preferable. Study of angle μ shows sufficient values in the low negative and 0 offset factor. An ideal value to transmit motion is 90 .

3. Conclusion

After conducted studies of the geometer-kinematic features of spherical gearing with intermediate ball elements, it is possible to conclude that low positive correction (x=0.2–0.4) is ineffective. The use of tear-shaped teeth (a high value of the correction factor: x=0.6–0.8) may be useful in kinematic drives, where there is no heavy loads, due to an attractive value of the contact ratio. These gears can smoothly transmit motion avoiding fluctuations in the velocity ratio and impact blows in case of teeth changeover. This kind of transmission possesses a whole number of positive qualities relevant to transmissions with free intermediate rolling elements. The approach developed by the authors provide engineers with a tool to make a justified choice of spherical gearing parameters. The contact conditions are better for separatorless gearing, but in this case, the closing surface is required to prevent loss of BIB.

References

[1] A E Beljev, An I- Kan, V V Gurin and L D Garanin, USSR. Patent No. 261,072
[2] A N Morozov, V N Kundel and N V Vasilchenko, USSR. Patent No. 1,173,099
[3] A E Beljev, An I- Kan, V I Risman and A N Morozov, USSR. Patent No. 658345
[4] An I-Kan, E Efremenkov and A Cheremnov, Russia. Patent No. 2,457,377
[5] An I-Kan, A S Il'in and A V Lazurkevich 2016 *IOP Conf. Ser.: Mater. Sci. Eng.* Vol 124 (England: IOP Publishing) 012003
[6] I Kenji, U.S. Patent No. 4,829,851A
[7] K Akiyoshi, K Sato, M Nakakoji, EP Patent No. 2,759,740 A1
[8] Gorbenko M V 2000 *Investigation of errors affect the kinematics and dynamics of a spherical ball-transfer intermediate bodies and the choice of rational parameters* (Tomsk: Tomsk Polytechnic University Press) p 123

Modeling a VR-type piston engine as the power plant

A A Gorshkalev, S S Kayukov, S S Korneev and V V Urlapkin

Samara National Research University, 34, Moskovskoe shosse, Samara, 443086, Russian Federation

E-mail: AGorsh@bk.ru

Abstract. This article describes existing methods of internal combustion engine calculation. The developed algorithm of diesel engine modeling in 'AVL FIRE ESE DIESEL' and 'LMS Imagine.Lab AMESim' software is presented. The algorithm includes description of model preparation, boundary condition setting and calculation execution. Obtained results of the modeling show significant enhancement in accuracy in contrast with analytical calculation. The difference between results obtained in different software is caused by a simplified combustion model in 'LMS Imagine.Lab AMESim' and a lack of a submodel describing a refrigeration system.

1. Introduction

Methods of the internal combustion engine in systems of automatized designing are described in [1-8]. Articles [9-15] deal with engine heat calculation in the 'AVL FIRE ESE DIESEL' software. They also deal with different approaches to boundary conditions setting and engine calculation settings. Data on the R-type and V-type piston engine 'LMS Imagine.Lab AMESim' simulation are presented in [16-20]. In the same time, information about designing of VR-type engine, which is a combination of V-type and R-type engines, was not found. This article deals with the modeling algorithm of VR-type piston engine. The engine, which is described in the given work, represents a four-cylinder, four-strokes, VR-type diesel engine with the V-angle equal to 15°, developed on the basis of the Volkswagen AAA VR–6 engine.

2. Problem to be solved

The first stage of the given research deals with heat calculation of the piston diesel internal combustion engine in the 'AVL FIRE ESE DIESEL' software [21]. Combustion chamber setting and calculation were carried out in the 'AVL FIRE ESE DIESEL' software. The type of the engine (Vee engine, which is a mark for V-type engine), the number of cylinders, the piston diameter (85 mm) and the compression ratio (18) were set in the menu 'General engine parameters'. Data about the crank radius (42.5 mm), the shaft length (150 mm) and cylinder axis displacement from the axis of the crankshaft (0 mm) were introduced in the menu 'Piston movement specification'. Models of the piston and the fuel sprayer were created in the 'Sketcher' window with parameters of the designed engine. Geometric parameters of the combustion chamber and the injector sprayer are represented in figure 1.

The next stage is creation of the mesh model of the combustion chamber by the integrated mesh generator. To check the mesh operation ability, it is displayed in different positions in relation to the crankshaft angle. The angle of calculation start (560°) and finish (900°), as well as the crankshaft frequency (3000 rounds per minute) are set in the 'Simulation parameters' chapter after mesh model generation. The next step is the choice of the boundary conditions in the corresponding menu. Air was chosen as a working fluid, and common boundary conditions were set: pressure (101325 Pa) and temperature (380 K). Turbulence model constants are also chosen at this stage. Then, the type of fuel is selected. Diesel fuel was chosen in the list of the 'Type of hydrocarbon fuel' tab. Equations for the process of burning and a 'k-zeta-f' turbulence model are selected in the 'Activate equations' tab. This model has a more suitable representation of the equation for elliptic functions and for boundary conditions near the wall and is also less sensitive to a wall non-uniformity. Advantages of this model are very important in the area of the nozzle sprayer where boundary conditions have a significant impact.

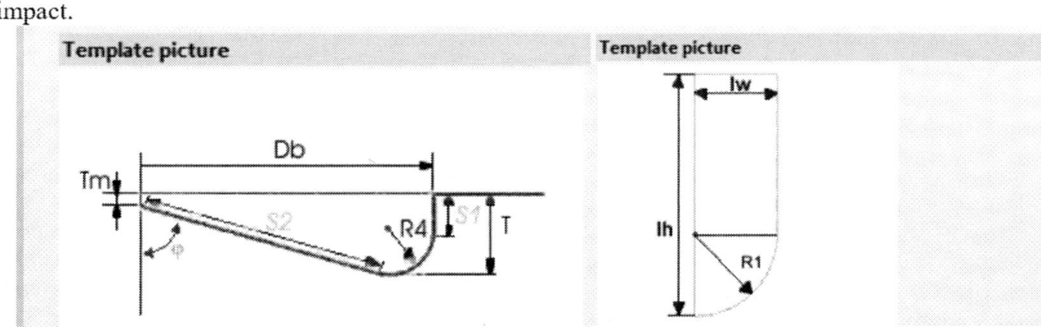

Figure 1. Geometric parameters of the combustion chamber and the nozzle sprayer

The combustion model 'Coherent Flame Model' was chosen in the 'Combustion models' chapter. This model is able to model burning of both uniform and non-uniform mixtures of air and fuel which is typical of diesel engines.

Nozzle parameters are also taken into account during the calculation due to the nozzle presence in the fuel-support system. The initial time, duration and the amount of fuel, injected in the combustion chamber, are selected in the 'Nozzles' chapter along with the nozzle geometry. The fuel injection duration is equal to 1.7 ms and the cyclic dose is equal to 24 mg.

A cycle of calculation with the initial time of injection of the varied parameter was carried out to determine an optimal variant of the engine operation. The variant with injection start on the crank angle equal to 703 ° was selected as optimal because maximal values of power and momentum were obtained in this regime without an increase of gas loads on engine details. The indicator diagram was plotted as a result of the calculation, and it is presented in figure 2. Maximal pressure reaches 9.4 MPa. Characteristics for the single cylinder are obtained: power is equal to 11.23 kW, and momentum — 35.75 N•m.

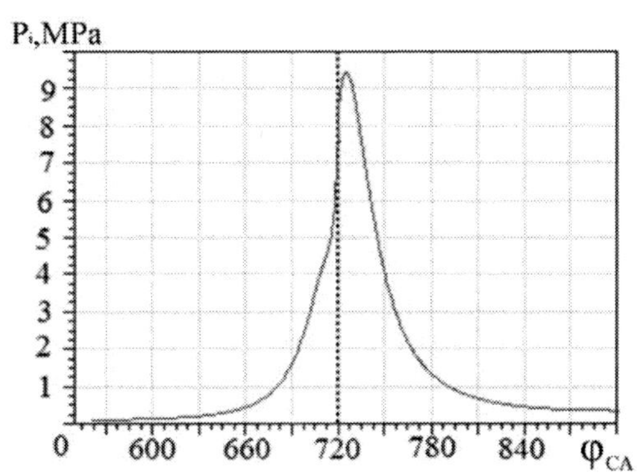

Figure 2. The indicator diagram

Pictures with fields of parameters distributions along the volume of the combustion chamber (figure 3) were generated in the 'Report generator' chapter.

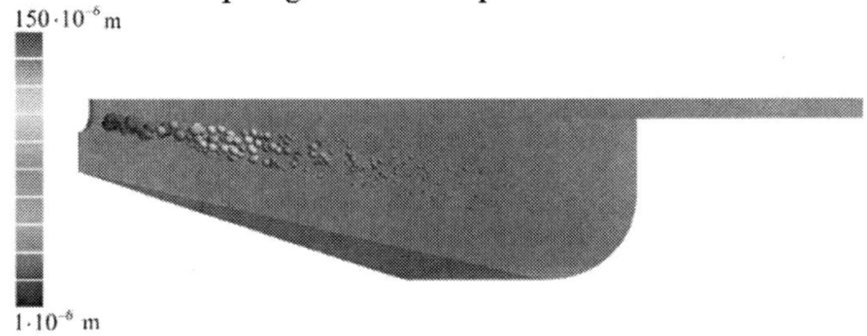

Figure 3. Distribution of the drops diameter in the torch in the combustion chamber (720°)

Figure 3 shows that given parameters of the nozzle provide acceptable distribution of the diameters of the drops in torch without their ricochet from the combustion chamber surface.

Data, which was obtained during the heat calculation, is differ from results of the numerical modelling. This is caused by the fact that polynominal equations, used during heat calculation were derived by the authors in the experimental and statistical way for a limited number of engines which decreases the accuracy of calculation during designing of modern engines.

Heat, kinematic and dynamic calculation of the piston diesel engine were carried out in the 'LMS Imagine.Lab AMESim' software at the second stage [22].

The engine model was composed in this program in the Sketch mode from separate submodels (figure 4), stored in the program libraries.

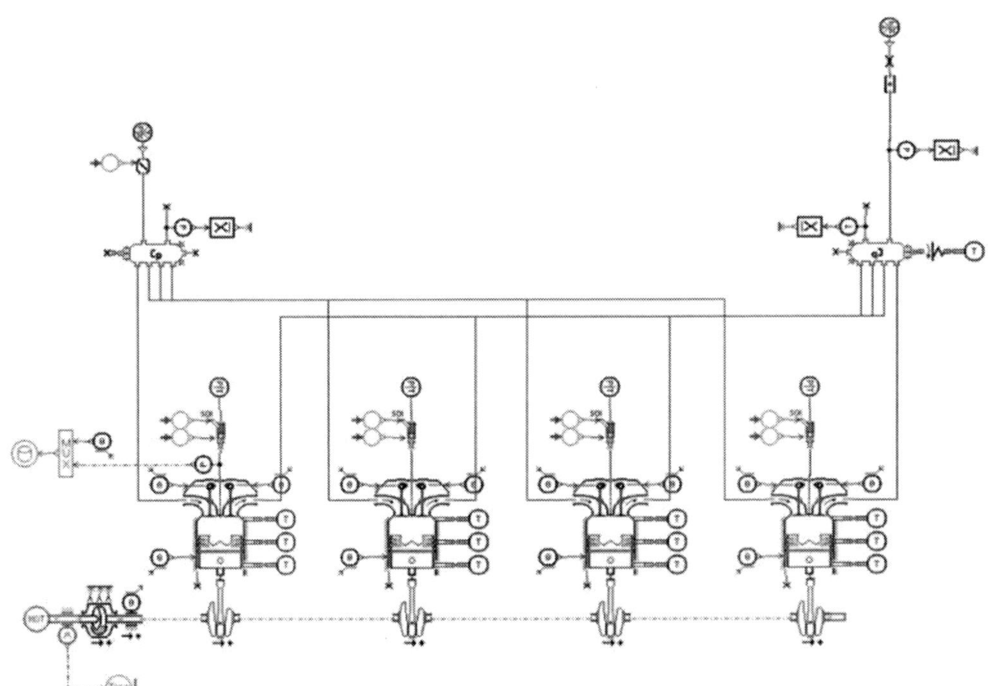

Figure 4. The ICE model

Submodel parameters were set in the parameter mode. There are two options of parameter setting in 'LMS Imagine.Lab AMESim': for every submodel separately and for several models simultaneously through the chapter 'Global parameters'.

Then, main data about the engine were inserted in the 'engine definition' chapter: engine architecture (V), the number of cylinders (4), the V-angle (5), the type of a crank-shaft mechanism (simple), the piston area of heat transfer (7032 mm2), the firing sequence (1-3-4-2), the heat exchange model (Annand). The chosen model (Annand) allows one to determine the heat transfer coefficient for constant mean gas velocity equal to mean piston velocity. The given model is simpler in comparison with other models which take into account the increase of the velocity in the cylinder.

Geometric parameters of the crank-shaft mechanism, the compression ratio (18), the advance angle of input valve opening (12°) and the delay angle of output valve closing (10°), the ICE operation regime, injection duration (1.7 ms), the injection advance angle (17 before the top dead centre), fuel supply pressure created by the low-pressure fuel pump and the type of the fuel are stated in the 'Global parameters' chapter. Geometric dimensions of valves, their number (1 input and 1 output ones), angles of the angular position countdown during the start of the rise (31,5°) and the fall (224°) of the valve are set in the "Gas distribution mechanism" chapter.

Conditions of the entrance and of the exit of the engine are also set: air pressure (0.124 MPa), air temperature (642 K), volumes of input and output ducts (2.545·10-3 m3).

Pressure distribution in the cylinder along the cycle was obtained as a result of the calculation and is represented in figure 5. Maximal effective pressure in the cylinder is equal to 9.51 MPa.

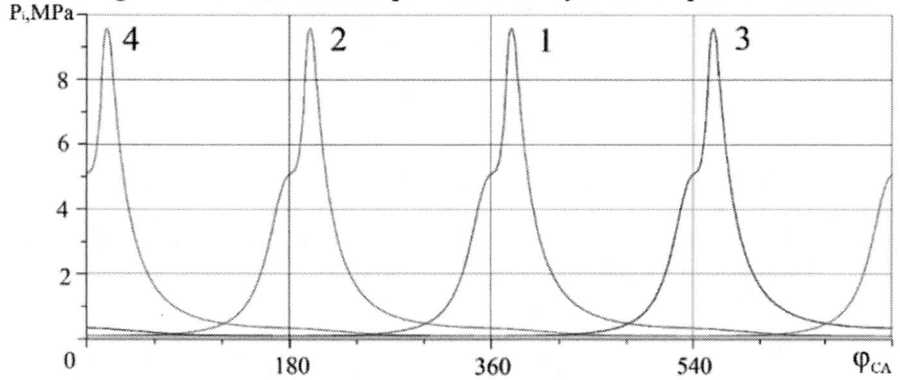

Figure 5. The pressure diagram in all cylinders

The feature of this calculation of the diesel engine in 'LMS Imagine.Lab AMESim' is the fact that engines submodels do not take into account heat exchange; therefore, there is no refrigeration system in this model. However, 'LMS Imagine.Lab AMESim' has an opportunity to model this type of system.

3. Results and Discussion

This chapter deals with results of analytical calculation of the diesel internal combustion engine and two types of numerical calculations: in 'AVL FIRE ESE DIESEL' and in 'LMS Imagine.Lab AMESim' software. The comparison of calculation results is represented in figure 6 in form of indicator diagrams.

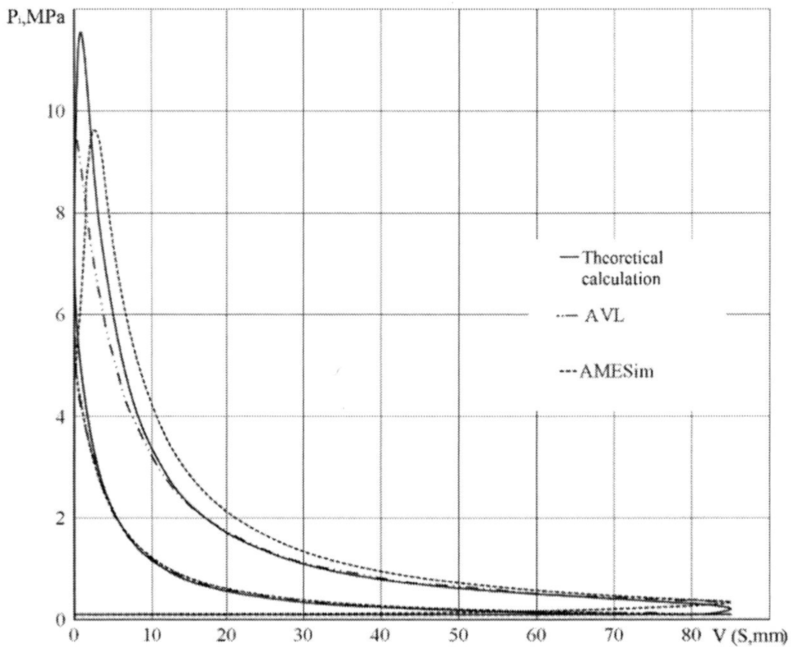

Figure 6. Indicator diagrams of every type of calculation

The values of indicator power and momentum are provided in table 1 for comparison of results of calculation by aforementioned methods. Data obtained from calculations in 'AVL' and 'AMESim' software differ from the analytical one. It is caused by the fact that polynomial equations, used during heat calculations, do not provide enough accuracy. Also, the 'AMESim' software does not consider heat transfer, so there is no refrigeration system.

Table 1. Comparison of calculation results

Calculation method	Power N_i (kW)	Momentum M_i (N·m)
Theoretical calculation	48,0	152,8
'AVL FIRE ESE DIESEL'	44.9 (-6.5 %)	143.0 (-6.5 %)
'LMS IMAGINE.LAB AMESIM'	41.5 (-13.5 %)	132.2 (-13.5 %)

4. Conclusion

A method for modeling of working processes in a VR-type engine in 'AVL FIRE ESE DIESEL' and 'LMS Imagine.Lab AMESim' is developed. The developed models allows one to calculate indicator parameters of working processes, which was compared to the results of the analytical research as a verification. 'AVL FIRE ESE DIESEL' allows modelling an injection from the nozzles to obtain a mixing and spraying quality in the cylinder. 'LMS Imagine.Lab AMESim' allows one to add different modules to the existing model such as a refrigeration system and external load to model the real behavior of the engine. Results deviations between numerical and analytical research varies from 5% to 10% which means that these models are suitable for engine designing.

Acknowledgements

This work was supported by the Ministry of Education and Science of the Russian Federation in the framework of the implementation of the Program 'Research and development on priority directions of the scientific-technological complex of Russia for 2014– 2020'

References

[1] Tavares F, Johri R and Filipi Z 2011 *J. of Engin. for Gas Turbines and Power* **133(12)** 463-476
[2] Almeida F L, Zoldak P, Wang Y, Sobiesiak A and Lacava P T 2014 *ASME 2014 Internal Combustion Engine Division Fall Technical Conference* (Columbus) vol **2** (New York: American Society of Mechanical Engineers) p 1044
[3] Biryuk V V, Gorshkalev A A, Kayukov S S and Uglanov D A 2014 *Open Mechanical Engin. J.* **8** 441-444
[4] Saigakov E A, Gorshkalev A A, Kayukov S S and Blagin E V 2014 *Research J. of Appl. Sciences* **9(10)** 669-673
[5] Falaleev S and Vinogradov A 2014 *ARPN J. of Engin. and Appl. Sciences* **9(10)** 1842-1848
[6] Salakhov R R, Yamalov D R, Gureev V M, Salakhov I R and Gattarov I F 2015 *Int. J. of Appl. Eng. Res.* **10(24)** 44974-44981
[7] Biryuk V V, Tsapkova A B and Shimanov A A 2016 *Key Engin. Materials* **685** 153-157
[8] Kozlov V E, Chechet I V, Matveev S G, Titova N S and Starik A M 2016 *Int. J. of Hydrogen Energy* **41(5)** 3689-3700
[9] Chen Y and Lv L 2014 *Appl. Thermal Engineering* **73(1)** 1332-1339
[10] Fridriksson H, Sundén B and Hajireza S 2010 *WIT Transactions on Engin. Sci.* **68** 177-188
[11] Sheikh S, Gokhale N, Thatte V, Gandhi N, Deshmukh B, Aghav Y, and Babu M K 2011 *SAE Technical Papers* **2011** 256-261
[12] Wu J, Wang H and Zhu L 2013 *Int. Conf. on Advanced Mechatronic Systems ICAMechS* (Luoyang) vol 1 (Luoyang: Institute of Electrical and Electronics Engineers) p 417-421
[13] Wu J, Wang H M, Zhu L L and Hua Y 2014 *Appl. Mechanics and Materials* **541-542** 763-768
[14] Zhang B, Mazlan S K, Jiang S and Boretti A 2015 *SAE Technical Papers* **2015** 391-398
[15] Zhang W, Shu G, Zhao W, Shen Y and Liang Y 2011 *Int. Conf. on Electric Information and Control Engin., ICEICE 2011* (Wuhan) vol 5 (Wuhan: Institute of Electrical and Electronics Engineers) p 134-5137
[16] Jing C, Wu W and Yuan S 2009 *Xitong Fangzhen Xuebao/J. of System Simulation* **21(23)** 7681-7685.
[17] Han W, Xu S, Deng X, Zhou Y and Zhang X 2011 *ICCRD2011 - 2011 3rd Int. Conf. on Computer Research and Development* (Shanghai) vol 2 (Shanghai: Institute of Electrical and Electronics Engineers) p 419-422
[18] Chen D G, Sun R Y, Wu Y D and Wang B 2013 *Adv. Mat. Res.* **724-725** 1317-23
[19] Payri R, López J J, Salvador F J and Martí-Aldaraví P 2013 *Int. J. of Engin. Education* **29(2)** 533-547
[20] Quan X H, Quan J J and He H 2014 *Appl. Mechanics and Materials* **496-500** 760-763
[21] https://www.avl.com/
[22] https://www.plm.automation.siemens.com/

Modeling of the egress of a drilling liquid from the nozzle of a drill bit with Ansys Fluent

D V Smorkalov[1], A V Tyutyaev[2], A M Shterenber[2] and A A Gorshkalev[3]

[1] LLC 'RosFin', 4, Moskovskoe shosse, Samara, 443013, Russian Federation
[2] Samara State Technical University, 244, Molodogvardeyskaya street, Samara, 443100, Russian Federation
[3] Samara National Research University, 34, Moskovskoe shosse, Samara, 443086, Russian Federation

E-mail: AGorsh@bk.ru

Abstract. A 3D model was built for the cylindrical outer nozzle of a drill bit which was in an immersed space at the distance of four diameters from the formation. A tetrahedral lattice was applied to the space filled with liquid, and a hexahedral lattice was applied to a section of the 3D model which imitated the rock (formation). A boundary layer was built near the walls. As a result of the calculations, the impact of the jet on the rock has been demonstrated, and it is similar to the impact of a drilling liquid jet on the bottom of a well during drilling. Distribution of the pressures, as well as the vector of velocities, the change in the volume ratio of a rock and the depth of penetration of the jet have been understood.

1. Introduction

Modern methods of designing rock destruction tools are quite advanced, however, in order to achieve the maximum performance of a drill bit when drilling oil and gas wells, one should account for a number of factors [1]. One of the factors is hydrodynamic parameters of a drilling liquid during the drilling process. Since it is rather difficult to perform the analysis of processes during well drilling due to extensive depths, by solving this problem, it would be possible to model the physical processes of drilling using the finite elements method. Currently, engineers worldwide utilize a great many tools for CFD computations, however, many years of international experience in running the software of this type point to the advantage of the ANSYS Fluent software package for a particular problem. An important factor for this study is that the ANSYS Fluent software package has the functionality to model two-phase and multicomponent operating environment including discrete granulated particles which are formed as a result of rock destruction. In order to perform calculations of this type, one requires significant computational capabilities, however, the use of average parameters of the operating environment enables us to determine the optimal ratio between the accuracy of results achieved and the time invested in the calculation, as well as to develop the methodology of calculation of the structure of drill bits at the stage of designing the relevant rock destruction tools, as well as the system of washing of drill bits, which enables a significant improvement of drilling KPIs, such as a drilling rate and a sinking advance per drill bit.

2. Setting of the problem

In this problem, we modeled the egress of a drilling liquid from the nozzle of a drill bit into the bottom of a well with an ablation effect. We used the parameters of water as a drilling liquid at the start of our study. The problem's solution is based on the numerical solution of Navier-Stokes equation which was implemented in ANSYS Fluent [2], while the moment of transfer from laminar conditions to turbulent conditions is determined by Reynolds criterion [3] which determines the ratio of viscous and inertia forces in a flow.

$$R_e = \frac{\rho \vartheta D_\Gamma}{\eta}, \tag{1}$$

where ρ – the density of the operating environment, D_Γ – the hydraulic diameter, ϑ - the velocity of the liquid flow, η – the dynamic viscosity of the operating environment.

In order to model the flow hydrodynamics, we used a two-phase model, where one phase is discrete granulated particles of the mine rock, and another phase is a drilling liquid. In order to describe changes in properties of the rock, we used empirical data from our works [5-8].

The change in viscosity of the granulated particles is described by Gidaspow equation [6]:

$$\mu_{s,kin} = \frac{10\rho_s d_s \sqrt{\Theta_s \pi}}{96\alpha_s(1+e_{ss})g_{0,ss}}\left[1+\frac{4}{5}g_{0,ss}\alpha_s(1+e_{ss})\right]^2 \alpha_s. \tag{2}$$

The volume viscosity of the granulate material determines resistance in case of expansion and compression. It can be demonstrated by the following ratio [7]:

$$\lambda_s = \frac{4}{3}\alpha_s \rho_s d_s g_{0,ss}(1+e_{ss})\left(\frac{\Theta_s}{\pi}\right)^{\frac{1}{2}}. \tag{3}$$

In the dense flows and at low shift velocity with a high level of concentration of solid particles, the stress is mostly formed due to friction between the particles. The viscosity of solid particles cannot be calculated by default; therefore the following equation is used to determine the viscous friction [8]:

$$\mu_{s,f,r} = \frac{\rho_s \sin\varphi}{2\sqrt{I_{2D}}}, \tag{4}$$

where p_s – the pressure of solid particles, φ – the internal friction angle, while I_{2D} – the second invariant of the deviator of a stress tensor.

In the flows with a high volume share of solid substances, an instantaneous collision is less important. The application of kinetic theory is not necessary for such flows, as the particles are in contact, and, therefore, the resulting friction must be taken into account.

In this section, the study began with computation of soft rock, 1-2 as per the IADC code which corresponds to a very soft and soft formation with the angle of internal friction of 14°-23°.

3. Results and Discussion

In order to perform CFD computation, we built a grid-scale model (Figure 1, a) which is composed of tetrahedral and hexahedral grids on two domains connected with a penetrable interface. The grid dimensions were 7.5 million elements with the maximum tapering of 0.84. We also built a prismatic borderline layer with a 1-mm thickness (Figure 1, b) [9-10].

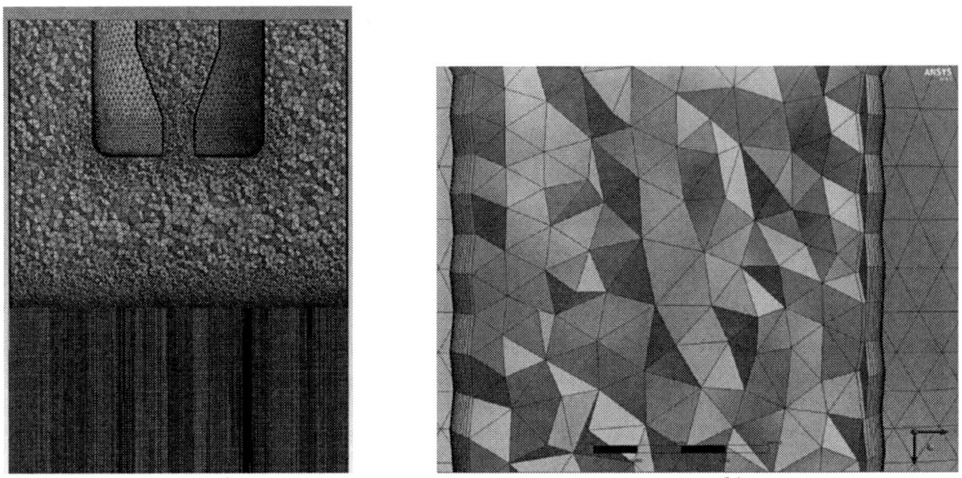

a) b)

Figure 1. The grid model
a) of the entire computation area, b) of the channel and borderline layer.

The problem was solved in an unconventional setting, using turbulence model k-ε. It was determined that in case of consumption of 1 kg/s of the drilling liquid via a submerged cylindrical nozzle with a 20 mm diameter, the flow reaches the established mode in 1 second (Figure 2).

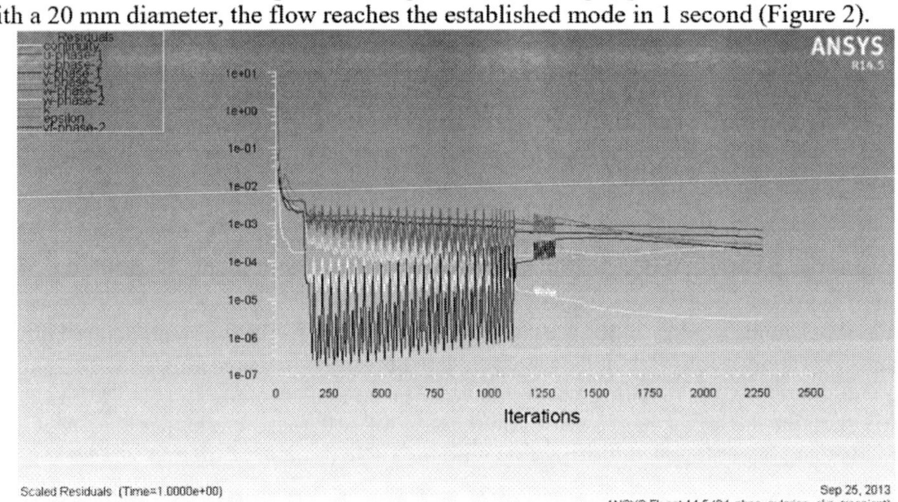

Figure 2. The diagram of convergence of the problem solutions.

As a result, we have the patterns of distribution of the parameters by the model cross sections, which enables us to receive the data necessary for the relevant verification (Figures 3-4). Figures 3 a, b, c, d show the change of the volume ratio of the rock over time.

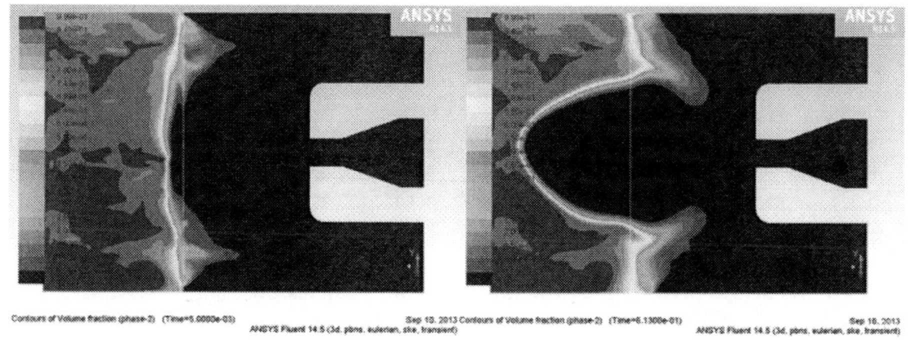

a) Computation time — 0.005 s b) Computation time — 0.26 s.

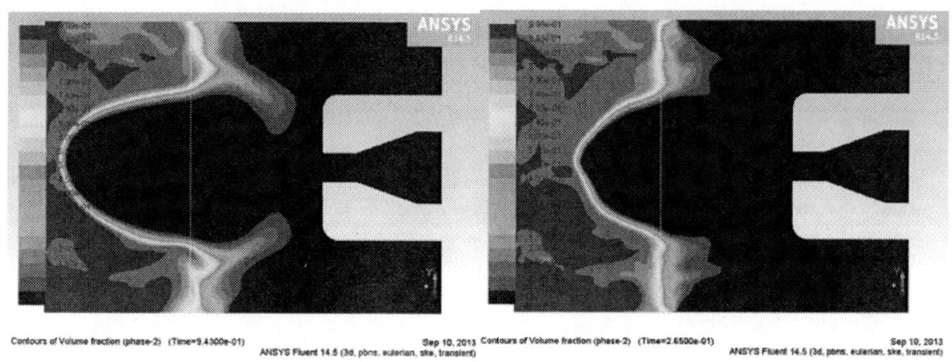

c) Computation time — 0.613 s d) Computation time — 0.943 s.

Figure 3. Distribution of the volume ratio of a hard rock formation during different time intervals.

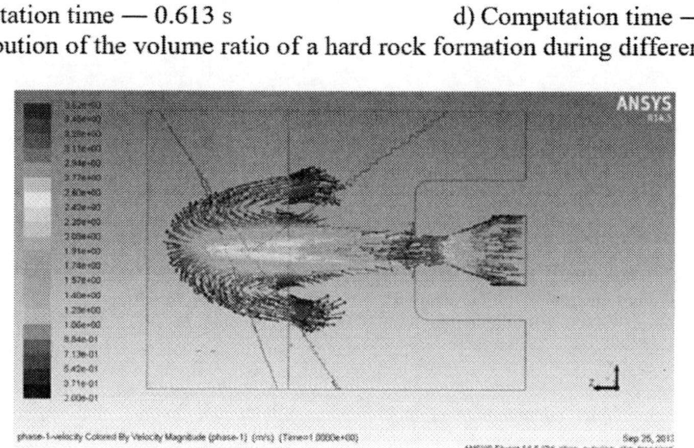

Figure 4. Vectors of velocities of the egress of a drilling liquid.

4. Conclusion

Based on the results of the calculations, it is demonstrated that in this problem setting for soft and very soft rock formation, the depth of penetration of the drilling liquid in the rock was around four diameters of the nozzle hole. This indicates to the degree of approximation of this numerical simulation to the reality [1]. However, in order to continue the studies, it is necessary to verify the model and the calculation for other types of formation. Also, it is required to account for the stressed-deformed state of the well bottom and unevenness of properties of the formation due to this. This methodology of calculation also enables us to evaluate the cutting transport which is formed as a result of the impact of the drilling liquid jet on the well's bottom, and its quantity and areas of maximum concentration in the annular space which may indirectly point to packing formation.

Acknowledgements
This work was supported by the Ministry of Education and Science of the Russian Federation in the framework of the implementation of the Program 'Research and development of priority directions of the scientific-technological complex of Russia for 2014– 2020'

References
[1] Shatsov N I 1961 *Drilling of oil and gas wells* (Moscow: State Scientific and Technical Publishers of Oil and Mining and Fuel Literature)
[2] Baturin O V, Popov G M, Goryachkin E S and Novikova Y D 2015 *Proc. 5th Inter. Conf. on Simulation and Modeling Methodologies, Technologies and Applications* (Vienna) vol 1 (Vienna: Institute of Electrical and Electronics Engineers) p 227-232
[3] Kramchenko V V and Savichev O G 2009 *Hydraulics* Part 2 (Tomsk: Tomsk Polytechnic University Publishers)
[4] Armsfield S and Street R 1999 *J. of Computational Physics* **153** 660-665
[5] Ding and Gidaspow D 1990 *AIChE J.* **36(4)** 523-538
[6] Gidaspow D and Huilin L 2003 *Chemical Engin. Sci.* **58(16)** 3777-3792
[7] Lun C K, Savage S B, Jeffrey D J and Chepurniy N 1984 *J. Fluid Mech.* **140** 223-256
[8] Schaeffer D G 1987 *J. Diff. Eq.* **66** 19-50
[9] Biryuk V V, Tsapkova A B and Shimanov A A 2016 *Key Eng. Mat.* **685** 153-157
[10] Saigakov E A, Gorshkalev A A, Kayukov S S and Blagin E V 2014 *Res. J. of Appl. Sci.* **9(10)** 669-673

Capacity calculation of the electrotechnical scheme of discharge gap replacement of the ozonizer in the COMSOL environment

Yu N Isaev[1], V A Kolchanova[1], A P Maltsev[1] and V P Aksenov[2]

[1] Tomsk Polytechnic University, 30, Lenina ave., Tomsk, 634050, Russia
[2] V E Zuev Institute of Atmospheric Optics SB RAS, 1, Academician Zuev square, Tomsk, 634055, Russia

E-mail: ni105@mail.ru

Abstract. The investigation of electrostatic field distribution of ozonator electrode gap caused by inhomogeneous permittivity has been done. The paper presents calculation of electrostatic field energy of ozonator electrode gap. From the energy of electrostatic field distribution the information about electrode system capacity has been extracted. For calculation of the electrostatic field and capacity of electrode system the software integrated environment COMSOL has been used.

1. Introduction

The application and physical phenomenon of barrier discharge is one of the most interesting and important direction for the fields like gas discharge of modern physics, gas electronics, physical chemistry and ozone synthesis. Currently, the barrier discharge is considered as discharge caused by electrodes voltage drop which takes place in a gas, where at least one of electrodes must be covered with dielectric.

The fields of technical application of barrier discharge have stimulated an interest in this type of discharge and physical phenomena of corresponding processes. These include application of barrier discharge for water purification, plasma technology, etching, semi-conducting materials processing, integrated circuits growing.

The successful ozone formation reaction in the barrier discharge is one of the few plasma-chemical processes utilize an industrial scale. The application of powerful ozonized installations is gradually growing and even 1MW capacity devices have been made.

However, strong spatial inhomogeneity and short duration of the physical processes which take place in the barrier discharge are causing difficulties for investigation of this phenomenon. Most recently, through the use of modern physical methods of research, modern methods of mathematical modeling and the growing capabilities of computer technology, we have been able to get a certain idea of the nature and sequence of the processes which take place in the discharge gap, but the full picture of the phenomenon still remains incomplete.

It is necessary to study in detail the development dynamics of the gas electrical breakdown in the barrier discharge in order to solve the optimization problem of plasma technologies and electrosynthesis of ozone technology.

The existing methods of mathematical modeling and interpretation of the electrical phenomena are mainly directed to the study of the processes which take place in the AC circuit. This is due to the fact that the alternating current can use well-developed mathematical technique of electrical engineering. However, the shape of the applied voltage in the circuit significantly affects the physical phenomenon of the processes occurring in the discharge gap, and as a consequence increases the energy contribution to the process. In some cases it is preferable and economical to use a pulse voltage. The processes which take place under the influence of a pulsed voltage include piecewise constant voltage areas. Within these areas, the voltage drop can be considered as a constant value, and the field in the electrode gap is considered as the electrostatic field.

Many authors suggest that the most successful way of barrier discharge description is an electrical equivalent circuit. The equivalent circuit diagram is a circuit diagram that has circuit elements which are chosen so that the current and voltage of the circuit have the same values as a discharge gap of the ozonizer. Thus, the information about the capacity of the equivalent circuit is extracted from the calculation of the electrostatic field of the discharge gap. Therefore, the aim of the article is the calculation of the electrostatic field of the discharge gap, followed by extraction of information about the capacity of the energy gap of the electrostatic field [1-6].

2. An Experimental Setup

The experimental setup that utilizes the method of water purification and disinfection is considered in various papers [5].

In the process of treatment the water droplets about 1-3 mm in diameter are falling on the electrode system which is illustrated in Figure 1. The discharge is generated between cylindrical electrodes, which are isolated from each other by dielectric barriers. Barriers are used to limit the discharge current through the gap and volume combustion of the discharge, when the uniform processing of the gap volume takes place.

In the case of a low permittivity of a barrier (a tube of quartz glass has permittivity $\varepsilon = 4$) the discharge exists in a form of a large number of low-voltage micro sparks, nonequilibrium plasma of which is an effective source of active species which are involved in the oxidation processes and degradation of contaminants in the treated water. The discharge is powered by a high-voltage pulse generator. The pulse duration is about $300 \div 500$ ns, and the amplitude of the voltage is – $20 \div 30$ KV.

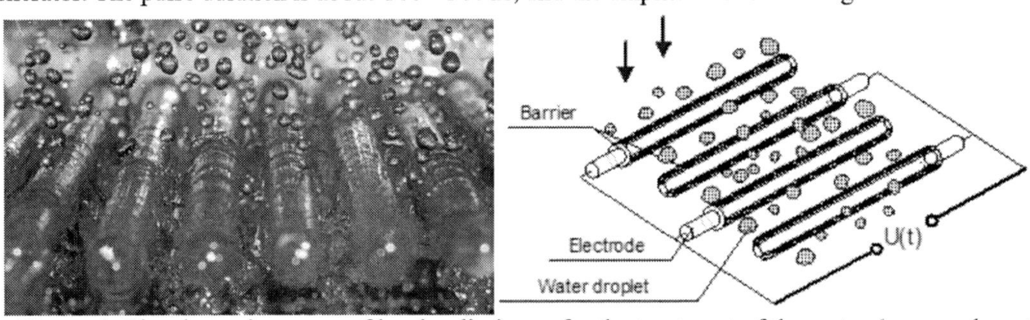

Figure 1. The electrode system of barrier discharge for the treatment of the water (arrows show the direction of the water flow).

3. Calculation of the Electrostatic Field

As a study of the system, let us consider the configuration of an electrode used for water purification as the group of round, equidistant metal cylinders with the diameter of 5 mm. These cylinders are arranged in parallel between the grounded planes (Figure 1, the sizes are given in millimeters). The geometry of the electrodes adopted in the calculations reproduces the real electrode system. The metal cylinders are inserted in 1 mm thick tightly fitting glass flasks (quartz glass barriers ε =4, Figure 2). Each cylinder is supplied with voltage $U = 10 \div 15$ kV with the alternating sign while the electrode potential difference is about $U = 20 \div 30$ kV. The electrodes are fed by voltage in the form of a pulse with the duration of about 0.6 microseconds. In the vertical direction, the various fractions of water droplets are passed through the electrode system. The size of each water droplet does not exceed the size of the air gap.

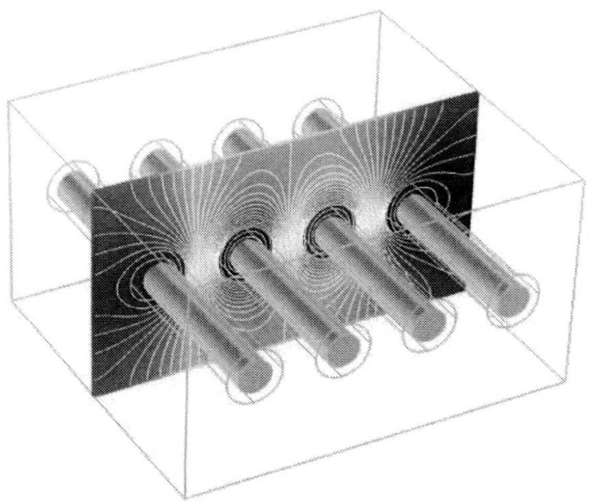

(a)

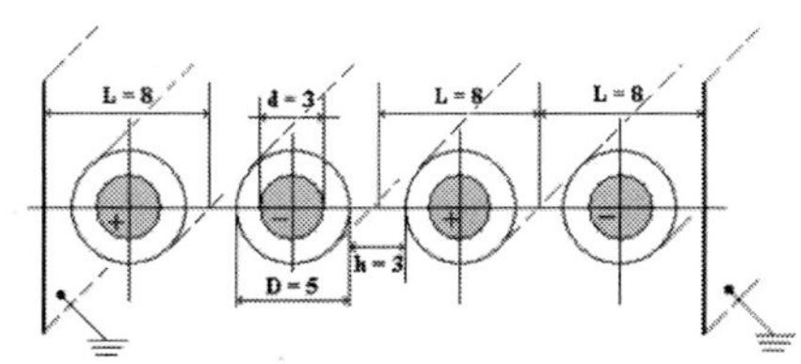

(b)

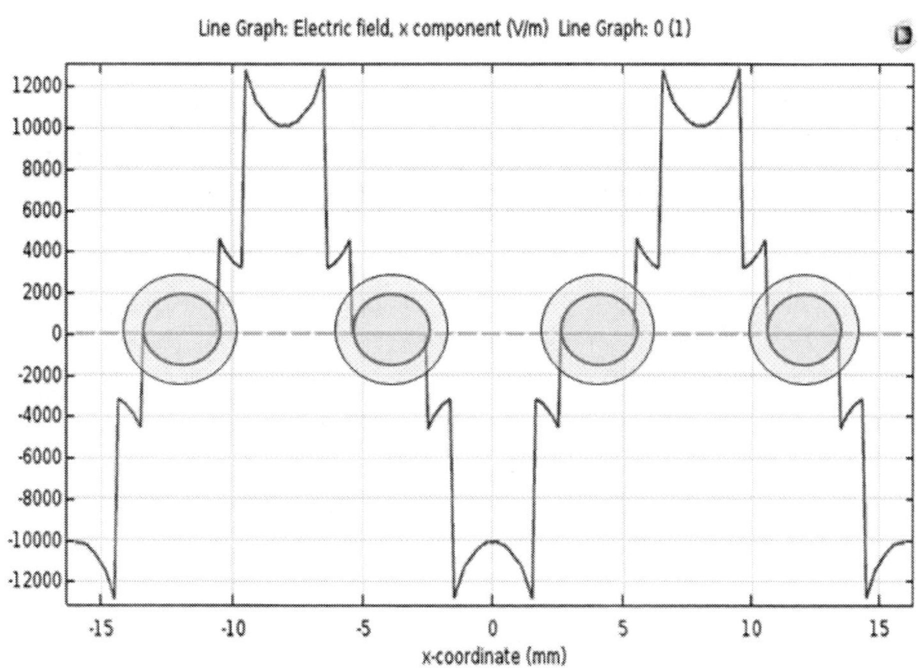

(c)

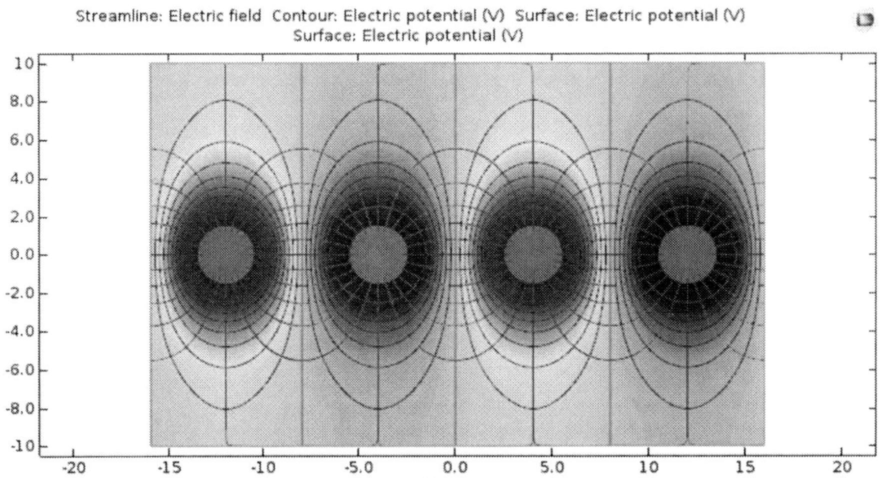

(d)

Figure 2. The picture of electrostatic field distribution of the electrode system which is presented in Figure 1

Since the electrode system has a complex form, the field of which is not uniform in the presence of barriers, the media between electrodes have boundaries and each separate part of media has different values of permittivity ε .The electrode system is not closed and the field distribution calculation for this electrode system is a challenging task [5].

The paper describes a method of electric field calculation of the electrode system with complex geometry by using software integrated environment COMSOL. The method simplifies the calculation of the complex system of electrodes. Calculation was done by using Electrostatics interface AC / DC module. After the geometry model is defined and assigned values of material properties, it is assumed that the conductor is an equipotential surface. The modeling domain is surrounded by a region of infinite elements, which is a way to truncate a domain which stretches to infinity. Although the thickness of the infinite element domain is finite, it can be thought of as a domain of infinite extent [8-10]. Then the physical parameters have been defined. On the first pair of electrodes positive voltage has been set U and on other pair of electrode negative voltage has been set $-U$.

The capacity calculation of the system was carried due to the calculation of the electric field on the assumption that the voltage supplied to each cylinder $U = 0.5$ V. In this case, the electrode potential difference was $U = 1$ V. The capacity of the electrode system is defined by a known ratio:

$$W = C\frac{U^2}{2} \rightarrow C = 2W .$$

From the simulation results, the information about the density of the energy distribution of the electrostatic field has been extracted. Then the calculation of the energy of the entire system as a volume integral has been done. In this case, it is necessary to take into account the geometrical dimensions of the system. The last stage was the calculation of full energy of the field as a value proportional to the capacity.

In the COMSOL environment, these actions were carried out in the following order: Surface Integraration / Energy and power/ Energy density.

The value of the capacitance obtained due to the simulation is a good estimation of its analytical counterpart obtained in [4] by the hybrid technique which involved the method of the integral equation and the method of Green function $C = 1.74 \cdot 10^{-10}$F/m. Let us note that the symmetric distribution of the electrostatic field allows us to determine the local capacities. To determine the local capacitance between two nearest electrodes, it is necessary to use the following ratio:

$$\frac{1}{C_1}+\frac{1}{C_1}+\frac{1}{C_1}+\frac{1}{C_1}=\frac{1}{C} \rightarrow C_1=4C.$$

Knowing the capacity of one section, we can determine the capacity of the gap filled with glass and the capacity of the gap filled with air.

4. Conclusion

The result of simulation presented above allows us identify the most intense areas of the electrostatic field. These areas are the boundaries between glasses and air (Figure 2). Thus, a strong field on the surface of conductive electrodes is pushed to the boundaries of the glass - air areas. Consequently, in the barrier discharge, the most intense impact of discharge on the water droplets will occur near the glass-air boundaries.

The calculation of the electrostatic field allowed us to determine the field energy, and the field energy is proportional to the value of capacitance of the electrode system. The symmetrical configuration of the electrostatic field helped us to determine the resulting capacitance of the electrode system as well as the local capacitance between two nearest electrodes.

The obtained capacitances can be used in the equivalent circuit of the barrier discharge.

References

[1] Samoylovich V I, Gibalov K V and Kozlov V K 1997 *Physical Chemistry of the Barrier Discharge* (Deutscher Verlag für Schweisstechnik)

[2] Bredow M, Rumyantsev V V and Toptygin I N 1985 *Classical electrodynamics* (Moscow: Science)

[3] Wagner H E , Brandenburg R, Kozlov K V, Sonnenfeld A, Michel P and Behnke J F 2003 *Biophy Research* (Marseille: France Vacuum 71) pp 417–436

[4] Kornev J, Yavorovsky N, Preis S, Khaskelberg M B, Isaev U and Chen B N 2006 *Ozone: Science and Engineering* **28(4)** 207-215

[5] Kolchanova V A, Isaev Yu N, Tarasenko S S and Filkov A I 2014 *Proc. Int. Conf. on Mechanical Engineering, Automation and Control Systems* (Institute of Electrical and Electronics Engineers Inc.) 6986868

[6] Isaev Y N, Kolchanova V A, Tarasenko S S and Tikhomirova O V 2015 *Proc. Int. Conf. on Mechanical Engineering, Automation and Control Systems* (Institute of Electrical and Electronics Engineers Inc.) 7414894

[7] Zhang X; Lee B J; Im H G and Cha M S, *IEEE Transactions on Plasma Science* 2601246

[8] Vasileva O V, Budko A A, Lavrinovich A V and Filkov A I 2014 *Proc. of 2014 Int. Conf. on Mechanical Eng., Automation and Control Systems* (Institute of Electrical and Electronics Engineers Inc.) 6986867

[9] Vasileva O V, Budko A A and Lavrinovich A V 2016 *IOP Conf. Ser.: Mater. Sci. Eng.* (Tomsk) vol 124 (England: IOP Publising) 012107

[10] Kuleshova E O, Plyusnin A A, Shandarova E B and Tikhomirova O V 2016 *IOP Conf. Ser.: Mater. Sci. Eng.* (Tomsk) vol 124 (England: IOP Publising) 012069

The single-zone numerical model of homogeneous charge compression ignition engine performance

E A Fedyanov, E M Itkis, V N Kuzmin and S N Shumskiy

Volgograd State Technical University, 28, Lenin ave., Volgograd, 400005, Russia

E-mail: tig@vstu.ru

Abstract. The single-zone model of methane-air mixture combustion in the Homogeneous Charge Compression Ignition engine was developed. First modeling efforts resulted in the selection of the detailed kinetic reaction mechanism, most appropriate for the conditions of the HCCI process. Then, the model was completed so as to simulate the performance of the four-stroke engine and was coupled by physically reasonable adjusting functions. Validation of calculations against experimental data showed acceptable agreement.

1. Introduction

Homogeneous Charge Compression Ignition (HCCI) is currently under widespread investigation. To limit the speed of combustion, lean mixtures are used. If the mixture is too rich, the pressure rises too rapid and generates knock-related problems. On the other side, lean equivalence ratios of the fuel/air mixture inducted into the cylinder provide low nitric oxides (NO_x) and particulate emissions, low fuel consumption. The HCCI process seems promising when running the engine using gas as fuel, especially using natural gas.

However, several problems with the HCCI combustion concept reappear throughout the majority of investigations – two combustion control aspects. First, the ignition event and the ignition timing are not controlled. Second, the rate of heat release is not controlled too. It occurs due to insufficient understanding of HCCI technology. In conjunction with extensive experimental researches, computational models provide a good basis for exploring the HCCI combustion phenomena. Yet a direct numerical simulation of fluid mechanics combined with the detailed chemical kinetic model is computationally complex. The typical method of the solution of the problem is simplification either in fluid mechanics or in the chemistry part of the model. An example of the latter approach is the single-zone model, developed in Volgograd State Technical University. The simulation of chemical kinetics is based on the law of mass action and the Arrhenius rate expression. This approach enables one to explore numerically the influence of the inlet air-fuel ratio, the initial mixture temperature and the in-cylinder pressure on the ignition timing that is directly dependent on the pre-flame chemical reactions.

In this paper, we present the investigation arranged as follows. The first section is the development of the first stage of the ignition event model – self ignition in the constant volume adiabatic reactor. The results obtained made it possible to choose the most acceptable kinetic reaction mechanism of methane burning. The second section is the development of the one-zone numerical model of the HCCI running process incorporating the kinetic model mentioned above. The third – the analyses of the preliminary results of the study, related to the validation of the model predictions against measurements, enabled us to propose a physically determined method to define heat release and, thus, the basic constants of the model more accurately.

2. Modelling description

2.1. Selection of the kinetic reaction mechanism

At the first step of the study, the computational model of ignition and combustion in the constant volume adiabatic reactor [1] is developed. Thus, we excluded the influence of such factors as variable volume and wall heat transfer. The model represents the set of equations including the law of mass action and the Arrhenius rate expression to define the rates of three types of reactions:

$$\frac{dc_i}{d\tau} = \sum_j W_{ij}, \tag{1}$$

$$W_{ij} = \pm k_j c_\alpha, \tag{2}$$

$$W_{ij} = \pm k_j c_\alpha c_\beta, \tag{3}$$

$$W_{ij} = \pm k_j c_\alpha c_\beta c_M, \tag{4}$$

where W denotes the reaction rate of chemical specie i in reaction j; α, β stands for i values of species involved in mono-, bi- and three molecular reactions; τ – the time; c_i – the molar concentration of the specie; c_M – the molar concentration of active centers; k_j – the reaction rate constant of the j-th reaction. Equation (2) is used for calculating the rates of monomolecular reactions; equation (3) – for bimolecular reactions; and equation (4) – for three molecular reactions.

2.2. The model of HCCI performance

The next step was the study of heat release evolution in the variable volume reactor using the developed model of the kinetic reaction mechanism. The model also includes the set of equations: the energy conservation balance and the ideal gas equation:

$$\begin{cases} \dfrac{dQ}{d\varphi} + \dfrac{\alpha_W (T_W - T)}{\dfrac{\pi n}{30}} A = p\dfrac{dV}{d\varphi} + c_v m \dfrac{dT}{d\varphi}, \\ pV = mRT, \end{cases} \tag{5}$$

where p, V, T denote the in-cylinder pressure, the volume and the temperature in the combustion chamber;

φ – the crank angle; m, R – are respectively the mass and the individual gas constant of the air-methane mixture; c_v – the specific heat capacity of the mixture at constant volume; Q – heat release during the combustion process (calculated by the model depicted above); α_W – the heat transfer coefficient; T_W – the cylinder wall mean temperature; A – the heat transfer area of the combustion chamber.

To date, the calculation of heat transfer coefficient α_W is based on Woschni's formula [2]. However, taking into account the results of investigations presented in [3], Woschni's formula was modified to enhance the accuracy of heat loss prediction for the HCCI engine operation:

$$\alpha_W = 819.5 \cdot \frac{p^{0.8} \cdot w^{0.8}}{T^{0.73} \cdot D^{0.2}}, \tag{6}$$

$$w = 2.28 \cdot c_m, \tag{7}$$

$$w = 2.28 \cdot c_m + 3.24 \cdot 10^3 \cdot \frac{V_s \cdot T_1}{6 \cdot p_1 \cdot V_1} \cdot (p - p_0), \tag{8}$$

where p – the in-cylinder pressure; w – the mean bulk speed; D – the engine bore; T – the mean in-cylinder gas temperature ; c_m – the mean piston speed; p_1, T_1, V_1 – the pressure, the temperature and the volume in the compression beginning, respectively; V_s – the displaced volume; p_0 – the in-cylinder pressure under motoring conditions. Two formulas are used for calculating: (7) for a compression process, (8) for combustion and expansion.

3. Results and discussion

3.1. A kinetic mechanism

In our opinion, the combustion process in HCCI engines can be presented as three sequential stages: the stage of pre-flame reactions, the stage of active burning and the post-burning stage. For overcoming the combustion control problems mentioned above, the main attention should be paid to studying of the first stage.

By means of the model, the evolution of ignition and burning processes of the air-methane homogeneous mixture in the constant volume adiabatic reactor was explored. The aim of the study was to determine the choice of a kinetic reaction mechanism that shows best agreement of calculated pre-flame stage duration with corresponding experimental data. The necessity of these investigations is based on the fact that the majority of chemical kinetics models observed do not take into account the peculiarities of the HCCI process.

The experimental data that serve as the baseline to compare the simulation results are generalized in reference [4]. The validation criterion of the model predictions against experiments is pre-flame stage timing τ_I. Goldsborough [5] has shown that this stage is over when the magnitude of current pressure exceeds the initial by one more than 5 %.

In figure 1, the effect of the initial mixture temperature on duration of the pre-flame reactions stage is shown, which was obtained from the experimental data [6] (lines 1–5) and from calculations with two versions of the kinetic reaction mechanism: one — by V. Ya. Basevich [6] (line 6) and the other is developed in San Diego University [7] (line 7). The latter chemical kinetics model is often used to simulate the combustion processes in the flame front of the air-methane mixture. Both experimental and computed data are taken for a stoichiometric mixture (α = 1.0) at initial pressure p_0 =15 bar. Initial mixture temperature T_0 varied from 800 to 2000 K, corresponding to the conditions in the combustion chamber when the piston reaches TDC. Line 1 reveals the experimental data obtained by D. W. Walker et al [8] for air-methane mixture self ignition while the initial temperature varied from 800 K to 2000 K. Line 2 corresponds to the experimental data of self ignition delay of mixtures of pure methane, oxygen and argon for the initial temperature of 1250 K to 2500 K obtained by K. S. Krishman et al [9]. Lines 3 and 4 show the experimental data for self ignition delay of the oxygen-methane mixture including ballast gases (N_2, Ar or He_2) obtained by E. L. Petersen et al [10]. In the experiments mentioned, shock pipes were used.

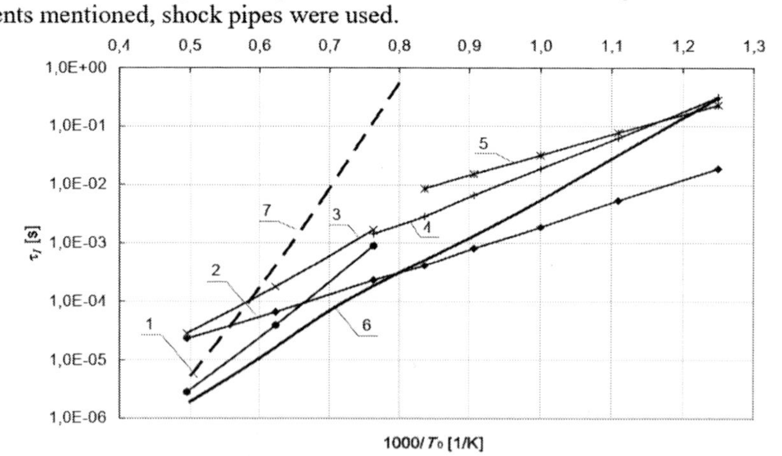

Figure 1. Comparison of calculations results with experimental data.

Obviously, values of pre-flame timing obtained by the Basevich kinetics mechanism show good agreement with measurements for the observed initial temperature range, while those by San Diego University diverge dramatically at temperatures less than 1700 K. Hence, the kinetics mechanism developed by Basevich was used in further calculations.

3.2. Simulation of HCCI performance

Numerical simulation of compression and combustion events of the HCCI-engine 'Volvo TD100' [11] was performed. The experimental data involved in the model were as follows. The engine was run at the speed of n = 1000 rpm using the air-methane mixture at fuel-air ratio α = 3.12. The initial temperature and pressure at IVC were obtained by extrapolation of those from [11] at 60 CAD BTDC. The results of calculations and experimental data are presented in figure 2.

As can be seen from figure 2, the model simulates the indicator process only qualitatively. Similar divergence of experimental data and calculations occur in some other researches [12, 13]. Experimentally observed burning rates of lean air-fuel mixtures are significantly lower than the calculated ones. Moreover, as the experimental data show, this divergence follows the increase of burnt gas fraction in the reacting mixture.

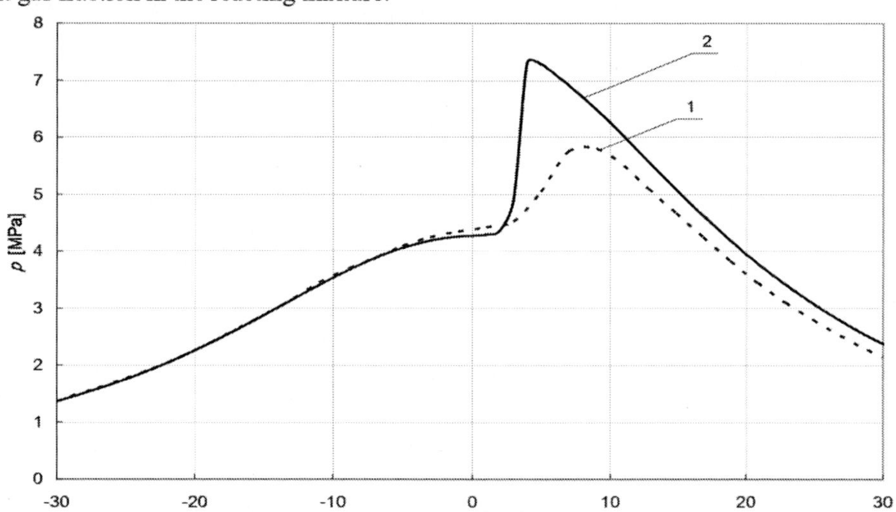

Figure 2. Experimental (line 1) and calculated (line 2) pressure profiles.

In our opinion, the reason of the mentioned divergence is the specifics of HCCI performance that is burning of extra lean fuel/air mixtures. We can point out two main factors that have influence on the burning rate. The first – comparison of calculated and experimentally observed pre-flame reactions timing leads to the conclusion that this divergence follows the increase of α. The latter shows that the surplus air in the air-fuel mixture reduces the rate of burning reactions at the first and second burning stages. The second factor that decreases burning rates is the increase of the burnt gas fraction in the reacting mixture during active burning and post burning stages. In the next section, the modification of the model taking into account phenomena mentioned is presented.

3.3. Model modification

It is known [14] that duration τ_I of the pre-flame reactions stage when the self ignition of the air-fuel mixture occurs depends exponentially on the reciprocal value of temperature T and activation energy E_a of burning reaction:

$$\tau_I \sim \exp\left(-\frac{E_a}{R_\mu \cdot T}\right), \tag{9}$$

where R_μ is the universal gas constant.

So the impact of the air-fuel ratio on the pre-flame stage timing can be taken into account by varying the activation energy magnitude by including adjusting function k_α in the argument of the exponential function. We suppose that the adjusting procedure of the activation energy should be implemented for all calculations of pre-flame and subsequent burning reactions, that is for all elementary reactions of the kinetic scheme. As the magnitude of the pre-flame reactions rate depends almost linearly on the air-fuel ratio and k_α is a multiplier to the argument of the exponential function, its correlation with the air-fuel ratio is assumed to be logarithmic:

$$k_\alpha = a_1 \cdot \ln(\alpha) + 1. \tag{10}$$

It provides, in particular, $k_\alpha = 1$ for $\alpha = 1$. Using the experimental data mentioned above, the magnitude of $a_1 = 0.1341$ is calculated.

The slowdown of fuel burning as a sequence of the increase of the burnt gas concentration is also considered by the adjusting function. Function k_C simulating the burnt gas fraction impact is involved in the Arrhenius rate expression as an additional pre-exponential multiplier. The relationship between k_C and the instantaneous burning products fraction in the reacting mixture is represented as a power function:

$$k_C = 1 - \left[\frac{C_b}{C_{b_{end}}}\right]^x, \tag{11}$$

where C_b, $C_{b_{end}}$ denote the instantaneous and final burning products fraction both calculated for the instantaneous state; x – the exponent. The analysis of the experimental data determined the value of exponent x as 0.16.

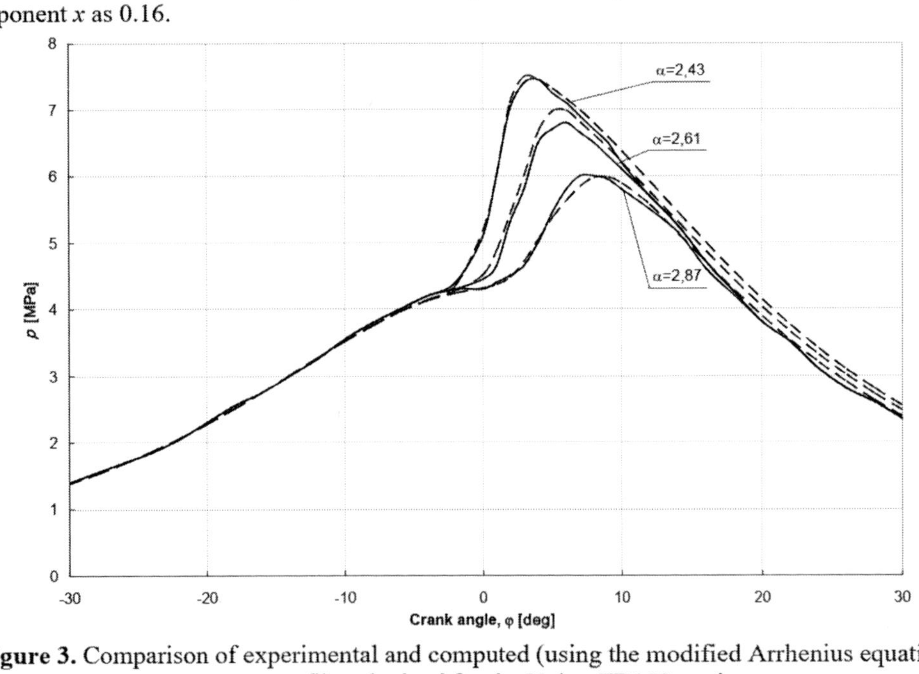

Figure 3. Comparison of experimental and computed (using the modified Arrhenius equation) pressure profiles obtained for the Volvo TD100 engine.
———————— – Experimental − − − − − − − − – Computed.

Thus, taking into account both adjusting functions, the rate constant for each chemical reaction incorporated in the kinetics mechanism is computed by the modified Arrhenius equation:

$$k = k_C \cdot A \cdot T^n \cdot \exp\left(-\frac{k_\alpha \cdot E_a}{R_\mu \cdot T} \right). \tag{12}$$

In figure 3, the experimental pressure profiles obtained on the Volvo TD100 engine [11], and the results of calculations with the modified Arrhenius equation are presented as an example. Both experimental and computed profiles are obtained for 1400 rpm and three air-fuel ratios α (2.43; 2.61; 2.87). As shown in the figure, the calculated profiles are in good agreement with experimental ones.

4. Conclusion

By means of the developed mathematical model numerical researches were performed which enabled the choice of the kinetic reaction mechanism of methane-air mixture burning, which is most appropriate for the conditions of the HCCI process. Being completed to simulate the performance of the four-stroke engine, the model must be coupled with physically reasonable adjusting functions to obtain acceptable quantitative agreement of calculations with experimental data.

References

[1] Jun D, Ishii K and Iida N 2003 *Trans. Jap. Soc. Mech. Eng. B.* **69(681)** 1282-89
[2] Woschni G 1967 *SAE Technical paper* **09** 31
[3] Chang J, Güralp O, Filipi Z, Assanis D N, Kuo T W, Najt P and Rask R 2004 *SAE Technical paper* **01** 2996
[4] Samuelsen S, McDonell V, Chen J and Jermakian V 2003 (Irvin: University of California)
[5] Goldsborough S S 2005 (Philadelphia: The Combustion Institute)
[6] Basevich V Ya, Vedeneev V I and Arutyunov V S 1994 *Comb., Explosion and Shock Waves* **30(2)** 140-146
[7] Hewson J C and Bpllig M 1996 (San Diego: University of California)
[8] Walker D W, Diehl L A, Strauss W A and Edse R 1969 *USAF Report AFAPL-TR* 69-82
[9] Krishnan S and Ravikumar R 1980 *Combustion Science and Technology* **24(5-6)** 239-245
[10] Petersen E L, Davidson D F and Hanson RK 1999 *J of Propulsion and Power* **15(1)** 82-91
[11] Amneus P, Nilsson D, Mauss F, Christensen M and Johansson B 1998 (COMODIA98) 567-572
[12] Soyhan H S and Andrae J 2008 *Turkish J. of Eng. and Environmental Sciences* **31(6)** 383-390
[13] Maiwald O, Schießl R, Böttiger J, König K, Schubert A and Maas U 2002 *Proceedings of the German-Russian Workshop on Reactive Flows* (Berlin, German)
[14] Heywood J B 1988 (New York: Mcgraw-hill)

Simulation of switching overvoltages in the mine electric power supply system

D I Ivanchenko and N G Novozhilov

Department of Electrical Engineering, Saint-Petersburg Mining University, 2, 21st Line, St Petersburg, 199106, Russia Federation

E-mail: dan87332@mail.ru

Abstract. Overvoltages occur in mine power supply systems during switching off consumers with high inductive load, such as transformers, reactors and electrical machines. Overvoltages lead to an increase of insulation degradation rate and may cause electric faults, power outage, fire and explosion of methane and coal dust. This paper is dedicated to simulation of vacuum circuit breaker switching overvoltages in a mine power supply system by means of Simulink MATLAB. The model of the vacuum circuit breaker implements simulation of transient recovery voltage, current chopping and an electric arc. Obtained results were compared to available experimental data.

1. Introduction

Switching off the overvoltage is one of the most common types of electric power supply system problems. These overvoltages are mainly caused by switching off the high inductive and capacitive or heavy resistive load, such as power transformers, reactors, capacitor banks and electrical machines. Those overvoltages are especially relevant to ungrounded neutral mine power supply systems.

The problem for the underground coal mine electric power supply system comes from availability of long 6 kV cable lines with significant capacitance supplying heavy inductive load which consist of clustered powerful electric drives. Overvoltage causes the degradation of insulation, and those are the major causes of outages in 6-10 kV cable lines. For instance, almost 50% of ground faults in 6-10 kV cable lines are caused by switching overvoltage. It is also possible that these electric faults would cause both power supply outages and accidents with ignition and explosion of methane and coal dust [1, 2].

A simplified structure of the mine section electric power supply is represented in Figure 1. Main consumers are electric drives of conveyor lines, coal plough-machines, ventilators and pumps. Mine is supplied by the 100 kV aerial line. Electric energy is distributed from the main transformer station of 110/6 kV to surfaced and underground facilities. Underground consumers are supplied by two cable lines through mine section transformers rated as 6/0.6 kV. The mine section transformer station supplies three asynchronous drives of the conveyor rated as 500 kW each, the cutting-loading machine rated as 536 kW and the pumping station rated as 1000 kW. There are two vacuum circuit breakers QF at the beginning and the end of the line. Vacuum circuit breakers are used because of their reliability and fire-explosions safety replacing oil circuit breakers, despite having a higher overvoltage ratio [3].

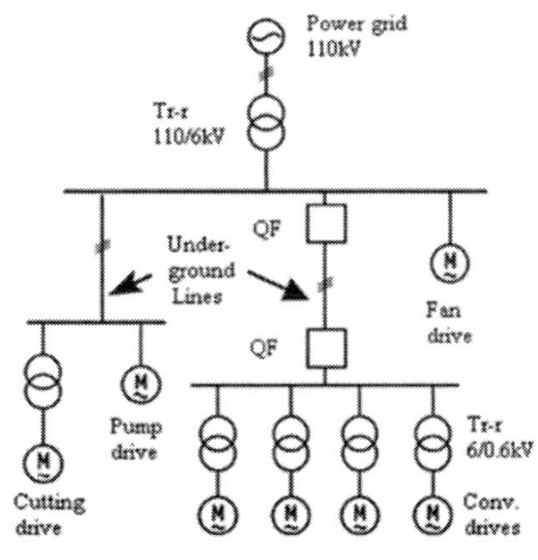

Figure 1. The diagram of the mine power supply system.

In order to address possible power supply problems caused by breaker switching, it is required to estimate possible overvoltages. As it is hard to analyze these processes experimentally, the mathematical and computer simulation of the circuit breaker operation could be suggested. The model of the circuit breaker must adequately represent the switching process by means of modeling the arc ignition, the existence of the chopping current electric arc and arc re-ignition in case recovery voltage is higher than momentary dielectric strength.

There are models of vacuum breakers [4, 5] in the EMPT/ATP environment, but there are no similar solutions for Simulink MATLAB. The authors of papers [6, 7] provided the simulation of transient recovery voltage of vacuum breakers, but did not simulate the electric arc. Also, available simulations of the electric arc in Simulink MATLAB [8, 9, 10] are based on the controlled current source model and are not applicable for breaker simulation. This paper represents an attempt to simulate all mentioned processes in the vacuum circuit breaker for switching overvoltage simulation.

2. Vacuum breaker modeling

Modeling of the vacuum breaker switching is complicated by the difficulty of physical processes and their mathematical description. In order to represent them in SimPowerSystem Simulink MATLAB, two models were made: the model of an arc interrupter of the vacuum circuit breaker simulates a dynamic arc current-voltage characteristic; the breaker control model simulates transient voltage recovery and chopping current.

2.1. Transient voltage recovery simulation

After physical separation of breaker's contacts, the electric arc is ignited. Arc exists until current drops to the chopping current value of I_c. It varies from 3A to 15A depending on a contact surface material and is approximately 2÷5 A for Cu-Cr contacts [6]. After arc extinguishing, recovery voltage U_r is formed between the contacts of the breaker:

$$U_r = U_g - U_l, \quad (1)$$

where U_g – electric grid voltage, U_l – load voltage.

Load voltage U_l depends on magnetic energy which is stored in inductive elements in the switched off part of the circuit and can be calculated according to formula

$$W = \frac{L_0 I_c^2}{2}, \quad (2)$$

where L_0 – inductance of the switched off part of the circuit, I_c – chopping current.

Stored energy is transferred from inductance L_0 to capacitance C_0 and back causing oscillations with frequency f_0:

$$f_o = \frac{1}{2\pi\sqrt{L_0 C_0}}. \quad (3)$$

Maximum overvoltage corresponds to peak value U_p of oscillating load voltage U_l. It can be calculated as:

$$U_p = \sqrt{\frac{L_0 I_c^2}{C_0}}. \quad (4)$$

If recovery voltage exceeds discharge voltage between contacts of the breaker, the electric arc will re-ignite and the process will be repeated multiple times. The recovery voltage increases in accordance with the contacts movement speed and depends on availability of electrical charges in the arc extinguishing chamber [11, 12].

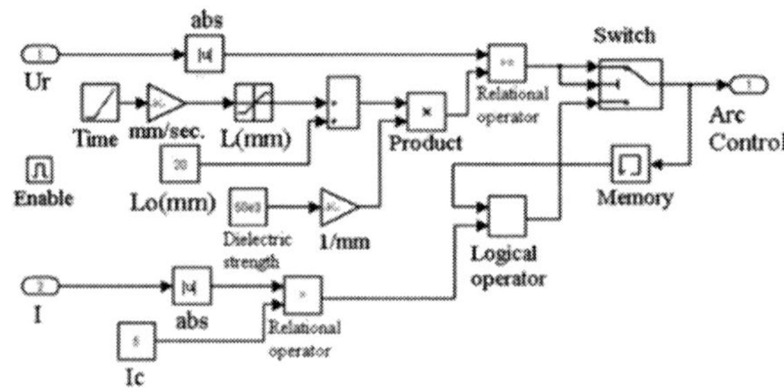

Figure 2. The circuit breaker state control subsystem.

This process was modeled in Simulink Matlab as a breaker state control subsystem represented in Figure 2. It is activated for the closing operation when a *Breaker control* singular signal is recieved and formulates an *Arc control* signal in accordance with measured recovery voltage, breaker current and the previous state of the breaker. The breaker state control subsystem generates a singular *Arc control* signal in two cases: recovery voltage is higher than the discharge voltage; recovery voltage is lower than the discharge voltage, but arc current is higher than chopping current. In any other cases this signal is equal to 0.

In order to simulate this process, several assumptions were made. It was considered that the contacts movement and dielectric strength of the gap are constant so discharge voltage increases linearly. Initial dielectric withstand was modeled by variable initial gap *Lo* between contacts. Electric strength of the gap between contacts was considered to be 50kV/mm, chopping current was considered to be equal to 5A.

2.2. *Electric arc simulation*

There are different approaches to electric arc modeling. Thermodynamic based models of the arc were proposed by various authors such as Mayer, Cassie, Schwarz, Schavemaker, Habedank and many others [8, 9, 10].

All models implement the arc with non-linear electric conductance, where momentary arc conductance is a function of power, supplied through the arc plasma channel, but are based on different concepts of the arc column. As the Mayr model represents a low current arc and conductivity reduction better, it was considered to be more suitable for the vacuum circuit breaker simulation case.

Simulation is based on the generalized model of the process in the electrical arc, described at statics by equation (9):

$$u_{arc} = R_{st}(i_\theta)i + U_c = \frac{U_{arc\,st}(i_\theta)}{i_\theta}i + U_c, \quad (5)$$

where u_{arc} – arc column voltage, i - arc column current, i_θ – arc state current, R_{st} – static arc column resistance, $U_{arc\,st}\,(i_\theta)$ – the static arc current-voltage characteristic, U_c – the voltage drop at contactors.

The low current arc voltage loss at contacts can be neglected and the current-voltage characteristic can be generally approximated to the static process as follows:

$$U_{arc}(i_\theta) = U_0\left(\frac{i_\theta}{I_0}\right)^n, \qquad (6)$$

where U_0, I_0 are reference points of the static arc current-voltage characteristic, n – the parameter based on the type of the arc characteristic.

For the case of the static arc with characteristic equation (6) for the Mayer arc model, the arc conductance is assumed to vary with time at low currents, in accordance with

$$\frac{1}{g}\frac{\mathrm{d}g}{\mathrm{d}t} = \left(\frac{u_{arc}i}{P} - 1\right)\frac{1}{\tau}, \qquad (7)$$

where τ – the Mayer arc time constant, P – arc power losses, g – momentary conductivity of the arc.

In order to simulate the breaker with arc ignition in Simulink SimPowerSystem, the model represented in figure 3a was developed in accordance with equation (7).

The model is based on the controlled voltage source and represents a voltage drop under closed, open and arc conditions of the vacuum breaker. The breaker switch is triggered by the *Breaker control* signal and turns the breaker from the closed state to the open state if the control signal is 0. The arc switch is triggered by the signal from the *Arc control* subsystem and turns the breaker from the closed state to the arc state if the control signal is 1. Arc inertia is modelled by means of the Transfer Fcn block. In order to avoid singularity at current zero crossing, the viscous friction function was used.

Developed models were integrated into the singular three-phase vacuum circuit breaker model, represented in figure 3b.

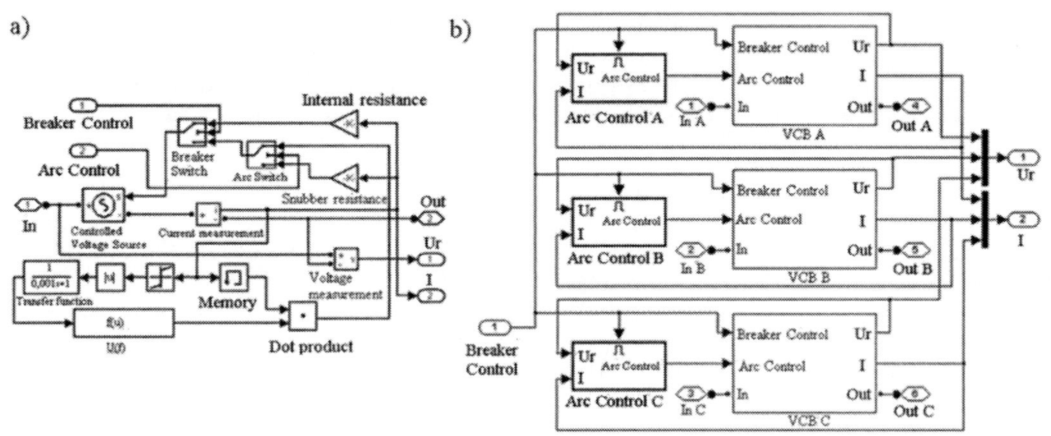

Figure 3. The circuit breaker model with electric arc implementation – a); the model of a three-phase vacuum circuit braker – b)

The model consists of three one-phase breakers and breaker control subsystems, one per phase. It is activated by the *Breaker control* signal and provides measurement of recovery voltage U_r and breaker current I.

IOP Conf. Series: Materials Science and Engineering **177** (2017) 012070 doi:10.1088/1757-899X/177/1/012070

3. Breaker Simulation Results

The developed model of the vacuum circuit breaker was tested in the electric circuit of the underground mine section (figure 4) simulated by means of SimPowerSystems Simulink MATLAB.

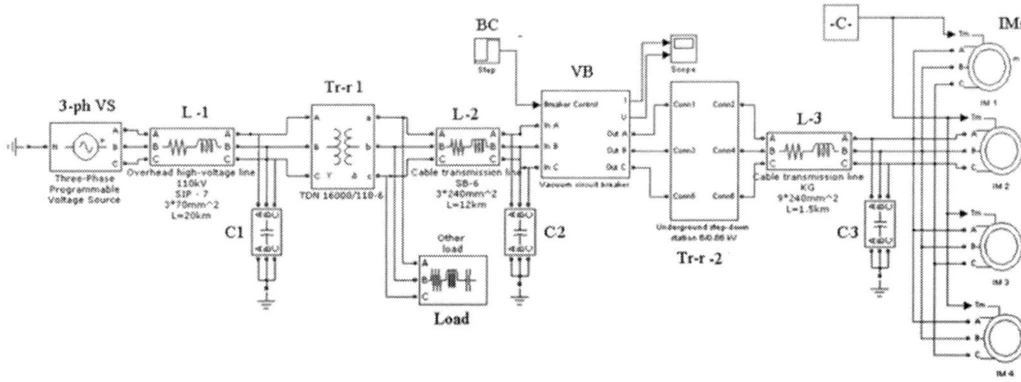

Figure 4. The model of the mine power supply system.

The purpose of the experiment was to investigate processes in the circuit breaker switching off. Disconnection of the mine section with the belt conveyor shown in figure 1 was taken as an example. The deenergized part of the circuit consists of underground transformer station Tr-r2 with three transformers 6/0.66 kV rated as 630 kW each, cable line 0.66 kV L-3, and motors (IM) of the conveyor electric drive. All other consumers were represented by equivalent RLC load (Load). The line section is switched off by the vacuum circuit breaker (VB) at the command of the breaker control block (BC). Phase currents and circuit breaker voltage measurements are presented on the oscilloscope (Osc.). Model parameters are represented in table 1.

The model also includes a three-phase power source, the aerial line of 110 kV (L-1), the main power transformer of 110/6 kV (Tr-r1) and the cable line of 6 kV(L-2). Capacitance of the cable and aerial lines was modeled by conductance C1, C2, C3.

Table 1. Parameters of Simulink model elements.

3-ph VS	L-1	Tr-r 1	L-2	Tr-r 2	L-3	IM	Load
$U_{l-l} = 110$ kV;	$R_0 = 0.414$ Ohm/km;	110/6kV $(Y - D_{11})$	$R_0 = 0.0845$ Ohm/km;	6/0.66 kV $(Y - D_{11})$	$R_0 = 0.0405$ Ohm/km;	$P_n = 250$ kW;	$P_n = 6000$ kW;
$f = 50$ Hz.	$L_0 = 9 \cdot 10^{-4}$ H/km;	$R_1 = 1.9745$ Ohm;	$L_0 = 22 \cdot 10^{-5}$ H/km;	$R_1 = 0.199$ Ohm;	$L_0 = 113 \cdot 10^{-}$ H/km;	$V_n = 0.6$ kV.	$Q_n = 2764$ kVAr.
	$C_0 = 10^{-8}$ F/km;	$L_1 = 0.1256$ H	$C_0 = 8{,}7 \cdot 10^{-9}$ F/km;	$L_1 = 0.0031$ H	$C_0 = 8{,}7 \cdot 10^{-9}$ F/km;		
	$l = 20$ km.	$R_2 = 0.01758$ Ohm;	$l = 12$ km.	$R_2 = 0.0072$ Ohm;	$l = 1.5$ km.		
		$L_2 = 0.00111$ H		$L_2 = 0.00013$ H.			

At 0.4 s, the breaker control block sends a switching off signal. Figure 5 represents oscillograms of the breaker voltage and current in relation to nominal line voltage and current. Figure 5c and figure 5b provide oscillograms of transient voltage in the vacuum breaker with simultaneous phase switching and without it.

As figure 5 shows, in case of nonsimultaneous switching, overvoltage on the circuit breaker exceeds

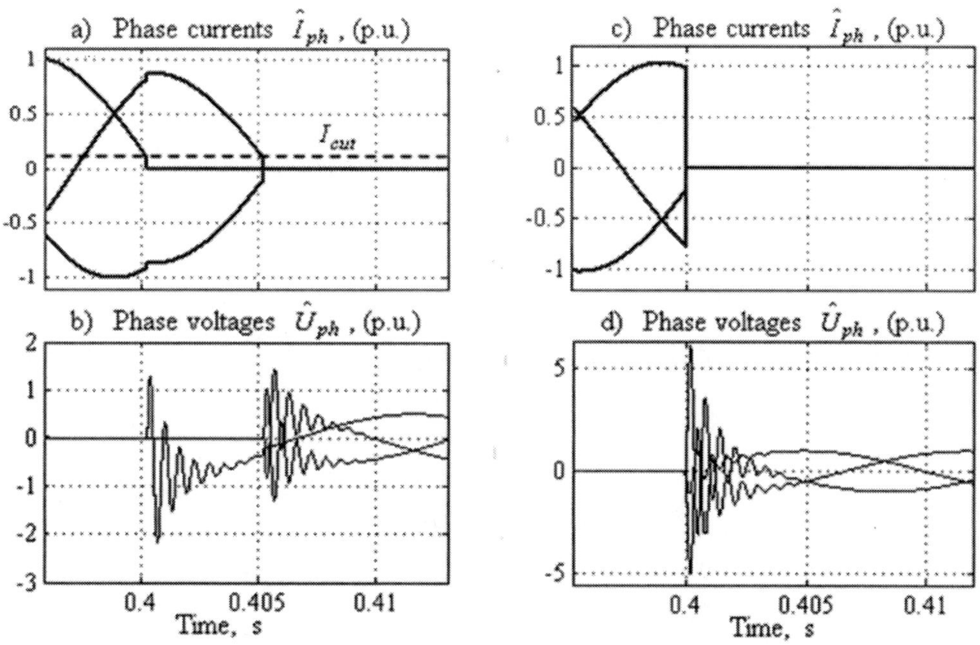

Figure 5. Simulation of vacuum breaker nonsimultaneous switching off phase currents (a) and voltage (b) and simultaneous switching off phase currents (c) and voltage (d).

the nominal line voltage 2 times. For the case of non phase-controlled switching, the overvoltage is 6 times higher than the nominal voltage. A high level of overvoltage is caused by switching of high power motors with low nominal voltage and requires the usage of surge arresters and other means of limitation. Multiple arc reignitions did not occur. The results of the simulation correspond to experimental data provided in [10, 13-15].

4. Conclusion

This paper presented modeling of switching overvoltages in the mine power supply system caused by the vacuum circuit breaker. The model allows the simulation of switching overvoltages, caused by the existence of chopping current in vacuum breakers; the ignition of the electric arc with a nonlinear current-voltage characteristic; the electric arc re-ignition in case the transient recovery voltage is higher than the discharge voltage between contacts of the breaker.

The vacuum circuit breaker model was tested for the case of switching off the mine section circuit, including high powered conveyor electric drives. Simulation results indicate a possibility of high ratio overvoltages in mine power supply and correspond to available experimental data.

The developed model could be applied for vacuum circuit breaker switching simulations in order to estimate a possible effect on the power quality in mine power supply systems.

References

[1] Volotkovsky S, Razumny Y and Pyvnyak G 1984 *Power supply of coal mines.* (Moscow: Nedra)
[2] Plashanskiy L A 2006 *Principles of energy supply in mining industry* (Moscow: MSMU publ)
[3] Gavrilova E V 2011 *Improvement of methods of assessment, prediction and means of limiting switching overvoltage in power supply system 6 - 10kv* Master thesis (Krasnoyarsk: SFU) p 21
[4] Borghetti A, Napolitano F, Nucci C, Paolone M, Sultan M and Tripaldi N 2011 *Proc. of Int. Conf. on power Systems Transient* (Delft: the Netherlands) 14-17

[5] Navalkar P, Gajjar G 2011 *Proc. of Eighth Int. Conf. on Switchgear & Controlgear* (Mumbai: India) pp 258-262
[6] Kuczek T, Florkowski M 2013 *Zeszyty Naukowe Wydziału Elektrotechniki i Automatyki PG* **36** 529
[7] Chystyakov G, Glushkin E and Zybkov N 2012 *Electro* **6** 46-50
[8] Yuan L, Sun L and Wu H 2013 *Energy and Power Engineering* **5** 833
[9] Marković N, Bjelić S, Živanić J and U. Jakšić 2013 *Przegląd Elektrotechniczny*, R. **89** 529
[10] Nitu S, Nitu C, Mihalache C, Anghelita P and Pavelescu D 2008 *J. of Optoelectr. and Adv. Mat.* **10** 1192-96
[11] Vereshago E, Kostuchenko V 2014 *Electrical engineering* **6** 36-42
[12] Varetsky Y 2010 *Proc. of Modern Electric Power Systems 2010 symposium* pp 1-4
[13] Singh V 2012 *Examination of Electric Arc Behavior in Open Air* Master thesis (Aalto university: Finland)
[14] Sima W, Zou M, Yang Q, Yangand M and Li L 2016 *Energies* **9(88)** 1-14
[15] Bazavluk A, Sarin L, Mikhailovsky G, Naumkin I and Gogolyk V 2011 *Energy expert* **2** 27-32

Fracture of cylindrical ampoules during solid phase synthesis under explosive loading

S A Zelepugin[1,2], O V Ivanova[2], A S Yunoshev[3] and A S Zelepugin[1,2]

[1] National Research Tomsk State University, 36, Lenina ave., 634050, Tomsk, Russia
[2] Tomsk Scientific Center of Siberian Branch of the Russian Academy of Sciences, 10/3 Akademichesky ave., 634055, Tomsk, Russia
[3] Lavrentyev Institute of Hydrodynamics of Siberian Branch of the Russian Academy of Sciences, 15, Lavrentyev ave., 630090, Novosibirsk, Russia

E-mail: bliz3@yandex.ru

Abstract. The paper presents the numerical and experimental study of the behavior of porous reactive mixtures (Al-S and Al-Tf) under explosive loading using a multicomponent medium model. The initial dispersion of the reactive mixture is varied to evaluate its contribution to the fracture of the cylindrical ampoule. The results have shown that the high rate of heat release during the chemical reaction in the bottom part of the ampoule and a sharp increase in pressure lead to the formation of the gas phase, which causes the ampoule fracture. It is found that the higher the initial dispersion of the mixture components is, the more intensive the ampoule failure is.

1. Introduction

By now, the solid-phase synthesis of materials placed in cylindrical ampoules under explosive loading has been insufficiently studied and, therefore, it is of great interest for further research. This direction has not yet reached the level of technology due to the lack of experimental data and numerical techniques correctly describing the process. During shock-wave compression, there is a significant energy release due to exothermic reactions in reacting mixtures, which, on the one hand, can lead to a self-sustaining propagation of chemical reactions in mixtures and, on the other hand, to the failure of ampoules, as well as the complete fracture [1-3]. It is nessesary to mention the investigations directed to the study of solid-phase chemical reactions in the detonation regime [4-6]. If the reaction rate is high enough, then it can lead to the conditions which will allow the solid phase (or gasless) detonation to occur.

For the analysis of these processes, it is necessary to develop mathematical models with the introduction of additional parameters and equations describing the kinetics of chemical reactions and considering the initial parameters and characteristics of reactive components, as well as the appropriate computational algorithms. The developed numerical models in combination with the available experimental data will provide the possibility to receive reliable information on the behavior of reactive media, including the mechanisms and kinetics of physical and chemical transformations and ways to form the new states of substances.

The goal of this work is to study the problems and peculiarities which occur during the solid-phase synthesis in the cylindrical ampoules subjected to explosive loading.

2. Formulation of the problem

The governing equations of continuity, momentum, and energy for the nonstationary adiabatic motion of each component in some fixed volume V of a compressible mixture bound by surface S can be written in the following form [7]:

$$\frac{\partial}{\partial t}(\alpha_i\rho_i)+\nabla\alpha_i\rho_i\upsilon_i=0,\ (i=1, 2, \ldots, N),$$

$$\alpha_i\rho_i\frac{\mathrm{d}_i\upsilon_i}{\mathrm{d}t}=\nabla\sigma_i+\alpha_i\sum_{j=1}^{N}\alpha_j\mathbf{R}_{ji},\ (i=1, 2, \ldots, N),$$

$$\alpha_i\rho_i\frac{\mathrm{d}_i E_i}{\mathrm{d}t}=\sigma_i\varepsilon_i+\alpha_i\sum_{j=1}^{N}\alpha_j F_{ji},\ (i=1, 2, \ldots, N),$$

$$\text{where } \frac{\mathrm{d}_i}{\mathrm{d}t}\equiv\frac{\partial}{\partial t}+\upsilon_i^k\frac{\partial}{\partial x^k}.$$

Here, t is the time, ρ_i is the density of the i-th component equal to the mass of the i-th component per unit volume of the i-th component, υ_i is the velocity vector, E_i is the internal specific energy, ε_i is the strain rate tensor, $\sigma_i=-P_i\delta_i+S_i$ is the stress tensor, P_i is pressure, S_i is the stress deviator, R_{ji} is the intensity of the momentum exchange between the j-th and i-th components, F_{ji} is the intensity of the energy exchange between the j-th and i-th components, N is the number of components.

Volume fractions of the mixture occupied by each component are given by:

$$\alpha_1+\alpha_2+\ldots+\alpha_N=1,\ (\alpha_i\geq 0),\ \alpha_i=\rho_i^*/\rho_i,$$

where ρ_i^* is the reduced density (mass of the i-th component per unit volume).

The mixture components interact with each other by exchanging momentum $\boldsymbol{R}_{ji}$, energy F_{ji}, and mass J_{ji} in the presence of chemical reactions within the framework of the multicomponent medium model.

In modeling chemical reactions under shock wave loading, we use the zeroth-order kinetic relation characterized by a constant rate of chemical transformations [1, 2, 8]:

$$J_{ji}=\frac{\mathrm{d}\eta}{\mathrm{d}t}=\begin{cases}0, & \text{if } \eta=1 \text{ or } (T_i<T_\eta \text{ and } P<P_\eta)\\ f(P_\eta), & \text{if } \eta<1 \text{ and } (T_i\geq T_\eta \text{ or } P\geq P_\eta)\end{cases},$$

$$f(P_\eta)=\begin{cases}K_0, & \text{if } P<P_\eta\\ K_p K_0, & \text{if } P\geq P_\eta,\end{cases}$$

where T_i is the temperature, P is the matched pressure of components, and T_η, P_η, K_p, K_0 are the constants, η is the conversion degree.

Studying the deformation of multicomponent media, it is necessary to take into account the state and response of each component, and, in contrast to a homogeneous mixture, not only the displacement of the external boundaries of the selected volume, but also the displacement of components in the selected volume of the mixture. Here, we consider the equal pressures of the components to be a condition for joint deformation of components during the interaction in the mixture, which determines the volume concentrations of the components:

$$P=P_i(V_i, E_i)=P_j(V_j, E_j)=\ldots=P_N(V_N, E_N).$$

3. Experimental results

To study experimentally the explosive synthesis, the powder mixture of aluminum and fluoroplastic (Al/Tf) was used as a loaded material that was placed in a cylindrical ampoule. The powders were mixed in an AGO-2U planetary mill. The mass fractions of components were 30/70 (Al/Tf), which was close to the stoichiometry of the aluminum fluoride (AlF_3) formation. The initial porosity of the loaded mixture components was varied in the experiments. In the first case, the initial porosity of the mixture components was 0.5 (the ratio between the volume of pores and the total volume) and in the second case, the initial porosity was 0.045. The pressed mixture was placed in a steel cylindrical ampoule (Figure 1). The height of the cylindrical ampoule was 95 mm, the external diameter was 20 mm. The thickness of the lateral wall of the ampoule was 3 mm. The ampoule was closed with lids 10 mm in sickness.

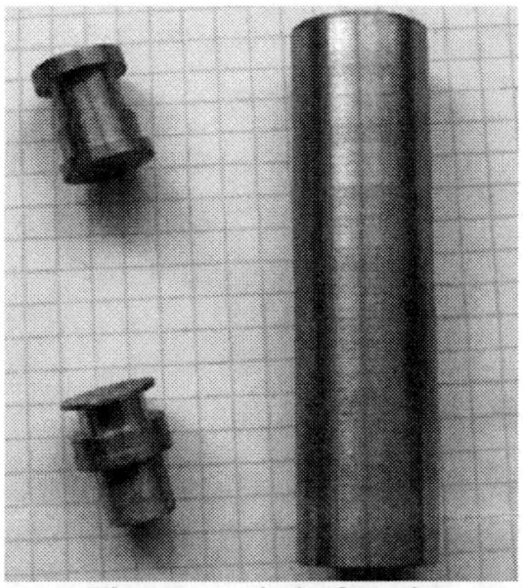

Figure 1. The ampoule before the experiment.

Figure 2 shows the structure of the Al/Tf sample surface before the experiment.

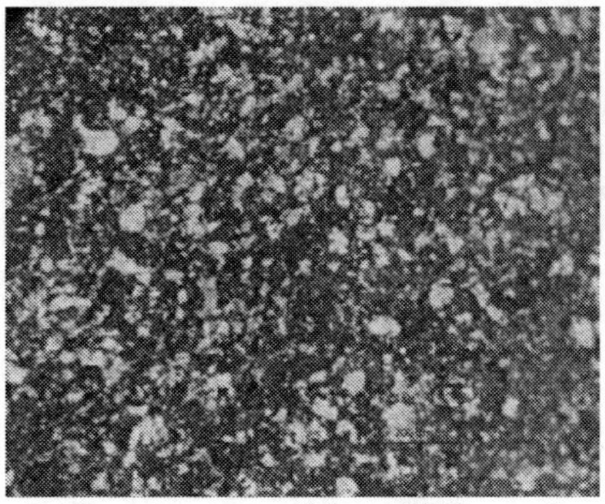

Figure 2. The structure of the Al/Tf sample surface.

The ampoule was loaded by an explosive (ammonite 6GV). The external diameter of the explosive was 45 mm. The density of the explosive was 1.0 g/cm^3, and the detonation velocity was 3.85 km/s.

The first ampoule was broken into large fragments (Figure 3).

Figure 3. Fragments of the ampoule after explosive loading.

The second ampoule was damaged along the whole length of the material. Only the upper layer of the sample, which is 5 mm in thickness, was saved after explosive loading. There are no traces of the reaction in this part of the sample according to the X-ray diffraction analysis.

4. Numerical results

To study the explosive synthesis, the numerical computations were carried out by using the model of a multicomponent medium and the finite element method [9-10]. We numerically solved an axisymmetric problem of a cylindrical steel ampoule containing a porous Al–S mixture under explosive loading. The mass fractions of the components were as follows: Al – 0.35; S – 0.65, which corresponded to the stoichiometry of the aluminum sulfide (Al_2S_3) formation after initiation of a chemical reaction.

The height and the diameter of the cylindrical mixture sample were 65 mm and 14 mm, respectively. The thickness of the lateral wall of the ampoule was Δh = 3 mm; the thickness of the lids was 20 mm. The external diameter of the ampoule was 20 mm. In the computations, the action of the explosive surrounding of the ampoule was modeled as the pressure of the explosion products acting on the horizontal and vertical layers of the ampoule during propagation of the detonation front. The detonation velocity was 3.3 km/s and corresponded to the experimental evaluations.

In the work, we assume that a chemical reaction initiation criterion (pressure criterion) and a reaction rate depend on the initial value of the mixture dispersion. For the aluminum and sulfur powders with low dispersion, we chose the reaction initiation criteria in terms of temperature, $T_\eta = 717$ K (sulfur evaporation temperature), and pressure, P_η= 2.0 GPa. Reaction rate K_0 was taken to be 301.6 GJ/(kg s), and K_p = 2.0. For the aluminum and sulfur powders with high dispersion, we chose $T_\eta = 717$ K and P_η= 1.0 GPa.

Figure 4 demonstrates the distribution of pressure profiles in the Al–S low dispersion mixture placed in a cylindrical ampoule at different moments of time, describing the dynamics of the shock wave process during the explosive synthesis of aluminum sulfide.

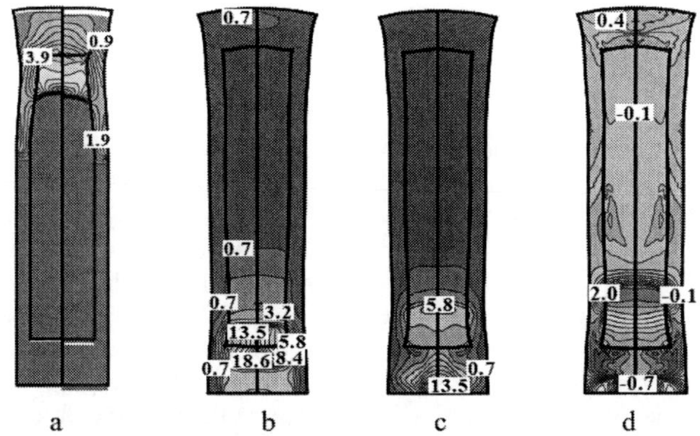

Figure 4. Distribution of pressure profiles (GPa) for the low dispersion reactive mixture (Al-S) in an axial section of the ampoule at different moments of time: (a) 10 μs, (b) 26 μs, (c) 27 μs, and (d) 28 μs.

A shock wave caused by the explosion products propagates along the sample. It should be noted that the wave front propagating in the sample is behind the front propagating along the ampoule due to collapsing the pores in the sample. When the shock wave is reflected from the bottom of the ampoule in the form of a compression wave, the pressure at the bottom of the sample increases by several times (Figure 4b) and is greater than 18 GPa.

Figure 5 demonstrates the distribution of pressure profiles in the Al–S high dispersion mixture placed in a cylindrical ampoule at different moments of time, describing the dynamics of the shock wave process during the explosive synthesis of aluminum sulfide.

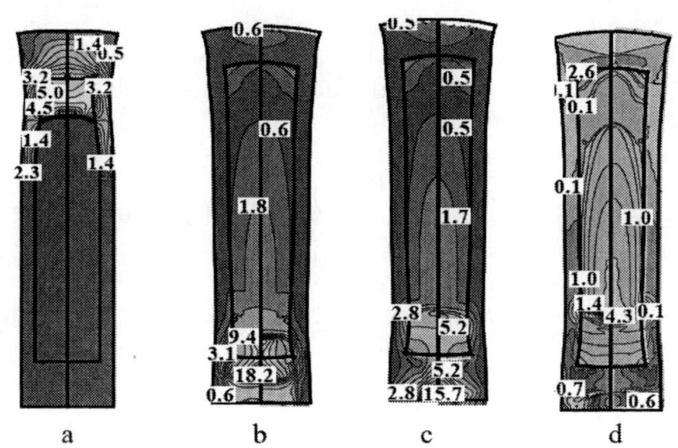

Figure 5. Distribution of pressure profiles (GPa) for the high dispersion reactive mixture (Al-S) in an axial section of the ampoule at different moments of time: (a) 10 μs, (b) 26 μs, (c) 27 μs, and (d) 28 μs.

During the propagation of the shock wave, the higher pressure is observed in the high dispersion sample compared to the low dispersion one, which indicates a faster chemical reaction and its contribution to the increase in the pressure and the temperature. The shock wave reflected from the bottom of the ampoule encounters the shock compression wave propagating in the sample, which leads to a sharp increase in the pressure at the bottom of the sample (Figure 5b) and is greater than 18 GPa. After the encounter, the shock waves start propagating in opposite directions (Figure 5c). The high residual pressures are observed in the sample and reach more than 4 GPa for the high dispersion mixture (Figure 5d) and 2 GPa — for the low dispersion one (Figure 4d).

5. Conclusion

Based on the analysis of numerical and experimental results, we can assume the following dynamics of the shock wave propagation during the solid-phase synthesis in cylindrical ampoules subjected to explosive loading. During the propagation of the shock wave, an exothermic reaction is initiated in the sample and the higher pressures are observed for the high dispersion sample compared to the low dispersion one. The shock wave, reflected from the bottom of the ampoule in the form of a compression wave, encounters the shock wave propagating in the sample, which leads to a sharp increase in the pressure and the rate of chemical reactions at the bottom of the sample. The high rate of heat release during the chemical reaction in the bottom part of the ampoule and a sharp increase in pressure lead to the formation of the gas phase, which causes the ampoule fracture.

The initial dispersion of the reactive mixture was varied to evaluate its contribution to the fracture of the cylindrical ampoule. The experimental and numerical results allowed us to find that the higher the initial dispersion of the mixture components was, the more intensive the ampoule failure was. In some experiments, there was a complete fracture of the ampoule.

Acknowledgments

The reported study was funded by the Russian Foundation for Basic Research, according to research project No. 14-03-00666.

References

[1] Zelepugin S, Ivanova O, Yunoshev A and Zelepugin A 2015 *Letters on materials* **5(4)** 468-472
[2] Zelepugin S A, Ivanova O V, Yunoshev A S and Zelepugin A S 2016 *IOP Conf. Series: Mat. Sci. and Eng.* **127(1)** 012057
[3] Batsanov S S 1994 *Effects of explosions on materials* (New York: Springer-Verlag)
[4] Dolgoborodov A Yu 2015 *Combustion, Explosion, and Shock Waves* **51(1)** 86-99
[5] Dolgoborodov A Yu, Makhov M N, Streletskii A N, Kolbanev I V, Gogulya M F and Fortov V E 2004 *Russian J. Chemical Physics* **23(9)** 85-88
[6] Jetté F-X, Higgins A J, Goroshin S, Frost D L, Charron-Tousignant Y, Radulescu M I and Lee J J 2011 *Journal of Applied Physics* **109** 084905
[7] Nigmatulin R I 1991 *Dynamics of Multiphase Media* vol 1 (New York: Hemisphere) p 507
[8] Shteinberg A S, Berlin A A, Denisaev A A and Mukasyan A S 2011 *Int. J. of Self-Propagating High-Temperature Synthesis* **20(4)** 259-265
[9] Gorel'skii V A, Zelepugin S A and Sidorov V N 1994 *Int. Appl. Mech.* **30(3)** 193-198
[10] Johnson G R 2011 *Int. J. Impact Eng.* **38(6)** 456-472

IOP Conf. Series: Materials Science and Engineering **177** (2017) 012072 doi:10.1088/1757-899X/177/1/012072

The use of magnetic marks in steel wire ropes

A S Khoroshev[1], A V Pavlenko[1], A A Korotky[2], D A Tchoutchkin[1], V S Puzin[1] and E V Khorosheva[1]

[1] Platov South-Russian State Polytechnic University (NPI), 132, Prosveschenia str., Novocherkassk, 346428, Russia
[2] Don State Technical University, 1, Gagarin sq., Rostov-on-Don, 344000, Russia

E-mail: vskych@gmail.com

Abstract. Various methods of marking wire ropes are not devoid of certain disadvantages. In this paper, the authors consider the theoretical possibility of using integrated magnetic marks. The use of mathematical modelling (GMSH + GetDP) assessed the degree of shielding magnetic field of a mark by rope wire. Through the simulation, the authors have determined that the using of a cylindrical mark with the magnetization direction along the rope has qualitative advantages over other forms of marks, or the transverse magnetization direction in terms of its detection by the external detecting device. This type of mark can be easily embedded in the polymeric core of the rope.

1. Introduction

With the development of the cable transport, a need for marking the different portions of the rope arises. For example, it becomes necessary to analyse rope slacking, registering its speed or measuring the distance, which the section of rope or the structure, rigidly connected with it (cabin, column), passed.

Currently, there are several ways of marking mining ropes and armored cables. Thus, the marking of the mining rope or cable can be accomplished by coating ferromagnetic powders on the outer surface of the armor or strands with their subsequent magnetization. The resulting marks have a low life due to mechanical abrasion of the mark caused by friction and reversal magnetization by external magnetic fields. Another method of marking is through the magnetized portions of the armor mining rope [1]. Magnetic marks are pretty quickly demagnetized, because the armor of mining cables is made of soft magnetic steel. Furthermore, the application of such marks to a rope or armor of the moving rope or cable, for example during the manufacture of the product, leads to the longitudinal deformation of the mark. Longitudinal deformation of the mark is caused by a displacement of the rope towards the magnetizing device for the time needed to magnetize the rope part.

There is also a well-known method for making magnetic marks consisting in creating capsules with a ferromagnetic fluid, enclosed into a flexible magnetic material of the core rope [2]. In this case, the material fills the space (gap) between the strands and the length of the capsule which is equal to a strand pitch. As the identifier marks RFID tags [3] are applied, their use may be complicated by interference in reading data from the mark caused during the work of the cableway devices or other devices.

In paper [4], it was proposed to implement magnetic marks based on the permanent magnets in the core of the rope, which will greatly increase their service life in comparison with the external marks (ferromagnetic powders or the obtained magnetization portion of the rope or armor). The major problem is that in most of cases, mild unalloyed steel is used for rope manufacturing, which largely shields the magnetic field. In order to investigate and to study the possibility of using magnetic marks that are integrated into the core of the rope, the numerical simulation of the electromagnetic field distribution in the part of the rope with a permanent magnet fitted in the core, is performed. This design can be implemented on the ropes with a polymeric core that can be replaced by rubber ferrite.

2. The model and simulation

For the analysis, the authors used the method of finite elements with the implementation of it in the software package 'GMSH + GetDP' [5, 6]. The computational domain is the portion of the rope with six strands (the geometric parameters correspond to those listed in The State Standard 3062-80) [7] which are homogeneous areas (Figure 1). The length of the portion of the rope is set to be 70 mm. The magnetic properties of the rope model correspond to the magnetic properties of steel AISI 1010 (St 10). The model rope [8] takes into account the lateral gap between the strands, the presence of which corresponds to the state of a rope not worn out [9].

The magnetic mark is installed into the core, made of a nonmagnetic material having a minimum gap size of 0.5 mm from the inner wires of the rope. The rope core used in a model simulates airspace. The magnetic mark is represented by characteristics corresponding to hard-magnetic material ANI-8 (the value of the residual induction – 0.53 T, coercive force – 790 kA/m).

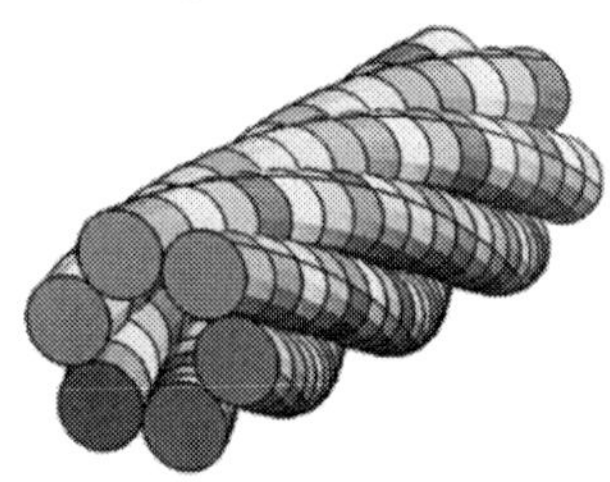

Figure 1. The 3D model of the rope.

When the magnetization takes place along axes Y and Z, the magnetic mark is performed in the form of a rectangular parallelepiped (dimensions ranging from 8x8x4 to 8x3x2 mm (Figures 2a, 2b)). For magnetizing along the X axis, the cylindrical mark was used (mark length is 4 mm, the diameter varied from 2 to 6 mm (Figure 2c)). When conducting the research, the rope diameter and the diameter of the strands ranged within 12-30 mm and 4-10 mm, respectively.

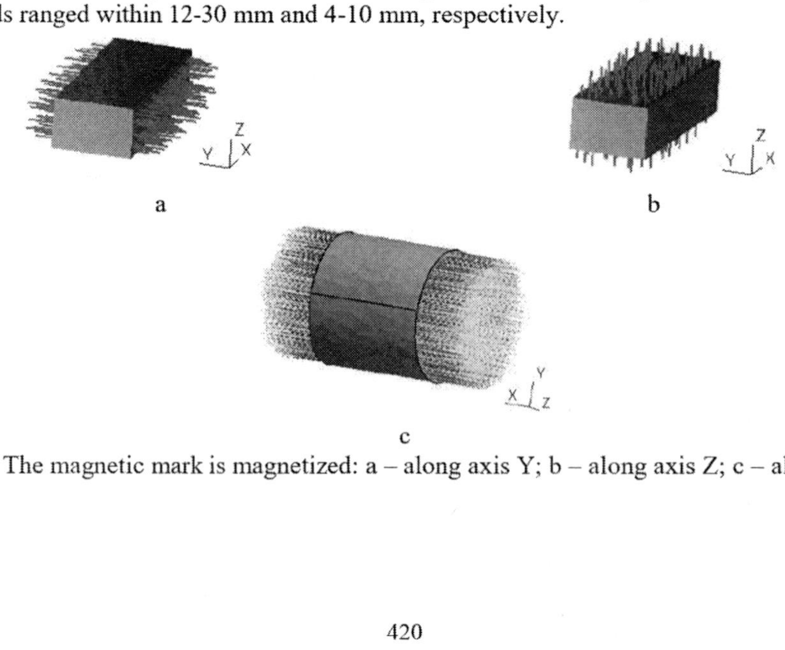

Figure 2. The magnetic mark is magnetized: a – along axis Y; b – along axis Z; c – along axis X.

The picture of the magnetic field was obtained using the finite element method. The computational domain (a three-dimensional model of the rope with the magnetic mark and the magnetic concentrator) is divided into a finite element mesh with built-in GMSH package [10]. The simulation has found that the strands of the rope largely shield the magnetic field generated by the magnetic mark: the magnetic field induction on the surface of the rope along the line of magnetic mark arrangement is evenly distributed (Figure 3).

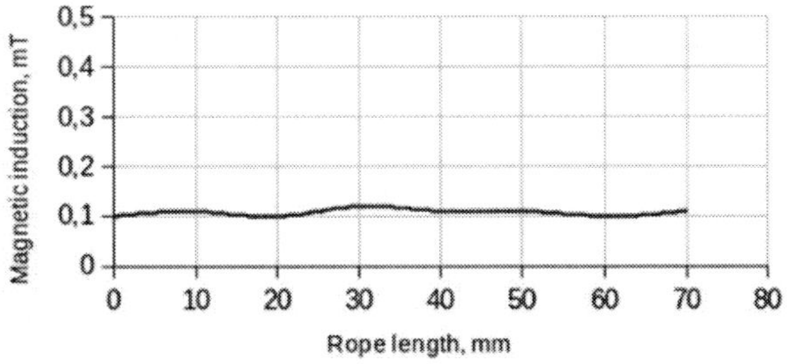

Figure 3. Distribution of the magnetic field in the rope without a magnetic concentrator.

For pronounced inhomogeneity of the magnetic field, a horseshoe-shaped ferromagnetic concentrator (Figure 4), partially covering the cable, is placed at the location of the magnetic mark. The gap between the ferromagnetic concentrator and the rope is 1 mm, the thickness of the concentrator is constant along its length and is equal to 3 mm. The radius of the curvature varies depending on the diameter of the rope to ensure the air gap between the rope and the concentrator is equal to 1 mm. The width of the magnetic concentrator located directly above the magnetic mark equal to the length of the permanent magnet. The magnetic properties of the material of the magnetic concentrator correspond to structural steel AISI 1008 (St 08). In case of using a cylindrical mark, the magnetic concentrator has a U-shape and is located on the lateral surface of the rope, the length of the concentrator is 12 mm (Figure 5).

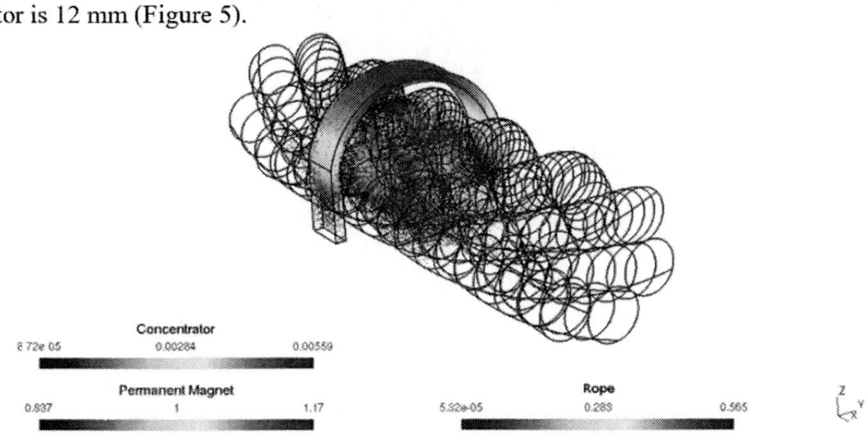

Figure 4. Distribution of the magnetic field with the magnetization of the magnetic mark directed along the Y-axis.

With the magnetization of the magnetic mark is directed along axes Y and Z (across the rope), the magnetic induction in the gap of the concentrator depends on the angle of rotation of the mark (rope with a mark) around axis X. Modulus values of the magnetic induction will be maximal if the vector of magnetic induction of the magnetic mark is directed from one pole to the other pole of the magnetic concentrator, i.e. along axis Y.

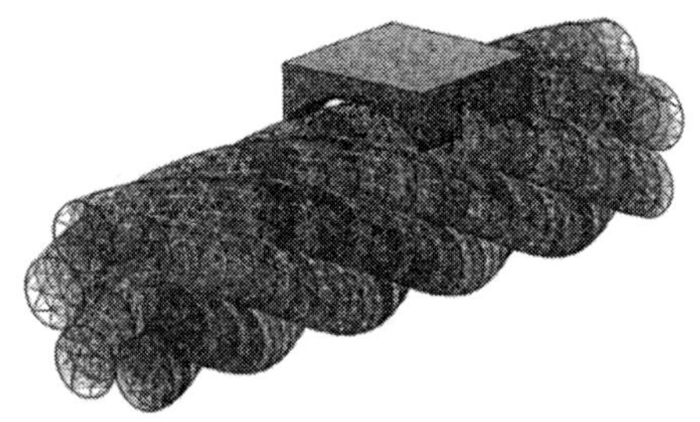

Figure 5. The3D model of the rope with the mark and the concentrator is divided into finite elements.

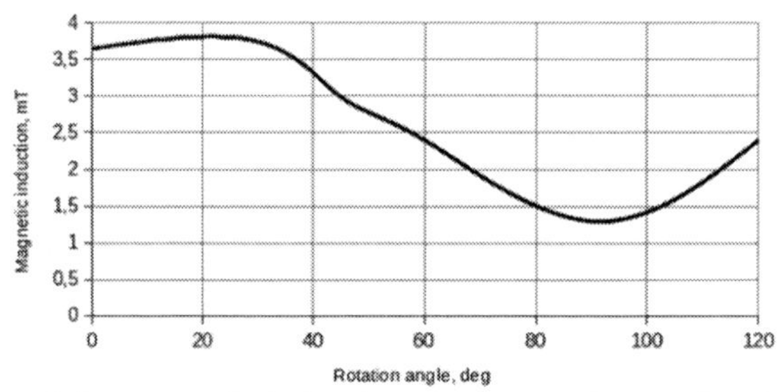

Figure 6. Dependence of the magnetic induction on the rotation angle of the rope relative to the stationary concentrator for the magnetization mark transversely the rope.

With increasing the angle of rotation of the mark, the value of the magnetic induction in the gap of the concentrator is lowered and reaches a minimum when the mark rotational angle is 90°, which corresponds to its magnetization along axis Z. The magnitude of the magnetic induction is 20-25% of the initial value (the rotation angle is 0°). The dependence of the modulus magnetic induction on the rotation angle is shown in Figure 6.

Thus, the authors concluded that magnetization of the mark along axes Y and Z (transversely the rope) has a drawback. It is a variation of the value of magnetic induction in the gap of the magnetic concentrator depending on the angle of rotation of the mark. The magnetic induction in the gap of the concentrator during magnetizing the magnetic mark along the X-axis does not change during the rotation of the rope, since the direction of the magnetic induction vector in this case is not changed.

3. The resolution

To determine the resolution when reading the signal, the dependence of the magnetic induction in the concentrator gap on the value displacement of the magnetic mark relative the concentrator along the rope axis has been studied. The geometrical dimensions of the model were not different from those mentioned above. Zero displacement corresponds to the location of the concentrator above the magnetic mark. The magnetic induction in this case reaches its maximum. The simulation was carried out for both rectangular and cylindrical marks. The value of the air gap between the concentrator and the rope was 0.5 and 2 mm.

When using a cylindrical mark, a sharp decrease of the magnetic induction in the gap of the concentrator with increasing displacement occurs. The 5-time reduction of the magnetic induction is achieved with the displacement of the magnetic mark by 16 mm (four lengths of the mark) in case of a cylindrical mark, and by 28 mm (seven lengths of the mark) in case of a rectangular mark (Figure 7).

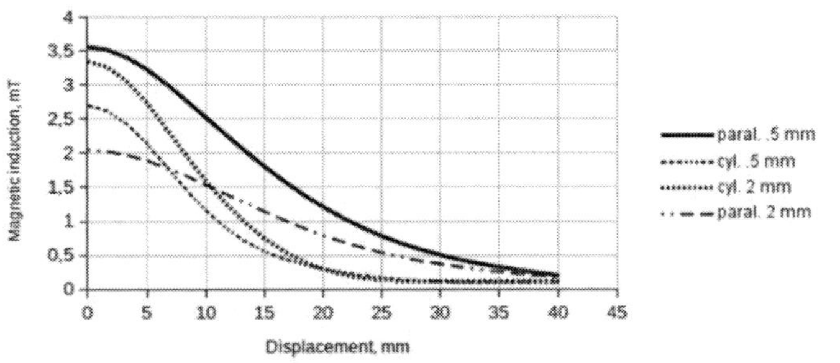

Figure 7. The dependence of the magnetic induction in the gap of the concentrator on the displacement of the cylindrical and parallelepipedic magnetic marks with the corresponding magnetization direction and the shape of the concentrator.

The authors concluded that the more reliable detection of the magnetic mark, possibly by determining the place of changing the sign of the normal (in the direction of the radius of the rope) component of the magnetic induction vector. The sign of the normal component of the magnetic induction vector abruptly reversed itself above the magnetic mark (hereinafter the mark is cylindrical). Increasing the length of the mark leads to a significant increase in the difference between the maximum (positive) and minimum (negative) value of the normal component of the magnetic induction vector, which has a positive effect on the quality of the potential detection of the magnetic mark. The negative effect is a somewhat decrease in the rate of the change in the value of the normal component, for example, during passage of the recording device above the mark (Figure 8).

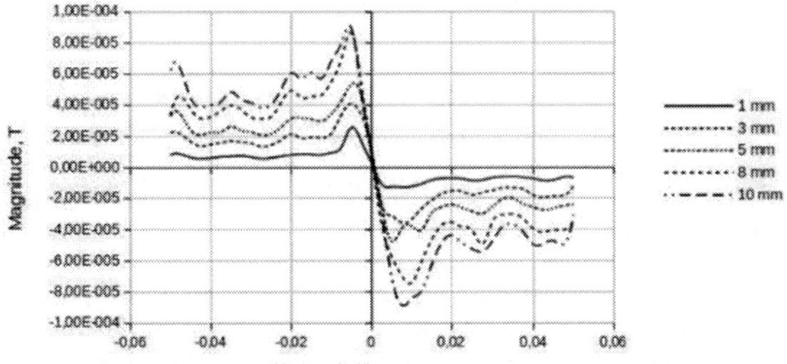

Figure 8. Changing the value of the normal component along the length of the rope for different values of the length of the magnetic marks.

The resolution will be determined by changing the sign of the normal component of the number of times corresponding to the number of marks. To estimate the resolution in determining the mark by the aforementioned method, the numerical simulation of the rope part with the length of 200 mm was performed with two cylindrical magnetic marks with the length of 5 mm. The distance between the adjacent ends of the marks was varied from 1 to 40 mm. Simulation results are presented in Figure 9. The value of the normal component corresponds to the values of the magnetic induction along a line parallel to the axis of the rope at a distance of 1 mm from the outer surface. The small distance between magnetic marks prevents from identifying the two marks. Thus, a double change in the sign of the normal component of the vector of magnetic induction recorded only at distances between the magnetic marks about 15 mm and more, which exceeds the diameter of the rope (11.5 mm).

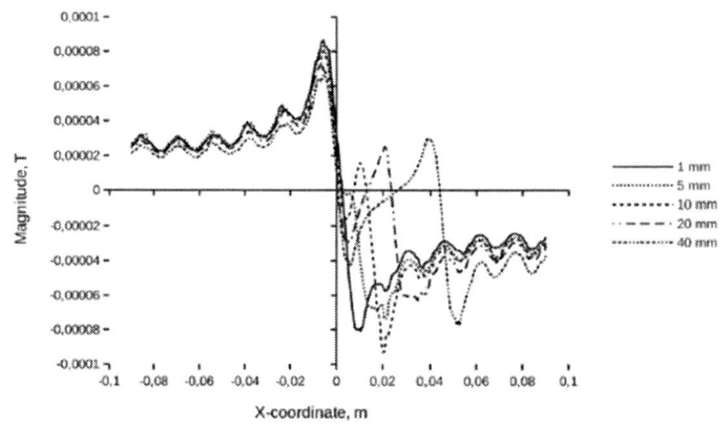

Figure 9. Changing the value of the normal component along the length of the rope for different values of the distance between the magnetic marks.

At shorter distances, the identification of two marks by the aforementioned method is not possible.

4. Conclusion

Analysis of research results leads to the following conclusions.

- The results confirm the possibility of mathematical modelling and the efficient use of permanent magnets in marking the rope with six strands.
- It was determined that for display a maximum variation of the magnetic field in a zone of the mark in the reading device, it is advisable to install a concentrator composed of the ferromagnetic material.
- Analysis of the magnetic field, generated by the permanent magnet, which form a magnetic mark, confirms the effectiveness of the registration normal component of the magnetic field induction vector.
- Using the permanent magnet with the magnetization direction along of the rope reduces the influence of the angle of rotation about the axis relative to the concentrator on the value of the induction in the concentrator measuring gap.
- Increasing the length of the permanent magnet performed as a magnetic mark increases the magnitude of the amplitudes of the normal component of the magnetic field induction vector on the relevant section of the rope that allows improving the quality of the reading signal of the magnetic mark.
- The distance between the individual adjacent marks should be more than the rope diameter for reliable identification.

Acknowledgments

The reported study was funded by RFBR according to research project No. 16-38-60078.

References

[1] G I Krivolapov, RU Patent No. 2,172,832 (24 March 1999)
[2] O G Mostovoy, N A Popov, V M Ivanovsky and B I Belov, SU Patent No. 1,803,486 (25 December 1990)
[3] D A Zirolf, RU Patent No. 2,295,140 (26 April 2002)
[4] A A Korotky, A V Pavlenko, A S Khoroshev, V M Prikhodko, G V Kustarev, V B Maslov and D A Korotky, RU Patent No. 116,861 (05 May 2011)
[5] Gmsh: A three-dimensional finite element mesh generator with built-in pre- and post-processing facilities *http://gmsh.info/*
[6] GetDP: A Generator Environment for the Treatment of Discrete Problems *http://getdp.info/*

[7] The State Standart 3062-80 1996 One lay rope type LK-O construction 1x7 (1+6). Dimensions *Publisher of standards*
[8] Khoroshev AS, Pavlenko AV, Puzin VS, Medvedev VV and Tchoutchkin DA 2012 *Proceedings of the universities. North-Caucasian region. Technical science* **1** 40-45
[9] A A Korotky, V B Maslov, D A Korotky, V A Tytskiy and A V Panfilov, RU Patent No. 2,489,542 (12 March 2012)
[10] Geuzaine C and Remacle J-F 2009 *Int. J. Numer. Meth. Engng* **79(11)** 1309-1331

Mathematical model for the power supply system of an autonomous object with an AC power transmission over a cable rope

V M Rulevskiy[1], V G Bukreev[2], E B Shandarova[2], E O Kuleshova[2], S M Shandarov[3] and Yu Z Vasilyeva[2]

[1] Research Institute of Automation and Electromechanics, Tomsk State University of Control Systems and Radioelectronics, 53, Belinsky Str., Tomsk, 634034, Russia
[2] Tomsk Polytechnic University, 30, Lenina Ave., Tomsk, 634050, Russia
[3] Tomsk State University of Control Systems and Radioelectronics, 40, Lenina Ave., Tomsk, 634050, Russia

E-mail: bukreev@tpu.ru, shandarovaelena@mail.ru, kuleshova@tpu.ru

Abstract. A modeling problem of the power system, which provides an AC power transmission to a submersible device over the conducting rope, was considered. The power supply system units and their parameters are described. The system multi-dimensional mathematical model in the variables state space with regard to the nonlinear characteristic of system elements is proposed.

1. Introduction

A power supply system (PSS) of an autonomous object with the power transmission over a cablerope should be improved. For this purpose, it is possible to apply the optimal controllers for efficient stabilization of voltage at consuming loads and minimization of weight and size characteristics for parametric and external disturbances [1-4]. Creating these regulators assumes the use of synthesis methods of complex objects modern control theory based on the description of dynamic processes in the variables state space [5, 6]. It should be noted that the system of differential equations in the Cauchy form is the most adequate mathematical model in the variables state space of many objects. It reflects the 'physics' of occurring phenomena and processes under certain assumptions. The number of differential equations must fully reflect elements properties of the power supply system representing an equivalent circuit with lumped parameters.

Figure 1 shows the structure of the PSS model, which provides an AC power transmission to a submersible device over the cable rope.

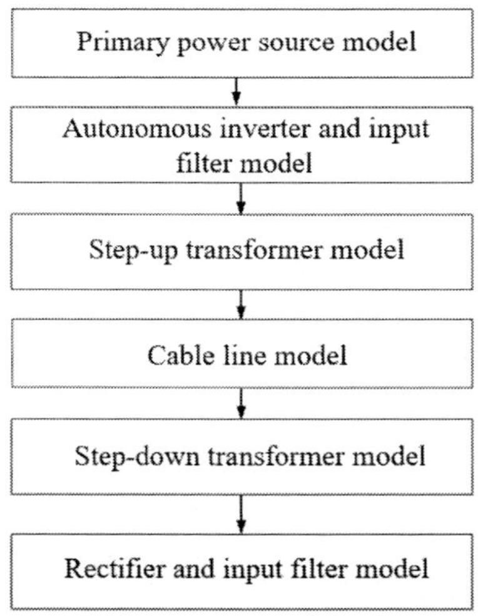

Figure 1. The structure of the PSS model.

It is assumed that parameters of the PSS equivalent circuit, shown in Figure 2, are the same for each phase in preparing mathematical models. It means that the system is symmetrical, and calculation can be carried out on one phase; other phases currents and voltages can be obtained by rotating through an angle of 120°.

The power supply is an autonomous synchronous generator, and the rectifier connected through the output to an *L*-shaped *LC*-filter. The power supply model is not considered in this paper.

Supply voltage U_s (Figure 2) is assumed to be constant and devoid of pulsations as a result of a smoothing filter action in the preparation of the PSS mathematical model.

An independent voltage inverter (IVI) is a static inverter of DC voltage U_s to AC voltage via semiconductor switches, which are used as metal–oxide–semiconductor field-effect transistors (MOSFET-transistors). A power converter is controlled by the method of Pulse Width Modulation (PWM). An output voltage curve is formed as a series of high frequency pulses, the duration of which changes under the law of the modulating signal; clock frequency f_{ref} is equal to 48 kHz, and the modulation frequency – 1 kHz.

2. The mathematical model

The mathematical model made for the idealized inverter, assuming supply voltage U_s is constant and keys — ideal. Control pulses, supplied to the power transistor base, are formed as a result of comparing a modulated low-frequency three-phase voltage, having in its part the first and the third harmonic of the following form:

$$u_{Amd}(t) = \frac{\sin(1000t) + 0.14 \cdot \sin(3000t)}{\cos(30^\circ)}; \quad u_{Bmd}(t) = \frac{\sin(1000t - 120^\circ) + 0.14 \cdot \sin(3000t)}{\cos(30^\circ)}$$

$$u_{Cmd}(t) = \frac{\sin(1000t + 120^\circ) + 0.14 \cdot \sin(3000t)}{\cos(30^\circ)}$$

with the reference voltage of a sawtooth shape, shown in the model as Fourier series:

$$u_{ref}(t) = \frac{2}{\pi} \cdot \left(\sin(\omega_{ref} t) - \frac{1}{2}\sin(2\omega_{ref} t) + \frac{1}{3}\sin(3\omega_{ref} t) - \frac{1}{4}\sin(4\omega_{ref} t) + \ldots \right),$$

where $\omega_{ref} = 2\pi f_{ref}$ is an angular clock frequency.

Application of the third harmonic premodulation increases the maximum achievable ratio of the first harmonic amplitude to power supply voltage U_s and then leads to a marked reduction in harmonic distortion.

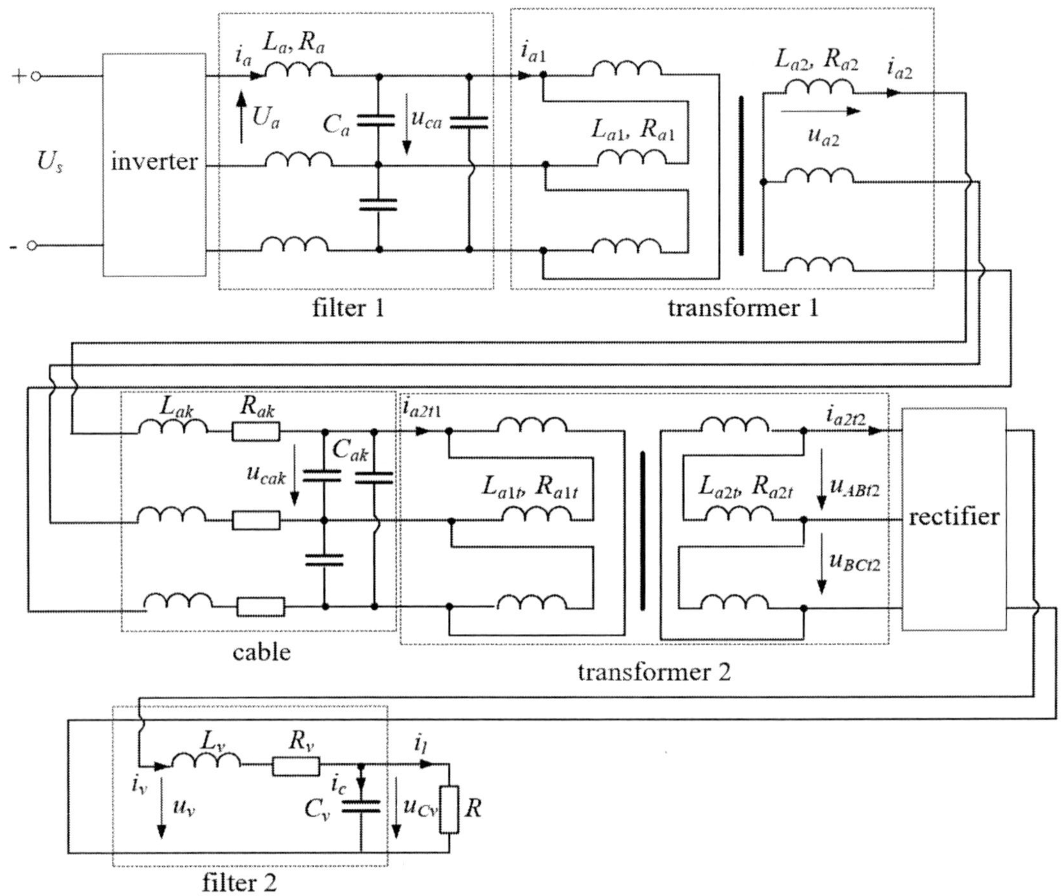

Figure 2. The PSS equivalent circuit.

The difference function is calculated for each phase according to the following formulas:

$$f_{Adif}(t) = u_{Amd}(t) - u_{ref}(t);\ f_{Bdif}(t) = u_{Bmd}(t) - u_{ref}(t);\ f_{Cdif}(t) = u_{Cmd}(t) - u_{ref}(t).$$

The phase switching function is recorded on the basis of the abovementioned formulas:

$$K_A(t) = \begin{Bmatrix} 1 \text{ if } f_{Adif} \geq 0 \\ 0 \text{ if } f_{Adif} < 0 \end{Bmatrix};\ K_B(t) = \begin{Bmatrix} 1 \text{ if } f_{Bdif} \geq 0 \\ 0 \text{ if } f_{Bdif} < 0 \end{Bmatrix};\ K_C(t) = \begin{Bmatrix} 1 \text{ if } f_{Cdif} \geq 0 \\ 0 \text{ if } f_{Cdif} < 0 \end{Bmatrix}.$$

The total commutation function autonomous inverter is written as:

$$K(t)=\begin{cases}-1/6 \text{ if } K_A(t)=1 \wedge K_B(t)=-1 \wedge K_C(t)=1\\ 1/6 \text{ if } K_A(t)=1 \wedge K_B(t)=-1 \wedge K_C(t)=-1\\ -1/6 \text{ if } K_A(t)=1 \wedge K_B(t)=1 \wedge K_C(t)=-1\\ 1/6 \text{ if } K_A(t)=-1 \wedge K_B(t)=1 \wedge K_C(t)=-1\\ -1/6 \text{ if } K_A(t)=-1 \wedge K_B(t)=1 \wedge K_C(t)=1\\ 1/6 \text{ if } K_A(t)=-1 \wedge K_B(t)=-1 \wedge K_C(t)=1\\ 1/2 \text{ if } K_A(t)=-1 \wedge K_B(t)=-1 \wedge K_C(t)=-1\\ -1/2 \text{ if } K_A(t)=1 \wedge K_B(t)=1 \wedge K_C(t)=1\end{cases}$$

The mathematical model of the *L*-shaped output filter (Figure 2) for phase *A* can be written using such obtained switching function as:

$$\frac{di_a(t)}{dt}=\frac{-R_a}{L_a}\cdot i_a(t)+\frac{-1}{L_a}\cdot u_{Caz}(t)+\left(\frac{K_A(t)}{2L_a}+\frac{K(t)}{L_a}\right)\cdot U_s;\ \frac{du_{Caz}(t)}{dt}=\frac{1}{C_{az}}\cdot i_a(t),$$

where R_a [Ohm], L_a [mH], C_{az} [uF] (capacities have a star connection) – are parameters of the IVI output filter, which is shown as filter 1 in Figure 2, $u_{Caz}(t)$ - is a voltage across capacities, at the star connection.

The basis of the three-phase step-up and step-down transformer model in Figure 2 is a T-equivalent circuit [7, 9].

The transformers parameters are windings active resistance (R_{a1}, R_{a2}, R_{a1t}, R_{a2t}) and leakage inductance (L_{a1}, L_{a2}, L_{a1t}, L_{a2t}), taking into account the leakage flux influence. A transfer characteristic is modeled by the branch, containing active magnetic resistance R_{m1} for the first transformer and R_{m2} — for the second one (it simulates active losses in the core), and, respectively, L_{m1} and inductance L_{m2} (it simulates reactive core loss). The core saturation can be taken into account via the transfer characteristic, which is described by a piecewise linear approximation of the relation between the flux and magnetizing current (hysteresis was not considered). A saturation can also be considered as an introduction of saturated parameters to the model in the static modes simulation.

It should be noted that the special transformers were designed to operate in the considered power supply system.

Active-inductive load with the star-connection and parameters R_{eqv} [Ohms], L_{eqv} [mH] was connected to the secondary winding of the step-down transformer for recording the mathematical model. As it is seen from Figure 2 the step-down transformer is a transformer 2. Load parameters are calculated as the equivalent for the second filter and load at its output according to the known methods: if the parameters of the filter and load are known, the circuit complex resistance is calculated at the rectifier fundamental frequency. Active and reactive components of the complex resistance parameters allow calculating an equivalent load [8].

Cable line design features must be taken into account when designing their models. Cable line main parameters are: phase active resistance, distributed internal inductance of phases, phase distributed capacity relative to the screen, and the interphase mutual inductance and capacitance. In general case, the listed parameters are different for each plot. It is due to variation of manufacturing parameters, insulation depreciation, different qualities of the electrical connections in couplings [9]. Besides, the cable parameters are different in various areas due to the wide range of ambient temperature change depending on the depth of submersible device suspension.

Conduction currents were not taken into account when simulating the mathematical model of the cable line, because the value of the active conduction current through the cable insulation was less than 0.01%, and the value of the capacitive conduction current — less than 0.4% of the cable rated current [10].

A cable line can be simulated as a line with distributed and lumped parameters. The model with lumped parameters, which is shown in Figure 2 as «cable» unit, was chosen as the mathematical model because of its simplicity and clarity.

The three-phase bridge rectifier, which is represented in Figure 2 as « rectifier» unit, is shown in the transfer function model, where rectified voltage U_v at intervals equaling to one-sixth of the period defined by difference-phase transformer EMF $u_{ABt2}, u_{BCt2}, u_{CAt2}$:

$$u_v = \begin{cases} u_{ABt2} - u_{BCt2} \text{ if } u_{ABt2} > u_{BCt2} \wedge u_{ABt2} \geq u_{CAt2} \wedge u_{BCt2} < u_{ABt2} \wedge u_{BCt2} \leq u_{CAt2} \\ u_{ABt2} - u_{CAt2} \text{ if } u_{ABt2} \geq u_{BCt2} \wedge u_{ABt2} > u_{CAt2} \wedge u_{CAt2} < u_{ABt2} \wedge u_{CAt2} \leq u_{BCt2} \\ u_{BCt2} - u_{CAt2} \text{ if } u_{BCt2} \geq u_{ABt2} \wedge u_{BCt2} > u_{CAt2} \wedge u_{CAt2} < u_{BCt2} \wedge u_{CAt2} \leq u_{ABt2} \\ u_{BCt2} - u_{ABt2} \text{ if } u_{BCt2} \geq u_{CAt2} \wedge u_{BCt2} > u_{ABt2} \wedge u_{ABt2} < u_{BCt2} \wedge u_{ABt2} \leq u_{CAt2} \\ u_{CAt2} - u_{BCt2} \text{ if } u_{CAt2} \geq u_{BCt2} \wedge u_{CAt2} > u_{ABt2} \wedge u_{ABt2} < u_{CAt2} \wedge u_{BCt2} \leq u_{CAt2} \\ u_{CAt2} - u_{ABt2} \text{ if } u_{CAt2} \geq u_{ABt2} \wedge u_{CAt2} > u_{BCt2} \wedge u_{ABt2} < u_{CAt2} \wedge u_{ABt2} \leq u_{BCt2} \end{cases}.$$

In the rectifier simulation, the following assumptions were taken: switching processes do not affect the shape of the output voltage and current; valves are ideal.

The PSS mathematical model was compiled for one phase (phase A was selected) to reduce the system order taking into account that the system is symmetrical. In this case, all parameters with a Δ-connection were transformed into a Y-connection (index z in the equations). The equation in an integro-differential form according to the Kirchhoff laws was made up for the scheme shown in Figure 2, where state variables are expressed, for drawing up a system of differential equations in the Cauchy form. Thus, state variables currents of the first and second filter (i_a, i_v), line currents of primary and secondary transformers windings ($i_{a1}, i_{a2}\ i_{a2t1}, i_{a2t2}$), and the voltage across the capacitive element (u_{Caz}, u_{Cakz}) were selected.

As a result, the PSS mathematical model is written as follows in the state space:

$$\begin{aligned} &\frac{di_a}{dt} = a_{11} \cdot i_a + a_{12} \cdot u_{Caz} + a_{13} \cdot \left(\frac{K_A(t)}{2L_a} + \frac{K(t)}{L_a} \right) \cdot U_s; \\ &\frac{di_{a1}}{dt} = a_{21} \cdot i_{a1} + a_{22} \cdot i_{a2} + a_{23} \cdot u_{Caz} + a_{24} \cdot u_{Cakz}; \\ &\frac{di_{a2}}{dt} = a_{31} \cdot i_{a1} + a_{32} \cdot i_{a2} + a_{33} \cdot u_{Caz} + a_{34} \cdot u_{Cakz}; \\ &\frac{di_{a2t1}}{dt} = a_{41} \cdot i_{a2t1} + a_{42} \cdot i_{a2t2} + a_{43} \cdot u_{Cakz}; \\ &\frac{di_{a2t2}}{dt} = a_{51} \cdot i_{a2t1} + a_{52} \cdot i_{a2t2} + a_{53} \cdot u_{Cakz}; \\ &\frac{du_{Caz}}{dt} = \frac{1}{C_{az}} \cdot i_a - \frac{1}{C_{az}} \cdot i_{a1}; \\ &\frac{du_{Cakz}}{dt} = \frac{1}{C_{akz}} \cdot i_{a2} - \frac{1}{C_{akz}} \cdot i_{a2t1}, \end{aligned} \tag{1}$$

where $L_{\Sigma 1} = L_{a1z}\left(L_{m1}k_1^2 + L_{a2} + L_{ak}\right) + L_{m1}\left(L_{a2} + L_{ak}\right)$; $L_{\Sigma 2} = L_{m2}\left(L_{a1tz}k_2^2 + L_2\right) + L_2 L_{a1tz}$;

k_1, k_2 – the ratio of the secondary winding phase voltage to the primary winding phase voltage for the first and second transformers, respectively;

$R_2 = R_{a2tz} + R_{eqv}$; $L_2 = L_{a2tz} + L_{eqv}$;

$$a_{11}=-\frac{R_a}{L_a};\ a_{12}=-\frac{1}{L_a};\ a_{13}=-\frac{1}{L_a};\ a_{21}=\frac{-L_{a2}\left(R_{m1}+R_{a1z}\right)-L_{ak}\left(R_{m1}+R_{a1z}\right)-L_{m1}k_1^2R_{a1z}}{L_{\Sigma1}};$$

$$a_{22}=\frac{-L_{m1}k_1\left(R_{a2}+R_{ak}\right)+k_1R_{m1}\left(L_{a2}+L_{ak}\right)}{L_{\Sigma1}};\ a_{23}=\frac{L_{a2}+L_{ak}+L_{m1}k_1^2}{L_{\Sigma1}};\ a_{24}=\frac{-L_{m1}k_1}{L_{\Sigma1}};$$

$$a_{31}=\frac{k_1\left(R_{m1}L_{a1z}-L_{m1}R_{a1z}\right)}{L_{\Sigma1}};\ a_{32}=\frac{-L_{m1}\left(R_{a2}+R_{ak}\right)-L_{a1z}\left(R_{a2}+R_{ak}+R_{m1}k_1^2\right)}{L_{\Sigma1}};$$

$$a_{33}=\frac{L_{m1}k_1}{L_{\Sigma1}};\ a_{34}=\frac{-L_{a1z}-L_{m1}}{L_{\Sigma1}};\ a_{41}=\frac{-L_2\left(R_{m2}+R_{a1tz}\right)-L_{m2}k_2^2R_{a1tz}}{L_{\Sigma2}};\ a_{42}=\frac{k_2\left(R_{m2}L_2-L_{m2}R_2\right)}{L_{\Sigma2}};$$

$$a_{43}=\frac{L_2+L_{m2}k_2^2}{L_{\Sigma2}};\ a_{51}=\frac{k_2\left(R_{m2}L_{a1tz}-L_{m2}R_{a1tz}\right)}{L_{\Sigma2}};\ a_{52}=\frac{L_{a1tz}\left(-R_{m2}k_2^2-R_2\right)-L_{m2}R_2}{L_{\Sigma2}};\ a_{53}=\frac{L_{m2}k_2^2}{L_{\Sigma2}}.$$

The phase voltage at the first and second transformer output is calculated through the state variables:

$$u_{a2}=b_{11}\cdot i_{a1}+b_{12}\cdot i_{a2}+b_{13}\cdot u_{Caz}+b_{14}\cdot u_{Cakz};\quad u_{At2}=b_{21}\cdot i_{a2t1}+b_{22}\cdot i_{a2t2}+b_{23}\cdot u_{Cakz},$$

where

$$b_{11}=\frac{L_{ak}k_1\left(R_{m1}L_{a1z}-L_{m1}R_{a1z}\right)}{L_{\Sigma1}};$$

$$b_{12}=\frac{L_{m1}\left(L_{a2}R_{ak}-L_{ak}R_{a2}+R_{ak}k_1^2L_{a1z}\right)+L_{a1z}\left(L_{a2}R_{ak}-L_{ak}R_{a2}-L_{ak}k_1^2R_{m1}\right)}{L_{\Sigma1}};$$

$$b_{13}=\frac{L_{m1}k_1L_{ak}}{L_{\Sigma1}};\ b_{14}=\frac{L_{a2}\left(L_{m1}+L_{a1z}\right)+L_{m1}k_1^2L_{a1z}}{L_{\Sigma1}};\ b_{21}=\frac{L_vk_2\left(R_{m2}L_{a1tz}-L_{m2}R_{a1tz}\right)}{L_{\Sigma2}};$$

$$b_{22}=\frac{L_{m2}\left(L_2R_v-L_vR_2\right)+L_{a1tz}\left(L_2R_v-L_vR_2+k_2^2\left(L_{m2}R_v-L_vR_{m2}\right)\right)}{L_{\Sigma2}};\ b_{23}=\frac{L_{m2}L_vk_2}{L_{\Sigma2}}.$$

The voltage and the current of the rectifier load, which is shown in Figure 2, are calculated according to formulas:

$$\frac{di_v}{dt}=\frac{R_v}{L_v}\cdot i_v+\frac{1}{L_v}\cdot u_v+\frac{1}{L_v}\cdot u_{Cv};\quad \frac{du_{Cv}}{dt}=\frac{1}{C_v}\cdot i_C, \tag{2}$$

where $i_C=i_v-\frac{1}{R}\cdot u_{Cv}$ and $i_l=\frac{1}{R}\cdot u_{Cv}$.

As a result, the mathematical model of PSS in the space of state variables is described as the system of equations (1) and (2).

3. Conclusion

The mathematical model for the power supply system of an autonomous object with an AC power transmission over a cable rope is described in the paper. The differential equations, describing a symmetrical three-phase power supply system, are presented in the Cauchy form. Thus, the model was developed, which allows determining the currents of input and output filters, linear currents of the primary and secondary transformer windings, the output voltage of the rectifier. The mathematical model simulation was carried out in MathCAD software tools. The advantage of this model is the ability to obtain values of the abovementioned voltages and currents, both in a steady state and during transients. Thus, this model allows one to reset and release a load current.

References

[1] Xiang X, Niu Z, Lapierre L and Zuo M 2015 *Hong Kong Institution of Engineers* **22** 103-116
[2] Song B, Zhang B, Jiang J 2016 *J. of Huazhong Univ. of Science and Techn.* **44** (7) 52-56
[3] Maalouf D, Creuze V, Chemori A, Tamanaja I T, Mercado E C, Muñoz J T, Lozano, R, Tempier O 2015 *Int. J. of Adv. Robotic Systems* **12** 1-15
[4] Cui R, Zhang X and Cui D 2016 *Ocean Engineering* **123** 45-54
[5] Fang M-C, Hou C-S and Luo J-H 2007 *Ocean Engineering* **34** (8-9) 1275-89
[6] Shilin A A. and Bukreev V G 2014 *Thermal Eng.* **61** (10) 741-746
[7] Vasileva O V, Budko A A and Lavrinovich A V 2016 *IOP Conf. Ser.: Mater. Sci. Eng.* (Tomsk) vol 124 (Bristol: IOP Publishing Ltd.) 012107
[8] Koran A, Labella T, Lai J-S 2014 *IEEE Transact. on Power Electr.* **29** 1285-1297
[9] Bychkov P N, Zabrodina I K, Shlapak V S 2016 *IEEE Transact. on Dielectrics and Electr. Ins.* **23** 288-293
[10] Kuleshova E O, Plyusnin A A, Shandarova E B, Tikhomirova O V 2016 *IOP Conf. Ser.: Mater. Sci. Eng.* (Tomsk) vol 124 (Bristol: IOP Publishing Ltd.) 012069

Method of technological control of proportional electromagnets for control systems of their production

A M Lankin, M V Lankin and M Y Lankina

Federal State Educational Institution of Higher Professional Education 'South-Russian State Technical University of the M I Platov', 132, Prosvesheniya str., Novocherkassk, 346428, Russia

E-mail: lankinjohn@yandex.ru

Abstract. This article describes the algorithm of the method of technological control of proportional solenoids to control their production process. An important element in the process of production of proportional solenoids is to identify deviations from the nominal mode of its occurrence. This situation shows the relevance of developing a method of technological control of proportional solenoids providing information on process control in a short time. The proposed algorithm consists of four steps: measuring the dynamic characteristics of the magnetization of electromagnet testing, the use of principal component analysis to reduce the dimension of the analyzed information, the classification of products by the condition and sub-standards, the definition of the numerical values of the manufacturing process variations using a calibration method. This approach allows us to determine the type and the degree of impairment of the mode of the process that will lead to effective management of the production process of the electromagnetic actuators.

1. Introduction

Identification of deviations from the nominal flow regimes of the process of proportional solenoid production is an integral part of enhancing the quality of products. This requires an effective method of process control, which allows determining the place and the degree of deviation from the nominal values of the process conditions for further management [6].

An important element of the production process is the management system of proportional solenoids production, it was the system that makes up the management modes of production equipment. We will consider the application of the developed method of process control as part of production process control system. The proposed block diagram of the process control system of production of proportional solenoids is shown in Figure 1. It contains dynamic characteristics of the magnetization determination device (MDD) [5, 8] and the information processing apparatus for generating the control action.

The detection device measures the dynamic characteristics of the magnetization of each electromagnet DCM and transmits the measurement result to the information processing apparatus and generates the control action. This in turn selects articles having deviations in operating parameters, and thus carries the information about the violation of the manufacturing process. These deviations are detected by means of a mathematical model obtained by the calibration method that connects the operating parameters with the parameters of the electromagnet and its manufacturing process. In addition, the control action processing unit collects and analyzes the information on the change of technological modes in case of the appearance of sustainable trends in their change process, which generates a control signal to eliminate the possibility of visiting lots of sub-standard products.

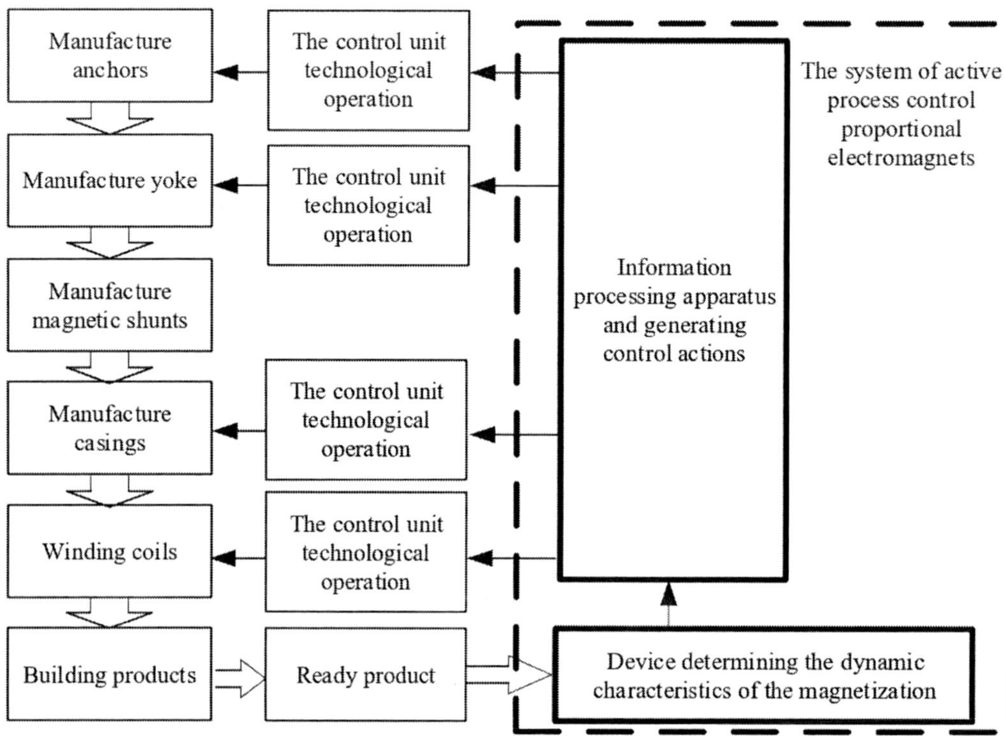

Figure 1. Block diagram of process control system production of proportional electromagnets.

Controllable process parameters are the deviation of the magnetic characteristics of the magnetic material parts of the electromagnet (casing, armature, yoke), arising from violations of their modes of machining and the number of operating winding turns. They are controlled by adjusting the process operating modes of winding equipment, and equipment machining workpieces annealing or adding operations.

2. Methods and results

The proposed method of process control involves four basic steps. The first step is measuring DCM of the electromagnetic drive, because it contains latent information on the majority of the electromagnetic parameters of the drive. The fact that the dynamic characteristics of the magnetization of the electromagnetic actuator has a complex character is not unique. In the second step, we reduce the dimension of the analyzed data using the projection method. In the third step, we carry out classification of groups of measured characteristics to highlight products that carry information on deviations of sustainable production process of proportional solenoids. In the final step, the numerical values are determined by deviations from the normal process with the help of calibration.

Application of DCM in an array of pairs of points (magnetic flux current) for further processing is difficult, because the use of large amounts of data (DCM measured with an accuracy of ± 3%, contains about 15,000 pairs of points) places high demands on the processing performance of funds. In this connection, it is advisable to obtain the measurement result in the form of approximating expressions, allowing one to reduce the processing time measurement information.

Let us consider obtaining step DCM. A mathematical model of the electromagnet is based on the method of harmonic balance [7]. The model is based on the equation of the circuit with a nonlinear inductance which approximate DCM to the degree of the polynomial:

$$\begin{cases} u = iR + \dfrac{d\Phi}{dt} \\ i(\Phi) = \sum_{1}^{m} k_q \Phi^q \end{cases},$$

where u – the voltage applied to the operating coil of the electromagnet, R– the resistance of the operation of the electromagnet coil, Φ – the value of the magnetic flux, k – approximation polynomial coefficients of the power, m – the maximum degree of the approximating polynomial.

The algorithm of natural-model tests [1-4, 9, 10] in relation to the problem of determining the proportional electromagnets of DCM is as follows: during operation, the electromagnet coil is energized, obtains a predetermined shape, and current versus time $i_{meas}(t)$ is measured. Electromagnet voltage is introduced into the mathematical model, and it generates output current $i_{model}(t)$. With the optimization algorithm (simplex optimization method) approximation coefficients k_q changes achieving coincidence of the measured $i_{meas}(t)$ and formed model of current electromagnet $i_{model}(t)$. The level of current differences $i_{meas}(t)$ and $i_{model}(t)$ determined by the value of functional J is:

$$J = \frac{\int_0^{T/2} (i_{model}(t) - i_{meas}(t))dt}{\int_0^{T/2} i_{meas}(t)dt}.$$

At each iteration, the optimization condition is tested:

$$J \le \varepsilon, \tag{1}$$

where ε – current measurement accuracy $i_{meas}(t)$.

Let us consider the adaptation in the method of a submission form of harmonic balance current. We turn to the discrete representation of functional form J:

$$J = \frac{\sum_{m=1}^{n} \left| I_{(2m-1)_{model}} - I_{(2m-1)_{meas}} \right|}{\sum_{m=1}^{n} I_{(2m-1)_{meas}}},$$

where $I_{(2m-1)_{meas}}$ and $I_{(2m-1)_{model}}$ – harmonic measurements ($2m$-1) and calculations of the current model in the coil proportional solenoid.

If condition (1) is not satisfied, then the parameters of the mathematical model change, defining the shape of dynamic characteristics of the magnetization, newly defined relationship $i_{model}(t)$ and performed functional calculation J. If condition (1) is satisfied, the dynamic characteristic of the magnetization proportional solenoid is found.

Due to the fact that an electromagnet is proportional to DCM complex hysteresis, the second step is to reduce the dimension of the analyzed information. For this, a projection applied approach uses principal component analysis (PCA). The essence of the method of the principal component consists of a transition from the original variables to new values - the principal components (PC), which are linear combinations of the original variables, with the maximum possible dispersion.

The application of the PCA to the group comprising k DCM of proportional solenoid actuators. On each original DCM, n fixed values of magnetic flux $\Phi_d = d \cdot \Delta\Phi$ are selected:

$$\Delta\Phi = \Phi_{max}/n,$$

where Φ_{max} – the maximum possible flow value for all studied characteristics, n – the selected number of fixed flow values, d – the point number.

Determined current values $i_j(\Phi_d)$, which are the elements of vector-column matrix **I** dimensionality $n \times k$, where n is the number of recorded pixels, k is the number of the test curves:

$$\mathbf{I} = \begin{bmatrix} i_1(\Phi_1) & i_2(\Phi_1) & \cdots & i_j(\Phi_1) & \cdots & i_k(\Phi_1) \\ i_1(\Phi_2) & i_2(\Phi_2) & \cdots & i_j(\Phi_2) & \cdots & i_k(\Phi_2) \\ \vdots & \vdots & & \vdots & \vdots & \vdots \\ i_1(\Phi_d) & i_2(\Phi_d) & \cdots & i_j(\Phi_d) & \cdots & i_k(\Phi_d) \\ \vdots & \vdots & & \vdots & & \vdots \\ i_1(\Phi_n) & i_2(\Phi_n) & & i_j(\Phi_n) & & i_k(\Phi_n) \end{bmatrix}.$$

Application of polynomial approximation of DCM allows us to calculate the missing values by interpolation vectors $i_j(\Phi_d)$, if Φ_{jmax} less than Φ_{max}.

The connection between original matrix **I** and matrix values of principal components **Z** performed by a matrix of eigenvectors **B** has the form:

$$\mathbf{Z} = \mathbf{B}^T \cdot \mathbf{I}. \quad (2)$$

Projection procedure is to find the PC matrix with the largest variances. To do this, we find the matrix of variances **D** and **Z**:

$$\mathbf{D}(\mathbf{Z}) = \mathbf{B}^{\mathrm{T}} \cdot \mathbf{D}(\mathbf{I}) \cdot \mathbf{B}.$$

Considering that the dispersion of n random variables is their covariance, the previous expression can be written as follows:

$$\mathbf{D}(\mathbf{Z}) = \mathbf{B}^T \cdot S \cdot \mathbf{B}.$$

To find the covariance S vector of calculated sample means, rows $\bar{I}_d$, and deviations from the mean for each case are summarized in matrix **F**, which has formulated the elements as follows:

$$f_{dj} = I_j(\Phi_d) - \bar{I}_d, \qquad d=1 \div n, j=1 \div k.$$

Then, the estimate of covariance S is:

$$S = \frac{\mathrm{F} \cdot \mathrm{F}^{\mathrm{T}}}{k-1}.$$

Not all PC calculated according to expression (2) are significant. Based on the evaluation matrix dispersion values of **D**, we leave only (n-l) PC with the largest variances, whereupon the matrix **Z** is:

$$\mathbf{Z} = \begin{bmatrix} z_{11} & z_{21} & \cdots & z_{k1} \\ z_{12} & z_{22} & \cdots & z_{k2} \\ \vdots & \vdots & \cdots & \vdots \\ z_{1n-l} & z_{2n-l} & \cdots & z_{kn-l} \end{bmatrix}.$$

The dimension of matrix **Z** is significantly less than that of original matrix **I**, (n-l) =1÷3. As a result of this step, we get a new space of principal components, each of which is a point of particular products DCM.

The second step is the classification of manufactured products by conditions and products with deviations of parameters as conditioning products do not carry information about deviations in the process. To do this, let us turn to the DCM, projected into space PC method discrimination - formally independent modeling of class analogies (FIMCA).

The application of FIMCA to the resulting matrix is in previous step **Z**. Let us find magnitude h (difference between the highest and lowest values of observations), each column vector of matrix **Z**:

$$h=\sum_{1}^{n-l}\left(\frac{z_i}{\sqrt{\lambda_i}}\right)^2,$$

where λ_i – eigenvalues **Z**, z_i – elements of **Z**.

Using training set **Z**, we will find the p range of $h_1,\ldots,h_i$ and $v_1,\ldots,v_i$. As it is possible to estimate the corresponding average values:

$$h_0=\frac{1}{p}\sum_{1}^{p}h_i \qquad v_0=\frac{1}{p}\sum_{1}^{p}v_i,$$

and the corresponding values of the dispersions:

$$S_h=\frac{1}{p-1}\sum_{1}^{p}(h_i-h_0)^2 \qquad S_v=\frac{1}{p-1}\sum_{1}^{p}(v_i-v_0)^2.$$

The numbers of degrees of freedom N_h, and N_v:

$$N_h=\frac{2h_0^2}{S_h} \qquad N_v=\frac{2v_0^2}{S_v}.$$

Accounting ratio f, over which ranking is calculated as:

$$f=N_h\frac{h}{h_0}+N_v\frac{v}{v_0}.$$

The critical factor f is given by:

$$f_{crit}=\chi^{-2}(\alpha,N_h+N_v)$$

where $\chi^{-2}(\alpha,n)$ – α-quantile of the chi-square distribution with n degrees of freedom.

The classification rule in FIMCA: A sample is taken as belonging to the class, if:

$$f<f_{crit}.$$

In the next step of the method of process control for stable detection of deviations of the process, we get information about the numerical values of the deviations of process parameters of the manufacturing process of products. For this calibration method, we use a regression to latent structures (RLS). The method consists in obtaining a model linking matrix **Z** with DCM coordinates of points in a space projected PC and matrix parameters **Y**, characterizing the process of manufacturing modes of proportional electromagnets.

The RLS model is based on the training set ($\mathbf{Z_o}$, $\mathbf{Y_o}$), for which there is a matrix of multiple regression coefficients **A**:

$$\mathbf{A}=f(\mathbf{Z_o}\mathbf{Y_o}).$$

Determination of process parameters $\mathbf{Y}_t$ for the next electromagnet with DCM, projected into the space of the PC $\mathbf{Z}_t$, we use mathematical models of RLS:

$$\mathbf{Y}_t=\mathbf{Z}_t\mathbf{A}.$$

3. Conclusions

Thus, the developed method of process control involves four basic steps. In the first step, the dynamic characteristic are measured by magnetization of the electromagnet, since it contains the information on the most latent electromagnetic drive parameters. The dynamic characteristics of the magnetization of the electromagnetic actuator has a complex character are not unique; in the second step, we reduce the dimension of the analyzed data using the projection method. In the third step, we carry out classification of the groups of measured characteristics to highlight products that carry information on sustainable production process deviations proportional to solenoids. In the final step, the numerical values are determined by deviations from the normal process with the help of calibration. Application of the proposed method of technological control of proportional solenoids control systems as part of their production will improve the quality and quantity of proportional electromagnets.

Acknowledgment
The results obtained with the support of the grant RFBR № 15-38-20652 "Development of the theory of predictive methods for sensorless control and diagnosis of electric drives" using equipment center for collective use "Diagnosis and energy-efficient electrical equipment" (NPI).

References

[1] Bakhvalov Y A, Gorbatenko N I and Grechikhin V V 2015 *Measurement Techniques* **58** 336-340
[2] Borovoi V V, Gorbatenko N I and Grechikhin V V 2015 *Measurement Techniques* **56** 614-617
[3] Gorbatenko N I, Grechikhin V V and Shaikhutdinov D V 2015 *Measurement Techniques* **56** 609-613
[4] Gorbatenko N I and Lankin M V 2004 *Russian Electrical Engineering* **8** 63-66
[5] Gordon A V and Slivinskaya A G 1960 Electromagnets DC power (Moscow: State Energy Publishing)
[6] Il'enkova S D, Il'enkova N D and Mhitaryan V S 2000 Management of quality (Moscow: UNITY)
[7] Lankin A M, Lankin M V, Grechikhin V V and Shayhutdinov D V 2015 *Research J. of Appl. Sci.* **10** 690-695
[8] Slivinskaya A G 1972 Electromagnets and permanent magnets (Moscow: Energy)
[9] Shaykhutdinov D V, Gorbatentko N I, Akhmedov Sh V, Shaykhutdinova M V and Shirokov K M 2013 *Life Science J.* **10** 2698-2702
[10] Shaikhutdinov D V, Gorbatenko N I and Shirokov K M 2015 *Measurement Techniques* **56** 618-620

Approximation of dynamic characteristics of the magnetization for electromagnets

A M Lankin, M V Lankin and M Y Lankina

Federal State Educational Institution of Higher Professional Education 'South-Russian State Technical University named after M.I. Platov', 132, Prosvesheniya Str., Novocherkassk, 346428, Russia

E-mail: lankinjohn@yandex.ru

Abstract. The article describes the use of piecewise-polynomial approximation and approximation in the form of a power polynomial to the dynamic characteristics of the magnetization. The feasibility of this approach consists in the reduction of the amount of data used to describe the characteristics. For clarity, only the ascending branch of the dynamic characteristics of the magnetization was used. The error, arisen during approximation of the characteristic which according to the results of experimental studies did not exceed the required 5%, was estimated. As a result of the studies, it is possible to conclude about the appropriateness of approximating expressions application to describe the dynamic characteristics of the magnetization. Approximation defined as a power polynomial with the same error allows fewer approximating coefficients.

1. Introduction

Process inspection of the proportional solenoids manufacturing begins with the obtainment of the characteristic, containing information about the technical parameters depending on the mode of the process. The definition of such characteristics should be carried out based on the fully assembled product, and must be low-cost in terms of time and technical resources. This integrated feature is the dynamic characteristic of the magnetization (DCM) [6, 10] (Figure 1). It is possible to calculate the number of characteristics regulated by GOST: static and dynamic traction $F=f(\delta)$ and $F_d=f(\delta)$, respectively; armature movement characteristics $\delta=f(t)$ and changing the current in the winding with time $i=f(t)$ [1-4, 8, 9].

After the electric power supply, the coil current reaches the pick-up current, which corresponds to point 1. At this instant, the armature is in motion, during which the working air gap is reduced, the winding inductance increases and the current drops until the armature is attracted to the core, to which corresponds point 2. During the movement of the anchor, the relationship between flux F and current I is determined by curve 1-2.

At the end of the motion of the armature, current starts to increase again, reaching a steady-state value in point 3. After disconnecting the power supply of the electromagnet, the current in the coil falls, and when reaching the value of drop-out current I_{rel}, (point 4), the anchor is set in motion, and the working air gap is increased from minimum δ_{min} to maximum δ_{max}. Point 5 corresponds to the end point of the motion armature. Then, the current drops to zero, which corresponds to point 6. Because of the residual magnetization of the magnetic conductor, the flux corresponding to point 6 is above zero.

The dynamic characteristic of the magnetization of the electromagnetic actuator carries information about the magnetic, electrical, traction and dynamic properties of proportional electromagnet, and for its obtainment, it is possible to use so-called 'non-sensor' devices. It does not require disassembly of the test product and the currents, equivalent to the nominal ones, flow along the working coil [5].

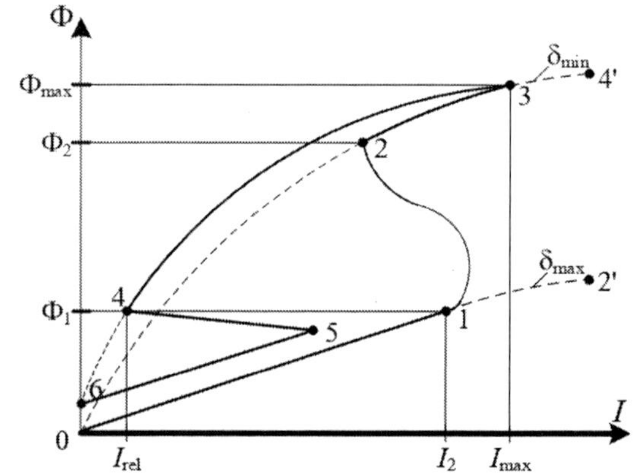

Figure. 1. DCM of the electromagnetic drive.

The use of the DCM in the form of an array of points pairs of the magnetic flux current for further processing is difficult. A large amount of data (one DCM measured with an accuracy of up to 3% contains about 15.000 pairs of points) places high demands for the performance of processing facilities and increases the computation time. In this connection, it is advisable to use this approximation characteristics.

2. Materials and methods

Let us consider the use of two of the most optimal approximations for the description of DCM: the piecewise-polynomial approximation and the approximation by power polynomial [7]. For clarity, we will continue to consider only the ascending branch of the DCM, i.e. plots 0-1, 1-2, 2-3. Because of the ambiguity of DCM (one value of current response may take several flux values), let us consider not dependence $\Phi(i)$, but reverse characteristic $i(\Phi)$.

The method of piecewise-polynomial approximation consists in the replacement of sections 0-1, 1-2, 2-3, 3-4, 4-5, 5-6 DCM with approximating curves described by polynomials of the following type:

$$i^{(q)} = \sum_{1}^{m} k_n^{(q)} \Phi^n + b^{(q)},$$

where k_n – coefficients describing the slope and the bends of the curve, b – the coefficient describing the displacement of the curve with respect to the x-axis, m – the maximum power of the polynomial, q – the number of the approximated area.

The first and second areas of DCM can be described by the first degree polynomial. The first comes from the coordinate origin. Therefore, factor $b^{(1)}$=0:

$$i^{(1)} = k_1^{(1)} \Phi .$$

The second area of DCM looks like:

$$i^{(2)} = k_1^{(2)} \Phi + b^{(2)} .$$

The third portion may be described by a polynomial of the second or third degree:

$$i^{(3)} = k_1^{(3)} \Phi + k_2^{(3)} \Phi^2 + b^{(3)}, \qquad i^{(3)} = k_1^{(3)} \Phi + k_2^{(3)} \Phi^2 + k_3^{(3)} \Phi^3 + b^{(3)}$$

Thus, to describe the ascending branch of the DCM using piecewise polynomial approximation takes six or seven approximation coefficients and the four values of the flow and the current at the switching points of approximation segments.

The approximation method by the power polynomial consists in the replacement of DCM throughout its whole period by the power polynomial of the *n-th* degree of the type:

$$i = \sum_{1}^{m} k_n \Phi^n .$$

The advantage of this method is to use a smaller number of values required to describe the DCM. For example, for the same DCM, with the approximation error not more than 5%, for approximation by power polynomial, six coefficient values are required, and for a piecewise polynomial approximation — eleven coefficient values.

3. Results

We conducted research on the approximation of the selected type of DCM $i(\Phi)$, shown in Figure 2.

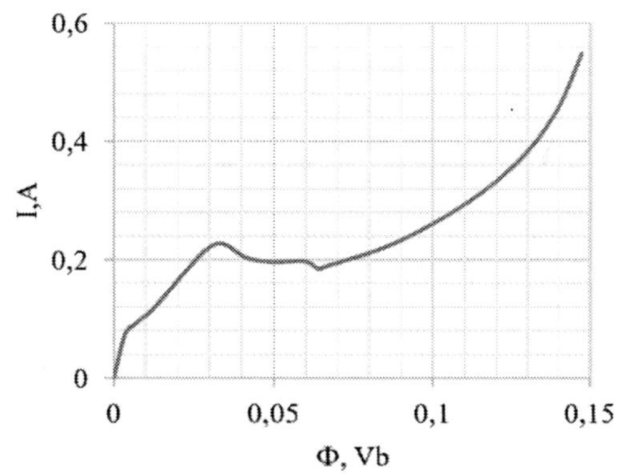

Figure. 2. DCM of $i(\Phi)$ type of proportional electromagnet

We used the piecewise polynomial approximation using approximations of the third area with the second and third degrees (Figure 3).

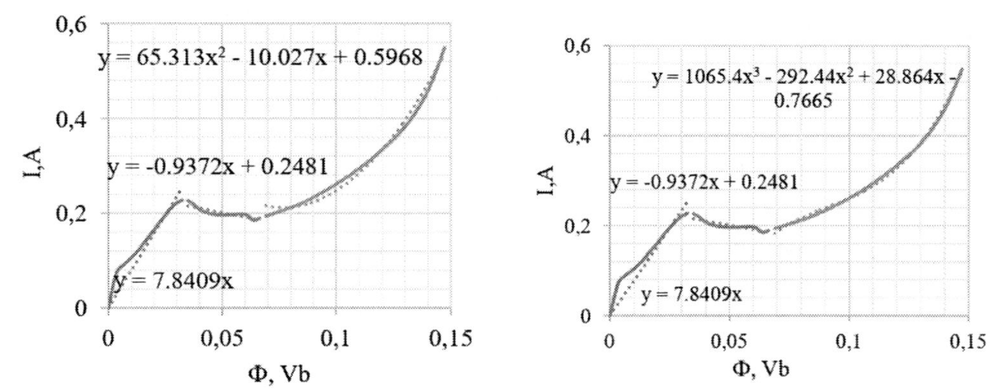

Figure 3. Piecewise-polynomial approximation of DCM of $i(\Phi)$ type

The error in the application of a second degree of the polynomial was 8%, and when using the third degree of the polynomial – 5% , which meets the requirements for magnetic measurements.

Let us consider the use of a polynomial approximation shown in Figure 2. We use a power polynomial of the sixth degree for describing DCM (Figure 4).

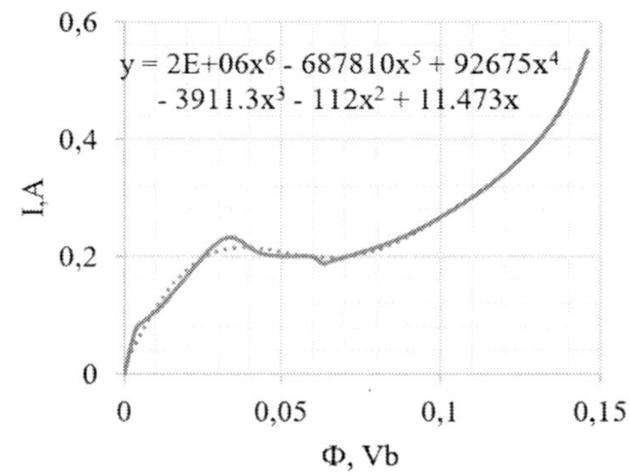

Figure 4. Polynomial approximation of DCM of $i(\Phi)$ type

When using a sixth-degree polynomial power, the error did not exceed 5%. However, when using a power polynomial for description of DCM, we need a less number of parameters than when using a piecewise polynomial approximation with the same accuracy. Thus, we can conclude about the possibility of using both types of approximation to describe the DCM.

4. Conclusions

The application of approximating expressions for description of dynamic characteristics of magnetization was studied. The studies allow concluding on the appropriateness of approximating expressions to describe the dynamic characteristics of the magnetization. The use of this approach will reduce the amount of measurement information without its losses, which will positively affect the speed of data processing. Both studied approximations demonstrated the error of the description of DCM of no more than 5%. However, a more efficient is the use of the power polynomial for describing DCM, since it allows using a small number of parameters than when using piecewise polynomial approximation.

Acknowledgment

The results obtained with the support of the project № 2.7193.2017 / BP performed within the base of the state task using equipment center for collective use "Diagnosis and energy-efficient electrical equipment" (NPI).

References

[1] Bakhvalov Y A, Gorbatenko N I and Grechikhin V V 2015 *Measurement Techniques* **58** 336-340
[2] Borovoi V V, Gorbatenko N I and Grechikhin V V 2015 *Measurement Techniques* **56** 614-617
[3] Gorbatenko N I and Lankin M V 2004 *Russian Electrical Engineering* **8** 63-66
[4] Gorbatenko N I, Lankin A M, Lankin M V and Shayhutdinov D V 2015 *Int. J. of Appl. Eng. Res.* **3** 6509-19
[5] Gorbatenko N I, Grechikhin V V and Shaikhutdinov D V 2015 *Measurement Techniques* **56** 609-613
[6] Gordon A V and Slivinskaya A G 1960 Electromagnets DC power (Moscow: State Energy Publishing)
[7] Lankin A M, Lankin M V, Aleksanyan G K and Narakidze N D 2015 *Int. J. of Appl. Eng. Res.* **3** 6489-98
[8] Shaykhutdinov D V, Gorbatentko N I, Akhmedov Sh V, Shaykhutdinova M V, Shirokov K M 2013 *Life Sci. J.* **10** 2698-2702
[9] Shaikhutdinov D V, Gorbatenko N I and Shirokov K M 2015 *Measurement Techniques* **56** 618-620
[10] Slivinskaya A G 1972 Electromagnets and permanent magnets (Moscow: Energy)

Modelling of flame propagation in the gasoline fuelled Wankel rotary engine with hydrogen additives

E A Fedyanov, E A Zakharov, K V Prikhodkov and Y V Levin

Volgograd State Technical University, 28, Lenin Avenue, Volgograd, 400005, Russia

E-mail: tig@vstu.ru

Abstract. Recently, hydrogen has been considered as an alternative fuel for a vehicles power unit. The Wankel engine is the most suitable to be adapted to hydrogen feeding. A hydrogen additive helps to decrease incompleteness of combustion in the volumes near the apex of the rotor. Results of theoretical researches of the hydrogen additives influence on the flame propagation in the combustion chamber of the Wankel rotary engine are presented. The theoretical research shows that the blend of 70% gasoline with 30% hydrogen could accomplish combustion near the T-apex in the stoichiometric mixture and in lean one. Maps of the flame front location versus the angle of rotor rotation and hydrogen fraction are obtained. Relations of a minimum required amount of hydrogen addition versus the engine speed are shown on the engine modes close to the average city driving cycle. The amount of hydrogen addition that could be injected by the nozzle with different flow sections is calculated in order to analyze the capacity of the feed system.

1. Introduction

The main advantages of the rotor-piston engine, also known as the Wankel engine, include higher power-to-weight and power-to-size ratios than those of traditional engines. It is simpler, more compact and has less components, almost balanced and has lower vibration [1, 2].

Thereby Wankel engines may be considered as an alternative to the petroleum-fueled internal combustion engines (ICE) such as two-stroke ones for boats or four-stroke ones for general aviation [3] and hybrid vehicles [4-6].

However, the Wankel engine has a higher level of unburned hydrocarbons and fuel consumption. One of the reasons is a combustion incompleteness of the fuel-air mixture near the rotor's T-apex due to one direction flow. The twin spark plug ignition system allows improving combustion, but it cannot ensure that all of the mixture is burnt [3, 7, 8].

It can be assumed that increasing the flame speed in the combustion chamber could improve burning of the air-fuel mixture. One of the ways to increase the flame speed of the conventional fuel is addition of hydrogen [9, 10].

Hydrogen has the highest combustion heat and flame speed than any other hydrocarbon fuels. Main properties of hydrogen in comparison with other fuels are listed in table 1 [9-13]. In addition, it has a small quenching distance so it would be burned near the apex of the rotor. Moreover, many researchers point that the Wankel engine is more suitable for hydrogen feeding than the piston engine due to less probability of pre-ignition or backfire [6, 14]. However, hydrogen has low density, so the use of pure hydrogen is limited by difficulties of on board storage or production [15-17]. It seems that addition of a hydrogen portion to the main fuel could resolve those problems [14].

Table 1. Properties of fuel

Property	Hydrogen	Gasoline	LPG	Methane
Density [kg/m^3]	0.089	750	2.1	0.72
Autoignition temperature [C]	510	350-450	500	436
Maximum laminar flame speed [cm/s]	275	30	46	37
Lower heating value (LHV) [Mj/kg]	120	44	46	50
Stoichiometry [kg air/kg fuel]	34.3	14.9	15.1	17.2
Diffusion coefficient [cm^2/s]	0.63	0.06	0.08	0.16
Quenching distance [mm]	0.64	2.51	1.95	2.03

2. Model details

Theoretical investigation of combustion incompleteness has been carried out with multizone modeling of flame propagation.

The combustion chamber of the Wankel engine has a compound shape, and, as a rule, it is equipped with two spark plugs, so we have to divide the inner space into five zones (figure 1): three of them are with the unburned mixture (marked as u_1, u_2 and u_3) and the rest ones – with burned gases (marked as b_1 and b_2). The zone of burned gases has expanded by the moving of flame front through the unburned mixture.

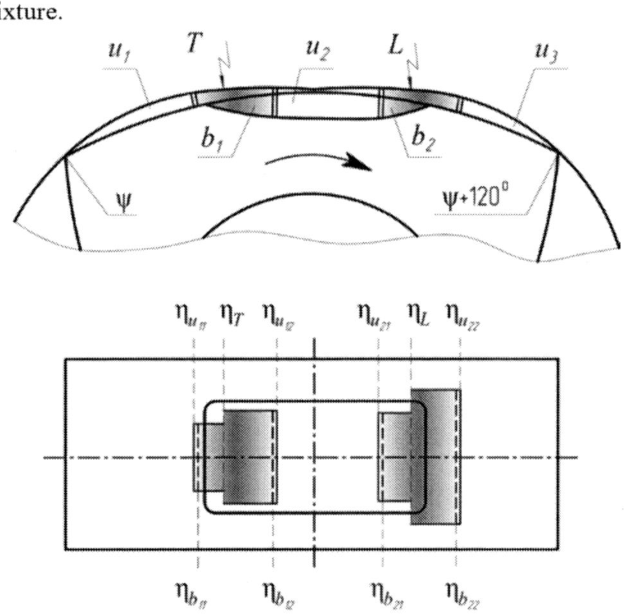

Figure 1. The plane of flame distribution in the Wankel engine: ψ – a T-apex angular coordinate; η_T, η_L – spark kernel angular coordinates; η_{u11}, η_{u12}, η_{u21}, η_{u22} – flame front angular coordinates; η_{b11}, η_{b12}, η_{b21}, η_{b22} – angular coordinates of burned gaseous zones.

It is considered that the flame is expanded from the spark plug in both directions: in the longitudinal one and in the cross one. Moreover, each of zones is subdivided into two subzones that are bordered by the flame front. One of them moves to the T-apex, another one moves to the L-apex.

The model is based on the following system of equations:

The energy-conservation equation for the initial zone with unburned gases:

$$\frac{k_{ui}}{k_{ui}-1}p\frac{dV_{ui}}{d\psi}+\frac{1}{k_{ui}-1}V_{ui}\frac{dp}{d\psi}=c_{pui}\cdot T_{ui}\frac{dm_{ui}}{d\psi}-\frac{dQ_{Wui}}{\omega}. \quad (1)$$

The energy-conservation equation for the zone with combusted gases:

$$\frac{k_{bj}}{k_{bj}-1}p\frac{dV_{bj}}{d\psi}+\frac{1}{k_{bj}-1}V_{bj}\frac{dp}{d\psi}=\left(Q_{LHV}-Q_D\right)\frac{dm_{bj}}{d\psi}-\frac{dQ_{Wui}}{\omega}+$$
$$+c_{puj}\cdot T_{uj}\frac{dm_{bj1}}{d\psi}+c_{pu(j+1)}\cdot T_{u(j+1)}\frac{dm_{bj2}}{d\psi} \quad (2)$$

The equations of the state for the initial zone with unburned gases:

$$p\cdot V_u=m_u\cdot R_u\cdot T_u, \quad (3)$$

and the zone with combusted gases:

$$p\cdot V_b=m_b\cdot R_b\cdot T_b. \quad (4)$$

The volumes conservation equation is

$$\sum_1^3\frac{dV_{ui}}{d\psi}+\sum_1^2\frac{dV_{bj}}{d\psi}=\frac{dV_{\Sigma(\psi)}}{d\psi}, \quad (5)$$

and the mass conservation equation is

$$\sum_1^3 m_{ui}+\sum_1^2\sum_1^2 m_{bji}=m_a, \quad (6)$$

where i and j – is the zone index; p – is the pressure; T_{ui}, V_{ui}, m_{ui}, R_u and T_{bi}, V_{bj}, m_{bj}, R_b – is the temperature, volume, mass and specific gas constant for the initial zone with unburned gases and the area with combusted gases, respectively; Q_{LHV} and Q_D – combustion heat and the heat of dissociation, respectively; Q_W – heat losses in the walls of the combustion chamber; k and c_p – the adiabatic exponent and specific heat at constant pressure, respectively; ψ – the angle of rotation of the rotor; ω – the angular speed of the rotor; $V_{\Sigma(\psi)}$ – the instant volume of the combustion chamber; m_a – the mass of the fresh mixture in the combustion chamber.

To take into account the influence of hydrogen additives, thermal properties of equations of unburned and burned gas were adapted. The lower heating value of fuel was calculated as

$$Q_{LHV}=Q_{CH}\cdot(1-g_{H_2})+Q_{H_2}\cdot g_{H_2}, \quad (7)$$

where Q_{CH}, Q_{H_2} - lower heating values of gasoline and pure hydrogen, respectively; g_{H_2} - the mass fraction of hydrogen.

The stoichiometric air-fuel ratio was calculated as

$$AF=AF_{CH}\cdot(1-g_{H_2})+AF_{H_2}\cdot g_{H_2}, \quad (8)$$

where AF_{CH}, AF_{H_2} - the stoichiometric ratios of gasoline and pure hydrogen, respectively.

The system of equations (1) - (6) could be solved with the Runge-Kutta method. Moreover, the system of the equations is complemented by equations for calculation of the shape of the combustion chamber, turbulent characteristics and the combustion duration of turbulent moles dipped in the flame front. The growth of the combusted gases volume and hence unburned gases volume decrease is determined by simulating the turbulent flame moving. The model called "dipping" has been used to describe flame front propagation [8, 18]. This model suggests that the movement of the flame front depends on turbulent characteristics of the incoming unburned mixture flow. Combustion of the mixture occurs in a great number of laminar fronts with the surfaces.

Experimental data [11, 12] of the laminar flame speed of gasoline-air and hydrogen-air mixtures are also used. To determine laminar flame speed w_{nm} of the gasoline-air mixture with hydrogen addition, the following equation is used:

$$w_{nm}=w_{nCH}\cdot(1-g_{H_2})+w_{nH_2}\cdot g_{H_2}, \quad (9)$$

where w_{nCH}, w_{nH_2} – the flame speed of gasoline and pure hydrogen, respectively.

3. Results and discussion

3.1. Flame propagation

The combustion chamber of the Wankel engine VAZ-311 was used in calculation. The engine technical specifications are listed in table 2.

Table 2. The technical specifications of Wankel rotary engine VAZ-311

Engine manufacturer	Volga Automobile Plant, Russia
Number of rotors	Single rotor
Cooling	Water cooled
Ignition source	Two spark plugs per rotor
Working volume of each chamber	654 cm^3
Compression ratio	9.3
Width of rotors	80 mm
Eccentricity	15 mm
Maximum power output	52 kW at 6000 rpm
Maximum torque	95 Nm at 4000 rpm

The mode of the engine is a full throttle at a speed of the eccentric shaft equal to 2000 rpm and the equivalence ratio is 1. The spark advance is 26 deg for the L-spark plug and 30 deg — for the T-spark plug.

Theoretical research shows that the hydrogen additive allows fixing combustion completeness near the T-apex both in the stoichiometric and poor mixtures. Figure 2 shows the flame front propagation ignited by two spark plugs versus the amount of hydrogen. The hydrogen mass fraction in the total fuel mass varies from 0% to 40%. The additive of 20% of hydrogen causes the reduction of the combustion duration. Additives of 30% and more of hydrogen have allowed the flame front to achieve the T-apex before the power stroke is finished. Let us note that the additive of hydrogen has a negligible influence on the flame velocity in the direction to the L-apex. From our point of view, it depends on the mixture flow speed in the combustion chamber.

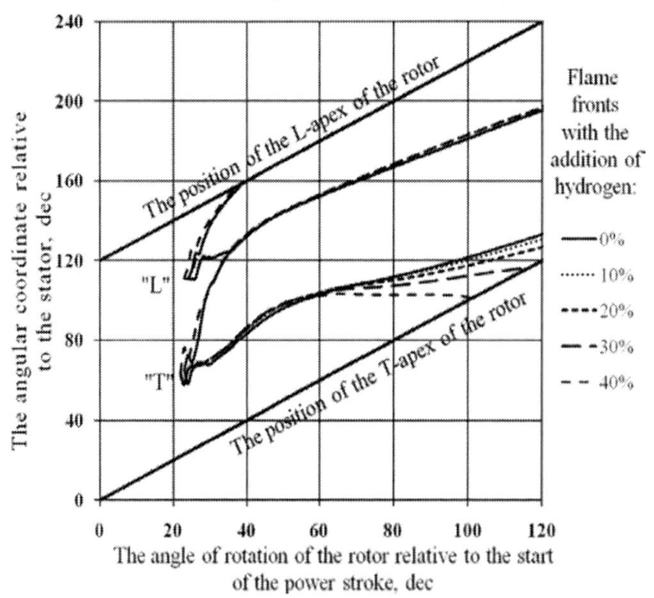

Figure 2. Propagation of the flame front versus hydrogen additives (λ=1; n=2000 rpm).

Hydrogen feeding in the Wankel engine can be realized by means of distributed injection into the inlet manifold [14, 15, 19, 20]. It is obvious that the amount of hydrogen addition needed for perfect combustion is caused by the mode of operation of the engine. It is illustrated in figure 3 where the relations of the minimum required amount of hydrogen addition versus the engine speed are shown on the engine modes close to the average urban driving cycle ($n \leq 3000$ rpm; $\varphi_{th} \leq 30\%$). It is noticed that the engine speed has a stronger influence than the load mode does.

We assume that it is due to the fact that the average speed of the mixture relative to the stator is increased with the engine speed so the flame front propagation on the rear apex of the rotor is hampered.

Hydrogen has the least density than other gases, so the implementation of the engine operation with the hydrogen additives is connected with special requirements for the feeding system [15]. The feed system must provide supply of the required quantity of hydrogen in the whole range of engine modes up to a full throttle. The last one is distinguished as a mode with the minimum duration of the intake, while hydrogen addition must be close to maximum. So, the amount of hydrogen addition that could be injected by the nozzle with different flow sections is calculated in order to analyze the capacity of the feed system.

3.2. Injection time

In figure 4, both the theoretical injection duration of hydrogen through the nozzle with a 1.2 mm^2 flow section and the duration of the intake are plotted versus the speed of the eccentric shaft. An excessive pressure is taken to be equal to 300 kPa that corresponds to the vehicle commercial onboard generator of hydrogen [15, 17].

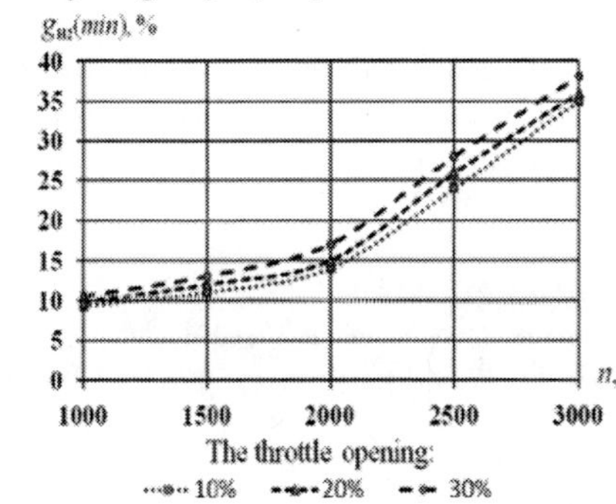

Figure 3. The minimum required amount of hydrogen addition in different driving modes.

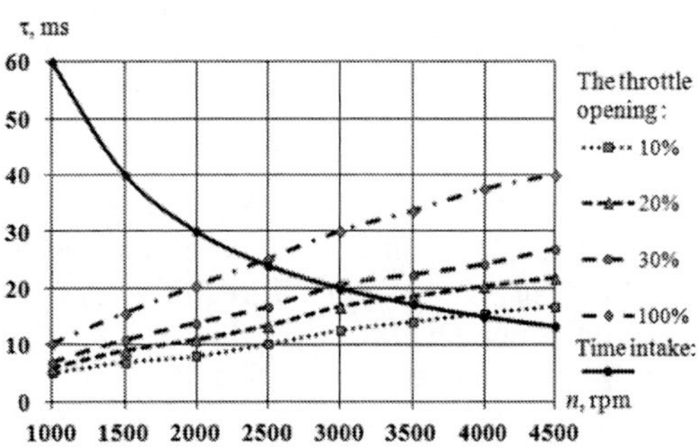

Figure 4. The intake duration and duration of hydrogen injection versus operation modes.

As we can see, the duration of the intake decreases significantly as the engine speed increases so it leads to complication of injection of a great amount of hydrogen. The nozzle that we use allows injecting the required quantity of hydrogen in a wide range of operation modes from almost idle load to the average urban driving mode. The intersections of the intake duration line with each of the lines of the hydrogen injection duration point on the limiting engine speed provide the required quantity of hydrogen and perfect combustion of mixtures. For example, in a 30% throttle opening mode, the implemented nozzle provides minimum hydrogen addition at an engine speed of up to 3000 rpm.

In some operation modes, the required hydrogen injection duration is shorter than that of the intake duration. In this case, the hydrogen injection must be started closer to the end of the intake stroke. It would increase the concentration of hydrogen near the T-apex where the incompleteness of combustion is observed.

It was calculated that the nozzle with the flow section of 4 mm^2 could provide necessary hydrogen feeding in all the operation modes of the investigated engine. However, the use of this nozzle is linked with the problem of the dispensing accuracy of the injection in the idle mode and with low loads. So, the solution is the implementation of the twin nozzles feeding system. One of them would operate in all modes, and the other one would be joined only in a high speed mode and a full throttle.

4. Conclusion

The obtained results show that hydrogen additives to gasoline-air mixtures can improve completeness of combustion in the quenching area near the T-apex.

Hydrogen addition can be realized by means of injection into the intake manifold through the twin-nozzle feeding system. During the study, it turned out that a number of factors, such as the operation mode, the nozzle capacity, the rate of the minimum hydrogen addition and the duration of the intake process should be taken into consideration.

After the stock-list of nozzles had been looked through, it was found out that the nozzle for LPG or methane – could be used to fuel the Wankel engine with the same parameters by hydrogen.

References

[1] Chen H, Pan C, Xu X, Zhang X and Xu H 2015 *Conf. on Ind. Electr. and Appl.* (London: Atlantis Press) p 180
[2] Thompson G J, Wowczuk Z S and Smith J E 2003 *SAE Technical Paper* **01** 3206
[3] Boretti A 2015 *Advances in aircraft and spacecraft science* **2(4)** 445-467
[4] Ribau J, Silva C, Brito F P and Martins J 2012 *En. Conversion and Management* **58** 120-133
[5] Butti A and Site V D 1995 *Society of Automotive Engineers (United States)* p 292
[6] Salanki P A and Wallace J S 1996 *SAE Technical Paper* **02** 32-35
[7] Ohkubo M, Tashima S, Shimizu R, Fuse S and Ebino H. 2004 *SAE Technical Paper* **01** 1790
[8] Danieli G A, Keck J C and Heywood J B 1978 *SAE Technical Paper* **04** 16-21
[9] Wang S, Ji C and Zhang B 2010 *Energy* **35(12)** 4754-60
[10] Shivaprasad K V, Raviteja S, Chitragar P and Kumar G N 2014 *Procedia Techn.* **14** 141-148
[11] Gelfand B E, Silnikov M V, Medvedev S P and Khomik S V 2012 *Springer Science & Business Media* p 326
[12] Glassman I, Yetter R A and Glumac N G 2014 *Academic press* p 619
[13] Heywood J B 1988 (New York: Mcgraw-hill) p 930
[14] Amrouche F, Erickson P, Park J and Varnhagen S 2014 *Int. J. of Hydrogen Ener.* **39(16)** 8525-34
[15] Verhelst S and Wallner T 2009 *Progress in Energy and Combustion Science* **35(6)** 490-527
[16] Cipriani G, Di Dio V, Genduso F, La Cascia D, Liga R, Miceli R and Galluzzo G R 2014 *Int. J. of Hydrogen Ener.* **39(16)** 8482-94
[17] Basile A and Iulianelli A 2014 *Elsevier* p 557
[18] Blizard N C and Keck J C 1974 *SAE Technical Paper* **01** 91-96
[19] White C M, Steeper R R and Lutz A E 2006 *Int. J. of Hydrogen Ener.* **31(10)** 1292-1305
[20] Zhao Z Y and Liu F S 2010 *In Optoelectronics and Image Processing* **1** 530-533

IOP Conf. Series: Materials Science and Engineering **177** (2017) 012077 doi:10.1088/1757-899X/177/1/012077

Topology optimization of a gas-turbine engine part

R N Faskhutdinov[1,2], A S Dubrovskaya[1,2], K A Dongauzer[1,2] P V Maksimov[1] and N A Trufanov[1]

[1] Perm National Research Polytechnic University, 29, Komsomolsky ave., Perm, 614990, Russian Federation
[2] Aviadvigatel, 93, Komsomolsky ave., Perm, 614990, Russian Federation

E-mail: dubrovskaya-as@avid.ru

Abstract. One of the key goals of aerospace industry is a reduction of the gas turbine engine weight. The solution of this task consists in the design of gas turbine engine components with reduced weight retaining their functional capabilities. Topology optimization of the part geometry leads to an efficient weight reduction. A complex geometry can be achieved in a single operation with the Selective Laser Melting technology. It should be noted that the complexity of structural features design does not affect the product cost in this case. Let us consider a step-by-step procedure of topology optimization by an example of a gas turbine engine part.

1. Introduction

For aerospace industry, the reduction in the part weight allows for fuel savings and the pollutant emissions reduction. The reduction in the parts weight and consequently the reduction in the engine weight bring down the cost of its life cycle. Thus, the main task in this industry is designing parts with lower weight while maintaining their utility. Up-to-date software allows for the part geometry to be optimized under known loading conditions, which results in effective weight reductions. However, manufacturing of optimized structures was impractical and often impossible in the past, as it is labour intensive and involves considerable investments in order to provide a complex geometry with conventional production methods of aircraft parts. Subsequent machining further increases a finished part price. The method of selective laser melting (SLM) allows providing a structure (as well as a bionic one) with a complex profile, while the geometry being complex does not influence the prime cost of a finished product. Besides, for SLM, minimum machining is required.

The ideas exchange between biology and engineering is an important trend in development of the optimization methods [1]. Principles of live forms organizations embodied into engineering devices were named biomimetrics. Many mathematical algorithms of topological optimization are now based on the equivalent ‘bionic‘ principles, with these algorithms allowing for the structure topology and shape with the best characteristics to be designed under the given forms of this structure loading. Thus, the ‘bionic design’ operation will be thought of as parts design based on the methods and algorithms of topological optimization.

2. Topology optimization of a structure

2.1 Target Functions

Optimization shall mean a process of choice of the best option imaginable. The choice of the better decision or comparison of the two alternative decisions is performed with some dependent values (a function) defined by the design data. This value is called a target function. While the optimization task being solved, the values of design data were found at which a target function has a minimum. It worth noting that a target function may not always be formulary. Sometimes it can take only some of the values and be defined in a tabular form, etc. In all the cases, it must be a single function of the design data. There may be several target functions. For example, the aircraft engines design requires that maximum reliability, minimum materials consumption and maximum payload volume (or load-lifting capacity) be provided simultaneously. Some target functions may appear to be incompatible. In such cases, a priority of one target function or another shall be incorporated.

2.2 Essence of the Approach

Topology optimization includes identification of the number, the shape and locations of the cavities in the solid structure, as well as the rules for cohesiveness targeting in a structure. Topological optimization allows finding an optimal material distribution in a given design area under the certain loads and boundary conditions. Loads, boundary conditions, the volume of the created structure and, possibly, some additional restrictions, such as targeted locations, and sizes of cavities and areas full of solid material, are targeted for topological optimization. The topology, the shape and the size of the structure are not expressed in terms of standard parameter functions. For the advantages and capabilities of the additive manufacturing being used in the most efficient way, designing of the parts using topological optimization is necessary [2].

A key property for the task solution by means of methods of topological optimization is material density distribution. If there are some elements left after optimization in a computation region, having the density value equal either to initial density or to zero, this result is then called 'black-and-white' topology (or 0-1 topology). However, regions with intermediate density, so-called 'grey' material zones, will always appear [3, 4]. One of the ways to eliminate 'grey' topology is SIMP-methods (methods involving Solid Isotropic Materials with Penalization), where material density for each element acts as a final element characteristic, with density related to an elastic module with a power-law dependence [5-6]. Under the corresponding values of the index greater than 1, such type of a dependence acts as a penalty function that does not allow the elements with intermediate density to be generated during optimization and results in a 'black-and-white' design of the part designed [7].

2.3 Schematic Usage of the Bionic Design Concept

The technology for the design of reduced-weight parts for gas-turbine engines using the bionic design concept includes the following stages:

- Importing of the part geometric model into the developed software package. The input data for the bionic design operation are 3D electronic solid-body geometric models of gas-turbine engine parts generated with common design CAD-means;
- Assigning of physical and mechanical material properties of the part;
- Formalization of the physical-and-mechanical and operational requirements to a redesigned part; strength, reliability, definition of a loading diagram and load variation range, dynamic characteristics of the part, etc. (dependent upon the redesigned part type);
- Formalization of the technology requirements for a redesigned part; identification of the areas and surfaces staying unchanged during redesign; formalization of the part topology taking into account its location within a finished product (the size and the location of the fasteners, part dimensions, etc.), as well as capabilities and characteristics of the SLM process being taken into account;
- Widening of the calculated area within the given dimensions, simplification of the initial topology of the designed part;

- Generation of a finite element representation of the part which is acceptable for topological optimization performance;
- Performance of the part topological optimization using up-to-date software;
- Smoothing of the redesigned part surface. Creation and rework of the 3D solid-body geometric model of the part after topological optimization being performed;
- Generation of detailed finite element representations of the parts, which are acceptable for performance of verifying engineering analysis on using up-to-date software packages;
- Performance of computing experiments in order to verify compliance of properties of the part with a new topology with initially targeted physical-and-mechanical and operational specifications. In case the redesigned part is incompatible with given specifications, a repeated redesign of the part is necessary with changed input parameters of the topological optimization algorithm;

For a part with a new shape and topology, compatible with the given requirements, a standard procedure of the part design is to be performed with the necessary design documents being issued.

3. Practical usage of the bionic design concept for gas-turbine engine parts

3.1 Task Setting

A bracket for drawing bars of a reverse device, manufactured in Aviadvigatel JCS, was selected as a modelling task to test the topological optimization procedure, Figure 1.

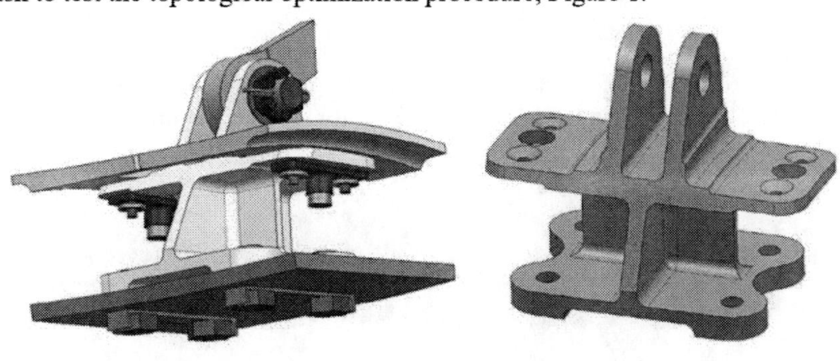

Figure 1. The geometry of a source model of the part under investigation: a – assembly; b – single.

This part takes up an aggregate complex of loadings in the different modes. The full load set was provided by the Engineering Department. The approach included two load steps.

The maximum stiffness was selected as a target function, namely minimization of the strain energy sum in the model to combine two selected static load sets. The decrease in the weight of the bracket simplified structure up to 35% was chosen as a limit. Thus, it results in a new bracket with an optimal geometry, with its weight being equal to 80% of that of the initially designed part. As a result, stresses at the selected load steps are expected to decrease in the optimized version of the bracket in comparison with the initially designed part.

3.2. Topological Optimization Performance

In order to solve the optimization task, the part geometry was maximally simplified by building it up with the additional material. The limiting condition was the capability to install the simplified part into the attaching lug. The fastening elements which must be targeted unchanged were kept as separate bodies, as boundary conditions are applied to them. As the result, the part weight increased five times, Figure 2.

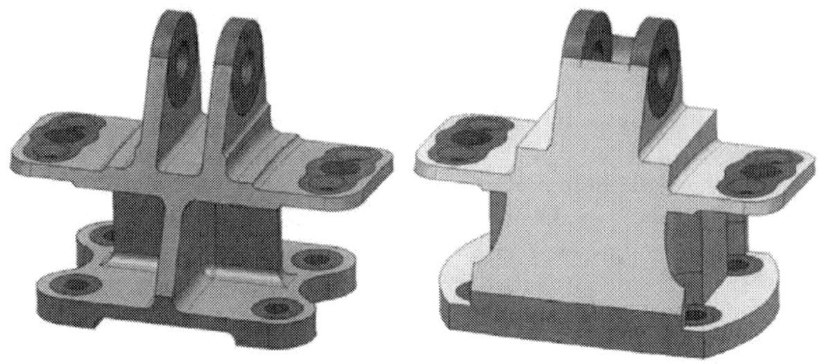

a b

Figure 2. Simplification of the part geometry: a – selection of unchanged fastening elements; b – simplified part.

The independent CATOPO software was selected as a basis for introduction of the topological optimization method. It permits to perform analysis of the stress-strain state of the part caused by application of the loads in the form of concentrated and distributed loads, thermal effects as well as its own natural oscillation.

3.3. Optimization Results

Figure 3 shows the geometry which was identified as the result of topological optimization of the selected part of a gas turbine engine. The results of the topological optimization are summarized in Table 1.

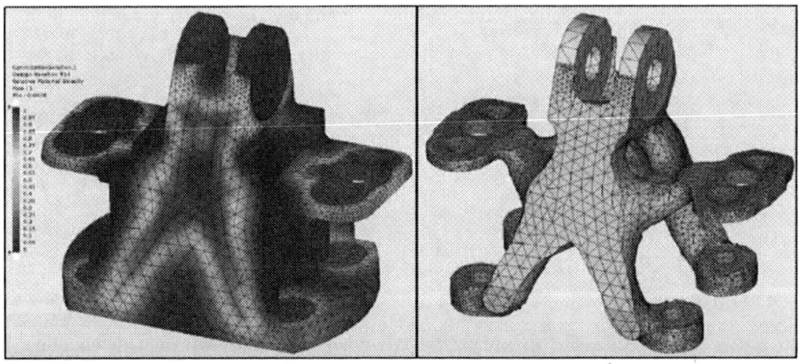

a b

Figure 3. The studied part with the optimized topology: a – density distribution; b – optimised part.

Table 1. Topological optimization results.

Version	Weight, kg	von Mises stress (step No. 1), MPa	von Mises stress (step No. 2), MPa
Initial	0.138	77	44
Simplified	0.687	34	22
Optimized	0.111	30	28

Accordingly, the use of the bionic design technology allowed not only for weight reduction of the fastening element under investigation, but also allowed for the significant reduction in the stresses at the investigated load steps. At that, strain-stress state calculations for the part showed that when the bracket weight reduced by 20%, the main von Mises stresses in it decreased by 48% on the average in comparison with the initial model.

3.4. Practical Use

The part under investigation was manufactured from steel powder by the SLM method in two versions: with the initial geometry and with the optimized geometry, Figure 4.

a b

Figure 4. The parts produced with the SLM method by Aviadvigatel JCS: a – initial geometry; b – optimal geometry.

The weight of the part decreased by 20%. The parts are now ready for the mechanical tests.

4. Conclusion

The use of the bionic design technology and topological optimization for the weight reduction of the engine parts allows for the final weight of the engine to be reduced with all other properties being maintained. The weight reduction for some parts can reach 50%.

The use of the bionic design technology is essentially related to the additive manufacturing methods due to the geometric shapes being complex. The similar technologies have been successfully introduced by the leading engines corporations, and more recently, they have been developed at Aviadvigatel's facility. The bionic design of aero engines offers vast prospects for the aerospace industry.

Acknowledgments

The paper has been executed with the financial support of the Ministry of Education and Science of the Russian Federation (contract No. 02.G25.31.0168 dated 01.12.2015) within implementation of 'The order of the Government of the Russian Federation' No. 218, 'About the measures of the state support of development of cooperation of the Russian educational organizations of higher education, the public scientific institutions and the organizations implementing complex projects on creation of high technology production'.

References

[1] Vincent J 2007 Adaptive structures – some biological paradigms *Adaptive Structures. Engineering Applications* (Chichester, West Sussex, England: John Wiley & Sons Ltd) pp 261-283
[2] Kobayashi M 2010 *Comm. on Nonlinear Sci. and Numerical Simulation* **15** 787-802
[3] Bendsoe M and Kikuchi N 1988 *Computer Methods in Applied Mech. and Eng.* **17** 197-224
[4] Bendsoe M 1995 *Optimization of Structural Topology, Shape, and Material* (Berlin: Springer)
[5] Bendsoe M 1989 *Structural Optimization* **1(4)** 193-202
[6] Rozvany G, Zhou N, Sigmund O 1994 *Adv. in Design Optimization* 240-299
[7] Brackett D, Ashcroft I, Hague R 2011 *22nd Annual Iint. Solid Freeform Fabrication Symp.* 348-362

Calculation by iterative method of linear viscoelastic plate under biaxial tension

A A Svetashkov[1], J Miciński[2], K K Manabaev[1] and A A Vakurov[1]

[1] Tomsk Polytechnic University, 30, Lenina Ave., Tomsk, 634050, Russia
[2] University of Warmia and Mazury in Olsztyn Faculty of Animal Bioengineering Department of Cattle Breeding and Milk Evaluation, 5, Oczapowskiego Ave., Olsztyn, 10-957, Poland

E-mail: mkk@tpu.ru

Abstract. In this paper, we used the iterative solution algorithm, proposed in the work of Pavlov and Svetashkova. This algorithm results in a complete separation of spatial and temporal variables, if we set up the boundary loads and (or) volumetric forces in the same kind. In this paper, we have examined the stress-strain state of a viscoelastic plate, and the results of the calculation displacements, stresses are given. In addition, we made a comparison of the calculation indices rate of convergence for the iterative process with their theoretical values.

1. Introduction
The problem of algorithmization for solving boundary tasks of linear viscoelasticity in case the constitutive equations are given in an integral form is actual. Storage of the full temporal history of changing the stress-strain state previous to time t is essential in solving this problem when using traditional approaches [1-3]. Thus, the obtainment of information about stresses and strains at time t requires the information on their changes in prior times $\tau < t$.

The application of the iterative algorithm in the calculation of tasks linear viscoelasticity allows circumventing the given problem, because implementation of this method involves the separation of spatial and temporal variables. This, in turn, eliminates epy necessity ща storage of information about the history of loading the construction.

Implementation of the new computational method of the iterative algorithm in the environment of the software complex of the finite element method is the aim of this work. The method is implemented by the example of solving a boundary problem about biaxial loading of the viscoelastic plate. We compare the indicators of rate convergence which were obtained by the theoretical and calculation methods. circumvent

2. The problem about loading plate
We consider the solution of the problem of a biaxial stretching linear viscoelastic plate under the action of the uniformly distributed loads applied to the ends of T_3, T_4 and directed along axes x and y. The plate has been fixed at sides T_1, T_2, figure 1.

The equilibrium equations during displacements of flat problem linear viscoelasticity have the form:

$$\left(K_0+\frac{1}{3}G^*\right)\theta_{,\alpha}+G^*\Delta u_\alpha=0,\ \alpha=1,\ 2 \tag{1}$$

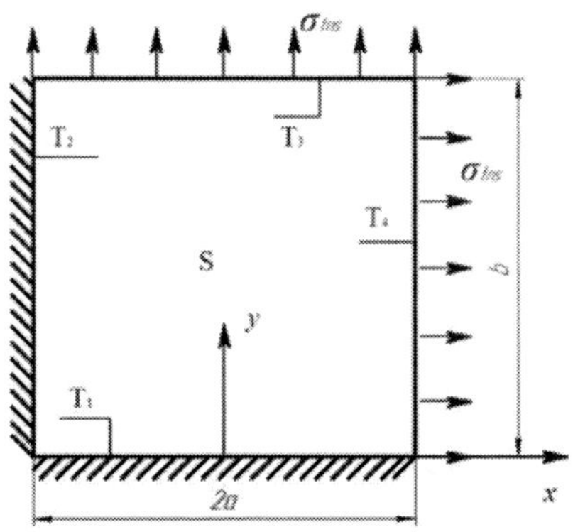

Figure 1. Scheme of loaded plate.

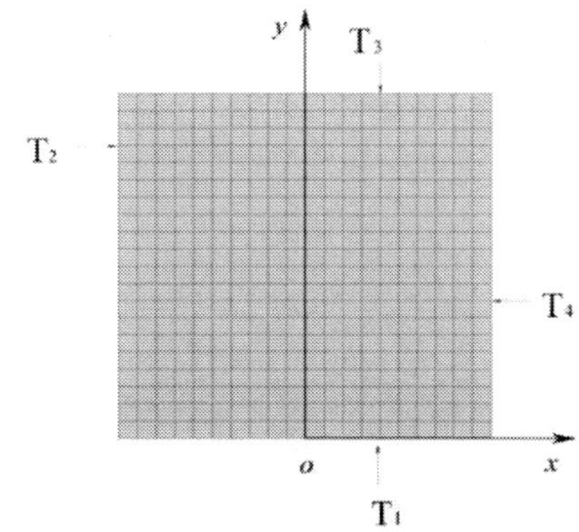

Figure 2. Image of finite – element model plate.

The boundary conditions in stresses are:

$$\begin{cases}\sigma_{11}l+\tau_{12}m=X_n\\ \sigma_{22}\,\mathrm{m}+\tau_{21}l=Y_n,\end{cases} \tag{2}$$

here, $\theta=u_{1,1}+u_{2,2}$ – volumetric deformation; comma before the index corresponds to differentiation with respect to the spatial coordinate (for example, $u_{1,1}=\frac{\partial u_1}{\partial x}$); $u_\alpha,\alpha=1,2$ – components of displacement vector $\bar{u}$; Δ – Laplace operator; $\sigma_{\alpha\beta}$ – components of the stress tensor; l,m – direction cosines of the normal to the boundary contour; X_n,Y_n – boundary loads; G^* – integral operator of relaxation.

$$G^*\varepsilon_{\alpha\beta}\equiv\int_0^t R(t-\tau)d\varepsilon_{\alpha\beta}(\tau), \tag{3}$$

where $R(t)$ – function of the shear relaxation, $R(0)=G_0$; G_0, K_0 – elastic-instantaneous modules of shear and volumetric compression; $\varepsilon_{\alpha\beta}$ – components of the deformation tensor. The physical equations of linear viscoelastic body have the form:

$$\sigma_{\alpha\beta}=\left(K_0-\frac{2}{3}G^*\right)\theta\delta_{\alpha\beta}+2G^*\varepsilon_{\alpha\beta},\ \delta_{\alpha\beta}=\begin{cases}1,\ \alpha=\beta,\\ 0,\ \alpha\neq\beta.\end{cases} \tag{4}$$

We will solve system integral-differential equations (1), (2) by the iterative method, presented in the work of Pavlov and Svetashkov [4]. For this purpose, we introduce the auxiliary constitutive equations:

$$\sigma_{\alpha\beta}=G^*\left[\left(k_S-\frac{2}{3}\right)\theta\delta_{\alpha\beta}+2\varepsilon_{\alpha\beta}\right] \tag{5}$$

where k_S is a constant, $k_S\in\left[1,K_0/G_0\right]$.

Constitutive equations (5) correspond to a viscoelastic material with proportional operators of volumetric and shearing relaxation. In this case, as Christensen [5] shown, the stress-strain state of the structure varies in proportion to a function of time.

We will solve boundary problems (1), (2) by replacing constitutive equations (4) with equations (5), and we will represent the discrepancy as fictitious volumetric and surface forces. As a result, in each iteration, we will obtain the iterative type equations:

$$G^*\left[\left(k_S+\frac{1}{3}\right)\theta^{n+1}_{,\alpha}+\Delta u^{n+1}_{\alpha}\right]=\left(k_S G^*-K^*\right)\theta^{n}_{,\alpha},$$
$$\sigma^0_{\alpha\beta}\left(\overline{u}^{n+1}\right)n_\beta=S^0_\alpha+\left[\sigma^0_{\alpha\beta}\left(\overline{u}^{n}\right)-\sigma_{\alpha\beta}\left(\overline{u}^{n}\right)\right]n_\beta,\ (\alpha,\ \beta=1,2),\ n=0,1,... \tag{6}$$

Here $S^0_1=X_n$, $S^0_2=Y_n$. Then, for the n-th approximation, we will have:

$$u^{(n)}_\alpha(x,y,t)=\sum_{i=1}^{n}\tilde{u}^{(i)}_\alpha(x,y)\varphi_{i-1}(t),$$
$$\varphi_i(t)=\left[\bar{G}^{*-1}\left(k_S G^*-K_0\right)\right]^i\varphi_0(t), \tag{7}$$
$$\varphi_0(t)=\bar{G}^{*-1}H,\ \bar{G}^*=G^*/G_0.$$

Stresses in viscoelastic body are defined as:

$$\sigma^{(n)}_{\alpha\beta}=\sigma^{(n-1)}_{\alpha\beta}+\tilde{\theta}^{(n)}\left(K_0-\frac{2}{3}G^*\right)\varphi_{n-1}\delta_{\alpha\beta}+2\tilde{\varepsilon}^{(n)}_{\alpha\beta}G^*\varphi_{n-1},\ \mathrm{n}=1,2,... \tag{8}$$

Functions $\tilde{u}^{(i)}_\alpha(x,y)$ are defined as the solution of elastic problems.

3. Analysis of the results

The numerical calculation of the elastic problem has been implemented in software package ANSYS, the calculated area of the finite – element model is presented in Figure 2. The constants of material are: $G_0=2$ MPa, $K_0=10$ MPa. The boundary loads equal $X_n=Y_n=\sigma_{ms}=15$ MPa.

According to (7), the results of the viscoelastic solution of the problem for the average speed of relaxation and specified parameters $\lambda=0.04$, $\gamma=0.01$, $k_S=4.5$, as well as when parameter η that determines the ratio of the resiliently instant module to prolonged module $\eta=G_0/G_\infty$ takes a value equal to 5, $t=100$ min, are shown in Table 1.

Table 1. Results $u^{(i)}_\alpha(x,y,t)$ of plate calculation

n	$\tilde{u}^{(i)}_1$, cm	$\tilde{u}^{(i)}_2$, cm	$\varphi_{n-1}(t)$	$u^{(i)}_1$, cm	$u^{(i)}_2$, cm
1	1.621E-02	1.692E-02	4.107	0.066	0.069
2	3.613E-03	3.723E-03	-3.147	-0.011	-0.011
3	8.504E-04	8.618E-04	2.372	0.002	0.002
4	2.108E-04	2.095E-04	-1.748	-3.669E-04	-3.669E-04
5	5.406E-05	5.314E-05	1.245	6.726E-05	6.612E-05
6	1.437E-05	1.392E-05	-0.840	-1.204E-05	-1.172E-05
7	3.880E-06	3.754E-06	0.514	2.000E-06	1.932E-06
8	1.070E-06	1.038E-06	-0.252	-2.694E-07	-2.590E-07

For assessment of coefficient k_S impact on the rate of the convergence process, we consider the following cases of its assignment:

$$1)\ k_S=0.7\tilde{k}_S;\ 2)\ k_S=0.85\tilde{k}_S;\ 3)\ k_S=0.9\tilde{k}_S, \tag{9}$$

where $\tilde{k}_S=K_0/G_0=5$.

In the future, we will consider the calculated indicators of convergence speed, which are determined by the formula:

$$q_n = \frac{W_n\left(u^{n+1} - u^n\right)}{W_n\left(u^n - u^{n-1}\right)}, \quad u^n = \left(u^n, v^n\right), n = 1, 2, \ldots$$

Here $W_n\left(u^n\right)$ – specific potential energies of stresses and deformations.

$$W_n\left(u^n\right) = \frac{1}{2}\left(\sigma^n_{\alpha\beta}\varepsilon^n_{\alpha\beta}\right)$$

Functions $W_n\left(u^n\right)$ have been calculated at each time point for the selected element of the grid calculated model of the plate. Next, we determine the total calculated indicator of degree compression for n iterations

$$q = \prod_{k=1}^{n} q_k \tag{10}$$

This indicator expresses the ratio of convolution, calculated for the difference of n and (n-1) approximations, to the respective convolution, calculated for difference of the first and zero approximation.

We have given Figure 3 for the assessment of impact k_S on the rate of convergence solution. The logarithmic dependences for the total indicator of degree compression q, defined by (10), depending on the time and iterations, are also presented in Figure 3

The logarithmic dependences of the theoretical $\overline{q}$ and calculated q (10) indicators of the convergence rate of the iterative solution depending on number n of steps for procedure solution are shown in Figure 4.

The results of solving the problem are given for the average rate of relaxation and the set parameters: $\lambda = 0.04$; $\gamma = 0.01, \eta = 5$, $k_S = 0.9\tilde{k}_S$. The dashed curves are obtained by the iterative method, the continuous curves – theoretically. The time interval of 50 minutes corresponds to curves 1, 5; 100 min –curves 2, 6; 1000 min – curves 3, 7; 1500 min – curves 4, 8.

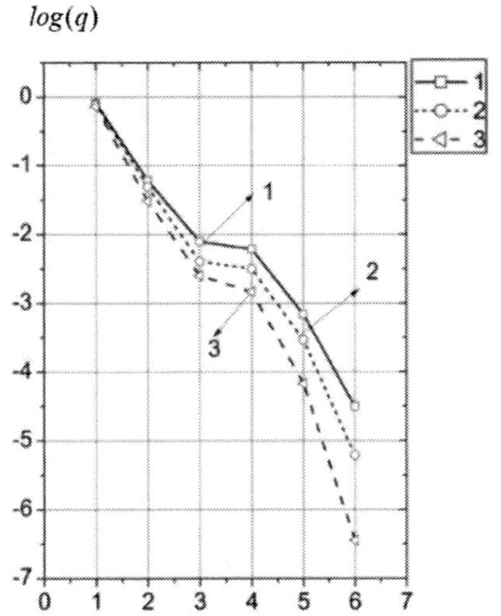

Figure 3. The logarithmic dependences for q from the number of iteration, $t = 1000$ min

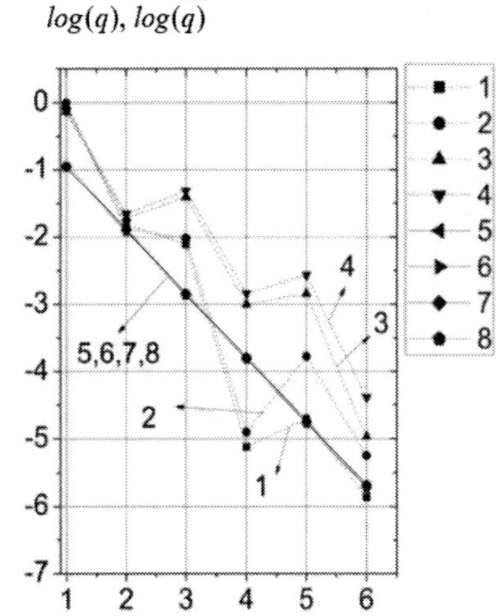

Figure 4. The dependence of indicators for the rate of convergence and n.

The curve of stresses σ_{11}, σ_{22} (biaxial tension by equal efforts) depending on the time is shown in Figure 5.

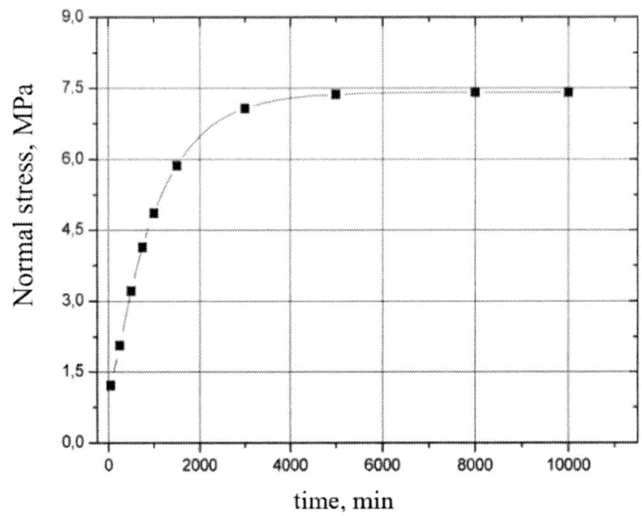

Figure 5. The dependence of stresses σ_{11}, σ_{22} on time.

The curve of stresses values σ_{22} depending on the number of iteration is presented in Figure 6. The time was chosen to be 200 minutes.

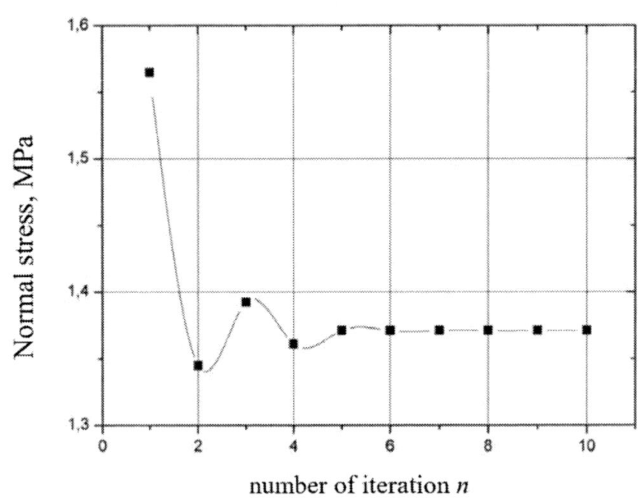

Figure 6. The curve of convergence stress σ_{22} depending on the number of iterations.

4. Conclusion

The numerical solution of the plane problem of linear viscoelasticity theory [8 – 10] as applied to the calculation rectangular plate (biaxial tension) is presented in the work. The following results were obtained:

a) the iterative algorithm which is used in the work, gives a fairly rapid convergence (for stresses, the full convergence is achieved after ten iterations, for displacements after eight);

b) the outlays of machine time are the same for any preassigned moment of the loading history plate. This fact is an advantage for the used iterative method;

c) comparison of the calculated and theoretical indicators of convergence for the iterative algorithm showed their convergence within the limits of 10%.

d) we investigated the effect of arbitrary numerical parameter k_s included in the auxiliary physical equations of the communications law of viscoelastic stresses and deformations on the convergence of the iterative process.

References

[1] Chazal C, Pitti R M 2001 *J. of Theor. and Appl. Mech.* **49** 1029-1048
[2] Matveenko V, Trufanov N 2013 *J. of Eng. Math.* **78** 119-129
[3] Ataoglu S 2009 *Struct. Eng. and Mech.* **32** 407-427
[4] Pavlov S M and Svetashkov A A 1993 *Russ. Phys. J.* **36** 400-406
[5] Christensen R M *1979 Mechanics of composite materials* (NewYork: AcademicPress)
[6] Svetashkov A A, Kupriyanov N A and Manabaev K K 2013 *Russian Physics J.* 209-211
[7] Kupriyanov N A, Simankin F A and Manabaev K K 2016 *IOP Conf. Ser.: Mat. Sci. and Eng.* (Tomsk) vol 124 (IOP Publishing) 012100
[8] Svetashkov A A, Vakurov A A 2016 *IOP Conf. Ser.: Mat. Sci. and Eng.* (Tomsk) vol 124 (IOP Publishing) 012099
[9] Svetashkov A A 2000 *Mechanics of Composite Materials* **36** 37-44
[10] Svetashkov A A, Kupriyanov N A and Manabaev K K 2016 *Key Eng. Mat.: Sci. J.* **685** 206-210

Simulation modelling of tribotechnologies system and its parametric reliability assessment on tribotechnical parameters of the joints of sliding friction

M N Nagorkin, V P Fyodorov and V V Nagorkina

Bryansk State Technical University, 7, 50-letiya Oktyabrya bul., Bryansk, 241035, Russia

E-mail: nagorkin_mn@mail.ru

Abstract. The article discusses the formation of tribotechnical characteristics of cylindrical sliding friction joints by means of simulation. Experiments were conducted on physical models of joints. The parametric reliability of tribotechnological parameters of burnish operation were determined for the studied tribotechnological system.

1. Introduction

It is known that up to 80% of the technical devices fail due to wear and tear of the joints of parts. Improvement of wear resistance of joints is one of the most urgent and fundamental problems of modern engineering, and its solution is largely to be found in improving the processing technology of contacting surfaces of triboelements. Thus, the solution of the problem of effective tribotechnology methods development of the formation of functional surfaces of triboelements providing the desired quality characteristics with high reliability is so important [1 – 6].

The article considers the questions of formation of tribotechnical characteristics of burnishing process of the cylindrical friction pairs 'Steel 45 – bronze CuSn5Zn5Pb5' or 'Steel 45 – antifriction cast iron grey ACiG-1'. Special attention was paid to the process of formation of tribotechnical characteristics depending on the processing conditions of the contacting surfaces and of some features of operation.

2. Materials and methods

The formation of the quality parameters determining the wear resistance of rubbing parts surfaces was carried out at the final stage of processing in the three-stage combined technological system (TS) of processing, that includes three main stages (Figure 1):

– Stage 1: ensures specified accuracy of part machining and the required geometrical parameters of surface quality for high-quality coating of materials for forming the running-films; main operation are finishing turning O_{11} or grinding O_{12};

– Stage 2: the coating on the parts surface of materials to obtain a soft running-film by friction O_{21} or by chemical methods O_{22} is provided;

– Stage 3: the microprofile is formed, excluding the process of microtcut and providing the required physico-mechanical and operational properties of the surface (e.g., durability). Processing can be carried out by methods of surface plastic deformation (SPD), by diamond burnishing O_{31}; by roll burnishing O_2 etc.

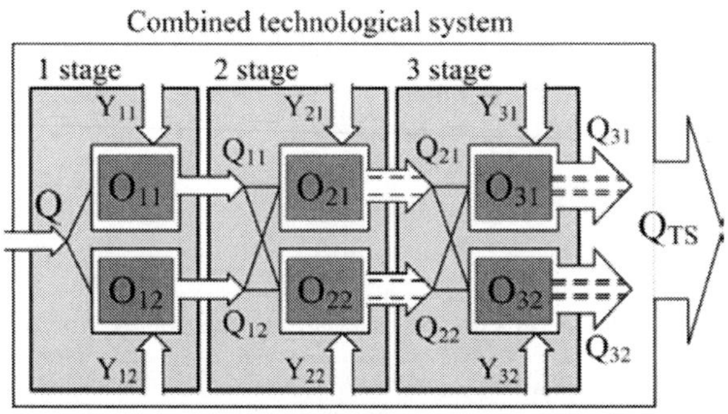

Figure 1. Diagram of three-stage combined technological treatment system of details: Q – the quality of the surface layer (QSL) obtained in the previous process steps; Q_{11}, …, Q_{mn} – the parameters of QSL in intermediate stages of the technological process; Q_{TS} – QSL after treatment; Y_{mn} – conditions of processing for technological operation; O_{mn} – the type of process operation; m – the processing stage; n – the processing method

In the study, in stage 1 parameters of parts surfaces quality were formed on a lathe finish turning with standard cutting inserted pieces, the cutting part of which is equipped with a composite 10 (geksanit R). In stage 2 of formation of the running-of films the finishing antifriction non-abrasive treatment (FANAT) of the work piece surface or chemical plating was applied.

The FANAT process was carried out using friction brass plating surfaces of parts with the use of surfactants (glycerin). The pressure in the contact area of the brass rod with the treated surface is 30 MPa; the processing speed is 20 m/min; the feed is 0.05 mm/rev.

In the process of a chemical copper plating the solution of copper chloride $CuCl_2$ in a mixture of 70% acetic acid and glycerin was applied to the on the degreased surface of the part rotating with a speed of 20 m/min. Uniform copper film was formed on the surface after 10 min of exposure of a part with constant speed of rotation.

At stage 3 of considered combined technologies the diamond burnishing by tools with elastic action with indenters, equipped with synthetic diamonds ASPK (carbonado, PSD) with a corner radius of 3.5 mm was used.

We studied the samples of steel 45 with hardness HRC 48...50, get by volume hardening. Samples were prepared in the form of step noses or rings. Models of liners were made of antifriction cast iron ACiG-1 or of bronze CuSn5Zn5Pb5. The unit model of sliding friction installed on the test bench is shown in Figure 2.

The parts surfaces were processed on a thread-turning machine 16K20 and on lathe and milling machining center mod. EX-308 (TAKISAWA) in accordance with the plan of the experiment that is a regular fractional replica of type 2^{12-8} of full factorial experiment of the type 2^8. The variation levels of the experiment factors are presented in table 1.

Table 1. Factorial planning of experiment for formation of tribotechnical characteristics study in tribotechnological system.

Fact or code	Factors											
	X_1	X_2	X_3	X_4	X_5	X_6	X_7	X_8	X_9	X_{10}	X_{11}	X_{12}
	V_T	S_T	t	j	M_f	Q_{DB}	S_{DB}	V_{DB}	M	V_R	P_R	ΔP
–	65	0.05	0.1	2.5	1.5	100	0.075	65	2	10	30	0.15
+	200	0.15	0.25	1.62	3	300	0.15	100	5	50	50	0.25

Eight technological factors of processing were varied affecting to the formation of tribotechnical characteristics of friction pairs:

– modes of finish turning: V_T – speed of steel 45 turning with composite 10, m/min.; S_T – tool feeding, mm/rev; t – cutting depth, mm;

– the rigidity of technological system j at stages 1 and 3 of treatment, kN/mm;

– *M* – the material of soft running-film on the surface of the shaft (for friction brass plating (FBP) the code is 3,0; for chemical copper plating (CCP) – 1,5);

– modes of diamond burnishing: Q_{DB} – the force of diamond burnishing, N; V_{DB} – the speed of diamond burnishing, m/min; S_{DB} – the feed of diamond burnishing, mm/rev.

As factors, taking into account the operating conditions were adopted:

– M_l – the liner material: for antifriction cast iron the code is 2; for bronze – 5;

– V_R – the speed of the relative sliding of triboelements, m/min;

– P_R – the value of nominal linear load on the tribojoint during running, N/mm;

– ΔP – the variation percentage of nominal linear load on the tribojoint during running when modeling dynamic operating conditions, %.

Studies of tribotechnical characteristics of joint were carried out on the installation, scheme of which is shown in Figure 2.

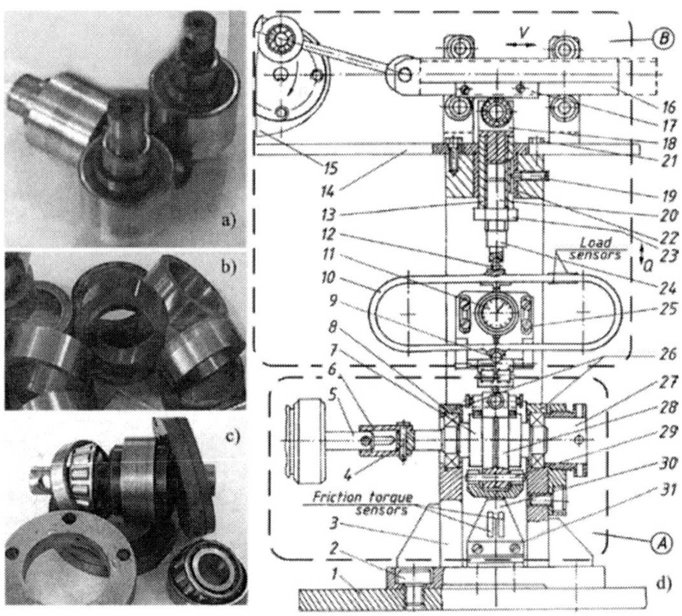

Figure 2. Installation for tribotechnical tests: a – shafts models; b – rings models; c – model of assembly of a friction slip; d – installation diagram: 1 – plate, 2 – screw; 3 – body; 4 – pin; 5 – lead shank; 6 – bushing; 7 – shaft; 8 – bearing; 9, 12 – balls; 10 – dynamometer; 11 – indicator; 13 – nut; 14 – plate; 15 engine to convert rotary motion into reciprocating motion; 16 – slider; 17 – cam mechanism to transfer the load to the sample; 18 – bracket; 19 – pin; 20 – screw; 22 – bushing; 23 – locknut; 24 – stock; 25 – screws; 26 – cover; 27 – bushing; 29 – pin; 30 – cantilever beam

In general, the experimental installation includes three modules:

1) main module A, in which a tested model of the friction pair, the drive, adjustable by speed, and dynamometer to create a static load are located;

2) module B to create a variable programmable load carrying dynamic loading according to the law: $P = P_{Rated}(a + b\sin\omega t)$,

where a, b – the factors used to change the load P according to a given harmonic law ($0 \leq a \leq 1$, $0 \leq b \leq 1$).

3) module C of electronic count of the number of loading cycles, that is controlled by electric probe (on Figure 2 is not conventionally shown).

The running-in process of the samples was carried out in the conditions set forth in the plan of experiment, however, in the early running-in and after it the friction coefficient was measured at various speeds (V_R = 10 – 90 m/min) and loads (P_R = 10 – 150 N/mm). The registration of the measured characteristics envisaged the use of computer measuring system 'Oscilloscope', that is connected to the output of the strain gauge amplifier or the use of loopback oscilloscope and two-coordinate recorder.

3. Results and Discussion

The following tribotechnical properties were studied:

– f_1, f_0 – friction coefficients at the beginning and at the end of the running-in process;
– h_{01}, h_{02} – values of an initial wear of surface of the shaft (01) and the liner (02), μm;
– L_{01}, L_{02} – running-in path for the shaft and the liner, m;
– I_1, I_2 – wear rate of the shaft and the liner surfaces.

Graphs of linear wear of the shafts and the friction coefficient depending on friction path are given in Figure 3. The graph shows the bounds for the 80% confidence intervals. In addition, micrographs of the topography of the shaft surfaces before and after running-in are showed. Color shades on all shades photomicrographs prove the presence of the running-film that in the conditions of the experiment remained on the surface both after diamond burnishing and after completion of running-in process.

Considering the dependence of the friction coefficient of running-in path, it should be borne in mind that their initial points ($L \rightarrow min$) correspond to the first registration. Naturally, that friction coefficient at L = 0 has a specific value close to registered, but not infinite as it may seem when considering curves 2, 4 (Figure 3).

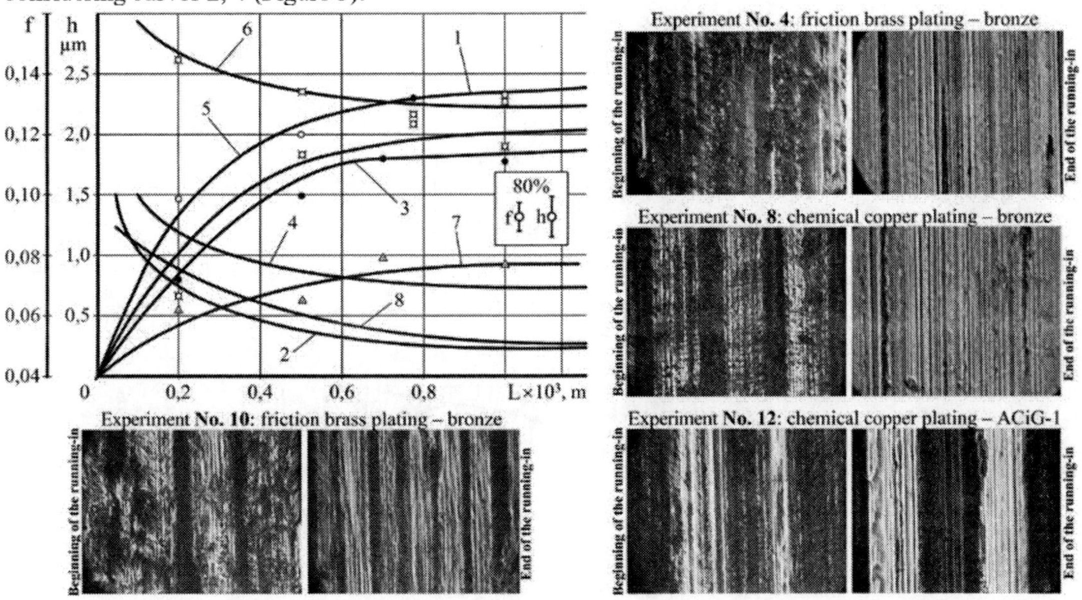

Figure 3. Dependencies of linear wear h of shafts (odd numbers) and friction coefficient f (even numbers) from the way of friction L for experiments conducted in accordance with the experiment plan and photomicrographs of the friction surfaces of the shafts at the beginning and end of the running-in: curves 1, 2 – experiment 4; curves 3, 4 – experiment 8; curves 5, 6 – experiment No. 10; curves 7, 8 – experiment No. 12

As a result of processing of experimental data by the method described in [7], adequate physical-statistical simulation models of formation of tribotechnical characteristics (Cobb-Douglas models) were built:

$$Y_i = b_0 K_1 K_2 V_T^{b_1} S_T^{b_2} t^{b_3} j^{b_4} Q_{DB}^{b_6} S_{DB}^{b_7} V_{DB}^{b_8} V_R^{b_{10}} P_R^{b_{11}} \Delta P^{b_{12}}, \quad (1)$$

where Y_i – i-th tribotechnical characteristics; K_1, K_2 – coefficients taking into account the type of the running-film (FBP or CCP) and liner material (bronze (Br) or antifriction cast iron (ACi); b_0, b_k – model coefficients.

Models parameters (1) and calculated values of Fisher criterion F_{calc} are presented in Table 2. F_{calc} compared with the table value at the significance level α = 0,05 (F_{tab} = 6,59) proves the adequacy of models.

Analysis of factors influence of considered tribotechnological system on main tribotechnical characteristics (f_1, f_0, h_{01}) can be performed using Pareto charts (Figure 4). It should be noted that such factors of diamond burnishing, as Q_{DB} and S_{DB} have a strong positive impact on reducing the friction coefficient f_1. From the point of view of friction coefficients reduction f_1 and f_0, the running-in process should be running on forced mode, that is, at maximum permissible speed of relative sliding.

Table 2. Parameters of simulation models of formation of tribotechnical characteristics.

Coef.	b_0	K_1		K_2		b_1	b_2	b_3	b_4	b_6	b_7	b_8	b_{10}	b_{11}	b_{12}	F_{calc}
		FBP	CCP	Br	ACi											
f_1	2.0	1.31	1.10	1.11	1.05	-0.06	0.04	-0.05	0.02	-0.07	0.47	-0.05	-0.32	-0.09	–	1.3
f_0	2.82	1.64	1.20	0.92	0.96	0.03	0.20	0.19	0.11	-0.14	0.28	-0.03	-0.35	0.03	0.62	0.5
h_{01}	3.74	1.20	1.07	1.23	1.09	0.31	0.15	0.02	0.04	-0.38	0.32	0.68	0.05	-0.74	-0.19	5.6
h_{02}	8.08	1.10	1.04	1.28	1.11	0.07	0.27	0.09	-0.07	-0.16	-0.04	0.05	0.04	0.10	-0.40	4.7
$I_1\times10^{-12}$	316	1.77	1.23	0.87	0.94	0.08	0.22	0.21	0.13	-0.17	0.43	-0.05	-0.43	-0.10	0.76	2.8
$I_2\times10^{-11}$	147	1.80	1.24	0.94	0.97	0.08	0.22	0.14	0.14	-0.13	0.25	-0.06	-0.4	0.02	0.64	2.1

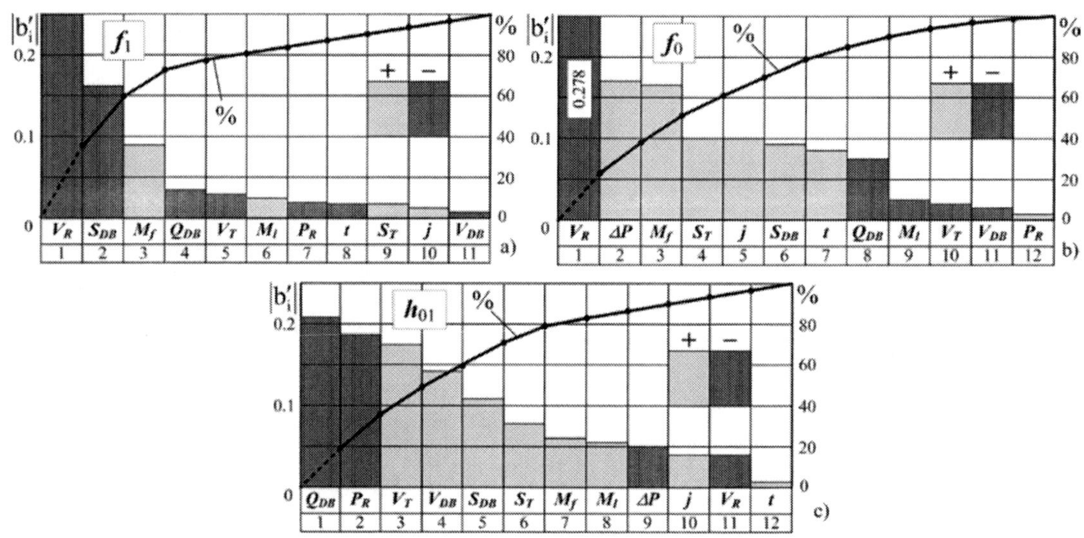

Figure 4. Pareto charts of tribotechnological system factors effects on: a – friction coefficient f_1; b – friction coefficient f_0; c – value of shaft initial wear h_{01}

Detailed analysis of Pareto chart allows to identify a number of features, in particular, the increasing of the rigidity j of TS within experiment leads to the increase of all of the characteristics that is impractical. This, apparently, can be explained by the fact that machining of parts of TS with high stiffness leads to the surfaces, the quality parameters of which are higher than the required operating conditions. In this case, the running-in process must match the decrease in the level of these parameters to the operating.

Thus, the provision of the considered tribotechnical characteristics due to the factors of studied tribotechnological system is possible. At that the limits of support intervals for individual characteristics are shown in Figure 5 a, and the reliability P of their provision [7] in the symmetric relative δ-interval is shown in Figure 5 b.

The most reliable is the ratio f_1/f_0 (at $\delta = 0.1$ $P = 0.99$), and less reliable is the wear intensity I_1 и I_2 (at $\delta = 0.1$ $P = 0.32$ и 0.45, respectively). The high reliability has a friction coefficient after burnishing at $\delta = 0.1$ $P = 0.76$.

Intensities of wear I_1 and I_2 have low reliability of technological provision, but their absolute value is quite low, so considered pairs refer to pairs of sliding friction with high abrasion class. The separation of influence factors on tribotechnical properties on technological and operational characteristics in this case can be considered as inappropriate because the uniform tribotechnological system is considered.

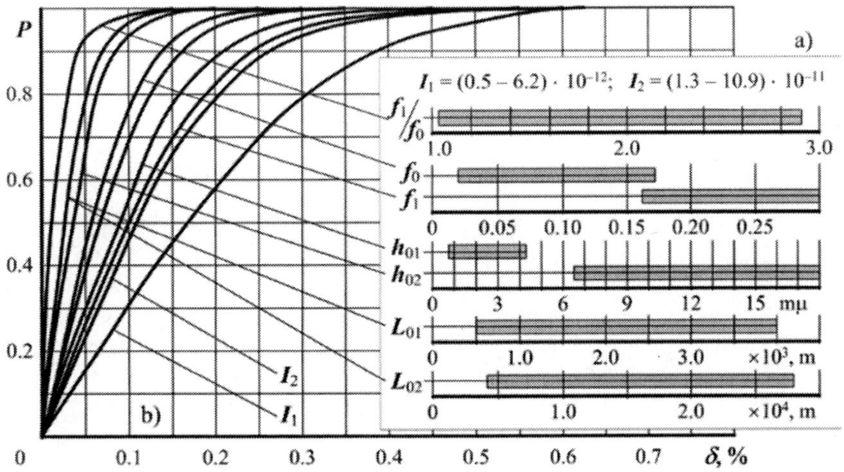

Figure 5. Diagram of variation (a) and the reliability provision P (b) of tribotechnical characteristics depending on values δ

4. Conclusion

An effective combined anti-friction technology for sliding friction surfaces treating of shafts of hardened steel, combining the finishing methods of machining, application of soft running-in film containing copper by friction or chemical means and the subsequent processing of the SPD is proved and developed. As a result of comprehensive research of tribotechnological system (surface treatment and conditions of joints of sliding friction burnish), a number of regularities of formation of parameters of wear resistance is established. The complex of adequate physical-statistical models 'mode – property' for tribotechnological system suitable for practical use is obtained.

References

[1] Gorlenko A O and Davidov S V 2016 *IOP Conf. Series: Mat. Sci. and Eng.* (Tomsk) vol 124/1 (IOP Publishing) 012154
[2] Kirichek A V, Altukhov A Yu, Soloviev D L and Silantiev S A 2015 *J. of Nano and Electronic Physics* **7(4)** 04019
[3] Kirichek A V and Solov'ev D L 2008 *Russian Eng. Res.* **28(3)** 277–279
[4] Suslov A G and Medvedev D M 2010 *Russian Eng. Res.* **30(11)** 1172–1175
[5] Suslov A G and Petreshin D I 2010 *Russian Eng. Res.* **30(4)** 418–423
[6] Fyodorov V P, Averchenkov V I and Nagorkin M N 2016 *Procedia Eng. 150 (2016): 2nd Int. Conf. on Industrial Eng.* (Chelyabinsk) (Elsevier Ltd.) p 899–905
[7] Fyodorov V P, Nagorkin M N and Totai A V 2016 *IOP Conf. Series: Mat. Sci. and Eng.* (Tomsk) vol 124/1 (IOP Publishing) 012053

Simulation of the plain milling process

A Yu Nikolaev

Irkutsk National Research Technical University, 83, Lermontov str., Irkutsk, 664074, Russia

E-mail: AndrNikolajev@gmail.ru

Abstract. The article considers the dynamics of the plain milling process. The model takes into account the time-lag effect due to the algorithm of new surface formation during material cutting-off. A reduced model based on the results of a modal analysis of the milling cutter is applied to describe the dynamics of the tool. The results of the work are presented in the comparison of the microprofiles obtained as a result of processing and simulation of surfaces.

1. Introduction

Metal milling is a significant part of the production process. The study of the machining process aims to increase productivity and quality of processing [1-7]. The authors use different methods for research of metal processing. In paper [3], it is proposed to evaluate residual stresses in the surface layers of steels by x-ray diffraction and Barkhausen noise. One of the research methods is simulation. Development of the model of a correct machining process is the issue of the day. The authors of many publications present the simulation as an effective method for optimization of cutting conditions [8-10]. The works [9, 10] propose the method based on the study of vibrations that occur during metal processing.

When selecting simulation as the main research method, it is primarily required to create an adequate model, which clearly reflects the dynamics of the object. To obtain this model, it is necessary to describe the processing surface, workpiece, tool and laws, according to which the cutting force will change. These are basic, but not always sufficient blocks that should be present in the model. This model will not be accurate, and, therefore, it is not suitable for determining the optimal cutting conditions and correction of the NC machine tool control program. The problem of accuracy is solved by the complexity of the model, adding to it such additional part as the deviations correction. Further refinement of the model is possible when we introduce additional extra factors obtained by correlation with experimental data or derived ab initio. The creation of model that will do for simulating of regenerative vibration during cutting is basic object of research. The model may enable us to receive a clear picture of the machining process.

2. Materials and methods

In the case of plain milling simulation, a mathematical model depending on its complexity can take into account a number of mechanisms of vibration excitation which occurs during the cutting process [11-13]. For example:

1. Resonance of the chip breaking-off;
2. Self-oscillations due to the joint action of the cutting and friction forces;

3. Self-oscillations due to a nonlinear descending character of the cutting forces depending on the cutting speed;
4. Self-oscillations due to the differences in the cutting forces at the entrance and exit of the cutter;
5. A mechanism of coordinate relation;
6. Self-oscillation during cutting the machined surface – a time-lag mechanism.

In the work, we applied the approach of 'Simulation modeling' [14-15]. It consists in the numerical integration of equations of the system motion. For this purpose, the algorithm of geometric modelling with the determination of the instantaneous values of the chip thickness is given as the time-lag. It may be possible owing to retaining of the surface formation history. The time-lag mechanism underlies the developed model. The system time-lag is formed because of regeneration of the surface. Figure 1 is a schematic diagram of cutting process dynamics.

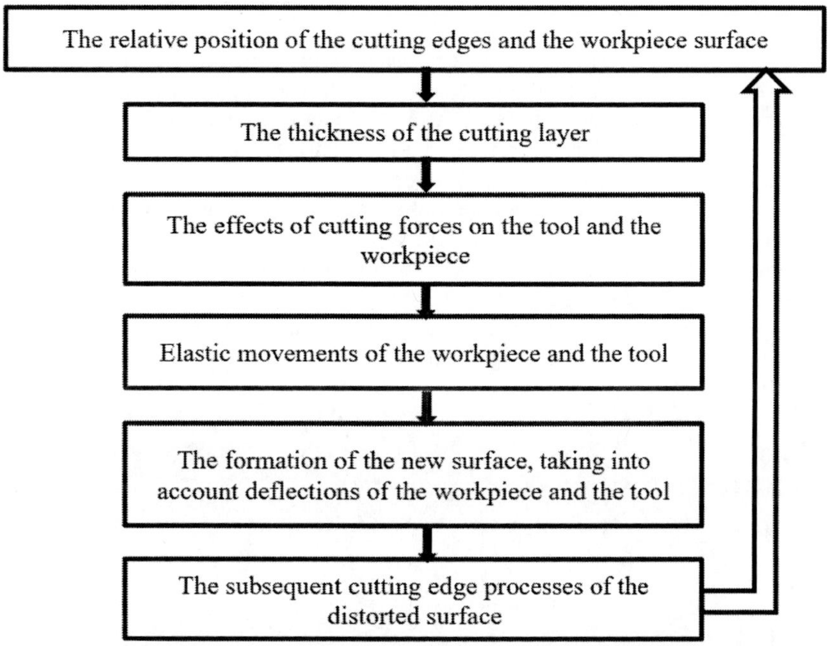

Figure 1. The scheme of cutting process dynamics.

To exclude errors during formation of the machined surface, it is necessary to consider the moments when the tool is entering and exiting the workpiece. For this purpose, in the algorithm, before the calculation of the chip thickness and cutting forces, we should take into consideration the relative position of cutting edges and the workpiece surface, thereby checking whether milling cutter teeth are in the metal or not.

During passage of the milling cutter tooth, the elastic movements of the tool are calculated (with allowance for specified parameters of damping, rigidness and mass) and formed by the machined surface. The next tooth passes on the newly formed surface. This approach allows one to take into account the variable chip thickness, and consequently to simulate regenerative vibration during the machining process. Later, that allows us to get an adequate picture of the machining process. The structure of the mathematical model is presented in figure 2.

Designations of the equation: M, C, K - matrices that define the dynamic model of the milling cutter; F[P, H, G, V(t), V(t-T)] – cutting forces; P – parameters of the processed material; H – parameters of the processing conditions; G – parameters of the milling cutter geometry; V(t) - dynamic movements of the milling cutter; V(t-T) - dynamic movements of the milling cutter with time-lag. To obtain tool dynamic characteristics used in the simulation, a modal analysis is performed.

IOP Conf. Series: Materials Science and Engineering **177** (2017) 012080 doi:10.1088/1757-899X/177/1/012080

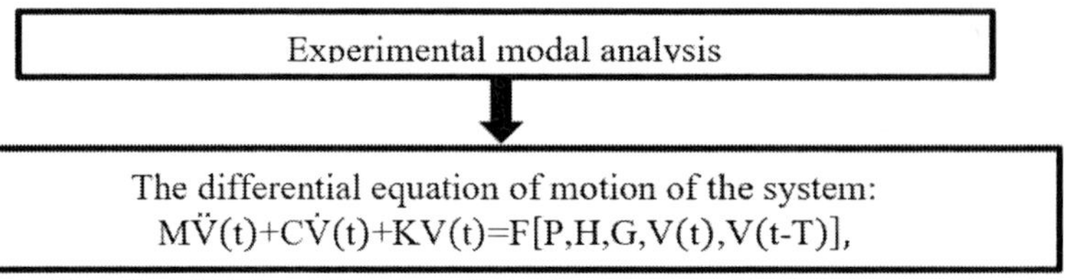

Figure 2. The structure of the mathematical model.

3. An experimental part. Modal analysis

The tool setup clamped in the spindle is presented in the form of a combined body which has mass, elasticity, resiliency and the ability to change the shape for a short time (Figure 3, a). Vibration of the setup, which has all these characteristics, can be divided into three main categories: free, forced and self-excitation.

Forced vibrations arise as a result of continuous external influences on the system. While the free vibrations are often represented in the time domain, the forced vibrations, as a rule, are analyzed in the frequency domain. This emphasizes the amplitude and phase dependence on frequency and permits to determine the natural frequencies. Vibrations with a relatively high amplitude occur when frequency of forced vibrations are close to the natural frequency.

The result of modal analysis is the stable cutting depth dependence on the spindle speed (Stability lobes). Cutting conditions, which reach within the area above the blue line, are unstable and accompanied by vibrations, modes located below the line theoretically should be stable by reducing the vibration component of the cutting forces.

According to the results of the performed modal analysis of finishing milling cutter AZ-3D16R6L65, made of hard alloy, the diagram of stability is built (figure 3, b). Modal analysis has been performed in the system 'Cutpro'.

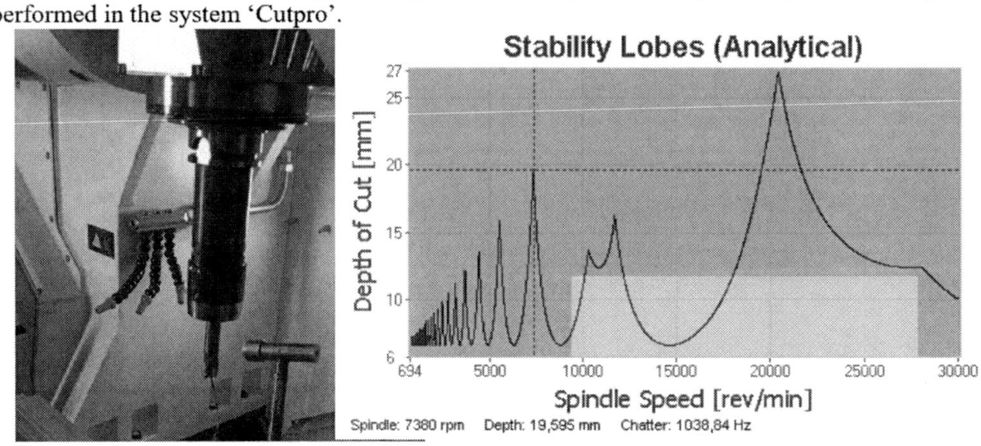

a) b)

Figure 3. Modal analysis of cutter AZ-3D16R6L65: a – studied instrumental setup; b – stability diagram.

Table 1 presents the machining process parameters, which are selected in accordance with the stability diagram.

Table 1. Machining process parameters

Tool	Cutting depth, a_p, mm	Cutting width, a_e, mm	Spindle speed, s, min^{-1}	Feed, f_z, mm/tooth
AZ-3D16R6L65	20	0.5	15000	0.08

The machined surface has been measured on contact profilometer Taylor Hobson Form TalySurf i200. The results are presented in figure 4.

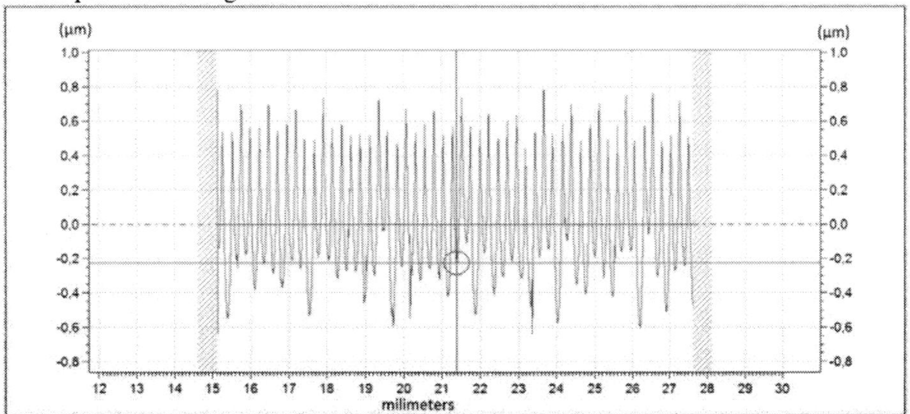

Ra	0,2432 (µm)		
Rp	0,7432(µm)	Rv	0,5842 (µm)
		Rz	1,3274 (µm)

Figure 4. The surface obtained after the machining process.

As we can see from figure 4:
R_{max}=1.4 µm, R_a=0.24 µm, R_z=1.33 µm.
Figure 5 shows the simulation result of the profile of the machined surface.

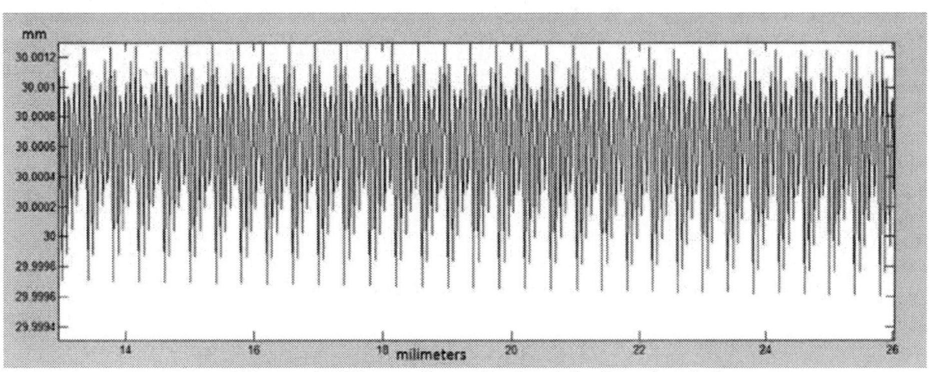

Figure 5. The surface obtained during simulation

In the program, the height of the workpiece was set to be equal to 50 mm; taking into account a_p=20 mm, we obtain the surface profile with:

$$R_{max}=|(30.0012-30)|+|(29.9998-30)|=0.0014 \text{ mm}=1.4 \ \mu\text{m}.$$

That correlates with the experimental data.

4. Conclusion

The convergence of calculation and experiment is evidence of the correct reaction of the math model to the changing machining process parameters. The time-lag mechanism underlying the developed model allows us to obtain the surface with set-up parameters of the machining process quite accurately. That gives us the opportunity to adjust the machining process conditions to obtain the best quality. The use of 'Simulation modeling' allows us to simulate regenerative vibration during cutting that gives the opportunity to receive a clear picture of the machining process.

Acknowledgments
This research was supported by the government of Russia within Project No. 2012-218-03-120. The authors are grateful to I A Kiselev and the staff of the High Productivity Machining laboratory of Irkutsk State Technical University, Alexey Pyatykh, Sergey Timofeev and Dmitry Paykin, for supporting this research and their contribution to conducting the experiments.

References
[1] Ponomarev B and Paykin D 2014 *Int. J. of Mech. and Mechatronics Eng.* **14** 1-5
[2] Rodygina A E 2013 *Int. J. of Eng. and Technol.* **5** 5140-5154
[3] Nikolaeva E P 2016 *Structure Investigation of the Constructional Steel St3ps after Argon-Arc Plasma Treatment* (Materials Science Forum vol **870**) ed. Radionov A A (Chelyabinsk: SUSU) pp 500-506
[4] Zamashchikov Y I 2015 *Int. J. of Machining and Machinability of Mat.* **17** 233-258
[5] Zamashchikov Y I 2011 *Int. J. of Advances in Machining and Forming Operations* **3** 141-152
[6] Zamashchikov Y I 2006 *Materials and Manufacturing Proc.* **21** 551-566
[7] Zamachtchikov Y I, Breaban F, Vantomme P, Deffontaine A 2002 *Lasers in Engineering* **12** 27
[8] Ahmadi K and Savilov A V 2015 *Int. J. of Machine Tools and Manufacture* **89** 208-220
[9] Ahmadi K and Altintas Y 2013 *Int. J. of Machine Tools and Manufacture* **68** 63-74
[10] Roukema J C and Altintas Y 2007 *Int. J. of Machine Tools and Manufacture* **47** 1455-1473
[11] Schmitz T L and Smith K S 2009 *Machining dynamics* (New York: Springer)
[12] Altintas Y 2012 *Metal cutting mechanics, machine tool vibrations, and cnc design* (New York: Cambridge University Press)
[13] Minis L, Yanushevsky R, Tembo A 1990 *CIRP Annals – Manufacturing Technol.* **39** 459-462
[14] Balachandran B and Zhao M X 2000 *Meccanica* **35** 89-109
[15] Campomanes M L, Altintas Y 2003 *J. of Manufacturing Science and Eng.* **125** 416-422

IOP Conf. Series: Materials Science and Engineering **177** (2017) 012081 doi:10.1088/1757-899X/177/1/012081

The frequency-domain method of calculation for the pulsed electromagnetic field in a conductive ferromagnetic plate

G V Nosov[1], E O Kuleshova[1], S Lefebvre[2], A A Plyusnin[1] and D M Tokmashev[1]

[1] Tomsk Polytechnic University, 30, Lenina ave., Tomsk, 634050, Russia
[2] Conservatoire National des Arts et Métiers, École Normale Supérieure de Cachan, 61 Av. Du President Wilson, Paris, F94235, France

E-mail: nosov@tpu.ru, kuleshova@tpu.ru

Abstract. The technique for parameters determination of magnetic skin effect on ferromagnetic plate at a specified pulse of magnetic field intensity on the plate surface is proposed. It is based on a frequency-domain method and could be applied for a pulsing transformer, a dynamoelectric pulse generator and a commutating inductor that contains an imbricated core. Due to this technique, such plate parameters as specific heat loss energy, the average power of this energy and the plate temperature raise, the magnetic flux attenuation factor and the plate q-factor could be calculated. These parameters depend on the steel type, the amplitude, the rms value, the duration and the form of the magnetic field intensity impulse on the plate surface. The plate thickness is defined by the value of the flux attenuation factor and the plate q-factor that should be maximal. The reliability of the proposed technique is built on a common frequency-domain usage applicable for pulse transient study under zero boundary conditions of the electric circuit and the conformity of obtained results with the sinusoidal steady-state mode.

1. Introduction

Many of today's electrophysical devices contain an imbricated core run under the pulsed electromagnetic field (EMF). Pulsing transformers, dynamoelectric pulse generators and commutating inductors provide examples [1-5, 10]. It is necessary to take uneven distribution of the pulsed magnetic flow in ferromagnetic plates of the imbricated core (magnetic skin effect) into account. To calculate its parameters, it is possible to solve EMF equations by a numerical technique. However, this technique is quite cumbersome so it makes determination of the pulsed EMF influence on imbricated cores too complicated [6-8]. On the other hand, the using of the frequency-domain method simplify computations and allows us to consider the shape, amplitude and duration of pulsed EMF at its diffusion in the ferromagnetic plate [9]. Consequently, such task as calculation of the magnetic skin effect in the ferromagnetic plate due to the frequency-domain method and choosing the plate thickness with respect to the pulsed EMF shape, duration and amplitude seems to be a vital one and reflects the purpose of the paper.

2. Assumptions

The following assumptions were used to obtain calculating formula.

1. Isolated from each other, ferromagnetic plates of the imbricated core at the same initial

temperature of θ_0 (°C) have the same values of absolute permeability μ_a (H/m), specific conductivity γ (1/Ω·m), volume density ρ (kg/m^3) and specific heat capacity C_T (J/kg°C).

2. Quite short EMF pulse of τ (s) heats ferromagnetic plates only by eddy currents (hysteresis neglected). It is adiabatic heating and heat transfer with the environment that has been neglected. Moreover, identical temperature rise ϑ (°C) of plates due to the pulsed EMF influence is minor in comparison with initial temperature θ_0 and does not lead to the change of μ_a, γ, ρ, C_T.
3. Thickness d (m) of ferromagnetic plates is much smaller than its height h (m) and length l (m), namely $d << h$ and $d << l$, so we will proceed from the conception of the plane one-dimensional electromagnetic wave penetrating from the opposite sides to conductive ferromagnetic plates along the X axe (Figure 1) perpendicular to plate lateral surfaces.
4. The EMF pulse is prescribed as an identical magnetic field intensity on plates lateral surfaces $H_S(t)$ – i.e. pulse time function t with duration τ and directed along the Z axe (Figure 1).
5. According to the frequency-domain method, the definable magnetic field in conductive ferromagnetic plates has zero initial conditions [9] that is the magnetic field in plates is completely absent (magnetic field intensity and induction are equal to zero).

3. The calculation method

Within these assumptions, let us consider the EMF influence on one ferromagnetic plate. Let us suppose that the plate lateral surface is an $x0y$ plane of Cartesian coordinates as it is shown in Figure 1. Then, E-vector $\overline{E}$, H-vector $\overline{H}$ and current density vector $\overline{\delta}$ of the plane one-dimensional electromagnetic wave have unit components each of which is dependent on the x coordinate and the t time [9]:

$$\overline{E} = E_y(x,t)\cdot\overline{1}_y;\ \overline{H} = H_z(x,t)\cdot\overline{1}_z;\ \overline{\delta} = \delta_y(x,t)\cdot\overline{1}_y,$$

where $\overline{1}_y, \overline{1}_z$ – lie in y and z axis unit vectors, respectively (Figure 1).

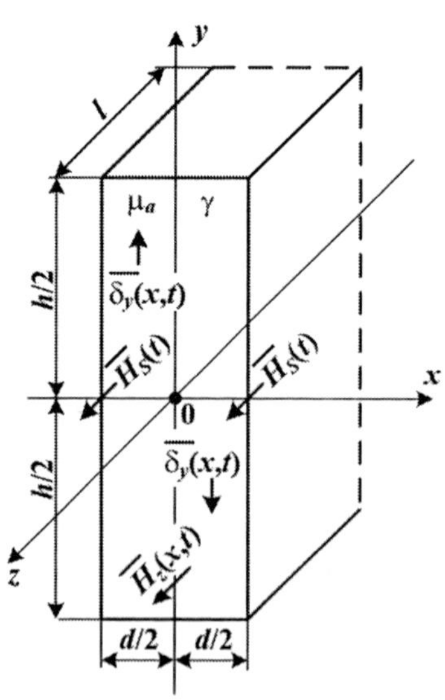

Figure 1. The conductive ferromagnetic plate: $H_S(t) = H_z(\pm d/2, t)$ – the magnetic field intensity on the plate surface.

In this case, EMF diffusion in the plate is described by the following equations [10]:

$$\frac{\partial^2 H_z(x,t)}{\partial x^2} = \mu_a \gamma \frac{\partial H_z(x,t)}{\partial t}; \ \delta_y(x,t) = \gamma E_y(x,t) = -\frac{\partial H_z(x,t)}{\partial x}, \tag{1}$$

where boundary and zero conditions are

$$H_y(\pm d/2, t) = H_S(t); \ H_y(x,0) = 0. \tag{2}$$

Let us find a spectral function of the magnetic field intensity on the conductor surface by the spectral (frequency-domain) method of unilateral forward Fourier transformation:

$$H_{Sm}(j\omega) = \int_0^\infty H_S(t)\cdot \exp(-j\omega t)\,dt = H_{Sm}(\omega)\cdot \exp\left[j\varphi_S(\omega)\right]; \ j = \sqrt{-1}, \tag{3}$$

where $H_{Sm}(\omega)$ and $\varphi_S(\omega)$ – the amplitude and the phase response of the magnetic field intensity pulse on the conductor surface as a function of cyclic frequency ω.

Then, equation (1) becomes:

$$\frac{d^2 H_{zm}(x,j\omega)}{dx^2} = j\omega\mu_a\gamma H_{zm}(x,j\omega); \ \delta_{ym}(x,j\omega) = -\frac{dH_{zm}(x,j\omega)}{dx}, \tag{4}$$

where $H_{zm}(x,j\omega)$ and $\delta_{ym}(x,j\omega)$ – the complex amplitude of the magnetic field intensity and the eddy current density as a function of the x coordinate and cyclic frequency ω.

Considering conditions (2, 3), the solutions of equations (4) will be:

$$H_{zm}(x,j\omega) = \frac{H_{Sm}(j\omega)}{\cosh\left[0.5\cdot p(j\omega)\cdot d\right]}\cdot \cosh\left[p(j\omega)\cdot x\right]; \tag{5}$$

$$\delta_{ym}(x,j\omega) = -\frac{p(j\omega)\cdot H_{Sm}(j\omega)}{\cosh\left[0.5\cdot p(j\omega)\cdot d\right]}\cdot \sinh\left[p(j\omega)\cdot x\right]; \ p(j\omega) = \sqrt{j\omega\mu_a\gamma}. \tag{6}$$

Next, it is possible to determine the magnetic field density and the eddy current density in the plate:

$$H_z(x,t) = \frac{2}{\pi}\int_0^\infty \left\{\mathrm{Re}\left[H_{zm}(x,j\omega)\right]\cdot \cos(\omega t)\right\} d\omega; \tag{7}$$

$$\delta_y(x,t) = \frac{2}{\pi}\int_0^\infty \left\{\mathrm{Re}\left[\delta_{ym}(x,j\omega)\right]\cdot \cos(\omega t)\right\} d\omega. \tag{8}$$

As a result, specific heat loss energy W_T (J/kg), the average power of its energy P_T (W/kg) and the plate temperature rise caused by eddy currents (8) will be:

$$W_T = \frac{1}{\gamma\rho d}\int_0^\infty \left[\int_{-0.5d}^{0.5d} \delta_y(x,t)^2\,dx\right] dt; \ P_T = W_T/\tau; \ \vartheta = W_T/C_T. \tag{9}$$

In virtue of (7), let us find the pulsed magnetic flux in plate $\Phi(t)$ and its rms value Φ (Wb/m):

$$\Phi(t) = \mu_a \int_{-0.5d}^{0.5d} H_z(x,t)\,dx; \ \Phi = \sqrt{\frac{1}{\tau}\int_0^\tau \Phi(t)^2\,dt}, \tag{10}$$

and calculate the values in the absence of eddy currents and the rms value of density on plate surface H_S that are then used for determination of magnetic flux attenuation factor K_Φ with regard to eddy currents:

$$\Phi_S(t) = \mu_a H_S(t) d; \ H_S = \sqrt{\frac{1}{\tau}\int_0^\tau H_S(t)^2\,dt}; \ \Phi_S = \mu_a H_S d; \ K_\Phi = \frac{\Phi}{\Phi_S}. \tag{11}$$

Taking into account (7, 9), we will define the average specific magnetic field energy in plate W_M (J/kg) and the ratio of W_M and W_T that we will call plate q-factor K_W:

$$W_M = \frac{\mu_a}{2\rho\tau d}\int_0^{\tau}\left[\int_{-0.5d}^{0.5d} H_z(x,t)^2\,dx\right]dt\;;\; K_W = \frac{W_M}{W_T}. \tag{12}$$

Let us examine the sinusoidal steady-state mode for comparison under:

$$H_S(t) = H_m \sin(\omega_0 t + \varphi), \tag{13}$$

when the complex amplitude of magnetic field intensity $\dot{H}_{zm0}(x)$ and eddy current density $\dot{\delta}_{ym0}(x)$ with respect to (5, 6) when $\omega = \omega_0$ are rated as follows:

$$\dot{H}_{zm0}(x) = \frac{H_m e^{j\varphi}\cdot\cosh\left[x\sqrt{j\omega_0\mu_a\gamma}\right]}{\cosh\left[0.5d\sqrt{j\omega_0\mu_a\gamma}\right]} = \left|\dot{H}_{zm0}(x)\right|\cdot\exp\left[j\psi_H(x)\right]; \tag{14}$$

$$\dot{\delta}_{ym0}(x) = -\frac{\sqrt{j\omega_0\mu_a\gamma}\cdot H_m e^{j\varphi}\cdot\sinh\left[x\sqrt{j\omega_0\mu_a\gamma}\right]}{\cosh\left[0.5d\sqrt{j\omega_0\mu_a\gamma}\right]} = \left|\dot{\delta}_{ym0}(x)\right|\cdot\exp\left[j\psi_\delta(x)\right], \tag{15}$$

In other words, magnetic field intensity $H_z(x,t)$ and eddy current density $\delta_y(x,t)$ will be:

$$H_z(x,t) = \left|\dot{H}_{zm0}(x)\right|\cdot\sin\left[\omega_0 t + \psi_H(x)\right];\; \delta_y(x,t) = \left|\dot{\delta}_{ym0}(x)\right|\cdot\sin\left[\omega_0 t + \psi_\delta(x)\right]. \tag{16}$$

Due to substitution (14-16) into (9-12), let us determine EMF integral characteristics in the sinusoidal steady-state mode.

4. Calculation results

Calculation of the magnetic skin-effect for the electrical steel plate of type 2411 and thickness d=0.5 (mm) with the intensity pulse of magnetic field equal to:

$$H_S(t) = \begin{cases} H_m \sin^n\left(\dfrac{\pi}{\tau}t\right) \text{ at } 0 < t < \tau; \\ 0 \text{ at } t > \tau \end{cases} \tag{17}$$

using formulas (3, 5-17) and replacement of integrals by sums have been carried out. Steel parameters at $\mu_0 = 4\pi\cdot 10^{-7}$ (H/m) are shown in table 1. Calculation results under various magnetic field intensity amplitudes H_m (17) are listed in tables 2, 3, 4. In addition, the sinusoidal steady-state mode at n=1; $\omega_0 = 2\pi f = \pi/\tau$ (tables 2, 3) and the pulse mode at $\tau = 0.1; 1$ (ms); $n = 0; 1$ (table 4) have been computed. Specific time dependence of pulsed intensity $H_S(t)$ (17) and magnetic field flux $\Phi(t)$ (10) for the steel plate of type 2411 at $H_m = 1000$ (A/m); $\tau = 0.1$ (ms); d=0.5 (mm); $\Phi_m = \mu_a H_m d = 0.685$ (mWb/m) are shown in Figure 2.

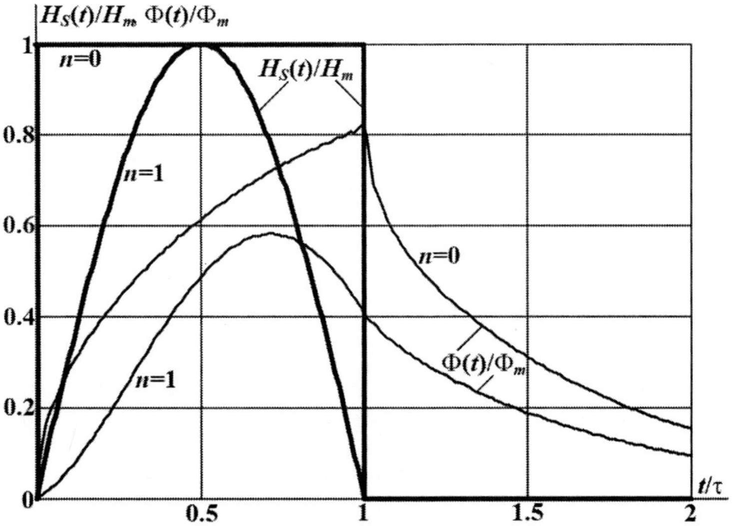

Figure 2. Specific time dependence of pulsed intensity $H_S(t)$ and magnetic field flux $\Phi(t)$ for the steel plate of type 2411: $H_m = 1000$ (A/m); $\tau = 0{,}1$ (ms); d=0.5 (mm); $\Phi_m = \mu_a H_m d = 0.685$ (mWb/m)

Table 1. Cold-rolled isotopic electric steel of type 2411 (silicon 2.8-3.8%)

H_m, A/m	$\mu_r = \mu_a/\mu_0$	B_m=$\mu_a H_m$, T	θ_0, °C	γ, 1/Ω·m	ρ, kg/m^3	C_T, J/kg°C
1000	1090	1.37				
2500	477.6	1.50	20	$2\cdot10^6$	7650	472.3
10000	135.3	1.70				

We can see that such characteristics as specific heat loss energy W_T, average power of this energy P_T and plate temperature ϑ grow with the amplitude H_m growth and the decline in duration τ of magnetic field intensity (17) (tables 2, 4). It is obviously that the use of the steel plate of type 2411 of thickness $d = 0.5$ (mm) is most likely for the pulse mode at such H_m, n, H_S and τ values when K_Φ and K_W are maximal (table 4): $H_m = 10000$ (A/m); $n = 1$; $\tau = 1$ (ms).

Table 2. The sinusoidal steady-state mode of the steel plate of type 2411 and thickness d=0.5 (mm)

H_m, A/m	f/τ, Hz/ms	W_T, J/kg	P_T, W/kg	ϑ, °C	K_Φ	K_W
1000	500/1	0.051	50.56	$1.07\cdot10^{-4}$	0.988	0.862
	5000/0.1	0.097	970.74	$2.06\cdot10^{-4}$	0.474	0.156
2500	500/1	0.063	62.99	$1.33\cdot10^{-4}$	1	1.951
	5000/0.1	0.334	3340	$7.07\cdot10^{-4}$	0.751	0.230
10000	500/1	0.082	81.56	$1.73\cdot10^{-4}$	1	6.876
	5000/0.1	0.760	7604	$1.61\cdot10^{-3}$	0.977	0.698

Table 3. The sinusoidal steady-state mode of the steel plate of type 2411 at frequency f=5000 (Hz)

H_m, A/m	K_Φ / K_W				
	d=0.05 (mm)	d=0.15 (mm)	d=0.25 (mm)	d=0.35 (mm)	d=0.5 (mm)
1000	1 / 8.534	0.992 / 0.956	0.895 / 0.362	0.714 / 0.213	0.474 / 0.156
2500	1 / 19.476	1 / 2.167	0.984 / 0.789	0.921 / 0.416	0.751 / 0.230
10000	1 / 68.748	1 / 7.640	1 / 2.753	1 / 1.408	0.977 / 0.698

Table 4. The pulse mode of the steel plate of type 2411 and thickness d=0.5 (mm)

H_m, A/m	τ, ms	H_S/H_m, A/m	n	W_T, J/kg	P_T, W/kg	ϑ, °C	K_Φ	K_W
1000	1	0.5	4.837	0.053	53.49	$1.13\cdot10^{-4}$	0.963	0.390
		0.707	1	0.043	43.15	$9.14\cdot10^{-5}$	0.994	1.021
		1	0	0.177	176.51	$3.74\cdot10^{-4}$	0.971	0.476
	0.1	0.5	4.837	0.047	468.74	$9.92\cdot10^{-5}$	0.518	0.167
		0.707	1	0.068	682.48	$1.44\cdot10^{-4}$	0.603	0.274
		1	0	0.148	1485	$3.14\cdot10^{-4}$	0.614	0.249
2500	1	0.5	4.837	0.079	78.84	$1.67\cdot10^{-4}$	1	0.772
		0.707	1	0.058	58.10	$1.23\cdot10^{-4}$	1	2.133
		1	0	0.468	468.02	$9.91\cdot10^{-4}$	0.997	0.517
	0.1	0.5	4.837	0.178	1778	$3.76\cdot10^{-4}$	0.724	0.201
		0.707	1	0.224	2240	$4.74\cdot10^{-4}$	0.816	0.379
		1	0	0.487	4873	$1.03\cdot10^{-3}$	0.807	0.335
10000	1	0.5	4.837	0.107	107.33	$2.27\cdot10^{-4}$	1	2.615
		0.707	1	0.079	78.63	$1.66\cdot10^{-4}$	1	7.190
		1	0	1.812	1812	$3.84\cdot10^{-3}$	1	0.621
	0.1	0.5	4.837	0.742	7421	$1.57\cdot10^{-3}$	0.944	0.338
		0.707	1	0.629	6286	$1.33\cdot10^{-3}$	0.985	0.855
		1	0	2.207	22070	$4.67\cdot10^{-3}$	0.959	0.462

5. Conclusion

1. The technique for parameters determination of the magnetic skin effect on the ferromagnetic plate at a specified pulse of magnetic field intensity on the plate surface is proposed. The technique is based on a spectral method and could be applied for a pulsing transformer, a dynamoelectric pulse generator and a commutating inductor that contains an imbricated core.
2. Due to this technique such plate parameters as specific heat loss energy, the average power of this energy and the plate temperature raise, the magnetic flux attenuation factor and the plate q-factor could be calculated. These characteristics depend on the steel type, the amplitude, the rms value, the duration and the form of magnetic field intensity impulse on the plate surface.
3. In virtue of the minor temperature rise of the given type of the plate under one impulse, its considering is necessary only under numerous EMF impulses.
4. The plate thickness is governed by the value of the flux attenuation factor and the plate q-factor that should be maximal.
5. The reliability of the offered methodology is built on a common frequency-domain usage applicable for pulse transient study under zero boundary conditions of the electric circuit and conformity of the obtained results with the sinusoidal steady-state mode.

References

[1] Spann M L, Bird W L, Weldon W F and Woodson H H 1983 *4th IEEE Pulsed Power Conf.,*

Albuquerque (NM, USA) 675-678
[2] Walls W A 2002 *IEEE Conference Record of Power Modulation Symposium* 33-38
[3] Rasilo P, Salem A, Abdallh A, Dupré L, Melkebeek J A 2015 *IEEE Transact. on Energy Conversion* **30** 736-744
[4] Lukutin B V, Shandarova E B, Matukhin D L, Makarova A F, Fuks I L 2016 *IOP Conf. Ser.: Mater. Sci. Eng.* **124** 012070
[5] Blume S, Jaritz M, Biela J 2015 *IEEE Transact. on Plasma Science* **43** 3385-3391
[6] Petrun M, Steentjes S, Hameyer K, Ritonja J, Dolinar D 2015 *Int. J. for Computation and Math. in Electr. and Electr. Eng.* **34** 710-723
[7] Rasilo P, Singh D, Belahcen A, Arkkio A 2013 *IEEE Transact. on Magnetics,* **49** 2041-2044
[8] Fuzi Janos 1995 *Periodica Polytechnica, Electr. Eng.* **39** 131-143
[9] Nosov G V, Kuleshova E O, Vassilyeva Yu Z and Elizarov A I 2016 *IOP Conf. Ser.: Mater. Sci. Eng.* **124** 012081
[10] Isaev Y N, Kolchanova, V A, Tarasenko S S, Tikhomirova O V 2015 *Proc. Int. Conf. on Mechanical Engineering, Automation and Control Systems* 7414894

The simulation of the workpiece and saw teeth heating during the high-speed cutting process

L M Gurevich, A I Bannikov, D V Pronichev, R E Novikov and A G Serov

Volgograd State Technical University, 28, Lenin Ave., Volgograd, 400005, Russia

E-mail: NovikovRome@gmail.com

Abstract. The paper presents the result of the simulation of the workpiece and saw teeth heating during high-speed cutting using the Deform 3D software package. The distribution of temperature and stressing fields during the high-speed cutting process was obtained. The investigation was confirmed by metallographic analysis, and it shows good convergence of simulation results with practical results for the temperature fields.

1. Introduction

The process of metal cutting proceeds under the condition of intense plastic deformation and chipping. The mechanical energy of the process is converted into heat, which affects plastic properties of the cut-off layer and its deformation, cutting forces, wear and tool life [1]. The existing studies contain little information on the high-speed cutting processes at high temperatures. There are some examples of works describing the influence of temperature on strength characteristics of the material at the high-speed cutting process [2–4]. The cutting process simulation can be performed via the finite element method (FEM) because the method considers a large number of material's properties and is suitable for multi-physics tasks. FEM allows one to reduce costs of new products development, to reduce or eliminate the necessity in field experiments; besides, it is one of the most informative and illustrative method among other modern research methods.

The present study aims to verify the suitability of FEM for the cutting process simulation. Temperature fields in the processed material as well as in the saw teeth during high-speed cutting of cylindrical workpieces are obtained.

2. Materials and methods

FEM simulations of the cutting process of the cylindrical steel workpiece was carried out using the Deform 3D software package. The longest time contact of the saw teeth with the hot workpiece (fig. 1) was considered via the simulation of the evolution of the saw teeth passing through the middle cross-section of the workpiece.

Model parameters: the diameter of the workpiece and the saw is 350 mm and 2500 mm, respectively, and the thickness of the workpiece and the saw is 100 mm and 12 mm, respectively. Saw tooth dimensions are shown in Figure 2. To simulate the heat transfer process and the stress-strain state, the finite element tetrahedral mesh was applied (fig. 3). When deformation exceeds 20%, the mesh is automatically rebuilt. The initial temperature of the workpiece was 1200°C while the initial temperature of the saw and the surrounding air was considered 20°C. The heat transfer coefficient between the workpiece and the saw – 30 W/(m^2×C), between the air and the parts – 0.002 W/(m^2×C). Side ends of the workpiece are fixed. The saw rotation speed is 1070 RPM, the feed rate is 3.5 m/sec.

Figure 1. The general view of the assembled model.

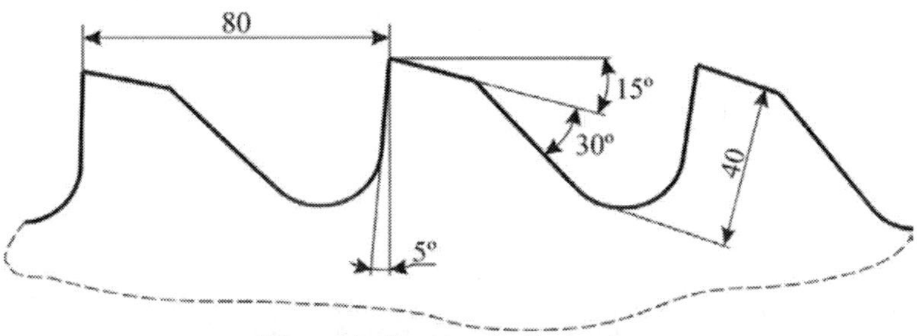

Figure 2. Saw tooth dimensions.

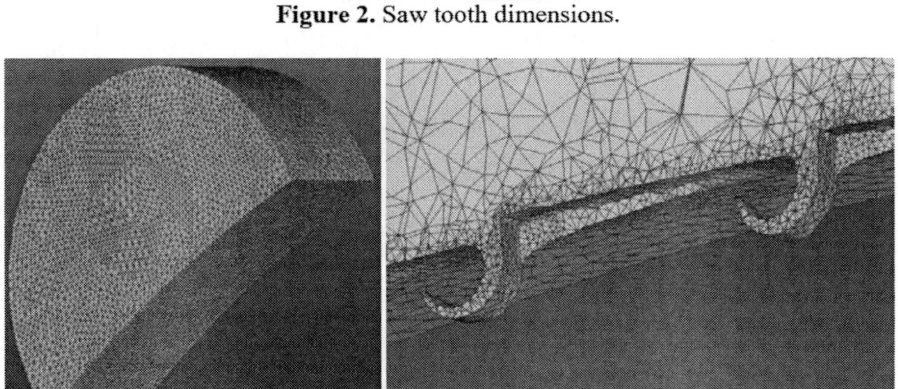

a b

Figure 3. The original mesh (a) and remeshing during chip formation (b, the section along the median plane).

The workpiece material was steel 0.08%C, while the saw material – AISI 4140. 3D deformation has an inbuilt base of temperature and strain dependent functions of Young's modulus, flow stress (graphs for steel AISI 4140 are shown in fig. 4a, b, c), heat capacity and thermal conductivity (graphs for both steels are shown in fig. 4d, e) for these steels. For steel 0.08%C, the strength properties are set using the Johnson-

Cook model [5] (table 1). As a criterion for the destruction of both steels, the Cockcroft-Latham model was used [6].

Table 1. The coefficients for the Johnson-Cook model for steel 0,08% C

A, MPa	B, MPa	m	n	$\dot{\varepsilon}_0$, sec^{-1}	T_m, K	T_r, K
350	275	1	0.36	1	293	1790

To verify the simulation results metallographic studies of the saw teeth, the structure and microhardness research was carried out. Optical images of the saw teeth were investigated by the AnaliSYS software for the Olympus BX61 microscope. The microhardness distribution was measured using a standard microhardness tester (PMT-3) with the Vickers indenter, carried out with loads of 50 g for 15 sec.

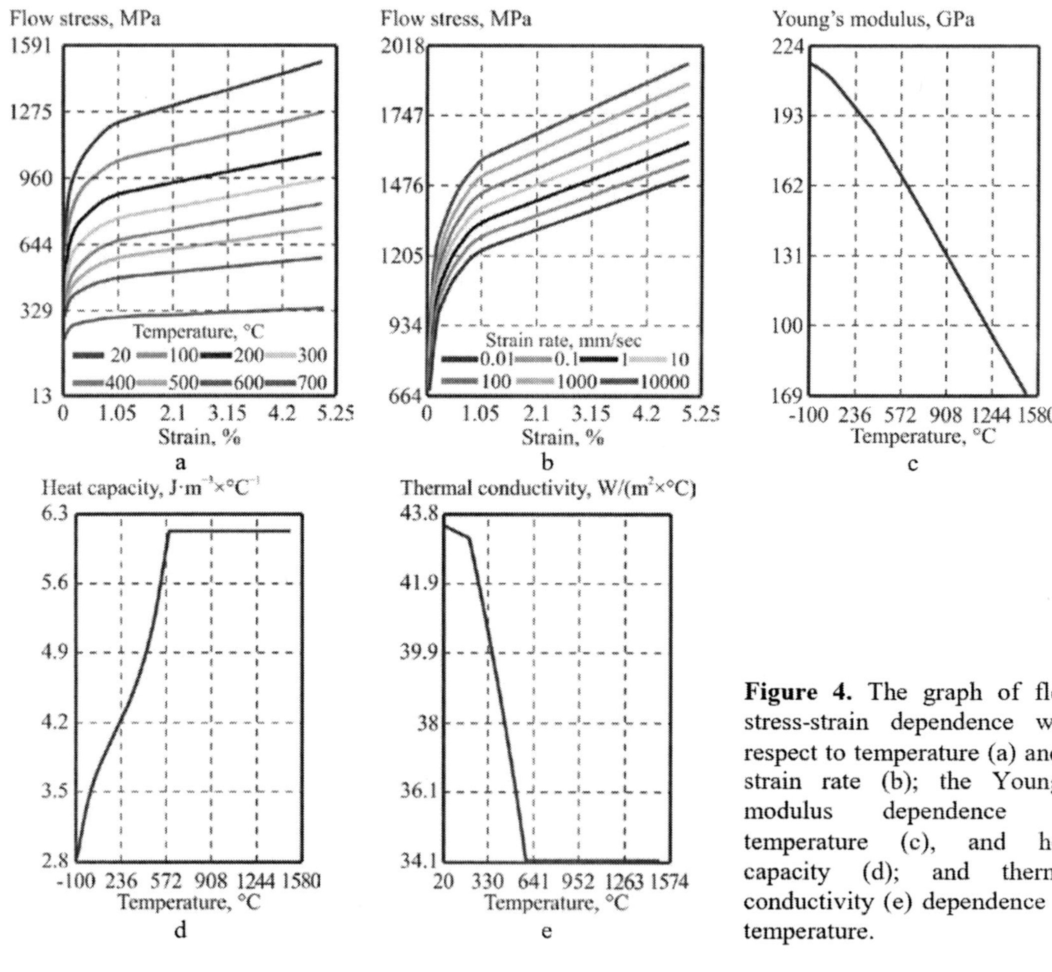

Figure 4. The graph of flow stress-strain dependence with respect to temperature (a) and a strain rate (b); the Young's modulus dependence on temperature (c), and heat capacity (d); and thermal conductivity (e) dependence on temperature.

3. Results and discussion

The simulation of temperature fields in the workpiece, the chip and the saw teeth during high-speed cutting showed that during intensive plastic deformation of the workpiece a thin layer is significantly heated along the plane of contact with the saw tooth of about 1500°C. The depth of heating the workpiece metal is approximately 5 mm (fig. 5a, b). Significant heating of the metal workpiece is confirmed by the high-speed movie of the cutting process on the seamless pipes cutting line [7].

The heated volume of saw teeth localized in the contact zone with the workpiece. The maximum heating temperature of the saw teeth was about 1220°C. The depth of the zone heated up to temperatures of phase transformations in steel (> 727°C) does not exceed 150 μm. During cutting of a workpiece, the deformation occurs both along and perpendicular to the cutting direction. Therefore, the cutting width is slightly greater than the thickness of the saw, and hence almost has no heating through the tooth sides (fig. 5c).

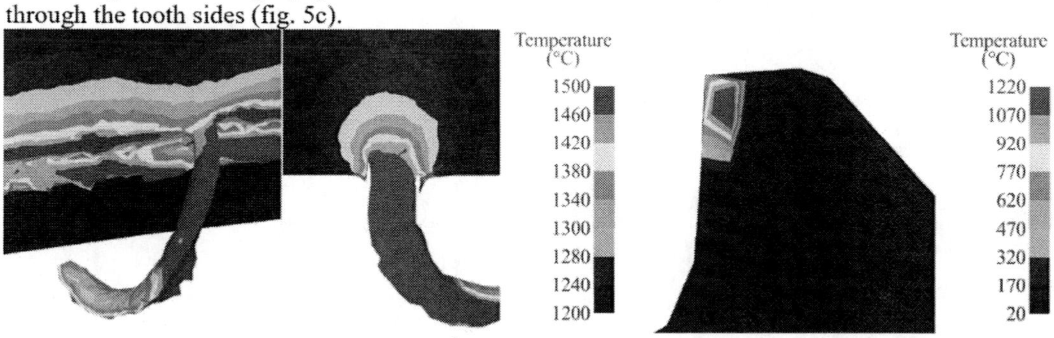

Figure 5. The temperature distribution along (a) and transverse (b) the section along the median plane of the workpiece, and the section along the median plane of the saw (c).

Maximum Mises stress values in the workpiece are located in the area of the contact with the top of the tooth (fig. 6a, b), and in the tip and the gullet of the tooth for the saw (fig. 6c).

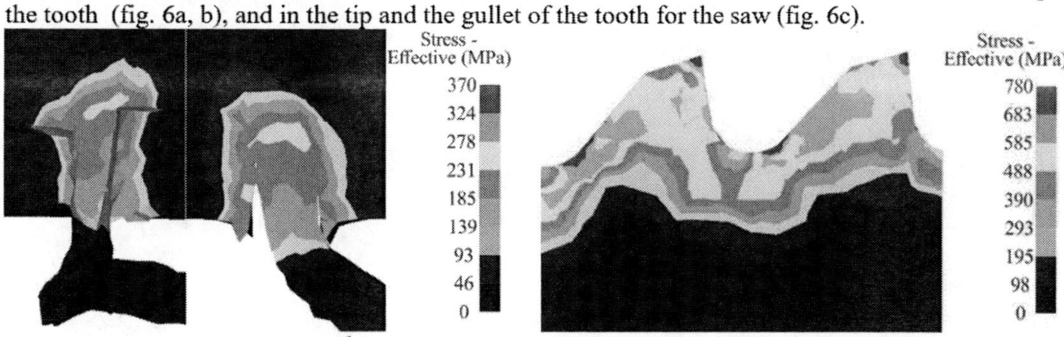

Figure 6. Stress distribution along (a) and transverse (b) the section along the median plane of the workpiece, and the section along the median plane of the saw (c).

To verify the simulation results, the metallographic examination of the saw tooth structure before and after the cutting process was carried out. Changes in the structure allowed us to determine the approximate depth of the saw metal heating during the cutting process up to the temperature above the temperature of phase transformations.

Initially, the microstructure of the saw tooth metal is heterogeneous etched tempered sorbite (fig. 7a). The heterogeneity of the tooth steel structure led to the variance of the measured microhardness from 2.5 to 3.8 GPa (fig. 7b) (the average value of microhardness was 2.9 GPa).

Metallographic examination of the tooth after the exploitation revealed the emergence of the metal layer in the workpiece-saw contact area. The layer formed as a result of hot metal adhesion [8]. The microhardness of the formed layer was 4.7-5.5 GPa (fig. 8). High speed of deformation restricts recrystallization processes even at temperatures of 1200°C. The formed layer gives off heat in the tooth and cools down rapidly after exiting the cutting zone. The high-etched zone (fig. 8a) consisting of structureless martensite and carbides formed during rapid cooling after pulse heating above the A_{C3} of the thin layer of tooth metal was found near the top of the working part of the tooth. The thickness of this zone is 130-180 μm, and the microhardness is ranged from 6 to 3.5 GPa by the distance from the surface. Figure 8b shows the distribution of microhardness from the surface into the depth in the working and non-working parts of the tooth. The structural change leading to the increase of the microhardness was found only in the working part of the tooth which is in contact with the hot workpiece. The depth of structural transformations is in good agreement with the results of temperature fields simulation.

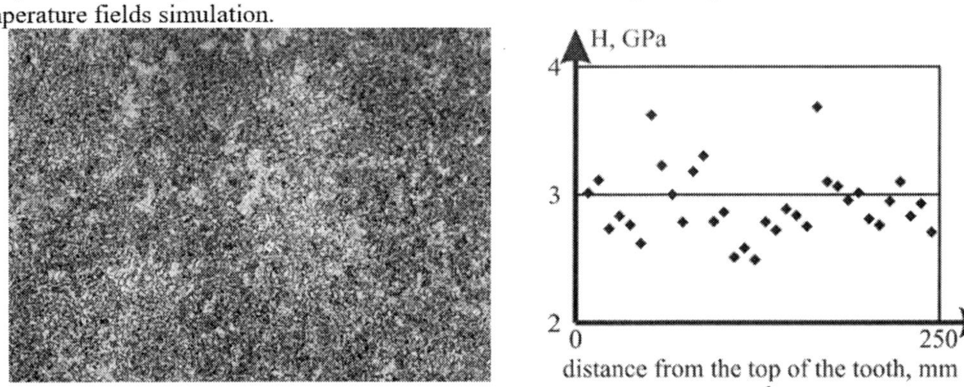

a b

Figure 7. The tooth microstructure (a, ×500) and distribution of microhardness (b).

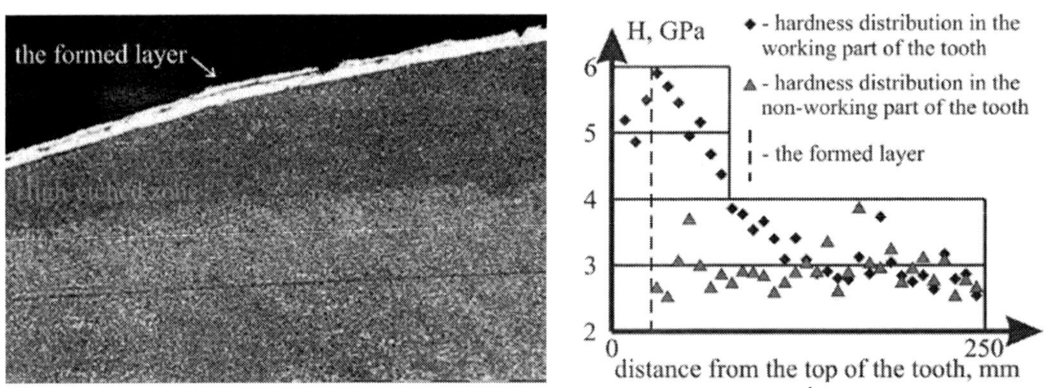

a b

Figure 8. The microstructure of the tooth surface layer after exploitation (a, ×500) and distribution of microhardness (b).

4. Conclusion

FEM simulation of the high-speed cutting process of the hot steel cylindrical workpiece using the Deform 3D software package was carried out. The simulation has allowed defining the distribution of both temperature and stressing fields during the high-speed cutting process.

Metallographic analysis of the saw tooth structure for high-speed hot cutting has revealed the hardened layer in the surface area formed due to the high thermal exposure. The thickness of this layer confirms the simulation results of heat transfer processes in the Deform 3D software package.

Acknowledgments

The reported study was funded by the Russian Science Foundation according to research project No.-29-00158.

References

[1] Edward M T and Paul K W 2000 *Metal Cutting* (Woburn: Butterworth-Heinemann)

[2] Sutter G and Ranc N 2007 *Int. J. of Machine Tools and Manufacture* **47(10)** 1507-17
[3] Wang C, Ding F, Tang D, Zheng L, Li S and Xie Y 2016 *Int. J. of Machine Tools and Manufacture* **108** 13-26
[4] Chen G J, Kong L G and Liu X L 2014 *Research of Dynamic Cutting Force in Hardened Steel Precision Turning Process* (Materials Science Forum) vol 800-801 pp 221-226
[5] Johnson G J and Cook W H 1983 A constitutive model and data for metals subjected to large strains, high strain rates and high temperatures *In proceedings of the seventh Int. symposium on Ballistics* **21** 541-547
[6] Cockcroft M G and Latham D J 1968 *Journal of the Institute of Metals* **96** 33-39
[7] Bannikov A I, Bannikov A A, Makarova O A and Permyakov I L 2014 *Russian Eng. Res.* **34(12)** 832-833
[8] Bannikov A I, Bannikov A A, Kurchenko A I, Dyatlov N A and Permyakov I L 2011 *Russian Eng. Res.* **31(1)** 88-90

Implementation of inverse calculation method in diagnostics of digital electric substations

R G Oganyan, N D Narakidze, D V Shaykhutdinov, N I Gorbatenko and S O Maksuta

Platov South-Russian State Polytechnic University (NPI), 132, Prosveshenia str., Novocherkassk, Rostov region, 346428, Russia

E-mail: roman.work18@gmail.com

Abstract. Diagnostics of the equipment of a substation is an important element of effective work of the machine-building industry in view of the fact that engineering plants need continuous and high-quality power supply. The article presents the solution for the accelerated analysis of the complex technical system equipment condition by the example of a digital substation. It is based on Boolean algebra and logic-numerical polynomic models. This approach can be applied to state estimation of the power equipment and digital devices of the substation.

1. Introduction

Precipitant development of modern information technologies, microprocessor equipment allows one to move to a new level of automation of electric substations – to digital substations. To accomplish this transition, a lot of research related to digital substations are performed [1-10].

The digital substation represents the complex technical system (TS) which is subject to an operating mode change, so, continuous monitoring of its condition is necessary. The change of a reliability status of the TS by an example of a digital substation is connected with the transition of the TS to a condition of refusal or with transition of the TS to a pre-failure condition. Accordingly, considering specificity of the technical system, the requirements for agility control are imposed to it.

The power equipment, used on the substation, is delivered by the companies which are engaged in the mechanical engineering area. An important question is obtaining information on the current state of such equipment during its operation for the purpose of its diagnostics because the main functions of the substation are assigned to this equipment. In addition, any machine-building plant needs uninterrupted power supply, therefore, there are high requirements to the reliability of the power supply.

An electrical power supply system of machine-building plants consists of supplying, distributive, transformer and converting substations, power line and current distributors of high and low voltage. The system of power supply should be reliable, convenient and safe in service and provide necessary quality of energy and uninterrupted electric power supply in normal and post-emergency conditions.

Thus, diagnostics of the power equipment and substations in general, which feed machine-building plants, is an important element of effective work of the machine-building branch.

2. Materials and methods

System of Boolean functions, tree of malfunctions, mathematical model operation on the basis of application of logical-numerical models.

3. Results and Discussion

For ensuring uninterrupted power supply, it is necessary to carry out the fast control of a condition of substations. One of the systems, necessary for the work of the digital substation, is an information system of monitoring (ISM), namely an SCADA-system. The structure of ISM includes the software (the interpreter of values of the TS controllable parameters into the values characterizing conditions of the TS), allowing one to carry out the analysis of a condition of the system.

Management of processes on the digital substation depends on the reaction time. If the reaction time of t_r ISM and TS state change big in magnitude, this change can be missed by ISM (figure 1).

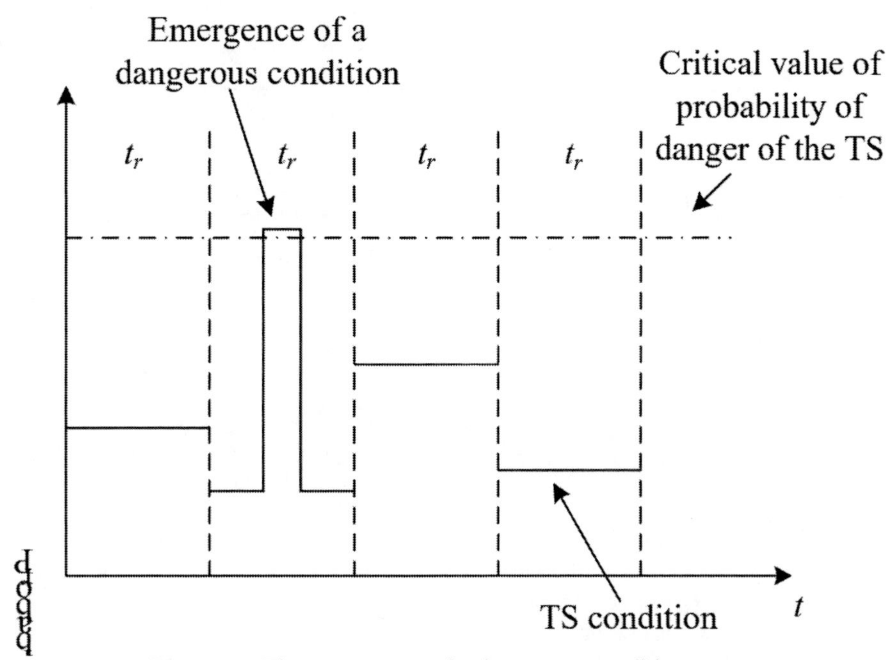

Figure 1. The emergence of a dangerous condition.

However, the time of reaction consists of several temporary intervals, such as time t_1 necessary for data collection, the time necessary for the analysis of this data t_2 and the time t_3 necessary for providing information characterizing a TS condition:

$$t_r = t_1 + t_2 + t_3.$$

But, with these time intervals, time $t_{в}$ for control action is also needed. Accordingly, the duration of a cycle of management of the TS is defined by the sum of times t_r of the reaction and times t_{ca} of realization of the control action.

For reduction of time t_r of the reaction, it is necessary to reduce the time of the analysis of condition, t_2. As a part of ISM, there should be models and control methods of a condition of the TS. One of the methods of representation of models is creation of the logic functions (LF) of failures of the TS. For designation of working ability (1) or not working state (0) of each of n TS elements, Boolean variables x_i ($i = \overline{1, n}$) are used. Values of the function on each of sets of values of variables allow one to draw a conclusion what condition a TS is in.

One of standard methods of creation of such LF is the method of creation a technical system malfunctions tree. The tree of malfunctions comprises a set of the initiating events (a deviation of values of parameters from admissible values, refusals of the TS elements), expressed by Boolean variables, as well as the hierarchy of nodes, to each of which Boolean function is compared. By means of a tree of malfunctions, it is possible to track cause-effect relationship between simulating events. During creation of a tree of malfunction for the TS, having an essential quantity of elements, there are difficulties connected with a large number of communications between a TS elements.

Let us make the tree of malfunctions of the TS by the example of the digital substation represented in figure 2.

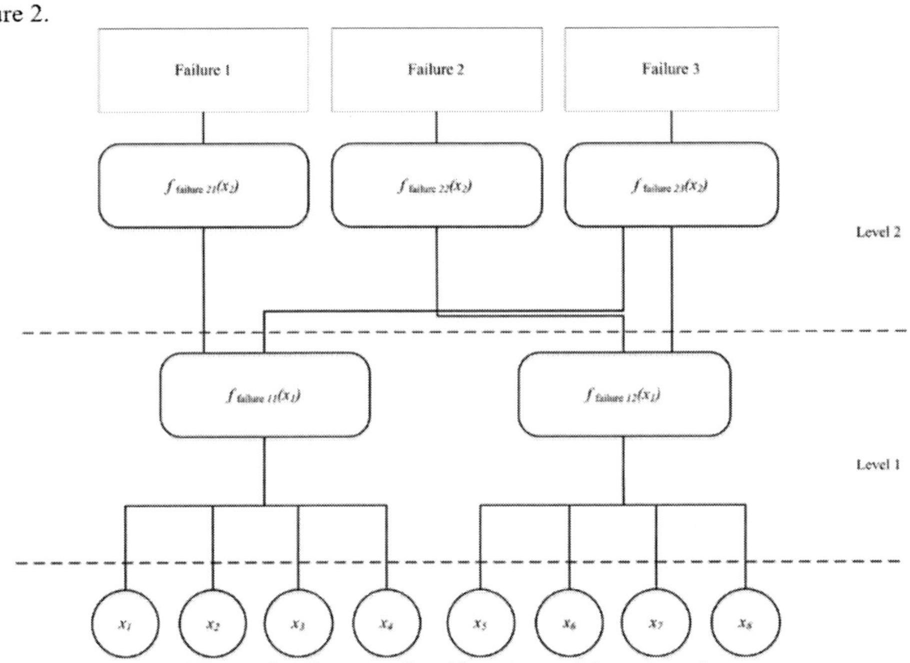

Figure 2. The tree of malfunctions of the substation.

For the substation, following malfunctions are defined:

- Failure of the power equipment;
- Failure of networks of data transmission;
- Failure of the power equipment and data-communication network.

One of logic functions 'AND', 'OR', 'ONDITION', 'EXCLUDING OR' and 'PRIORITY AND' corresponds to each node of the malfunction tree. The nodes of the tree, being at the lowermost level of the hierarchy, correspond to the initiating events changing conditions of the TS which are entered by logical variables.

A Boolean function (BF) can be corresponded to each node of the malfunction tree. For each level of the tree of malfunctions of the TS a system of Boolean functions (SBF) can be associated. SBF is used for the description of processes of TS in a failure condition transition.

Following SBF correspond to tree of malfunctions represented in figure 2,:

$$\text{Level 1:} \begin{cases} f_{failure11} = x_1 \wedge x_2 \vee x_3, \\ f_{failure12} = x_4 \wedge x_5 \vee x_6, \end{cases} \quad (1)$$

$$\text{Level 2:} \begin{cases} f_{failure21} = f_{failure11}, \\ f_{failure22} = f_{failure12}, \\ f_{failure23} = f_{failure11} \wedge f_{failure12}, \end{cases} \quad (2)$$

where $f_{failureij}$ – BF of failure of the node j, located on the i level of the tree of malfunctions considered by the TS;

$x_1 \ldots x_6$ – the Boolean variables describing existence of initiating events for transition considered by the TS under conditions of failure (value "0" of Boolean variable x_i corresponds to lack of the initiating event, and values "1" corresponds to event existence), symbols $\wedge$, $\vee$ have designated operations of logic addition and multiplication.

However, in case of a large number of the initiating events and/or additional functions, malfunctions analysis of the TS assumes the creation of SBF of high complexity. The described fact testifies to complexity of SBF, storage and realization of BF corresponding to a logic structure of the tree of malfunctions that in turn reduces productivity of the information system of monitoring of the technical system.

Parallel processes of the modern TS causing nessesarity of simultaneous realization of several BF – i.e. SBF. The effective solution of this task is application of a logical-numerical polynom (LNP) [11] which allows creating SBF in one arithmetic expression.

Using expressions:

$$x_1 \wedge x_2 = x_1 x_2,$$

$$x_1 \vee x_2 = x_1 + x_2 - x_1 x_2,$$

it is necessary to transform systems (1) and (2), using presented expressions:

$$\text{Level 1:} \begin{cases} f_{failure11}(x_1) = x_1 x_2 + x_3 - x_1 x_2 x_3, \\ f_{failure12}(x_1) = x_4 x_5 + x_6 - x_4 x_5 x_6, \end{cases}$$

$$\text{Level 2:} \begin{cases} f_{failure21}(x_2) = f_{failure11}(x_1), \\ f_{failure22}(x_2) = f_{failure12}(x_1), \\ f_{failure23}(x_2) = f_{failure11}(x_1) f_{failure12}(x_1). \end{cases}$$

Then, it is necessary to multiply level-by-level expressions on relative to them weighting factors: $2^0, 2^1, 2^2, \ldots, 2^{s-1}$ (s – quantity of BF entering into SBF at each level):

$$\text{Level 1:} \begin{cases} f'_{failure11}(x_1) = 2^0 (x_1 x_2 + x_3 - x_1 x_2 x_3), \\ f'_{failure12}(x_1) = 2^1 (x_4 x_5 + x_6 - x_4 x_5 x_6), \end{cases}$$

$$\text{Level 2:} \begin{cases} f'_{failure21}(x_2) = 2^0 f_{failure11}(x_1), \\ f'_{failure22}(x_2) = 2^1 f_{failure12}(x_1), \\ f'_{failure23}(x_2) = 2^2 f_{failure11}(x_1) f_{failure12}(x_1). \end{cases}$$

For receiving LNP $D_i(x_i)$ for each of the levels of the tree of malfunctions, summation of factors of an arithmetic polynom on each level is performed:

$$D_1(x_1) = x_1 x_2 + x_3 - x_1 x_2 x_3 + 2x_4 x_5 + 2x_6 - 2x_4 x_5 x_6,$$

$$D_2(x_2) = f_{failure11}(x_1) + 2 f_{failure12}(x_1) + 4 f_{failure11}(x_1) f_{failure12}(x_1).$$

Let's consider following abnormal situation example.

The initiating events x_1 - transformer temperature increase; x_2 – the increase of the current proceeding through the transformer; x_3 - turn-to-turn short circuit; x_4 - duplicative stream supply on the information network; x_5 - overload of the information network; x_6 - time server malfunction.

Let's simulate the situation as follows: $x_1 = [0\ 0\ 1\ 1\ 0\ 1]$. Then, the values of polynom $D_1(x_1)$ during substitution will be equal to $D_1(x_1) = 3_{(10)}=11_{(2)}$. Thus, values of polynom $D_1(x_1)$ will allow us to write down value $x_2 = [1\ 1]$. Then values of polynom $D_2(x_2)$ during substitution it will be equal to $D_2(x_2) = 7_{(10)}=111_{(2)}$, then analyzing binary representation of the value of polynom $D_2(x_2)$ it is possible to draw a conclusion that the emergence of three refusals will occure on the set of values of variables x_1,. This conclusion can be traced graphically on a basis of figure 2.

4. Conclusion

Thus, the used mathematical support based on LNP, allowing to increase productivity of ISM-difficult TS by an example of the digital substation, as well as to estimate the current status of the equipment of the substation. This approach is effective, because the digital substation represents the system with big amount of elements and connections and the analysis of a condition of the digital substation is a critical time parameter. Providing the fastest analysis of a condition of the substation will allow to take actions on elimination of the arisen malfunctions and the fast restoration of power supply that is an important element of the machine-building area.

Acknowledgments

The results were obtained using the equipment center for collective use 'Diagnosis and energy-efficient electrical equipment' (NPI).

References

[1] Rybakov S N, Ukhov V I, Slabospitssky S R, Kovtsova I O 2014 *J. Releyshchik* **1** 32-39
[2] Dubrov V I, Oganyan R G, Shaykhutdinov D V, Kiriyevsky E V, Kruglova T N and Narakidze N D 2016 *J. Basic researches* **9** 16-20
[3] V I Dubrov, R G Oganyan, D V Shaykhutdinov, E V Kiriyevsky, N I Gorbatenko and N D Narakidze 2016 *J. Modern High Techn.* **9** 36-40
[4] Drozdova T V, Elov N E and Morozov A P 2016 *J. Energy of a Uniform Network* **3** 54-61
[5] Xin J, Liao Z and Wen F 2010 *Int. Conf. on Power System Technology* (Tongji Univ., Shanghai) pp 1-5
[6] Jingbo H and Longhua M 2006 *Int. Conf. on Power System Technology* (Tongji Univ., Shanghai) pp 1-5
[7] Cipriani G, Cascia D La, Dio V Di, Lipari A and Miceli R 2014 *16th Int. Power Electronics and Motion Control Conf. and Expos.* (Antalya) pp 1300-05
[8] Ling Z, Chaoran Y and Yanxiang H 2011 *Int. Conf. on Mechatronic Sci., Electric Eng. and Computer* (Power Univ., Beijing) pp 1374-77
[9] Peng P, Liao Z, Wen F and Huang J 2013 *SCI 465* (Springer-Verlag Berlin Heidelberg)
[10] Zhao S, Jia X, Li B and Yuan J 2004 *Int. Conf. on Power System Technology* vol 2 pp 1179-82
[11] Sharay V A, Burangulova O S and Andriuts M V 2011 *J. News SFU. Technical Sci.* **12** 35-49

Diagnostic features of relief formations on the nanostructured titanium VT1-0 surface after laser shock-wave treatment

I V Lytvynenko[1], S A Lupenko[1], P O Maruschak[1], S V Panin[2,3] and Yu I Hats[1]

[1] Ternopil Ivan Pul'uj National Technical University, 56, Ruska st., Ternopil, 46001, Ukraine

[2] Institute of Strength Physics and Materials Science RAS, 2/4, Akademicheskii Ave., Tomsk, 634021, Russia

[3] National Research Tomsk Polytechnic University, 30, Lenin Ave. Tomsk, 634050, Russia

E-mail: maruschak.tu.edu@gmail.com

Abstract. A new class of diagnostic features for conducting morphological analysis of relief formations induced by laser shock-wave treatment on the surface of the nanostructured titanium VT1-0 alloy is proposed. They are the coefficients of series expansions of statistical estimates for the orthogonal basis of Chebyshev, Laguerre, Kravchuk discrete polynomials and trigonometric functions. Based on the criterion of the minimum number of the diagnostic features in the above-mentioned bases, the Chebyshev one was selected as the most appropriate to solve this problem.

1. Introduction

The relationship between structural features of materials and constructions with their physical-mechanical properties, particularly strain resistance, can be established on the basis of the concept of structural levels of deformation, which belongs to the area covered by the physical mesomechanics of materials [1].

Practical applications of this approach are of importance when analyzing complex deformation processes that take place in structure-sensitive materials and constructions [2]. In doing so the material surface is considered as an indicator to characterize processes of damage accumulation in the form of shear and rotational displacements of elementary structural elements – grains, or rarely their conglomerates. Since this process possesses an expressed stage and a monotonic pattern, it makes it possible to use the data on the surface strain induced relief to evaluate a (diagnostic) state of the object. Let us consider the behavior of the surface characteristics of the material in more detail [3].

The surface is treated as a complex, hierarchically organized system that under loading induced accumulation of micro-damages evolves and adapts to the force effect [4-6]. This self-organization results in the formation of strain induced relief which might be taken as an informative parameter of the system mechanical state. There are some literature data on attempts to propose a mathematical description of the relief formations. They allow us to establish the relationship between deformation processes that develop at the macro-, meso- and microscale levels [7-9]. Also, a number of papers devoted to the experimental analysis of the surfaces state in a number of materials under static, cyclic and shock-wave loading are available. However, in general, mathematical processing of the results and particularly procedures of their numerical automated analysis requires their further development. Design of modern systems for the processing and simulation of cyclic attributes of the relief formation allows automating and substantially intensifying the process of their analysis, diagnosis and prognosis, as well as opens the possibility of carrying out computer simulations during their modeling.

The aim of this study is to substantiate the required minimum set of diagnostic parameters that are sensitive to the relief formation process and might be employed as the informative features/attributes for automated diagnostic systems.

2. Physical reasoning for relief formation attributes analysis

Plastic deformation at the mesoscale level in the surface layers of nanostructured materials is developed with low energy dissipation. It is this fact that allows one to reveal nonlinear waves of localized plastic flow in nanostructured surface layers [10]. Regularities of nonlinear wave formation are associated with an ordered distribution of stresses and self-organization which take place in the nanostructured material under loading.

The data obtained with the use of the interferential profilometer 'Micron-alpha' were employed to investigate micro-localizations of relief formations on the surface of the nanostructured VT1-0 titanium after laser shock-wave treatment. The obtained experimental data make it possible to reveal distribution of relief formations that exhibit a "chessboard" pattern in two-dimensional observation where the regions experiencing compressive stress are alternated with tensile stressed ones [1, 2]. Stress relaxation can take place through the processes of heat and mass transfer being initiated in the tensioned regions towards the compressed ones. In so doing, a folded relief is formed in places where the maximum compressive stresses are reached.

Nanostructured titanium specimens were subjected to laser shock-wave treatment in various environments: air, water and ink. Specimens with the 80×80×5 mm gauge length were treated with the use of the GOS-1001 laser with the LiF gate in a Q-switched mode. It is suggested that the material surface "feels" deformation manifestations that are accompanied by the generation of deformation fields which spatial structure reflects the functional state of the material. The using of the non-contact profilometry of high resolution makes it possible to obtain a clear numerical characteristic of the surface state being covered with strain induced relief formations. The data obtained were regarded as the attributes which can be analyzed using statistical methods [11]. This predetermines the fundamental possibility of increasing the accuracy, informativeness of automated (computer-aided) processing of such attributes and determination of the surface state condition through the analysis of the relief formations. This approach can ensure standardization of automated processing and simulation of relief formation attributes at various stages of structures life-time assessment, and improve the accuracy and completeness of the technical diagnosis resulting from the use of the similar diagnostic features.

3. Methodological approach

The estimates of the obtained probability characteristics [9] might be used as diagnostic features. However, they possess the substantial drawback – a large number of samples that are necessary to be set (samples of the estimate for the expectation and the variance). This makes it difficult to use them in constructing training sets in the systems for analysis of the diagnostic attributes.

In order to reduce the dimension of the diagnostic feature space under analysis of the obtained probability estimates, it is suggested to use the coefficients of orthogonal representation estimates of the mathematical expectation and the variance in the series of orthogonal bases as the diagnostic attributes. If a system of orthonormal discrete functions is given as $\left\{\psi_n(l),\ n \in \boldsymbol{Z},\ l=\overline{0,L}\right\}$, discrete function $f(l),\ l=\overline{0,L}$ can be written as a series of

$$f(l)=\sum_{n=0}^{\infty} c_n \cdot \psi_n(l),$$
$$c_n=\sum_{l} f(l)\cdot \psi_n(l)\cdot \rho(l), \tag{1}$$

where c_n – the coefficients of the series; $\rho(l)$ – the weight function which selection depends upon the type of the approximating function: the weight function should reach a maximum in the area where the best approximation is required.

Complex $\{c_n,\ n \in \mathbf{Z}\}$ is called function range $f(l),\ l = \overline{0,L}$ in orthonormal basis $\left\{\psi_n(l),\ n \in \mathbf{Z},\ l = \overline{0,L}\right\}$.

It is known that the same function can be expanded in a series of different orthonormal bases. Let us represent function $f(l),\ l = \overline{0,L}$ as a series (2) as well as one in the basis of other orthonormal functions $\left\{\varphi_k(l),\ k \in \mathbf{Z},\ l = \overline{0,L}\right\}$

$$f(l) = \sum_{k=0}^{\infty} d_k \cdot \varphi_k(l), \tag{2}$$

where d_k – the serial coefficient.

Let us multiply both sides (1) by $\varphi_k(l)$ and sum over l, where function $f(l),\ l = \overline{0,L}$ is given:

$$\sum_l f(l) \cdot \varphi_k(l) = \sum_l \sum_{n=0}^{\infty} c_n \cdot \psi_n(l) \cdot \varphi_k(l) = \sum_{n=0}^{\infty} c_n \sum_l \psi_n(l) \cdot \varphi_k(l). \tag{3}$$

The left side of this equation is the k-th coefficient of d_k of the normalized number for function $f(l),\ l = \overline{0,L}$ while the sum of the right side (denoted by Ψ_{nk}) is a ratio of the same series for basic functions $\psi_n(l)$. Here,

$$d_k = \sum_{k=0}^{\infty} c_n \cdot \Psi_{nk}. \tag{4}$$

Coefficients Ψ_{nk} are independent of $f(l)$. Thus, expression (4) establishes a relationship between the spectra of expansion coefficients in series of function $f(l)$ in different orthogonal bases.

Basic functions $\left\{\psi_n(l),\ n \in \mathbf{Z},\ l = \overline{0,L}\right\}$ can possess quite complex spectrum at their expansion in a series over the system of other orthogonal functions $\left\{\varphi_k(l),\ k \in \mathbf{Z},\ l = \overline{0,L}\right\}$.

This proves the possibility of a system of basic functions to exist among certain systems of basic ones that allow approximating function $f(l),\ l = \overline{0,L}$ with the same accuracy by the smallest number of coefficients in comparison with all other terms of this set.

A study on decomposition of different assessments of probability characteristics for variety of orthogonal bases was conducted. Within this research, we have limited it by the expansions of only a few orthogonal bases.

The most well-known and widespread polynomials with well-studied properties of an orthogonal basis are discrete argument polynomials of Chebyshev, Kravchuk, Laguerre [12], and discrete trigonometric functions. First, we have to determine the selection criteria for the orthogonal functions system in which a series expansion will be carried out being based on the statistical processing assessment results of expectation of the relief under analysis. In doing so, the criterion based on the minimum amount of the first coefficients of the orthogonal series in case of their equal percentage contribution to the total energy of the obtained probabilistic assessments was used. It is the so-called energy criterion.

The expansion of the expectation and the variance by the series for the four above-mentioned orthogonal bases was carried out. It is based on the criterion of a minimum number of the series coefficients, which contribute at least 95% to the total energy of the signal. It was found that the fewest number of coefficients is used during the expansion in series by Chebyshev orthogonal polynomials.

Thus, the coefficients of Chebyshev series are encouraged to use as the diagnostic features during the diagnostic evaluation of probability characteristics. The expectation estimates which were expanded in series by the selected polynomials is shown in Figure 1, while their results – in Figures 2, 3, 4.

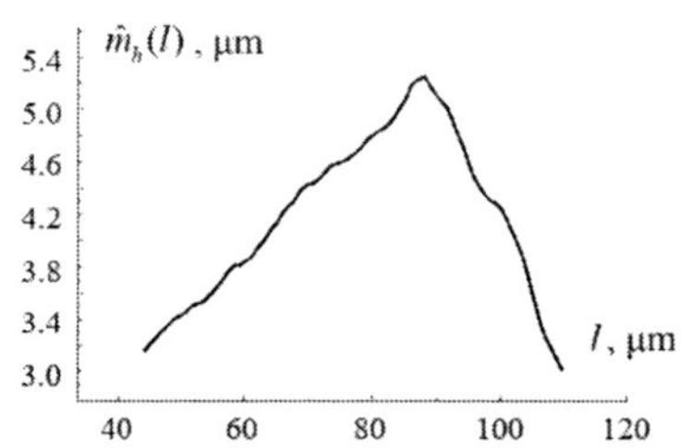

Figure 1. Estimation of the expectation for the relief formations width on the surface of nanostructured titanium after laser shock-wave influence in air [9].

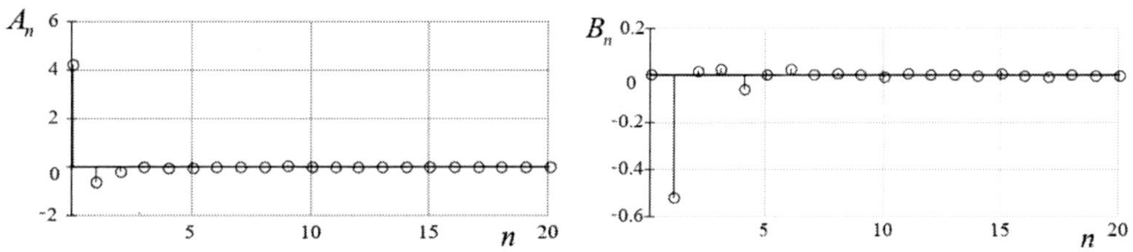

Figure 2. Expansion of estimates in series of trigonometric functions; two-component.

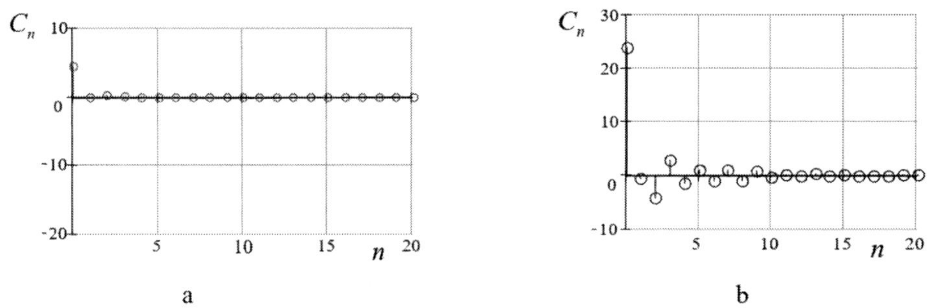

Figure 3. Expansion of estimates in Laguerre (a) and Chebyshev (b) series.

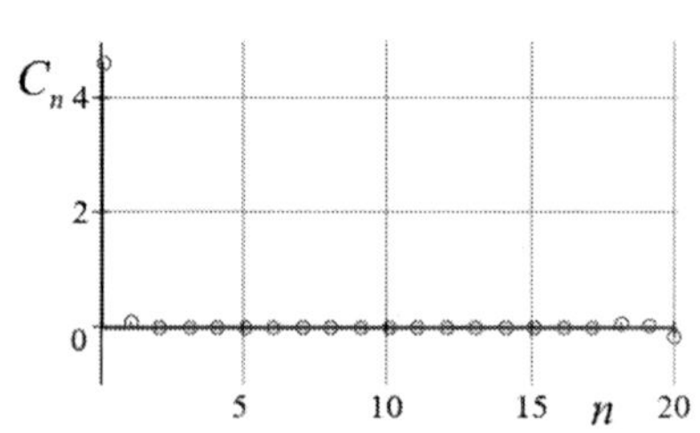

Figure 4. Expansion of estimates in Kravchuk series.

4. Discussion

Let us note that "the relief segments" formed on the surface of the material after the treatment are not equal. This is related to the local heterogeneity of the metal melting, crystallization and evaporation processes to take place on the material surface under laser irradiation [13]. Development of these processes depends on the pressure and duration of the shock pulse which are crucial during ordered relief formation in the area under modification.

The above-stated features of the processing of the analyzed cyclic signal allow us to cover all modes of nanostructured titanium modification. In doing so, they might be considered as a particular case in the framework of a united theoretical-methodological approach as well as take into account the wide range of possible cyclic attributes (to reflect the recurrence of signals in the structure) and the substantial structural differences during relief variation. As a result of the expansion and the analysis, the block diagram is given in Fig. 5 which shows the percentage energy contribution of two series factors of the studied orthogonal polynomials. The greatest energy contribution is made by two coefficients of the Chebyshev series – 99.45 %, while the coefficients of the series of trigonometric functions are 99.56 %. However, in the case of the series expansion in the trigonometric functions, the four coefficients (A_1, A_2, B_1, B_2) are to be considered. In order to minimize the number of the diagnostic features, only two coefficients of Chebyshev series were selected.

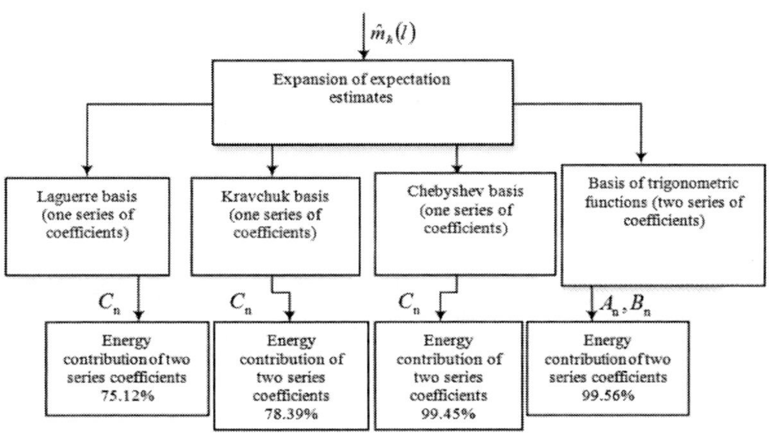

Figure 5. The block-diagram illustrating series expansion in orthogonal bases.

A series of experiments in which the expectation estimates were expanded in series of orthogonal Chebyshev, Kravchuk, Laguerre bases by the discrete trigonometric functions has been conducted. On the basis of the criterion of the minimum number of terms in the series that bring the total signal energy contribution not less than 95%, it is proposed to use the coefficients of the Chebyshev series of discrete argument as a diagnostic attributes.

The proposed mathematical approaches can be used to develop special software aimed at running surface diagnostics of the nanostructured titanium implants after the high-energy treatment [14]. The use of this approach allows us not only to develop an express method for diagnostics, but also to predict the mechanical state of the surface layer of modified materials.

5. Conclusion

A new class of diagnostic attributes in the form of expansion coefficients of statistical estimates (expectation estimates) of relief formation in a series of Chebyshev orthogonal polynomials for the problem of morphological analysis under different modes of the laser shock-wave treatment has been proposed The resulting expectation estimates can serve as a physical-mechanical basis for further diagnosis of the mechanical state of these materials surface layer through the analysis of the local region subjected to various types of the treatment/modification.

References

[1] Panin V E, Egorushkin V E and Panin A V 2010 *Phys. Mesomechanics* **13(5-6)** 215-232
[2] Panin V E and Egorushkin V E 2013 *Phys. Mesomechanics* **16(4)** 267-286
[3] Panin V E and Egorushkin V E 2011 *Phys. Mesomechanics* **14(5-6)** 207-223
[4] Balokhonov R R, Romanova V A and Martynov S A 2015 *Phys. Mesomechanics* **18(1)** 81-87
[5] Romanova V, Balokhonov R and Zinovieva O 2016 *Meccanica* **51(2)** 359-370
[6] Panin A V, Kazachenok M S, Kozelskaya A, Hairullina R R and Sinyakova E A 2015 *Mat. Sci. and Eng.: A* **647** 43-50
[7] Lytvynenko I V, Maruschak P O and Lupenko S A 2014 *Autom. Control and Comp. Sci.* **48(1)** 1-9
[8] Lytvynenko I V and Maruschak P O 2015 *Optoelectr., Instrumentation and Data Processing* **51(3)** 254-263
[9] Panin V E, Egorushkin V E, Panin A V and Chernyavskii A G 2016 *Phys. Mesomech.* **19(3)** 255-268
[10] Lytvynenko I, Maruschak P and Menou A 2013 *Proc. of the Intern. Symp. on Operational Res. and Appl.* (Marrakech, Morocco) pp 678-684
[11] Demyanchuk N and Lupenko S 2010 *Proc. of the X-th Int. Conf. 'Modern Problems of Radio Eng., Telecomm. and Comp. Sci.* (Slavske, Lviv) p 298
[12] Marchenko V B 1992 *Orthogonal to discrete argument functions and their application in geophysics* (Kiev: Naukova Dumka)
[13] Okipnyi I B, Maruschak P O, Zakiev V I and Mocharskyi V S 2014 *J. of Failure Anal. and Prevention* **14(5)** 668-674
[14] Allegrini S Jr, Yoshimoto M, Salles M B, de Almeida Bressiani A H 2014 *Int. J. Oral Maxillofac Implants* **29(1)** 63-70

Numerical simulation of electron beam welding with beam oscillations

D N Trushnikov and G L Permyakov

The department for Welding Production and Technology of Constructional Materials, Perm National Research Polytechnic University, 29 Komsomolsky ave., Perm, 614990, Russia

E-mail: gleb.permyakov@yandex.ru

Abstract: This research examines the process of electron-beam welding in a keyhole mode with the use of beam oscillations. We study the impact of various beam oscillations and their parameters on the shape of the keyhole, the flow of heat and mass transfer processes and weld parameters to develop methodological recommendations. A numerical three-dimensional mathematical model of electron beam welding is presented. The model was developed on the basis of a heat conduction equation and a Navier-Stokes equation taking into account phase transitions at the interface of a solid and liquid phase and thermocapillary convection (Marangoni effect). The shape of the keyhole is determined based on experimental data on the parameters of the secondary signal by using the method of a synchronous accumulation. Calculations of thermal and hydrodynamic processes were carried out based on a computer cluster, using a simulation package COMSOL Multiphysics.

1. Introduction

Currently, there have been widely known advances in the numerical simulation of electron-beam welding (EBW) and laser welding [1-3]. However, all results are related to modeling of welding without beam oscillations. A complete dynamic welding model, including beam oscillations, is still missing. The complex character, high speed of processes, high temperature gradients and multifactorial nature of the process make the direct numerical simulation of the conditions of periodic actions extremely difficult, even for modern computing resources. The main difficulty in modeling is determining the shape of the keyhole.

To solve this problem, a method for experimental determination of the keyhole shape using the parameters of the secondary signal in EBW with beam oscillation has been proposed [4]. Apart from the keyhole shape, this method provides additional information about the processes in the channel. In particular, this technique can be used to determine the distribution of the beam energy on the channel walls [5]. This approach eliminates the need to take into account all the complex factors that affect the formation of the keyhole.

The purpose of this study is to use these parameters as inputs to calculations to determine the effect of different beam oscillations and their parameters on the keyhole shape, the heat and mass transfer and the weld parameters, in the development of methodological recommendations. The calculations for the thermal and hydrodynamic processes were carried out based on a computer cluster, using a simulation package COMSOL Multiphysics. The calculations gave the geometry of the welds, the temperature fields and the flow velocity of the melt.

2. A mathematical model

Figure 1 shows a schematic of the welding model. The process parameters and physical parameters of steel are presented in Tables 1 and 2, respectively.

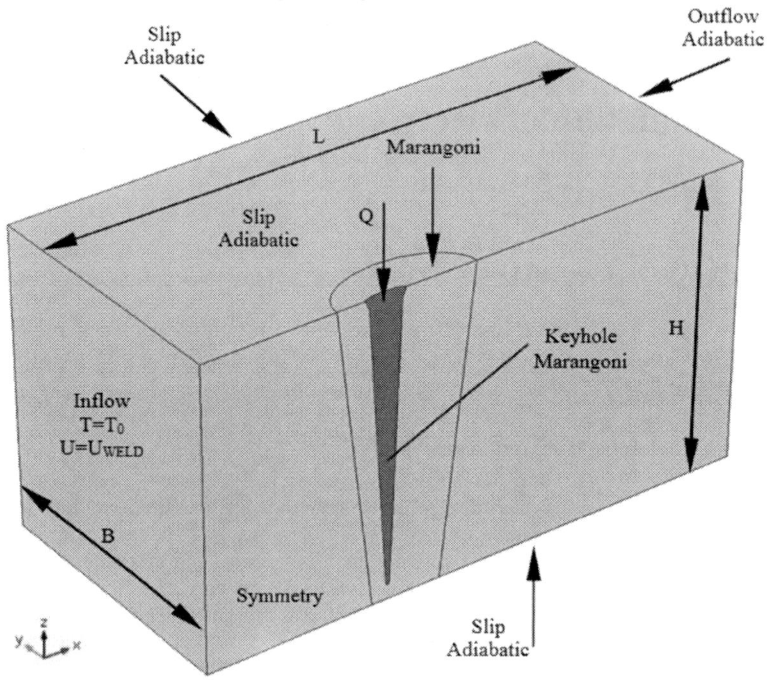

Figure 1. The schematic of the geometry of the model, initial and boundary conditions

Table 1. Process parameters.

Type of oscillations	Beam's parameters		Oscillation's parameters		Welding speed, U[mm/s]
	Power [kW]	Min radius [mm]	Amplitude [mm]	Frequency [Hz]	
Lengthwise	3	0.25	0.75	645	5

Table 2. Physical parameters of steel (0.15 wt. % C, 5 wt. % Cr and around 1 wt. % Mo).

Parameters	Symbol [Unit]	Value
Ambient temperature	T_0 [K]	293.15
Liquidus	T_L [K]	1730
Solidus	T_S [K]	1700
Fusion latent heat	L_f [J·kg^{-1}]	700
Dynamic viscosity	μ [Pa·s]	0.007
Density	ρ [kg·m^{-3}]	7750
Heat capacity	c [J·kg^{-1}·K^{-1}]	483
Thermal conductivity	λ [W·m^{-1}·K^{-1}]	37

The following simplifying assumptions were made when mathematically modeling the welding process:

- Flow is laminar in the weld pool.
- The power distribution in the beam is Gaussian.
- The heat source is linear along the x axis.

- The solid–liquid phase change is modeled by a jump in viscosity and the introduction of the latent heat of the phase transition.
- The shape and dimensions of the keyhole were determined from experimental data. The keyhole was approximated by an oblique elliptical cone with a spherical apex.

2.1 Reconstruction of a keyhole shape
The keyhole's shape is determined by the method of synchronous accumulation. The data of the reconstruction of the keyhole shape in EBW with oscillation along the joint [3, 4] were used for further calculations. These data are presented in Figure 2.

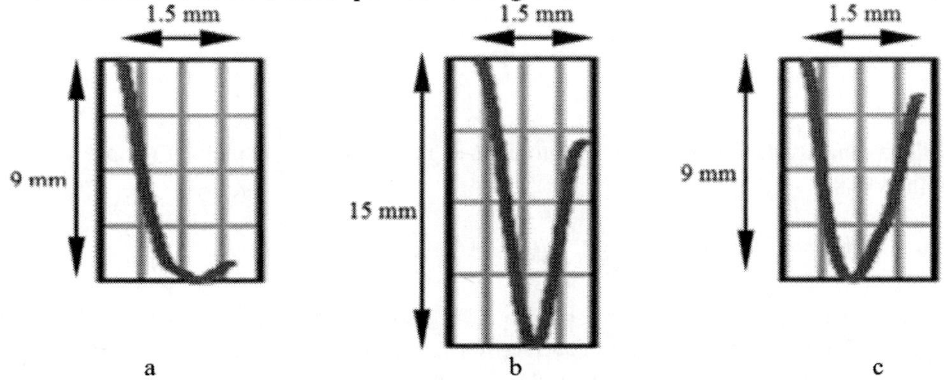

Figure 2. The result of synchronous accumulation for steel welding with oscillation along the joint for different focus regimes (longitudinal projection): a – under-focused; b – sharp focus; c – over-focused modes

2.2 Governing equations
The governing equations for the calculations are as follows [1, 6].
Mass Conservation:

$$\nabla \vec{u} = 0 \tag{1}$$

The equation of motion for an incompressible fluid:

$$\rho \frac{\partial \vec{u}}{\partial t} + \rho(\vec{u}\nabla)\vec{u} = \mu \cdot \nabla^2 \vec{u} - \nabla P + F , \tag{2}$$

where ρ is density, $\vec{u}$ is the melt flow rate vector, μ is dynamic viscosity, P is pressure, F is force term.

Force term:

$$F = \rho g \alpha (T - T_{ref}) - C\left(\frac{1 - f_L}{f_L^3 + B}\right)\vec{u} + \rho U \frac{\partial \vec{u}}{\partial x} \tag{3}$$

where g is gravitational acceleration, α is thermal expansion coefficient, T the absolute temperature, T_{ref} is reference temperature, taken as a solidus temperature (T_S), U is welding speed.

The liquid fraction:

$$f_L = \begin{cases} 1 & T > T_L \\ (T - T_S)/(T_L - T_S) & T_S \le T \le T_L \\ 0 & T < T_S \end{cases} \tag{4}$$

where T_L is liquidus, T_S is solidus.

The differential equation of energy transfer:

$$\frac{\partial T}{\partial t}+u\frac{\partial T}{\partial x}+\upsilon\frac{\partial T}{\partial y}+w\frac{\partial T}{\partial z}=a\cdot\left(\frac{\partial^2 T}{\partial x^2}+\frac{\partial^2 T}{\partial y^2}\right)+\frac{Q}{c\rho} \quad (5)$$

where $a = \lambda / C_{eff} \cdot \rho$ is the thermal diffusivity, u, v, w are velocity components along the axes, c is heat capacity, ρ is density, Q is input heat.

The latent heat of melting and solidification was taken into account by introducing effective heat capacity

$$C_{eff}=C_0+\frac{\exp[-((T-T_{melt})/(T_L-T_S))^2]}{\sqrt{\pi}(T_L-T_S)}H_f, \quad (6)$$

where C_0 is heat capacity depending on temperature, H_f is latent heat, T_{melt} is melting temperature.
The Maramgoni effect:

$$\mu\frac{\partial u}{\partial z}=f_L\frac{d\gamma}{dT}\frac{\partial T}{\partial x};\quad \mu\frac{\partial v}{\partial z}=f_L\frac{d\gamma}{dT}\frac{\partial T}{\partial y};\quad \mu\frac{\partial w}{\partial z}=f_L\frac{d\gamma}{dT}\frac{\partial T}{\partial z} \quad (7)$$

where u, v, w are velocity components along axes X, Y and Z, μ is dynamic viscosity, f_L is liquid fraction, $d\gamma/dT$ is temperature coefficient of surface tension, T is the absolute temperature.

This effect is applied to the model as a boundary condition at the top surface of the weld pool and walls of the keyhole. Initial and boundary conditions are presented in Figure 1.

3. Results and Discussion

Figure 3 presents the experimental and calculated cross sections of weld joints for different focus regimes.

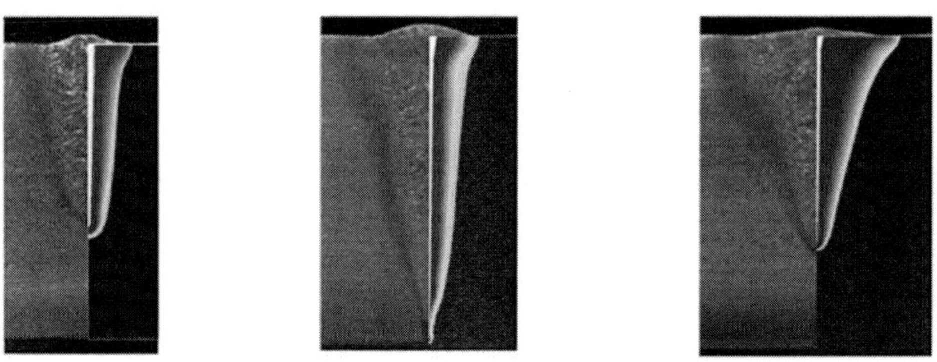

a b c

Figure 3 Experimental and calculated cross sections of weld joints for different focus regimes: a – under-focused; b – sharp focus; c – over-focused modes

The calculated and experimental data for sharp and over-focused modes are well correlated. Differences between simulation and experiment for the under-focused mode are caused by errors in determining the depth of the channel.

The analysis of the influence of various factors on the weld's geometry formation and on the probability of defects formation during electron beam welding with oscillation was carried out. Convective phenomena in the liquid phase are one of the reasons of a specific form of welds with the broadening upper part (the characteristic values of the Marangoni and Rayleigh numbers were $5\cdot10^3$

and $2 \cdot 10^3$ respectively). The intensity of these effects depends on the input energy distribution and on the focusing mode.

4. Conclusions

It is possible to analyze the effect of the oscillation parameters through studying the effect of channel geometry on heat and mass transfer in a liquid bath. The methodology proposed in [4, 5] can be used to determine the geometry of the keyhole. It was found that the intensity of convection caused by the Marangoni effect increases during the transition from the over-focused to the under-focused mode. The intensity of thermogravitational convection in this case is decreased.

Acknowledgements

This work was supported by a grant from the Russian Foundation for Basic Research, No. 14-08-96008, and with the financial support from the Russian Ministry of Education and Science at the base part of the state task (Project No. 01201460538).

References

[1] Rai R, Palmer T A, Elmer J W, Debroy T 2009 *Welding Journal* **88** 54-61
[2] W Sudniky, D Radajz and W Erofeew 1996 *J. Phys. D: Appl. Phys.* **29** 2811-2817
[3] W-I. Cho, S-J. Na, C Thomy, F Vollertsen 2012 *J. of Mat. Proc. Technology* **212** 262-275
[4] D Trushnikov, V Belenkiy, V Schavlev, A Piskunov, A Abdulin and G Mladenov 2012 *Sensors* **12**(12) 17433-17445
[5] D N Trushnikov, V Ya Belen'kii 2013 *Welding Int.* **27**(11) 877-880
[6] M Bachmann, V Avilov, A Gumenyuk, M Rethmeier 2014 *J. of Mat. Proc. Techn.* **214** 578-591

Estimation of the vibration loading vehicle with pneumohydraulic suspensions

A S Gorobtsov[1], S K Kartsov[2] and Yu A Polyakov[3]

[1] Volgograd State Technical University, 28, Lenina ave., Volgograd, 400131, Russia
[2] Moscow Automobile and Road Construction State Technical University (MADI), 64, Leningradskiy ave., Moscow, 630073, Russia
[3] National Research Technological University 'MISIS', 4, Lenina ave., Moscow, 119049, Russia

E-mail: polyakov_yu.a@mail.ru

Abstract. The spatial model of an all-terrain vehicle with pneumohydraulic suspensions, which allowed executing the estimation of a vibration loading construction taking into account large motions of solid bodies, exact kinematic of suspension elements and features of their mounting on a vehicle, is built. Comparison of charts of power spectral density of vertical accelerations shows quality accordance of calculation and experimental spectra for the examined modes of motion. The built dynamic model fully satisfies the requirements of adequacy to the real object and can be used for further calculations.

1. Introduction

The increase of efficiency of decision of the tasks related to the necessity of perfection of the vibroprotective systems of modern vehicles is indissolubly related to the necessity of search of new technical decisions, allowing us to use the different variants of management of resilient and damping descriptions of the suspension directly in the process of the motion of the vehicle. In this regard, there are significant prospects for the creation of mathematical models of vehicles with adaptive suspensions, giving a possibility of debugging the construction of the suspension at the stage of imitation computer modeling.

A distinctive feature of this work is the formulation of the problem in the extended interpretation, which implies rejection of the hypothesis of the small motions of bodies, traditionally used for similar calculations, and allows us to carry out the complete calculation chart of the vehicle of geometrical non-linearity of the motion of elements of the construction on the base of differential equations of large motions of bodies [1-6]. Under large motions here, we imply the term meaning exact description in equations of dynamics of the angular orientation of bodies, without the use of assumption about the smallness of the rotation angles.

This approach allows the creation of dynamic model of the vehicle more exactly taking into account such properties, as a spatial character of the motion, the structure of the dynamic model, different non-linearity of descriptions of resilient and damping elements, multidimensional determined and casual indignations, as well as providing a possibility to include the models of specific interactions, for example, woobling of wheel with elastic tyre.

2. The dynamic model of the all-terrain vehicle with pneumohydraulic suspensions

This article describes the multimass dynamic model of all-terrain vehicle with pneumohydraulic suspensions, taking into account the specific of kinematics of elements of independent suspensions and corresponding resilient and damping connections.

The dynamic model of the examined vehicle contains a frame, a cab, a van, a power aggregate, elements of independent pneumohydraulic suspensions, elements of the steering system, wheels connected to each other by connecting elements. The longitudinally located engine with a gear-box is set on a frame on two front and two back rubber mountings having identical resilient and damping descriptions. For a calculation scheme, shown in Figure 1, all bodies are accepted as absolutely solid. The operation of the transmission is not taken into account; therefore, a transfer box and cardan shafts are absent from a model.

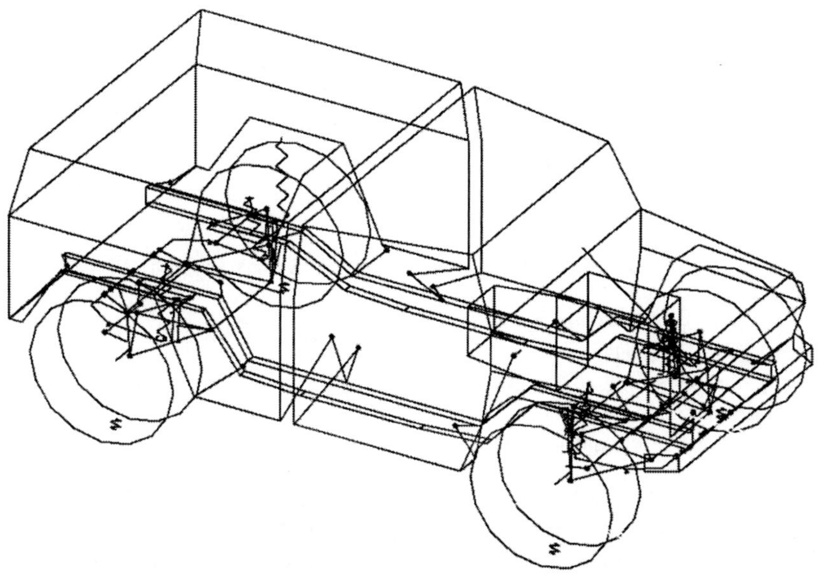

Figure 1. The calculation scheme of the all-terrain vehicle with pneumohydraulic suspensions.

The functions of the resilient and damping devices of suspension combine in themselves a pneumohydraulic element (Figure 2). Its underbody is a joint fastened to the axis of the lower arm suspension, and an overhead – to the bracket of the vehicle frame.

The advantages of pneumohydraulic elements include: non-linearity descriptions; compactness, conditioned by high pressure of gas and combining of the resilient element and the shock absorber in one knot; a possibility of adjusting the height of the basket above by the linen of the road. A defect is the high cost related to the necessity of producing details with high accuracy.

Control suspension arms of each wheel are presented by two arms swinging in a transversal plane. Every upper arm has a V-configuration, thus the heads of its fastening are located at the tops of a corresponding triangle. In contrast to the upper lever, the lower one of suspension of the ∀-configuration is provided with an axis for joint connecting to its lower head of pneumohydraulic element fastening.

Providing counteraction to the transversal rolls of the vehicle is achieved by plugging of front and back stabilizators of transversal stability in the model. Each of them appears to be consisting of two halves connected by an element with given torsional stiffness. The end part of each of these halves is connected to the lower arm of the suspension by means of an earring joint, and the middle part is fastened through rubber pillows to the bracket of the vehicle frame.

Researches of vehicle dynamics were conducted by means of the modeling system 'FRUND' [1, 2].

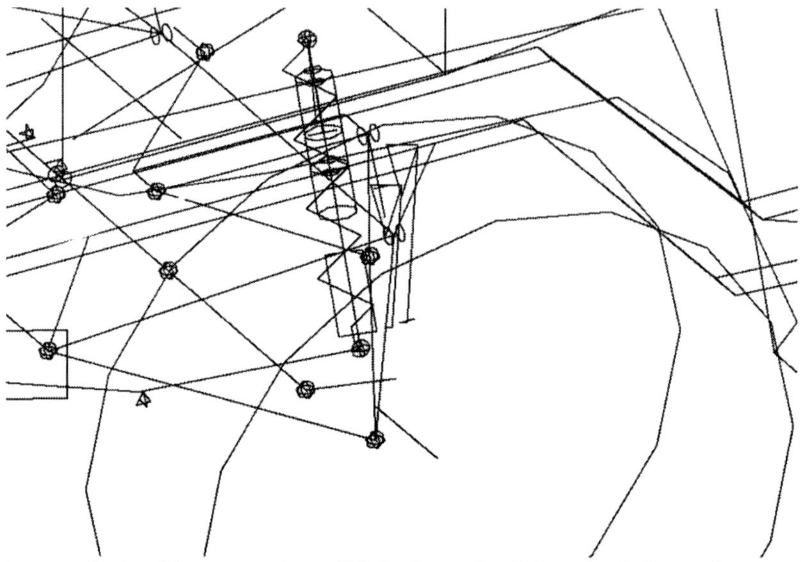

Figure 2. Pneumohydraulic suspension of left fore wheel in calculation scheme of all-terrain vehicle.

3. Results and Discussion

To verify the adequacy of the dynamic model, we compared the calculation results with experimental data, obtained during the road tests.

External road forcing was determined by two discrete, stationary and random processes of the change of ordinates' microprofile of the road surface, given separately on right and left wheels.

A rectilinear motion of the vehicle has been simulated on the random microprofile of the road surface with constant vehicle velocity. The standard types of the microprofile are used in the calculations: even a cobble (velocity 25…30 km/h) and dynamometric road (velocity 50…60 km/h).

For the estimation of vibration loading of the vehicle construction, the signals from sensors of vertical accelerations were written down in the following points:

- on the cab floor, under the driver seat (Figure 3, Figure 5);
- on the left frame girder, above the back left wheel (Figure 4, Figure 6).

Realizations of accelerations were processed by means of the programmatic system 'FRUND' [1, 2].

The calculation charts of power spectral density of vertical accelerations in indicated points were compared with those obtained experimentally during road tests.

The spectrum analysis of vertical accelerations in the indicated points of the vehicle construction shows, that the shape of the spectrum, even for the same transport vehicle and for the same loading, can be different depending on a level and spectrum indignations operating from the side of the road, and also from the position of the measuring point.

Since in experimental tests, it is difficult to provide a strict constancy of the vehicle velocity, it is required to adjust the calculation and experimental spectrums of vertical accelerations to the minimums conditioned by the effect of the wheel base.

As shown in Figure 3, during the motion of the vehicle along even cobble with a speed of 27, 30 km/h on the spectrums of vertical accelerations of the point located on the cab floor, under the driver seat, we clearly see three minimums from the first three harmonics of excitation of vertical vibrations corresponding to wave lengths L; $(1/2)L$; $(1/3)L$, where L is the wheel base of the vehicle. Such phenomenon is named after the effect of the wheel base [7]. The best coincidence with the experimental data is provided by the spectrums obtained during motion along even cobble at a speed of 27 km/h.

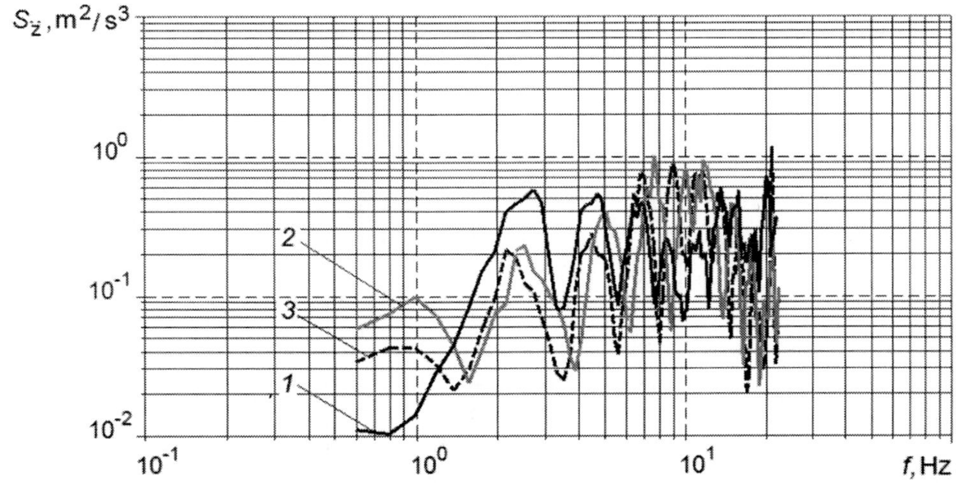

Figure 3. Adjustment of spectrums of vertical accelerations on the cab floor, under the driver seat, on the minimums conditioned by the effect of the wheel base (during the motion of the vehicle along even cobble): *1* – experiment, velocity is 25…30 km/h; *2* – calculation, velocity is 30 km/h; *3* – calculation, velocity is 27 km/h.

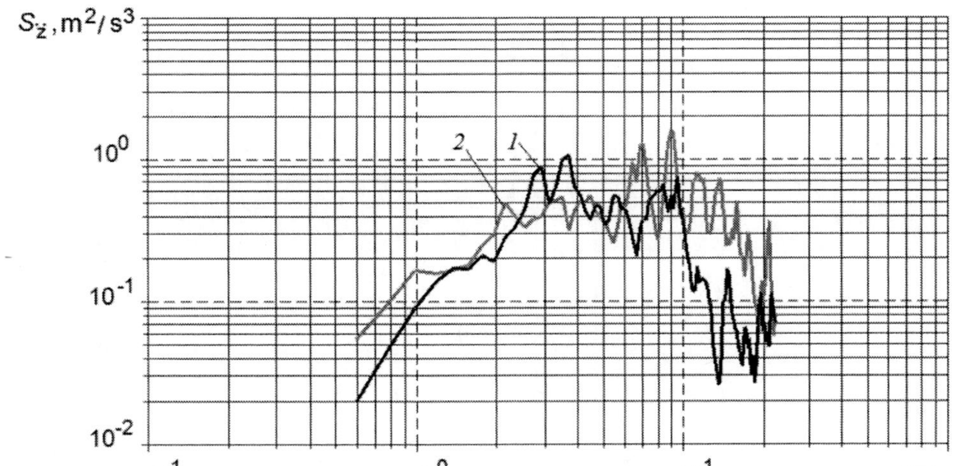

Figure 4. Charts of power spectral density of vertical accelerations on the left frame girder, above the back left wheel (during the motion of the vehicle along even cobble): *1* – experiment, velocity is 25…30 km/h; *2* – calculation, velocity is 27 km/h.

During the motion of the vehicle along a dynamometric road with speeds of 50, 60 km/h on the spectrums of vertical accelerations of the point located on the cab floor, under the driver seat, we can clearly see only one minimum on the frequency of the first harmonic of the effect of the wheel base (Figure 5), corresponding to wave length L.

Thus, the best coincidence with experimental data is provided by the spectrums obtained during the motion of the vehicle along an even cobble at a speed of 27 km/h (Figure 3) and on a dynamometric road at a speed of 60 km/h (Figure 5).

Comparison of charts of power spectral density of vertical accelerations shows the quality accordance of calculation and experimental spectrums for the examined modes of motion.

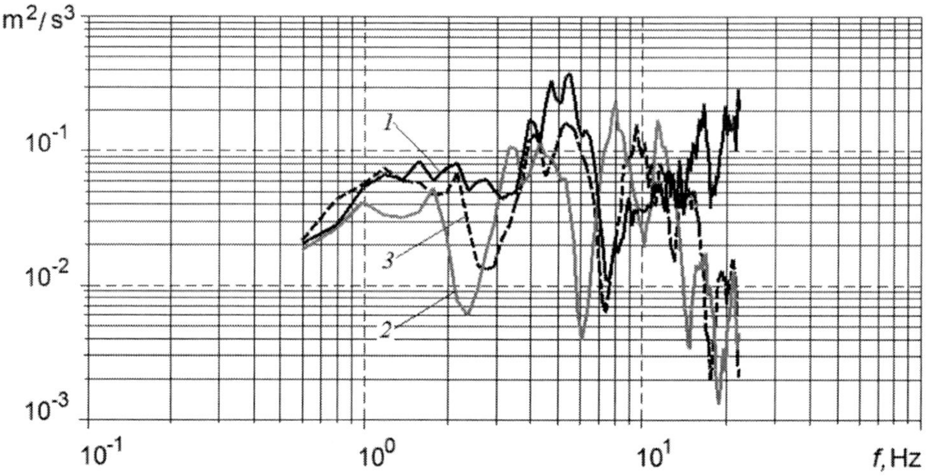

Figure 5. Adjustment of spectrums of vertical accelerations on the cab floor, under the driver seat, on the minimum conditioned by the effect of the wheel base (during the motion of the vehicle along a dynamometric road): *1* – experiment, velocity is 50…60 km/h; *2* – calculation, velocity is 50 km/h; *3* – calculation, velocity is 60 km/h.

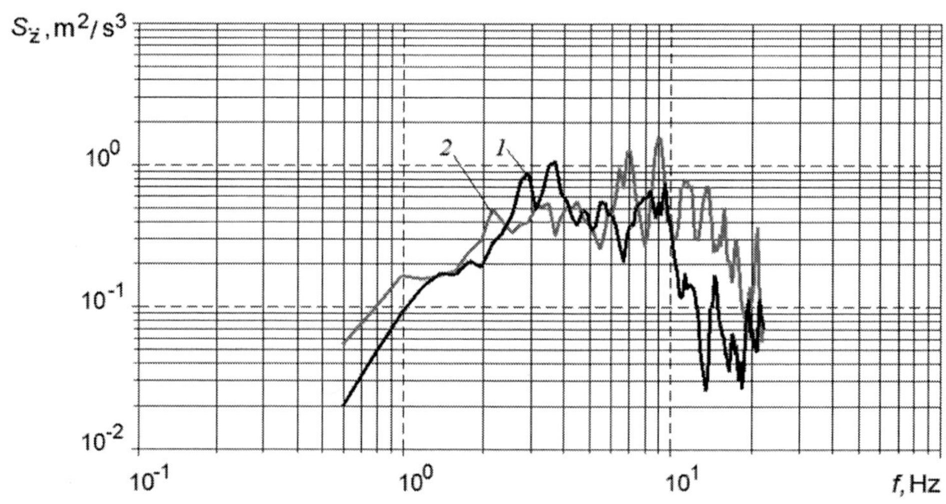

Figure 6. Charts of power spectral density of vertical accelerations on the left frame girder, above the back left wheel (during the motion of the vehicle along a dynamometric road): *1* – experiment, velocity is 50…60 km/h; *2* – calculation, velocity is 60 km/h

4. Conclusion

The spatial model of an all-terrain vehicle with pneumohydraulic suspensions, which allowed executing the estimation of a vibration loading construction taking into account large motions of solid bodies, exact kinematic of suspension elements and features of their mounting on a vehicle, is built.

Since in experimental tests, it is difficult to provide a strict constancy of the vehicle velocity, the adjustment of the calculation and experimental spectrums of vertical accelerations is required on the minimums conditioned by the effect of wheel base.

Comparison of charts of power spectral density of vertical accelerations shows quality accordance of calculation and experimental spectrums for the examined modes of motion. The built dynamic model fully satisfies the requirements of adequacy to the real object and can be used for further calculations.

References

[1] Gorobtsov A S, Kartsov S K, Pletnev A E and Polyakov Y A 2011 *Kompyuternye metody postroeniya i issledovaniya matematicheskikh modeley dinamiki konstruktsiy avtomobiley* (Moscow: Mashinostroenie Pablishing)
[2] Gorobtsov A S, Kartsov S K and Polyakov Y A 2013 *News of the Tula State University. Series 'Technical Sciences'* **6**(1) 102-115
[3] BauChau O A, Damilano G and Theron N J 1995 *Int. J. for Numerical Methods in Eng.* **38** 2727-51
[4] Bayo E, Garcia de Jalon and Serna M A 1988 *Computer Methods in Appl. Mech. and Eng.* **71** 183-195
[5] Bayo E, Garcia de Jalon, Serna M A and Cuadrano J 1991 *Computer Methods in Appl. Mech. and Eng.* **92** 377-395
[6] Shabana A A, Hwang Y L and Wehage R A 1992 *Int. J. for Numerical Methods in Eng.* **35** 1927-66
[7] Yatsenko N N 1972 *Kolebaniya, prochnost I forsirovannye ispytaniya gruzovykh avtomobiley* (Moscow: Mashinostroenie Pablishing)

Using a specialized Markov chain in the reliability model of disk arrays RAID-10 with data mirroring and striping

P A Rahman

Department of Automated Technological and Informational Systems, Ufa State Petroleum Technological University, 2, October ave., Sterlitamak, 453118, Russia

E-mail: pavelar@yandex.ru

Abstract. This paper deals with a specialized Markov chain for a reliability model of the fault-tolerant system with dual-type failures, non-synchronized repairs and full restore after reaching the failed state. A generalized iterative procedure for calculation of the system's stationary availability factor, the mean time of repair and the mean time of failure is introduced. An application of the specialized Markov chain in the reliability model of the fault-tolerant dual-level disk arrays 'RAID-10' and calculation of the mean time to data loss by using the generalized iterative procedure are also observed. Finally, calculation examples of mean time to data loss of the RAID-10 array are also given.

1. Introduction

Currently, the world deals with complex technical devices and systems, which are used in everyday life as well as in the industrial manufacturing processes. In addition to these parameters of technical systems, like performance, time latency, capacity, power consumption etc., the reliability parameters are also quite important, because they directly affect the stability, efficiency and security of technical systems.

There is a set of the academic books, dedicated to reliability theory [1-4], but these in most cases deal with the well-known reliability model of the repairable systems, based on Markov Birth-Death Chain [5, 6]. These models allow us to obtain calculation formulas for such complex and integral reliability parameters like mean time to failure and a stationary availability factor. However, in more sophisticated cases, with several types of failures and special kinds of repairs, well-known Markov Birth-Death Chain cannot be applied to the reliability model of these systems, and development of the specialized Markov Chains is required. In particular, the reliability model of the fault-tolerant dual-level disk array 'RAID-10' (RAID – redundant array of inexpensive disks) [7, 8] cannot be properly represented by Markov Birth-Death Chain and requires development of a specialized chain.

Over the past few years, within the ambit of the research work in the field of reliability models of data transmission, processing and storage systems [9, 10], the author developed a specialized Markov Chain for the reliability model of the fault-tolerant system with n elements, dual-type failures, non-synchronized repairs and full restore after reaching the failed state, and obtained a generalized procedure for calculation of the mean time to failure and a stationary availability factor.

Finally, the author applied the specialized Markov Chain to the reliability model of the fault-tolerant dual-level 'RAID-10' array and obtained a specialized procedure for calculation of the mean time to data loss of the 'RAID-10' array.

2. Specialized Markov chain for the reliability model of the system with dual-type failures, non-synchronized repairs and full restore after reaching the failed state

Let us introduce a system with a set of operable states $j = 0 \ldots n$ and one failed state F. State 0 is an initial fully operable state and states $j = 1 \ldots n$ are degraded operable states. In degraded states, the system is operable, however for larger j, the system is closer to the failed state, and system functioning performance may be lower.

Also, let us introduce the following scheme of transitions between the system states:

- From operable state $j = 0 \ldots n$, the system can pass to the next more degraded operable state, $j + 1$, if $j < n$, or to failed state F, if $j = n$, with given failure rate λ_j.
- From the operable state, $j = 1 \ldots n$, the system can pass back to the less degraded operable state, $j - 1$, with given repair rate μ_j.
- In each operable state, $j = 0 \ldots n$, some kind of critical failure may occur in the system, which transfers it from the operable state directly to failed state F with the given rate of σ_j.
- In the failed state, a full restore procedure for the system is used, which returns the system to initial operable state 0 with given rate γ.

Now, let us introduce Markov chain (figure 1), which represents graphically the reliability model, described above:

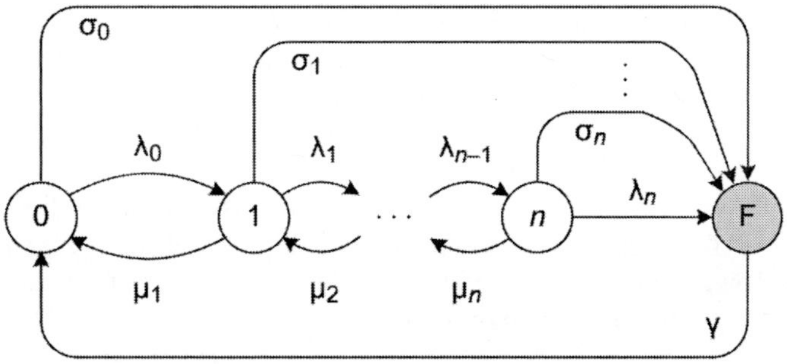

Figure 1. Specialized Markov chain for the reliability model of the system with dual-type failures, non-synchronized repairs and full restore after reaching the failed-state.

Accordingly, the system of Kolmogorov-Chapman equations for the stationary case is as follows:

$$\begin{cases} P_0 + P_1 + \ldots + P_n + P_F = 1; \\ -(\lambda_0 + \sigma_0)P_0 + \mu_1 P_1 + \gamma P_F = 0; \\ \lambda_0 P_0 - (\mu_1 + \lambda_1 + \sigma_1)P_1 + \mu_2 P_2 = 0; \\ \vdots \\ \lambda_{n-2}P_{n-2} - (\mu_{n-1} + \lambda_{n-1} + \sigma_{n-1})P_{n-1} + \mu_n P_n = 0; \\ \lambda_{n-1}P_{n-1} - (\mu_n + \lambda_n + \sigma_n)P_n = 0; \\ \sigma_0 P_0 + \ldots + \sigma_{n-1}P_{n-1} + (\lambda_n + \sigma_n)P_n - \gamma P_F = 0. \end{cases} \quad (1)$$

An availability factor of the system is equal to the sum of probabilities of all operable states:

$$K_S = \sum_{j=0}^{n} P_j .$$

Next, we can easily calculate the mean time to repair by taking into consideration the fact that the system from the failed state can transit only to the initial operable state with rate γ:

$$T_R = 1/\gamma .$$

Finally, mean time to failure can be derived by using the well-known in reliability theory identity for repairable systems $K_S = T_F/(T_F + T_R)$:

$$T_F = K_S/(\gamma(1-K_S)).$$

So, to obtain the stationary availability factor and mean time to failure, we need to solve the equations system and get all stationary probabilities.

It should be noted, that solving equation system (1) is a multi-level iterative process with cubic computational complexity ~$2(n+2)^3$. However, the author obtained a general analytic solution (for any given $n > 0$) in the form of the following matrix formula:

$$\Psi = \prod_{j=1}^{n} \begin{bmatrix} \lambda_{n+1-j} & 0 & 0 \\ 1 & \mu_{n+1-j} & \sigma_{n+1-j} \\ 1 & \mu_{n+1-j} & \lambda_{n+1-j} + \sigma_{n+1-j} \end{bmatrix};$$

$$\begin{bmatrix} U & 0 & 0 \\ V & 0 & W \\ M & 0 & D \end{bmatrix} = \Psi \times \begin{bmatrix} \lambda_0 & 0 & 0 \\ 1 & 0 & \sigma_0 \\ 1 & 0 & \lambda_0 + \sigma_0 \end{bmatrix}; \qquad (2)$$

$$K_S = \frac{\gamma M}{\gamma M + D}; \quad T_F = \frac{M}{D}; \quad T_R = \frac{1}{\gamma}.$$

A matrix formula requires multiplication of the $n + 1$ square matrixes with sizes 3 by 3, which contains all source reliability parameters of the system. The matrix formula has linear computational complexity ~$138(n+1)$. In general case, the matrix formula produces a result matrix with five non-zero coefficients U, V, W, M and D, and only two of them, M and D, are used for calculation of the stationary availability factor and mean time to failure.

By taking into the consideration the associative feature of matrix multiplication $(\mathbf{AB})\mathbf{C} = \mathbf{A}(\mathbf{BC})$, the author obtained the following recurrent computational scheme, which has also linear, but significantly less computational complexity ~$36n$ and provides fast calculation of the stationary availability factor and mean time to failure:

$$\begin{cases} U^{(1)} = \lambda_0; \quad V^{(1)} = 1; \quad M^{(1)} = 1; \\ W^{(1)} = \sigma_0; \quad D^{(1)} = \lambda_0 + \sigma_0; \\ r = 1 \ldots n; \\ \begin{cases} U^{(r+1)} = \lambda_r U^{(r)}; \\ V^{(r+1)} = \sigma_r M^{(r)} + \mu_r V^{(r)} + U^{(r)}; \\ M^{(r+1)} = \lambda_r M^{(r)} + V^{(r+1)}; \\ W^{(r+1)} = \sigma_r D^{(r)} + \mu_r W^{(r)}; \\ D^{(r+1)} = \lambda_r D^{(r)} + W^{(r+1)}; \end{cases} \\ M = M^{(n+1)}; \quad D = D^{(n+1)}; \end{cases} \qquad (3)$$

$$K_S = \frac{\gamma M}{\gamma M + D}; \quad T_F = \frac{M}{D}; \quad T_R = \frac{1}{\gamma}.$$

3. Specialized Markov chain in the reliability model of the dual-level disk array 'RAID-10'

The fault-tolerant dual-level disk array 'RAID-10' is a data storage system, which combines high reliability in the inner level due to the data mirroring in 'RAID-1' technology and high-performance in the outer level due to data striping in 'RAID-0' technology. In the inner level, there is a set of n independent dual-disk 'RAID-1' arrays. Data in each dual-disk 'RAID-1' array are always synchronized between two disks. In the outer level, a set of n dual-disk 'RAID-1' arrays are united to an array of arrays by using the 'RAID-0' technology (figure 2). Data in the 'RAID-0' array are distributed between blocks, located on the different 'RAID-1' arrays.

The 'RAID-10' array in total contains $2n$ disks ($2n \geq 4$). User data capacity is equal to 50% of total disk space because of data mirroring between two disks in each 'RAID-1' array.

In the best case, 'RAID-10' is operable even if n disks fail and all of them belong to different 'RAID-1' arrays, and there is no pair of disks, which belong to the same 'RAID-1' array. In the worst case, even failure of two disks can cause the whole 'RAID-10' failure and data loss, if both of them belong to the same 'RAID-1' array, because the 'RAID-0' technology does not provide any fault-tolerance and failure of any dual-disk 'RAID-1' array, which causes failure of the whole 'RAID-10'.

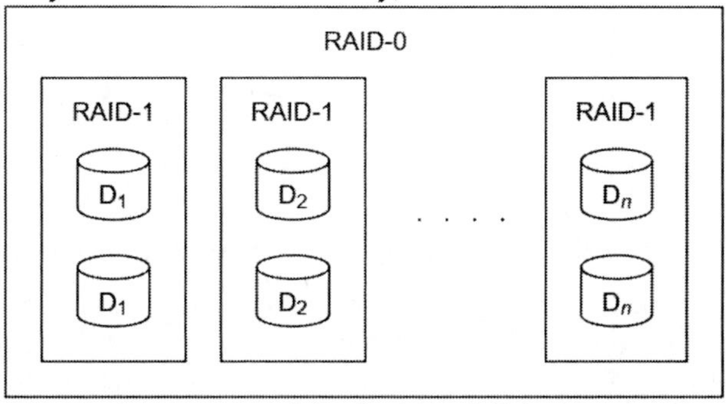

Figure 2. The structure of the dual-level disk array 'RAID-10'.

Let us now overview the offered reliability model of the data storage system, based on the dual-level disk array 'RAID-10' with $2n$ identical disks (the same capacity, manufacturer and model of all disks). The disk failure rate is λ. Disks can fail independently.

The initial state of the data storage system is 0 (all disks are operable). From this state, in case of failure of one of the $2n$ disks, the system passes to state 1 (one of non-fatal failures occurs).

From state 1, in case of failure of one of the $2n - 1$ disks, the system passes either to state 2 (two of non-fatal failures occurred) if the failed disk is the one of the $2n - 2$ disks, which do not belong to the 'RAID-1' array with the previously failed disk, or to failed state F if currently and previously failed disks belong to the same 'RAID-1' array.

From state 2, in case of failure of one of the $2n - 2$ disks, the system passes either to state 3 (three of non-fatal failures occurred) if the failed disk is the one of the $2n - 4$ disks, which do not belong to the one of 'RAID-1' arrays with one of the previously failed disks, or to failed state F, if current and one of the previously failed disks belong to the same 'RAID-1'.

Accordingly, in state n (n of non-fatal failures occurred), in each 'RAID-1' array, one of two disks will be failed. Thus, in state n, failure of any remaining n disks will transfer the system to failed state F because, in this case, one of the 'RAID-1' arrays will fail.

Next, let us assume, that 'RAID-1' arrays are independent for repairs, and in each of them, the rebuild process can be started independently after replacement of the failed disk. Completions of the rebuild processes in different 'RAID-1' arrays are also independent and not synchronized. The 'RAID-1' array rebuild rate is μ. We must mention that for simplification of the reliability model, we will consider the disk replacement time as negligible (assuming that the unlimited amount of the hot-spare disks is available). Completion of the rebuild process in the separate 'RAID-1' array returns the system from state $j = 1 \ldots n$ back to state $j - 1$.

Also it should be noted that for the rebuild process in each of the 'RAID-1' arrays, we will take into consideration additional disk read error rate ε. It should be added to the failure rate of the disk, from which data are copied to the replaced disk in the 'RAID-1' array during the rebuild process. Failure of the data source disk as well as the read error on it during the rebuild process can cause failure of the whole 'RAID-1' array and transfer the system to failed state F.

Next, let us take into consideration critical array controller errors with rate σ, which transfer the system from any operable state $j = 0\ldots n$ directly to failed state F.

Finally, let us assume that the system does not include any additional hardware and software to provide total backup of the user data. In case of reaching the failed state, all data will be lost. So, we will consider the rate of the system full recovery rate, which is zero, $\gamma = 0$.

Now, we may introduce Markov chain (figure 3), which represents graphically the reliability model, described above:

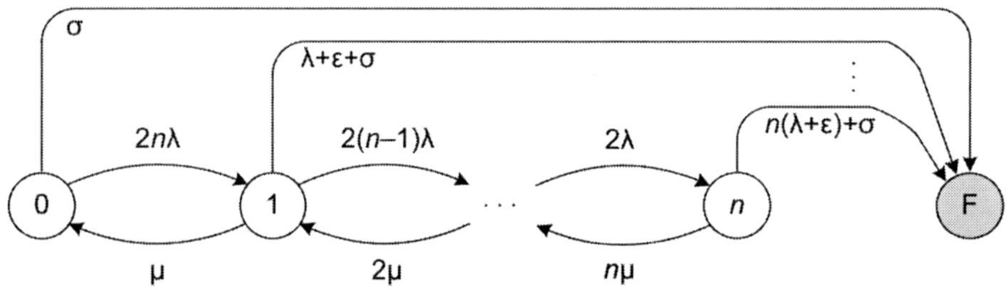

Figure. 3. The specialized reliability model for the dual-level disk array 'RAID-10'.

This model is a particular case of the specialized Markov chain, discussed above, with the following substitutions of the source reliability parameters:

$$\begin{cases} \lambda_0 = 2n\lambda; \quad \lambda_j = 2(n-j)\lambda; \\ \sigma_0 = \sigma; \quad \sigma_j = \sigma + j(\lambda+\varepsilon); \\ \gamma = 0; \quad \mu_j = j\mu; \\ j = 1\ldots n. \end{cases} \tag{4}$$

As the system full recovery rate is zero, $\gamma = 0$, the stationary availability factor is zero and mean time to recovery is infinite for the discussed 'RAID-10' array. However, mean time to data loss for 'RAID-10' is equal to the mean time to failure in specialized Markov chain. So, after substitution of the source reliability parameters into the above-mentioned recurrent calculation scheme (3), we obtain the following recurrent scheme for calculation of the mean time to data loss of the 'RAID-10' array:

$$\begin{cases} U^{(1)} = 2n\lambda; \quad V^{(1)} = 1; \quad M^{(1)} = 1; \\ W^{(1)} = \sigma; \quad D^{(1)} = 2n\lambda + \sigma; \\ r = 1\ldots n; \\ \begin{cases} U^{(r+1)} = 2(n-r)\lambda U^{(r)}; \\ V^{(r+1)} = (\sigma + r(\lambda+\varepsilon))M^{(r)} + r\mu V^{(r)} + U^{(r)}; \\ M^{(r+1)} = 2(n-r)\lambda M^{(r)} + V^{(r+1)}; \\ W^{(r+1)} = (\sigma + r(\lambda+\varepsilon))D^{(r)} + r\mu W^{(r)}; \\ D^{(r+1)} = 2(n-r)\lambda D^{(r)} + W^{(r+1)}; \end{cases} \\ M = M^{(n+1)}; \quad D = D^{(n+1)}; \\ T_{\mathrm{DL}} = M / D. \end{cases} \tag{5}$$

4. The calculation example for the mean time to data loss of the 'RAID-10' array

A 'RAID-10' array with total number of disks $2n$ is given. The disk failure rate is $\lambda = 1/120000$ hour^{-1}, the additional disk read error rate during the rebuild process is $\varepsilon = 1/112$ hour^{-1}, the 'RAID-1' array rebuild rate is $\mu = 1/9$ hour^{-1} and the array controller critical error rate is $\sigma = 1/1200000$ hour^{-1}. Let us calculate the mean time to data loss of the 'RAID-10' array for the total number of disks $2n$ = 2, 4, 6, 8, 10 and 12.

Calculations by the recurrent scheme (5) give us the following values, presented in table 1:

Table 1. Mean times to data loss of the 'RAID-10' array for different total numbers of disks.

$2n$	T_{DL} (hours)
4	302642
6	220282
8	173159
10	142645
12	121275
14	105473
16	93315

Calculation results show that the mean time to data loss of the 'RAID-10' array is moderately decreased with the increase in the total number of disks.

5. Conclusion

In this scientific paper, a specialized Markov chain for the reliability model of the fault-tolerant system with dual-type failures, non-synchronized repairs and full restore after reaching the failed state is discussed. A generalized iterative procedure for calculation of the system's stationary availability factor, mean time to repair and mean time of failure is also introduced. Finally, an application of the specialized Markov chain in the reliability model of the fault-tolerant dual-level disk arrays 'RAID-10' and calculation of the mean time to data loss by using the generalized iterative procedure is also observed.

Scientific results were used by the author in designing of fault-tolerant data storage systems based on the 'RAID-10' arrays for Moscow Power Engineering Institute, Nuclear Power Plant "Balakovo" and several other enterprises.

Acknowledgements

The author thanks professor I I Ladygin, Moscow Power Engineering Institute (MPEI), for scientific support, and S N Khorkov, chief network administrator of MPEI, for technical support.

References

[1] Cherkesov G N 2005 *Reliability of Hardware and Software Systems* (Saint-Petersburg: Piter)
[2] Polovko A M and Gurov S V 2006 *Basis of Reliability Theory* (Saint-Petersburg: BHV-Petersburg)
[3] Shooman M L 2002 *Reliability of computer systems and networks* (John Wiley & Sons, Inc.)
[4] Koren I and Krishna C M 2007 *Fault-Tolerant Systems* (Morgan Kaufmann Publishers)
[5] Anderson W J 1991 *Continuous-Time Markov Chains* (Springer-Verlag)
[6] Wang Z, Wang T and Yang X 1992 *Birth and death processes and Markov chains* (Springer-Verlag)
[7] Schmidt K 2006 *High-availability and disaster recovery* (Springer-Verlag)
[8] Nelson S 2011 *Pro data backup and recovery* (Apress)
[9] Rahman P A, Muraveva E A and Sharipov M I 2016 *Key Eng. Mat.* **685** 805-810
[10] Rahman P A and Bobkova E Yu 2016 *IOP Conf. Ser.: Mater. Sci. Eng.* (Tomsk) vol 124 (IOP Publishing) 012023

Analysis of the mean time to data loss of nested disk arrays RAID-01 on basis of a specialized mathematical model

P A Rahman

Department of Automated Technological and Informational Systems, Ufa State Petroleum Technological University, 2, October ave., Sterlitamak, 453118, Russia

E-mail: pavelar@yandex.ru

Abstract. In this scientific paper a specialized Markov chain for a reliability model of the fault-tolerant system with two types of failures, single batch repair and full restore after reaching the failed state is discussed. The generalized formulas for calculation of the system's stationary availability factor, the mean time to failure and the mean time to repair are also introduced. Application of the special type of the Markov chain in the reliability model of the nested disk array 'RAID-01' and calculation of mean time to data loss by using the generalized formulas are also discussed. The calculation examples of mean time to data loss of the 'RAID-01' array are also given.

1. Introduction

The modern world deals with complex technical devices and systems, which are used in the industrial manufacturing processes as well as in everyday life. In addition to the primary parameters of technical systems, like performance, capacity, power etc, the reliability parameters are also quite important, because they define efficiency, stability and security of technical systems.

The academic books, dedicated to the reliability theory [1-4], in most cases deal with the well-known simplified reliability model of the repairable systems, based on Markov Birth-Death Chain [5, 6], which allows us to obtain calculation formulas for such complex and integral reliability parameters as a stationary availability factor and mean time to failure. In more sophisticated cases with several types of failures and special kinds of repairs, the traditional Markov Birth-Death Chain cannot be used in reliability models. For these systems, development of a specialized Markov Chain is required. In particular, the reliability model of the nested disk array 'RAID-01' (RAID – redundant array of inexpensive disks) [7, 8] cannot be adequately described by Markov Birth-Death Chain and requires development of a specialized Markov chain.

Within the research work in the field of reliability models of data processing, transmission and storage systems [9, 10] the author developed a special type of Markov Chain for the reliability model of the fault-tolerant system with n elements, two types of failures, single batch repair and full restore after reaching the failed state, and derived the generalized formulas for calculation of the stationary availability factor and the mean time to failure.

Finally, the author applied the specialized Markov Chain to the reliability model of the nested disk array 'RAID-01' and obtained a special formula for calculation of the mean time to data loss.

2. Specialized Markov chain for the reliability model of the system with two types of failures, single batch repair and full restore after reaching the failed state

Let us introduce a system with a set of operable states $j = 0 \ldots n$ and one failed state F. State 0 is an initial fully operable state and the states $j = 1 \ldots n$ are degraded operable states. In degraded states, the system is operable, however for larger j, system's functioning performance may be lower, and the system is closer to the failed state.

Also, let us introduce the following scheme of transitions between the different states of system:

- From operable state $j = 0 \ldots n$, the system can pass to the next more degraded operable state $j + 1$, if $j < n$, or to failed state F, if $j = n$, with given failure rate λ_j.
- From the operable state $j = 1 \ldots n$, the system can pass back directly to initial operable state 0 with given repair rate μ_j by using a single batch repair procedure for one or more failed redundant parts of the system.
- In each operable state $j = 0 \ldots n$, some kind of critical failure may occur in the system, which transfers it from operable state directly to failed state F with given rate σ_j.
- In the failed state a full restore procedure for the system is used, and it returns the system to initial operable state 0 with given rate γ.

Now, we may introduce a Markov chain (figure 1), which represents graphically the reliability model, described above:

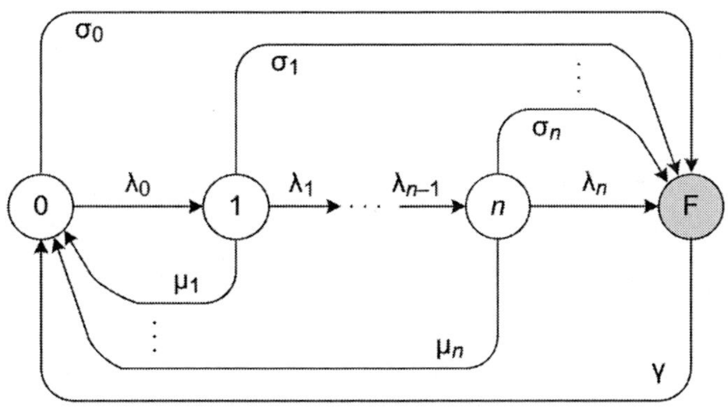

Figure 1. Special type of Markov chain for the reliability model of the system with two types of failures, single batch repair and full restore after reaching the failed-state.

Accordingly, the system of Kolmogorov-Chapman equations for the stationary case is as follows:

$$\begin{cases} P_0 + P_1 + \ldots + P_n + P_{\mathrm{F}} = 1; \\ -(\lambda_0 + \sigma_0)P_0 + \mu_1 P_1 + \ldots + \mu_n P_n + \gamma P_{\mathrm{F}} = 0; \\ \lambda_0 P_0 - (\mu_1 + \lambda_1 + \sigma_1)P_1 = 0; \\ \vdots \\ \lambda_{n-1} P_{n-1} - (\mu_n + \lambda_n + \sigma_n)P_n = 0; \\ \sigma_0 P_0 + \ldots + \sigma_{n-1} P_{n-1} + (\lambda_n + \sigma_n)P_n - \gamma P_{\mathrm{F}} = 0. \end{cases} \quad (1)$$

Availability factor of system is equal to the sum of probabilities of all operable states:

$$K_{\mathrm{S}} = \sum_{j=0}^{n} P_j .$$

Next, we can easily calculate the mean time to repair by taking into consideration the fact that, the system from failed state can transit only to initial operable state with rate γ:

$$T_{\mathrm{R}} = 1/\gamma.$$

Finally, mean time to failure can be derived by using the well-known in reliability theory identity for repairable systems $K_{\mathrm{S}} = T_{\mathrm{F}}/(T_{\mathrm{F}} + T_{\mathrm{R}})$:

$$T_{\mathrm{F}} = K_{\mathrm{S}}/(\gamma(1-K_{\mathrm{S}})).$$

So, to obtain the stationary availability factor and the mean time to failure, we need to solve the equations system and get all stationary probabilities.

Within the scientific research work, the author obtained the general analytic formulas (for any given $n > 0$) for calculation of the stationary availability factor and the mean time to failure:

$$\begin{cases} M = \sum_{q=0}^{n}\left(\frac{1}{\lambda_q}\prod_{j=1}^{n-q}\left(1+\frac{\mu_{q+j}+\sigma_{q+j}}{\lambda_{q+j}}\right)\right); \\ D = 1+\sum_{q=0}^{n}\left(\frac{\sigma_q}{\lambda_q}\prod_{j=1}^{n-q}\left(1+\frac{\mu_{q+j}+\sigma_{q+j}}{\lambda_{q+j}}\right)\right); \end{cases} \quad (2)$$

$$K_{\mathrm{S}} = \frac{\gamma M}{\gamma M + D}; \quad T_{\mathrm{F}} = \frac{M}{D}; \quad T_{\mathrm{R}} = \frac{1}{\gamma}.$$

We should mention that in our later discussion we will also need a special case with $\lambda_n = 0$, in case of which a division by zero problem is occurred in calculation of coefficients M and D. However, algebraic conversions show that, in fraction M / D, parameter $\lambda_n = 0$ can be eliminated. Thus, for case $\lambda_n = 0$, the author derived the specially converted formulas to avoid the division by zero in calculation of the coefficients M and D:

$$\begin{cases} M = 1+(\mu_n+\sigma_n)\sum_{q=0}^{n-1}\left(\frac{1}{\lambda_q}\prod_{j=1}^{n-1-q}\left(1+\frac{\mu_{q+j}+\sigma_{q+j}}{\lambda_{q+j}}\right)\right); \\ D = \sigma_n+(\mu_n+\sigma_n)\sum_{q=0}^{n-1}\left(\frac{\sigma_q}{\lambda_q}\prod_{j=1}^{n-1-q}\left(1+\frac{\mu_{q+j}+\sigma_{q+j}}{\lambda_{q+j}}\right)\right); \end{cases} \quad (3)$$

$$K_{\mathrm{S}} = \frac{\gamma M}{\gamma M + D}; \quad T_{\mathrm{F}} = \frac{M}{D}; \quad T_{\mathrm{R}} = \frac{1}{\gamma}.$$

3. Specialized Markov chain in the reliability model of nested disk array 'RAID-01'

The nested disk array 'RAID-01' is a data storage system, which combines high performance in the inner level due to data striping in 'RAID-0' technology and high reliability in the outer level due to the data mirroring in 'RAID-1' technology. In the inner level, there are two independent n-disk 'RAID-0' arrays. Data in each n-disk 'RAID-0' array is distributed between blocks, located on the different disks. In the outer level, the two 'RAID-0' arrays are united to an array of arrays by using the 'RAID-1' technology (figure 2). Data in the 'RAID-1' array is synchronized between two 'RAID-0' arrays.

The 'RAID-01' array in total contains $2n$ disks ($2n \geq 4$). User data capacity is equal to 50% of total disk space because of data mirroring between two 'RAID-1' arrays.

In the best case, the 'RAID-01' array is operable even if n disks fail, if all of them belong to the one of the 'RAID-0' arrays. In the worst case, even failure of two disks can cause the whole 'RAID-01' failure and data loss, if one of the disks belongs to first 'RAID-0' array, and the other one belongs to second 'RAID-0' array.

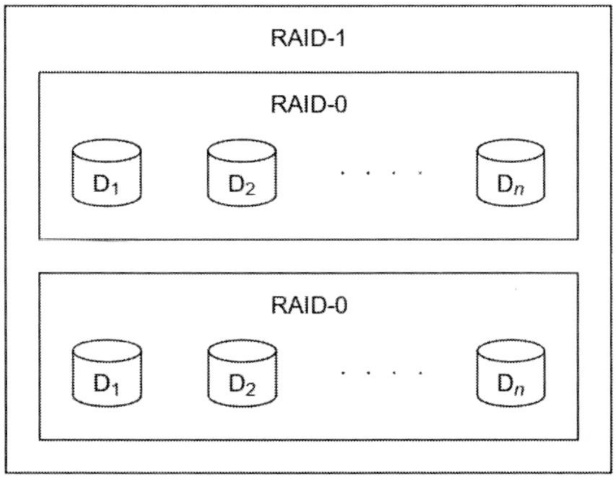

Figure 2. The structure of the nested disk array 'RAID-01'.

Let us now observe the offered reliability model of data storage system, based on the nested disk array 'RAID-01' with $2n$ identical disks (the same capacity, manufacturer and model of all disks). The disk failure rate is λ. Disks can fail independently.

The initial state of the data storage system is 0 (all disks are operable). From this state, in case of failure one of the $2n$ disks, system passes to the state 1 (one of non-fatal failure occurred). In this case one of the 'RAID-0' arrays, in which failed disk is located, fails, however, due to the second operable 'RAID-0' array, the whole data storage system still operable.

From the state 1, in case of failure of one of the $2n - 1$ disks, the system passes either to the state 2 (two of non-fatal failure occurred), if failed disk is the one of the $n - 1$ disks, which belong to the previously failed 'RAID-0' array, or to failed state F if failed disk belongs to the another 'RAID-0'.

From state 2, in case of failure of one of the $2n - 2$ disks, the system passes either to the state 3 (three of non-fatal failures occurred), if failed disk is the one of the $n - 2$ disks, which belong to the previously failed 'RAID-0' array, or to failed state F, if failed disk belongs to another 'RAID-0'.

Accordingly, in state n (n of non-fatal failures occurred) all n disks in of the 'RAID-0' arrays are failed, thus failure of any remaining n disks, all of which in this case belong to the another 'RAID-0' array, will transfer the system to failed state F, because in this case both of the 'RAID-0' arrays fail.

Next, let us assume, that in any of state $j = 1 \ldots n$, when all failed disks belong to one of the 'RAID-0' arrays, the 'RAID-1' technology launches a single batch rebuild process between the two 'RAID-0' arrays after replacement of the all failed disks in the failed 'RAID-0' array. After completion of the rebuild process system passes to initial state 0. The 'RAID-1' array rebuild rate is μ. If a disk failure occurs during the batch rebuild process and failed disk belongs to the previously failed 'RAID-0' array, then the rebuild process will be restarted after replacement of the failed disk. Rebuild rate in all states $j = 1 \ldots n$ has a same value and does not depend on number of the replaced disks, because rebuild process involves two whole 'RAID-0' arrays. Moreover, for simplification of the reliability model, we will consider the disk replacement time as negligible (assuming that the unlimited amount of the hot-spare disks is available).

We must mention, that during the rebuild process, data from the n disks of the operable 'RAID-0' array are copied to the n disks of the failed 'RAID-0' array, and we will take into consideration an additional disk read error rate ε, which should be added to the disk failure rate. A read error as well as failure of any disk on the operable 'RAID-0' array during the rebuild process will transfer the system directly to failed state F.

Next, let us take into consideration critical array controller errors with rate σ, which transfer the system from any operable state $j = 0 \ldots n$ directly to failed state F.

Finally, let us assume that the data storage system does not include any additional hardware and software to provide total backup of the user data, thus, in case of reaching the failed state all data will be lost. So, we will consider the rate of system full recovery rate, which is zero, $\gamma = 0$.

Now, we may introduce next Markov chain (figure 3), which represents graphically the reliability model, described above:

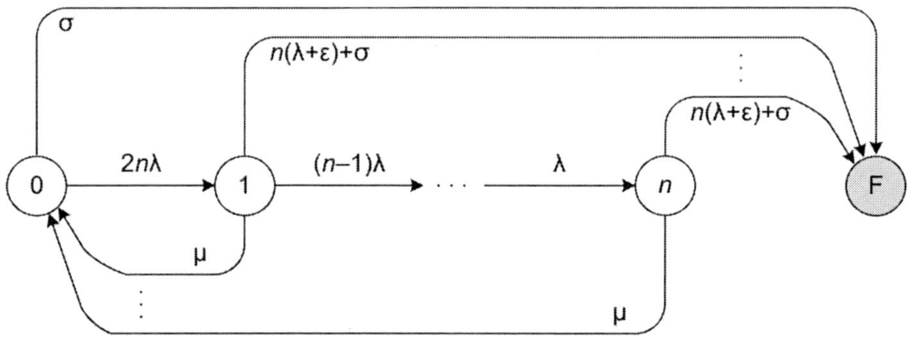

Figure 3. Special type of reliability model for the nested disk array 'RAID-01'.

This model is a particular case of the specialized Markov chain, discussed above, with the following substitutions of the source reliability parameters:

$$\begin{cases} \lambda_0 = 2n\lambda; \quad \lambda_j = (n-j)\lambda; \\ \sigma_0 = \sigma; \quad \sigma_j = \sigma + n(\lambda+\varepsilon); \\ \gamma = 0; \quad \mu_j = \mu; \\ j = 1 \ldots n. \end{cases} \tag{4}$$

As the system full recovery rate is zero, $\gamma = 0$, mean time to recovery is infinite and stationary availability factor is zero for the discussed 'RAID-01' array. However, the mean time to data loss for 'RAID-01' is equal to the mean time to failure in discussed above special type of Markov chain. So, after substitution of the source reliability parameters to the discussed above calculation formula (3), with taking into consideration $\lambda_n = 0$, we obtain the following formula for calculation of the mean time to data loss of the 'RAID-01' array:

$$\begin{cases} M = 1 + (\mu + \sigma + n(\lambda+\varepsilon)) \times \\ \times \sum_{q=0}^{n-1} \left(\frac{1}{((2-\min(1,q))n - q)\lambda} \prod_{j=1}^{n-1-q} \left(1 + \frac{\mu + \sigma + n(\lambda+\varepsilon)}{(n-q-j)\lambda} \right) \right); \\ D = \sigma + n(\lambda+\varepsilon) + (\mu + \sigma + n(\lambda+\varepsilon)) \times \\ \times \sum_{q=0}^{n-1} \left(\frac{\sigma + \min(1,q) n(\lambda+\varepsilon)}{((2-\min(1,q))n - q)\lambda} \prod_{j=1}^{n-1-q} \left(1 + \frac{\mu + \sigma + n(\lambda+\varepsilon)}{(n-q-j)\lambda} \right) \right); \\ T_{\mathrm{DL}} = M / D. \end{cases} \tag{5}$$

4. The calculation examples for the mean time to data loss of the 'RAID-01' array

A 'RAID-01' array with total number of disks $2n$ is given. The disk failure rate is $\lambda = 1/120000$ hour^{-1}, the additional disk read error rate during rebuild process is $\varepsilon = 1/112$ hour^{-1}, the 'RAID-1' array rebuild rate is $\mu = 1/9$ hour^{-1} and the array controller critical error rate is $\sigma = 1/1200000$ hour^{-1}.

Let us calculate mean time to data loss of the 'RAID-01' array for the total number of disks $2n$ = 2, 4, 6, 8, 10 and 12.

Calculations by the formula (5) give us the following values, presented in table 1:

Table 1. Mean times to data loss of the 'RAID-01' array for different total number of disks.

$2n$	T_{DL} (hours)
4	184042
6	95106
8	58827
10	40574
12	30060
14	23413
16	18918

Calculation results show that the mean time to data loss of the 'RAID-01' array is rapidly decreased with the growth of the total number of disks.

5. Conclusion

Within the scope of this scientific paper, a special type of Markov chain for the reliability model of the fault-tolerant system with two types of failures, single batch repair and full restore after reaching the failed state is discussed.

The generalized formulas for calculation of the system's stationary availability factor, the mean time of failure and the mean time to repair are also introduced. Finally, application of a special type of Markov chain in the reliability model of the nested disk array 'RAID-01' and calculation of the mean time to data loss by using the generalized formulas are also presented.

Obtained scientific results were used by the author in designing of the fault-tolerant data storage systems based on the 'RAID-01' arrays for Moscow Power Engineering Institute, Nuclear Power Plant 'Balakovo' and several other enterprises.

Acknowledgements

Author thanks S. N. Khorkov, chief network administrator of Moscow Power Engineering Institute (MPEI), for technical support and professor I. I. Ladygin, MPEI, for scientific support.

References

[1] Ostreykovsky V A 2003 *Reliability Theory* (Moscow: Vyshaya Shkola)
[2] Ushakov I A 2008 *Reliability Theory Course* (Moscow: Drofa)
[3] Rausand M and Holyand A 2009 *System Reliability Theory* (John Wiley & Sons)
[4] Abd-El-Barr M 2007 *Design and Analysis of Reliable and Fault-Tolerant Computer Systems* (London: Imperial College Press)
[5] Kijma M 1997 *Markov Processes for Stochastic Modeling* (London: Chapman & Hall)
[6] Anderson W J 1991 *Continuous-Time Markov Chains* (Springer-Verlag)
[7] Khurshidov A S 2001 *The Essential Guide to Computer Data Storage* (Prentice Hall)
[8] Schmidt K 2006 *High-availability and disaster recovery* (Springer-Verlag)
[9] Rahman P A and Bobkova E Yu 2016 *IOP Conf. Ser.: Mater. Sci. Eng.* **124** 012023
[10] Rahman P A, Muraveva E A and Sharipov M I 2016 *Key Eng. Mat.* **685** 805-810

Modelling the behaviour of ingot axial defects under open die forging

D V Rutskii[1], A N Stjagin[2], N A Zyuban[1] and S B Gamanyuk[1]

[1] Volgograd state technical university, 28, Lenin ave., Volgograd, 400005, Russia
[2] 'Red October' VMC, 110, Lenin ave., Volgograd, 400007, Russia

E-mail: tecmat@vstu.ru

Abstract. The paper reports findings on mathematical modelling of ingots of different geometry. It is established that V-shaped cracks concentrated in the ingot bottom are gradually removed through upset, while arch-shaped cracks located around the sedimentation cone top open up. A computer simulation made it possible to calculate efficient deformations in the workpiece axial zone. An alteration of the ingot geometry leads to an increase in efficient deformation and, consequently, a better treatment of axial areas and elimination of the axial defects of the ingot metal.

1. Introduction

Nowadays, there is an increasing need in the production of large forgings used to manufacture heavy duty large-sized products for nuclear power engineering, heavy engineering etc. The weight of the ingots required to produce these workpieces can achieve 850 tons. The solidification of a large mass of metal takes an extended period of time (1 to 7 days) and is accompanied with an intensive development of segregation and shrinkage phenomena. The natural shrinkage of the melt is equal to 4% of the solidifying volume, which leads to the development of shrinkage defects inside large ingots. The detailed studies [1-3] demonstrated that after solidification, large ingots have a well-developed heterogeneity of their inner structure. This heterogeneity may be of one of the three types (figure. 1):

- physical heterogeneity (shrinkage cavities, V-shaped cracks, porosity);
- chemical heterogeneity (macrosegregation, microsegregation, 'A'-segregation);
- structural non-homogeneity.

Currently, there are a number of works [1, 4-6] considering the problems of ingot solidification and the development of internal defects inside them. Computer modeling is the most advanced method which enables us to assess the effects of heating and pouring process parameters on solidification [7-11].

Computer modeling has made it possible to develop methods of improving the quality of the metal in ingots. Somehow, the proposed methods do not always make it possible to fully avoid the development of the aforementioned defects because extensive solidification leads to a considerable development of these defects.

After solidification, ingots undergo metallurgical treatment which includes open die forging, preliminary and final mechanical treatment and heat treatment. Some internal structure defects can be eliminated through these operations.

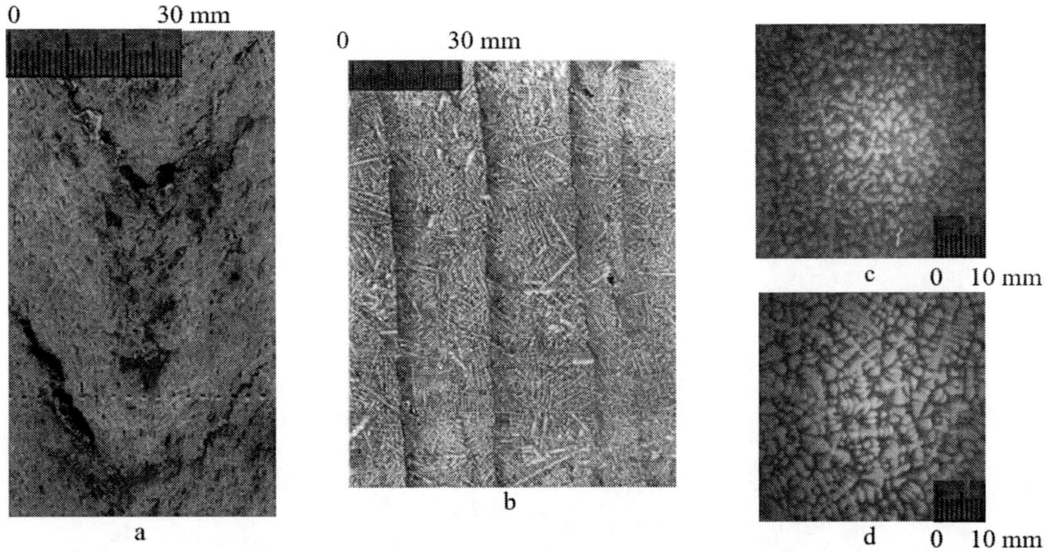

Figure 1. Defects of a large forging ingot: a – axial zone defects in the form of V–shaped cracks; b – 'A' segregation streaks; c– ingot upper part macrostructure; d – ingot bottom part macrostructure

Open die forging of ingots is one of the major stages of metallurgical treatment when inner defects can be eliminated. Structural heterogeneity is fully eliminated during hot deformation (figure 1c, d). Axial zone defects as well as the "A"-segregation zone noticeably reduce in size and length (figure 1a, 1b).

Somehow, if axial zone defects are well developed, they cannot be removed by forging and remain in the finished product, which is unacceptable. The authors [1, 3, 12] have established that the reason for the rejection of large-sized products is a noticeable concentration of discontinuities in forging axial sections identified by UT (figure 2). Thus, it is an important task to establish the features of ingot internal defect behaviour while forging. Upset is one of the efficient operations of open die forging which helps to eliminate internal axial defects. The paper contains a mathematical modeling of the upset of billets having various geometry.

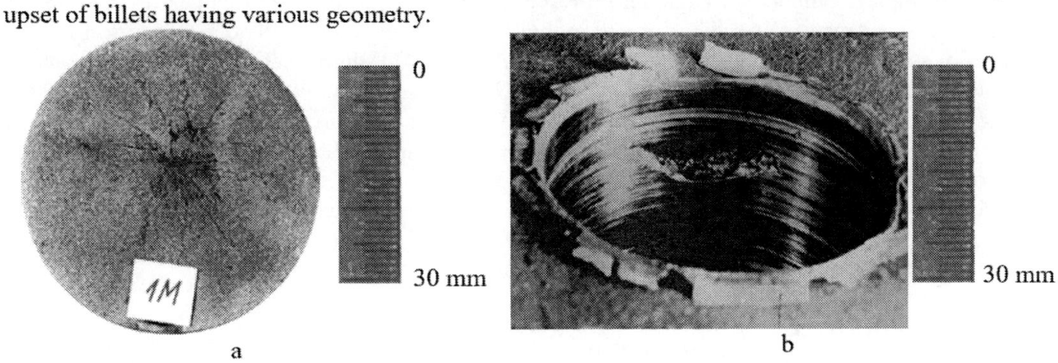

Figure 2. Defects identified in the axial areas of the forgings produced from large ingots: a – a defect in the axial area of a shaft forging; b – a defect identified while boring the axial channel in a hollow forging.

2. Materials and methods

The authors [1, 2, 9, 12] have carried out a detailed study of a poured structure which enabled to establish the location and extension of axial zone defects (V-shaped cracks, arch-shaped cracks) in ingots with conventional and altered geometry of the bottom part. While modeling the chill, axial zone defects were specified in the form of cavities with the dimensions and shape similar to real ingot defects identified when studying the ingot internal structure (figure 3).

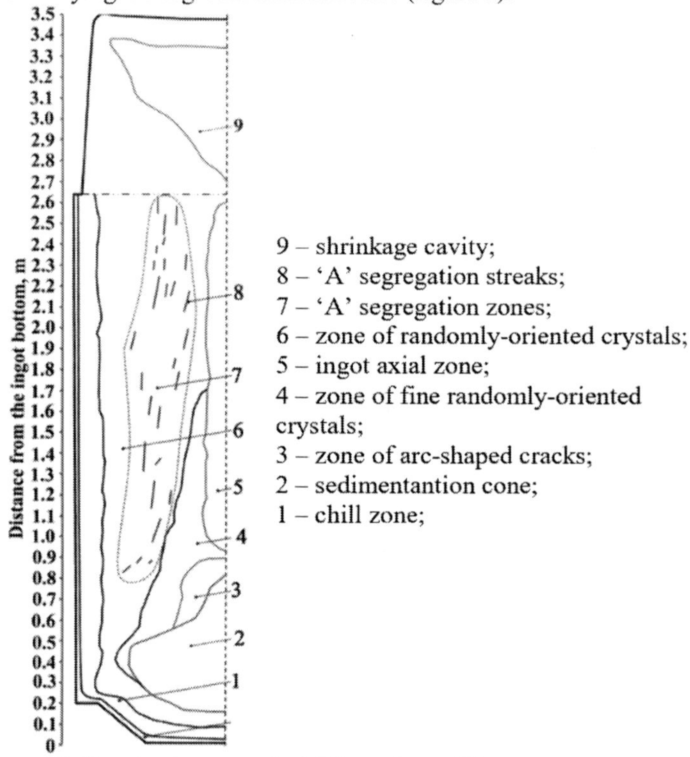

Figure 3. Structural zones of a 24.2 ton ingot [1].

Mathematical modeling was carried out with the Deform 3D application.

Geometry import, symmetry assignment, presetting border conditions and calculation parameters (time interval, stop time etc.) were performed in the ‘Preprocessor’ module. Tool deformation was not considered when modeling the upset operation; besides, only a 90° ingot section was modeled because the problem is fully symmetrical. In addition, steel 38ХН3МФА entry was made in the application in order to make simulation as real as possible. The forging temperature range in the simulation was 1073 K (800 °C) to 1373 K (1100 °C). Heat loss to the tool was calculated through an increased heat loss factor. Ambient temperature was equal to 293 K (30 °C).

The forging press speed was assumed to be unchanged and was equal to 0.5 mm/s. Calculation was stopped at the distance between the tools equal to 1200 mm. Estimated forging time was equal to 40 minutes which fully corresponds to a real production process.

3. Results and discussion

Temperature field calculation throughout the ingot under upset shows that upon upset completion, the temperature of the billet center was not lower than 1263 K (990 °C), which complies with forging temperatures for this grade of steel. The temperature throughout the axial zone is equal and it leads to equal plastic properties of steel throughout the billet and prevents crack formation when forging the ingot.

The application makes it possible to monitor deformation process (figure 5) thus enabling to establish the time of the complete removal of defects thus avoiding additional loading and, consequently, additional costs.

The calculation results provided in figure 7 demonstrate that axial zone cracks are gradually removed through upset.

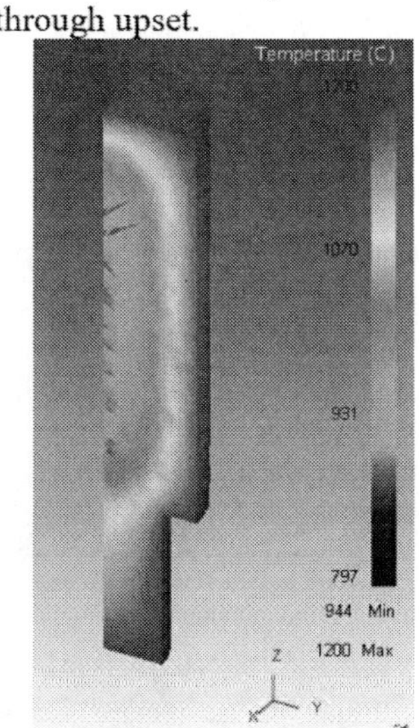

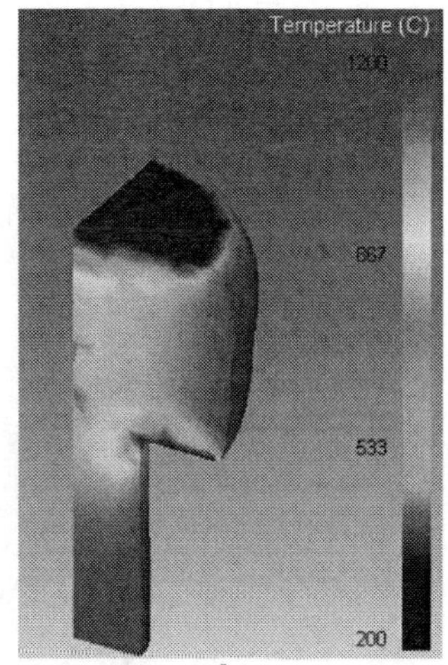

b

a

Figure 4. Temperature distribution field in the conventional shape ingot: a – before upset; b – after upset

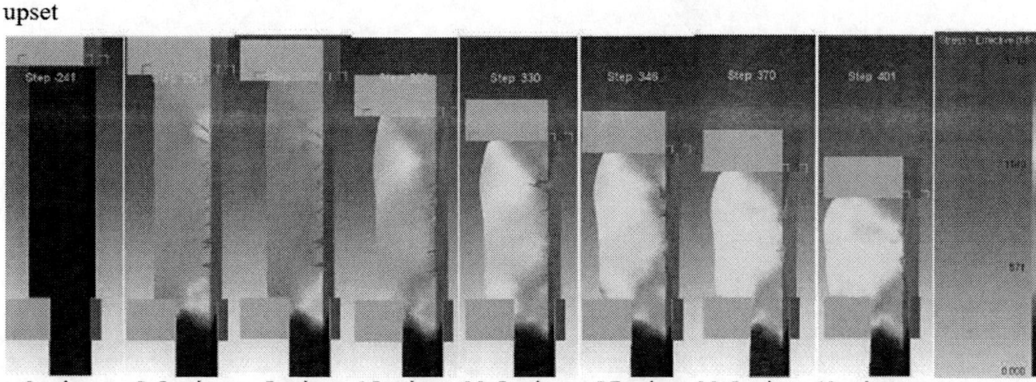

0 min 2.5 min 5 min 15 min 22.5 min 27 min 32.5 min 40 min

a

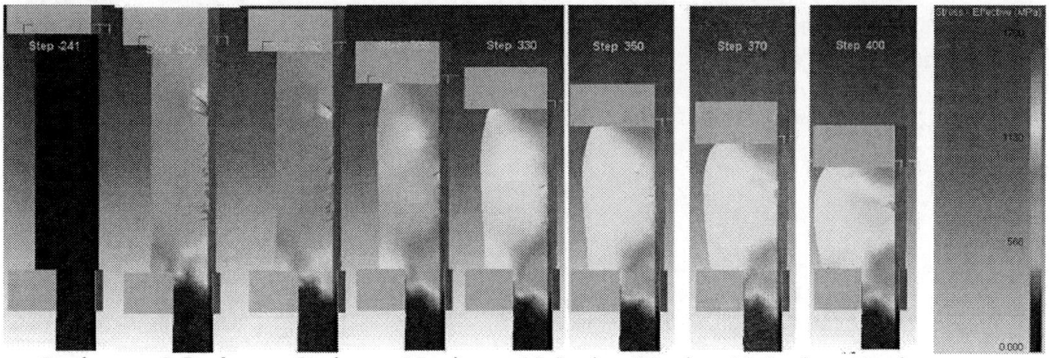

0 min 2.5 min 5 min 15 min 22.5 min 27 min 32.5 min 40 min

b

Figure 5. Shape and strain changes throughout the ingot during hot working: a – the conventional geometry ingot; b – the altered geometry ingot.

During minute 15 of upset, the geometry of cracks starts to change intensively; as this takes place, V-shaped cracks are gradually removed by forging while arch-shaped cracks around the sedimentation cone top open up by minute 23 (figure 5). This behaviour of cracks is typical for both ingots.

A mathematical calculation of efficient deformations in the axial zone of the workpiece manufactured from the altered geometry ingot demonstrates that cracks are eliminated more efficiently in the ingot due to their advantageous orientation in the original ingot and a difference in dendrite structure density.

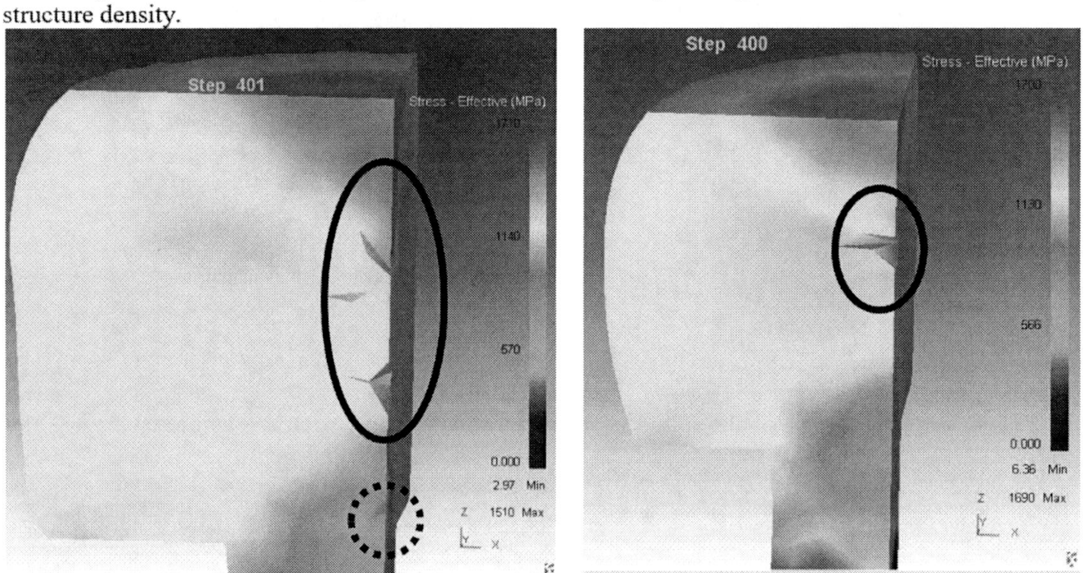

a b

Figure 6. Distribution of total strains after the hot working of ingots: a – the conventional geometry ingot; b – the altered geometry ingot; ——— deformed metal section with crack remnants in the axial zone; ••••••••• deformed metal section with defect remnants of the arch-shaped crack zone.

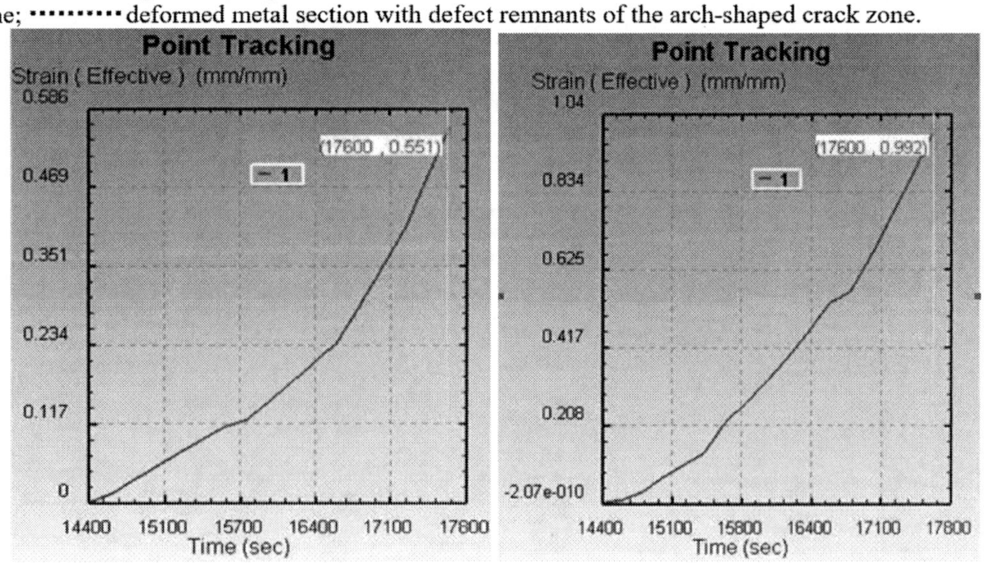

a b

Figure 7. Efficient deformation behavior in the course of time in the ingot axial zone: a – the conventional geometry ingot; b – the altered geometry ingot.

Somehow, even an alteration of the ingot geometry does not allow one to get rid of crack development in the central section of the forging. Moreover, some cracks develop to a great extent (figure 6) because the direction of working coincides with the crack inclination angle thus favoring the crack growth.

We can clearly see from figure 7 that efficient working along the axis of the altered geometry ingot is notably higher than that in the conventional geometry ingot. It can be explained by a higher density of metal in the axial zone; this higher density was obtained during pouring and teeming. An increased density improves the degree of deformation of the axial areas throughout the workpiece thus contributing to the removal of defects.

4. Conclusion

A computer modeling of ingot upset demonstrates that axial zone cracks behave differently during upset: V-shaped cracks are removed by forging while arch-shaped cracks open up. Cracks are not completely removed by the end of upset.

The computer modeling of hot working also proves that during the upset of the conventional geometry ingot, most axial zone defects are not removed as it was believed before, but continue to develop.

Axial zone defects do not develop so rapidly in the altered geometry ingot, cracks are eliminated in the course of hot working (upset) and efficient deformation is twice as high as that in the conventional shape ingot.

Acknowledgments

The reported study was funded by the RFBR, research project No. *HK 15-08-08098\15*

References

[1] Zyuban N, Rutskii D, Konovalov S and Gamanyuk S 2014 *Metallurgical and Mat. Transactions A: Phys. Metallurgy and Mat. Sci.* **45** 6200-6206
[2] Poslamovskaya Yu A and Zhul'ev S I 2008 *Metallurgist* **52** 253-258
[3] Shamrei V A and Zhul'ev, S I 2007 *Metallurgist* **51** 617-623
[4] Romashkin A, Dub V, Tolstykh D, Ivanov I and Ekhvaya G 2016 *Metallurgist* **56** 1163-1172
[5] Liu H, Fu P, Kang X and Ma X 2014 *China Foundry* **11** 46-51
[6] Dub A V, Dub V S, Makarycheva E V and Ivanov I *Russian Metallurgy* **7** 560-564
[7] By Sang-Hun OH, Jung Namkung, Seog-Ou, Cho, and Dong-Hee Lee 2011 *18th Int. Forgemasters Market and Technical Proc.* p 179-182
[8] Tu W, Zhang X and Shen H 2014 *China Foundry* **11** 52-58
[9] Jiaqi W, Paixian F, Hogwei L, Dianzhong L and Yiyi L 2012 *Materials and Design* **35** 446-456
[10] Lan P and Zhang J 2014 *Ironmaking & Steelmaking* **41** 598-606
[11] Liu D, Sang B, Kang X, Li D, Oh J, Yun E, Golkovski M and Lee S 2009 *Acta Phisica Sinica* **58** 104-111
[12] Rutskii D V, Zyuban N A and Titov K E 2009 *Metallurgist* **53** 585-591

IOP Conf. Series: Materials Science and Engineering 177 (2017) 012090 doi:10.1088/1757-899X/177/1/012090

Simulation and optimization of wind and diesel power supply systems

B V Lukutin[1], E B Shandarova[1], D L Matukhin[1], A A Igisenov[1] and S M Shandarov[2]

[1] Tomsk Polytechnic University, 30, Lenina ave., Tomsk, 634050, Russia
[2] Tomsk State University of Control Systems and Radioelectronics, 40, Lenina ave., Tomsk, 634050, Russia

E-mail: shandarovaelena@mail.ru

Abstract. The paper proposes an algorithm to optimize the structure and the choice of capacity of wind and diesel power units of the combined power plant, depending on the wind energy potential and electricity consumption of electrified facility. The algorithm is based on mathematical models of technical and economic characteristics of wind and diesel power plants as well as an optimization method of coordinate descent. The algorithm takes into account the structure of the combined power plant, changing modes of its operation, construction and operation costs of the power facility. The objective function of the algorithm is to minimize the cost of electricity generated.

1. Introduction

A large number of papers are devoted to the issues of simulation and optimization of the power supply system (PSS), using renewable energy sources (RES). Let us note the works [1-3], which significantly contributed to designing software packages for automatic optimization. The most common are as follows: HOGA – Hybrid optimization by the genesis algorithm (Spain), HOMER – Hybrid optimization modeling software (USA), HYPORA – Hybrid power optimized for rural / remote areas (USA).

Mathematical models of optimization software packages are implemented by various means of computer programming: C ++, MS, Excel, JAVA, etc.

The advantage of these optimization software packages is the possibility to solve optimization problems of power facility with an optimal ratio of generating units having different physical properties. The presented software packages feature a convenient interface, which simplifies their use.

Disadvantages of this software refer to the following restrictions:

- the wind energy potential is assessed based on the average annual or average monthly wind speed without taking into account its random origin;
- the lack of correlation between the parameters of the environment and the parameters of power equipment;
- inability to perform calculations and measurements of electric power losses in the power supply system;
- inability to perform investment detailing during the construction and operation of PSS;

- inability to change the structure of the combined power supply system and supplement it with new elements.

These restrictions do not allow optimizing the structure and parameters of wind and diesel PSS subject to a number of important characteristics of the system. These features include a smart interaction algorithm of PSS elements as well as technical and economic performance detailing of the system over the estimated period of operation [4].

The aim of the study is to develop a methodology in order to optimize the structure and the technical and economic characteristics of wind-diesel power plants.

2. Simulation and optimization of wind and diesel power supply systems

The process of choosing the best option - the minimum value of the optimization criterion, e.g., the cost of electricity, can be carried out by various methods, for example, the method of coordinate descent [5].

Let us assume that we need to find the smallest value of objective function $u = f(M) = f(x_1, x_2 \ldots x_n)$. Here, M denotes the point of the n-dimensional space with coordinates $x_1, x_2 \ldots x_n$: $M = x_1, x_2 \ldots x_n$. Let us choose starting point $M_0 = (x_1^0, x_2^0 \ldots x_n^0)$ and consider function f with the fixed values of all variables, except for the first: $f(x_1, x_2^0, x_3^0 \ldots x_n^0)$. Then it becomes a function of one variable x_1. Changing this variable, we move from starting point $x_1 = x_1^0$ in the direction of decreasing the function, until we reach its minimum at $x_1 = x_1^1$, after which it begins to increase. A point with coordinates $(x_1^1, x_2^0, x_3^0 \ldots x_n^0)$ will be denoted by M_1, while $f(M_0) \geq f(M_1)$.

We fix variables: $x_1 = x_1^1$, $x_3 = x_3^0 \ldots x_n = x_n^0$ and consider function f as a function of one variable x_2: $f(x_1^1, x_2^2, x_3^0 \ldots x_n^0)$. Changing x_2, we move from initial value $x_2 = x_2^0$ again in the direction of the decreasing function, until we reach a minimum at $x_2 = x_2^1$. The point with coordinates $(x_1^1, x_2^1, x_3^0 \ldots x_n^0)$ will be denoted by M_2, while $f(M_1) \geq f(M_2)$.

We draw the same minimization of the objective function with variables $x_3, x_4 \ldots x_n$. Reaching variable x_n, let us return to x_1 and continue the process. This procedure fully justifies the name of the method. Based on it, we design a sequence of points $M_0, M_1, M_2 \ldots$, which corresponds to a monotonic sequence of function values $f(M_0) \geq f(M_1) \geq f(M_2) \ldots$ Breaking it at some point k, we can approximately assume the value of function $f(M_k)$ to be its smallest value in this area.

Let us note that this method reduces the problem of finding the smallest value of the function of several variables to a multiple solution of one-dimensional optimization problems. The method is quite simple; however, taking into account the discrete parameters of power equipment, its capacity is sufficient for the given field of power engineering.

The main criteria of economic efficiency of wind and diesel power plants include the reduced annual costs of 1 kW of installed capacity of a power plant and the costs of 1 kWh of electricity. These criteria are determined by the following expressions:

$$C = \frac{P_n \cdot K + E}{P}, \tag{1}$$

where P – installed capacity of power facility (kW); K – total investment, which can be calculated using the formula:

$$K = K_{eq} + K_{de} + K_{con},$$

where K_{eq} – costs of complete plant equipment; K_{de} – design costs; K_{con} – costs of construction and installation of power plants (substations); $P_n = 1/T$ – norm coefficient of profitability, T – economic life of the equipment (years).

In formula (1), E – total annual operating costs, which can be calculated from the formula:

$$E = E_{op} + E_{main} + E_{fu} + E_{tr},$$

where E_{op} – annual costs of power supply system operation; E_{main} – annual maintenance costs; E_{fu} – annual fuel costs; E_{tr} – annual costs of fuel transportation.

The prime cost of 1 kWh of electricity is:

$$E_{el} = \frac{P_n \cdot K + E}{W},$$

where W – total amount of electrical energy generated by power plants during the year.

Economic analysis of the stand-alone power supply, under existing economic conditions in Russia in this regard, can be aimed in many cases at reducing the subsidies cost. Accordingly, electricity production in decentralized power systems is often unprofitable. It is advisable to determine the norm coefficient of profitability in these situations based on the life of the main power equipment. Renewable energy sources and the smart energy system, under certain conditions, can improve the technical and economic characteristics of the stand-alone power supply systems [6, 7]. Given that the vast majority of existing power supply systems include diesel power plants (DPP), the possible increase of technical and economic indicators should start with an analysis of DPP operation modes.

The aim of optimizing the composition and operation modes of DPP is to reduce the fuel component in the electricity costs. Optimizing DPP load modes will increase the service life of diesel generators, and possibly reduce their installed capacity, which will decrease capital costs. The introduction of the wind power plant to stand-alone diesel PSS, under certain conditions, reduces the cost of electricity produced. Capability study for renewable energy integration into the stand-alone power supply system should start with analyzing the energy potential of renewable energy sources and, then, choosing a combined power plant structure [8]. Basically, combined power plants can be constructed both for separate operation of diesel and renewable parts or their parallel operation to supply stand-alone electricity consumers. A simplified block diagram of this facility is shown in Figure 1.

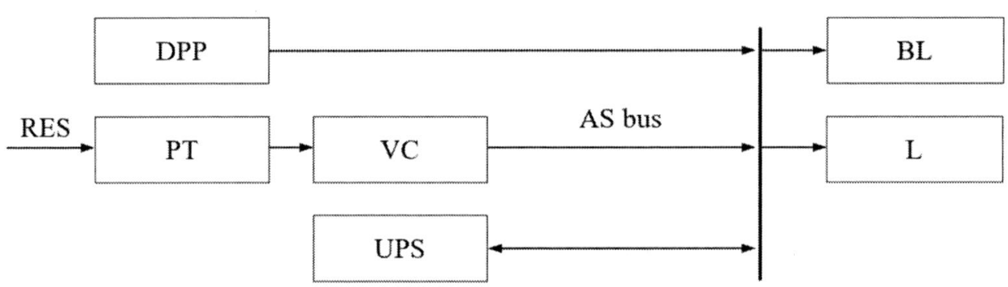

Figure 1. The simplified block diagram of the wind and diesel power plant.

Figure 1 shows the following abbreviations: DPP – diesel power plant; PT – primary transducer of wind energy RES; VC – voltage converter; UPS – uninterruptible power supply; L – load; BL – ballast load. The combined operation mode does not require the use of UPS and BL, and is characterized by the low use of wind energy. The installed capacity of each independent energy source of the combined power plant with the separate operation mode of WPP and DPP should be capable of covering the electrical load of the system. Consequently, the available power of each energy source shall comply with the maximum power of the consumer. Available DPP power usually ranges from 0.7 to 0.8 of its installed capacity. The available capacity of wind power plants, in addition to the installed capacity, is determined by the wind energy potential.

Obviously, for the approximation of available WPP capacity to its installed capacity, it is necessary to choose a wind turbine with the rated wind speed close to the mean wind speed in the area of the proposed windfarm installation. Increasing the installed capacity utilization factor of wind power improves its economic performance and, accordingly, technical and economic characteristics of the combined power supply system. For example, the cost of electricity generated by WPP depending on the average wind speed, other things being equal, is characterized by the following values: 7.16 m/s - 4.8 cents per kWh; 8.08 m/s – 3.6 cents per kWh; 9.32 m/s – 2.6 cents per kWh. It should be noted that the rated wind speed for wind turbines is commonly in the range of 10 – 15 m/s.

Therefore, combined wind and diesel power plants with separate operation of generating units are suitable for areas with high wind power potential to provide a large part of electricity generation for the consumer [8].

To ensure the quality of power supply and the maximum wind energy in these systems, uninterruptible power supply and ballast load are used. Stand-alone inverters are used for inversion of wind power parameters. From an economic point of view, WPP supplement to diesel power systems, designed to cover the electrical load for most of the time, requires considerable additional costs for the design and construction works as well as WPP equipment. This results in a fold increase of overall capital costs. If we consider that the amount of electricity produced in the stand-alone power supply system remains approximately unchanged, economic characteristics of the power system with wind generation deteriorate.

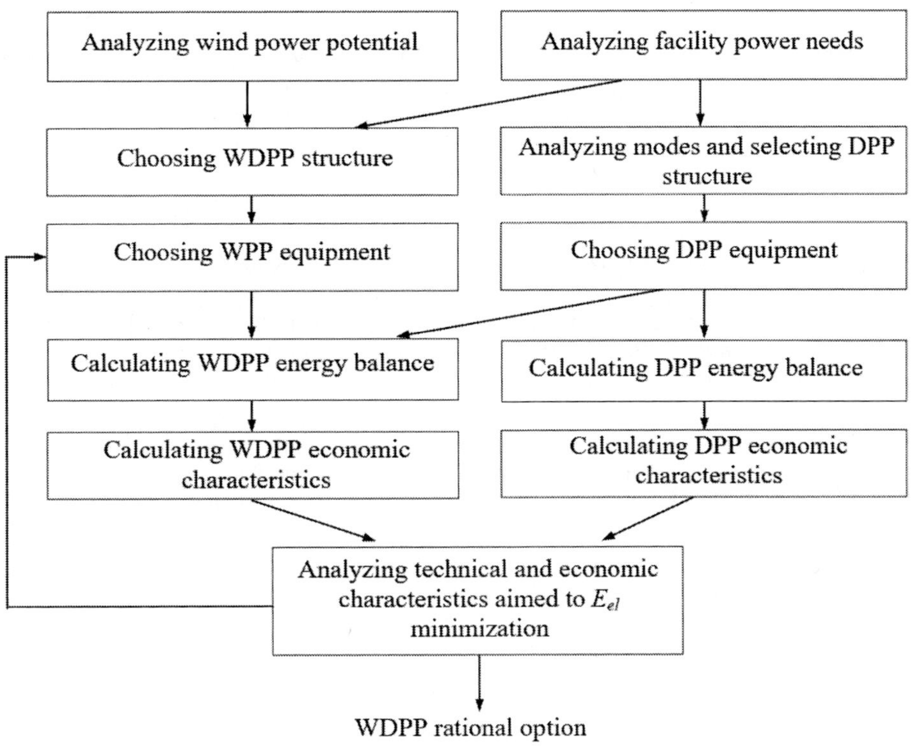

Figure 2. The block diagram of methodology of designing a rational option of the wind and diesel power supply system.

On the other hand, partial substitution of the diesel generation for wind generation saves expensive fuel and, reducing the operation time of diesel generators, increases their service life [9, 10]. Wind power equipment features longer life as compared to diesel generators, which also helps to improve the economic performance of the combined power facility.

Therefore, the decrease in profitability of norm coefficient P_n and operating costs of wind-diesel power contribute to the improvement of its economic performance. Obviously, economic feasibility of the construction of stand-alone wind and diesel power facility with separate operation modes of wind and diesel parts is achieved if the savings per annum in operating costs exceed the cost of compensation for the increased capital costs. These opportunities are more typical of stand-alone diesel power systems with high cost of electricity, determined primarily by the cost of fuel. Stand-alone systems with the main energy source – the continuously operating diesel power plant and the wind power plant, connected to it in parallel, according to the current wind energy potential, are less efficient. Typically, the wind power plant is connected to DPP through the grid tie inverter, which limits the maximum wind power at the level of 30-40% relative to the power of the diesel power system. The economic viability of these systems is only possible with very expensive diesel fuel.

3. Results and Discussion

The methodology of designing a rational, based on technical and economic parameters, option of the wind-diesel power system is shown as a block diagram in Figure 2. The input information for analyzing possible options of designing a stand-alone power supply system is wind power potential in the area of electrified facility and consumer's energy needs. High wind energy potential, characterized by the average annual wind speed of about 5 m/sec or more, allows choosing the structure of the wind-diesel power supply system with separately operating generating units.

At lower wind speeds, it is possible to design a more rational structure of combined power facility with the parallel operation of wind and diesel generating units, where diesel generation prevails.

In accordance with the structure of the chosen WDPP, wind power equipment is determined as follows: installed capacity of main generating units, an inverter type, availability and capacity of energy storage equipment. The DPP structure and composition are determined by the facility energy needs, recommendations to ensure energy-efficient modes of diesel generators and practically does not depend on the wind power plant. It should be noted that WDPP structure recommendations, depending on the average wind speed, can be adjusted according to the achieved economic project indicators.

Economic indicators of stand-alone wind and diesel power systems are determined by the overall capital investment and maintenance costs, and the volume of electricity produced and supplied to the consumer. Capital investment includes the equipment, design, construction and installation costs. We shall note that capital investment in the wind power plant sufficiently exceeds the costs of the diesel power plant construction.

Operating costs for the wind farm, due to the lack of fuel cost, operational staff and a long life of the power equipment is much less than those of DPP.An optimal option of wind-diesel power systems can be found by calculating the minimal prime cost of electricity produced. An optimization target function, in this case, should be expressed as $E_{el} \leq \min$. As the input information, we should take into account the stability of the wind power potential and facility power consumption. The installed capacity of DPP is unchangeable as it is determined by the maximum load power. The cost of diesel fuel remains unchanged or projected to change based on price fluctuations in recent years.

4. Conclusion

To sum up, optimization of the wind and diesel power plant can be reduced to a one-dimensional problem with the independent variable – WPP installed capacity.

The initial value of the WPP installed capacity shall be chosen based on the combined power plant structure and the interaction algorithm between wind and diesel parts. During separate operation of generating units, power ratio restrictions of DPP and WPP virtually do not exist. During combined operation of wind and diesel generators, DPP usually becomes the leading energy source. WPP available capacity, in this mode of operation, should not exceed 30-40% of the power generated by DPP.

Based on these recommendations, we shall choose the initial option of WPP installed capacity and calculate the cost of electricity. Further, according to the method of coordinate descent, we can change the value of the optimization variable – the WPP installed capacity, taking into account available capacities of wind turbines with suitable characteristics and find a new value. Based on the difference sign, we shall adjust the direction to the minimum point and carry on calculating power equipment options for the combined power plant.

References

[1] Can L K, Shek J K H, T and Mueller M A 2016 *Energ. Convers. Manage* **126** 446-462
[2] Das D, Aditya S K and Kothari D P 1999 *Int. J. of Electrical Power and Energy Systems JEPE* **21**(3) 183-189
[3] Shilin A A and Bukreev V G 2014 *Thermal Eng.* **61** (10) 741-746
[4] Elhadidy M A and Shaahid S M 2005 *Energ. Convers. Manage* **46** 2501-13
[5] Wang X, Palazoglu A and El-Farra N H 2015 *Applied Energy* **143** 324-335
[6] Nosov G V, Kuleshova E O, Vassilyeva Yu Z and Elizarov A I 2016 *IOP Conf. Ser.: Mater. Sci. Eng.* (Tomsk) vol 124 (Bristol: IOP Publising) 012081
[7] Bansal R C 2006 *IEEE Transactions on Industrial Electronics* **53** (4) 1116-26
[8] Bansal R C, Bhatti T S, Kothari D P 2004 *Int. J. of Modelling and Simulation* **24** (1) 1-7
[9] Lukutin B V, Sarsikeyev Y Zh, Surkov M A and Lyapunov D Yu 2014 *Int. Conf. on Environment and Electrical Engineering EEEIC* 2014 (Krakow) (USA: IEEE Computer Society) 459-463
[10] Guo T and Wu Q 2016 *CCDC* 2016 7532110

The choice of the principle of functioning of the system of magnetic levitation for the device of high-performance testing of powder permanent magnets

D V Shaykhutdinov, N I Gorbatenko, N D Narakidze, A S Vlasov and I A Stetsenko

Platov South Russia State Polytechnic University (NPI), 132, Prosveschenija str., Novocherkassk, 346400, Russia

E-mail: shaykhutdinov@npi-tu.ru

Abstract. The present article focuses on permanent magnets quality control problems. High-performance direct-flow type systems for the mechanical engineering production processes are considered. The main lack of the existing high-performance direct-flow type systems is a completing phase of movement of a tested product when the movement is oscillatory and abrupt braking may be harmful for high fragility samples. A special system for permanent magnets control is offered. The system realizes the magnetic levitation of a test sample. Active correction of the electric current in magnetizing coils as the basic functioning principle of this system is offered. The system provides the required parameters of the movement of the test sample by using opposite connection of magnetizing coils. This new technique provides aperiodic nature of the movement and limited acceleration with saving of high accuracy and required timeframe of the installation in the measuring position.

1. Introduction

Electric motors production is one of the most important areas of electrical engineering. DC motors have high energy efficiency and huge torque. An important element of most of electromagnetic drives is permanent magnets. Recently, permanent magnets made of ferromagnetic materials powder [1-7] have been widely used. However, their quality must be controlled. For implementation of production control of permanent magnets, direct-flow type magnetic systems are used [8]. These systems carry out high-performance movement, orientation through a magnetic texture and fixing of the test sample in the measuring position with electromagnetic field. The main lack of these systems is the completing phase of the movement of the test sample (figure 1). In the completing phase, the movement has an oscillatory nature, and braking of the test sample is rather abrupt. These factors do not allow one to use magnetic systems of direct-flow type for testing products with the increased fragility.

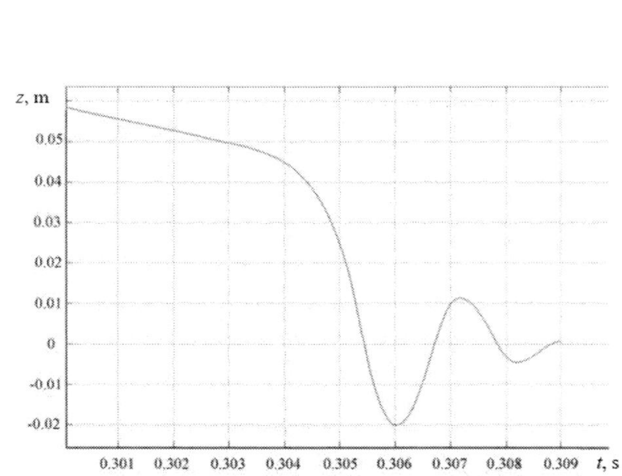

Figure 1. The trajectory of movement of the test sample in the direct-flow type magnetic system.

The objective of this research is implementation principles development of the control system based on magnetic levitation [9, 10] of the test sample by active correction of the current in magnetizing coils of the magnetic system for ensuring the set parameters of the sample movement: aperiodic nature of the movement and acceleration restriction and saving of high accuracy and timeframe of the installation in a measuring position.

2. Materials and methods

The control system of test sample magnetic levitation is offered as the solution of this task. The function chart of the control system is presented in figure 2.

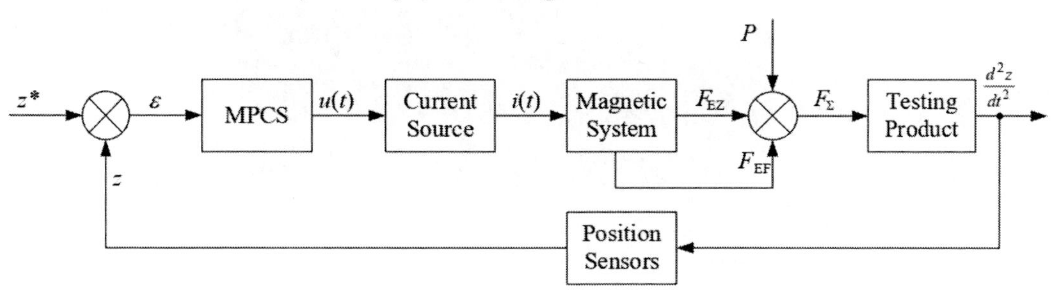

Figure 2. The block diagram of the control system of parameters of examinee levitation of the product.

Parts of the system (figure 2) are: the microprocessor control system (MPCS), the managed current source, the magnetic system, the situation sensor.

The law of the test sample movement with predetermined geometrical sizes in the direct-flow type magnetic system is described by a second order differential equation:

$$m\frac{d^2z}{dt^2} + P + F_z(z,I) + F_x(z,I)\cdot k_F \cdot sign(\frac{dz}{dt}) = 0, \quad (1)$$

where z – the distance between centers of the test sample and the interpolar space; m, P – respectively, the weight and the gravity of the test sample; F_z, F_x – tangential and normal components of the electromagnetic force affecting the test sample, k_F – the friction coefficient of the test sample of the product against guide rails, I – the current in magnetizing coils.

3. The implementation and research of the magnetic system

Two options of current forming in magnetizing coils of the magnetic system have been considered. With the coordinated connection of magnetizing coils, the direction of vector tangential force F_z has been coordinated with gravity *P*. When the velocity vector of the test sample is directed to the center of the system, this force does not limit free fall acceleration, and, on the contrary, creates additional acceleration that leads to "flight" of the test sample through the installation position. Restriction on acceleration in this case is only imposed by the normal component force of F_x (1), which will be transformed to friction force F_{EF} (figure 2). There is the second shortcoming: the arisen friction force of F_{EF} results in considerable mechanical wear of guides and partial damage of the test sample.

The assumption is made that the use of opposite connection of magnetizing coils will allow one to reduce friction force and to limit acceleration of an object. To confirm that, the model of the magnetic system has been developed in the environment of *GMSH GetDP*. The model is presented in figure 3. In figure 3, the following designations are accepted: *MC* - magnetic core, *TP* - testing product, *MC* - magnetizing coils. The value of the magnetomotive force in the model was set as *Iw*=20 kA/m. The received schedules of tangential and the normal components of electromagnetic forces as dependences on the distance test sample to the center of magnetizing coils are represented in figure 4. Schedules of dependence of forces are symmetric concerning the center of the magnetic system interpolar space.

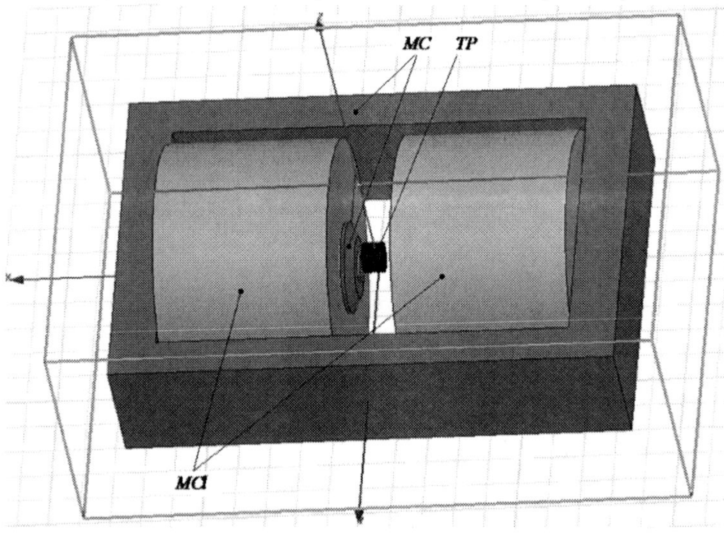

Figure 3. The model of the magnetic system in the environment of *GMSH GetDP*.

In case of oppositely connected magnetizing coils, the specified value of magnetomotive force, *Iw*=20 kA/m, owing to the centered position of the test sample in the magnetic system. The value of the normal component of electromagnetic force is F_x =1.83 N, and the value of the tangential force component is F_z =0. For retention of the test sample in the centered position, the value of the amount of operation forces has to satisfy the condition: $F_x \cdot k_F = P$ or $F_x \cdot k_F = mg$.

To solve the problem of the position control of the permanent magnet, it is necessary to develop a mathematical model of the movement process. The following prior restraints on various subjects of permanent magnets *D* (diameter) and *h* (height) were raised: $18 \le D \le 22\ mm; 11 \le h \le 14.7\ mm$. To construct a mathematical model of the movement process, the apparatus of orthogonal polynomials and multi-level experiments planning is applied. The graphs in figure 4 show that in the studied area of position *z*, tangential $F_z(z)$ and normal $F_x(z)$ components of electromagnetic forces can not be described even by linear and quadratic functions. Similar preliminary studies indicated that dependences $F_z(h)$, $F_z(D)$, $F_x(h)$, $F_x(D)$ can be described by a linear function.

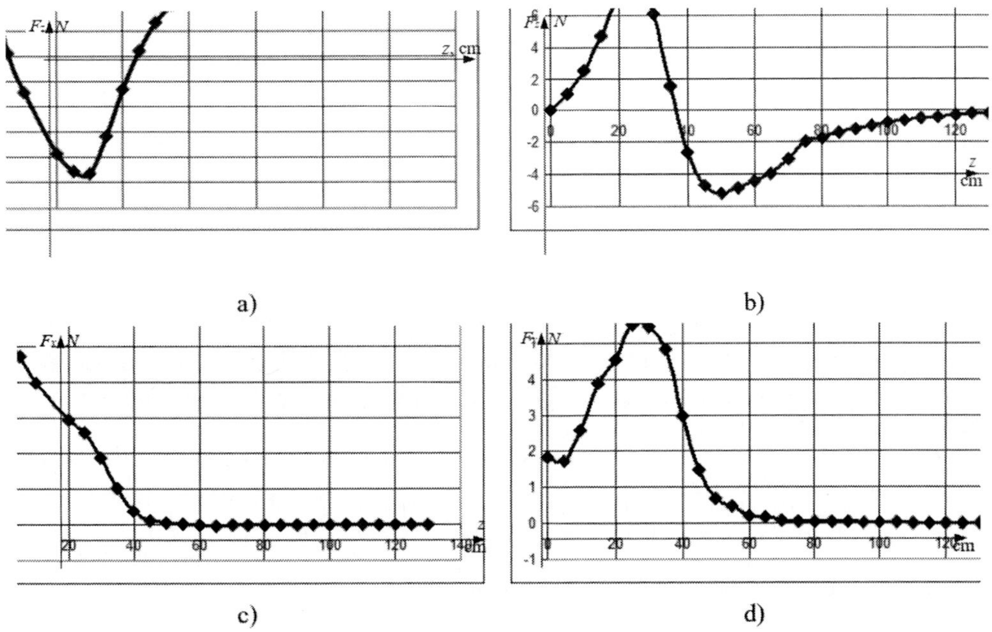

Figure 4. Schedules of the electromagnetic forces operating on the test sample depending on the distance between the center of the sample and the center of the magnetic system.

The aim of the experiment is to build the polynomial mathematical model of the movement process in the form of the regression equation. For polynomial modeling, the orthogonal polynomials are selected as a basis function. The possibility of constructing the model of a particular type depends on the number and location of the experimental points in the factor space. In this case, a complete factorial experiment was conducted. It consisted of 96 different sets of F_z and 84 different sets of F_x, which define a four-factor space, respectively, 96 and 84 experimental points. The models include orthogonal polynomials in four variables; the degree of the polynomial older for any variable is less than the number of varying levels.

In addition to single variables, the model introduced a twin, triple and quadruple product of polynomials of different variables. The complete F_z model includes: 33 pairs of works, 37 triple, 14 quadruple. The complete F_x model includes: 29 pairs of works, 32 triple, 12 quadruple. The model does not include products of polynomials of the same variable. General expression patterns are not recorded due to its bulkiness.

Experimental design for F_z includes 96 experimental points and for F_x — 84 experimental points. Since the experiment was performed with the help of a computer model, the randomization and repeated experiments were not pursued. Processing of the results consists of three operations: the calculation of a *b*-coefficient of the regression mathematical model, determination of their statistical significance, check of the adequacy of the model.

To find the values of the *b*-coefficient, it is necessary to make a structural matrix with the order of 96 for F_z and with the order of 84 for F_x. To test the adequacy of the resulting model, the predicted responses values were calculated. Then, there is the dispersion of the adequacy calculated by formula:

$$S_{ad}^2 = \sum_{u=1}^{N} \frac{\varepsilon_u^2}{f_{ad}},$$

where $f_{ad} = N\text{-}l$ - number of degrees of freedom; l - number of significant b-coefficients.

For the tangential component of electromagnetic force, F_z $f_{ad,z}$=N_z-l_z=96-50=46, and the dispersion of adequacy was $S^2_{ad,z} = 697.2544$.

The value of the reproducibility variance was calculated when determining the significance of the coefficients: $S^2_{res,z} = 10242$. Therefore, the calculated value of the F-criteria was $F^z_{calc} = 0.0681$.

For the normal component of electromagnetic force, F_x: $f_{ad,x}$=N_x-l_x=84-35=49, and the dispersion of adequate was $S^2_{ad,x} = 1210.4$.

The value of the reproducibility variance was calculated when determining the significance of the coefficients: $S^2_{res,x} = 13285$. Therefore, the calculated value of the F-criteria was $F^x_{calc} = 0.0911$.

The tabular value F-criteria for the significance level of α=0.05 and the number of degrees of freedom f_{ad}=46 and f_{res}=96 is $F^z_{tb} = 1.00$. For α=0.05 and the number of degrees of freedom f_{ad}=49 and f_{res}=84, the tabular value F-criteria is $F^x_{tb} = 1.00$. Since $F^{z,x}_{calc} < F^{z,x}_{tb}$ in both cases, both models should recognize the adequacy of existing statistical data. The resulting adequate equation for models of F_z and F_x does not condition their complexity.

The body weight, which the magnetic system is capable to hold by means of the managed electromagnetic forces of m_{max}, makes:

$$m_{\max} = \frac{F_x \cdot k_F}{g}.$$

The scheme used for determination of size k_F is presented in figure 5.

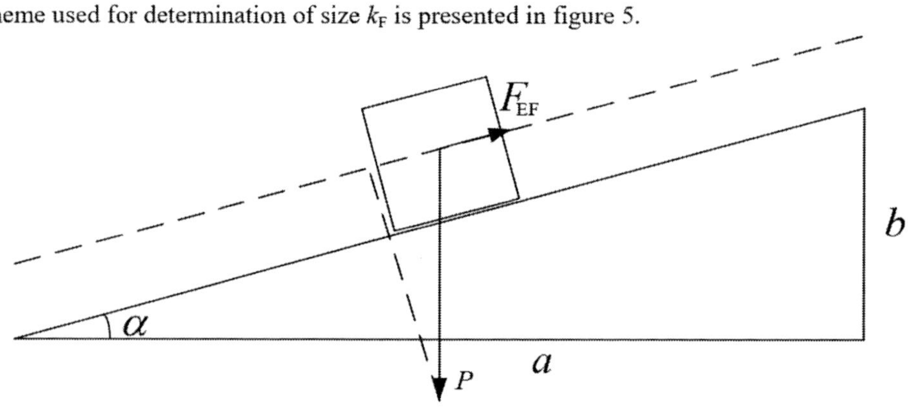

Figure 5. Scheme of k_F definition.

At some value of the slope angle of plane α on which the test product is placed, the body begins the movement. We will determine the size and the tangent of angle α by value k_F. For this purpose, we will fix values of a and b. The received friction coefficient was: $k_F = 0.25 \pm 0.02$.

Admission (± 0.02) implements a possibility of friction of the test product about the directing systems of tests by different parts. In this case, $m_{\max} = 0.046$ kg.

4. Conclusion

From the analysis of charts in figure 4 that treat shortcomings of oppositely connected magnetizing coils in the magnetic system, we can conclude the following.

1. There is a need for creation of a considerably bigger value of magnetomotive force affecting the test sample.

2. The vector of the tangential component of electromagnetic force F_z is not directed opposite to the force gravity vector at all values z. When it is directed opposite to the gravity vector at z=(37÷120) mm, the test product "is involved" in the center of the magnetic system.

3. There is a significant value of normal component F_x at z=(20÷40) mm. The situation when the test sample does not reach the center of interpolar space of the magnetic system is possible.

These imperfections can be solved in the system parameters calculation phase.

The advantages of oppositely connected magnetizing coils are:

1. the direction of vectors of forces affecting the test sample at oppositely connected magnetizing coils has matched the expected. The vector of electromagnetic forces is directed conversely relatively gravity P.

2. The value of forces for oppositely connected magnetizing coils is much less than values of forces during the coordinated turning of magnetizing coils. The advantage of opposite connection, thus, is a considerable decrease in friction force of the test sample between guide rails, which leads to a considerable decrease in resources consumption of the system.

In general, it is possible to draw a conclusion that with oppositely connected magnetizing coils of the magnetic system, it is possible to create counteracting magnetic force in the required timeframe and to achieve necessary parameters of the movement of the test sample in the direct-flow type magnetic system. There is aperiodic nature of the movement and a restriction on acceleration.

The mathematical models of F_z and F_x as the basis for the construction of high-performance permanent magnet control systems used in the manufacturing process of mechanical engineering were developed.

Acknowledgments

This article is based on results of project No. SP 4108.2015.1, implemented under the program 'Scholarship provided by the President of the Russian Federation for young scientists and graduate students engaged in advanced research and development in priority areas of modernization of the Russian economy' using the equipment of shared facility 'Diagnosis and energy-efficient electrical equipment' (NPI).

References

[1] Volodchenkov A D, Kodera Y, Garay J E 2016 *J. of Mat. Chem. C* **24(4)** 5593-5601

[2] Pedrosa F J, Rial J, Golasinski K M, Rodríguez-Osorio M, Salas G, Granados D, Camarero J, Bollero A 2016 *Royal Society of Chemistry Advances* **90(6)** 87282-87

[3] Niada R F, Bittencourt S D, De Martins R M, De Aquim P M, Dias M M, Schaeffer L 2015 *Ciencia e Tecnologia dos Materiais* **27(1)** 53-62

[4] Singh N, Shyam R, Upadhyay N K, Dhar A 2015 *IOP Conference Series: Materials Science and Engineering* (Kerala) vol **73** (Bristol: IOP Publishing) 012042

[5] Cui J, Choi J P, Li G, Polikarpov E, Darsell J, Kramer M J, Zarkevich N A, Wang L L, Johnson D D, Marinescu M, Huang Q Z, Wu H, Vuong N V, Liu J P 2014 *J. of Appl. Phys.* **115(17)** 17A743

[6] Liu W Q, Wang Z, Sun C, Yue M, Liu Y Q, Zhang D T, Zhang J X 2014 *J. of Appl. Phys.* **115(17)** 17A716

[7] Oniku O D, Bowers B J, Shetye S B, Wang N, Arnold D P 2013 *J. of Micromech. and Microeng.* **23(7)** 075027

[8] Gorbatenko N I 2001 *Full-scale-model tests of the ferromagnet materials products* (Rostov-on-Don: Publishing house SKNTs VSh) p 244-281

[9] Matheys C, Tumanova N, Leyssens T, Myerson A S 2016 *Crystal Growth and Design* **16(9)** 5549-5553

[10] Ozturk K, Kabaer M, Abdioglu M, Patel A, Cansiz A 2016 *J. of Alloys and Comp.* **689(25)** 1076-1082

Research of dynamic loading in a drivetrain by means of mathematical modeling

N S Sokolov-Dobrev, M V Ljashenko, V V Shekhovtsov and P V Potapov

Volgograd State Technical University, 28, Lenin ave., 400005, Russia

E-mail: pvicpotapov@gmail.com

Abstract. This paper describes the development of the initial full dynamic model of a caterpillar agricultural tractor 'Chetra 6C-315' drivetrain, the processes of the model reduction; the mathematical apparatus for defining loads acting on drivetrain shafting elements. The main results of computational researches of dynamic loadings of drivetrain elements in various tractor operation modes are presented.

1. Introduction

Despite the continuous improvement of modern tractors drivetrains constructions, dynamic loading of drivetrains remains at a high level. It results in accumulation of fatigue damage and premature failures of details subjected to the highest loading. It especially concerns machines which were tested insufficiently and their construction was not enhanced properly. For example, agricultural tractor Chetra-6C315 (drawbar category 6) was designed on the base of an industrial tractor (drawbar category 9) construction. At that, constructions of some units, particularly a gearbox and a steering mechanism, were taken from a 9-tonn machine to a 6-tonn one without any changes. So, the development of an adequate mathematical model of the drivetrain and the search of ways for decreasing dynamic loadings in the drivetrain of this tractor in various operation modes by the created model is a topical goal.

2. Development and specifics of the dynamic model

The structural scheme of the tractor drivetrain created on the base of analysis of technical documentation was used as the base for development of a dynamic model. The full initial model includes 98 lumped masses with elastic and friction links. Dynamic parameters of its elements were defined by computations according to the method described in [1].

As far as the initial dynamic model of the drivetrain has complicated the branched structure with the large number of masses, which have high partial frequencies (above 10000 Hz), simplification of the initial model was made. The initial model was changed to the equivalent model with fewer moving masses. For simplification, Rivin's method was used. This method is based on changing of a separate elementary two-mass oscillation system with the highest partial frequency to a one-mass system by combination of two masses into one and changing the links compliance of the combined mass.

The reduced dynamic model is presented and described in [2].

The advantage of the reduced model is providing analysis of:

- a mutual influence of torsional vibrations in drivetrain and longitudinal-angular vibrations of the tractor frame;

- the influence of elastic and inertial parameters of reactive elements of the drivetrain (tractor chassis mass and suspension stiffness, and also masses of separate units in a transmission case and its bearing stiffness) on the dynamic loading of the drivetrain from the main operation loads;
- the influence of elastic, inertial and damping parameters of each drivetrain elements on loading from the operational impact, and also the influence of these parameters on the process of transferring torsional vibrations through shafting;
- the character of loading changes on drivetrain areas as the result of installed dampers of torsional vibrations;
- Dynamical loadings of drivetrain elements on transient modes: when starting from stand-still and stopping, during gear changing under loading, during cornering, during slipping.

3. Development and research of the mathematical model

On the base of the dynamic model [2], the mathematical model of the drivetrain [3] was created in Simulink. The structural scheme of the general visual model of the tractor drivetrain is presented in figure 1. Main subsystems of the visual model are described by separate blocks.

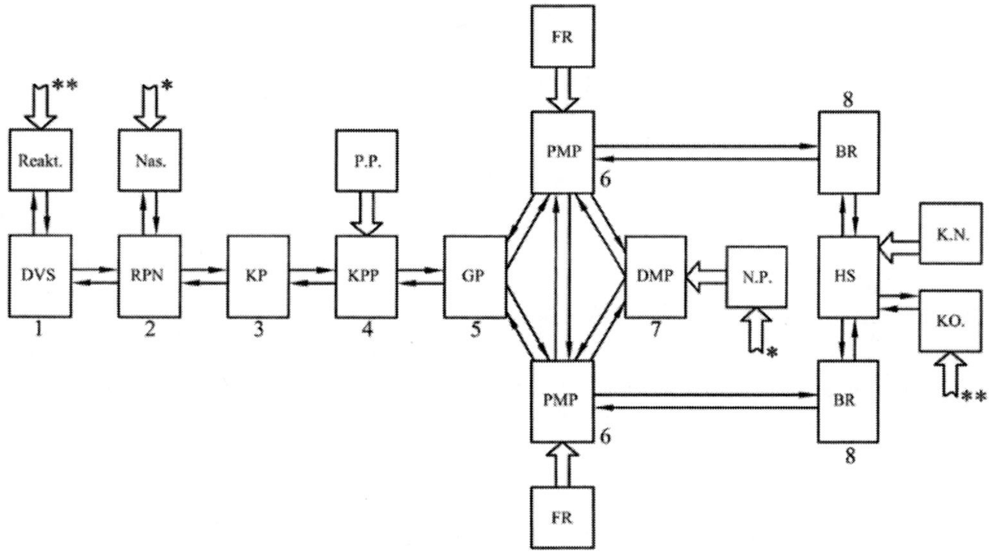

Figure 1. The structural scheme of the drivetrain for creating the general visual model in MATLAB/Simulink. Reakt – reactive mass of the engine frame; 1 – IC-engine; 2 – pump drive reduction gear; 3 – propeller shaft; 4 – gearbox; 5 – final gear; 6 – planetary steering mechanism; 7 – differential steering mechanism; 8 – side reducing gear; 9 – chassis; NAS – power take-off for pumps; PP. – dynamic impacts during gears changing in the gearbox; FR – dynamic impacts of clutches engaging during cornering; NP – impacts of the steering mechanism pump; KN – dynamic impacts of drawbar loading; KO – dynamic impacts of chassis vibration.

Main elements of subsystems are blocks corresponding to inertial masses of the system. These blocks are enumerated in accordance with figure 1. Other blocks represent external and internal impacts. The model provides researches of the drivetrain loading both by separate impact factors and by their mutual influence.

The feature of the model is a new quality describing of the planetary mechanism operation in which all laws of energy transferring and transmitting are observed, and inertial parameters of each mechanism element and elastic parameters of elements interactions are taken into account. The corresponding block provides researching of speed and power characteristics of power flow through the planetary mechanism both when any element is stopped, and during rotation of all elements with various angular speeds. Also, vibration processes generated between connected elements and damping characteristics of their link are taken into account.

4. Dynamic parameters of loading modes

The general loading mode of pull and transport machines could be presented as a sum of quasistatic and dynamic components of the torque [1-4]. A static component is simple to calculate – necessary data are engine power characteristic, conditions of a propulsor and work equipment interaction with the ground.

A dynamic component is considered by means of the coefficient of external dynamic loads k_{dv} calculated from relation:

$$k_{dv} = 1 + \frac{P_A}{P}, \tag{1}$$

where P – tangential load corresponding to the rated torque, P_A – average dynamic load [2]. Tangential load is calculated from expression:

$$P = \frac{2 \cdot M}{m \cdot z}, \tag{2}$$

where M – rated torque, m – average modulus, z – gear teeth number.

$$P_A = \sqrt{P_a \cdot (2 \cdot D - P_a)}, \tag{3}$$

$$P_a = \frac{P_1 \cdot P_2}{P_1 + P_2}, \tag{4}$$

where P_1 – force in gearing in case of absence of elastic deformation of teeth, shafts and other elements; P_2 – force in gearing during steady rotation of gears and bending of teeth in the range of a step errors value.

Dynamic factor D is calculated from expression:

$$D = \Delta \cdot b_c \cdot C \cdot \cos^2 \beta, \tag{5}$$

where Δ – rated factory inaccuracy of gear, C – specific teeth stiffness, b_c – average value of the rim, β – pressure angle.

The model research allows one to obtain oscillograms of toques on every areas of the drivetrain, to define maximal and minimal dynamic components, and the average value of the static component.

5. Research of dynamic loading of the drivetrain from the sprocket and caterpillar chain rewinding unevenness

The plan of the computational research is presented further. On the calculation start in dialog mode rated values of external and internal forces and torques, the character of loads are entered. In accordance with drawbar force calculation, the operation mode of the engine incase of the maximal torque is set [5, 6]. For this research, it was assumed that the drawbar load is constant in time and equal to the load on maximal possible drawbar power. Modelling time was set equal to 10 sec. During that time, all transient processes in the drivetrain will end. The range of the sprocket and caterpillar chain rewinding frequencies is defined by the range of tractor working speeds. The relation between these parameters is the following:

$$V = \frac{\omega \cdot 2\pi \cdot R_S}{z}, \tag{6}$$

where ω – frequency of sprocket and caterpillar chain rewinding, R_S – sprocket radii, z – teeth number of the sprocket.

Oscillograms of elastic torques on each area of the drivetrain were obtained as the result of modeling. An example of the oscillogram is presented in figure 2.

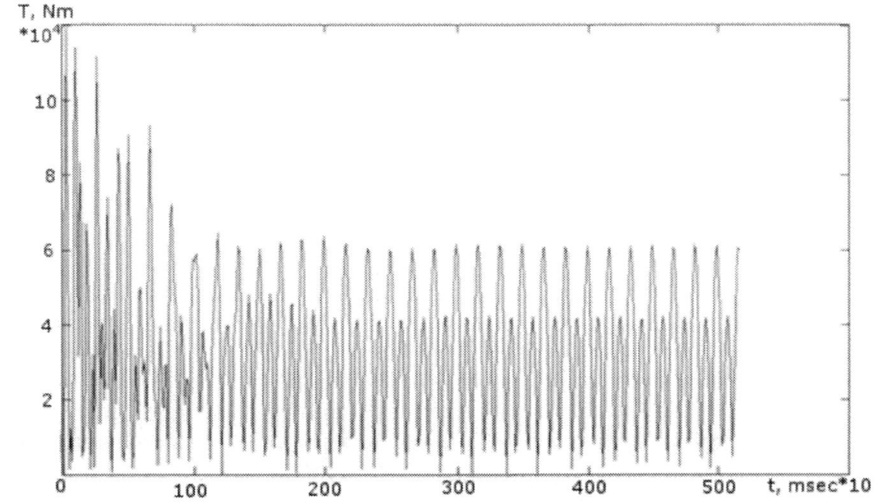

Figure 2. The oscillogram of the elastic torque on 20th area (planetary mechanism in differential steering mechanism) at the tractor speed 0.75 m/sec (sprocket and caterpillar chain rewinding frequency – 5 Hz).

Oscillograms were processed by means of the program: maximal M_{max} and minimal M_{min} values of the elastic torque were calculated, and the value of the average M_c torque on each area was calculated according to the following relation:

$$M_C = M_{min} + \frac{M_{max} - M_{min}}{2}. \tag{7}$$

Calculation of the dynamic factor is as follows:

$$k_d = \frac{M_{max}}{M_C}. \tag{8}$$

Research results are relations of loads dynamic factor variations on areas in the range of tractor working speeds. An example of those relations of the dynamic factor variation is presented in figure 3.

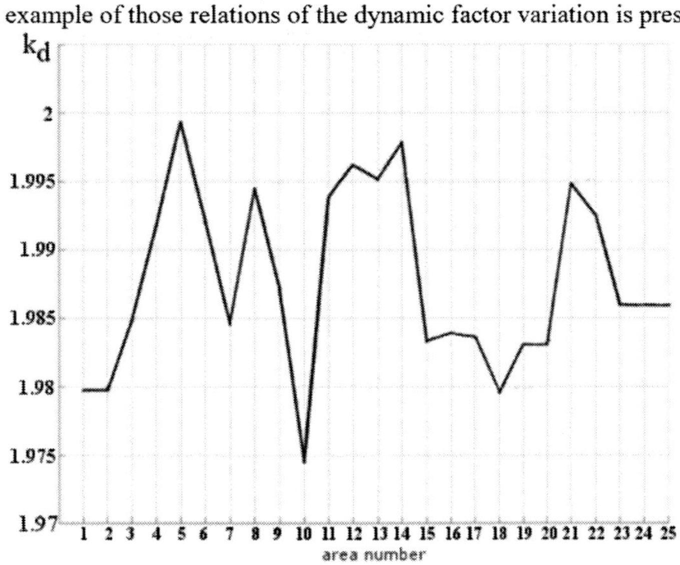

Figure 3. Values of the dynamic factor on areas at tractor speed V=2.26 m/sec.

Also, the range of computational researches for defining the pattern of the torsional vibrations transfer through shafting was performed. These torsional vibrations are generated by unevenness of operation loads and rewinding of the caterpillar chain.

6. Conclusion

- Analysis of the obtained graphics allows identifying areas with the highest dynamic loading on main tractor operation modes. Modes with the highest loads on areas lying near loads source (final drive and differential turn mechanism), and modes where the highest loading is on distant areas (propeller shaft, shafts and a gear of the gearbox) were found.
- Analysis of torsional vibration transferring through shafting shows that loads with frequencies of 0 – 12 Hz from the sprocket go through the whole shafting almost without amplitudes decreasing. Additional dynamic loading in this case is 50 – 120% of the activation torque. Additional loadings from caterpillar chain rewinding with frequencies 12-24 Hz on rear axle areas rises to 90-120% of the activation torque. Amplitudes of high-frequency loadings from engine harmonics are significant on areas which are close to the engine, but behind the gearbox and farther, these amplitudes are close to zero.
- Shafting is practically transparent for transferring of low-frequency vibrations. Construction measures are necessary for limiting (by means of installing elastic elements or vibration dampers) action of low-frequency vibrations on areas behind the final drive.

References

[1] Giancarlo G. 2009 Vibration dybamic and control (Springer science: Buisness media)
[2] Shekhovtsov V 2013 *The Archives of Automotive Eng.* **59** 85-97
[3] Shekhovtsov V, Sokolov-Dobrev N and Potapov P 2016 *Procedia Eng.* **150** 1239-44
[4] Popp K, Schiehlen W 2010 *Ground vehicle dynamic* (Berlin: Springer-Verlag Heidelberg) p 348
[5] Rajamani R 2012 *Vehicle dynamics and control* (Springer)
[6] Rill G 2012 *Road vehicle dynamics: fundamentals and modeling* (CRC press. Taylor and Francies group)
[7] Shevchuk V L 2013 *Modern High Technologies* **2** 44-49

IOP Conf. Series: Materials Science and Engineering 177 (2017) 012093 doi:10.1088/1757-899X/177/1/012093

GTE blade injection moulding modeling and verification of models during process approbation

I S Stepanenko and A I Khaimovich

Department of Aircrafts Engines' Construction Technologies, Samara National Resarch University, 34, Moskovskoye Hwy, Samara, 443086, Russia

E-mail: iliya.stepanenko@gmail.com

Abstract. The simulation model for filling the mould was developed using Moldex3D, and it was experimentally verified in order to perform further optimization calculations of the moulding process conditions. The method described in the article allows adjusting the finite-element model by minimizing the airfoil profile difference between the design and experimental melt motion front due to the differentiated change of power supplied to heating elements, which heat the injection mould in simulation. As a result of calibrating the injection mould for the gas-turbine engine blade, the mean difference between the design melt motion profile and the experimental airfoil profile of no more than 4% was achieved.

1. Introduction

The strength of polymer composite material (PCM) items featuring a combined reinforcement structure and short-fibre reinforcement is conditioned by the fibre distribution and direction. It is particularly important for the GTE blades, for which strict quality requirements are placed. The strengthener blade made based on the moulded multilayer PCM allows applying high-modular fibres. The control of the reinforcement structure i.e. laying-up angles allows obtaining the required anisotropic structure considering the effective load properties. At the same time, structures made based on the multilayer PCM have a relatively low interlayer strength and are susceptible to damages represented by layer separation and cracking in the transversal direction of the reinforcement [1]. This type of PCM damage is the most dangerous since it results in a considerable reduction of flexural and torsional rigidity, furthermore, it refers to invisible damages, i.e. it may be detected only with special means and methods.

In addition, forming of the blade root area by laying-up of multilayer prepregs is difficult due to its complex geometry. In this connection, the lock portion and the inner portion are reasonable to be manufactured by injection moulding (IM) of PCM by means of filling the forming cavities of the injection mould by the PEEK melt with the reinforcing discrete carbon fibre. Therefore, the blade is manufactured of the pre-formed prepreg shells, with pouring the PEEK PMC reinforced with short fibres using the IM method.

An important factor to obtain a high-quality moulding is foremoulding its structure, in particular, distribution of the reinforcing fibre which is ensured by the structure of the moulding injection mould and process modes.

In the context of the injection mould structure, the IM process parameters are optimized using the FE simulation modelling such as Autodesk Simulation Moldflow [2] or Moldex3D [3, 4]. In general, the modelling reliability depends on the FE model quality as well as the properly set boundary conditions. The optimized process parameters allow obtaining a precise moulding with the required anisotropy of properties.

The issues of precision analysis for the blade type products are described in detail in works [5, 6].

This experiment was aimed at the development and experimental verification of the simulation model for filling the mould in plant conditions in order to perform further optimization calculations of the moulding process conditions. Main stages of the research described below.

1. Development of the moulding simulation model considering the process peculiarities.
2. Preliminary optimization of moulding conditions. Injection mould filling process analysis and melt front motion modelling.
3. Experimental analysis of the melt front motion in the shell-free mould moulding. Verification and specification of the simulation model.
4. Modelling of the injection mould filling with the installed reinforcing shells. Reinforcing fibre orientation analysis.
5. The experimental verification of the simulation model based on the fibre orientation calculation.

The first three stages are described in this article.

2. Theoretical Aspects Of Melt Motion In Moulding

The reinforcing fibre orientation is conditioned by the melt motion. Theoretic melt motion models and their numerical implementation are described in works [7, 8].

In real conditions, the flow of polymeric melt is non-stationary, quasi-viscous, and non-isorthermal. Layer hardening is started near the mould walls with the material mixture distribution in the mould cavity. This process is non-linear and the material properties depend on rheological and thermal conditions.

In Moldex3D, a high-efficiency and precise finite element method for calculating the non-stationary melt flow in complex 3D cavities is implemented. In particular, the main advantage of the 3D-FVM is the possibility to implement the non-balanced gating system. The particular features of the 3D-FVM method are described in [9].

To trace the melt front motion, function f of the inclusion volume fraction is calculated by the method of 3D-FVM [2]. $f=0$ is to determine the air phase, $f=1$ is the polymer melt phase. Values within the range of $0<f<1$ show that the melt is in the FE-mesh cells, its motion over time is described by the transfer equation:

$$\frac{\partial f}{\partial t}+\nabla\cdot(uf)=0 \quad (1)$$

It should be noted that the flow rate or the injection pressure are predetermined at the injection mould inlet; the wall friction is not considered; the inlet boundary condition only is required for the hyperbolic differential transfer equation of volume fraction function.

To describe the polymer melt viscosity curve considering the shear rate, the Cross model version, namely Cross-Williams-Landell-Ferry model is used, where zero shift viscosity $\eta_0(T, p)$ depends on temperature T and pressure p. The Cross-WLF model is described by seven parameters [4]:

$$\eta(T,P,\gamma)=\frac{\eta_0(T,P)}{1+(\eta_0\gamma/\tau^*)^{1-n}};\quad \eta_0(T,P)=D_1\exp\left(\frac{-A_1(T-T_c)}{A_2+(T-T_c)}\right)$$

$$T_c=D_2+D_3P;\quad A_2=\tilde{A}_2+D_3P \quad (2)$$

In formula (2), n is the power dependence exponent; τ^* the parameter describing the transition interval from the zero shear to the viscosity curve exponential law. Parameters $D1$, $D2$, $D3$, $A1$, $A2$ are the additional model constants. According to the pVT diagram, a modified Tait model describes a complex thermodynamic condition by the dependencies as follows:

$$V = V_0[1 - C\ln(1 + P/B)] + V_t$$
$$V_0 = \begin{cases} b_{1S} + b_{2S}\bar{T}, \text{ if } T \le T_t \\ b_{1L} + b_{2L}\bar{T}, \text{ if } T > T_t \end{cases}; \quad B = \begin{cases} b_{3S}\exp(-b_{4S}\bar{T}), \text{ if } T \le T_t \\ b_{3L}\exp(-b_{4L}\bar{T}), \text{ if } T > T_t \end{cases}; \quad V_t = \begin{cases} b_{7S}\exp(b_8\bar{T} - b_9 P), \text{ if } T \le T_t \\ 0, \text{ if } T > T_t \end{cases}$$
$$\bar{T} = T - b_5, \; T_t = b_5 + b_6 P, \; C = 0.0894 \tag{3}$$

13 parameters of the modified Tait model are determined based on the reference data of material properties [10].

3. Stator Blade IM Simulation Modelling

The purpose of blade simulation modelling in Moldex3D is the development of a calibration method for the finite element model to determine the impact caused by the error when setting the boundary conditions for the moulding thermal condition with the simplified representation of the injection mould structure model.

Further analysis of the flow hydrodynamics in the injection mould filling is necessary to detect challenging areas during filling as well as to establish the adequate reinforcing fibre distribution pattern. Calibration was carried out by comparing the design and experimental data of the current hydrodynamics based on a model material (PA-MXD6 polyamide) in the moulding conditions ensuring compatibility according to the PEEK blade material viscosity condition.

Based on a CAD model, using the Moldex3D Designer tools, a FE-mesh was obtained (consisting of 6,466,708 elements) which reflected the characteristic areas of the part to the most precise extent. For the blade, the airfoil represents this type of characteristic area. The density of finite elements in possibly hindered melt flowing points was selected as the largest value. The geometric shape of the injection mould was preset as a simplified properly sized parallelepiped with the surface of the joint, melt supply gates and thermostabilization channels arranging the heating elements. Moulding process parameters were set in the Moldex3D pre-processor according to table 1.

Table 1 – Moulding process parameters

Parameter	Value	Parameter	Value
Injection pressure	180 MPa	Heating parameters: method	continuous
Injection rate	15 mm/s	Cooling medium temperature	100°C
Rate control - pressure control switching moment	98%	Extraction temperature	190°C
Pressure holding time	9.6 s	Cooling time	20 s
Melt temperature	280°C	Mould opening time	5 s
Injection mould temperature	100°C		

The injection time was calculated automatically depending on the injection rate and pressure.

Figure 1 and figure 2 shows the results of airfoil filling simulation modelling at the beginning of the process and at the second stage – after full filling of the root flange.

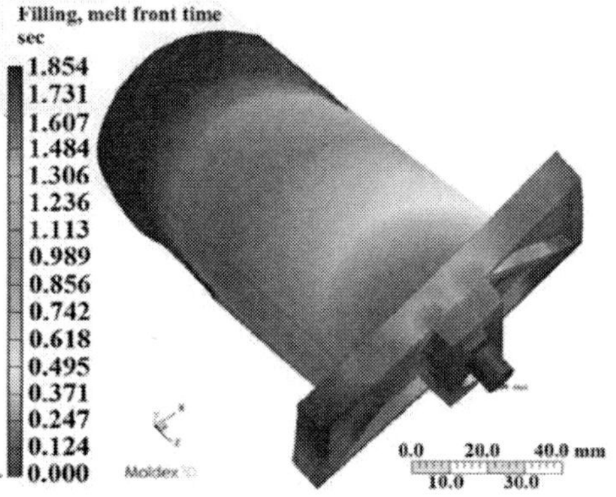

Figure 1. The melt front in the airfoil blade after the root flange filling

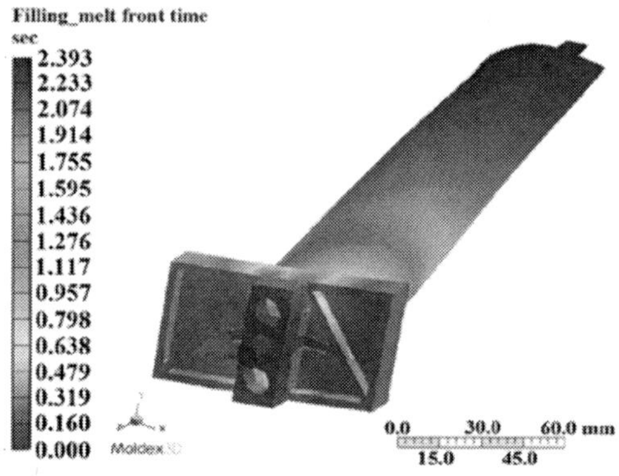

Figure 2. Filling of moulding cavities at the final stage before holding at constant pressure

As we can see, a minor underfill is observed in the most challenging zone, i.e. within the airfoil edge thinning area, which is eliminated with further filling of the mould. Verification of the IM simulation model in the melt filling was carried out following the results of the experiment using the algorithm as follows.

The preliminary calculation melt front motion during filling the root and airfoil section was carried out in Moldex3D. At a fixed moment of the injection time, volume V of the melt injected into the mould cavity and the gate system was determined. Further, a theoretical enveloping curve for the forward melt front was built. Experimentally, the blade sample was moulded under the design conditions with the same injected volume, V. The airfoil upper profile coordinates of the specially underfolded blade were measured. The relative mean square deviation between the theoretical and experimental curve was calculated by the following dependence:

$$\sigma = \sqrt{\frac{1}{n}\sum_{i=1}^{n}\left(1-\frac{y_{theor}}{y_{\exp}}\right)^{2}} \quad (4)$$

In formula (4), n is the number of reference points, $y_{theor.}$ is the coordinate of the theoretical flow front point, y_{exp} is the coordinate of the experimental flow front point.

4. Experimental Analysis of the Melt Front Motion in Moulding. Simulation Model Verification and Specification

The IM process was carried out using the Negri Bossi VE 1200 H-440 injection moulding machine equipped with a hot runner gate system mould. Injection mould (figure 3) consists of the movable 1 and stationary 2 parts.

The blade forming airfoil surfaces of parts 3 and 4 and the package of inserts 5, 6 forming the blade root are thermally insulated from flange plates 7, 8 and spacers 9, 10. The PCM melt is supplied to the mould cavity through central bush 11, plate 1 with the hot channel and heated nozzle 13. The melt temperature is controlled by the thermocouple installed at injection nozzle 13. The temperature of moulding plates 3, 4 and inserts 5,6 is automatically controlled and maintained by the controller (PLC) via changing the power of cylinder heaters located in holes 14, 15.

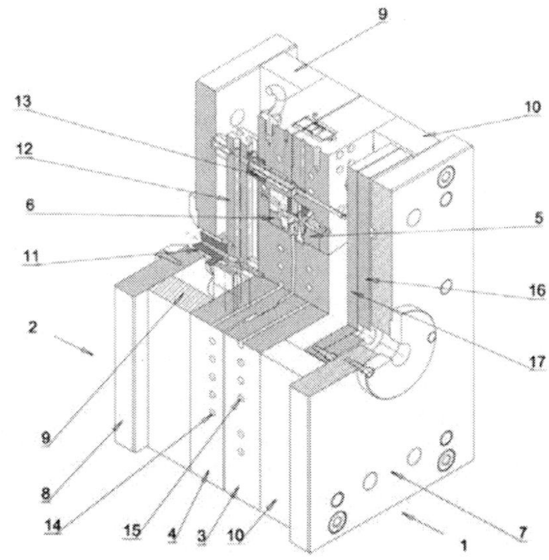

Figure 3. Injection mould for the blade IM

Upon completion of the IM cycle and the injection mould opening, the blade is taken from the injection mould (figure 4) jointly with inserts 5 and 6 while pusher plates 16 and 17 move.

Figure 4. An opened injection mould with the moulded blade before its removal jointly with inserts.

The melt front measuring results for the experimental blade with the unfinished airfoil and comparison of the results with the theoretical melt motion profile are given in table 2 and are represented in figure 5 a and 5 b.

6 points along the experimental ($T_{exp.}$) and theoretical ($T_{theor.}$) front lines of the melt flow were taken. After measuring the coordinates, the difference between them as well as the mean square error were calculated.

Table 2. Evaluation of the coincidence of the theoretical and experimental front

№	$T_{exp.}$	$T_{theor.}$	$\|\delta\|$	σ
P1	55.00	55.00	0.00	
P2	118.35	118.35	0.00	
P3	122.06	127.04	4.98	
P4	118.85	125.32	6.47	0.04
P5	105.44	110.13	4.69	
P6	55.00	55.00	0.00	
P’	98.28	97.45	0.83	

With the given calculation option, the melt flow profile difference within 4.69–6.47 is observed. The difference is caused by the fact that the actual surface condition (in terms of roughness) is not considered by the mathematical model and real moulding temperature values somewhat differ from the design values. Consideration of roughness is possible if the friction factors between the moulding cavity material of the injection mould and the cast part are known.

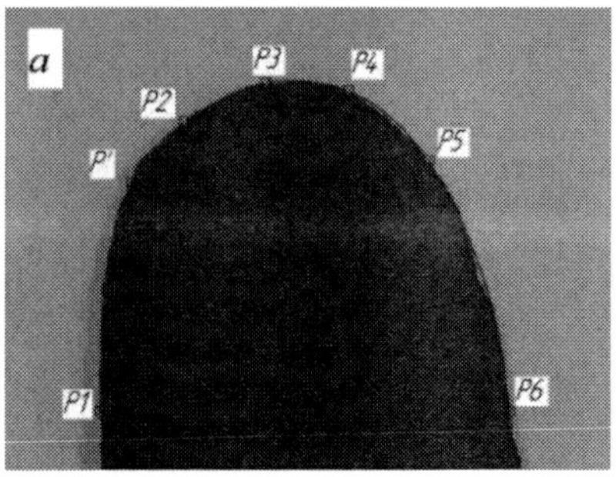

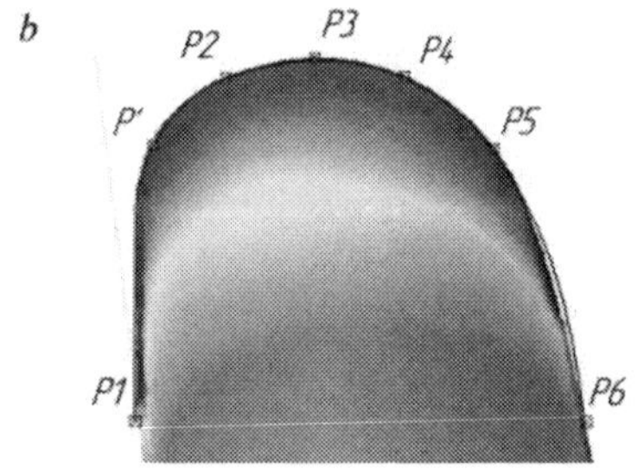

Figure 5. Melt motion front profile: a – experimental; b – theoretical

Since the melt motion in the injection mould is initially stipulated by its temperature gradient, the simulation model was adjusted with the target function by means of virtual adjustment of the heating elements power in the movable and static parts ensuring the required heat exchange with the melt.

5. Conclusions

Since a simplified representation of the mould design is used for the development of a finite-element model, the real pattern of the heat flow distribution differs from the modelled one. In addition, real values of the friction factor and the threshold shear value in the melt layer adjacent to the injection mould surface are unknown.

As a result, the hydrodynamics for melt flow within the injection mould may differ from the modelled one. Therefore, to perform correct calculations, the finite-element model shall be calibrated. The method described in the article allows performing the detailed adjustment by minimizing the profile difference between the design and experimental melt motion front due to the differentiated change of power supplied to heating elements, which heat the model representation of an injection mould. As a result of calibrating the injection mould for the blade airfoil, the mean difference between the design melt motion profile and the experimental profile of no more than 4% was achieved.

Acknowledgments
This work was supported by the Ministry of Education and Science of the Russian Federation in the framework of the implementation of the Program 'Research and development on priority directions of the scientific-technological complex of Russia for 2014-2020'.

References

[1] Feraboli P, Peitso E, Deleo F, Cleveland T, and Stickler P B 2008 *J. Reinf. Plast. Comp.* **10** 1191-1214
[2] Wang J and Jin X 2010 *Proc. of the Polymer Processing Society 26th Annual Meeting* (Banff)
[3] Foss P H, Tseng H C, Snawerdt J, Chang Y J, Yang W H and Hsu C H 2014 *Pol. Comp.* **4** 671-680
[4] Kitayama S, Onuki R, and Yamazaki K 2014 *Int. J. of Adv. Manuf. Tech.* **5-8** 827-838
[5] Pechenin V A, Bolotov M A, Ruzanov N V and Kondovin V A 2016 *J. Mach. Manuf. Reliab.* **2** 185-190
[6] Pechenin V A, Bolotov M A and Ruzanov N V 2016 *Key Eng. Mat.* **685** 334-339
[7] Vdovin R A and Smelov V G 2014 *Int. J. Eng. Tech.* **5** 2269-2275
[8] Nosova E A and Grechnikov F V 2016 *Key Eng. Mat.* **684** 366-370
[9] Chang R Y and Chiou S Y 1995 *Pol. Eng. Sci.* **22** 1733-1747
[10] Van Krevelen D W and Nijenhuis K Te 2009 *Properties of polymers* (Amsterdam: Elsevier)
[11] Ryazanov A I 2016 *Russian Eng. Res.* **36** 751-754

IOP Conf. Series: Materials Science and Engineering **177** (2017) 012094 doi:10.1088/1757-899X/177/1/012094

Algorithm of iterative transformation for effective modules of multicomponent isotropic composite

A A Svetashkov[1], J Miciński[2], M S Pavlov[1] and A A Vakurov[1]

[1] Tomsk Polytechnic University, 30, Lenina ave., Tomsk, 634050, Russia
[2] University of Warmia and Mazury in Olsztyn Faculty of Animal Bioengineering Department of Cattle Breeding and Milk Evaluation, 5, Oczapowskiego ave., Olsztyn, 10-957, Poland

E-mail: mkk@tpu.ru

Abstract. We consider the effective modules of Voigt, Reiss for isotropic elastic composites. We have reformed the method for constructing iterative transformation of the upper and lower estimates of fork (Voigt-Reuss) towards two-component composite in case of an arbitrary number of components. The method is based on the fact that effective modules of Voigt and Reuss can be regarded as elementary symmetric functions introduced by Gauss. The conditions, which the iteratively – transformed efficient modules must fulfill at every iteration, are shown.

1. Introduction

One of the basic problems in the mechanics of composite materials is the determination of effective properties for non-homogeneous elastic bodies. To solve this problem, we apply the theory of the effective module that represents the method of calculation of the elastic characteristics for a homogeneous comparison environment, in which the potential energy is as close as possible to the energy of the elastic composite. At this, the elastic modules of the basic material and the inclusions are taken as inputs to solve the problem.

The solution of this problem allows obtaining estimates of the stress state of structures, without the need for studies related to consideration of the structure of composites. This capability is important for designing the real constructive elements, for example [1, 2].

In addition to the theory of effective module, there is an approach based on the analysis of the geometry for inclusions of elastic composite. This approach is described in numerous publications. Let us name just some of them: [3-6].

As for the theory of effective modules, the models of Voigt and Reuss (VR) [7], Hashin and Shtrikman (HS) [8, 9], as well as model Krishera [10], developed in relation to the problems of thermal conductivity, are the basic mathematical models for definition effective elastic modules of the composite which does not take into account the geometry of the inclusions. The iteratively - converted effective modules for bi-component elastic composites were first introduced in [11, 12]. Despite the fact that the mathematical model has not received enough physical basis, calculations of parameters of the stress-strain state showed greater accuracy of the new model compared to the classical approach.

The main purpose of this paper is to reform the results [11, 12, 14-16] of iterative conversion of the effective modules of Voigt and Reuss for the case of multi-component composites.

2. An algorithm for constructing iterative conversions of the effective modules of isotropic multicomponent composite

An algorithm of constructing the new effective modules for the two-component elastic composite.
Iterative transformations of the upper and lower bounds of Hashin-Shtrikman (H-S) modules were constructed in [11].

We will explain the essence of the method using of the effective modules by Voigt and Reuss. Let us suppose that we have the two-component composite with shear modules $G_1 > G_2$ and volume contents γ_1, γ_2, and $\gamma_1 + \gamma_2 = 1$.

The expressions of Voigt (V) and Reuss (R) modules have the form:

$$G_V = \gamma_1 G_1 + \gamma_2 G_2, \tag{1}$$

$$G_R = \left(\frac{\gamma_1}{G_1} + \frac{\gamma_2}{G_2} \right)^{-1}. \tag{2}$$

The easily verifiable inequalities take place between modules G_1, G_2 and G_v, G_R:

$$G_1 \ge G_V \ge G_R \ge G_2. \tag{3}$$

Thus, model V-R, defined by relations (1), (2), produces modules which are located inside fork $G_1 > G_2$. The question is as follows: can we get a new compression of the fork by applying iterative transformation?
Let us introduce the new modules of such type as Voigt and Reuss:

$$G_V^1 = \gamma_1 G_V + \gamma_2 G_R, \quad G_R^1 = \left(\frac{\gamma_1}{G_V} + \frac{\gamma_2}{G_R} \right)^{-1}. \tag{4}$$

We can easily show that the inequalities are performed as follows:

$$G_1 \ge G_V \ge G_V^1 \ge G_R^1 \ge G_R \ge G_2 \tag{5}$$

Thus, expressions (4) produce compression already in the two forks: fork $G_1 > G_2$ and fork $G_V > G_R$.
Relationships (4) can be easily generalized to the case of *n* steps of iterative transformations:

$$G_V^n = \gamma_1 G_v^{n-1} + \gamma_2 G_R^{n-1}, \quad G_R^n = \left(\frac{\gamma_1}{G_V^{n-1}} + \frac{\gamma_2}{G_R^{n-1}} \right)^{-1}. \tag{6}$$

We can prove that the chain of inequalities is performed:

$$G_1 \ge G_V \ge G_V^1 \ge \ldots G_V^n \ge G_R^n \ge \ldots \ge G_R^1 \ge G_R \ge G_2. \tag{7}$$

The restrictions which are superimposed on the model of iterative transformation.
When building type sequences (6), we use an approach that is applied in determining the elementary symmetric Gaussian functions [13]. Let us suppose we have $a_0 \ge b_0 \ge 0$. Next, let us find:

$$a_{n+1} = \frac{1}{2}(a_n + b_n), \; b_{n+1} = \sqrt{a_n b_n}, \; n = 0,1,\ldots.$$

Then, the inequalities of types (5), (7), are valid for a_n, b_n:

$$a_0 \ge a_1 \ge \ldots \ge a_n \ge \ldots \ge b_n \ge \ldots \ge b_0$$

The limit of sequences $\{a_n\}$, $\{b_n\}$, [14], is equal to some function $M(a_0, b_0)$.
For the three positive numbers $a_0 \ge b_0 \ge c_0$, we have:

$$a_{n+1} = \frac{1}{3}(a_n + b_n + c_n), \; b_{n+1} = \sqrt{\frac{a_n b_n + a_n c_n + b_n c_n}{3}}, \; c_{n+1} = \sqrt[3]{a_n b_n c_n}.$$

The authors of [14] argue that

$$\lim_{n\to\infty} a_n = \lim_{n\to\infty} b_n = \lim_{n\to\infty} c_n = M(a_0, b_0, c_0).$$

Thus, the tasks of transforming sequences of positive numbers and sequences of effective modules in many ways have common patterns. The difference lies in the fact that the effective modules comprise a quantity of positive numbers each of which is multiplied by some weight multiplier of less than 1. Constructions of convergent sequences have the general requirements:

- positivity of each component at each step;
- fulfillment of the inequalities of kinds (5) and (7) at each step of iterative transformations.

In this case, obviously, the convergence of sequences effective modules to the limit, which depends only on the initial value of the sequences (from $\gamma_1 G_1, \gamma_2 G_2$) for the case of a two-component composite, can be proved.

3. An algorithm of constructing the new effective modules for three-, four-, five-component composites

Let us consider a generalization of the mathematical model (6) for the case of multi-component elastic isotropic composites. We will implement all reasoning by the example of the V-R effective modules.

Let us consider the inequality of the form:

$$G_V \geq \sqrt{G_V G_R} \geq G_R \tag{8}$$

For a three-component composite, along with the effective modules of Voigt and Reis:

$$G_V = \gamma_1 G_1 + \gamma_2 G_2 + \gamma_3 G_3,$$

$$G_R = \left(\frac{\gamma_1}{G_1} + \frac{\gamma_2}{G_2} + \frac{\gamma_3}{G_3}\right)^{-1}.$$

We will introduce a module that can be called the geometric mean:

$$G_a = \sqrt{G_V G_R} \tag{9}$$

Next, we will construct a sequence of the iteratively – reformed effective modules for three-component composite. The first iteration will look like:

$$G_V^1 = \gamma_1 G_V + \gamma_2 G_a + \gamma_3 G_R,$$

$$G_R^1 = \left(\frac{\gamma_1}{G_V} + \frac{\gamma_2}{G_a} + \frac{\gamma_3}{G_R}\right)^{-1}, \tag{10}$$

$$G_a^1 = \sqrt{G_V^1 G_R^1}.$$

It is easy to verify the fulfillment of inequalities:

$$G_1 \geq G_V \geq G_V^1 \geq G_a^1 \geq G_a \geq G_R^1 \geq G_R \geq G_3$$

For n iterations, we have:

$$G_V^n = \gamma_1 G_v^{n-1} + \gamma_2 G_a^{n-1} + \gamma_3 G_R^{n-1},$$

$$G_R^n = \left(\frac{\gamma_1}{G_V^{n-1}} + \frac{\gamma_2}{G_a^{n-1}} + \frac{\gamma_3}{G_R^{n-1}}\right)^{-1}, \tag{11}$$

$$G_a^n = \sqrt{G_V^n G_R^n}.$$

For the four-component composite, we need the expressions of the two geometric mean modules. Let us suppose that we have:

$$G_1 \geq G_2 \geq G_3 \geq G_4,$$

$$G_V = \gamma_1 G_1 + \gamma_2 G_2 + \gamma_3 G_3 + \gamma_4 G_4,$$

$$G_R = \left(\frac{\gamma_1}{G_1} + \frac{\gamma_2}{G_2} + \frac{\gamma_3}{G_3} + \frac{\gamma_4}{G_4}\right)^{-1}.$$

Let us introduce:

$$G_a = \sqrt{G_R G_V},\ G_b = \sqrt{G_a G_R}. \tag{12}$$

Let us prove the fulfillment of inequalities:

$$G_V \geq G_a \geq G_b \geq G_R.$$

We have:

$$\frac{G_V}{G_a} = \frac{G_V}{\sqrt{G_V}\sqrt{G_R}} = \sqrt{\frac{G_V}{G_R}} > 1,$$

$$\frac{G_a}{G_b} = \frac{\sqrt{G_V G_R}}{\sqrt{G_R\sqrt{G_R G_V}}} = \frac{\sqrt{G_V}}{\sqrt[4]{G_V G_R}} = \sqrt[4]{\frac{G_V}{G_R}} > 1,$$

$$\frac{G_b}{G_R} = \frac{\sqrt{G_R\sqrt{G_R G_V}}}{G_R} = \sqrt[4]{\frac{G_V}{G_R}} > 1.$$

We will write down the expressions of the iterated effective modules for the n-th iteration:

$$G_V^n = \gamma_1 G_V^{n-1} + \gamma_2 G_a^{n-1} + \gamma_3 G_b^{n-1} + \gamma_4 G_R^{n-1},$$

$$G_R^n = \left(\frac{\gamma_1}{G_V^{n-1}} + \frac{\gamma_2}{G_a^{n-1}} + \frac{\gamma_3}{G_b^{n-1}} + \frac{\gamma_4}{G_R^{n-1}}\right)^{-1}, \tag{13}$$

$$G_a^n = \sqrt{G_R^n G_V^n},\ G_b^n = \sqrt{G_a^n G_R^n}.$$

It is easy to prove the fulfillment of inequalities:

$$G_V^n \geq G_a^n \geq G_b^n \geq G_R^n. \tag{14}$$

We will consider a five-component composite. Let us suppose that we have an environment which is characterized by the five modules of shift:

$$G_1 > G_2 > G_3 > G_4 > G_5,$$

$$G_V = \gamma_1 G_1 + \gamma_2 G_2 + \gamma_3 G_3 + \gamma_4 G_4 + \gamma_5 G_5,$$

$$G_R = \left(\frac{\gamma_1}{G_1} + \frac{\gamma_2}{G_2} + \frac{\gamma_3}{G_3} + \frac{\gamma_4}{G_4} + \frac{\gamma_5}{G_5}\right)^{-1}.$$

In addition, we introduce three modules (geometrical mean) of the next kind:

$$G_c = \sqrt{G_V G_R},\ G_a = \sqrt{G_V G_c},\ G_b = \sqrt{G_c G_R}.$$

It is easy to prove the inequalities:

$$G_V \geq G_a \geq G_c \geq G_b \geq G_R.$$

For n iterations, we have:

$$G_V^n = \gamma_1 G_V^{n-1} + \gamma_2 G_a^{n-1} + \gamma_3 G_c^{n-1} + \gamma_4 G_b^{n-1} + \gamma_5 G_R^{n-1},$$

$$G_R^n = \left(\frac{\gamma_1}{G_V^{n-1}} + \frac{\gamma_2}{G_a^{n-1}} + \frac{\gamma_3}{G_c^{n-1}} + \frac{\gamma_4}{G_b^{n-1}} + \frac{\gamma_5}{G_R^{n-1}}\right)^{-1}, \tag{15}$$

$$G_c^n = \sqrt{G_V^n G_R^n},\ G_a^n = \sqrt{G_V^n G_c^n},\ G_b^n = \sqrt{G_c^n G_R^n}.$$

Similarly, we can construct the sequences of the iterated effective modules for n components, where $n > 5$.

4. The discussion of the results

The construction of the sequences of iteratively converted effective modules (EM) of the V-R model, carried out in this paper, includes the following necessary steps.

a. The localization of new effective modules having properties similar to the symmetric Gaussian functions. As shown in item 2, the construction of the inertial sequences of two variables requires two symmetric functions. For this, we can take the expressions of effective modules (V and R). And we must have three symmetric functions in case of three variables, etc. Since the tasks of building of the inertial sequences of some positive rational numbers are the same as the problem of constructing iterative sequences of EM, for the case of three-component composite in addition to the EM of V and R, we need to have one more additional effective module (G_{add}), which should satisfy the inequality:

$$G_V \geq G_{add} \geq G_R$$

For the four- and five-component composites we must have already 2 or 3 additional EM. The modules from the family of the geometric mean act as additional EM. Formulas (10), (11) and (14) give their visual representations.

b. At each stage of the introduction of new additional EM, we have to check the inequalities, which are similar to the inequalities that are performed when constructing of transformations of the symmetric Gaussian functions. These inequalities are shown in the complexes of (11), (13) and (14). The proof of their fulfillment is displayed, too.

The property of convergence of the various sequences to the same limit is one of the most remarkable properties of the iterated effective modules. So, the two-component effective modules of V and R model are reduced to the limit, which can be found by the numerical and analytical ways. The analytical limit of sequences for EM of V- R was found and analyzed by the authors (the mathematical calculations, obtained in deriving the results, are very bulky and this material was not included in this work). It turned out that, first, the limits of analytical and numerical sequences EM V and R coincide completely. Secondly, the calculation of the analytical limit confirmed reasoning of Gauss that the limit of the sequences of symmetric functions and sequences of EM depend only on their initial values (that is, on values $\gamma_1 G_1$, $\gamma_2 G_2$ which characterize the two-component composite). Third, it was found that the analytical limit of two sequences of Voigt and Reuss EM equals to the geometric mean of G_V and G_R which are defined by (4). For the case of three or more components, finding the analytical limit of sequences for iterative-converted EM is a more complicated mathematical problem. However, the latter does not mean that its solution cannot be found.

5. Conclusion

Expressions of the new effective characteristics of multicomponent elastic composites, presented in this paper, are constructed in accordance with the requirements for the iterative transformation of a symmetric mean [14]. They have the following properties:

a. The new efficient characteristics are positive for each step of iterative transformations;

b. The values all of iterated modules lay within the fork, which is formed by the maximum and minimum elastic shear modules of phases.

Recurrence relations, which determine the number of geometric mean modules needed to build iterative transformations, have been obtained. For the three-component composite , we have to have one geometric mean module for the four-component composite – two geometric mean modules, for the five-component composite - three, and so on. Numerical analysis of convergence for sequences of the iterated effective modules and calculations of the stress-strain state of structures with new effective characteristics are given in the second part of this work.

References

[1] I-Kan An, Ilin A S and Lazurkevich A V 2016 *IOP Conf. Ser.: Mat. Sci. and Eng.* (Tomsk) vol 124 (IOP Publishing) 012003

[2] I-Kan An, Ilin A S and Lazurkevich A V 2016 *IOP Conf. Ser.: Mat. Sci. and Eng.* (Tomsk) vol 124 (IOP Publishing) 012004

[3] Rasool A, Böhm H J *2012 Int. J. Eng. Sci.* **58** 21 – 34

[4] Klusemann B, Böhm H J and Svendsen B 2012 *J. Mech. A. Solids.* **38** 21 – 37

[5] Shermergor T D 1977 *Elasticity Theory microinhomogeneous bodies* (Moscow: Nauka)

[6] Sendetski G P 1974 *Composite Materials, Academic Press* 45

[7] Christensen R M *1979 Mechanics of composite materials* (NewYork: AcademicPress)
[8] Hashin Z 1962 *J. Appl.* **29** 143-150
[9] Hill R 1965 *J. Mech.Phys. Solids* **13** 213-222
[10] Budiansky B 1965 *J. Mech. and Phys. Solids* **13** 223-227
[11] Svetashkov A A, Kupriyanov N A and Manabaev K K 2013 *Russian Physics J.* 209-211
[12] Pobedrya B E 1984 *Mechanics of Composites* (Moscow: MGU)
[13] Kupriyanov N A, Simankin F A and Manabaev K K 2016 *IOP Conf. Ser.: Mat. Sci. and Eng.* (Tomsk) vol. 124 (IOP Publishing) 012100
[14] Pavlov S M and Svetashkov A A 1993 *Russ. Phys. J.* **36** 400-406
[15] Svetashkov A A 2000 *Mechanics of Composite Materials* **36(1)** 37-44
[16] Svetashkov A A, Kupriyanov N A and Manabaev K K 2016 *Key Eng. Mat.* **685** 206-210

Algorithm of constructing hybrid effective modules for elastic isotropic composites

A A Svetashkov[1], J Miciński[2], N A Kupriyanov[1], V N Barashkov[3] and A V Lushnikov[1]

[1] Tomsk Polytechnic University, 30, Lenina ave., Tomsk, 634050, Russia
[2] University of Warmia and Mazury in Olsztyn Faculty of Animal Bioengineering Department of Cattle Breeding and Milk Evaluation, 5, Oczapowskiego ave., Olsztyn, 10-957, Poland
[3] Tomsk State University of Architecture and Building, 2, Solyanaya sq., Tomsk, 634003, Russia

E-mail: mkk@tpu.ru

Abstract. The algorithm of constructing of new effective elastic characteristics of two-component composites based on the superposition of the models of Reiss and Voigt, Hashin and Strikman, as well as models of the geometric average for effective modules. These effective characteristics are inside forks Voigt and Reiss. Additionally, the calculations of the stress-strain state of composite structures with new effective characteristics give more accurate prediction than classical models do.

1. Introduction

In this paper, we consider the problem of constructing new constitutive equations of elastic isotropic two-component composites. Most of modern publications focus on mathematical models that take into account the geometry of heterogeneities in an elastic body, while the creation of mathematical models of effective modules is neglected. In this paper, we used a model based on the formulation of effective relations for the hybrid module. For the first time, effective modules, called hybrid, were introduced in [1] as a Crisher model [2]. In the present paper, this model was further developed: implemented new types of hybrid modules, in particular the geometric average effective modules and their combinations with upper and lower bounds of the models of Voigt, Reiss, Hanshin and Strikman. Numerical and graphical analysis of the new effective characteristics have been implemented. Also, to address the problems in effective modules used [6-11].

2. Analysis of the Crisher model

The research work of Krisher [2] can be attributed to the number of the first jobs, in which hybrid models of the effective modules elastic composites have been received and have been used. We will give the basic idea of this model as an example.

The next shear module of the two-component composite was introduced by Krisher during modeling of the effective characteristics:

$$G_{eff} = \left(\frac{1-a}{G_V} + \frac{a}{G_R} \right)^{-1}. \tag{1}$$

According to Crisher, a – experimentally determined coefficient, G_V, G_R – effective modules of Voigt (V) and Reiss (R)

$$G_V = \gamma_1 G_1 + \gamma_2 G_2,\ G_R = \left(\frac{\gamma_1}{G_1} + \frac{\gamma_2}{G_2} \right)^{-1}. \tag{2}$$

G_1, G_2 – phases shift modules, γ_1, γ_2 – volume content.

If the ratio of (1) is replaced by $1-a\ \ \gamma_1$, we get:

$$G_{eff} = \left(\frac{\gamma_1}{G_V} + \frac{\gamma_2}{G_R} \right)^{-1}. \tag{3}$$

A new expression of G_{eff} allowed us to obtain more accurate performance characteristics in comparison with the forecast defined by relations (2).

Let us analyze (3). In fact, this is a ratio of a new effective module of Reiss type, wherein classic modules Voigt and Reiss are used as phase modules. Thus, a new module (3) prepared as a function of two classic modules Voigt and Reiss. If the new module of Voigt is built in a similar way, we obtain:

$$G_V^1 = \gamma_1 G_V + \gamma_2 G_R. \tag{4}$$

The question arises, whether all functions are applicable for getting more accurate effective characteristics? It turns out [3], [4] that functions are applicable, if they produce compressive transformations, in which new modules satisfy the inequalities:

$$G_1 \ge G_V \ge G_V^1 \ge G_{eff} \ge G_R \ge G_2. \tag{5}$$

For transformations (3), (4) as we can easily verify, these inequalities are fulfilled.

3. Mixed-type hybrid modules

Let us consider the effective shear modules of the Hanshin, Shtrikman models (H-S) [5] of a two-component elastic composite.

$$\begin{aligned} G^{''} &= G_2 + \frac{\gamma_1 (G_1 - G_2)}{1 + \gamma_2 (G_1 - G_2)/(G_2 - G_U)}, \\ G^{'} &= G_2 + \frac{\gamma_1 (G_1 - G_2)}{1 + \gamma_2 (G_1 - G_2)/(G_2 - G_L)}. \end{aligned} \tag{6}$$

Here:

$$G_L = \frac{3}{2}\left(G_2 + \frac{10}{9/K_2 + 8/G_2} \right)^{-1},\ G_U = \frac{3}{2}\left(G_1 + \frac{10}{9/K_1 + 8/G_1} \right)^{-1}.$$

K_1, K_2 – the modules of volume elasticity of the phases. For $G^{''}$, $G^{'}$, fork H-S is performed:

$$G^{''} > G^{*} > G^{'}.$$

We form new effective shear moduli obtained as a superposition of effective modules V-R and H-S:

$$\tilde{G}_V = \gamma_1 G^{''} + \gamma_2 G^{'},\ \tilde{G}_R = \left(\frac{\gamma_1}{G^{''}} + \frac{\gamma_2}{G^{'}} \right)^{-1}, \tag{7}$$

and let us consider the following ratio:

$$\begin{aligned} \tilde{\tilde{G}}^{*} &= G_R + \frac{\gamma_1(G_V - G_R)}{1+\gamma_2(G_V - G_R)/(G_R - \tilde{\tilde{G}}_U)}, \\ \tilde{\tilde{G}}^{'} &= G_R + \frac{\gamma_1(G_V - G_R)}{1+\gamma_2(G_V - G_R)/(G_R - \tilde{\tilde{G}}_L)}. \end{aligned} \tag{8}$$

Here:

$$\tilde{\tilde{G}}_U = \frac{3}{2}\left(G_R + \frac{10}{9/K_R + 8/G_R}\right)^{-1}, \ \tilde{\tilde{G}}_L = \frac{3}{2}\left(G_V + \frac{10}{9/K_V + 8/G_V}\right)^{-1}.$$

For volumetric modules, K_V, K_R perform usual ratio V-R:

$$K_V = \gamma_1 K_1 + \gamma_2 K_2, \ K_R = \left(\frac{\gamma_1}{K_1} + \frac{\gamma_2}{K_2}\right)^{-1}.$$

As we can see from (7), (8), the new hybrid modules are produced by inserting the *H, S* modules in the classical *V, R* model instead of G_1, G_2 and by inserting the effective *V, R* modules in the *H, S* model.

It can be shown that for effective new modules (7), (8), the inequalities that provide compressibility conditions of the form are:

$$G_1 > G_V > \tilde{G}_V > \tilde{G}_R > G_R > G_2, \ G_1 > G^{''} > \tilde{\tilde{G}}_V^{''} > \tilde{\tilde{G}}_V^{'} > G^{'} > G_2.$$

Let us consider two more pairs of new effective hybrid modules. We will introduce into consideration geometric mean modules for pairs of upper and lower estimations of models (V-R) and (H-S):

$$G_{S,VR} = \sqrt{G_V G_R}, \ G_{S,\mathrm{HS}} = \sqrt{G^{'} G^{''}}. \tag{9}$$

Then, new effective characteristics will be:

a) based on model (V- R):

$$G_{V,S} = \gamma_1 G_V + \gamma_2 G_{S,VR}, \ G_{R,S} = \left(\frac{\gamma_1}{G_V} + \frac{\gamma_2}{G_{S,VR}}\right)^{-1}. \tag{10}$$

b) based on model (H-S):

$$\begin{aligned} G_{V,S}^{'} &= G_{S,HS} + \frac{\gamma_1(G_V - G_{S,HS})}{1+\gamma_2(G_V - G_{S,HS})/(G_{S,HS} - G_{L,S})}, \\ G_{V,S}^{''} &= G_{S,HS} + \frac{\gamma_1(G_V - G_{S,HS})}{1+\gamma_2(G_V - G_{S,HS})/(G_{S,HS} - G_{V,S})}. \end{aligned} \tag{11}$$

Here: $G_{L,S} = \frac{3}{2}\left(G_{S,HS} + \frac{10}{9/K_{S,HS} + 8/G_{S,HS}}\right)^{-1}$, $G_{V,S} = \tilde{\tilde{G}}_V$, $K_{S,\mathrm{HS}} = \sqrt{K^{'} K^{''}}$.

4. The discussion of the results

As we noted earlier, O Krisher's work was the first productive experience of using the hybrid effective modules in the mechanics of composites. Here the expressions of the upper and lower estimations of the classical model of Voigt and Reuss were introduced into the expression of the effective module of Reuss type instead of the modules shear phases. In this paper, the given method of constructing effective hybrid characteristics was further developed.

The first hybrid model is structured as follows. The expressions are taken for the upper and lower evaluations of models Hashin-Shtrikman and Voigt-Reuss. Then, the upper and lower assessments of model Hashin-Shtrikman are inserted in the ratios of Voigt and Reuss model instead of phase shift modules. As a result, we obtain (7). The second hybrid model has been obtained by reverse substitution – the effective modules of Voigt and Reuss have been put into the expressions of model Hashin-Shtrikman instead of phase shift modules. This model is represented by expressions (8).

The following hybrid modules were obtained by the superposition of estimates Hashin and Shtrikman, Voigt-Reuss jointly with the geometric mean effective modules, have been introduced in [6]. These models are represented by relations (9-11).

The requirement of compressibility of the function that produces a transition from one pair of effective modules to another is the criterion for the correctness of the procedure construction new effective characteristics. Mathematically, this requirement is expressed as an inequality of type (5).

For the purpose of clarity, a graphical representation of relations for the hybrid modules is shown separately for each pair of found effective characteristics in comparison with modules of Voigt and Reuss or with modules of Hashin and Shtrikman. The hybrid effective modules (7) are shown in figure 1. The graph shows that the values of the new effective characteristics fully lay within bounds of the upper and lower estimates of the model of Voigt-Reuss. Similarly, the values of new effective characteristics of model (9) lay within bounds of fork Voigt-Reuss (figure 2).

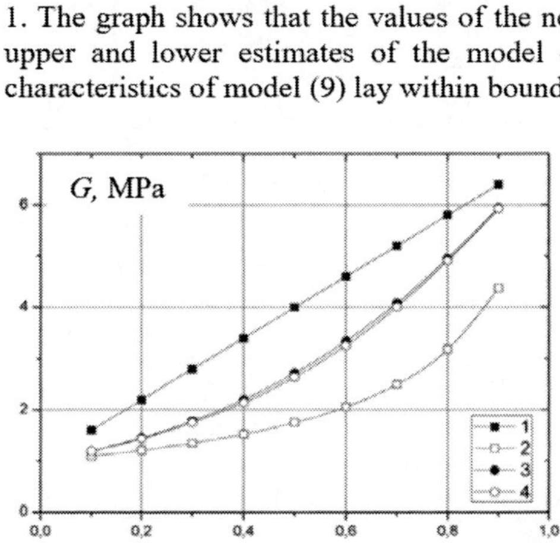

Figure 1. Curves of values of effective characteristics depending on γ_1. 1, 2 – modules Voight and Reuss; 3, 4 – $\tilde{G}_V$, $\tilde{G}_R$.

γ_1

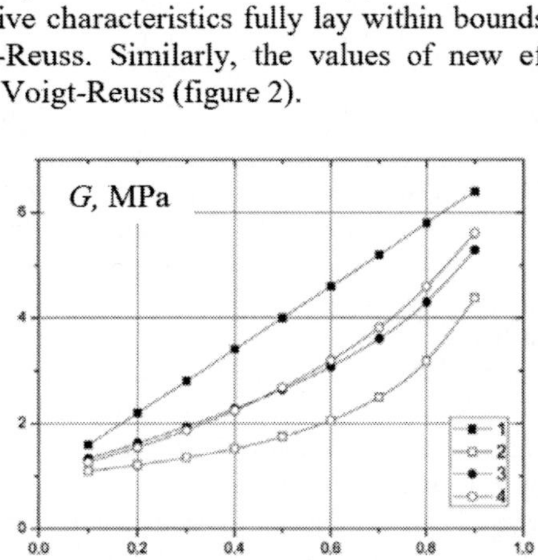

Figure 2. Effective characteristics depend on γ_1.1, 2 –modules Voight and Reuss, 3, 4 – $G_{S,VR}$, $G_{S,HS}$.

γ_1

The hybrid effective modules (11) are shown in figure 3. The graph shows that the values of the new effective characteristics fully lay within bounds of the upper and lower estimates of the model of Hashin-Shtrikman. Similarly, the values of new effective characteristics of model (9) lay within bounds of the fork Voigt-Reuss (figure 4).

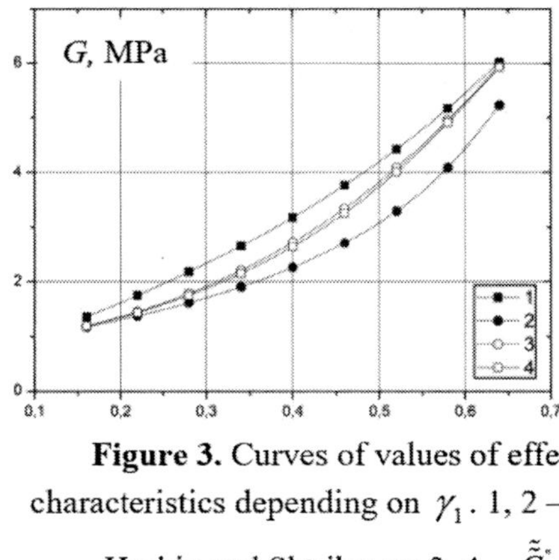

Figure 3. Curves of values of effective characteristics depending on γ_1. 1, 2 – modules Hashin and Shtrikman; 3, 4 – $\tilde{\tilde{G}}''$, $\tilde{\tilde{G}}'$.

γ_1

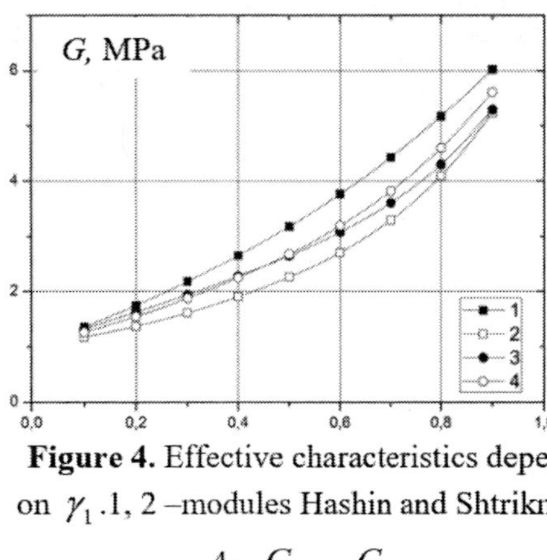

Figure 4. Effective characteristics depending on γ_1.1, 2 –modules Hashin and Shtrikman, 3, 4 – $G_{S,VR}$, $G_{S,HS}$.

γ_1

Table 1 gives comparison of the values for effective modules (9). The table shows that the values of effective characteristics of models (9) are practically the same in the range of 5-7 %.

Table 2. Effective characteristics depending on volume of fraction γ_1

γ_1	G_V ,MPa	G_R , MPa	$G^{''}$, MPa	$G^{'}$, MPa	$G_{S,VR}$, MPa	$G_{S,HS}$, MPa
0.1	1.621	1.094	1.357	1.172	1.346	1.291
0.2	2.214	1.207	1.748	1.373	1.719	1.659
0.3	2.887	1.346	2.178	1.613	2.138	2.106
0.4	3.415	1.522	2.652	1.902	2.622	2.638
0.5	4.254	1.750	3.178	2.258	3.185	3.258
0.6	4.696	2.059	3.765	2.708	3.843	3.960
0.7	5.288	2.500	4.425	3.294	4.591	4.728
0.8	5.841	3.182	5.170	4.089	5.420	5.531
0.9	6.416	4.375	6.021	5.229	6.266	6.315

Only practical calculation of a structure made of composite material with the help of a particular model and a comparison of this calculation with the exact solution can serve as a criterion for the selection of a type for new effective characteristics

5. Conclusion

Some new characteristics of effective two-component elastic composites had performed. Couples of new effective characteristics are defined by (7)–(11) and, so, we have 10 efficient modules expressions. Numerical analysis showed that the values of all found effective characteristics are within the scope of external evaluations Voigt-Reiss models. The values of effective characteristics of the models (8-11) are practically the same in the range of 7 %.

References

[1] Svetashkov A A, Kupriyanov N A and Manabaev K K 2016 *Key Eng. Mat.* **685** 206-210
[2] Budiansky B 1965 *J. Mech. and Phys. Solids* **13** 223-227
[4] Sendetski G P 1974 *Composite Materials, Academic Press* 45
[5] Hashin Z 1962 *J. Appl.* **29** 143-150
[6] Kupriyanov N A, Simankin F A and Manabaev K K 2016 *IOP Conf. Ser.: Mat. Sci. and Eng.* (Tomsk) vol 124(1) 012100
[7] Svetashkov A A, Vakurov A A 2016 *IOP Conf. Ser.: Mat. Sci. and Eng.* (Tomsk) vol 124 (IOP Publishing) 012099
[8] Ziyakaev G R 2016 *Key Eng. Mat.* **685** 441-444
[9] Ziyakaev G R, Simankin F A 2016 *IOP Conf. Ser.: Mat. Sci. and Eng.* (Tomsk) vol 124 (IOP Publishing) 012072
[10] Pavlov S M and Svetashkov A A 1993 *Russ. Phys. J.* **36** 400-406
[11] Svetashkov AA 2000 *Mechanics of Composite Materials* **36** 37-44

IOP Conf. Series: Materials Science and Engineering **177** (2017) 012096 doi:10.1088/1757-899X/177/1/012096

Methods of determining the margin of cyclic crack resistance of metal structures for hoisting machinery

V Yu Antsev[1], A S Tolokonnikov[1], A V Vorobev[1] and V I Sakalo[2]

[1] Tula State University, 92, Lenina ave., Tula, 300012, Russia
[2] Bryansk State Technical University, 7, 50-letiya Octyabrya bul., Bryansk, 241035, Russia

E-mail: anzev@yandex.ru

Abstract. In the course of manufacturing and operation, there can appear cracks in load-carrying metal structures which are hard to be detected till a definite moment. That is why in the course of designing metal structures, it is necessary to ensure their durability during possible existence of cracks up to the next routine inspection of a structure. For the prognostication of the durability of a structure with a crack, a research of coefficients of stress intensity on end-element models with crack-like defects of different length is carried out. As a result of this, a procedure for the definition of crack-resistance resource allowing the supposition of safe operation of a metal structure with a crack within the limits of recurrent routine works is developed.

1. Introduction

Both technological and fatigue cracks can appear in parts of load carrying metal structures in hoisting machines during production or operation processes. The resistance of metal to structure crack propagation is called fracture toughness.

The known concept of secure damage is based on the assumption that in any metal structure, even in new, there are always defects that may remain long undetected. Dimensions of detected defects are determined by resolution of applicable testing aids. All defects which are beyond the scope of testing aids remain undetected and, when present in the metal structure, they are potentially dangerous for metal structure operation.

To assess the vitality and the remaining life of metal structures, as well as to assess the impact of technological and operational defects on their load-bearing capacity, we must have a calculation device to predict behavior of metal structures with cracks under operational loads, which includes methods of fracture mechanics using the stress intensity factor (SIF) as the main characteristics of the stress and strain state of the material near the crack tip. The value of SIF may be found, for example, by reference data [1, 2] or by numerical methods. However, SIF guides do not contain practically any solutions to the cracks in the spatial objects of box-section typical for parts of metal structures of hoisting machines. In the paper presented this problem is solved on the basis of the end-element modeling of structures with crack defects with the purpose of the assessment of metal structure durability at the stage of designing.

2. Simulation of the stress and strain state at the tip of the fatigue crack

Many scientists were engaged in the solution of problems concerning cracking and fracture of load carrying metal structures of machinery, with the use of the finite element method as well [3-7]. In this study, finite element models of metal structures with box-section design were built to calculate the stresses in the vicinity of the crack tip and SIF in the automated design system and in the analysis of automated workstations WinMachine. These constructions are typical for the type of bridge cranes, which are the most common means of mechanization of cargo handling, transport and warehousing operations during various types of production and warehouse complexes [8]. In the bearing section models, simulated crack-like defects of different lengths were simulated.

When constructing models and studying the stress and strain state, some measures were foreseen. They allowed one to improve the accuracy of the solution, namely, small discretization was carried out during the crack tip partition into small elements; at points located at different distances r from the crack tip, the values of stress intensity coefficient K_I were determined by calculation; diagrams of K_I dependence on r were built. Required value K_I corresponding to the considered crack is defined as a result of extrapolation when $r \to 0$. When carrying out extrapolation, we exclude values K_I, corresponding to the points which are most closely located to a crack.

In the current study, the values of SIF were found for normal separation cracks by extrapolating the graph of function $K_I(r) = \sigma(r)\sqrt{2\pi r}$ at point $r = 0$, where $\sigma(r)$ is stress at distance r from the crack tip. The linear extrapolation is carried out within the zone of accumulation of elastic-plastic fractures [9], where the stress partially exceeds yield stress σ_T.

As a result of numerical experiment, stress values and SIF have been determined under rated stresses in the considered section of the metal structure equal to 50, 75 and 100 MPa, respectively.

The obtained SIF values were approximated by polynomial functional dependencies $K_I(\alpha)$ (figure 1):

- at 50 MPa: $K_I(\alpha) = 4.1908 + 181.1681 \cdot \alpha - 136.3966 \cdot \alpha^2$;

- at 75 MPa: $K_I(\alpha) = 10.4058 + 122.5162 \cdot \alpha + 442.216 \cdot \alpha^2$;

- at 100 MPa: $K_I(\alpha) = 12.0978 + 210.3875 \cdot \alpha + 225.2415 \cdot \alpha^2$,

where α – incremental dimension of the crack.

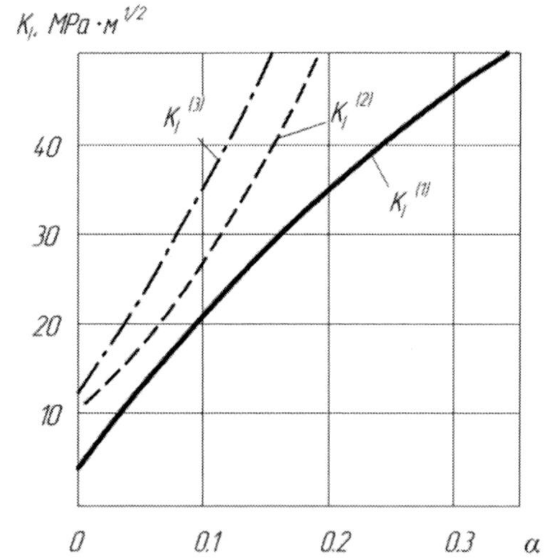

Figure 1. Experimental functions

$K_I^{(1)}(\alpha)$ – at 50 MPa,
$K_I^{(2)}(\alpha)$ – at 75 MPa,
$K_I^{(3)}(\alpha)$ – at 100 MPa

The diagram of SIF changes for different lengths of crack-like defects at a nominal level of the current stresses of 50 MPa is presented in figure 2.

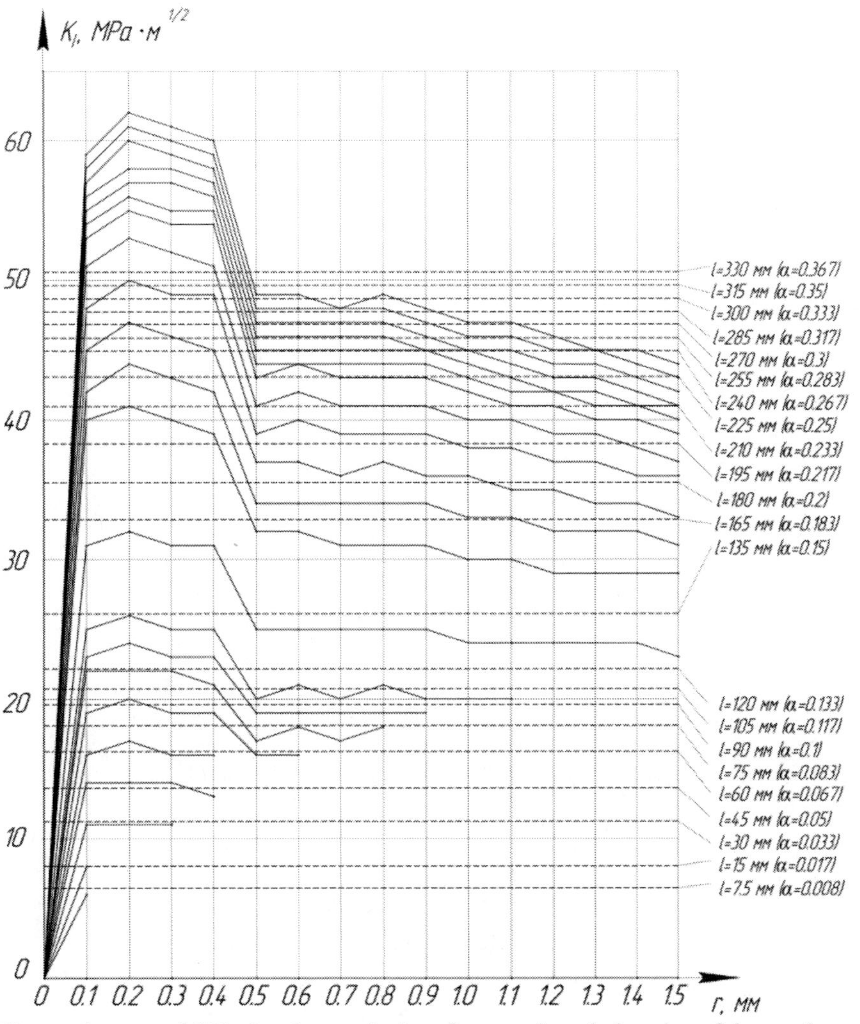

Figure 2. Charts changes of SIF: l – the cracks length, α – the relative size of the crack, r – the distance from the crack tip.

The margin of crack toughness of a metal structure under cyclic loading is characterized by the number of cycles required for the crack growth from initial size a_0 to critical a_c. If the cycle of stress changes in the crack location is of constant signs, then modeling and planning of the process, operational load of a metal structure are exercised by the method described in [10].

3. Forecasting methods of structure durability with a crack

Engineering assessment of the crack toughness margin is performed in the framework of the settlement system at the ultimate limit states, so we must enter the relevant factors that will ensure the reliability of this assessment. The number of crane working cycles up to failure Z_a when the crack growth occurs to a critical size is defined by the formula:

$$Z_a = \frac{\gamma_{dN}\left(\gamma_n \gamma_m \Delta K_0\right)^q \cdot P}{\zeta_{ea} V_0 \cdot K_I(\alpha)^q} \int_{a_0}^{a_c} \frac{d\alpha}{K_I(\alpha)^q \cdot \alpha^{q/2}} \quad (1)$$

where γ_{dN} is the coefficient of reliability of calculation methods; γ_m is safety factors as an intended design or a part of the structure; γ_m is safety factors according to characteristics of the material; ΔK_0, q, V_0 are parameters of the Paris's equation (ΔK_0 – amplitude of the stress intensity factor (SIF); $\Delta K_0 = 0{,}005\sigma_B - 9$; q, V_0 are empirical constants ($q = 3$, $V_0 = 10^{-7}$ м); P is a sectional perimeter with a crack; ζ_{ea} is a cyclic loading rate; $K_I(\alpha)$ is a function of SIF changes from the relative crack length; a_0, a_c is the initial and relative size of the crack.

By results of modeling and calculation, dependence graphs of durability of a metal design from a relative length of a crack are constructed (figure 3).

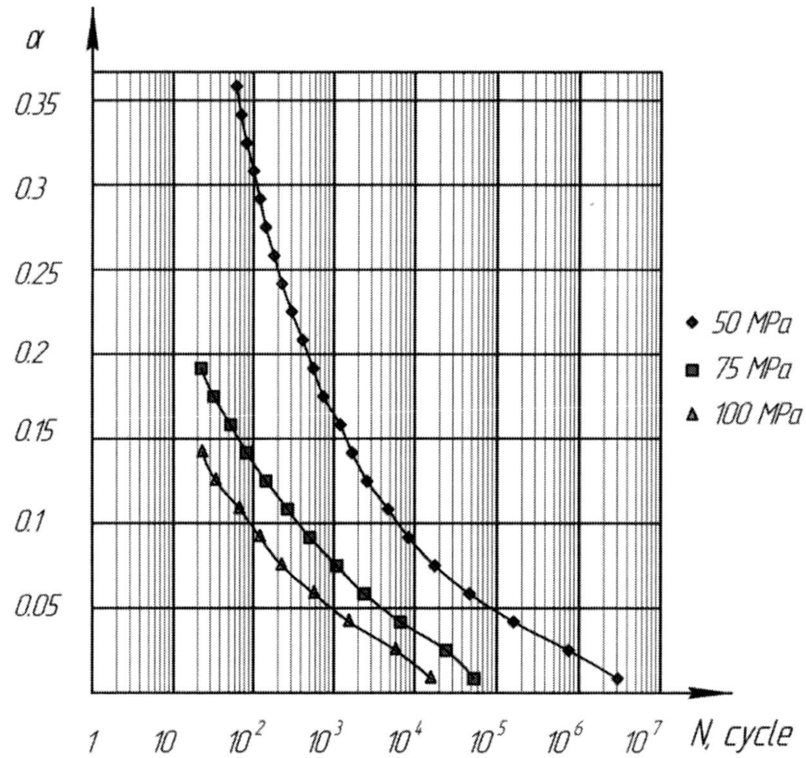

Figure 3. The dependence graphs of durability of the relative crack length.

The condition of safe operation of the crane for the period between its complete engineering certification (CEC), met once per every three years within the limits of standard service life, looks like:

$$Z_a \geq Z_d = n_c n_s n_d N, \quad (2)$$

where Z_d – number of loading cycles for the period between complete technical examinations; n_c – number of cycles per shift; n_s – number of shifts per day; n_d – number of working days a year; N = 3 years – the period of operation of the crane between complete technical examinations.

4. Conclusion

According to the study results, the following method of forecasting of the crane safe operation can be offered for a period of up to the next full technical inspection:

1. Current stresses at different load levels are computed.
2. Depending on the level of nominal stress in the most dangerous section, this or another dependence $K_I(\alpha)$ is accepted
3. According to formula (1), the value of the crack toughness margin is calculated, Z_a.
4. The margin of crack toughness Z_a is compared with the permitted number of cycles Z_d (the interval between the crane complete engineering certification (CEC)). If $Z_a \geq Z_d$, so the crack toughness condition is met, which means that even by an undetected crack at the technical inspections of metal structure of the bridge crane will serve Z_a cycles. And if $Z_a < Z_d$, then we need to make structural changes into the metal structure to ensure the fulfilment of condition (2).

References

[1] Murakami Y 1987 *Stress intensity factors handbook* (New York: Pergamon Press)
[2] Tada H, Paris P C and Irwin G R 1985 *The stress analysis of cracks handbook. Second ed.*(St. Louis, Mo: Paris Productions)
[3] Nagai M, Miura N and Shiratori M 2015 *Int. J.l of Pressure Vessels and Piping* **131** 2
[4] Sih G C, Paris P C and Irwin G R 1965 *Int. J.l of Fracture Mechanics* **1 (3)** 189
[5] Newman J C Jr, Raju I S 1981 *Eng. Fracture Mechanics* **15 (1-2)** 185
[6] Leach A M, Daniewicz S R and Newman Jr. J C 2007 *Eng. Fracture Mechanics* **74 (8)** 1233
[7] Boukharouba T, Tamine T, Niu L, Chehimi C, and Pluvinage G 1995 *Eng. Fracture Mechanics* **52 (3)** 503
[8] Shadskii G V, Antsev V Yu and Koveshnikov V A 1989 *Soviet Eng. Res.* **9 (6)** 66
[9] Seliverstov G V, Sorokin P A and Tolokonnikov A S 2004 *Heavy-duty production* **1** 14
[10] Murakami Y 1992 *The Rainflow Method in Fatigue: The Tatsuo Endo Memorial Volume* (Oxford: Butterworth-Heinemann)

Parameters assessment of the inductively-coupled circuit for wireless power transfer

Yu N Isaev[1], O V Vasileva[1], A A Budko[1] and S Lefebvre[2]

[1] National Research Tomsk Polytechnic University, 30, Lenina ave., Tomsk, 634050, Russia
[2] Conservatoire National des Arts et Métiers, École Normale Supérieure de Cachan, Paris, France

E-mail: vasileva.o.v@mail.ru

Abstract. In this paper, a wireless power transfer model through the example of inductively-coupled coils of irregular shape in software package COMSOL Multiphysics is studied. Circuit parameters, such as inductance, coil resistance and self-capacitance were defined through electromagnetic energy by the finite-element method. The study was carried out according to Helmholtz equation. Spatial distribution of current per unit depending on frequency and the coupling coefficient for analysis of resonant frequency and spatial distribution of the vector magnetic potential at different distances between coils were presented. The resulting algorithm allows simulating the wireless power transfer between the inductively coupled coils of irregular shape with the assessment of the optimal parameters.

1. Introduction

Wireless power transfer technologies are of great practical interest especially in the development of intelligent systems [1]. The purpose of this study is the optimum parameters assessment of the inductively-coupled circuit, such as inductance, coil resistance and self-capacitance for a wireless power transfer.

Statement of the research problem resides in estimation of the optimum parameters of the inductive coil corresponding to wireless power transfer with maximum efficiency at a distance not exceeding the size of the room [2]. The frequencies range is given with step f=10^5 Hz from initial value f_0=100 Hz to finite value f_k=600 MHz, because the radio frequency (until 300 GHz) electromagnetic field effect is not dangerous for human health [3]. The generator wavelength at a finite value of frequency f_k is equal to $\lambda = c/f_k = 0.5$ m, where c – light velocity.

2. Technique

Parameters will be estimated in the system of two inductively-coupled helical coils having N=4, N - number of turns, which are the source and device coils, respectively (Figure 1). In the simulation, we present a helical coil in the form of four rings for the calculation convenience. Sine voltage with amplitude V=1 V is applied to the inductive coil of the source. Both coils are made of cooper wire with radius r=0.3 m, the internal radius of the helical coil smaller turn is equal to R=0.25 m, the distance between inductive coils varies in the range of 0.5-2 meters maintaining the power transfer to the distance exceeding the radius of coils by the factor of 8.

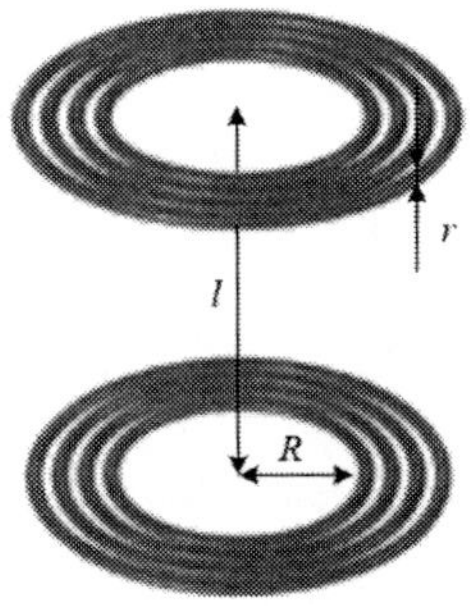

Figure 1. The modeled system of two inductive coupled coils: l – distance between coils; r – radius of cooper wire; R – radius of coil.

3. Mathematical model description

When voltage is applied to the high-frequency range of the transmitter, coil high-frequency currents will flow, so we use Maxwell equations system as a basis for calculation of coil parameters [4].

The equation for direct current, connecting flux density **B** and magnetic vector potential **A**, is the following:

$$\mathbf{B}=\nabla\mathbf{A}=\begin{pmatrix}\mathbf{i} & \mathbf{j} & \mathbf{k}\\ \frac{\partial}{\partial x} & \frac{\partial}{\partial y} & \frac{\partial}{\partial z}\\ A_x & A_y & A_z\end{pmatrix}=\mathbf{i}\left(\frac{\partial}{\partial y}A_z-\frac{\partial}{\partial z}A_y\right)+\mathbf{j}\left(\frac{\partial}{\partial z}A_x-\frac{\partial}{\partial x}A_z\right)+\mathbf{k}\left(\frac{\partial}{\partial x}A_y-\frac{\partial}{\partial y}A_x\right). \quad (1)$$

The equation for direct current, connecting magnetic force **H** and current density vector **J**, is the following:

$$\nabla\mathbf{H}=\begin{pmatrix}\mathbf{i} & \mathbf{j} & \mathbf{k}\\ \frac{\partial}{\partial x} & \frac{\partial}{\partial y} & \frac{\partial}{\partial z}\\ H_x & H_y & H_z\end{pmatrix}=\mathbf{i}\left(\frac{\partial}{\partial y}H_z-\frac{\partial}{\partial z}H_y\right)+\mathbf{j}\left(\frac{\partial}{\partial z}H_x-\frac{\partial}{\partial x}H_z\right)+\mathbf{k}\left(\frac{\partial}{\partial x}H_y-\frac{\partial}{\partial y}H_x\right)=\mathbf{J}. \quad (2)$$

Using (1) and (2), we can receive Poisson equation for magnetic vector potential **A** through current density vector **J** as follows:

$$\Delta\mathbf{A}=\frac{\partial^2}{\partial x^2}\mathbf{A}+\frac{\partial^2}{\partial x^2}\mathbf{A}+\frac{\partial^2}{\partial x^2}\mathbf{A}=\mathbf{J}. \quad (3)$$

For alternative current, using Helmholtz equation and (3) we can receive the following expression:

$$\left(j\omega\sigma-\omega^2\varepsilon_0\varepsilon\right)\mathbf{A}+\nabla\mathbf{H}=\mathbf{J}, \quad (4)$$

where: $\omega=2\pi f$ – angular frequency; σ – wire conductivity; ε – dielectric permeability; ε_0 – permittivity of vacuum; **A** – magnetic vector potential; **H** – magnetic force; $\mathbf{J}=\nabla\times(\mu^{-1}\nabla\times\mathbf{A})$ – current density vector.

Helmholtz equation (4) can be rewritten through voltage V, applied to the coil:

$$(j\omega\sigma-\omega^2\varepsilon)\mathbf{A}+\nabla\times(\mu^{-1}\nabla\times\mathbf{A})=\frac{\sigma V}{2\pi r}. \quad (5)$$

The current in the coil of wire, using (5), can be expressed as:

$$I=\pi\int_0^R\mathbf{J}r^2dr.$$

Magnetic field energy is given by [2]:

$$W = 2\pi \iint \mathbf{A} r dr \, .$$

Circuit parameters, such as inductance, coil resistance and capacitance were defined as follows:

$$L = \frac{2W}{|I^2|}, \ R = \mathrm{Re}\left(\frac{V}{I}\right), \ C = \frac{2W_e}{|V^2|}, \qquad (6)$$

where: W_e – electrical field energy.

4. Experimental part

The direct current is assumed to be I=1 A for inductance calculation. Distribution of vector magnetic potential **A** of the inductive coil with N=1 was calculated in COMSOL Multiphysics and is shown in Figure 2.

Inductance calculation with N=1 has been checked on the basis of [5], using $L = \mu_0 R\left(\ln\frac{8R}{r} - 2\right)$.

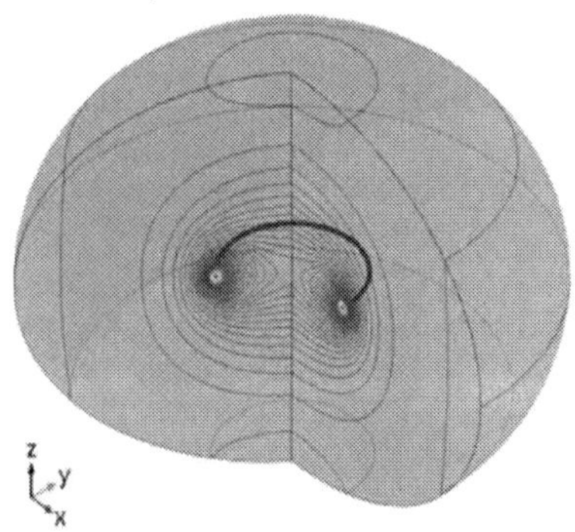

Figure 2. Distribution of vector magnetic potential **A** for the inductive coil with N=1

The inductance and the resistance of the irregular shape coil with N=4 as a function of the frequency were calculated according to the same algorithm (Figure 2), using (5) and (6) by the finite-element method in COMSOL Multiphysics.

Calculation results are presented in Figure 3.

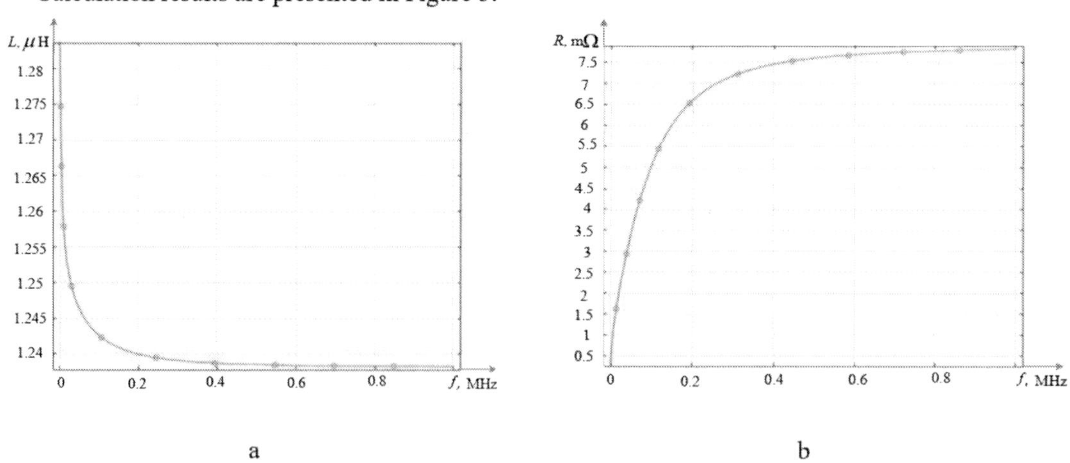

Figure 3. The frequency characteristic for the inductive coil: a) inductance $L(\omega)$; b) coil resistance $R(\omega)$

Figure 3, *a* shows that coil inductance decreases sharply and has a constant value in a high-frequency band, which is L=1.238 μH according to (6) due to saturation of the material. Figure 3, *b* shows that coil resistance increases and has a constant value in a high-frequency band, which is R=8 mΩ.

Poisson equation for the electric field in Cartesian coordinates for coil self-capacitance calculation is given by:

$$\Delta\varphi = \frac{\partial^2}{\partial x^2}\varphi + \frac{\partial^2}{\partial y^2}\varphi + \frac{\partial^2}{\partial z^2}\varphi = -\frac{\rho}{\varepsilon\varepsilon_0}. \tag{7}$$

The direct voltage is assumed to be V=1 V for self-capacitance calculation, which is independent of frequency. Self-capacitance calculation with N=1 has been checked on the basis of [6], using $C_0 = 4\pi^2\varepsilon R\cdot\left(\ln\left(\frac{8R}{r}\right)\right)^{-1}$. The simulation result in COMSOL Multiphysics coincided with the reference data. The self-capacitance of the irregular shape coil with N=4 was calculated through electric field energy according to (6) and (7), which is C_0=0.21 pF.

Resonant frequency was defined as $f_0=1/2\pi(LC)^{-1}$ and is equal to f_0=312 MHz. Thus, the resonant wavelength is λ_0 =0.96 m.

Spatial distribution of current per unit $I(\omega,k)$ for the inductive coil as a function of angular frequency ω and coupling coefficient k for analysis of resonant frequencies range f_0 may be presented as follows:

$$I(\omega,k) = \varpi(\omega)\cdot\frac{1+\frac{\omega_0^2\cdot k^2\cdot L(\omega)^2}{R(\omega)^2}}{\left(1+j\cdot\frac{X(\omega)}{R(\omega)}\right)^2+\frac{\omega^2\cdot k^2\cdot L(\omega)^2}{R(\omega)^2}}, \tag{8}$$

where: $\varpi(\omega)=\frac{\omega}{\omega_0}$ – relative angular frequency; ω_0 – resonant angular frequency; $k = M(\omega)/\sqrt{L_1(\omega)\cdot L_2(\omega)} = M(\omega)/L(\omega)$ – coupling coefficient; $M(\omega)$ – mutual inductance; $X(\omega)=\omega\cdot L(\omega)-\frac{1}{\omega\cdot C}$ – equivalent reactive resistance [7].

Expression (8) can be rewritten through general detuning ξ, which is absolutely dependent on quality factor Q:

$$I(\omega,k) = \varpi(\omega)\cdot\frac{1+(k\cdot Q)^2}{(1+j\cdot\xi)^2+(k\cdot Q)^2}, \tag{9}$$

where: $\xi = \frac{X(\omega)}{R(\omega)} = Q\cdot(\varpi(\omega)-\frac{1}{\varpi(\omega)})$ - general detuning; $Q=\frac{\omega_0\cdot L(\omega)}{R(\omega)}$ – quality factor.

Expression (9) shows that the effect of circuit parameters on the form of resonance curve $I(\omega,k)$ is taken into consideration fully by quality factor Q [8]. Spatial distribution of current per unit $I(\omega,k)$ for the inductive coil is presented in Figure 4.

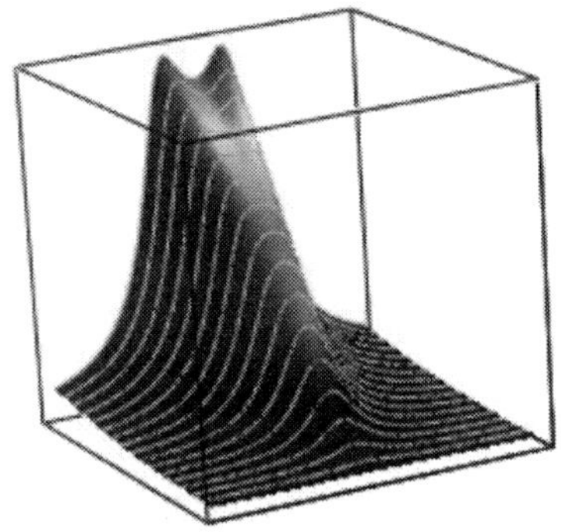

Figure 4. Spatial distribution of current per unit $I(\omega,k)$ for the inductive coil

Figure 4 shows a possibility to estimate the role of losses in the circuit and to select only natural resonance frequency from the entire frequencies range in $k \cdot Q << 1$ (weak-coupling regime) or two resonance frequencies in $k \cdot Q >> 1$ (strong-coupling regime). Optimal parameter estimation of the inductively-coupled circuit has allowed one to establish which part of the power is transferring to the receiving coil when the circuit is tuned to resonance [9, 10].

5. Simulation of the wireless power transfer in COMSOL Multiphysics

The system is modeled and calculated in software package COMSOL Multiphysics on the basis of Helmholtz equation (4). Sine voltage with amplitude V=1 V and resonant frequency f_0 was supplied on the source inductive coil. Results of simulation for different distance of the coils between each other as a function of the resonant wavelength λ_0 are shown in Figure 5.

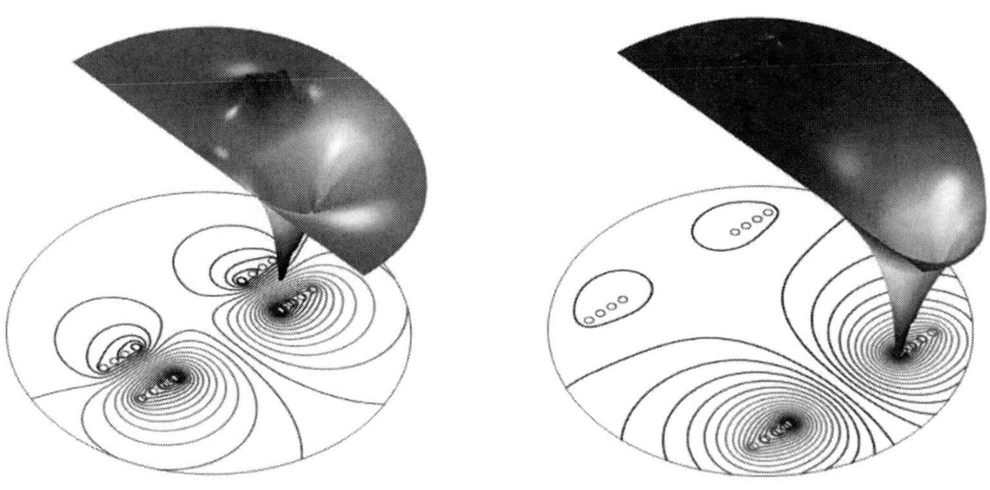

a b

Figure 5. Spatial distribution of vector magnetic potential **A**: a) $l=\frac{\lambda_0}{2}$; b) $l=\frac{4 \cdot \lambda_0}{2}$

Figure 5 shows that the field pattern is symmetric, so we can plot half of the field pattern of vector magnetic potential spatial distribution **A** relative to the central axis of the coil. It may be noted that if the distance between coils increases, the power transfer decreases (Figure 5, *b*) [11].

Thus, a research transition phase about a wireless power transfer was presented.

The power is the energy change in time, so power equation $P(\omega)$, transferring to the receiving coil (Figure 5), is given by:

$$P(\omega)=\frac{dW(\omega)}{dt}=\frac{dW(\omega)}{d\omega}\frac{d\omega}{dt}=-\omega^2\frac{dW(\omega)}{d\omega}. \tag{9}$$

We believe that the optimum parameters of the inductively-coupled circuit and coupling coefficient k, which depend on the distance between coils l, relationship of diameters of the primary and secondary coils, and also quality factor Q, can appreciably improve the power (9) transfer distances [12]. The suggested algorithm could be useful in practical applications for further study.

6. Conclusion

The model for wireless power transfer research through the example of inductively-coupled coils of irregular shape in software package COMSOL Multiphysics was suggested.

Circuit parameters, such as inductance, coil resistance and self-capacitance were defined through electromagnetic energy by the finite-element method. The coil inductance was determined on the basis of calculation of the magnetic vector potential succeeded by calculation of magnetic field energy. The self-capacitance was calculated through electric field energy.

Spatial distribution of current $I(\omega,k)$ per unit in dependence to frequency ω and coupling coefficient k for analysis of resonant frequency f_0 and spatial distribution of vector magnetic potential $\mathbf{A}$ at different distances l between coils were presented. The study was carried out according to Helmholtz equation.

The resulting algorithm allows us to simulate the wireless power transfer between the inductively coupled coils of irregular shape with the assessment of the optimal parameters for improving the efficiency of resonant energy transfer.

References

[1] Huang L, Hu A P, Swain A K and Su Y 2016 *IEEE Transactions on Power Electronics* **31**(11) 7556-63

[2] Reza Khan S and Choi G 2016 *Microwave and Optical Technology Letters* **58**(8) 1861-66

[3] Vasileva O V, Budko A A and Lavrinovich A V 2016 *IOP Conf. Ser.: Mater. Sci. Eng.* (Tomsk) vol 124 (Bristol: IOP Publishing) 012107

[4] Isaev Y N, Kolchanova V A, Tarasenko S S and Tikhomirova O V 2015 *Proc. Int. Conf. on Mechanical Engineering, Automation and Control Systems* (Institute of Electrical and Electronics Engineers Inc.) 7414894

[5] Kalantarov P L and Tseytlin L A 1986 *Inductance Calculation* ed P L Kalantarov and L A Tseytlin (Leningrad: Energy-atom Publishing) p 488

[6] Iosel Yu Ya, Kochanov E S and Strunskii M G 1981 *Electrical Capacity Calculation* ed Yu Ya Iosel, E S Kochanov et al (Leningrad: Energy Publishing) p 288

[7] Kuleshova E O, Plyusnin A A, Shandarova E B and Tikhomirova O V 2016 *IOP Conf. Ser.: Mater. Sci. Eng.* (Tomsk) vol 124 (Bristol: IOP Publishing) 012069

[8] Bychkov P N, Zabrodina I K and Shlapak V S 2016 *IEEE Transactions on Dielectrics and Electrical Insulation* **3**(1) 288-93

[9] Bordunov S V and Galtseva O V 2016 *J. Phys.: Conf. Ser.* (Tomsk) vol 671 (Bristol: IOP Publishing) 012009

[10] Pritulov A M, Usmanov R U, Gal'Tseva O V, Kondratyuk A A, Bezuglov V V and Serbin V I 2007 *Russ. Phys. J.* **50**(2) 187-92

[11] Campi T, Dionisi F, Cruciani S, De Santis V, Feliziani M and Maradei F 2016 *Asia-Pacific Int. Symp. on Electromagnetic Compatibility* (Institute of Electrical and Electronics Engineers Inc.) 544-47

[12] Ozana S, Pies M and Docekal T *AIP Conf. Proc.* (Greece) vol 1738 (New York: AIP Publishing) 370006

The power supply system model of the process submersible device with AC power transmission over the cable-rope

V M Rulevskiy[1], V G Bukreev[2], E O Kuleshova[2], E B Shandarova[2], S M Shandarov[3] and Yu Z Vasilyeva[2]

[1] Research Institute of Automation and Electromechanics Tomsk State University of Control Systems and Radioelectronics, 53, Belinsky str., Tomsk, 634034, Russia
[2] Tomsk Polytechnic University, 30, Lenina ave., Tomsk, 634050, Russia
[3] Tomsk State University of Control Systems and Radioelectronics, 40, Lenina ave., Tomsk, 634050, Russia

E-mail: bukreev@tpu.ru, shandarovaelena@mail.ru, kuleshova@tpu.ru

Abstract. A practical problem of power supply system modeling for the process submersible device with AC power transmission over the cable-rope was considered. The problem is highly relevant in developing and operation of submersible centrifugal pumps and submersibles. The results of modeling a symmetrical three-phase power supply system and their compliance with the real data are given at the paper. The obtained results in the mathematical and simulation models were similar.

1. Introduction

Nowadays, there is a problem of voltage regulation systems at the remote dynamic object [1-4]. Creating regulators of voltage assumes the use of synthesis methods of complex objects in modern control theory based on the description of dynamic processes in the variables state-space [5, 6]. To solve the problem, it is necessary to create a mathematical model that most accurately describes the operating modes of the object power supply system (PSS) and solves system differential equations [7, 8]. Various methods can be used to create a mathematical model [9]. At the primary stage of creating the mathematical model, it is assumed that the three-phase system is symmetrical. Thus, calculations can be performed only for one phase. Figure 1 shows the PSS scheme, reduced to a single phase. The pulse inverter connected to a DC power source with a nominal voltage of U_s = 540 V. A model with concentrated parameters was selected as a mathematical model of the cable line; see Figure 1 (cable). When the cable-rope length is 6 km, equivalent circuit parameters are as follows: active resistance - $R_{ak} = 19.2$ Ohm, inductance - $L_{ak} = 2.4$ mH, capacitance - $C_{ak} = 0.496$ uF.

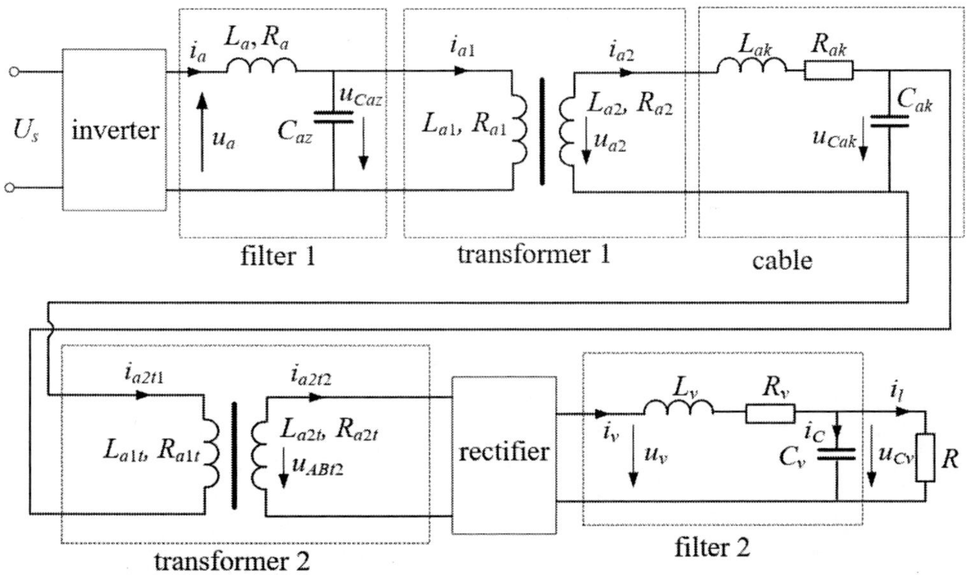

Figure 1. The power supply scheme, reduced to phase A.

The following parameters of the equivalent circuit are accepted:

1. For the input filter: $R_a = 1$ Ohm, $L_a = 0.4$ mH, $C_{az} = 6$ uF.
2. It should be noted that the special transformers were designed to operate in the considered power supply system. The parameters of these transformers, calculated from the results of short circuit and open circuit experiments, are shown in Table 1.

Table 1 . The transformers parameters

Step-up transformer (Fig.2, transformer 1)	
Rated power, windings voltage, frequency	$S_{n1} = 30$ kVA, $U_{n1} = 660$ V, $U_{n2} = 1830$ V, $f = 1000$ Hz
Winding connection	star-delta
Primary winding parameters	$R_{a1} = 0.866$ Ohm, $L_{a1} = 0.067$ mH
Secondary winding parameters	$R_{a2} = 2.28$ Ohm, $L_{a2} = 0.17$ mH
Magnetizing branch parameters	$R_{m1} = 494$ Ohm, $L_{m1} = 0.255$ H
Step-down transformer (Fig.2, transformer 2)	
Rated power, windings voltage, frequency	$S_{n2} = 60$ kVA, $U_{n1} = 1830$ V, $U_{n2} = 400$ V, $f = 1000$ Hz
Winding connection	delta-delta
Primary winding parameters	$R_{a1t} = 0.84$ Ohm, $L_{a1t} = 0.586$ mH
Secondary winding parameters	$R_{a2t} = 0.04$ Ohm, $L_{a2t} = 0.03$ mH
Magnetizing branch parameters	$R_{m2} = 373$ Ohm, $L_{m2} = 0.889$ H

2. Results of mathematical modeling

The rectifier (Figure 1, rectifier) is connected to the secondary winding of the step-down transformer (Figure 1, transformer 2) to obtain a constant load voltage.

Load parameters are calculated as a complex circuit impedance, $\underline{Z}_{eqv} = R_{eqv} + j\omega_v L_{eqv}$, at the fundamental frequency of the rectifier as a result of the equivalenting second filter and the load on its output. Active and reactive components of the complex impedance allow calculating the equivalent load parameters [10].

The power supply system, shown in Figure 1, is described by a system of differential equations in the Cauchy form [7, 8]. The mathematical model and the system of differential equations are given in 'Mathematical model for power supply system of autonomous object with AC power transmission over cable-rope'.

Figure 2 shows plots of the input current after inverter $i_a(t)$ and currents of the primary $i_{a1}(t)$ and secondary $i_{a2}(t)$ winding of the first transformer. As we can see from the plots, the input filter considerably smooths ripple of the inverter current. A transient time is not more than one period.

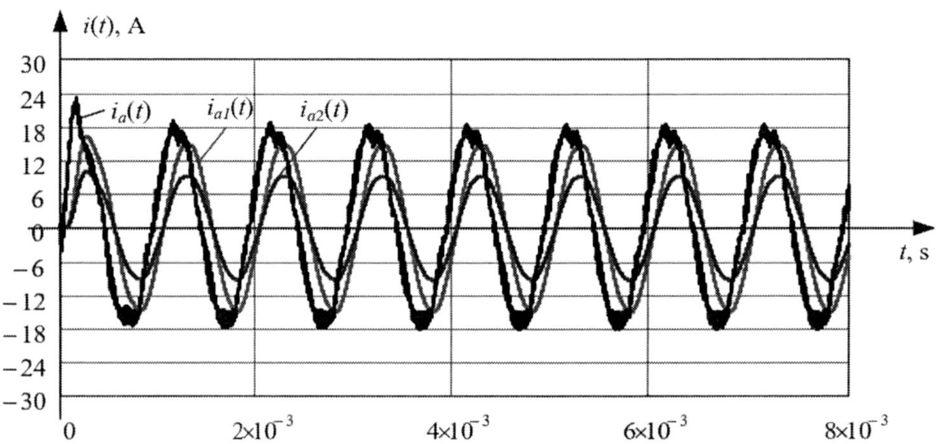

Figure 2. Input current $i_a(t)$ after inverter and linear currents of primary $i_{a1}(t)$ and secondary $i_{a2}(t)$ windings of transformer 1.

Primary u_{Caz} and secondary u_{a2} phase voltage of transformer 1 are shown in Figure 3. The turn ratio is equal to k_{f1}=0.624.

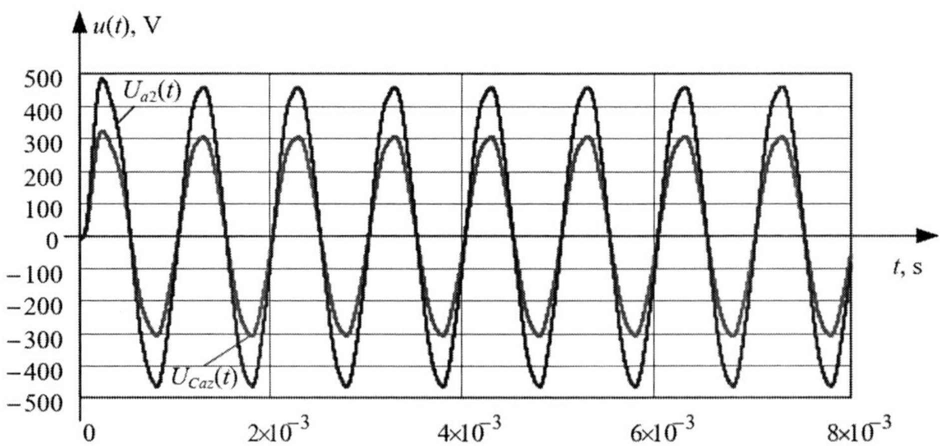

Figure 3. Primary and secondary phase voltage of transformer 1.

Figure 4 shows currents of the primary i_{a2t1} and secondary i_{a2t2} windings of transformer 2 with turn ratio $k_{f2} = U_{f2}/U_{f1} = 0.222$, where $U_{f1} = 925$ V is rated voltage of the primary winding, and $U_{f2} = 205$ V is rated voltage of the secondary winding. The currents have a harmonic form.

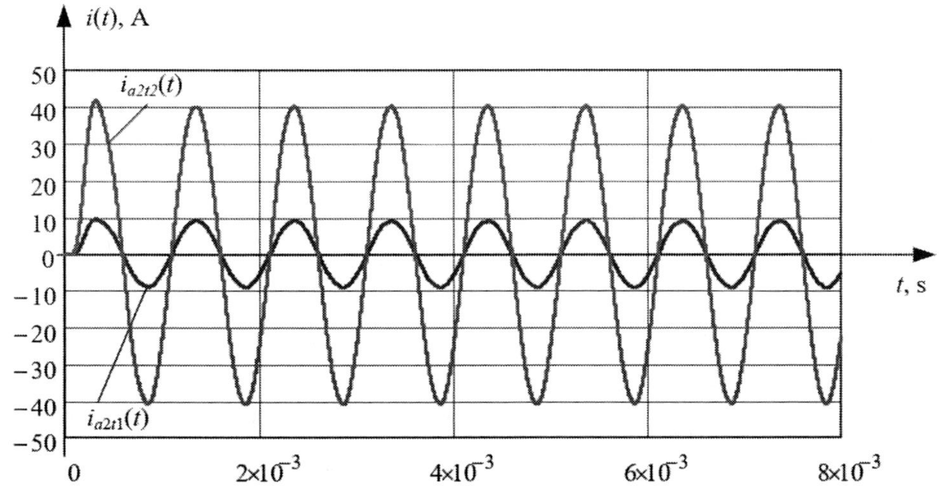

Figure 4. Currents of the primary i_{a2t1} and secondary i_{a2t2} windings of transformer 2.

Voltages on the primary and secondary windings of transformer 2 are different in turn ratio k_{f2} = 0.222. The voltage has a harmonic form.

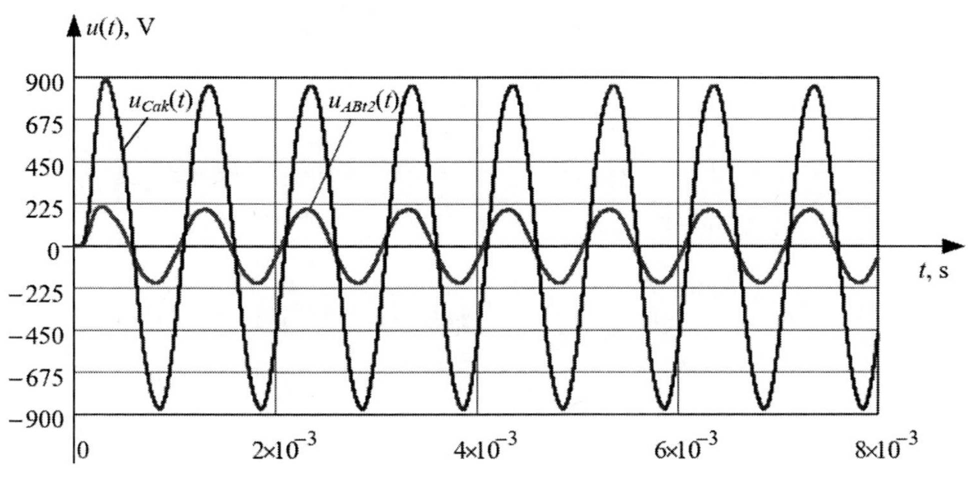

Figure 5. Voltage on the primary and secondary windings of transformer 2.

A transient process for currents i_d, i_C, i_{nag} is obtained due to the presence inductive and capacitive elements in the output filter. The transient process ends approximately at 0.12 s. The load current is equal to i_{nag} = 69 A in the steady-state mode. The capacity current approaches zero (Figure 6).

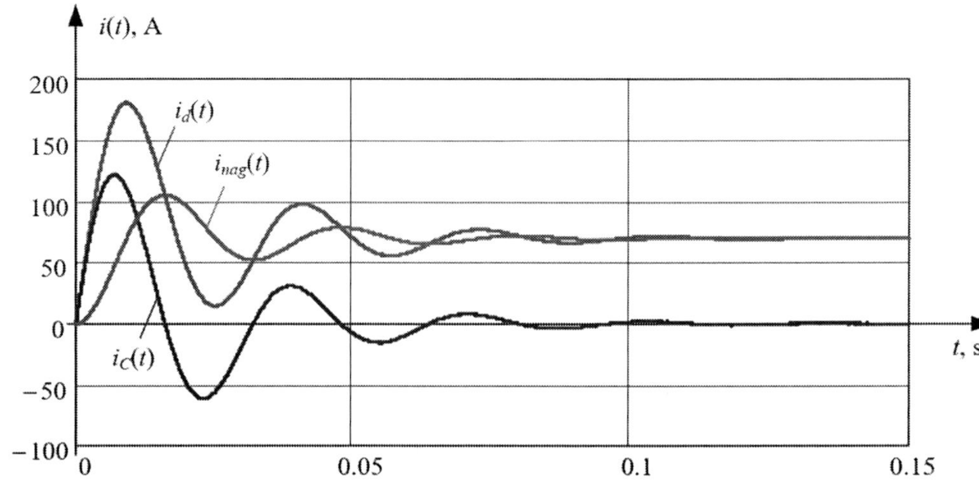

Figure 6. The current after rectifier i_d, filter capacity current i_C and load current i_{nag}.

Figure 7 shows the curve of the rectifier output voltage. As we can see from the figure, the rectified voltage has a six-phase ripple.

To smooth pulsations, the *LC*-filter is connected to the rectifier output. *LC*-filter parameters are: $R_v = 1\,\text{Ohm}$, $L_v = 10\,\text{mH}$, $C_v = 2240\,\text{uF}$. The PSS load is modeled by active resistance, which is accepted to be equal to $R = 4.7$ Ohm in the nominal operation mode.

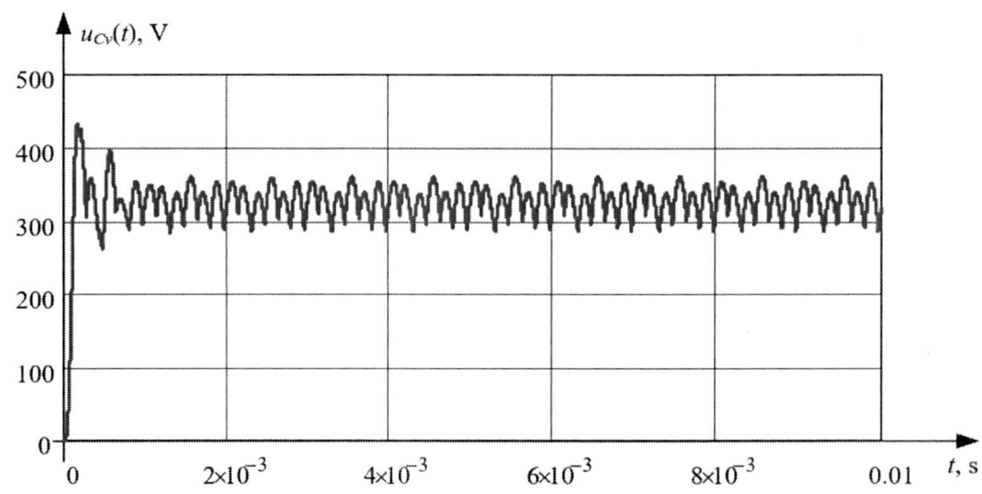

Figure 7. The curve of the rectifier output voltage.

There are smoothing pulsations after the filter. Experimentally, the voltage on the load of submersible device u_{os} was obtained by $329 \pm 1.5\%$ (Figure 8). The transient process ends at time *t*=0.05s.

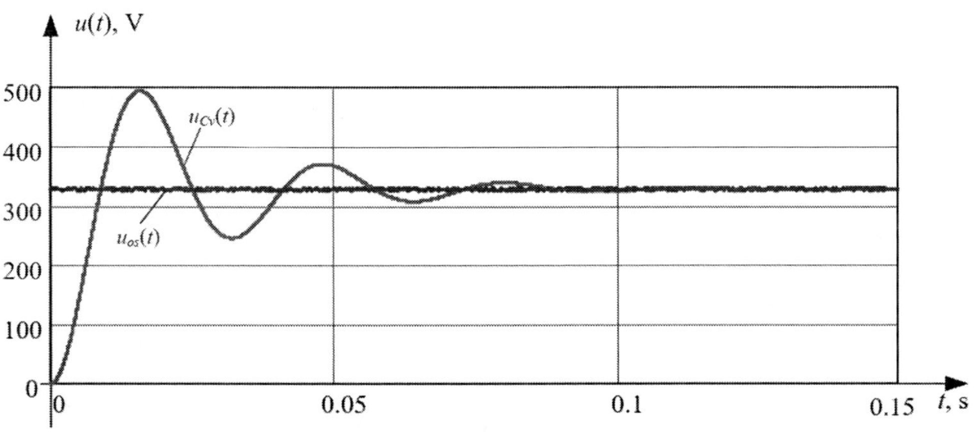

Figure 8. Calculated and experimental load voltage.

Figure 8 shows the plot of the voltage under load u_{CV}. The voltage was obtained by mathematical modeling of the power supply system of the remote dynamic object. Power transmission is ensured by a cable-rope on alternating high frequency current. The mathematical model simulation was carried out in MathCAD software tools. The calculated results were compared with experimental results (Figure 8). The comparison showed that the difference in the steady-state mode is not more than 5%. Thus, these results confirm the adequacy of the developed model.

3. Conclusion

Thus, in this paper the mathematical model for the power supply system of the remote dynamic object with AC power transmission over the cable-rope was developed using MathCAD software tools. The calculated results were compared with experimental results.

The resulting mathematical model allows investigating static and dynamic modes of the power supply system of the submersible device.

On the basis of the resulting mathematical model, it is planned to develop an optimal control for the effective voltage stabilization under the load.

References

[1] Xiang X, Niu Z, Lapierre L and Zuo M 2015 *HKIE Transact. Hong Kong Institution of Engineers* **22** 103-116
[2] Song B, Zhang B, Jiang J 2016 *J. of Huazhong Univ. of Science and Techn.* **44**(7) 52-56
[3] Maalouf D, Creuze V, Chemori A, Tamanaja I.T, Mercado E.C, Muñoz J.T, Lozano, R, Tempier O 2015 *Int. Journal of Advanced Robotic Systems* **12** 1-15
[4] Cui R, Zhang X, Cui D 2016 *Ocean Engineering* **123** 45-54
[5] Fang M-C, Hou C-S, Luo J-H 2007 *Ocean Engineering* **34**(8-9) 1275-1289
[6] Shilin A A. and Bukreev V G 2014 *Thermal Eng.* **61** (10) 741-746
[7] Mamis M S, Kaygusuzi A and Koksal M 2014 *Turk J Elec Eng & Comp Sci* **18**(1) 31-42
[8] Kuleshova E O, Plyusnin A A, Shandarova E B, Tikhomirova O V 2016 *IOP Conf. Ser.: Mater. Sci. Eng.* (Tomsk) vol 124 (Bristol: IOP Publishing Ltd.) 012069
[9] Goheen K R 1991 *J. of Robotic Systems* **8**(3) 295-317
[10] Koran A, Labella T, Lai J-S 2014 *IEEE Transact. on Power Electr.* **29** 1285-1297

Studies in electromagnetic compatibility of optical and digital current and voltage transformers

V D Lebedev and A A Yablokov

Ivanovo State Power Engineering University, 34, Rabfakovskaya str., Ivanovo, 153003, Russia

E-mail: AndrewYablokov@yandex.ru

Abstract. The use of microprocessor devices for relay protection and automation, devices for measuring and determining electric power quality permits the introduction and development of digital circuits leading from primary current and voltage transformers. The use of optical channels for digital circuits addresses the problem of offsetting electromagnetic interference and voltage shift. Creating optical and digital current and voltage transformers with digitization of signals already at high measuring transformer voltage is the best solution for the above problem of mechanical engineering in the field of electromagnetic compatibility of high voltage measuring transformers. However, a difficulty arises in ensuring electromagnetic compatibility of sensors and microelectronic equipment adjoining live parts. The present study is devoted to examining the impact of electromagnetic fields on sensors and solving the problems of electromagnetic compatibility of optical and digital current and voltage transformers.

1. Introduction

The development of electric power involves introduction of modern microprocessor devices for relay protection, automation, diagnostic and measurement means etc. [1-3] With all the above devices, information on current and voltage has traditionally been communicated via secondary circuits from primary measuring converters of current and voltage transformers. Microprocessor devices are the most sensitive elements in the above circuit. No transitional electromagnetic processes in current and voltage measurement transformers are permissible for sensitive microprocessor devices.

To reduce the effect of primary converters there exists the notion of safety coefficient, which, however, does not protect microprocessor equipment against secondary circuit interference. A solution to this problem of mechanical engineering might be to use optical transformers or digital transformers with digitization of signals received from the primary converter at the place of the transformers' installation [4-6]. The above signals are transmitted via an optic cable as light signals unexposed to the effect of electromagnetic interference resulting from industrial and radio frequency current fields. The most promising approach is digitization of the signals immediately at high potential with simultaneous provision of high voltage insulation [7, 8].

It should be noted that electronics (ADCs, microprocessors, power units, optical transmitters) subjected to high voltage can in real conditions be exposed to intense electromagnetic action. It is therefore important to fulfil measures to ensure conditions of electromagnetic compatibility in accordance with the functional resilience of the microelectronics located next to and even inside live parts. Providing these conditions is related to theoretical research based on studies of electromagnetic fields, to choosing optimal locations for electronic equipment and to developing shielding harness.

2. Studying electromagnetic compatibility of optical voltage transformers

In recent times, optical current and voltage transformers have gained wide publicity. In optical voltage transformers, the sensors (optical crystals), which are located in the insulating column body (figure 1), respond to an electrical field, rotating the light polarization plane, measured with the aid of optical sensors. The electrical field in the transformer's insulator body and the actual crystals is determined by the voltage applied to the transformer, and also depends on a number of other factors, such as the nearby presence of high-voltage live parts and earthed units of electrical equipment metal structures. All these elements distort the electrical field not only around but also inside the transformer's insulator body.

To evaluate the impact of high-voltage conductors we performed analytical calculations to determine the electrical field at the location of optical voltage converters. These calculations were based on the following assumptions: dielectric permittivity of the support structure of the transformer's insulator being equal to one; the electrical field's sources being the endless conductors located in the upper part of the transformer (figure 1), which permits us to regard the field as plane-parallel.

Taking into account the above assumptions, we may determine the values of electrical field voltage along the axis of maximum transmitter sensitivity using the method of image charges, and also on the basis of Maxwell's postulate, assuming earth dielectric permittivity to be equal to infinity, according to the following formulae:

$$E_y^{\mathrm{M}} = \sum_{i=1}^{n} E_y^{i\mathrm{M}}, \tag{1}$$

where i – conductor number, n – quantity of conductors; E_y^{iM} – electrical field intensity created by i-conductor at point M and determined according to the following formulae:

$$\begin{aligned} E_y^{1\mathrm{M}} &= \left|\vec{E}^{1\mathrm{M}}\right|;\ E_y^{2\mathrm{M}} = \left|\vec{E}^{2\mathrm{M}}\right| \cdot \frac{a_{1\mathrm{M}}}{a_{2\mathrm{M}}};\ E_y^{3\mathrm{M}} = \left|\vec{E}^{3\mathrm{M}}\right| \cdot \frac{a_{1\mathrm{M}}}{a_{3\mathrm{M}}}; \\ E_y^{4\mathrm{M}} &= \left|\vec{E}^{4\mathrm{M}}\right|;\ E_y^{5\mathrm{M}} = \left|\vec{E}^{5\mathrm{M}}\right| \cdot \frac{b_{1\mathrm{M}}}{b_{2\mathrm{M}}};\ E_y^{6\mathrm{M}} = \left|\vec{E}^{6\mathrm{M}}\right| \cdot \frac{b_{1\mathrm{M}}}{b_{3\mathrm{M}}}. \end{aligned} \tag{2}$$

The module of electrical field intensity of an endless line conductor is determined by the formula:

$$\left|\vec{E}_i\right| = \frac{\tau_i}{2\pi\varepsilon\varepsilon_0 r_{i\mathrm{M}}} \tag{3}$$

where τ_i – conductor charge, C; r – distance from i-conductor to point M, m; ε – dielectric medium permittivity.

Conductor charges may be determined according to the second group of Maxwell equations:

$$\begin{aligned} \tau_1 &= \beta_{11}\varphi_1 + \beta_{12}\varphi_2 + \beta_{13}\varphi_3; \\ \tau_2 &= \beta_{21}\varphi_1 + \beta_{22}\varphi_2 + \beta_{23}\varphi_3; \\ \tau_3 &= \beta_{31}\varphi_1 + \beta_{32}\varphi_2 + \beta_{33}\varphi_3, \end{aligned} \tag{4}$$

where β_{11}-β_{33} – volume coefficients related to potential coefficients α_{11}-α_{33} of the first Maxwell equation group as:

$$[\alpha] = [\beta]^{-1}. \tag{5}$$

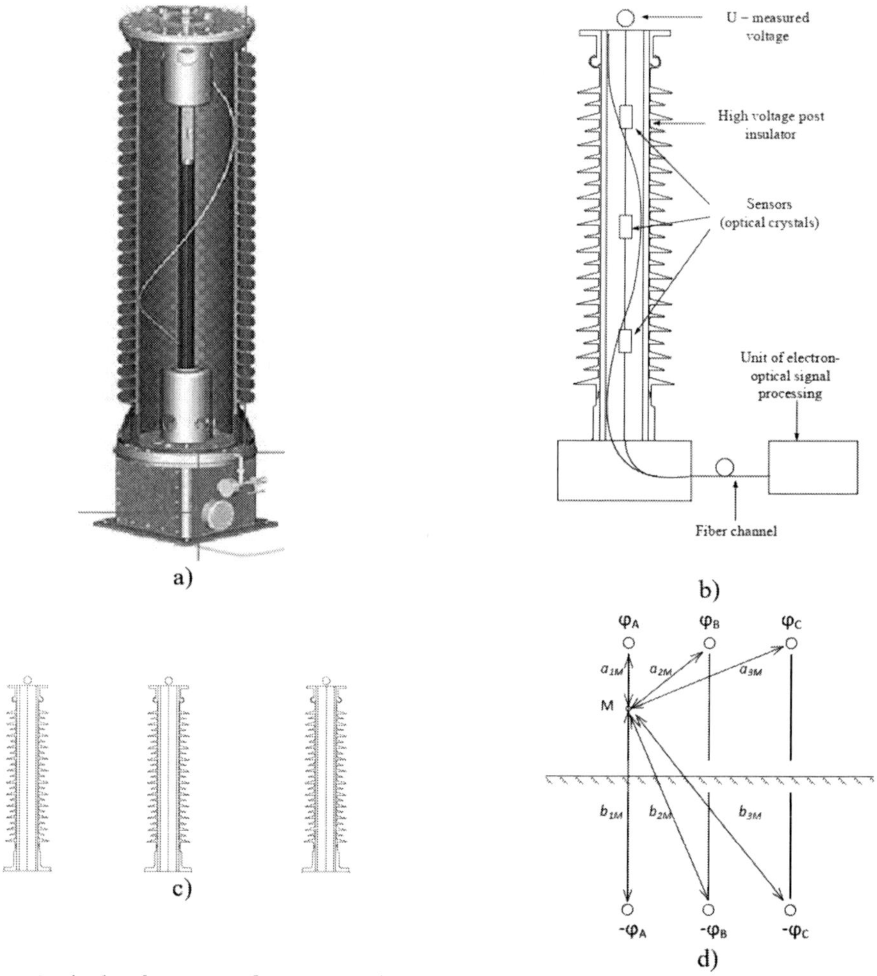

Figure 1. Optical voltage transformer: a – three-dimensional model; b – structure; c – location on open switchgear; d – diagram for calculating sensor electrical field.

Potential coefficients are determined using formulae obtained by the image charges method (figure 1, d):

$$\alpha_{km} = \frac{1}{2\pi\varepsilon\varepsilon_0}\ln\frac{b_{km}}{a_{km}}, \quad \alpha_{kk} = \frac{1}{2\pi\varepsilon\varepsilon_0}\ln\frac{2h}{r}, \tag{6}$$

where h – distance from conductor to earth's surface, m; r – conductor radius, m.

Analysis of the impact of live parts is performed for standard equipment layout at a substation's open switchgear. Research shows that the greatest impact on transformer sensors is caused by bus conductors under high voltage of their own phase (at which voltage is measured) and neighbouring phases.

Calculations made according to the above analytical formulae show that the level of interference from neighbouring phases is rather high, and may reach 10% of the useful signal.

For more detailed study, we developed two-and three-dimensional mathematical models based on differential equation solutions using the finite element method with MATLAB and COMSOL Multiphysics software.

The computed electrical field pattern for positioning of three voltage transformers is shown in figure 2.

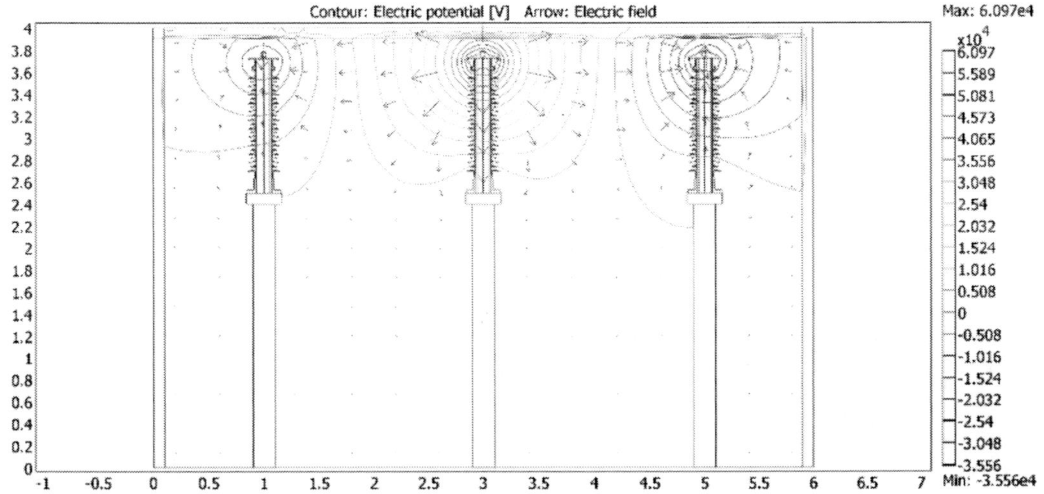

Figure 2. Distribution of electrical potential (contours) and electrical field (arrows).

The results of our calculations enable us to determine electrical field values in a measuring transformer's sensors caused by both applied voltage and neighbouring phase voltage (figure 3). Studies based on the developed mathematical models confirmed earlier results obtained with analytical formulae.

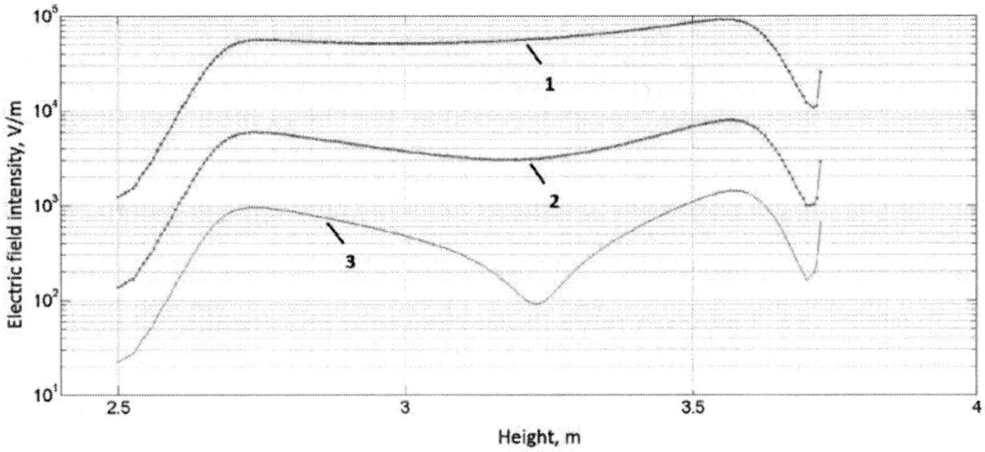

Figure 3. Electric field intensity created by first conductive: 1 – field distribution in the first voltage transformer; 2 – field distribution in the second voltage transformer (interference); 3 – field distribution in the third voltage transformer (interference).

Table 1 shows the results of examining the dependence of optical voltage transformer error due to interference from neighbouring phases on the number and location of Pockels cells.

The impact of interference on measuring accuracy may be reduced by:

- using metal shields with sufficient insulation gaps;
- using materials with high dielectric permittivity;
- algorithmic processing of signals, taking into account neighbouring phase voltage.

IOP Conf. Series: Materials Science and Engineering **177** (2017) 012099 doi:10.1088/1757-899X/177/1/012099

None of these solutions can entirely preclude the shortcomings described earlier, being technically difficult to implement [9]. Furthermore, external factors (temperature, pressure and vibration) have an impact on the output signal of optical transformers, which adversely affects the accuracy of measurement [12-14].

It should be noted that this problem does not exist in conventional electromagnetic and capacitor transformers, as the electric field is formed by the voltage divider created by single transformer cascades in electromagnetic transformers, or by capacitor cascades in capacitor transformers.

Table 1. Dependence of optical voltage transformer error on number and location of Pockels cells

Number of cells	Location of cells*	Amplitude error, %	Phase error, degrees
1	top 1	5.929	3.134
	top 2	2.596	1.586
	centre	3.392	2.046
	bottom 2	3.841	2.257
	bottom 1	4.106	2.377
2	top 1, centre	4.487	2.420
	top 1, bottom 1	5.713	3.031
	centre, bottom 1	3.469	2.082
	top 2, centre	2.827	1.718
	top 2, bottom 2	2.879	1.736
	centre, bottom 2	3.580	2.134
3	top 1, centre, bottom 1	4.330	2.342
	top 2, centre, bottom 2	3.002	1.810
5	top 1, top 2, centre, bottom 2, bottom 1	1.986	1.110

* Top1 presupposes cell position at height of 1.113 m above support level; Top 2 – 0.86 m; Centre – 0.613 m; bottom 2 – 0.363 m; bottom 1 – 0.113 m

The authors' design for digital voltage transformers uses low-power voltage dividers with sufficient intrinsic conductance to ensure the target accuracy class, taking into account the impact of electromagnetic field interference [10, 11].

3. Examining electromagnetic compatibility of digital current and voltage transformers

Digital current measuring transformers (DCT) and combined digital current and voltage transformers (DCVT) contain electronic units on the primary side i.e. under the primary conductor's potential in the zone of intense magnetic fields.

Digital transformers envisage arrangement of microprocessor equipment with ADC inside the conductive shield used directly as a conductor (figure 4, a).

Arranging electronic equipment inside the cylindrical conductor is the most effective option, as demonstrated by our studies. In accordance with the law of total current

$$\oint_l \vec{H}d\vec{l} = i, \tag{7}$$

and bearing in mind axial symmetry, there is no magnetic field inside an infinitely long live cylindrical conductor. Numerical calculation of electromagnetic field in a real conductor (figure 4,b) also confirms minimal field inside the conductor.

For initial evaluation of electromagnetic impact, the field may be resolved into two components – longitudinal and lateral.

Let us consider a cylindrical conductor in a longitudinal magnetic field. Here the external magnetic field $\vec{H}_0$ causes electromagnetic induction in the electromotive force conductor, determined by the equation:

$$\oint_l \vec{E}d\vec{l} = -\frac{d\vec{\Phi}}{dt}. \tag{8}$$

For sinusoidally varying magnetic fields the equation may be written as:

$$\oint_l \dot{E}dl = -\frac{j\omega \dot{H}_i R}{2}. \tag{9}$$

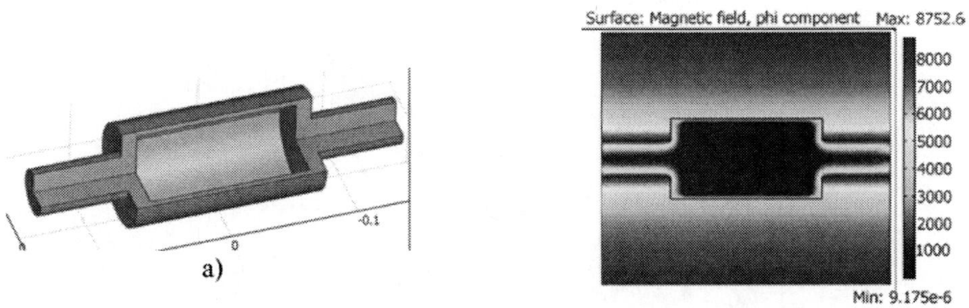

Figure 4. Geometry of a current conductor with internal space to accommodate microelectronics (a) and calculation of magnetic field intensity [A/m] at primary current of 1000 A (b).

The field inside H_i is determined by external field H_0taking into account the compensating effect of a field created by current along the circumference of the cylinder, the density of which is determined by the equation $\dot{\delta} = \dot{E} \cdot \gamma$, where γ– material conductance. It is not difficult to show that the higher the conductance, the higher the shielding constant, because material conductance is a finite value and it will therefore be impossible to absolutely exclude magnetic field inside the tube. Moreover, the shielding effect depends on frequency and is entirely absent with magnetic field that is constant in time (table 2).

Magnetic permittivity of the cylinder significantly influences emergence of the magnetic field pattern, though as a rule, magnetic field strength inside the tube is not reduced.

Screen attenuation b_Q for high and low frequencies may be approximately determined by the following formula Dyakov *et al* [15]:

$$b_Q = \ln\left|\frac{\vec{H}_0}{\vec{H}_i}\right| \approx \begin{cases} \frac{1}{2}\ln\left[1+\left(\frac{\omega\mu_0\gamma Dd}{2m}\right)^2\right] & for\, d < \delta, \\ \frac{d}{\delta} + \ln\frac{\mu_0}{\mu}\cdot\frac{D}{2\sqrt{2}m\delta} & for\, d > \delta, \end{cases}. \tag{10}$$

where D – tube inner diameter, d – wall thickness, $\delta = \left(\frac{2}{\omega\mu\mu_0\gamma}\right)^{1/2}$ - equivalent depth of penetration, m – coefficient, for cylindrical shields equal to two.

Lateral magnetic field is also attenuated inside the cylindrical conductor (figure 5). With low frequency electromagnetic impact a significant factor is the field's magnetic component. In this case, the shielding effect is significantly influenced by the magnetic permittivity of the material μ (as distinct from the shielding effect in a longitudinal magnetic field). For a cylindrical shield with internal tube radius μ, external radius r_2 and relative magnetic permittivityof the shield metal μ the shielding coefficient (constant) for a homogeneous field with intensity H_0 is determined by the formula

$$k_s = \frac{H_0}{H_i} = 1 + \frac{(\mu - 1)}{4\mu}\left(1 - \frac{r_1^2}{r_2^2}\right). \tag{11}$$

Electromagnetic shielding predominates at high frequencies. The field's induced electrical component along the conductor (figure 5) results in the emergence of currents which in turn create a compensating magnetic field inside the cylinder (Table 2).

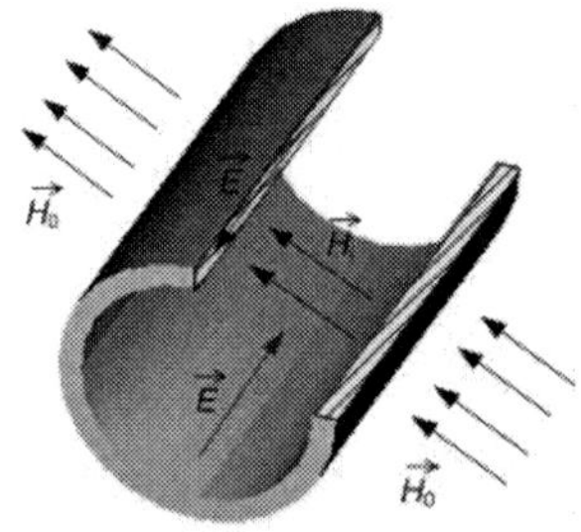

Figure 5. Cylindrical shield with lateral magnetic field.

The above resolution into the conducting tube's own field and external homogeneous longitudinal and lateral fields permits evaluation of the shielding effect. However, examination of the impact of the magnetic field of a digital transformer's live parts requires proper three-dimensional field simulation, taking into account the live parts total geometry. For this purpose we created a field model, the geometry of which is shown in figure 6, a, enabling us to study magnetic field distribution both within and outside the measuring and converting parts of a digital current transformer.

For our simulation we used the following space-frequency equations:

$$\begin{gathered} -\nabla\left((j\omega\gamma)\vec{A} + (\gamma)\nabla U\right) = 0, \\ (j\omega\gamma)\vec{A} + \nabla \times \left(\mu_0^{-1}\mu_r^{-1}\nabla \times \vec{A}\right) + (\gamma)\nabla U = 0. \end{gathered} \tag{12}$$

To solve this equation (12) requires addition of the necessary border conditions. The task was fulfilled with the aid of COMSOL Multiphysics software, in which the obtained three-dimensional patterns (figure 6) enable us to accurately determine levels of electromagnetic activity, select the material and dimensions for additional shielding elements.

a)

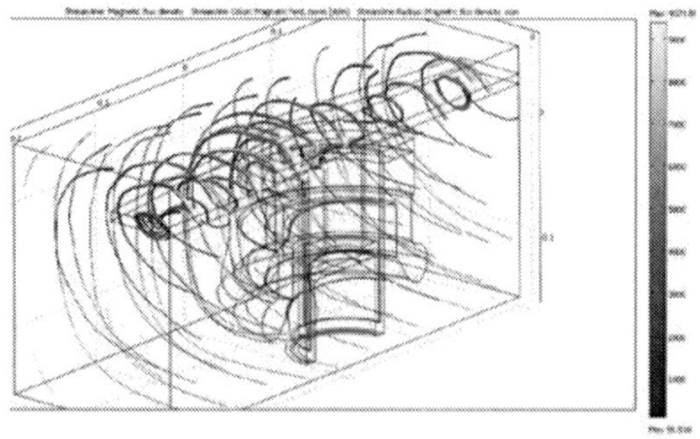

b)

Figure 6. Geometry of a three-dimensional model of a digital transformer's live part (a), and the magnetic field pattern (b).

Table 2. Magnetic field power lines pattern

Freq., Hz	Relative magnetic permittivity μ= 1	μ= 10	μ= 100
	Longitudinal external field		
0	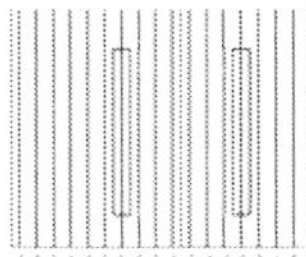	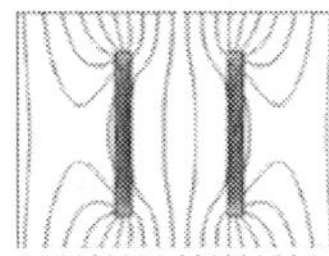	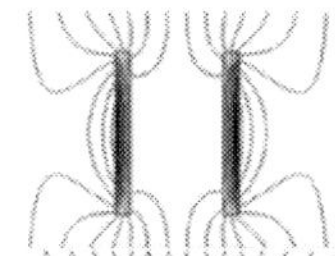
10	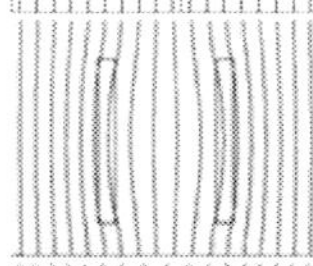	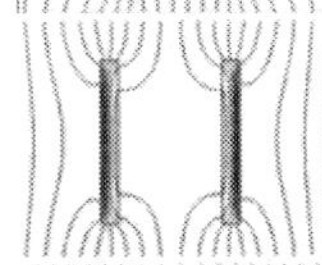	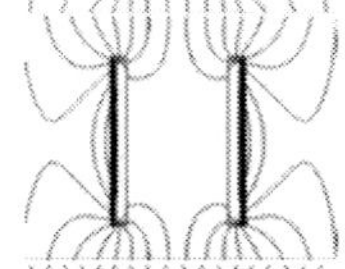
50	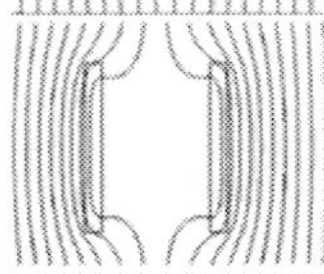	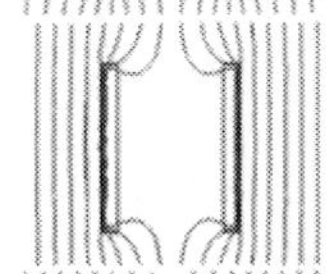	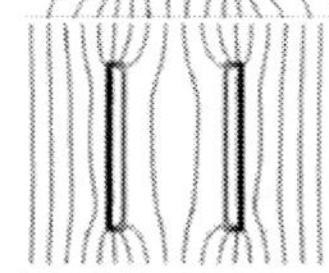
100	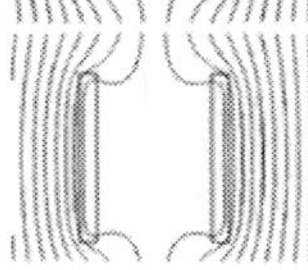	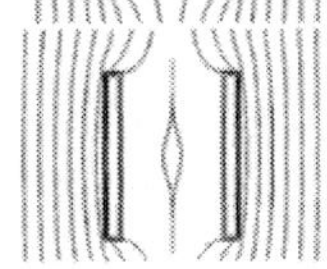	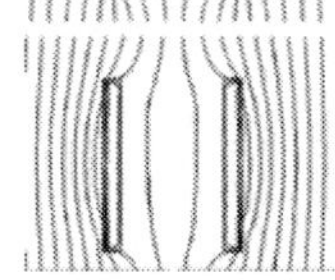
	Lateral external field		
0	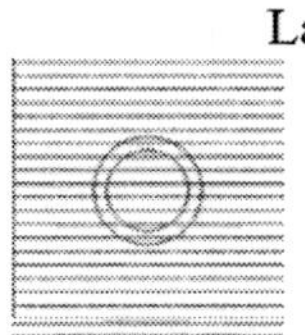	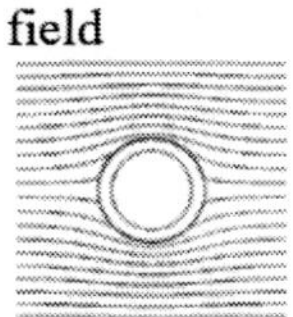	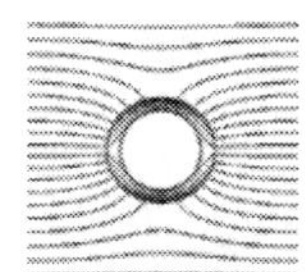
10	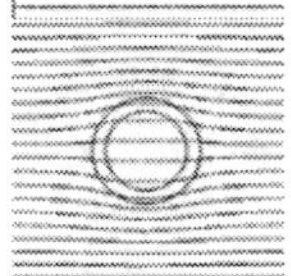	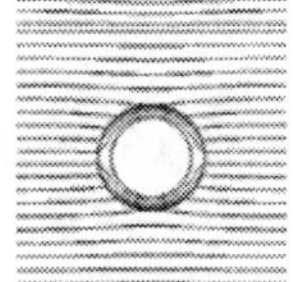	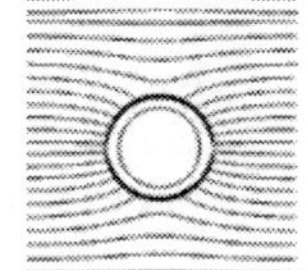
50	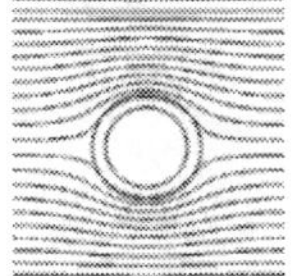	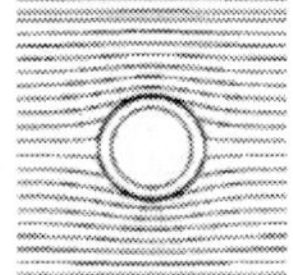	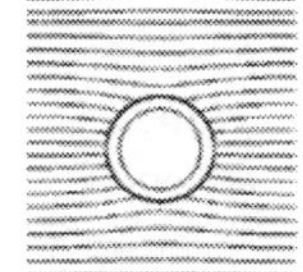
100	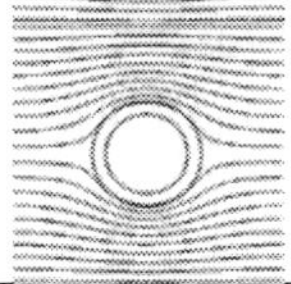	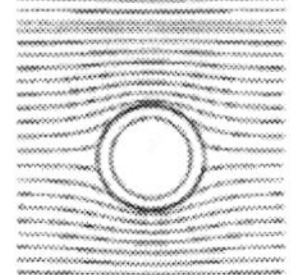	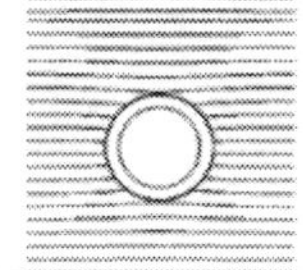

4. Conclusion

The results of our research show significant interference from neighbouring phases and, as a result, significant optical voltage transformer errors.

Our research shows that arranging digital transformer microelectronics inside a cylindrical conductor enables us to exclude the main electromagnetic field component, created by the transformer's intrinsic current. In addition, the cylindrical conductor is also a shield for fields induced by neighboring conductors. Furthermore, the shielding effect may be evaluated according to approximate formulae, and more accurately on the basis of computer mathematical simulation.

Acknowledgments
This research was conducted with financial support of the Ministry of Education and Science of the Russian Federation at Ivanovo State Power Engineering University within the framework of a federal target programme, 'Research and Development in Priority Areas of Developing the Russian Scientific and Technological Complex in 2014-2020' concerning 'Development and Research Relating to Digital 110 V Transformers Based on Basic Physical Laws with Optoelectronic Interface for Metering Electric Power in an Intelligent Electric Power System with a Smart Grid' (Agreement No. 14.574.21.0072 on Granting Subsidies, dated 27 June 2014, Unique Identifier for Applied Scientific Research (Projects) RFMEFI57414X0072).

References

[1] Buhagiar T, Cayuela J P, Procopiou A and Richards S 2016 *Proc. of 13th Int. Conf. on Development in Power System Protection* (Bangkok) vol 1 (Bang Kruai: AESIEAP) pp 4-9
[2] Ma J, Wang T, Wu J and Wang Z 2010 *Proc. of 5th Int. Conf. on Critical Infrastructure* (Beijing) vol 1 (Beijing: NCEPU) pp 1-5
[3] Huang Q, Jing S, Li J, Cai D,Wu J and Zhen W 1999 *IEEE Transactions on Power Delivery* 1-6
[4] Igarashi G, Santos J, Junior S and Pellini E 2015 *Proc. of 2015 IEEE PES Innovative Smart Grid Technologies Latin America* (Montevideo) vol 1 (New York: IEEE) pp 893-7
[5] Thomas R, Vujanic A, Xu D, Sjödin J, Salazar H, Yang M and Powers N 2016 *Proc. of 2016 IEEE/PES Transmission and Distribution Conf. and Exp.* (Dallas) vol 1 (New York: IEEE) pp 1-5
[6] Chatrefou D 2009 *Proc. of 8th Int. Conf. on Adv. in Power System Control, Operation and Management* (Hong Kong) vol 561 (Stevenage: IET) pp 456-461
[7] Wang H and Guan Y 2015 *Proc. of 2015 Int. Conf. on Intelligent Transportation, Big Data and Smart City* (Halong Bay) vol 1 (New York: IEEE) pp 767-770
[8] Grechukhin V 2006 *Ivanovo State Power Engineering University J.* **4** 1-9
[9] Chen J, Li H, Zhang M and Zhang Y 2012 *Metrology and Measurement Syst.* **3** 611-16
[10] Lebedev V, Zhukov V and Yablokov A 2015 *IOP Conference Series: Materials Science and Engineering* (Tomsk) vol 1 (Bristol: IOP Publishing) pp **93** 1-6
[11] Lebedev V and Yablokov A 2015 *Appl. Mech. and Mat.* **792** 220-9
[12] Christensen L 1995 *IEEE Transaction on Power Delivery* **10(3)** 1332-7
[13] Xiao X, Xu Y and Dong Z 2015 *Sensors* **15** 7125-35
[14] Weina G and Wenjie H 2010 *Proc. of 2010 Asia-Pacific Conf. on Power Electronics and Design* (Wuhan) vol 1 (New York: IEEE) pp 47-50
[15] Dyakov A, Maximov B, Borisov R, Kuzhekin I and Zhukov A 2003 *Electromagnetic Compatibility in Electric Power and Electronics* (Moscow: Energoatomizdat)

Modelling of dredging process with vibrating digging bucket under subzero temperatures

S A Zenkov, D Y Kobzov and N A Balahonov

Bratsk State University, 40, Makarenko str., Bratsk, 665709, Russia

E-mail: mf@brstu.ru

Abstract. A mathematical model of ground motion on the vibrating surface of the digging bucket under subzero temperature is suggested. High-frequency oscillations influence on the ground sliding resistance force on the surface under subzero temperature is analyzed. This analysis proves that shear strength is defined by the pressure magnitude of ultrasonic waves affecting the ground from the sliding surface and significantly depends on acoustic properties of the ground. The ground sliding resistance force is stated to decrease under the exposure of ultrasonic oscillations mainly due to a decrease of cohesive forces and ground friction on the metal.

1. Introduction

The research objective is to define the ground shear strength along the sliding surface under the exposure of ultrasonic oscillations and subzero temperature. The presented model of ground motion on the sliding surface, which emits high-frequency fluctuations, is designed using the approximate physical modelling method with the linear scales distributed to reference axes [1-5].

The approximate physical modelling method with the linear scales distributed to reference axes allows one to carry out the research without modifying environment conditions. The main requirement is creating the conditions for the propagation of oscillatory waves in the model which are identical to real-life conditions.

To define the propagation process of ultrasonic and audio-frequency range oscillations, we use the general equation of elastic waves propagation in a medium [1]:

$$\left(\frac{d}{dt}\right)\left[(\lambda+2G)\frac{d^2u}{dx^2}\right]=\frac{\gamma_r d^3u}{gdt^3}+\frac{\gamma_r d^2u}{gt_2dt^2}+\frac{kd^2u}{gt_2dx^2} \tag{1}$$

where: λ, G – Lame constants; u – path; γ_r– medium bilk density; k – volume compression modulus; t_2 – relaxation time (for the given process $t_2 = G/\eta'$); η' – medium viscosity under the exposure of high-frequency oscillations.

2. Formulation of the problem and method of solution

Let us consider the digging process. The problem of reducing the digging strength while reducing the friction on the inner bucket surface can be considered under the following assumption: let us presume that at the final stage of filling the bucket with ground, while the chip is moving along the front wall and the bottom, it also surmounts the inner and outer friction forces. The friction between chips and bucket walls is ignorable. The mathematic expression of the cut bank motion for the given case is arrange in accordance to the calculation model presented in figure 1.

The forces applied to the bank are given in figure 2. The ground inside the bank is assumed to be incompressible.

The cut volume of the ground moves along the curved surface of the work tool. The ground is affected by ultrasonic exposure from the sliding surface at the same time. Ground particles under the exposure of the ultrasonic wave are oscillating around their equilibrium point. The impact of high-frequency oscillations (with frequencies from the audio and ultrasonic range) results in changing the viscous friction forces, which causes thixotropy effect under some conditions.

The cut elementary volume of ground is affected by the following forces: weight force dg, inertial force (dP_{IN}–tangential component, dN_C–normal component), pressure forces $dP_{1,2}$, friction forces $dF_{1,2}$, resisting force dN_{BEND} caused by bending of the bank, ultrasonic pressure force dP_{US} applied normally to the ground sliding surface and friction force dF_{US} caused by dP_{US}.

Force P_R is applied to the ground elementary volume to conform to the D'Alembert principle. Therefore, the sum of force projections in the radial direction is defined as follows:

$$dP_2 + (P_R + dP_R)sin\frac{d\varphi}{2} + P_R\,sin\frac{d\varphi}{2} - dN_C - dP_1 - dN_{BEND} + dg\left(\varphi - \frac{\psi}{2}\right) + dP_{US} = 0 \quad (2)$$

where φ – the angle defining the element position on the surface; ψ – the central angle of the sliding surface.

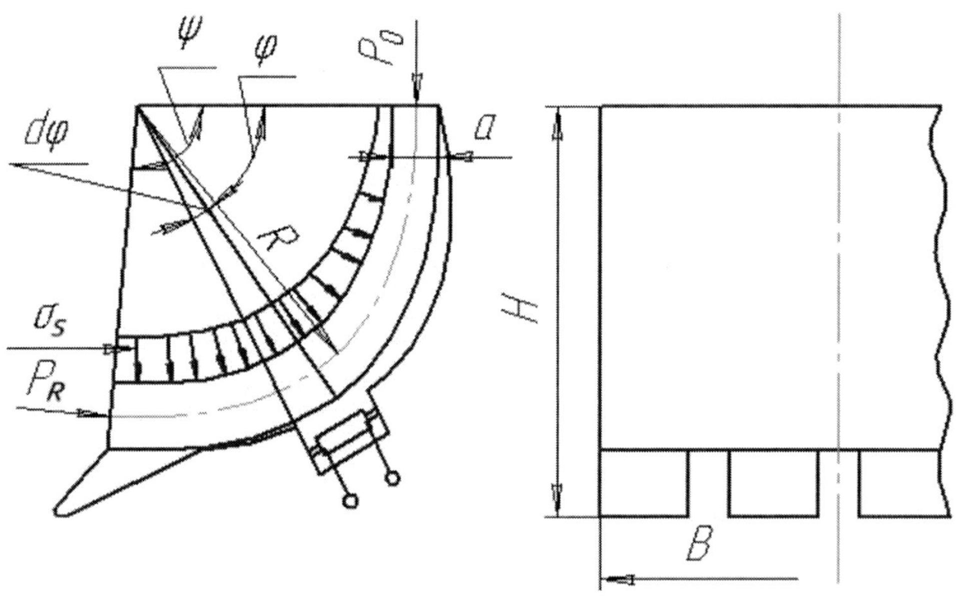

Figure 1. The plan of the ground chip motion on the curved surface of the work tool (bucket) emitting high-frequency oscillations: P_0 – the force preventing bank movement; H – the altitude of the ground ascending on the sliding surface; $σ_s$ - direct stress on the considered surface; a – soil layer thickness; B – bucket width.

Bearing in mind that $dP_1 \approx dP_2$; $sin\frac{d\varphi}{2} \approx \frac{d\varphi}{2}$; $dP_R\frac{d\varphi}{2} \to 0$ we have:

$$dN_C + dN_{BEND} + dP_{US} = P_R d\varphi + dg\,sin\left(\varphi - \frac{\psi}{2}\right) \quad (3)$$

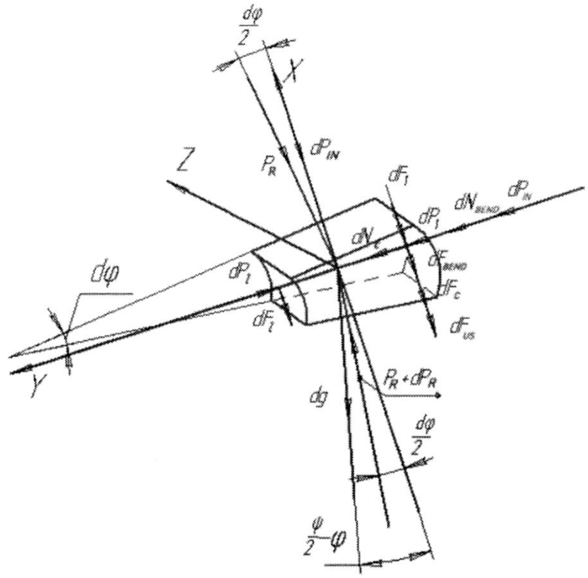

Figure 2. The force diagram for the element of the ground bank.

The sum of force projections on the normal to the radius is given by:

$$dF_1 + dF_2 + dF_{BEND} + dF_C + P_R\cos\frac{d\varphi}{2} + dP_{IN} + dg\cos\left(\varphi - \frac{\psi}{2}\right) - \left(P_R + dP_C\right)\cos\frac{d\varphi}{2} + dF_{US} = 0 \quad (4)$$

The weight of an elementary ground volume is:

$$dg = \gamma_r aRd\varphi dz \quad (5)$$

where R – sliding surface curve radius; a – chip thickness.
The tangential component of the inertial forces is:

$$dP_{IN} = \frac{\gamma_r}{g}dxdydzd\cdot\frac{dv_x}{dt} \quad (6)$$

where v_x – ground chip movement velocity on the sliding surface.

Since $\cos\frac{d\varphi}{2}\approx 1$, we take the friction forces to be $dF_1 \approx dF_2 \approx dF_\tau$ as first approximation.

dF_τ can be defined as the function of tangential stresses:

$$dF_\tau = \tau_\pi Rd\varphi dz \quad (7)$$

where $\tau_\text{п}$ - tangential stress in the plane of sliding.
Taking into account that [6, 7]

$$\tau_\pi = \frac{fP + f_1 p_{AD}S}{S} = f_{SFF}\frac{P}{S} \quad (8)$$

where S – area of contact; f_{SFF} – superficial friction factor; p_{AD} – specific adhesion force; P – resultant of contact normal pressure force; f – coefficient of proportionality for the deformation component of friction; f_1 – coefficient of proportionality for the adhesion component of friction.
Therefore,

$$dF_{\tau} = f_{SFF} \frac{dPRd\varphi}{dx} \tag{9}$$

Centrifugal inertial force dN_C, bank bending force dN_{BEND} and ultrasonic pressure force dP_{US} cause the tangential friction forces, respectively dF_C,dF_{BEND}, dF_{US}.

The sum of these forces is:

$$dF_C + dF_{BEND} + dF_{US} = tg\delta'(dN_C = dN_{BEND} + dP_{US}) \tag{10}$$

where δ' - angle of wall friction along the sliding surface under the ultrasonic exposure.

The pressure force of the ultrasonic wave can be defined as the product of wave pressure and the area [7]:

$$dP_{US} = dP_{US} Rd\varphi dz = 2\pi f_{US} dA_{US} C \frac{\gamma_r}{g} Rd\varphi dz \tag{11}$$

where f_{US}– oscillations frequency; A_{US}– oscillations amplitude; C – velocity of wave propagation in the ground.

With reference to (3), we have:

$$dF_C = dF_{BEND} + dF_{US} = P_R tg\delta' d\varphi + dgtg\delta' sin\left(\varphi - \frac{\psi}{2}\right) \tag{12}$$

Having implemented in (4) the corresponding substitutes, we get:

$$\begin{aligned} dP_R - P_R tg\delta' d\varphi = 2 f_{SFF} \frac{dP_R d\varphi}{dx} + \gamma_r dtg\delta' sin\left(\varphi - \frac{\psi}{2}\right) Rd\varphi dz \\ + \gamma_r cos\left(\varphi - \frac{\psi}{2}\right) + Rd\varphi dydz + \frac{\gamma_r dv_x}{gdt} dxdydz \end{aligned} \tag{13}$$

As known [1], the influence of the curve radius on changing formation resistivity can be neglected if $R/H \geq 9$. For the case of formation movement on a flat surface along the X axis, equation (13) takes the form:

$$dP_R = 2 f_{SFF} dP + \frac{\gamma_r dv_x}{gdt} dxdydz \tag{14}$$

where P_R – resisting force to ground movement in the bucket surface.

Taking into account the charts available from experiments, ground motion resistance [7] in the bucket surface can be defined:

$$dP_R = 2\begin{pmatrix} 4.42 + 0.15P + 0.2W + 0.58T - 0.07t + 330.5D - 24DT + 59Dt + \\ 0.01PW - 0.01PT - 0.03WT - 19 \cdot 10^4 D^2 - 0.01P^2 - 0.02W^2 - 0.01t^2 \end{pmatrix} dxdz + \frac{\gamma_r dv_x}{gdt} dxdydz \tag{15}$$

where P – ground pressure, kPa; T – zone area temperature,°C; W – ground relative wetness, %; t – contact time, sec.; D – ground dispersion, mm.

The continuity equation for moving ground bank with elastic wave propagation is:

$$\frac{d\gamma_r}{dt} + \gamma_r\left(\frac{dv_x}{dx} + \frac{dv_y}{dy}\right) + v_x \frac{d\gamma_r}{dx} + v_y \frac{d\gamma_r}{dy} = 0 \tag{16}$$

where v_x, v_y – ground motion velocity projections.

Boundary conditions on the surface of chip sliding on the ground and metal take the form:

for $\varphi = 0$, $P_R = P_0$; for $\varphi = \psi$, $P_R = P_{Rmax}$;

for $t = 0$, adhesion coefficient $C'_W = C_W$, internal friction angle $p' = p$, wall friction angle $\delta' = \delta$;

for $t = t$, $C'_W = C_{WUS}$, $p' = p_{US}$, $\delta' = \delta_{US}$.

According to the boundary conditions, ultrasonic exposure on the ground starts when $t \neq 0$.

Therefore, expression (15) defines the mathematical model of ground moving on vibrating surface of the work tool for calculating ground motion resisting force under the influence of ultrasonic oscillations.

3. Results and Discussion

Analysis of high-frequency oscillations effect on ground motion resistance forces on a sliding surface under subzero temperatures proves that bank motion resistance is defined by the pressure of ultrasonic waves, affecting the ground from the sliding surface, and to a great extent depends on acoustic properties of the ground. Resisting force to ground bank motion on a sliding surface under ultrasonic oscillations can be reduced essentially by means of reducing adhesion forces and friction forces on metal. Quantitative assessment of the effect under investigation will be calculated through experiments.

The conditions of approximate physical modelling for the ground movement process on a surface that emits the audio range and ultrasonic oscillations are defined by basic similarity criteria [1]. The approximate physical modelling for the ground movement process on a surface that emits the audio range and ultrasonic oscillations is reasonable to conduct with scale division to the coordinate axis. This allows one to carry out the research without modifying conditions of the medium and parameters of intensifying equipment. Methods of multi factor planning were used for experiments.

The effect was assessed through the magnitude of the conditionally momentary freezing distribution coefficient, which is approved to be shear stress U. It conforms to the start of ground element motion against the working surface which is defined as follows:

$$U = \frac{P_R}{S}, \tag{17}$$

where P_R – load required for the metal surface to move against the ground, H; S – working area of the iced ground element, m^2. Shear force U was calculated without external influence (Uwi), but with ultrasonic exposure (Uus).

To carry out an active experiment, the rotatable central composite design was chosen for the five-factor model with plan linear kernel on 2^{5-1} half-replicate and the total number of points N = 32. The plan has star points with the axial distance of α=±2. Planning the matrix and the results of the multifactor experiment with the ultrasonic effect of the PMS-6M emitter are outlined in [7].

Parameters of ultrasonic equipment (rational values) were not modified during the experiment: frequency is 21.8 kHz, amplitude is 0.005 mm, exposure time is 10 sec, emitting surface temperature is 25 °C.

Experimental research [7] processing using the computer program 'MNKLUX' resulted in the regression equation that approximates experimental data:

– under ultrasonic exposure in coded representation:

$$Y_{T.A.B} = 7.72 + 1.61X_2 + 0.49X_3 + 1.96X_4 + 0.38X_5 - 0.48X_1X_4 - 1.18X_1X_5 + 0.72X_2X_3 + \\ 1.09X_2X_4 + 1.26X_3X_4 - 0.76X_1^2 - 1.01X_2^2 - 0.55X_3^2 - 0.45X_2^2 \tag{18}$$

– in natural representation:

$$\tau_{T.A.B} = 4.42 + 0.15p_p + 0.2W + 0.58T - 0.07t + 330.5D - 24DT + 59Dt + \\ 0.001p_pW - 0.01p_pT - 0.03WT - 19\cdot 10^4 D^2 - 0.01p_p^2 - 0.02W^2 - 0.01t^2 \tag{19}$$

4. Conclusion

The equation (17) analysis for optimum showed that this point is located in the negative area of response. All factors within the investigated range mainly facilitate an increase in adfreezing strength. In ascending order of adhesion, they make up a series: *D, t, W, P, T.*

The increase of pressure impact under the ultrasonic exposure can be explained by fading cementing links of ice under the influence of thermal energy (melting) and the transition of its water into an open-grained and free condition.

The experiments proved the 25-time reduction of adfreezing strength of the ground and the metal surface under ultrasonic exposure in average.

References

[1] Balovnev V I 1981 *Modelling of road-building machines work tools interacting processes with the medium* (Moscow: Vysshaya shkola) p 335
[2] Sharma V K, Drew L O, Nelson L 1977 *Transactions of the ASAE* **20** 46-51
[3] Tong J 1993 *Study on reducing adhesion and resistance of soil to soil engaging components of machinery for land locomotion by bionics.* A Ph.D. Dissertation, Jilin University of Technology,Changchun, China (in Chinese)
[4] Tong J, Ren L, Yan J, Ma Y and Chen B 1999 *Int. Agricultural Eng. J.* **8** pp 1-22
[5] Wang X L, Ito N, Kito K and Garcia P P 1998 *J. of Terramech.* **35** pp 87-101
[6] Zenkov S A, Ignatyev K A, Filonov A S and Lkhanag D 2013 *Materials of the 8th Int. Forum on Strategic Technology (MUST 2013* vol 1*)* pp 729-731
[7] Zenkov S A 2009 *Systems Methods Technologies* **1** 59-64

IOP Conf. Series: Materials Science and Engineering **177** (2017) 012101 doi:10.1088/1757-899X/177/1/012101

Piercing mandrel strengthening by surfacing with nickel aluminide-based alloy

I V Zorin, Yu N Dubtsov, G N Sokolov A A Artem'ev, V I Lysak and S N Elsukov

Volgograd State Technical University, 28, Lenina ave., Volgograd, 400005, Russia

E-mail: kewa991@yandex.ru

Abstract. Electrode composite wire (CW) was used for argon-arc surfacing of a thermal-resisting nickel aluminide-based alloy (Ni-Al-Cr-W-Mo-Ta system) on the butt-end surface of the non-water-cooled piercing mandrel. It was shown that multipassing surfacing forms a defect-free deposited metal based on the γ'-Ni_3Al phase of various structural origins. Using high-temperature sclerometry and thermal fatigue testing methods, the metal deposited with CW containing ultrafine particle of 0.3–0.4 % wt. WC carbide features increased resistance to thermal and force effects at temperatures up to 1200 °C.

1. Introduction

It is known that in the temperature range of 1100-1200 °C, the most efficient alloys are heat-resistant γ'-Ni_3Al phase-based alloys produced by casting with high-gradient directed crystallisation with a specified structure throughout the whole workpiece [1-3]. This approach is not always possible or efficient when it is necessary to strengthen the most stressed areas of the metallurgical tools (piercing mandrels, pressing tools). Their intensive wear is due to the metal deformation in temperature-force cycling up to 1200 °C. Such operation conditions are the most common for large non-water-cooled mandrels, piercing the workpieces of difficult-to-form heat-resistant and corrosion-resisting steels and alloys [4, 5]. This increases the relevance of the technologically flexible processes of surfacing of such mandrels with γ'-Ni_3Al-based materials, featuring high deformation resistance of the deposited metal at temperature up to 1200 °C. One way to increase the strength properties of Ni_3Al-based metal is microalloying with rare-earth (La, Y, Ce) metals (REM) [6, 7]. This makes it possible to stabilise the structure with nanosized (up to 70 nm) endogenous phases due to their low solubility in $\gamma'+\gamma$ solid solutions [7]. At the same time, due to their surface activity, REMs decrease the melt surface tension and influence the energy required for nucleation centers, which increases their amount [8]. Such a mechanism can also be employed with refractory thermally stable compounds when introducing them into the weld pool melt in the form of exogenous ultrafine particles [9], as well as with REMs with high surface energy.

The objective of this paper is to increase the deposited metal wear resistance under the thermal-force action using CW containing ultrafine components, as well as to study their effect on the structure formation of nickel aluminide γ'-Ni_3Al-based metal.

2. Materials and methods

The composite wire was produced by an accumulative mechanical 4x drawing through the hard-alloy die of the shell that was made of nickel strand and filler; components were metal wires of aluminium, tungsten, molybdenum as well as tantalum foil and chrome metal powder. The composite wire produced in that way represented a leak-free shell structure 2.5 mm in diameter, ensuring the stable arc [10]. The rated chemical composition ensured the production of the deposited metal on the basis of alloyed nickel aluminide: (wt. %) 0.3 C; 3.0 W; 2.5 Mo; 4.2 Cr; 1.9 Ta; 11 Al; and Ni-remaining element. The CW composition also included 0.3-0.4 % wt. of nickel micropowder with implanted ultrafine (less than 80 nm) particles of WC carbide. The nickel micropowder and ultrafine carbide ratio was 1 : 1.

The deposited metal structure was studied with the inverted digital microscope Zeiss Axiovert 40 MAT and the scanning electron microscope Versa 3D (FEI Company, USA).

The deposited metal deformation resistance within a temperature range of 800-1200 °C was appraised by the sclerometer tests of experimental samples 50×5×8 mm in size, which was produced from the metal deposited on the mandrel. During the testing in argon, the Rockwell indenter travelled on the polished surface of the sample, and heated with the passing current at the rate of 0.8-1.0 mm/sec, and with load on the indenter 0.55 N (Pat. RF 87018). The criterion of the metal's resistance to large deformation at a normal load was k indicator, inversely related to the total volume V_d of the metal deformed by the spherical surface of the Rockwell cone. The specified volume was calculated on the basis of the analysis (with Image Analysis software) of the geometrical parameters (width and height) of the 10 mm tracks' cross-sectional profiles. The tracks' profiles were acquired by scanning the deformed metal surface with the scanning probe microscope Solver Pro (NT-MDT, Russia).

The deposited metal was tested for resistance to the formation of thermal fatigue cracks by the developed procedure-heating the sample with the plasma spray up to 1200 °C and subsequent water cooling up to 50-60 °C at the rate of 140-150 °C/sec. The length of one heating-cooling cycle was 1 min. These testing conditions created higher thermal stresses in the Ni_3Al-based metal, and the time to the appearance of the first fatigue crack was reduced in comparison with the iron-nickel alloys. The criterion of the deposited metal resistance to thermal fatigue was the amount of heating-cooling cycles before the cracks were visible at 4x optical magnification appeared on the sample surface.

3. Technological characteristics of surfacing

The surfacing object was a non-water-cooled large mandrel of the piercing mill 156 mm in diameter, made of 34CrNiMo6 steel (EN 1.6582). In order to reduce the influence of the base metal carbon on the rated chemical composition of the deposited metal, the 2.5 mm high undercoat of the nickel alloy HN60VT grade (Russian grade abbreviation GOST 5632-2014; 0.1 C; 26 Cr; 16 W; 1.0 Ti+Al; Ni-remaining element) was first formed on the mandrel butt-end surface.

The automatic argon-arc surfacing was conducted with two separate beads with a lateral motion of the electrode CW at the rate of ~ 108 m/h. This helped to level the thermal field in the mandrel, which prevented the formation of deposited metal cracks due to the low thermal relaxation capability of nickel aluminide. To prevent the mandrel butt-end surface overheating and the related defects of the deposited metal formation, the duration of surfacing for each passing was limited to 4-5 seconds. The surfacing was performed with the direct current of reverse polarity. The key characteristics of the arc surfacing mode (Table 1) ensured the high-quality formation of the deposited metal and the absence of thermal cracks in the range of the surfacing energy input rate of 800–850 kJ/m. The mandrels were first heated to 400 °C.

Table 1. Key characteristics of the arc surfacing mode

Welding current rate (A)	Arc voltage (V)	Welding speed (m / h)	Oscillation range (mm)
270-290	26-28	25-30	20-25

4. The study of the structure and properties of deposited metal

The study of well-formed 10 mm high metal (Figure 1 a) deposited on the mandrel butt-end surface revealed that its fir-tree structure is represented by the solid solution on the γ'-phase, produced as a result of the eutectic and peritectic reactions involving the γ-solid solution. This corresponds to the stoichiometry area for the Ni_3Al intermetallic compound on the Ni-Al state diagram. The transition area between the alloy HN60VT undercoat and the nickel aluminide-based deposited metal is small (20-30 μm) (Figure 1 b). It does not have any carbide line precipitations or diffusive and crystallisation layers. The basic phase in the transition area structure are dendrites based on the γ-solid solution which contains γ'-Ni_3Al phase, the size of which smoothly decreases through the deposited metal layer height (Figure 1 b), and the alloying elements' content in this area corresponds to their rated values.

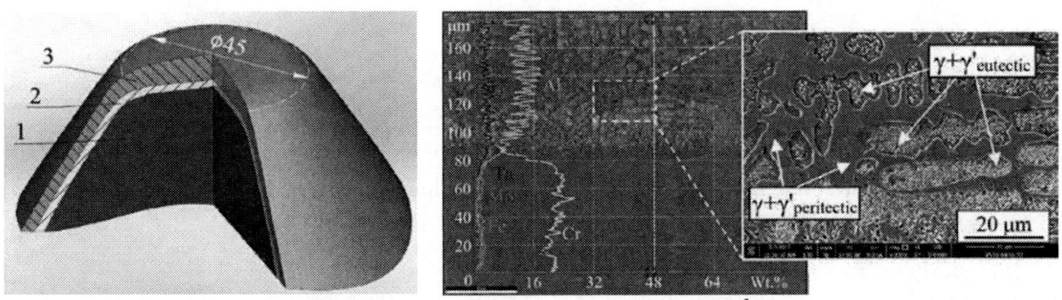

a b

Figure 1. The structure of the deposited butt-end surface of the mandrel (a) and the transition section (b) between the deposited alloy HN60VT (2) and the alloy on the basis of alloyed nickel aluminide (3): (1) – 34CrNiMo6 steel.

It is demonstrated Figure 2 a that the alloy deposited with CW containing microgranules with ultrafine particles has the highest deformation resistance at temperature up to 1200 °C. One possible reason for this is the formation of new crystallisation centre due to WC nanoparticles, which results in the redistribution of molybdenum, tantalum and tungsten in the volume of γ' and γ phases of the forming dendrite; this helps with the strengthening of the solid solution of the γ'-phase with these elements. The specified peculiarities of the γ'+γ structure formation determine the 1.5-1.7 times increase of the *k* indicator of large deformation resistance of the metal in the temperature range of 800-1200 °C against the alloy deposited without ultrafine particles Figure 2. The data provided for comparison on the large deformation resistance of alloy HN60VT shows that despite the content of up to 16 % wt. of tungsten and up to 26 % wt. of chrome, the operating temperature range of this deposited metal type is limited to 800-900 °C.

The study of the deposited metal thermal resistance revealed its high sensitivity to the thermal stress level, which manifests itself in a relatively small number of cycles before the first fatigue crack appears Figure 2 b. The lowest resistance to the thermal fatigue cracks under the considered conditions is characteristic of metal on the basis of high nickel alloy. A considerable increase of the nickel aluminide-based metal thermal resistance is attributable to the higher thermal stability of its structure provided by a high concentration (up to 85 % wt.) of γ'-phase of various structural origin. In the metal deposited with ultrafine WC particles, the fatigue crack nucleation is less intensive, and its opening width is 4-6 times less, and the crack development is blocked in the concentration area of the $\gamma'_{eut.}$-phase particles.

Obviously, the formation of the fatigue cracks in the metal deposited without ultrafine particles can be connected with the critical amount of the TCP phases of the form (Ni, Ta) (W, Mo, Cr) attained during the thermal cycling. In the metal alloyed with the ultrafine particles, during the thermal cycling these phases assume cubic shape, which is less active stress concentrator than the plate-like one.

Provided that the coherent bond with the alloy matrix is preserved, the cubic particles are able to hinder the dislocation migration. At high temperature this factor will prevent the deposited metal softening.

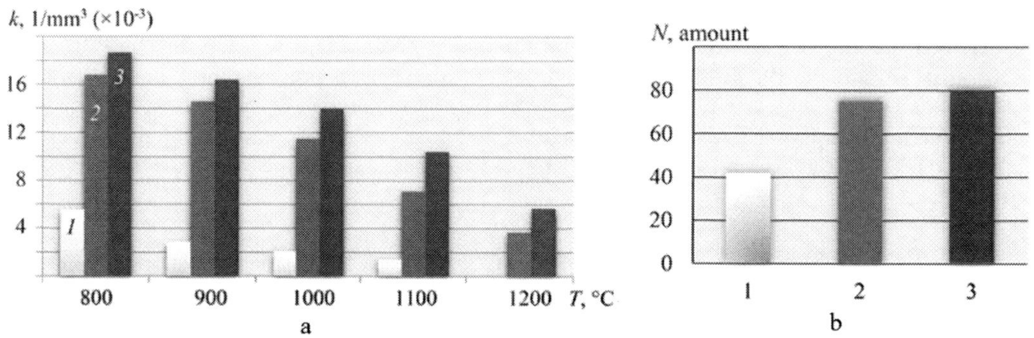

Figure 2. Comparative analysis of the indicator k of the deposited metal's plastic resistance at sclerometry (a) and thermal fatigue tests (b): 1 – HN60VT alloy; nickel aluminide-based metal alloyed with carbide WC ultrafine particles (3) and without them (2). N – amount of heating-cooling cycles before the crack.

5. Conclusion

The high-quality formation of multiple-bead deposited metal on the butt-end surface of the piercing mandrel is ensured under the conditions of argon-arc surfacing, with the electrode composite wire featuring a small energy input rate and process duration.

The obtained results allow us to consider microalloying with ultrafine refractory particles as one possible way to modify the alloys with a $\gamma'+\gamma$ structure in order to improve their strength and thermal fatigue properties at temperature up to 1200 °C.

Acknowledgments

The reported study was funded by the Russian Foundation of Fundamental Research (RFBR 16-08-01276_a; 16-38-00764_mol_a) and a grant from the President of the Russian Federation No. MK-4713.2016.8.

References

[1] Hirano T, Mawari T, Demura M and Isoda Y 1997 *Mater. Sci. Eng. A* **239** 324-329
[2] Li H, Zheng L, Zhang Huarui, Li S, Zhang Hu and Gong S 2012 *Procedia Eng.* **27** 1187-1192
[3] Bazyleva O A, Arginbaeva E G, Zaytsev D V and Fesenko T V 2015 *Met. Sci. Heat Treat.* **57(3)** 133-137
[4] Romantsev B A, Matyko O K, Goncharuk A V, Aleshchenko A S and Polivets A V 2008 *Steel Transl.* **38(11)** 897-899
[5] Vavilkin N M and Bodrov D V 2011 *Russ. Metall.* **1** 25-28
[6] Zexi Yuan, Shenhua Song, Faulkner R G, Zongsen Yu 1998 *J Mater. Sci.* **33(2)** 463-469
[7] Kazanskaya N K, Drozdov A A, Bazyleva O A, Kostina M V, Antonova A V and Morozov A E 2011 *Russ. Metall.* **1** 47-54
[8] Povarova K B, Bondarenko Yu A, Drozdov A A, Bazyleva O A, Antonova A V, Morozov A E and Arginbaeva E G 2015 *Russ. Metall.* **1** 43-50
[9] Sokolov G N, Zorin I V, Artem'ev A A, Litvinenko-Ar'kov V B, Dubtsov Yu N, Lysak V I, Kharlamov V O, Samokhin A V and Tsvetkov Yu V 2015 *Inorg. Mater. Appl. Res.* **6(3)** 240-248
[10] Dubtsov Yu N, Zorin I V, Sokolov G N, Artem'ev A A and Lysak V I 2015 *Key Eng. Mat.* **685** 495-499

Preparation of the highly dispersed powder of titanium carbonitride by SHS azide technology with previous partial nitriding

A P Amosov, Yu M Markov, R A Dobrovolskaya and E N Nikolaeva

Samara State Technical University, 244, Molodogvardeyskaya str., Samara, 443100, Russia

E-mail: egundor@yandex.ru

Abstract. It is shown that the powder of very hard refractory titanium carbonitride ($TiC_{0.5}N_{0.5}$) is the basis of tungsten-free hard alloys which are prospective for application as inexpensive cutting tools. The finer the powder of titanium carbonitrideis, the moreenhanced properties of hard alloys,sintered from the powder,are. An opportunity to reduce the particle size of the titanium carbonitride powder obtained by energy-saving azide technology of self-propagating high-temperature synthesis at the cost of reducing the particle size of the initial titanium powderwas investigated. To ensure the safety of the grinding process of the initial metal titanium powder, it was offered to nitride a Ti powder partially into a $TiN_{0.2}$ compound. Such partial nitriding was performed by the azidetechnology with lack of sodium azide (NaN_3) as a nitriding reagent. After intensive grinding in the planetary ball mill, the $TiN_{0.2}$ powder turned into a superfine powder with an ultrafine structure. This powder was capable of nitriding and carburizing in the azide technology with formation of superfine pure powder agglomerates which are composed of ultrafine and nano-particles of $TiC_{0.5}N_{0.5}$.

1. Introduction

Hard alloys are widely used in mechanical engineering for manufacture of cutting and punching tools, wear parts, operating in abrasive and corrosive environments [1, 2]. The most common hard alloys are based on tungsten carbide, but due to the shortage of tungsten, its high cost, widening application areas, the problem of the development, production and application of tungsten-free hard alloys becomes more urgent. In this respect, hard alloys, based on refractory compounds of titanium, which is much more common in the earth crust and less expensive than tungsten, are very attractive. First, hard alloys based on titanium carbide were developed, but cutting tools made from these alloys were not widely used because of high brittleness [1, 2]. To reduce the brittleness, the titanium nitride (TiN) was added to the hard alloy composition, and then the titanium carbonitride ($TiC_{1-x}N_x$) was used in place of carbide and nitride of titanium [3]. The titanium carbonitride is a solid solution of carbide and nitride of titanium with a wide homogeneity range and combines their best properties: a high melting point, hardness and wear resistance of TiC with the low friction, chemical and oxidation resistance of TiN [4]. Hard alloys based on titanium carbonitride are fine-grained, with a uniform distribution of the binder of Ni-Mo, with low porosity and high values of hardness and bending strength. They are characterized by high scale resistance, heat resistance, a low coefficient of friction on steel, which ensures good cutting properties approaching those of tungsten-containing hard alloys. Hard alloys based on titanium carbonitride are applicable for semi-finish and semi-rough turning and milling of steels, cast iron, nonferrous metals, and also for metal forming by pressure without sticking of the formed metal on the forming tool made from these alloys. In connection with the above-mentioned, the hard alloys based on titanium carbonitride are considered as promising for the replacement of tungsten-containing alloys in their application as cutting and forming tools and wear parts [1-5].

There are several methods of producing titanium carbonitride powders which are used for sintering hard alloys and for spraying wear resistant coatings [1, 3, 5, 6]. The powders are produced from the mixture of powders of TiC and TiN due to high temperature diffusion for several hours at 1600-1800°C and under the pressure of 30-300 MPa, followed by prolonged grinding. The high temperature gas nitriding (1700°C, 2 hours) mixture of powders of Ti and TiC leads to the synthesis of the titanium carbonitride, which turns into a powder during less prolonged grinding. Industrial production of titanium carbonitride powders is based on the carbothermal reduction of titanium from cheap titanium oxide in an atmosphere of nitrogen (1400-1800°C, 2 h): $2TiO_2 + 2(3\text{-}x)C + xN_2 = 4CO + 2TiC_{1\text{-}x}N_x$. This method allows yielding the powders with a particle size of more than 0.5 μm that are bonded in the strong agglomerates, which requires subsequent grinding. Additives of powders of some metals oxides, added to the mixture of TiO_2-C, lead to the synthesis of titanium carbonitride nano-powders by the carbothermal method [7]. Methods for the synthesis of the titanium carbonitride powders at relatively low temperatures and pressures have been also developed; they include the thermal decomposition of $TiCl_4$ complexes with organic amines (800-1200°C, 1-2 h) [8], the reduction of the solution of $TiCl_4$ in CCl_4 with magnesium (850-1050°C) in a nitrogen atmosphere [9] or with sodium azide NaN_3 [6], the use of exchange reactions at 450°C between reagents of $TiCl_4$, CaC_2 and NaN_3 [10]. These methods allow producing highly dispersed powders of the titanium carbonitride including nano-powders, but they are complex and involve the use of costly and toxic components. We should also mention a method of synthesis of titanium carbonitride nano-powders during high-energy milling of powders of carbon and titanium in a nitrogen atmosphere [11].

The above-listed methods of producing the titanium carbonitride powders are energy consuming, long-term, and are implemented on the basis of complex equipment at high temperatures, which lead to the formation of large particles or strong agglomerates of fine particles and the subsequent need for prolonged grinding. The powders often contain impurities of oxygen, free carbon and a material of grinding bodies [6, 9]. In this connection, it is of interest to use the method of self-propagating high-temperature synthesis (SHS) with low energy consumption, simple equipment, a short duration of the synthesis process, high purity products and great possibilities to control the particle size of the synthesized powders [12].

Direct synthesis of the titanium carbonitride of a different composition is possible when burning a mixture of powders of Ti and C in the nitrogen atmosphere only when this mixture is diluted with the TiN powder and at increased pressure of nitrogen (0.6-1.65 MPa) [13, 14]. A finer and less sintered powder can be obtained by the method of azide SHS (SHS-Az) when burning mixtures of Ti and C with sodium azide (NaN_3) and halide salts [15]. In this case, the combustion temperature is relatively low, a lot of gaseous products, which separate the powder particles, are generated; the particle size of the titanium carbonitride powder corresponds to the size of the particles of the initial titanium powder (less than 40 μm), and the content of the basic substance is 96.5-98.5 wt. %. Recently, the SHS method was used to prepared a fine powder of titanium carbonitride, when the titanium reacted with the zinc dicyanide, $4Ti + Zn(CN)_2 \rightarrow 4TiC_{0.5}N_{0.5} + Zn$, followed by removal of the zinc by dilute hydrochloric acid [6]. The temperature of proceeding of this reaction was approximately 2600°C and the titanium carbonitride powder was synthesized in the form of particles of a rounded shape with a size of about 1 μm which were not agglomerated. When three moles of zinc were added to the original mixture, the reaction temperature decreased to about 1000°C, the particles became of a submicron size, but the final product contained the impurity in the amount of 0.86 wt. % Zn.

Increased dispersion of the titanium carbonitride powder, that is a reduced size of its particles, is an important task, as the smaller and purer the powder is, the stronger the hard alloys, sintered from it, are [9]. Along with this, to expand the use of the titanium carbonitride powder, it is necessary to reduce its cost [5]. The aim of this work was to study the possibility of increasing the dispersion of the $TiC_{0.5}N_{0.5}$ powder prepared by the SHS azide technology due to the increase of dispersity of the initial titanium powder.

2. Materials and methods of investigations

The following initial powders were used in the experimental study: the titanium powder of the PTM (small powder of titanium) brand (purity ≥ 98.9 wt. %, particle size < 45 μm), carbon black of the P804T brand (99.5%, 0.2 μm), sodium azide NaN_3 (98.7%) and ammonium fluoride NH_4F (98.5%). All the powders were sifted through a sieve with a cell of 80 μm on a shaking machine of the WS-2 type. Intense grinding with mechanical activation of powders was performed in a planetary mill 'Activator 28L' with a centrifugal acceleration of 30 g. Two drums with a volume of 250 ml contained (each) 375 g of cemented carbide grinding balls with a diameter of 7-11 mm and 37.5 grams of the powder to be milled. Experimental investigations were conducted in the SHS-Az reactor of constant pressure with a volume of 4.5 l. The arrangement of this reactor and methods of work with the measurements of the linear velocity and the maximum combustion temperature by thermocouple method using tungsten-rhenium thermocouples TR 5/20 were described in detail in [16]. The cylindrical samples (40 mm diameter, 50 mm height) of the initial powders mixture of bulk density were burned. The phase composition of the powdery combustion product was determined on an automated X-ray ARL X TRA diffractometer 'ThermoScientific'. Shooting X-ray spectra was performed using Cu-radiation with continuous scanning in the range of angles $2\theta = 20°\div80°$ at a speed of 2 deg./min. The obtained spectra were processed using a special software package 'WinXRD'. The study of the morphology and the size of powder particles was performed on the scanning electron JSM-6390A microscope 'Jeol'.

3. Results and discussion

Our experience shows that a high dispersity of the initial titanium powder can be achieved by intensive grinding in the planetary mill 'Activator 28L', which allows reducing the particle size from 45 μm to a submicron size. The use of this finely ground powder instead of the standard titanium powder of the PTM brand inspires the hope that it is possible to prepare titanium carbonitride powder by SHS-Az process of the same small size as the milled titanium. However, we must keep in mind that if we make a grinding pure metallic titanium in a planetary mill, it may present a fire hazard at high dispersity, especially if it is mixed with oxidants, when its fire and explosion hazard increases many times. To reduce the activity of the highly dispersed powder of titanium is possible if the powder is passivated by paraffin during milling in the planetary mill. Our research in this direction has shown that effective passivation of the submicron powder is achieved at a large amount of paraffin ranging from 5 to 10 wt. %. This paraffin decomposes to carbon and hydrogen in the combustion wave and participates as a carbon reactant in the synthesis of titanium carbonitride. However, paraffin reduces the temperature and velocity of combustion at the expense of losing a part of the combustion heat for the decomposition of paraffin, and also due to the fact that the major portion of carbon is emitted when the combustion wave passes and free carbon remains in the target product. Because of this, using paraffin, it is difficult to achieve the desired ratio of carbon and nitrogen in the carbonitride and to prepare a pure product without free carbon.

In this connection, to reduce the activity of titanium and to enhance the safety of grinding, prior partial nitriding of Ti and transformation into $TiN_{0.2}$ compound were used in this study instead of passivation by paraffin. Titanium nitride of such composition of $TiN_{0.2}$ should maintain its ability to participate in the reactions of carburizing and nitriding in the SHS-Az process and to ensure fire safety of grinding and preparation of powder mixtures. In addition, the partially nitrided titanium, in contrast to pure titanium, is more brittle and is dispersed in the mill better.

$TiN_{0.2}$ was obtained by the SHS-Az method with the limited content of sodium azide according to the following equation:

$$Ti + 0.066NaN_3 + 0.066NH_4F \rightarrow TiN_{0.2} + 0.066NaF + 0.033N_2 + 0.132H_2. \quad (1)$$

To increase the efficiency of the SHS-Az process, a halide salt of ammonium fluoride NH_4F was added to the mixture as it is usually done [15, 16]. The process was carried out in the reactor in argon atmosphere at a pressure of 4 MPa instead of nitrogen atmosphere to avoid additional nitriding of titanium with gaseous nitrogen. Synthesized powder $TiN_{0.2}$ was washed with distilled water for removal of byproduct salt of NaF and then was dried.

Synthesized powder $TiN_{0.2}$ was intensively ground with mechanical activation in the planetary mill 'Activator 28L' within 8 min. An SEM image with different magnifications of the ground $TiN_{0.2}$ is presented in figure 1.

(a) (b)

Figure 1. SEM magnifications ofthe ground $TiN_{0.2}$:
(a) x 5000; (b) x 20000.

As can be seen from this figure, the size of the ground particles of the $TiN_{0.2}$ is in the range from 100 nm to 3 µm, but these particles are combined into agglomerates of a micron size. This powder can be qualified as a superfine powder with an ultrafine structure.

This $TiN_{0.2}$ powder has been used for the synthesis of the powder of titanium carbonitride by the SHS-Az process with the application as reagents powders of carbon, sodium azide and ammonium halide according to the stoichiometric equation:

$$TiN_{0.2} + 0.5C + 0.1(NaN_3 + NH_4F) \rightarrow TiC_{0.5}N_{0.5} + 0.1NaF + 1.08N_2 + 1.3H_2. \quad (2)$$

To reduce sintering of the target product during the synthesis process and to obtain a very fine powder, the oxidizing agents of NaN_3 and NH_4F were introduced into the initial charge in excess as compared with $0.1(NaN_3 + NH_4F)$. They yield a large amount of gaseous products and thereby divide the particles of carbonitride without letting them merge and become stronger. In addition, the presence of an excess of oxidants reduces the temperature of combustion of powder mixtures, thereby also reducing the size of the synthesized product particles [12]. An excess of oxidizing agents is equal to 100% for $0.2(NaN_3 + NH_4F)$, 200% - for $0.3(NaN_3 + NH_4F)$, and so on. To determine the optimal excess of oxidizing agents, the dependence of combustion temperature T_c, °C, and combustion velocity U_c, mm/s, on the value of the excess of oxidants has been investigated. The results of these studies are as follows: 1400 °C and 6.2 mm/s at a 100% excess of oxidants, 1350 °C and 4.1 mm/s — at 200%, 1200 °C and 3.2 mm/s — at 300%, 900 °C and 2.3 mm/s — at 400%, and no combustion — at 500%. Thus, with an excess of the oxidants over 400 %, the combustion is impossible. The optimum excess can be considered as 300%, when the temperature is sufficient for the synthesis of titanium carbonitride, but its sintering is prevented.

Figure 2 shows a XRD pattern of combustion products of the mixture with a 300% excess of sodium azide and ammonium fluoride. As can be seen, X-ray analysis of the washed powder confirmed the presence of titanium carbonitride only in the SHS-Az product.

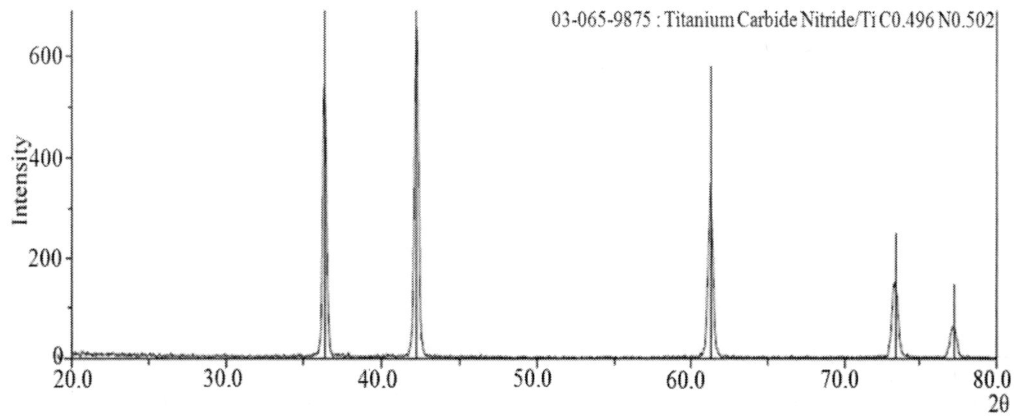

Figure 2. The XRD pattern of the combustion product of the mixture with a 300% excess of oxidants.

(a) (b)

Figure 3. SEM magnifications of the combustion product of the mixture with a 300% excess of oxidants:
(a) x 20000; (b) x 80000.

Micrographs of the synthesized powder of $TiC_{0.5}N_{0.5}$ are presented in figure 3 with different magnifications. These pictures show that the particle sizes of the target powder of $TiC_{0.5}N_{0.5}$range from 40 to 400 nm, but these particles are combined into agglomerates of a micron size. Thus, synthesis of titanium carbonitride, using partially nitrided titanium of composition $TiN_{0.2}$ activated in a planetary mill, confirmed that the use of the initial mixture of $TiN_{0.2}$ + 0.5C + 0.4(NaN_3 + NH_4F) with a 300% excess of oxidizing agents made it possible to store and even slightly reduce the initial particle size of $TiN_{0.2}$ when turning it into carbonitride. Therefore, using a superfine partially nitrided titanium powder with an ultrafine structure, it is possible to obtain superfine pure titanium carbonitride agglomerates which are composed of ultrafine- and nano-particles of $TiC_{0.5}N_{0.5}$ in a safe way.

4. Conclusion

A method of passivation of a pure metallic titanium powder is proposed by partially nitriding it by the SHS azide technology into $TiN_{0.2}$ to ensure fire safety and effectiveness of grinding in a planetary ball mill till obtainment of a superfine powder with an ultrafine structure. It is shown that the partially nitrided and milled superfine powder of $TiN_{0.2}$ can be then used in the simple energy-saving azide SHS technology to produce the superfine agglomerated powders with size less than 3 μm which are composed of $TiC_{0.5}N_{0.5}$ ultrafine- and nano-particles of size from 40 to 400 nm instead of a much larger powder of titanium carbonitride with a particle size of 40 μm [15].

Acknowledgements
The authors thank A Yu Illarionov for assistance in SEM investigations and V A Novikov for capturing diffraction patterns for X-ray phase analysis of the products of SHS-Az.

References

[1] Lengauer W 2000 *Handbook of Ceramic Hard Materials*, ed R Riedel (editor) (Weinheim: Wiley-VCH) p 202-52
[2] Panov V S, Chuvilin A M and Falkovsky V A2004 *Technology and Properties of Sintered Hard Alloys and Articles manufactured from Them* ed 2 (Moscow: MISIS) (in Russian)
[3] Pastor H 1988 *Mater. Sci. Eng.***A 105/106** 401-409
[4] Pierson H O 1996 *Handbook of Refractory Carbides and Nitrides* (Westwood, New Jersey: Noyes Publications)
[5] Gong X, Sun J, Quan K and Meng Y 2013 *Adv. Mater. Res.* **634-638** 2373-7
[6] Dyjak S, Wasilewska M, Cudzilo S 2015 *Centr. Eur. J. Energ. Mater.* **12** (2) 249-59
[7] Matsuda T, Matsubara H 2014 *Int. J. Refr. Met. Hard Mater.* **42** 1-8
[8] Lichtenberger O, Pippel E, Woltersdorf J and Riedel R 2003 *Mater. Chem. Phys.* **81** 195-201
[9] Lee D-W, Yu J-H and Jang T 2007 *Solid State Phen.* **124-126** 1225-8
[10] Feng X and Shi L-Y 2005 *Chem. Let.* **34** 1002-3
[11] Yin F, Zhou L, Xu Z, Xue B and Jiang X 2009 *J. Alloy Compd.* **470** 369-74
[12] Amosov A P, Borovinskaya I P, Merzhanov A G and Sytchev A E 2005 *J. Self-Prop. High-Temp. Synth.* 2005 **14** 165-86
[13] Eslamloo-Grami M and Munir Z A 1994 *J. Mater. Res.***9** 431-5
[14] Yeh C L and Chen Y D 2005 *Ceram. Int.* **31** 719-29
[15] Amosov A P, Bichurov G V, Bolshova N F, Erin V M, Makarenko A G and Markov Y M 1992 *J. Self-Prop. High-Temp. Synth.***1** (2) 239-45
[16] Shiganova L A, Bichurov G V, Titova Yu V, Ermoshkin A A and Bichurova P G 2011 *Rus. J. Non-Ferr. Met.* **52** (1) 91-5

The study of mechanical effect on the characteristics of the PZT-system piezoelectric elements

V I Berg, A E Brand, V A Petryakov and A G Zakirzakov

Federal State Budgetary Educational Institution of Higher Education 'Tyumen Industrial University', 38, Volodarsky str., Tyumen, 625000, Russia

E-mail: Allbert-@mail.ru

Abstract. The paper examines the characteristics of the piezoengine by the method of glancing collisions under the conditions of the moving scheme with mobile connections. The research results of the values' changes of the PZT materials' electrophysical parameters and their structural composition are represented. In the course of the research, the voltage boundary values, occurring during glancing collision of the inclined piezoceramic transducer, have been obtained.

1. Introduction

There is an active development of power engineering in the past 30 years in the world. At the end of the XX century a great interest of the humanity was focused on a large amount of electricity generation to cover the increasing needs of population and industry. However, with the advent of the new Millennium, there is an active development of small generations: microelectronics, automated manufacturing, robotics, precision engineering, where an important role is given to the accuracy of movement, assembly, etc. Special attention should be devoted to the high- frequency piezoelectric transducers' region on the basis of which the piezoengines, used as the industrial drives of linear displacement, are constructed, [1-3]. The constructions and materials, applied in piezoelectric motors do not always correspond to the movement accuracy and quality requirements.

Piezocelectric transducers are based on the use of the phenomenon of the direct and inverse piezoelectric effect and are made from materials based on lead zirconate-titanate (PZT) with various structural modifications. When the direct piezoelectric effect is used, a polarization charge on the piezoelectric surface is observed; the transducers have a mechanical input and an electrical output. Electrical polarization, associated with the surface charge, increases linearly relative to mechanical stress as a first approximation [4, 5]. In the process of the inverse piezoelectric effect, the change of the piezoelectric length is observed, but the transducer has an electrical input and mechanical output.

The most important element of a piezoengine is a piezocelectric element, the choice of which should be based on the maximum efficiency achievement requirement. The application of the scheme with direct and glancing collision, the velocity of which is a fraction of a second, is absolutely actual in precision engineering. When these schemes are used, the quality of piezoelectric elements is characterized by the following main parameters, adopted in Russia and abroad [6, 7]:

K^T_{33} (e^T_{33}/e_0) – relative permittivity;

tg d – loss-angle tangent at frequency of 1 kHz in weak fields;

T_c (T_k) – Curie temperature point;

K_p, K_{33}, K_{31}, K_{15} – electromechanical coupling coefficients;
d_{33} -d_{31} d_{15} – piezoelectric modules;
g_{33}, g_{31}, g_{15} – electric voltage factors;
Y^E_{11}, Y^E_{33} – Young's moduli;
N_L, N_T, N_R – frequency constants;
S^E_{11}, S^E_{33} – the elasticity parameter;
r – density;
Q_m – the mechanical quality factor.

The most important parameters are dielectric permittivity K^T_{33} (e^T_{33}/e_0), as a measure of the piezoelectric element capacity, and piezoelectric moduli d_{33} -d_{31}, d_{15}, determining the value of the electric charge on the piezoelectric electrodes when the force is applied [8,9].

The aim of the article is to study the influence of the mechanical effect on the characteristics of the piezoelectric element and to determine the optimal material for piezoelectric motors.

2. Materials and methods

To conduct the study, concerning an assessment of the mechanical effects impact, the authors of the article have chosen the most common piezoelectric materials, based on PZT: PZT-19, PZT-19M, PZTtBS-1. Special disc blanks with a diameter of 15 mm and a thickness of 2 mm were made from the abovementioned materials. The mechanical effect was performed on the laboratory setup, depicted in figure 1.

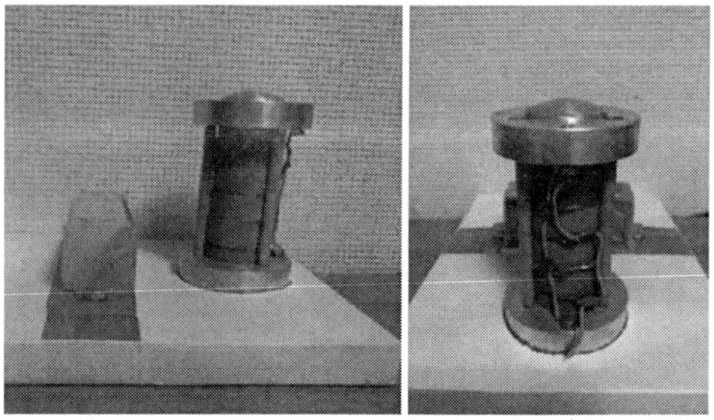

a b

Figure 1. The laboratory setup: a – the side view, b – the back view

Figure 2 shows a scheme of the investigated piezoelectric motor with thickness δ with electrodes deposited on two parallel faces [8-12].

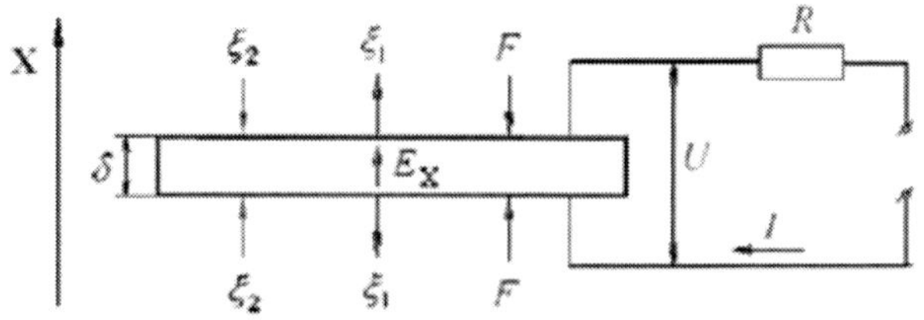

Figure 2. The piezoelectric motor scheme: ξ_1, ξ_2 – plate faces displacement under tension and compression; F – external forces; E_x– electric field intensity; U – electrodes' plate voltage; I - current, R – load.

The inverse piezo effect equation for the longitudinal fluctuations under the voltage control has the following view [12-16]:

$$S_x = d_x E_x(t) + \varepsilon_x T_x, \quad (1)$$

where S_X – the piezoplate strain rate in direction X; ξ – x plate section displacement; d_{33} – the piezoelectric modulus under the longitudinal plate polarization.

The force equation, influencing the piezoplate, is represented in the following way:

$$T_x S_0 = F + \frac{m \partial S_x}{\partial t}, \quad (2)$$

where F – external force, applied to the piezoelectric element; m – piezoengine's moving mass.

The dynamic load with a pressure range from 2 to 5 N/cm^2 was applied to the element from the selected materials to assess the mechanical effect influence upon the piezoelectric element characterictics. Also, in the process of the study, the electrophysical properties at the temperature change from '-60^0' to '+60^0' have been estimated.

3. Results and Discussion

As a result of mechanical effect, the following data, represented in table 1, 2 and in figure 3 have been obtained.

Table 1. Electrophysical parameters of piezoelectric materials

Parameter	Symbol	PZT-19	PZT-19M	PZTtBS-1
Relative permittivity	$e^T{}_{33}/e_0$	1650	1850	4100
Loss-angle tangent at frequency	tg	0.0300	0.0250	0.0270
Electromechanical coupling coefficients	$K_p/K_{31}/K_{33}$	0.56/0.29/0.64	0.62/0.35/0.72	0.65/0.38/0.73
Piezoelectric modules	D_{31}/d_{33}	155/360	184/410	289/600
Mechanical quality factor	Q_m	90	90	50
Curie temperature point;	T_c	290	300	170
Density	r	7.60	7.60	7.30

When there is a change of the piezoelectic element temperature, the following electrophysical parameters have been obtained. The results are given in table 2.

Table 2. The change temperature and electrophysical parameters of piezoelectric materials

Material	Effecting factor		Electrophysical parameters change, %		
	Lower temperature of the agent (limiting)	Elevated temperature of the agent (limiting)	$\varepsilon^T{}_{33}/\varepsilon_0$	d_{31}	$V^E{}_1$
PZT-19			± 10.0	± 10.0	± 2.0
PZT-19M	- 60^0	+ 60^0	± 10.0	± 10.0	± 2.0
PZTtBS-1			± 10.0	± 10.0	± 2.0

The temperature gradient and electrophysical parameters changes comparison has shown that material PZT-19 is much less susceptible to the temperature fluctuations and is suitable for the use in the middle and Northern latitudes.

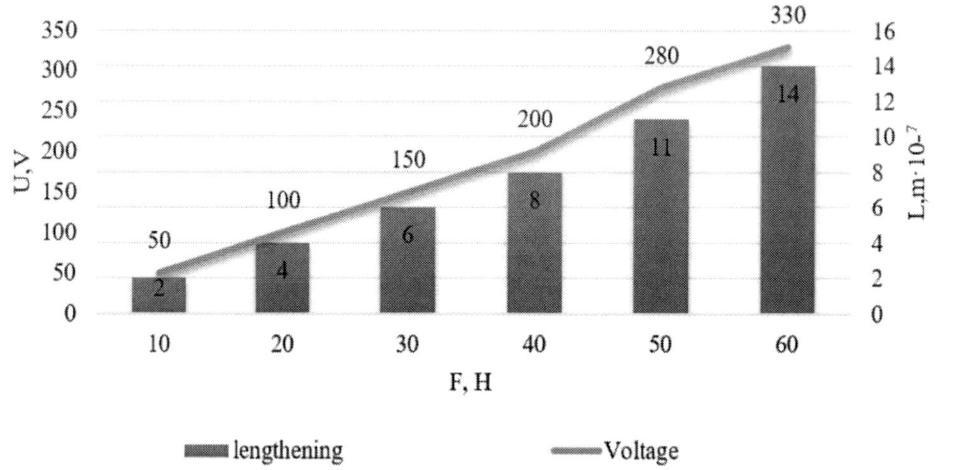

Figure 3. Extension and the external load for different values of the resulting voltage.

The graph shows that the resulting voltage at the ends of the electrodes increases almost linearly with the increasing of the applied load. The extension of the piezoelectric element being tested also increases almost linearly.

The comparative analysis of the electrophysical parameters, given in figure 4, shows the presence of significant advantages of the PZT-19M.

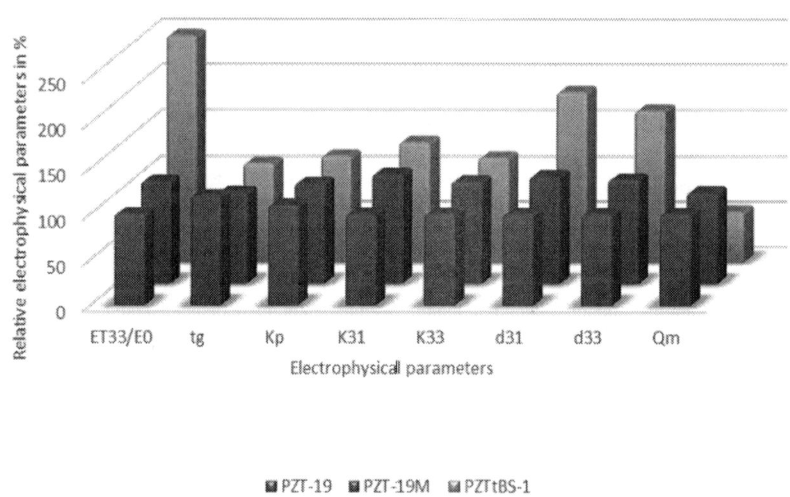

Figure 4. Comparative analysis of electrophysical parameters.

4. Conclusion

The calculation of the piezoelectric motor characteristics was carried out for the scheme of the movable link by the glancing collisions method. As a result, a range of the impact forces values that occur under the conditions of vibrations of the high frequency piezoelectric transducer has been received. Thus, the conducted researches allowed us to reveal the dependence of the output power of the piezoelectric motor on the material under low mechanical loads and the behavior trends of the piezoelectric constants. The best material for piezoelectric motors is PZTtBS-1.

References

[1] Haiyan C, Xiaobo G and Zhongyan M 2002 *J. Materials Chemistry and Physics* **75** 202
[2] Iula A and Pappalardo M 2003 *J. Ultrasonics* **42** 291
[3] Karimli T I 2016 *Inter. Collec.of Scien. Pap.* **1(52)** 69
[4] Vullers R J M, Schaijk R, Doms I, Hoof C V and Mertens R 2009 *J. Solid-State Elect.* **53** 684–693
[5] Schwankl M, Rübner M, Singer R F and Körner C 2013 *J. Proc. Mater. Sci.* **2** 166
[6] Komarov V P, Peresadchenko A N, Bushnaya A N and Khrebtov A O 2014 *J. Scien. works of DonNTU* **2(23)** 60
[7] Semenistaya T V, Petrov V V and Ladigina A A 2014 *J. Izvestiya SFedU. Eng. Sc.* **4** 213
[8] Neubauer M and Wallaschek J 2013 J. *Mech.l Sys.s and Sign. Proces.* **36(1)** 36
[9] Rupitsch J S, Ilg J, Sutor A and Lerch R 2011 *J. Proc.Eng.* **25** 1441
[10] Ivanov A A 2014 *J. Trans. of NSTU n.a. R.E. Alexeev* **5(102)** 61
[11] Jiao Q, Shifeng H, Yuesheng X, Hudong J, Dongyu X and Xin C 2014 *J. Ceram. Inter.* **40(8)** 13019
[12] Du G, Liang R, Wang J, Wang L, Zhang W, Wang G and Dong X 2013 *J. Ceram. Inter.* **39 (8)** 9299
[13] Cross J S, Shinozakia K, Yoshiokaa T, Tanakaa J, Kimb S H, Moriokac H and Saito K 2010 *J. Mater Sci. and Eng: B* **173 (1)** 18
[14] Wang D, Rocks S A and Dorey R A 2009 *J. of the Europ. Ceram. Soc.* **29 (6)** 1147
[15] Wang Z, Miao J and Chee W T 2009 *J. Sens. and Actuat. A: Phys.* **149(2)** 277
[16] Takechi S, Onishi T, Minami S, Miyachi T, Fujii M, Hasebe N, Mori K, Nogami K, Ohashi H, Sasaki S, Shibata H, Iwai T, Grün E, Srama R and Okada N 2008 *J. Planet. and Spac. Sci.* **56 (9)** 1309

Metal of cavitation erosion of a hydrodynamic reactor

A G Zakirzakov, A E Brand, V A Petryakov and E F Gordievskaya

Federal State Budgetary Educational Institution of Higher Education 'Tyumen Industrial University', 38, Volodarsky str., Tyumen, 625000, Russia

E-mail: Allbert-@mail.ru

Abstract. Cavitation erosion is a major cause of the petroleum equipment hydraulic erosion, which leads to the metal weight loss of the equipment and its breakdown, which can be followed by the full stop of the plant or company work. The probability of the metal weight loss and equipment failure can be reduced by the use of special protective coatings or rivets, made of the sacrificial metals, the use of which significantly increases the service life and the production equipment reliability. The article investigates the cavitation erosion effect, occurred under the condition of the advanced hydrodynamic cavitation on the hydrodynamic cavitation reactor. This article presents the results of the experiments and recommendations for increasing the operational resource.

1. Introduction

The problem of the machine parts and mechanisms service life increase in the aggressive environments and at high loads has always been relevant to all areas of the industry. In the Russian Federation, the most important parameter is the pipelines and the pumping units operating life in the petroleum industry, which is the principle of the federal budget taxable base. The machine parts and mechanisms operating life can be prolonged as at the manufacturing stage, by applying the impurities and additional protective coatings, and at the operational stage. In the context of the techno-economic aspect, the pipelines and pumps service life extension at the manufacturing stage is more beneficial.

Considering the increasing dynamics of the hydrocarbon production and oil and gas companies' technical and economic difficulties, the cavitation processing technology increasingly finds its application [1, 2]. It improves the oil rheological properties, reduces the pipeline wall buildup, increases the efficiency of the viscous and highly-viscous oil transportation, increases the light cut yield, etc [3-7]. However, the cavitation technology application is associated with the high risks of the operating equipment breakdown, caused by the cavitation erosion effect and the pressure fluctuations.

The analysis of the scientific literature has shown that there are a large number of hypotheses for the explanation of the cavitation mechanism and the physical nature and its effect on the metals destruction, which have both their advantages and disadvantages [7-10]. So, the results of the research have identified two stages of cavitation development (including the cavitation erosion), however the study implies the existence of four stages (periods) of cavitation erosion [11-14]. There is also difficulty in the analysis when the researchers are experimenting with the application of the simulative equipment that is inconsistent with the real process. Thus, often, in the process of the cavitation erosion investigation, the ultrasonic dispersant or the magnetostriction vibrator with the acoustic vibrations concentrator are used. They are inferior by their effect parameters to the hydrodynamic cavitation reactors, which develop stronger cavitation phenomena. Namely the cavitation erosion, which occurs under the condition of the advanced hydrodynamic cavitation, is the most dangerous for the pump equipment and the pipeline transport.

A wide range of methods is used for the protection against the cavitation erosion, including the use of the erosion resistant steel in the constructions base, the inside construction carbidic coatings application, application of multilayer vacuum coatings and rivets made of the sacrificial metals, the use of chemical and thermal processing and others.

The aim of this article is the investigation of the cavitation erosion effect, occurred under the condition of the advanced hydrodynamic cavitation on the hydrodynamic cavitation reactor walls and the investigation of the anti-erosion coating effect on the reactor operating life.

2. Materials and methods

To conduct the research, the hydrodynamic cavitation reactor with the following parameters was used: length – 100 mm, diameter – 40 mm, confuser inlet angle – 40', diffusor outlet angle – 50', channel contraction ratio – 3. The reactor is made of X39Cr13 steel, having the following chemical composition distribution: Fe – 84.1%, Cr – 14.2%, C – 0.41%, Mn – 0.48%, Si – 0.33%, Ni- 0.31%. The remainder of the chemical composition is represented by P, Al, Co, Cu and other substances. It was decided to cover the reactor inner surface with the titanium, nickel and copper mixture to explore the ways of the service life improvement. The model and the sample are presented in figures 1 and 2.

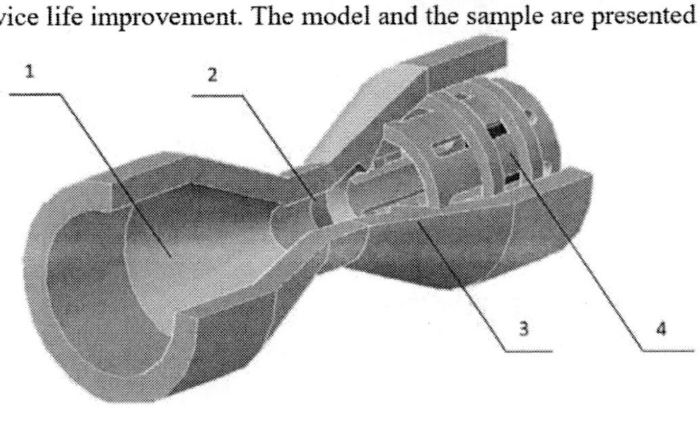

Figure 1. The hydrodynamic cavitation reactor model: 1 – confuser, 2 – processing chamber, 3 – diffusor, 4 – slit-type cylinder arrangement.

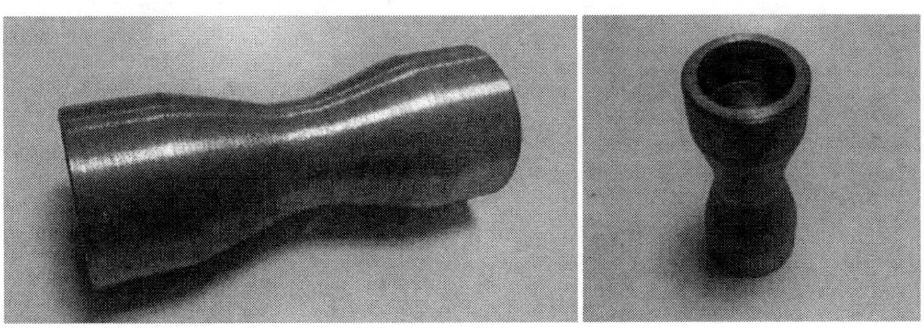

Figure 2. The hydrodynamic cavitation reactor sample.

The experiment was carried out on the testing bench, shown in Figure 3, which consisted of the pumping equipment, a pipeline that pumped the thick liquid, the cavitation equipment and the temperature and pressure sensors.

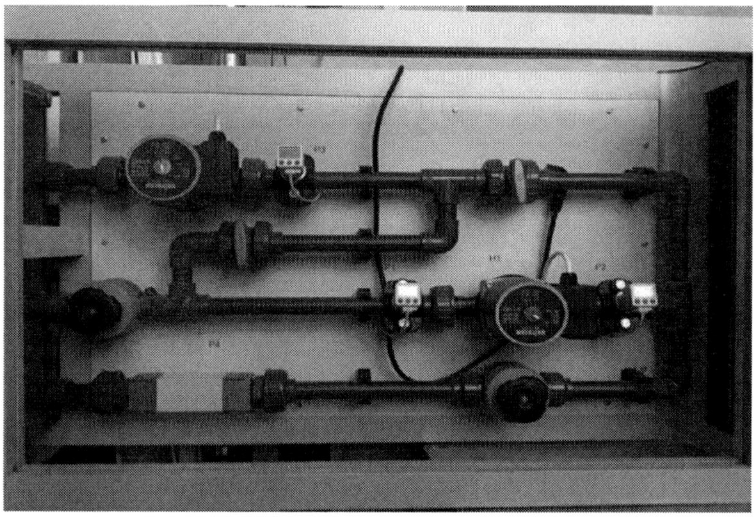

Figure 3. The laboratory unit.

The cavitation reactor metal weight loss was measured to evaluate the cavitation erosion effect and the coating efficiency.

3. Results and Discussion

Figure 4 and table 1 show the metal weight loss of the cavitation reactor with the protective coating (2) and without it (1) and its dependence on the time.

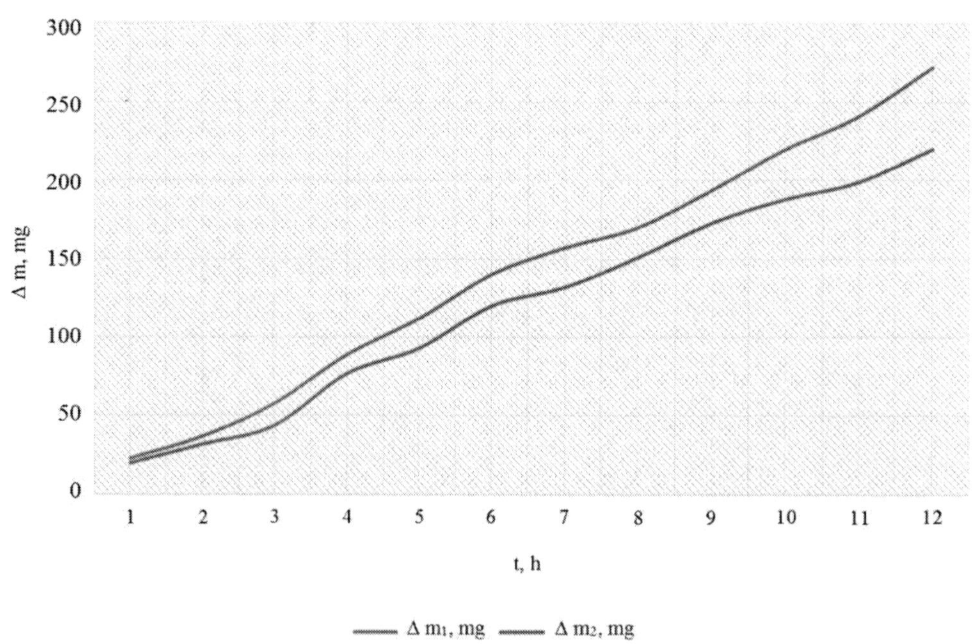

Figure 4. The metal weight loss of the cavitation reactor with protective coating (2) and without it (1) the and its dependence on the time.

Table 1. The cumulative difference in metal weight loss

Time, h	$\sum(\Delta m_1\text{-}\Delta m_2)$, mg
0-2	2
3-4	16
5-6	41
7-8	83
9-10	129
11-12	192

Figure 4 shows that the coating increases the erosion resistance of the construction and reduces the metal weight loss. It can be observed that the reactor with the protective coating has the highest corrosion resistance since the lowest weight was lost during the exposure period. After 2 hours of test, the reactor inner surface starts presenting plastic deformation, after 4 hours, the plastic deformation becomes more severe. The cracks are initiated from the internal defects of the material and along the weakest spots. After 8 hours surface cracks join each other, increase erosion rate.

Figure 5 and table 2 represents the dependence of metal loss on the periods of cavitation for the uncoated and coated reactor.

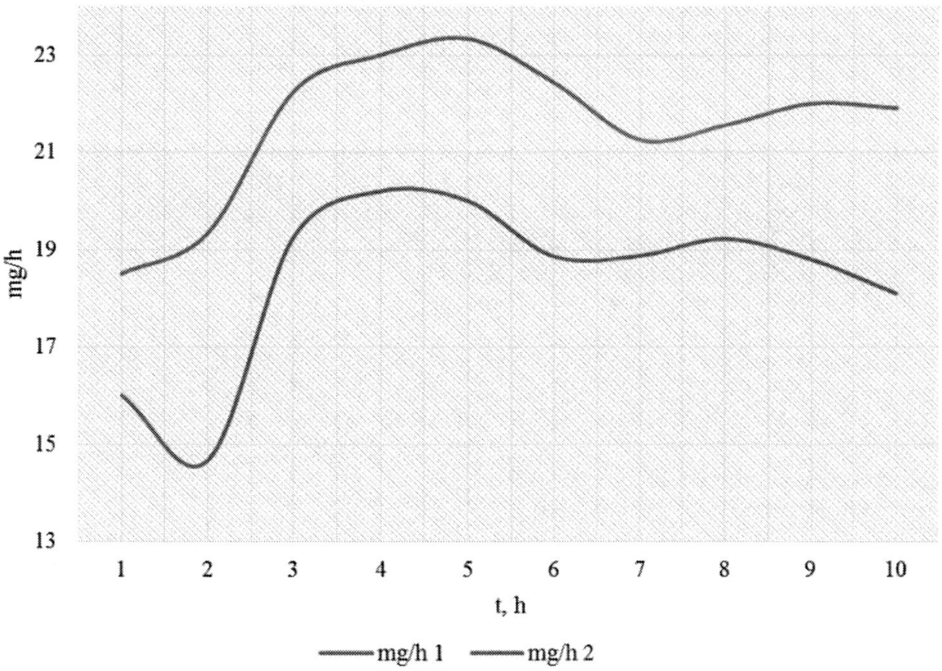

Figure 5. Metal weight loss according to the periods of cavitation.

The figure illustrates that the metal weight loss rate decreases and thereafter increases to a given value, and then again decreases. The coating increases the incubation (first) period time, during which the minimum metal weight loss and the plastic deformations accumulation occur. These results are in good agreement with other conducted researches [15-18].

Table 2. The periods of cavitation

Period	Time, h
1	0-2
2	2-6
3	6-10

The metal weight loss results according to the periods of cavitation can be roughly divided into 3 stages: an incubation (first) period where little weight loss occurs, a transition period with the increasing weight loss rate and finally a stable period during which the weight loss rate is approximately constant.The figure demonstrates that the coating application has a positive effect on the service life. On the basis of the conducted experiment and scientific papers analysis, the optimum coating composition is the mixture of titanium - 55%, nickel 25% and copper 20%.

4. Conclusion

The service life and the processes reliability are the most important technical parameters in the industry. The careful approach to the selection of the construction material brand, used in the cavitation treatment, and the coating materials, protecting the base metal from the rapidly developing cavitation erosion, can enable us to achieve the high rates of performance. The results obtained in the experiment and the scientific literature analysis have shown that the most optimal way is the usage of the titanium, nickel and copper mixture. The given results can be used in the engineering design of the pumps and pipelines by the research and design institutes in the petroleum industry.

References

[1] Brand A E. Vershinina S V, Vengerov A A and N A Mostovaya 2015 *IOP Conf. Ser.: Mater. Sci. Eng.* **93** 012005
[2] Bordeasu I, Popoviciu M O, Mitelea I, Ghiban B, Ghiban N, Sava M, Duma S T and Badarau R 2014 *IOP Conf. Ser.: Mater. Sci. Eng.* **57** 012006
[3] Khvatov B N 2002 *J. Trans. of the TSTU* **8(3)** 507
[4] Tôn-Thât L 2012 *8th Inter. Sympos. on Cavitation Singapore*
[5] Vakulenko K V, Biblik I V, and Kazak I B 2014 *J. Aerospace Eng. and Tech.* **7 (114)** 116
[6] Linderov M L and Merson D L 2010 *J. Vektor Nauki of TSU* **3(13)** 43
[7] Shestoperov V Yu 2014 *J. Trans. of NSTU n.a. R.E. Alexeev* **5 (102)** 79
[8] Shestopyorov V Yu 2014 *J. Trans. of NSTU n.a. R.E. Alexeev* **5 (107)** 361
[9] Vyas B and Preece C 1977 *Metal. and Mater. Trans. A* **8(6)** 915
[10] Gireń B G and Frączak J 2016 *J. Wear* **364** 1
[11] Hattori S, Ishikura R and Zhang Q 2004 *J. Wear* **257** 1022
[12] SangKi Chi, JinHwan Park, MinYoung Shon 2015 *J. of Indust. and Eng. Chem.* **26** 384
[13] Dular M, Stoffel B and Sirok B 2006 *J. Wear* **261(5)** 642
[14] Matevž D, Bernd S and Brane Š 2006 *J. Wear* **261** 642
[15] Koukouvinis P, Bergeles G and Gavaises M 2015 *J. of Hydrodyn., Ser. B* **27(4)** 579
[16] Mitelea I, Bordeas I, Pelle M, and Craciunescu C 2015 *J. Ultrason Sonochem* **23** 385
[17] Santa J, Blanco J, Giraldo J and Toro A 2011 *J. Wear* **271** 1445
[18] Wood R J K, Wharton J A, Speyer A J and Tan K S 2002 *J. Tribol. Int.* **35** 631

The connection analysis between the dilution of the deposited Fe-Cr-V-Mo-C layer by the basic metal and the parameters of its microstructure

A S Degterev and S F Gnusov

National Research Tomsk Polytechnic University, 30, Lenina ave., Tomsk, 634050, Russia

E-mail: Degterev@tpu.ru

Abstract. In this work, the structure of the Fe-Cr-V-Mo-C coatings received by plasma transferred arc cladding was investigated. Coatings were deposited on plates with a thickness of 10 mm and made from constructional steel (steel 20). The correlation analysis of relationships between dilution of the deposited layers by the basic metal and the parameters of their microstructure was carried out. The parameters were as follows: volume fraction, a size, a shape factor, the distance between particles, the number of particles of vanadium carbide, volume fraction of the eutectic on the basis of carbide M_7C_3 and the distances between its colonies, as well as the volume fraction of the α-phase in the alloy matrix.

1. Introduction

Notwithstanding the cladding methods, their regimes should provide the minimum penetration of a processing detail. It provides the preservation of the initial chemical content of a powder composition in the deposited layer. On the other hand, the intention to minimize the dilution of the deposited layers by the basic metal (*D*) is restricted by necessity of realization of groups of important requirements for coatings. Among these are: the quality of fusion of deposited layers with the basic metal (substrate); the provision of the demanded coatings geometry, the provision of minimum porosity, absence of cracks and ruptures; high efficiency of the cladding process; and the low-loss adding material. For the foregoing reasons, the optimization task of a plasma transferred arc (PTA) cladding regime can be solved exclusively by searching for a compromise solution.

In [1] for the formation of qualitative hard-facing Fe-Cr-V-Mo-C layers, the recommendations for choosing a PTA cladding regime are given. The demand of a limitation of dilution (*D*) of the deposited layer by the basic metal at a level of 10 % is rooted in the recommendations. However, the reasons of this limitation *D* are not explained by the authors.

Articles [2-5] underline the necessity of a choice of the cladding regime of the hardened layer taking into account the concrete data about the influence of the regime on its microstructure. The classical orientation to dilution, the macro sizes and the absence of imperfections cannot be universal for all varieties of composite coatings. The realization approach to the choosing of a cladding regime, along with knowledge of its influence on a microstructure, demands labor-intensive research.

The aims of this work are to estimate the correlation force between dilution of the deposited Fe-Cr-V-Mo-C layer, which was generated by the PTA method, and the parameters of its microstructure.

This will allow defining the parameters of a microstructure which demands the individual account for the search of an optimum cladding regime.

2. Materials and experimental procedures

Coating formations were made by the plasma transferred arc welding method, using the UPN-303 standard equipment. The industrial powder 'PR-Cr18VNiMo' (PN-AN2) was used as an adding material. The chemical compositions were (wt. %): 2.1-2.4 C, 17-19 Cr, 2-3 Ni, 7-8 V, 2-2.6 Mo, Fe - the rest. The fraction of the powder was used in the range from 50 to 200 μm. The basic material (substrate) were plates with a thickness of 10 mm. The plates were made of steel 20 (wt. %: 0.17-0.24 C, 0.17-0.37 Si, 0.35-0.65 Mn).

For the coatings production of varied dilutions, two groups of experiments were performed. Group A – cladding without plasmatron lateral oscillation, and group B – cladding with plasmatron lateral oscillation. In experiments of group A – arc current (I) was varied within the limit from 220 to 260 A, cladding speed ranged (V_s) from 4.5 to 13 m/h, the rate of powder feed (V_p) was from 1.1 to 2.2 kg/h. In experiments of group B – arc current was established in a range from 220 to 260 A, V_s equaled 4.5 m/h, V_p was changed in limits from 0.7 to 2.2 kg/h, the peak-to-peak amplitudes of plasmatron oscillation (R) ranged from 10 to 30 mm, and their frequency (v) — from 0.25 to 0.5 Hz. The working distance between the plasmatron and the item was 10-13 mm, the plasma gas flow rate was 2-3 l/min, carrier gas — 13-17 l/min and up to 17 l/min of the shielding gas. Argon was used as the enumerated gases. In this paper, microsections preparation was performed in the traditional fashion, through mechanical grinding and mechanical polishing on the diamond paste of various dispersibility.

A microstructure of deposited layers was investigated in section I (figure 1(a)) through the use of an optical inverted microscope, Olympus GX51.

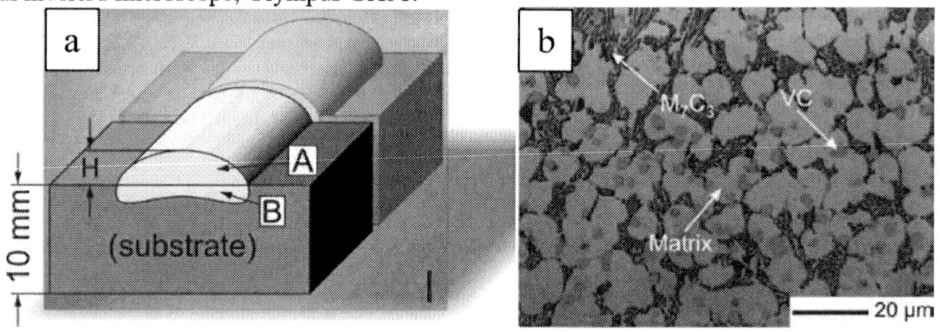

Figure 1. The specimen cutoff scheme (a) and the microstructure of coatings (b).

Macroetching of samples were executed with a 4 % solution of HNO_3 in C_2H_5OH at room temperature during 15-30 s. Dilution (D) was computed according to the following equation [6]:

$$D(\%) = \frac{B}{A+B}100 \quad (1)$$

In the above-stated equation, B is the area of penetration, mm^2, A is the deposit area without penetration, mm^2.

Particles of vanadium carbide were determined by relief polishing with the following application of microscopy methods: dark-field illumination and differential interference contrast (DIC). To detect eutectic carbide of the M_7C_3 type, the following etching agent was applied: 8.3 g $K_3[Fe(CN)_6]$, 8.3 g KOH, 83.4 ml H_2O. The etching was performed by dipping samples at room temperature for 60-90 sec. Martensite of a matrix was determined by means of a reactant: 20 ml HCl, 1 g $CuCl_2$, 15 ml H_2O, 65 ml C_2H_5OH (etching time from 2 to 20 min). The quantitative analysis of structural elements was performed with application of the analyzer 'SIAMS 700'. The following parameters were measured: the average volume fraction (V_{VC}), the size (d_{VC}, diameter), the number of particles (N_{VC}) related to the unit of the area of the section (10^4 μm^2), the shape factor (f_{VC}), the distance between particles VC (H_{VC}), the eutectic volume fraction on the basis of carbide M_7C_3 (V_{eut}) and the average distances between its colonies (H_{eut}). The shape factor was computed as a ratio of an orthogonal maximum projection of a particle to the maximum projection of a particle. For the definition of the average rates of parameters under consideration of a microstructure on the transverse microsection from the fusion border to the top, the track of pricks (markers) was pierced by a diamond pyramid with a 300 μm step. To the left and to the right of each marker, a microstructure was analyzed. For an estimation of linear dependences, a Pearson's coefficient of correlation was defined according to the following equation:

$$r_{xy} = \frac{\sum_{i=1}^{n}\left(X_i - \overline{X}\right)\left(Y_i - \overline{Y}\right)}{\sqrt{\sum_{i=1}^{n}\left(X_i - \overline{X}\right)^2 \sum_{i=1}^{n}\left(Y_i - \overline{Y}\right)^2}}. \quad (2)$$

In the above-mentioned equation, $\overline{X}$ and $\overline{Y}$ are the arithmetical means of two datasets: $(X_1...X_n)$ and $(Y_1...Y_n)$. Variable *X* was *D*, and variable *Y* was each of the microstructure parameters. Correlation force was estimated according to the Cheddok's scale: $0.1 \le r_{xy} < 0.3$ – weak correlation; $0.3 \le r_{xy} < 0.5$ – moderated correlation; $0.5 \le r_{xy} < 0.7$ – appreciable correlation; $0.7 \le r_{xy} < 0.9$ – high correlation; $0.9 \le r_{xy} < 1.0$ – very high (strong) correlation.

In researching the phase composition of coatings, X-ray diffraction analysis was carried out by a Shimadzu XRD 6000 diffractometer with CuK$_\alpha$ radiation in a scan mode in the range of corners 2θ from 30 to 110° with step 0.1°. Well-known card files were used during qualitative phase analysis. Integral intensity rates of the diffractional maximum (111) γ-phase and (110) α-phase were used in the quantitative phase analysis. Hardness of samples was measured by the Rockwell C method (HRC in TK-2M equipment).

3. Results and discussion

The microstructure of the coatings, according to optical microscopy data and X-ray diffraction analysis, consists of disperse particles of vanadium carbide and eutectic skeletal colonies on the basis of carbide M_7C_3 (figure 1 (b)). The matrix in deposited layers is presented by α - and γ - solid solutions (figure 2).

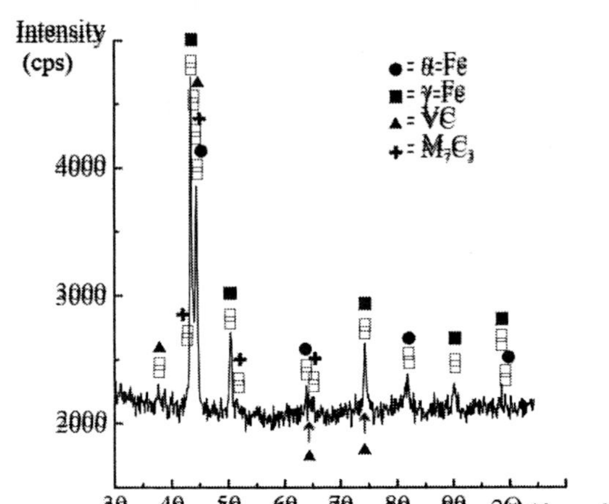

Figure 2. X-ray diffraction pattern of the Fe-Cr-V-Mo-C coating.

In figure 3 (a), there is a dependence of the α-phase volume fraction in the alloy matrix (V_α) on dilution (*D*). Its analysis allows emphasize 3 characteristic areas for the groups of experiments A and

B. In area I, the α-Fe volume fraction in the matrix decreases from 45 to 20 % with the increase of dilution ($1.5 \leq D \leq 18$ %). In area II ($18 < D \leq 28$ %), V_α does not vary. The further increase of D to 50 % is accompanied by an intense increase of the α-Fe volume fraction in the alloy matrix to 75 %. The difficult nature of communication of the rates is explained by the behavior of two competing processes. The first process is the increasing of the degree of eutectic carbide and vanadium carbide dissolution in a molten pool with the growth of heat input in the work part. As a result, it provides a high content of Cr, V and C in the austenite coating after molten pool crystallization, which greatly reduces martensite start temperature (M_s) and martensite finish temperature (M_f). The second process is the increasing of dilution with an increase of the heat input. It provides the common concentration reduction of doping elements in a coating and, consequently, the rise of the M_s temperature. In the area I, the dissolution process makes a greater contribution to change V_α than the dilution process does. In area II, the dissolution and dilution processes balance each other out, and in area III, the dilution process dominates.

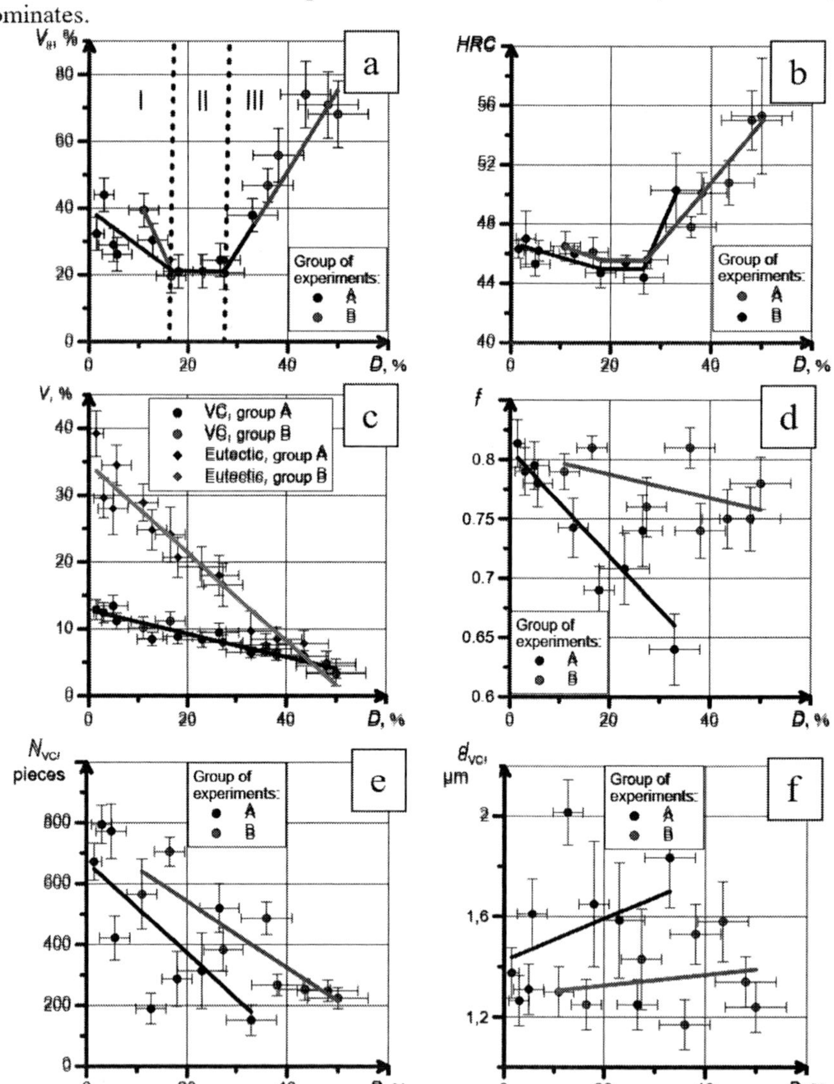

Figure 3. Dependences of V_α (a), HRC (b), V_{VC} and V_{eut} (c), f_{VC} (d), N_{VC} (e) and d_{VC} (f) on dilution (D).

With the D growth, the volume fraction of inclusions of VC and eutectic on the basis of M_7C_3 decreases from 13.5 to 3.5 % and from 39.0 to 3.5 % (figure 3 (c)). This decrease is explained by the two processes described above. The connections of V_{VC} and V_{eut} with D, according to the joint data of experiments of A and B groups, are well described by the first degree polynomials of the following types:

$$V_{VC} = 12.81993838 - 0.1752292615D, \tag{3}$$

$$V_{eut} = 34.66331883 - 0.6607581353D. \tag{4}$$

The intensity of the decrease of an eutectic volume fraction on the basis of M_7C_3 with growth D is considerably higher than the intensity of decrease of V_{VC}. According to the data [7], M_7C_3 is completely dissolved in austenite in the temperatures range of 950-1150°C. Vanadium carbide is dissolved restrictedly (1.5-2 %) in austenite at 1100-1150 °C. However, in the alloy under consideration, M_7C_3 is less refractory and contains up to 15 % of V. M_7C_3 saturates the hard solution with vanadium in the process of dissolution. In liquid steel it can be dissolved up to 3 % of VC. This information indicates that the reduction of the volume fraction of the eutectic on the basis of M_7C_3 is carried out both at the expense of the dissolution process and the concentration reduction of doping elements in the coating. The V_{VC} reduction occurs mainly at the expense of dilution.

Correlation coefficients take values of 0.95 and 0.97 for the first and the second equations, accordingly. This proves the very high correlation of the VC volume fraction and the eutectic volume fraction on the basis of carbide M_7C_3 with dilution. The equations can be used for prediction of the above-stated parameters of the microstructure coatings, generated by plasma-cladding of powder PR-Cr18VNiMo and other similar Fe-Cr-V-Mo-C alloys. The hardness of deposited layers depends a lot on a α-Fe volume fraction in their matrix (a correlation coefficient is 0.92) (figure 3 (b)).

Figures 3 (d, e, f) and 4 (a, b) present the graphs of dependences f_{VC}, d_{VC}, N_{VC}, H_{VC}, H_{eut} on D, and table 1 shows the estimations of the correlation force. The average diameter of carbide vanadium particles does not depend on dilution. The average shape factor and the average number of VC particles in comparison with V_{VC} and V_{eut} correlates with D comparatively weakly.

Table 1. The estimation of the correlation force of microstructure parameters with dilution (D)

VariableY	Correlation coefficient r_{xy}			The correlation force
	The group of regimes			
	A	B	A+B	
V_{VC}	0.90	0.97	0.95	Very high
V_{eut}	0.94	0.98	0.97	Very high
f_{VC}	0.91	0.52	0.32	Moderated
N_{VC}	0.70	0.88	0.68	Appreciable
d_{VC}	0.33	0.20	0.06	Absent
H_{VC}	0.78	0.89	0.81	High
H_{eut}	0.92	0.92	0.91	Very high

The morphology and the number of vanadium carbide particles are defined by the amount and correlation of C and V in an alloy [8, 9], and by the degree of molten pool supercooling [10, 11]. With the growth of supercooling, the critical radius of a nucleus reduces, as well as the mean diameter of VC particles crystallized from the melt first decreases and the number of particles grows. Also, the disperse VC particles (the diameter is less than 1 μm) which are situated in the adding powder in great numbers (more than 1500 pieces per 10000 μm^2) and which have not been dissolved less in a molten pool coagulate as a result of the decrease of residence time at a high temperature. The shape of VC particles with the growth of supercooling approaches to the equiaxed one. The reduction of the V and

C concentration in a coating as a result of its dilution leads to a decrease of number of VC particles and their elongation. The PTA method allows the formation of the coatings during various thermal cycles. Thus, at the same (*D*), the degree of molten pool supercooling can vary considerably.

The distances between VC particles and eutectic colonies on the basis of carbide M_7C_3 increased with the growth of dilution (figure 4). Approximating polynomials are the following:

$$H_{VC} = 7.346399084 + 0.2566490277D, \quad (5)$$

$$H_{eut} = 3.232293933 + 0.2928848323D. \quad (6)$$

Correlation coefficients for the first and second equations take values of 0.81 and 0.91. The strength of the positive correlation is high.

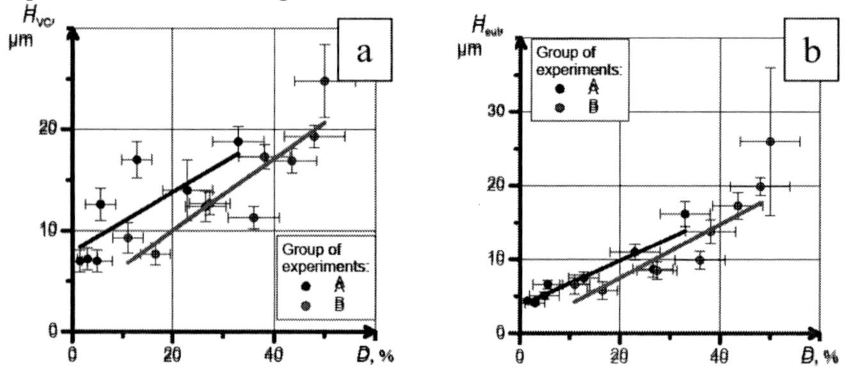

Figure 4. Dependences of the distances between VC particles (H_{VC}) (a) and eutectic on the basis of M_7C_3 (H_{eut}) (b) on dilution (*D*).

Distinction in correlation coefficients of particle number N_{VC} (0.68) and distance H_{VC} (0.81) with *D* is explained by changing a matrix's grains size, which mainly has VC spreading along them, as well as by changing this spreading character. With the growth of the dilution, the matrix's grains size increases, the ratio of the number of highly rigid inclusions located along the grains boundary to inclusions located inside of grains grows. The regime of PTA cladding is capable of providing the disperse VC particles obtained in a comparatively great amount even at greater *D*, however the distance between these particles will be considerable as well.

4. Conclusions

Correlation between the average volume fraction of VC particles or the eutectic volume fraction on the basis of carbide M_7C_3 and the dilution of the deposited layer by the basic metal (*D*) is linear, negative and very high. Correlation of the average distances between VC particles or eutectic colonies M_7C_3 with *D* is positive and high. The number, the shape, and the diameter of VC particles, as well as the α-phase volume fraction in a matrix are weakly correlated with the dilution. These rates are defined by a thermal welding cycle. HRC Hardness in the group of alloys under consideration depends a lot on the α-phase's content in their matrix.

References

[1] Perepletchikov E F , Ryabtsev I A and Gordan G M 2003 *J. Avtom. Svarka* **3** 21-5
[2] Schevchenko O I 2003 *J. Weld. Int.* **17(2)** 139-46
[3] Keränen M 2010 *Effect of Welding Parameters of Plasma Transferred Arc Welding Method on Abrasive Wear Resistance of 12 V Tool Steel Deposit* (Helsinki: Aalto University) p 166
[4] Ilo S, Just C and Xhiku F 2012 *J. Mater. and Des.* **33** 459-68
[5] Gruzdys E and Meškinis Š 2011 *Mater. Sci. Medz* **17(2)** 140-4
[6] Balasubrarnanian V, Lakshminarayanan A K, Varahamoorthy R and Babu S 2009 *J. of Iron and*

Steel Res. Int. **16(1)** 44-53
[7] Bilek P, Sobotova J and Jurci P 2011 *J. Mater. in Tehn.* **45(5)** 489-93
[8] Wei S, Zhu J and Xu L 2005 *J. Mater. Science and Eng. A* **404** 138-45
[9] Wei S, Zhu J, Xu L and Long R 2006 *J. Mater. and Des.* **27** 58-63
[10] Yan F, Xu Z, Shi H and Fan J 2008 *J. Mater. Charact.* **59** 592-7
[11] Luan Y K, Song N N, Bai Y L, Kang X H and Li D Z 2010 *J. Mat. Process. Technol.* **210** 536-41

Study of the Ti-5Al-5Mo-5V-1Cr-1Fe titanium alloy grain structure uniformity after bending and annealing

A V Balaykin, E A Nosova and N V Galkina

Samara University, Russia, 34, Moskovskoye Shosse, 443086, Samara, Russia

E-mail: nata12_92@mail.ru

Abstract. The uniform grain structure of titanium alloys is required to enhance the product manufacturability and precision. For the research, flat blanks cut out of the BT-22 titanium alloy forged bar were used. Bending tests with the subsequent annealing were carried out at the temperature of 800°C during 1–5 h. The grain size change and its uniformity depending on the annealing duration were analysed. The possibility of obtaining the uniform fine-grained structure after deformation and annealing for 4 hours was established.

1. Introduction

The stability of the manufacturing and machining process for metal alloy parts depends on the continuity and uniformity of the grain structure [1, 2]. Recently, titanium alloys have become increasingly applicable in the deformed condition owing to the developments connected with the thermal and deformation processing technologies [3]. When comparing between a cast and the deformed structure and the structure after the selective laser sintering of powder metallic materials [4], the deformed structure appears to be more advantageous. Such defects as porosity, non-uniformity of grains, sweating can be observed in the cast structure, while porosity, cracks, lack of penetration are characteristic of the selective laser sintering [5-8]. The structural change of the deformed semi-finished products as well as provision of the necessary structure depend on the coordinated combination of the deformation and thermal processing conditions [9, 10].

This work is aimed at analysing the grain structure uniformity in bending of forged Ti-5Al-5Mo-5V-1Cr-1Fe titanium-alloy blanks with the subsequent annealing at the temperature of 800 °C and the duration of 1, 2, 3, 4 and 5 hours.

2. Methods and materials

Samples with the dimensions of 2x10x30 mm were cut out of the bar (AMTU 451-67) using the electroerosion cutting method. The chemical composition of the alloy according to GOST 90013-81 [11] is given in table 1.

After cutting out, the samples were subjected to the bend deformation in the tool-making die with the angle at the vertex of about 90 ° and the rounding radius of 10 mm. After bending, the samples were annealed at the temperature of 800 ° C for 1, 2, 3, 4 and 5 h [12, 13].

Table 1. The chemical composition of Ti-5Al-5Mo-5V-1Cr-1Fe

Element' wt. %											
Fe	C	Si	Cr	Mo	V	N	Ti	Al	Zr	O	H
0.5 – 1.5	Up to 0.1	Up to 0.15	0.5 – 2	4 – 5.5	4 – 5.5	Up to 0.05	78.485 – 86.6	4.4 – 5.9	Up to 0.3	Up to 0.2	Up to 0.015

Later, microsections were manufactured of the samples obtained. Etching was performed using the following mixture: 180 ml of H_2O, 180 ml of HCl, 120 ml of HNO_3, 30 ml of HF. The duration of etching was 1-2 min; washing was carried out using running water followed by drying with filter paper. After etching, the microstructure was analysed using the metallographic METAM LV-32 tool-maker microscope. After photographing the microstructure, the uniformity of the grain structure was evaluated. To this end, 10 adjacent grains were detected in the figure along with measuring their smallest and largest size (figure 1).

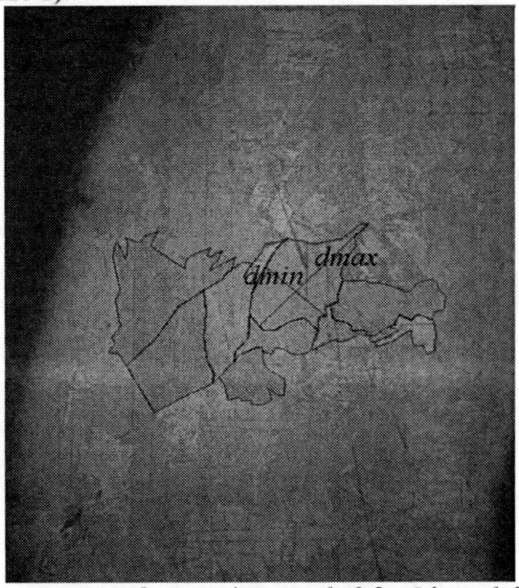

Figure 1. The microstructure of a sample annealed for 5 h, and the grain selection and size measurement pattern.

Then, using the Microsoft Excel features, the maximum and the minimum values of the average grain size as well as their standard deviation were determined. Based on the received data, the number of intervals and the pitch were calculated. After that, the probability density distribution graphs for the grain size were built. As a result, the probability density vs the medium-size interval graph was obtained (figure 2).

Using the obtained data, the variety of the grain size was calculated using the following formula: $\beta = \frac{\Delta d_{gr}}{d_{prob} \cdot P}$, where Δd_{gr} is the half-height grain size change, d_{prob} – the most probable grain size, P – the probability value.

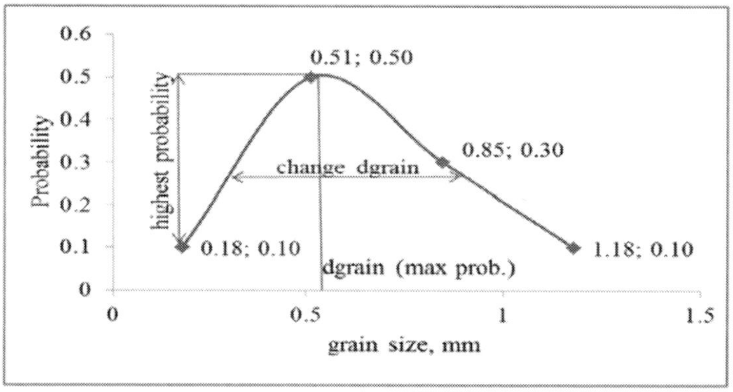

Figure 2. The probability density vs the interval for the sample annealed for 5 h

3. Results and discussion

The diagram of the grain size and grain structure non-uniformity vs annealing time is given in figure 3.

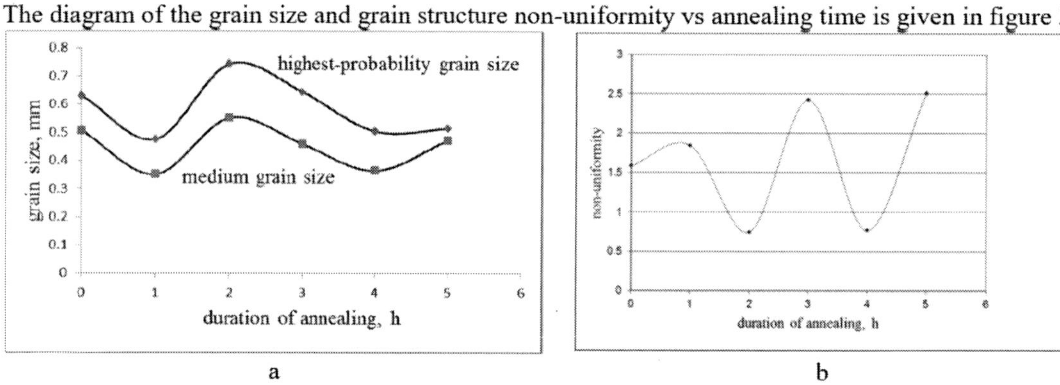

Figure 3. Grain structure change dependency: a – grain size, b – non-uniformity

Medium and highest-probability grain sizes have a similar nature of changes with the increased annealing duration. At the initial annealing stage, a minor grain structure size reduction takes place followed by the increase after the 2-hour holding. Further holding time extension results in the grain structure size reduction, which reaches the minimum after 4-hour holding. Then, a minor grain size increase is observed.

The non-uniformity and annealing time graph has a wavy form of changing. The change is insufficient within the interval from 0 to 1 h, an approximate two-fold non-uniformity reduction occurs within the interval of 2 h. The non-uniformity increases 3 times within the interval of 2-3 h; then it decreases 3.3 times after 4 h again, and increases 3 times for 5 h.

As a rule, the grain structure uniformity change is connected with the recrystallization processes. The data provided by the sources [14-16] indicate that the recrystallization threshold of titanium alloys largely depends on the extent of the preceding deformation. Before the beginning of recrystallization all the processes connected with the structure change take place inside grains, therefore its grain size and uniformity shall not change. The reduced non-uniformity may be connected with polygonal cells appearing inside the grains, which may be observed as particular grains.

Formation of new equiaxed grains on the background of a deformed structure shall result in the grain size variety, which is seen in figure 3 with the holding time of 2 h and more. Annealing for 3 h resulted in the highest grain structure non-uniformity, which becomes minimal after 4 h of holding. This may evidence the initial recrystallization completion. Holding during annealing for 5 h increases the grain structure non-uniformity again, which may represent a secondary recrystallization manifestation.

Rebuilding of the grain structure may cause changes of the residual macro- and microstresses. As we see from figure 1, 6-8 grains fit within the plate thickness of 1 mm, which is indicative of a rather large grain, and the effect of recrystallization on both residual micro- and macro- stress. Change of residual microstress, represented in the study [12], shows that the annealing lasting for up to 3 hours at the temperature of 800 °C causes the increase of the residual compression stress both in the tension and compression zone. Increasing the holding time up to 4 h results in the decreased compressing microstress. This may result from the growth of new non-deformed grains, which reduces the moving stress for the grain boundary movement. Macrostress increases with holding within 2 to 3 h, which is supposed to be connected with the beginning of initial recrystallization. The holding time extended further will not cause critical changes in macrostresses at the beginning of the subsequent recrystallization stage as well; furthermore, the change of the grain structure non-uniformity allows evaluating time intervals of each stage.

4. Conclusions

Evaluating the non-uniformity allowed establishing the sequence of structural changes occurring during the annealing. This study allowed establishing the fact that when annealing the Ti-5Al-5Mo-5V-1Cr-1Fe titanium alloy at the temperature of 800 °C during 4 h, the smallest grain size, which possesses the minimal non-uniformity, is formed, i.e. the most preferable structure is created.

Acknowledgements

These studies were conducted using the equipment of CAM in the center of collective usage (RFMEFI59314X0003). This work was supported by the Ministry of Education and Science of the Russian Federation in the framework of the implementation of the Program 'Research and development in priority directions of the scientific-technological complex of Russia for 2014-2020'.

References

[1] Salischev G A 1996 *Formation of submicrocrystalline structure in titanium at plastic deformation and its effect on the mechanical properties* No 4 (Metals) pp 86-91

[2] Wang Y B, Liao X Z, Zhao Y H, Lavernia E J, Ringer S P, Horita Z, Langdon T G and Zhu Y T 2010 *Mater Sci Eng. A.* **527** 4959-4966

[3] Il'in A A, Skvortsova S V and Mamonov A M 2008 *Mater. Sci.* **44**(3) 341

[4] Smelov V G, Sotov A V, Agapovichev A V and Tomilina T M 2016 *IOP Conf. Series: Mater. Sci. Eng.* (Yurga) **142** (Yurga: Institute of Physics Publishing) p 115

[5] Smelov V G, Sotov A V and Agapovichev A V 2016 *Ferrous Metals* **9** 61-65

[6] Gu D and Shen Y 2006 *Mater Sci & Eng A.* **435** 54-61

[7] Zlenko M A, Popovich A A and Mutylina I N 2013 *Additivnye tekhnologii v mashinostroenii (Additive technologies in mechanical engineering)* (Saint Petersburg: Publishing House of Polytechnical University) p 210

[8] Smelov V G, Vdovin R A and Golanov S P 2015 *App. Mech. Mater.* **770** 223-228

[9] Zherebtsov S, Salishchev G, Galeyev R and Maekawa K 2005 *Plastic Deformation Materials Transactions* **46** 2020-2025

[10] Vdovin R A, Smelov V G, Bolotov M A and Pronichev N D 2016 *IOP Conf. Series: Mater. Sci. Eng.* (Yurga) **142** (Yurga: Institute of Physics Publishing) p 242

[11] GOST 90013-81. Alloys of titanium. Moscow, 1999

[12] Balaykin A V, Nosova E A and Galkina N V 2016 *IOP Conf. Series: Mater. Sci. Eng.* (Yurga) **142** (Yurga: Institute of Physics Publishing) p 142

[13] Nekhoroshev M V, Pronichev N D and Smirnov G V 2014 *Open Mech. Eng. J.* **8** pp. 436-440

[14] Leyens C and Peters M 2003 *Titanium and Titanium Alloys* (Copyright WILEY-VCH Verlag GmbH & Co. KGaA, Weinheim) pp 26-27

[15] Kolbasnikov N G, Matveev M A and Mishnev P A 2016 *Met .Sci. H. Tre.* pp 1-7
[16] Gorelik S S 1981 *Recrystallization in metals and alloys* (Moscow: Mir) pp 25–30

The effect of cooling speed on the structure and properties of the heat affected zone in welded compounds of ferrite-austenitic steels

I L Gonik, D N Gurulev and O P Bondareva

Volgograd State Technical University, 28, Lenina ave., Volgograd, 400005, Russia

E-mail: rct2005@yandex.ru

Abstract. Such parameters as the maximum heating temperature, duration of stay at high temperatures, the rate of cooling influence greatly the structure and properties of the heat-affected zone of welded joints of steels and alloys. In the present work, the effect of different cooling speed upon the impact of the thermal cycle of welding on the structure, the fine structure and toughness of ferrite-austenitic steels is investigated. It is established that the cooling speed after welding has a great influence on the shock impact toughness, the phase composition and the structure of the zone of ferrite-austenitic steels.

1. Introduction

The parameters of the thermal cycle of welding have great influence on the structure and properties of the heat-affected zone of welded joints of steels and alloys: the maximum temperature, the duration of stay at high temperatures, the speed of cooling. The scope of application of ferrite-austenitic steels is narrower than that of chromium-Nickel austenitic steels due to their embrittlement at welding temperatures and exploitation temperatures. Under the influence of the heat-affected zone, the intensive growth of ferritic grains is observed in the area of overheating as well as changing of the ratio of the phases and disintegration of unstable structures [1-3].

2. Materials and methods

The influence of cooling rates on the structure, the fine structure, and toughness of ferrite-austenitic steels Fe-0.08%C-22%Cr-6%Ni-1%Ti and Fe-0.08%C-18%Cr-8%Mn-2%Ni-1%Ti of industrial melting according to GOST 5632–72 under the heat-affected zone impact was investigated.

The study was carried out on the samples cut from the parent metal in the delivery condition, with subsequent simulation of the thermal cycle of welding on the installation of high-speed electric heating. Simulation of the thermal cycle of welding was conducted on the samples sized as 12 x 12 x 60 mm.

The sample was rigidly fixed by current-carrying copper water-cooled grips of the installation and was heated by passing current through the sample. The maximum heating temperature was 1300 °C, the heating rate was 100 °C/s, the cooling speed was 2, 10, 40, 100 °C/s. During the heating process, the heat affected zone of a welded joint was imitated, performed by automatic welding with submerged arc and electroslag welding.

The temperature at the center of the sample was controlled with the aid of chromelalumel thermocouple calked in the sample. The entry mode was carried out using a potentiometer 'KSP-4' included in the control circuit. The speed of cooling was controlled via additional heating and in the air. The samples with shock type 11 [4] were produced out of the specimens with the heat affected zone. The incision was performed at the place of fastening of the thermocouple in the center of the sample.

Diffractometer DRF-2 was used for structural quantitative analysis of the metal. Hardness was determined according to Vickers (TPP-2) (GOST 2999-75) and microhardness of structural components was found out on the device 'PMT-3' according to (GOST 9450-76). The microstructure of the samples was studied on the microscope 'MIM-8'. Fractographic study was carried out using electron microscopy in the fractures percussion samples in a transmission microscope 'UEMB - 100K'.

The structure was revealed by electrolytic etching in oxalic acid. To separate the phases, staining reagent of Groesbeck was used. The ratio of α- and γ- phases was determined by metallographic and x-ray diffraction analysis according to standard techniques [5]. Fractographic investigation of the surface of impact specimens tested at a temperature of minus 40°C, was carried out at the maximum and minimum cooling rates of 100 and 2°C/s.

3. The study of the structure of steels Fe-0.08%C-22%Cr-6%Ni-1%Ti and Fe-0.08%C-18%Cr-8%Mn-2%Ni-1%Ti

The microstructure of the samples of steel Fe-0.08%C-22%Cr-6%Ni-1% Ti under the delivery condition and after simulation of the thermal cycle of welding is shown in Figure 1.

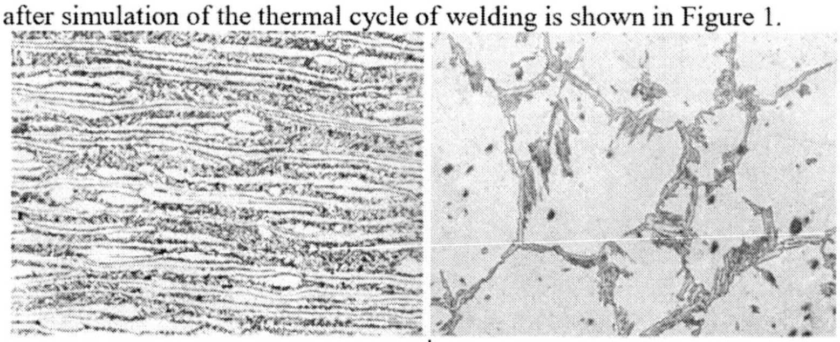

a | b

Figure 1. The microstructure of steel Fe-0.08%C-22%Cr-6%Ni-1%Ti under the delivery condition (a) and the heat affected zone, the cooling rate of 40 °C/s (b), ×100.

Microfractogrammes of steel specimens Fe-0.08%C-22%Cr-6%Ni-1%Ti after exposure to the thermal cycle of welding with different cooling rates are shown in Figure 2.

a | b

Figure 2. The fractures of steel Fe-0.08%C-22%Cr-6%Ni-1%Ti after exposure to the thermal cycle of welding at cooling of 2 °C/s (a), and cooling to 100 °C/s (b), x4500.

The results of the tests for impact toughness are presented in table 1. The fractogrammes of the samples cooled at a rate of 2°C/s are characterized mainly by ductile fracture. There are major foil allocations of carbides, clusters of carbides of dendritic form, a large selection of carbonitrides in the fracture.

Table 1. Impact toughness of steels Fe-0.08%C-22%Cr-6%Ni-1%Ti and Fe-0.08%C-18%Cr-8%Mn-2%Ni-1%Ti after the influence of the heat-affected zone (T_{max} = 1300 °C)

Type of Steel	Cooling Speed under heat affected zone, °C/s	Shock impact toughness, KCV, MJ/m^2 at temperatures, °C		
		- 40	- 20	20
Fe-0.08%C-22%Cr-6%Ni-1%Ti	base metal	0.84	1.15	1.35
	2	0.15	0.18	1.10
	10	0.18	0.50	1.20
	40	0.30	0.61	1.30
	100	0.32	0.84	1.40
Fe-0.08%C-18%Cr-8%Mn-2%Ni-1%Ti	base metal	0.89	0.94	1.10
	2	0.16	0.21	1.00
	10	0.19	0.40	1.00
	40	0.24	0.65	1.30
	100	0.32	0.87	1.55

The cooling speed has a significant influence on the impact strength of the investigated steels. With the increase in cooling speed, shock impact toughness also increases.

The minimum values of the impact strength were obtained at the cooling speed of 2 °C/s, characteristic of the electroslag welding method at all the temperatures tested. At this cooling speed, the toughness of both steels in the temperature test of minus 40 °C, minus 20 °C is less than 0.20 MJ/m^2. The increase in the cooling speed of up to 10 °C/s increases shock impact toughness up to ~0.50 MJ/m^2 for steel Fe-0.08%C-22%Cr-6%Ni-1%Ti and up to 0.40 MJ/m^2 — for steel Fe-0.08%C-18%Cr-8%Mn-2%Ni-1%Ti at the test temperature of minus 20 °C. The values of shock impact toughness remain low – less than 0.20 MJ/m^2 for both steels at the test temperature of minus 40°C. When exposed to the heat-affected zone with the cooling speed of 40°C/s, shock impact toughness rises a bit at all the temperatures tested.

The highest shock impact toughness for steels Fe-0.08%C-22%Cr-6%Ni-1%Ti and Fe-0.08%C-18%Cr-8%Mn-2%Ni-1%Ti was obtained at cooling speed 100°C/s. The duration of stay of the metal in the high temperature region increases with the decreasing cooling speed, i.e. the time for the occurrence of diffusion processes increases. Chromium from ferrite diffuses, forming chromium carbides. In the areas depleted of chromium, there is a transformation of δ - ferrite into austenite. The percentage content of austenite reaches a maximum of 25 % in steel Fe-0.08%C-22%Cr-6%Ni-1%Ti and 47 % — in the Fe-0.08%C-18%Cr-8%Mn-2%Ni-1%Ti steel at a cooling speed of 2 °C/s, which is due to the long duration of stay at high temperatures. With the increase of the cooling speed, the amount of austenite in the studied steels decreases, reaching a minimum value of 20 % in steel Fe-0.08%C-22%Cr-6%Ni-1%Ti and 35% — in Fe-0.08%C-18%Cr-8%Mn-2%Ni-1%Ti at a cooling speed of 100 °C/s. At the cooling speeds of 2, 10, 40 °C/s, a part of the austenite is transformed into martensite, the formation of which is confirmed by the increase in the microhardness of the γ' - phase (table 2), as well as the increase of the lattice parameter of the α - phase and the microstresses of the second kind (table 3).

Table 2. The results of metallography research of the destruction surface of the samples tested on a shock bend at minus 40 °C

Type of Steel	Cooling Speed under heat-affected zone, °C/s	Ferrite grain diameter, μm	The ratio of the phases, %			Microhardness, P = 20g		Hardness, HV
			γ′-phase	α-phase	δ-ferrite	δ-ferrite	γ′-phase	
Fe-0.08%C-22%Cr-6%Ni-1%Ti	2	420	25	10	65	243	384	310
	10	410	23	5	72	241	320	291
	40	400	22	5	73	244	310	281
	100	380	20	no	80	238	274	270
Fe-0.08%C-18%Cr-8%Mn-2%Ni-1%Ti	2	390	47.1	15	47.9	224	368	300
	10	381	36	10	54	218	351	280
	40	380	30	10	60	215	347	273
	100	360	35	no	65	220	342	251

Table 3. The effect of the cooling speed under the influence of the heat-affected zone (T_{max} = 1300 °C) on parameters of a thin structure of steels Fe-0.08%C-22%Cr-6%Ni-1%Ti, Fe-0.08%C-18%Cr-8%Mn-2%Ni-1%Ti

Type of Steel	Cooling Speed under heat-affected zone, °C/s	α - phases		γ′ - phases	
		Lattice Parameter, nm	The true width of the line $(211)^{\alpha}$, radian 10^{-3}	Lattice Parameter, nm	The true width of the line $(311)^{\beta}$, radian 10^{-3}
Fe-0.08%C-22%Cr-6%Ni-1%Ti	base metal	0.2858	5.1	0.3628	2.8
	2	0.2856	-	0.3626	2.1
	10	0.2856	3.7	0.3628	2.7
	40	0.2889	4.8	0.3628	2.8
	100	0.2889	5.9	0.3630	3.05
Fe-0.08%C-18%Cr-8%Mn-2%Ni-1%Ti	base metal	0.2869	6.7	0.3591	3.09
	2	0.2874	-	0.3594	1.9
	10	0.2874	3.7	0.3595	2.2
	40	0.2875	5.1	0.3596	2.4
	100	0.2876	5.6	0.3599	2.9

The amount of martensite increases with the increase of the cooling speed. The metallographic examination of the microstructure showed that at high cooling speeds of ~100°C/s, in steel Fe-0.08%C-22%Cr-6%Ni-1%Ti, secondary austenite in the form of needles is formed mainly at the boundaries of ferritic grains. In steel Fe-0.08%C-18%Cr-8%Mn-2%Ni-1%Ti, the formation of secondary austenite inside the grains is observed [6-7].

With the decrease of the cooling rate, as it has been already noted, the number of austenite increases, and at a cooling speed of 2 °C/s, separate grains of ferrite are completely transformed into austenite. The austenite in the form of a layer is situated along the grain boundaries. We measured the maximum diameter of a ferritic grain which achieves δ - ferrite in the weld heat until it starts to separate secondary austenite. With the increase of the cooling speed, the diameter of the grain of δ-ferrite in the studied steels slightly decreases: in steel Fe-0.08%C-22%Cr-6%Ni-1%Ti – from 420 μm to 380 μm, and in steel Fe-0.08%C-18%Cr-8%Mn-2%Ni-1%Ti — from 390 μm to 360 μm. The lattice parameter of the γ′ - phase and the α - phase of steels Fe-0.08%C-22%Cr-6%Ni-1% Ti and Fe-0.08%C-18%Cr-8%Mn-2%Ni-1%Ti decreases at a cooling speed of 2°C/s compared to the base metal, which indicates the process of carbide formation (table 3).

At a cooling speed of 40 and 100 °C/s, the lattice parameter of the α - phase of steels Fe-0.08%C-22%Cr-6%Ni-1% Ti and Fe-0.08%C-18%Cr-8%Mn-2%Ni-1%Ti will increase to 0.2889 nm and 0.2876 nm, respectively. The lattice parameter of the γ′ - phase with the increasing cooling speed increases in both steels as the process of carbide formation is impeded and the content of alloying elements, primarily chromium, in the solid solution increases.

The fractographic investigation of the surface of the tested samples at a temperature of minus 40 °C was carried out at the maximum and minimum cooling speed of 100 °C/s and 2 °C/s. The fractography samples cooled at a speed of 2 °C/s are characterized by mainly ductile fracture. There is large foil emission of carbides and clusters of carbides of dendritic form in the fracture.

4. Conclusion

At a cooling speed of 100 °C/s, in the fractures, the total number of precipitated carbides and carbonitrides is much smaller, and they are more fine. There are no coarse precipitates of dendritic type observed. Along with the ductile fracture, there are areas of brittle fracture.

Taking everything into account, the cooling speed after welding has a great influence on the shock impact toughness of the phase composition and the structure of the zone of ferrite-austenitic steels.

References

[1] Hubrich M A, Salnikov G A, Bondareva O P and Zhukova I I 1986 *Chemical and Petroleum Engineering* **3** 23-25
[2] Hubrich M A, Salnikov G A, and Lepilina J A 1986 *Automatic Welding* **4** 21-33
[3] Kakhovsky N I, 1975 *Welding of high-alloyed steels* 375
[4] GOST RF 9454-78, Metals. Method for testing the impact strength at the low, room and high temperature, Moscow, 1999
[5] Trykov Y P, Arisova V N, Gurevich L M, Trudov A F, Gurulev D N and Volobuev S A 2002 *Welding International* **11** 890-893
[6] Bondareva O P, Sedov E V and Gonik I L 2015 *Izvestia Volggtu. Ser. Problems of materials science, welding and strength in machine building* **5(160)** 139-142
[7] Bondareva O P, Sedov E V and Gonik I L 2016 *Chemical and petroleum engineering* **7** 45-48

Tribological properties of hydraulic fluids modified by peat-based additives

V V Ionov[1], S A Larionov[1], Ju S Sarkisov[1], N O Kopanica[1], A V Gorchkova[1], N P Gorlenko[1], N A Tzevtkov[1] and K V Ikonnikova[2]

[1] Tomsk State University of Architecture and Building, 2, Solyanaya sq., Tomsk, 634003, Russia
[2] National Research Tomsk Polytechnic University, 30, Lenina ave., Tomsk, 634050 Russia

E-mail: 038103@sibmail.com

Abstract. The paper presents physicochemical investigations of the structure and properties of a nano-modifier synthesized from peat, the local raw material subjected to pyrolysis in air-free conditions. This nano-modifying additive is a combination of various forms of nanocarbon and polar and non-polar adsorbing materials such as silica (SiO_2), calcium carbonate ($CaCO_3$) and carbon (C). Different nanocarbon forms (nanotubes, fullerenes, nanodiamonds, nanofiber, nanodispersed carbon) used in different proportions with micro and macro peat components give multifunctional properties to the synthesized nano-modifier and the ability to positively change tribological properties of hydraulic fluids and oil lubricants. Test results of type TMT-600 show that its different percentage is required to modify tribological properties of the steel tribocouple under different loading conditions. At 0.5 wt.% content of this nano-modifier, stabilization of the friction ratio and an increase of seizure load are observed.

1. Introduction

Currently, nanomodifiers are being widely used in different fields of science, technology, and production [1-6]. However, in spite of their high efficiency, the use of nanomodifiers is often restricted by synthesis and expensiveness. Therefore, today the nanomodifier synthesis methods are not only the relevant but also important and significant problem of practical and theoretical interest, which should satisfy the criteria for technological availability and efficiency, energy and resource saving, ecological safety, natural balance, biocompatibility, and economic feasibility[7-9].

In-situ resources of the Siberian region allow the researchers to develop peat-based modifiers for the property control of dry mix mortars and other composite materials. The stock of this partially renewable resource available in Tomsk region achieves 29.3 billion tons per year per 40% humidity. In terms of this factor, Tomsk region ranks second after Tyumen region.

During the thermal treatment of peat, the larger amount of organic substances is subjected to destruction. The modifiers obtained due to the thermal treatment are divided into four types depending on the treatment temperature. These types are organic modifiers (200°C); organomineral (400°C) with prevailing organic substances; mineral-organic (600°C) with prevailing mineral substances; and mineral (800°C and higher) with a low amount of organic substances [10-11].

2. Materials and methods

The synthesized TMT-600 additive is used in construction industry to control the properties of cement systems. The comprehensive analysis provides the information on the structure and properties of this additive. Pyrolized peat in the amount of 100 mg was put into a container with 5 ml of 96% ethanol and then dispersed in ultrasonic bath ST-400D at 60 W. The obtained suspension was settled for 5 min. During this period, the coarse particles precipitated on the bottom of the container, while the suspension divided into two visible layers. Two or three drops of the suspended matter were pipetted from each of the three layers (including the lowest sediment layer) and then put onto the clear surface of the specimen glass. After drying of the suspended matter, the carbon film was sprayed onto the specimen glass and then torn from it by gelatins. The obtained specimens were investigated on the transmission electron microscope (TEM) 'JEM-2100' at 200 kV of accelerating voltage and in different magnification ranges.

3. TEM investigations

TEM investigations show that the particle size varies from tenths of a micrometer to several micrometers (Figure 1).

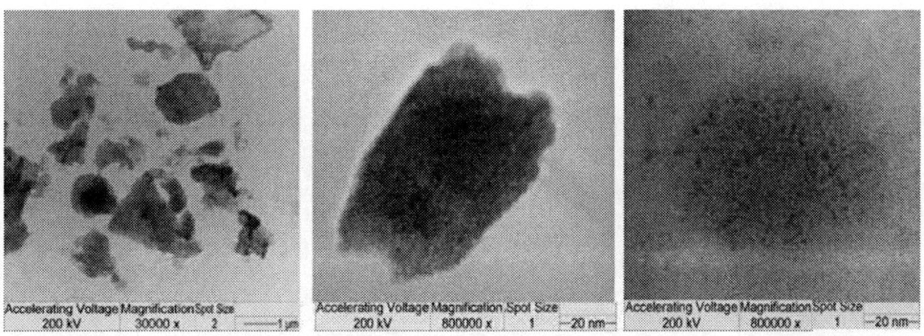

(a) (b) (c)

Figure 1. TEM images of particle formation: *a* – in sediment; *b, c* – clusters of fine particles.

Along with the formation of coarse particles, we observed the clusters of fine particles with the size of less than 5 nm (Figure 2 b, c) and elongated particles or nanofibers having the size of 200 nm and larger at their cross-section of 20 nm and less (Figure 2).

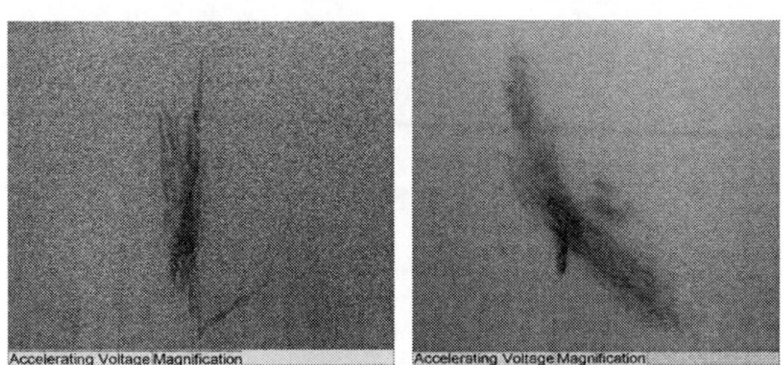

Figure 2. TEM images of nanofibers.

Isolated particles are formed in the central and upper layers of the suspension similar to those presented in Figure 1. Moreover, nanofibers 200 nm long having 2-4 nm in diameter are present in this suspension. Also, there are 10-20 nm particles with an irregular shape which adhere to these nanofibers (Figure 3).

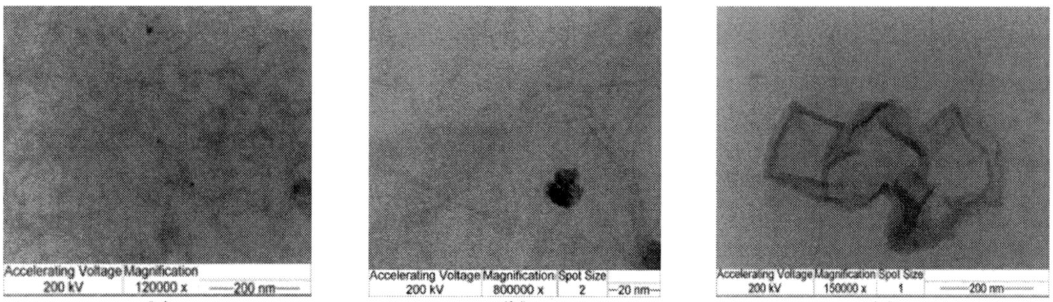

(a) (b) (c)

Figure 3. TEM images of nanofibers: *a, b* – nanoparticles; *c* – particles with an irregular shape.

In all suspension layers, the observed particles possess the clearly visible periodic structure (Figure 4) represented by nanotubes. The size of the periodic structure ranges from 1.0–1.2 nm to 4.7–4.8 nm.

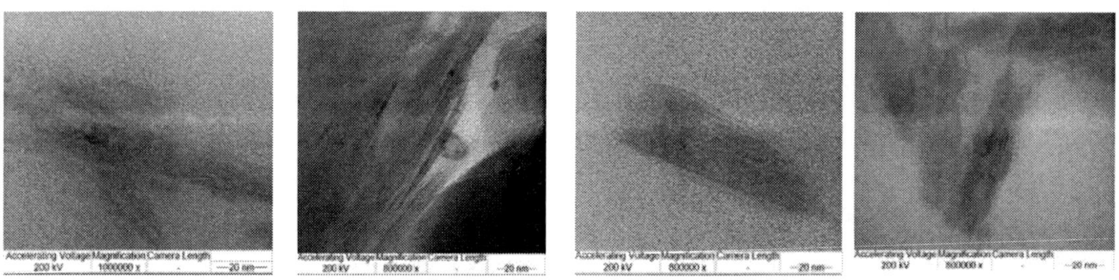

Figure 4. TEM images of nanoparticles with the periodic structure.

According to physicochemical investigations of the TMT-600 additive, it is the efficient mixture of different forms of nanocarbon including micro and macro peat components. The quantitative content of this additive is given in Table 1.

Table 1. Quantitative content of the TMT-600 modifier

Composition	Size (nm)	Content (%)
SiO_2	>400	43.81
$CaCO_3$	180	47.99
C (graphite)	15	7.77
$CaAl_2Si_2O_8$	100	0.31
C60	10-20	0.12

Carbon-based nanomodifiers have already found their use in engineering industry. As it is known, one of the dominant requirements for machining fluids is their lubricating capacity which prevents mechanisms from increased wear and frictional seizure in tribocoupling [12-13].

Nanostructure materials are characterized by the ability to modify the tribological properties of lubricants. Of great interest is the research carried out into the service properties of hydraulic oils modified by nanostructure materials.

Lubricants are estimated by such parameters as a frictional ratio, a loading capacity (at which frictional seizure of specimens occurs), and wear resistance.

The miltifactorial experiment includes the rotating roller–stationary shoe tribocoupling with immersion in oil.

The I-20-type hydraulic oil was the original material subjected to friction and wear tests. New shoes were produced for each experiment for the roller–shoe friction pair. The roller was cut from BelAZ 7540 wrist-pin made of the 15X steel type, while the shoe material was the 20 steel type. The rollers were made of 15X steel type having 60 Rockwell hardness and 0.63 surface finish. The contour area of contact was 0.55 cm^2. The experimental matrix is given in Table 2.

Table 2. Experimental matrix.

No.	Hydraulic oil	Friction pair	Test type	TMT-600 concentration in base oil (%)
1	I-20	Steel 15X – Steel 20	Antifrictionality, score resistance	
2	I-20+ TMT-600	Steel 15X – Steel 20	Antifrictionality, score resistance	0.01; 0.1; 0.5 1.0; 5.0

The antifriction tests were conducted under the dead load and the following conditions:

- 2.5 m/s sliding speed in I-20 and I-20+TMT hydraulic oils;
- 200 N normal load on each specimen for I-20 hydraulic oil;
- 60 min test duration for I-20 and I-20+TMT hydraulic oils.

Tribological tests were conducted on the original test bench 'TK-2' produced in the research laboratory of Tomsk State University of Architecture and Building. Wear resistance, antifriction properties, and loading capacity of oil additives were estimated by the wear-in of specimens in pure and modified oil compositions until the friction ratio achieved its stationary value. The wear-in process was carried out at a step load not exceeding 0.75 N. The scuffing load capacity of hydraulic oils was detected under the step loading of specimens up to their frictional seizure.

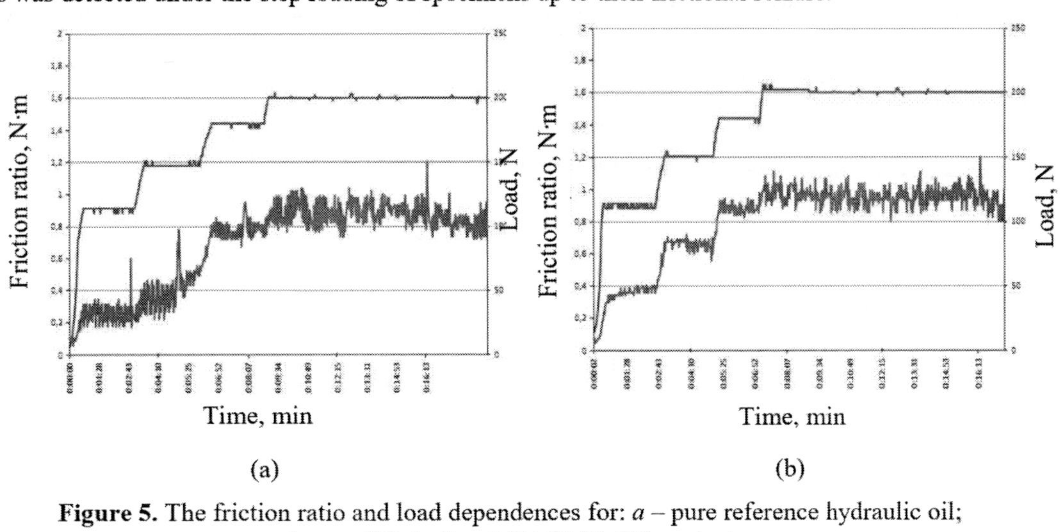

Figure 5. The friction ratio and load dependences for: *a* – pure reference hydraulic oil; *b* – modified oil at 0.5% TMT-600 content.

Some of the experimental findings, obtained in accordance with Table 2, are presented in Figure 5 and Table 3 in the real-time mode.

Table 3. Ultimate loads for investigated lubricants

TMT-600 content (%)	Load ratio (%)
0.01	No change
0.1	No change
0.5	+36
1.0	+53
5.0	-41

The studies of tribological properties of hydraulic oils modified by the TMT-600 additive will be further explained in our future works.

3. Conclusion

The experiments showed that the peat-based modifier improved the efficiency of hydraulic fluid lubricity. The modification of tribological properties of hydraulic fluids is of great theoretical and practical interest, since the peat dust has a negative effect on the friction and wear processes. Since the size of the modifier particles lies within the nanoscale level and that of the peat dust – within the micro and millimeter range, the nanodisperse state plays the important part in the reduction of friction and wear energy in machine parts and units.

References

[1] Li Z Wang H, He S, Lu Y and Wang 2006 *Mater. Lett.* **3** 356-359
[2] Nazari A and Riahi H 2010 *Mat. Res.* **4** 1-13
[3] Nazari A and Riahi H 2011 *Sadhana* **36**(3) 371-391
[4] Zhao Y, Qi X, Dong Y, Ma J, Zhang Q, Song L, Yang Y and Yang Q *Tribolog International* 2016 **103** 599-608
[5] Zhao W, Mo Y, Pu J and Bai M *Tribol. International* 2009 **42** 828-835
[6] Li B, Wang X, Liu W and Xue Q 2006 *Tribol. Lett.* **22** 79-84
[7] Zhigal'skii A A, Ikonnikova L F, Minakova T S, Mukhachev V A and Troyan P E 1996 *Russian Phys. J.* **39(6)** 576-578
[8] Ikonnikova K, Ikonnikova L and Koltunova E 2016 *Key Eng. Mat.* **683** 301-305
[9] Ikonnikova K, Ikonnikova L and Koltunova E 2016 *AIP Conf. Proc.* **1772** 020010
[10] Li B, Wang X, Liu W and Xue Q 2006 *Tribol. Lett.* **22** 79-84
[11] Sarkisov Yu S, Kopanitsa N O and Kasatkina A.V. UMP Rus. Fed. 107151 Publ. 10.08.2011
[12] Ku B C, Han Y C, Lee J E, Lee J K, Park S H and Hwang Y J 2010 *Eng. Manuf.* **11** 607-611
[13] Maharaj Y, Bhushan B and Iijima S 2013 *J. Colloid Interf.* 147-160

The structure and phase composition of hard alloys of the Cr_3C_2-Ti system produced by explosive compacting of powders

V O Kharlamov, A V Krokhalev, M A Tupitsin, S V Kuz'min and V I Lysak

Volgograd State Technical University, 28, Lenin ave., Volgograd, 400005, Russia

E-mail: harlamov_vo@mail.ru

Abstract. The work presents the findings of theoretical and experimental studies by scanning electron microscopy and energy-dispersive electron microprobe analysis of the phase composition of hard alloys produced by explosive compacting of the powders of chromium carbide Cr_3C_2 with titanium. It was found that when the powder mixture is heated in shock waves to 660 °C, the phase composition of hard alloys corresponds to that of the initial components of the powder mixture. With the increasing intensity of the explosive compacting, formation of secondary carbides is observed on the border of the initial components. A further increase in temperature results in a local melting and formation of new fine phases. With the subsequent temperature rise in the shock waves, a transition to the calculated equilibrium composition is observed.

1. Introduction

Explosive compacting is one of the promising types of impulse compaction, which significantly expands the capabilities of powder metallurgy technology in the creation of new materials [1]. The explosive processing enables achievement of the pressure sufficient for uniform compaction of powders to a practically pore-free state, and at the same time, the temperature required for consolidation (welding) of the structural components of powder material [2]. A short-term exposure to high pressure and temperature makes it possible to prevent the secondary interaction between the components of alloys that allows for using titanium, a metal reactive in high temperature, as a binder [3].

To analyse the processes occurring during both the shock-wave action on powder mixtures and the subsequent heating of hard alloys, it is very important to know the nature of the interfacial interactions between their components. We usually lack this knowledge when using new, little-known compositions of hard alloys. In these circumstances, we have to resort to computer thermodynamic simulation followed by direct experiments.

2. Materials and methods

We used hard alloy samples of the Cr_3C_2-Ti system, containing from 22 to 40 wt. % of titanium, produced by explosive compacting of the mixture of pure chromium carbide powders Cr_3C_2 KHNP-1 (TU 14-22-28-90) with PTS powders of titanium (TU 14-22-57-92) on the substrate of steel 09G2S according to the scheme of a normal-incidence detonation wave [4].

Chromium carbide (Cr_3C_2) was chosen as the alloy base because it is the most promising carbide for operation in oxidizing medium at temperatures of 300–400°C [5, 6], which correspond to the conditions typically achieved during the operation of friction assemblies of pumps for the transfer of overheated distillate. Titanium was used as the metallic binder due to its low acoustic stiffness, which ensures good compressibility of the powder mixture during explosive processing [7].

Compaction according to a similar scheme is accompanied by the propagation of the incident shock wave through the powder mixture and wave reflection from the surface of the monolithic substrate and the intermediate spacer. Simultaneously, the temperature increase in the powder material typically occurs in the incident wave travelling through the undisturbed powder, and the maximum pressure of its compression is achieved within the reflected shock waves, which is optimal for pressure welding.

The maximum pressure of the shock wave compression was calculated using the (P, u) diagram method based on the step-by-step determination of the characteristics of the incident and reflected waves by analysis of the intersection of shock adiabats of the spacer, powder, base, and detonation products in 'pressure–mass velocity' coordinates by applying the specially developed software. The powder heating temperature was calculated from the increase in its internal energy behind the front of the incident shock wave, which was determined by the mass velocity of the powder particles [8].

During the explosive compaction, the powder mixture in the shock waves was heated to 650–1300 °C, with the maximum compression pressure of 12 to 15 GPa, which ensured the obtaining of a monolithic hard alloy of the Cr_3C_2-Ti system at the compaction stage [9].

Thermodynamic calculations of the phase state were performed using the Thermo-Calc software package, based on the numerical simulation of phase equilibrium CALPHAD [10, 11].

Studies of the microstructure and the phase composition of the obtained materials were carried out using optical metallographic microscope Carl Zeiss Axiovert and a dual beam scanning electron microscope (SEM) 'FEI Versa 3D', equipped with gas-injection platinum deposition system GIS, transmitted electrons detector STEM that forms images in a dark-field (DF) and bright field (BF) transmission mode with high contrast to the crystallographic planes and the grain structure, accordingly. The integrated energy-dispersive electron microprobe analysis system 'EDAX Apollo X' allows a determination of the elemental composition with a high degree of localization on a specially prepared sample – foil of about 50–100 nm thick [12].

The foil preparation procedure involved the microsection preparation on a semi-automatic grinding-and-polishing machine with microprocessor control TegraPol-11. The sample was then placed in an electron microscope chamber, wherein a platinum layer, required to protect the surface from the ion beam effect, was deposited on the location of interest on the sample, using the gas-injection system. After that, cutting on the sides and etching with a focused ion beam of Ga^+ is carried out from above and below the platinum layer, thereby a special sample piece – a foil 20–25 μm long, 5–7 μm high, and 1–2 μm thick – is obtained for transmission microscopy. Then, a needle - manipulator is brought and soldered to the obtained plate by platinum deposition. At the next stage, the foil sample piece is transferred onto a special holder – a copper grid, and fixed to it. Next, the foil is thinned with an ion beam on both sides until it becomes electron-light (thickness of about 50–100 nm). Now, the copper grid with the fixed foil is placed in a special holder of the STEM detector.

3. Results and Discussion

Thermodynamic calculations (Figure 1) demonstrated that equilibrium phases in the alloys of chromium carbide and titanium in the investigated range of concentrations (up to 40 wt. % Ti) may be: starting chromium carbide Cr_3C_2, titanium carbide TiC, chromium carbides bound with carbon Cr_7C_3 and $Cr_{23}C_6$, and solid solution Ti in Cr.

Therefore, the exposure of a chromium carbide and titanium mixture to a rather high temperature should theoretically cause a change in the phase composition of the material. It can be expected that the interaction of the starting components in this case will result in the formation on the interface surface between Cr_3C_2 and Ti, of the alternating layers of stable phases TiC and Cr_7C_3, $Cr_{23}C_6$ or Cr, the boundaries of which will move towards the initial phases up to the disappearance (full or partial) of the latter ones.

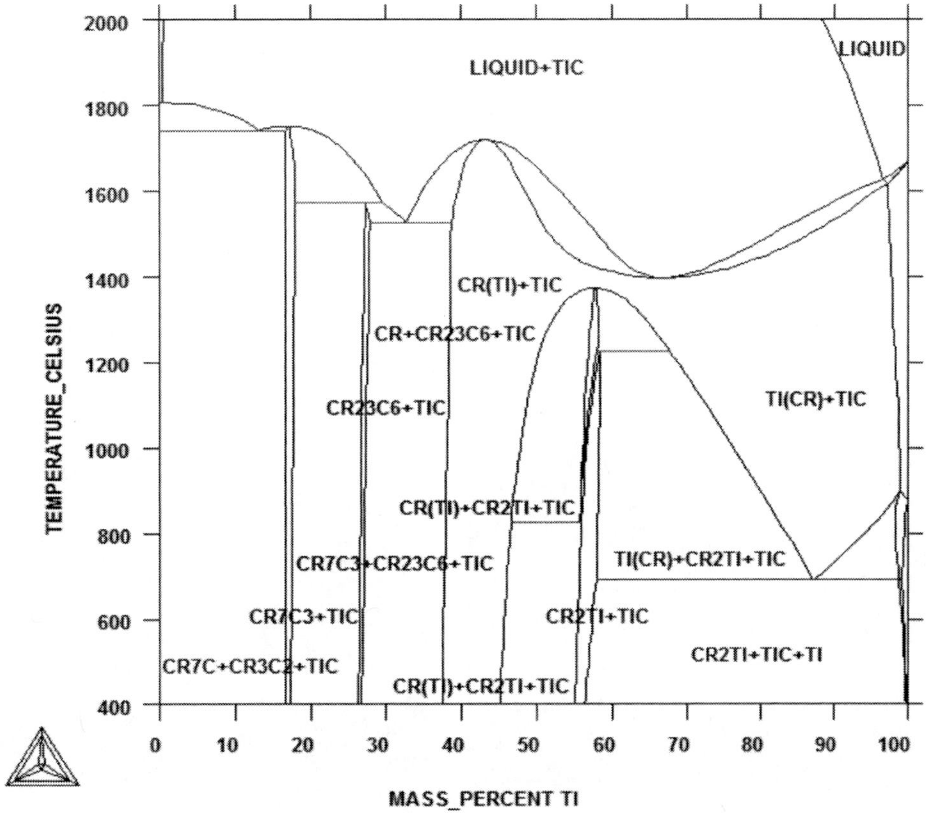

Figure 1. The quasi-binary cross-section of Cr_3C_2-Ti system Cr-C-Ti (Thermo-Calc software).

However, the study of the microstructure and line profile of the elements distribution between the phases of the hard alloy produced by explosion at a maximum compression pressure of 12 GPa, with simultaneous heating of powder chromium carbide mixture Cr_3C_2 with the titanium binder in the amount of 22 wt.% in the shock waves up to 660°C (Figure 2 a) showed no traces of chemical interaction between the components of the base powder mixture: all the titanium remained a part of the metal phase of the alloy, and chromium, and carbon – of the carbide phase.

The revealed fact can only be explained by an extremely short duration of the material exposure to pressure and temperature, within which the diffusion processes could not change the phase composition of the material in any noticeable degree.

With increasing explosive compacting pressure and a corresponding rise in the average heating temperature for the powder mixture in the shock waves, we nevertheless managed to find changes in the microstructure and nature of the elements distribution between the phases in the alloys containing 40 wt.% of titanium (Figure 2 b, c, d).

A double layer formed on the 'carbide – binder' border (Figure 2 b). On the outer side of the carbide particle, we detected a bright layer about 700 nm thick, and on the side of the material – a dark layer about 400 nm thick, which is indicative of a change in the average atomic number of the studied area [13].

The intensity distribution of the characteristic radiation of the carbon atoms along the scan line shows that as we approach the carbide particle boundary, we can see a slight decrease in the carbon content in the near-surface layer of the chromium carbide particle, followed by an abrupt change at the electron probe exit on the boundary layer of the material binder, that can be explained by the diffusion of carbon in the boundary layers of the chromium carbide particles towards the titanium binder, with formation of the titanium carbide layer on its border.

With the increasing alloy heating temperature during explosive processing, we can see the signs of melting on the surface of the initial particles of chromium carbide and the material matrix [14], which causes the appearance of three fine phases in the metal binder which evolved during the melt crystallisation (Figure 2 c): of a presumably solid solution based on chromium, titanium carbide and Laves phase $TiCr_2$, which can thermodynamically emerge (Figure 1) because the content of titanium in the metal binder in the case of incomplete dissolution of the starting chromium carbide may greatly exceed 45 wt.%. These phase changes are manifested in the appearance of alternating peaks on the chromium and titanium percentage curves (Figure 2 c).

A further increase in the explosive loading rate results in a complete transition of a hard alloy in the equilibrium state: (Figure 2 d) chromium carbide particles Cr_3C_2 disappear from its structure, and particles of a new phase appear instead. They are round-shaped and sized from 500 nm to 1.2 μm, uniformly arranged in the matrix.

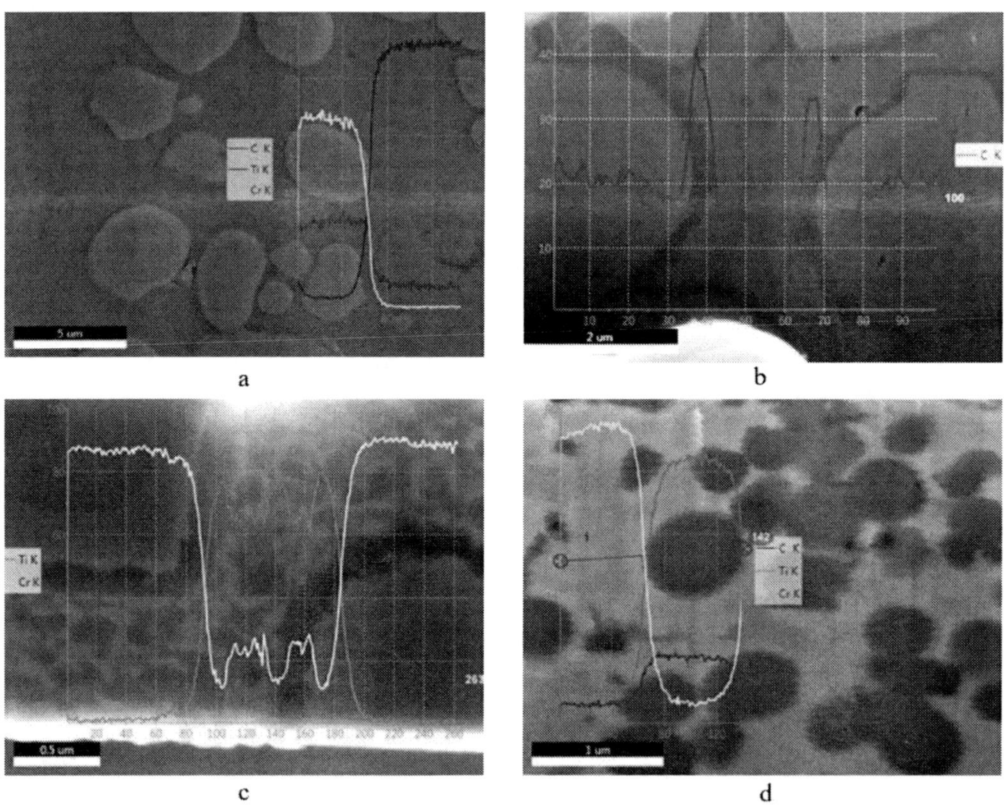

Figure 2. The microstructures of hard alloys of the Cr_3C_2-Ti system, at various stages of interfacial interaction, SEM Versa 3D: a – the structure corresponding to the phase composition of the starting components; b – the stage of the double boundary layer; c – the melting stage with the appearance of fine phases; d – the equilibrium structure of hard alloy Cr_3C_2-Ti after explosive compaction.

The electron-probe test along the scan line indicates that particles of titanium and carbon have titanium and carbon in their composition in this case, whereas the alloy matrix consists essentially of chromium. This corresponds to the calculation of phase equilibria in the studied system (Figure 1), under which the hard metal carbide phase should be titanium carbide TiC, and a binder – a solid solution of titanium in chromium.

4. Conclusions

The phase composition of hard alloys of the Cr_3C_2-Ti system, produced by explosive compacting of powders on a steel base in the modes, providing heating of the powder mixture in the shock waves of up to 660°C, corresponds to the composition of the starting components of the powder mixture.

With the increasing intensity of the explosive loading, we observe interfacial interactions between the hard alloy components which mainly take place in the liquid phase, and result in the disappearance of the initial constituents and appearance of the new ones, thus the alloy becomes equilibrated: titanium carbide becomes the major carbide phase, and chromium - the metal binder.

Acknowledgments

This work was supported by the grant of the President of the Russian Federation for young scientists, MK-4189.2015.8

References

[1] Gourdin W H 1986 *Prog. Mater Sci.* **30** 39-80
[2] Prummer R A 1989 *Materialwiss. Werkstofftech.* **20** 410-415
[3] Lysak V I, Kuz'min S V, Krokhalev A V and Grinberg B A 2013 *Phys. Met. Metallogr.* **114** 947-952
[4] Prummer R A, Balakrishna B T., Siva Kumar K and Hokamoto K 2006 *Explosive compaction of powders and composites* (Enfield: Science Publishers)
[5] Kayuk V G, Masljuk V A and Kostenko A D 2003 *Powder Metall. Met. Ceram.* **42** 257-261
[6] Hussainova I, Jasiuk I, Sardela M and Antonov M 2009 *Wear* **267(1)** 152-159
[7] Murr L E, Staudhammer K P and Meyers M A 1986 *Metallurgical applications of shock-wave and high-strain-rate phenomena* (New York: Marcel Dekker)
[8] Mamalis A G, Vottea I N, Manolakos D E 2001 *J. Mater. Process. Technol.* **108** 165-178
[9] Krokhalev A V, Kharlamov V O, Kuz'min S V and Lysak V I 2013 *Russ. J Non-Ferr. Met.* **54** 522-526
[10] Lukas H L, Fries S G and Sundman B 2007 *Computational thermodynamics. The Calphad method* (Cambridge: University Press)
[11] Raghavan R, Hari Kumar K C, Murty B S 2012 *J. Alloys Compd.* **544** 152-158
[12] Newbury D E, Scott J H, Wight S and Small J A 2005 *Handbook of microscopy for nanotechnology* vol 12, ed N Yao and Z L Wang (Boston: Kluwer academic publishers) p 361
[13] Reimer L 1998 *Scanning electron microscopy: physics of image formation and microanalysis* (New York: Springer-Verlag)
[14] Songa J, Kostka A, Veehmayer M, Raabe D 2011 *Mater. Sci. Eng. A* **528** 2641-47

IOP Conf. Series: Materials Science and Engineering **177** (2017) 012110 doi:10.1088/1757-899X/177/1/012110

The structure and the phase composition of copper aluminide-based coatings formed on the aluminum surface under contact melting in the Al-Cu system joints

V G Shmorgun, O V Slautin and V P Kulevich

Volgograd State Technical University, 28, Lenina ave., Volgograd, 400005, Russia

E-mail: mv@vstu.ru

Abstract. A contact melting mechanism in the Al-Cu system is proposed and experimentally verified. The impact of contact melting thermal and temporal conditions on the phase composition, structure and properties of coatings produced on aluminum surfaces is investigated.

1. Introduction

Wear-resistant coatings based on intermetallic compounds are used to increase the wear resistance of the electrically conductive material surface [1-5]. Complex technologies are employed to obtain these surfaces, and they include a preliminary coating of a component surface with a layer of the metal which can form intermetallic joints with the base metal accompanied by a subsequent diffusion annealing of the composite material produced [6].

A significant increase of diffusion interaction between dissimilar metals can be achieved through contact melting at the metal interface [7-9]. The process consists in the transition of contacting dissimilar hard materials to the liquid state at temperatures below their melting points. This phenomenon is typical of eutectic systems and the systems which form solid solutions with a minimum on the liquidus curve.

The use of explosion welding in this technology is substantiated by a number of advantages which this method has for joining dissimilar metals. The main advantage is the production of a composite with a firm-and-impervious joint of layers with the thickness required.

The paper aims at establishing the regularities in structure formation and the properties of aluminide copper-based coatings which form on the aluminum surface under contact melting in the Al-Cu system joints produced by explosion welding.

2. Materials and methods

Explosion welded specimens of grade AD1 of aluminum with grade M1 of copper (16 + 0.1-0.5 mm) were used as experimental material. The specimens were heat treated in the SNOL heater ‘8.2/1100’ at temperatures of 560°C and 580°C. A metallographic study was performed with the Olympus BX-61 metallographic microscope. The interaction zone phase composition was evaluated by comparing the data obtained with the DRON-3 diffractometer and the Versa 3D DualBeam electronic microscope. Microhardness measurements were performed using the PMT-3M instrument with the 100 gram indentor load.

3. Results and discussion

The Al-Cu system (Figure 1) belongs to heavily degenerated eutectic systems with the eutectic point in the phase diagram shifted in the direction of Al. Hence, the decisive role in the liquid phase formation in the Al-Cu contact should be played by the melting point depression [7]. The analysis of the Al-Cu phase diagram made it possible to suggest the following sequence of interaction zone formation in contact melting at the interface of explosively welded bimetal AD1+M1.

The contact melting process in the Al-Cu system starts at temperature above 548 °C (Figure 1) and is initiated by the reciprocal diffusion which leads to the formation of the Al(Cu)-based solid solution of maximum concentration at the interface of the explosively welded metals; this solution deliquesces easily.

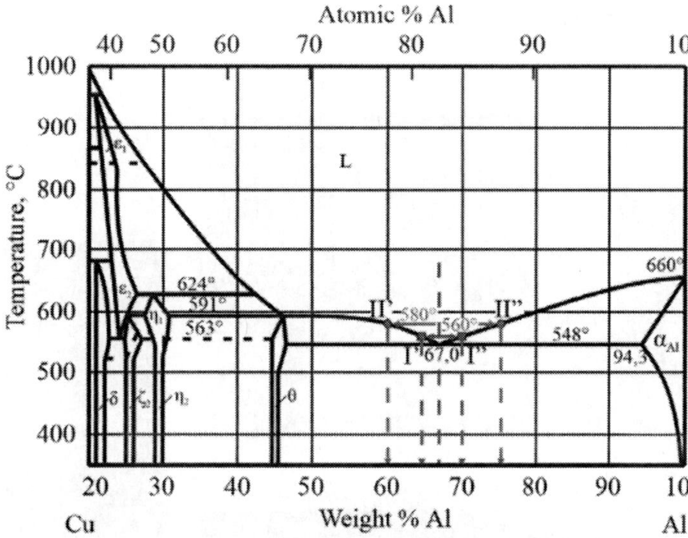

Figure 1. A portion of the Al-Cu system phase diagram [10].

Upon the achievement of a certain critical Cu concentration in Al(Cu), separate small areas of the solid solution turn out to be "blocked" from each other by Cu atom 'partitions' as a result of breaking Al-Al interatomic bonds. It leads to the formation of nano-sized clusters where melting point depression shows up [7].

Further, the chaotically distributed clusters merge upon the contact and form a liquid matrix with separate Al(Cu) solid-phased particles (microcrystals). With time, the solid particles, surrounded with the liquid, dissolve and the entire solid solution area turns into a thin liquid film.

The liquid interlayer thickness growth is accompanied by a decrease in the Cu concentration as per the liquidus line in the Al-Cu phase diagram. In doing so, the copper concentration falls in the interlayer part adjoining the Al-based solid solution (points 'I' and 'II'). The concentration growth is observed in the interlayer part adjoining the Cu-based solid solution (points 'I' and 'II').

The nonequilibrium crystallization in the interaction zone may be followed by the formation of Al(Cu) and $CuAl_2$ subsequent phases. In case of a prolonged heat treatment, a complete Cu dissipation is to be followed by the liquid phase coming into equilibrium as per the points 'I' and 'II'.

In order to experimentally verify that the interaction zone structure complies with the predictions, based on the phase diagram, the real structure obtained under heat treatment as per the contact melting regime at hold times which provide the presence or absence of the non-reacted Cu layer; heat treatment of a set of specimens was performed at 560°C and 580°C (Figures 2, 3).

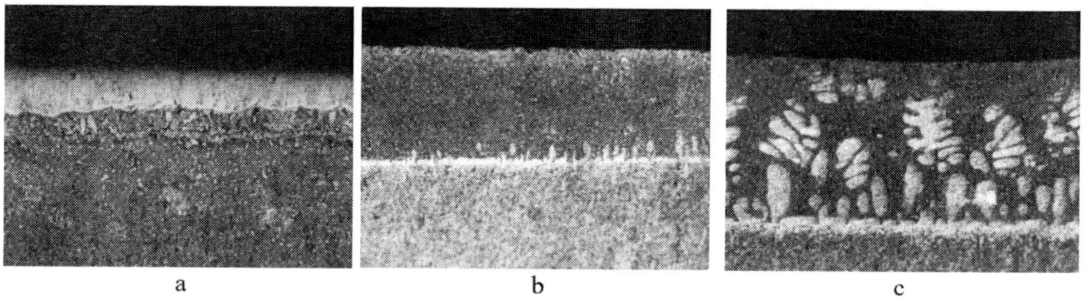

Figure 2. The interaction zone microstructure after heat treatment at 560°C within 5 (a) x200, 10 (b) and 20 min (c) x100

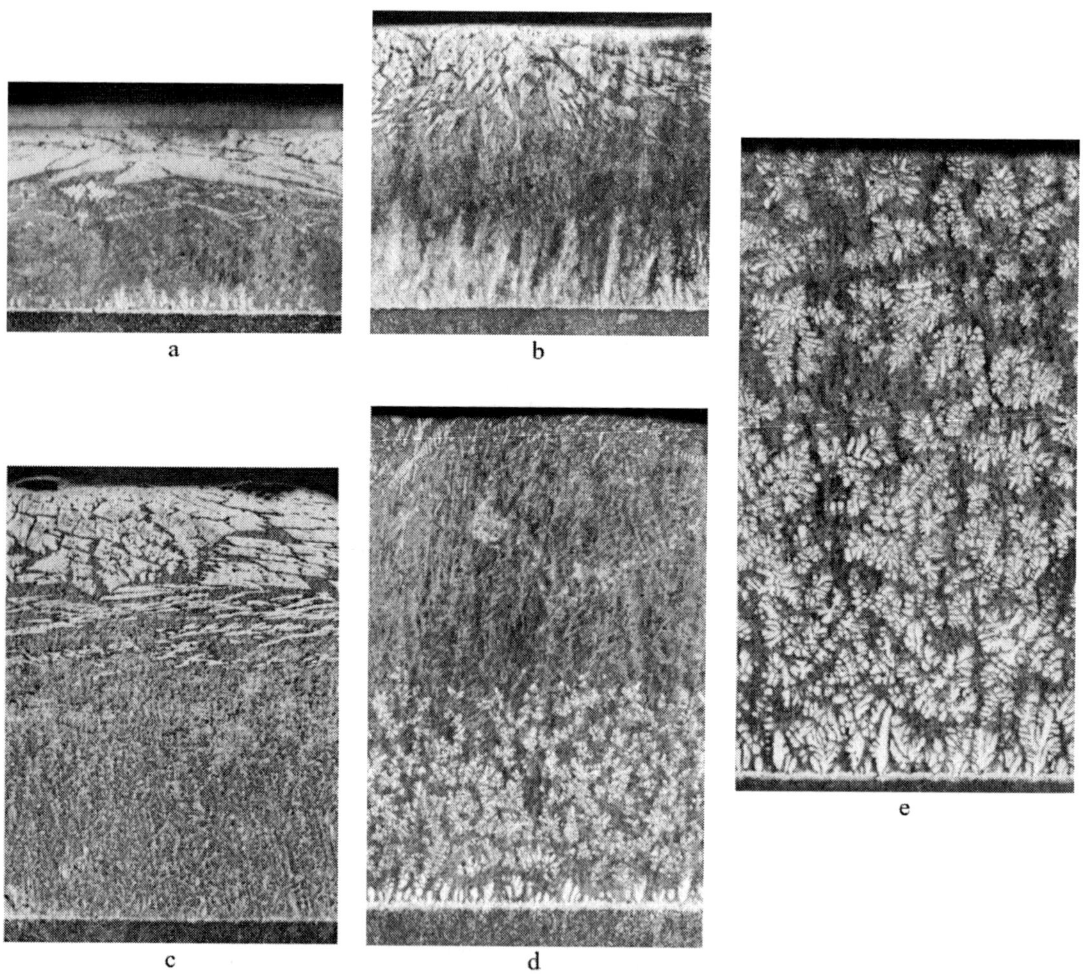

Figure 3. The interaction zone microstructure after heat treatment at 580°C within 5 (a), 10 (b), 15 (c), 20 (d) and 50 minutes (e) x50.

It was established that the interaction zone thickness (Figure 2), formed at the AD1+M1 interface (16 + 0.1 mm) after heat treatment at 560°C, changes from 55 μm (5 min) to 690 μm (20 min). The quantity of Al and Cu in the interaction zone, expressed on a percentage base and calculated by the amount of the components which took part in the reaction (thickness loss compared with the initial values after welding considering component densities), ranges from 30 to 42 wt.% for Cu and 58 to 70 wt.% for Al. For the hold time of 5 min, which is insufficient for complete Cu melting (Figure 2, a), the copper concentration in the interaction zone (42 wt.%) somewhat exceeds the eutectic concentration (33 wt.%). The hold time increase to 10 min ensures Cu melting in the interaction zone and reduces the copper concentration to the eutectic value (Figure 2, b). When the hold time equals 20 min, the Cu concentration continues to decrease and is accompanied with the emergence of Al(Cu) inclusions in the interaction zone structure (Figure 2, c).

The heat treatment temperature increase facilitates the contact melting velocity increase. The interaction zone thickness after heat treatment at 580°C (copper layer thickness is 0.5 mm) changes from 1.6 (5 min) to 5.2 mm (50 min (Figures 3, 5, a). The quantity of Al and Cu in the interaction zone expressed on the percentage basis is within the range of 59 to 74 wt. % and from 26 to 41 wt.% respectively. For hold times insufficient for Cu complete melting (from 5 to 10 min), the Cu concentration is somewhat higher (38 to 41 wt. %) than the eutectic one. An increase in the hold time after Cu complete melting leads to the coagulation of the structural components and the formation of flake-like homogenous insertions in the coating (Figures 3, d, 3, e).

The analysis of the interaction zone surface XRD patterns, the study of chemical element distribution throughout the interaction zone thickness and point energy-dispersion microanalysis (Figure 4, Tables 1, 2) made it possible to conclude that after heat treatment at 560°C, the interaction zone major structural components are structurally free intermetallic $CuAl_2$ and eutectic Al(Cu)+$CuAl_2$. Intermetallic Cu_2Al_3 was additionally found at the Cu boundary after heat treatment at 580°C.

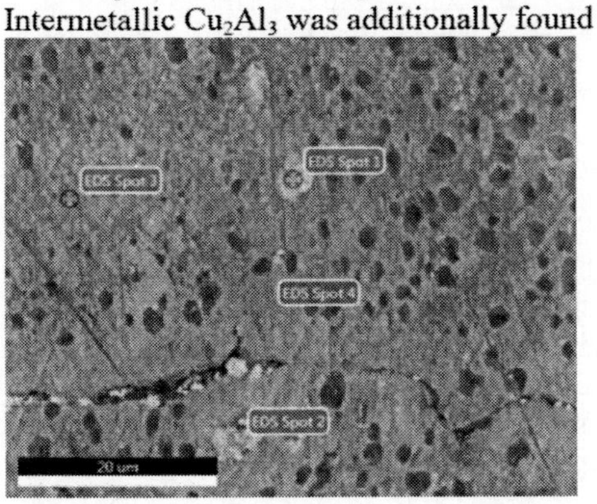

a

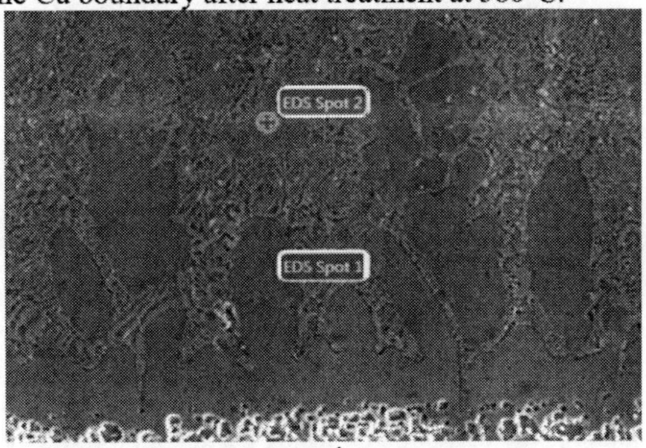

b

Figure 4. The SEM image of the coating structure (heat treatment at 580°C, 5 min) from the copper side (a) and from the aluminum side (b) with point location for element composition analysis.

Table 1. Point energy dispersion microanalysis results (Figure 4, a)

point	element	wt. %	at. %	error, %	phase
1	Al	5.78	12.62	10.35	Cu(Al)
	Cu	94.22	87.38	2.02	
2	Al	1.55	3.57	13.57	Cu(Al)
	Cu	98.45	96.43	1.96	
3	Al	22.53	40.64	8.46	Al_2Cu_3
	Cu	77.47	59.36	2.06	
4	Al	22.71	40.89	8.47	Al_2Cu_3
	Cu	77.29	59.11	2.09	

Table 2. Point energy dispersion microanalysis results (Figure 4, b)

point	element	wt. %	at. %	error, %	phase
1	Al	93.89	97.31	2.3	Al(Cu)
	Cu	6.11	2.69	5.23	
2	Al	53.16	72.77	6.77	$CuAl_2$ + эвт
	Cu	46.84	27.23	2.23	

The interaction zone microhardness after heat treatment at 560°C is approximately 1.5 GPa. Microhardness mean values in the interaction zone obtained at 580°C diminish with the Cu concentration gradient from 2.7 GPa (5 minutes) to 1.4 GPa (50 minutes) (Figure 5, b).

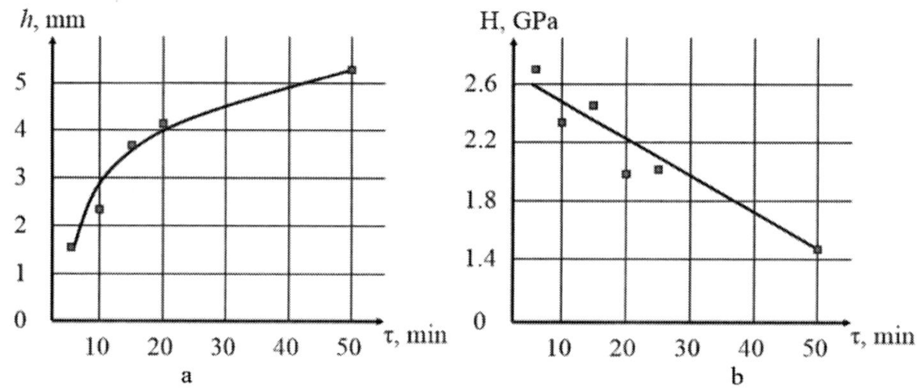

Figure 5. Variations of interaction zone thickness (a) and microhardness mean values (b) depending on the hold time at 580°C.

4. Conclusions

The contact melting process in the Al-Cu system is initiated by reciprocal diffusion which leads to solid solution formation in Al, and this solution deliquesces easily. Diffusion interaction in the liquid phase presence results in intermetallic coating formation on the Al surface. The coating structure and the phase composition are conditioned by thermal and temporal heat treatment regimes and the weight content of the components which participated in the reaction.

The major structural components of the system Al-Cu coating obtained on the Al base after heat treatment as per the contact melting regime which provides total melting of the copper layer are $CuAl_2$+Al(Cu) eutectics and Al(Cu) solid solution.

An increase in heat treatment temperature leads to an increase of the contact melting velocity and the interaction zone thickness while an increase in the hold time brings about the coagulation of the structurally free solid solution and a decrease in the coating hardness.

Acknowledgments
The reported study was funded by the Russian Science Foundation according to research project No. 14-29-00158

References

[1] Kobelev A E, Kuznetsov V E and Titlyanov A E 2000 *Mat. Sci. and Heat Treatment of Metals* **12** 30-33
[2] Titlyanov A E, Radyuk A G, Kozlov S N and Purekhov A N 2002 *Higher Educational Institutions News. Non-Ferrous Metallurgy* **2** 54-55
[3] Radyuk A G, Titlyanov A E and Ukraintsev A E 2007 *Non-Ferrous Met.* **5** 95-97

[4] Yakoev A G, Titlyanov A E, Radyuk A G and Pedos S I 2002 *Non-Ferrous Met.* **8** 70-73
[5] Radyuk A G, Titlyanov A E and Filatov S V 2006 *Non-Ferrous Met.* **8** 106-111
[6] Shmorgun V G, Gurevich L M, Slautin O V, Arisova V N and Evstropov D A 2016 *Metallurgist* 1-6
[7] Ahkubekov A A, Orkvasov T A and Sozaev V A 2008 *Kontaktnoe plavlenie metallov i nanostruktur: monografija* (Moscow: Fizmatlit Publ.)
[8] Afashokov V Z, Ahkubekov A A, Ahkubekova S N and Baysultanov M M 2009 *Bulletin of the Russian Academy of Sciences: Physics* **73(3)** 422-424
[9] Akhkubekov A A, Akhkubekova S N, Bagov A M, Zubkhadzhiev M A and Mamaeva, Zh M 2010 *Bulletin of the Russian Academy of Sciences: Physics* **74(5)** 644-649
[10] Ljakisheva N P 1999 *Diagrammy sostojanija dvojnyh metallicheskih sistem* vol. 3 book 1 (Moscow: Mashinostoenie Publ.)

Non-destructive testing of nanomaterials by using subminiature eddy current transducers

S F Dmitriev[1], A V Ishkov[2], A O Katasonov[1], V N Malikov[1], A M Sagalakov[1] and L I Shevtsova[3]

[1] Altay State University, 61, Lenin str., Barnaul, 656049, Russia
[2] Altay State Agrarian University, 98, Krasnoarmeyskiy ave., Barnaul, 656049, Russia
[3] Novosibirsk State Technical University, 20, Karka Marksa ave., Novosibirsk, 630073, Russia

E-mail: osys@mail.ru

Abstract. A sensor for studying nanomaterials has been developed on the basis of a transformer-type eddy-current transducer. The basic technical data are stated (the number of windings is 100-400 turns, and the value of the initial permeability of the core is $\mu_{max} = 36000$). Measurements technique which allows high-precision measuring the electrical conductivity in thin film Ce-Nb. The electrical conductivity of Niobium-Selenium varies from $3.3 \cdot 10^5$ to $3.7 \cdot 10^5$ MS/m in the ratio of four from 4 to 60 nm.

1. Introduction

Advances in modern micro- and optoelectronics are inextricably linked with the intensive introduction of thin-film elements. The structural state and the topology of nanocrystalline materials, which include thin metal films, are one of the most important problems of modern solid-state physics. In connection with this, experimental studies of the similar materials are of fundamental importance for the understanding of the regularities of formation of nanostructures. Particularly relevant in this case is the development of new methods and devices, allowing the analysis of the structure and topology of thin films. T A de Assis and el study the effects of time-dependent substrate/film temperature in the deposition of a mesoscopically thick film using a statistical model that accounts for diffusion of adatoms without lateral neighbors which coefficients depend on an activation energy and temperature [1]. Effects of depositing atom kinetic energies and atomic composition are studied in order to predict the evolution of morphologies and the atomic structure of molecular dynamics (MD) grown thin films [2]. Recently, both the charge density wave and the superconductivity in monolayer $NbSe_2$ have been investigated [3-6]. In the construction industry, Nb plays an important role in the development of more resistant steels, which can contribute to reduction of total project costs of infrastructure, even if they become increasingly complex. On the other hand, the demand of the construction industry will continue to be increasingly growing due to urbanization, the population growth and replacement of obsolete infrastructure, so the need for lighter structures results in a greater use of high quality steel containing Nb [7, 8].

2. Specifications, material choices and design

While working on the improvement of the devices that implement the eddy current testing method [9, 10], subminiature eddy-current transducers which led to a significant increase in the localization of the magnetic field were created. Reached locality allows the study of nanostructures, including nanometer thin films.

The subminiature eddy-current transducer (ECT) is connected to the sound board of a computer with special software that controls the voltage in the transformers exciting winding and also reads the voltage in the measuring winding (in arbitrary units). On the basis of preliminary calibration, these readings are converted to values of the electrical conductivity. The exciting winding (diameter D_1 = 0.12–0.13 mm) of the superminiature eddy-current transformer consists of ten turns of copper wire (cross-sectional area is 5 μm^2). The measuring winding (diameter is 0.05–0.08 mm) consists of 130 turns of copper wire (cross-sectional area is 20 μm^2). To minimize the influence of the exciting winding on the final signal, the circuit includes a compensation winding that consists of 20 turns of copper wire (cross-sectional area is 5 μm^2), connected to the measuring winding in such a way that the voltage of the exciting winding is subtracted from the result.

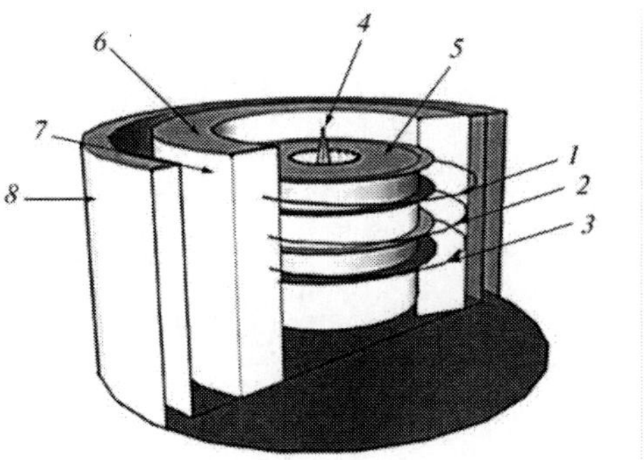

Figure 1. The scheme of the eddy current transducer.

The eddy-current transducer (Figure 1) is a transformer with measuring (1), exciting (2), and compensation (3) windings and a magnetic circuit (4), which is located inside cylindrical platform (5) with tracks intended for windings that are cut on the external side. The platform is impregnated with compound 6 at a temperature of 200°C to prevent the disintegration of the windings when ferrite screen 7, which is intended for the localization of the electromagnetic field on the tested object, is put in place. From the outside, the transducer is included in corundum washer 8, which protects core 4 from contacting the tested object.

The developed device allows, in particular, determining the electrical conductivity, which allows drawing some conclusions about the properties of the test films. The purpose of this study was to develop non-destructive testing methods for thickness, electrical conductivity and oxidative stability of nanofilms alloy Nb-Ce, intended for the use as contact and barrier materials in low-temperature superconducting electronic devices. Nb-Ce nanofilms were prepared by pulsed (200-250 ms), high-vacuum (10^{-3} -10^{-4}Pa) deposition of alloy (0.005-0.008 g) at temperatures of 3000-3200°C. The alloy was evaporated from the surface of the Pt-Pd-strip-heater (1×15×0.05 mm) on the pre-annealed quartz or Pt-Pd-substrates in a vacuum chamber of the ALA-TOO installation of IMASH-275 type. Conductivity (σ) nanofilms found contactless eddy currents developed by us with the use of the subminiature eddy-current transducer.

Measurement of volume electrical resistivity of films (R_x) produced by a bridge on the E7-11 device using benchmark resistance class 1 accuracy. Resistance of samples (Rx), measured with device E7-11, was calculated by the formula:

$$R_X = R_N \cdot \frac{\alpha_{X_1} - \alpha_{X_{11}} + U_{X_1}}{\alpha_{N_1} - \alpha_{N_{11}} + U_{N_1} + U_{N_{11}}},$$

where: R_X – resistance of the sample, α_{X_1}, $\alpha_{X_{11}}$, α_{N_1}, $\alpha_{N_{11}}$ – indications on block AK of device E7 – 11, R_N – benchmark resistance of the sample made of copper M0, U_{X_1} – R_X voltage,

U_{N_1} – R_N voltage, $U_{N_{11}}$ – R_N voltage for a change in the direction of the current bridge, R_N, calculated according to formula $R_N = \frac{l}{\sigma \cdot s}$, where σ is a known value.

3. Experimental results

Results of measurements electrophysical characteristics of nanofilms alloys Ce: Nb - 1:4 and 4:1, obtained by pulsed high-vacuum sputtering of different amounts on the Pt-Pd-substrate are shown in Table 1.

Table 1. Electrophysical characteristics of nanofilms alloys Ce: Nb (N=5, P=0.95).

Composition of alloy, mass. %		Electrophysical characteristics	
Ce	Nb	R_x, Ohm	σ, MS/m
20	80	1.04±0.03	2.2±0.2
80	20	$(1.02\pm0.4)\cdot10^{-2}$	$(3.5\pm0.2)\cdot10^{5}$

Calculation of the film thickness (λ) of (2) based on the data in Table 1 was carried out using the resistance measurement circuit section assuming a constant specific volume resistance of the alloy:

$$l = \frac{\sigma \cdot S}{2} \cdot (R_x - R_{Lin.}), \quad (2)$$

where: R_x – resistance of the sample, S – the area of the film, σ – the film electrical conductivity, $R_{Lin.}$ – the reactance line including a resistance wire, resistance of the two Pt-Pd-substrates and the amount of contact resistance at boundaries Cu/Pt and Pt-Pd/Pt-Pd (for our installation, $R_{Lin.}$. was $(3.5 \pm 0.1)\ 10^{-3}$ Ohm).

The calculated thickness of the alloy film, Niobium-Selenium, in the ratio of four to one of 60 nanometers and the Niobium-Selenium alloy in the ratio of one to four nm.

As it has been intended to make the deposition of nanofilms of Ce:Nb alloys, state, adhesion and stability of which considerably depend on the condition of the substrate surface, then the process of their vacuum annealing followed by atmospheric gases and vapors adsorption has been previously studied. Substrates annealing has been carried out by a passage of current through them from 10 to 50 A at a voltage of 4 V during 2-3 s to red heat, and then from 50 to 100 A during 250 ms up to white heat of platinum (3000-3200°C). This mode of substrates annealing has caused reduction of their weight from 1.7 to 4% depending on the initial mass of a heater(in absolute terms, the greatest reduction in weight of the substrate – 0.0012 g, and the smallest reduction – 0.0003 g) that is attributable to removal of a layer of adsorbed gases, vapors and dust-like particles from chemically inert surface that would prevent further sedimentation of a nanofilm.

Investigation of the kinetics oxidation of nanofilms oxygen in the air at 25 °C was carried out through the contactless visually-optical method by photographing them at different time intervals and determining the decreasing area of the film. Then, the conversion substance (α) can be calculated from the formula (3) for reducing the area of the alloy film:

$$\alpha = \frac{S_0 - S_t}{S_0}. \quad (3)$$

Here, S_0 and S_t – the original areas of the film and the film in time t.

The film area was determined with the help of the programs for ECM Analizer и FracDim, allowing the users to search for and to identify the edges in bitmap images, to perform their black-and-white contrasting and to calculate the closed loops areas and the fractal dimension of the edges in the images. The kinetic data processing was conducted according to the Kolmogorov-Erofeev equation (4)

$$\alpha = 1 - \exp(-kt^n)\cdot, \qquad (4)$$

where k- the kinetic constant; t- the decomposition time; n- the kinetic parameter.

The kinetic oxidation curves of two Ce-Nb 1:4 alloy nanofilms, differentiated by the specific electrical conductivity, in the air at temperature of 25°C are shown in Figure 2. 1 – film with electroconductivity 3.3×10^5 MS/m, 2 – film with electroconductivity 3.7×10^5 MS/m. As an example, in the same place, original video frames of the second film, oxidizing in the air during 3, 10, 20 and 60 minutes, respectively, are provided. Similar kinetic curves were obtained for Se-Nb 1:4 as well. The kinetic equation parameters (4) presented in Table 2 were determined from the kinetic curves linear anamorphoses by the OLS method. For films having different conductivity and the composition, kinetic constant k differed by more than 100 times, the kinetic parameter n – almost 4 times.

The obtained results testify the quickest oxidation and, consequently, the least stability of the Ce 80%: Nb 20% alloy nanofilms in the air, the film characterized by higher resistance and less thickness oxidizes quicker than a thicker one does. The Ce 20% : Nb 80% alloy films oxidize 10 to 100 times slower with the common pattern being broken; a thinner film (with higher resistance) oxidizes slower than a thicker one. Depending on the thickness and the nature of the film, the process mechanism changes as well, which is manifested by the alteration in the topochemical reaction order.

Table 2. Kolmogorov-Erofeev equation of kinetic equation parameters

Composition of alloy, mass. %		σ, MS/m	Kinetic equation parameters	
Ce	Nb		$k\cdot10^{-2}$	n
20	80	$3.3\cdot10^5$	0.07	0.98
		$3.7\cdot10^7$	3.2	0.48
80	20	2.1	12.5	0.26
		2.5	0.6	0.56

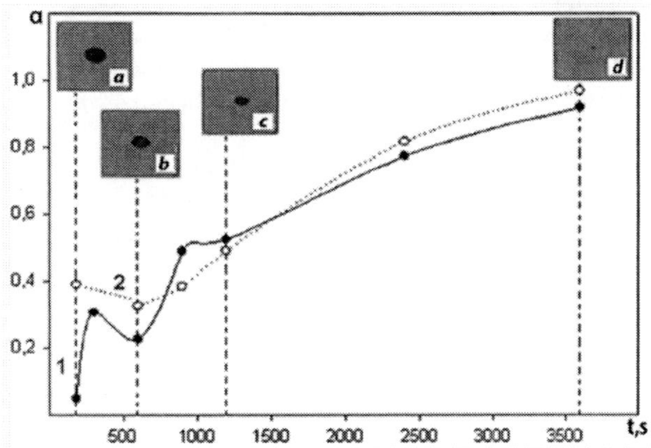

Figure 2. Kinetics of alloy nanofilms Ce-Nb.

It is found that the substance oxidation level is considerably changed in the process of oxidation. Upon that, the time it takes for achievement of a certain oxidation level considerably changes depending on conductivity of a substance and a proportion of elements.

For the first alloy with a Ce:Nb 1:4 proportion for the first film with specific electrical conductivity of $3.3\cdot10^5$ MS/m, the oxidation level is: for 3 minutes of oxidation – 5%, for 10 minutes of oxidation – 24%, for 20 minutes of oxidation – 50%, for 60 minutes of oxidation – 90%. The kinetic constant is $0.07\cdot10^{-2}$, the kinetic parameter is of 0.98.

For the second film with specific electrical conductivity of $3.7 \cdot 10^5$ MS/m, the oxidation level is: for 3 minutes of oxidation – 39%, for 10 minutes of oxidation – 40%, for 20 minutes of oxidation – 51%, for 60 minutes of oxidation – 92%. The kinetic constant is $3.2 \cdot 10^{-2}$, the kinetic parameter is of 0.48.

For the second alloy with a Ce:Nb 4:1 proportion for the first film with specific electrical conductivity of 2.1 MS/m, the oxidation level is: for 3 minutes of oxidation – 38%, for 10 minutes of oxidation – 48%, for 20 minutes of oxidation – 54%, for 60 minutes of oxidation – 65%. The kinetic constant is $12.5 \cdot 10^{-2}$, the kinetic parameter is 0.26.

For the second film with specific electrical conductivity of 2.5 MS/m, the oxidation level is: for 3 minutes of oxidation – 10%, for 10 minutes of oxidation – 19.5%, for 20 minutes of oxidation – 27%, for 60 minutes of oxidation – 44.5%. The kinetic constant is $0.6 \cdot 10^{-2}$, the kinetic parameter is of 0.56.

4. Conclusions

While alloying the Ce and Nb mixtures within a wide range of the PT mixing ratio in full vacuum vials, the alloys are formed; the latter have the form of solid solutions of γCe in niobium, which is verified by the chemical and X-ray diffraction analysis data. The pulsed high-vacuum deposition of the alloys with the Ce:Nb 1:4 and 4:1 (Ce 20% : Nb 80% and Ce 80% : Nb 20%) mixing ratio onto pre-annealed Pt-Pd-templates led to the obtainment of the nanofilms equal to 4 and 60 nm, respectively. The specific electrical conductivity of the nanofilms amounted to 2.1 – 2.5 (for the Ce:Nb 1:4 alloy) and $(3.3\text{-}3.7) \cdot 10^5$ (for the Ce:Nb 4:1 alloy) mS/m. The air oxidation of the nanofilms obtained in the Ce-Nb system at the temperature of 25°C studied by means of a contactless optical method is a topochemical reaction, described by the Erofeev-Kolmogorov equation with the following constants: k $(0.07\text{-}12.5) \cdot 10^{-2}$ and n 0.26-0.98 depending on the film specific electrical conductivity.

References

[1] T A de Assis and F D A Aarão Reis 2013 *J. of Statist. Mech.: Theory and Exper.t* **92** 1-10
[2] Xie L and BedraL 2014 *J. of Physics D: Appl. Phys.* **47**(22) 1-17
[3] Xi X and Mak K F 2015 *Nature Nanotechnol.* **10** 765-769
[4] Ugeda M M, and Riss A 2016 *Nature Phys.* **12** 92-97
[5] Xi X and Mak K F 2016 *Nature Phys.* **12** 139-143
[6] Cao Y and Khestanova E 2015 *Nano Lett.* **15** 4914-4921
[7] Grill R and Gnadenberger A 2006 *Int. J. of Refr. Metals & Hard Mat.* **24**(4) 275–282
[8] Tither G 2001 *The International Symposium Niobium* (Orlando, Florida) vol 1 p 845
[9] Malikov V N and Sagalakov A M 2015 *Int. Conf. on Mechan. Eng., Automation and Control Systems* (Tomsk) vol 1 (IEEE) pp 1-4
[10] Malikov V N and Sagalakov A M 2016 *Russian Eng. Res.* **8** 626-629

Development and implementation of the NMR-spectrometer on the basis of the National Instruments technologies

N D Narakidze, D V Shaykhutdinov, K M Shirokov, N I Gorbatenko and S G Yanvarev

Platov South Russia State Polytechnic University (NPI), 132, Prosvescheniya str., Novocherkassk, 346400, Russia

E-mail: narakidze@npi-tu.ru

Abstract. The quality of lubricating oil in mechanical engineering, technology of creation of units, in particular in equipment of transmission gears, is a factor which considerably defines reliability and safety of the whole propulsion system or the greased constructive components. There are many soluble oil additives such as, for example, different additives for extreme compression conditions or additives against wear. Additives are used with mineral oils, products from mineral oils or synthetic oils for lubricant action or chemical properties improvement. The most exact way of definition of the chemical composition of a substance at the moment is the method of nuclear magnetic resonance (NMR). In the first section of this article, a brief and very simplified review of the NMR basic principles using classical physics is provided. The second section is focused on the description of the hardware solutions and the architecture of the NMR spectrometers. The software developments (LabVIEW programs) of the data-acquisition and signal processing techniques are presented in the third section. At the end, results of measurements are provided.

1. Introduction

The quality of lubricating oil in mechanical engineering, technology of creation of units, in particular in the equipment of transmission gears, is a factor which considerably defines reliability and safety of all propulsion systems or the greased constructive components. There are many soluble oil additives such as, for example, different additives for extreme compression conditions or additives against wear. Additives are used with mineral oils, products from mineral oils or synthetic oils for lubricant action or chemical properties improvement. The difference in lubricating oil production technologies is a decisive criterion in competitive fighting. The experience of transmission gears operating shows that even best oils degrade and are subject to change. Thus, there is a gradual transition from planned intervals of oil change to oil change terms depending on oil condition. A criterion is the classical analysis that determine physical and chemical properties of lubricating oil. For example, for wind-power installations, there is a standard instruction according to which it is necessary to take samples of oil and check quality in a laboratory. If parameters of oil worsen, lubricating oil is replaced for safety reasons. A well equipped analytical laboratory and exact sampling time are necessary to determine the corresponding time of oil change.

The most exact way of definition of a substance chemical composition at the moment is the method of nuclear magnetic resonance. Typically, high field NMR systems have many advantages, such as a

IOP Conf. Series: Materials Science and Engineering 177 (2017) 012112 doi:10.1088/1757-899X/177/1/012112

high signal-to-noise ratio (SNR), resolution and high image quality. Actually, in many cases for particular purposes, only NMR spectrometers or MR imagers with a subset of some standard commercial features are needed [1]. In addition, using of medium fields (below 0.5 T) is more sufficient in some cases. The cost of the system can then be reduced since these medium fields - with, sometimes, relatively poor performances requirements - could be easily produced. Moreover, the use of medium fields simplifies the design and realization of compact NMR systems which could be especially appreciated for the in situ applications.

A number of groups have worked to develop dedicated NMR systems by using compact MR magnets. For example, a home-built and fully digital system working at 0.1 T (resonance frequency is of about 4.25 MHz) [2]. The system was based on the use of a high-performance Digital Signal Processor (DSP), a direct digital synthesizer (DDS) and a digital receiver. Based on this work, Shen [3] proposed another system working at 0.3 T and allowing larger imaging sizes than in [2]. Another work, carried out in [4], was focused on the realization of a wideband receiver for a home-built NMR spectrometer working at 55.84 MHz (high field).

Some groups have NMR systems for the specific application of measurement of the polarization for the NMR of hyperpolarized gases (129Xe, 3He etc.). Most of these systems were actually developed by modifying high frequency and high cost commercial spectrometers. One research group has, however, developed its own NMR system [5]. This system was used for monitoring the polarization of hyperpolarized helium (3He) at 3 mT. It was a fully analog system where authors performed a phase-sensitive detection of the NMR signal. They used then an oscilloscope for signal visualization. Despite the great merit of the original and elegant electronic solutions developed in [5], detection of hyperpolarized 3He signals was relatively not a hard task since their levels were quite high (at least 10 mV). Actually, this spectrometer did not allow the sufficient dynamic range to detect the NMR signal of the proton (1H) in such field. In any case, dedicated NMR systems are still far from the experience of most NMR groups. Some groups developed low-field NMR spectrometers that allow detection of the 1H NMR signals at 4.5 mT [6, 7]. These systems are based on National Instruments data-acquisition boards (NI DAQ). These boards are adequate at low frequencies. Moreover, the increased performance and the related software (LabVIEW) are user-friendly and quite easy to use. In these new NMR spectrometers, many analog electronic components were replaced by NI DAQ boards and software. It was shown that the use of advanced data-acquisition and signal processing techniques allow 1H NMR signals detection at 4.5 mT.

The aim of this research is to present advances of standardized National Instruments board and software LabVIEW in the development of medium-field NMR systems. One of the underlying ideas of this research is to make these systems versatile and easy-to-replicate so as to help developers and research groups to realize NMR spectrometers with flexibility, low cost and minimum development time. This includes both hardware design and software developments.

In the first section of this article, a brief and very simplified review of the NMR basic principles is provided using classical physics. The second section is focused on the description of the hardware solutions and the architecture of the NMR spectrometers. This architecture is mainly based on the use of the signal generator and boards from National Instruments. The software developments (LabVIEW programs) of the advanced data-acquisition and signal processing techniques are presented in the third section. The last section will concentrate on results and discussions. The use of the developed system for the measurement will be particularly illustrated.

2. Materials and methods

When a sample, consisting of NMR-sensitive nuclei (^{1}H, ^{3}He, ^{129}Xe; ^{23}Na, etc.), is subjected to uniform static magnetic field B_0, net macroscopic magnetization M of the sample appears in the same direction of B_0. This magnetization is proportional to this polarizing field B_0, to the density of nuclei within the sample and to characteristic gyro-magnetic ratio γ of the nucleus being studied. In a typical one-pulse experiment, the sample is subjected to a short pulse (called excitation pulse) of radiofrequency magnetic field B_1, applied perpendicularly to B_0 and at characteristic Larmor frequency

f_0. For the proton (^{1}H nucleus), this frequency is about 42.25 MHz at B_0 = 1 T and about 21.125 MHz at 0.5 T.

The effect of the excitation pulse is that the magnetization, M, is 'tipped' or rotated from its initial direction (or from its thermal equilibrium state) by angle α. This angle is called a "*flip angle*". It is proportional to field B_1 and to its duration, τ, according to the equation: $\alpha=\gamma\cdot\tau\, B_1$.

At the end of the excitation pulse, the NMR signal – called also the Free Induction Decay (FID) – is received at the same frequency f_0. This signal, which is proportional to magnetization M (then to B_0) and to γ, is processed for usage while obtaining the studied nucleus.

The flip angle can be set through the adjustment of the amplitude, B_1, or/and the duration, τ, of the excitation pulse. For a one-pulse sequence, the maximum NMR signal level is obtained at flip angles of 90° and 180° [8]. One should also know that a variety of parameters contribute to the signal-to-noise ratio (SNR). Firstly, and roughly speaking, the SNR is proportional to the square of the static magnetic strength. The SNR, the image quality and spectral resolution are enhanced at the high field. This is one of the main reasons for which NMR experiments are usually performed at high fields. Secondly, the SNR is proportional density of the nuclei within the sample being studied and it depends on the nucleus of interest (through gyro magnetic ratio γ).

3. The implementation of the hardware structure for medium-field NMR systems

Based on these simplified principles, Figure 1 illustrates the general hardware architecture of the developed low-field NMR systems.

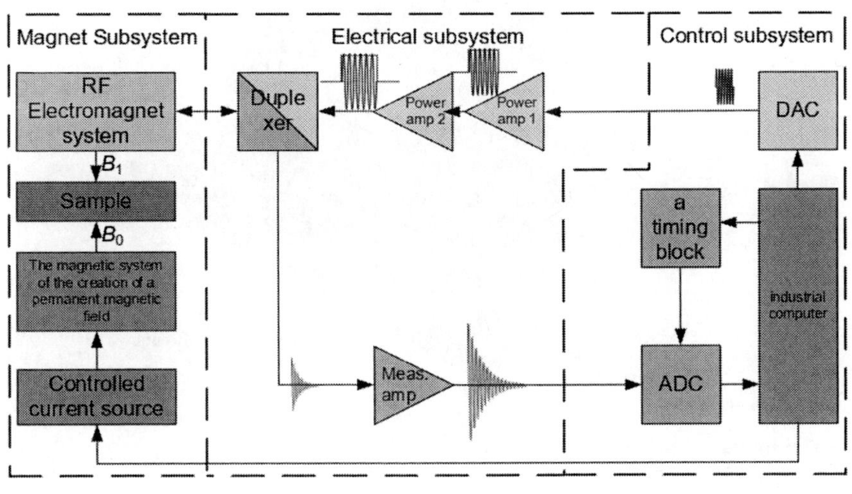

Figure 1. The block scheme of the developed spectrometer NMR.

The system is based on industrial computer NI PSIe-1078. Static field B_0 of about 0.5 T is produced by a permanent magnet and magnetization coils on the magnet core from electric steel. The excitation pulse (at about 21.125 MHz for ^{1}H) is generated by the Digital-to-Analog Converter (arbitrary waveform generator board NI 5442 of National Instruments). This pulse is amplified by a power amplifier and sent to the well-tuned coil (at the working frequency of 21.125 MHz for the ^{1}H) which generates excitation field B_1. At the end of the excitation pulse, this same tuned coil detects the weak NMR signal. This signal is transmitted to a low-noise preamplifier. The amplified signal is then received by the receiving board (A digitizer board NI 5154 from National Instruments) for digitalization and processing. A monostable-based circuit generates TTL control and synchronization signals from a single and very short (about 10 ns) TTL pulse ('*Marker*') that could be generated from the NI 5442. At least, two signals are necessary. Since the same coil is used for both transmitting and receiving (i.e. a transmit-receive coil), a 'blanking signal' is required to control the duplexer. This signal 'blanks' the preamplifier input during the excitation pulse and it isolates the transmitting section from the receiving one during the NMR signal detection. Another control signal (trigger signal) is necessary for triggering the signal acquisition with the receiving board.

In NMR systems, the static magnetic fields, B_0, could be produced using a variety of magnet categories and structures. The choice of a category and a structure is strongly related to the application, and it depends on many considerations such as the value of the magnet field, the desired performances (field stability, spatial homogeneity), the cost and complexity of realization as well as the ease-of-use. However, these magnets can be divided into two categories: permanent magnets and electromagnets. The main advantage of permanent magnets is that they do not use any power supply. However, these magnets could not offer a good stability of the field because of the temperature-dependence of their magnetization [9, 10]. Another disadvantage, which is the imperfections of the magnetic materials, may increase complexity of implementation. Electromagnets [11] can offer an alternative solution. Typically, the obtained field strength could be as high as 0.5 T. The hybrid system of creation of the magnetic field based on the use of permanent magnets and additional magnetizing coils is offered. Magnetizing coils are applied on the pole core of 10x10 cm. The model of the magnetic system is presented in figure 2.

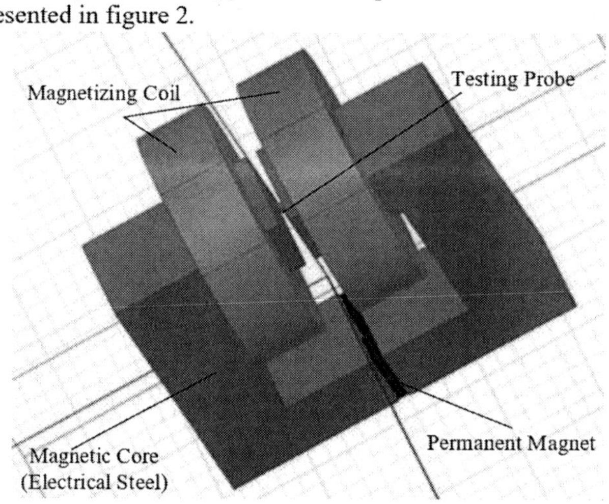

Figure 2. The model of the magnetic system.

For the solution of the problem of creation of an adequate model of the magnetic system, the effective approach of natural model trials provided in [12] has been used. The developed strategy of the solution of the return task has allowed solving the problem of identification of the magnetic system, optimization and implementation of its parameters.

4. Development and realization of the software: LabVIEW program

The '*NMR Spectrometer*' program was developed using LabVIEW and associated instrument drivers (NI-FGEN and NI-SCOPE) of the NI 5442 and the NI 5154 devices. The architecture of the program is open which allows users to build their own modulus if needed. The main panel of the Graphical User Interface (GUI) is shown in Figure 3. The program allows investigating a sample by '90° impulses' and '180° impulses' techniques.

IOP Conf. Series: Materials Science and Engineering **177** (2017) 012112 doi:10.1088/1757-899X/177/1/012112

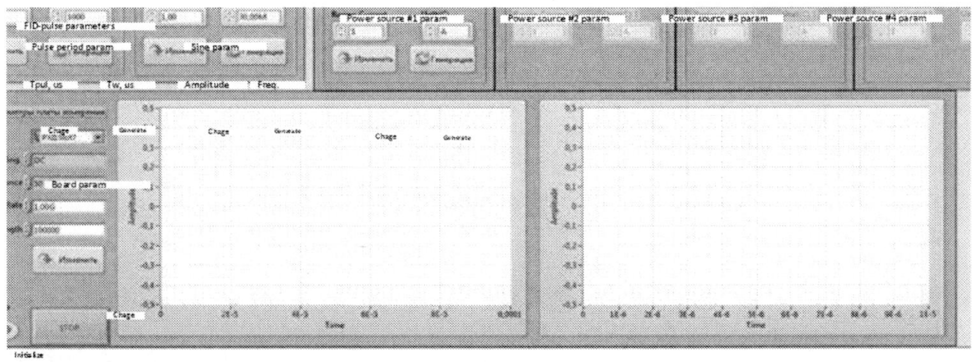

Figure 3. The main panel of the Graphical User Interface.

The algorithm of the program work is shown in figure 4.

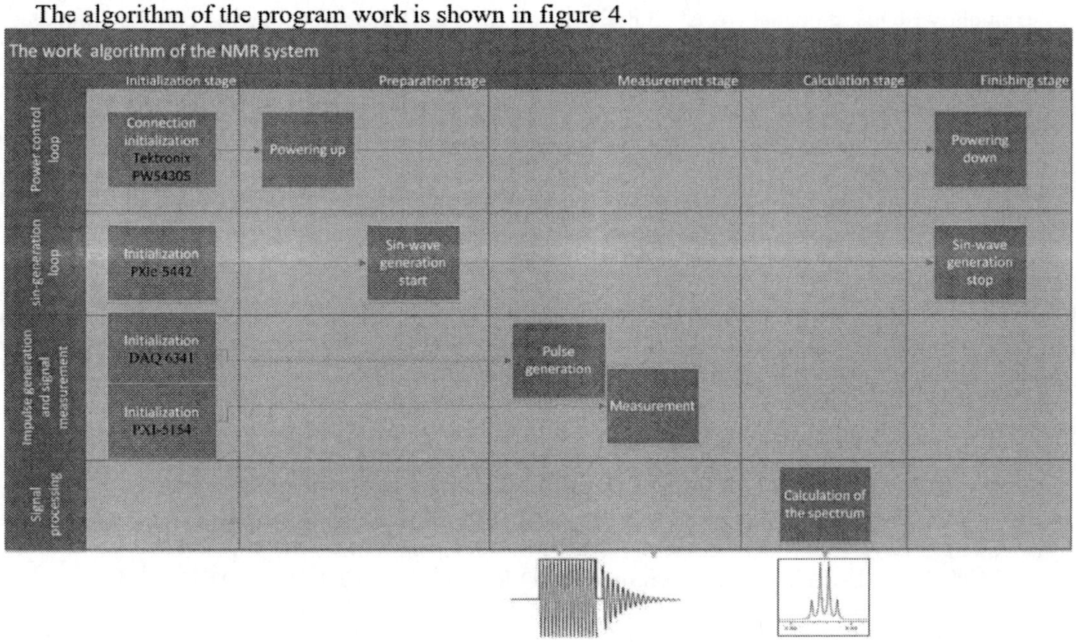

Figure 4. The algorithm of the program.

The user could choose the frequency, amplitude, and duration of the excitation pulse as well as the repetition time (TR) for a one-pulse NMR sequence. Hardware configurations of the NI 5442 and the NI 5154 are not available in the main front panel, but they could be modified if required in the LabVIEW diagrams.

In figure 5, the results of research of the developed system, where a 180° impulse use allow one to exclude influence of electromagnetic properties of the FID coil, are presented by the example of isopropanol studiyng.

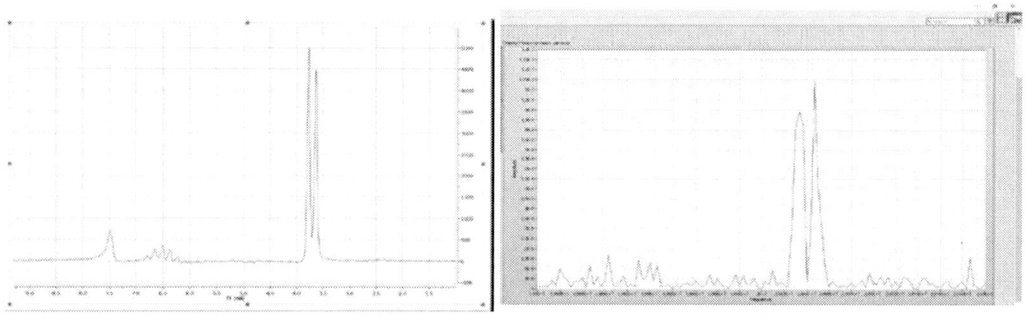

a b

Figure 5. The result of isopropanol spectral measurement: a – received by means of NMR on the basis of superconducting-coil electromagnets of high resolution; b – 0 received by means of the developed NMR device.

The peak of frequency is the share of 20.033 MHz and 20.047 MHz. It should be noted that charts of exemplary FID isopropanol (received by NMR on the basis of superconducting-coil electromagnets of high resolution) and ours are very similar.

5. Conclusion

A National Instruments boards and LabVIEW-based medium field NMR spectrometer working were presented. This spectrometer allows the detection of NMR signals of ^{1}H at 0.5 T. The flexibility of the system allows its use for a variety of NMR applications without (or with minor) hardware and software modification.

Acknowledgments

The results obtained with the support of grant RFBR No. 15-38-20652 ‘Development of the theory of predictive methods for sensorless control and diagnosis of electric drives’ using equipment of shared facility ‘Diagnosis and energy-efficient electrical equipment’ (NPI).

References

[1] Gengying L J, Xiaolong Y and Yun J 2002 *Rev. of Scientific Instr.* **72(12)** 105101-08
[2] Raoof K, Asfour A and Fournier J-M 2002 *Proc. of the IMTC 2002 19th IEEE Instrumentation and Measur. Techn. Conf. (Anchorage)* vol 1 (Piscataway: IEEE Service Center) 341-345
[3] Shen J, Xu Q, Liu Y and Gengying L 2005 *Rev. of Scientific Instr.* **76(10)** 105101-08
[4] Michal C A, Broughton K and Hansen E 2002 *Rev. of Scientific Instr.* **73(2)** 453-458
[5] Saam B T and Conradi M S 1998 *J. of Magnetic Resonance* **134(1)** 67-71
[6] Asfour A, Hyacinthe J N and Leviel J L 2006 *Proc. of the IMTC 2006 IEEE Instrumentation and Measurement Techn. Conf. (Sorrento)* vol 1 (Piscataway: IEEE Service Center) 1839-43
[7] Asfour A 2008 *Proc. of the IMTC 2008 IEEE Int. Instrumentation and Measurement Techn. Conf. (Victoria)* vol 1 (Piscataway: IEEE Service Center) 697-701
[8] Saam B T and Conradi M S 1998 *J. of Magnetic Resonance* **134(1)** 67-71
[9] Meethan W, Thomas I, Thomas C 2014 *Chiang Mai Univ. J. of Natural Sciences* **13(1)** 57-66
[10] Chen Q Y, Zhang G C, Xu Y J, Chang Y, Wang H and Yang X D 2014 *Appl. Mech. and Mat.* **543-547** 509-513
[11] Xia Y, Jeffrey K R, Callaghan P T 1992 *Magnetic Resonance Imaging* **10(3)** 411-426
[12] Shaykhutdinov D, Shurygin D, Aleksanyan G, Grushko I, Leukhin R, Stetsenko I, Oganyan R, Bondarenko V, Leukhin V 2016 *Asian J. of Information Technology* **15(9)** 1443-46

Effect of heat treatment conditions on structure and properties of high-speed steel

E P Nikolaeva[1] **and D B Vlasov**[2]

[1] Irkutsk National Research Technical University, 83, Lermontov str., Irkutsk, 664074, Russia
[2] IRGIREDMET, 38, Gagarin bul., Irkutsk, 664025, Russia

E-mail: nicolaeva-ep@yandex.ru

Abstract. High-speed steels made by powder and traditional metallurgical technology have been investigated. These steels are used for manufacturing of metalworking tools. Powder high-speed steels have a large safety margin and a high level of mechanical properties. Mechanical properties and the structure of steels S390 MICROCLEAN® and Fe-9%W-4%Mo-8%Co were compared after their hardening heat treatment. The composition of the carbides was determined by x-ray phase analysis. Residual stresses of the finished cutters were determined. It was shown that the mechanical properties of steels have been determined by the phase composition and the degree of carbides dispersion. The dependence of the microstructure and the high-speed steels phase composition on heat treatment conditions was identified. The conclusion about structure control effectiveness in the adjustment of heat treatment conditions of high speed steels was made.

1. Introduction

The metal cutting is an important method of components obtaining in aircraft engineering. In many publications, great attention is paid to the study of milling, drilling, turning, etc., from various points of view [1-7]. For example, A V Savilov and K Ahmadi [3] investigated the vibrations that occur in cut-ting tools. The works of Y I Zamashchikov are devoted to the study of residual stresses in the treated material [5-7]. The article [8] describes methods that were used to study the structure of carbon steel. The same methods (determination of residual stresses by the x-ray method, the study of steel by Barkhausen noise) are successfully used to study high-speed steel.

Currently, most of tools are made of high-speed steel. High demands are placed on the material of milling cutters, which are used for machining of titanium. It is known that the production technology (smelting or extrusion, etc.) affects the structure of high-speed steel. Powder steel is preferable for tool manufacture, as they are not carbide heterogeneity.

The composition and the morphology of carbides are completely formed during the smelting. Subsequent heat treatment of the tool forms the phase composition of the steel and generates secondary phases – carbides, intermetallics, etc. The most successful heat treatment of the tool is carried out in vacuum furnaces. Vacuum furnaces are not always available in companies, because of their high cost. The impossibility of quality heat treatment restrains the wide application of powder high-speed steel.

Steels heat-treated in resistance chamber furnaces and salt baths have been investigated. We studied the sensitivity of powder steels to disturbances of the heat treatment conditions. The aim of the authors was to find out how the heat treatment effects the phase composition of high-speed steels.

2. Materials and methods

High-speed steels S390 MICROCLEAN® and Fe-9%W-4%Mo-8%Co were investigated. Powder steel has a high content of carbon, vanadium, molybdenum, tungsten and chromium. The content of the main alloying elements of these steels is given in table 1.

Table 1. Chemical composition of steels Fe-9%W-4%Mo-8%Co and S390

Steel grade	**Weight content of element, %**					
	C	W	Mo	Cr	V	Co
Fe-9%W-4%Mo-8%Co	1.05	9.00	4.00	3.20	2.30	8.00
S390	1.60	10.50	2.00	4.80	5.00	8.00

Heat treatment was carried out for steels hardening. Mechanical properties were determined.

Determination of the phase composition of high-speed steels was carried out on the Shimadzu X-ray diffractometer (XRD) with Cu-K_α radiation. Obtained diffractograms were compared with diffractograms of standard samples.

X-ray diffractometer Xstress 3000 G3/G3R was used. The principle and methodology of measurement are described in works [8-10].

3. Results and discussion

3.1. Study of heat treatment effect on mechanical properties

The tempering has been performed to relieve stresses. Annealing temperature is 600 °C.

Steels were gradually heated and kept at the temperatures of 500 °C (for 1.5 h) and 860 °C (for about 6 min). The third preheating before quenching was excluded for steel S390 (1050 °C), but the exposure of workpieces at the second stage of heating was increased. The second preheating (860 °C) and final preheating to a quenching temperature were performed in salt baths. Three sample parties were heated up to three different temperatures. Steels were cooled in the salt bath, and then in the oil. Triple tempering (560 °C) was carried out after quenching in chamber furnaces.

Testing of steel for hardness (HRC), impact viscosity (KC, J/cm^2) and transverse bend (Rbm N/mm^2) was carried out after the tempering. The results of the tests – the fluctuation ranges of values HRC, KC and Rbm – are shown in figures 1 and 2.

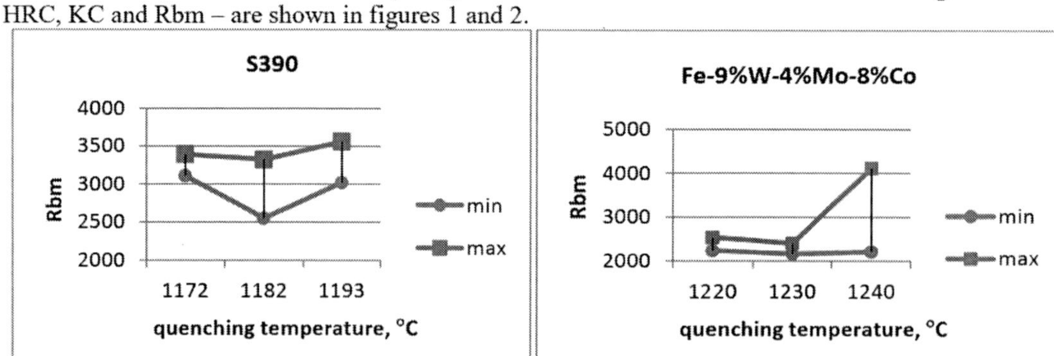

a b

Figure 1. Changes of transverse bend of high-speed steel S390 (a) and Fe-9%W-4%Mo-8%Co (b) depending on quenching temperature.

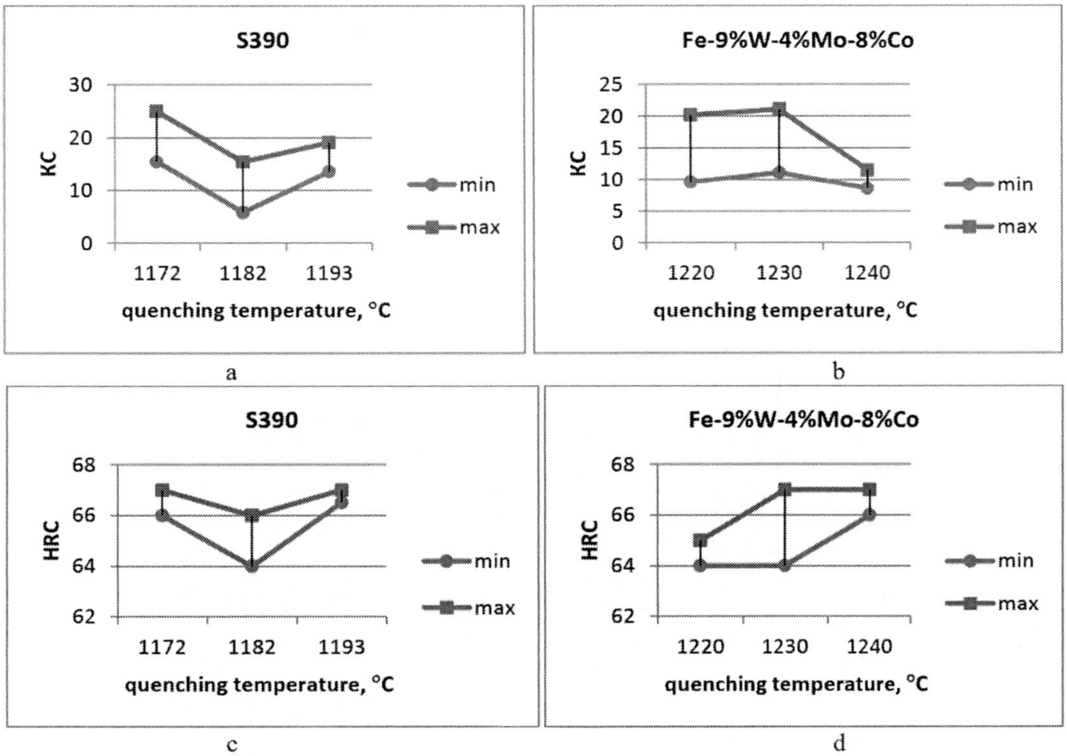

Figure 2. Changes of impact viscosity (a. b) and hardness (c, d) depending on quenching temperature.

Powder steel demonstrates sensitivity to the lack of 3rd preheating before quenching. The decrease of the mechanical properties S390 is observed at tempering temperature from 1182 °C.
Properties of steel Fe-9%W-4%Mo-8%Co change ambiguously.

There is a large range of values in the samples of the same party. Steel Fe-9%W-4%Mo-8%Co hardened at the temperature from 1240 °C has higher hardness, but its range of tensile strength (Rbm) increases, and its impact viscosity (KC) reduces.

3.2. Study of the microstructure
Investigations of the microstructure were performed on samples intended for destructive testing. After the surface preparation, the samples were studied at magnification from 250 to 1.000 times. The microstructure of steels is represented by tempered martensite and carbides (figure 3).

Very fine carbides are homogeneously distributed throughout the volume of the powder steel microstructure. Carbide heterogeneity is marked in the microstructure of steel Fe-9%W-4%Mo-8%Co: here the large carbides of different shapes, which form clusters and chains, are visible. The quenching temperature increase causes the grain growth of austenite and formation of large laminated martensite in steel Fe-9%W-4%Mo-8%Co.

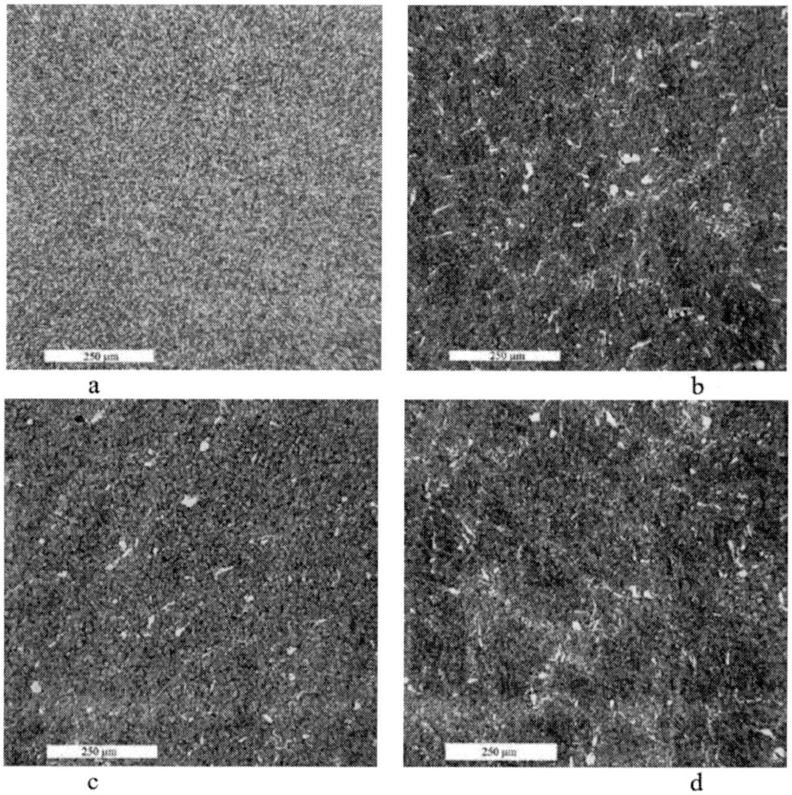

Figure 3. Microstructure of steels: a – steel S390; b, c, d – steel Fe-9%W-4%Mo-8%Co, quenching temperature, 1220 °C, 1230 °C, 1240 °C, respectively.

3.3. Determination of the high-speed steels phase composition

Figures 4 and 5 show diffractogram of steels Fe-9%W-4%Mo-8%Co and S390.

Studies have shown that most of the designated chemical elements (Cr, Ni, Co) are in the solid solution based on iron (in martensite). Martensite is a basis of high-speed steels. Alloying elements saturation of martensite depends on treatment conditions.

There is a difference in the phase composition of steels Fe-9%W-4%Mo-8%Co and S390. For steel Fe-9%W-4%Mo-8%Co, we observe a decrease of vanadium carbide V_8C_7 (angles 2θ = 36-41°) and an increase of the diffuse peak (angles 2θ = 76-80°) corresponding to the group of Me_xC_y carbides. Complex tungsten (Fe_3W_3C) and molybdenum ($Co_6Mo_6C_2$) carbides are identified in both steels. In steel S390, cobalt is composed of complex carbides.

At the quenching temperature of 1240°C of steel Fe-9%W-4%Mo-8%Co at the angles of reflection approximately equal to 2θ = 77.5° diffused peak corresponding to carbides Cr_2C, Mn_3ZnC etc. appears. The composition of carbides corresponds to the general formula of MexCy; an increase of the diffraction peak is observed for angles 2θ = 44-45°, which corresponds to an increase of proportion of solid solution based on iron – alloyed martensite. The diffraction peak of Cr_2C carbide (2θ = 56.5°) also appears.

In steel S390, the amount of carbide Me_xC_y decreases after quenching from 1193 °C, which is associated with a transition of carbon in vanadium carbide V_8C_7 and an increasing of its proportion in steel. The amount of carbide $Co_6Mo_6C_2$ increases.

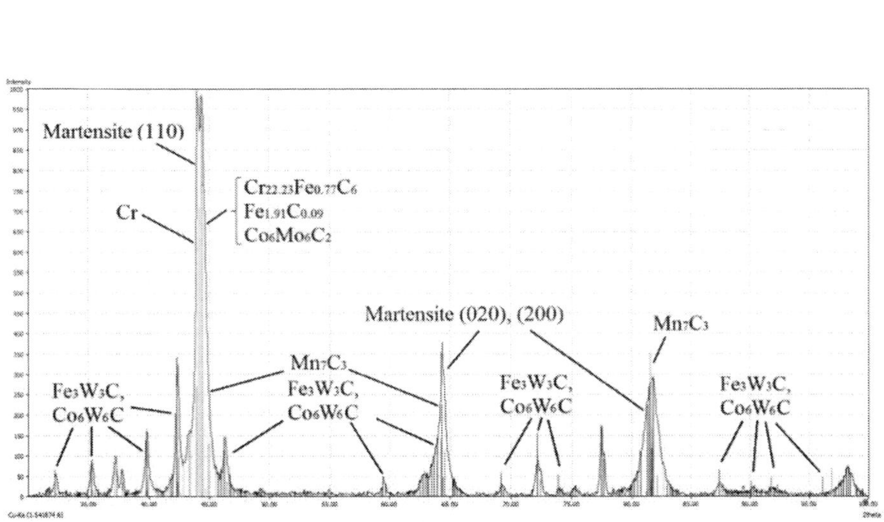

Figure 4. Diffractogram of steel Fe-9%W-4%Mo-8%Co (quenching temperature is 1220 °C).

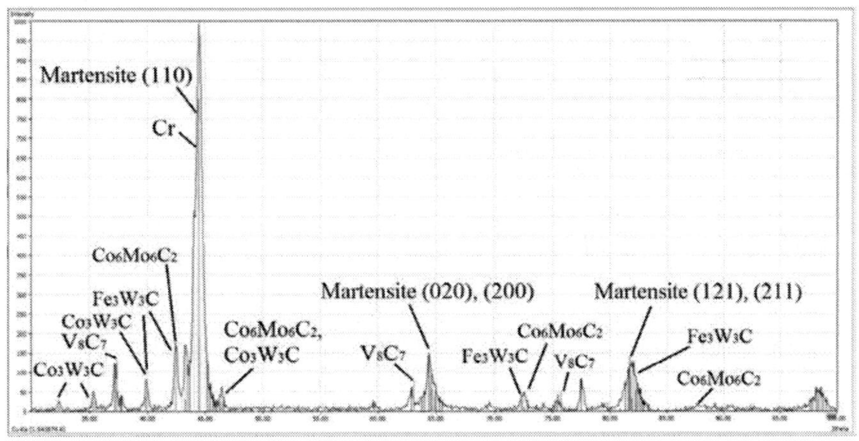

Figure 5. Diffractogram of steel S390 (quenching temperature is 1182 °C).

3.4. Determination of residual stresses

Finish milling cutters were investigated before their use. Heat treatment of the tool was carried out according to the conditions described above. Residual stresses were determined in different sections of cutters.

Measurement parameters of residual stresses: Cr-Kα radiation; the angle of diffraction – 156.4; the plane of reflection (211). Young's moduli – 217000 MPa (for S390) and 242000 MPa (for Fe-9%W-4%Mo-8%Co) and Poisson's ratio equal to 0.3 were set for the automated calculation of stresses in the program 'Xtronic'. Table 2 shows the residual stresses that are defined in different sections of the finished cutters.

Table 2. Residual stress on cutting tool surface

Steel grade	**Residual stresses, MPa**		
	Cutting edge	Shank	Milling cutter neck
S390	-1016…-1174	-610…-668	-709…-790
Fe-9%W-4%Mo-8%Co	-1189…-1300	-651…-682	-726…-821

As we can see from the table, the compressive residual stresses in the cutting edge reach high values. An excessively stressed state may be the cause of crack nucleation and premature fracture of the tool.

4. Conclusion

Our studies have shown that during heat treatment of high speed-steel in resistance chamber furnaces and in salt baths, are not possible to obtain stable mechanical properties. High-speed steels are very sensitive to changes of heat treatment conditions.

The resulting structure affects mechanical properties negatively. Eventually, the unfortunate combination of phases and the size of the structural elements are the cause of low endurance of the cutting tool.

Obtained results allow us to conclude about the important role of steel structure studying. Research findings give the opportunity to adjust the heat treatment conditions

Acknowledgments

The research was supported by the Government of the Russian Federation according to research project 2012-218-03-120. Authors are grateful to the laboratory staff of destructive and other types of testing OJSC 'IrkutskNIIhimmash' for supporting this research.

References

[1] Ponomarev B and Paykin D 2014 *Int. J. of Mech. and Mechatronics Eng.* **14** 1-5
[2] Rodygina A E 2013 *Int. J. of Eng. and Technol.* **5** 5140-5154
[3] Ahmadi K and Savilov A V 2015 *Int. J. of Machine Tools and Manufacture* **89** 208-220
[4] Ahmadi K and Altintas Y 2013 *Int. J. of Machine Tools and Manufacture* **68** 63-74
[5] Zamashchikov Y I 2014 *Int. J. of Machining and Machinability of Mat.* **16** 187-211
[6] Zamashchikov Y I 2007 *Int. J. of Machining and Machinability of Mat.* **2** 378
[7] Zamachtchikov Y I, Breaban F, Vantomme P, Deffontaine A 2002 *Lasers in Eng.* **12** 27
[8] Nikolaeva E P 2016 *Mat. Sci. Forum* (Chelyabinsk) vol 870 (Chelyabinsk: SUSU) pp 500-50
[9] Hauk V 1997 *Structural and residual stress analysis by nondestructive method* (New York: Elsevier)
[10] Suominen L, Khurshid M, Parantainen J 2013 *Procedia Eng. 5th Fatigue Design Conf.* vol 66 (France: Senlis) pp 181-191

Reasons for crack nucleation in welded joints of main gas-pipelines after a long-term operation

P O Maruschak[1], **R T Bishchak**[2], **L S Shlapak**[2] **and S V Panin**[3,4]

[1] Ternopil Ivan Pul'uj National Technical University, 56, Ruska str., Ternopil, 46001, Ukraine
[2] Ivano-Frankivsk National Technical University of Oil and Gas, 15 Karpatska str., Ivano-Frankivsk, 76000, Ukraine
[3] Institute of Strength Physics and Materials Science RAS, 2/4, Akademicheskii ave., Tomsk, 634021, Russia
[4] National Research Tomsk Polytechnic University, 30, Lenin ave. Tomsk, 634050, Russia

E-mail: Maruschak.tu.edu@gmail.com

Abstract. A crack of operational origin in the welded joint of the main gas pipeline is analyzed. The reasons for its nucleation and impact on technological microdefects that were formed earlier during the welding are found. Micromechanisms and stages of nucleation and propagation of the crack are investigated.

1. Introduction

The main gas pipeline 'Urengoy - Pomary – Uzhgorod' was commissioned in 1984 to supply natural gas from the deposits in the north of Western Siberia to Central and Western Europe [1]. The total length of the pipeline is 4451 km, and on the territory of Ukraine, its length is equal to 1160 km. So, in 2016, its service life will make up 32-33 years. Diagnostic measures of its current mechanical state allow detecting a significant amount of defects [2, 3].

Of particular importance under main gas pipelines exploitation is the fact that along with operating defects, they often have defects of welded joints, which are formed under a building stage (results from imperfect production technology). Typically, the fault repair requires a significant amount of work, cutting out and replacing defective parts of the pipeline. In addition, it is important to understand the real mechanisms of the weld damage, the conditions of its degradation and fracture [1-3]. We analysed a fragment of the pipeline with a diameter of 1420 mm, which was built in 1982-1983. The pipeline made of imported steel X70 is designed to work under pressure of 7.4 MPa. Pipe billets for the pipeline were produced at Khartsyzsk Pipe Plant.

The objective of this paper is to analyze the reasons for cracking in welded joints of main gas pipelines.

2. Experimental

To study the reasons of defect nucleation in welded joints as well as crack propagation from them (I), Figure 1a, b, comprehensive studies of a fragment of the main gas pipeline were conducted. They included mechanical testing (to evaluate mechanical properties of the metal in the weld zone), visual inspection, metallographic analysis (with the use of the AXIOVERT 40- MAT microscope). Thin sections were cut from a fragment of the pipe, which were then ground with sandpaper with gradual reduction in the grit. Finally, it was polished with the diamond paste. The prepared surface of the thin section was etched in the 4 % nitric acid solution in ethanol.

Mechanical testing of the specimen cut along the welded joint of the pipeline with an operating defect was performed under static uniaxial tension with the use of hydraulic testing machine R100 (Figure 1c, d). Fractographic analysis of the specimen's fracture surface was performed with the help of scanning electron microscope REM-106.

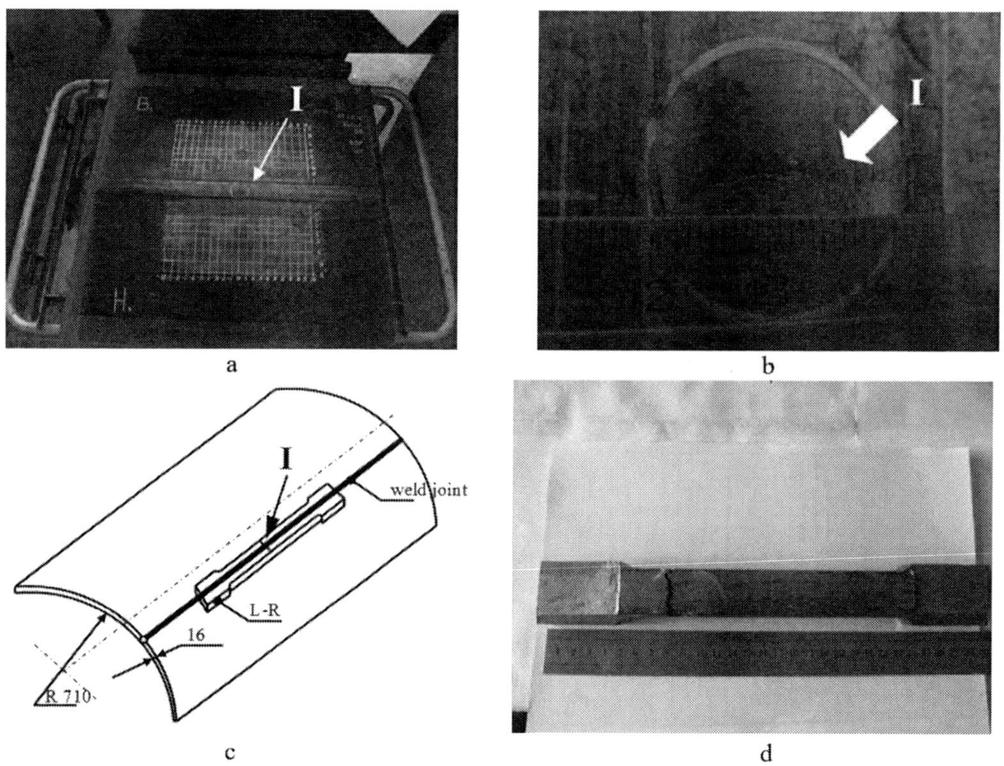

Figure 1. The general view of the pipe fragment (a) and the welded joint with a crack (b), the specimen cutting scheme (c) and the fractured specimen (d); I – defect.

3. Analysis of the fracture surface of the welded joint

The analysis of the specimen fracture surface has shown that it is covered with a layer of oxide scale without a metallic luster typical of cold cracks. In our opinion, the defect is a typical cold crack, Figure 2a. It was formed in the zone adjacent to the outer surface of the welded joint.

Based on the analysis of the fracture surface in the vicinity of the macroconcentrator with the use of the scanning electron microscope, the areas of nucleation (A) and growth (B) of the defect were revealed. It should be emphasized that almost the entire surface inside the defect was covered with a layer of corrosion, Figure 2c. However, its presence does not preclude observation of a 'chevron pattern', which 'rays' are oriented at the angle of 70-80 degrees to each other, Figure 2c. Given that the edges of the chevron pattern are always oriented towards the direction opposite to the crack propagation, the concentrator which initiates fracture behind the 'convergence' of the chevron pattern was revealed. This allows us to conclude that the crack occurred in the subsurface layer of the welded joint and propagated from the outer to the inner surface.

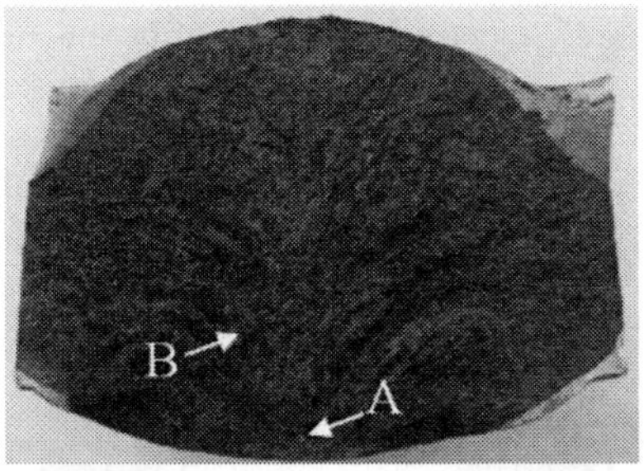

a

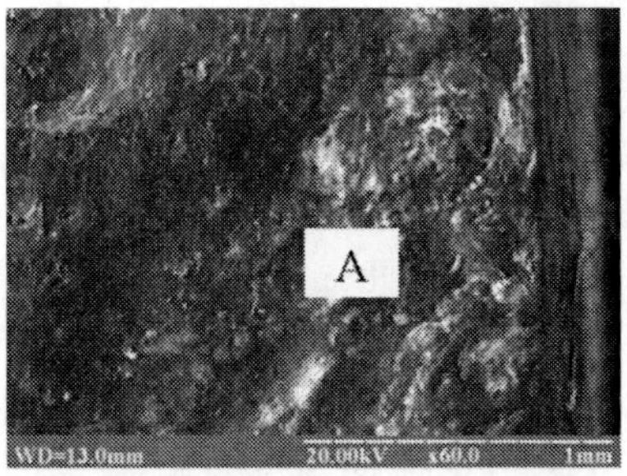

b

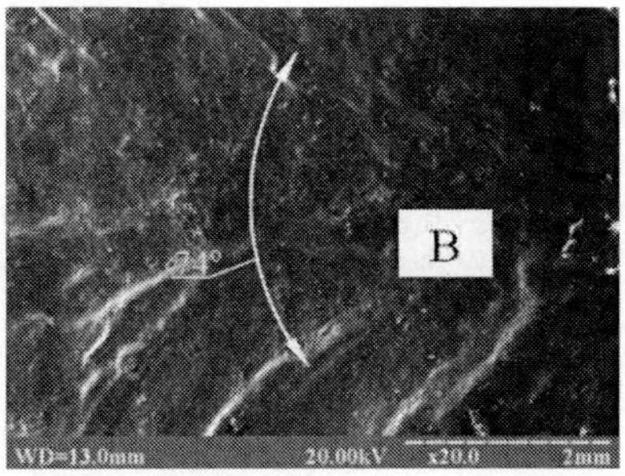

c

Figure 2. The general view of the opened crack in the welded joint, and the zone of its start (b) and propagation (c)

4. Metallographic analysis of the welded joint and discussion

The microstructure of the pipe X70 steel of the gas pipeline 'Urengoy- Pomary-Uzhgorod' cut out in different directions was studied in [1]. It is established that the microstructure consists of ferrite, fine pearlite and belongs to the ferritic-pearlitic class. The grain size corresponded to the 9-11 rank according to GOST 5639-82.

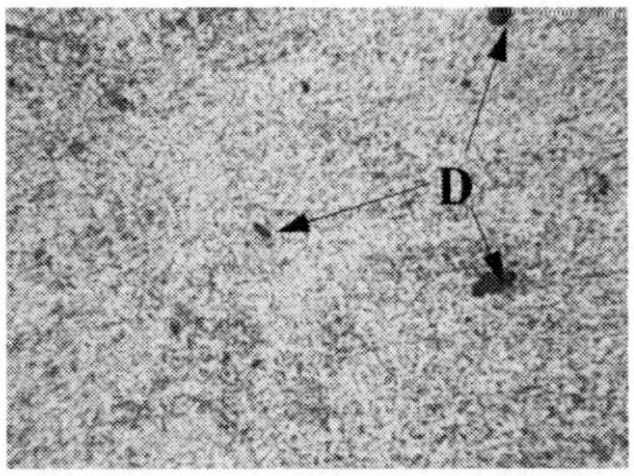

a

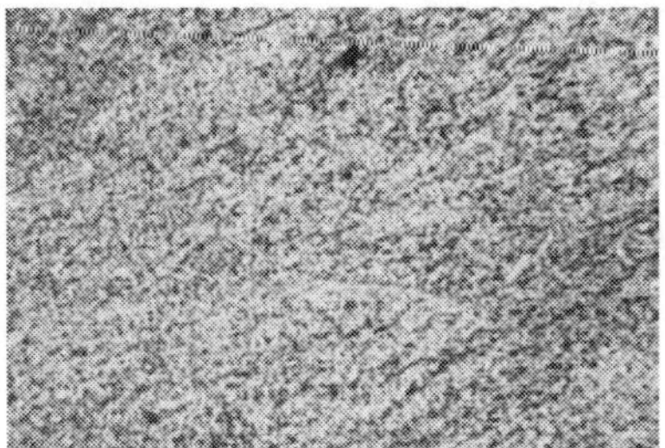

b

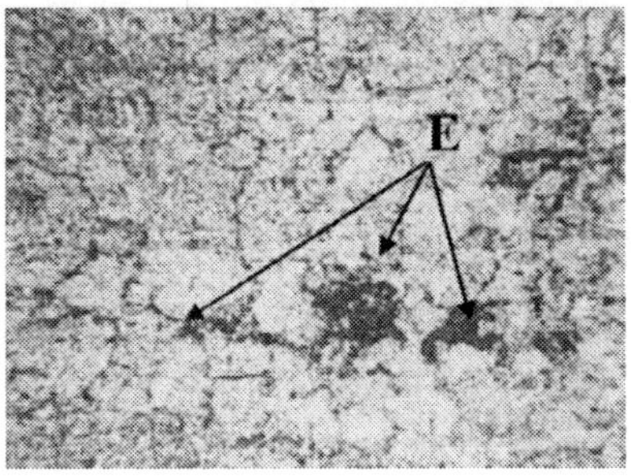

c

Figure 3. The microstructure of the welded joint – ferrite-bainite mixture (×100); D – microdefects, E – damaging at the boundaries between 'coarse grains'.

Fine crystallites of the 'ordered' structure were found in the structure of the welded joint. These crystallites can serve as an indicator [2, 4] of the direction of crystallization of the weld metal from a liquid state, Figure 3a. The structure of the weld metal represents a ferrite-bainite mixture, Figure 3b. In the vicinity of the area, where the crack nucleation took place, a weld overheating region was found, where 'coarse grains' were found. Their shape accounted for a significant length of weakened boundaries, and contributed to the formation of laminations during the crack formation process, Figure 3c. Metallographic studies conducted on etched thin sections allowed revealing small micropores in the area of crack propagation, which may serve as evidence of hydrogen impact on the weld material [5]. Defects of this type are known, which were found in the vicinity of a through transverse crack in the pipeline with a diameter of 1420 mm. The estimated technological reason for their formation is very rapid and uneven cooling of the welded joint. While analyzing metallographic images taken from the area of the welded joint, it was found that most probably the crack emerged on the surface of the welded joint in the areas with coarse grains. Numerous micropores with a diameter of a few micrometers and cracks oriented along the boundaries between several grains served as additional factors [6], which contributed to the formation of a macrodefect.

In our opinion, cracks were formed in the weld zone under the influence of damage accumulated in the grain boundaries [7, 8]. This damaging took place in the first hours after welding, however, it was scattered and could not coalesce, because the level of welding stress at the site was inadequate for its growth. After mounting the pipe in the pipeline, the local growth and coalescence of microdefects took place under the influence of operational factors. To our mind, this technological effect could be the 'imposition and summation' of welding and technological stresses, resulting in the localization of microdeformation processes. The additional impact could result from hydrogenation of the welded joint. Therefore, from the analysis of the damage, we can conclude that the defect was formed due to adverse structural and mechanical factors: the size of austenitic grain, a significant gradient of the grain structure in the weld zone, and the 'consolidation' of grains.

5. Conclusions

The analyzed crack of the welded joint was of the operational origin; however, its nucleation is a result of the impact of the earlier created technological microdefects that were formed during the welding. Fractographic and metallographic studies allowed revealing the area of crack nucleation and micromechanisms of crack propagation. The study of microstructural features of the weld zone and its comparison with the crack growth mechanisms made it possible to reveal the reasons of its origin, stages of nucleation and the growth of a macrodefect.

Acknowledgments

The reported study was funded by the Ministry of Education and Science of Ukraine, according to research project No. 0116U006908.

References

[1] Krasovskii A Ya, Lokhman I V and Orynyak I V 2012 *Strength of Materials* **44** 129-143
[2] Tsyrul'nyk O T, Voloshyn V A, Petryna D Yu, Hredil' M I and Zvirko O I 2011 *Mat. Sci.* **46** 628
[3] Tsyrul'nyk O T, Kryzhanivs'kyi E I, Petryna D Yu, Taraevs'kyi O S and Hredil' M I 2004 *Mat. Sci.* **40** 844-849
[4] Khotinov V A, Farber V M, Morozova A N, Valov M A and Sharipov G A 2014 *Metal Sci. and Heat Treatment* **56** 315-319
[5] Rybakov A A, Goncharenko L V, Filipchuk T N, Lokhman I V and Burak I Z 2014 *The Paton Welding J.* **3** 49-52
[6] Shirinzadeh-Dastgiri M, Mohammadi J, Behnamian Y, Eghlimi A and Mostafaei A 2015 *Eng. Failure Analysis* **53** 78-96
[7] Bordbar S, Alizadeh M, Hashemi S H 2013 *Mat. and Design* **45** 597-604
[8] Marushchak P O, Salo U V, Bishchak R T, Poberezhnyi L Ya 2014 *Chem. and Petr. Eng.* **50** 58-61

Development of SHS azide technology of silicon carbide nanopowder

Y V Titova[1], A Yu Illarionov[1], A P Amosov[1], D A Maidan[1] and K S Smetanin[2]

[1] Samara State Technical University, 244, Molodogvardeyskaya str., Samara, 443100, Russia
[2] Vyksa Metallurgical Plant, 45, Br. Batashevykh str., Vyksa, Nizhegorogskii region, 607060, Russia

E-mail: titova600@mail.ru

Abstract. The possibility of increasing the purity and yield of the SiC nanopowder using an intermediate synthesis of silicon nitride (Si_3N_4) by azide technology of self-propagating high-temperature synthesis based on the use of a powder of sodium azide (NaN_3) as a nitriding reagent was investigated. As a result of combustion of the initial mixture of powders '19Si + $6NaN_3$ + $(NH_4)_2SiF_6$ + 20', a powdery product was obtained, consisting almost entirely of SiC (about 90%) with an admixture of Si_3N_4 and Si (about 5% each), and representing nanoparticles (80-150 nm) combined into agglomerates with the size up to 50 μm.

1. Introduction

Silicon carbide (SiC) is an important ceramic nonoxide material used in mechanical engineering, because of the unique set of properties: low density, high hardness, high-temperature strength and high chemical resistance [1]. For many years, the silicon carbide powder has been used as an excellent abrasive 'carborundum' in the free state and in the various abrasive and cutting tools in the bound state [2, 3]. The SiC powder is used also to sinter high-quality technical ceramics for a structural purpose with very good mechanical properties. This ceramics is used to produce two groups of components: wear – and corrosion-resistant components (mechanical seals, bushings, shafts and bearings of pumps) and high-temperature components (combustion chambers, rotors and blades of gas turbines, heat exchangers and fans) [1]. Recently, much attention has been paid to the development of nanostructured ceramics, which is sintered from nanopowders and its properties markedly exceed the properties of traditional technical ceramics [4].

The main part of SiC is produced by a simple and cheap Acheson method based on the reduction of SiO_2 by carbon in core-type electric resistance furnaces at 2200-2700°C with a process time of about 40 h. The resulting blocks of crystals larger than 10 mm are then ground and sieved, but the product is not enough quality in composition and structure and requires additional treatment [5]. SiC manufactured by this method generally has a coarse grain, with a mean particle size of 5 μm at best [1]. These coarse powders are mainly used as abrasives, for the manufacture of electric heaters, refractories and structural ceramics, partially. Smaller and pure SiC powders, necessary for the production of structural ceramics, can be obtained by the Acheson method from the more homogeneous mixture of small and pure Si and C powders at a low temperature of 1500-1800°C. Since the eighties of the last century, it is known that the smallest submicron and nanosized SiC powders (less than 0.1 μm) can be synthesized by methods of vapour phase reactions of SiH_4 or $SiCl_4$ with hydrocarbons CH_4 and C_3H_8 or thermal decomposition of complex chemical compounds CH_3SiCl_3, $(CH_3)_4Si$ or carbosilane, as well as the evaporation of solid-phase reagents Si and C in the atmosphere of a mixture of inert gas with H_2 or N_2 [1]. Then these methods were supplemented by the sol-gel method, plasma and microwave methods [6-8]. However, all these methods are high energy-consuming, prolonged, complicated, resulting in high cost of nanopowders, therefore, attempts are made to date to develop new methods for the preparation of SiC nanopowders.

Among the new methods, the method of combustion synthesis or the method of self-propagating high-temperature synthesis (SHS) of silicon carbide is of undeniable interest, because it is characterized by low energy consumption, short duration of the process, high purity products, simple equipment, wide possibilities of regulation of dispersed structure of powders: from single crystal grains to nanosized particles [9]. However, the problem is that the reaction of direct synthesis of silicon carbide from elements

$$Si + C = SiC + 73 \text{ кJ/mole} \quad (1)$$

does not have sufficiently high heat liberation, and, thus, has a relatively low adiabatic combustion temperature (1173 K). Hence, this binary system (1) fails to implement a self-sustaining combustion process. Therefore, for the synthesis of SiC in the SHS mode, different types of the 'activation' process are used, among which are pre-heating the initial mixtures [10], addition of chemical activators [11], the use of intermediate high-exothermic synthesis of Si_3N_4 by combustion of a mixture of Si+C in a nitrogen atmosphere or air [12, 13], mechanical activation of the initial mixture [14].

Since 1970, in Samara State Technical University, the azide technology of self-propagating high-temperature synthesis (SHS-Az) has been developed, which allows us to obtain micro- and nanopowders of nitrides and compositions on their basis with the use of a powder of sodium azide (NaN_3) as a nitriding reagent and halide salts [15]. The SHS-Az method was used for the preparation of nanopowder of SiC through intermediate synthesis of Si_3N_4 using energy additive of aluminum [16]:

$$14Si + 3NaN_3 + (NH_4)_2SiF_6 + 15C + Al = 15SiC + Na_3AlF_6 + 5N_2 + 4H_2. \quad (2)$$

As the result of burning this mixture of powders, a powdery product was obtained based on silicon carbide nanopowder of composition: β-SiC – 48.6 wt. %, α-Si_3N_4 – 27.0%, β-Si_3N_4 – 5.8%, Na_3AlF_6 – 18.6%. The particle size of silicon carbide was in the range of 50-150 nm, and whiskers of silicon nitride had a diameter of about 100 nm. Although the yield for the silicon carbide was about 50%, this composite nanopowdery product was successfully used for modification of aluminum alloy [17].

This work is a continuation of the research initiated in [16], with the aim of increasing the purity and yield of SiC nanopowder during the synthesis by SHS azide technology without the use of energy additive of aluminum.

2. Materials and methods of investigations

For preparation of more pure silicon carbide and elimination of Na_3AlF_6 impurity, it was necessary to exclude the addition of aluminum to the initial mixture of powders and to investigate the influence of the carbon content in the initial mixture by bringing it to 20 moles, when all initial silicon Si is bound into silicon carbide SiC, and the nitride Si_3N_4 is not contained in the final product:

$$19Si + 6NaN_3 + (NH_4)_2SiF_6 + 5C = 5Si_3N_4 + 5SiC + 6NaF + 4H_2, \quad (3)$$

$$19Si + 6NaN_3 + (NH_4)_2SiF_6 + 8C = 4Si_3N_4 + 8SiC + 6NaF + 4H_2 + 2N_2, \quad (4)$$

$$19Si + 6NaN_3 + (NH_4)_2SiF_6 + 11C = 3Si_3N_4 + 11SiC + 6NaF + 4H_2 + 4N_2, \quad (5)$$

$$19Si + 6NaN_3 + (NH_4)_2SiF_6 + 14C = 2Si_3N_4 + 14SiC + 6NaF + 4H_2 + 6N_2, \quad (6)$$

$$19Si + 6NaN_3 + (NH_4)_2SiF_6 + 17C = Si_3N_4 + 17SiC + 6NaF + 4H_2 + 8N_2, \quad (7)$$

$$19Si + 6NaN_3 + (NH_4)_2SiF_6 + 20C = 20SiC + 6NaF + 4H_2 + 10N_2. \quad (8)$$

For a preliminary analysis of the temperature of combustion of the starting components and the composition of the synthesis products, the thermodynamic calculations were performed using the 'Thermo' program [18].

The following powders were used in the experimental study as starting raw materials: silicon powder of KR0 brand (purity ≥ 98.8 wt. %, particle size < 10 μm), carbon black of P701 brand (88%, 0.2 μm), sodium azide NaN_3 (98.7%), ammonium hexafluorosilicate $(NH_4)_2SiF_6$ (99.0%). The methodology of experimental investigations in reactor SHS-Az of constant pressure with a volume of 4.5 l was described in [15]. The cylindrical samples (30 mm diameter, 45 mm height) of initial powders mixture of bulk density were burned. Phase composition of the powdery combustion product was determined on automated X-ray diffractometer ARL X TRA 'Thermo Scientific'. Shooting X-ray spectra was performed using Cu-radiation with continuous scanning in the range of angles $2\theta = 20°\div80°$ at a speed of 2 deg/min. The obtained spectra were processed using a special software package WinXRD. Quantitative phase analysis was carried out by full-profile analysis (Rietveld method) using the software PDXL 1.8.1.0 using the crystallographic open database (COD). The study of the morphology and chemical element composition of powder particles was performed on the scanning electron microscope JSM-6390A 'Jeol' with the EDS attachment JED-2200 'Jeol'. The specific surface of powders was determined by the adsorption method on the device 'Sorbtometr'.

3. Results and discussion

Figure 1 presents the results of thermodynamic calculations showing values of the adiabatic combustion temperature, the enthalpy change of the system in the reaction and the composition of the synthesis products with different contents of carbon in the initial mixture.

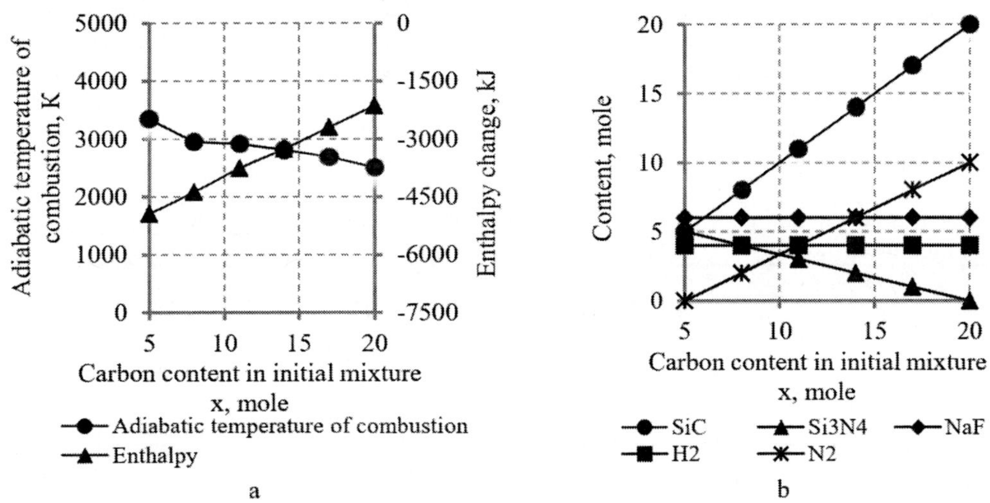

Figure 1. The results of thermodynamic calculations of combustion parameters (a) and composition of synthesis products (b).

The significant reduction of heat liberation and adiabatic temperature of combustion is seen with the increasing content of carbon black in the initial powder mixture (see figure 1 (a)). Referring to figure 1 (b), when the carbon content is near 5 moles, the reaction products contain silicon carbide, silicon nitride, sodium fluoride and hydrogen. With increasing content of carbon black to 20 moles, the products of synthesis become free of silicon nitride. The obtained results of thermodynamic analysis are consistent with equations (3)-(8). Thus the results of thermodynamic calculations show that the initial mixture '$19Si+6NaN_3+(NH_4)_2SiF_6+20C$' is optimal for the synthesis of silicon carbide by SHS azide process because the combustion of this mixture gives rise to the target silicon carbide and by — products (nitrogen, hydrogen and sodium fluoride), which are easily removed and do not contaminate the final product.

The results of experimental determination of temperature and velocity of combustion is presented in figure 2.

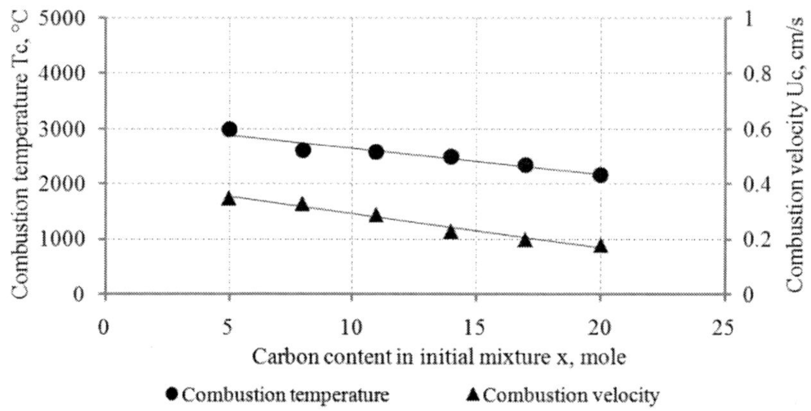

Figure 2. Combustion parameters of mixture $19Si + 6NaN_3 + (NH_4)_2SiF_6 + xC$.

The figure shows that the temperature and velocity of combustion reduce with increasing carbon black content (x) in the initial mixture which is consistent with the results of thermodynamic calculations. Let us note that the experimental values of the combustion temperature are below the theoretical ones. This is because, the heat of combustion is partially consumed for heating the environment. In addition, the real composition of the combustion products differs somewhat from the composition in accordance with the thermodynamic calculations, that is seen in table 1 and figure 3 from the results of X-ray phase analysis of the products of chemical reactions (3)-(8).

Table 1. The phase ratio in washed products of combustion.

Carbon content in initial mixture x, mole	Composition of combustion product, wt. %				
	SiC	α-Si_3N_4	β-Si_3N_4	Si	SiO_2
5	8.40	66.56	23.80	0	1.24
8	10.80	62.10	26.00	1.10	0
11	50.30	27.40	21.40	0.90	0
14	64.70	19.50	14.00	1.80	0
17	70.70	24.70	4.60	0	0
20	89.40	5.50	0	5.10	0

It follows from the presented data that the increase in carbon content in the starting material leads to the change of the ratio of the phases in the products of combustion. The washed products of the combustion of mixtures with a carbon content up to 17 moles contain three main phases: β-SiC, α-Si_3N_4, β-Si_3N_4 and impurities of Si or SiO_2 in small proportion. (Sodium fluoride (NaF) due to its good solubility is completely removed as a result of washing in distilled water.) The maximum yield of silicon carbide of about 90% is observed with 20 moles of carbon black, as a result of the combustion; combustion products also contain side products: about 5% Si_3N_4 and 5% Si. An increase of the content of carbon black above 20 moles did not make sense, since this would have led to the increase in the content of free silicon in the final product of combustion.

Figure 4 shows the morphology and the results of EDS analysis of the powder particle synthesized from two mixtures with 5 and 20 moles of carbon black. Figure 4 (a) demonstrates the shape of columnar crystals with a diameter of from 100 to 200 nm which is typical for silicon nitride. The presence of carbon in these crystals marked with numbers 1-6 can be explained by the fact that the penetration depth of the electron beam is larger than a single crystal of Si_3N_4 (table 2). Silicon carbide is synthesized in the form of equiaxed particles in the size range of 80 to 150 nm, combined into agglomerates with the size up to 50 μm. The specific surface of the powder of silicon carbide synthesized from '$19Si + 6NaN_3 + (NH_4)_2SiF_6 + 20C$' mixture is 19.16 m^2/g.

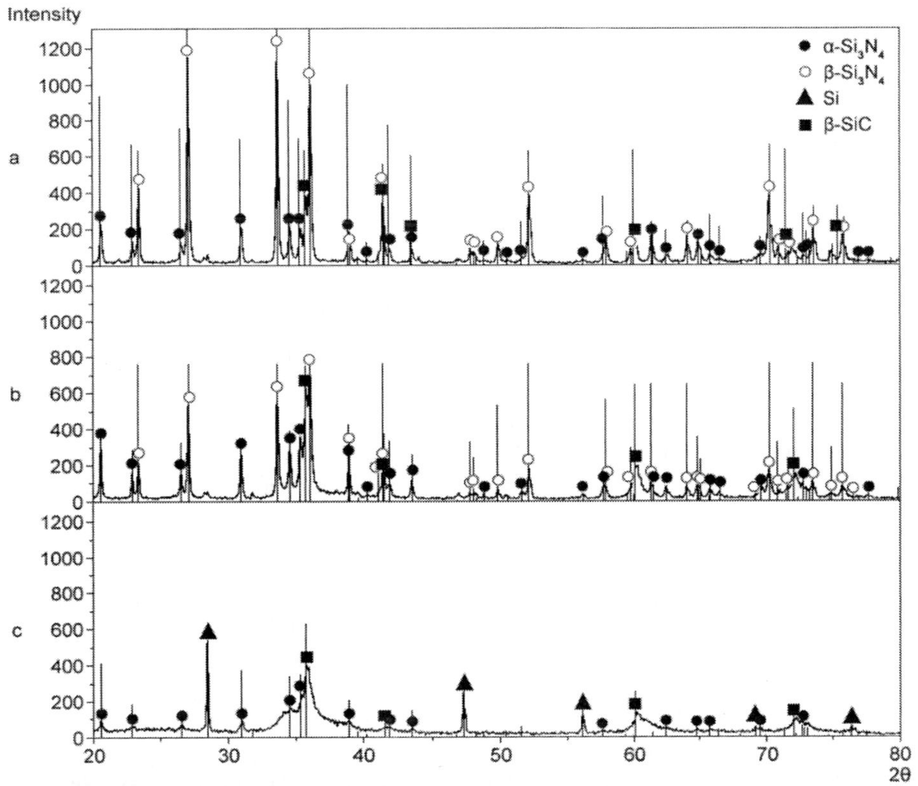

Figure 3. XRD patterns of combustion products of mixture 19Si + $6NaN_3$ + $(NH_4)_2SiF_6$ + xC after washing: (a) x = 5 moles, (b) x = 11 moles, (c) x = 20 moles.

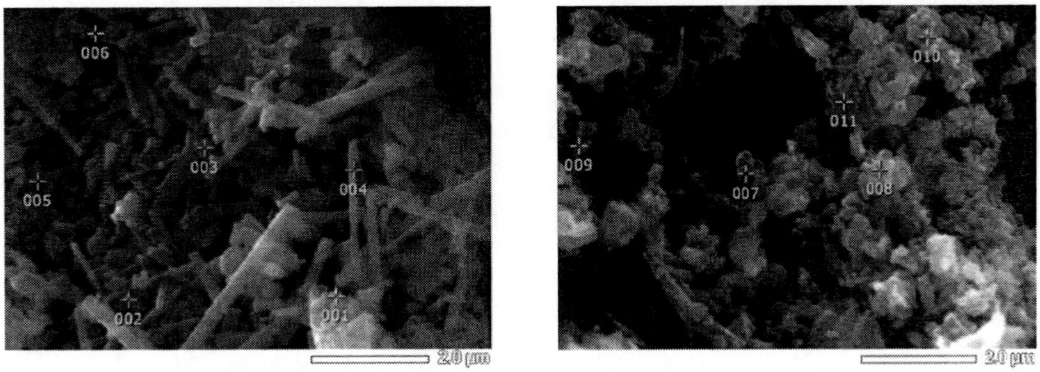

a b

Figure 4. SEM image and EDS analysis of powders synthesized from mixtures: (a) 19Si + $6NaN_3$ + $(NH_4)_2SiF_6$ + 5C, (b) 19Si + $6NaN_3$ + $(NH_4)_2SiF_6$ + 20C.

Table 2. Content of chemical elements at points marked in figure 4, wt. %.

Chemical element	Marker number										
	001	002	003	004	005	006	007	008	009	010	011
C	34.62	19.47	27.72	17.07	18.44	10.39	59.21	67.51	58.12	75.54	66.63
N	–	23.32	8.37	13.77	20.23	36.40	9.32	–	12.47	–	–
F	0.48	0.62	0.51	0.27	0.47	0.53	–	–	–	–	–
Na	0.29	–	0.31	0.08	0.19	0.13	1.09	1.32	0.84	0.70	1.08
Si	64.61	56.59	63.10	68.81	60.67	52.54	30.38	31.17	28.47	23.75	32.29

4. Conclusion

Application of azide technology of SHS has allowed us to obtain from the initial powder mixture '19Si + $6NaN_3$ + $(NH_4)_2SiF_6$ + 20C', a powdery product consisting almost entirely of β-SiC (89.4%) with an admixture of α-Si_3N_4 (5.5%) and Si (5.1%), and representing nanosized particles (80-150 nm) combined into agglomerates with the size up to 50 μm.

Acknowledgements

The authors thank V A Novikov for capturing diffraction patterns for X-ray phase analysis of the products of SHS-Az.

References

[1] Yamada K and Mohri M 1991 *Silicon Carbide Ceramics – 1: Fundamental and Solid Reaction*, ed S Somiya and Y Inomata (London and New York: Elsevier) p 13-44
[2] Jackson M J and Hitchiner M P 2013 *High Performance Grindingand Advanced Cutting Tools (*New York: Springer)
[3] *Ceramic Cutting Tools: Materials, Development, and Performance* 1994, ed E Dow Whitney (New Jersey: Noyes Publications)
[4] Basu B and Balani K 2011 *Advanced Structural Ceramics* (Hoboken, New Jersey: WILEY)
[5] Inomata Y 1991 *Silicon Carbide Ceramics – 1: Fundamental and Solid Reaction*, ed S Somiya and Y Inomata (London and New York: Elsevier) p 1-12
[6] Najafi A, Fard F G, Rezaie H R and Ehsani N 2012 *Powder Technol.* **219** 202-10
[7] Lin H, Gerbec J A, Sushchikh M and McFarland E W *Nanotechnology* 2008 **19**(32) 325601-8
[8] Mukhopadhyay A K, Datta S and Basu D 2009 *Bull. Mater. Sci.* **32**(1) 1-13
[9] Mukasyan A S 2011 *Properties and Applications of Silicon Carbide,* ed R Gerhardt (Janeza Trdine: In Tech) p 389-409
[10] Pampuch R, Stobierski L, Liz J and Raczka M 1987 *Mat. Res. Bull.* **22** 1225-31
[11] Nersisyan G A, Nikogosov V N, Kharatyan S L and Merzhanov A G 1991 *Combust. Explos. Shock Waves* **27**(6) 720-4
[12] Yamoda O, Hirco K, Koizumi M, et. al. 1989 *J. Am. Ceram. Soc.* **72**(9) 1735-8
[13] Kata D, Lis J, Pampuch R, Stobierski L 1998 *Int. J. Self-Propag. High-Temp. Synth.* **4**(7) 475-85
[14] Mukasyan A S, Lin Y-C, Rogachev A S and Moskovskikh D O 2012 *J. Am. Ceram. Soc.* 1-7
[15] Bichurov G V 2015 *Nitride Ceramics: Combustion Synthesis, Properties, and Applications*, ed A A Gromov and L N Chukhlomina (Weinheim: Wiley-VCH) pp 229-263
[16] Titova Y V, Amosov A P, Ermoshkin A A, Markov Y M, Khusainova T N and Popova A V 2014 *Rus. J. Non-Ferr. Met.* **55**(6) 620-6
[17] Amosov A P, Titova Y V, Maydan D A, Ermoshkin A A and Timoshkin I Y 2015 *Rus. J. Non-Ferr. Met.* **56**(2) 222-8
[18] Mamyan S S 2002 *Key Eng. Mat.* **217** 1-8

Contact stresses modeling at the Panda-type fiber single-layer winding and evaluation of their impact on the fiber optic properties

Yu I Lesnikova, O Yu Smetannikov, A N Trufanov and N A Trufanov

Perm National Research Polytechnic University, 29, Komsomolsky ave., Perm, 614990, Russia

E-mail: ulesig@gmail.com

Abstract. The impact of contact transverse forces on the birefringence of the single-mode polarization-maintaining Panda-type fiber is numerically modeled. It has been established that with a single-row power winding on a cylindrical mandrel, the fiber tension at winding is the principal factor that influences birefringence. When coiling the fiber based on the local defect microbending, the birefringence at the microbending point differs from that of the free fiber by 1.3%.

1. Introduction

Polarization-maintaining optical fibers are widely used for the production of interferometric sensors, fiber-optic gyroscope sensors and in the monitoring systems as distributed sensors of thermomechanic stresses [1-3]. Performance parameters of these fibers are, first of all, provided by the internal residual stress fields that are formed in the manufacturing process [4], and secondly, by the level and type of fibers mutual effects and interaction with the mandrel during power wiring under technological tests and sensing circuit coil manufacturing. As a result, the stress state of the polarization-maintaining fiber and its optical properties are determined by a combination of factors which are connected with the properties of silica fiber materials and polymer protective and strengthening coatings (PSC), manufacturing conditions, and external stresses of contact nature [5]. Papers [6, 7] consider the effect of power winding on the optical fiber characteristics disregarding PSC and contact fiber – mandrel interaction is not considered for simplicity. The present paper is focused on numerical mathematical models that enable description of technological stresses development regularities considering all the above factors.

The paper sets the boundary problems of quarts heterogeneous fiber system interaction in the two-layer linear viscoelastic polymer protective-strengthening coating with the elastic cylindrical mandrel and with neighboring fibers along the lateral surfaces under single-layer power winding. Contact boundary conditions are formulated without friction. Finite element procedure for the numerical solution of linear viscoelasticity contact problems for the fiber power winding on the mandrel has been developed and implemented. Basing on the numerical polyvariant studies, the contact stress state evolution regularities in the optic fiber have been investigated considering the residual technological stresses, tensile force at wiring, the mandrel diameter, and fiber orientation in relation to the mandrel. Residual stresses were predicted as described in [4].

2. Dependence of optomechanical characteristics of the Panda-type fiber with the protective-strengthening coating on the power effects at a single-layer winding on the aluminum coil

The finite element three-dimensional analog of the investigated structure is presented in Figure 1. The calculation was made for two fiber section orientations in relation to the radial direction of the coil – parallel (α =0°) and normal (α =90°) ones, as well as for two coil diameters – 44 and 100 mm. Fiber tension F_n changed from 0.2 to 1 H.

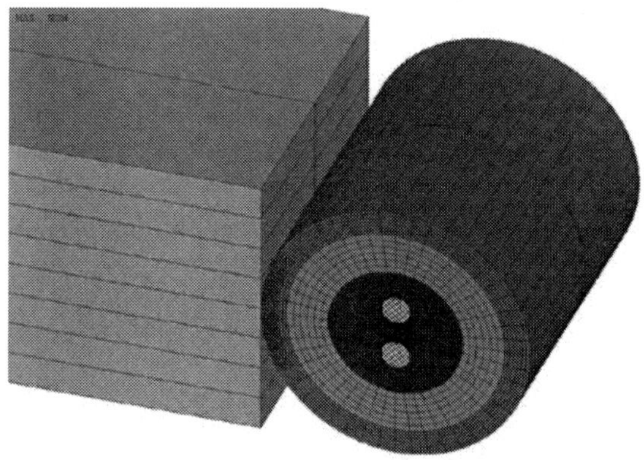

Figure 1. The finite-element model of contact fiber - coil interaction under power winding for $\alpha = 90^\circ$ and D_k =44mm.

Figure 2 illustrates first principal stress fields in the structure for two tension values F_n=0 H and F_n=0.2 H. The maximum first principal stress rises from 173 to 202 MPa. The figures also clearly demonstrate the kink effect on the fiber stress-strain state. It should also be noted that the stress level in PSC is considerably lower than in silica fiber, which results from Young's modulus of the protective coating inner cladding material, which is lower by several digits.

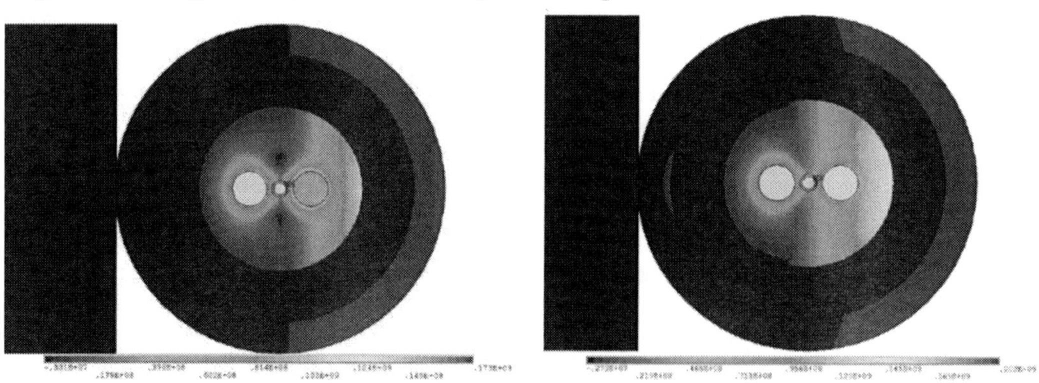

a b

Figure 2. First principal stress σ_1, Pa: a – F_n=0, α=0°, b – F_n=0.2, α=0°

The transverse contact forces change the principal stress difference fields distribution pattern in the optical fiber (Figure 3), which eventually leads to the modal birefringence change. The effect of viscoelastic properties of the protective-strengthening coating is manifested in some evolution of stresses, and, as a result, of birefringence (Figure 4), however, it is clear that the effect is negligible and the birefringence value stabilizes rapidly enough (within some minutes after winding).

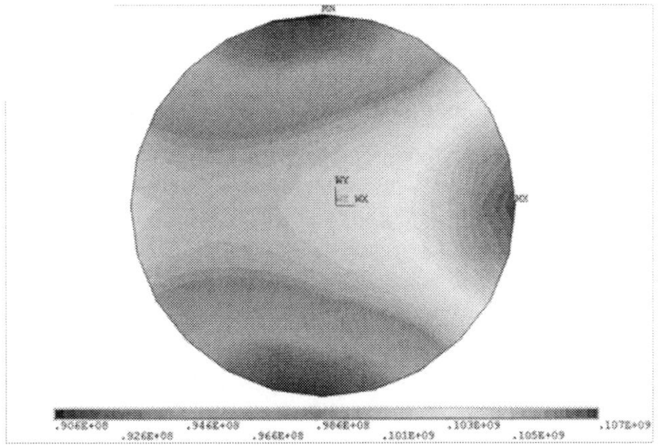

Figure 3. F_n =0.2 H. Difference of principal stresses in the optic fiber core.

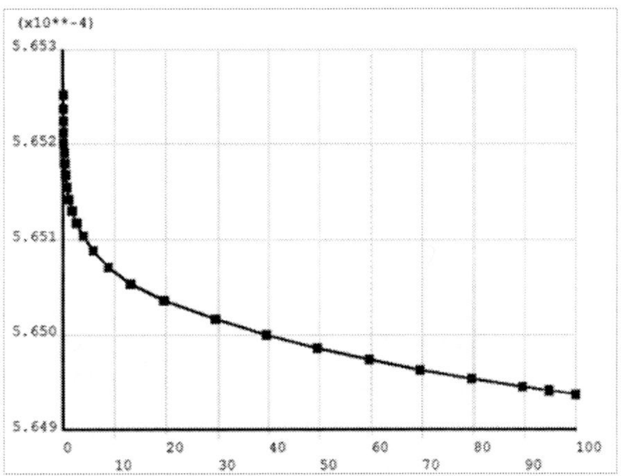

Figure 4. F_n =0.2 H. Modal birefringence time evolution.

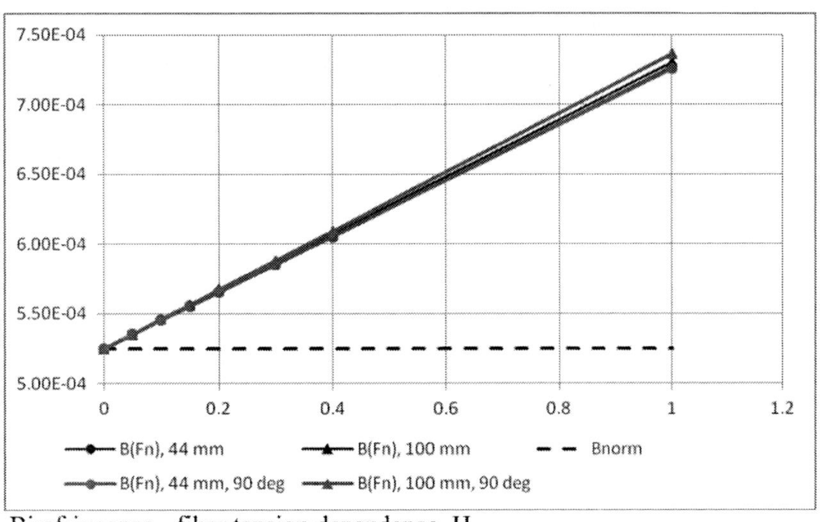

Figure 5. Birefringence - fiber tension dependence, H.

Figure 5 shows summarized data on the optical fiber modal birefringence B dependence on tension for two coil diameters and two orientations. The graph analysis demonstrates that the principal factor that affects the B value is the fiber tension during winding which forms contact stresses along the fiber – coil surface interaction boundary. As in [6], obtained dependence $B(F_n)$ is linear, moreover, birefringence does not practically depend on the coil diameter and fiber section orientation. For the maximum force studied F_n=1 H, birefringence increases by 38%. The effect of other factors (orientation and coil diameter) varies within 0.8% of average values.

3. Effect of 167 micron fiber microbending

To study the microbending effect on the fiber opto-mechanical properties, the mandrel-fiber system stress-strained state problem was solved with tension-coiled fiber and transverse kink through the 167 micron fiber lengthwise stuck to the mandrel. Such local imperfection leads to additional transverse contact forces which influence stress-strain state in the fiber core and, as a result, optical characteristics.

Isoline and stress intensity patterns (Figure 6.) show the effect of external forces on the stress pattern: fields are vertically weakly asymmetrical, which is induced by the transverse contact forces and kinking. The defined impact does not extend to the optical fiber core area. Significant redistribution of localized contact forces is provided by the protective inner shell from the highly elastic low mode incompressible material.

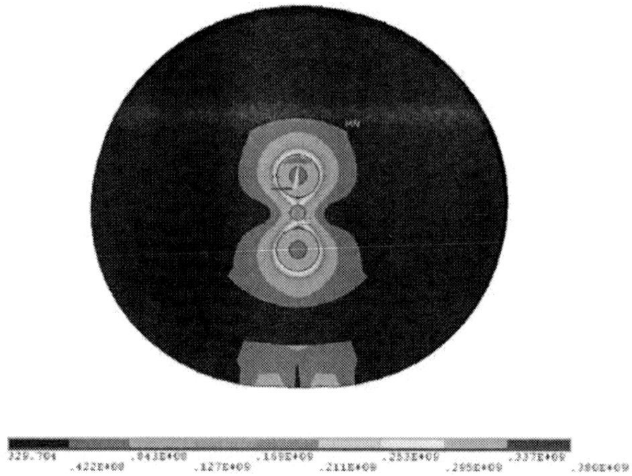

Figure 6. Stress tensor intensity in the cross-sectional plane of the PSC fiber

The numerical investigation ascertained that the birefringence value in the optic fiber core immediately above the microbending point differs from that of the free fiber (equally strained) no more than by 1.3%. The farther from the defect, the more birefringence values approach those of the free (not kinked) fiber (Figure 7).

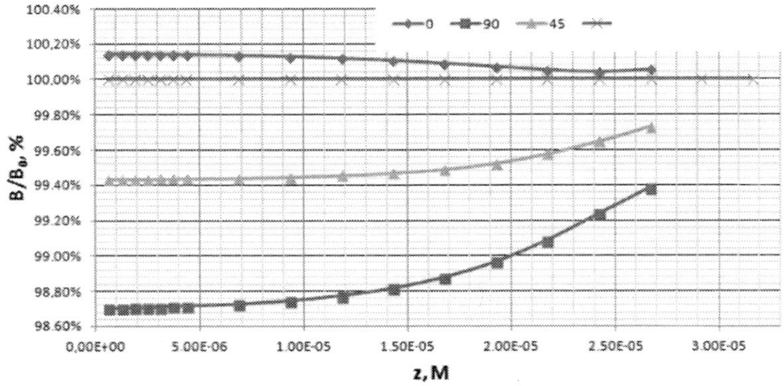

Figure 7. Relative change of modal birefringence along the fiber for three cross-section orientations (α =0°, α =90°, α =45°).

4. Conclusion

As a result of the research, a mathematical model has been constructed and its numerical realization procedure has been developed. The obtained birefringence dependence on fiber tension during its winding on the mandrel $B(F_n)$ is linear, which correlates with the results published earlier [6]. It has been determined that birefringence does not practically depend on the coil diameter and fiber section orientation. It has been established that there is modal birefringence in the section above; the microbending point differs from that of the free fiber by no more than 1.3% (at winding tension F_n =0.2 H). The farther from the defect, the closer birefringence values are to those of the free fiber (Figure 6).

Acknowledgments

The work was carried out with the support of the Russian Foundation for Basic Research, RFBR grant No. 16-48-590660 p_a.

References

[1] Lee B 2003 *Opt. Fiber Technol.* **9** 57-79
[2] Bao X, Chen L 2012 *Sensors* **12** 8601
[3] Karimi M, Sun T, Grattan K T V 2013 *IEEE Sens. J.* **13** 4459-4464
[4] Trufanov A N, Trufanov N A, Semenov N V 2016 *Opt. Fiber Technol.* **31** 83-91
[5] Guan R, Wang X, Wang X, Huang D, Liu S 2005 *Chin. Opt. Lett.* **3** 42-45
[6] Ulrich R, Rashleigh S 1982 *IEEE J. Quantum Electron* **18** 2032-2039
[7] Bløtekjær K. 1993 *Opt. Lett.* **18** 1059

IOP Conf. Series: Materials Science and Engineering **177** (2017) 012117 doi:10.1088/1757-899X/177/1/012117

Influence of stress concentrator shape and testing temperature on impact fracture regularities of pipeline steel

I V Vlasov[1,3], **S V Panin**[1,3], **P O Maruschak**[2], **D D Moiseenko**[1] **and F Berto**[4]

[1] Institute of Strength Physics and Materials Science RAS, 2/4, pr. Akademicheskii, Tomsk, 634021, Russia
[2] Ternopil Ivan Puluj National Technical University, 56, str. Ruska, Ternopil, 46001, Ukraine
[3] National Research Tomsk Polytechnic University, 30, Lenina ave. Tomsk, 634050, Russia
[4] NTNU, Department of Engineering Design and Materials, 7491, Richard Birkelands vei 2b, Trondheim, Norway

E-mail: maruschak.tu.edu@gmail.com

Abstract. The structure and impact toughness of the pipeline 17Mn1Si steel have been studied. The main attention was paid to the analysis of various conditions of stress concentration under dynamic loadings. The process of strain localization with increasing stress state stiffness at the tip of the concentrator with decreasing testing temperature was investigated. Impact loading diagrams for specimens with various stress concentrator shapes were registered and analyzed.

1. Introduction

Increasing service life time of structural elements is one of the most difficult and actual requirements for transportation engineering. Recently, an intensive building and renovation of existing oil and gas pipelines takes place. During operation, the pipe material experiences pressure drops, strain aging, corrosion, abrasion, etc. [1] Large oil and gas deposits to be excavated in the north conditions make the problem more complex due to operation at negative temperatures [2]. In this regard, the actual problem is to improve the operational properties of pipe steel by providing higher mechanical properties [3]. One of the most common and well-standardized methods for mechanical properties certification of pipe steel is the impact testing [4, 5].

Pipe steels are characterized by high values of impact toughness. This is related to the fact that one should diminish the risk of accidental failures of the pipe under pressure as well as reduce the impact of the mechanical properties dispersion and presence of manufacturing defects induced during the steel production [6, 7]. Thus, the assessment of the influence of defects and structure heterogeneity of pipe steel by varying the shape of stress concentrators, especially its radius is of particular importance [5-8]. The data of this kind will allow taking into account the embrittlement factors that influence the stability of pipe steel to impact deformations.

An important task is also the study of fracture mechanisms at different stress stiffness values. Its importance is governed by practical applications associated with the interpretation of elastic-plastic characteristics of fracture resistance of structural materials in the presence of service defects [1].

This paper studies the influence of the notch shape on the impact fracture of 17Mn1Si steel at different, first of all, negative temperatures.

2. Experimental

Specimens for investigation were spark cut out of a pipe steel sheet of 30 mm thickness produced at the Machine Building Factory of Yurga (Russia). A batch of specimens with V-, U- and I-shaped notches of equal depth (2 mm) was machined. V- and U-notches were made by standard milling cutters; I-notches were made by electroerosion. Notch tip radius: U-1.0 mm; V-0.25 mm; I-0.1 mm. Before low-temperature impact loading the specimens were kept in a refrigerator (cooling chamber) 'Lauda rp870' for 10 minutes in the temperature range from –60 °C to 0 °C. The time interval between the specimen extraction from the cooling chamber and further testing did not exceed 5 seconds. At least three specimens of each type were tested at temperatures 20, 0, –20, –40 and –60 °C on an impact pendulum Instron 450MPX with recording the impact diagram, Table 1. The data processing program divides the specimen fracture energy into two constituents, such as load-time (*P-t*) and load-curvature (*P-s*). The deformation and fracture mechanisms were studied by fracture surface images using a scanning electron microscope LEO EVO 50 (Zeiss, Germany).

3. Results and Discussion

Metallography. Specimens for testing had the ferrite-pearlite structure with an average grain size of 14.7 ± 2.5 μm (figure 1). Ferrite grains had predominantly globular shape which is typical for 17Mn1Si steel. The high content of ferrite phase provides greater fracture toughness that is of particular importance under dynamic loading. Higher strength of the pearlite phase ensures possessing appropriate mechanical properties being required for the pipe operation under exploitation pressure.

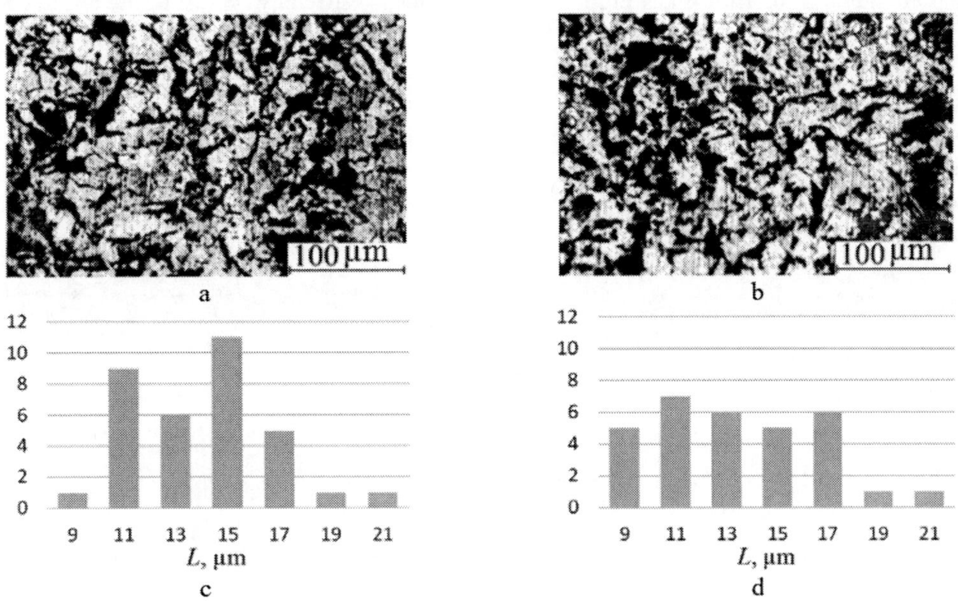

Figure 1. Optical micrographs of the 17Mn1Si steel in the supplied state (a, b); grain size distribution histograms along vertical (c) and horizontal (d) directions

Fracture macromechanisms. The test temperature dependence of impact toughness (figure 2, a, Table 1) exhibits several portions [9]. On the first portion (from T = –60 °C to T = –40 °C), the specimen fracture is brittle, without pronounced signs of plastic deformation. The test temperature growth up to –20 °C triggers combined fracture mechanisms. The third portion (from –0 °C to 20 °C) corresponds to the region of ductile fracture characterized by plastic flow on both the micro- and macroscale. The literature data [9, 10] indicate that each portion corresponds to certain fracture micromechanisms.

Let us note that the form of the obtained impact toughness dependences for specimens with sharp V- and I-notches almost coincide within the entire studied temperature range. An exception is the temperature –20 °C at which the impact toughness of V-notched specimens is slightly higher than that of I-notched specimens. On the whole, the impact toughness of U-notched specimens is about 3 times higher in the entire test temperature range than that of V-notched specimens (figure 2). It can be assumed in a certain approximation that the impact toughness value for specimens with all three types of notches linearly decreases with the decreasing test temperature.

The obtained dynamic loading curves of the specimens corroborate their sensitivity to changes in the macrofracture localization conditions which are associated with the changed concentrator shape (figure 2). It can be said in general that the curve shape is typical of ductile fracture within the entire studied temperature range from 20 °C to –60 °C [11], as well as for specimens with sharper V- and I-notches. The form of the impact diagrams of the V- and I-notched specimens is almost the same, which indicates that crack initiation and growth occur at similar stages. Let us consider them in more detail.

V-notch. Specimens of 17Mn1Si steel fracture in a ductile manner at test temperatures from 20 °C to –20 °C, which is evidenced by gently ascending and descending curve portions (figure 2, b–d, curve *KCV*). The microscale deformation mechanisms allow effective stress relaxation [12]. The crack initiation and growth take place in a "classical" manner, i.e., consecutively, without interruption and abrupt bifurcation transitions. The material shows high ductility and crack resistance. At the test temperatures –40 °C and –60 °C, the fracture diagram narrows and peaks. The form of the descending curve portion points to brittle crack propagation [12]. This is especially typical of the test temperature $T = -60$ °C at which the maximum load decreases down to $P_{max} = 7$ kN, which is indicative of a partial loss of the bearing capacity of the material [13].

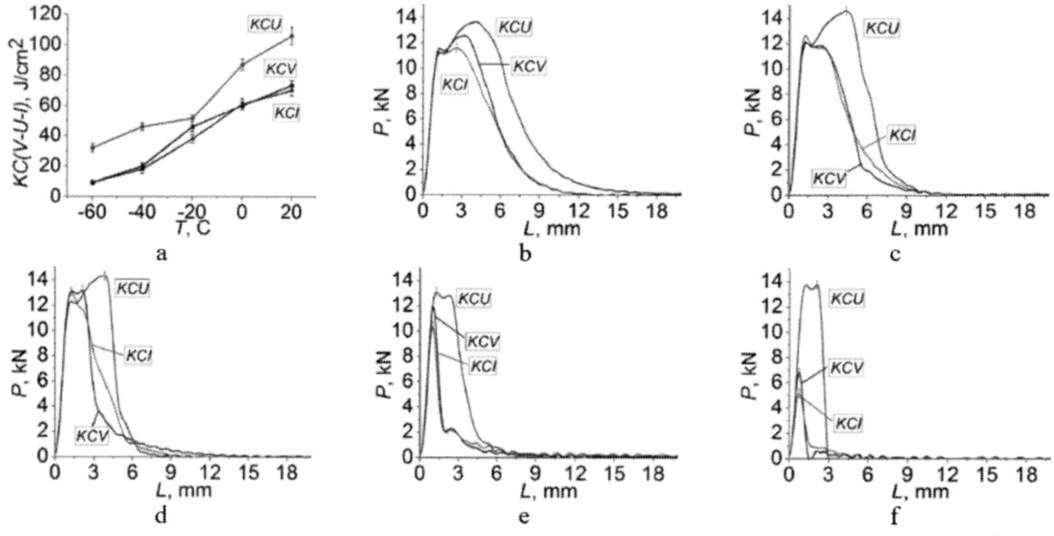

Figure 2. Test temperature dependence of impact toughness (a); impact diagrams in the load/displacement coordinates at test temperatures: 20 °C (b); 0 °C (c); –20 °C (d); –40 °C (e); –60 °C (f) for V-, U- and I-notched specimens.

I-notch. The form of the loading diagram (figure 2, b–f, curve *KCI*) is similar to that of the V-shaped concentrator (figure 2, b–f). Hence, with the sharpest concentrator (compared to the V-notch; notch tip radius is 0.25 and 0.1 mm, respectively) the fracture behavior of the material does not appear to be more brittle, even in the case of such a sharp notch similar in shape to a fatigue crack. In our opinion, this is due to the fact that plastic deformation ahead the tip of the main crack during its propagation in the studied steel provides for partial stress relaxation.

The influence of the test temperature reduction for the I-notched specimens is similar to the influence described above for the V-notched specimens. At $T = -60$ °C the maximum load value decreases even more abruptly down to P_{max} = 5 kN. This is testimony to a more pronounced embrittling effect of the concentrator, which corresponds well to the minimum notch tip radius.

U-notch. The impact fracture of notched 17Mn1Si steel specimens with the maximum notch tip radius is accompanied by ductile deformation. The observed macroscopic deformation behavior of the material bears witness to the activation of relaxation processes, which leads to an increase in the height and width of the impact diagram in the entire studied temperature range.

As one can see, the maximum load value decreases at room temperature (figure 2, b) with the growing stress concentration at the notch tip (in transition from the U- to I-notch). The height of the yield point increases in this case and its peak in the I-notched specimen corresponds to maximum load P_{max}. Thus, an increase in the material volume involved in deformation resistance, like for the U-notch due to its larger area and larger radius, increases resistance to macrocrack initiation and growth. However, this occurs at lower values of load P at the stage of elastic deformation as compared to the case of the I-shaped notch.

The yield plateau in the impact diagrams of I-notched specimens is little pronounced, 'degenerated'. This is testimony to the occurrence of local hardening processes on early deformation stages, which is accompanied by a decrease in the resistance of steel to strain localization in bending (this is dangerous for pipelines).

Integral estimation of impact diagrams. From the viewpoint of experimental data application, one of the most important issues in post-critical deformation models is the determination of the point of transition to the post-critical stage of specimen deformation, i.e., the stage of crack growth. Table 1 contain the force and energy parameters of impact toughness for 17Mn1Si steel. Like in our previous papers [1, 12], we may assume the maximum load in impact testing Pmax as the macrocharacteristic that reflects the material strength.

Table 1. Maximum load values on impact diagrams.

T, °C	V-notch		U-notch		I-notch	
	KCV, J/cm^2	P_{max}, kN	*KCU*, J/cm^2	P_{max}, kN	*KCI*, J/cm^2	P_{max}, kN
–60	9.3±1.1	6.8±0.2	32±2.8	13.8±1.7	9.4±1.3	5.2±0.1
–40	20±2.3	12±1.03	46±2.5	13.1±1	18±2.7	10.4±0.8
–20	46±3.4	13.2±1.01	51.6±2.2	14.3±0.6	38±2.6	13.1±1.3
0	60±2.6	12.2±1.02	87±3.9	14.6±1.4	61±3.7	12.7±1.2
20	73.3±3.2	12.6±2.3	106±5.7	13.7±1.3	70±3.4	11.6±1.5

These data are very important for understanding the effect of stress stiffness on the crack resistance of 17Mn1Si steel because *P*max corresponds to the point of transition from macrocrack initiation to its propagation. The sharper is the concentrator, the lower is the value of parameter *P*max at which the crack grows, which is indicative of a stiffer stress state at the crack tip [12].

Macroanalysis of fracture surface morphology. Analysis of laboratory and full-scale pneumatic test data has shown that the capability to arrest extended ductile fracture is determined by the volume and intensity of metal plastic deformation ahead the propagating crack tip [4]. In so doing, the volume of plastically deformed metal ahead the crack tip on the stage of stable ductile crack growth remains almost unchanged [4]. The larger volume is involved in plastic deformation near the propagating crack and the higher is the deformation intensity, the shorter is the fracture length. Thus, an increase in the fracture toughness, first of all, its constituent responsible for the crack growth energy, as well as the presence of plastic strains contractions along the crack front in Charpy specimens are additional factors attesting to high mechanical properties of pipe steel [14].

T = 20 °C (for V-, U-, I-notches). The crack initiation zone (zone I) is characterized by transition from the notch at an oblique angle, which points to the presence of macro- and microplastic strains and is confirmed by the shape of shear lips (figure 3, a–c). The fracture surface (zone II) has the form typical of shear rupture. This bears witness to the formation of a plasticity zone ahead the notch tip and to a heterogeneous stress-strain state in the specimen at crack initiation and growth. In this case, fracture is determined by the inhomogeneity of energy absorption in different specimen regions, particularly, in the zone of shear lips (zone III), which is manifested in the generation of different fracture growth mechanisms and in final specimen rupture (zone IV).

T= –20 °C (for V-, U-, I-notches). The influence of the plastic deformation processes decreases, which affects the external view of the specimen fracture surfaces (figure 3, d–f). The crack initiation and crack growth zones (zones I and II) formed in a ductile manner, with local regions of plastic deformation. The fracture surface profile takes on the form typical of cleavage fracture (direct fracture). The fracture surface structure exhibits cleavages that lead to plastic strain localization through multiple necking, and thus the volume of plastically deformed metal decreases, including the zones of shear lips (zone III) and final specimen rupture (zone IV).

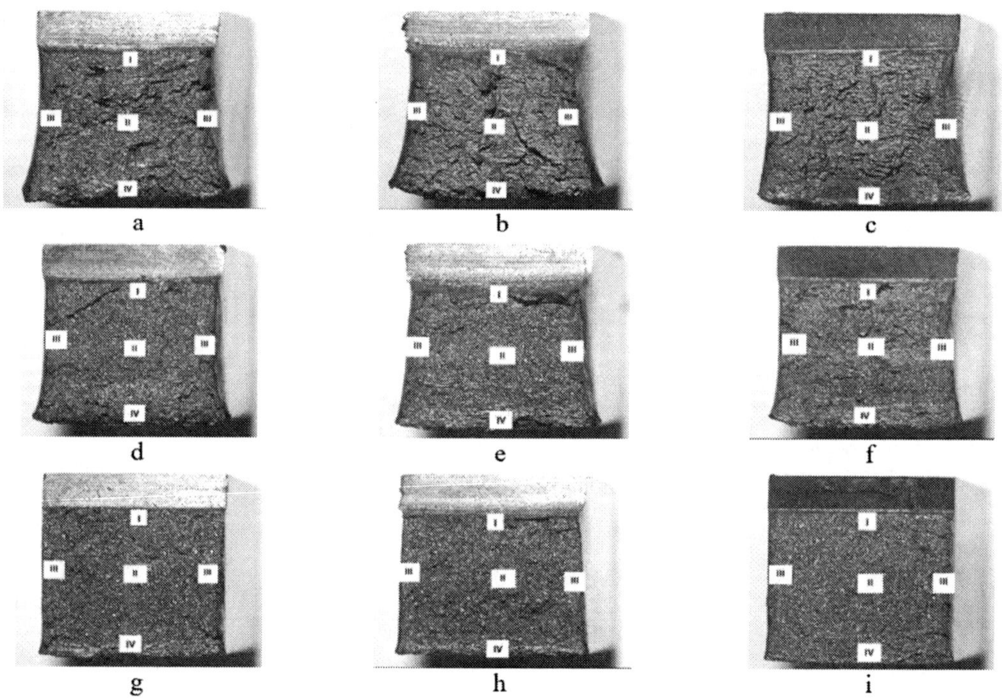

Figure 3 Macrofractures of impact test specimens at different temperatures with the designation of zones: I) – crack initiation; II) - crack growth; III) – shear lips; IV) – final rupture; a), b), c) T=20 °C; d), e), f) T=-20 °C; g), h), i) T=-60 °C; a), d, g) V-notch; b), e), h) U-notch; c), f), i) I-notch.

T = –60 °C (for V-, U-, I-notches). The regions of crack initiation (zone I) and growth (zone II) exhibit flat surfaces without visible signs of material tear outs at the macroscale. This means that crack propagation along the entire front occurred in stationary conditions, i.e., in nearly the same stress-strain state along the entire front (figure 3, g–i). In this case, crack initiation corresponded to the transition of material to the nonequilibrium state. Fracture in this zone is brittle. The strain rates, shear strain and tangential stress values exerted almost no effect on fracture micromechanisms. We shall notice that the fracture surface of all concentrators in specimens is flat (on the macroscale) with a small zone of shear lips (zone III). This indicates that the stress and strain fields were symmetric relative to the notch and initiated crack as well as during crack propagation and final rupture (zone IV).

The performed tests have shown that with the increasing stress stiffness induced by temperature reduction and strain localization [15], the material approaches a more brittle state and hence its fracture energy decreases.

4. Conclusion

The structure of the 17Mn1Si steel as well as micro- and macroscale mechanisms of its impact fracture in the temperature range of -60 °C – 20 °C have been studied. The main regularities of the notch shape influence onto impact failure energy as well as the form of the loading diagram were revealed. The discussion and generalization of the obtained results have been conducted.

Acknowledgments

The work was performed in the framework of fundamental research projects of the Russian State Academies of Sciences (2013–2020), with a partial support of RFBR Grant No. 15-08-05818_a, and Project of the Headquarter of the Russian Academy of Sciences on Artic research.

The authors are grateful to Shared Use Center 'Nanotech' of ISPMS SB RAS for the assistance in fractographic investigations.

References

[1] Maruschak P, Bishchak R, Prentkovskis O, et al 2016 *Adv. in Mech. Eng.* **8(4)** 1-8

[2] Kontorovich A E, Ermilov O M, Laperdin A N 2013 *Herald of the Russian Academy of Sci.* **83**(5) 400-410

[3] Arabei A B, Pyshmintsev I Yu, Farber V M, Khotinov V A, Struin A O 2012 *Steel in Translation* **42(3)** 212-218

[4] Hwang B, Kim Y G, Lee S, Kim N J, Yoo J Y 2005 *Metall. Mater. Trans.* 36A 371-87

[5] Kudrya A V, Kuz'ko E I, Sokolovskaya E A 2014 *Russian J. Metallurgy* **10** 837-845

[6] Oh C-K, Kim Y-J, Baek J-H, Kim Y-P, Kim W 2007 *Int. J. of Mech. Sci.* **49(12)** 1399-1412

[7] Pyshmintsev I Yu, Arabei A B, Farber V M, Khotinov V A, Lezhnin N V 2012 *The Physics of Metals and Metallography* **113(4)** 411-417

[8] Hashemi S H 2009 *Int.l J. of Pressure Vessels and Piping* **86 (8)** 533-540

[9] Kotrechko S A, Meshkov Yu Ya 2001 *Strength of Materials* **33(4)** 356-361

[10] Hashemi S H, Mohammadyani D, Pouranvari M, Mousavizadeh S M 2009 *Fatigue & Fracture of Engin. Materials & Structures* **32** 33-40

[11] Venkatsurya P K C, Misra R D K, Mulholland M D, Manohar M, Hartmann J E 2014 *Metallurgical and Materials Transactions A* **45(5)** 2335-2342

[12] Maruschak P O, Danyliuk I M, Bishchak R T, Vuherer T 2014 *Central European J. of Eng.* **4(4)** 408-415

[13] Sih G C, Tzou D Y 1986 *Theoretical and Applied Fracture Mech.* **5(3)** 189-203

[14] Xiao-Long Yang, Yun-Bo Xu, Xiao-Dong Tan, Di Wu 2014 *Materials Sci. and Eng.: A* **607(23)** 53-62

[15] Kazberuk A, Savruk M P, Chornenkyi A B 2016 *Int. J. of Solids and Structures* **85(86)** 134-143

Influence of soaking time of modifier in melt on microstructure of Al-12%Si alloys

N V Martyushev, V S Bashev and A P Zykova

Tomsk Polytechnic University, 30, Lenina ave., Tomsk, 634050, Russia

E-mail: nikitoid@mail.ru

Abstract. The influence of the soaking time of the silumin melt with a modifying nanopowder W on the microstructure and the impact toughness of the Al-12%Si silumin is studied in the paper. It was established that when introducing 0.1 mass. % of W and soaking it for 10 min, up to 30% of the inassimilable nanopowder W remains in the crucible after pouring. At that, the microstructure of castings is refined, and the impact toughness increases by 12 % compared to the initial alloy. It was established that the powder remaining in the crucible after melting consists of the remnants of metallic W, WO_3 and W_3O. Nanopowder W is completely assimilated by the melt if the soaking time is increased by 120 min; in this case, the impact toughness reduces, which is connected with the increase of dendrites and silicon wafers sizes.

1. Introduction

At present, silumins are one of the most wide spread alloys among non-ferrous metals. Silumins are widely used in aircraft, shipbuilding, automobile production and for daily living needs. Their wide application is caused by their increased fluidity, low tendency to shrinkage formation during pouring, a possibility of soldering and welding, low cost. Eutectic and hypereutectic silumins possess enhanced foundry properties because of low shrinkage and high fluidity. Eutectic silumins contain about 12 % Si and at room temperature, they consist of eutectic – a mixture of the solid solution of aluminium containing up to 1.65 % Si and pure silicon wafers. This type of the phase composition of eutectic silumins in the solid state determines their low mechanical properties. Low mechanical properties of eutectic silumins limit the field of application of this material.

To enhance the mechanical characteristics of silumins is possible owing to [1, 2]:

1) changing the morphology and reducing the crystals sizes of primary and eutectic silicon.

2) refining structural components of the alloy including predominantly ferrous-based intermetallides insoluble in matrix.

3) inhibiting dendrite crystallisation by means of dispersion and transformation of dendrites of the solid solution into equiaxed crystals.

One of the efficient ways to realise these mechanisms is an increase of the number of nucleation centres during the crystallisation process by means of modifying silumins. A widespread method influencing the uniform formation of the microstructure and enhancement of strength properties of Al-Si alloys is modification of super- and nano-dispersed particles of different chemical compositions [3-7]. The basic nanodimensional modifiers used for silumins are the powders of metal oxide and nitrides [3, 4, 8]. They possess high melting temperature and high inertia, which allows them not to dissolve in the melt and to serve as crystallisation centres. The particles of refractory metals (W, Ti, Mo and others) act in a similar way. When introducing the powder particles into the melt, it is necessary to provide their wettability in order to achieve the modification effect.

The effect of the particles of chemical compounds differs from the effect of pure metals. Metals of powders (Ti, Mo, Cr, Sb etc.) interact properly with the silumin melt [9]. According to the state diagram of Al-W, the solubility of W in the α-Al solid solution amounts to 0.16-0.20 mass. % [10]. Therefore, the particles of nanopowder W, introduced into the melt in the amount of up to 0.5 mass.

%, will dissolve completely in the α-Al solid solution. In the process of formation of silumin structures, refractory wolfram particles being additional crystallisation centres will influence the refinement of structural components of castings.

This paper is aimed at the study of the influence of the soaking time of the silumin melt with the modifying nanopowder W on the microstructure and the impact toughness of the Al-12%Si alloy.

2. Methods and materials

The Al-12%Si alloy was chosen as a material of the study. The pig silumin, manufactured at a metallurgical plant, with the chemical composition according to Table 1 was used as a mixture for melting.

Table 1. Chemical composition of Al-12%Si alloy used in this work (mass. %)

Si	Fe	Cu	Mn	Zn	Ti	Mg	Ca	Pb	Ni
11.2	0.44	0.29	0.28	0.08	0.04	0.05	0.04	0.01	0.03

The wolfram powder obtained by the electric blasting method with an average particle size of 240–380 nm and a specific surface area of 1.7-2.4 m^2/g was used for alloy modification. The average size of wolfram particles amounts to $d_{av} = 1.32$ μm. On the surface of wolfram particles, there are nanoparticles, the average size of which is $d_{av} = 0.07$ μm (Figure 1).

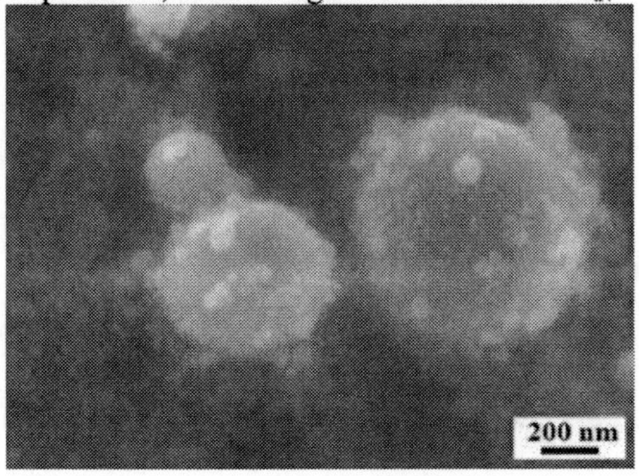

a

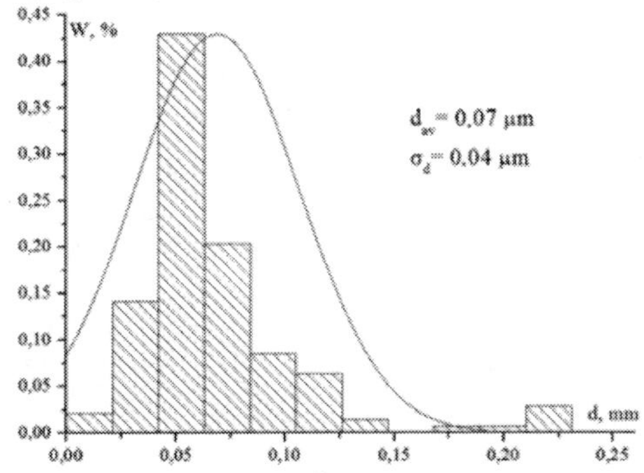

b

Figure 1. Nanoparticles on tungsten particles surface (a) and histogram of nanoparticles distribution by sizes (b).

Silumin melting was carried out in a muffle furnace. For speeding up the melting, the furnace was, first, heated up to 800°C, and then a steel crucible with a batch of the mixture was placed into the furnace. The mass of one melting amounted to 500 g. The mass of the charged metal and the mass of the powders intended for modification were measured by means of an analytical balance 'Shinko HJR-620CE' with an accuracy of up to 0.01 g. After silumin melting, the crucible was withdrawn from the furnace, the mirror of the melt was cleaned from the oxide film and then the modifier was introduced. The amount of the introduced modifier was 0.1 mass. %. After the introduction of the modifier, the melt was soaked in the furnace at a temperature of 800°C during 10, 60, 120 min. The pouring was carried out at a rate of 0.06-0.09 l/s. Per each soaking period, there were 3 meltings to accumulate statistics.

Granulometric analysis of nanopowders W was conducted by the method of laser diffraction (particle analyzer '*SALD-7101*'). The chemical and phase compositions of the wolfram powder were studied by the methods of optical emission spectroscopy (*PMI-MASTER Sort*) and X-ray diffraction analysis (*XRD-7000S*). The chemical analysis of silumins is determined by an average value of five measurements. Before measuring, the surface of samples was milled at a depth of 5 mm. The measurement error of the device amounted to 0.01%.

3. Results and discussion

The results of the conducted experiment showed that after conducting the melting and soaking of the melt containing the powder in the furnace for 10 min at 800 °C, 25-30 % of the powder remains on the walls of the crucible after pouring of the melt. The increase of the soaking time of the crucible with the nano-powder, introduced into the melt, up to 60 min results in the fact that the amount of the powder remaining in the crucible after pouring reduces to 10-15 %. When increasing the soaking time to 120 min, no powder remains on the crucible walls.

The studies showed that the powder, remaining on the crucible walls after melting, differs from the initial one by the composition. According to the data of XRD, the initial wolfram powder contains metallic wolfram (85 vol. %) and wolfram oxide W_3O (15 vol. %) (Figure 2 a). According to XRD, the powder remaining in the crucible after melting consists of the remnants of metallic W and its oxides of different modifications (WO_3, W_3O) (Figure 2 b). After melting, the volume fraction of the metallic W amounts to 72 vol. %, W_3O – 8 vol. % and WO_3 – 20 vol. %.

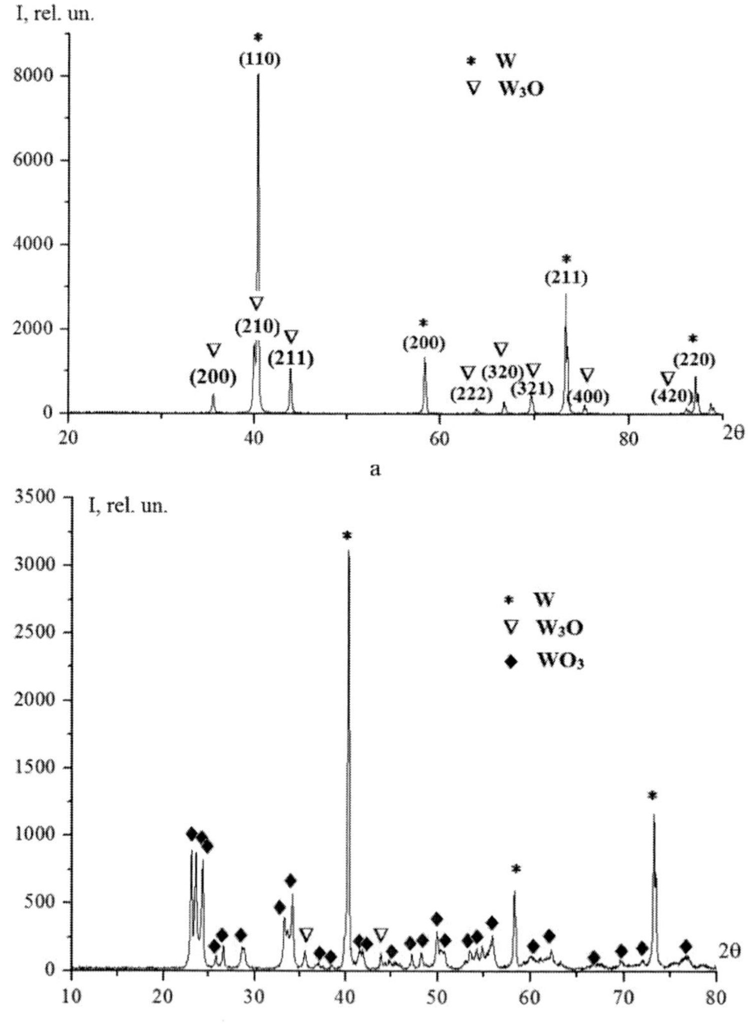

Figure 2. X-ray pattern of nanopowder W: a – initial state; b – remaining in crucible after melting (soaking of the melt containing powder in the furnace during 60 min).

The variation in the time of soaking of the modified melt in the furnace leads to a fairly significant change in the microstructure. A ten-minute soaking of the melt significantly refines dendrites of the solid solution. Short axes of the first order with a length of less than 200 μm and short axes of the second order (with a length of not more than 30-40 μm) are formed. Silicon wafers are subjected to significant refinement. When the melt with the nanopowder W is soaked in the furnace for 60 min, developed dendrites of the solid solution of silicon in aluminium are formed in the castings structure. According to the data of optical microscopy (Figure 3 c), it is obvious that the dendrites of the solid solution have developed axes of the first order of a 500 μm length, as well as quite developed axes of the second order. When increasing the soaking time up to 120 min (Figure 3 d), the length of the first-order axes reduces substantially. Similarly, the length of the second-order axes reduces as well, and they become less developed (of less length and thickness). Changes occur in the silicon wafers. Thus, the increase of the soaking time in the furnace up to 120 min reduces the average length of silicon wafers by 15-25%. In this case, the thickness of wafers increases.

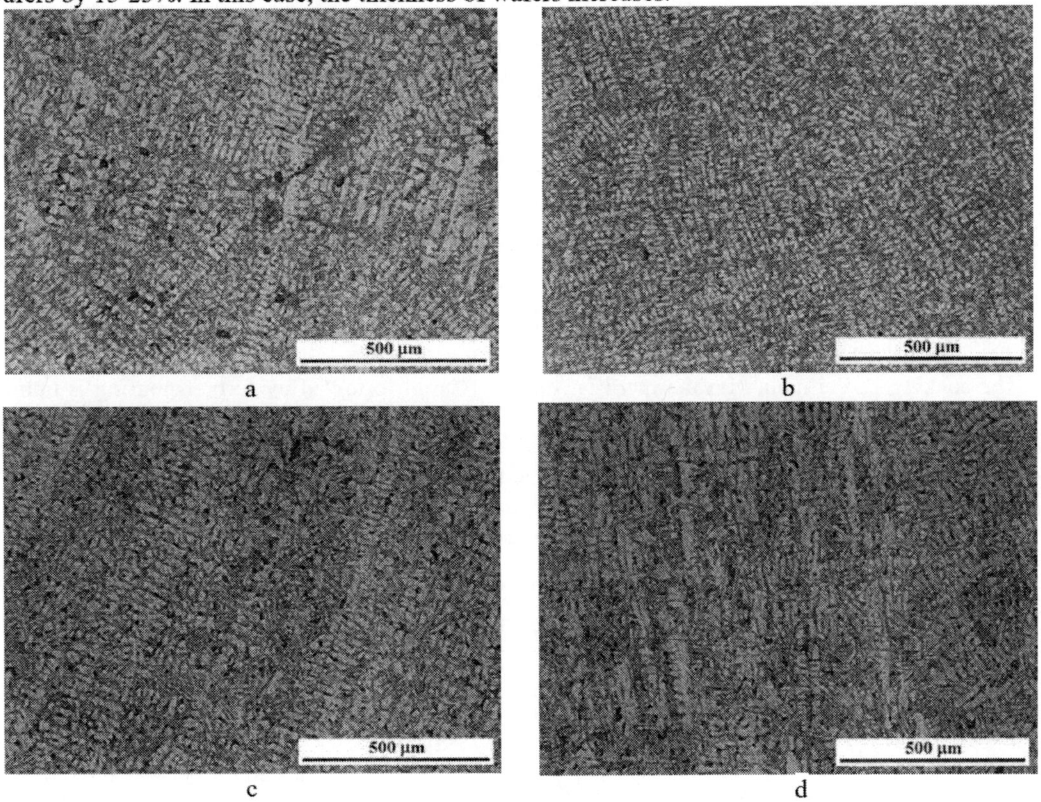

Figure 3. Microstructure of Al-12%Si alloy: a – initial state, soaking time is 10 min; b – 0.1 mass. % W, soaking time is 10 min; c – 0.1 mass. % W, soaking time is 60 min; d – 0.1 mass. % W, soaking time is 120 min.

These variations in the structure affect the properties significantly. The performed testings of samples obtained during different soaking periods with the introduced modifier in the furnace demonstrated a significant influence on the impact toughness (Figure 4). These data demonstrate that the impact toughness of unmodified samples amounts to KC 15.6. Addition of the powder and soaking of the melt with the powder in the furnace during 10 min facilitates the increase of the impact toughness up to KC 18. A further increase of the soaking time of the melt in the furnace leads to the reduction of the impact strength up to KC 13.4 (Figure 4).

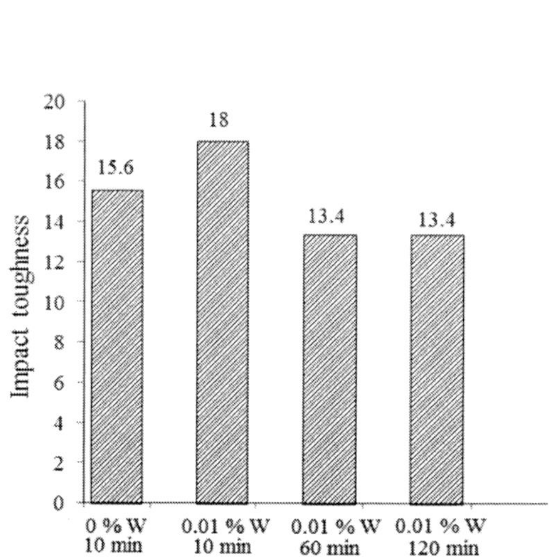

Figure 4. The impact toughness of Al-12%Si alloys with different soaking of the melt of nanopowder W.

4. Conclusion

On the basis of the carried out experiment, it is possible to conclude that the introduction of the nanopowder W into melt Al-12%Si produces a significant modifying effect and allows changing the microstructure. In this case, a significant influence on the effect of powder modification will be exerted by the time of soaking the alloy with the nanopowder in a heated state. In case of the nanopowder introduction, soaking during 10 min and the pouring of the melt, up to 30 % of the inassimilable powder remains in the crucible. It is possible to avoid it by increasing the presence time of the melt with the powder in the crucible up to 120 min.

The soaking of the melt with the modifier during 10 min followed by further pouring is optimal from the viewpoint of impact toughness. This is in good agreement with the data of metallographic analysis. Samples, obtained during a 10 min soaking period, possess refined dendrites with an undeveloped structure and short silicon wafers. The increase of the melt soaking period in the furnace leads to the increase of the dendrites and silicon wafers sizes even in comparison with the unmodified structure. Therefore, the impact toughness of these samples reduces.

Acknowledgments

The reported study was funded by RFBR, according to research project No. 16-38-60146 mol_a_dk.

References:

[1] Kuztsova V Z, Nosko O A and Sherstobitova A S 2008 *Russian J. Starodubovskiye reading* 1-6
[2] Petrov S S, Prigunova A G, Prigunov S V and Klyuchnik D N 2015 *Metal Phys. and Adv. Techn.* **37**(3) 367-78
[3] El-Mahallawi I S, Shash A Y and Amer A E 2015 *Metals* **5** 802-821
[4] Zykova A P, Kazantseva L A and Kurzina I A 2016 *AIP Conf. Proc.* (Tomsk) vol 1772 (AIP Publishing LLC.) 1-7
[5] Chen C, Liu Z, Ren B, Wang M, Weng Y and Liu Z 2007 *Trans. Nonferr. Met. Soc. China.* **17** 301
[6] Li J H, Wang X D, Ludwig T H, Tsunekawa Y, Arnberg L, Jiang J Z and Schumacher P 2015 *Acta Materialia* **84** 153
[7] Li Q, Xia T, Lan Y, Zhao W, Fan L and Li P 2013 *J. of Alloys and Comp.* **562** 25-32
[8] Nowak M, Bolzoni L and Hari Babu N *J. of Alloys and Comp.* **641** 22-29
[9] Belov N A, Savchenko S V and Khvan A V 2007 *Phase composition and structure of silumins: Reference Book* (Moscow: MISIS)
[10] L F Mondolfo 1979 *Structure and properties of aluminium alloys* (Moscow: Metallurgiya)

Research of the possibility of using an electrical discharge machining metal powder in selective laser melting

A A Golubeva, A V Sotov, A V Agapovichev, V G Smelov and V N Dmitriev

Samara University, 34, Moskovskoe sh., Samara, 443086, Russia

E-mail: agapovichev5@mail.ru

Abstract. In this paper the research of a Ni-20Cr-10Fe-3Ti (heat-resistant) alloy metal powder conducted for use in a selective laser melting technology. This metal powder is the slime after electric discharge machining. The technology of cleaning and melting the powder discussed in this article. As a control input of the powder, immediately before 3D printing, dimensional analysis, surface morphology and the internal structure of the powder particles after the treatment were examined using optical and electron microscopes. The powder granules are round, oval, of different diameters with non-metallic inclusions. The internal structure of the particles is solid with no apparent defects. The content of the required diameter of the total volume of test powder granules was 15%. X-ray fluorescence analysis of the powder materials carried out. The possibility of powder melting was investigated in the selective laser melting machine 'SLM 280HL'. A selection of the melting modes based on the physical properties of the Ni-20Cr-10Fe-3Ti alloy, data obtained from similar studies and a mathematical model of the process. Conclusions on the further investigation of the possibility of using electric discharge machining slime were made.

1. Introduction

Modern industrial production has a tendency to search new materials and methods for their preparation. Currently, additive technologies introduced in all areas of production. Today, it is one of the most dynamically developing directions of "digital" manufacture. These technologies accelerate research and development, solve the problems of pre-production, and in some cases actively applied for the production of a functional product [1-6].

One of the directions of additive technology is the selective laser melting (SLM) of a metal powder composition. The development of this technology area has stimulated the development of technologies for producing metal powders. Today, the nomenclature of metal compositions has a wide range of materials based on Ni and Co (Co-Cr-MO, Inconel, Ni-Cr-Mo), on Fe (tool steel: 18%Ni300, H13; stainless steel: 316L), on Ti (Ti6-4, CpTigr1) and on Al (Al-10%Si-Mg, Al-12%Si).

Overall, the situation concerning the introduction of additive technologies in the Russian industry is extremely unfavorable. The aerospace production is in need of the introduction of this technology, because difficult-to-cut, refractory materials and alloys are used in the production of parts for rockets and aircraft engines.

The basic problem is production of powder materials for additive technologies [7]. Installations for the production of metal powders existing today in Russia are using different methods. The most widely used method is spraying a plasma jet of the rapidly rotating rod of the corresponding alloy in an argon atmosphere. Such installations are expensive, and the possibility of the use of such powders in additive technology has not been studied.

In this paper, the treatment technology of electric discharge machining (EDM) of slime made of graphite, copper and mineral oils has been developed, the basic technological properties of the purified powder were studied and the use of EDM slime in SLM process has been investigated by the example of the Ni-20Cr-10Fe-3Ti alloy.

2. The methodology of the researches

The object of the research is the Ni-20Cr-10Fe-3Ti alloy in the form of EDM slime. EDM slime is a difficult-to-cut material [8-10]. This slime consists of micro powder, which consists of spherical particles with impurities of graphite, copper, and mineral oil. For cleaning the slime, the method for removing contaminants has been developed.

Dimensional analysis, the surface morphology and the internal structure of the powder particles after treatment have been investigated using electron and optical microscopes. These researches are necessary as the input of the powder should be controlled immediately before the SLM process.

For a detailed analysis of the chemical composition of the purified Ni-20Cr-10Fe-3Ti powder, spectral analysis using the optical emission spectrometer with inductively coupled plasma (PlasmaQuant PQ9000) conducted. This is a high-resolution spectrometer designed for high-precision multi-element analysis of liquid samples of various types with the minimum requirements for sample preparation.

The spectrometer has a range of 0.002 nm, which allows using the device for solving the most complex analytical problems. Unique optics and built-in database allow defining more than 75 elements. The possibility of melting the particles of the metal Ni-20Cr-10Fe-3Ti powder alloy was used on the basis of 3D machine SLM 280HL. The installation uses an infrared fiber laser with a wavelength of 1075 nm and a maximum power of 400 W, operated in a continuous mode. This machine has an open system that allows using its own production powders for the manufacture of parts.

3. Results and discussion

Since the additive manufacturing of metal powders sets sufficiently high quality requirements [1, 11], in particular, the requirements for morphology (spherical, the allowable number of defects), the grain size (the upper and lower limit of the particle size range is 0.03-0.05 mm), the requirements for process properties, the requirements for the chemical composition of the gas and impurities, the powder input control was conducted. In addition, the internal structure of the powder particles in the presence of pores investigated. Figure 1 shows an image of the Ni-20Cr-10Fe-3Ti alloy powder with pollution included therein.

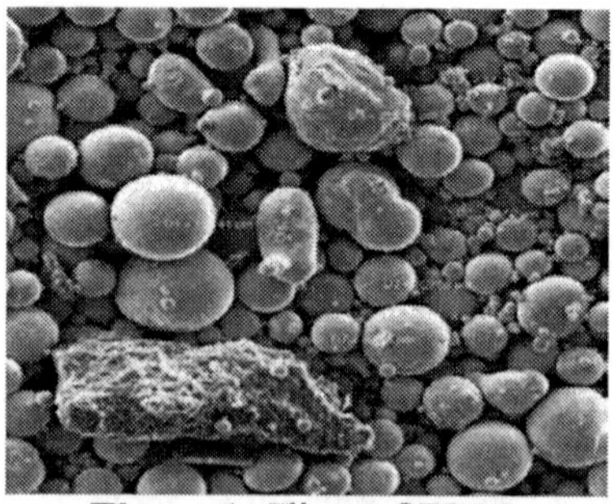

Figure 1. Slime of EDM.

The preliminary analysis showed that the powder granules are round, oval and of different diameters with nonmetallic inclusions. The internal structure of particles is solid with no apparent defects. The content of the granules of the required diameter is 15 % of the total amount of the investigated powder.

For obtaining a pure powder from slime, the clean technology of mineral oil, graphite and copper was developed and applied. The technology is based on the knowledge of chemical properties of the Ni-20Cr-10Fe-3Ti alloy metals and derivatives of EDM. Distilled water, baking soda, technical soap used as a chemical reagent. The centrifuge, the drying cabinet, the screen were used as equipment.

After cleaning the slime, the obtained powder materials sent to the X-ray fluorescence analysis (XRF). This method of analysis is one of the modern spectroscopic methods for studying the substance in order to obtain its elemental composition. It allows detecting various elements beginning from beryllium (Be) up to uranium (U). An XRF method is based on the collection and subsequent analysis of the spectrum upon irradiation of X-ray radiation of the material. The results of this study are shown in Figures 2 and Table 1 and 2. Figure 3 shows an image of the surface morphology of the particles of the Ni-20Cr-10Fe-3Ti alloy powder.

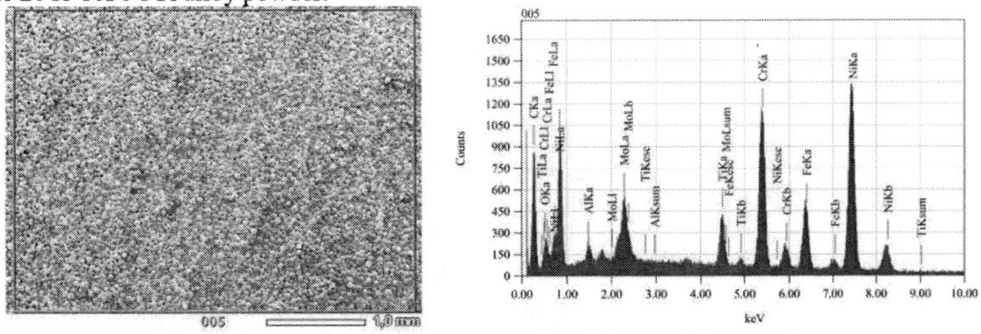

Figure 2. Analysis of some quantity of the particles powder.

Table 1. Standardless thin film. Standardless quantitative analysis (Fitting Coefficient: 0.6333).

Element		ke, V	Mass, %	Counts	Error, %	Atom, %	K
C K*		0.277	14.89	6311.02	0.01	45.05	0.8603
O K*		0.525	0.99	1103.47	0.21	2.25	0.3273
Al K*		1.486	0.64	937.85	0.42	0.86	0.2476
Ti K*		4.508	5.01	3650.23	0.11	3.80	0.50002
Cr K*		5.411	19.35	11579.05	0.04	13.52	0.6092
Fe K*		6.398	10.77	5093.45	0.08	7.00	0.7707
Ni K*	(Ref.)	7.471	38.29	13963.46	0.03	23.70	1.0000
Mo K*		2.293	10.07	3348.89	0.13	3.81	1.0964
Total			100.00				

Table 2. ZAF Method. Standardless quantitative analysis (Fitting Coefficient: 0.6640).

Element	ke, V	Mass, %	Error, %	Atom, %	K
C K*	0.277	26.97	2.05	63.56	2.1505
Al K*	1.486	1.09	0.91	1.14	0.1773
Ti K*	4.508	3.18	0.47	1.88	1.7552
Cr K*	5.411	15.81	0.62	8.61	9.0372
Fe K*	6.398	7.03	0.70	3.56	4.0189
Ni K*	7.471	41.22	1.08	19.87	22.3395
Mo K*	2.293	4.69	1.27	1.38	1.8602
Total		100.00		100	

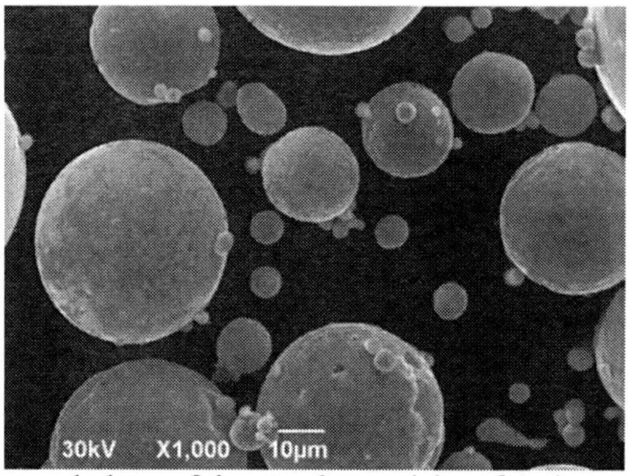

Figure 3. The morphology of the powder particles after the surface cleaning.

For a detailed chemical composition of the purified powder, a unique PlasmaQuant PQ9000 spectrometer used. The spectrograph deals with the solutions of any substance. To obtain the solution, the powder boiled in the hydrochloric acid solution (figure 4). The resulting solution filtered and loaded into the spectrograph. The chemical composition identified after analysis of the solution of the purified powder (table 3).

Table 3. The chemical composition of the purified powder.

Chemical element, mass, %									
Cr	Ni	Co	Cu	Zn	Mn	Fe	Al	Ti	Mo
15.82	65.34	0.19	0.14	0.02	0.01	11.45	1.28	1.47	0.17

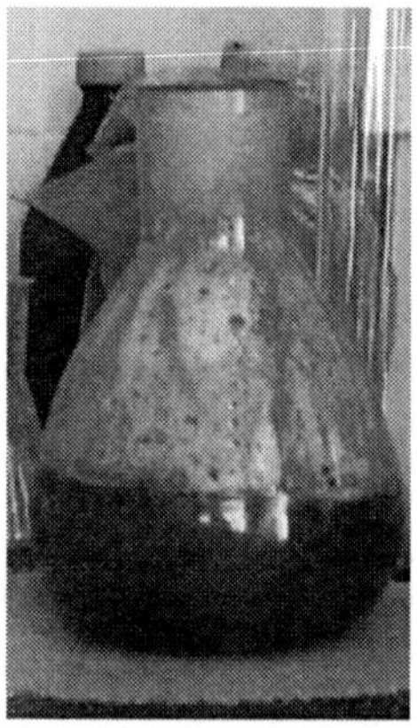

Figure 4. The Ni-20Cr-10Fe-3Ti powder dissolved in the acid solution.

The precipitate (the powder remained after boiling in the hydrochloric acid) was dissolved in nitric acid. Then, the bulk of powder was analyzed with a spectrograph. The composition of the slime appeared to be similar to that of the bulk powder. The mass fraction of minor elements is not more than 0.01 %. Thus, a certain chemical composition of a purified powder should not prevent melting of the particles in the SLM machine.

After cleaning and researching of the powder, the melting process was conducted with the use of 3D machine SLM 280HL. The empirical search of the laser processing modes, which provide the necessary structural and phase composition, is extremely difficult and labor intensive. In fact, the required quality layers formed in a very narrow range of laser modes. The selection of melting modes was carried out based on the physical properties of the Ni-20Cr-10Fe-3Ti alloy.

The results of melting were unsatisfactory. The synthesized material was sufficiently fragile, leading to detachment from the building platform (Figure 5). Selected laser modes were to ensure the melting of the powder particles, and this suggests an idea that the oxide film is present on the surface of the particles.

For the analysis of oxides, the additional X-ray analysis, which will identify the crystal structure of the powder inside and a classification of relations between the elements of the crystal lattice, is required. The results of X-ray diffraction are expected in the near future. The nitrogen as a protective gas could interfere with melting. Further melting will be carried out using argon as a protective gas.

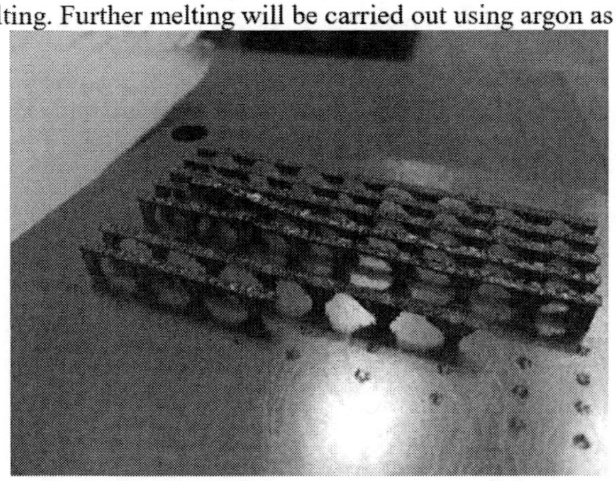

Figure 5. The melting results of the powder in the SLM machine.

4. Conclusion

In this paper, the research on the possibility of using EDM slime in the selective laser melting technology showed in general a positive result. As a result of this work, we can conclude the following:

1. The treatment technology of EDM slime made of graphite, copper and mineral oils has been developed. Distilled water, baking soda, technical soap were used as chemical reagents. The centrifuge, the drying cabinet and the screen were used as equipment.
2. The surface morphology of powder investigated after cleaning EDM slime. The results of analysis showed that the granules of powder have a round shape with different diameters. The content of the required diameter granules (25 ... 35 microns) of the total powder volume was 15%. The mass fraction of minor elements is not more than 0.01 %.
3. After analyzing of the properties, the metal powder was melted in the SLM machine, where the synthesized material was sufficiently fragile. These data suggest the presence of the oxide film on the powder surface, preventing melting.

Acknowledgements

These studies were conducted on the basis of the equipment of CAM Technology Center (RFMEFI59314X0003). This work was supported by the Ministry of Education and Science of the Russian Federation in the framework of the implementation of the Program "Research and development on priority directions of the scientific-technological complex of Russia for 2014-2020".

References

[1] Zlenko M A, Popovich A A and Mutylin I N 2013 *Publishing house of the Polytechnic University* (Saint Petersburg)

[2] Vdovin R A and Smelov V G 2014 *Int. J. of Eng. and Tech.* **6(5)** 2269-2275
[3] Smelov V G, Sotov A V and Agapovichev A V 2016 *Key Eng. Mat.* **684** 316-322
[4] Grjaznov M U, Shoshin S V and Chuvildeev V N 2012 *Bulletin of the Nizhny Novgorod University named after N. Lobachevsky* **5(1)** 45-50
[5] Smelov V G, Sotov A V, Agapovichev A V, Tomilina T M 2016 *IOP Conf. Series: Materials Science and Engineering* **142**
[6] Sufiiarov V Sh, Popovich A A, Borisov E V and Polozov I A 2015 *Tsvetnye Metally* **1** 79-84
[7] Gu D D, Meiners W, Wissenbach K, Poprawe R 2012 *Mat., Proc. and Mech. Int. Mat. Rev.* **57(3)** 133-164
[8] Ivanova I I, Serebrovskiy V I, Kolmykov V I 2012 *Southwestern State University* (Kursk)
[9] Fominskiy L P, Levchuk M V, Tarabrina V P 2010 *Powder metallurgy* **4** 1-6
[10] Fominskiy L P 2016 *Electrophysical and electrochemical methods of processing* **8** 6-8
[11] Kruth J P, Froyen L, Van Vaerenbergh J, Mercelis P, Rombouts M and Lauwers B 2004 *J. of Mat. Proc. Techn.* **149** 616-622

Influence of the direction of selective laser sintering on machinability of parts from 316L steel

V P Alexeev, A V Balyakin and A I Khaimovich

Samara University, 34, Moskovskoye Shosse, Samara, 443086, Russia

Email: alexeev_v.p@mail.ru

Abstract. This work presents the results of research of the impact of layer-by-layer growing of workpieces made of 316L steel on their machinability. The results of determination of residual stresses and measurement of hardness of the workpieces grown have been demonstrated. A series of experimental studies has been performed in order to determine the cutting force which occurs in the process of machining. The microstructure of the workpieces grown has been examined. It has been shown that the workpieces machined using Selective Laser Melting technology have the microstructure which is a totality of 'microwelded seams', which have a significant influence on the behavior of deformation processes in case of machining. The studies have shown that in case of lateral milling of the horizontally grown workpiece, the codirectional microwelded borders prevent any significant deformation of the misalignment which increases the cutting force by up to 10% as compared with milling of the vertically grown workpiece.

1. Introduction

The development of new technologies for fast manufacture of goods and products is the most promising area in modern machine building. One of the methods which enable the reduction of the manufacture time of parts and components is the technology of Selective Laser Melting (SLM) of a powdered material based on a three-dimesinoal model [1]. Selective laser melting opens up new possibilities in the manufacture of superlight metallic structures of irregular shapes and reduces the time of manufacture, while also enabling one to avoid manufacture of labor-intensive tooling (cast or extruded), which is normally expensive, and is undoubtedly interesting for the development and test deployment of the manufacturing process for new products [2].

Austenitic stainless steels (ASSS) are widely used due to their high resistance to corrosion and oxidizing. However, these materials are considered to be hard-to-machine as they feature high mechanical- and microstructure responsivity to deformation and deformation speed. 316L steel has been extensively used in medicine due to its high corrosion resistance and biocompatibility.

There have been studies [7, 8] of the conditions for formation of the microstructure and evaluation of properties of grown workpieces. There are also several works on the formation of materials and products with the use of the methods of layer-by-layer laser melting [3-8], however, until now, a number of issues relating to subsequent machining of parts made using SLM technology have not been properly studied, particularly the issue of mechanical treatment. The purpose of this work is to research the impact of layer-by-layer growing of workpieces made of 316L steel on their machinability.

2. Materials and Methods

The method of investigation of the influence of the direction of layer-by-layer laser growning of workpieces on their machinability includes the following steps: acquisition of workpieces with different direction of growning out of 316L material using SLM mehod with SLM 280HL machine and studying their microstructure; studying the condition of the surface layer by hardness measurement and determination of residual stresses in the subsurface layer of the grown workpieces; set up and carrying out of in-situ tests for determination of the cutting force.

Workpieces with dimensions (H×W×D) 6×24×40 mm (Figure 1) were prepared for the study of the impact of layer-by-layer growing. They were grown out of the powdered metal of 316L stainless steel using the SLM 280HL machine. The average size of the particles is 15…60 μm. For powder melting within the machine an IR fiber laser with the wavelength of 1075 nm and peak power of 400 Wt working in the continious regime is used. So as to prevent powder particles from oxidizing and inflammation during the process of monolayers formation, the chamber is filled with noble gas, while the oxygen content does not exceed 0.2 %. The chemical composition of the powder is shown in Table 1.

Table 1. The chemical composition 'mass. %' of 316L

Grade	C	Si	Mn	P	S	Cr	Ni	Mo
AISI 316 L	≤0.030	≤0.75	≤2.0	≤0.045	≤0.030	16.0–18.0	10.0–14.0	2.0 –3.0

Workpieces were placed on a growth platform horizontally (Figure 1 a) and vertically (Figure 1 b). When growing workpieces using the 316L stainless steel powder on the SLM 280HL machine, 1-mm thick supports were additionally grown across the perimeter of the lower surface. These supports served as a heat sink (Figure 1 c).

a, b c d

Figure 1. Workpieces synthesized from the powdered metal of 316L stainless steel on the growth platform: a – horizontal placement of the workpiece; b – vertical placement of the workpiece; c, d – method of placement of the workpieces with additional supports and without supports.

Contraction of the workpieces due to temperature stresses (Figure 1 d) was observed during the growth process in case there is no additional heat sink via the supports grown.

3. Measurement of Rockwell hardness of the grown workpieces

In order to determine the impact of the direction of the growth on hardness, the metals and alloys were tested for hardness in accordance with Rockwell method using the TK-2M hardness gage. Indenters with a steel ball of 1.588 mm in diameter were used under the load of 100 kgf. The hardness value was determined as an arithmetic average of three measurements in 6 points which were evenly distributed across the surface. The results of the hardness measurements are shown in Table 2.

Table 2. Results of the measurement of hardness of the workpieces made of 316L stainless steel.

Point #	Rockwell Hardness Number					
Workpieces	1	2	3	4	5	6
Vertically grown workpiece	92	89	91	88	92	87
Horizontally grown workpiece	91	91	87.5	88.5	92	92

Table 2 shows that hardness of the workpieces is identical within the limits of imprecision.

4. Determination of residual stresses in the core of the rectangular section

Determination of the residual stresses was performed using the automated system 'ASB-1' using the method of measuring deflection of the thin workpieces with preliminarily etched out surface layers which were cut out of the workpieces grown. The distribution of residual stresses in the workpiece grown vertically is shown in Figure 2a, and the distribution of residual stresses in the workpiece grown horizontally is shown in Figure 2b.

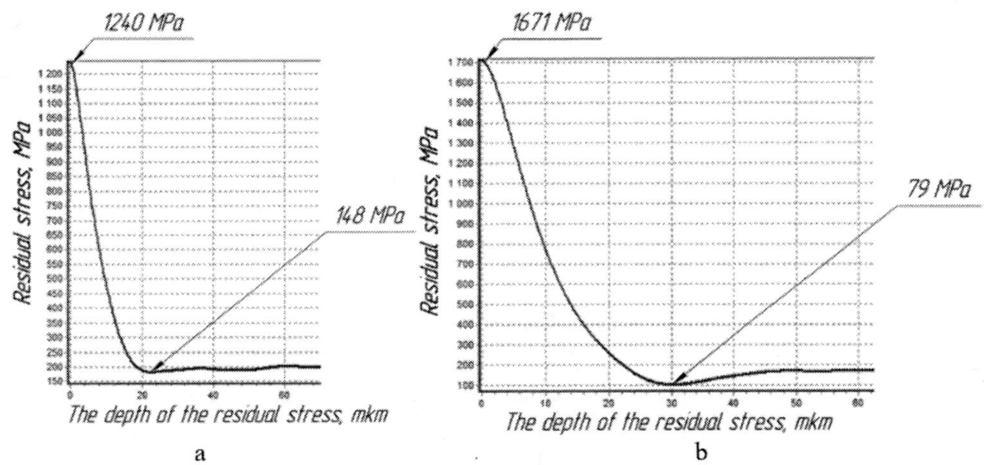

Figure 2. Distribution of residual stresses workpiece grown a – vertically, b – horizontally.

Chemical milling of the workpieces was carried out in the electrolyte: 1.7 l of orthophosphoric acid (density is 1.56 g/cm^3); 0.3l of sulpharic acid (density is 1.84 g/cm^3); 100g of chromium anhydrite. The speed of chemical milling was 1.15 μm/min. In Figure 2a,b it is shown that the maximum value of residual stresses in the vertically grown workpieces is by 431MPa less than that in the workpiece grown horizontally. However, the minimum value of the residual stresses is by 69 MPa greater for the vertically grown workpiece, than that for the workpiece grown horizontally. Furthermore, the depth of formation of residual stresses in the workpiece grown horizontally is by 10 μm greater. This may be explained by a different area of heat removal and the direction of the microwelded seam in the process of growing, as the chemical milling of the workpieces was carried out along the 40 mm side, and the microwelded seam will be either horizontal or vertical for each workpiece correspondently.

The graphs show that the depth of formation of residual stresses in the workpiece grown horizontally is greater than that in the workpiece grown vertically. This may be explained by the different conditions of heat removal from the workpieces in the process of growing them and the time of their exposure to laser irradiation. Heat removal and formation of thermal stresses should influence the differences in microstructure of the workpieces grown. This was the aim of the following study of microstructure.

5. Studies of the microstructure of the workpieces grown

The workpieces were etched using the following solution 'vol. %': HCl – 22; HNO_3 – 15; HF – 41; H_2O – 22. The microstructure was examined using the 'TESCAN VEGA' scanning electronic microscope. The microstructure of the workpieces (Figure 3) is a mixture of layers: the light gray grains of the hard solution are based on chrome; the gray areas – ferrite, and the dark areas – austenite.

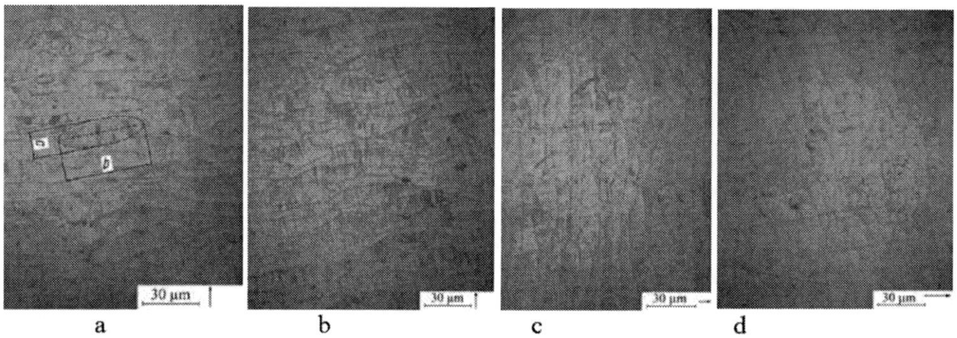

Figure 3. The microstructure of the workpieces grown: a, b – horizontally; c, d – vertically; ↑ – direction of the layer-by-layer workpiece growing process.

In Figure 3, one can clearly see the borders of the section of the border type 'microwelded seams' (hereinafter – Microwelded Seams Borders). These layer-type borders which are in the perpendicular direction towards the melting are significantly different than the borders of conventional grains in terms of crystallography.

There are no traces of recrystallizing and growth of grains in the workpieces grown. The volume of all the grains is determined by the particle size of initial powder. Let us denote the value of the unit of the surface energy of the microwelded seam with σ_{surf} and inscribe the average grain in the spheroid with semi-axes a, b ($a \le b$) (Figure 3a), equal to the grain volume V, then the boundary energy U_{surf} of the total area S_{surf} can be described as follows:

$$U_{surf} = S_{surf}\sigma_{surf} = 2\pi a^2\left(1+\frac{e^2}{\sqrt{e^2-1}}\arcsin\left(\frac{\sqrt{e^2-1}}{e}\right)\right)\sigma_{surf} \quad (1)$$

Relation $e = b \cdot a^{-1}$ characterizes the form of the spheroid and determines the boundary surface energy. Therefore, taking into consideration that the boundary belongs to two neighbor grains, the boundary energy per the unit of grain volume will be determined as follows:

$$u_{surf} = \frac{U_{surf}}{2V} = \frac{3}{4b}\left(1+\frac{e^2}{\sqrt{e^2-1}}\arcsin\left(\frac{\sqrt{e^2-1}}{e}\right)\right)\sigma_{surf} \quad (2)$$

The greater is e in (2), the greater is the specific energy of the boundary u_{surf}. The minimum value of u_{surf} is for the grain, the form of which is similar to a sphere ($e \to 1$).

The boundary specific energy of spherical shape grain of equal volume will be expressed as:

$$u_{spher} = \frac{3\sigma_{surf}}{2r},\ r = b / e^{2/3} \quad (3)$$

See Table 3 for the comparison values of boundary specific energy for the workpieces with different direction of growing.

Table 3. Connection between the boundary specific energy and direction of workpiece growing

Workpieces grown	b (μm)	$e = b/a$	u_{surf} / σ_{surf} (μm^{-1})	$u_{spher} / \sigma_{surf}$ (μm^{-1})	u_{surf} / u_{spher} (%)
Vertically	49	3.55	0.089	0.072	19
Horizontally	50	3.76	0.091	0.072	20

According to Table 3, boundary specific energy of the grown workpieces is commensurable and does not exceed 20% of the specific energy of the grain spherical shape of the equal volume. Crystallization of the melted alloy begins with the seal border, due to which the seam border is significantly different from the grain material in terms of the chemical composition (surface segregation). High energy of the borders makes them a preferred place for nucleation and allocation of the second phase [11]. Second phase segregation is carried out at the boundaries (figure 3) corresponding to the direction of b-axis of the spheroid. Study [7] pointed out the influence of the 'microwelded seam' type borders on the mesostructural age hardening of 316L alloy.

6. Setting up and conducting of full-scale tests to determine the cutting force

In order to determine the cutting force, we used the methodology of previous experimental research [9]. The cutting instrument used was an end-cutting carbide mill [10], with 12 mm diameters, and the number of teeth is 4. The workbench for the experiments included the 'ALZMETALLBAZ 15 CNC' production center equipped with a Kistler dynamometric table which served to register changes in the components of cutting force F_x, F_y, F_z in real time. For this test, the cutting speed varied from 60 – 100 m/min, and a transfer to a tooth from 0.05 mm - 0.2 mm/tooth, lateral removal t = 0.1 mm – 1 mm. The surface of each workpiece was preliminarily machined so as to ensure that roughness of all the workpieces had the same Ra 1.25. Table 4 shows the results of measurement of the cutting force when machining in the following modes: S = 0.2 mm/tooth, t = 1 mm.

Table 4. Values of the cutting force when machining in modes Sz = 0.05 mm/tooth; t = 1 mm.

	Cutting speed (m/min.)		
	60	80	100
Cutting form when processing the vertically grown workpiece, H	97.42	99.11	100.44
Cutting form when processing the horizontally grown workpiece, H	108.45	106.9	105.7
Relative difference in the cutting force, %	11.3	7.9	5.2

7. Discussion of the Results of the Study

The formation of sheet-like phases of austenite in the grain located perpendicular to the border of the microwelded seam results in hardening of the material because partition distortions of the crystalline lattice, close to the borders of the layer, are the reason of the occurrence of volumetric residual stresses (residual stresses of the first kind). Due to the enhanced heat removal when growing the workpieces horizontally, versus the workpieces grown vertically, there is a difference in the kinetics of crystallization of the grain of the melted layer. The horizontally grown workpieces feature more intensive extraction of the second phase (Figures 3 a, b). This conclusion can also be indirectly validated by the results of measurement of residual stresses (Figures 2). Residual stresses in the horizontally grown workpieces near the surface at the depth of 10 μm reach relatively higher values as compared with those of the vertically grown workpieces – 750 MPa and 400 MPa, correspondingly. Thereafter, at the depth of 60 μm, the values of stresses level out up to 200 MPa. Microwelded borders have significant influence over the behavior of deformation processes as they are effective barriers for the movement of dislocations, which additionally contributes to age hardening of the material.

Another important result which requires discussion is the effect of hardening which can be observed when milling a horizontally grown workpiece. At the macro level, chip formation during cutting is a process of consecutive deformations of dislocation at the time of application of compression stress. Accumulation of local deformations of dislocation results in micro-segmentation of chips, which means significant deformations of dislocation of the macrovolume – chip segments. In case of the lateral milling of a horizontally grown workpiece, the co-directional microwelded borders prevent any significant deformation of the dislocation, which, according to the test results (Table 4), increases the cutting force by up to 10% as compared with milling of a vertically grown workpiece. The comparative data analysis (Table 4) shows that as the cutting speed increases, the difference in the cutting force in case of lateral milling of the workpieces grown in different directions decreases. This fact can be explained by the thermal effect of plastic deformation. Due to increased localization of the temperature near the cutting edge as the speed of cutting increases, the movement of dislocations, which provide for the plastic deformation, overcomes the barrier formed by the microwelded seam. The softening action of the temperature has varying effects on the vertically- and horizontally grown workpieces. Whereas in case of the former, increased viscosity of the material along with the increased cutting speed result in the insignificant rise of the cutting force, in case of the latter it leads to its reduction due to the observed effect of the reduced action of the hardening effect of the microwelded seam as temperature increases.

8. Conclusion

1. The structure and physical and mechanical properties of workpieces made of 316L alloy were tested. These workpieces were manufactured using the technology of layer-by-layer laser melting. Because of more efficient heat removal during the process of horizontal growing of the workpieces as compared with the vertically grown workpieces, a difference was observed in the kinetics of crystallization of the grain of the melted layer, which influences its age hardening and the value of residual stresses. Residual stresses in the horizontally grown workpieces near the surface at the depth of 10 μm reach relatively higher values as compared with those of the vertically grown workpieces – 750 MPa and 400 MPa, correspondingly. Thereafter, at the depth of 60 μm the values of stresses level out up to 200 MPa.
2. It was demonstrated that the workpieces manufactured using SLM technology feature the microstructure which is a totality of 'microwelded seams' that have a significant influence on behavior of deformation processes during milling, as they are effective barriers for movement of the dislocations which additionally contributes to resistance of the material to the cutting force.
3. It was established that in case of lateral milling of a horizontally grown workpiece, the co-directional microwelded borders prevent any significant deformation of the dislocation, which, according to the test results, increases the cutting force by up to 10% as compared with milling of a vertically grown workpiece. As the cutting speed of horizontally grown workpieces increases, the hardening effect diminishes, which is stipulated by the stalling effect of the microwelded seam in terms of movement of the dislocations.

Acknowledgments
This work was supported by the Ministry of Education and Science of the Russian Federation in the framework of the implementation of the Program 'Research and development on priority directions of scientific-technological complex of Russia for 2014– 2020'.

References

[1] Agapovichev A V, Balaykin A V and Smelov V G 2015 *Modern Applied Science* **9**(4) 335-343
[2] Shishkovsky I V 2016 *The basics of high definition additive technologies* (St Petersburg)
[3] Abe F, Osakada K, Shiomi M, Uematsu K and Matsumoto M 2001 *J. of Mat. Proc. Tech.* vol 111 (Osaka: Osaka University) pp 210-213
[4] Mullen L, Stamp R C, Brooks W K, Jones E and Sutcliffe C J 2009 *J. of Biomed. Mat. Res. A* **12** 264-270
[5] San-Juana M, Martina O, Tiedraa M, Santosa F J and Lopeza R, Cebriana J A 2015 *Procedia*

Eng. vol 132 (Valladolid: Universidad de Valladolid) pp 500-506
[6] Shishkovsky I V 2009 *Laser synthesis of functionally-graded mesostructures and three dimensional products* 248-265
[7] Gryaznov M Yu, Shotin S V and Chuvildeev V N 2012 *The Herald of N.I. Lobachevsky Nizhny Novgorod University* **5**(1) 43-50
[8] Sufiiarov V S, Popovich A A, Borisov E V and Polozov I A 2015 *Tsvetnye Metally* **1** 79-84
[9] Alekseev V P and Khaimovich A I 2015 *Key Eng. Mat.* **684** 13-20
[10] Skuratov D L, Zhidyaev A N and Sazonov M B 2016 *Res. J. of Appl. Sci.* **9**(11) 767-770
[11] Orlov A N, Perevezentsev V N and Rybin V V 1980 *Metal grain boundaries* (Moscow: Mettalurgiya)

The investigation of the deformation wave hardening effect on the strength of the medium and low alloy steels

A V Kirichek[1,a], **S V Barinov**[2,b], **A V Yashin**[2,b] **and S E Kolontsov**[2,b]

[1] Bryansk State Technical University, 7, 50-letiya Octyabrya bul., Bryansk, 241028, Russia
[2] Murom Institute (branch) of Vladimir State University, 23, Orlovskaya str., 602209, Russia

E-mail: [a]avk.57@yandex.ru, [b]box64@rambler.ru

Abstract. The article presents the data on the effect of wave deformation hardening on the strength of the 45, 40X and 35 HGSA steel. To improve the strength of these steels, it is proposed to create structured areas consisting of alternating solid and viscous-plastic sections in their surface layer. The evenness of arrangement of the sections is characterized by the overlap factor. The studies found that wave deformation treatment of the samples, made of the 45, 40X 35HGSA steel, made it possible to increase the tensile strength by 8, 4.2 and 13%, the values of elastic deformations – by 37, 81 and 51% during their hardening with overlapping coefficients 0.7; 0.9 and 0.7, respectively.

1. Introduction

The surface destruction due to the strength loss is critical for a large number of products in different fields. The increase of the metal strength while maintaining sufficient ductility and viscosity increases the reliability and durability of machines (structures) and reduces the consumption of the metal used for their fabrication due to the reduced cross section of machine parts that has important economic value.

Strength can be enhanced both by applying different hardening technologies and by creation of composite materials [1-10]. The use of composite materials combining properties of the relatively soft matrix and durable high modulus (up to 4900 MPa and more) fibers, giving small deformation under loads, allows increasing the specific strength of the material by 20-40% [2].

One of the most promising directions of modern hardening technologies development is the formation of the heterogeneous structure of the modified surface layer of solids, but not the gradient material that, on one hand, has no clearly defined areas boundaries with changed properties, and on the other hand, is similar in the properties to the composite material with the soft matrix and hard inclusions.

The heterogeneous structure generated in a homogeneous material is mainly applied to enhance wear resistance and durability under contact cyclic loading. The examples of heterogeneous structures are shown in Figure 1. To obtain the heterogeneous structure, the well-known methods of hardening were used such as heat treatment and chemical-heat treatment, and, more recently, with the advent of the wave strain hardening, the surface plastic deformation has been widely used [1, 2].

Increasing the surface strength by plastic deformation is more advantageous compared to other hardening methods, as it has a number of advantages. First, with its small energy-output ratio, it is possible to provide a greater hardening depth with smooth transition from a hardened surface to a non-hardened one, that is especially important for materials working in conditions of static load. Second, there is the possibility of parts machining of any sizes and configurations. Third, due to its manufacturability and ease of implementation, it is easy to achieve the mechanization and automation of the process. Fourth, there is the possibility of local hardening of details parts.

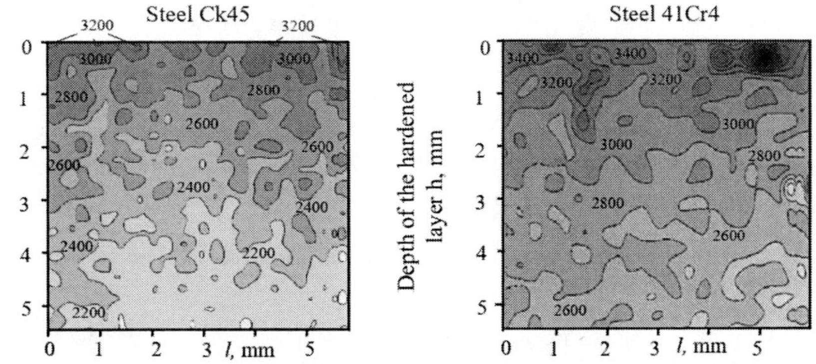

Figure 1. Examples of uneven microhardness distribution (MPa) in the surface layer with a heterogeneous structure obtained by hardening by the deformation wave.

The creation of the heterogeneous structure by surface plastic deformation (SPD) in a homogenous material to improve the strength was not used. One of the reasons for this is the high energy consumption and low precision of creation of the hardened surface layer of large thickness with SPD methods, the lack of research of uniformity of the hardened cold-worked layer and its influence on the strength properties of parts of different steels.

The goal of the prospective research is to establish the deformation wave hardening effect on the strength of medium and low alloy steels.

2. Materials and methods

To solve this problem, it is proposed to use the wave deformational hardening that is a modern method of surface plastic deformation. It has wide technological possibilities for the formation of the hardened surface layer and allows one to create the layer with a greater depth (up to 6...10 mm) and a high hardening degree (up to 6500 MPa) [1]. The peculiarity of the method consists in deformational waves generation by the shock system with an intermediate member and their transfer to the deformation zone to strengthen responsible surfaces of machines details. The method effectiveness is achieved by the full use of the wave deformation energy during hardening. Depending on required hardening parameters of the surface layer, elements of the shock system are selected that generated the crash pulse of the required amplitude and duration. As a result of the deformation wave parameters control, there is an opportunity to form not only a uniformly hardened surface layer, but also areas that are heterogeneously reinforced. The uniformity of plastic overlapping of imprints is evaluated using coverage factor K. The range of the K change is from 0 to 1: when $K = 0$, prints do not overlap, print edges are contiguous with each other; when $K = 1$, there is a complete overlap of prints, strikes are in the same place. Figure 1 shows that as a result of microhardness measurements, it is found that in processing modes when $0 \leq K \leq 0.5$, a pronounced heterogeneous structure of the hardened surface layer is formed, that is characterized by alternating of hard and soft areas. The examples of these areas are presented in Figure 1. In processing modes when $0.5 < K < 0.8$, the upper surface layer is hardened enough evenly, and unevenly reinforced areas are displaced in the subsurface layer. In modes when

$0.8 \leq K < 1$, the almost completely uniformly hardened surface layer is formed, and microcracks appear on the surface as a result of over cold hammering [1, 2].

To research the deformation wave hardening effect on the strength, flat samples Ck45, 41Cr4 and 35HGSA with a thickness of 7mm were used. The choice of samples material is based on the fact that steels 41Cr4 and 35HGSA are quite often used for the manufacture of machine parts working in severe conditions, and increased requirements for strength are imposed on them. Steel Ck45 is a reference in engineering. When hardening, the heterogeneous structure with uniformity of $0.2 < K < 0.9$ was created. The rod roller with a width and diameter of 10 mm was used as a tool. The energy of strikes was 25 joules. The studied samples were hardened with the deformation wave only on one surface.

In addition, the investigation of the influence of the directions of tool strikes' prints on the processed material obtained for the same modes of wave deformation hardening was conducted. To do this, due to rotation of the hardening tool (rod roller) at a specified angle in the direction of its feeding, which coincides with the direction of the gap, the rod roller strikes ' prints of different directions are applied to the processed surface. The examples of the rod roller strikes' prints of different directions are shown in Figure 2.

Strength characteristics of investigated samples were determined on a universal electromechanical machine with computer control WDW100E series.

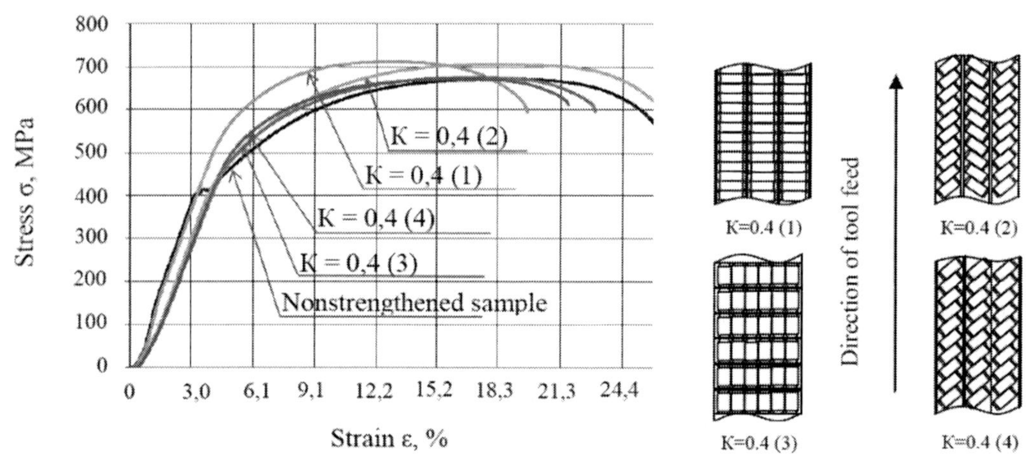

Figure 2. The diagram of stretching of Ck45 steel samples, hardened with overlap K=0.4, for different types of application of impact prints of the tool for hardening relative to the direction of its feed.

3. The investigation of the deformation wave hardening effect on the strength of the medium and low alloy steels

As a result of strength tests, it was found that due to the location of prints on type 1 and 2 shown in Figure 2 on the samples surface with the same strengthening modes, it was possible to increase the tensile strength by 4-5%. The prints of type 1, in comparison to 2, tend to be located in the surface layer, with relatively equal values of tensile strength, of more pronounced elastic properties. Therefore, only this type was used for hardening of all samples.

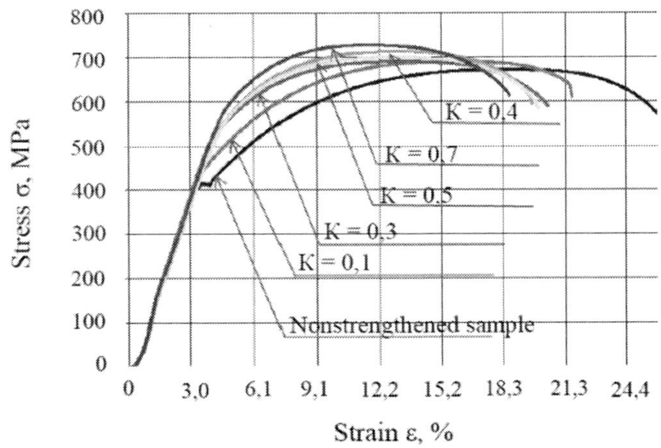

Figure 3. The diagram of straining of Ck45 steel samples, hardened with the deformational wave with varying uniformity (K).

As we can see from Figure 3, as a result of Ck45 steel wave deformation hardening, the greatest increase in strength was 8% attained at K=0.7, while elastic deformations were increased by 43%, while the ductility was decreased by 37% compared to non-hardened samples. In addition, it was found that during hardening of steel Ck45 with the increase of the overlap ratio from 0.1 to 0.7, the tensile strength increases from 2.6% to 8%, the ductility decreases from 22 to 37%, and elastic deformations increase from 8.6% to 43%.

As it can be seen from Figure 4a, in case of 41Cr4 steel hardening, the greatest increase in strength was 13% attained at K=0.9, while elastic deformations were increased by 81%, while the ductility was decreased by 38% compared to non-hardened samples. When hardening steel 41Cr4 with the increase of the overlap ratio from 0.1 to 0.9, the tensile strength increases from 3.5% to 13%, the ductility decreases from 23 to 38%, and elastic deformations increase from 39% to 81 %.

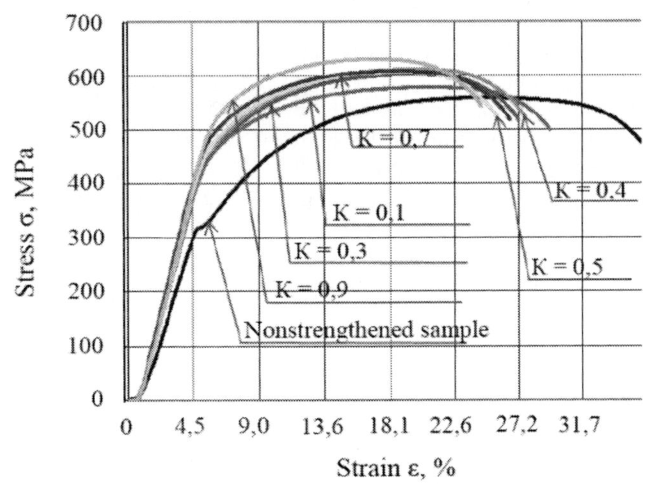

a)

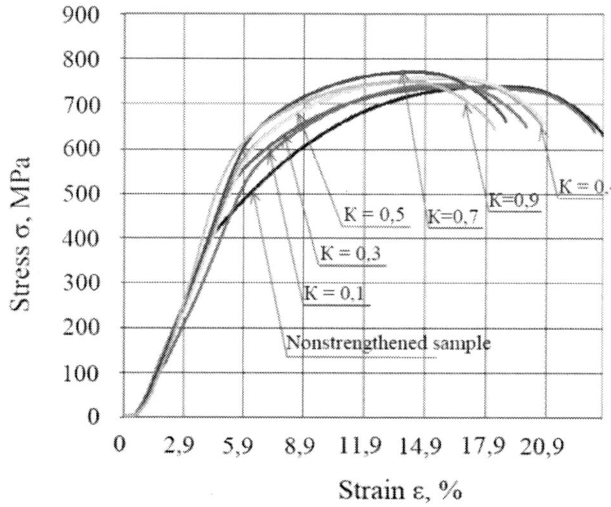

b)

Figure 4. The diagram of specimens strain reinforced with different overlap (K) of steel: a) 41Cr4; b) 35HGSA.

As it can be seen from Figure 4,b, the hardening by the deformation wave of 35HGSA steel allowed to achieve the greatest increase in strength only by 4.2% attained when K=0.7, while elastic deformations increased by 51%, while the ductility decreased by 28% compared to non-hardened samples. During hardening of steel 35HGSA with the increase of the overlap ratio from 0.1 to 0.7, the tensile strength increases from 0.4% to 4.2%, the ductility decreases from 1 to 28%, and elastic deformations increase from 39% to 51%. The sample with K=0.9 demonstrates a decrease in the tensile strength and ductility due to supercooled hammering in comparison with the sample with K=0.7.

4. Conclusion

1. It was found that the prints of the strikes of the tool which is perpendicular to the direction of feed, formed in the surface layer, can improve the tensile strength of the material by 5%, compared with the prints of directions obtained in the same modes of hardening by deformational waves.

2. It was found that through the application of wave deformation hardening of samples with a thickness of 7 mm and made of the 45, 40X, 35HGSA steel, it is possible to increase the tensile strength by 8, 13 and 4.2%, respectively, and the values of elastic deformations — by 37, 81 and 51%.

Acknowledgements

The research has been performed with the assistance of the Russian Foundation for Basic Researches, grant No. 16-38-00698 mol_a.

References

[1] Kirichek A V, Barinov S V, Yashin A V 2015 *J. of Nano and Electronic Physics* **7 (4)** 04019

[2] Kirichek A V, Barinov S V, Yashin A V 2016 *IOP: Conf. Ser. Mater. Scien. and Engin.* (Tomsk) vol **124** (England: IOP Publishing) p 1-6

[3] Panin V E, Egorushkin V E 2013 *Physical Mesomechanics* **16(4)** 267

[4] Dmitriev A I, Zolnikov K P, Psakhie S G, Panin V E and Goldin S V 2005 *Theoretical and Applied Fracture Mechanics* **43(3)** 324

[5] Valiev R Z, Krasilnikov N A and Tsenev N K 1991 *Mater. Science and Engin.* **137** 35

[6] Ivanov G P, Kartonova L V, Khudoshin A A 1997 *Stroit. Dorozhn. Mashiny* **1** 33

[7] Kositsyna I I, Sagaradze V V, Kopylov V I 1999 *Physics of Metals and Metallogr.* **88(5)** 493
[8] Panin V E and Deryugin E E 2003 *The Physics of Metals and Metallogr.* **96** 2
[9] Valiev R Z, Korznikov A V and Mulyukov R R 1993 *Mater. Science and Engin.* **168** 141
[10] Zhu Y T, Huang J Y, Gubicza J, Ungár T, Wang Y M, Ma E and Valiev R Z 2003 *J. of Mater. Res.* **18(8)** 1908

Research of the process of axisymmetric forming of thin-walled flat blanks into the conical parts with minimal thickness variation

E G Demyanenko[a], **I P Popov**[b] **and V S Menshikov**[c]

Samara University, 34, Moskovskoye roadway, Samara, 443086, Russian Federation

E-mail: [a]e-dem@mail.ru, [b]igr_popov@mail.ru, [c]me-vs@yandex.ru

Abstract. In this paper, the method based on the process of forming a flat blank into the thin-walled conical part is proposed. The forming method is based on two stages. At the first stage, forming of the free portion of the blank is carried out until it fully contacts the conical surface of the punch. During this stage, thickness unevenness is created, which can be compensated during the second stage by the action of active friction forces. The shaping of the workpiece during the forming of the conical part is performed by thinning. The thinning increases with the increase of the penetration depth of the punch, i.e. the final part will be thinner, but its thickness variation will be reduced. This method allows achieving minimal thickness variation in the walls of the part.

1. Introduction

Thin-walled conical parts with the $S_{blank}/D\langle 0.008$ (where S_{blank} – thickness of a blank; D – larger edge diameter of a part) thickness ratio are widely used in aircraft and rockets manufacturing, especially for liquid rocket engines (LRE) [1 – 3]. Various methods of manufacturing of these parts [4 – 20], based on welding of conical or flat ring-shaped blanks, are known. Conical blanks are used for manufacturing tall tapering parts, flat blanks are used for relatively low parts ($H/R \le 0.25$, where R – the larger edge diameter of a part, H – the height of a part). Both cases demonstrate significant thickness variations leading to reduced specific strength of the part or a necessity of additional mechanical processing. In this paper, the method (shown in Figure 2) of manufacturing of these parts is investigated.

The forming method is based on two stages. At the first stage, forming of the free portion of the blank is carried out until it fully contacts the conical surface of the punch. During this stage, thickness unevenness is created, which can be compensated during the second stage by the action of active friction forces. In the maximum thickness area near the clamp, additional tensile stresses are created, leading to thinning of that area and equalization of the thickness in the entire volume of the conical portion of a part. Height h of the emerging cylindrical-like shaped portion depends on deformation during the second stage. The determination of the thinning value during the second stage in the vicinity of radial transition of the punch is the primary goal under given parameters (geometric: conic angle, D/r ratio; mechanical: relative elongation, anisotropy coefficient). Both stages provide forming by thinning of a blank. As the thinning increases, the achieved height of a part increases as well. This depends primarily on mechanical properties of the material being used in the process (such as δ_U – relative uniform elongation; anisotropy of the material and the number of stamping steps).

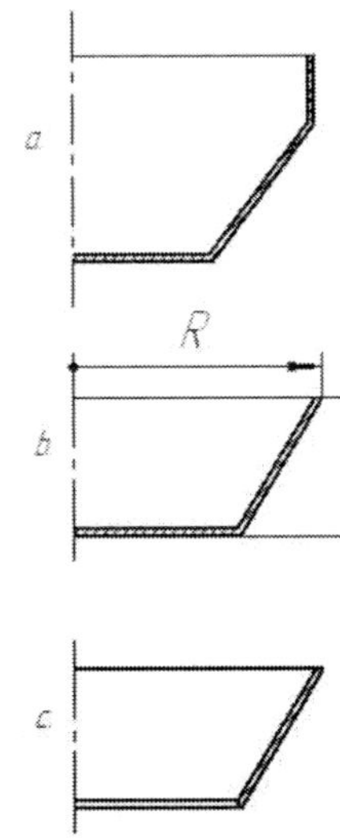

Figure 1. Typical forms of conical parts.

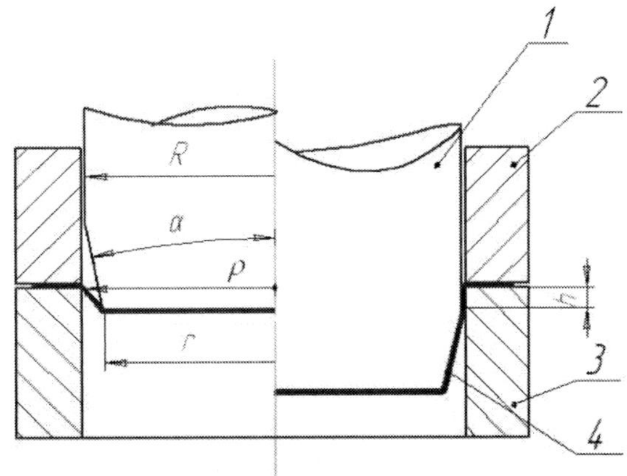

1 – punch; 2 – clamp; 3 – die; 4– part

Figure 2. The device for flanging of the ring blanks

2. Theoretical foundations

The equation for the conical shell blanks is as follows [21]:

$$\rho(d\sigma_\rho / d_\rho) + \sigma_\rho - \sigma_\theta - \sigma_\theta f ctg\alpha = 0, \tag{1}$$

where σ_ρ and σ_θ – stresses in meridian and tangential directions, accordingly;

f – the friction coefficient;

α_{part} – the cone angle of the part;

ρ – the independent variable.

In the immobile portion of the blank, between the die and the clamp, tangential deformation equals zero, and from stress-strain equation, the following ratio takes place:

$$\sigma_\theta / \sigma_\rho = \mu \tag{2}$$

where, μ – the coefficient of a transversally isotropic body.

In the flat bottom portion of the workpiece, stress equals:

$$\sigma_\theta / \sigma_\rho = 1. \tag{3}$$

By applying the linear relation of stresses due to the minimal length of the free conical portion, we obtain:

$$\frac{\sigma_\theta}{\sigma_\rho} = 1 - \frac{1-\mu}{\overline{R}-1}\left(\overline{\rho}-1\right). \tag{4}$$

Solution (1) with consideration of (4) for the first stage takes the following form:

$$\sigma_{\rho 1} = \sigma_{\rho 1}^{*}\left(\overline{\rho}\right)^{\frac{-\mu+1}{\overline{R}-1}} \exp\left[\left(1-\overline{\rho}\right)\frac{1-\mu}{\overline{R}-1}\right], \tag{5}$$

where, $\sigma^*_{\rho 1}$ – boundary stress on the smaller edge of the workpiece during the end of the first stage, when $\overline{\rho} = \frac{\rho}{r}$, $\overline{R} = \frac{R}{r}$

Considering that at the second stage, the free portion is formed and condition (3) is applied to the whole conical region, and solution (1) takes the following form:

$$\sigma_{\rho 2} = \sigma^*_{\rho 2} \left(\overline{\rho}\right)^{fctg\alpha}, \quad (6)$$

where, $\sigma^*_{\rho 2}$ – boundary stress on the smaller edge of the conical portion at $\overline{\rho} = 1$ during the second stage.

Thickness values can be calculated by assuming that, according to plasticity condition [22]:

$$\sigma_S^2 = \sigma_\rho^2 + \sigma_\theta^2 - 2\mu\sigma_\rho\sigma_\theta,$$

the following ratio takes place:

$$\sigma_\rho = \sigma_S / \left(1-\mu^2\right). \quad (7)$$

At the second stage,

$$\sigma_\rho = \sigma_S / \sqrt{2(1-\mu)}. \quad (8)$$

The thickness deformation is take as the intensity of deformation $e_i = e_S$. The linear hardening condition is [23]

$$\overline{\sigma}_S = \overline{\sigma}_{TO} + e_S = \overline{\sigma}_{TO} + 1 - \overline{S}. \quad (9)$$

Applied to the first stage,

$$S_1 = \left(\overline{\sigma}_{TO} + 1 - \overline{\sigma}_{\rho 1}\sqrt{1+\mu^2}\right)S_{blank}. \quad (10)$$

And the second stage,

$$S_2 = \left[\overline{\sigma}_{TO} + 1 - \overline{\sigma}_{\rho 2}\sqrt{2(1-\mu)}\right]S_1, \quad (11)$$

where the following is accepted:

$$\overline{\sigma}_{TO} = \sigma_{TO}/\Pi; \overline{\sigma}_{\rho 1} = \sigma_{\rho 1}/\Pi; \overline{\sigma}_{\rho 2} = \sigma_{\rho 2}/\Pi,$$

where S_1, S_2 – thicknesses of a workpiece at the first and the second stages;

σ_{T_0} – extrapolated yield stress;

Π – the hardening modulus.

In Figure 6, the thicknesses curves, including the one of the first stage, are shown.

It is possible and advisable to regulate the thickness of the conical portion during the second stage. The condition of providing minimal thickness variation can be written as:

$$\int_1^{\overline{R}} \left(S_{part} - S_2\right)^2 d\rho \to \min, \quad (12)$$

where:

$$S_{part} = \left[\overline{\sigma}_{TO} + 1 - \overline{\sigma}^*_{\rho 2}\sqrt{2(1-\mu)}\right]\left[\overline{\sigma}_{TO} + 1 - \overline{\sigma}^*_{\rho 1}\right]S_{blank}.$$

This means that the part thickness is determined by the minimal value in the area of radial transition of a punch.

If we assume $\overline{\sigma}^*_{\rho 2} = \sigma^*_{\rho 2}/\Pi$ as a variable parameter, which determines thinning of the workpiece, after derivation, we get:

$$\partial/\partial x = 0, \int_1^{\overline{R}} \left[(a-xc)b - \left(a - x\rho^k c\right)\left(a - \rho^p \cdot e\right)\right]\cdot\left[-cb + \rho^k c\left(a-\rho^p \cdot e\right)\right]d\overline{\rho} = 0, \quad (13)$$

where

$$x = \overline{\sigma}^*_{\rho 2}; a = \overline{\sigma}_{TO} + 1; b = \overline{\sigma}_{TO} + 1 - \overline{\sigma}^*_{\rho 1}; c = \sqrt{2(1-\mu)}; d = \sqrt{1-\mu^2}; e = \overline{\sigma}^*_{\rho 1}\sqrt{1-\mu^2};$$
$$\rho^k = \rho^{-fctg\alpha}; \rho^p = \rho^{\frac{1-\mu}{\overline{R}-1}} \exp\left[\left(1-\overline{\rho}\right)\frac{1-\mu}{\overline{R}-1}\right]. \tag{14}$$

After replacing the integrals with the finite sums, let us find:

$$\overline{\sigma}^*_{\rho 2} = x = \frac{ac\sum_{i=1}^{n} E_i \overline{\rho}_i^k - ec\sum_{i=1}^{n} E_i \rho_i^{\rho+k} - cb\sum E_i}{cb\sum_{i=1}^{n} M_i - ca\sum_{i=1}^{n} M_i \rho_i^k + ec\sum_{i=1}^{n} M_i \rho_i^{k+p}}, \tag{15}$$

where $E_i = ab - a^2 + ae\rho_i^p; M_i = -cb + ac\rho_i^k - ce\rho_i^{k+p}$.

In order to use expression (15), not only mechanical property values, but also stress values in the radial transition area of the punch (boundary conditions) must be known. To establish boundary conditions at the first stage, we use the volume constancy condition. In addition, let us assume that the bottom part of the workpiece is thinned equally.

Then, the thickness value equals:

$$S_{Bot} = S_{blank}\left(\overline{\sigma}_{TO} + 1 - \overline{\sigma}^*_{\rho 1}\sqrt{1-\mu^2}\right) = S_{blank} b. \tag{16}$$

The emerged conical part has its thickness complying with the linear law, and its volume can be calculated using the average thickness of the conical portion.

Thus, we get:

$$\pi R^2 S_{blank} = \pi r^2 S_{Bot} + \pi \frac{R^2 - r^2}{\sin\alpha} S_{med}, \tag{17}$$

where $S_{med} = (S_{BOT} + S_k)/2$, $S_k = S_{blank}\left(\overline{\sigma}_{TO} + 1 - \overline{\sigma}^*_{\rho 1}\sqrt{1-\mu^2}\, \overline{R}^{\frac{1-\mu}{\overline{R}-1}} \exp\left[\left(1-\overline{R}\right)\frac{1-\mu}{\overline{R}-1}\right]\right)$ (18)

From expressions (17,18), let us find:

$$\overline{\sigma}^*_{\rho 1} = \frac{\overline{\sigma}_{TO} + 1 - \overline{R}^2 + \frac{\overline{R}^2 - 1}{2\sin\alpha}\cdot\left(\overline{\sigma}_{TO} + 1\right)}{\sqrt{1-\mu^2}\left[1 + 0.5\frac{\overline{R}^2 - 1}{2\sin\alpha}\left(1 + \overline{R}^{\frac{1-\mu}{\overline{R}-1}} e^{\mu - 1}\right)\right]}. \tag{19}$$

Considering $\overline{\sigma}^*_{\rho 2}$, found earlier, the thickness value of the bottom portion, calculated using expression (11), should not exceed the boundary value. In case stresses ratio is $\sigma_\rho/\sigma_\theta = 1$, thinning should not exceed [24]:

$$e_S = (S_{blank} - S_{Bot})/S_{blank} \le 2n, \tag{20}$$

where $n = \delta_u$ – the exponential function of hardening rate, equal to the relative uniform elongation.

From (20), we can find:

$$\overline{S}_{Bot} = 1 - 2\delta_u. \tag{21}$$

According to the found $\overline{\sigma}^*_{\rho 2}$ value for the second stage, let us calculate the part thickness, considering it as a constant value. After that, we calulate the height h value of the free cylinder-like portion using the volume constancy condition. To the expression (17) summand, representing the volume of the cylindrical portion, we must add:

$$\pi R^2 S_{blank} = \pi r^2 S_{blank} + \pi \frac{R^2 - r^2}{\sin\alpha} S_{part} + 2\pi R S_{blank} h,$$

where $S_{blank} = S_2$.

Thus, we get:

$$\bar{h} = (\bar{R}^2 - (\bar{R}^2 - 1)/\sin\alpha)/(2\bar{R} \cdot \bar{S}_{part}). \quad (22)$$

3. Experimental studies

The experimental technique.

For conducting the experiment according to the presented schematic (Figure 2), the experimental dies were used (Figure3).

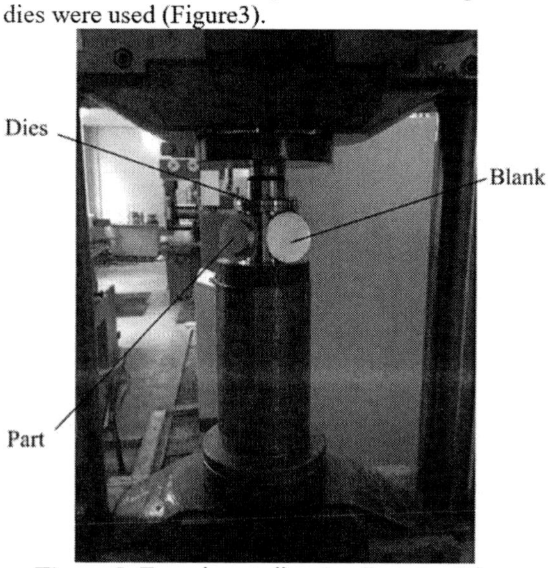

Figure 3. Experiment dies: 1 – Part; 2 – Dies; 3 – Blank.

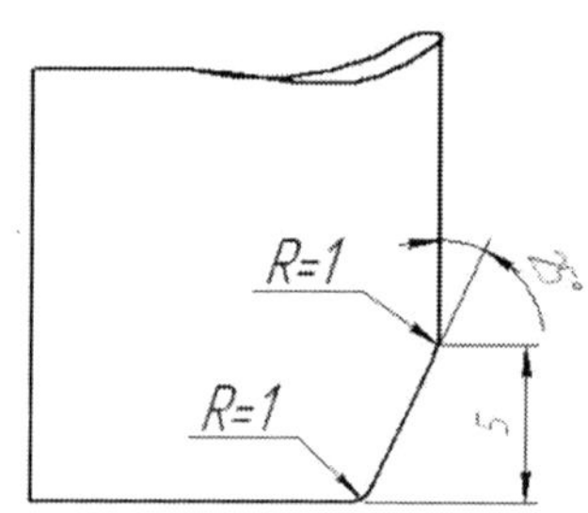

Figure 4. The geometry of the punch.

Fixation of the flanges of a blank was performed using bolts, with enough force to render flanges immobile during the stamping process. Working surfaces of the die and the holder clamp must be strictly parallel to conduct equal fixation of the flange on the whole surface, and rigid enough to prevent significant elastic strains caused by the forces applied to the fixation bolts, which are placed around the blank. The holder clamp and the die are made of U7 Steel, which then was thermally treated to hardness HRC = 40 – 45. The height of the conical portion of the punch is 5.0 mm with the radial transition of 1 mm (Figure 4).

Deformation of the blank was carried out using hydraulic press CDMPU-10.

The blank was obtained using the sheet of the aluminum AD0 alloy with 0.5 mm of thickness rolled on rolling mill Quarto K220/75-300 DUO up to a 0.325 mm thickness. After cutting out the blank with the 58 mm diameter, we annealed it at $420° \pm 5° C$ for 30 min, and cooled in the furnace. The samples for estimating the transversally isotropic body anisotropy coefficient were prepared for 3 dimensions: along, across and at a $45°$ angle to the rolling direction [25].

The transversally isotropic body anisotropy coefficient was calculated using the following expression:

$$e_{width} = \ln|b_k / b_0|, \; e_{length} = \ln|l_k / l_0|, \quad (23)$$

where b_k – the width of the sample after the tensile test;

b_0 – the width of the sample before the tensile test;

l_k – the calculated length after the tensile test;

l_0 – the calculated length before the tensile test.

$$\mu_{along} = -\left(\frac{e_{width}}{e_{length}}\right),\ \mu_{across} = -\left(\frac{e_{width}}{e_{length}}\right),\ \mu_{45^0} = -\left(\frac{e_{width}}{e_{length}}\right),\ \mu = \frac{\mu_{along} + \mu_{across} + \mu_{45^0}}{3}. \quad (24)$$

Measurements of the absolute thickness and width values for the calculation were performed using the electronic tool with a 0.001 mm accuracy and the micrometer with a 0.01 mm accuracy. The mean square error value did not exceed 1%

The mechanical properties of the samples were evaluated by the tensile tests according to GOST7855-55 and GOST1497-61, using the Tinius Olsen H5KT machine. After the series of experiments, the results were processed using the PC software, stored and printed out as charts. The obtained diagrams of linear elongation of the material (force – time dependence) were used to calculate the constant value in expression (9) and evaluate the mechanical properties, such as: $\sigma_{0.2}, \sigma_u, \delta_u$.

Using obtained values of stresses and strains, the linear relation was found:

$$\Pi = \left(\sum_{i=1}^{k}\sigma_i e_i - \frac{\sum_{i=1}^{k}\sigma_i \sum_{i=1}^{k} e_i}{\sum_{i=1}^{k} e_i}\right) \Bigg/ \left(\sum_{i=1}^{k} e_i^2 - \frac{\left(\sum_{i=1}^{k} e_i\right)^2}{\sum_{i=1}^{k} e_i}\right) \quad (25)$$

$$\sigma_{T_O} = \left(\sum_{i=1}^{k}\sigma_i - \Pi \cdot \sum_{i=1}^{k} e_i\right) \Big/ k, \quad (26)$$

where e_i – the intensity of deformation at the considered interval of deformation.

The forming process was conducted in multiple steps with a 1 mm penetration depth of the punch at each step up to a 5.5 – 6 mm depth. The height of the cylinder-like shaped portion was in accordance to the value, calculated using expression (22). At the moment of complete contact with the workpiece at 4 – 4.5 mm of depth along the conical part, the thickness measurement for the first stage was performed. The next measurement was conducted after increasing the penetration depth by 1 – 1.5 mm; then, the height of the part reached 5.5 – 6 mm (the second stage). The conical surface along the generatrix was marked on 5 circular sections from a smaller to larger diameter (Figure 5). Rings were marked on 4 sections, two of which were markers along the rolling direction of the sheet, and two were marked perpendicular to the first two. Thickness measurements were performed using the electronic measurement tool with a 0.001 mm accuracy. The average thickness values for the four sections of each ring were calculated. According to the values of relative thicknesses after the measurements, the regression equation of the first degree was build, using the PC software. For comparison, the equations of relative thicknesses for the first and second stages, found analytically, are presented in Figure 6. The movement distance of the punch was determined by the scale, attached to the press.

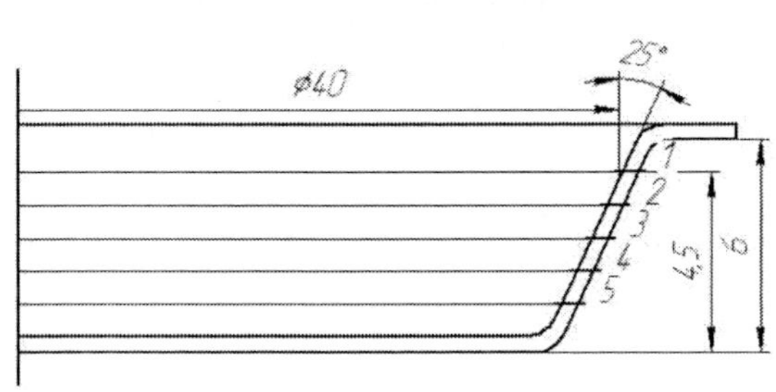

Figure 5. The measurements scheme.

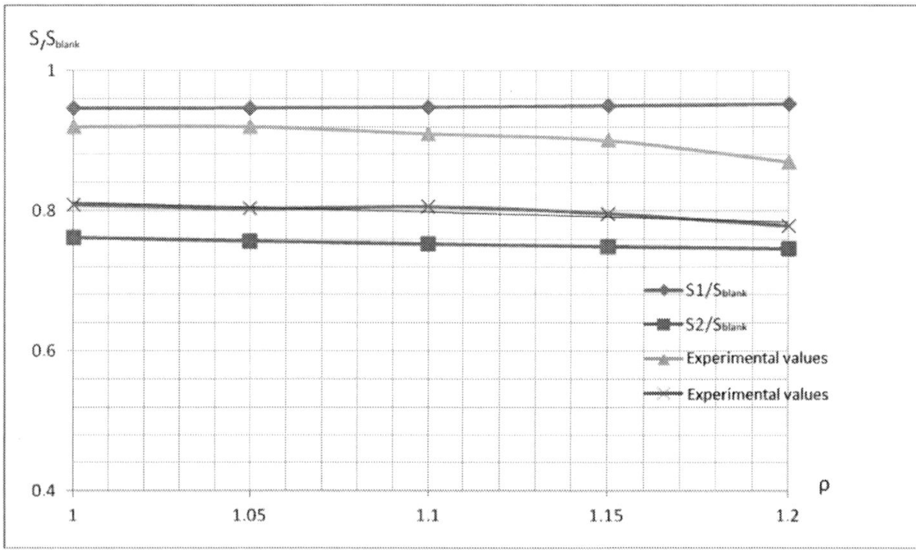

Figure 6. Thickness values of the workpiece.

4. Results of the experiments

Table 1. Mechanical properties of the tested material.

Material	δ_u	μ	σ_u (MPa)	$\sigma_{0.2}$ (MPa)	Thermal treatment (C°) – t (min)
Aluminium AD0	0.22	0.44	167.9	61.3	410 – 20
$S_{blank} = 0.325$ m					

In Table 2, the thickness values for the first and second stages are presented.

Table 2. Thickness values per stage

Stage	Average thickness values per circular portions (mm)				
	1	2	3	4	5
I	0.272	0.274	0.266	0.263	0.26
II	0.263	0.261	0.262	0.259	0.253

5. Analysis and conclusion

The shaping of the workpiece during the forming of the conical part takes place due to thinning. With the increase of the penetration depth of the punch, thinning also increases, i.e. the final part will be thinner, but its thickness variation will be reduced. At the I stage, the thickness variation value was 4.6 %, at the II stage – 3.9 %. With the exclusion of the ring elements on the borders of the curvature radii and the conical portion (circular portions 1 and 5), the average thickness values demonstrate the thickness variation of 4.1% at the first stage, and 0.77% — at the second stage. A sharp decrease of the thickness unevenness occurs on the conical surface.

Using the formation of the flat blank into the conical part, we demonstrate the possibility of obtaining the parts with relative height $H/R \approx 0.1 \div 0.15$ and relatively small conical angle $\alpha \approx 25° \pm 1°$ and thickness variations less than 1%, which in combination with the other modern improvements [26] could benefit to better and possibly cheaper manufacturing processes of thin-walled conical parts for aviation and rocket applications.

References

[1] Popov V G and Yaroslavcev N L 2001 *Liquid rocket engines* (Moscow: 'MATI' Publishing and typographical center – Tsiolkovsky KTU)
[2] Sinyarev G B and Dobrovolsky M B 1955 *Liquid rocket engines: Theory and designing* (Moscow: Oborongiz)
[3] Alisov V E 1980 *Rocket engines theory* (Moscow: Mashinostroenie)
[4] Chistyakov V P, Popov I P, Ovchinnikov A G, Maslov V D et al, USSR Patent No. 1050778 A (1983)
[5] Chictyakov V P, Popov I P, Maslov V D and Gostev A I, USSR Patent No. 1349836 A1 (1987)
[6] Chistyakov V P, Popov I P, Maslov V D et al, USSR Patent No. 1391773 A1 (1988)
[7] Asvobodinov B A, Shashkov V I and Filatova E A, USSR Patent No. 1447480 (1988)
[8] Popov I P, Karpuhin A N and Maslov V D, USSR Patent No. 1454545 A1 (1989)
[9] Chumadin A S, Ershov V I, Kovalev A D and Deberdeev E M, USSR Patent No. 1465152 (1989)
[10] Shipilov A N and Shulga S A, USSR Patent No. 1563813 (1990)
[11] Popov I P, Maslov V D and Karpuhin A N, USSR Patent No. 1748905 (1992)
[12] Chistyakov V P, Popov I P, Ovchinnikov A G and Maslov V D, USSR Patent No. 633643 (1978)
[13] Moiseev V K, Schegolevatykh V F, Komarov A D, Goliusov T P and Galperin V E, USSR Patent No. 755378 (1980)
[14] Averkiev A Y 1996 *Russian J. Vestnik of engineering* **2** 33-37
[15] Alyshin Y A and Shipilov A N 1987 *Russian J. Mashinovedenie* **1** 98-104
[16] Gluschenkov V A 2008 *Special kings of stamping: textbook. Part 1. Stamping using elastic mediums* (Samara: Samara state aerospace university) p 72
[17] Yakovlev S S, Kukhar V D, Remnev K S and Talalaev A K 2012 *Izvestiya TulGU. Technical Sci.* **2** 148-155
[18] Yakovlev S S, Kukhar V D and Remnev K S 2012 *Izvestiya TulGU. Technical Sci.* **1** 185-191
[19] Yakovlev S S, Remnev K S and Tregubov V I 2012 *Izvestiya TulGU. Technical Sci.* **4** 98-102
[20] Khaimovich I N 2014 *Russ. Aeronaut.* **57(2)** 169-174
[21] Storozhev M V and Popov E A 1971 *Theory of pressure working of metals* (Moscow: Mashinostroenie)
[22] Grechnikov F V 1998 *Deformation of anisotropic materials* (Moscow: Mashinostroenie)
[23] Dmitriev A M and Vorontsov A L 2004 Approximation of hardening curves of metals *KSP OMD* **1** 23-26
[24] Tomlenov A D 1968 *Plastic flow of metals* (Moscow: Nauka)
[25] Erisov Y A, Grechnikov F V and Surudin S V 2016 *Structural Eng. and Mech.* **58 (4)** 677-687
[26] Haimovich I N, Frolov M A 2016 *Key Eng. Mat.* **684** 487-507

Creation of wear-resistant frictional surfaces by implanting materials based on tungsten carbide

S V Davidov and A O Gorlenko

Bryansk State Technical University, 7, 50-letiya Oktyabrya bul., Bryansk, 241035, Russia

E-mail: davidov69@tu-bryansk.ru

Abstract. The influenceof the implanted tungsten carbide on the formation of wear-resistant structures, formed in the process of implementation of the combined electro-processing technology in the friction surfaces, is studied. It has been shown that during the thermal force influence in the deformation zone, there is intensive austenizationof the steel with the dissolution of the tungsten carbide powder and the subsequent formation of the composite nanostructures as a result of decomposition of the supercooled austenite, supersaturated with tungsten. The results of tribologicaltestings of cylindrical samples by the normalized method are presented.

1. Introduction

The tungsten compounds with carbon, which are W_2C and WC tungsten carbides, are essential from a practical point of view. These carbides have a high level of hardness, wear resistance and refractoriness, which is an exceptional combination of properties to create wear-resistant and heat-resisting alloys.

In the steel, tungsten forms carbide W_6C, which under austenitization is partially transformed into a solid solution, thus providing obtainment of martensite alloyed by tungsten, which prevents martensitedisintegration during heating, thus providing the required red hardness of the steel. The undissolved part of the W_6C carbide leads to an increase of wear resistance of steel.

On the basis of tungsten carbide (WC, W_2C), the most effective tool hard alloys, containing 85-95 % of WC and 5-14 % of Co, are created. Heat resistant and wear-resistant alloys-stellites (3-5 % W, 25-35 % Cr, 45-65 % Co) are applied to the surfaces of machine parts, subjected to significant wear, with the help of welding.

The main objective of this research is the development of technology of saturation of the surface layer of carbon steel with tungsten carbide in order to achieve high wear resistance of the surface layers of machine parts, working under different conditions of friction.

2. Materials and methods

Currently, synergy-technologies are being intensively developed [1]. These technologies include not only the Additive Technology classification ASTM standard [2], but also traditional hardening techniques such as induction heating, plasma deposition, surface modification, ion implantation, and others [1]. A special feature of these technologies is the use of highly concentrated energy sources.

The main objective is to provide the surface phenomena of synergy-technologies in sequential formation of structures of functional surface layers of different materials or gradient structures and management of the properties of these structures with different physical and high-force action. At the same time, a key element of surface phenomena of self-organization is a sustainable and stable formation of a layer of a certain thickness with a given structure and phase composition, providing the required set of service properties. One of the key issues is the question of structure formation processes in the emerging layer.

Taking into account the experience gained during implantation of detonation synthesis [3] of nanodiamonds in order to create wear-resistant surface layers on friction surfaces of machine parts, the combined electric and mechanical treatment technology (hereinafter – CEMTT) [4], presupposing the formation of the layers implanted with tungsten carbides, with further electric and mechanical hardening of the treated surface, has been considered.

The hardening effect during EMT is achieved due to the fact that high heating and cooling rates are provided, and a high degree of the austenite grain fragmentation produces the fine-grained structure of the surface layer hardening, having high physical and mechanical and performance properties [3-5].

The technology is implemented with a specific installation, which is a technological complex, consisting of the multi-purpose machine (used for mechanical treatment of workpieces) with the appropriate tools and devices to fix the workpiece to be treated; the supply of electric current of high strength and low voltage; the power unit for converting the industrial electric power; the control unit treatment mode; switching tools and supply of lubricating and cooling technological environment; the block for connection with the PC.

Implantation of tungsten carbides into the surface layer during electric and mechanical treatment is performed on the basis of certain modes. They are applied to the surface prior to such treatment as coating, previously stirred with the consistent graphite lubricant in a certain ratio (for better current conductivity). Tungsten carbide particles are embedded in the surface layer reinforcing it. Then electric and mechanical treatment (EMT) is performed on the same surface in reinforcing modes [3,4].

Austenization of the steel surface layer in the contact zone takes place during a high-temperature plastic deformation at high temperatures and pressures. Carbon included into the coating, consisting of graphite and tungsten carbides, diffuses into the surface layers, increasing the carbon content in the austenite during solid phase saturation. When conducting investigations, the friction surfaces of cylindrical samples were treated with the tungsten carbide powder (Table 1), obtained at the plant of refractory metals, CJSC 'Wolfram 'Company' (Unecha, Bryansk Region), according to the specification of the German company 'Element Six GmbH'.

Table 1. The brand of the tungsten carbide powder, the particle size, the oxygen and carbon content

Brand of tungsten carbide powder	Average size of particles according to Fisher Scale, μm	Oxygen content, wt % (max)	General Carbon (typical content),wt %	General Carbon (minimal content), wt %	Free Carbon, wt %
*WC*1,0	0.8…1.3	0.20	6.13…6.18	6.08	0.06
*WC*1,5	1.3…1.8	0.20	6.13…6.18	6.08	0.06
*WC*2	1.8…2.5	0.10	6.13…6.18	6.08	0.05

The tungsten carbide powder is applied to the surface of cylindrical samples made of steel 45 (Figure 1). As a result of treatment, a multilayer structure, consisting of a hardened layer with the thickness of (180 ... 220) μm (Figure 1, layer 1), the first lower sublayerwith the thickness of (200 ... 250) μm (Figure 1, layer 2), the second lower sublayer with the thickness of (20 ... 40) μm (Figure 1, layer 3) and a matrix, consisting of normalized steel 45, are formed on the surface. The microstructure research was conducted usingthe auto-emission scanning election microscope of ultrahigh resolution 'Zeiss Ultra plus' on the basis of 'Ultra 55', Germany ('Zeiss Ultra plus' Field Emission Scanning Electron Microscope).

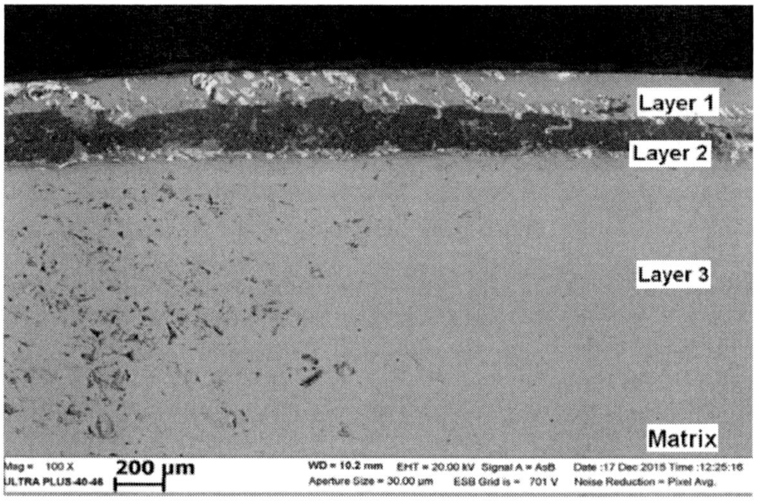

Figure 1. The surface structure of the steel sample hardened with tungsten carbide.

Vickers microhardness was measured with the microhardness meter 'PMT-3M', with a load of 200 g. Measuring the diagonals of prints was conducted with the metallographic inverted microscope 'Metam LV-34' with the use of the automated analysis system 'Micro-Analysis View'. The Vickers microhardness values (Figure 1) for the reinforced multilayer structure (MPa) are the following: layer 1 — 741…846; layer 2 — 546…633; layer3 — 431…525; matrix — 304…332. The Vickers hardness of the R18 steel, heat-treated at elevated temperatures, is the following: 200^{0}C– 815MPa; 400^{0}C – 755MPa [6]. Comparison of the hardness data shows that, in fact, the surface layer of carbon steel 45 is a gradient structure of the P18 tool steel, and it is close to the steelby its properties.

3. Results and discussion

Figure 2 shows the microstructure of layers 1 and 2 in Table 2, which represents the chemical composition of the phase. Layer 1 is a mixture of tungsten carbide WC particles on a steel basis. Therefore, in the process of implanting, the plastic mixture of tungsten carbide in a volume of steel transformed into the solid state.

Table 2. Energy-dispersive analysis of Figure 2, wt %

Points	C	Al	Fe	Co	Cu	W	In total
1	16.8	-	4.0	-	-	79.2	100.0
2	7.6	-	6.4	14.7	-	71.3	100.0
3	3.3	-	84.2	-	2.9	9.6	100.0
4	3.6	-	83.6	-	2.7	10.1	100.0
5	15.1	0.6	71.5	-	4.0	8.8	100.0
6	11.8	0.7	76.9	-	2.1	8.5	100.0
7	11.3	0.6	88.1	-	-	-	100.0

IOP Conf. Series: Materials Science and Engineering **177** (2017) 012123 doi:10.1088/1757-899X/177/1/012123

Layer 2 (Figures 2, 3) is ferrite undersaturated with tungsten. Along the boundaries of its grains, the grid of tungsten carbide is located. Therefore, in the process of implantation, the maximum amount of energy is emitted in layer 2, which pushes the system into a metastable state, followed by formation of the cellular structure (Figure 3).

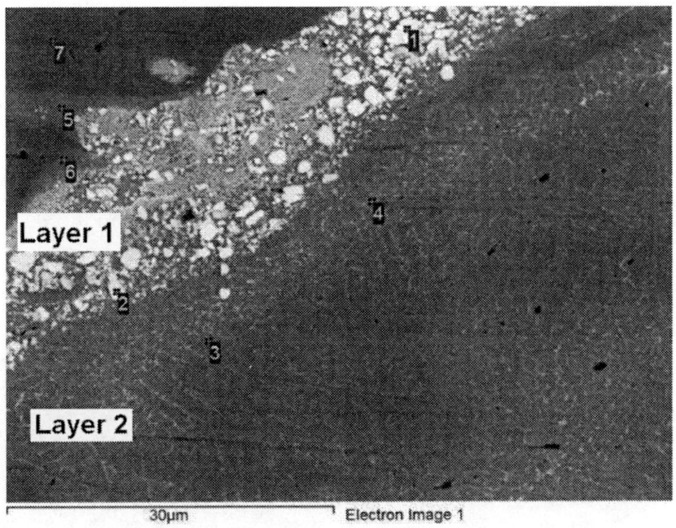

Figure 2. The microstructure of layer 1 (tungsten carbide particles) and layer 2 (the grid of tungsten carbide along the borders of austenite grains)

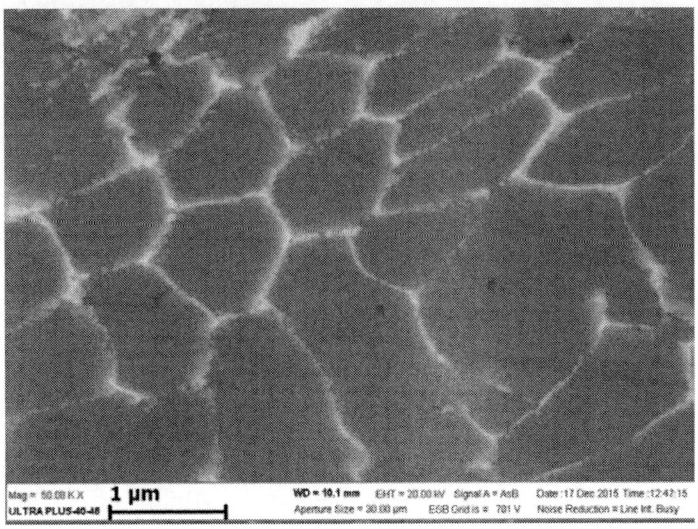

Figure 3. The cellular structure of overcooled austenite in layer 2 (the grid of tungsten carbide along the boundaries of austenite grains).

In accordance with known diagram Fe - W, the temperature of layer 2 is greater than temperature 1060°C, corresponding to the peritectoid reaction. Under this temperature, polymorphic transformation of iron with austenite formation and tungsten carbide dissolution occurs in austenite.

For example, about 8% of W is dissolved in steel austenite R_6M_5 with optimum heating temperature during tempering (1200 ... 1230) °C. About 8% of W also dissolves in steel R18 in the austenite, but it takes place at temperatures of (1270 ... 1290) °C. Consequently, the maximum limit of steel heating during thermal power exposure through implantation can be considered to be the limit of up to 1300°C.

Upon cooling, the austenite solution becomes oversaturated with tungsten and thus stabilizes. Upon further cooling, the structure of stable overcooled austenite is created and the excess tungsten is released in a tungsten carbide grid along the austenite grain boundaries, slightly coated with tungsten. In this case, the resulting structure is identical to that of cast high-speed steel, corresponding to complex carbide eutectic, resembling ledeburite, and is located along the grain boundaries.A similar structure is observed (Figure 4) in the transition oflayer 2 into layer 3, which is much thinner than layer 1, but is also saturated with the particles of tungsten carbide (Table 3).

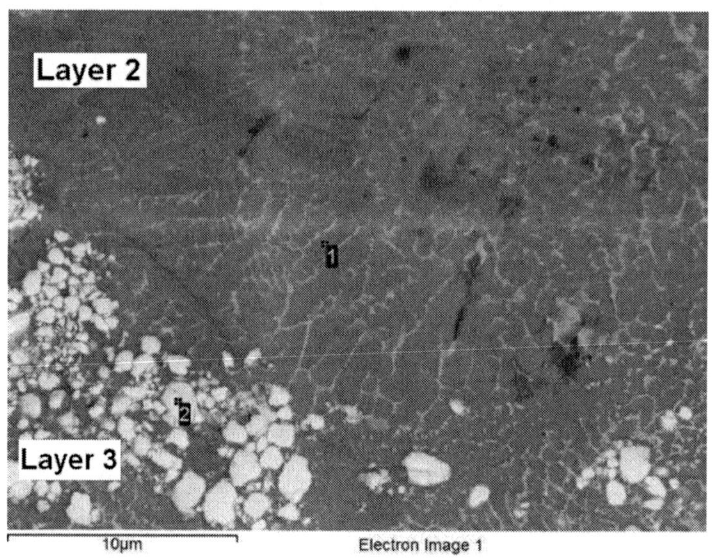

Figure 4. The microstructure of layer 2 (tungsten carbide grid) and layer 3 (tungsten carbide particles)

Table 3. Energy-dispersive analysis of Figure 4, wt. %

Points	C	Fe	Co	Cu	W	In total
1	13.5	68.3	-	6.2	12.0	100.0
2	14.5	9.1	4.2	3.2	69.0	100.0

For complex comparative durability tests, the outer cylindrical surface of the samples produced of steels P18 and 45 was investigated using the following techniques:
- bulk hardening, low temperature tempering (P18 steel);
- bulk hardening, low temperature tempering (steel P18), plus coating with solid-solution hardening of connections with different types of atomic bonds of the Ti-Al-N system with a thickness of 3 μm

applied using PVD technology (in vacuum — using an arc plasma source and separation of the plasma flux);

- bulk hardening, low temperature tempering (steel P18), plus coating with multiphase system structure Mo-Cr-N with a thickness of 3 μm applied using PVD technology (in vacuum — using an arc plasma source and separation of the plasma flux);
- combined electric and mechanical treatment (CEMT) represents formation of the tungsten surface layer implanted with carbides on the surface of steel 45, which was not subjected to thermal treatment and subsequent electric and mechanical hardening of the treated surface.

Samples with coatings based on Ti-Al-N and Mo-Cr-N were manufactured by the LLC 'NPF 'Plazmatsentr' (Saint-Petersburg). Comprehensive comparative tests of the samples were carried out on the automated installation, created on the basis of the friction machine 'Mi-1M' with a normalized method.

According to the analysis of the recorded parameters, the following indicators of tribological properties were determined: running-in time t_0, *h*; running wear h_0, μm; the average value of the friction factor during normal wear *f*; ratio of the maximum value of the friction factor during running, f_0 to *f*; the average value of the wear intensity during normal wear $I_h = (h - h_0) / (L - L_0)$, where *h* is the total value of sample depreciation during the test, μm; *L* is the path of friction, completed by the surface of the sample during the test, μm; L_0 is the path of friction, completed by the surface of the sample during handling; the value of the wear rate within the total time of testing is $I_{h\Sigma} = h / L$.

Choosing the best method of hardening for friction surfaces according to the parameters of tribological tests can be carried out on the basis of the determination of the above-mentioned criteria and their comparison. Thus, the friction surface is more wear-resistant with smaller values of wear intensity, the duration of handling, the friction factor, as well as with the falling characteristic of the friction factor curve change in time.

Testing of the samples was carried out under the following conditions: the sliding speed is υ = 1 m / sec; the normal loading force is N = 100 ± 0,5%, H (which corresponds to the pressuresof about 150 MPa, calculated according to Hertz); the kind of initial contact is a saturated plastic one; the type of lubrication is a boundary-type one; the type of lubrication is a plunging one; the leading type of wear is a fatigue one; the lubricant is industrial oil 'I — 20A' (GOST 20799 - 88); the indenter material is hard alloy BK8; the total time of each sample testing is 8 hours.Test results of the samples treated with the above-mentioned technology onthe modernized installation 'MI-1M'are shown in Table 4.

Comparison of the results of tribological tests of samples with a normalized method has shown that the minimum values of the friction factor, the running-in time and wear are characteristic of the tungsten surface layer implanted with carbides, followed by electric and mechanical hardening.

Table 4. The results of tribological tests onthe modernized installation 'MI-1M'

Tribo-engineering property	Value	Factorvalue for a sample			
		Steel P18	Steel P18 + Ti-Al-N	Steel P18 + Mo-Cr-N	Steel 45 + CEMTT
Running-in ability	t_0, ч	1.12	0.58	0.75	0.45
	h_0, μm	7.5	1.30	1.50	1.70
	f_0/ f	1.61	1.42	1.46	1.19
Antifrictionality	f	0.31	0.32	0.25	0.24
Wear-resistance	h, μm	16.1	9.70	9.60	4.6
	$I_h \cdot 10^{-10}$	3.44	3.12	3.09	1.44
	$I_{h\Sigma} \cdot 10^{-10}$	5.55	3.35	3.32	2.03

Wear curves for samples coated with Ti-Al-N and Mo-Cr-N systems are characterized by the rapid increase of wear after the wear of the coating. The wear curve for the sample subjected to CEMTT is more stable. The results of tribologicaltests have shown that the wear rate of the samples with a layer, implanted by tungsten carbides, and the subsequent electric and mechanical hardening during the normal wear period is less compared with:

- heat-treated samples - 2.4 times;
- samples subjected to PVD (coating of the Ti-Al-N system) - 2.2 times;
- samples subjected to PVD (coating of the Mo-Cr-N system) - 2.1 times.

Chichinadze A.V.introduced the system [7] of ten classes of wear resistance, whereby the wear resistance of steel R18 with different coatings according to the test results (Table 4) relates to class 7, and the wear resistance of steel with a gradient of 45 — to the structure of the 8th class.

Thus, high levels of wear resistance of the gradient structure layer to steel 45implanted with tungsten carbide are a consequence of the composite hardening due to the formation of fine carbide structures, based on tungsten carbide (less than 1 μm) of different morphology (mesh net, thread, grains), and as a consequence of higher hardness values, compared to the layer with a hardness of typical tool steel R18.

4.Conclusions

1. For the first time,it has become possible to apply the coatings on the basis of tungsten carbide by plastic deformation
2. In fact, the surface layer of carbon steel 45is a gradient structure of tool steel P18, and as a result, carbide nanocomposite hardening phases with higher hardness of different morphology are distinguished.
3. The gradient structure is a honeycomb supercooled austenite, which is stabilized and reinforced with the tungsten carbide grid consisting of aggregated nanoscale filamentary and rounded tungsten carbide particles.
4. The presence of a gradient structure having an entirely metallic matrix base material provides a monolithic adhesion hardened layers without discontinuity in the process of wear with considerable dynamic loads, which is also confirmed by the tribological tests.
5. Modification of the friction surface of steel 45 by implanting and strengthening nanocomposite powder tungsten carbide by the IKEM method can significantly increase the wear of the friction surfaces, as evidenced bytribological tests.

References

[1] Kheifets M L 2016 *Sci.Intensive Techn. in Mech.Eng.***4** 3-9
[2] Grigoriev N and TarasovaT 2015*Met Sci. Heat. Treat.***10 (724)** 5-10
[3] Davydov S V Gorlenko O A Scancem V M and Kurakin M Y 2014 *Metall. and Heat Treatm. of Metals* **5(707)** 46-50
[4] Gorlenko A O and Gorlenko O A 2011 *Sci. Intensive Tech. in Mech. Eng.* **6(06)** 21-25
[5] Stanski D 2015 *Met Sci. Heat. Treat.***7 (721)** 77-83
[6] Geller Y A 1983 *Tool steel* (Moscow: Metallurgy)
[7] Chichinadze A V 2001 *Fundamentals of tribology (wear, friction, lubrication)* (Moscow: Mashinostroenie)

The influence of surface finishing methods on touch-sensitive reactions

M S Kukhta[1], **A P Sokolov**[1], **P Y Krauinsh**[1], **A D Kozlova**[1] **and C Bouchard**[2]

[1] Tomsk Polytechnic University, 30, Lenina ave., Tomsk, 634050, Russia
[2] Ecole nationale superieure d'arts et metiers, 151, boulevard de l'Hopital, 75013, Paris, France

E-mail: eukuh@mail.tomsknet.ru

Abstract. This paper describes the modern technological development trends in jewelry design. In the jewelry industry, new trends, associated with the introduction of updated non-traditional materials and finishing techniques, are appearing. The existing information-oriented society enhances the visual aesthetics of new jewelry forms, decoration techniques (depth and surface), synthesis of different materials, which, all in all, reveal a bias towards positive effects of visual design. Today, the jewelry industry includes not only traditional techniques, but also such improved techniques as computer-assisted design, 3D-prototyping and other alternatives to produce an updated level of jewelry material processing. The authors present the specific features of ornamental pattern designing, decoration types (depth and surface) and comparative analysis of different approaches in surface finishing. Identifying the appearance or the effect of jewelry is based on proposed evaluation criteria, providing an advanced visual aesthetics basis is predicated on touch-sensitive responses.

1. Introduction

The relevancy of the paper is associated with the fact that the role of jewelry in our information-oriented society is changing from day to day. Possible updated surface finishing techniques and non-traditional materials determining the visual aesthetics of jewelry are considered. The characteristic aspects of jewelry forming in terms of new visual aesthetics requirements are being investigated. It is these aspects that do develop the information-oriented society itself, digital technology, Inter Web and the virtual content of a new culture.

Nowadays, the range of materials used in jewelry is expanding. The existing raw source is becoming a less important jewelry classification factor as newly-applicable materials can not be defined as ‘precious’ ones. Raw feature becomes less decisive in the classification of jewelry as already used materials and is far removed from the concept of ‘precious’ [1]. It is possible that in the near future, different ornaments and decorations would be considered as pieces of jewelry if they involved such jewelry work methods as engraving, filigree, artistic enameling.

2. Materials and methods

The research included engineering methods for handling processes of complex surface finishing via updated techniques, (computer configuration INTANT i7025, 3D printer WANHAO Duplicator 4, 3D printer Alaris 30), methods of analyzing perceptual quality (portable ECG, equipment for video shooting, SMI iView X HED — a device for video recording of eye movements in the infrared radiation range, portable ECG) and comparative analysis methods to compare similar visual counterparts.

3. The influence of surface finishing techniques on forming processes

Traditionally, the jewelry crafting technique was handmade. Nowadays, it is either mechanized or automated. One of the tasks of art-design is the distribution of unique items. This involves two widely-used processes involving robot-manipulators and 3D-printers. Due to the modern technology development level, many items that were only handmade 50 years ago can now be produced. In view of this fact, the jewelry life-cycle is focused more on the product design than on its own design. Consequently, the most relevant issue is to study all the factors influencing visual aesthetics of jewelry design. The main factor is the application of computer technology. [2, 3] In this case, as quickly as possible, it is necessary to build on the technological chain, which is close to computer-aided design and actual production levels. Improving the production process software is focused on creating decorative surfaces for jewelry.

Considering the fact that modern techniques have increased freedom for a designer in forming highly-aesthetic jewelry, the jewelers are paying more and more attention to the decoration itself. To define the above-mentioned tendencies, the concept "decoration" should be classified into 2 types: depth decoration and surface decoration. Depth decoration is achieved by changing the material structure. This typically embraces traditional techniques, involving the development of 3D-printing. Surface decoration is ornamenting the surface of jewelry items. This tendency is rapidly developing, as well as the so-called "plating techniques", i.e. the finishing technique, the form of which is similar to thin sheets. Samples for surface decoration are extremely wide-ranging: from macro to micro-world forms. There is an atlas of decoration samples — illustrated photos of iron-carbon alloy structures. Designing the decorative pattern involves the selection of a corresponding sample from the atlas. This matching is so exact that it is possible to design such an item embracing this or that appreciable mood. For example, a rounded-shape sample creates a sense of peace. If a designer wants to create a sense of dynamics or expressiveness, then, in this case, the most suitable samples would be those generated by relevant material finishing techniques. In this case, one could use enlarged images of micro-splinters and micro-cracks on the surface of ceramics. Today, the range of decoration means is replenished by updated techniques which either reduplicate natural forms or imitate natural forming processes [4,5,6].

Modern information culture forwards new requirements applicable to the designing of jewelry forms. Intensive integration of computer technology and the Internet transformed the design activities in general. We can observe morphologically advanced processes which emerge at the interface between reality and virtuality and is connected with new techniques and updated materials. This made it possible to automatize the processes in producing complex forms, which previously could be only handmade. However, automated designing of complex-formed items requires sophisticated software. The technology based on 3D-printers excludes the application of sophisticated software. On the other hand, a limited range of applied materials restricts the application of these techniques.

Consequently, the development of new techniques and materials furthers the jewelry development trends, which, in its turn, expands the possibilities of fine surface finishing and involves the design of new forms.

Developing modern jewelry techniques embrace not only the production of new alloys, but also the implementation of sophisticated technological processes, including an alternative IT environment. Different software tools are used in designing and modeling jewelry form variations.

4. The effect of forms and surface characteristics on the quality of visual aesthetics

Alternatives of sophisticated technology in material surface finishing determined the specific features of the culturally industrial jewelry design. The era of computerization resulted in the substitution of the perceptual jewelry concept as a jewelry masterpiece, traditionally being a cultural indicator of wealth and the holder's status and often empowering some magical power, but now, in present-day world, only serving some functions. The first and foremost function is its unique jewelry design concept. If previously, the creation of new jewelry forms originated in a designer's workshop, now, in the information culture, this process is accomplished by continuous production variations [7–9]. The innovative methods in developing forms and surface characteristics via sophisticated techniques of jewelry surface finishing introduced a new concept of status based on the combination of design, material and techniques. The array of modern jewelry materials also include non-traditional materials-valuable wood and different plastic materials. We can consider a similar example-application of ebony wood in jewelry design, having density of 1300kg/m³ at 15% humidity. Ebony exhibits the following properties: it does not float on water; requires specific drying conditions (its volume decreases after drying); it contains a significant amount of ether oil being resistant to the environmental effect and temperature and a humidity drop; it does not rot and is resistant to insect damage. Due to the ebony fibre structure and its hardness, a jeweler should be experienced in processing this wood. Another important factor is that ebony rapidly and easily blunts any wood-carving tools, so it is necessary to have special-purpose wear resistant tools. The cost of ebony is comparable to that of commonly used jewelry metals which are truly undergoing revolutionary changes. Being applied in jewelry design, fragrant and fibrous ebony is often encased with gems. A wide range of techniques in material surface finishing is applied in modern jewelry design [5, 6, 10]. The reticulated sample of 800 silver is illustrated in Figure 1. Many jewelers apply such technique as riffles or ripples on metal plates as it exhibits an aesthetic effect.

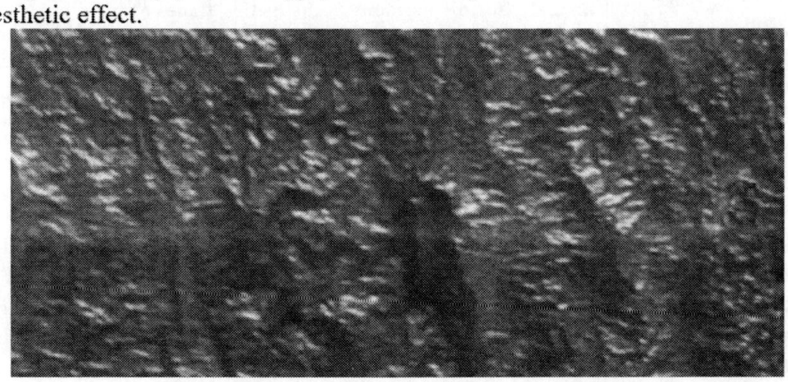

Figure 1. The reticulated sample of 800 silver.

The wide selection of materials and techniques provides an opportunity to create a variety of jewelry forms which would satisfy even the most refined taste. The jewelry-ware of Christine Skobina (Irkutsk) was selected to study the perception effect on different materials. These included ammonites and rock crystal encased with silver (Figure 2).

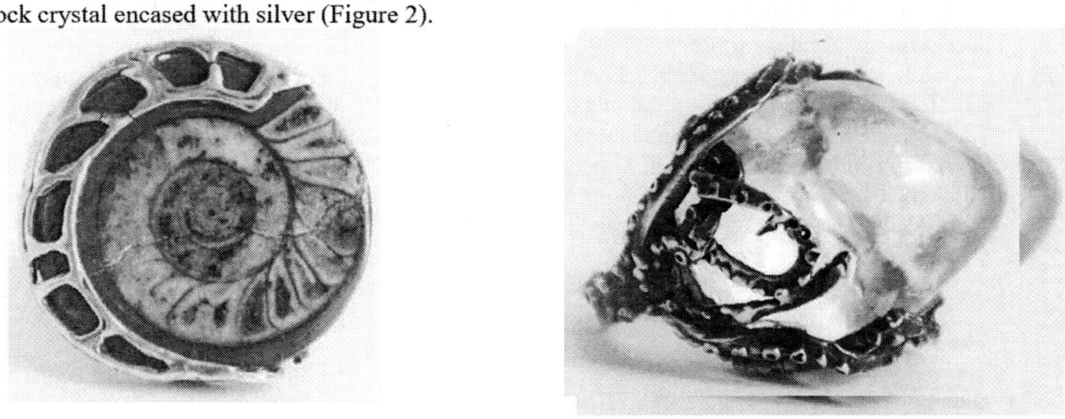

Figure 2. Design of the rings: left – the ring encased with ammonites; right – the ring encased with rock crystal. Designer: C. Skobina (Irkutsk).

Perception of objects is quite different and governed by human mental state laws responding in a different way to materials of varied surface characteristics, forms and colors. The described analysis is based on the authors' proposed method which measures attention delay time, reveals specific perception effect on different natural materials (ammonites, rock crystal). This analysis proved the fact that jewelry–ware, matching unfinished natural materials with jewelry-forming finished metal or alloys, is in great demand. Abandoning the traditional jewelry forms furthered the possibility of focusing more attention on the techniques and materials, i.e. in the sense of 'feeling' (tactile/touch sensing level), the quality of such things as conchite or stones. Out–of–style–related, each jewelry transmits this or that visual legend and transforms into a source of new senses unique in itself involving a touch-sensitive contact for the holder (Table 1).

Table 1. Perception of jewelry designed from different materials

Specific perception features	Jewelry from ammonites	Jewelry from rock crystal
Tactile (touch) sense of material surface	smooth, warm surface	cold, intricate structure surface
Visual perception of material surface	non-transparent, matte surface, natural form harmony, distinct perfection	intricate transparent contour, surface with "traces of time"
Perception of jewelry form	simplicity, laconism, geometrical purity (spirals, circles of natural ammonite shells and rhythmic repeat of mounted silver)	jewelry form resembles archaeological artifacts, creating visual legends; jewelry attains the status of historical rarity
Tactile sense	wave fragrance	practically no
Usability	high level	low level

Ammonites and rock crystal are practically not exposed to finishing, and their natural forms are preserved in their primeval features by the designer, which makes it possible to feel the vigorous bond with the Earth energy. Touch, visual, tactile perception of ring forms transmits the richest pallet of senses which was confirmed by the results of the finished material tested on the quality of touch–sensitive responses and presented in Table 1.

5. Conclusion

Jewelry design in our modern society is connected with the search for new forms which are determined by emerging visual aesthetics based on existing alternatives of the information culture. This, in its turn, enables the human to penetrate beyond the boundaries of the visible world, to intrude into the microstructure, to observe the images created by Nature and to re–create them in ornamental patterns on the jewelry surface.

A wide spectrum of proposed techniques for surface finishing of jewelry is connected with two basic trends: introducing non–traditional materials in jewelry and finding new finishing techniques.

The priority trend is the combination of finished and natural materials. The designer selects the material finishing technique which determines the jewelry form. The diversity of proposed techniques is governed by the demand in creating different touch–sensitive responses, which would provide a varied level of emotional feelings.

Computer technology automatizes not only the process of producing visually defined jewelry forms, but also quality engineering of touch-sensitive responses.

Acknowledgments
This work is done under the financial support of the Ministry of Education and Science of the Russian Federation by contract No 02.G25.31.0190

References

[1] McGrath J 2005 *The Jeweler's Directory of Decorative Finishes: From Enameling and Engraving to Anodizing and Hokum Gene.* (London : Bloomsbury Publ.)
[2] Sokolov A P, Kukhta M S, Pelevin E A 2014 *Proceedings of the International Conference, Tomsk* (National Research Tomsk Polytechnic University) pp 1-4
[3] Kukhta M, Sokolov A, Pelevin E 2015 *IOP Conf. Ser.: Mat. Sci. and Eng.* (Tomsk) vol **93(1)** (IOP Publishing) pp 1-5
[4] Chernyh M M, Rudenko M J 2013 *Proc. of the Academy of Industrial Design* **1** 6-10
[5] Galanin S I, Viskovaty I V, Gladiy J P 2016 *Surface Eng. and Appl. Electrochemistry* **52(3)** 6-15
[6] Galanin S I, Viskovaty I V 2015 *Surface Eng. and Applied Electrochemistry* **51**(4) 332-338
[7] Rao Chepuri R K, Trivedi D C *Coordination Chemistry Reviews* **249** 613-631
[8] Gardan M 2012 *Jewelry Technology Forum* 1-11
[9] Wastiels L, Schifferstein H N, Wouters I and Heylighen A 2013 *Int. J. of Design* 7**(2)** 31-41
[10] Saltykova N A 2003 *J. of Mining and Metallurgy* **39(1-2)** 201-208

Application of the geometric immersion method based on the Castigliano variational principle for the axisymmetric problems of elasticity theory

Y S Kuznetsova, N A Vorobyev and N A Trufanov

Perm National Research Polytechnic University, 29, Komsomol ave., Perm, 614000, Russia

E-mail: Suhodolchik@mail.ru

Abstract. The fundamentals of the geometric immersion stress method for axisymmetric problems of elasticity theory have been stated. The geometric immersion method involves reduction of the initial problem for Clapeyron's free form elastic body to an iteration sequence of elasticity theory problems on some canonical domain. The iteration procedure for the variational equation of the geometric immersion method has been stated, as well as its discrete analog construction technique suggested with the finite-element stress method for the axisymmetric problem of the elasticity theory in the cylindrical coordinates. A practical application of the method has been demonstrated by a test problem. A reasonably good fit between the stress fields definition and the numerical solution by the traditional finite-element displacement method has been obtained.

1. Introduction

The boundary problem of the elasticity theory can be traditionally stated as displacement or stress, which under variational formulation implies the principle of minimum potential energy of the system (Lagrange principle) and the principle of the least work (Castigliano principle), respectively. Approximative and numerical methods of the elasticity theory such as the finite-element method (FEM) and the variational difference method, which implement extreme principles, in most cases, base on the displacement formulation as it is more convenient for choosing basis functions that should satisfy rather simple requirements for kinematic acceptability, that is fulfillment of boundary displacement conditions and existence of space variable first derivatives. Well-known shortcomings of numerical displacement solutions consist in low precision in strain and stress fields definition, weakly compressible material analysis problems, and impossibility to calculate incompressible bodies. Therefore, different mixed formulations based on Reissner, Hu-Washizu, etc. variational principles, which are more difficult to realize, but, to a certain extent, free from the given shortcomings, have been developed sufficiently widely.

Approximate and numerical stress solutions based on the Castigliano variational principle have found a fairly limited application, primarily, due to the construction problems with basis functions which, in this formulation, must be statically acceptable, i.e. satisfy the equilibrium equations and static boundary conditions in the domain [1-5]. Generally, solutions for such elasticity theory problems have been obtained in the canonical form domains.

There are some methods to reduce the elasticity theory boundary problem for an arbitrary shape domain to the canonical domain: the fictitious domain method, perturbation technique, etc. The paper offers theoretical justification and practical examples of the geometrical immersion method (GIM) for the elasticity theory stress problem. The appeal of this approach is connected with the possibility in principle to obtain stress fields in the numerical realization of the GIM with higher precision and to solve incompressible and weakly compressible material problems. Moreover, displacement and stress solutions enable construction of lower and upper variational boundaries for the guaranteed exact solution.

2. Problem statement

Let us consider the functional of complementary work of an elastic body [6]:

$$V(\hat{\sigma})=\frac{1}{2}\int_{D}\sigma_{ij}\varepsilon_{ij}(\hat{\sigma})dD-\int_{S_u}t_i(\hat{\sigma})U_i dS_u\,, \tag{1}$$

where $\hat{\sigma}$ is symmetrical stress tensor with components σ_{ij}; $\hat{\varepsilon}$ – symmetrical strain tensor with components ε_{ij}; U_i – displacement vector components specified on boundary S_u, $t_i=\sigma_{ij}n_j$ – forces on surface S_u, n_i – components of the boundary S unit outward the normal vector. Functional (1) is defined in domain D on the set of statically acceptable stress fields σ_{ij} which satisfy equilibrium equations and static boundary conditions on boundary S_σ; $S=S_\sigma\cup S_u$.

From the functional (1) stationarity condition, given that $\delta\sigma_{ij}$ variations are also statically acceptable stress fields, we have:

$$\delta V(\hat{\sigma})=\int_{D}\delta\sigma_{ij}\varepsilon_{ij}(\hat{\sigma})dD-\int_{S_u}t_i(\delta\hat{\sigma})U_i dS_u=0\,. \tag{2}$$

Hence, the realization of compatibility equations in the Beltrami-Michell form in the D domain and boundary conditions in displacements on S_u follow:

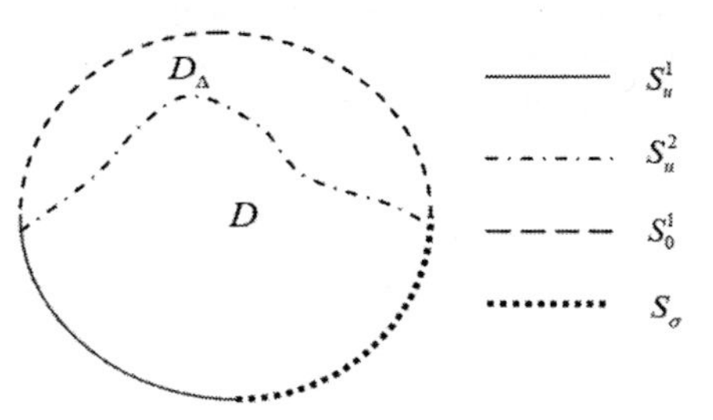

Figure 1. Initial domain D, complement domain D_Δ and boundary delineation.

According to the geometric immersion procedure, we will consider canonical domain D_0 ($D_\Delta=D_0\setminus D$, Figure 1) and redefine our functional (1) on the canonical domain:

$$\begin{aligned}V(\hat{\sigma})&=\frac{1}{2}\int_{D}\sigma_{ij}\varepsilon_{ij}(\hat{\sigma})dD-\int_{S_u}t_i(\hat{\sigma})U_i dS_u=\\&=\frac{1}{2}\int_{D}\sigma_{ij}\varepsilon_{ij}(\hat{\sigma})dD-\int_{S_u}t_i(\hat{\sigma})U_i dS_u+\frac{1}{2}\int_{D_\Delta}\sigma_{ij}\varepsilon_{ij}(\hat{\sigma})dD_\Delta-\frac{1}{2}\int_{D_\Delta}\sigma_{ij}\varepsilon_{ij}(\hat{\sigma})dD_\Delta=\\&=\frac{1}{2}\int_{D_0}\sigma_{ij}\varepsilon_{ij}(\hat{\sigma})dD-\frac{1}{2}\int_{D_\Delta}\sigma_{ij}\varepsilon_{ij}(\hat{\sigma})dD_\Delta-\int_{S_u}t_i(\hat{\sigma})U_i dS_u\end{aligned} \tag{3}$$

By varying the stress fields, we come to the following GIM variational equation which is defined in canonical domain D_0:

$$\delta V(\hat{\sigma}) = \int_{D_0} \delta\sigma_{ij}\varepsilon_{ij}(\hat{\sigma})dD - \int_{D_\Delta} \delta\sigma_{ij}\varepsilon_{ij}(\hat{\sigma})dD_\Delta - \int_{S_u} t_i(\delta\hat{\sigma})U_i dS_u = 0. \quad (4)$$

Functional analysis methods have determined requirements for functions according to which functional (3) is defined in domain D_Δ that complements initial domain D up to canonical D_0; established a correspondence between the stress fields that minimize functionals (1) and (3); shown the correspondence of variational equations solutions (2) and (4) in the D domain; and obtained differentional formulation of the elasticity theory boundary stress problem in domain D_0, which corresponds to variational formulation (4). Necessary conditions for the functional (4) stationarity are realization of compatibility equations in the Beltrami-Michelle form in domain D_0 and kinematic boundary conditions on boundaries S_u^1 and S_u^2. It has been theoretically obtained that on transition via boundaries S_u^2 and S_u^1, a jump of the displacement vector is possible. Immersion of the initial domain into the canonical one evidently leads to the appearance of new boundary S_0^1 on which the boundary conditions should be formulated. Three possible types of such boundary conditions, which enable the selection of the most acceptable boundary condition type for the current problem, have been ascertained.

It is suggested that equation (4) be solved with the following iteration procedure:

$$\int_{D_0} \delta\sigma_{ij}^{(k+1)}\varepsilon_{ij}\left(\hat{\sigma}^{(k+1)}\right)dD_0 = \int_{D_\Delta} \delta\sigma_{ij}^{(k+1)}\varepsilon_{ij}\left(\hat{\sigma}^{(k)}\right)dD_\Delta + \int_{S_u} t_i\left(\delta\hat{\sigma}^{(k+1)}\right)U_i dS_u,$$
$$\sigma_{ij}^{(0)} \equiv 0, \quad k = 1,2,3..., \quad (5)$$

The iterative process convergence theorem in the geometric immersion method (5), irrespective of the initial D domain difference from the canonical D_0 one, has been proved.

In a number of cases, the GIM allows analytical realization, but in most cases, one has to resort to numerical methods. Numerical realization presupposes that variational difference or finite element methods are applied, which enable the construction of a discrete analog of the GIM variational iterative procedure [7-10]. The present paper expands on the finite element realization of the method by the example of axisymmetrical elasticity problems.

3. Exemplifications

As an example of the described GIM procedure, let us consider an axisymmetrical elasticity problem: determination of the stress state of the finite hollow cylinder with a rigid rectangle in the cross-section and a circular inclusion under the action of uniformly distributed internal pressure P (Figure 2). On the inclusion contour, both displacement vector components are zero. A cylinder without inclusion, completely covering the real domain, was chosen as the canonical domain.

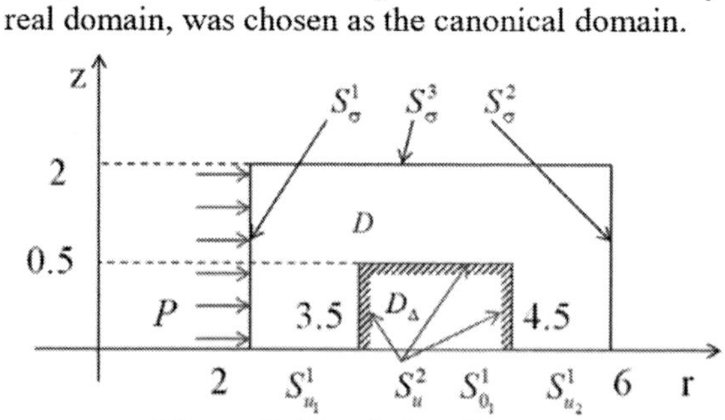

Figure 2. The design diagram.

The problem solution by the geometric immersion method is reduced to the variational-iterative procedure (5). To ensure static admissibility of the stress fields, the following boundary conditions should be satisfied:

$$\sigma_{rr}(2,z)=-P,\quad \tau_{rz}(2,z)=0,\quad z\in S_{\sigma}^{1},$$
$$\sigma_{rr}(6,z)=0,\quad \tau_{rz}(6,z)=0,\quad z\in S_{\sigma}^{2},$$
$$\sigma_{zz}(r,2)=\tau_{rz}(r,2)=0,\quad r\in S_{\sigma}^{3},$$
$$\tau_{rz}(r,0)=0,\quad r\in S_{u_1}^{1}\cup S_{u_2}^{1}.$$

Kinematic boundary conditions are:

$$u_r(r,z)=u_z(r,z)=0,\quad (r,z)\in S_{u}^{2},$$
$$u_z(r,0)=0,\quad (r,z)\in S_{u_1}^{1}\cup S_{u_2}^{1}.$$

During immersion, new boundary $S_{0_1}^{1}$ is also formed, for which a boundary condition should be stated; in this case, it is convenient to extend symmetry conditions, such as in $S_{u_1}^{1}$ and $S_{u_2}^{1}$:

$$\tau_{rz}(r,0)=u_z(r,0)=0,\quad (r,z)\in S_{0_1}^{1}.$$

Material parameters are: $E=2.3E11$ Pa, $\nu=0.3$. Pressure - $P=10$ Pa.

In order to establish the GIM iterative convergence on the discretization of the canonical domain, a series of calculations of a sequence of uniformly condensing meshes was made (Figure 3). Results obtained by the finite-element displacement method (discretization degree is 6088) were taken as the reference solution.

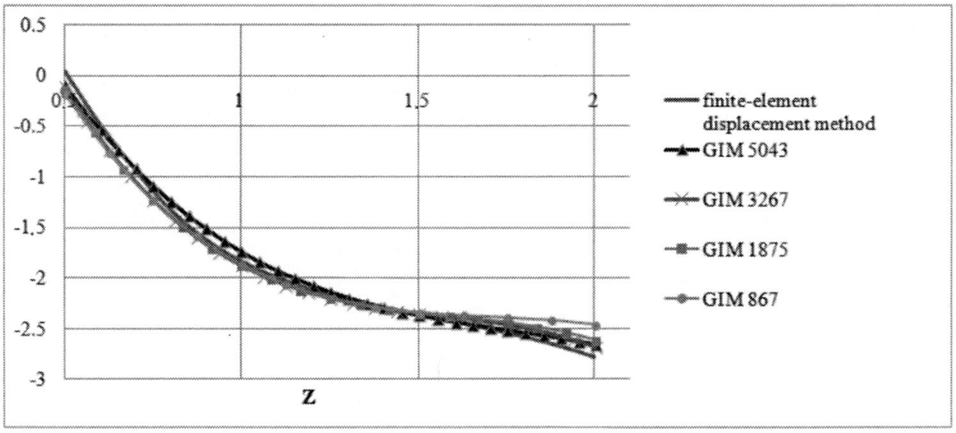

Figure 3. Distribution σ_{rr}, *Pa* depending on the degrees of freedom on the cylinder section at $r=4$m.

It was established that the 3267 nodal unknown mesh will be sufficient for the problem. Further condensing of the mesh does not result in a significant change of the numerical solution.

In order to determine the iterative convergence of the number of iterations, let us consider a change in the relative residuals at each solution step in Figure 4.

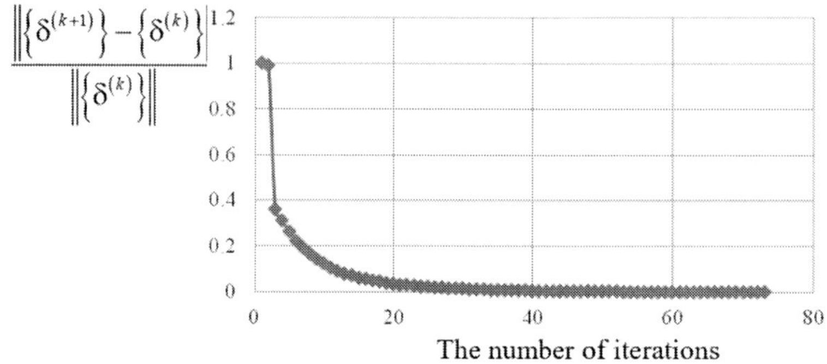

Figure 4. Change in the relative residual on the iteration number.

Figure 4 shows that we need about 40 iterations to obtain a quality solution. If we increase the iteration number further, relative residual $\frac{\left\|\{\delta^{(k+1)}\}-\{\delta^{(k)}\}\right\|}{\left\|\{\delta^{(k)}\}\right\|}$ will be less than 0.0001, which will change results insignificantly.

Since an undisputable advantage of the numerical approaches based on the Castigliano variational priciple is calculation of structures from incompressible materials, let us consider practical application of GIM in the investigation problem of the stressed state of the rubber-metal shock absorber in the form of shaped bushing designed to compensate for the longitudinal impact (Figure 5(a)). Steel elements are supposed to be absolutely rigid and continuously bonded with rubber. Figure 5(b) shows a simplified design model and indicates complement domain D_Δ and new boundary S_0^1 , which appeared as a result of geometric immersion.

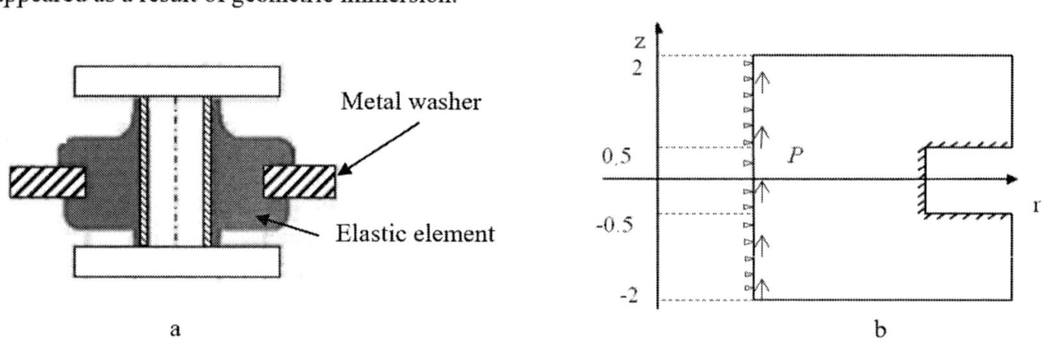

Figure 5. The rubber-metal shock absorber in the form of shaped bushing.

The given problem was also explored for convergence; the results are presented as a stress intensity pattern in the whole shock absorber in Figure 6.

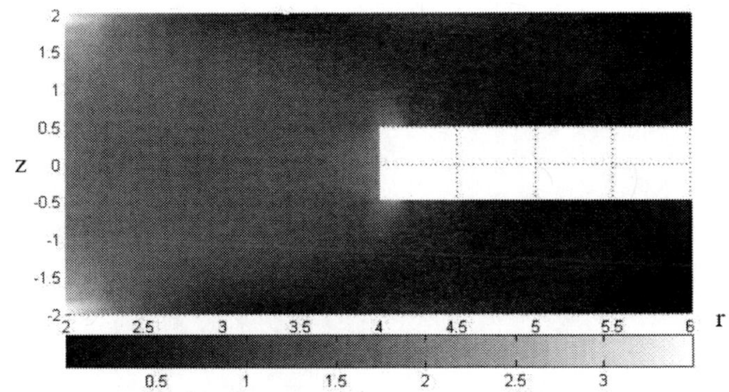

Figure 6. Stress intensity reduced to load intensity.

4. Conclusion

Thus, the paper has demonstrated practical application of the geometric immersion method based on the Castigliano variational principle for the solution of axisymmetrical elastic problems, including problems for incompressible elastic materials. The iteration procedure for solving the variational equation by the geometric immersion method has been formulated, as well as the procedure for its discrete analog construction by the finite-element stress method for the axisymmetric elasticity problem in the cylindrical coordinates has been suggested. The considered model axisymmetric problems have demonstrated the GIM efficiency. A sufficiently good agreement of the stress fields determination in comparison with the numerical solution by the traditional finite-element displacement method has been obtained.

Acknowledgments

The work was supported by the Government of the Russian Federation (Resolution No. 220 of April 9, 2010), contract No. 14.B25.310006 of June 24, 2013.

References

[1] Girija Vallabhan C V, Muluneh Azene 1982 *Int. J. Numer. Meth. Eng.* **18(2)** 291-309
[2] Sarigul N, Gallagher R H 1989 *Int. J. Numer. Meth. Eng.* **28(7)** 1577-98
[3] Watwood V B Jr, Hartz B J 1968 *Int. J. Solids Struct.* **4(9)** 857-73
[4] Fraeijs de Veubeke B 2001 *Int. J. Numer. Meth. Eng.* **52(3)** 287-342
[5] Pin Tong, Pian Theodore H H 1969 *Int. J. Solids Struct.* **5(5)** 463-72
[6] Lurie A I 2005 *Theory of elasticity* (Berlin: Springer)
[7] Gallager R 1975 *Finite element method. Fundamentals* (Englewood Cliffs, N.J.: Prentice Hall)
[8] Norrie D, de Vries J 1978 *An introduction to finite element analysis* (New York: Academic Press)
[9] Zienkiewicz O C, Taylor R L, Zhu J Z 2005 *The Finite Element Method: Its Basis and Fundamentals* (Oxford: Elsevier Butterworth-Heinemann)
[10] Spilker R L, Pian T H H 1978 *Computers and Structures* **9** 273-9

The use of the rotating electromagnetic field for hardening treatment of details

V A Lebedev[1], A A Kochubey[1] and A V Kiricheck[2]

[1] Don State Technical University, 1, Gagarin Sq., Rostov-on-don, 344000, Russia
[2] Bryansk state technical University, 7, 50-letiya Oktyabrya bul., Bryansk , 241035, Russia

E-mail: va.lebidev@yandex.ru

Abstract. The article discusses energy aspects of details' hardening with convective flows of freely moving indenters under the conditions of the rotating electromagnetic field. Results of theoretical studies of the kinetics of the movement of the ferromagnetic indenters are presented and the energy model of the state of the rotating magnetic liquefied layer is proposed, formed under the influence of the rotating electromagnetic field.

1. Introduction

Despite the variety of methods to increase details durability, the issue of strengthening treatment of long parts with a relatively small cross-sectional area and inner surfaces of the hollow thin-walled parts is still incompletely resolved. To resolve this problem, the magneto-dynamic method is seen as promising, which in its physical-technological essence and the nature of the shock-pulse impact on the surface is a kind of dynamic methods meant to increase the strength of details with the convective flows of freely moving indenters. The most famous representatives of this group of methods are: vibro-impact [1], pneumodynamic and centrifugal rotary processes [2].

In [3, 4], it is shown that if non-equiaxed ferromagnetic particles are placed in a rotating electromagnetic field of sufficient strength, they come in a complex, chaotic movement, creating a kind of the rotating magnetic liquefied (RML) layer. For creation and practical use of the RML layer of ferromagnetic freely moving particles, the last are loaded into the device, creating a rotating electromagnetic field (REMF). The basis for these devices (Figure 1) is inductor 2, placed in housing 1. The inductor is a modified stator of an asynchronous motor. Cylindrical tube 3 of a non-magnetic material is installed into a stator boring, the inner surface of which limits the working area of the device. Tube 3 serves as the working area. To increase the resistance of the working area of the pipe, replaceable insert 5 is inserted into it, that is, replaced when worn. Cylindrical particles 4 serve as the working environment that is made of the ferromagnetic material. Devices can be made with the water and air cooling.

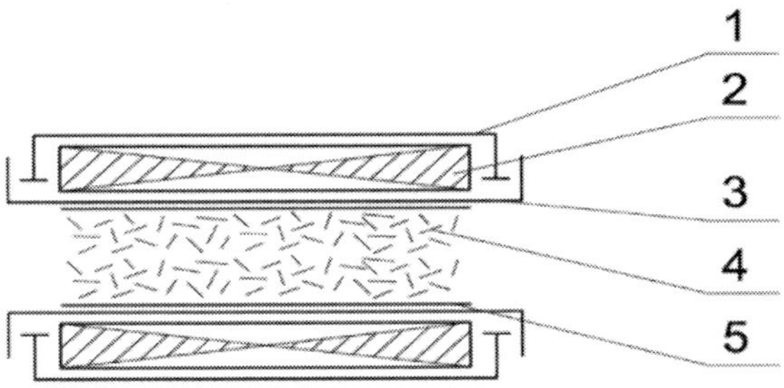

Figure 1. The diagram of the device with the rotating electromagnetic field.

Creation of the RML layer in devices under the influence of the rotating electromagnetic field, allows one to intensify a number of technological processes due to the complex flow of the physico-chemical effects in the working area such as stirring and dispersion, acoustic and electromagnetic phenomena, friction, high local pressures, electrolysis and shock-pulse interactions. Due to their relatively simple design, devices with REMF have been widely applied in technological processes of waste processing for electroplating, petroleum refineries, processing of animal waste, in powder metallurgy, oil refining, production of paints and polymers, in construction, and so on.

In this connection, the use of the rotating electromagnetic field is not without interest to improve the efficiency of technological processes of manufacturing parts [5]. Based on preliminary experimental studies, we have established that the chaotic motion of particles in the rotating electromagnetic field leads to a large number of collisions between particles, particles with the treated products, accompanied by both direct and glancing blows. The number of such collisions reaches 10^3–10^4 by every particle per second. As the result of impact-impulse collision, there is a local elastic-plastic deformation of the surface of the parts in the working area of the device. Thus, we can conclude: devices with REMF can be quite successfully used to harden the surface layer of parts. To justify the technological possibilities of machining under conditions of REMF, we will analyze the main patterns of the dynamics of the RML layer.

2. The kinetic state of ferromagnetic indentors in the rotating magnetic liquefied layer

If bodies are placed in the electromagnetic field, then forces applied to the electric charges of these bodies will act on them. Thus, the ferromagnetic non-equiaxed indentors, becoming magnetized to saturation, can be considered as the elementary magnetic dipoles.

Let us define the kinematic characteristics of the magnetic indenter with mass m and magnetic moment p_m, moving in a magnetic field formed by the superposition of a homogeneous constant field, and an inhomogeneous alternating field.

The value of the moment of couple of forces tending to turn the magnetic indentor in the direction of an external alternating electromagnetic field can be expressed as follows:

$$M_r = p_m B \sin\varphi \tag{1}$$

where ω is the angular velocity of REMF;

p_m is the magnetic moment of the ferromagnetic indentor;

$\vec{B} = \vec{B_c} + \vec{B_v} \cos \omega t$ - REMF induction;

φ – the angle between the vector of the magnetic moment of the indentor and the vector of electromagnetic induction.

In addition, in an inhomogeneous field, the biasing (transportation) force acts on the magnetic indentor:

$$F_p = \chi\, UB \frac{dB}{dy}, \tag{2}$$

where χ – magnetic susceptibility of the indentor material; U – the volume of the indenter; $U = \pi d^2 l/4$, where d and l – the diameter and the length of the indenter, accordingly; $\frac{dB}{dy}$ – the field gradient.

Under the action of the rotating moment, the ferromagnetic indenter rotates in a magnetic field with absolute angular speed ψ_a, that in the fixed coordinate system is equal to the sum of the relative and transportation speeds, i.e.:

$$\psi_a = \psi_r + \psi_t \,. \tag{3}$$

On the basis of numerous experimental studies in [6], it is shown that with a high degree of correlation, the transportation speed of rotation of the magnetic particle is almost equal to the speed of rotation of the external magnetic field. This allows one to record $\psi_t = \omega$.

To calculate the relative angular speed of ferromagnetic indentors in the electromagnetic rotating field, the following formula is received:

$$\psi_r = \varphi_1' = \frac{d^2 l \vec{B} \vec{B}_v}{I \omega K_\mu} \sin\varphi_0 \sin\omega t, \tag{4}$$

Where $K_\mu = \mu \left(1 + (\mu-1) \dfrac{\frac{\lambda}{\sqrt{\lambda^2-1}} \ln\left(\lambda + \sqrt{\lambda^2-1-1}\right)}{0.785(\lambda^2-1)(\mu-1)} \right)$; I – the moment of inertia of the particle μ – magnetic permeability of the material of the indentor; $\lambda = l/d$ – the ratio of the length of the cylinder to its diameter.

Based on (9) and (10), and considering the fact that ferromagnetic indentors are magnetized to full saturation, the equation of the absolute rotation speed of the ferromagnetic indentors in the RML layer will take the form:

$$\psi_a = \omega + \frac{\pi d^2 l \vec{B} \vec{B}_v (\mu-1)}{4 I \omega \mu} \sin\varphi_0 \sin\omega t \,. \tag{5}$$

From this equation it follows that the non-equiaxed ferromagnetic indentor in a rotating electromagnetic field at a constant angular speed moves with a variable angular speed, fluctuating relative to the induction vector of the magnetic field with a frequency depending on the magnitude of the moment of inertia of the indentor, its magnetic moment, the induction of the external magnetic field and the angle between the vector of the magnetic moment of the indentor and the magnetic field induction vector. To take into account all conditioning factors for the movement of indentors in the RML layer and to get admissible estimated analytical dependences is a very complex task from a mathematical point of view. So we will use the most versatile energy approach to the description of these processes and to the energy state estimate of ferromagnetic indention in the RML layer.

3. The energy balance of the ferromagnetic indentor of the rotating magnetic liquefied layer

The proposed model is based on the following provisions:

- in a steady state of developed magnetic liquefaction, the energy roll-in to indentors from the external field is compensated by energy dissipation due to indentors interaction with the processed environment and as a result of inelastic collisions [4,7];
- changing the orientation of the indentors magnetic moments during the time between successive collisions is relatively small, which makes it possible to linearize the equation of motion of a ferromagnetic indentor in an external field;
- the energy roll-in from the field is provided through both progressive and rotational degrees of freedom;
- inter-indentor interactions lead to the establishment of an effective balance between progressive and rotational degrees of freedom of the balanced indentors;
- indenters of cylindrical shape have frozen-in magnetic moments.

To determine the energy imparted by the external alternating field with magnetic induction $\vec{B} = \vec{B}_c + \vec{B}_v \cos\omega t$ to rotational degrees of freedom of indentors, possessing a magnetic moment constant by modulus $\vec{p}_m$, let us consider the equation of indentor rotation relative to the axis perpendicular to the plane $(\vec{B}, \vec{p}_m)$:

$$\varphi_1' \cong \frac{p_m B_v}{I\omega} \sin\varphi_0 E(\omega t;1) \cong \frac{p_m B_v}{I\omega} \sin\varphi_0 \sin\omega t. \tag{6}$$

Let us assume the following equalities:

$$\varphi' = \varphi_0' + \varphi_1'(t) - \varphi_1'(t_0). \tag{7}$$

The average amount of energy, transmitted by the external field to rotational degrees of freedom to one indenter during time $\frac{T}{2}=\frac{\pi}{\omega}$, will be after averaging $\frac{1}{2}I(\varphi'-\varphi_0')^2$ at $t=t_0+\frac{T}{2}$ with respect to initial values φ_0 and t_0. Value $\frac{T}{2}$ is selected as the time between collisions in the assumption that the collisional processes are implemented at the extreme points of the trajectory of reciprocation.

After conversion, taking into account equality (7) and the following correlation $\langle\sin^2\varphi_0\rangle=\langle\sin^2\omega t_0\rangle=\frac{1}{2}$, the energy transmitted by the external field to ferromagnetic indentors at a time at the rotational-oscillative motion will be written in the following expression:

$$E_p = \frac{N}{T/2}\left\langle\frac{1}{2}I\left(\varphi'^2-\varphi_0'^2\right)\right\rangle = \frac{N}{2\pi}\frac{\left(p_m B_v\right)^2}{I\omega}, \tag{8}$$

where *N* is the numeric concentration of indentors in the RML layer.

Let us determine the energy imparted by an inhomogeneous external field with a gradient of induction $\frac{\partial B_v}{\partial y}$ to translational degrees of freedom of indentors. Not taking into account the resistance to progressive-oscillative motion, we will write the equation of motion:

$$m y_1'' = p_m \frac{\partial B_v}{\partial y} \sin\varphi_0 \cos\omega t. \tag{9}$$

We integrate the equation of motion after transformation energy, transmitted by the field to indentors per time unit during progressive-oscillative motion, which can be written as:

$$E_n = \frac{N}{2\pi}\frac{p_m^2}{m\omega}\left[\frac{\partial B_v}{\partial y}\right]^2 \tag{10}$$

The total energy transmitted by the external field to indentors per time unit is equal to

$$E_p + E_n = \frac{N}{2\pi}\frac{p_m^2}{\omega}\left[\frac{B_v^2}{I}+\frac{1}{m}\left[\frac{\partial B_v}{\partial y}\right]^2\right]. \tag{11}$$

Let us consider the energy dissipation of indentors due to the resistance to deformation of the treated medium in the process of impact-pulse interaction. To simplify the supposed resistance, we shall consider it as linear to relevant speeds that is true neglecting effects of attached mass. The equation for the energy, dissipated by the indentor in the rotating magnetic liquefied layer per time unit, can be represented as

$$E_{dtm}^{-} \simeq 6\varepsilon N\left(\frac{\lambda}{m}+\frac{\sigma}{I}\right), \tag{12}$$

where ε is the average energy of one degree of freedom, and the average is taken over the distribution for the corresponding linear or angular speed; λ and σ are, respectively, the friction coefficients of progressive and rotational motion.

In general, λ and σ depend on internal friction η of the processed material, the size and the shape of indentors, the Reynolds number for progressive and rotational motion, and the volume concentration of indentors in the working area covered by the rotating electromagnetic field are measured by parameter *P*. In [8,9], it is shown that for spherical particles of the r radius interacting with the continuous phase at Reynolds numbers less than or about one, coefficients λ and σ can be determined from ratios:

$$\lambda = 6\pi\eta\, r f(P), \quad \sigma = 8\pi\eta r^3 f(P), \quad f(0)=1, \tag{13}$$

where f(P) is the function that takes into account the influence of the pressure of indentors movement on resistance experienced by them.

For cylindrical indentors having $l/d>1$, taking into account (13), we will write:

$$\frac{\lambda}{m}+\frac{\sigma}{I}=\frac{9}{2}\frac{\eta f}{\rho r_e^2}+15\frac{\eta f}{\rho r_e^2}=\frac{39\eta f}{2\rho r_e^2},$$

where r_e is the equivalent radius of the indentor, (under the condition of equality of the masses of the cylinder and the sphere) equal to $r_e=\sqrt[3]{\frac{3}{16}d^2 l}$.

From the obtained expression, we can see that the viscous dissipation of energy of the rotational motion is more than three times greater than the dissipation of the progressive motion.

To account the dissipation due to inelastic collisions, we need to solve the problem of a single collision of two freely rotating indentors taking into account the surface friction, plastic deformation and speed changes of indentors during their convergence due to the dipole interaction. For this, we will limit ourselves by a semi-empirical approach, introducing the effective coefficient of elastic collisions γ, describing the conservation of kinetic energy of colliding indentors in a single act, but we will present an expression of the energy dissipation due to collisions per volume unit and time [10] unit in the form of

$$E^{\cdot}_{di}\simeq 6(1-\gamma)\varepsilon N. \tag{14}$$

Thus, the energy dissipation of indentors in the RML layer will be equal to

$$E^{\cdot}_{dtm}+E^{\cdot}_{di}\simeq 6\varepsilon N\left(1+\frac{\lambda}{m}+\frac{\sigma}{I}-\gamma\right). \tag{15}$$

For the sustainable process of magnetic liquefaction of the rotating layer, the dissipation and conducted energy should be equal. Therefore:

$$\frac{N{p_m}^2}{2\pi\omega}\left[\frac{B_v^2}{I}+\frac{1}{m}\left(\frac{\partial B_v}{\partial y}\right)^2\right]=6\varepsilon N\left(1+\frac{\lambda}{m}+\frac{\sigma}{I}-\gamma\right). \tag{16}$$

The equality (16) is valid under the assumption of the dominance of the viscous mechanism of dissipation and for this case allows determining the threshold value of the field induction gradient providing a stable mode of magnetic liquefaction of the rotating layer:

$$\left(\frac{\partial B_v}{\partial y}\right)^2=\frac{12\pi\omega m}{{p_m}^2}\varepsilon\left(1+\frac{\lambda}{m}+\frac{\sigma}{I}-\gamma\right)-\frac{m}{I}B_v^2. \tag{17}$$

As follows from (17), the sustainable magnetic liquefaction of the rotating layer depends on the induction of the rotating electromagnetic field, the magnetic moment of the indentor, its mass and moment of inertia, the frequency of the alternating magnetic field and the resistance coefficients to progressive and rotational motions.

4. Assessment of the energy state of the rotating magnetic liquefied layer at phase transitions

The ferromagnetic indentor of the RML layer has a reserve of kinetic energy, defined by expression (16), as well as potential energy related to the presence of a constant component of induction B_c of the external field. Thus, the total energy per unit of volume of the RML layer per time unit can be written in the form of

$$E_{te}=\frac{Np_m}{2\pi\omega}\left[\frac{p_m}{I}B_v^2+\frac{p_m}{m}\left(\frac{\partial B_v}{\partial y}\right)^2-\omega^2 B_c\cos\alpha\right], \tag{18}$$

where α – the average value of the angle between p_m and B_c for the period.

According to [11], the transition of the RML layer from the magnetic liquefied state in the "solid" phase will occur under the condition that ferromagnetic indentors stop the reciprocating motion. This process is realized when increasing the constant field induction to values satisfying the following relation:

$$\frac{N}{2\pi}p_m\omega\, B_c\cos\alpha\geq\frac{N\,{p_m}^2}{2\pi\, m\omega}\left(\frac{\partial B_v}{\partial y}\right)^2. \tag{19}$$

From the last inequality, we can obtain the expression for the induction of the constant field corresponding to the transition disorder – order:

$$B_c\geq\frac{p_m}{m\omega^2}\left(\frac{\partial B_v}{\partial y}\right)^2. \tag{20}$$

In inequality (20), the $\cos\alpha=1$ is adopted that can be justified by the assumption about the collinear orientation of vectors $\overrightarrow{p_m}$ and $\overrightarrow{B_c}$ in the ferromagnetic indentor transition from the magnetic liquefied state to the "solid" phase.

The induction of the constant field can be defined, at which the mechanical motion of indentors in the structure of the "solid" phase stops, wherefore the energy of the rotational motion of the indenter will be equated to the energy of inter-indentors interaction. In the approximation of the nearest neighbors and collinear orientation of magnetic dipoles, let us write:

$$\frac{1}{2}\frac{p_m{}^2 B_v^2}{I\omega^2} = \frac{\mu_0}{4\pi}\frac{p_m{}^2}{4r_e^3} + p_m B_c \quad ; \tag{21}$$

and for the required value of the induction, we will receive:

$$B_c = \frac{1}{2}\frac{\rho_m B_v^2}{I\omega^2} - \frac{\mu_0}{4\pi}\frac{\rho_m}{r_e^3}. \tag{22}$$

5. Conclusion

1. The proposed magneto-dynamic method of treatment based on creating a convective flow of ferromagnetic indentors under the conditions of the rotating electromagnetic field rightly expands the theological possibilities of dynamic methods to increase the strength of details and has its own specific technological purpose associated with its use in operations to increase the strength details with expressed paramagnetic properties. These parts include thin-walled tubes, stringers, belts, spars, that are widely used for aircraft manufacturing.

2. From the obtained results of theoretical studies, it follows that the kinematic characteristics of the ferromagnetic indentors are determined by the parameters of the field – induction of constant and alternating fields, frequency and the gradient of induction of the alternating field, and magnetic and inertial characteristics of indentors.

3. The optimum ratio between the magnetic field parameters and inertial and magnetic characteristics of indentors proposed based on the energy model of the RML layer, allows one: to justify the energy potential of ferromagnetic indentors, giving the effect of elastic-plastic deformation of the treated medium, leading to its strengthening; to define the threshold value of the magnetic field gradient providing a stable mode of magnetic liquefaction of the rotating layer; to justify the induction of the constant field in which the mechanical movement of indentors in the structure of the "solid" phase stops.

References

[1] Babichev A P and Babishev I A 2008 *Fundamentals of Vibration Technology* (Rostov-upon-Don: Publishing Center of DSTU)
[2] Tamarkin M A, Tishchenko E E, Korol'kov Y V and Rozhnenko O A 2009 *Russian Eng. Res.* **29**(5) 518-521
[3] Logvinenko D D Shchelyakov O P 1976 *Intens. of Tech. Proc. in Devices with Vortex Layer* (Kiev: Technics)
[4] Vernigorov Y M, Egorova S I, Egorov I N and Plotnikov D M 2009 *Russian J. of Non-Ferrous Metals* **50(4)** 400-403
[5] KoSung-Lim -L, Baron Y M, Park J I 2007 *Key Eng. Materials* **329(1)** 237-242
[6] Buevich Y A, Syutkin S V and Tetyukhin V V 1984 *Magnetic Hydrodynamics* **4**-11
[7] Vernigorov Y M, Egorov I N and Egorova S I 2005 *Euro PM2005 Powder Metallurgy Cong. and Exhib.: Proc.* (Prague, Czech Republic) vol **1** p 451
[8] Lebedev V A, Kirichek A V, Sokolov V D 2016 *Int. Conf. on Industrial Eng., ICIE 2016, Proc. Eng.* **150** 775-781
[9] Kirichek A V, Soloviev D L, Altukhov A Y 2014 *J. of Nano-and Electronic Phys.* **6(3)** 03069
[10] Hall W F, Busenberg S N 1969 *Y. Chem. Phys.* **51(1)** 137-144
[11] Lourie A and Khojayev K S 1976 *Papers on Eng. Mech.* (Moskow Higher School) **6** 72-81

Thermodynamic aspects of the coating formation through mechanochemical synthesis in vibration technology systems

S U Shtyn[1], V A Lebedev[1] and A O Gorlenko[2]

[1] Don State Technical University, 1, Gagarin Sq., Rostov-on-Don, 344000, Russia
[2] Bryansk State Technical University, 7, 50-letiya October bul., Bryansk 241035, Russia

E-mail: va.lebidev@yandex.ru

Abstract. On the basis of thermodynamic concepts of the process, we proposed an energy model that reflects the mechanochemical essence of coating forming in terms of vibration technology systems, which takes into account the contribution to the formation of the coating, the increase of unavailable energy due to the growth of entropy, the increase in the energy of elastic-plastic crystal lattice distortion as a result of the mechanical influence of working environment indenters, surface layer internal energy change which occurs as a result of chemical interaction of the contacting media. We proposed adhesion strength of the local volume modified through processing as a criterion of the energy condition of the formed coating. We established analytical dependence which helps to obtain the coating strength of the material required by operating conditions.

1. Introduction

The process of forming of mechanochemical coatings is a set of simultaneous chemical and physico-chemical phenomena occurring in the surface layer of metal under the influence of external forces. A distinctive feature of vibration mechanochemical coatings (VMCC), which distinguishes them from other types of coatings, is that their formation occurs under conditions of the vibration shock-pulse impact of indenters, moving freely in the closed volume of the working chamber. As a result of the dynamic interaction of the indenters with the surface of the part, chemical and physicochemical processes occurring in the surface layer are activated and a change in its geometrical and physical and mechanical properties occurs [1, 2].

From the thermodynamic point of view, all elementary acts of the mechanochemical process, taking place in the local micro-volumes at the ‘coating-substrate’ interface, are divided into two distinctive groups, characterized by different kinetic laws [3, 4]. The first group of elementary acts of atomic and molecular rearrangements is associated with the establishment of adhesive bonds between contacting substances and therefore they cause accumulation in the local volume of stored energy. Elementary acts of the second group are connected with establishment and cancellation of connections as a result of the transformation (conversion) of irreversible mechanochemical thermal energy, which results in the form of a thermal effect. Therefore, in accordance with the basic law of thermodynamics - the law of energy conservation during its transformation, the equation of the energy balance of the process of the local micro volume of the vibration mechanochemical coating at the "coating-substrate" interface can be written as follows:

$$A=\Delta U+q \tag{1}$$

$$\Delta U=\Delta U_{pm}+\Delta U_{pch}+\Delta U_{hm}+\Delta U_{hch}$$

where, ΔU_{pm}, ΔU_{pch} – changes in the mechanical and chemical potential component of the local volume of internal energy; ΔU_{hm},ΔU_{hch} – a part of heat energy expended to increase its energy state; q – a part of thermal energy dissipated into the substrate body and the environment.

2. The energy model of mechanochemical synthesis of VMCC

Considering mechanochemical processing (VMCP) generated in the course of vibrating at the interface 'coating-sublayer", we can present the modified local microvolume as the open thermodynamic system under established mechanochemical conditions in the ratio of local balance. It is possible to present a condition of its formation in the shape of:

$$\Delta U=U_0, \tag{2}$$

where ΔU,U_0- change of mole internal energy of the system, and its mole energy, accordingly, defining conditions of formation of the modified local microvolume.

The physical sense of this condition lies in the fact that when the left part, corresponding to an increase of the internal energy of system ΔU, will reach the right U_0, estimated by the fundamental power characteristic, the modification process of the local microvolume will reach its completion (balance).

The concept of the surface layer as the phase layer possessing thickness and volume is taken as a basis of thermodynamic approaches to the surface layer description. It allows using the description the same thermodynamic device for the surface layer as that for three-dimensional phases [5]. According to the method of surplus values of Gibbs [6], in compliance with it, the used thermodynamic parameters are not absolute but superfluous in comparison with the corresponding parameters of the macroscopical phases, the fundamental equation for the energy of surface layer energy U at the interface of the phases separation looks like:

$$\Delta U=T\Delta S+p\Delta V+\sum \mu_i\Delta\omega_i, \tag{3}$$

where T – absolute temperature; p – pressure; S,V,μ_i – mole values of entropy, the volume and the chemical potential; $\Delta\omega_i$ –mole fraction of the i-component of the system.

Consideration of each component under conditions of VMCP has shown that the first summand is a mole excess of connected energy ΔS which irreversible growth during any non-equilibrium processes is postulated by the second law of thermodynamics. According to Boltsman's interpretation, the change of mole entropy is described in the shape of $\Delta S=R\ln\frac{W_2}{W_1}$,

where W_2, W_1- thermodynamic probabilities of the initial and final condition of the system. If we assume that entropy manufacture in the system at VMCP is estimated by the duration of mechanochemical synthesis of the process of coating formation, then the thermodynamic weight of the system can be estimated from the correlation:

$$\frac{W_2}{W_1}=\frac{t}{t_0}, \tag{4}$$

where t – the time of the system period under the condition of the mechanochemical synthesis; t_0 – the period of thermal deviation of atoms.

The contribution of mechanical forces to an excess of internal energy of the system is composed of potential energy E_{es} , caused by external and internal (residual) elastic stress, and energy ΔE_{pd}, accumulated by plastic deformation, which means $pdV=E_{es}+\Delta E_{pd}$. The change of the mole potential caused by elastic deformations makes:

$$E_{es}=V_m 10^{-6}\frac{\sigma^2}{2E}, \tag{5}$$

where V_m– the mole volume, mm^3/mole; σ – effective stress, MPa; E – the stress module, MPa.

Elastic energy is not accumulated in a material eventually; therefore, current value E_{es} is a function of the instant stresses value. Besides, it is necessary to note that the experimentally established fact proves efficiency of VMCP. In spite of the fact that during elastic deformations, elastic energy is not accumulated in a material eventually, E_{es} is not equal to zero as microplastic deformations take place.

Unlike E_{es}, the work spent on the plastic deformation, eventually accumulates in a material, and is transferred in this way to the thermodynamic system being at the interface 'coating – sublayer'. The mechanical energy is equal to mole work of rigid forces, which is equivalent to the square under a true curve 'stress – deformation':

$$\Delta E_{pd}=V_m 10^{-6}\sigma\Delta\varepsilon, \tag{6}$$

where $\Delta\varepsilon$ - relative change of the plastic deformation of the system under the condition of mechanochemical influence t.

The relative change of the plastic deformation in the course of processing by dynamic methods of surface plastic deformation of free moving indenters (characterized by a small degree of deformation $\Delta\varepsilon \leq 0.3\Delta\varepsilon$), which variety is vibrational-striking processing, can be estimated by correlation $\Delta\varepsilon=\frac{D}{d}$, where d — the diameter of a plastic print formed on a surface under the impact of a particle of a working environment of diameter D.

Partial transition of deformation energy in the heat can be considered by means of the factor of accumulation of mechanical energy k_a which shows a share of reserved energy regarding all spent work (k_a=0,2÷0,3). In this case:

$$\Delta E_{pd}=k_a V_m 10^{-6}\sigma\frac{d}{D}. \tag{7}$$

Taking into account (4-7), we receive:

$$pdV=\Delta E_{es}+\Delta E_{pd}=V_m 10^{-6}\frac{\sigma^2}{2E}+k_a V_m 10^{-6}\sigma\frac{d}{D}=V_m 10^{-6}\left(\frac{\sigma^2}{2E}+k_a\sigma\frac{d}{D}\right). \tag{8}$$

Experimental researches of the formation processes of the basic groups of the vibration mechanochemical coating have shown that for obtaining of VMCP of the first group, considerable power influences on working environment indenters are required, providing plastic deformation of the surface layer with the result that there is a penetration of dispersed microparticles of a material of a coating into a surface. As a consequence, we have a coating formation. In this case, the energy of elastic stresses can be neglected and the first summand in equation (8) is not to be considered. When obtaining VMCP of the second group, the defining role in their formation should belong both to elastic deformations, and to plastic ones. The former activate oxide films, the latter destroy them and strengthen adhesive links of the coating with the substrate. For VMCP formation of the third group, it is quite enough to provide mechanical work of indenters of a working environment at the level of elastic deformations as coating formation is carried out on the account of chemical reactions in the surface layer of a material, and the task of mechanical forces is to activate this process. At designing of VMCP technology, the mechanical component of the process of application of a demanded coating is controlled by the choice of a material and the indenters sizes of the working environment.

The third summand in equation (3) having dimensional energy is characterized by the contribution of non-mechanical reversible forces to accumulation of internal energy of the system, caused basically at VMCP by the course of material chemical and diffusion processes in the surface layer. This summand is presented in the form of the sum of the members containing chemical potentials of components, participating in the formation of a coating and describing chemical work spent on the change of the system structure, in a general view:

$$A_{ch}=\sum \mu_i \Delta w_i \quad (9)$$

If we consider chemical work as a work of an actually chemical reaction [7], then in non-equilibrium chemical reactions, processes in which some chemical forms of substance turn into other, chemical potentials define thermodynamic affinity - the integrated set of operating chemical forces characterizing transformation of the reaction to an equilibrium state. According to Lewis's method, the expression describing the change of chemical potential looks like

$$\mu_i=\mu_i^0+RT\ln a_i\ , \quad (10)$$

where μ_i – the standard chemical potential of the system component; $\mu_i^0=\frac{H}{N_a\mu_i^0}=\frac{H}{N_a}$; H – component enthalpy; N_a- Avogardo's number; a_i- relative activity of the component in the system; $a_i=y_i c_i$; y_i- activity factors; c_i- component concentration.

From expression (10), it follows that activity, as well as chemical potential, is the measure of a tendency of the system component to turn into another phase. Herewith, chemical potentials are divided into two components - a constant one (standard) (depending only on temperature) and a variable one anyhow depending on the concentration of the substance in the system. At the same time, in work [8], it is shown that both chemical potentials and standard affinity can be presented through standard enthalpy and standard entropy of a reaction, in other words, through enthalpy change ΔH^0 and entropy ΔS^0in a chemical reaction of the substances which are in standard conditions. Both those and other changes are defined, accordingly, by partial (and for pure substances as specific) mole enthalpies and entropies of participants of the reaction in standard conditions. This circumstance allows chemical reaction proceeding in the surface layer at VMCP and an increase of potential internal energy of local volume modified in the course of VMCP at the interface 'coating – sublayer' expressed through the change of free energy of Gibbs, kJ/mole:

$$A_{ch}=-\Delta G^0=-(\Delta H^0-T\Delta S^0), \quad (11)$$

where $\Delta H^0=\sum v_i h_i^0$; $\Delta S^0=v_i s_i^0$; v_i – the stoichiometric factor; h_i^0 – enthalpy of the component mole; s_i^0 – mole entropy of the component.

The relation between enthalpic and entropic members in formula (11) at different temperatures defines the influence of temperature on the position of the reaction equilibrium. The opposite directedness of influence of enthalpic and entropic components on standard affinity is obvious, hence, on a direction of a chemical reaction as well. With the fall of the temperature, the role of the entropic component decreases essentially.

Taking into account (4, 8, 11), we have obtained the model uncovering the mechanism of mechanic-chemical synthesis of coating forming in the conditions of vibrational technological systems and defining the power condition of local microvolumes at the interface 'coating – underlayer' modified in the course of VMCP:

$$TR\ln\left(\frac{t}{t_0}\right)+V_m 10^{-6}\left(\frac{\sigma^2}{2E}+k_a\sigma\frac{d}{D}\right)-\Delta G^0=U_0. \quad (12)$$

Power model (12) accounts for the contribution to the process of VMCP formation of the surface layer of a raise of connected energy at the cost of the growth of entropy (the first summand), the increase of energy of elastic-plastic distortions of a crystalline grid as a result of mechanical action working environment indenters (the second summand), modifications of interior energy of the surface layer which was modified as a result of chemical interaction of contacting media. The role of each of them in kinetics of formation of classification groups of vibrational mechanochemical coatings according to the offered model (12) is schematically illustrated in figure 1, from which it is visible that the basic contribution of VMCP formation of the first group (figure 1a) to the process is made by a deformed summand monotonically growing in the course of time. VMCP formation of the second group (figure 1b) is the result of deformation and chemical summands. During VMCP formation of the third group (figure 1c), the leading part belongs to a chemical component of the model.

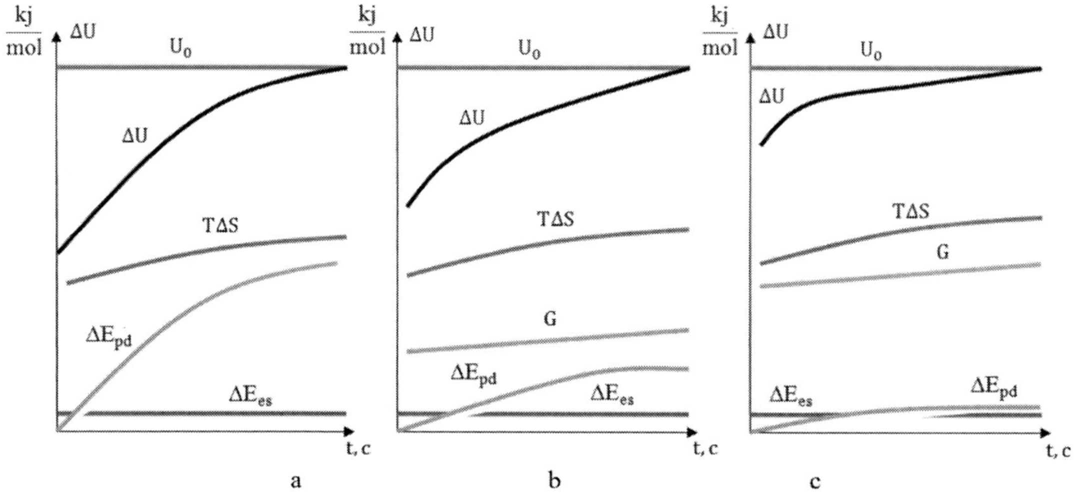

Figure 1. The diagram of energy accumulation in the local volume of the surface at the interface 'coating – substrate' by VMCP.

3. The energetic condition of VMCP durability provision.

The basic indicator of quality of any coating of metal applied to a metal surface is adhesion which characterizes the grasping durability between two materials, the connection occurrence between surface layers of two heterogeneous contacted substances. The basic performances of adhesion are energy of adhesion, the grasping durability and the force of adhesion. In work [9], devoted to the research of adhesion performances of contacting materials with coatings, it is established that adhesion energy and the grasping durability of a 'coating-substrate' are connected to each other by the following relation:

$$\tau_0=\frac{1}{\sqrt{\pi h}}\sqrt{E_a E}\,, \tag{13}$$

where, h- the thickness of a coating,

And the association between the force of adhesion and the grasping durability can be defined by the formula:

$$F_a=\pi\tau_0 r^2\,, \tag{14}$$

where, r- the indenter radius at the top of a scleroscope needle.

Thus, having defined the energy of adhesion by one of the above-stated techniques in section 1, it is possible to calculate the grasping durability and the force of adhesion of the SP method by formulas (13, 14).

Quantitative performance of adhesion is adhesion work W_a- the work necessary for reversible isothermal separation of two condensed phases given in the contact of the condensed phase on a square of the unit section. It is possible to present it in a following aspect:

$$W_a=E_a N, \tag{15}$$

where, E_a – the average energy of a communication unit providing adhesion of a modified on the basis of mechanochemical synthesis of the local microvolume at the interface of contacting environments; N- the number of links accounted for a contact square unit of a 'coating-substrate'.

If it is necessary to equate the mole energy defining conditions of formation of the modified local microvolume at the interface 'coating – underlayer' to the average energy of a communication unit providing its adhesion $U_0=E_a$ on the basis of power model (12), we will receive an analytical dependence, providing coating generation of a material demanded by service conditions of durability on the surface:

$$TR\ln\left(\frac{t}{\tau_0}\right)+V_m 10^{-6}\left(\frac{\sigma^2}{2E}+k_a\sigma\frac{d}{D}\right)-\Delta G^0=E_a. \quad (16)$$

The left side of equation (16) represented in the form of the time function is the VMCP kinetic equation which at $t<t_p$, where t_p— the processing duration, describes the growth process of the internal energy of the forming coating. At moment $t=t_p$, when plastic deformation becomes limitary for the given material, that is, $d=d_l$, and the activity of system components reaches thermodynamic constant *K* (according to [1] $\prod_{i=1}^{m}(a_i)^{v_i}=K$), the forming process of the vibrational mechanochemical coating completes. Having expressed t_p, we receive a generalized formula for the assessment of the VMCP process duration, providing VMCP formation:

$$t_p=t_0\exp\left[\frac{E_a-V_m 10^{-6}\left(\frac{\sigma^2}{2E}+\sigma\frac{d_l}{D}\right)+\Delta G^0}{RT}\right] \quad (17)$$

4. Conclusion

The obtained power model (12) can be taken as a management principle of mechanochemical synthesis of the formation process of coatings in vibrational technological systems as it discovers the mechanism of influence of the technological environment and conditions of the VMCP course on kinetics of internal energy changes of local microvolumes of the surface layer modified on a surface during activating the strike-impulse impact of free moving (metallic or nonmetallic) indenters.

The offered analytical dependence (16) allows solving a number of optimizing problems connected with a choice of technological conditions of VMCP on quality maintenance of VMCP and operational properties of a surface of parts at a stage of technological preparation of manufacture.

Acknowledgments

The research has been carried out by the finance support of the Russian Fund of Fundamental Researches within scientific project No 16-38-00276.

Reference

[1] Ivanov V V, Lebedev V A and Pinahin I A 2014 *J. of Friction and Wear* **35(4)** 339-342
[2] Lebedev V A, Kirichek A V and Sokolov V D 2016 *Int. Conf. on Industrial Eng. Proc. Engin.* **150** 775-781
[3] Rusanov A I 2005 *Source of the Document Surface Science Reports* **58(5-8)** 111-239
[4] Boldyrev V V 2000 *J. of Materials Synthesis and Proc.* **8(3-4)** 121-132
[5] Vitkovsky I V, Konev A N and Shorkin V S 2009 *Source of the Doc. Tech. Physics* **54**(2) 170-175
[6] Kokotov Yu A 1999 *Source of the Doc.t Solvent Extraction and Ion Exchange* **17(4)** 1083-1131
[7] Romashin S N, Presnetsova V Yu, Frolenkova L Yu and Shorkin V S 2015 *Int. Conf. on Mechanics - Seventh Polyakhov's Reading* **1** 71
[8] Boldyrev V V 2014 *8th Int. Conf. on Mechanochemistry and Mechanical Alloying, INCOME 2014* (Krakow; Poland) 111995
[9] Frolenkova L J, Shorkin V S 2013 *Method of Calculating the Surface and Adhesion Energies of Elastic Bodies* (Orel:State University-Education-Science-Production Complex) **1** 235-259
[10] Bull S J, Belde K 2008 *J. of Adhesion Sci. and Tech.* **22**(2) 121-132
[11] Fedorov A V, Van T R, Vellinga W P 2007 *Progress in Organic Coatings* **58(2-3)** 180-186

Investigation of cutting process and process of surface quality formation during machining of fragile components

O N Fedonin and D V Levy

Bryansk State Technical University, 7, 50-letiya Oktyabrya bul., Bryansk, 241035, Russia

E-mail: dims78-l@yandex.ru

Abstract. This article presents the process of the material destruction and formation of roughness during the edge cutting machining of brittle materials. Also, this article presents the theoretical equations of macro-chips and micro-chips formation, as well as equations of relationship of processing conditions with the surface quality parameters, which allow providing the desired roughness of parts.

1. Introduction

One of the most urgent tasks of engineering production is to increase the efficiency of processing methods, which means to provide the required quality of parts at minimal cost. Graphite parts (inserts, bearings, electrodes, crucibles, etc.) are widely used in engineering production. A production process of brittle materials is different from that of other structural materials because of specific properties (low hardness, lack of plasticity, high brittleness, and porosity). Reference books do not contain data on possible edge cutting machining to provide perfection factors of the surface made of brittle materials and to reduce their production costs. That is why, it is very urgent to study the ways to increase the efficiency of edge cutting machining of brittle parts.

The aim of this paper is to provide the efficiency of parts made of brittle materials with minimum production costs during edge cutting machining.

2. Theoretical description of the chip formation process in cutting edge machining of brittle materials

The process of cutting brittle materials (graphite, glass, ceramics, etc.) compared to the cutting of plastic materials has a number of features due to their properties. This leads to the cyclical nature of chip formation, which is accompanied by the formation of macrochipping (pulverized) and microchipping (fracture) (Figure 1). When we move the blade of the tool from position I, in which microchipping 1 was formed, towards the work piece cut layer of zone 2, there will be normal (σ) and shear loads (Figure 1b), the magnitude of which will increase up to σ material, which will inevitably lead to the contact destruction under compression with the formation of small dust-like shavings. This process will continue until the contact area on the blade front surface increases to a certain critical value, at which again there will be spall and microchipping formation (tool blade position II (Figure 1a)).

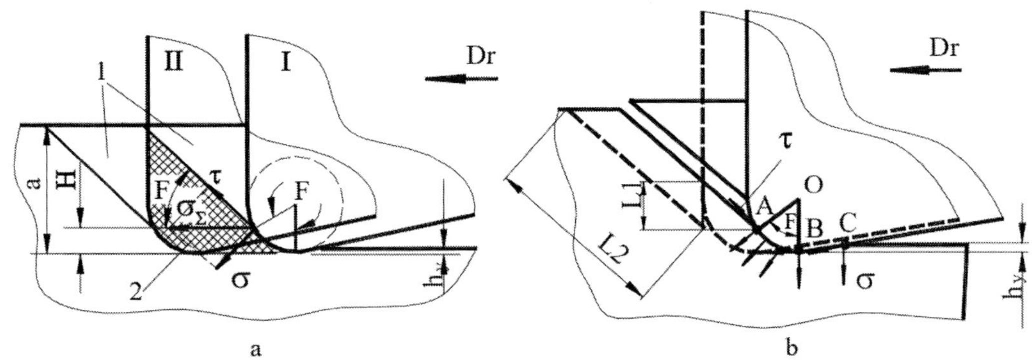

Figure 1. The scheme of the chip formation zone: a – sequence of chip formation cycles; b – conditions of microchipping formation; 1 – macrochipping; 2 – the material turns into microchipping; I, II – positions of the blade tool in forming macrochipping.

Having considered the treatment process (Figure 1) and making the assumption that a force acts on the material from the tool that causes shear and normal stresses, we obtain the following equation of forces acting on the cut layer:

$$\tau \cdot \mathrm{b} \cdot \mathrm{L2} = \sigma \cdot \mathrm{b} \cdot \mathrm{L1},$$

where b is the width of the area, which is under the influence of shear stresses and normal stresses; L1 is the length of the area, which is under the influence of normal compressive stresses; L2 is the length of the area which is under the influence of shear stresses.

$$L2 = \frac{(\mathrm{a} - H)}{\sin F},$$

where a – the slice thickness; N – the value of the brittle fracture layer with a radius of rounding the cutting edge; f – conditional shear angle ($F = arctg \frac{\sigma}{[\tau]}$).

If shear stresses exceed ultimate shearing strength [τ], it will spall; if not — it will be brittle fracture with the formation of dust-like chips due to exceeding temporary strength by normal stresses σ.

Considering all mentioned above, we get the conditions of the chip formation when machining brittle blade materials:

$$L1 \le \frac{[\tau]}{\sigma} \frac{(\mathrm{a} - H)}{\sin F}$$ – the conditions for microchipping formation;

$$L1 > \frac{[\tau]}{\sigma} \frac{(\mathrm{a} - H)}{\sin F}$$ – the conditions for macrochipping formation.

Defining the nature of chipping formation with processing blade, one can create a theoretical picture of roughness formation during edge cutting machining of brittle materials.

3. Theoretical description of the part surface quality formation made of brittle materials with a processing blade

When manufacturing parts, their quality is assessed by accuracy of dimensions and the quality of the surface layer. From the quality parameters set of the surface layer [1-4], only roughness parameters are standardized. Therefore, in many cases, the surface layer quality is estimated according to their size.

To predict the magnitude of high-rise roughness parameters is possible by the dependence [3]:

$$R_z = h_1 + h_2 + h_3 + h_4,$$

where h_1 is a component of the surface roughness caused by the geometry of the working part of the tool (blade) and its working motion kinematics relative to the treated surface; h_2 is a component of the surface roughness caused by oscillatory movements of the tool relative to the treated surface; h_3 is a component of the surface roughness caused by plastic deformation of the workpiece material in the zone of contact with the working tool; h_4 is a component of the surface roughness caused by roughness of the working part of the tool

However, this linear connection is valid for the plastic materials, during the processing of which the plastic edging of the material is possible by means of the auxiliary cutting edge [5, 6, 10].

And during processing of fragile materials the pattern of surface roughness formation will be different. Its formation (as well as during the machining of plastic materials) will be determined by the geometry and movement kinematics of working parts of the tool relative to the treated surface (h_1) by means of tool oscillations (h_2). However, the component h1 is partially reduced due to the cut of the roughness tip during the shift of the material by means of the secondary cutting edge by value h_3 (Figure 2). It is natural that plastic edging of the material caused by its brittleness during the formation of roughness will not occur during the processing.

The roughness of the cutting edge at the top of the cutter will further increase formed roughness up to value $h_4=Rz_{to}$.

When processing by a multi-point tool such as a milling cutter, the runout of the teeth will affect formed roughness $h_5=\Delta te$. Possible porosity of the material h_6 can also influence the roughness height in the machining process.

Thus, in the general case, the average height of the roughness profile will be determined by the equation with the blade machining brittle materials:

$$R_z = h_1 + h_2 - h_3 + h_4 + h_5 + h_6 . \quad (1)$$

The roughness profile formation when processing single-lip cutters is presented in Figure 2.

The dependence (1) for accuracy takes the following form:

$$R_z = h_1 + h_2 - h_3 + h_4 + h_6 . \quad (2)$$

Components h_1, h_2 can be calculated according to known dependences [1, 2]. The formation of component h_3 will be influenced by compressive stress and shear stress. The equilibrium condition will be as follows:

$$[\tau]\cdot b\cdot c = \sigma \cdot \sin\chi_1 \cdot c \cdot l_{AB},$$

where c is the width of the area, which is under the influence of compressive stress and shear stress; b is the length of the area, which is under the influence of shear stress; $l_{AB}=\rho\cdot F$ is the length of the area, which is under the influence of compressive stress, where ρ is the radius of the cutting edge rounding, and F is the phase angle; χ_1 is the contact angle of the blade top with processed surface along the auxiliary cutting edge (Figure 2).

The following condition must be met in order to ensure the cut at the roughness tip:

$$b < \frac{\sigma}{\tau}\sin\chi_1 \cdot \rho \cdot F,$$

and this dependence must substitute the following values of $\chi 1$:

1) when $\varphi \geq \arcsin\frac{S}{2r}$ and $\varphi_1 \geq \arcsin\frac{S}{2r}$, $\chi_1 = \arcsin\frac{S}{2r}$;

2) when $\varphi \geq \arcsin\frac{S}{2r}$ and $\varphi_1 < \arcsin\frac{S}{2r}$, $\chi_1 = \varphi_1$ (auxiliary angle of the tool);

3) when $\varphi < \arcsin\frac{S}{2r}$ and $\varphi_1 < \arcsin\frac{S}{2r}$, $\chi_1 = \varphi_1$.

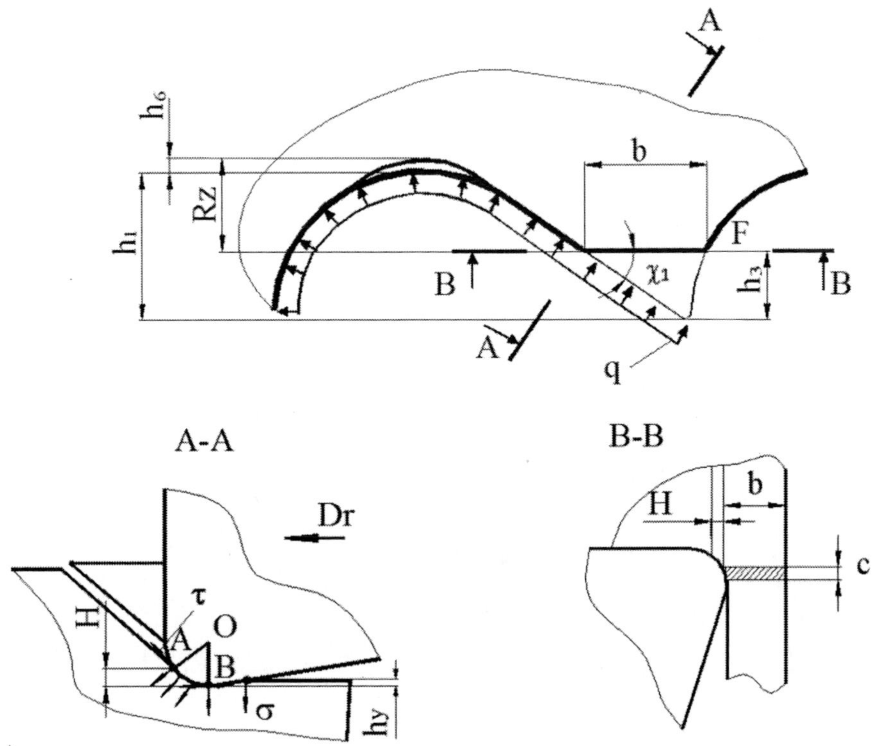

Figure 2. The formation of the roughness profile with the blade machining brittle materials.

Correlation equations for b and h_3 values, which are the sides of the triangle chip ridge, can be obtained under the following processing conditions:

1) when $\varphi \geq \arcsin\frac{S}{2r}$ and $\varphi_1 \geq \arcsin\frac{S}{2r}$ (Figure 3a)

$$h_3 = \frac{btg\chi}{2} \qquad \left(\chi = \chi_1 = \arcsin\frac{S}{2r}\right),$$

where χ – the contact angle of the blade top with a machined surface on the main cutting edge;

2) when $\varphi \geq \arcsin\frac{S}{2r}$ and $\varphi_1 < \arcsin\frac{S}{2r}$ (Figure 3b)

$$h_3 = \frac{b}{\left(\frac{1}{tg\chi_1} + \frac{1}{tg\chi}\right)} \qquad \left(\chi_1 = \varphi_1, \chi = \arcsin\frac{S}{2r}\right);$$

3) when $\varphi < \arcsin\frac{S}{2r}$ and $\varphi_1 < \arcsin\frac{S}{2r}$ (Figure 3c)

$$h_3 = \frac{b}{\left(\frac{1}{tg\chi_1} + \frac{1}{tg\chi}\right)}, \quad (\chi_1 = \varphi_1, \chi = \varphi).$$

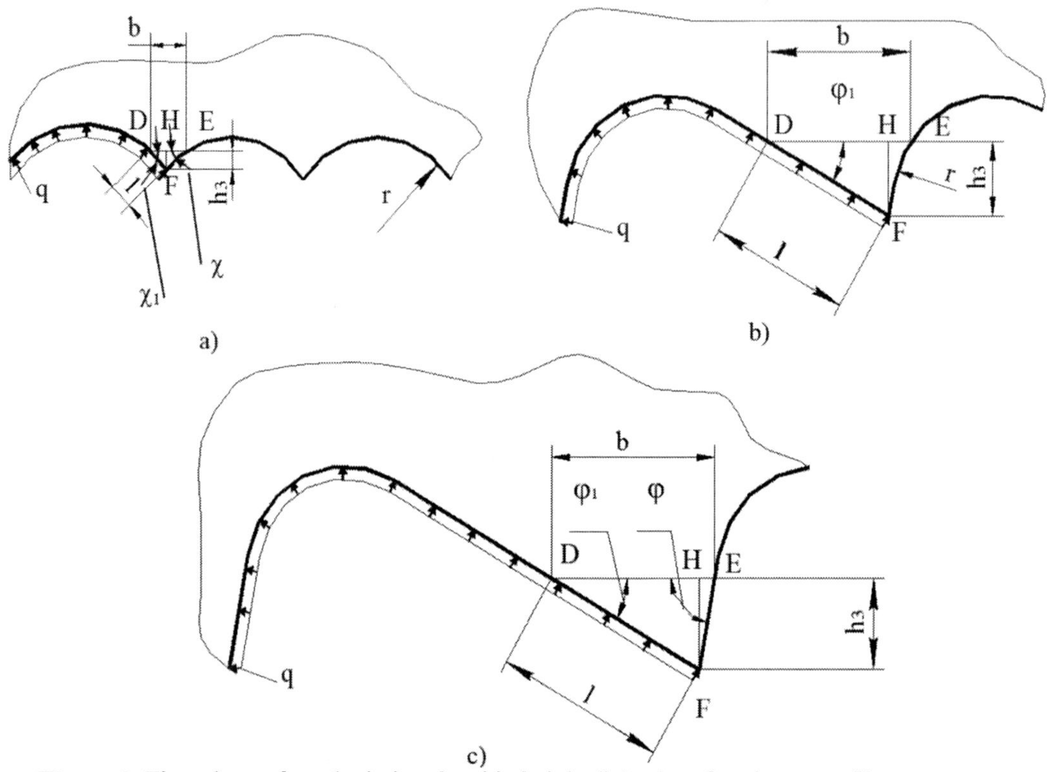

Figure 3. The scheme for calculating the chip height (h_3) when forming a profile:
a – by the radial part of the tool blade; b – by the radial part of the blade and support cutting edge of the tool; c – by the radial part of the blade, the main and auxiliary cutting edges of the tool.

The proposed model was tested in machining graphite (grade MG1) on the machine model '16K20' by a cutter with brazed plate WK8. Given that the magnitude of cutting force fluctuation ΔPy in processing graphite is much less than in processing structural steel, we can neglect this value, thus, h_2=0. Thus, equation (2) in case of turning graphite will be as follows:

$$R_z = h_1 - h_3 + h_4 + h_6 .$$

4. Conclusion

Taken surface profilograms confirm the roughness formation picture (Figure 4), so the derived dependence can be used to predict the cutting process and to define high-rise parameters of roughness parts made of brittle materials [5, 7-9]. The purpose of the processing modes by the proposed dependencies will improve the performance and reduce the cost of manufacturing parts due to the absence of test moves and reducing the number of defective parts.

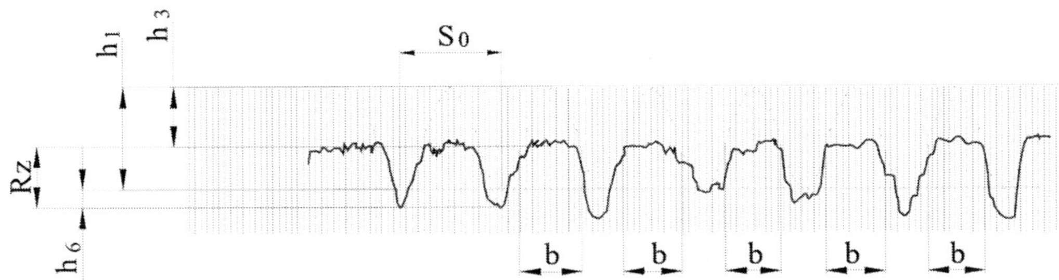

Figure 4. The surface profilograms treated by machining (top radius r = 0.4 mm, cutting depth t = 1 mm, cutting speed V = 200 m/min, feed S_0 = 0.7 mm/rev, angles $\varphi = \varphi_1 = 45°$).

References

[1] Suslov A G, Oleynik A V 2015 *Sci. Intensive Techn. in Mechanical Eng.* **1(43)** 3-7
[2] Suslov A G 2008 Surface Engineering (Moscow: Mashinostroenie)
[3] Gashev E A, Muratov K R *Russian Eng. Res.* 34 **4** 268-271
[4] Rastorguev G A 2014 *Reference Eng. J. with the App* **3** (204) 32-43
[5] Levyj D V 2013 *Modern trends in science and education collection of scientific works on materials of the Int. Scientific-Practical Conf.: in 26 parts.* (Tambov) vol 5 (Tambov: Consulting company Ucom) pp 82-83
[6] Fedonin O N, Levyj D V 2009 *Vestnik of Bryansk State Technical Univ.* **4** 13-16
[7] Balykov A V 2015 *materials In journal: Scientific works of the IV Int. Sci. Conf. Fund. Res. and Innovative Techn. in Mech. Eng.* (Moscow.) vol 1 (Moscow: LLC ID SPEKTR) pp 55-58
[8] Syanov S Yu 2008 *Vestnik of Bryansk State Technical Univ.* **1** 14-19
[9] Syanov S Yu 2014 *Sci. Intensive Techn. in Mechanical Eng.* **6** 24-29
[10] Scherbakov A N 2006 *Vestnik of Bryansk State Technical Univ.* **2** 85-92

Dependence of the diffusion wear of the hard alloy surface on its fractal dimension

V P Nesterenko, A A Kondratyuk, E B Belousova and I A Shulepov

Tomsk Polytechnic University, 30, Lenin ave., Tomsk, 634050, Russia

E-mail: nest2004@bk.ru

Abstract. The paper presents the research results in the synergies between the wear resistance of carbide cutting tools of P-group applicability and fractal dimension of the wear surface occurring on the rake face of the tool when processing the material, which causes intensive diffusion wear. It was found that the resistance of carbide cutting tools increases as the fractal dimension of their wear surface reduces.

1. Introduction

Austenitic, perlitic and ledeburitic steels exposed to the P-group cutting tools treatment are used to manufacture details and components for nuclear-power reactors. High operational performance of the tools provides high quality standard of the output product and operational safety. Therefore, the accurate prediction of the cutting tool wear during machining is very important.

While manufacturing and operating, the surface structure of the P-group hard alloys undergoes oxygenation and nitrogenization when exposed to the surrounding gaseous atmosphere [1]. The intensity of the oxygen interaction process with a hard alloy composition depends on the presence of a tungsten constituent, and that of nitrogen compounds is determined by the titanium constituent. The surface oxycarbide structures, formed at high temperatures, effectively prevent the insert contact surfaces from the devastating effect of the most active workpiece elements and phases. The increase of the titanium carbide constituent in the hard alloy composition results in the decreased thickness of the oxycarbonitride film and the increased screening capacity, which limits the mass and velocity of the work piece material diffusion into the boundary layers of the hard alloys interfaces. This occurs due to densification of the chemically inactive nitride compounds in the composition of oxycarbonitride structures.

When cutting, the structure of the contact surface is continuously subjected to the destruction and regeneration due to interaction with the elements of gaseous atmosphere. As a rule, the wear surface microrelief of the P-group cutting inserts exhibits its poor development and an insignificant fractal dimension value. As the titanium constituent in the hard alloy composition increases, its density and screening capacity, tend upwards. Simultaneously, the fractal dimension of the wear surface, which causes an intensive chemical interaction in contact zones, decreases. The radiation preprocessing of carbide cutting inserts with a gamma-quantum promotes the intensification of the hydrogen desorption process from the hard alloy structure at the stage of its operation. This process results in high oxidation characteristics of the contact surfaces, an increase in the wear microrelief development and fractal dimension, and some decrease in the wear resistance.

2. Materials and methods

The investigation of the wear resistance of the P-group carbide cutting inserts was carried out with the help of a turning lathe of model 163. Steel 50, which causes intensive diffusion wear, was used as a processed material [2]. Machining was performed by the carbide cutting inserts of the industrial TZ0K4grade related to the above specified group. The cutting conditions were as follows: cutting speed = 180 m/min (and it was optimal as it provided the minimum wear rate), feed rate = 0.23

mm/rev and cutting depth = 1.5 mm. The wear resistance was estimated in minutes of the tool operating time up to the specified criterion of tool dulling that was 1.0 mm. The treatment of tools by the gamma quantum was performed in a special radiation unit. The radiation fluence was 10^{17} gamma-quantum/cm^2.

The morphology of the wear surface was investigated with the help of non-contact three-dimensional profilometer MICROMEASURE 3D STATION by STIL. The wear surface testing was carried out with the pulsed light beam, generated by the halogen lamp with the frequency of 10Hz. This frequency provided the more exact simulation of the wear lines. The dimensions of the microrelief peaks and dimples were determined from the difference between the incident and reflected light intensities.

The roughness of the surface profile was estimated based on its measurements in the interval of 0.5 mm. The profiles were processed using a special computer program, a part of the profilometer equipment. The fractal dimension value was identified automatically using the roughness parameters of the measured profile. Roughness control and fractal dimension measurement were performed on the flank land located in the central part of the insert. For all the worn inserts, the running coordinates of the flank land were similar.

3. Results and Discussion

Materials, which cause intensive diffusion wear are usually processed by the P-group carbide cutting tools. At high temperatures, these materials have a tendency to oxygenation and nitrogenization. The aeration process of the hard alloys takes place not only at the stage of preparation of different components and their sintering, but also during the use of manufactured cutting tools. Surface and subsurface oxide and oxycarbide structures occurring while cutting steels act as a screen that prevents chemical solid-phase interaction between the workpiece and hard alloy materials. Testing shows that the microrelief of the wear surface has fractal properties [3]. The main characteristic of fractal surfaces is their fractal dimension. The level of the wear surface roughness, fractal dimension, correlates with the material wear resistance and recovery properties of the damaged objects.

Figure 1 shows the dependence of cutting inserts operating time up to the specified criterion of tool dulling on the fractal dimension value. The data in the figure indicate that the decreased roughness development of the flank wear leads to the increase in the wear resistance of the cutting inserts. From this, it can be expected that the higher chemical, thermal and mechanical stability of the oxycarbonitride film structure, occurring on the contact surfaces, the less the possibility of the formation of the developed wear microrelief, and the lower the value of fractal dimension, the higher the expected wear resistance of the cutting tool. The key factors that influence the formation of the fractal dimension of the wear surface are the mechanical and chemical characteristics of the surface film structure. The experiments showed that when the mechanical and chemical stability, occurring in the hard alloy material during the cutting process of the surface structure, grows, the fractal dimension of the wear microrelief reduces, and the operating time of the cutting inserts up to the specified criterion of tool dulling increases.

Generally, the composition of the P-group hard alloys is presented by titanium carbide, the solid solution of tungsten carbide in titanium carbide, titanium carbide and cobalt binding. The base of these hard alloys is titanium carbide [4]. As the titanium constituent in the composition of hard alloy increases, its brittle strength decreases, and the chemical stability grows. However, brittle strength does not affect the wear resistance of the cutting tools since the materials intended for the processing of these alloys do not cause intensive adhesion.

The overall thermodynamic stability of these hard-alloys depends on their bulk and surface structure. In its turn, the thermodynamic stability of the surface structure depends on the composition of the formed thin oxide, oxycarbide and oxycarbonitride films. The following film structures shield the diffusion processes that develop at high temperatures between the workpiece and tool materials, and reduce the likelihood of the solid-phase reactions [5]. Eventually, the resultant working capacity of the cutting tool depends on the high temperature chemical stability of both surface and bulk structures. Thus, the stability of the occurring surface structures greatly depends on the composition of the bulk structure including the gas elements content and the kinetics of the mass transfer of these elements from the bulk to the contact zone. This depends on the operating conditions of the cutting tools. As a result, it can be assumed that the fractality of the cutting tool contact surface occurs in the course of self-organization of various processes. In addition, the directivity of the self-organization processes can be controlled with the help of a composition and a structure. To provide a low fractal dimension of contact surfaces, it is necessary to create the conditions for the formation of surface structures with high mechanical, thermal and chemical stability.

The film structure is formed from separate areas of tungsten, cobalt and titanium oxides. Simultaneously, the surface relief may be affected by the degree of interaction of hard-alloy individual components and gaseous atmosphere. As a result, the surface consists of peaks and dimples. When the film thickness grows, the difference between peaks and dimples reduces. Under the wear conditions, separate areas of the surface microrelief perform special functions. At specific correlation between the surface peaks and dimples, the operational effectiveness becomes maximal.

The intensity of the interaction of hard alloy deep layer components and the elements of gaseous atmosphere decreases. At the same time, while heating, the boundary layers actively interact with gas elements of the bulk structure. It should be noted that the composition of the environment gas and the gaseous atmosphere of the hard alloy vary widely. As a result, the properties of the occurring film structures will significantly differ within the distance from the surface to the depth. It was found that when the temperature grows, the content of titanium and cobalt oxides in the composition of the surface film increases. The concentration of the tungsten constituent is observed to decrease insignificantly. The results of the phase analysis showed that the probability of the oxycarbonitride titanium compound formation in the composition of the surface film is considerably low. At high temperatures, the probability of formation increases. Therefore, the increased content of titanium oxide in the composition of the surface film structure leads to the increase in the tool wear resistance while cutting perlitic and ledeburitic steels.

The process of the surface film structure formation depends on the stoichiometry of the carbide systems in the hard-alloy content. It can be assumed that increased carbide nonstoichiometry increases the probability of oxide and nitride structures formation. As a result, the mechanical properties of the surface and bulk structures deteriorate. Hence, the fractal dimension of the formed wear surface increases.

At higher temperatures, the intensity of the contact surfaces oxidation changes. As a result, the balance between the separate phases of the surface polyoxide structure changes, and new phases or phase compounds occur. In this case, oxidation spreads deeper. The transformation from the carbide bulk to the transitional oxy-carbide and then to the surface oxide structure is more regular. The diffusion resistance of the surface structure is higher. However, after formation of this surface structure, its mechanical properties deteriorate. Due to adhesive wear, occurring in the areas of intensive oxidation and directed into the bulk structure, formed drag lines can differ in their height and development. As a result, the fractal dimension of the wear surface increases.

During oxide films formation, compressive and tensile stresses develop in their structure. The type and the stress value are closely related to the specific aspects of the hard alloy oxidation. In oxide structures, the actual stress impacts diffusion resistant and mechanical characteristics of films and the formed microrelief of the wear surface in particular.

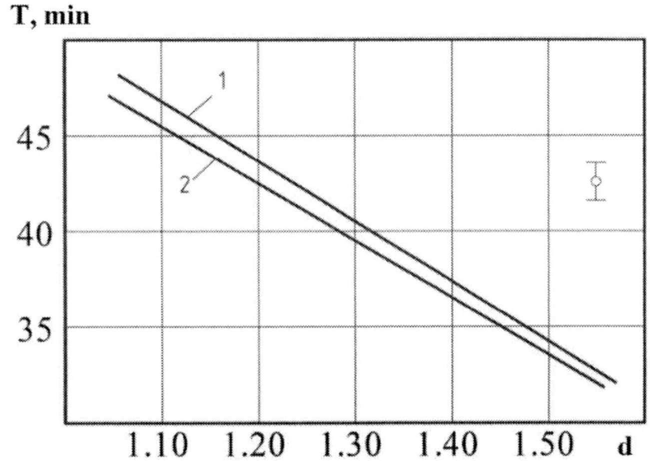

Figure 1. The dependence of wear resistance of the T30K4 grade hard alloy on the value of the fractal dimension of the flank land: 1 – controlled inserts; 2 – radiation pretreated inserts.

The increase of tensile stress in the film structure encourages the formation of a porous structure at phase boundaries. It results in the intensification of diffusion processes, solid-phase reactions and oxidation processes in this area. As the result, the possibility of the developed structure formation on the hard alloy surface increases and the tool wear resistance decreases.

The functional efficiency of the cutting tools generally depends on the hard alloy composition as well as surfaces film composition that influences mechanical, chemical and thermal characteristics. The bonding strength between the surface film and the base material is a very important fact. This property fully depends on the coherence between the film and the base material. In its turn, the degree of coherence depends on many factors; one of them is the presence of impurity oxide and oxide compounds in the material composition. It is notable that the creation of the reinforced contact surfaces by means of different modifications is related to the increased adhesion strength between the base material and the surface thin-film structure.

While using the cutting tool, the degree of coherence between the film and the base material is continuously changing due to the wear processes and the elements diffusion of the processing material, hard alloy base and gaseous atmosphere into the film structure. An optimal intensity rate of the processes provides the higher quality of the forming surface structure. The minimal fractal dimension of the forming wear surface corresponds to the highest wear resistance of the tool.

The conformance of the lattice parameters of the forming film and the base material also depends on the temperatures and load bearing characteristics (stresses) in the contact zones. When the temperature increases, the chemical reaction of contact zones with the gaseous atmosphere and gas constituents of hard alloy intensifies. At the same time, the processing material stresses on the forming film structure decreases, and the closeness of the interfacing elements grows. In this case, the surface structure is formed due to the interaction of the tool contact zones with gas elements of the hard alloy bulk. The formed surface film has a more homogeneous structure, and its crystallographic parameters coincide with the base material parameters.

The value and the nature of the strain-stress state in the film structure is a factor of the reaction order and the oxidation intensity of hard-alloy components. The surface film structure includes tungsten oxide, titanium oxide and cobalt oxide. The concentration of oxycarbide compounds in the film structure decreases when its formation temperature grows. At the same time, the formation of the nitride compounds is possible.

If the balance in this system is upset due to the alteration of the wear rate, temperature fields, interface stresses or due to gas flows rearrangement, another form of coherence between the film and the base material can be observed. Through this process the wear microrelief, the fractal dimension value and the wear rate of the wiping surface of the insert change. It is possible to consider that the processes of formation, destruction and regeneration of the oxide and oxycarbide surface structures follow the laws of self-regulation.

Processing of materials, which cause an entirely intensive diffusion wear, is generally performed by the carbide cutting inserts that under the optimal conditions of operation must set up a high correspondence between the forming oxide film structure and the crystallographic boundaries. In this case, the surface film plays the role of a diffusion-resistant and heat protection shield, and the formed wear microrelief exhibits low fractal dimension.

The process of oxide film formation in the contact zones of the cutting insert starts with the capture of gas elements from the gaseous atmosphere [6]. More often, open and closed pores of the hard-alloy surface structure, formed as a result of tribological interaction of the workpiece and tool materials are more efficient areas of gas capture. Different types of adsorption may occur in the contact zones of the cutting insert depending on the available and forming cavities, their surface and gas access rate. The capture of gas elements by the juvenile surface of the forming microcavity is usually followed by the chemical adsorption. The high concentration of juvenile surfaces speeds up the process of the film structure formation. When the gas elements are captured by the juvenile pore surface, physical adsorption takes place first. In this case, the process of the surface film formation slows down even at high concentration of the cavities. In practice, the process of the film structure formation is caused by both physical and chemical adsorptions with the predominance of one of them in each particular case. As a consequence of oxygen adsorption, the sections of oxide structure occur, expand and gradually unite into large areas.

The energy of thermal activation necessary for the formation of the oxide compound in a juvenile pore space is significantly lower than that in the preliminarily contaminated pore. After adsorption of gas molecules, the surface needs some time and energy consumption for binding the reacting atoms and the oxide compound synthesis.

Physical adsorption of molecules does not always end with decomposition into atoms, taking part in the formation of new compounds. However, the molecule physically adsorbed by the cavity, which is not susceptible to active decomposition into atoms, can be exposed to desorption and adsorbed again by the pore formed nearby with the more active internal surface.

The compounds formed in the internal spaces of juvenile and non-juvenile pores can grow and form the so-called island structure [7]. The bonding between the islands and the surface is the factor of the internal space diameter and the adsorption history on the pore walls at the initial stage. When islands grow, they overlap and form a solid surface. The formation of the oxide and oxycarbide structures and their regeneration due to inevitable damage continue throughout the whole period of the tool life, and their microrelief undergoes the continuous transformation. Because of layering, the surface structure can be assumed as mobile and its composition and properties — as inhomogeneous. First of all, it depends on the level of uniformity of the hard-alloy composition, the temperature pattern and actual stresses on the contact surface. Consequently, the chemical activity of different parts of the cutting insert interfaces and the aeration velocity will vary.

When increasing the uniformity of thermal and mechanical patterns, occurring on the interfaces the stay-put feature of the surface oxide and oxycarbonitride structures, under the conditions of intensive diffusion wear, the fractal dimension of the wear roughness decreases. This process results in improvement of the cutting tool wear resistance.

The least developed rough surface occurs on the insert rake face at some optimal temperature corresponding to an optimal cutting speed. At temperatures (speeds) lower than optimal, the high development of the wear surface microrelief and the high value of fractal dimension are specified by the considerably low intensity of formation of the surface oxide and oxycarbonitride compounds in hard alloy. This leads to the formation of low quality film structures with low screening capacity to resist diffusion processes.

The failure of the oxide structures also results from the low level of stoichiometry. Due to oxygen deficiency, the low activity of the desorption processes can be observed. At temperatures higher than optimal, the development of the wear surface microrelief and the value of fractal dimension are higher than that caused by the intensive oxide and oxycarbonitride film destruction due to the intensification of their sublimation. This is due to the high adsorption activity, the high oxidation rate and the high transition rate of oxides and oxycarbonitride into a gaseous state. This process is accompanied by the intensification of the diffusion wear.

4. Conclusion

The radiation pre-treatment of the carbide cutting tools causes an intensive desorption of impurity hydrogen that decreases the formation of the oxide and oxycarbide structures on their surfaces and subsurfaces [8]. This involves an increase in the fractal dimension of the flank wear and a decrease in the wear resistance of the cutting tool inserts. It is possible to conclude that the wear resistance of the carbide insert is specified by the optimal composition of the surface and subsurface oxide, the oxycarbide structures, and the decreased fractal dimension of the wear surface results in the increased wear resistance of the cutting insert.

References

[1] GOST 3882-74. Firm alloys sintered. Stamps, Moscow, 1974
[2] Guo S, Li J, Cheng H, Yang J, Zhong M and Zhu E 2013 *Rare Metals and Cem. Carbides* **41 (6)** 67-69
[3] Roldugin V I 2011 *Physicochemistry of surface* (Moscow: Intellect)
[4] Adaskin A N 2015 *Tool materials in mechanical engineering* (Moscow: Info-M)
[5] Bowden F P and Tabor D 2011 *The friction and lubrication of solids* (Oxford: University Press)
[6] Adamson W 2011 *Physical chemistry of surfaces* (Moscow: Mir)
[7] Kolesnikov V I, Bulgarevitch S B, Boyko M V and Feyzova V A 2011 *Wear and Tear* **(4)** 233-243
[8] Bondarenko G G 2016 *Radiation physics, the structure and strength of solids* (Moscow: Binom)

Effect of technological heredity on the fatigue strength in the manufacture of gas turbine engine blades

G V Smirnov, N D Pronichev and M V Nekhoroshev

Samara University, 34, Moscow Hwy, Samara, 443086, Russia

E-mail: maxnogood@ssau.ru

Abstract. In the study, the task of researching of the finishing-strengthening machining stage of gas turbine engine compressor blades manufactured of titanium and nickel-chromium alloys in order to extend their service life was solved. The application of electrochemical pulse machining as a technological heredity barrier was substantiated since this method allows a considerable decrease of the residual stress and surface layer work hardening. To ensure the extended service life of blades, the conditions for the subsequent finishing-strengthening machining were identified.

1. Introduction

Modern aircraft engines have a rather complex design and demonstrate a large degree of manufacturing labour-intensity. Strict operating requirements are imposed on them including their operational reliability. Therefore, in serial manufacture of engines, much attention is paid to the issues of reducing labour input and improving the manufacturing quality to enhance the operational reliability and service life of the gas turbine engines while reducing their self-cost. An important feature in the manufacture of engines is the intent of extending service life for serialized products by implementing design and technological methods. Furthermore, the blade is one of the most loaded parts of the gas turbine engine, and, to a large extent, it produces a great impact on the service life and operational reliability of the engine. Therefore, improving the blade service life will reasonably favour extending the engine service life. Basic service life-defining properties of the blade surface layer are developing in the course of the final machining; therefore particular attention is paid to the finishing machining technology applied for the GTE compressor blades.

The shortcoming of the airfoil final machining methods causes necessity of airfoil profile manual correction to ensure the desirable blade geometry. However, the worker's professional competence (a subjective factor) produces a great impact on the blade surface quality. The occurrence of burn marks on the blade airfoil surface after the correction bench work may be stated to illustrate the unfavourable influence of the subjective factor. The decisive effect on burn makes is produced by the workman's qualification, labour skills and state, as well as by the applied efforts, which are necessary in grinding. All this causes irregular quality characteristics of the blade airfoil surface layer. Therefore, the decreased scope of manual blade airfoil correction works is favorable both for decreasing of the labour intensity in manufacturing the engines and extending their service life.

With a certain degree of convention, rotor blades may be divided into three standard sizes: large, medium and small. Low pressure compressor (LPC) rotor blades with the airfoil length of 300...500 mm, the blade chord of 100...250 mm may be referred to the large-size blades. They are manufactured of titanium alloys. The medium-size blades have the airfoil length of 100…200 mm with a chord of 50…80 mm. All the rest are referred to the small-size ones. As a rule, these are high pressure compressor blades with the maximum airfoil length of 15…35 mm and the chord size of 11…30 mm. The most typical of them may be found in modern two- and three-shaft gas turbine engines, where high air compression rates are implemented. Therefore, the last stages of high pressure compressors (HPC) are equipped with blades of the maximum airfoil length of 15 mm and the chord of 11 mm. Since the HPC blades operate at high temperatures, nickel-chromium steels and alloys are applied for their manufacture. The design features of these airfoils are defined by strict roughness requirements for all the path surfaces (R_a=0.32...0.08 μm), the thin profile (C_{max}=1.1 mm) and the maximum radius of the leading and tailing edge being 0.1 mm. Based on the operating conditions of axial compressors, the blades shall comply with the requirements as follows:

- The blade material shall conform to the heat-resistant and hot-temperature stability requirements in the effect of high temperatures, pressure drops and revolutions;
- The weight and sections of blades shall be minimal and their dimensions shall be highly precise thus ensuring the parametrical stability and high compressor efficiency;
- Blade path surfaces shall demonstrate a low roughness indicator, since it helps reducing the losses in a gas flow and increasing the fatigue resistance;
- The labour effort of the technological process shall be kept to a minimum.

Success achieved in the application of low-allowance workpieces raises the topicality of the issue of developing a technology allowing one to process openwork pieces of hard-to-treat materials. It should be noted, that small blades are deprived of the developed base surfaces. In these conditions, circular (airfoil, shroud and edges simultaneously) electrochemical dimensional machining (ECM) of the airfoil profile from the finished root becomes efficient.

However, this is true only for the case, when the ECM is used as a final shaping operation to ensure compliance with the drawing's requirements. The enhanced ECM precision of the airfoil profile and path surfaces allows abandoning the subsequent manual profile correction.

A special feature of medium-size blades resulting from their operating conditions is that, firstly, they may be manufactured both of titanium and of iron-chromium-nickel alloys.

Secondly, smaller dimensions, as compared to the LPC blades, allows using low-allowance workpiece manufacturing methods, e.g. high-velocity forming. Thus, the minimum allowance value along the airfoil will be decreased to 0.3...0.5 mm and its irregularity will be 0.6...1.3 mm. That is, with a minor allowance, its irregularity is rather considerable. The allowance may not be removed in two or three operations, as compared to the LPC blade airfoil. The airfoil machining using edge and abrasive tool is problematic due to the considerable impact and thermal effect produced on the workpiece openwork airfoil. High requirements for the airfoil profile precision and its displacement relative to the root reduce them to small-size blades. All the above-mentioned makes the ECM a priority airfoil machining method which may reliably ensure a non-stressed allowance removal to exclude a considerable impact and thermal effect, caused on an airfoil, and ensure the uniform distribution of minimal surface residual stresses in all the airfoil sections after machining provided the limited dimensions of basing surfaces.

2. Materials and methods

Presently, the ECM is used not only for the airfoil machining, but also in treating of bladed discs; in particular, the work [1] provides the experimental ECM data for the Ti60 titanium alloy which is used in the manufacture of bladed discs for the gas turbine engine compressor. The machining results for interblade areas of the blisk of the given alloy were obtained ensuring high precision of the blade profile (the profile error of 0.05...0.07 mm).

Including the airfoil ECM operation in the rotor blade manufacturing process is expedient since it ensures the smallest level of residual stresses in the airfoil surface layer as compared to other mechanical machining types. In addition, multiple studies have shown that the blades, which have passed the electrochemical machining and the respective strengthening machining, possess the improved strength properties as compared to those ones which undergone the mechanical machining and the subsequent work hardening. The durability and fatigue strength is improved up to 20%. For example, titanium alloys after the electrochemical machining demonstrate the improved endurance characteristics at an optimal performance level. A defective layer from the preceding machining is removed during the ECM. Furthermore, the removal is not accompanied by the power effect on the surface being treated. Therefore, the electrochemical airfoil dimensional machining represents a special technological barrier of the negative technological heredity (TH).

It should be noted that the analysis of technological machining processes from the TH standpoint has been employed by Minsk, Moscow and Samara technological schools since the 80s, and recently this approach has become widespread in analysing various technologies:

- Metal working;
- Machining of composite materials and itermetallic compounds;
- Volumetric deformation technologies;
- Assembly.

Thus, in work [2] the authors propose studying the manufacturing technology for metal structures as a sequence of macroscopic, microscopic and submicroscopic transformations characteristic of the structure initiation from the TH standpoint to forecast the product quality. As an example of such approach, a method of continuous deformation nanostructuring is employed, allowing the authors to define the evaluation strategy basis for various multi-stage material machining processes to achieve the required set of properties.

The work [3] contains a comparative analysis of TH factors for the parts manufactured both of metal alloys and polymer materials. During the analysis, the factors that have the greatest influence on the heredity of properties were identified. Technological connections ensuring the heredity transfer from one technological operation to the other were considered.

The authors of the work [4] also employed the TH theory in analysing the hydraulic cylinder manufacturing technology; in addition, the mathematical statistic device was involved to estimate the efficiency of production operations. All these allowed the authors to find the most acceptable hydraulic cylinder machining option, ensuring the suppression of negative heredity.

In addition, the researchers [5] analysed the technology for manufacture of a cylindrical gearwheel with involving the TH theory and a multi-factor correlation analysis. As a result, a flow diagram for the machining error heredity in manufacturing the cylindrical gearwheel was developed.

In work [6], an approach from a TH perspective in analysing machine working of casts and the respective assembly of the item jointly with the use of the functional gradient structure model of grain boundaries was employed. This allowed determining the influence of cutting rates on formulating the structural stress raisers during finishing cast milling.

In article [7], following the rough-turning and finish turning example and the subsequent surface plastic deformation, the methods of surface quality control methods for the machine parts and their performance characteristics considering the TH were analysed.

The authors of work [8] employed the TH principle for a comparative analysis of the bearing ring working technologies ensuring the minimal adhesion wear in the subsequent operation.

In article [9], the authors, relying on the TH theory, studied the workpiece hot volumetric deformation process, which allowed them to efficiently suppress the negative heredity, improving the plasticity of high-strength aluminium alloys as well as discovering implementability of phase transitions in manufacturing semi-finished products within thixotropic conditions close to superplastic deformation.

In work [10], the influence of TH on the strength and service life of the welded joint between the header and the heat exchange tube was studied. The residual stresses in the header assembly were determined depending on the main technological pressing parameters and strength characteristics of the tube and header materials.

Within this work, aimed at introducing the airfoil electrochemical pulse dimensional machining into the blade manufacturing technology as a finishing machining method, we have performed complex studies using the sample blades and actual blades to search for the optimal electrolytes and machining conditions. In addition, the studies included the subsequent strengthening machining and fatigue testing of blades with the airfoil treated under the optimal ECM conditions. Also, tests are mandatory when introducing changes to the existing technological part machining process stipulating the engine service life [11].

This work is a continuation of the authors' cycle of works on the ECM information modelling [12].

3. The comparative studies of the final stage of the gas turbine compressor blade manufacturing technology

The final stages of manufacturing technologies for the titanium alloy LPC rotary blade and the chromium nickel HPC blade were chosen as the subjects of research. The airfoils of the mentioned blades underwent finishing machining using the electrochemical pulse dimensional machining in electrolytes under conditions ensuring the preset precision parameters. The residual stresses in the airfoil surface layer after the pulse ECM at the optimal performance were below 100-150 MPa, thus demonstrating the negative value with the occurrence depth of less than 100 μm for all the blade materials. In other words, the ECM has produced almost no stress in the surface airfoil layer thus representing a negative technological heredity barrier for the stressed condition of the surface layer.

Further, strengthening machining using different methods was applied to the blades according to the options as represented in table 1.

Table 1. The options of surface plastic machining of the blade airfoil after the ECM

For the rotary blade of the EP718VD alloy	ECM + glass micro-ball strengthening (glass-MB) + vibrogrinding ECM + glass-MB strengthening ECM + steel micro-ball strengthening (steel MB) + vibrogrinding
For the rotary blade of the VT-8 alloy	ECM + glass-MB strengthening ECM + glass micro-ball strengthening + vibrogrinding
For the guide blade of the EP718VD alloy	ECM (initial option) ECM + manual polishing ECM + steel -MB strengthening

The blades treated according to these blade diagrams underwent the standard fatigue testing with a basic number of loading cycles being $N \cdot 10^6$.

The experimental data obtained were subjected to the linear regressive analysis resulting in determination of the fatigue strength and stress-cycle relationship values for the failure probability, P=50%.

The analysis results are given in figure 1.

The dispersion analysis determined the statistical significance of the technological factors' influence according to the strengthening options adopted in the manufacturing process. As a result of the unifactor dispersion analysis, it was established that the testing results of the working blades manufactured of the EP718VD alloy treated with ECM + micro-ball strengthening, both with the vibrogrounding and without it, belongs to the same universe general population and may be considered as a consolidated option.

The selection of hardening bodies (glass and steel micro-balls) is a considerable factor, therefore the results of testing HPC blades of the EP718VD with the ECM machining + steel micro-ball strengthening + vibrogrinding form a separate population. It was also established that the testing results of blades manufactured using the ECM method and steel micro-ball strengthening with the preliminary airfoil polishing and without polishing shall belong to the same general population and may be considered as a joint option.

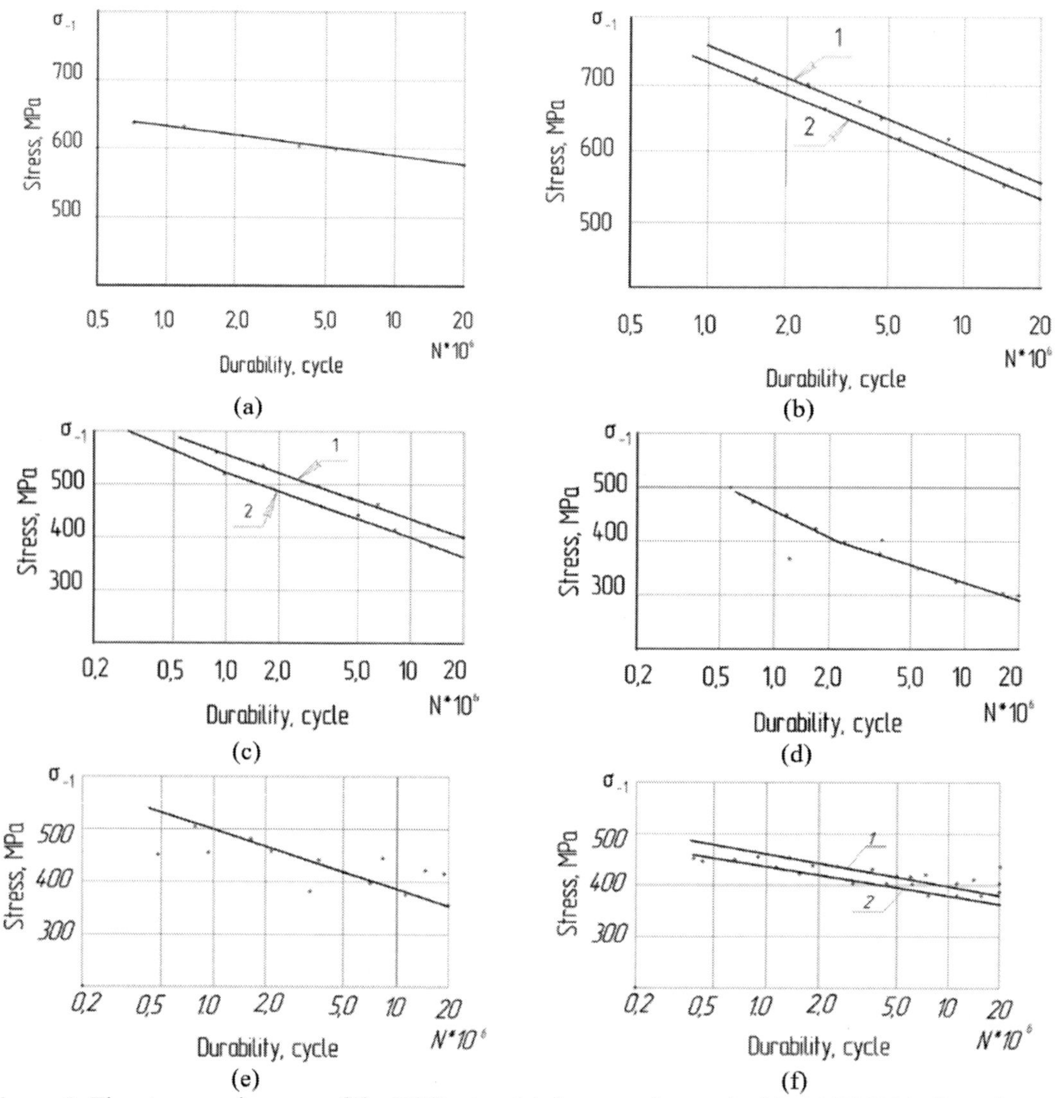

Figure 1. The stress-cycle curve of the HPC rotary blades manufactured of the EP718VD alloy after:
(a) ECM + steel micro-ball strengthening + vibrogrinding;
(b) 1 – ECM + glass micro-ball strengthening + vibrogrinding; 2 – ECM + glass micro-ball strengthening;
(c) 1 – ECM + glass micro-ball strengthening + vibrogrinding; 2 – ECM + glass micro-ball strengthening;
(d) ECM;
(e) ECM + manual polishing;
(f) 1 – ECM + steel micro-ball strengthening; 2 – ECM + manual polishing + steel micro-ball strengthening.

The dispersion analysis of testing results for blades forming options without micro-ball strengthening using manual polishing and without it has demonstrated that the testing results for these two options belong to different populations.

The dispersion analysis of other combinations of options was meaningless due to the obvious difference of testing results. The linear regressive analysis was repeatedly performed for the generalized options. Therefore, following the results of strength testing, the following values of fatigue strength were obtained, which are given in table 2.

Previous studies for the EP718VD blades, which airfoils were treated by the double milling and the subsequent manual polishing as well as subjected to steel microball strengthening and vibrogrinding, demonstrated in the fatigue testing that their fatigue resistance value was lower and amounted to σ_{-1}=512 MPa. For the glass microball strengthening option, it is σ_{-1}=470 MPa. This means a considerable increase of the fatigue strength for the blades treated by ECM with the subsequent strengthening machining.

For airfoils which failed during the strength testing, metallographic studies were carried out which confirmed the correspondence to the metal quality specifications. The condition analysis of the blade surfaces was performed with a MBS-2 binocular microscope with the magnification of 12,5 and x25.

The surface of blades of EP718VD alloy option 1 is grey, matte, demonstrating particular traces of knocking with glass micro-balls and their slipping. The option 2 blade surface is light grey, matte, demonstrating traces of glass microball machining. The option 3 blade surface is grey, matte, demonstrating traces of knocking with glass micro-balls and their slipping.

Table 2. The values of fatigue strength σ_{-1}, MPa after the strength tests performed

Working blade of the EP718VD alloy	
ECM + glass micro-ball strengthening (with/without vibrogrinding)	540
ECM + steel micro-ball strengthening + vibrogrinding	580
4th stage rotary blade of the VT9 alloy	
ECM + glass micro-ball strengthening (with/without vibrogrinding)	370
6th stage guiding blade of the EP718VD alloy	
ECM	280
ECM + manual polishing	360
ECM + steel micro-ball strengthening (with/without preliminary polishing)	480

The microanalysis performed in fractional blade sections established that the structure of blade material is normal for the EP718VD alloy solid solution + strengthening phases. The grain size for all the blades corresponds to points 5-6 of the grain scale under GOST 5639-82 and complies with the requirements of technical specifications for blades. To determine the blade surface strengthening level for all the options, the micro-hardness was checked using the PTM-3 device with the load of 50 g. The micro-hardness measurement results are given in table 3. The micro-hardness measurement results demonstrate that the increase of surface micro-hardness corresponds to the work hardening level 31 (39%).

As a result of the research, an optimal technology for the finishing-strengthening stage of the HPC blade airfoil treatment after pulse ECM was offered. The employment of this technology decreases the influence of negative technological heredity on the type II size precision. It enables us to form the optimal surface layer quality.

Table 3. The measurement results of micro-hardness after different machining methods

Machining option	Measuring point	Average impression diameter, μm	Micro-hardness, MPa
1 option	surface	44	4724
1 option	core	50.3	3580
2 option	surface	45	4510
2 option	core	50	3650
3 option	surface	42.8	4990
3 option	core	50.3	3580

Furthermore, the improved fatigue resistance is ensured with retaining the increased long-term strength as in case of general pulse ECM. After electrochemical machining, the residual stresses are minimal in the surface layer, the work hardening is absent.

4. Conclusion

Therefore, in testing various blade airfoil machining options, it was demonstrated that the electrochemical pulse dimensional machining is the best method to comply with the requirements for the finite machining technique with the subsequent micro-ball strengthening, representing the barrier for the negative technological heredity. In this event, a considerable growth of long-term strength (approximately 10-15%) is ensured as compared to the traditional mechanical working of the airfoil and the consequent strengthening, which is greatly in line with the results of studies performed earlier for the medium- and large-sized blades.

Acknowledgments

The authors of the article are grateful for the assistance in the research to the employees of engine building companies of Samara region, the Russian Federation.

References

[1] Xu Z, Chen X, Zhou Z, Qin P and Zhu D 2016 *18th CIRP Conf. on Elec. Phys. and Chem. Mach.* **42** 125-130
[2] Korchunov A, Polyakova M, Gulin A and Konstantinov D 2014 *Appl. Mech. and Mater.* **555** 401
[3] Baurova N I, Zorin V A and Prikhodko V M 2015 *Polymer Sci.: D* **8(3)** 219-222
[4] Kryvyi P D, Dzyura V O, Tymoshenko N M and Krupa V V 2014 *ASME 2014 Inter. Manuf. Science and Engin. Conf.* **2** V002T02A037
[5] Ajoy Kumar Das 1999 *J. of Mat. Proc. Techn.* **91(1-3)** 66-74
[6] Kusyi Y, Kuzin O and Kuzin N 2016 *East. Eur. J. of Enterprise Techn.* **1(5)** 39-47
[7] Fyodorov V P, Averchenkov V I and Nagorkin M N 2016 *Proc. Eng.* **150** 899-905
[8] Dubovska R and Majerik J 2015 *Proc. Eng.* **100** 730-736
[9] Lisunets N L, Smirnov O M and Tsepin M A 2008 *Mat. Sci. Forum* **575-578(2)** 1134-1138
[10] Stepanov G V and Kharchenko V V 1998 *Strength Problems* **2** 66-86
[11] Kozhina T D and Kurochkin A V 2016 *IOP Conf. Series: Mat. Sci. and Eng.* **127(1)** 012044
[12] Nekhoroshev M V, Pronichev N D and Smirnov G V 2014 *Open Mech. Eng. J.* **8** **A436** 436-440

Experimental and theoretical study of the hydriding behaviour in the pulse ecm of titanium alloys

G V Smirnov, N D Pronichev, M V Nekhoroshev and V I Bogdanovich

Samara University, 34, Moscow Hwy, Samara, 443086, Russia

E-mail: maxnogood@ssau.ru

Abstract. The titanium alloy hydriding is a negative factor since it may result in a slow-action destruction of parts operating at small loads. The introduction of pulse electrochemical machining requires consideration and evaluation of this factor's influence onto the operating performance of components. Since electrochemical machining is performed at small gaps, and hydrogen release is very intensive on electrodes, favourable conditions for the hydriding process are developed. The work describes a profound theoretical study of this process with proposing technological methods to reduce hydriding.

1. Introduction

An important task of the aircraft engine building is enhancing its competitive ability by virtue of the improved quality, operational reliability and extended service life along with reducing the self-cost of the major engine components. These components conditioning the level of the engine building product are primarily the compressor and turbine blades. Solving this task as applied to these components is possible within the framework of the finishing improvement.

One of the methods in improving the finishing machining of compressor blades is the use of pulse electrochemical machining (ECM) at a final stage as the most appropriate finishing airfoil forming method. Minimal power and thermal effect on the low-hardness airfoil is ensured by the ECM when removing the stock, which is especially important in the absence of the developed technological bases, a small stock and its considerable irregularity. The introduction of ECM is restricted by the technological heredity factors, which partially impair the surface layer condition, for example, by hydriding the machined surface of a titanium-alloy blade airfoil.

It is obvious that when using the electrochemical pulse dimensional machining (ECM) of the blade airfoil as a finishing forming operation, the importance of hydriding as a negative technological heredity factor will be considerably intensified. This is due to the fact that the surface layer formed in the ECM will be definitive of the component's service life.

The reason of surface hydriding is high diffusion mobility of hydrogen atoms, which is conditioned by its small size. The problems of hydrogen interaction with metals are diverse and they include adsorption, diffusion, dissolution and formation of hydrides. In addition, the negative impact of components' surface hydriding affects their operational performance as a result of the hydrogen brittleness.

2. Materials and Methods

Surface hydriding occurs in various engineering processes and is the subject of the detailed study. So, in work [1], when studying the titanium alloy laser welding process, it was detected that if the hydrogen concentration is exceeded by more than 0.2 %, the superplastic ductility of the welded joint will decrease, the deformation ratio non-uniformity will grow. In article [2], when studying the stability of Ti_3SiC_2 base plate in hydrogen at the temperature of 1200–1400°C during three hours, the hydrogenation mechanism using the combined use of x-ray diffraction, scanning electron microscopy and the diffused light spectrometry was clarified. The authors observed separation of the base plate layer at the temperature of above 1300°C due to the increased hydrogen diffusion along the basal plane. In work [3], the impact of hydrogen on the structural and phase state of the fine-grained and submicrocrystalline two-phase Ti-6Al-4V titanium alloy was studied by electronic microscopy and by the x-ray diffraction analysis. It was established that hydriding results in minor structural and phase changes both in the fine-grained and submicrocrystalline samples. The use of electron-beam irradiation combined with heating for hydrogen release in Ti-6Al-4V alloy resulted in the decreased degasation time and the reduced hydrogen concentration down to the values adopted in the engineering standards for the Ti-6Al-4V alloy.

In work [4], comparative studies of the effect caused by 0.002–0.12 % hydriding on the structure and phase composition of the submicrocrystalline and coarse-grained Ti-6Al-4V alloy were described. The evolution of deformation processes in hydrogenated alloys within the extension conditions at the temperature of 293 K was studied depending on the hydrogen content and the condition of alloy. The presence of hydrogen in solid-solution Ti-6Al-4V nanostructured alloys was established, which resulted in their reduced yield as well as the improved tensile strength.

Work [5] studied the peculiarities of the ultrafine-grained structure formed in the Ti-6Al-4V alloy through the use of the method combining the intensive plastic deformation and the reversible hydrogen treatment. It is shown that martensitic and $\alpha \leftrightarrow \beta$ phase transformations, due to the presence of hydrogen, favour the improved structure during the intensive plastic deformation of the alloy. This allows reducing the deformation rate necessary to obtain the ultrafine-grained structures with the component dimensions of 0.1–0.5 μm. It was demonstrated that the formation of ultrafine-grained structures in the Ti-6Al-4V alloy results in the growth of its strength properties of about 1.5 times, the increase of hydrogen brittleness resistance and the reduction of superplastic condition transition temperature by 200-300 K.

In article [6], thermo-emf and the electrical resistivity of hydrogenated titanium were measured. Furthermore, the hydrogen concentration in titanium has changed from 0.005 to 0.25%. The methods applicable to their sensitivity to thermo-emf and electrical resistivity change during the titanium hydrogenation process were compared. Precision of the results obtained was checked by comparing them with the results for palladium. It was shown that the specified method is sensitive to hydrogen concentration in titanium at different alloy hydrogenation depths. Based on the measurements carried out, the authors proposed a graphic method for detecting the titanium structure change.

A large number of works are devoted to the issue of reducing hydrogenation during the titanium alloy ECM, where authors attempted to find the best composition of electrolyte and selecting the optimal machining mode capable of ensuring the low surface roughness value and the small surface layer hydrogen concentration along with the high machining precision. So, for example, the compositions of electrolytes were developed, where the surface smoothing effect is achieved at regular as well as at low current densities. According to the data [7], with the standard ECM conditions, high process performance is achieved in chloride-nitrate electrolyte. For the chloride-bromide electrolyte, the reduced hydrogenation level is observed in proportion to the decreasing surface roughness.

The author selected the optimal electrolyte for the VT3-1 alloy. Mixture 14%NaCl+11%NH_4NO_3 was preferred, where the metal removal rate was 0.85 g/min instead of 0.7 for the sodium chloride solution (14 % NaCl) and the current density of 100 A/cm^2. High removal rates were obtained in similar chloride-nitrate electrolytes for VT9 and OT4-2 alloys [8]. At the same time, work [9] showed that the removal rate for titanium alloys was lower in the chloride-nitrate solutions as compared to the chloride solutions. According to [10], surface hydriding in electrolytes based on NH_4NO_3 is over the limit. According to other results of the study [11], hydriding of titanium alloys in chloride-nitrate electrolytes is within tolerance. In comprehensive work [12], describing the extensive study of the machinability for titanium alloys, the prospects of chloride-nitrate electrolytes for machining compressor blades were verified in the context of the achieved high machining rate, as well as low surface layer roughness and hydriding values.

In article [13], the authors observed generation of the hydrided layer on the surface of the titanium alloy influenced by high temperature developed in the course of milling. Hydriding produced a considerable impact on the plastic properties of the alloy.

To sum up, a suggestion can be made that pure nitrates are of little use for titanium alloy ECM due to their high hydriding, but when combined with other components, they provide the enhancement of the technological characteristics. Among these electrolytes, sodium chloride and sodium nitrate (ammonium)-based electrolytes may be called as the most universal ones since the improved process performance as well as the reduced surface roughness and hydriding are observed in them. As opposed to multicomponent electrolytes, they comply with the operating requirements.

The issue of selecting a chloride-nitrate mixture for the pulse ECT has not been sufficiently studied until present regardless of its prospects. The results of the enhanced process performance study for these solutions are sparse and contradictory. The issue of the possibility to improve the precision, to reduce hydriding and to enhance surface roughness has not been addressed sufficiently.

3. Developing A Mathematical Model for The Hydriding Process in The Titanium Alloy ECM

The analysis was aimed at searching for the optimal electrolytes and pulse dimensional electrochemical machining conditions for the VT9 alloy as well as developing the theoretical surface hydriding model at the pulse electrochemical dimensional machining as well as development of the theoretical surface hydriding model in the pulse ECM.

The experimental studies of the surface hydriding during electrochemical machining were carried out on the experimental bench by employing cylindrical samples of the VT9 titanium alloy with the area of 100 mm^2. In the initial condition, the hydrogen content in samples was 0.005% and the surface roughness – R_a=2.5 μm. The following electrolytes were used for the analysis: sodium chloride, sodium nitrate, ammonium nitrate. The selection of electrolytes was limited by the most perspective electrolytes among the ones that were initially detected as ensuring the lowest gas emission, minimal hydriding, responding to the process selectivity requirements and ensuring low surface roughness with the pulse current ECM. It was shown that with a decrease of the gap down to 0.04–0.06 mm (within the range of ultra-small gaps, where the blade ECM bears the risk of short circuiting, and the increased process selectivity is possible at the same time) a tendency to the decreased hydriding is noted. That is, within the gap range where the ECT may really act as a finishing method, minimal hydriding may be achieved. Hydriding grows with the increased current pulse duration and pauses between the pulses. In the pulse ECM, the surface layer generation behaviour stipulating the hydriding during machining, will considerably differ from the direct current machining. It was established that the titanium alloy hydriding source in nitrate and chloride-nitrate solutions is hydrogen, which is generated in the anode. The hydriding mechanism was proposed based on the adsorption of anode atomic hydrogen and its penetration into the surface layer. The minimal contents of hydrogen in the surface layer may be obtained via the pulse ECM of titanium alloys in nitrate-based electrolytes. When using chloride-nitrate mixtures in optimal conditions, hydriding is ensured without exceeding the permissible limits. Adding superficially active materials (SAS) to electrolytes in the amount of 0.1...0.15 is useful for the pulse titanium alloy ECM. According to the analysis of the effect caused on the hydriding value by major ECM factors, it was defined that major factors are the duration and the rate of pulse, an electrode gap, total electrolyte concentration and electrode current. At the subsequent stage of the analysis, using the experiment planning, we determined the dependence of hydriding based on the above-mentioned factors. A fractional factorial experiment was performed with the use of the half-replicate 25-1 (16 experiments). The factor variation range is given in table 1.

Table 1. The range of conditions under study

Factors	Code factor symbol	Natural factor levels corresponding to the coded ones.	
		Lower -1	Upper+1
Pulse duration t_i, ms	x_1	0.06	2.0
Pulse rate, q	x_2	2	6
Gap a, mm	x_3	0.04	0.2
Electrolyte concentration, %	x_4	5	25
Voltage U, V	x_5	12	25

As a result of the above-mentioned analysis, the regression equation was obtained as follows: $y = 0.0527 + 0.00025X_1 + 0.000875X_2 + 0.00264X_3 + 0.00025X_4 - 0.00501X_5$.

The equation adequacy as well as the significance of factors were checked. Using this dependence, it is possible, firstly, to determine the expected level of hydrogen content in the surface with the selected machining conditions in the context of the airfoil forming precision. Secondly, it allows adjusting the machining conditions towards lower hydriding values if they do not comply with the maximum permissible limit of hydrogen contents.

To obtain the universal generalized conclusions on the principal hydriding reduction directions, which would suit various alloy structures, conditions and electrolytes, we developed the mathematical model of the hydriding process in the titanium alloy ECM considering the joint effect caused by all the factors acting during machining and built with the adequate phenomenological model. In work [14], the authors studied the diffusion hydriding in conditions of the direct current ECM, and it was established that when increasing the surface dissolution rate, the density of the hydrogen diffusion flow directed inside the metal would grow and the quasi-stationary distribution of bound hydrogen would be set on the moving dissolution front. After the process termination, hydrogen accumulated in the surface layer continues diffusing into the depth of the metal, where it is transformed to the bound state.

For the pulse ECM, the large number of pulses ~ 10^5–10^7 equal in their intensity and supplied during the stock removal, are inherent. Small thickness of the layer dissolved within one pulse h~ 10^{-8}-10^{-4} mm, short pulses and pauses with the order of ~ 10^{-4}–10^{-3} as compared to the characteristic relaxation times (the time for establishment of quasi-stationary moving front (~3–5 s) and the time of hydrogen transition into the bound state (~60 min)) allow considering this process as a sequence of momentary focused pulses.

Solving the task on determining the hydrogen concentration in a surface layer after the pulse ECM is reduced to determining the concentration balance on the moving dissolution surface. With the momentary removal of the layer with thickness h, the amount of hydrogen $N_p = h(C_h + C_p)$ is generated on a newly formed surface, where C_h is the background hydrogen concentration in metal, C_p is a reactive component stipulated by the anode dissolution electrochemical reaction. Furthermore, when the layer is dissolved, the excessive hydrogen in the amount of $N_c = \int_0^h \overline{C}(z)dz$ contained in it is generated, where $\overline{C}(z)$ is the excessive hydrogen concentration established in the ECM. A fraction of this quantity of hydrogen equal to $\alpha(N_p + N_c)$, where 0<α<1, is the mass-transfer rate, enters the metal; the rest is transferred to the electrolyte solution, where a uniform concentration of reagents is maintained by mixing.

In turn, $\overline{C}(z)$ may be represented as a sum of free hydrogen $C'(z)$ and bound hydrogen (in the form of hydride) $C_x(z)$: $\overline{C}(z) = C'(z) + C_x(z)$. When considering a single flat pulse of power v on surface z=+0, the diffusion equation will take the form of:

$$\frac{\partial C}{\partial t} = D\frac{\partial^2 C}{\partial z^2}(C - C_x) + \nu\delta(z,t), \tag{1}$$

where ν is a specific amount of hydrogen generated as a result of the layer dissolving of thickness *h*, δ(z,t) is the pulse function. The rate of free hydrogen flowing into the bound conditon is determined by the first-order chemical reaction kinetics equation:

$$\frac{\partial C_x}{\partial t} = A_\kappa(C - C_x), \tag{2}$$

where A_κ is the averaged reaction constant for free hydrogen transformation into different bound conditions. If we pass on to $C'(z)$ in equations (1) and (2), the result will be as follows:

$$\frac{\partial C'}{\partial t} = D\frac{\partial^2 C'}{\partial z^2} - A_\kappa C' + \nu\delta(z,t) \tag{3}$$

$$z>0,\ t>0 \text{ with limiting condition } \left.\frac{\partial C'(z,t)}{\partial z}\right|_{z=0} = 0. \tag{4}$$

The equation of form (3) is reduced to the parabolic normal-form equation by substituting for $C'(z,t) = u(z,t)\exp[-A_\kappa t]$. Consequently, u(z,t) corresponds to solving the 2nd type heat conduction boundary problem for the half-line with pulse source ν at the end:

$$u(z,t) = \frac{\nu\cdot\exp\left[-\frac{z^2}{4\cdot D\cdot t}\right]}{\sqrt{\pi\cdot D\cdot t}}. \tag{5}$$

When multiplying (5) by $\exp[-A_\kappa t]$, the required free hydrogen concentration during the passage of one pulse will be obtained:

$$C'(z,t) = G_i(z,t) = \frac{\nu\cdot\exp\left[-\frac{z^2}{4\cdot D\cdot t} - A_\kappa\cdot t\right]}{\sqrt{\pi\cdot D\cdot t}}, \tag{6}$$

where G_i is the function of the surface source - solution of equation (3) for the half-line. The concentration of bound hydrogen during the passage of one pulse will amount to:

$$C_x(z) = A_\kappa\int_0^\infty G_i(z,t)dt = \nu\sqrt{\frac{A_\kappa}{D}}\exp\left[-z\sqrt{\frac{A_\kappa}{D}}\right]. \tag{7}$$

During the passage of the following pulse, after time *T*, a new surface hydrogen layer will be generated along with the disappearance of the previous one together with the dissolved layer with leaving traces:

$$S' = G_i(z+h,t) = \frac{\nu\cdot\exp\left[-\frac{(z+h)^2}{4\cdot D\cdot t} - A_\kappa\cdot t\right]}{\sqrt{\pi\cdot D\cdot t}} \tag{8}$$

$$S_x = A_\kappa\cdot\int_0^T G_i(z+h,t)dt =$$

$$= \frac{1}{2}\cdot\nu\cdot\sqrt{\frac{A_\kappa}{D}}\left(\exp\left[-(z+h)\sqrt{\frac{A_\kappa}{D}}\right]\cdot erfc\left[\frac{z+h}{2\cdot\sqrt{D\cdot T}} - \sqrt{A_\kappa\cdot T}\right] -,\right. \tag{9}$$

$$\left. -\exp\left[(z+h)\sqrt{\frac{A_\kappa}{D}}\right]\cdot erfc\left[\frac{z+h}{2\cdot\sqrt{D\cdot T}} + \sqrt{A_\kappa\cdot T}\right]\right)$$

where $erfc\xi = \frac{2}{\sqrt{\pi}}\int_{\xi}^{\infty}\exp\left(-x^2\right)dx^2$, $S'(z)$ is the concentration of free hydrogen from the previous pulse at the moment of the subsequent one, $S_x(z)$ is the concentration of bound hydrogen. $S'(z)$ is the initial condition for the 2nd boundary value problem with the zero right side. Consequently, the concentration field of the momentary surface hydrogen source will take the form as follows:

$$F_1(z,t) = \int_0^{\infty} S'(\xi,t)\frac{\exp\left[-\frac{(z-\xi)^2}{4\cdot D\cdot t} - A_\kappa\cdot t\right] + \exp\left[-\frac{(z+\xi)^2}{4\cdot D\cdot t} - A_\kappa\cdot t\right]}{2\sqrt{\pi\cdot D\cdot t}}d\xi, \quad (10)$$

where F_1 is the first-order phantom source. Then the concentration of bound hydrogen after two pulses as well as the full hydrogen concentration may be found:

$$\overline{C}_s(z) = C_H + \alpha(C_p + C_H)\frac{\sigma\exp\left[-z\sqrt{\frac{A_\kappa}{D}}\right]}{\sqrt{1+\sigma^2} - 2\alpha}, \quad (11)$$

where $\sigma = \frac{2\sqrt{A_\kappa DT}}{h}$. The total excessive amount of hydrogen: $N(\sigma) = \int_0^{\infty}(C(z) - C_H)dz = \alpha\sqrt{\frac{A_\kappa}{D}}(C_p - C_H)\frac{\sigma}{\sqrt{1+\sigma^2} - 2\alpha}$. Function $\phi(\sigma) = \frac{\sigma}{\sqrt{1+\sigma^2} - 2\alpha}$ has its maximum $\sigma^* = \frac{\sqrt{1-4\alpha^2}}{2\alpha}$ in point $\phi^* = \frac{1}{\sqrt{1-4\alpha^2}}$. From here, $\alpha = \frac{1}{2\sqrt{1+\sigma^{*2}}}$. The maximum hydriding shall be expected at the dissolution rate of $V_{\kappa p} = \frac{h}{T} = \frac{4\alpha\sqrt{A_\kappa D}}{\sqrt{1-4\alpha^2}}$. So as to evaluate the hydrogen intensity, the total excessive quality of hydrogen in terms of the per unit area of the dissolution surface is used:

$$N(\sigma) = \int_0^{\infty}(C(z) - C_H)dz = \alpha\sqrt{\frac{A_\kappa}{D}}(C_p - C_H)\frac{\sigma}{\sqrt{1+\sigma^2} - 2\alpha}. \quad (12)$$

These characteristics allow determining the maximum performance of the technology and the direction of hydriding reduction as well as estimating mass-transfer rate value α. Thus, it follows from boundedness condition N at small σ that $\alpha<1/2$. The developed model reflects the hydrogen surface saturation condition at the direct finishing completion moment, while, in practice, the described situation will always be distorted by the natural hydrogen diffusion inside the metal. The model indicates a way to achieving the minimal initial starting concentration of hydrogen during natural diffusion upon the ECM completion.

We have developed a method for determining the expected surface layer hydriding with the pulse ECM depending on the properties of the material being machined as well as machining conditions. It comprises the steps as follows:

1. A number of pulse ECM experiments are carried out for samples with the predetermined electrolyte in a wide range of machining conditions for the material being analysed in order to determine the metal dissolution rate; and the surface hydriding curves are defined. The evaluation of blade airfoil surface hydriding is proposed to be performed using the spectral method according to the procedure developed by the All-Russian Research Institute of Aviation Materials (VIAM).
2. A factor of σ and the value of total excessive amount of hydrogen N are determined for each experimental curve obtained.

3. Then, chart $\varphi=N(\sigma)$ is built. Function $\phi(\sigma)=\dfrac{\sigma}{\sqrt{1+\sigma^2}-2\alpha}$ has its maximum at point $\sigma^*=\dfrac{\sqrt{1-4\alpha^2}}{2\alpha}$ with the value of $\phi^*=\dfrac{1}{\sqrt{1-4\alpha^2}}$, allowing us to experimentally determine the mass-transfer rate value: $\alpha=\dfrac{1}{2\sqrt{1+\sigma^{*2}}}$. The value of kinetic constant A_κ is determined as follows. By selecting random point z_0 on the experimental curve (figure 1) $C(z)$, the following dependence shall be built:

$$\psi(z)=\frac{C(z)-C_H}{C(z_0)-C_H}. \qquad (13).$$

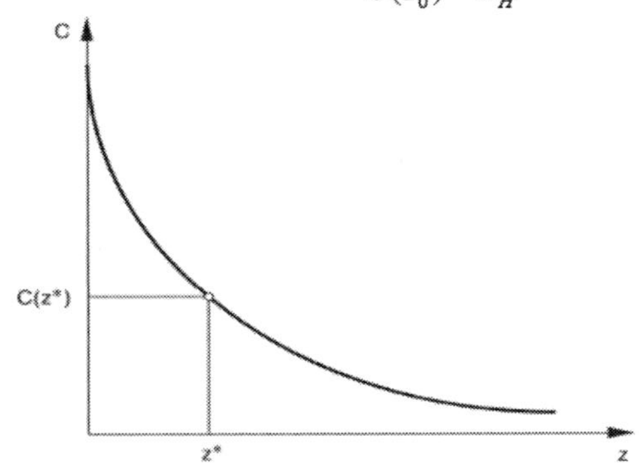

Figure 1. The dependence of hydriding along the depth of the surface layer on random point z*

When substituting (11) in (13), the following result is obtained: $\psi(z)=\exp\left[-\sqrt{\dfrac{A_\kappa}{D}}(z-z_0)\right]$. After taking logarithms of the last relation and making necessary transformations $\sqrt{\dfrac{A_\kappa}{D}}=\dfrac{1}{z_0-z}\ln\left(\dfrac{C(z)-C_H}{C(z_0)-C_H}\right)$, from which

$$A_\kappa=D\left[\frac{1}{z_0-z}\ln\left(\frac{C(z)-C_H}{C(z_0)-C_H}\right)\right]^2, \qquad (14)$$

where A_κ is the average value at experimental points of curve C(z). Value A_κ is a constant for the material grade and its working type before the ECM, which has little dependence on the ECM mode with the absence of the intercrystalline etching to the depth of diffusion.

The last of remaining parameters C_p is determined by calculated values σ, α, A_κ, $N(\sigma)$:

$$C_p=\frac{N(\sigma)\cdot\sqrt{A_\kappa}\cdot(\sqrt{1+\sigma^2}-2\cdot\alpha)}{\alpha\cdot\sigma\cdot\sqrt{D}}-C_H. \qquad (15)$$

The results of these studies and the hydriding mathematical model are planned to be used in developing a database for the EDM tool automated profiling module [15].

4. Conclusion

As a result of studies carried out, a dependence of BT-9 alloy ECM hydriding on the principal factors influencing the process (i.e. pulse duration, pulse rate, electrode gap, concentration of electrolyte and electrode current) was established. To define the universal dependence for the surface hydriding in titanium alloy ECM applicable to different structures of alloys, conditions and electrolytes, a mathematical model was developed for the surface layer saturation process in the pulse ECM. Based on the model, the method for determining the expected surface layer hydriding during pulse ECM depending on the properties of material being processed as well as machining conditions was developed. Also, the program for determining the expected surface layer hydriding in the pulse ECM depending on the properties of the material being machined as well machining conditions was developed and adjusted.

Acknowledgements
The authors of the article express their gratitude to the employees of engine-building companies in Samara region, the Russian Federation, for the assistance in the research.

References

[1] Fan Z, Cheng D, Chen Y and Hu D 2016 *Hanjie Xuebao/Transactions of the China Welding Institution* **37(4)** 57-60
[2] Chen C, Li F-Z, Xu C-H, Zhang H-B, Peng S-M and Zhang G-J 2016 *Adv. in Appl. Ceramics* **115(5)** 288-293
[3] Stepanova E N, Kudiiarov V N, Sypchenko V S, Lider A M, and Liu G 2016 *Key Eng. Mat.* **683** 187
[4] Stepanova E N, Grabovetskaya G P, Zabudchenko O V and Mishin I P 2011 *Russian Phys. J.* **54(6)** 690-696
[5] Grabovetskaya G P, Stepanova E N, Ratochka I V, Naidenkin E V and Lykova O N 2013 *Inorganic Materials: Appl. Res.* **4(2)** 92-97
[6] Lider A M, Larionov V V and Syrtanov M S 2016 *Key Eng. Mat.* **683** 199-202
[7] Klopova S V 1968 *Research machinability of titanium alloys VT3-1, VT8 and OT4-1 blades and the compressor operating mode of these alloys by electrochemical method. Abstract of PhD thesis* (Moscow: IPCE RAS) p 22
[8] Zhuravsky A K 1968 *On the question of increasing the accuracy of electrochemical machining parts* (Ufa: The improvement of machine technology) vol 2 p 6
[9] Saushkin B P 1986 Improving the efficiency of the electrochemical machining of titanium alloys. Systems approach (Tula: TPI) pp 209-229
[10] Radchenko V G and Grigorieva I V 1979 Effect of electrolyte composition on the ECM titanium alloy VT3-1 (Ufa: NTO Mashprom) pp 21-24
[11] Shmanev V A 1975 Influence of quality of the surface layer after the electrochemical machining on the fatigue strength of titanium alloys (Tula: TPI) vol 1 p 198
[12] Pronichev N D 1997 *Development of methodology and design principles of the electrochemical processing technology heavy duty gas turbine engine parts. Doctoral Dissertation* (Samara)
[13] Khaimovich A I and Balaykin A V 2014 *Int. J. of Eng. and Techn.* **6(5)** 2184-90
[14] Filimoshin V G, Shulepov A P and Mulkidzhanyan V N *Determination of the temperature on the electrode surface electrochemically in an electrolyte flow* (Kuibyshev: Electrochemical machining) vol 63 p 24
[15] Nekhoroshev M V, Pronichev N D and Smirnov G V 2014 *Open Mech. Eng. J.* **8 A436** 436-440

Surface profiling in mating parts by combined nonabrasive finishing

EV Smolentsev[1], ON Fedonin[2] and VP Smolentsev[1],

[1] Voronezh State Technical University, 14, Moskovsky ave., Voronezh, 394026, Russia
[2] Bryansk State Technical University, 7, 50-letiya Octyabrya bul., Bryansk, 241035, Russia

E-mail: smolentsev.rabota@gmail.com

Abstract. Nonabrasive finishing of precision mating surfaces in locking devices with the use of a combined erosion-chemical process at the first stage of the processing and with the use of anodic dissolution by alternating low-voltage current at the final stage of a refinement operation till gapless joints obtaining is considered. It is shown that the application of electro-erosion, electrochemical and combined nonabrasive finishing in mating parts opens up a possibility to ensure stable impermeability in locking devices on a macro- and micro-level through the method of a substantiated purpose of technological modes. A procedure is created for the development of such modes, and on their basis technological processes for the obtaining of gapless mating surfaces meeting the performance requirements for locking devices are developed. For this purpose, qualitative devices resistant to hostile environment are manufactured that is urgent for the mechanical engineering including repetition work for the equipment of petrochemical industry, transport and household machinery.

1. Introduction

Mating units with different systems of impermeability are used in the equipment which excludes falls of the flows of liquid and fluid media at small pressure at the expense of elastic seals (rubber and plastic rings, stuffing-boxes and so on), and washers. But with the pressure increase at the input, the sealing element wear increases, and the reliability and life of sealing elements decrease. Therefore, if pressure difference reaches tens of megapascals, then to ensure impermeability one begins to use seals with discharging chambers of the type shown in [1]. But the field of their application is rather limited, and for many locking devices with high pressure (particularly in aggressive flowing media), they are ineffective.

In the majority of locking devices used, it is necessary to ensure high accuracy and indices of a surface layer in mating movable parts. For high-pressure devices, a mutual processing of these surfaces without abrasive use which can charge the places of mating, cause increased wear, create gaps and cause the leaks of pumped media is required. At the same time, the errors in mating areas must be commensurable with a height of micro-roughness, and a width of micro-defect opening of a surface layer must not exceed tens of nanometers.

2. Analysis of mechanical finishing operations

In Table 1, the technological values and requirements for mechanical finishing operations are shown in [2].

Table 1. Technological potentialities of mating surfaces finishing.

Stages of finishing	Technological potentialities and requirements		
	Allowance, μm	Error, μm	Roughness (Rz), μm
Preliminary	20-50	3-5	3.2-0.8
Medium	5-20	1-3	0.8-0.4
Final	2-5	0.5-1.0	0.4-0.1
Fine	2-0.2	0.1-0.5	0.1-0.02

The analysis of Table 1 allows justifying a choice of a rational technological method of processing for each stage of precise surfaces finishing including mating ones.

One of the ways for a controlled removal of imperfections in the areas of locking elements consists in the use of a contactless processing with the superposition of an electric field. But technological problems arise here since electro-erosion finishing ensures obtaining the roughness (Rz) not less than 0.8-1.0 μm, which is unacceptable for many locking devices.

An anodic process with a recommended voltage range not lower than 5-6 V also does not make it possible to obtain the required accuracy (within parts of the micrometer) and roughness (minimum value Rz is about 0.1μm). Furthermore, before the beginning of a finishing operation, each of mating surfaces may have an error of 5-10 μm which requires allowance removal with the value not less than a limit value of tolerance for a blank at the first stage of finishing. For this purpose, the use of electro-erosion processing with 'soft' technological modes increases considerably labour-intensiveness of an operation and forms a cellular surface on each mating element unsuitable for locking devices. Besides, the low voltage, used in this case, violates the process stability and causes the appearance of local micro-defects and imperfections with a height greater than it is allowed by requirements of a drawing. A contactless anodic dilution does not allow ensuring accuracy in the areas of mating and at recommended voltages below breaking ones [3], it is impossible to reduce an initial height of roughness, that is, to obtain surfaces with required indices after finishing at a nano- and microlevel.

The works carried out have shown that during nonabrasive finishing of mating movable surfaces it is expedient to use two stages of processing: the first stage is with the use of a combined erosion-chemical process with direct current with a polarity periodical change, the second stage - at low-voltage anodic dissolution of micro-imperfections of alternating current with mechanical depassivation of the area of parts mating. In the latter case, the voltage on electrodes decreases up to 3-4 V at the expense of losses during the removal of a passivating film removed partially by a mechanical impact of a moving opposite mating surface of a locking device.

3. Mating surfaces geometry

In the equipment, mating surfaces of several types are used; some of them are shown in Figure 1.The process flow-sheets of the locking device grinding have significant differences. For the flat units (Figure 1 a), there is a typical possibility to control a pressing force of contact elements (gate 3 and nipple face 4) by means of changing elastic properties of part 5 (for instance, a spring). Besides, a swinging movement ('A' in Figure 1 a) allows supporting a constant position of an operation track on a gate during finishing and stabilizing a rate of a swinging movement of contact elements.

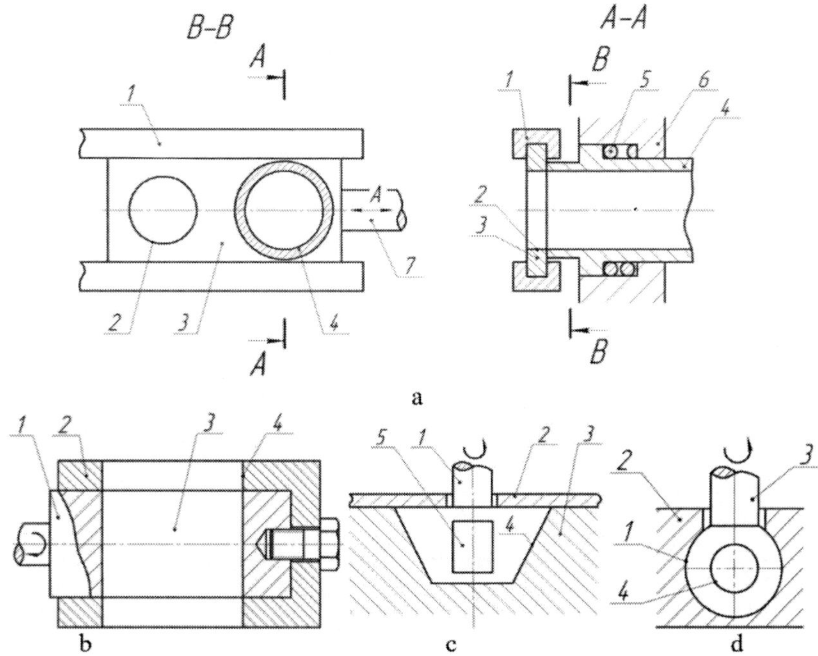

a – flat type, 1– guiding elements; 2 – hole; 3 – shutoff gate; 4 – pipe nipple; 5 – elastic element; 6 – locking device case; 7 – shutoff gate drive; b – cylindrical type, 1 – rotation part; 2 – case; 3 – groove in a rotation part; 4 – groove in a case; c – tapered type, 1 – rotation part; 2 – holder; 3 – case; 4 – bevel coupling; 5 – groove; d – spherical, 1 – ball coupling; 2 – case; 3 – rotation part drive; 4 – hole in asphere.

Figure 1. Basic types of mating surfaces in locking devices.

These surfaces, as a rule, are subjected to chemical-thermal treatment (for instance, nitration), that is why an allowance for their finishing should not exceed 50 μm. Similar locking devices can have face contact elements closing a saddle. In case of high pressures and expenses, in such elements, hydraulic impacts, breaking a flow of working environment, can arise; that is why, they are used mainly in hydraulic networks with the pressure up to 20 MPa.

In case of cylindrical locking devices (Figure 1 b), in the course of the preliminary finishing, there are gaps formed between a rotation part (1) and a case (2) which are required to use at this treatment stage of a technological rotation part with its substitution for a standard part at the final stage of treatment. At this stage of finishing, there arise complexities with the adjustment of operation clearances and their support in the course of maintenance, particularly in a pumping hostile environment, considerable temperature differences between flowing liquid and a case of a locking arrangement.

Cone structures (Figure 1 c) are more convenient for finishing in our case including those without using abrasive tools, as they allow adjusting a pressing force in a bevel coupling (4) between a rotation part (1) and a case (3), and also replacing the reciprocating travels of contact elements (Figure 1 a) by a turning motion of rotation part 1 in case 3. But their use is limited by the environment consumption passing through a device.

Ball valves (Figure 1 d) are rather complicated in manufacturing, though they widen considerably operational potentialities of locking devices complementing them by the functions of mixers and some other capacities. Here, new constructs are required for the adjustment of gaps and expensive equipment for manufacturing spherical contact surfaces. During their finishing, it is difficult to exclude the application of abrasive tools.

4. Mating parts finishing

All locking devices shown in Figure 1 can be produced by finishing without using abrasives by means of a two-stage micro- and macro removal of allowance by a combined method with the superposition of an electric field.

At the first stage, an erosion-chemical removal of macro- and micro-imperfections with the removal of projections occurs at the expense mainly of anodic dissolution of metal from tips of imperfections of both mating surfaces (Figure 2).

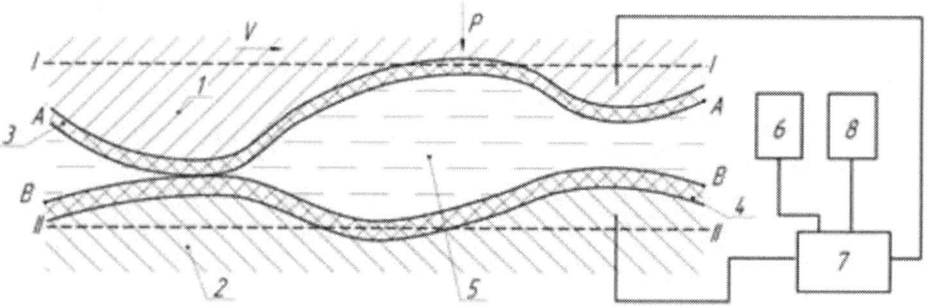

1; 2 – mating parts; 3; 4 – oxide film; 5 – low-conducting liquid medium; 6 – RC pulse generator; 7 – polarity switch; 8 – low-voltage current source; I-I; II-II – boundaries of mating surfaces after the first stage of processing; A-A; B-B – external surface of an oxide film; P – pressure upon mating surfaces; V – rate of mutual displacement of mating surfaces.

Figure 2. The scheme of mating surfaces processing.

In the environment of the weakly-conducting liquid (5) when approaching parts 1 and 2 up to the contact of oxide films 3 and 4, an electro-erosion microremoval occurs to boundaries I-I and II-II which can ensure the mating of components at the nano-level of accuracy. It is well-known [4] that during the combined treatment, the removal takes place mainly from an anode. If an initial roughness and error of one of mating surfaces is considerably higher than the other, then it is possible to carry out finishing with the polarity at which the first surface will be constantly an anode and the removal from the other surface can be at the expense of the current reverse half-wave. But in most cases of finishing the allowances are provided for both mating parts that is why a current polarity is changed from time to time by switch 7.

The boundaries of an oxide film (A-A and B-B) change the position of a contact area of mating parts, where pressing is controlled by pressure P. Depending on the pressure of parts 1 and 2 the thickness of a film in the areas of a contact becomes smaller. Here the resistance drops and an increased removal of imperfections at the expense of an electro-erosion constituent of the process take place. Simultaneously, during the combined treatment, an anodic dissolution of micro-projections takes place by means of the current of constant-current source 8. To ensure the continuity of a finishing process and a periodic replacement of operation environment in the area of treatment, parts 1 and 2 are moved with the velocity 'V'.

Under the influence of electrical discharges upon the surface of mating parts arise imperfections in the form of spherical holes (Figure 3).

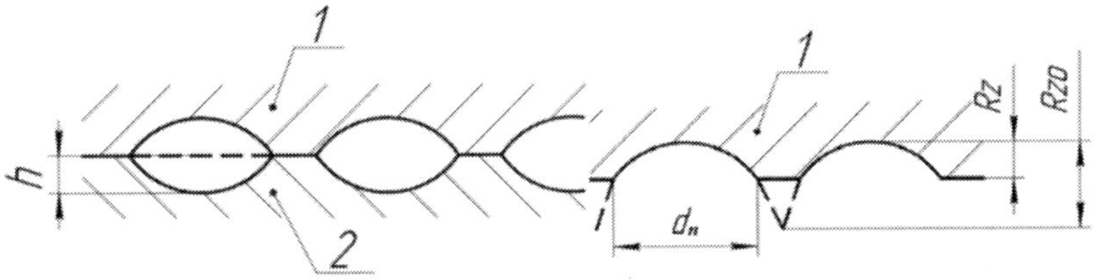

a b

a – position of mating parts 1 and 2; b – change of micro-hole shape at the expense of anodic dissolution of imperfection projection; h – depth of micro-hole; Rz – height of micro-imperfection after a set of erosion discharges; Rzo – height of imperfections at the end of the first stage of treatment; d_n – diameter of a hole.

Figure 3. Surface micro-geometry after first stage of treatment.

Some of them allow removing geometrical errors of a profile, other form micro-imperfections defining a surface smoothness (Rz in Figure 3). An electro-chemical constituent of the combined process contributes to a micro-profile smoothing and reduction of imperfections up to a nano-level (at the final stage of smoothing). As it is shown in [3], the correlation between a diameter (dh) and a depth of hole h depends upon modes of working. And at the same time, the depth (h) defines the reserve of the liquid working medium and the duration of anodic process passing controlled by the rate of mating parts travel.

At the second stage of finishing, only the anodic dissolution is used at a low alternating voltage from source 8 (Figure 2) with primary removal from the place of holes mating formed after the first stage of treatment (Figure 3). Then, there is a surface formed with the accuracy and macro-imperfections at a nano-level. A mechanical depassivation of the area of processing and the low voltage from a current source allows avoiding the intercrystalline etching of a working area and reduces friction [5] between contacting surfaces.

5. Substantiation for finishing modes

In [3], a dependence is offered for the estimation of an area diameter of the discharge effect which for electro-erosion finishing can be approximated by the diameter of hole d_n:

$$d_n = K_0 \cdot \left(\frac{A_u}{\tau_u} \right)^m \tau_u^n, \tag{1}$$

where

$$K_0 = \frac{L}{S(m - 0.5n) + 0.5n}. \tag{2}$$

Here L, K_o, m, n – empirical coefficients; L – measuring, mm; S – inter-electrode gap measurable with the thickness of an oxide film; τ_u – pulse duration; A_u – pulse energy. For the RC diagram:

$$A_u = \eta_1 \frac{CU^2}{2},$$

where η_1 – coefficient of energy efficiency (According to [3], $\eta_1 \leq 0.4$; according to [4], $\eta_1 \leq 0.2$); C – capacitor capacitance of pulse generator RC of the diagram; U – voltage on electrodes (at very small gaps U=10-20 V).

$$\tau_u = \frac{1}{qf}, \tag{3}$$

where q – on-off time ratio of pulses assumed as the ratio of a repetition period to their activity changes within the range of 2-10 (specified experimentally); f – frequency of pulse sequence. Approximately:

IOP Conf. Series: Materials Science and Engineering 177 (2017) 012132 doi:10.1088/1757-899X/177/1/012132

$$f = 0.837/RC,$$

where R – adjustable resistance of a RC -generator.

In Table 2, there are data for a choice of coefficients

Table 2. Numerical values of empirical coefficients.

Material of mating surfaces	Numerical values		
	L	m	n
Steel	0.08-0.1	0.75-0.78	
Copper alloys	0.036-0.04	0.65-0.68	0.28-0.3
Aluminum alloys	0.097-0.1	0.7-0.75	0.3-0.35
Cast iron	0.18-0.2	0.8-0.82	0.35-0.4

For the definition of pulse energy, it is necessary to determine a value of a minimum spark gap ('S' in (2)) which depends upon a thickness of an oxide film. According to [4], such films are formed under the impact of current and represent materials close to semiconductors. The film thickness (H) depends upon properties of work material, and in a classical comprehension it is expressed through the energy of discharge (A_u), Boltzmann constant R and temperature (T):

$$H = K_H \cdot e^{-Au/RT} \cdot \tau_u, \quad (4)$$

where K_H – dimension factor taking into account a rate of the film increase depending on material properties.

A limiting thickness of oxide films for nickel-containing alloys amounts to 15-30 μm according to [4], and in titanium materials - up to 50-60 μm. Hence, an inter-electrode gap (S) will be close to these values for obtaining a pulse breakdown voltage, even less than 10 V is enough. Oscillography has shown that with such voltages, the appearance of glow discharges is possible, which do not arouse a hole formation. Therefore, a mutual travel of mating surfaces under pressure allows thinning a film up to obtaining local areas with the conductivity during which micro-discharges can be formed [6;7].

The computation through formula (1) allows estimating the depth of a hole which, according to [3], equals:

$$h = K_h \cdot d_n, \quad (5)$$

where K_h – coefficient. According to [3], for steel K_h = 0.25-0.35. The value 'h' is limited with the allowance (Zo) for finishing:

$$h < Zo - \delta, \quad (6)$$

where δ– error of a surface after the first stage of finishing. According to [2], this value amounts to 3-4 μm. From this, it follows that the allowance for the final stage will be:

$$Zo = h + \delta,$$

and does not exceed 5-6 μm.

Through the depth of a hole, it is possible to determine value Rz (Figure 3) upon the dimension of which an anodic constituent of the process has an influence, and the appearance of nano- and micro-particles of the material was developed by an electroerosion method in the space between mating surfaces (Figure 4). The degree of their accumulation depends upon the area of mating areas, the speed of contact elements travel and the modes of erosion working.

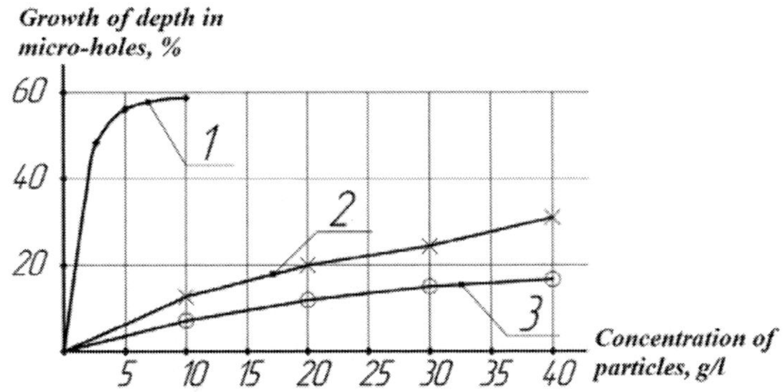

1 – aluminum alloys; 2 – copper alloys; 3 – steels.

Figure 4. Impact of products of electro-erosion machining upon changes in micro-hollows in surfaces under grinding.

The analysis of Figure 4 shows that for alloys based on iron and copper, the law of variation in a hole value is close to a linear one, and with the limiting concentration of particles, growth Rz can be considered by constant coefficients K_1 and K_2, taking into account particles conductivity and environment viscosity (for steels K_1 and K_2=1.15-1.16, for copper alloys – 1.25-1.3, aluminum alloys - 1.55-1.6). The existence of treatment products can change not only surface roughness, but also corrosion resistance of parts [9; 10].

The anodic dissolution (Figure 2) contributes to the smoothing of a microsurface in finishing modes of a combined process and allows decreasing an imperfection height by 20-30% which can be taken into account by coefficient K_2=0.7-0.8.

Value Rz_0 obtained without taking into account the influence of particles and an anodic constituent of the combined process (Figure 3) can be computed through the formula similar to that shown in [2;8]:

$$Rz_0 = \frac{\beta^2 \left(\frac{d_n}{2}\right)^3 K_h}{3 d_h^2 \left(K_h^2 + \frac{1}{2}\right)}, \tag{7}$$

where β – coefficient of hole covering.

$$\beta = \frac{d_n}{l},$$

where l – distance between neighboring holes. At the finishing stage of the treatment, β = 0.7-0.8.

Taking into account the impact of nano- and micro-particles and also an anodic constituent of the process, the height of micro-imperfections (Rz in Figure 3) can be estimated through the dependence:

$$Rz = Rz_0 \cdot K_1 \cdot K_2 = \frac{\beta^2 K_h d_n}{12\left(K_h^2 + \frac{1}{2}\right)} \cdot K_1 \cdot K_2. \tag{8}$$

From this it follows that it is possible to find a value of the capacitance of capacitors (C) during the machining with a RC- generator:

$$C = \left(\frac{R_z}{K_3 U}\right)^3, \tag{9}$$

where K_3 – coefficient, the values of which are shown in Table 3 and used for the combined treatment.

Table 3. Coefficients K_3 (μm/μF·V) for different materials.

Work material	Value K_3 at	
	Electro-erosion machining	Combined machining
Carbon steel	0.4-0.45	0.25-0.3
Alloyed steel	0.3-0.4	0.2-0.25
Copper alloys	0.7-0.8	0.3-0.4

A minimum value for the rate of mating surfaces travel (V in Figure 2) depends upon an initial profile of contact areas and grows according to imperfections removal as a contact area increases and products of machining begin to impede the deionization of an inter-electrode space. The size (L_d) of a contact surface along the direction of a travel depends upon a product design, and the rate of the accumulation of particles in an inter-electrode space - upon energy and frequency of pulses. With a certain reserve for a finishing electro-erosion constituent of machining, we can assume that with the limiting fill-up of volume V_n of a hole room (Figure 2) with products of machining, they must be removed with liquid working medium at the expense of a mutual movement of mating elements in a contact device.

If we assume a form of nano- and micro-particles as a spherical one, then during the filling of space Vn between electrodes with the concentration 'φ' time (τ_0) of a cycle of space purification:

$$\tau_0 = \frac{Vn \cdot \varphi}{(a+bC)UK_1K_2K_4}, \tag{10}$$

where a, b, K_4 - coefficients. Here, K_4 takes into account machining conditions (working medium, a treatment procedure, a sequence frequency of working pulses). φ -characterizes the degree of filling a hole with depth h by products of machining. For spherical particles, φ = 0.6-0.7. For RC, the procedure of machining numerical values a, b, K_4 are shown in Table 4:

Table 4. Values of coefficients a,b, K_4.

Work material	Values of coefficients		
	a	b	$K_4 \cdot 10^{-4}$
Carbon steel	0.23-0.25	0.7-0.72	6.7-6.8
Alloyed steel	0.2-0.22	0.65-0.7	5.8-5.9
Copper alloys	0.5-0.7	0.7-0.75	7.8-8.0

The period of polarity changing (τ_1) for an electro-erosion and electrochemical constituent of the combined process depends upon time of purification of an inter-electrode space and a height of imperfections of each mating surface before grinding. If these indices are close, then:

$$\tau_1 = \frac{1}{2}\tau_0 = \frac{Vn \cdot \varphi}{(a+bC)UK_1K_2K_4}. \tag{11}$$

So the rate of contact surfaces movement essential for the removal of machining products can be estimated through the dependence:

$$V \geq \frac{L_d(a+bC)UK_1K_2K_4}{Vn \cdot \varphi}. \tag{12}$$

The pressure of clamping (P in Figure 2) must be controlled according to the value of current between contact surfaces. For the stable procedure of an electroerosion treatment process, it is necessary to have an average current density (j) not lesser than 0.01 A/mm^2. In the area of contact surfaces (F_k), a control parameter (current I) makes:

$$I = j \cdot F_k. \tag{13}$$

For the electro-erosion and electro-chemical constituent of the combined process at the first stage of finishing, the polarity is equal. Here, an anodic impact is directed mainly to the decrease of a hole depth (hence, to the reduction of the height of micro-imperfections). At the expense of the adjustment of clamping pressure and the rate of the mutual movement of contact surfaces, the anodic dissolution of projections can be carried out at voltages of a low-voltage source (8 in Figure 2), 3-4 V, which earlier was considered to be unrealizable because of large impedances in an inter-electrode gap with current passing through an oxide film. The combined process offered at low voltage allows clearing short circuits and simplifying considerably the designs of plants for the precision finishing of contact devices.

The final stage of finishing is carried out through anodic dissolution with alternating current with voltage 3-4 V from source 8 (Figure 2). The dimensions of the particles of machining products in finishing modes do not exceed 100-1000 μm, and their capacity is 150-200 times larger than that of the metal removed.

If we assume that the volume of machining products (hydrates, gases and liquid vapors) in the space between holes during their confrontation is equal to 2 V_n, then before a final stage of finishing, the volume (V_M) of the metal removed from a contact surface close to a metal hole cannot exceed:

$$V_M = \frac{2V_n}{\varphi_1}, \quad (14)$$

where φ_1 - ratio of volumes of machining products in a hole to the volume of the metal removed close to a hole (φ_1 = 150-200).

In the process of micro-surface smoothing (Figure 3), the volumes of holes will decrease and the height of imperfections (Rz) will be reduced up to 20-30 μm (Table 1). It follows from (14) that the rate of anodic dissolution at the end of finishing will be measured by nano-meters in unit time, and a check parameter - duration of a process (τ_f) — for both mating surfaces will make:

$$\tau_f = \frac{\gamma_M \varphi_2 (Rz_H + \delta_H)[\varphi_2 (Rz_H + \delta_H) + 2Rz_H]}{\eta \alpha \chi (U - \Delta U)}, \quad (15)$$

where γ_M — density of work material; φ_2 — coefficient taking into account the rate of surface smoothing at an uneven allowance. For the final stage of finishing, φ_2 = 1.1-1.2. Rz_H - height of imperfections; after the first stage of finishing (according to Table 1, Rz_H = 0.8-3.2 μm); δ_H - error at the beginning of the second stage of finishing (0.1-1.0 μm according to Table 1); η - output on current. For very small gaps, according to [4], η is by 10-20% lower than it is shown in recommendations [3]; α - electro-chemical equivalent of work material (reference data, for example, of those shown in [3]); χ - specific conductivity of working medium. If working is carried out in industrial water, then electroconductivity may be chosen according to [3] for the concentration of salts of about 1%. U - voltage on electrodes. In the case under consideration, U = 3-4 V. ΔU - voltage losses (mainly, ohmic ones in an inter-electrode space). On the basis of [3], [4], at small gaps voltage, losses make ΔU = 1-2.2 V. The period of the final stage of finishing can make tens of minutes (Figure 5), whereas the duration of the first stage does not exceed 1-2 min.

During finishing on a mating part with a larger area, there can be formed a projection between an area of machining and an inoperative part of a surface with the value of up to 2-5 μm, which is commensurable with the roughness of a blank (3.2 μm), has a smooth transition (at the expense of the current dispersion during anodic dissolution of the boundary of a finishing area). A projection does not affect the capacity for work of locking equipment, but the narrow boundaries of tolerances (less than 1 μm in Table 1) at the final stage of machining can be a sign of a reject. Therefore, it is expedient under technical conditions regarding a mating unit to point at a permissibility of a local transfer area with a height drop within the limits of 5-10 μm.

6. Conclusion

1. The method for a nonabrasive finishing of mating surfaces in metal locking devices with the formation of contact areas having an error within the limits of 20-100 μm by means of two-stage combined machining is offered.

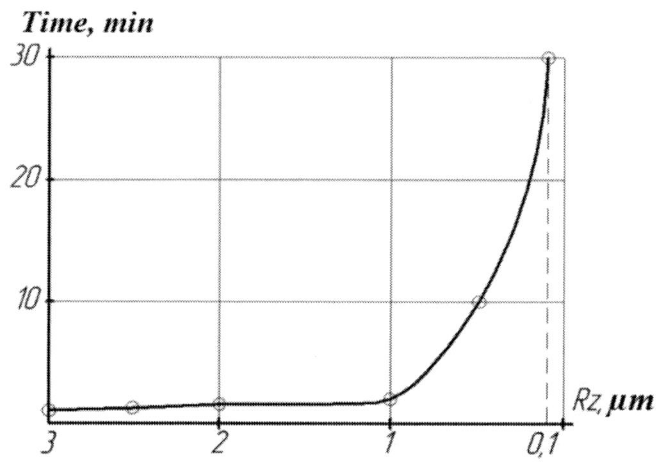

Figure 5. Labour-intensiveness at final stage of contact surfaces finishing (area 100 mm^2).

2. It is shown that at the first stage, without abrasive use, it is possible to obtain a micro-profile by means of an erosion-chemical method where geometrical deviations of mating surfaces are close to the height of microimperfections of the finishing stage in the mechanical refinement. It allows obtaining the imperfection height of a nano-value at the final stage of finishing by means of anodic dissolution which ensures the impermeability of locking equipment operation at high pressure differences, in hostile environment and with a high discharge of working environment.
3. There are developed operating practices taking into account local removals of the material with the allowance at the level of nano- and micro-values which allowed ensuring a required accuracy in mating surfaces without the use of abrasive environment causing a charging of contact areas in the area of finishing.

References

[1] Bondar A V 2002 Patent RF No 2183825
[2] Suslov A G 2005 *Mechanical Engineering. Encyclopedia* vol. III-3 (Moscow: Mashinostroenie)
[3] Smolentsev E V 2005 *The planning of electrical and combined processing methods* (Moscow: Mashinostroenie)
[4] Smolentsev G P and Koptev I T 2000 *Theory of Electro-chemical Working in Transient Mode* (Voronezh: VSTU)
[5] Chichinadze A V, Brown E D and Busheet N A 2001 *Fundamentals of Tribology (Friction, Wear, Lubrication)* (Moscow: Mashinostroenie)
[6] Klimenchenkov A A 2013 Non-abrasive Precision Machining of Mating Surfaces in Locking Devices *SSP-2013* pp 183-185
[7] Klimenchenkov A A and Sharov YuV 2013 *High Techn. in Mech. Eng.* **6** 33-36
[8] Klimenchenkov A A *Nano-engineering* **7** 8-15
[9] Fedonin O N 2012 *High Techn. in Mech. Eng.* **9** 15-17
[10] Baron Y M 1986 *Magnetic abrasive and magnetic processing of products and cutting tools* (Leningrad: Mashinostroenie)

The application of finite element analysis during development of the Integral Strain Gauges calibration method for the study of the welded construction

K V Syzrantseva[1], V N Syzrantsev[1] and M V Dvoynikov[2]

[1] Tyumen Industrial University, 38 Volodarskogo str., Tyumen, 625000, Russia
[2] National Mineral Resources University (Mining University), 2, 21st Line, St Petersburg, 199106, Russia

E-mail: kv.syzr@gmail.com

Abstract. Destruction of welds in metalwork is caused, in most cases, by low-cyclic fatigue. Among the wide spectrum of existing fatigue gauges, the Integral Strain Gauges are widely adopted in practice of experimental researches of different machine parts and carrying systems of vehicles. The paper proposes a new method based on finite element analysis of calibrating dependence establishment for ISG placed on the weld material by the use of few welded specimens of special geometrical shape providing the obtainment of several points of the calibrating curve in the tests process. Calibrating dependences allow determining the stress value σ_x in places of accumulated fatigue damages concentration which is used for serviceability estimation of welded joints by traditional techniques.

1. Introduction

It is known that destruction of welds in metalwork is caused, in most cases, by low-cyclic fatigue. Up to the present time, a wide range of the experimental methods allowing us to estimate the quality of welds is developed for prevention of emergencies [1-3]. However, these methods provide only detection of welds defects, without solution of the main problem: how detected defects will affect the fatigue durability of a construction. Among the wide spectrum of existing fatigue gauges [4-6], the Integral Strain Gauges [7-8] were widely adopted in practice of experimental researches of different machine parts and carrying systems of vehicles [8-9].

Operation, necessary for the use of ISG, is their calibration which is carried out by cyclic deforming of specimens with the pasted ISGs under the known stress-strain condition of specimens [8]. As a result of ISG calibrating, the dependence connecting the amplitude of the cyclic stresses (strains) with the number of loading cycles in accordance with one criterion of the ISG reaction (first 'dark spots' on the ISG surface, the relative area of 'dark spots' and others) is established. When using ISG for accumulated fatigue damages control of welds, it is necessary to get the calibrating dependences not only on the main material, but also on zone of thermal influence and on weld zone because the fatigue processes in these zones differ.

ISG calibrating can be carried out on standard plane specimens with a cross weld under conditions of cyclic tensile stress and compression. However, in this case, after testing of one specimen, only one experimental point of calibrating dependence will be established. And for obtaining the whole set of points, it is necessary to test several specimens under load varying.

This paper offers a new method of calibrating dependence establishment for ISG placed on the weld material by the use of few welded specimens of special geometrical shape providing the obtainment of several points of the calibrating curve in the tests process.

2. The task of the choice of an optimum geometrical shape of the specimen

At the first stage, the task of determination of an optimum geometrical shape of the test specimen is solved. Table 1 presents both the results of the stress-strain conditions calculation of different variants of specimens and the stress distribution diagram on the weld length.

In middle section (where the weld is placed) of the test specimen of traditional shape, the stress distribution diagram is close to uniform. To obtain the irregular distribution law of stress, the different concentrators were made on specimens 2 … 6.

Table 1. Influence of concentrators on the stress distribution law in the welded specimen

№	Stress von Mises distribution picture	Stress distribution diagram (σ_x) along the weld, MPa	Tensile force, kN
1	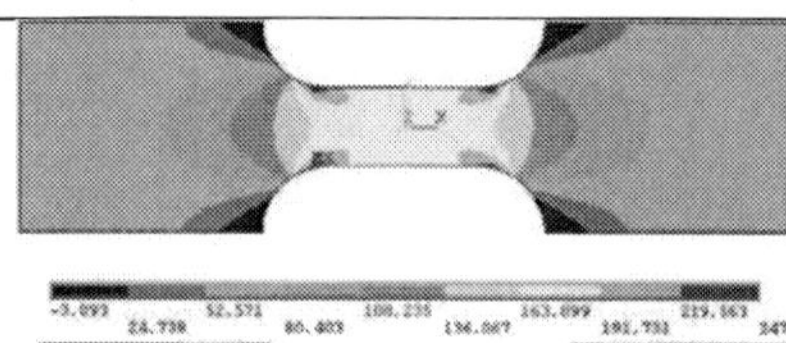		90
2	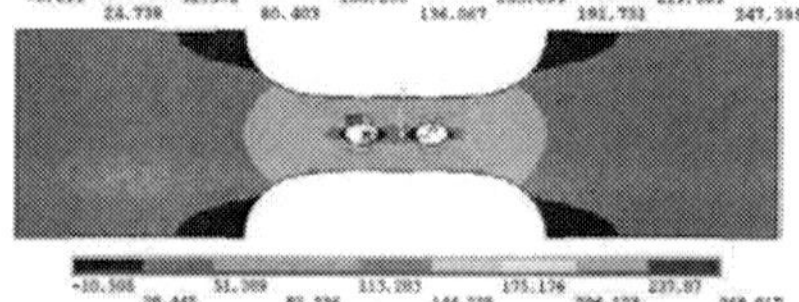		37
3	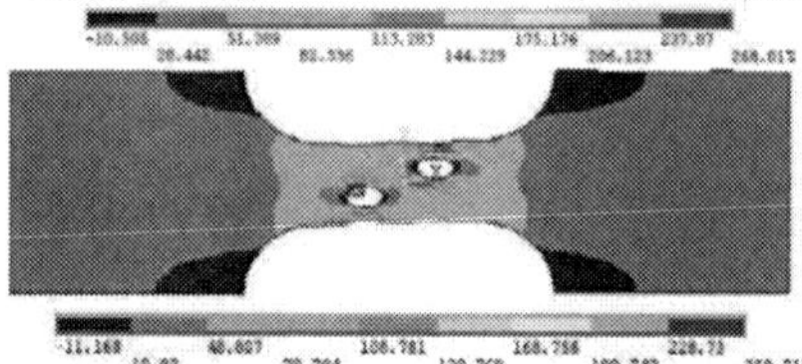		30
4	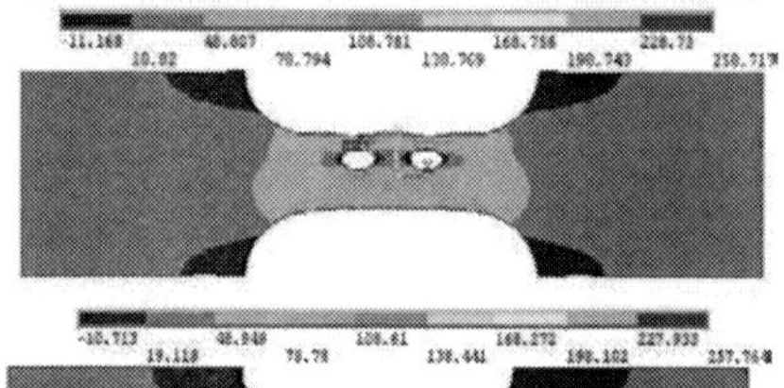		33
5	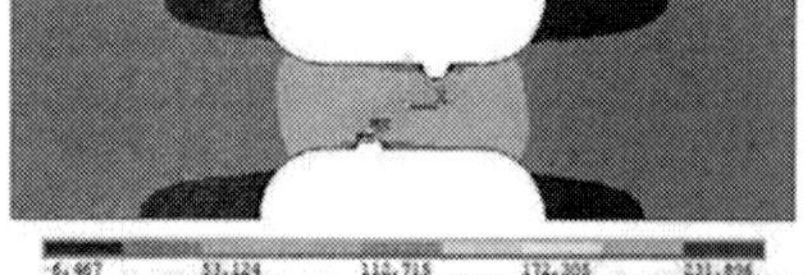		39
6	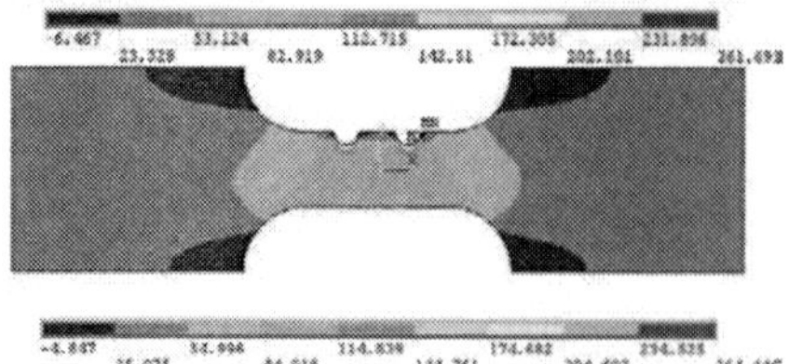		40

Analyzing results of stress calculation in the place of a weld presented in Table 1 , it is easy to see that concentrators significantly change the picture of stress distribution. Among researched specimens, specimen № 5 is of the greatest interest, because it is characterized by the optimum stress distribution law (close to parabolic) along the weld. The value of the maximum stress (88 MPa) exceeds the level of the minimum stress (39 MPa) approximately twice. It allows one to carry out the calibration of ISG in this stress range by using only one test specimen.

3. Analysis of the stress-strain condition of a three-dimensional welded specimen

At the second stage, computer simulation for the stress-strain condition calculation of the calibrating specimen of the geometrical shape chosen at the first stage (specimen №5) is carried out by Finite Element Method [1, 6, 10-11]. The finite element structured mesh [12] of the calibrating specimen is shown in Figure 1. The calibrating specimens material [9] was described by steel 14HGS, Russian State Standard 19282-73 (yield strength is 343.4 MPa, tensile strength is 490.5 MPa). For the analysis, the elastic material model was used. For the analysis of the influence of different boundary conditions on experiment results, the authors have considered 7 loading cases which are characterized by different boundary conditions (Table 2).

For quality estimation of the finite element mesh, the method described in works [8, 13] was used. Relative accuracy between nodal and element results was less than 7% that testifies to sufficient quality of the finite element mesh of the researched model.

The executed calculations have shown that the most rational loading cases for establishment of calibrating dependences for ISG are loading cases № 5, 6, 7 (Figures 3 and 4): tensile force varying at a fixed value of the vice lock on the pulsating stand. These test modes will allow carrying out the calibration of ISG in three ranges of stress variation by using only one calibrating specimen.

Figure 1. The finite element model of calibrating specimen №6

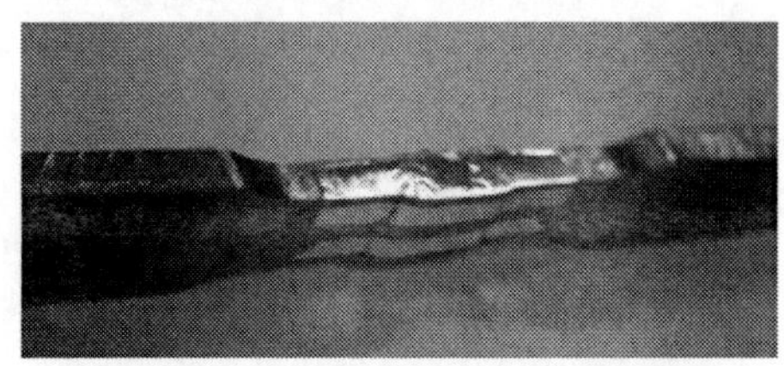

Figure 2. Calibrating specimen №1 with placed ISG

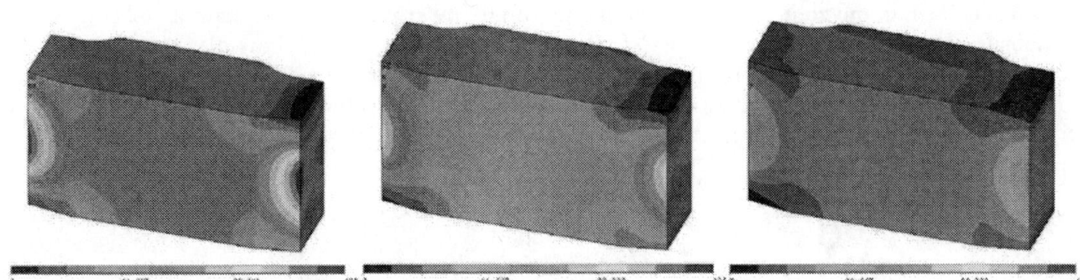

Figure 3. The stress von Mises distribution law in the weld material for loading cases №5, 6, 7

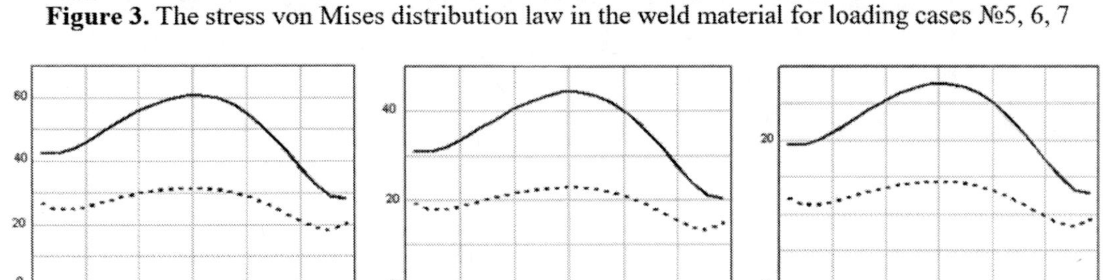

Figure 4. The stress von Mises distribution diagram along the strengthening edge of the weld

for loading cases № 5, 6, 7 (solid curve – weld upper surface, doted curve – weld lower surface)

Table 2. Boundary conditions for different loading cases of specimens

Loading case	Type of specimen	Tensile force F, N	Value of vice lock on pulsating stand L, mm
1	Welded	10000	40
2	Solid	10000	40
3	Welded	9700	10
4	Welded	13000	80
5	Welded	11000	50
6	Welded	8000	50
7	Welded	5000	50

4. The method of establishment of calibrating dependence

We will illustrate the realization of the developed method of calibrating dependence establishment for a welded specimen corresponding to the fixed number of loading cycles.

The welded specimens (Figure 2) made of a welded pipe were object of the experiment. The pipe material: 14HGS. The specimen on the pulsating stand [14] was subjected to deforming under the conditions of stretching at constant amplitude (F_{min} = 6 000 N, F_{max} = 35 000 N) during 80 000 cycles [9]. After N=20 000 cycles, the crack was registered and the experiment was interrupted. Using the photo camera, the digital pictures of the ISG reaction were obtained at different places of the specimen. Figure 5 illustrates the ISG reaction on the face surface of the welded specimen (a – the face crack, b - the upper weld in the weld zone with the main metal, c - the lower weld in the zone of thermal influence, d - the lower weld in the weld zone with the main metal). Figure 6 shows the ISG reaction on the lower surface of the weld (a - on the weld boundary, b, c, d – different zones of the main metal) [9]. The analysis of the achieved information gives evidence of the fact that the intensity of the ISG reaction in various zones of the weld differs significantly. The greatest ISG reaction was registered on a surface of the lower weld on the weld boundary (Figure 6a). The maximum ISG reaction testifies to stress (strains) concentration in the researched place and, accordingly, the greatest accumulated fatigue damage. Confirmation of the described fact is a fatigue crack initiation on the welded specimen in the zone of the weld boundary of the lower weld with the main metal (Figure 5a). The obtained photos of the ISG reaction (similar to the data of work [6]) show that stress distribution (accumulated during cyclic deforming of the specimen) on the surface of the welded specimen in its different zones (weld, zone of thermal influence, main metal, weld boundary) is very heterogeneous. Difficulty of serviceability forecasting of welded joints and constructions is explained by this circumstance.

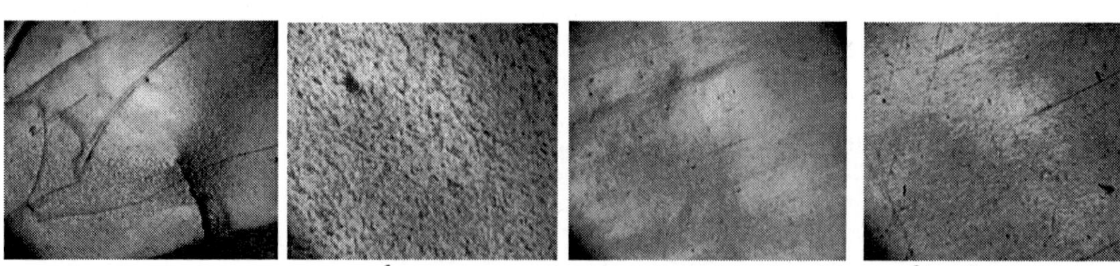

a *b* *c* *d*

Figure 5. The ISG reaction in different face regions of the welded specimen after 20000 cycles, ×200

For interpretation of the ISG reaction, the stress-strain condition estimation of the researched welded specimen was carried out by the finite element method during its stretching by axial force F =35000N. Values of the calculated stresses for 8 regions of the welded specimen are presented in the

IOP Conf. Series: Materials Science and Engineering **177** (2017) 012133 doi:10.1088/1757-899X/177/1/012133

last column of Table 3, and the results obtained during processing of photos of the same ISG regions by the histogram method with the help of the special software [8] are shown in the second and third columns of Table 3.

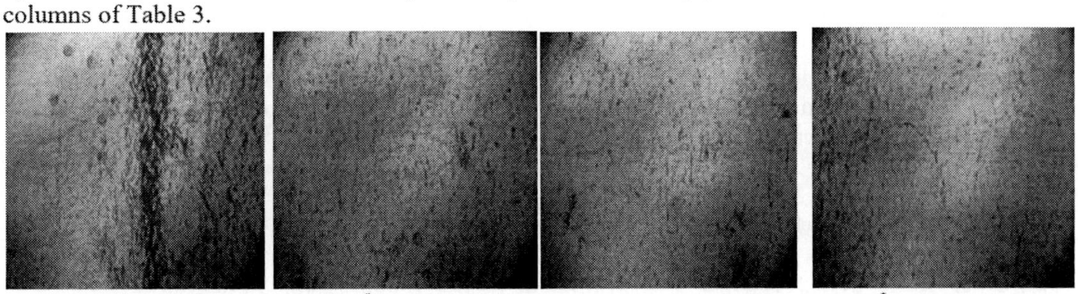

a *b* *c* *d*

Figure 6. The ISG reaction on the lower surface of the weld after 20000 cycles, ×200

Using the value of criteria (D and δ) of the ISG reaction estimation and calculated stress value, we shall obtain experimental points of calibrating dependence for the loading cycles number, N=20000, in coordinates D-σ_x (Figure 7) or δ-σ_x. After approximation of experimental data by the linear model, the required calibrating dependence can be presented as function:

$$D\,(\sigma_x, N{=}\text{const}) = 22.21\cdot\sigma_x - 4.16\cdot 10^3 \qquad (1)$$

or $$\delta(\sigma_x, N{=}\text{const}) = 1.539\cdot\sigma_x - 293.895\ . \qquad (2)$$

The use of dependences (1) and (2) for experimental researches of welded joints with the help of ISG consists in the following. ISG are placed on the surface of welds, on the weld boundary and on the main metal; the specimen is subjected to cyclic deforming during N=20000 cycles. After that, the regions of ISG with the greatest reaction are photographed and the obtained information is processed by the developed software [8].

Table 3. The results of computer processing of the ISG reaction

Photo of ISG reaction	Relative area of 'dark spots' δ, %	D-criteria (dispersion) of ISG reaction	Stress σ_x, MPa by FEM
Figure 5a	53.4	833	220.5
Figure 5b	29.0	517	212.33
Figure 5c	28.3	556	212.33
Figure 5d	27.4	464	212.33
Figure 6a	38.2	598	215.2
Figure 6b	13.2	327	199.67
Figure 6c	15.5	332	199.67
Figure 6d	16.9	282	199.67

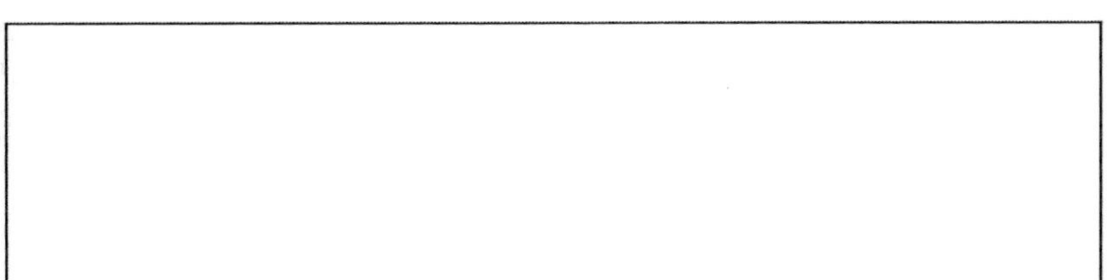

Figure 7. Experimental points and calibrating dependence on the base of the linear model in accordance with D-criteria of the ISG reaction.

5. Conclusion

Values D or δ obtained as a result of processing of calibrating dependences (1) or (2) allow determining the stress value σ_x in places of accumulated fatigue damages concentration which is used for serviceability estimation of welded joints by traditional techniques. It is easy to see that similar calibrating dependences can be established for any other numbers of loading cycles.

References

[1] Nadimi S, Khoushehmehr R J, Rohani B and Mostafapour A 2008 *J. Appl. Sci.* **8** 1014-20
[2] Joy Varghese V M, Sumanth M R, Suresh M R 2012 *Proc. Eng.* **38** 2913-21
[3] Ni Y-Q, Xia Y-X 2016 *Int. J. of Struct. Stability and Dynamics* **16**(4) 1640027
[4] Krishnamurthy S, Badcock R A, Machavaram V R, Fernando G F 2016 *Sensors* **16**(6) 777
[5] Černý I 2014 *51st Annual Int. Sc. Conf. on Experimental Stress Analysis EAN 2013* (Czech Republic) vol 486 58-61
[6] Fried M and others 2014 *2nd European Symp. on Superalloys and Their Appl. EUROSUPERALLOYS 2014* (France) vol 14 16001
[7] Syzrantsev V N, RU Patent 2,209,412 (2003)
[8] Syzrantsev V N, Golofast S L, Syzrantseva K V 2004 *Diagnostics of loading and resource of parts of machines transmissions and carrying systems according to integral strain gauges indications* (Novosibirsk: Nauka)
[9] Syzrantsev V, Syzrantseva K 2016 *IOP Conf. Series: Mat. Sci. and Eng.* vol 127 (IOP Publishing) 012051
[10] Tinsley Oden J 2010 *Finite element method* (Scholarpedia) vol 5(5) 9836
[11] Wittbrodt E, Szczotka M, Maczy'nski A, Wojciech S 2013 *Rigid finite element method in analysis of dynamics of offshore structures* (Springer, Berlin)
[12] Amir Javidinejad 2012 *Mechanical Engineering* 168941
[13] Liu Y and Glass G 2013 Effects of mesh density on finite element analysis (SAE Technical Paper) 2013-01-1375
[14] Fowler S, Toumpis A and Galloway A 2016 *Int. J. Adv. Manuf. Tech.* **84**(9) 2659-69

The study of the chemical and phase composition of the diffusion interaction zone in layered composite Cr20Ni80-AD1

V G Shmorgun[1], L D Iskhakova[2], A I Bogdanov[1], A O Taube[1] and R P Ermakov[2]

[1] Volgograd State Technical University, 28, Lenina ave., Volgograd, 400005, Russia
[2] Fiber Optics Research Center RAS, 38, Vavilova str., Moscow, 119333, Russia

E-mail: mv@vstu.ru

Abstract. The paper presents the research of chemical and phase composition of the zone of diffusion interaction in the layered composite Cr20Ni80-AD1. It is shown that interlayer border of bimetal Cr20Ni80-AD1 is a complex mixture of phases with the main phases in its composition in solid-phase interaction: Al_3Ni_2 and $Al_{14.687}Cr_{3.443}Ni_{0.87}$, and in the liquid phase - Al_3Ni_2, $Al_{0.987}Cr_{0.017}$, $Al_{14.687}Cr_{3.44}Ni_{0.87}$ and $Al_{45}Cr_7$. It is shown, that conditions of holding time of the heat treatment do not affect the phase and chemical composition of the diffusion zone.

1. Introduction

Papers [1, 2] describe the technological process of production of heat-resistant laminated coating system Al-Cr-Ni, including the explosion of the welding operation, the pressure treatment and the heat treatment (HT). The basis (semi-finished) for the preparation of these coatings are explosion welded layered composite materials with composition alloy Cr20Ni80 - aluminum AD1. Since the phase and chemical composition of the laminate coating, which determines its durability, is formed at the stage of diffusion annealing of composite Cr20Ni80 - AD1, the aim of this work was to study the effect of temperature and time on the phase and chemical composition of the zone of diffusion interaction (ZDI), formed on the interlayer boundary of explosion welded bimetal Cr20Ni80-AD1.

2. Materials and methods

The object of study - bimetal Cr20Ni80-AD1 obtained by explosion welding in the optimal mode, guarantees a minimum level of structural and chemical micro-inhomogeneity in the heat-affected zone: no voids, lack of fusion, fused metal portions.

HT of bimetal was carried out in two modes:
- solid phase - at 600 °C, not exceeding the melting point of aluminum (660 °C), for 1 to 50 h;
- the liquid phase at temperatures of 660 and 700 °C, above the melting point of aluminum, for 1-10 h.

HT carried out in the presence of a liquid phase allows due to the intensification of diffusion processes to form the greater thickness of ZDI in less time. ZDI thickness formed on both the first and the second modes was 120-150 μm.

The task of determining the phase and chemical composition of the ZDI was solved by a set of methods for electron microscopy, energy dispersive analysis (EDA) and X-ray diffraction (XRD).
The microstructure of the samples was investigated with a scanning electron microscope (JSM-

5910LV, JEOL) by the reflected electrons in the Z-contrast mode, in which brightness increases with the phase average atomic numbers on the analyzed area. Elemental analysis was carried out by energy dispersive analysis using analytical system AZtecENERGY, Oxford Instruments.

X-ray studies were performed using a powder X-ray diffractometer 'Bruker D2 Phaser' (CuK_α = 1.54056Å). The recording was made in the plane of the sample with the guidance of the X-ray beam on the surface of the ZDI and on the ZDI surface after stratification of bimetal. Processing of the results was performed using the TOPAS 4.2 software.

3. The results and discussion

The results of X-ray analysis in ZDI in unfibered samples after HT, providing solid-phase interaction, are difficult to interpret uniquely. Only two main phases have been discovered because of the small thickness of the ZDI: aluminum AD1 and phase with the Ni3Cr type structure - alloy Cr20Ni80 (figure 1).

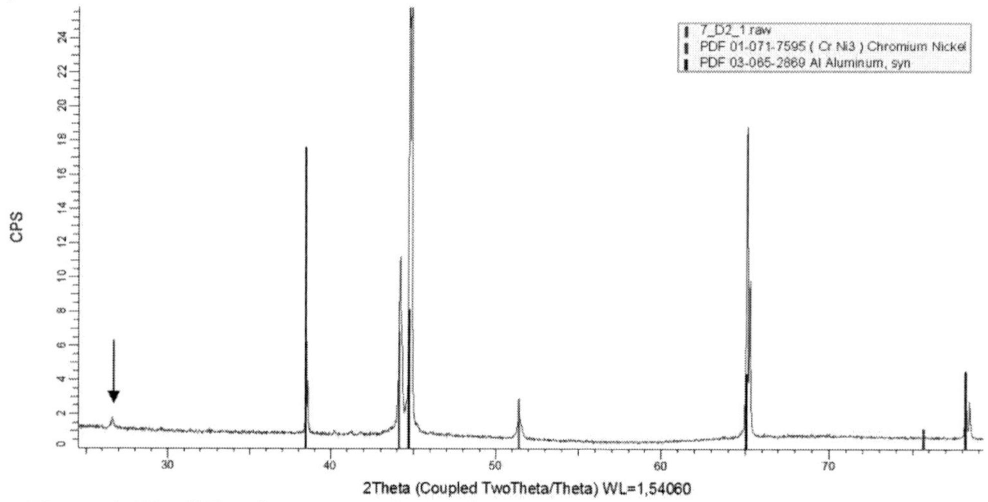

Figure 1. The diffraction pattern of ZDI after HT at 600 °C for 10 h (unfibered sample). The arrow marked reflex with the unidentified phase.

In this connection, the X-ray analysis was made for stratified samples. As a result, diffraction patterns were taken from the surface of the ZDI, remaining on Cr20Ni80 alloy (figure 2a) and aluminum AD1 (figure 2b).

As we can see from figure 2, the main phase in the ZDI is Al3Ni2. Radiometric characteristics of other phases identified therein are given in Table 1.

Table 1. Radiometric characteristics of phases in ZDI

ICDD card number	Chemical composition	Space group	The lattice parameters, Å
03-065-3454	Al_3Ni_2	P-3m1	a = 4.0363; c = 4.9004
00-058-0277	$Al_{0.64}Cr_{0.36}$	R3m	a = 12.77; c = 7.958
01-074-5256	$Al_{0.14}Ni_{0.86}$	Fm-3m	a = 3.5486
01-073-9456	$Al_{14.687}Cr_{3.443}Ni_{0.87}$	P63	a = 17.674; c = 12.516

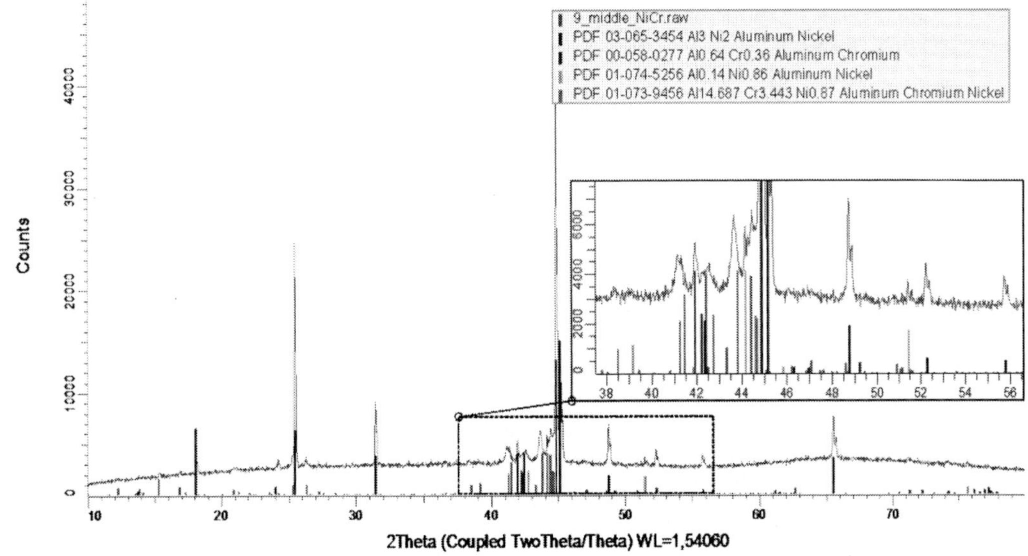

a

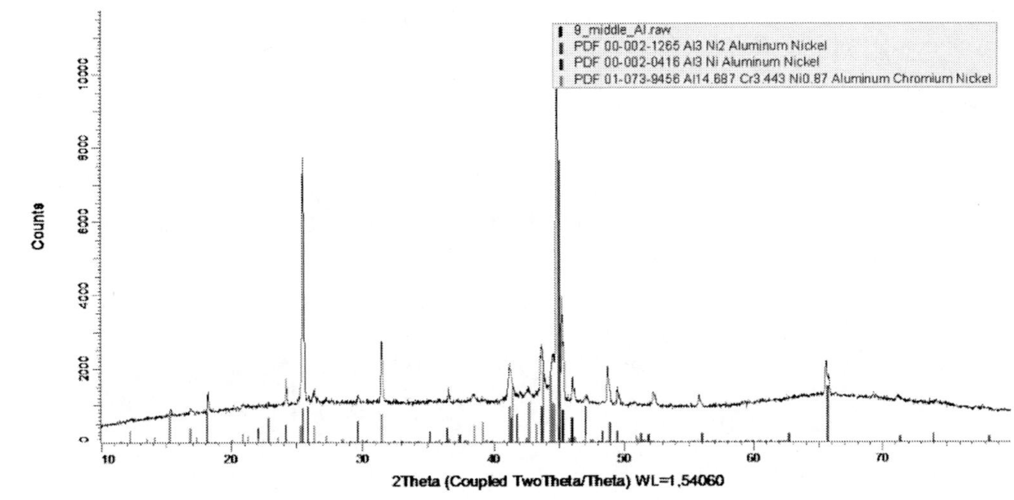

b

Figure 2. The diffraction patterns taken from the surface of the ZDI, remaining on Cr20Ni80 (a) and AD1 (b) after the HT at 600 °C for 10 h (stratified sample).

During the EDA, statistic data on the chemical composition of areas were recruited (at least five measurements) with the same microstructure and brightness on micro-images, which also showed the proximity of the elemental composition of the areas. These data are difficult to read and they are not shown not to overload the text.

The complex of received X-ray and EDA data allowed us to conclude that the main phases found in ZDI, formed at 600 °C (regardless of the holding time) are Al_3Ni_2 phase and $Al_{14.687}Cr_{3.443}Ni_{0.87}$. Phase Al_3Ni probably exists on the border between ZDI and AD1 and the $Al_{0.64}Cr_{0.36}$ phase is formed at the boundary of the ZDI and Cr20Ni80.

HT, carried out in the presence of a liquid phase, leads to a greater enrichment ZDI by chromium, which is confirmed by the presence on the diffraction patterns of additional reflections from $Al_{0,987}Cr_{0,017}$ type phase and θ-phase - $Al_{45}Cr_7$ (figure 3).

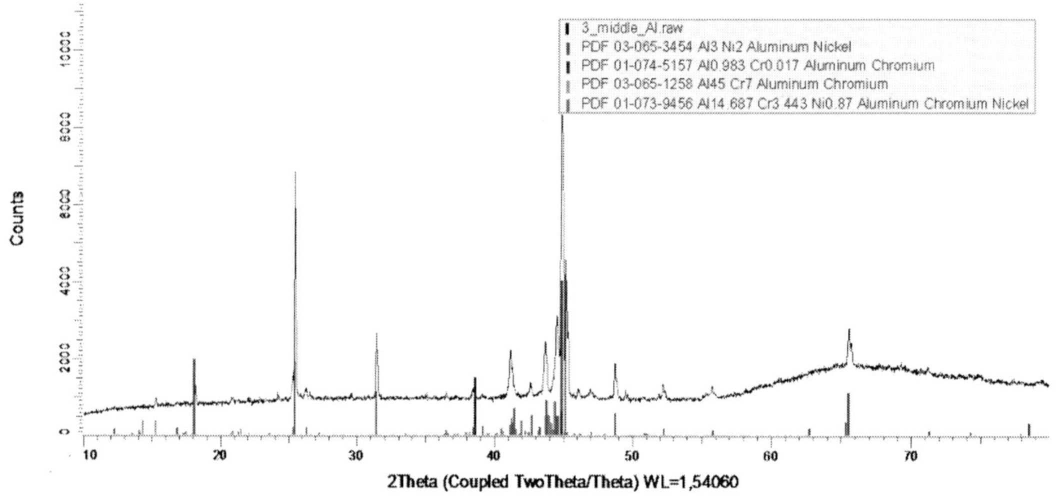

Figure 3. The diffraction pattern taken from the surface of the ZDI, remaining on the aluminum AD1 layer after HT at 700 °C for 10 h (stratified sample).

As we can see from figure 3, ZDI contains phases with the structures of Al_3Ni_2, $Al_{0.987}Cr_{0.017}$, $Al_{14.687}Cr_{3.44}Ni_{0.87}$ and θ-phase $Al_{45}Cr_7$.

As a result of EDA, in the chemical composition of ZDI, a significant change in the ratio of Al/(Cr+Ni) and Ni/Cr was recorded. For example, at various analysis points within the region marked by a square in figure 4, the ratio of Al/(Cr+Ni) varied from 0.78 to 2.63, and Ni/Cr - from 0.87 to 3.32, which indicates, that the phase and chemical composition of the ZDI are complicated.

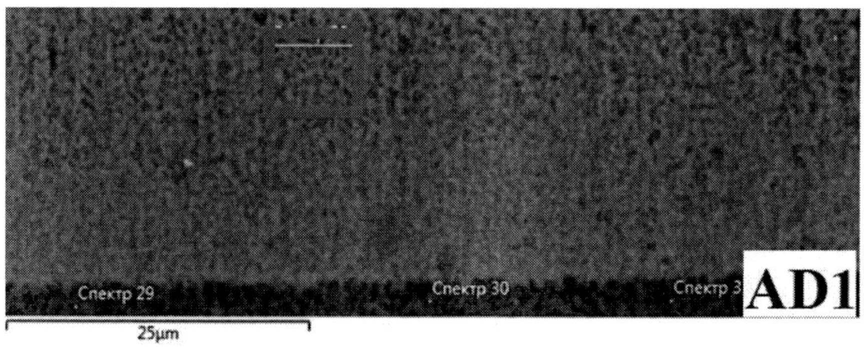

Figure 4. The microstructure (in the Z-contrast) of ZDI, remainig on the surface of the AD1 layer after the sample stratification, heat-treated at 700 °C for 10 h. The EDA area is marked by a square.

In conclusion, it should be noticed that most of diffraction patterns in these studies can not be uniquely interpretated due to several objective reasons:

- a series of intermetallic phases in the studied systems has similar diffraction patterns, resulting in the imposition of appropriate reflexes;
- a number of weak reflections exist on diffraction patterns, and it is not possible to identify them because of their low intensity;

- analysis of published data indicates the absence of full clarity in the composition of intermetallic compounds based on aluminum, chromium, and nickel, as different phases of the composition have the same radiometric characteristics.

4. Conclusion

1. The zone of diffusion interaction on the interlayer border of bimetal Cr20Ni80-AD1 is a complex mixture of phases with a significant change in the ratio of Al/(Cr+Ni) and Ni/Cr. The main identified phases in its composition in solid-phase interaction are Al_3Ni_2 and $Al_{14.687}Cr_{3.443}Ni_{0.87}$, and in the liquid phase - Al_3Ni_2, $Al_{0.987}Cr_{0.017}$, $Al_{14.687}Cr_{3.44}Ni_{0.87}$ и $Al_{45}Cr_7$.
2. Holding time conditions of the diffusion annealing do not affect the phase and chemical composition of the zone of diffusion interaction.

Acknowledgments
The investigation was performed based on the grant from the Russian Science Foundation (project No. 14-29-00158).

References

[1] Shmorgun V G, Bogdanov A I, Taube A O 2014 *Proc. of the Univ. Ferrous Metallurgy* **5** 64-65
[2] Shmorgun V G, Bogdanov A I, Taube A O and Serov A G 2016 *Proc. of the Univ. Powder Metall. and Funct. Coat.* **1** 51-59
[3] Swain M, Singh S, Basu S, Bhattacharya D, Gupta M 2014 *J. Appl. Phys.* **116** 222208
[4] Okamoto H 2004 *J. of Phase Equilibria and Diffusion* **25(4)** 394-399
[5] Dupin N, Ansara I, Sundman B 2001 *Calphad* **25** 279–298
[6] Chen H-L, Doernberg E, Svoboda P, Schmid-Fetzer R 2011 *Thermochem. Acta.* **512** 189–195
[7] Grushko B, Kowalski W, Pavluchkov D, Mi S, Surowiec M 2009 *J. Alloys&Comp.* **485** 132
[8] Progrushchenko A V, Lebedev Yu M 1969 *Ukrainian Journal of Physics* **14(2)** 282-286
[9] Hara Y, O'Handley R C, Grant N J 1986 *J. Magnetism & Magn. Mat.* **54** 1077-1078
[10] Mohan Rao P V 1992 *Phys. Stat. Sol. A.* **133(2)** 231-235
[11] Mohan Rao P V 1993 *J. Appl. Crystallography* **26(5)** 670-676

Analysis of cavitation on the surface of steel under the ultrasonic cleaning

I K Tsybry and I L Vyalikov

Don State Technical University, 1, Gagarin sq., Rostov-on-Don, 644000, Russia

E-mail: vialikov@mail.ru

Abstract. The article presents: the results of studies on the effects of cavitation on the surface of steel during processing in a liquid medium; the analysis of changes of the microstructure and the microhardness of steel under the effect of cavitation; the features of hardening and destruction of the surface layers of steel with a structure of granular pearlite and the martensitic steel structure. Comparative analysis of the structural transformations of the surface layers of steels 12H17 and U8 allows us to select the optimum time of ultrasonic treatment for a complete cleaning of the surface without destroying it.

1. Introduction

It is known that ultrasonic treatment in a fluid is widely used for cleaning metal surfaces from technological, biological and other types of contamination. Under the influence of ultrasonic cavitation in fluid, the plastic deformation processes are proceeding in the surface layers of the metal [1-4]. The intensity of these processes is determined by the structural features of the processed material [5-8]. Therefore, the efficiency of cleaning by using ultrasonic vibrations is determined by the choice of rational conditions under which the maximum degree of purification with a minimum negative impact of cavitation may be provided on the surface. This article presents a comparative analysis of the structural transformations of the surface layers of steels 12H17 and U8 under the process of ultrasonic treatment in the mode of developed cavitation.

2. Materials and methods

For research of cavitation resistance of steel under the process of ultrasonic treatment, the following steels were used: a 12H17 steel which structure is spheroidized carbides in the chromium alloyed ferritic matrix and a U8 steel with the martensite structure and a residual austenite content of 5-8 %.

After grinding, polishing, pickling by standard methods for carbon and alloy steel and application of the model pollution (preservative oil), the investigated samples were subjected to ultrasonic treatment in distilled water. To eliminate corrosion of the surface, sodium nitrite was added as a passivator. The intensity of the ultrasonic fluctuations of 57 kHz was chosen based on the possibility of obtaining a pre-cavitation mode (0.5 W/cm^2), and the developed cavitation mode (1.5 W/cm^2).

The microstructure of the surface layers was investigated under the microscope 'Neophot-21', the micro-hardness was measured with hardness tester PMT-3M.

The sequence of the operation in accordance with [9, 10] is the following: the intensity of the ultrasound of 0.5 W/cm^2 is set; a prepared sample is immersed in the ultrasonic bath and is held for 10 minutes; the microhardness of the treated surface is measured, and its microstructure is recorded.

Then, the process is repeated as long as no obvious signs of surface fracture are specified; similarly, the processing of the subsequent samples at 1.5 W/cm^2 ultrasound intensity is conducted.

3. Discussion of the experimental results

Changing of the microstructure of the surface of steels 12H17 and U8 under the influence of the developed cavitation is shown in figures 1 and 2, respectively. Microstructures in figures 1a and 2a are obtained after the ultrasonic cleaning during the time required to remove a layer of pollution. The microstructures in figures 1b and 2b correspond to the processing time, when signs of stable cavitation effects on the metal surface become visible. The microstructures in figures 1c and 2c illustrate the beginning of the surface fracture.

The results of measuring of microhardness of the treated surface of the sections in the mode of the developed cavitation are shown in table 1, and the change of the microhardness of the frontal section of the samples at maximum hardness on the surface – in table 2.

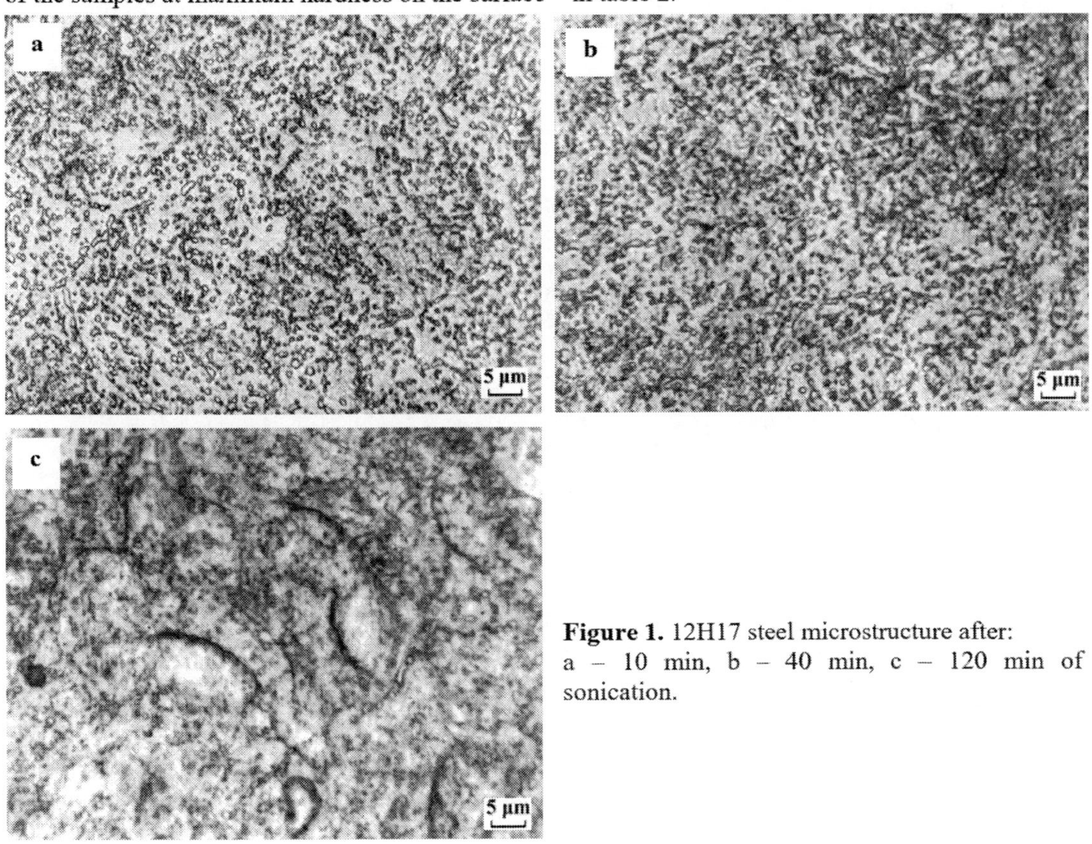

Figure 1. 12H17 steel microstructure after: a – 10 min, b – 40 min, c – 120 min of sonication.

The experimental results showed that during the ultrasonic treatment in the pre-cavitation mode, no changes in the microstructure of the surface of both investigated steels were revealed. The level of the micro-hardness of steel U8 also remained constant. The microhardness of steel 12H17 was slightly increased after 140 minutes of sonication. The main process determining destruction of the material in the process of cavitation is the mechanical action of pulsating vapor-gas bubbles in the liquid, during the closing, the surface in contact with the liquid receives the hydrodynamic shocks. The pressure at the impact point can reach 126-250 MPa and the temperature – 230-720 °C.

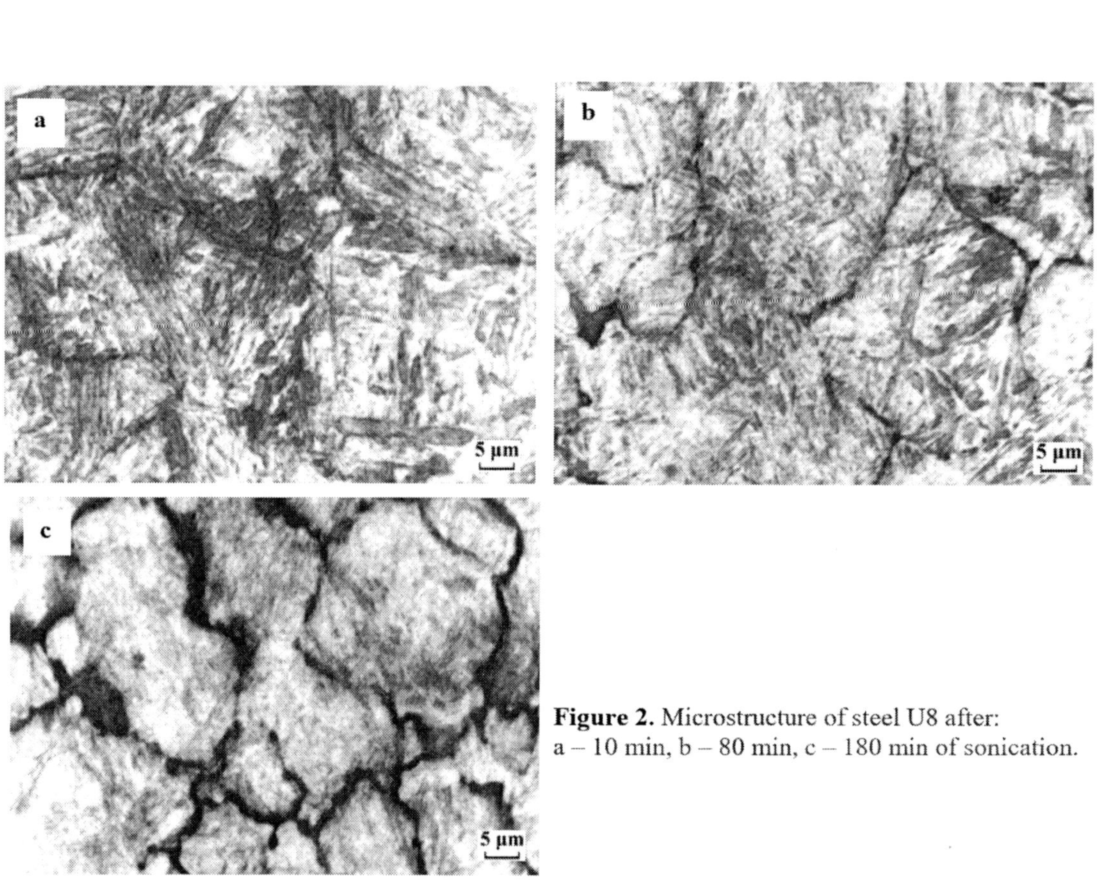

Figure 2. Microstructure of steel U8 after: a – 10 min, b – 80 min, c – 180 min of sonication.

The developed cavitation mode for both steels is characterized by the following periods:
– an incubation period, including cleaning the surface from contamination and the time during which cavitation has no apparent effect on the surface state;
– a period of strain hardening is accompanied by an increased microhardness and transformation of the surface structure (twinning, grain crush to subgrains, the structural transformations, etc.);
– a period of the softening and destruction of the surface layers of the processed material.

The duration of each period, and the intensity of the processes of hardening and destruction is determined by the structure of the investigated steels.

Thus, in the process of cavitation impact on steel 12H17, the incubation period is 35-40 min. Then, a period of plastic deformation of ferrite starts, accompanied by the display of a relief on the surface, broadening the interface between the ferrite matrix and carbides, the display of the boundaries of ferritic grains, and the micro-hardness increased by 25 % compared with the original one. Microhardness in the frontal section of the samples at a maximum degree of the surface hardening, decreases exponentially, reaching the original value at about 75 µm.

The hardening effect of the ferrite matrix is caused by formation of the subgrain structure, increasing of the dislocation solidity and complexity of the dislocation structure under the influence of the cavitating liquid bubbles.

Table 1. Changes in microhardness of surface of 12H17 and U8 steels in mode of developed cavitation.

Processing time, min	Microhardness of surface of steel 12H17, MPa	Microhardness of surface of steel U8, MPa
0	1810	8400
20	1810	8400
40	1930	8400
60	2170	8410
80	2430	8430
100	2470	8450
120	2330	8520
140	2250	8500
160	-	8430
180	-	8390

Table 2. Measurement of microhardness in face section of samples.

Microhardness, MPa	Distance from surface, μm			
	10	25	50	75
12H17 [9]	2460	2070	1900	1800
U8	8500	8400	8400	8400

The subsequent period of weakening and fracture of the surface starts after two hours of the cavitation impact after a fall of micro-hardness, the formation of deep dents in the thin section as a consequence of the plastic flow of the ferrite base, and in this process, the carbide grains are involved. There are also signs of erosion of the surface areas.

U8 steel is characterized by a long incubation period, due to the high initial strength and hardness of martensite. Microhardness in the samples' frontal section at the maximum degree of surface hardening remains practically unchanged. A slight increase in the surface microhardness of 3-4 % of the initial value starts 60 minutes after, and the visible traces of plastic deformation of the martensite start to appear in the microstructure after 80 minutes of cavitation. On microsections, the reorientation of martensitic needles and a blur relief, the emergence of micro-cracks at the boundaries of the prior austenite grains are observed. Hardening effect of the martensite can be due to an increase of the dislocation density or collapse of a residual austenite with formation of a martensite deformation.

The subsequent weakening and destruction along the boundaries of prior austenite grains appear under the mechanism of brittle fracture.

4. Conclusion

The experiments and analysis of their results suggest the following conclusions:

– cleaning efficiency with application of ultrasonic vibrations is dependent on the intensity and the time of action of the ultrasonic vibration;

– 12H17 and U8 steel processed with ultrasound in the pre-cavitation mode during experiment practically did not change the state of the samples' surface;

– at the same, intensity of ultrasonic vibrations in the mode of developed cavitation, the cavitation resistance of the material is determined by its structure;

– the deformation hardening of the ferrite of steel 12H17 peaks in 100 minutes of the treatment and is accompanied by an increase of the microhardness of 25% compared with the initial following gliding fracture of the surface;

– ultrasonic treatment of steel U8 with the structure of the martensite showed that long time cavitation had no effect on the state of the surface layer, a slight increase of the microhardness in the surface layers and subsequent brittle fracture of the surface are observed;

– comparative analysis of the structural transformations of the surface layers of steels 12H17 and U8 allows one to select the optimum time of ultrasonic treatment for a complete cleaning of the surface without destroying it, and if necessary, to set the optimal mode of steel 12H17 hardening .

Acknowledgments

The article is supported by the Russian Fund for Basic Research (grant №16-38-00276) 'Obtaining of wear-resistant and corrosion-resistant coatings based on energy managing model of mechanochemical synthesis process of their formation uder conditions of the shock-pulse, vibrowave impact of free moving indenters'.

References

[1] Hashish M 1991 *J. of Eng. Mat. and Techn.* **3** 354-358
[2] Buravova S N 1998 *Technical Physics* **68(9)** 110-114
[3] Karakulin E A 2010 *Mathematical Models and Computer Simulations* **6(2)** 790-799
[4] Sirotyuk M G 2008 *Acoustic Cavitation* (Moscow: Nauka)
[5] Chudina O V, Aleksandrov V A, Fatyuhin D S 2011 Hardening Techn. and Coating 2 3-6
[6] Khachaturyan V 2014 Effect of ultrasonic treatment steel surface. Avtomobil'. Doroga. Infrastruktura **2**(2)
[7] Alexandrov V A, Fatyuhin D S 2012 *Deformation and Fracture of Mat.* **10** 38-40
[8] Fedorov A V, Rymkevich A I, Davydova N V, Zubchenko A S and Bazhenov V V 2014 *Welding Production* **11** 31-39
[9] Tsybry I K Vyahireva E V and Kuharenko I S 2010 *Vestnik Don State Techn. Univ.* **6(49)** 90-96
[10] Tamarkin M A, Tishchenko E E and Vyalikov I L 2015 *Russian Eng. Res.* **35(10)** 740-744

The study of oscillations excitation patterns in the process of milling with portable equipment

A M Kozlov[1], E V Kiryuschenko[2] and A V Khandozhko[3]

[1] Lipetsk State Technical University, 30, Moskovskaya str., Lipetsk, 398600, Russia
[2] Novolipetsk Metallurgical Company, 2, Metallurg sq., Lipetsk, 398040, Russia
[3] Bryansk state technical University, 7, 50-let Oktyabrya bul., Bryansk 241035, Russia.

E-mail: kam-48@yandex.ru

Abstract. This paper examines the mechanism of excitation of vibrations in the milling process with low-rigid metal-cutting equipment. The results of theoretical and practical experiments are presented. The mechanism of self-adjustment of the technological system and the possibility of damping of its oscillations, not associated with increased stiffness of equipment or tool, are examined.

1. Introduction

With the development of mechanical engineering, the use of compact metal-cutting equipment becomes the more urgent task. Its most common representatives are compact multi-axis milling machines (they are also called 3D-milling) and portable milling equipment for the processing of large parts (when the machine is installed on a part). The peculiarity of the use of such equipment is the presence of the low-rigid technological system.

The well-known fact that during milling, the mechanism of occurrence of oscillations is quite complicated due to the intermittent nature of the cutting process, and the major cause of their increase is a regenerative excitement of the track on the cutting surface. The vibrations arising in the process of milling, are a powerful deterrent to increase its productivity. Therefore, the special attention is paid to the study of the mechanism of this type of excitation, as well as the possible impact on it.

2. Theoretical study

The system for theoretical studies of the mechanism of the regenerative vibration excitation includes one reduced mass having one degree of freedom in the direction perpendicular to the cutting speed (Figure 1). For it, the parameters of the tool subsystem of the milling machine in the direction of the feed axis were adopted, as this subsystem is most exposed to the trace of the cut surface [1, 2].

The variability of the slice thickness due to ongoing vibration movements and vibrations of the track is taken as the main reason of the oscillations excitation of the adopted system. In order not to take into account changing of the slice thickness versus cutter turning during the rotation (because they do not influence the regenerative process of excitation of oscillations), the treatment process was 'expanded' in linear: the cutting surface in its nominal position (without wave-shaped track) was adopted as flat and continuous, the movement of the teeth relative to it as straight and forward. It was also accepted that teeth are located in parallel and secured in a common unit (as in the gang processing). It has been accepted that the inertial properties of the oscillating system are concentrated in the cutter block, and they are represented in the diagram by weight *m*.

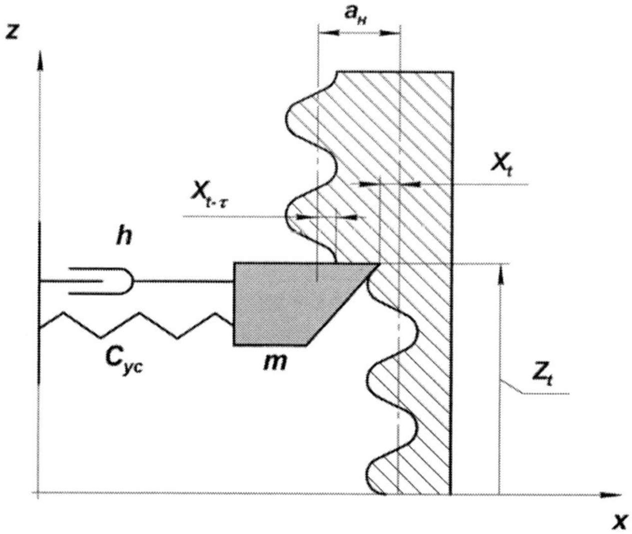

Figure 1. The oscillatory system scheme: X_t – mass displacement at a given time, m; $X_{t-\tau}$ – track oscillation on the surface of the cutting, m; C_{yc} – stiffness of the oscillating system, N/m; h – damping of oscillating system N·s/m; $a_н$ – nominal slice thickness, m; Z_t – coordinate characterizing the position of the tooth along axis z, m.

Teeth are displaced along the x-axis relative to each other by the amount equal to the slice thickness, and along the z-axis – by the amount equal to circumferential distance between them in the tool. Impact on the system formed by the tooth is determined by the thickness of the shear layer by the hardness of the cutting:

$$F_t = K_{pt} \cdot a = K_{pt} \cdot \left(a_н + X_t - X_{t-\tau}\right), \quad (1)$$

K_{pt} – hardness of the cutting in the direction of main movement, N/m; a – instantaneous slice thickness, m; F_t – tangent component of the cutting force, N.

To ease the calculation, it was assumed that the cutting stiffness does not depend on the thickness of the shear layer. Other components of the cutting force were expressed through coefficients:

$$F_r = K_r \cdot K_{pt} \cdot a = K_r \cdot F_t,$$

$$F_0 = K_0 \cdot K_{pt} \cdot a = K_0 \cdot F_t.$$

The equation of system motion according to Figure 1 can be written as:

$$m \cdot \ddot{X}_t = -C_{yc} \cdot X_t - K_p \cdot \left(X_t - X_{t-\tau} + a_н\right) - h \cdot \dot{X}_t, \quad (2)$$

K_p – hardness of the cutting, N/m.

The mass in the direction of the x-axis is influenced only by a radial force, so:

$$K_p = K_{pt} \cdot K_r.$$

Let us mark the tooth, leaving the trace on the cutting surface as the 'tooth 1', and the tooth, cutting this track — as 'tooth 2'. Let these teeth move along the z-axis at constant speed v_0, and the equations of motion in the direction of cutting movement of the first and second teeth are, respectively:

$$Z_1(t) = v_0 \cdot t;$$

$$Z_2(t) = Z_{12} + Z_1(t), \quad (3)$$

t – current time, s; $Z_1(t)$, $Z_2(t)$ – position of respectively the first and second teeth along axis z at time moment t, m; v_0 – velocity of cutting movement, m/s; Z_{12} – distance from tooth 1 to tooth 2 in the direction of the main movement of the cutting, m; $Z_{12} < 0$.

When moving along the surface of the cutting, tooth 2 is sequentially positioned at some positions, characterized by coordinate $Z_2(t)$. Preceding tooth 1 was also positioned at these positions some time ago, and left a trace on the surface. When the considered tooth passes through this position, it cuts off this trace, i.e. the so-called 'amount of lag' τ_{21}, is determined by the time between the moments when two consecutive teeth pass the same position along the direction of the main movement (i.e. along the z-axis). At constant velocity v_0 along the z-axis, value τ_{21} is determined by path length Z_{21}. In this case, path length Z_{21} does not contain the whole number of waves on the surface of cutting:

$$Z_{21}=-Z_{12}>0;\ Z_{21}=L\cdot(k+i);\ L=v_0/f_x=\text{const};\ \tau_{21}=(k+i)\cdot T_x,$$

L – length of one period of oscillations on the surface of the cutting, m; k – number of oscillation periods that fall entirely within the length of the path Z_{21}; i – ratio of the rest of the period of oscillation that does not fit in to the oscillation period of a single-mass system along axis x; f_x — vibration frequency of the system along axis x, Hz; T_x – period of oscillation of the system along axis x, s.

A condition to determine the magnitude of the lag is:

$$Z_2(t)=Z_1(t-\tau_{21}).$$

And taking into account (3):

$$\tau_{21}=-\frac{Z_{12}}{v_0}=\frac{Z_{21}}{v_0}.$$

Thus, at a constant speed of main movement v_0 the time delay is a constant value. The value of the phase difference corresponds to the time delay - this is the same interval but measured in phase (i.e. in phase angles):

$$\varphi_{21x}=\omega_x\cdot\tau_{21}=\omega_x\cdot(k+i)\cdot T_x=(k+i)\cdot 2\pi, \qquad (4)$$

φ_{21x} – the phase difference between the oscillations of the system along axis x when passing two adjacent teeth through the given point in space, rad; ω_x – angular frequency of oscillations along axis x, rad/s.

Thus, the value of phase difference φ_{21x} may include different numbers of oscillation periods. The whole part of this number determines the number of full oscillations and fractional remainder $2\pi i$ – the phase shift [1–3].

3. Results of the practical experiment

A feature of the milling process is the presence of periodic inputs-outputs of the teeth from contact with the workpiece. This leads to intense torsional vibration of the main movement drive, therefore, to the change of the cutting speed. When machining at a variable cutting speed, the phase shift is changing constantly, and the level of vibration should change depending on the instantaneous value of this angle.

Figure 2. Equipment for the experiment.

To confirm the amplifying effect of the regenerative effect on the level of vibrations when milling with portable equipment, a series of experiments was made. The experiments were carried out at the facility, which is a small milling machine with the CNC system, mounted on non-rigid supports (Figure 2). The processing was conducted in various modes by the end-cylindrical milling cutter with a diameter of 8 mm and teeth number 5 according to procedures [4]. Vibration measurements were carried out using device *VibXpert II*. Data processing was carried out in software *OMNITREND*.

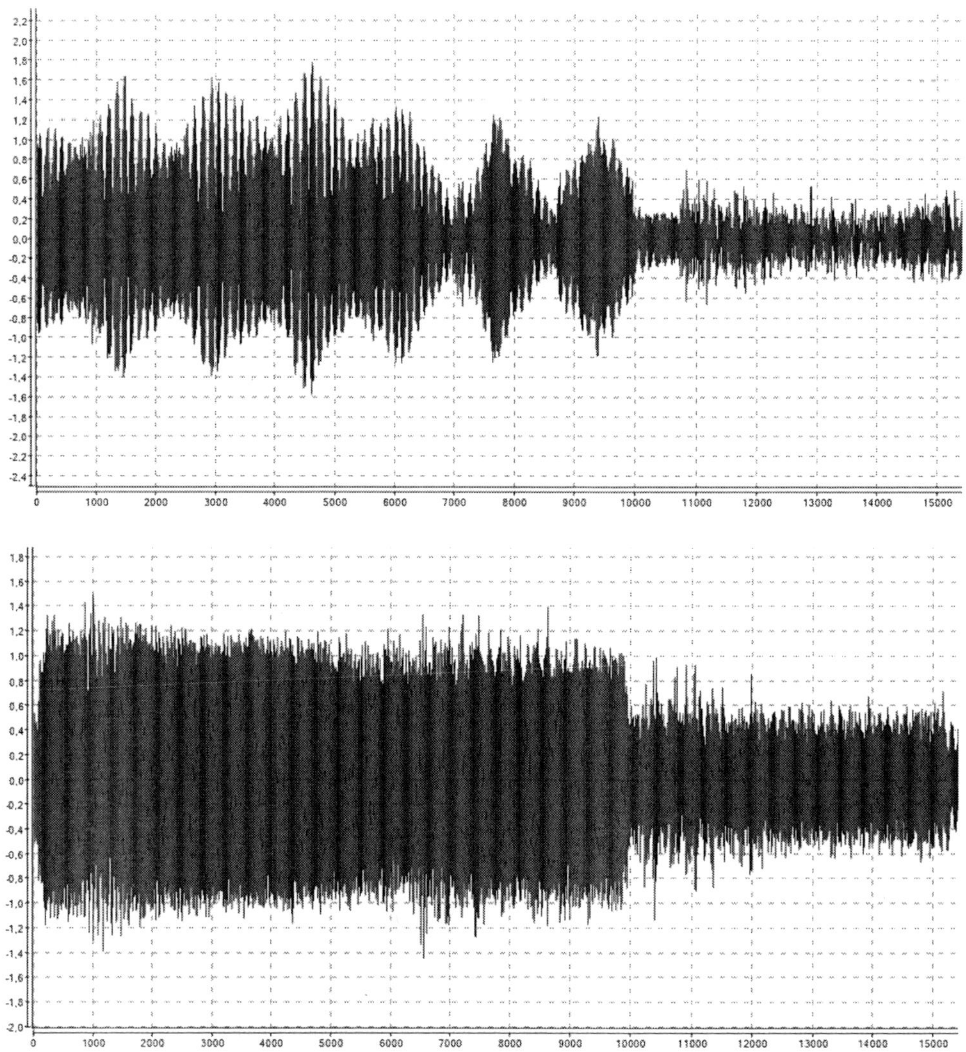

Figure 3. Wave form of system vibration in the direction of z-axis for various treatment modes.

A parameter that does not depend on phase change, but affects the level of regenerative oscillation, can be the lag time. Another parameter influencing the level of oscillation is the phase shift in the beginning of cutting on the trace. The minimum level of vibration occurs when the initial shear is equal to $\pi/2$, and is maximum at $-\pi/2$. Data analysis of vibrodiagnostics (Figure 3) shows that if the initial phase shift is different from $\pi/2$, the part of the vibration energy is spent on setting up the system, and the fluctuations are damped quite rapidly. If the initial phase shift equals $-\pi/2$ or is close to this value, the system immediately adjusts to the correct phase shift, and steady oscillations occur in the system. Due to the presence of friction in the system, these oscillations are gradually damped and the decay time depends on the initial phase shift.

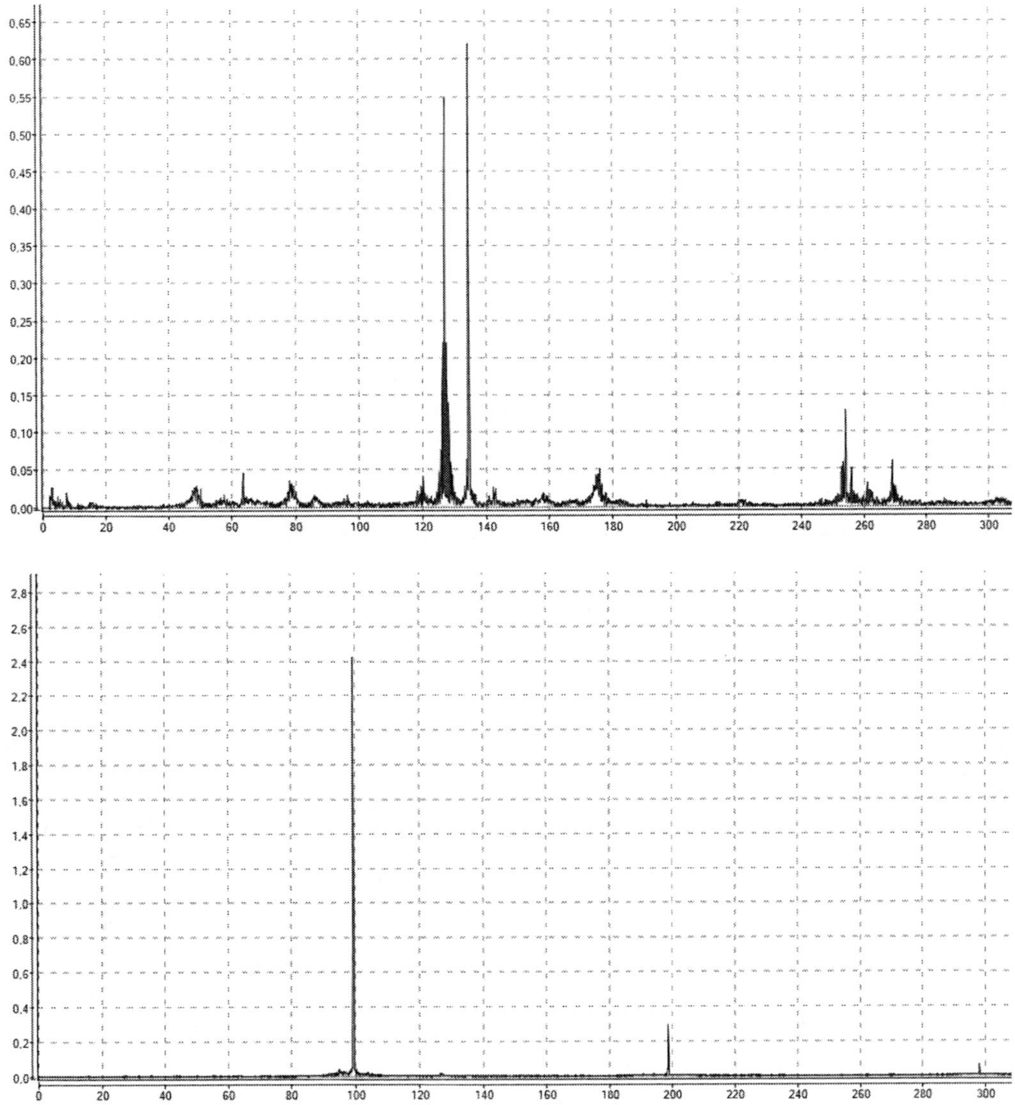

Figure 4. Vibration spectra of the experimental installation in the z-axis direction at different cutting modes.

Figure 4 shows the most characteristic spectra of the oscillatory process for different values of the phase shift between the track and system oscillations. It should be noted that the oscillatory process, in this case, has several components. There are forced resonant oscillations in the system, characterized by pronounced pulses from the entry and exit of the teeth and by the intense transition process. The level of these oscillations is quite high, and the presence of regenerative effect leads to a resonant increase in the amplitude of the harmonic with frequency corresponding to the natural frequency of the system. As a result, there are two frequencies that are in dominated in the spectrum (Figure 4): the frequency of the driving force and multiplied by its frequency is close to the frequency of the system;

in addition, the spectrum contains many other harmonics that are multiples to the frequency of the driving force. There are also self-oscillations that occur in the form of a beat. Therefore, the frequency of the driving force and the natural frequency of flexural vibrations of the instrument subsystem shifted up due to the influence of the stiffness of cutting are also present in the spectrum (Figure 4). The level of vibration velocity from self-oscillations is lower than in case of forced resonant oscillations.

Figures 3 and 4 show that the system tends in all cases to a state, when vibration velocity fluctuations are behind a trace at the angle of $\pi/2$ (i.e., $2\pi i=-\pi/2$), so they are always $i=-1/4$ or $i=3/4$.

If we consider the trace on the cutting surface as driving force with frequency p, the phase of the forced oscillations of the system should lag behind the phase of the perturbing force for the value:

$$2\pi \cdot i_p = -\arctan\frac{2\cdot\varepsilon\cdot\gamma}{1-\gamma^2},$$

ε – dimensionless damping coefficient; γ – ratio of the frequency of driving force p to its own frequency f_x.

In case of equality of its own and driving frequencies, the phase shift does not depend on ε and can only be equal to $\pi/2$. More specifically this value of phase shift corresponds to the conditions of resonance occurrence. Hence, the increase in the level of oscillations in the perturbation on the trace in the first place is due to resonance or near-resonance phenomena due to the coincidence of the own frequency of the system with the wave frequency on the cutting surface.

Consequently, the value of phase difference φ_{21x} calculated in equation (4) determines its value only at the initial moment of cutting in trace because after a while the phase of oscillations will be self-set with a new value. At the same time, the time of phase adjustment does not depend on its initial value and is equal to several periods of oscillation of the driving force. This self-adjusting of the system is due to the change of the oscillation period of the system itself, T_x. It follows that time τ_{21} always includes the whole number of T_x periods, determined by the own frequency of the system or the duration of the transition, i.e. $i=0$, and at this value of phase shift, the effect of regenerative excitation is minimal [1, 5]. Thus, the limit of stability of the cutting process does not depend on the disturbance on the trace.

In the analysis of vibration spectra (Figure 4), there is a soft excitement of regenerative vibrations. This implies that the system is stable when cutting on the trace but with a small stiffness coefficient; and it may become unstable when cutting on the trace, when the stiffness ratio exceeds a certain value [6–8].

4. Conclusion

Thus, if the magnitude of the impact force is small, the increase in vibrations level is caused by resonance phenomena with a trace on the cutting surface and, over time, due to the absence of the external periodic source, these vibrations are damped. If the magnitude of the impact force exceeds a certain value, there is a sharp increase in the amplitude of the oscillations which take the form of continuous beating. These fluctuations can not be attributed to the forced resonance, because they occur only for sufficiently large force action and outwardly behave as self-oscillation. This self-adjusting of the system always occurs at phase shift $\boldsymbol{\pi/2}$. As a result, the part of the oscillation energy is consumed not to unbalance the system but to adjust it. The more energy will be spent on the adjustment, the less of it will remain on the excitation of vibrations. It follows that one of the effective ways to damp regenerative oscillations can be to create such conditions of its operation in which it would be necessary to perform this reconfiguration constantly. Thus, if the wave period for the cutting surface will not coincide with the period of self-oscillations, the resonance will not occur, and the vibration level will be much less [9, 10]. The search for such conditions and determination of their interrelationships with its parameters is a goal for further research.

References

[1] Radulescu R A 1993 *General Cutting Process Model for High Speed Machining Dynamic and Thermal Considerations: PhD Thesis* (University of Illinois at Urbana-Champaign) p 257

[2] Sastry S, Kapoor S G, DeVor R E, Dullerud G E 2001 *ASME J. Eng. Indus.* **123** 753-756
[3] Svinin V M 2005 *Metal working* **3(28)** 28-30
[4] Kozlov A M, Kiryuschenko E V 2013 *Reference Book. Eng. J.* **1** 30-34
[5] Kozlov A M, Kiryuschenko E V 2012 *Fund. and Eng. Probl. of Technics and Techn.* **3** 65-73
[6] Gorodetskiy UI. 2004 *Collected papers of the VII Int. Scientific-Technical Conf.* (Saratov: Saratov state technical university) 161-165
[7] Gorodetskiy U I, Strebiliaev S N, Mayorova U E 2005 *Bull. of Scientific-Technical Dev. Nat. Tech. Group* **9(25)** 91-95
[8] Vasin S A 2006 *Prediction of tool vibration resistance in turning and milling.* (Moscow: Mashinistroienie)
[9] Kiryuschenko E V 2012 *Reporter of Voronezh State Technical Univ.* **10(8)** 87-93
[10] Kozlov AM, Kiryuschenko EV, Kuznetsov SF 2013 *Modern Problems of Mech. Eng.: Collection of Scientific Papers of VII Int. Sci.-Techn. Conf.* (Tomsk: Tomsk polytechnical university) 295-299

Friction self-oscillation decrease in nonlinear system of locomotive traction drive

D Ya Antipin, V I Vorobiyov, O V Izmerov, S G Shorokhov and D A Bondarenko

Bryansk state technical university, 7, 50-letiya Octyabrya bul., Bryansk, 241035, Russia

E-mail: adya24@rambler.ru

Abstract. The problems of the friction self-oscillation decrease in a nonlinear system of a locomotive traction drive are considered. It is determined that the self-oscillation amplitude decrease in a locomotive wheel pair during boxing in traction drives with an elastic linkage between an armature of a traction electric motor and gearing can be achieved due to drive damping capacity during impact vibro-damping in an axle reduction gear with a hard driven gear. The self-oscillation amplitude reduction in a wheel pair in the designs of locomotive traction drives with the location of elastic elements between a wheel pair and gearing can be obtained owing to the application of drive inertial masses as an anti-vibrator. On the basis of the carried out investigations, a design variant of a self-oscillation shock absorber of a traction electric motor framework on a reduction gear suspension with an absorber located beyond a wheel-motor unit was offered.

1. Introduction

In a locomotive boxing mode, the dynamic loads in a traction drive are caused by friction self-oscillations. Friction self-oscillations arise in systems with nonlinear dry friction in the presence of a falling area in friction characteristics, to which the dependence of the cohesion between a wheel and a rail upon a slip velocity corresponds. As applied to train crews, the friction oscillations were investigated in a number of works ([1, 2] and others). As appears from these works, the self-oscillations in a locomotive traction drive differ in:

- the multi-modiness (there are known five basic forms of oscillations: oscillations of a wheel pair, oscillations of an engine of the suspension, oscillations of a traction drive, vertical and horizontal oscillations of a bogie);
- the competition of modes (during boxing, one of the oscillation forms dominates which can be changed by other forms when changing external conditions).
- the hard predictability of self-oscillation development during each separate case of boxing, in view of the dependence of a constant of friction upon many external conditions subjected to a spontaneous change (rail pollution, contact temperature of a wheel and a rail and so on).

It is emphasized in [1, 2] that self-oscillations can result in breakage of traction drive parts; at the same time, the most dangerous failure is the breakage of a wheel-set axle which can result in a train crash. The experience of the operation of the DS3 electric locomotive shows that the application of traction induction motors (TIM) does not exclude a boxing mode. Thus, the problem of data systematization of methods for the decrease of self-oscillation amplitudes and criteria for the choice of these methods during locomotive designing arises. This paper is an attempt to solve this problem.

It is experimentally established that the process of the self-oscillation onset and development with increasing a slip velocity of wheels on a rail can be divided into the following phases: the clutch failure and slippage beginning; the self-oscillation development with the self-oscillation amplitude increase as the velocity increases; the beginning of the breakdown of self-oscillations when their amplitude ceases to grow; the breakdown of self-oscillations when the further velocity increase results in the amplitude decrease.

These phases are characteristic for any forms of friction self-oscillations during boxing which could be seen in Figures 1a and b, where the process of self-oscillation development during boxing of the 2TE121 type diesel locomotive according to two forms is shown: self-oscillations of a wheel pair as a rigid solid, in case of shafting compliance of a traction drive (Figure 1 a) and self-oscillations of a wheel pair in case of the compliance of its axle, when wheels oscillate in an anti-phase (Figure 1 b). It is possible to see in both figures that the basic influence upon the limit value of the self-oscillation amplitude is performed by a speed at which the power growth input to an oscillating system per one cycle of oscillations stops and the further increase of the wheels slip on a rail results in a quick decrease of the amplitude.

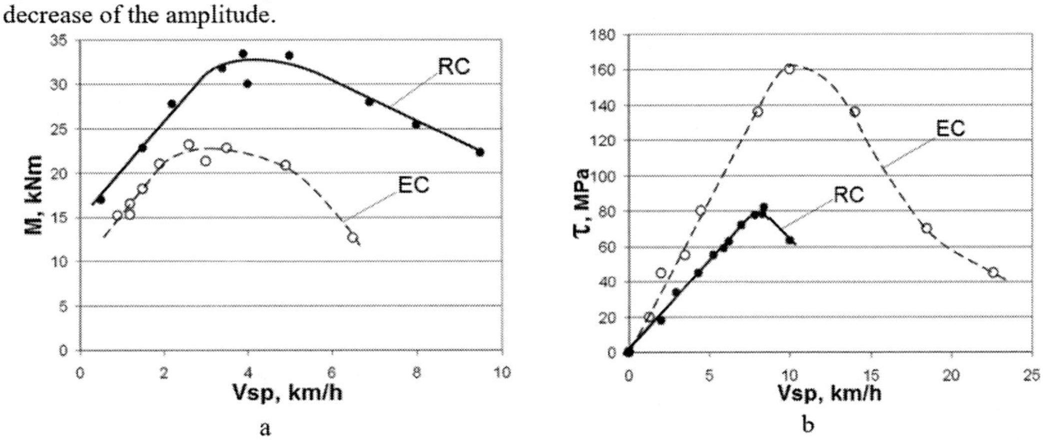

Figure 1. Examples of torque diagrams of a high-speed shaft of an axial reducing gear at self-oscillations of a traction drive (a) and tangential stresses in the axle of a wheel pair (b) in the boxing mode of diesel locomotive 2TE121: EC – variant with an elastic cog-wheel, RC – variant with a rigid cog-wheel.

A conclusion appears from the fact that for the most critical unit of a diesel locomotive traction drive, a wheel pair, tangential stresses during its self-oscillations in the mode of boxing could be decreased considerably at the expense of power diffusion in a traction drive which is also shown in work [1]. For self-oscillation absorbing, it is possible to introduce an additional oscillatory unit with an elastic-dissipative connection in a dynamic system as it is offered in [3], or to use the existing design elements of a drive. Let us consider first a possibility of self-oscillation power dispersion in a traction drive owing to the use of its design elements.

The appearance of new locomotives and road machines with an elastic link between a traction motor armature and a wheel pair (elastic cog-wheel, elastic clutch, torsion) poses a problem of the creation of a friction self-oscillation absorbing theory by elements of a traction drive and a crew part of a locomotive. Let us consider factors affecting the capacity of a drive to damp self-oscillations of a wheel pair and required dynamic properties of a traction drive ensuring a sufficient damping capacity.

2. Materials and methods

The main purpose of the work is the analysis of existing and search for new options for locomotive traction drive design in order to reduce impact of dynamic loads. In work use analytical methods to research the dynamics of locomotives, the results of which were confirmed by field experiments.

3. Impact oscillation damping in the axial reducing gear

During the testing time of a traction drive of diesel locomotive 2TE121 with a rigid cog-wheel of an axial reducing gear, during boxing, the stress in an axle of a wheel pair reached 85...90 MPa, and incase of an elastic cog-wheel – 140...160 MPa. A dynamic constituent of a torque on an input shaft of a reducing gear in both cases amounts to 6...7 kN/m. Inasmuch as in this case, the self-oscillation frequency in a wheel pair during boxing (49 Hz) is several times higher of the frequency of armature oscillations of a traction motor (TM) making 4-6 Hz, the armature oscillations in drive 2TE121 did not affect considerably dynamic processes in a drive, and the observable effect could be explained only by an impact damping of oscillations in a drive with a rigid cog-wheel. In the absence of self-oscillation damping in a wheel pair (a driven wheel is located in the oscillation unit [1], or a wheel pair is separated from a gearing by an elastic clutch [2]), stresses in an axle reach 160 MPa. Assuming 160 MPa as a level of stresses reached in the absence of considerable power scattering in elements of a drive, it is possible to draw a conclusion that the effect of power scattering in shock-absorbers of an elastic cog-wheel, elastic joints of the reducing gear suspension and friction in gearing is negligible in comparison with the effect of vibroimpact damping except when strong resonance oscillations develop.

As a rule [1], dynamic properties of a traction drive with non-linear elastic-dissipative properties were estimated on the basis of linearized models. At the same time, it was considered that the 'presence of considerable non-linearity (random kind) of drive elastic elements does not affect in any way the changes in the development and stabilization of friction self-oscillations'.

For a more specific problem of the analysis of power waste in a vibroimpact system we will consider not the process of self-oscillation itself but pulse interactions of system joints accompanying it. As it is shown in [4], in a symmetric two-mass dynamic system (Figure 2) at one concussion, the amount of released energy T is equal to

$$T = \frac{(1-R^2)}{2} \cdot \frac{Mm}{M+m} \cdot V^2, \tag{1}$$

where R – the coefficient of restoration at a blow; M – the mass of the deformable body (of a wheel center with a cog-wheel); m – the mass of an absorber (gear with a flange or a rotor of a traction engine); V – the relative velocity of bodies at the moment of the concussion.

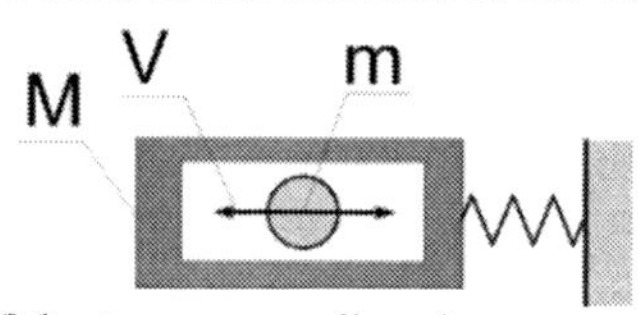

Figure 2. The scheme of the two-mass vibro-impact system of the traction drive.

For the traction drive of diesel locomotive 2TE121 in the mode of boxing it was revealed that on the body of a traction gear box close to a suspension bracket, the intensive accelerations arise both with the frequency of a basic tone of wheel pair self-oscillations and with double frequency. The phenomenon is noted both for the variant with a rigid rim, and for that with an elastic one of a cog-wheel of the axle reducing gear and it indicates that during one cycle of wheel pair self-oscillations in gearing, two concussions of teeth occur. A rigid rim of a cog-wheel impacts directly the axle blows, and an elastic one weakens them and, in this way, decreases the efficiency of impact vibro-absorbing. The connection rigidity of a cog-wheel hub in a reducing gear with the closest wheel center (for a drive with spur gears) is many times higher than that with other wheel center, and a moment of inertia of a driven cog-wheel does not exceed 20...25% of the moment of inertia of the wheel center. The results of experimental researches did not reveal a considerable influence of the cog-wheel place upon the location of an oscillation unit as the moment of inertia of wheel masses of a wheel pair is several times higher than the moment of inertia of masses of a cog-wheel (for diesel locomotive 2TE121 – 5 times higher). It substantiates the reason to suppose that an oscillation system of a wheel pair represents a symmetric two-mass system with an oscillation unit in the middle of an axle. On the basis of this, in [5], the authors offered a way to estimate efficiency of vibro-absorbing through a conditional parameter being the ratio of power used in the course of a cycle to the maximum possible power stored in the system in the absence of damping:

$$K_b = \left(1 - R^2\right)\frac{D}{2}\left(\frac{V}{V_s}\right)^2 \frac{J_b}{J_k + J_b}, \tag{2}$$

where V_s – slip velocity; D – the diameter of the wheel center; J_k – the moment of inertia of the wheel center, J_b – the moment of inertia of the driving gear and the input shaft with half coupling moved to the axle of the wheel pair.

A practical value of parameter K_b consists in the definition with its aid the design measures which allow decreasing stresses in the axle during boxing in the course of developmental testing. For a specific drive, when reaching a critical slip velocity in set mode V/V_s=const, it can be assumed that:

$$(1 - R^2)\frac{D}{2}\left(\frac{V}{V_s}\right)^2 = C_k, \tag{3}$$

where C_k – constant depending on design parameters of the drive.

A vibro-impact mode begins to finction at the fulfilment of the condition

$$V_s \geq \frac{M_t D}{2\omega J_b}, \tag{4}$$

where M_t – the motor traction moment applied to the axle of a wheel pair; ω – circular frequency of self-oscillations.

Thus, the shown inertia moment of a half-coupling flange defines a critical velocity at which a vibro-impact dissipation of self-oscillation power occurs. To increase the vibro-damping efficiency and the dynamic stress decrease in the axle of a wheel pair, it is necessary to increase ratio $J_b/(J_k+J_b)$, which could be carried out at the expense of the half-coupling inertia moment increase at the input shaft of a reducing gear.

In the mentioned designs of the drive, we did not observe large-scale failures of wheel pairs in contrast to, for instance, industrial electric locomotive 14KR1 [1], in spite of the boxing mode appearance during starting off. The presence of the rigid gearing in a traction drive, as a rule, ensures a sufficiently low level of tangential stresses during torsional self-oscillations of a wheel pair without the additional increase of the inertia moment in a half-coupling of a driving shaft.

It is confirmed also by the experience of the development and operation of electric locomotives 'RM' and 'RC4' of 'ASEA' Co., which have a support-frame drive with an axial reducing gear and the rigid gearing [6]. It is well-known that a large-scale appearance of cracks in wheel axles is observed at the amplitude of aletrnating stresses equal to 160 MPa [1, 8]. In [7], it is recommended for shafts of a traction drive to admit a lower limit of a safety margin equal to 1.4…1.6, which allows assuming tangential stresses in an axle defined during tests, 100…115 MPa, as a limit. In narrow-gauge locomotives MD54 and TU6, the tangential stresses in an axle during the tests reached 115 MPa [1], and a fact of a reliable operation of axles in these locomotives confirms the correctnes of safety margin chosen.

4. The use of inertia masses of the drive as a shock absorber

Elastic elements in a cog-wheel reduce sharply the efficiency of impact vibro-absorbing. The installation of an elastic self-installing cog-wheel (IESCW) in a traction drive of diesel locomotive 2TE116 with a support-axle drive did not cause large-scale failures in axles. It was defined experimentally that tangential stresses in an axle during self-oscillations do not exceed 40…60 MPa (the drive of diesel locomotive 2TE10L with a rigid cog-wheel – 30 MPa [1]).

For the IESCW of a diesel locomotive, the rigidity characteristic during the realization of traction efforts close to utmost ones on coupling is non-linear [9], which can result in the appearance of resonance phenomena at frequences several times less than excitation frequency. During the tests of the support-axial traction drive with an elastic cog-wheel and a rubber-metal cross-arm [9, 10], self-oscillations of a wheelset during boxing with frequency f_0=80…85 Hz were followed by intensive subharmonic oscillations of an engine on a rubber-metal cross-arm with frequency (19…20 Hz) ~$f_0/4$, which was not observed in the course of testing the traction drive of diesel locomotive 2TE10L and diesel locomotives with hydrotransmission.

The phenomenon revealed cannot be explained by the highest harmonics of traction engine self-oscillations around the axle of a wheelset since it is typical that for this form of self-oscillations, the armature current fluctuations with frequency 8…12 Hz were not in tests. The appearance of sub-harmonic fluctuations can be caused only by fluctuations in the system of gyrating masses. It allows using the armature of a driving motor or a flange of a half-coupling of an axial reducing gear in a traction drive with an axial reducing gear as a resonance absorber (antivibrator) since the basic frequency of self-oscillations in a wheelset in these drives is usually far from frequencies of characteristic oscillations of drive elements. Sub-harmonic oscillations can become excited in case a twisting rigidity of shaftings of a support-frame drive with an axial reducing gear has a significant non-linear characteristic, for instance, in case a driving motor is connected with an axial reducing gear through a rigid compensating coupling (gear clutch, cam clutch, hinged-actuating coupling and so on), and shafting twisting compliance is ensured by an elastic cog-wheel of an axial reducing gear with a rigidity non-linear characteristic.

If in the considered range of oscillation frequencies of a traction drive the damping in a rubber element depends slightly upon frequency, then power waste during an oscillation cycle will amount to

$$T = C_r \sum_{i=1}^{n} \Delta_i , \qquad (5)$$

where C_r – power waste in athe rubber element of the absorber in the course of a cycle of oscillations at the amplitude of deformation equal to a unit determined experimentally or in a computation way; i – the number of oscillation harmonic; Δi – the amplitude of rubber element deformation at oscillations at frequency $f_i = f_0/I$; n – the number of analyzable partial frequencies of oscillations of a damping mass.

The efficiency of self-oscillation damping in a wheelset during sub-harmonic resonance depends upon the amplitude of relative deformation of elastic rubber elements, the value of which can be large enough in a resonance operating mode of an absorber. The growth of the deformation amplitude in rubber elements of a drive during boxing results in the decrease of their life. From the practical point of view, the most convenient way of damping with the aid of structure elements of a drive should be shock vibro-damping in the gearing. The use of inertia masses of a drive as a shock absorber (anti-vibrator) is useful only when an elastic joint in shaftings of a drive according to design ideas cannot be located just between a wheelset and a cog-wheel.

5. Search of design solutions for additional devices of self-oscillation damping

The use of special absorbers to reduce self-oscillation amplitudes in a traction drive is difficult in connection with the fact that the elements of a traction drive are located in a limited area of a bogie. In a diesel locomotive with a hydrotransmission, it is possible in principle to use silicon absorbers for the dissipation of oscillation power in the shafting, but in practice, it is impossible to carry out for diesel locomotives with power transmission and electric locomotives. Besides, during the influence of buffing loadings upon hydraulic absorbers arising in a traction drive, these buffing loadings can pass to a bogie frame.

In connection with this, the authors have offered the solution for a self-oscillation absorber of a driving motor (DM) on a traverse suspension with the absorber located beyond the dimensions of a wheel-motor unit and combined with a suspension unit (Figure 3).

With this arrangement, the unit of the driving motor suspension has suspension link 1 and some rubber or rubber-metal elements 2 arranged in alignment and preliminarily tightened. At the same time, the end of the suspension link has non-self-braking thread 3 on which there is nut 4 connected with the axle of speed regulator 5, the body of which is connected with the frame of bogie 6 through rubber elements 7.

During deformation of rubber elements 5 during driving motor oscillations on surface imperfections of a track or during self-oscillations when locomotive boxing suspension link 1 moves upright, thereupon nut 4 turns on non-self-braking thread 3 and rotates the axle of speed regulator 5. During oscillation, the amplitude of the velocity of axle rotation in speed regulator 5 increases that results in the friction increase in speed regulator 5, in the increase of resistance to the nut 4 rotation on non-self-braking thread 3, which increases the resistance of a driving motor to oscillations, reduces the oscillation amplitude in case of the resonance or self-oscillations at boxing at the expense of the power disspation increase in speed regulator 5, preventing in this way the appearance of large loads in parts. Rubber elements 7 protect speed regulator 5 from damages at shocks when passing track imperfections.

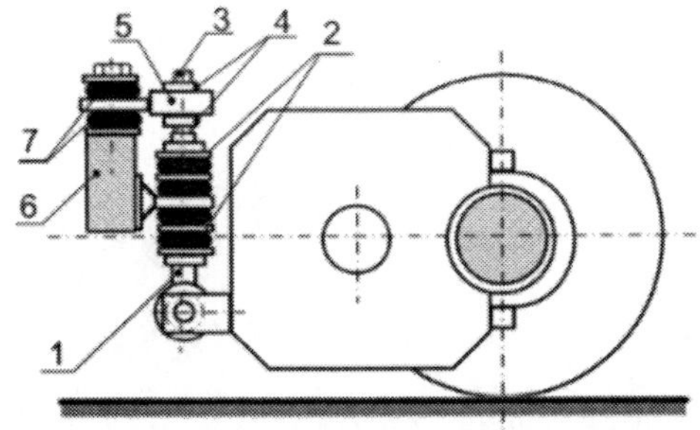

Figure 3. The version of friction self-oscillation absorber arrangement.

To solve effectively the problem of tangential stresses in the axle during self-oscillations is possible through the introduction of traction drive designs integrated with systems excluding a boxing mode at the expense of the control of a physical friction coefficient by means of the effect of the contact of a wheel and a rail of electric current and/or a magnetic field upon the area, and the creation of systems of boxing warning forecasting changes into a physical friction coefficient. This type of a drive allows also decreasing considerably power waste during wheel slipping on a track, which reaches 10-30% of power on traction which allows characterizing this sort of the drive as an energy-saving one [5].

6. Conclusion

1. The self-oscillation amplitude decrease in a wheelset during boxing in traction drives with an elastic link between the armature of a driving motor and gearing can be obtained at the expense of drive damping capacity during shock vibro-absorbing in an axial reducing gear with a rigid driven cog-wheel. At the same time, the efficiency of shock vibro-absorbing is small, which is critical to the change in design parameters of a wheel-motor unit and can be increased by the growth of an inertia moment of masses on a speed shaft of gearing.

2. In case of the elastic element location between a wheelset and gearing, the decrease of the self-oscillation amplitude in the wheelset can be obtained at the expense of using drive inertia masses as an anti-vibrator, in particular, at the expense of sub-harmonic resonance in a non-linear dynamic system of a drive, the presence of which was revealed in a diesel locomotive support-axial drive with an elastic cog-wheel.

3. Practically, on the basis of the rolling-stock operation, we can suppose that the value of tangential stresses in an axle during boxing must not exceed 100…115 MPa.

4. This paper offers a design of a self-oscillation absorber of the frame of the driving motor on the suspension of a reducing gear with an absorber located beyond the wheel-motor unit.

5. It is possible to solve the problem of the dynamic load decrease during self-oscillations through the realization of drive designs integrated with systems excluding boxing at the expense of control of a physical coefficient of friction with the effect upon the area of a wheel and a track contact of electric current and/or a magnetic field and to create boxing preventing systems, forecasting changes in the physical coefficient of friction.

References

[1] Sebesan I 2015 *Proc. of the Romanian Academy Series A* vol 16 (Bucharest: The publishing

house proceedings of the romanian academy of the romanian academy) pp 209-216
[2] Strelnikov V P 1973 Results of Wheelset Drive Dynamic Tests at Support-Frame Suspension of Driving Motor (Kolomna: Kolomensky zavod)
[3] Koropets P A and Sinyutin S A 2013 *Bulletin of the Rostov State University of Railway Transport* (Rostov: Rostov State University of Railway Transport) pp 58-63
[4] Anderson P M and Fouad A A 2002 *Power System Control and Stability* 2nd Edition (Hoboken: Wiley Interscience)
[5] Kosmodamiansky A S 2014 Development Concept of Energy-Saving Electro-Mechanical Systems (Orel: Orel State University)
[6] Lundberg K G 1979 *Electric locomotive of Rm series for Kiruna-Narvik ore transportation* (Moscow: World Railways) pp 25-28
[7] Zong J and Qian P 2012 *Trans Tech Publications* **516-517** 1935-38
[8] Belyaev A I and Bunin B B 1984 *Reliability Increase in Crew Section of Diesel Locomotives* (Moscow: Transport)
[9] Yevstratov A S and Kochergin A S 1987 *Crew Sections of Diesel Locomotives* (Mosow: Mechanical Engineering)
[10] Brunia S and Vinolasb J 2011 *Vehicle System Dynamics* **49** 1021-72

Multifactorial determination of the electric drive for the force compensating manipulator

G Ya Pyatibratov and A A Danshina

Platov South-Russian State Polytechnic University (NPI), 132, Prosveschenia str., Novocherkassk, Rostov Region, 346428, Russia

E-mail: da.angela@mail.ru

Abstract. The methodology of multifactorial determination of rational parameters of transmission gear and synchronous electric motor driving by permanent magnets for the system of the vertical freight movement of the force compensating manipulator is offered. An integrated approach to the selection of the power part of this manipulator takes into account: motor speed matching and an executive mechanism of the manipulator, operation of the electric drive with a minimum possible value of the maximum torque at the movement of freight with constant speed and with acceleration at different values of the freight mass. A reasonable radius of mechanism activation is determined from accepted values with application of the compromise approach enabling to consider at the same time the performance of all limiting conditions. The electromechanical module of the manipulator is selected when a value of the activation radius provides the minimum possible required motor torque.

1. Introduction

Development of mechanization and automation means of various productions is meant to improve the manipulators serving them and robotic complexes. Nowadays, during automation of processes related to the transportation of various goods and freights, the balanced manipulators are widely used. The feature of these manipulators is compensation of links weight and load grippers by means of counter balance or springs that allows one to improve power indicators of the drive operation [1]. To increase efficiency of loading and unloading operations and the level of their automation, it is necessary to improve electromechanical systems of the long-range balanced manipulators [2]. The promising direction of improvement of manipulators and special load grippers is application of force compensating manipulators (FCM) implemented with the use of control systems of efforts which make it possible to perform the required compensation of gravity not only links, but also the freight moved by a worker [3]. Herewith, it is reasonable to assign solving of additional tasks to decrease the impact of inertia forces of transfering masses, dry and viscous friction of mechanisms to an electric drive (ED) [4].

The feature of the manipulators operation is a necessity to transport freigts of various weights, which depends on the mass of the load gripper of manipulator Δm and freight m_0. Herewith, the minimum mass is defined by the mass of load grippers (hooks, electromagnetic, lever or pneumatic grippers and other devices) which can have the weight from 10 to 35% of the mass of the transferred freight. The maximum value of the weight is defined by the sum of masses $\Delta m + m_0$ of the load gripper and the mass of the transferred freight. According to the investigation, during design of power efficient FCM, it is reasonable to apply synchronous electric motors activated by permanent magnets with nominal rotation speeds from 4000 to 6000 rpm. Herewith, the problem of determination of rational parameters of the FCM mechanism taking into account the main requirements, conditions and performance features is urgent. For the effective solution of this task, it is reasonable to apply the methodology of multifactorial determination of necessary parameters of the mechanism and ED.

The purpose of work is adaptation of a technique of the multi-factorial combined selection of rational parameters of FCM and ED mechanisms during speeds matching of motor and freight movement, value minimization of required maximum motor torque, implementation of motor alive standing with given loading moment.

2. Task description of the research

In order to improve power and mass-dimensional indicators of FCM, it is necessary to apply ED providing fulfillment of the contradictory requirements imposed on it. For this purpose, it is offered to solve a number of the interrelated assignments wholistically: to provide necessary movement parameters of the executive mechanism, to compensate freight weight and links of the manipulator, to decrease the impact of friction forces and the inertia of moving masses, to provide the required accuracy of efforts control of executive mechanisms, to improve mass-dimensional performances. By means of ED, it is possible to reduce oscillations and the dynamic efforts caused by elasticity of mechanical gears and constructions, to provide long maintenance and smooth change of torques considerable in value on the motor shaft in the mode of its standing alive. During design of FCM taking into account the listed assignments, it is necessary to select reasonably rational parameters of mechanisms and ED.

During the solving of multifactorial tasks, they are reduced to optimization of some single combined parameter [5]. When designing FCM as the combined parameter, it is offered to implement the required reduction radius that connects speed of motor rotation Ω_M and freight transferring V_0 with kinematic parameters of mechanical gears:

$$\rho_R = V_0 / \Omega_M = d_D /(2i_P i_G) = d_{GD} /(2i_G), \tag{1}$$

where i_G, i_P – the gear ratio of the gearbox and the divisible factor of block and tackle; d_D, d_{GD} – drum diameters in the rope drive or gear wheel in the rack and th gear drive.

Let us define acceptable values of reduction radius ρ_{AV} providing coordination of movement speed of the executive mechanism and the motor; minimization of required maximum motor torque M_M [6], transportation of various mass freight with maximum values of speed and acceleration [7].

3. Problem solving

We wil solve the task, based on the loading capacity of manipulator m_0 and parameters of the freight movement — speeds V_0 and acceleration a_0. Let us consider FCM of average loading capacity $m_0 = 150$ kg, which construction contains supplementary elements with mass $\Delta m = 15 - 50$ kg. Let us introduce a concept of an actual mass of transferred freight m_A which, during the operation of FCM along with the executive mechanism in the form of the rope drive, can change from minimum value $m_{1A} = \Delta m = 15$ kg to maximum value $m_{2A} = \Delta m + m_0 = 200$ kg.

The analysis of papers [8-10] showed that in order to increase the productivity of the FCM operation, it is reasonable to have the maximum speed of freight movement equal to $V_0 = 0.5$ m/s. According to GOST 26057–84, the efforts, applied by the operator to freight, weighing up to 100 kg, should not exceed values $F_{E\min} = 50$ N, and weighing from 100 to 250 kg — should not exceed $F_{E\max} = 100$ N [11]. Taking into account these requirements, we will define possible maximum accelerations of the executive mechanism and freight:

$$a_{01} = \frac{F_{Emim}}{m_{1A}} = \frac{50}{15} = 3.3\,\text{m/s}^2 \quad \text{and} \quad a_{02} = \frac{F_{E\,max}}{m_{2A}} = \frac{100}{150+50} = 0.5\,\text{m/s}^2. \tag{2}$$

According to the methodology presented in [5], let us define the motor capacity which provides prescribed acceleration $a_{02} = 0.5\,\text{m/s}^2$ at maximum speed $V_0 = 0.5$ m/s of freight movement with total weight $m_{2A} = \Delta m + m_0 = 200$ кg:

$$P_{2R} = 2k_S(\Delta m + m_0)V_0(g + a_{02})\frac{1}{\eta_M} = 2 \cdot 1.2 \cdot (50+150) \cdot 0.5 \cdot (9.81+0.5) \cdot \frac{1}{0.8} = 3.1\,\text{kW}, \tag{3}$$

where $k_S = 1.2$ – the safety margin taking into account the possibility of noncoincidence of the actual reduction radius with the reduction radius providing minimum motor capacity; $g = 9.81\,\text{m/s}^2$ – the gravitational acceleration; $\eta_M = 0.8$ – the efficiency coefficient of the mechanism of the freight vertical movement system.

During the operation of FCM, the important requirement is to hold fixed hanging freight by means of ED. That is why, it is necessary to provide the mode of long standing of the motor alive at the maximum value of the steady load. According to the investigation among modern electric motors of alternating current, the synchronous motors 5DVM series produced in Russia have similar properties [12]. Selected for implementation of FCM, the high-speed motor of type 5DVM 115M has the following performance specifications: nominal capacity $P_N = 2.95$ kW, nominal rotation speed $\Omega_N =$ 6000 rpm, nominal torque of motor shaft $M_N = 2.35$ N·m, accepted torque of long standing of the motor alive M_0=4.7 N·m, maximum torque on motor shaft $M_{M\,max} = 11.75$ N·m and inertia torque J_M=0.00074 kg·m^2.

To solve the problem of multifactorial determination of the mechanism and ED CV, it is reasonable to accept reduction radius ρ_R as ф combined parameter [5]. To determine its rational values, we will review the impact of reduction radii which provide the required coordination of speeds of motor movement and freight ρ_Ω, define the minimum possible value of required maximum motor torque ρ_M, provide freight movement with constant speed ρ_S and with acceleration ρ_D.

To solve the described problem, it is necessary to define the value of required reduction radius ρ_R of the FCM mechanism which has to enter the range of its acceptable values ρ_{AV} taking into account conditions and restrictions presented in Table 1.

Table 1. Restrictive conditions.

№	Requirement	Condition	Restriction
1	Movement speed approval of the freight and the motor	$\Omega_{MR} \geq V_0 / \rho$	$\rho_R \geq \rho_\Omega$
2	Minimization of the required maximum motor torque	$M_M = M_S + M_D = \min$	$\rho_R = \rho_M$
3	Moving of load at a constant speed;	$M_S > M_N$	$\rho_R < \rho_S$
4	Moving of load with acceleration	$M_{DS} = M_S + M_{DM} + M_{DL} < M_{M\,max}$	$\rho_R > \rho_D$

Generally, component parts of load torques depend on the direction, speed and acceleration of freight movement. During solving the described problem, it is necessary to define the greatest required value of the motor torque which will be required for movement of freight with maximum speed and acceleration. Analytic forms for definition of motor load torques under various operating conditions of FCM can be found according to the following formulas:

$$M_S = m_A g\rho(1+f_{SF}) + f_{DF}V_0 \le M_N, \quad (4)$$

$$M_{ST} = m_A g\rho(1+f_{SF}) \le M_0, \quad (5)$$

$$M_{DS} = M_S + M_{DM} + M_{DL} < M_{M\,max}. \quad (6)$$

We will highlight a problem of multifactorial selection of the electric drive with application of the high-speed synchronous electric motor driven by permanent magnets of type 5DVM115M with the diagrams presented in Figure 1. In Table 1, in Figure 1 and in formulas (4) – (6), the following symbols are indicated: Ω_{MR} – the required speed of the electric motor; M_S and M_D – static and dynamic load torques in the researched system; M_{ST} – the load torque in the mode of long standing of the motor alive; M_N – the rating torque of the electric motor; M_N – the rating torque of the electric motor; M_{DS} – the total maximum torque taking into account static and dynamic loads of the electric motor specified by inertia torques of motor J_M and gearbox J_G: $M_{DM} = (J_D + J_G)a_0/\rho$ and freight weight $M_{DL} = m_A a_0 \rho$; $f_{SF} = 0.2$ and $f_{DF} = 0.05$, which are the coefficients of static friction and movement. The reduction radii are: ρ_Ω – takes into account the coordination of speeds of the motor and freight movement, ρ_M – provides a minimum value of the maximum motor torque, ρ_S – defines the motor torque during freight movement with constant speed ρ_D and with maximum acceleration, ρ_{AV} – the accepted range of required reduction radius ρ_{1R}. In Figure 1, the following symbols are given: the index '1' means freight weight m_{1A} = 50 kg, and the index '2' is equal to freight weight m_{2A} = 200 kg.

In Figure 1, dashed lines show the limiting values of reduction radii of the FCM mechanism during application of the selected electric motor, as well as areas of their acceptable values at the minimum and maximum of the moved mass which comply with the conditions given in Table 1.

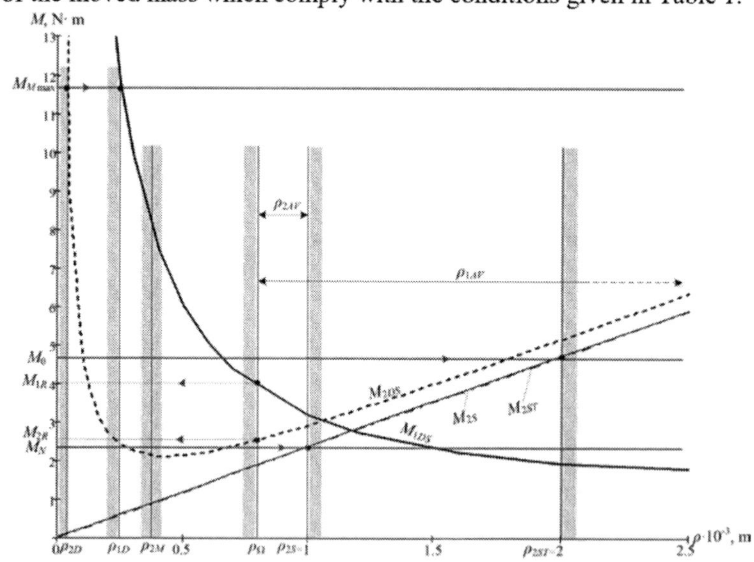

Figure 1. Definition of the required reduction radius of the FCM mechanism with the synchronous electric motor of type 5DVM115M.

4. Reviewing of obtained results

The review has revealed that conditions of long standing of the motor alive and during freight movement with constant speed of the considered electric motor during freight transferring are always fulfilled. Therefore, in Figure 1, motor torques and the reduction radii corresponding to them are not shown.

In case of transferring FCM of the freight of maximum weight m_{2A}= 200 kg, the comparison of the obtained limiting values of reduction radii showed that the acceptable range of acceptable values ρ_{2AV} is limited by coordination conditions of motor and freight speeds $\rho_{\Omega} = 0.8 \cdot 10^{-3}$ m and freight movement with constant speed $\rho_{2S} = 0.99 \cdot 10^{-3}$ m. Therefore, it is impossible to realize the ED and FCM executive mechanism with the motor of type 5DVM115M by the criterion of minimization of the required maximum motor torque equal to $M_{2M} = 2.14$ N·m because value $\rho_{2M} = 0.42 \cdot 10^{-3}$ m does not reach the range of acceptable values of reduction radius ρ_{2AV}. In the reviewed example, when $\rho_M < \rho_{\Omega}$, the defining factor at the selection of rational value ρ_R is the condition of coordination of motor and freight speeds. In this case, when $\rho_R = \rho_{\Omega}$, we will receive the minimum value of total motor torque $M_R = 2.5$ Nm. At the same time, it is possible to transport freight with accelerations a_{01} and a_{02}. And the maximum value of the total torque of the motor during transportation of the freight with weight m_{1A} = 50 kg and acceleration 3.3 m/s^2 will be 4 N·m; and with weight m_{2A}=200 kg and acceleration 0.5 m/s^2, the value of the torque will be 2.6 N·m that will be equal to 33% and 22% of the maximum acceptable motor torque, respectively.

5. Conclusion

Application of the methodology of the multifactorial selection of the electric drive and the mechanism of the force compensating manipulator with the synchronous electric motor of type 5DVM 115M allows making the following conclusions:

1. During development of FCM, the defining condition is specifying required reduction radius ρ_R that in case of application of the high-speed motor enables the reduction of the radius of coordination of motor and freight ρ_{Ω} speeds to value ρ_{2M} providing minimization of the required maximum motor torque.

2. In case of the given parameters of FCM, ρ_{2M} does not reach the range of accepted values of reduction radius ρ_{2AV} that make it impossible to obtain the lowest required values of the motor torque, however the obtained value at $\rho_R = \rho_{\Omega}$ of the required torque is much less than the maximum accepted motor torque.

3. During development of advanced FCM, it is recommended to apply high-speed synchronous electric motors driven by permanent magnets that enables a decrease of the required capacity of the electric motor and the mass of the electromechanical unit.

Acknowledgments

The paper results are obtained with the support of project № 2878 'Theory development and implementation of electro-technical systems of simulators complexes and mobile objects' executed within the framework of the basic part of state assignment № 2014/143.

References

[1] Vladov I L, Danilevskiy V N, Ionov P B and etc 1988 *Balanced Manipulators* (Moscow: Engineering)

[2] Sukhenko N A, Pyatibratov G Ya, Danshina A A and Altunyan L L 2016 Prospective electromechanical control systems of industrial manipulator efforts *IJPEDS* **7(2)** 416-21

[3] Sukhenko N A and Kravchenko O A 2003 *Rus. Electromech.* **5** 30-36
[4] Kravchenko O A, Pyatibratov G Ya, Sukhenko N A and Bekin A B 2013 *Un. News North-Caucasian Reg. Techn. Scienc. Ser.* **2** 32-35
[5] Pyatibratov G Ya, Danshina A A and Altunyan L L 2016 *The 2nd Int. Conf. on Ind. Eng. (ICIE)* vol 150, ed A A Radionov (Chelyabinsk: Procedia Engineering) pp 1403–09
[6] Myslivets N L and Sabinin Yu A 1975 *Unification Industrial Robots Actuator* (Leningrad:Electric Drive) **6(14)**
[7] Kochergin V V 1988 *Servo Systems with Direct Current Motor* (Leningrad: Energoatomizdat)
[8] Borovichi Experimental Engineering Plant 26.05.2016 Comparative table *BEEP* http://bomz.su/table/)
[9] Gorbel 27.05.2016 Intelligent lifting devices g-force®/easy arm *Gor.* https://www.gorbel.com/Resources/products/intelligentassistdevices/gforce/documents/g-forcebrochure.pdf
[10] Knight Global Inc 27.05.2016 Ergonomic material handling systems *KGI* http://www.knight-ind.com/brochures/knight_global_product_catalog_web.pdf
[11] GOST 26057–841985 1985 *Balanced Manipulators General Specifications* (Moscow:Standartinform Publ)
[12] Electrical Company 05.04.2016 BLDC (synchronous) motors 5DVM series *El. Equip.* http://www.elmotor.ru/dvg1300.html

Methods for rotational speed reduction for alternating current electric motors

A V Kagan and D Y Glukhanich

Saint-Petersburg Mining University, 2, 21st Line, St Petersburg, 199106, Russia

E-mail:kagan_55@mail.ru

Abstract. The analysis of rotational speed reduction methods for alternating current electric motors are given, assigned to low-speed electric drives of various power levels. The integrated classification of electric machines of well-known types is given, the rotational speed reduction method being used as the basis. The main advantages and disadvantages, defining perspectives for the application in various low-speed electric drives, are explained. The approximate bounds of engineering expediency of the applications of the motors are given for obtaining certain assessments in selection of a type of the drive motor.

1. Introduction

Growing demands for low-speed electric motor drives of various power levels, intended for various applications in machinery, have drawn special attention to up-to-date low speed a.c. machines. Although a great number of investigations have been devoted to research and development of the abovementioned machines, the working-out of recommendations on high-torque low-speed electric machine application were not considered in a proper way. So we will first consider the classification of the above-mentioned class of machines, as well as the prospects of application in the high-torque electric drive.

2.Classification

It has been found expedient to apply the principle of obtaining a reduced rotational speed of the output shaft as the basis for classification of low-speed high-torque a.c. machines (Figure 1).

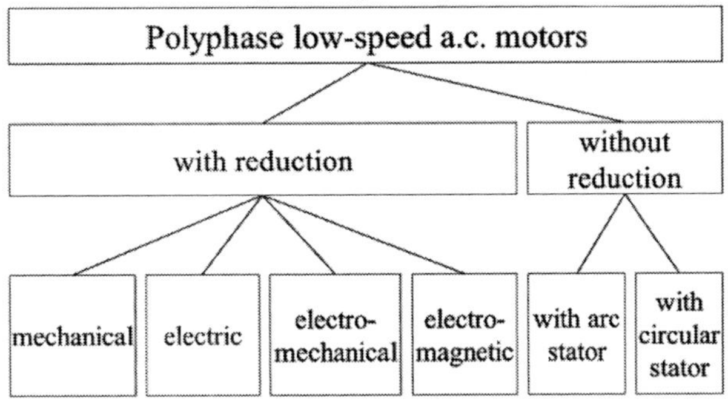

Figure 1. The classification diagram of low-speed high-torque electric motors.

In accordance with the given integrated classification, all the polyphase low-speed machines can be divided into two groups: geared motors and gearless motors.

Geared motors are specified by reduced angular velocity ω_2 of the output shaft, compared to angular velocity ω_1 of the main harmonic term of the magnetic field, the power being a source of conventional frequency. In connection with this, the reduction factor is $k = \omega_2 / \omega_1 > 1$.

Based upon the rotational speed reduction principle, geared electric machines are divided into motors with mechanical reduction, motors with electric reduction, motors with electromechanical reduction and motors with electromagnetic reduction.

Gearless low-speed electric machines ($k = 1$) are divided into circular stator motors and arc-stator motors [1], both types being fed from conventional frequency sources. The rotational speed reduction in arc-stator motors is achieved not through an increase of pole quantity but through the lay-out of the stator winding over a part of a circle. Hence, the power of the above-mentioned motors is substantially lower than the power of circular-stator motors, their low rotational speeds being equal. Therefore, arc-stator motors are basically designed for relatively low torques, and the application of the latter is effective in electric drives of the machines wherein the connection of the electric machine with its actuator is considered to be expedient. Hence, we consider exclusively arc-stator gearless motors as the most general case for multipolar electric machines.

It should be noted that the consideration of the above-mentioned low-speed electromechanical systems as cascaded ones, as well as the application of the multi-rotor motors, etc., presents a definite problem because of the variety of the effectiveness estimate criteria.

We will analyze the potentialities for classification principles of achieving the low rotational speed of the a.c motor output shaft (from the point of view of the potential use in high-torque motor drives).

3. Motors with reduction

A mechanical reduction gear extensively applied in a.c electric drives [2, 3] is the simplest way of achieving a low rotational speed for the actuator. In so doing, a reduction gear is built in the electric machine body or is located outside, the reduction factor of the above-mentioned motor being equal to reduction gear ratio i ($k = i$) and being varied within certain limits depending on a reduction gear type used.

The rotational speed reduction principle consists in feeding an electric machine from a certain dedicated electric power source with frequency f_0 (the latter being a connector between the motor and the power network), but not from a conventional frequency source. The necessity of a special source is a factor associated with the main disadvantage of a motors of the above-mentioned group and limiting their application.

Low frequency (in comparison to conventional frequency f) motors are excited by the reduced frequency current and have reduction factor $k = f / f_0$. The frequency converter transmits the required full power to the above-stated electric machine.

Well-known «double-powered» motors (slip-ring motors) are defined by reduction factor $k = f_1 / (f_1 - f_2)$ where f_1 and f_2– voltage frequencies feeding the stator and the rotor. Besides, contactless motor drives with slip rings are well known [4], two separate windings of different poles p_1 and p_2 being dropped in stator slots. The above motor reduction factor is defined by $k = f_1(p_1+p_2) / p_1(f_1-f_2)$.

The principle of electromechanical reduction of the rotational speed is based on the integration of a mechanical reduction gear box with the electromagnetic portion of the motor [5, 6]. The maximum reduction factor for the above-mentioned machines is estimated roughly as 100-5000. As we know, there are two main types of motors based on the above-mentioned principle.

In the rolling rotor machine, the rotor axis rotates about the stator axis of symmetry in synchrony with the field. The reduction factor is $k = D_2/(D_1 - D_2)$, where D_1 and D_2 are the stator bore diameter and the rotor bore diameter, respectively.

The wave-rotor motor (where a wave drive is physically integral with the electric machine) features reduction factor $k = z_2 / (z_2 - z_1)$, where z_1, and z_2– teeth numbers (teeth quantities) of rigid and flexible gear rings in the stator and the rotor. In spite of certain advantages, the above-mentioned motors are

IOP Conf. Series: Materials Science and Engineering **177** (2017) 012139 doi:10.1088/1757-899X/177/1/012139

currently not produced (with a relatively low torque), that is caused by design complexities and difficulties in manufacturing of an elastic rotor of the required core thickness.

The motors with electromagnetic reduction operate by the principle of magnetic flux amplitude modulation [7, 8]. Taken together, they are divided into excitation winding (active) motors and motors without the excitation winding (reactive motors).

The reduction factor for an active motor is $k = z_2 / p_1$, where z_2–the number of rotor teeth, p_1– the number of stator winding pole pairs.

The reduction factor for a reactive motor is defined by $k = z_2 / (z_2 - z_1)$, where z_1 and z_2 – the number of stator and rotor teeth, respectively.

The Megatorque high-torque a.c. machines with actuators produced by Motornetics (USA) are included into the above-mentioned class of low-speed electric machines. The above-mentioned motors comprise inner and outer stators with open poles (salient pole stators), the teeth being located over the pole pieces, and a thin ring rotor — with a low moment of inertia. Other designs are known as well, e.g. a low-speed machine [9].

The maximum reduction factor of motors with electromagnetic reduction is estimated roughly as 50-100. The motors based on two of the above-mentioned reduction principles are also available.

4. Gearless motors

The low rotational speed can be obtained at the expense of multipolar winding arranged over the stator. The versions of winding arrangements are thought to be the most practical in providing a maximum permissible low-speed: with slot numbers for a pole and for phase $q \leq 1$.

It is a fact that value q (according to the rules of symmetrical stator winding arrangement) cannot assume value $q \leq m-1$, where m – the number of phases. When$m-1 < q < 1$, the MMF winding curves include a wide range of relatively practically expressed harmonic components, the majority of the latter are of prohibitively high value, as research indicates. The above-stated fact causes a rise of differential dissipation, an increase of the complementary losses level, and thereby an efficiency decrease; it causes parasitic electromagnetic moments as well as noise and the vibration rise.

Our analysis shows that when $m-1 < q \leq 1$, the machine with stator windings at $q = 1$ and with slot skewing features the best properties. Yet the diameter of the machine is rather large, which is impossible to tolerate according to operating conditions. Thus, if the traditional design motor is unrealizable, it isadvantageous to direct attention to a polystator machine offering a certain advantage of raising its low-speed level at the required diameter. The design of the gearless multipolar machines demonstrates certain characteristic properties [10], which will be considered in the part given below.

5. Dimension relationships in multipolar machines

We will first consider the main overall dimensions in the multipolar motors. As theoretical investigations show, the Esson constant in multipolar motors is two times lower in comparison to high-speed electric machines and can be approximatly estimated by value $C \approx 2.5$ kVA·min/m^3, because of a relatively low induction value for a slot in a multipolar motor ($B_\delta < 0.55T$) and more complicated cooling conditions in comparison to high-speed electric machines. We proceed from the fact that the outer diameter is defined as $D_a = (1.1\text{-}1.25)D$, where D – the inner diameter of the core. Then the design length of the motor is determined by the following relationship:

$$l_\delta = \frac{(5.2\text{–}6.2)\cdot 10^{-4}}{D_a^2}\frac{S}{n}, \tag{1}$$

where S– the design power; n– the rotational speed.

The third main dimension-air gap ofδ is taken on the basis of providing an allowable dissipation in a multipolar motor as well as by manufacturing standards.

As it does so, δ satisfies the following relationship:

$$10^{-3}D < \delta < 1.3(D/p)^2, \tag{2}$$

where p – the number of pole pairs.

IOP Conf. Series: Materials Science and Engineering **177** (2017) 012139 doi:10.1088/1757-899X/177/1/012139

In certain cases of preliminary assessment, we need to define the maximum obtainable number of poles for the maximum allowable design outer diameter of the armature in a low-speed motor. That is done by means of relationship:

$$p \leq (3.2\text{–}3.7)\cdot 10^{-3}\sqrt{D_a}. \qquad (3)$$

The multipolar machine with a restricted diameter is characterized by relatively low pole pitch τ. In this situation, the core altitude of poles is to be extended for providing the required power level as well as design power factor $\cos\varphi$ in the synchronous electric motor. The required number of exitation winding turns is to be reduced in core slots, the conductor cross-sections being of a definite value. In this event, pole altitude h becomes objectionably high, the latter results in impermissible pole dissipation. Thus, the relative pole altitude should be restricted according to:

$$\frac{h}{\tau} \leq 0.3p\Big/\left(1+\frac{p^2}{100}\right). \qquad (4)$$

Relationships (1)-(4) have been approved by the practice of design and development of multipolar machines ($p = 12\text{-}60$) with certain limited cross-sectional dimensions ($D_a < 4m$), the power range being $S = 100\text{-}1000$ kVA, and the required quantity estimates can be done.

6. Conclusion

The diagram roughly estimating the bounds of engineering expediency of low-speed a.c. electric motors is presented, based on the torque developed and the rotational speed of the output shaft (Figure 2), and on the strength of the analysis of the given design relationships and statistical processing of research data. The given diagram provides obtaining assessments essential for a choice of a low-speed gear electric motor type. In the diagram, a point located in areas *a-e* (or outside these areas) corresponds to each pair of required torque values *T* and rotational speed *n*. The location of the above-mentioned point outside the areas marked by dashed lines (by shading) is evidence that in developing the motor with a combination of these parametres, there are certain engineering problems. In search of alternatives, the area (that is the reduction type), located at a lower level, is preferable. Thus, if we are to provide rotational speed $n = 100$ r/min at torque $T = 10$ kN·m, we get to the point overlapped with the *c*, *d*, *e* areas, simultaneously. Consequently, there is a scope for developing either an electric machine with mechanical reduction or an electric machine with electric reduction or a gearless multipolar machine. However, the principle of rotational speed mechanical reduction is more preferable from the point of view of maximum engineering and economical effectiveness in decision-making.

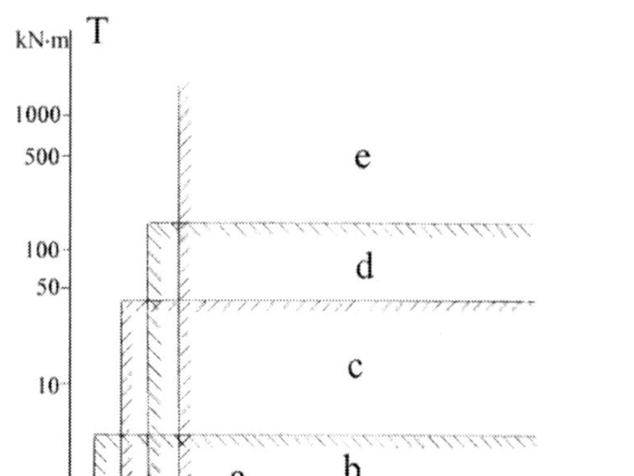

Figure 2. Statistically averaged fields of application for various types of low-speed electric machines: a – motors with electromechanical reductions; b – motors with electromagnetic reduction; c – motors with mechanical reduction; d – motors with electric reduction; f – gearless multipolar electric motors.

The data presented are averaged, but the allowance is made for the development of a low-speed electric drive on the basis of a high-torque motor, the limiting values of parameters lying outside the above-mentioned areas.

References

[1] Voldec A I and Popov V V 2007 *Electric Machines, Alternating Current Machines* (Saint-Petersburg: Piter)

[2] Bekeshev R F and Dementiev U N 2016 *Electrical drives* (Moscow: Urait)

[3] Kagan A V 2012 Construction of electrical machines (Saint-Petersburg: Saint-Petersburg Mining University)

[4] Broadway A and Burbridge L 2010 *Proc. IEEE (Piscataway)* vol 117 (Stevenage: IET) p 2160

[5] Borzyak Yu G, Zaikov M A and Nanii V P 2002 *Electric Motors with Rolling Motor* (Kiev: Technika)

[6] Green G W and Paul K J 2010 *Proc. IEEE (Piscataway)* vol 118 (Stevenage: IET) p1413

[7] Harris M, Finch J, Mallick J and Miller T 2008 *IEEE Trans. on Ind. Appl. (Greenfield)* vol IA-22 (Piscataway: IEEE) p 716

[8] Andrienko P D and Zhulovyan V V 2001 *Russian J. Electrotechnics* **11** 23-25

[9] Hashimoto I, Ishizaki K, Hayashida H, Miyashida K 1989 *Low-Speed High-Torque Motor* Patent US 4890024 A

[10] Kagan A V and Ryabukha V I 1988 *Russian J. Electromechanics* **6** 48-52

Forecasting of the electrical actuators condition using stator's current signals

T N Kruglova[1], I V Yaroshenko[2], N N Rabotalov[3] and M A Melnikov[3]

[1] Platov South-Russian State Technical University, 132, Prosveschenia str., Novocherkassk, Rostov Region, 346428, Russia
[2] Diagnostic Systems and Complexes L.L.C., 132, Warsaw ave., Moscow, 117519, Russia
[3] HV Measurement complexes and Systems L.L.C., 44/11 Balakovo, Trnavskaya str. Saratov Region, 413864, Russia

E-mail: tatyana.kruglova.02@mail.ru

Abstract. This article describes a forecasting method for electrical actuators realized through the combination of Fourier transformation and neural network techniques. The method allows finding the value of diagnostic functions in the iterating operating cycle and the number of operational cycles in time before the BLDC actuator fails. For forecasting of the condition of the actuator, we propose a hierarchical structure of the neural network aiming to reduce the training time of the neural network and improve estimation accuracy.

1. Introduction

Modern development of automation and robotics require the increased accuracy and reliability of the actuator operation. In this context, the BLDC motors are widely useful as actuator systems. A typical BLDC based actuator is an electromechanical system including an electrical motor and a reducer connected by a coupler. Long operation activity of the BLDC based actuator [1] under the large reversed loads can produce failures. Therefore, it is necessary to monitor periodically the BLDC actuator. It is equally important to determine the time and the cause of equipment failure. This problem can be solved using the forecasting technical condition method for BLDC actuators.

The main problem of the development of the condition estimation methods is the large amount of non-formalized qualitative information that cannot be used with conventional modelling methods in contrast to artificial intelligence methods, where these data can be managed.

2. Materials and methods

One of the simplest and the most accessible methods of diagnosing is the spectral analysis of stator's current signals, as it does not require additional setups and it is not time consuming. It can be performed directly on the working equipment.

This method allows estimating the technical condition of the electric motor and the mechanical devices connected with it. During the set time interval, the steady-state currents consumed by the motor are converted into the frequency domain using the Fourier transformation [2]. For the analysis, the current signal of a new serviceable motor, which is accepted to be the basic reference standard, is measured once before the long-time exploitation. When faults occur, there is a change in the common level and single amplitudes at the characteristic frequencies. The investigation of failures is carried out by comparison of the current spectrum with the basic reference spectrum done by means of computational intelligence. Failure search is performed by comparing the amplitudes of the current spectrum and a new serviceable the actuator.

Midrange current signal (1), which can be considered as a bias arising from the process noise, is obtained from all amplitudes of a current spectrum without characteristic frequencies, is computed as follows:

$$a_{mid} = \frac{\sum_{i=1}^{g} a_i - \sum_{j=1}^{h} a_j}{g-h}, \qquad (1)$$

where a_i – current signal amplitude ; i, j – frequency indices; g – frequencies of a spectrum interval; h – characteristic frequencies of diagnosing.

The following equation describing the spectrum analysis is restricted to the normalized characteristic frequencies under consideration:

$$k_i = \frac{A_i - A^0_i + \Delta a_{mid}}{a_{mid\ 0} + A_i^0}, \qquad (2)$$

where A_i – the amplitude of the analyzed spectrum on the i-th characteristic frequency; A_i^0 – the amplitude of a reference spectrum on the i-th characteristic frequency; $a_{mid\ 0}$ – the midrange current signal of the reference spectrum; $\Delta a_{mid} = a_{mid\ i} - a_{mid\ 0}$ – absolute deviation of the midrange current signal; $a_{mid\ i}$ – the midrange current signal of the analyzed spectrum.

If the analyzed spectrum is equal to the referential one, then normalized factor $k_i = 0$. If a fault occurred, then the change of the midrange current signal and amplitude at characteristic frequencies lead to a change of the normalized factor. These data are fed into the neuro-fuzzy troubleshooting model [3]. The received model output allows estimating a current condition of the object, having attributed it to one of the following classes: $\mathbf{F(x^*) = 1}$ – serviceable; $\mathbf{0 < \mathit{F}(x^*) < 1}$ – operative; $\mathbf{-1 \ll F(x^*) \ll 0}$ – corrupted.

If the BLDC actuator is operative, it is possible to forecast its technical condition. Forecasting is carried out at two stages. The first stage consists in predicting the next time values of the function. At the second stage, the outputs of the first phase are approximated and are determined by the state of the BLDC actuator at the next iteration.

For forecasting of faults, a simulation model based on the neural network such as perceptrons with three inputs and one output is used.

The network consists of three layers. The first and second layers have a sigmoid activation function. The third output layer has a linear activation function [4].

Initial data for the prediction values are defining functions in different time intervals $F_L(x_i^*),\ i \in [1, n]$.

The first input of the neural network is a vector of values defining function $F_1(x_i^*).....F_{L-3}(x_i^*)$. The second input has the same parameters, but with a shift in one parameter to the right — $F_1(x_2^*).....F_{L-2}(x_i^*)$. The third input has the parameters, which are shifted by one relative to the second entrance — $F_3(x_i^*).....F_{L-1}(x_i^*)$. The data form the input vector of the neural network represented as follows:

$$P_i = [F_1(x_i^*)...F_{L-3}(x_i^*);\ F_2(x_i^*)...F_{L-2}(x_i^*);\ F_3(x_i^*)...F_{L-1}(x_i^*)].$$

The purpose of vector H is to determine the value of the function from the fourth to the L-th:

$$H_i = [F_4(x_i^*),...F_L(x_i^*)].$$

Using input *P,* the neural network calculates the output value of defining function *Y*, which must comply with *H.* For the network learning algorithm, the back propagation approach is used [5,6]. To find the next value of the defining function, the input is represented as a column vector containing the last three values of the training sample:

$$C = \left[F_{L-2}(x_i^*); F_{L-1}(x_i^*); F_L(x_i^*)\right].$$

For the second phase of the forecasting process, a radial basis network (Figure 2) is simulated. The inputs of the network are fed to the output networks of the first stage. The mathematical minimum function is used for training the neural network.

If the network output of the second phase is positive, then the outputs networks of the first stage will be added to the training set, and the process repeats until the value of the output of the second phase become less than or equal to zero. The number of iterations equals the number of cumulative intervals up to the current time (Figure 1).

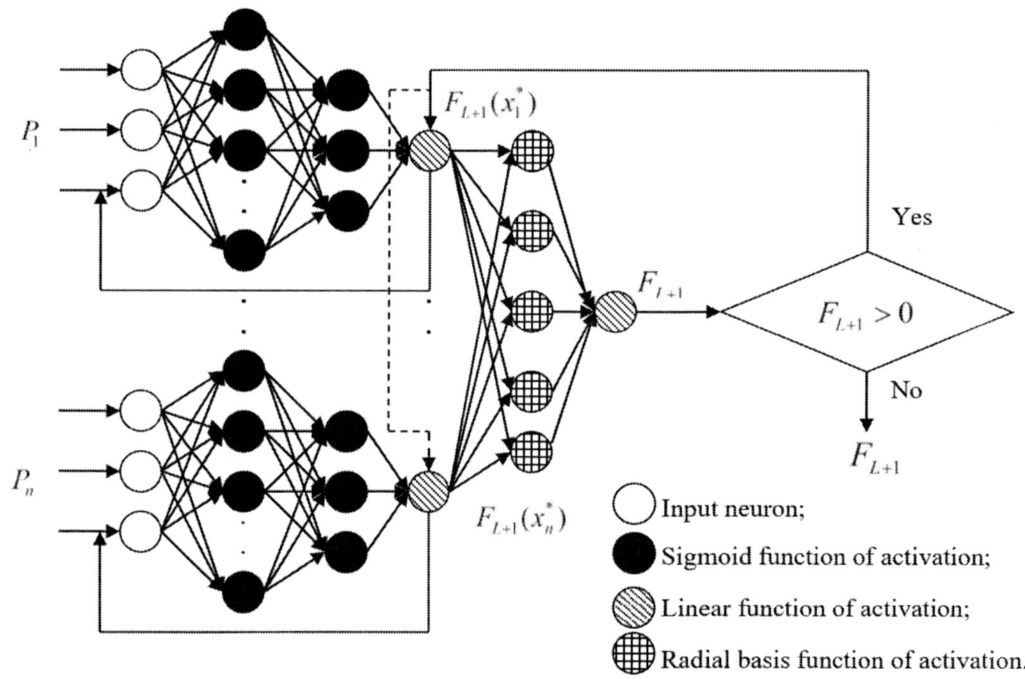

Figure 1. The neural network for the forecasting

3. Experimental results

The described method was used for BLDC motor condition estimation, where all the necessary readings were acquired from the coupling of the reducing gear connected to an actuation mechanism. Current measurement is carried out within a low level BLDC control, therefore, using a spectrum current, it is possible to determine only the technical condition of BLDC with the full load, which is directly connected with it. Reducing gear is a constant passive load hence there is no variable influence on a frequency spectrum of the current consumed by the motor. The given characteristic frequencies allow introducing the following classification illustrated in figure 2.

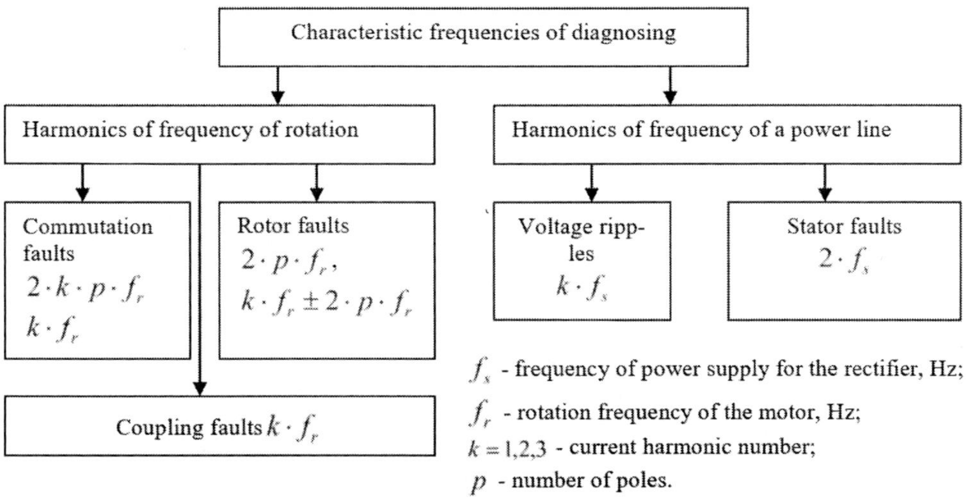

Figure 2. Classification of characteristic frequencies of diagnosing

The investigations of changes in the technical condition of the BLDC actuator are performed with a long-term drive. At the frequency of 15 Hz, six measurements of current in a motor with constantly velocity of rotating and an interval of 30 minutes are conducted. The obtained data diagnostics are shown in Table 1.

Table. 1 Diagnostic results of the BLDC actuator condition in the continuous mode operation

	Time of exploitation of the BLDC actuator, min					
	0	30	60	90	120	150
Commutation faults	0.877	0.705	0.96	0.9	0.21	0.998
Rotor faults	1	1	1	1	1	1
Voltage ripples	0.768	1	0.95	0.9	1	0.965
Coupling faults	-1	-1	-1	-1	-1	-1
Stator faults	1	1	1	0.7	1	1

From Table 1, it is evident that the dividing function, which characterizes the coupling state, reaches its minimal value. This indicates the coupling is failing. The changes in the coupling state can be estimated indirectly by the changes in the dividing function of others fault features. The changes in the coupling behavior impact the communication faults readings at overlapped characteristic frequencies. The forecasting of the time horizon, when the coupling damage appears, can be done by means of the evolution of dividing functions during the long-term actuator drive (Figure 3).

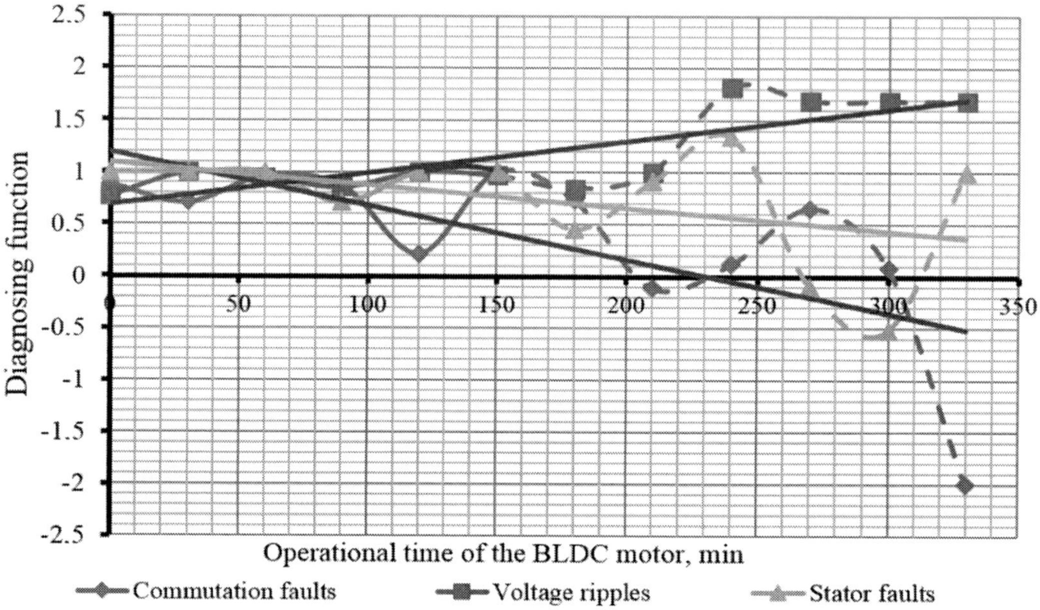

Figure 3. The forecast of failure development

From Figure 3, it is evident that the dividing functions progress periodically with a tendency approximated as linear. The coupling damage is possible at the time when one of the linear approximations cuts the time axis as first.

4. Conclusion

This paper proposed a forecasting method of the technical condition for the previous value of the current amplitude at specific frequencies to determine the time and cause of the drive fault based on the hierarchical neural network.

Using the trend line of previous current values, it is possible to perform short and long term condition estimation of the motor, the defects status and the forecasted time for equipment failure. The experimental research has confirmed the adequacy and accuracy of this method with reference to the theoretical findings and adopted technical solutions. The condition estimation reliability is 95%.

References

[1] Subhasis Nandi, Hamid A. Toliyat, and Xiaodong Li 2005 *IEEE Transactions on energy conversion (Vancouver)* vol 20(4) (IEEE Journals & Magazines) p 719
[2] T W Körner 1993 *Fourier Analysis* (New York: Press Syndicate of the Unuversuty of Cambrige)
[3] Kruglova T N 2015 *Procedia Engineering* **129** 219-224
[4] F Uossermen 1992 *The Neuro-Computer techniques: the Theory and practice* (New York: Van Nostrand Reinhold Co.)
[5] Guoqiang Zhang, B Eddy Patuwo, Michael Y Hu, 1998 *Int. J. of Forecasting* **14** 35-62
[6] Demin A V, Dmitrieva S 2015 *7th Intrenational Congress on Ultra Modern Telecommunication and Control Syttemand Workshops (Brno, Czech Republic)* p 14

Temperature control of power semiconductor devices in traction applications

A A Pugachev[1] and N N Strekalov[2]

[1] Bryansk State Technical University, 7, 50-letiya Octyabrya bul., Bryansk, 241035, Russia
[2] Moscow State University of Railway Engineering, 9, Obrazcova str., Moscow, 127994, Russia

E-mail: alexander-pugachev@rambler.ru

Abstract. The peculiarity of thermal management of traction frequency converters of a railway rolling stock is highlighted. The topology and the operation principle of the automatic temperature control system of power semiconductor modules of the traction frequency converter are designed and discussed. The features of semiconductors as an object of temperature control are considered; the equivalent circuit of thermal processes in the semiconductors is suggested, the power losses in the two-level voltage source inverters are evaluated and analyzed. The dynamic properties and characteristics of the cooling fan induction motor electric drive with the scalar control are presented. The results of simulation in Matlab are shown for the steady state of thermal processes.

1. Introduction

The power semiconductor module is a main source of heating in up-to-date frequency converters. Its cooling system is designed to provide the needed thermal behaviour limited by the maximum temperature of IGBT and diode junctions [1,2]. This task is the pivotal matter for the traction electric drives in railway applications, where the wide range of loads and speeds is a common condition. The significant currents flowing through the collector and the emitter during the start mode under the heavy freight or fast acceleration lead to the considerable power losses that, in its turn, cause the excessive heating. Otherwise, in the no-load mode, the traction drive could be operated in the mode close to the no-load. These load cycles depend on the profile of the track and, as a result, heating cycles of semiconductor devices vary very widely from almost a cold state to severe overheating. So, the best way to control this type of the condition is to apply an automatic temperature control system. This approach allows us to obtain the required level of air or liquid cooling flow to take away overheating in different modes. It also leads to the possibility of energy saving while the current approaches to the no-load mode. It appears to be an important matter for freight locomotives where the cooling system of the traction frequency converter consume a large amount of energy. For example, the 22 kW induction motor is used in the 2TE25A locomotive for the air cooling of the traction converter, the 55 kW induction motor is used in the VL85 electric locomotive for the same purpose.

The survey of locomotives used in the Russian railway shows that the induction motors of cooling fans are either controlled discretely or not controlled at all [3]. Meanwhile, the greatest effect in solving the problem of increasing the efficiency of cooling systems and providing the required temperature of semiconductor devices of a frequency converter can be achieved if the cooling systems will be equipped with controlled drive fans, allowing continuous and automatic change of cooling air or liquid depending on the temperature of the IGBT and diode junctions. So, the investigations of thermal behaviour of semiconductor devices used in traction applications and the developments of its control system are still relevant and important. This article presents the topology and the principle of operation of the automatic temperature control system of traction frequency converter semiconductor devices.

2. An automatic temperature control system of semiconductor devices

Despite the fact that there are a number of techniques [4-7] allowing one to reduce junction temperature and its fluctuation by means of changing the frequency switching, this approach does not get wide implementation in the traction drives due to the complex dependences of mechanical processes including adhesion, skid and slipping effects in starting and braking modes and the frequency of the torque and current oscillation. The forced air or liquid convection by means of a fan, a blower or a pump is much prefferable in these applications.

The scheme of the automatic temperature control system of semiconductor devices of the frequency converter, feasible for using in a locomotives, is given in Figure 1. It is necessary to analyse properties of each of the units of the scheme (Figure 1) for developing the whole block diagram of the control system. In the present article, the authors proposed the implication of the air fan driven by the induction motor drive with scalar control as the executive regulation device. The dynamic and static performances of the executive regulation device and semiconductor devices as an object of temperature control are investigated and discussed further in the sections below.

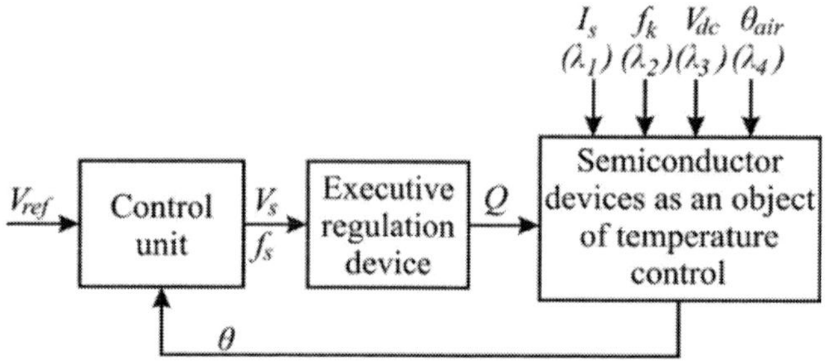

Figure 1. The scheme of the automatic temperature control system: θ – temperature of semiconductors, V_{ref} – temperature reference, V_s – stator voltage, f_s – stator current frequency, Q – air flow, λ_i – disturbances, I_s – stator current of a traction induction motor, f_k – switching frequency, V_{dc} – DC-link voltage, θ_{air} – air temperature

3. Semiconductor devices as an object of temperature control

The most interesting and difficult thing to be implemented in the suggested scheme is the feedback of the semiconductor temperature. It is obvious that the most thermally stressed part of any semiconductor is its junction and it is almost not available for the direct measurement. So, it is necessary to introduce some indirect techniques to measure or estimate the junction temperature. The technique allowing us to get come solution of this problem bases on the thermal equivalent circuit of the power semiconductor module [4,5]. The thermal equivalent circuit is presented by the multiple circuit of resistances and capacitors connected in series and parallel, the transient characteristics of it are given in the module datasheets as a thermal response in the single step reference or disturbance. The simplified and full thermal equivalent circuits of one leg of the two-level voltage source inverter (Figure 2) are shown in Figure 3.

In Figure 3, $Z_{j\text{-}c}$, $Z_{c\text{-}hs}$, $Z_{hs\text{-}air}$ denote the dynamic thermal impedance between the junction and the case, the case and the heatsink, the heatsink and the cooling air, respectively; θ_j, θ_c, θ_{hs} denote the temperature of the junction, the case and the heatsink of the semiconductor module, respectively; θ_{air} denotes the temperature of the cooling air, ΔP denotes the power losses. The power losses in Figure 3 are presented as a current source while the temperature of cooling air is presented as a voltage source. The dynamic thermal impedance can be described as follow:

$$Z_{x-y}(t) = R_{x-y}(1 - e^{-t/\tau_{x-y}}),$$

where $R_{x\text{-}y}$ denotes the thermal resistance between conjugated units of x and y; $\tau_{x\text{-}y}$ denotes the time constant of the thermal transient mode.

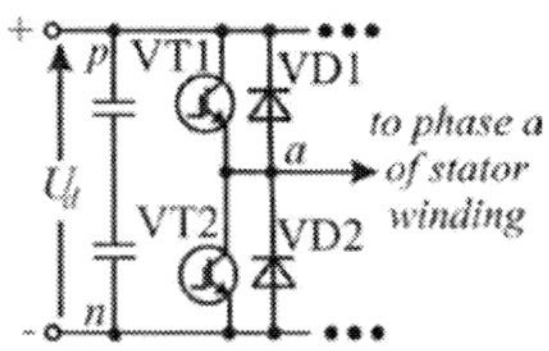

Figure 2. The equivalent circuit of one leg of the two-level inverter

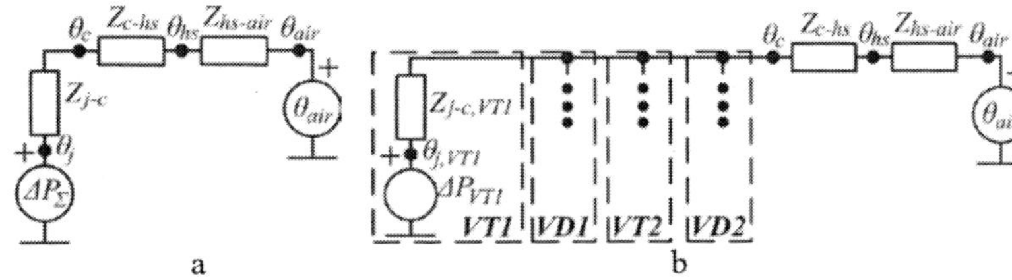

Figure 3. The simplified (a) and full thermal equivalent circuit of one leg of the two-level inverter (Figure 2)

The junction temperature in the steady state can be derived from the simplified equivalent circuit (Figure 3a) as follows:

$$\Theta_j = \Theta_{air} + \Delta P_\Sigma \Sigma R_{x\text{-}y}.$$

The equivalent circuit (Figure 3b) gives more detailed temperature distribution for the semiconductor module consisting of one leg of the two-level inverter (Figure 2). The junction temperature derived from the circuit (Figure 3b) is written as follow:

$$\Theta_j = \Theta_{air} + \Delta P_\Sigma R_{hs\text{-}air} + \Delta P_\Sigma R_{c\text{-}hs} + \Delta P_{VT1} R_{j\text{-}c}, \tag{1}$$

where ΔP_Σ denotes the total power losses in the semiconductor module; ΔP_{VT1} denotes the power losses in the *VT1*. It is obvious that Figure 3b gives more correct information for the junction temperature evaluation than Figure 3a does. It occurs due to the heat dissipation in the thermal resistance of $R_{j\text{-}c}$ caused by only power losses in the considered IGBT.

To use the equivalent circuits properly, it is necessary to take some assumptions for the investigation of temperature and power losses distribution in the semiconductor module. The main ones are as follows: the load current, i.e. the stator current, has the sinusoidal shape, the value of dead time of transistors is neglected, all the phases of voltage source inverter and induction motors are symmetric and even. Power losses in the power switches consist of conduction and switching losses. The conduction losses are mainly caused by the voltage drop on the junction under the forward current. The conduction losses depend on the stator current (as the load of the inverter), the temperature of the junction, the modulation type and its parameters, the power factor of the induction motor stator winding. The switching losses are caused by changing of the opened/closed state of transistors. The value of these losses depends on the load current, DC-link voltage, junction temperature, switching frequency and turn-on and turn-off energy. As some investigations show [6,7], the load current and switching frequency have the main influence on the total power losses.

With the numerical methods applied to the evaluation of electric drive dynamics, the following expressions are used for the power losses over one switching period T_k:

$$\Delta P_{cond,VT} = \frac{1}{2\pi}\int_0^{\pi} v_{ce} I_{\max}\sin(2\pi f_s t)\frac{1+\mu(t)}{2}d(2\pi f_s t) \tag{2}$$

$$\Delta P_{cond,VD} = \frac{1}{2\pi}\int_0^{2\pi} v_{ce} I_{\max}\left|\sin(2\pi f_s t)\right|\frac{1+\mu(t)}{2}d(2\pi f_s t) \tag{3}$$

$$\Delta P_{sw,VT} = \frac{1}{2\pi}\int_0^{\pi}\left(E_{on}(I_s)+E_{off}(I_s)\right)\frac{V_{dc}}{V_{dc,rat}}\frac{1}{T_k}d(2\pi f_s t) \tag{4}$$

$$\Delta P_{sw,VD} = \frac{1}{2\pi}\int_0^{2\pi} E_{rec}(I_s)\frac{V_{dc}}{V_{dc,rat}}\frac{1}{T_k}d(2\pi f_s t) \tag{5}$$

where $\Delta P_{cond,VT}$, $\Delta P_{cond,VD}$ denote the conduction power losses in IGBT and the diode, respectively; , $\Delta P_{sw,VT}$, $\Delta P_{sw,VD}$ denote the switching power losses in IGBT and the diode, respectively; v_{ce} denotes the collector-emitter voltage; E_{on}, E_{off} denote the turn-on and turn-off energy of IGBT, E_{rec} denotes the recovery energy of the diode; μ denotes the modulation coefficient; the subscript of *rat* denotes the rated value of some parameter.

The dependence of collector-emitter voltage v_{ce} from forward current I_s and junction temperature θ_j is approximated by the following equation [3]:

$$v_{ce}(I_s,\theta_j)= 1.21 + 9.93\cdot 10^{-3} I_s - 1.88\cdot 10^{-3} \theta_j - 2.54\cdot 10^{-5} I_s^2 + 3.22\cdot 10^{-5} I_s \theta_j. \quad (6)$$

The results of simulation of the electric drive for the power modules with thermal parameters $R_{c\text{-}hs,VT}$ = =0.085°C/W, $R_{c\text{-}hs,VD}$ = 0.13°C/W, $R_{j\text{-}c,VT}$ = 0.01 °C/W, $R_{j\text{-}c,VD}$ =0.015 °C/W, $R_{hs\text{-}air}$ =0.031 °C/W are shown in Figure 4.

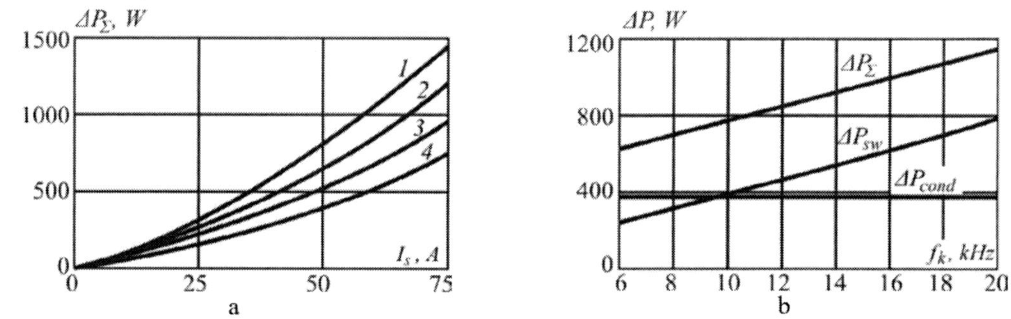

Figure 4. The total power losses vs. the stator current (a) (1 – f_k =20kHz, 2 – f_k =15kHz, 3 – f_k= =10kHz, 4 – f_k = 5 kHz) and the power losses vs. switching frequency (b) when stator current I_s = 65A

4. The cooling fan with the induction motor drive as an executive regulation device

The implication of the air cooling fan with the induction motor drive is proposed to cool the semiconductors. The most common features of electric drives of cooling fans for locomotives traction electric equipments and motors are highlighted from the survey [8,9] of operation modes of the auxuliary drives and machines. It can be listed as follows: shifting of speed to its lower values causes a considerable reduction of the load torque and the mechanical power on the motor shaft; the long period of time of the operation mode with permanent speed and load torque; the absence of the reverse mode; the absence of overloads, the value of overload may be determined in advance; the start-up time is not limited, the start is prefered to have small start-up time for restricting dynamic overloads. So, it is reasonable to apply a scalar control system of the induction motor because of some specific requirements the electric drive has to satisfy. To design the whole automatic temperature control system, the mathematical model and/or transfer functions of the induction motor and the electric drive have to be known. In this investigation, the well known equations [10] of an induction motor are used:

$$v_{s\alpha} = R_s i_{s\alpha} + \frac{d\lambda_{s\alpha}}{dt} - \omega_k \lambda_{s\beta},\ v_{s\beta} = R_s i_{s\beta} + \frac{d\lambda_{s\beta}}{dt} + \omega_k \lambda_{s\alpha}, \quad (7)$$

$$0 = R_r i_{r\alpha} + \frac{d\lambda_{r\alpha}}{dt} - (\omega_k - \omega)\lambda_{r\beta},\ 0 = R_r i_r + \frac{d\lambda_{r\beta}}{dt} + (\omega_k - \omega)\lambda_{r\alpha}, \quad (8)$$

$$\lambda_{s\alpha} = L_s i_{s\alpha} + L_\mu i_{r\alpha},\ \lambda_{s\beta} = L_s i_{s\beta} + L_\mu i_{r\beta},\ \lambda_{r\alpha} = L_r i_{r\alpha} + L_\mu i_{s\alpha},\ \lambda_{r\beta} = L_r i_{r\beta} + L_\mu i_{s\beta} \quad (9)$$

$$T = \frac{3}{2} p_n \left(i_{s\beta}\lambda_{s\alpha} - i_{s\alpha}\lambda_{s\beta}\right) \quad (10)$$

where α,β denote the coordinate axes; $v_{s\alpha}$, $v_{s\beta}$ denote the corresponding coordinates of the stator voltage; $i_{s\alpha}$, $i_{s\beta}$, $i_{r\alpha}$, $i_{r\beta}$ denote the corresponding coordinates of the stator and rotor currents; $\lambda_{s\alpha}$, $\lambda_{s\beta}$, $\lambda_{r\alpha}$, $\lambda_{r\beta}$ denote the respective coordinates of the stator and rotor flux linkages, R_s, R_r denote the stator and rotor resistances, L_s, L_r, L_μ denote the stator and rotor inductances and magnetizing inductance, respectively; ω_k denotes the speed of the reference; ω denotes the speed of the rotor shaft; T denotes the torque produced by the motor; p_n — the number of pole pairs.

The mechanical movement is described as follows:

$$T - T_L = J\, d\omega/dt \tag{11}$$

where T_L denotes the load torque, J denotes the moment of inertia.

5. Simulation of the automatic temperature control system

The scheme (Figure 1) adapted by the conducted research to be simulated is shown in Figure 5. To simulate the automatic temperature control system, the Matlab Simulink is used in this investigation. The executive regulation device is described by equations (6) – (11), the thermal behavior of the semiconductor devices is evaluated by equations (1) – (5). The cooling fan is simulated by the proportional gain. The PI-controllers are applied for both temperature (W_θ) and speed (W_ω) controllers to exclude the error in the steady state. As it was mentioned in section 4, the scalar control of the cooling fan induction motor is implemented. The coefficients of k_U and k_ω are chosen so as to provide the square dependence between stator voltage V_s and stator current frequency f_s. The algorithms of sinusoidal pulse-width modulation (*PWM*) is used to control semiconductors of the voltage source inverter (*VSI*). The presence of the forced air convection is taken into account by reducing thermal resistance $R_{hs\text{-}air}$ by the technique described in [11]. The results of the simulation in the steady state are shown in Figure 6 for I_s=65A. The junction of the IGBT is established to be the most thermal stressed unit of the semiconductor module.

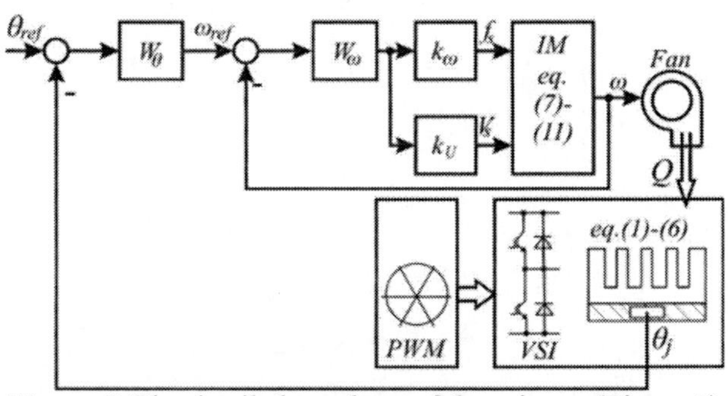

Figure 5. The detailed topology of the scheme (Figure 1) for simulation

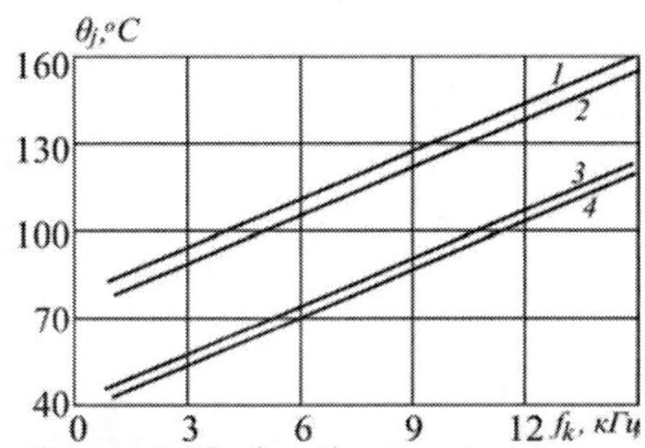

Figure 6. The junction temperature vs. switching frequency (1 – $\theta_{j,VT1}$, 2 – $\theta_{j,VD1}$ during free convection, 3 – $\theta_{j,VT1}$, 4 – $\theta_{j,VD1}$ during forced convection)

6. Conclusion

The investigation presented in the article shows the relevance and the necessity of designing the automatic temperature control system of the semiconductor modules of the traction frequency converter that allows one to get desired thermal behavior of the power switches at any time. The adequacy of the proposed topology and the operation principle of the control system is proved by means of Matlab Simulink. As the results of simulation show, the implementation of the induction motor electric drive for driving the cooling fan and the closed-loop control of junction temperature leads to the reduction of temperatures. Also, the results of simulation show that the most heated unit of the module is the IGBT junction.

References

[1] Thorsen O V, Dalva M 1995 *European Conference on Power Electronics And Applications* (EPE'95, Spain) vol 1 p 1033

[2] Lezana P, Poy J, Meynard T A, Rodriguez J, Ceballos S, Richardean F 2010 I*EEE Transactions on Industrial Electronics* **57(7)** 2207-18

[3] Kosmodamianskiy A S, Klyachko L M, Vorobiev V I and Pugachev A A 2014 R*ussian Electrical Engineering* **85(8)** 513-518

[4] Junke W, Luowei Z, Pengju S, Xiong D 2014 I*nt. Power Electronics and Appl. Conf. and Exposition* **1** 41-48

[5] Murdock D A, Ramos Torres J E, Connors J J, Lorenz R D 2006 IEEE Transactions on Industry Applications **42(2)** 552-558
[6] Huang X., Xizhimen W., Zhang L., Sun H., You X 2010 *IEEE Energy Conversion Congress and Exposition* 1666-71
[7] Blasko V [et al.] 1999 *Industry Appl. Conf.* **1** 1423-31
[8] Kosmodamianskiy A S, Vorobiev V I and Pugachev A A 2012 R*ussian Electrical Eng.* **83(12)** 667-671
[9] Krishnan V 2001 *Electric Motor Drives: modeling, analysis and control* (Virginia Tech, Blacksburg, VA)
[10] Nowotny D, Lipo T 1995 *Vector Control and Dynamics of AC drives* (Clarendon Press, Oxford, UK)
[11] Bachman P, Haiduk R *The effect of forced air cooling on heat sink thermal ratings* (white paper)

Assessment of duration of the drive operation in the mode of kinetic energy recovery under power supply voltage sags in electrical grids of mechanical engineering enterprises

O B Shonin and N G Novozhilov

Department of Electrical Engineering, Saint-Petersburg Mining University, 21st Line, 2, St Petersburg,199106, Russia

E-mail: ninosh_eltech@mail.ru, nikita_novozhilov@mail.ru

Abstract. Voltage sags in electric grids of mechanical engineering enterprises may lead to disconnection of important power consumers with variable frequency drives from the power grid and further interruption of the production process. The paper considers a sensorless V/f control system of epy induction motor drive under normal conditions and under voltage sags on the basis of a computer model of the drive and derivation of a formula for assessment of possible duration of the drive operation in the mode of controlled recovery of kinetic energy accumulated in rotating mass of the drive. Results of simulations have been used to validate results of calculations of the rotor velocity deceleration made in a closed form obtained from the equation reflecting the balance of torques. It is shown that results of calculations practically coincide with results of simulations in the range up to 5% of the velocity initial value. The proposed formula may be useful for estimation of the duration of the drive operation in the mode of recovery of kinetic energy depending on parameters of the motor and driven mechanisms.

1. Intoduction.

Variable frequency electric drive (VFD) is widely used in mechanical engineering industry because of its high level of reliability, controllability and flexibility – characteristics which are in great demand for numerous processing units [1, 2]. In spite of advantages, VFD is found to be susceptible to voltage sags, advent of which causes tripping of under-voltage relay in order to prevent the components of electric drive from damage due to inrush currents flowing at the recovery of normal power supply conditions [3-7].

Momentary power interuptions and voltage sags are normally caused by short curcuiting in transmittion lines and distribution lines. Parameters of voltage sags are known to depend on characteristics of protection relays tripping the faulted lines and topology of the network, that determines the area of propagation of voltage sags. Duration of voltage sags usually constitutes 0.1-3s and voltage magnitude is equal or less than 90% of nominal voltage value. Such events occuring 20-30 per year lead to restarting delays and disruption in technological processes and significant economic detriment because of lost productivity and distorted quality products [7-9]. Ensuring the reliability of VFD operation under voltage sags in order to keep the continuity of processes in industrial facilities is of great importance.

In order to run a facility through power interruptions and sags, either standby power is supplied or internal resourses of the drive are utilized in the form of maintaining the velocity of driven mechanism at the expence of controlled operation in the field weakening zone and/or recovering the kinetic energy accumulated in the driven mechanism in order to keep the voltage of DC link at predetermined level.

Feasibility and the cost of implementing each solution depend on practicality, economics and the losses incurred due to downtime[10, 11]. Running a facility through power interruptions and sags by means of internal resourses of the drive is preferable in most cases because of its simplicity and low cost of realization via modification of control system through software.

Algorithm-based strategies for improvements to voltage sag ride-through performance of VFD are normally implemented for the FOC electric drive. However, for widely spread v/f control systems realization of these kinds of strategies is also in great demand. For both types of VFD, it is essential to know possible duration of operation in the mode of recovery of kinetic energy available in rotating mass during a sag. Keeping the controls energized only solves a part of the problem and the actual motors and drives must also be analyzed.

The paper is devoted to building up a sensorless scalar conrol system of VFD with ability to be configured for realizing the ride-through strategy on the basis of maintaining the voltage of DC link by regulating an active component of the motor current. The second goal of the paper is to obtain a theoretical tool for assement of possible duration of the drive operation in a regenerative mode.

2. Sensorless v/f control system for IM drive under normal and voltage sags conditions.
The structure of a sensorless IM VFD with the two-loops v/f control system (figure 1) is built up on the basis of the rotor velocity observer [12] according to the principle of surbodinate regulation.

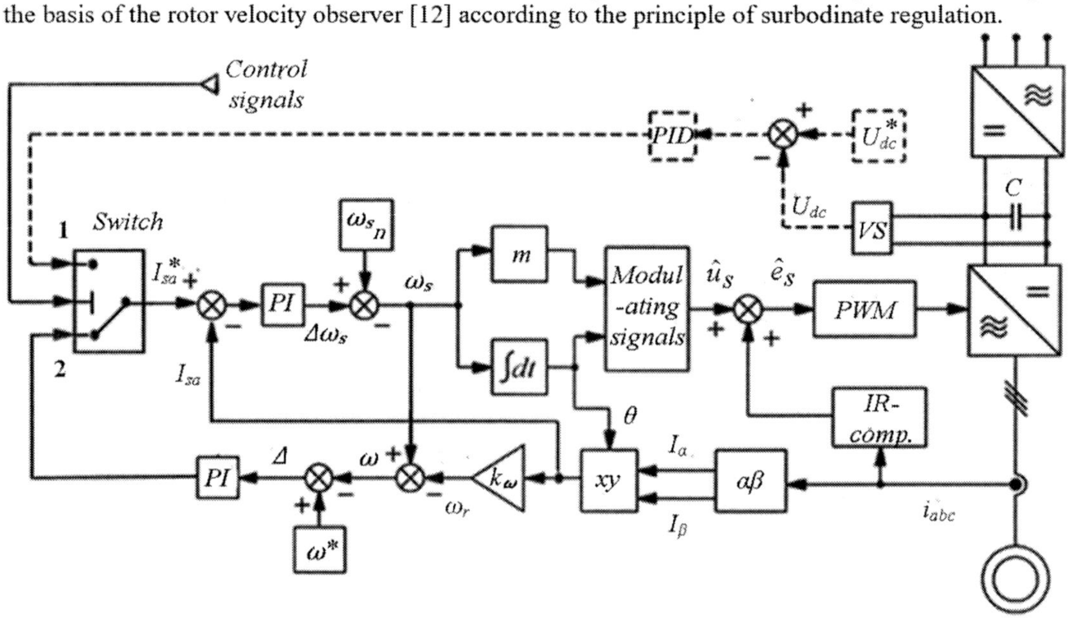

Figure 1. The sensorless v/f control system of IM VFD

The inner loop of regulating an active component of the stator current is under control of outer loop producing reference value I^*_{sa}. The control system comprises two outer loops depending on conditions of the drive performance – normal conditions when the rotor velocity is controlled to follow set-up reference value ω^* and under emergency conditions during AC voltage sags when the voltage of the DC link is kept at set value U^*_{dc} via charging the condensor from IM operting in regenerative mode which is controlled by varying angular velocity of rotating magnetic field in the air-gap of IM vs observed angular velocity of the rotor. Configuration of control system is changed via a virtual switch which state 1- normal mode or 2- emergency mode is controlled by signals from the block responsible for identification of AC voltage sags through processing the signals of input voltages.

In the normal mode of operation, the system under cosideration is functioning as follows. Measured stator currents i_a, i_b, i_c are converted into $\alpha\beta-$components of the current vector which then are projected onto the $x-$axis of the rotating frame (figure 2). These transformations allow determining the active component of the current as a sum of two projections:

$$I_{sa} = i_\alpha \cos\theta + i_\beta \sin\theta . \qquad (1)$$

Angle θ of fundamental harmonic of stator voltage $\vec{E}_s$ is set by VSI modulating vector $\bar{e}_s$, which is aligned along the $x-$axis. Slip frequency ω_r estimated according to linear relationship $\omega_r = k_\omega I_{sa}$, where $k_\omega = \omega_{r,nom} / I_{sa,nom}$ is used to calculate the rotor angular velocity as follows: $\omega^{est} = (\omega_s - \omega_r) / z_p$, where z_p – pole-pairs number.

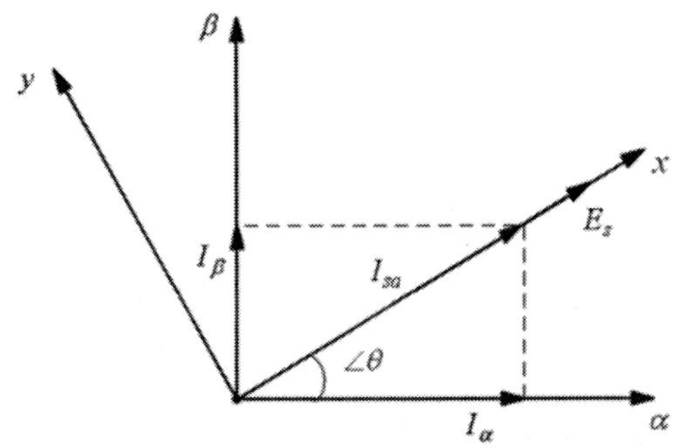

Figure 2. The stator current in stationary and rotating reference frames.

Comparison of estimated and set values of the the rotor velocities produces a signal $\Delta = \omega^* - \omega^{est}$ that is fed to PI-regulator, which outputs reference I^*_{sa} of active stator current for the inner control loop. The output of current PI-regulator $\Delta\omega_s$ is used to produce required frequency of PWM modulating signals $\omega_s = \omega_{s_{nom}} - \Delta\omega_s$, where $\omega_{s_{nom}}$ is a nominal value of the frequency. Obtained value ω_s is used to determine angle θ of generalized rotating vector $\bar{e}_s$ via the integrator and modulation index m, which for control law $E/\omega_s = const$ is calculated as follows: $m = \omega_s / \omega_{s_{nom}}$. This allows producing reference values for modulation signals $\hat{e}_a(t), \hat{e}_b(t), \hat{e}_c(t)$, which then are corrected with IR-compensation block. Comparison of calculated voltages $\hat{u}_a(t), \hat{u}_b(t), \hat{u}_c(t)$ with PWM carrier signals results in generating the switching functions S_a, S_b, S_c for IGBT modules of VSI.

If AC voltage dips are identified, the control system is reconfigured to implement the controlled recovery of kinetic energy stored in the drive. In this mode of operation, the reference value of active current I^*_{sa} is formed by a PID regulator as a reaction to input signal $\Delta U_{dc} = U^*_{dc} - U_{dc}$, where U^*_{dc} and U_{dc} – required and measured values, respectively.

3. Assessments of ride-through capability of proposed control system

3.1 Results of simulation

Investigation of IM VFD with the proposed control system in two modes of operation has been carried out via a computer model built up in MatLab/Simulink environment. The model includes squirrel-cage IM, inverter, fan load model, current sensors and sinusoidal PWM block. The motor parameters are as follows: P_{nom}=37kW; U_{ph}=220V; f=50Hz; z_p=7; s=0.02; R_s=0.084Ohm; $L_{s\sigma}$=0.0009H; R'_r=0.0564Ohm; $L'_{r\sigma}$=0.0011H; L_m=0.0109H; U_{dc}=540V; J=18 kg·m^2

In the normal mode, the rotor velocity constitutes 36.7 rad/s. Power supply interruption at time t_1=5s is reproduced by a controlled three-phase voltage source. Period of interruption was set up long enough to estimate maximum duration of the drive operation in the mode of controlled recovery of stored kinetic energy. Results of simulation (figure 3) represent time dependences of the rotor speed, DC-link voltage, electromagnetic torque and currents.

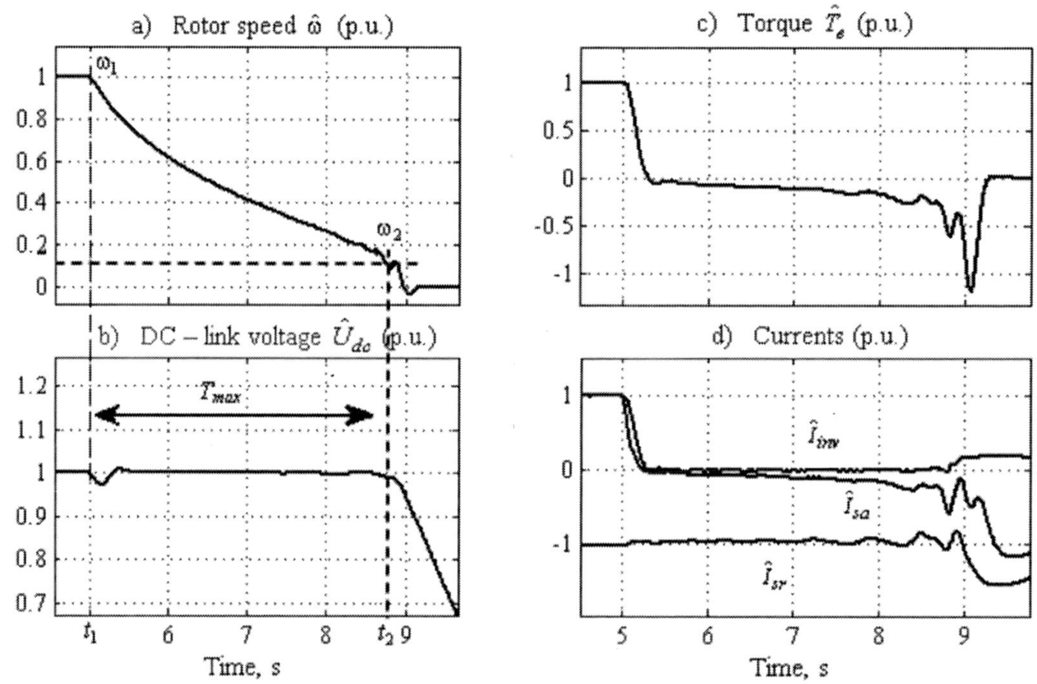

Figure 3. Characteristics of VFD under controlled recovery of kinetic energy.

Observation of submitted curves allows revealing following peculiarities of the processes under study. In the background of smooth deceleration of the rotor, transition from one mode to another is featured by a decrease in electromagnetic torque up to practically zero value, change of its sign and consequent increase in order to compensate for energy loss in the motor: $\Delta P=3\cdot R_s\cdot I^2_s$. Behavior of current component I_{sa} is similar to that of electromagnetic torque while reactive component I_{sr} stays practically unchanged which corresponds to the fact that energy loss during regenerative mode may be considered constant. This allows determining electromagnetic torque as a function of the rotor velocity as follows: $T_{em}=\Delta P\cdot\omega^{-1}$. In the normal mode, the input current of VSI $i_{inv}(t)$ is pulsated in respect to its mean value I_{inv}, which along with DC-voltage defines the power of DC-link. In the regenerative mode, offset value I_{inv} gets close to zero. As we can see from Figure 3b, DC-voltage is kept approximately equal to set value U^*_{dc} during interval of time $T_{max}=t_2-t_1$. Moment t_2 is defined as a moment when the rotor velocity slows down up the value of $\omega_2=0.1\ \omega_1$. In the range of variables variations $t \geq t_2$ and $\omega \leq \omega_2$, the drive goes to the braking mode at the expanse of growing electromagnetic torque.

3.2. Theoretical approach

Calculation of maximum duration of controlled recovery of kinetic energy T_{max} is possible to execute having a solution of equation, reflecting the balance of torques. For a driven mechanism with mechanical characteristic of the fan is possible to write down:

$$-\Delta P\cdot\omega^{-1}-K_f\omega^2=J\,d\omega/dt\,,\ K_f=P_{nom}\cdot\omega_{nom}^{-3}\,,\ \omega(0)=\omega_1 \tag{2}$$

Solution of differential equation (2) may be presented as follows:

$$t(\omega)=\frac{J/6}{\sqrt[3]{\Delta P\cdot K_f^2}}\left(\ln\frac{(\alpha^2\omega^2-\alpha\omega+1)}{(\alpha\omega+1)^2}+2\sqrt{3}arctg\left(\frac{2\alpha\omega-1}{\sqrt{3}}\right)\right)\Bigg|_{\omega}^{\omega_1},\quad \alpha=\sqrt[3]{\frac{K_f}{\Delta P}} \tag{3}$$

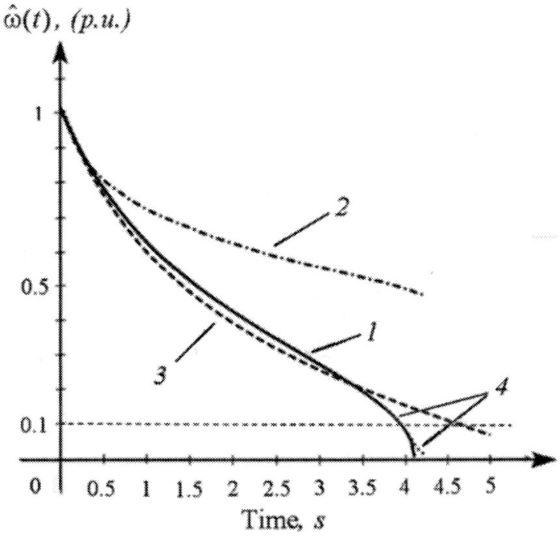

Figure 4. Characteristics of the drive under controlled recovery of kinetic energy.

Results of calculations obtained in accordance with formula (3) are presented in figure 4, where curve 1 relates to results of simulation, curves 2, 3, 4 – to results of calculations. Curve 2 depicts an uncontrolled deceleration of the motor in case of power supply interruption; curve 3 describes deceleration of the motor under condition that braking electromagnetic torque is constant, curve 4 takes into account dependency of electromagnetic torque on the rotor velocity.

As we can see from figure 4, results of simulation (curve 1) and calculations (curve 4) practically coincide that allows recommending formula (3) for assessment of maximum duration of VFD operation in the mode of controlled recovery of kinetic energy.

4. Conclusion.

Results of the paper may be summarized as follows.

The V/f control system of VFD with an observer of the rotor velocity has been elaborated to regulate the drive in two modes of operation: the main mode to secure required coordinates of the driven mechanism and the auxiliary mode to sustain the controllability of the drive during interruption of power supply and/or voltage sags. Stabilization of DC link voltage is achieved by extracting the kinetic energy in a regenerative mode of the motor operation. Control of the drive in both modes is executed in accordance with the principle of a subordinate regulation.

Simulation of the drive in Simulink/MatLab environment allows revealing basic peculiarities of the processes in the drive under voltage sags which were used to estimate possible duration of efficient extraction of kinetic energy T_{max} under condition $\omega(T_{max}) = 0.1\omega(0)$ and to substantiate the necessity of taking into account the contribution of braking electromagnetic torque in theoretical assessment of the process parameters. The use of proposed formula for calculation of time span T_{max} gives results coinciding with results of simulation in the range of the motor deceleration: $\omega(0) \le \omega \le 0.05\omega(0)$.

References

[1] Bose B K 2002 *Modern power electonics and AC drives* ed B Goodwyn and M Vincenti (Upper Saddle River: Prentice Hall) p 738
[2] Sokolovskii G G 2006 *Frequency controlled AC motor drives* (Moscow: Academy) p 265
[3] Petronijevi´c M, Jefteni´c B, Mitrovi´c N and Kosti´c V 2006 Ser.: Elec. Energ. **16** 231-7
[4] Djokic S Ž′, Stockman K, Milanovic J V′, Desmet J J M and Belmans R 2005 *IEEE Transactions on Power Delivery* **20(1)** 494-505
[5] Pietiläinen K, Harnefors L, Petersson and Nee H 2006 *IEEE Transactions on Industry Applications* **53(4)** 1261-68
[6] Pronin M, Shonin O, Vorontsov A and Gogolev G 2011 *Proc. of the 14-th Eur. Conf. on Power Electronics and Adjustable Speed Drives* (United Kingdom) vol **12** (Birmingham: United Kingdom) pp 1-8
[7] Holtz J, Lotzkat W 1994 *IEEE Transactions on Industry Applications* **30(5)** 1275-83
[8] Kornilov G P, Kovalenko A Yu, Nikolaev A A, Abdulveleev I R and Khramshin T R 2014 *Electrotechnical Systems and Complexes* **2(23)** 44-48
[9] Telichko L Y, Basov P M 2009 *Electrotechnical Complexes and Control Systems* **2** 16-20
[10] Jouanne A, Enjet P N, and Banerjee B 1999 *IEEE Transactions on Industry Applications* **35(4)** 908-16
[11] Deswal S S, Dahiya R and Jain D K 2009 *Int. J. of Comp. and Electrical Eng.* **1(2)** pp 141-7
[12] Shonin O B, Novozhilov N G and Kryltsov S B 2016 *Electrotechnical Systems and Complexes* **2(31)** 15-19

IOP Conf. Series: Materials Science and Engineering 177 (2017) 012143 doi:10.1088/1757-899X/177/1/012143

The experimental studies of operating modes of a diesel-generator set at variable speed

S G Obukhov, I A Plotnikov, M A Surkov and L P Sumarokova

Tomsk Polytechnic University, 30, Lenina ave., Tomsk, 634050, Russia

E-mail: masur@tpu.ru

Abstract. A diesel generator set working at variable speed to save fuel is studied. The results of experimental studies of the operating modes of an autonomous diesel generator set are presented. Areas for regulating operating modes are determined. It is demonstrated that the transfer of the diesel generator set to variable speed of the diesel engine makes it possible to improve the energy efficiency of the autonomous generator source, as well as the environmental and ergonomic performance of the equipment as compared with general industrial analogues.

1. Introduction

All industrial diesel generator sets (DGSs) work at a constant speed with a wide range of changes in electric load. This approach has been tested by years of practice and allows for use of sufficiently simple and reliable control systems. However, this strategy for controlling DGSs has a series of serious drawbacks: specific fuel consumption increases significantly in fractional load operating modes, the operating life of the diesel engines is lowered, and emissions of harmful substances increase. One of the more promising technical solutions to increasing the efficiency of autonomous power supply systems is changing the DGS operating mode from constant speed to variable speed, optimized according to the size of the electric load [1, 2].

When the DGS is switched to variable speed, the parameters of the output voltage, created by the generator, change [3, 4]. In order to guarantee supplied output voltage parameters, it is necessary to introduce power converters (controlling rectifiers and inverters) into the system, which, on the one hand, makes the system more complex and costly, but, on the other, creates a set of benefits: specific fuel consumption decreases significantly, the quality of generated electric power increases, the generator is discharged of idle currents, etc.

The goals of the research presented in this paper were to experimentally prove the efficiency of changing the DGS operating mode from constant to variable speed and to determine areas for regulating the unit's operating modes.

2. Diesel generator sets operating at variable speed

As stated above, the energy efficiency of a DGS can be increased significantly by changing the operating mode of the diesel engine from constant to variable speed and by using specialized regulators for controlling the plant [5-8]. A structural diagram of such a diesel generator set (DGS) is shown in Figure1.

As can be seen from Figure 1, the system must have several control loops in order for the variable speed DGS to operate. By acting on the actuating device of the high-pressure pump, the speed regulator sets the optimal shaft speed for the diesel engine based on the criteria for minimal fuel use and stable operation of the diesel engine. The synchronous generator field current regulator sets the current value based on criteria for loss minimization in electrical machinery. The power converter controller provides the necessary algorithms for switching semiconductor rectifiers to create parameters for the electricity provided to consumers in accordance with established standards.

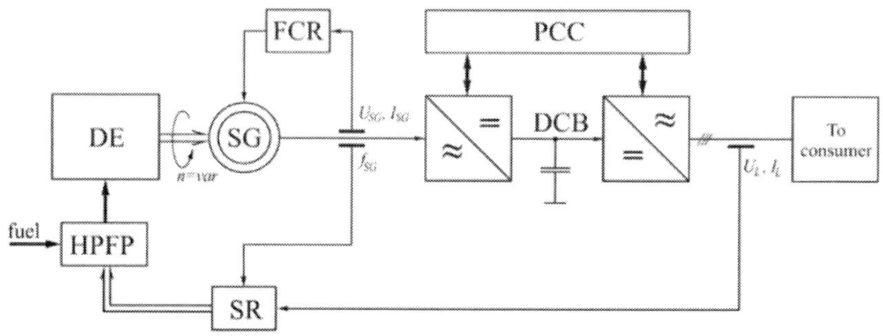

Figure 1. The variable speed DGS structural diagram: DE – diesel engine; PCC – power converter controller; SR – speed regulator; FCR – field current regulator; SG – synchronous generator; HPFP – high-pressure fuel pump; DCB – direct current bus.

While working at variable speed, the DGS is thus a rather complex, multichannel, non-linear, mobile automatic control system. The direct experiment, a common way to research them, makes it possible to obtain reliable results.

3. Laboratory bench

A laboratory bench was developed and manufactured in order to conduct experimental research on variable speed DGSs. A key development requirement was providing high level of automation for the control flow of bench components, and for control and recording of measured variables [9].

Development of the structural and functional design of the laboratory bench was conducted in view of the following technical requirements: the possibility of modeling static and dynamic DGS modes of operation for all main parameters of the operating mode; use of the modular construction principle, which allows for simple attachment/detachment of separate components from the power plant; satisfying all mandatory fire, environmental, and electrical safety regulations; convenience in use and maintenance; the possibility of further modernization of the bench.

4. Experimental studies

An experimental research program was developed that involved removing one of the primary characteristics of the DGS in standard mode (n=const) and at variable speed (n=var) and subsequently conducting a comparative analysis.

The following notation was used to characterize the position of the control rod adjusting screw: h_0 – the fuel pump control rod adjusting screw is in initial position (factory settings); h_4 – the fuel pump control rod has been displaced by 4 mm from its initial position in the direction of decreasing fuel delivery.

The DGS performance standards in standard operating mode were evaluated based on load characteristics. The experiments conducted have shown that, when the DGS load is measured within the range of idling to its nominal value, the frequency of output voltage is measured from 53.0 to 49.9 Hz. The all-regime shaft speed controller can regulate the fuel flow G_m, depending on the load size, in essentially linear fashion. However, the dependence of specific fuel consumption g on the load size is non-linear. In the entire range of electrical load changes, the DGS voltage standard regulating system stabilizes the size of output voltage within 230±5 V, and the field coil current changes from 0.85 to 2.45 A. The temperature of the engine cylinder block does not exceed 95^{0}C, the operating mode of the diesel engine is stable, and the vibrations do not exceed acceptable limits.

Mechanical loss in the diesel engine was determined by the dry motoring procedure and was equal, in the nominal mode, to P_m=5.47 kW, mechanical efficiency η_m=68%. The diesel engine temperature balance in the nominal load mode is shown in Figure 2. The net efficiency of energy conversion in the diesel engine at nominal load was equal to η_e=30.6%.

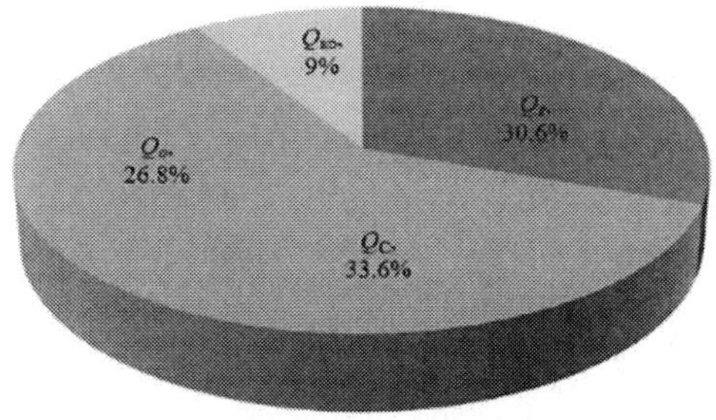

Figure 2. The temperature balance of the KM2V80 diesel engine in the nominal operating mode: Q_E – the amount of heat equivalent to effective operation; Q_G – the amount of heat lost with exhaust fumes; Q_c – the amount of heat removed by the cooling system; Q_{BD} – temperature balance discrepancy.

Figure 3 shows a series of regulatory characteristics of the KDE12EA3 DGS recorded at various positions of fuel feed control organ h. There is also a line of maximal outputs for partial speed modes 1, which determines the maximum accepted value of luminous flux declension coefficient K_{dop}, determined based on the experimental results. To ensure the most efficient diesel engine operational mode when in use, the resulting linear dependence must be displaced downward to the factory settings for the nominal capacity mode: $h\_0 - P_n = 9.5$ kW (line 2).

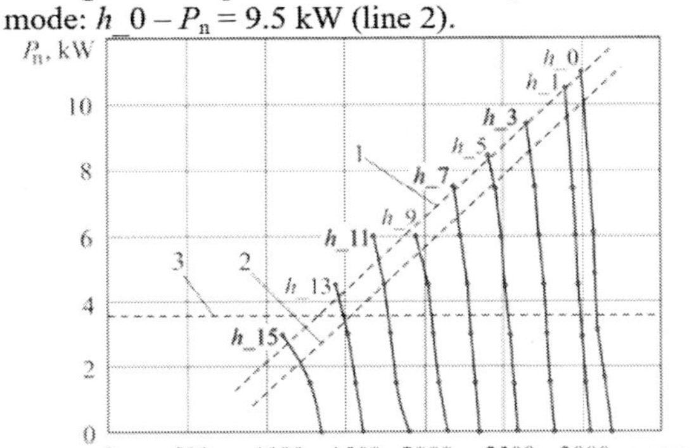

Figure 3. DGS KDE12EA3 regulatory characteristics.

In order to ensure stable operation of a diesel engine with small loads, its minimal operating shaft speed should be limited to 1600 rpm, which corresponds to a displacement of the fuel control rod by 12 mm from factory settings (line 3). Further lowering of the shaft speed is inadvisable, as the imbalance of fuel atomization and combustion increase significantly at low speeds, which leads to unstable engine operating modes.

The results of the experiments determine the displacement range of the KDE12EA3 DGS fuel pump control rod necessary for implementing the power station control mode in accordance with minimum fuel use criteria, which is equal to 12 mm from h_0 in nominal capacity mode to h_12 in small load mode, with an operating range of shaft speeds from 3000 to 1600 rpm (Figure 3). As a result, a functional relationship was obtained that connects the position of the fuel pump control rod with the capacity of the DGS electrical load at which specific fuel consumption will be at a minimum, Figure 4.

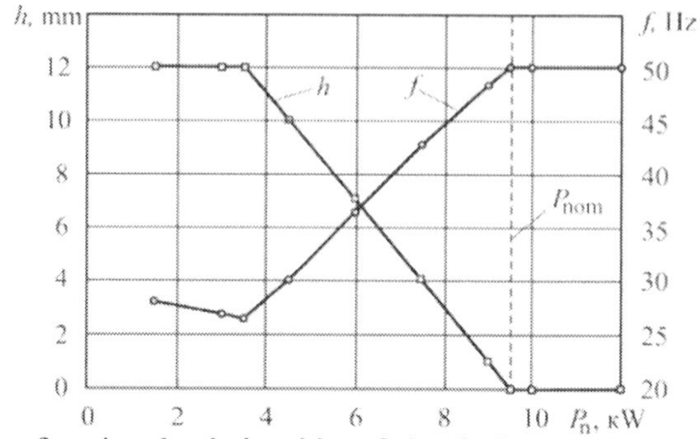

Figure 4. The functional relationship of the fuel pump control rod position with the KDE12EA3 DGS electrical load that guarantees minimum fuel use.

Figure 4 also shows the dependence of the frequency of output voltage *f* on the size of the DGS electric load which operating modes have been optimized for specific fuel consumption. The relationship is also linear, which makes it possible to use current shaft speed values as the feedback signal in an automated regulating system of the fuel feed control organ in inverter DGSs.

The empirical relation between the specific fuel consumption of the researched DGS operating in standard mode (n=const) and optimized for consumption (n=var) is presented in Figure 5. We can see from it that changing the DGS to variable speed saves fuel across the entire spectrum of operating modes, and the maximum savings are in small load modes.

Reducing specific fuel consumption in a variable speed DGS can be done by reducing mechanical loss in the diesel engine, increasing generator efficiency η_g and using efficient diesel fuel η_e.

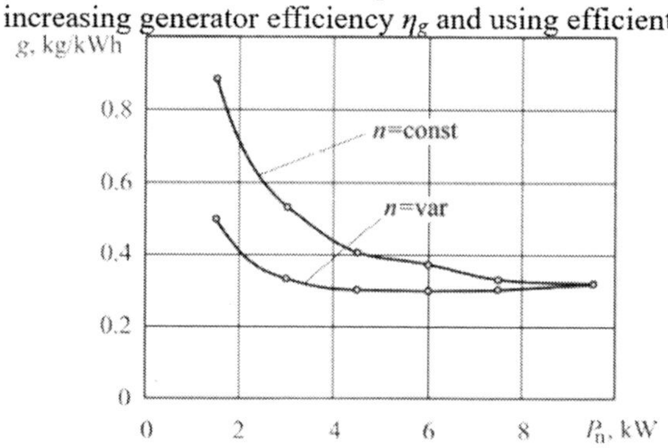

Figure 5. Dependence of specific fuel consumption on the size of electrical loads in variable and constant speed DGSs.

Figure 6 shows comparative DGS energy characteristics generated based on the results of the experiments conducted. The net efficiency of diesel fuel η_e in a variable speed DGS can be increased by using higher-quality air-fuel mixture in fractional load modes. The quality of the mixture is largely determined by excess air coefficient α which value in modern diesel engines varies between 1.2 and 1.8 (depending on the type of combustion chamber and pressurization). As a rule, manufacturers ensure optimal values for α in fractional modes, which are at about 80% of the engine's nominal capacity. A diesel engine without a pressurization system is used in most DGSs with a capacity of up to 60 kW.

Correspondingly, in DGSs with constant speed, the airflow rate remains practically unchanged for all fractional load modes, and the diesel engine is stabilized only by changing the fuel delivery, to which the size of α is inversely proportional. In a variable speed DGS, when the load size is reduced, the shaft speed decreases along with a change in fuel delivery, and, correspondingly, the airflow rate, which leads to slightly lower α values on fractional modes. A divided combustion chamber with a swirl chamber is used in the diesel engine being studied, and the best mixture conditions are created at values of $\alpha \approx 1.4$. During operation of a variable speed DGS, the α for most fractional load modes will be closer to optimal compared with general industrial units.

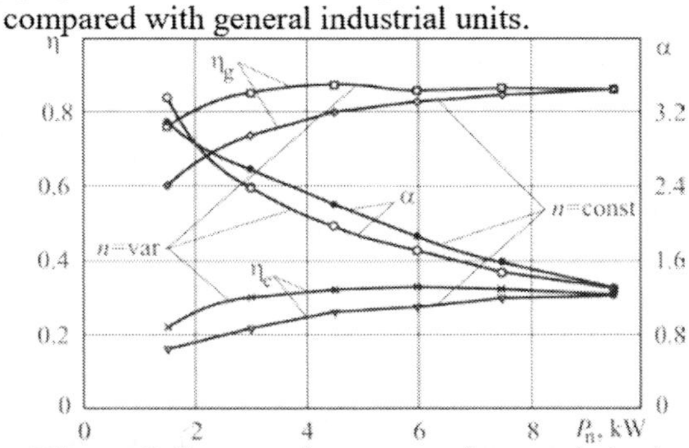

Figure 6. Comparative energy characteristics in constant and variable speed DGSs.

On the basis of the experiments, it can be concluded that changing the DGS operating mode to variable speed optimized for specific fuel consumption provides the following competitive advantages:

- lowers mechanical loss and increases efficiency of the diesel engine and generator for all fractional load modes except nominal mode;
- lowers specific fuel consumption for all fractional modes, saving up to 40% from the baseline scenario in low load modes.

Experiments were conducted on the basis of a specific model of a diesel generator set, yet the theory of similarity of internal combustion engines and electrical machinery, as well as analogous results obtained by other researchers [10-12] make it possible to extend the conclusions to other DGSs with the same capacity range up to 100 kW.

5. Conclusion

The main technical problem that arises when small-scale renewable power generation is used in practice is the necessity of coordinating production and energy use modes in an isolated energy system, with extreme temporal unreliability in primary energy carrier energy and random nature of the load of autonomous consumers. A guaranteed power source is necessary for such systems, and diesel generator sets are used as such in most practical scenarios.

Specific fuel consumption for the production of 1 kWh of electrical energy is a key indicator of DGS energy efficiency. In view of the fact that the largest portion of expenditures in prime cost of electrical energy produced by a DGS is taken up by fuel expenses, this factor is key.

As a result of the experiments, it has been empirically proven that changing a diesel generator set to variable speed optimized for the size of electric load lowers specific fuel consumption for the generation of electrical energy. The specific fuel consumption in a variable speed DGS is lowered by decreasing mechanical loss in the diesel engine, increasing the efficiency of the generator and efficiency of the diesel engine. For a typical load schedule of autonomous consumers, fuel economy amounts to 20-30%.

References

[1] Wang D, Nayar C and Wang C 2010 *2nd IEEE Int. Symp. on Power Electronics for Distributed Generation Systems* (USA: IEEE) p 1

[2] Chlodnicki Z, Koczara W and Al-Khayat N 2008 *J. Electr. Power Quality and Utilis.* **14** 13-24

[3] Iwanski G and Koczara W 2013 Variable Speed Power Generation *monograph: Power Electronics for Renewable and Distributed Energy Systems: A Sourcebook of Topologies, Control and Integration* (London:Springer Verlag) chapter 7 pp 309-378

[4] Lukutin B, Sarsikeyev Y, Surkov M and Lyapunov D 2014 *14th Int. Conf. on Environment and*

Electrical Engineering (Krakow: IEEE Computer Society) p 459
[5] Zhang Y, Li S, Lu G and Zhou Y 2012 *Int. Conf. on Systems and Informatics* (USA: IEEE) p 619
[6] Pichkalov I, Kyselova A, Kyselova O and Kyselov G 2013 *IEEE 33th Int. Scientific Conf. Electronics and Nanotechn.* (USA: IEEE) p 378
[7] Lukutin B, Shandarova E, Matukhin D, Makarova A and Fuks I 2016 *IOP Conf. Series: Mat. Sci. and Eng.* (Tomsk) vol 124 (Tomsk: Institute of Physics Publishing) p 4
[8] Lukutin B, Shandarova E, Makarova A, and Shvartsman I 2016 *IOP Conf. Series: Mat. Sci. and Eng.* (Yurga) vol 127 (Yurga: Institute of Physics Publishing) p 98
[9] Duraev N, Obukhov S and Plotnikov I 2014 Laboratory Stand for Research Mode Diesel-Generator Set for Variable Speed *Industrial ACS and controllers* (Moscow: Nauchtechlitizdat LTD) pp 33-40
[10] Leuchter J, Refucha V, Krupka Z and Bauer P 2007 *The 38th Annual IEEE Power Electronics Specialist Conf.* (USA: IEEE) p 2287
[11] Chlodnicki Z, Koczara W, Al-Khayatt and Brown N 2008 *13th Int. Power Electronics and Motion Control Conf.* (USA: IEEE) p 2150
[12] Beyerleyn E and Tyuteva P 2014 *15th Int. Conf. of Young Specialists on Micro/Nanotechnologies and Electron Devices* (USA: IEEE Computer Society) p 359

Structural heredity influence upon principles of strain wave hardening

A V Kiricheck[1,a], S V Barinov[2,b] and A V Yashin[2,b]

[1] Bryansk State Technical University, 7, 50-letiya Octyabrya bul., Bryansk, Russia
[2] Murom Institute (branch) of Vladimir State University, 23, Orlovskaya str., Russia

E-mail: [a]avk.57@yandex.ru, [b]box64@rambler.ru

Abstract. It was established experimentally that by penetration of a strain wave through material hardened not only the technological modes of processing, but also a technological heredity - the direction of the fibers of the original macrostructure have an influence upon the diagram of microhardness. By penetration of the strain wave along fibers, the degree of hardening the material is less, however, a product is hardened throughout its entire section mainly along fibers. In the direction of the strain waves across fibers of the original structure of material, the degree of material hardening is much higher, the depth of the hardened layer with the degree of hardening not less than 50% makes at least 3 mm. It was found that under certain conditions the strain wave can completely change the original structure of the material. Thus, a heterogeneously hardened structure characterized by the interchange of harder and more viscous areas is formed, which is beneficial for assurance of high operational properties of material.

1. Introduction

To improve performance properties, various methods of hardening are widely used in machinery [1-10]. The strain wave hardening is a relatively new method of surface plastic deformation. The peculiarity of the method is to generate the strain wave by a strike system, with an intermediate link and to post them into the deformation zone aiming at strengthening critical surfaces of machine parts. Efficiency of the process is provided by the full use of strain wave energy while hardening. This is achieved at the expense of a preliminary static approach of the system 'tool-waveguide' to the work surface and the subsequent dynamic loading of a hammer head. Static contraction provides regeneration of reflected strain waves. Deformation of the environment can occur at a depth of 6-8 mm, the degree of hardening can reach up to 150%. Depending on the desired parameters of hardening of the surface layer, selected elements of the strike system are selected which generate an impact impulse of the required amplitude and duration. By means of control of the strain wave parameters, it is possible not only to form a hardened surface layer uniformly, but also to get heterogeneously hardened areas [1, 2].

As a result of hardening of steel samples by the strain wave, it was determined that various diagrams of microhardness can be formed (Figure 3g) under the same processing conditions. The aim of this work is to identify the causes and effects of this phenomenon in relation to the original structural technological heredity.

2. Effect of structural heredity on the conformity of strain wave hardening

For research carrying out some samples were cut out of round rolled metal (steel 45) of a rectangular cylindrical shape and with the thickness of 10 to 20 mm (Figure 1 a, b). The use of round bars is due to a known fiber direction of the original heritable structure (Figure 2, a). One part of samples cut out of round rolled metal was hardened so that the strain waves passed across the fibers of the initial heritable structure during processing and another part of the samples – by the passage of the strain waves along the fibers. All test samples were reinforced by the same strain wave with the same modes: strikes energy was 25 J; impact frequency was 23.3 Hz; prints overlap coefficient was K = 0.4; the tool was a cylinder of 7 mm width and 10 mm in diameter.

Hardening efficiency of material by the strain wave was evaluated by a change of microhardness diagrams of the surface layer. Research samples were cut on a reinforced track symmetry axis in the feeding direction (Figure 1).

On the obtained polished specimen, in the transverse and longitudinal directions with a pitch of 0.3 mm, micro-hardness with a diamond pyramid by the method of Vickers hardness test was measured (Figure 1 b). Based on measuring results of microhardness in the Statistica program, the diagrams were formed according to its distribution in the surface layer.

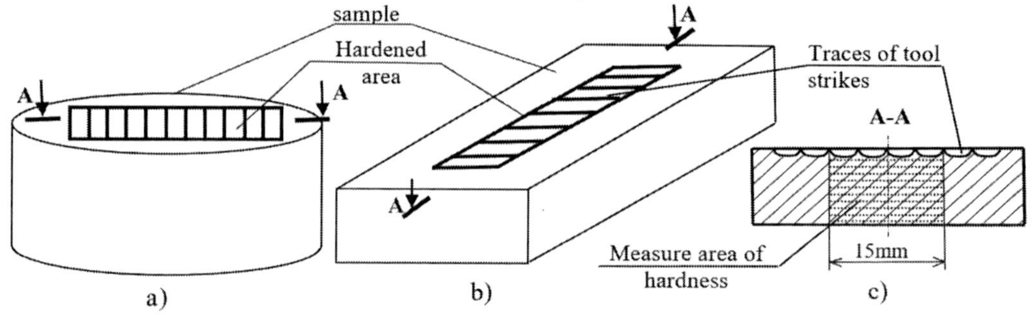

Figure 1. Diagrams of studied samples.

Microhardness diagrams of round bars of the original structure are presented in Figure 2. In the diagram (Figure 2a), we can see typical rolled fibers having a higher hardness as compared with basic metal. In figure 2b, we see the microhardness diagram of the sample with transverse arrangement of the original fiber structure, in Figure 2c, the structure is presented with longitudinal arrangement of fibers.

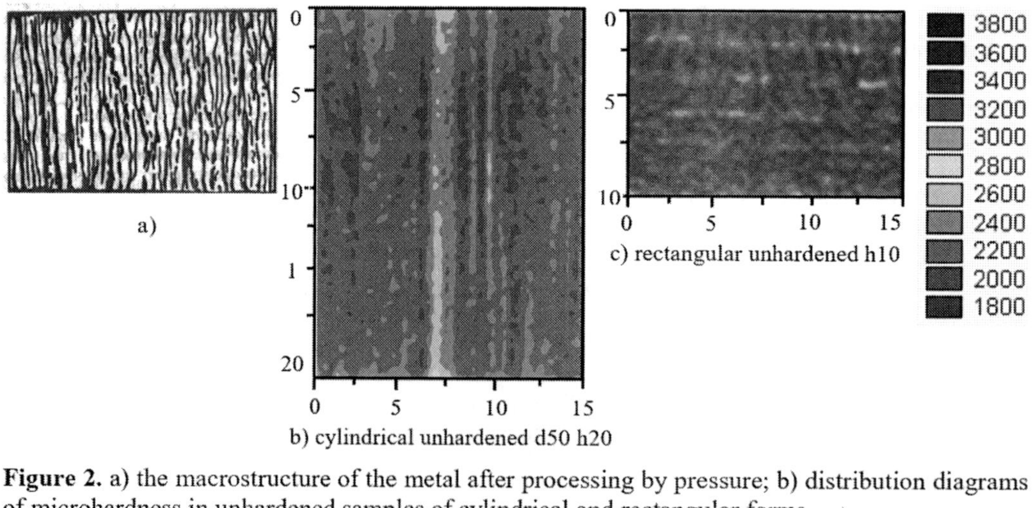

Figure 2. a) the macrostructure of the metal after processing by pressure; b) distribution diagrams of microhardness in unhardened samples of cylindrical and rectangular forms.

Microhardness diagrams in the samples hardened by a strain wave lengthwise and crosswise of the initial fibrous structure, are presented in Figure 3. The analysis of obtained diagrams has shown that in case of hardening of samples with a longitudinal arrangement of the hereditary fibrous structure (Figure 3, c-f), the strain wave owing to anisotropy of properties of the initial metal extends both on fibers and on the set of rods covering the whole sample. The material is more strongly hardened along fibers, however, the hardening degree is rather insignificant. Only in the surface layer for the depth of 2.5 mm, we can observe the fragmentary degree of hardening for more than 40 %. Among fibers and where the fibrous structure is expressed more poorly, the degree of hardening is less - no more than 20%.

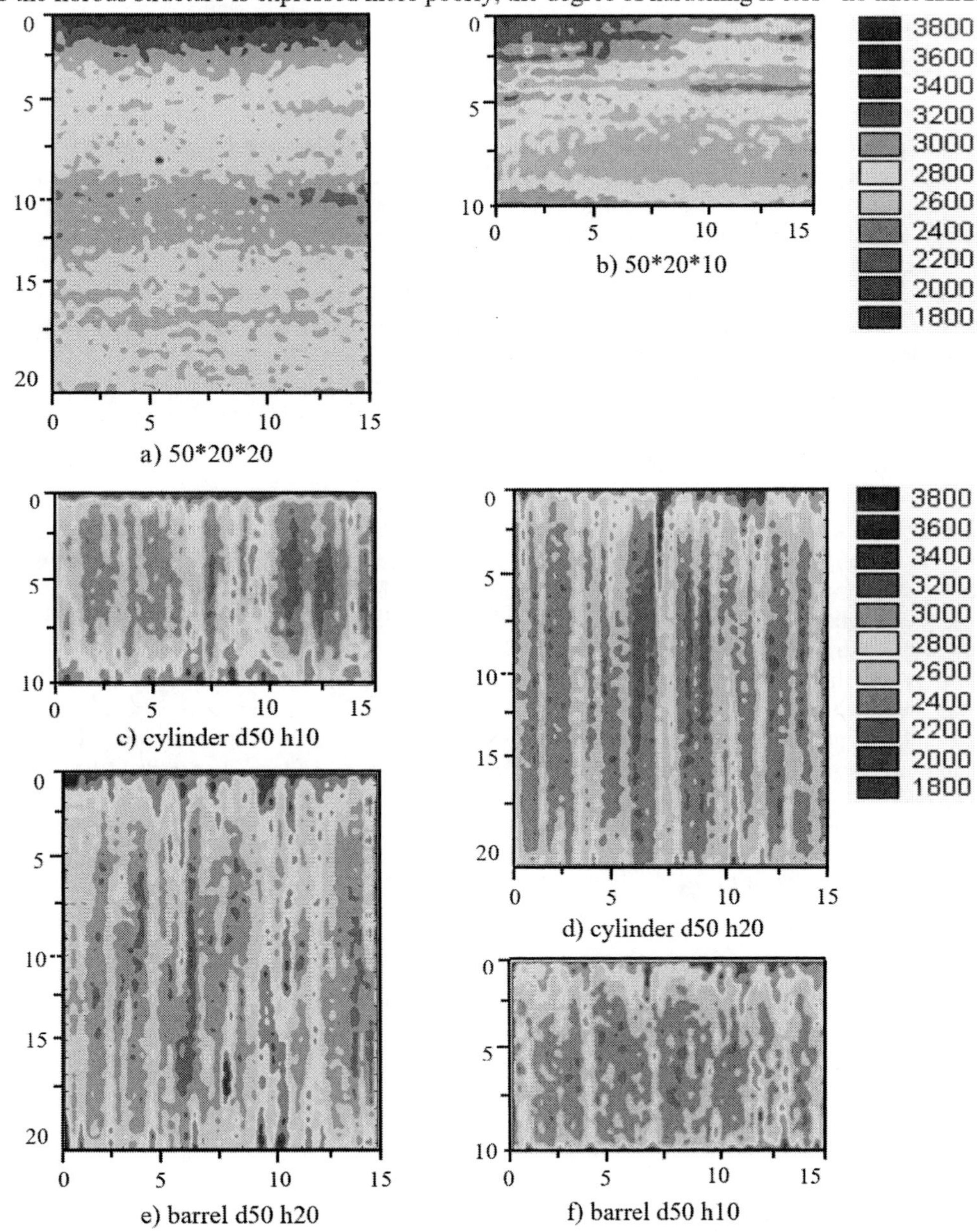

Figure 3. The microhardness diagram after hardening of the material by a strain wave.

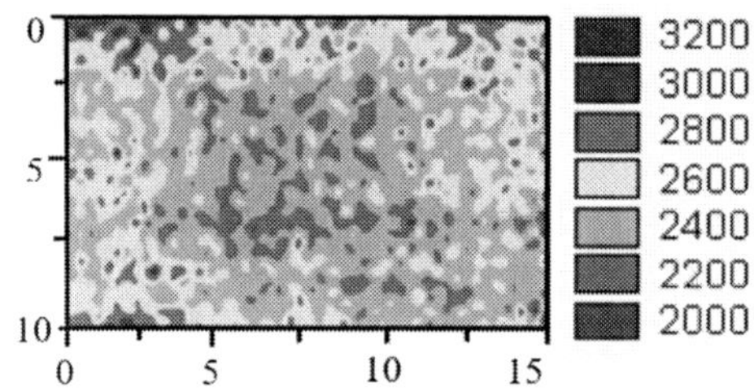

Figure 4. Obtaining of the strain wave of local alternations of firmly viscous areas in the surface layer.

In case of strain wave propagation across fibers (Figure 3, a, b), the degree of material hardening of the surface layer is much higher, and the depth of an essentially hardened layer is greater. Practically the whole surface layer at the depth of 3 mm has a degree of hardening of 50 % and more. Most often a sample gets a multilayered gradiently hardened structure without strongly manifested fibers. The least hardened zone is formed, as a rule, in the centre of the sample. However, under certain conditions and processing modes, strain wave hardening is capable of changing the inherited structure - of breaking fibers and generating the spotty heterogeneously hardened structure with alternating harder and more viscous areas (Figure 4). In this case, a more intensive increase in operational properties of a material is observed too.

It is necessary to notice that in all cases, the surface layer of a material both with contact and from the opposite side of the sample is hardened at the expense of the reflected strain waves.

3. Conclusions

1. It is established that technological heredity has a strong impact on efficiency of strain wave hardening - the fibers are generated at the stage of metal rolling.

2. It is established that for obtaining of great values of depth and degree of hardening of the surface layer the strain wave should be directed across fibers of initial structure.

3. It is established that for hardening of the whole sample, the strain wave should be directed along fibers of the initial structure.

4. It is established that under certain conditions the strain wave can completely change the initial fibrous structure (Figure 4) and provide alternation of much harder and more viscous local areas at considerable depth of the surface layer or on all section of the sample. The given type of the hardened structure has shown itself in the increase of operational properties of machines parts [2-9].

Acknowledgement

The researches have been performed with the assistance of the Russian Foundation for Basic Researches, No. 14-08-00112 A.

References

[1] Kirichek A V, Kuzmenko A P, Soloviev D L, Barinov S V, Altukhov A Y, Silantiev S A, Grechukhin A N, Myo Min Than and Dobromyslov M B 2015 *J. of Nano and Electronic Physics* **7** **4** 04023
[2] Kirichek A V, Barinov S V, Yashin A V 2016 *IOP: Proc. of the Conf. Mat. Sci. and Eng.* **124** 012158
[3] Dmitriev A I, Zolnikov K P, Psakhie S G, Panin V E and Goldin S V 2005 *Theoretical and Appl. Fract. Mech.* **43(3)** 324-334
[4] Ivanov G P, Kartonova L V, Khudoshin A A 1997 *Stroit. Dorozhn. Mashiny* **1** 33-34
[5] Panin V E and Deryugin E E 2003 *Phys. of Metals and Metallogr.* **96** 2-15
[6] Panin V E, Egorushkin V E 2013 *Physical Mesomechanics* **16(4)** 267-286
[7] Smelyanskii V M, Zemskov V A 2005 *Uprochn. Tekhnol. Pokryt.* **1** 27-35
[8] Valiev R Z, Krasilnikov N A and Tsenev N K 1991 *Mat. Sci. and Eng.* **137** 35-40

IOP Conf. Series: Materials Science and Engineering **177** (2017) 012144 doi:10.1088/1757-899X/177/1/012144

[9] Valiev R Z, Korznikov A V and Mulyukov R R 1993 *Mat. Sci. and Eng.* **168** 141-148
[10] Zhu Y T, Huang J Y, Gubicza J, Ungár T, Wang Y M, Ma E and Valiev R Z 2003 *J. of Mat. Res.* **18(8)** 1908-1917

Institute of Physics
Dirac House, Temple Back
Bristol BS1 6BE UK

ISSN: 1757-8981
ISBN 978-1-5108-3715-7
9 781510 837157

CURRAN ASSOCIATES INC.
proceedings
.com